Corrosion Resistance of Aluminum and Magnesium Alloys

WILEY SERIES IN CORROSION

R. Winston Revie, Series Editor

Corrosion Inspection and Monitoring. Pierre R. Roberge

Corrosion Resistance of Aluminum and Magnesium Alloys: Understanding, Performance, and Testing. Edward Ghali

Microbiologically Influenced Corrosion. Brenda J. Little and Jason S. Lee

Corrosion Resistance of Aluminum and magnesium Alloys. Edward Ghali

Corrosion Resistance of Aluminum and Magnesium Alloys

Understanding, Performance, and Testing

Edward Ghali

WILEY
A JOHN WILEY & SONS, INC., PUBLICATION

Published by John Wiley & Sons, Inc., Hoboken, New Jersey
Published simultaneously in Canada

Library of Congress Cataloging-in-Publication Data:

Ghali, Edward.
Corrosion resistance of aluminum and magnesium alloys : understanding, performance, and testing / Edward Ghali.
p. cm.
Includes index.
ISBN 978-0-471-71576-4 (cloth)
1. Aluminum alloys–Corrosion. 2. Magnesium alloys–Corrosion. 3. Corrosion and anti-corrosives. I. Title.
TA480.A6G45 2010
620.1′8623–dc22

2009014010

Printed in the United States of America

10 9 8 7 6 5 4 3 2 1

I am grateful to my wife, Helen, whose unfailing encouragement and support sustained me through the challenge of writing this book. There are no words that can express my heartfelt gratitude to members of my close happy family, in particular, Rafik and Sonia Ghali and their spouses, Isabelle Côté and Dominique Julien, who encouraged me enormously during the realization of this book.

Contents

Preface

This book is based on a one-semester graduate course on corrosion of Al and Mg alloys given at the Faculty of Science and Engineering at Laval University, Quebec City, Canada. Aluminum has a close relation to magnesium in regard to electronic configuration and at the same time both have active–passive behaviors in aqueous solutions. Although the mechanism of passivation is different for the two metals, it was justified to combine the discussion of corrosion for Al and Mg in one book. I believe that these two metals and their concerned alloys will play an important part in our modern societies, especially for environmental considerations.

The purpose of the first part of the book is to introduce the fundamentals of corrosion science while the second part covers the engineering performance and corrosion resistance of aluminum alloys. Part Three considers magnesium in the same way for another five chapters. In the absence of adopted standards of the definition of corrosion forms and types in aqueous media, I used an arbitrary approach, inspired from recent literature, and divided them into seven forms as a function of the corrosion mechanism and morphology of the corroded material.

The seven forms are treated in the following sequence: General corrosion (uniform and nonuniform), galvanic or dissimilar metal corrosion, localized corrosion, metallurgically influenced corrosion, microbiologically influenced corrosion, mechanically assisted corrosion, and environmentally induced corrosion. However, certain difficulties are encountered for some corrosion principles or history cases since some of these forms are intimately related whether at the beginning, during, or at the end of corrosion or fracture failure.

Corrosion prevention methods such as coatings for aluminum and magnesium are described in Part Four in two separate chapters. Part Five describes the different electrochemical methods of investigation in a general chapter followed up by two chapters on testing of aluminum and magnesium. These include testing and evaluation of forms and types of corrosion of some concerned alloys. Finally, Part Six has three appendixes to bring together the many symbols, equations, acronyms used throughout the book and to list some pertinent references.

During the preparation of this book, I came to realize the expanding roles of aluminum, magnesium, and their alloys in industrial practice, and the fascinating new service challenges, such as in nuclear applications, and research activities for new alloys and composites. A great effort has been made to cover these innovative uses for Al and Mg.

Edward Ghali

Quebec, Cananda

Acknowledgments

For forty years, I have been working intensively in the area of corrosion and electrochemical engineering as both a professor and a consultant. The Québec community has been very supportive of me and my work. I therefore acknowledge my fellow researchers and the students at the University of Quebec at Chicoutimi, and at Laval University (Québec), who have helped me to achieve a better understanding of the fundamentals of corrosion engineering.

My sincere thanks first to those who were involved directly in the preparation and writing of this book: Jean-Philippe Gravel, François Gilbert, Jean-Pierre Côté, and Carl Moniz, who were students at Laval University (Québec) in chemical and metallurgical engineering programs. The effort of Simon-Pierre Barrette, specialist in information science at the Bibliothèque Scientifique, Université Laval, is highly appreciated. I would like to express my gratitude to Dr. Winston Revie and Dr. Mimoun Elboujdaini, Materials Technology Laboratory, CANMET–Minerals and Metals Sector, Ottawa, and to Professor Karl-Ulrich Kainer, Dr. Carsten Blawert, Dr. Wolfgang Dietzel, and Dr. Norbert Hort, Members of the Center for Magnesium Technology, at GKSS-Forschungszentrum GmbH, Institute for Materials Research, Geesthacht, Germany. Special thanks go to Professor Andrej Atrens, Dr. Guangling Song (Research & Development Center, General Motors Corporation, Warren, MI, USA), and Dr. Nicholas Winzer from the University of Queensland, Brisbane, Australia.

Finally, I wish to offer my deep gratitude to Professor Réal Tremblay and Professor Dominique Dubé for their enthusiasm and numerous suggestions, and all the members of the research group working in the area of corrosion at the Department of Mining, Metallurgical and Materials Engineering, Faculty of Science and Engineering, Laval University, Québec, Canada.

E. G.

Part One

Electrochemical Fundamentals and Active–Passive Corrosion Behaviors

Chapter 1

Fundamentals of Electrochemical Corrosion

Overview

The thermodynamic tendency to corrosion and Gibbs free-energy change, $\Delta G = -nFE$, are discussed. The double layer is described and the Nernst equation is deduced to describe electrode potentials. Strong and weak electrolytes as well as Faraday laws are mentioned. The activity and conductance of the electrolyte are explained and the constant of dissociation and coefficients of dissociation and activity are defined.

Popular types of electrodes are examined and explained. The general approach to electrochemical cells is given and used to better explain the most frequently encountered corrosion cells. A classification of electrochemical cells and corrosion cells is then given. Electrochemical reactions are defined, such as those in which free electric charges, or electrons, participate. They are classified as micro- and macroelectrochemical cells (inseparable anode/cathode areas and separable anode/cathode areas). Corrosion chemical reactions, such as metal dissolution in liquid medium, can be described by the absence of charge transport in an electrolyte or a formed film at the interface in a metal–solvent reaction or at a metal–gas interface. A study of corrosion phenomena principally covers the corrosion product, the material, the medium, and the interface.

A. THERMODYNAMIC CONSIDERATIONS OF CORROSION

Corrosion of metals is mainly due to an irreversible oxidation–reduction reaction, where an oxidizing agent in an environment attacks the metal. An electrochemical reaction is a chemical transformation that implies charge transport at the interface from a metallic conductor (electrode) to an ionic conductor (electrolyte). These are dependent on the thermodynamic and physical properties of the electrode and the activities of the different species in solution at the interface. The properties of the interface are dependent on temperature variation, bulk solution properties, convection, diffusion, and so on. In addition, metallurgical and mechanical properties of the electrode and microbiological organisms in solution or at the interface can cause corrosion alone or increase corrosion rates by a synergetic effect with electrochemical corrosion. In other limited conditions of corrosion, chemical reactions, such as the dissolution in liquid metals and some solvent media and

Corrosion Resistance of Aluminum and Magnesium Alloys: Understanding, Performance, and Testing By Edward Ghali

metal–gas free interface reactions, can cause corrosion alone or assist an electrochemical reaction.

1.1. ELECTROLYTIC CONDUCTANCE

Electrons are the current carriers in solids while ions are the only means of charge transport in a solution. An ion is an atom that has lost or gained electrons; if it lost some, it is a positive ion and called a cation, while if it gained some, it is a negative ion and called an anion. An ion can be a simple ion or a complex one. However, ions can be monovalent, divalent, trivalent, and so on (e.g., Na^+, Mg^{2+}, Al^{3+}) or complex ions (e.g., $SO_4{}^{2-}$, $PO_4{}^{3-}$, $Fe(CN)_6{}^{4-}$). The strength of attraction between sodium and chloride ions inside the crystal is $F = qq'/r^2$ (q is the charge and r is the distance between two ions from center to center). Once the sodium chloride is added to water as a solvent, the strength of attraction between sodium and chloride ions becomes $F = qq'/Dr^2$, where D corresponds to the dielectric constant of water, which is close to 80 in centimeter-gram-second (CGS) units. The attraction therefore becomes 80 times smaller and the crystal separates into ions even at room temperature.

A solvated ion is an ion that fixes more closely one or several molecules of solvent. If the solvent is water, one says that the ion is hydrated. Many ions in aqueous solutions are hydrated such as given for HCl:

$$HCl + (n+m)H_2O \rightarrow H^+ \cdot nH_2O + Cl^- \cdot mH_2O$$

The values of n and m of H_2O are not exactly known. The hydrated ions can conduct current, while liquid hydrochloric acid and pure water are bad conductors. The properties of hydrated or solvated ions in general are critical to conduction and corrosion.

Molten salts can be good electrolytes, such as potassium chloride that undergoes thermal agitation till melting. As the temperature rises, the strength of crystalline cohesion goes down. Once the resulting liquid at the point of fusion is reached, the Cl^- and K^+ ions are relatively free to transport the charge and this can form a good electrolyte if the forces of attraction between differently charged ions do not dominate.

Under the effect of a sufficiently powerful outside electric field, one can produce a migration of ions of a solid salt in two opposite directions according to their signs. This is not the case of potassium chloride (KCl) and numerous other salts but this is the situation of several halides, such as silver bromide or iodide. AgI is considered a strong electrolyte but most solid electrolytes are weak. In the case of gas as an electrolyte, ions can form but their life span is very short and they recombine immediately to give the corresponding neutral molecules. It is necessary to use an outside energy source as the electric spark, and a plasma is helpful in separating and characterizing the ions separately for fundamental studies. Otherwise, there are many applications of plasma in the area of corrosion; for example, various plasma applications are emerging for anodizing (anodic oxidation) to protect Al or Mg alloys by forming their respective oxides.

All electrolytes are dissociated into ions independently of the passage of the current. Electrolytes can be divided into two groups:

1. *Strong electrolytes* dissociate almost completely in spite of the electrostatic attractions between oppositely charged ions. Their dissociation is almost total: $AB \rightarrow A^+ + B^-$.
2. *Weak electrolytes* are dissociated imperfectly or their dissociation is reversible: $AB \Leftrightarrow A^+ + B^-$.

In aqueous solution, with some exceptions, bases (e.g., NaOH) and salts (e.g., NaCl) are strong electrolytes. However, acids such as CH_3COOH and H_2S or bases such as NH_4OH are weak. In the molten state, one considers a salt like KCl to be dissociated completely, while some melted salts undergo the phenomenon of association because of thermal agitation and attraction between positive and negative ions driving the formation of a weak mobile complex or even nonionized molecule. In this case, the molten electrolyte is considered weak. Many compounds, notably organic, are not ionized or hardly ionized in the melted state. It is interesting to note that acetic and hydrochloric acids in the pure liquid state don't drive the current.

1.1.1. Faraday Laws

The following three laws of Faraday apply for electrolysis and corrosion:

1. In electrolysis, products of the electrochemical decomposition appear on the electrodes and not within the electrolyte.
2. The metals of salts and bases and the hydrogen of acids appear at the cathode; remainders of the molecule and/or its products of decomposition appear at the anode.
3. The mass m of deposited metal on the cathode by electrolysis is proportional to the quantity of current (I) crossing the cell and to the atomic mass of metal and inversely proportional to the valence of the metal:

$$m = \frac{1}{F} \times \frac{A}{n} \times I \times t$$

F is called the Faraday constant and is $\sim$96,500 coulombs/mole if the current I is expressed in amperes, the time t is given in seconds, the atomic mass is in grams/mole, and n is the electrovalence of an ion (sometimes expressed as z). The electrovalence of an ion should correspond to the number of electrons either gained or lost in the electrochemical reaction. Reactions of reduction occur at the cathode (gain of electrons) and reactions of oxidation occur at the anode (loss of electrons). The experience shows that a coulomb deposits 1.118 mg of silver. The atomic mass can be deposited by $107.88/0.001118 = 96{,}494$ coulombs, while the coulometer to iodine gives for 1 equivalent 96,514 coulombs (electrode of platinum and iridium using a solution of 10% KI). The quantitative law of Faraday applies to secondary reactions, that is, chemical reactions that follow electrochemical reactions.

Figure 1.1 shows a zinc–copper cell in a corrosive aqueous environment (exothermic reaction). By convention, in a cell, the polarity of the anode is negative and that of the cathode is positive (Figure 1.1). However, for electrolysis, the polarity of the anode is positive and that of the cathode is negative. In this case, the sign is imposed by the external electric current.

1.1.1.1. Primary and Secondary Reactions

The reaction that involves a gain or a loss of electrons is considered the primary reaction. Often the remainder that appears at the anode is not chemically stable in solution such as the OH product species of a primary reaction that leads to the evolution of oxygen by the

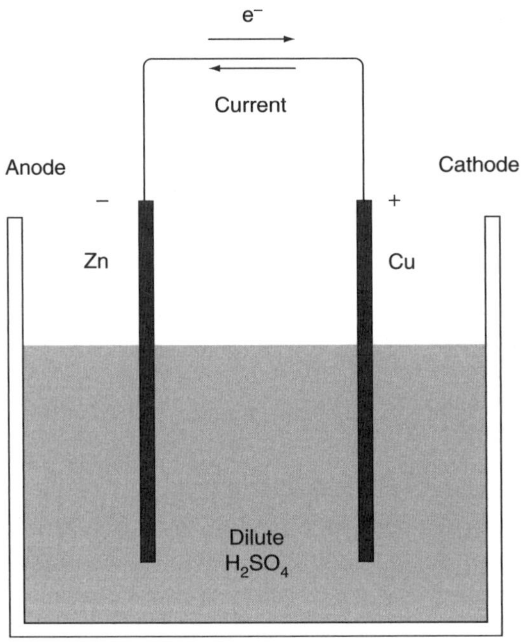

Figure 1.1 Zn–Cu cell in dilute sulfuric acid (H_2SO_4).

reaction; the same applies for SO_4 and SiF_6. This reaction is called a *secondary reaction.* The primary and secondary reactions are dependent on the nature, composition, and microstructure of the electrode, the composition of the electrolyte, and the properties at the interface. The secondary reactions could produce the electrochemically active ion by dissociation of a complex, for example, or involve the substances formed by the primary reactions. The secondary reaction mechanism is often composed of one or more hypotheses that are very frequently based on the properties of the metal or material, the electrolyte at the interface, thermodynamic considerations, and kinetic approaches. The existence of some transient species could also be proved experimentally [1]. The primary and secondary reactions are indicated in the following for oxygen evolution in acidic nitrate solution:

$$2NO_3^- \rightarrow 2NO_3 + 2e^- \qquad \text{Primary reaction}$$

$$\left.\begin{array}{l} 2NO_3 + H_2O \rightarrow 2HNO_3 + \tfrac{1}{2}O_2 \\ 2HNO_3 \rightarrow 2H^+ + 2NO_3 \end{array}\right] \text{Secondary reactions}$$

$$H_2O \rightarrow 2H^+ + \tfrac{1}{2}O_2 + 2e^- \qquad \text{Total reaction}$$

1.2. TENDENCY TO CORROSION

The change in the Gibbs free energy, frequently designated as free enthalpy, or Gibbs function can be calculated from the equation $\Delta G = \Delta H + T\,\Delta S$, where H is the enthalpy, T is the absolute temperature, and S is the entropy. The tendency for any chemical or electrochemical corrosion reaction is governed by the Gibbs free-energy change ΔG. The ΔG of a corrosion reaction can be considered in its reduction form for magnesium and aluminum:

$$Al^{3+} + 3e^- \rightarrow Al, \quad \Delta G = 481.374\ \text{kJ}$$

$$Mg^{2+} + 2e^- \rightarrow Mg, \quad \Delta G = 457.348\ \text{kJ}$$

$$(F = 96{,}487\ \text{C}, E^\circ(\text{Al}) = -1.663\ \text{V}, \text{ and } E^\circ(\text{Mg}) = -2.37\ \text{V})$$

The ΔG° (standard free-energy change) of the reduction reaction corresponds to the difference of enthalpy between the oxidized products and the reactants in the standard states. Since the law of physics or nature is that the most stable state is the one with the lowest free energy, the reaction is exothermic when ΔG is negative. The more negative the value of ΔG°, the greater the tendency to corrosion. The positive value of the reduction potential of a gold reaction indicates the stability of this metal in water saturated with atmospheric oxygen [2]. When ΔG° is equal to zero, there is no change in free energy of the reaction in each direction and this represents the equilibrium condition. It is assumed that every system reacts in a manner to offset any driving force and equilibrium is eventually obtained when $\Delta G^\circ = 0$. It should be emphasized that the tendency to corrode is not a measure of reaction rate. A large negative ΔG may or may not be accomplished by a high corrosion rate. If ΔG is negative, the reaction rate may be rapid or slow, depending on various kinetic factors and corrosion mechanisms. For chemical reactions it is convenient to make use of the van't Hoff equation: $\Delta G^\circ = -RT \ln K$, where K is the product of the activities of reactants (in moles) at the state of equilibrium.

The net electrical work performed by a reaction giving a potential E and supplying a quantity of electricity Q equals EQ, but $Q = nF$, where n is the number of equivalents, and so the net electrical work is nFE. However, any work by a cell can be accomplished only at the expense of a decrease in free energy occurring within the cell. The decrease in free energy must equal the electrical work done, and so $\Delta G = -nFE$. This equation of free energy is the bridge between thermodynamics and electrochemistry. Spontaneous reactions necessitate a positive potential and negative ΔG; nonspontaneous reactions require negative potential and positive ΔG. Every cell is composed of two individual single electrodes, such that their algebraic sum is equal to the total electromotive force (emf) of the cell [3]. In view of the electrochemical mechanism of corrosion, the tendency for a metal to corrode can also be expressed in terms of the electromotive force of the corrosion cells that are an integral part of the corrosion process. Since electrical energy is expressed as the product of volts and coulombs (joules, J), the relation between ΔG in joules and emf in volts, E, is defined by $\Delta G = -nFE$, where n is the number of electrons (or chemical equivalents) taking part in the reaction, and F is the Faraday constant (96,500 C/eq). The term ΔG can be converted from calories to joules by making use of the conversion 1 cal = 4.184 absolute joules. Thus the greater the value of E for any cell, the greater is the tendency for the overall reaction of the cell to proceed [2].

1.3. THE ELECTROCHEMICAL INTERFACE

The passage of a current in a metallic conductor means that there is circulation of electrons: that is, the transported electricity is always negative. It is necessary to underline that the free electrons don't penetrate in electrolytes. In an electrolytic solution, for example,

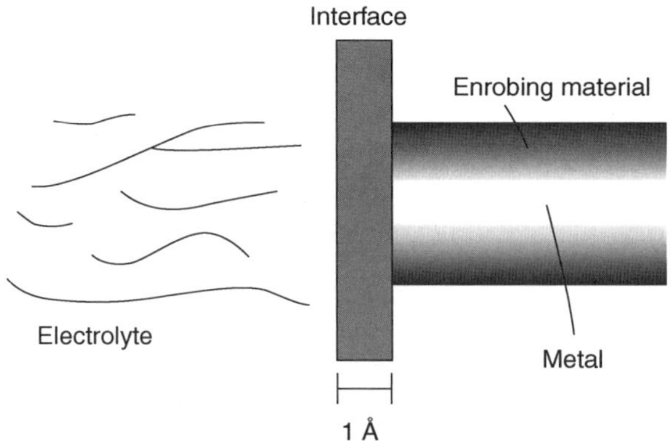

Figure 1.2 Schematic of a round metallic sample (1 cm^2) in contact with an electrolyte or corrosive medium held in an isolated resin and frequently polished and connected electrically for electrochemical studies.

an aqueous solution or a molten salt of sodium chloride, the passage of an electric current is done by the positive ions and the negative ions. At the solid–electrolyte interface, there is a zone of exchange of electrons. The products of electrochemical decomposition appear at the interface of the electrode and not in the mass of the electrolyte (first law of Faraday). This zone of interface is not thick and is on the order of 1 Å. It has the particular physicochemical properties that permit the exchange of electrons (Figure 1.2). This interface can vary rapidly because of its thickness and its dynamic change in chemical composition, properties and concentrations can control the kinetics of the reaction and sometimes can dictate the dominant reaction.

1.3.1. Electric Double Layer

A metal immersed in a solvent can have a tendency to pass into solution and the passage of ions takes place until their concentration at the neighborhood of the metal is large enough to cancel the tendency of metal ions to pass into solution and an equilibrium between the metal and its ions occurs:

$$M^{n+} + ze^- \Leftrightarrow M$$

where z (n) is the electrovalence of the considered ion. Excess electrons in the metal move to make an equal and oppositely charged layer to the one found in the solvent at the metallic interface. The atom or the crystal is electrically neutral, with equal positive and negative charges. In spite of their kinetic energy, electrons cannot move more than 2 nm from the crystal, because of the strength of attraction of the positive charge (Nernst double layer, internal and external Helmholtz planes).

However, unsymmetrical, polar H_2O molecules (H atoms positive, O atoms negative in the molecule) are attracted to the conductive surface, forming an oriented solvent layer, which prevents close approach of a charged species (ions) from the bulk solution. Charged ions also attract their own sheath of polar water-solvent molecules, which further insulate them from the conducting surface. The plane of closest layer of positively charged cations to the negatively charged metal surface is often referred to as the *outer Helmholtz plane*, as indicated in Figure 1.3 [4,5].

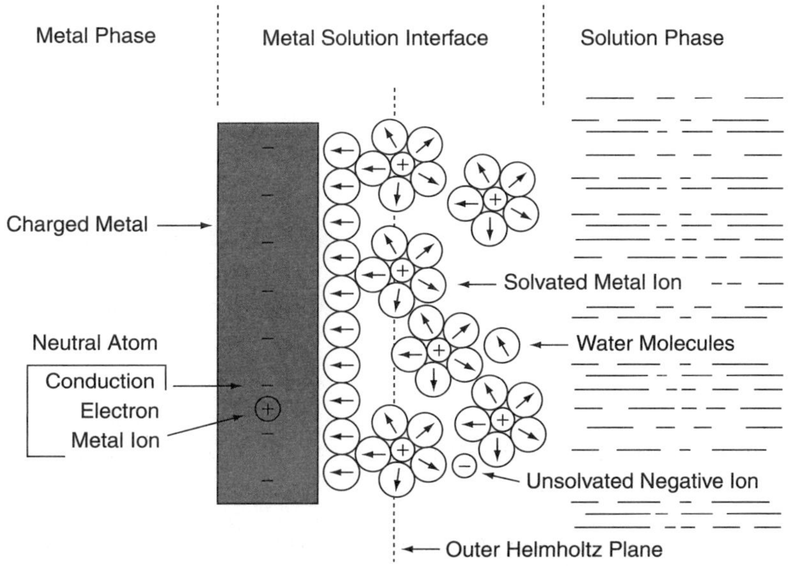

Figure 1.3 Schematic of the complete double layer of the metal–solution interface. (Adapted from Refs. 4 and 5.)

Because of the strength of attraction of solvated ions of opposite sign and the strength of repulsion of those with identical sign, there is a diffusion of cations between anions in the total mass of the solution in a state of equilibrium (diffuse layer). The ions do not occupy a stationary position in the plane of Helmholtz because of the thermal agitation; instead, they are arranged according to a Boltzmann distribution in a zone situated very close to the surface, called the double diffuse layer or Gouy–Chapman layer. The Stern model is a combination of the Helmholtz and Gouy–Chapman models and this represents the complete double layer (Figure 1.4) [6].

1.3.2. Equivalent Circuit of the Electric Double Layer

The electric double layer at the interface behaves as a charged capacitor connected in parallel with a resistance (Figure 1.5a) and this limits electrochemical reactions at the surface. The electric circuit indicates that a continuous current can cross the interface. This current, called the current of charge transfer or Faraday current, provokes an electrochemical reaction at the electrode–electrolyte interface. Some electrodes—called ideally polarizable electrodes—do not contain any reactive species and permit one to vary the potential on a large scale without producing an electrochemical reaction or giving a measurable Faraday current. This is the case for an electrode of mercury immersed in a weak solution of salt, such as NaCl or NaF. Mercury, which is liquid at ambient temperature, is particularly interesting for double layer studies (Figure 1.5b) [6].

1.4. NERNST EQUATION

Let's consider the following general reaction occurring in a cell:

$$l\mathrm{L} + m\mathrm{M} \rightarrow q\mathrm{Q} + r\mathrm{R}$$

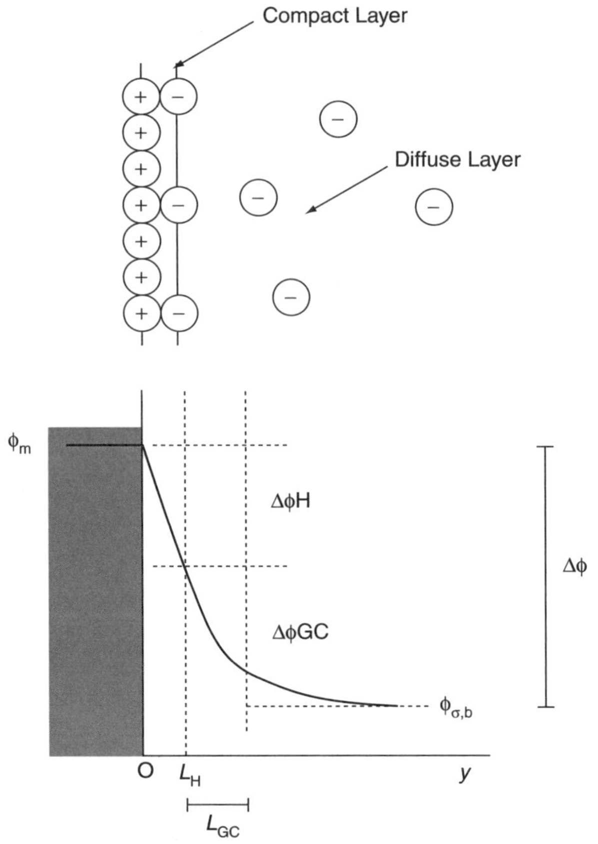

Figure 1.4 Stern model for the double layer [6] showing the potential evolution at the interface.

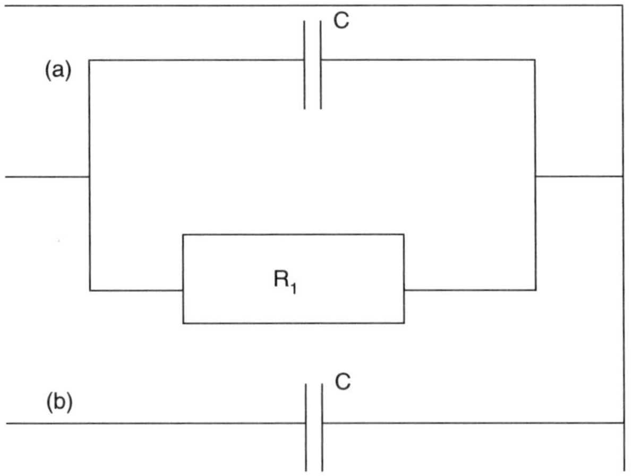

Figure 1.5 Equivalent circuit of the interface for (a) corrodible metal and (b) ideally polarizable metal [6].

where l moles of substance L and m moles of M react to form q moles of Q and r moles of R. The change of the free enthalpy, ΔG, for this reaction is given by the difference of the free energy between products and reactants:

$$\Delta G = (qG_{\mathrm{Q}} + rG_{\mathrm{R}}) - (lG_{\mathrm{L}} + mG_{\mathrm{M}})$$

At the arbitrary state of reference or equilibrium,

$$\Delta G^{\circ} = (qG_{\mathrm{Q}}^{\circ} + rG_{\mathrm{R}}^{\circ}) - (lG_{\mathrm{L}}^{\circ} + mG_{\mathrm{M}}^{\circ})$$

All concentrations or pressures of the different substances should be replaced by their activities considering the coefficient of activity in every case; the difference of the free energy of L in a certain state and its standard state is

$$l(G_{\mathrm{L}} - G_{\mathrm{L}}^{\circ}) = lRT \ln a_{\mathrm{L}} = RT \ln a_{\mathrm{L}}^{1}$$

where R is the perfect gas constant (8.31441 $\mathrm{J \cdot K^{-1} \cdot mol^{-1}}$) and T the absolute temperature in (K = °C + 273.16). By subtraction,

$$\Delta G - \Delta G^{\circ} = RT \ln \frac{a_{\mathrm{Q}}^{q} \times a_{\mathrm{R}}^{r}}{a_{\mathrm{L}}^{l} \times a_{\mathrm{M}}^{m}}$$

At equilibrium, the reaction is stationary and $\Delta G = 0$; so

$$K = \frac{a_{\mathrm{Q}}^{q} \times a_{\mathrm{R}}^{r}}{a_{\mathrm{L}}^{l} \times a_{\mathrm{M}}^{m}}$$

where K is the constant of equilibrium of the reaction. By substitution,

$$\Delta G^{\circ} = -RT \ln K$$

Since $\Delta G = -zFE$, by substitution we have

$$-zEF = -RT \ln K + RT \ln \frac{a_{\mathrm{Q}}^{q} \times a_{\mathrm{R}}^{r}}{a_{\mathrm{L}}^{l} \times a_{\mathrm{M}}^{m}}$$

$$E = \frac{RT}{zF} \ln K - \frac{RT}{zF} \ln \frac{a_{\mathrm{Q}}^{q} \times a_{\mathrm{R}}^{r}}{a_{\mathrm{L}}^{l} \times a_{\mathrm{M}}^{m}}$$

Then the Nernst equation is

$$E = E^{\circ} - \frac{RT}{zF} \ln \frac{a_{\mathrm{Q}}^{q} \times a_{\mathrm{R}}^{r}}{a_{\mathrm{L}}^{l} \times a_{\mathrm{M}}^{m}}$$

where z is the number of electrons considered in the electrochemical equation and F is the Faraday constant.

At 25°C, the coefficient of the logarithmic term is $2.303RT/zF = 0.0592/z$. Then

$$E = E^{\circ} + \frac{0.0592}{z} \log \frac{\text{Reduced form}}{\text{Oxidized form}}$$

where

$$E^{\circ} = -\frac{\Delta G^{\circ}}{zF} = \frac{RT}{zF} \ln K$$

For example, to calculate the potential of reduction of zinc, we have

$$Zn^{2+} + 2e^{-} \rightarrow Zn$$

$$E_{Zn} = E^{\circ}_{Zn^{2+}/Zn} - \frac{RT}{zF} \ln \frac{a_{Zn}}{a_{Zn^{2+}}}$$

$$E = E^{\circ}_{Zn^{2+}/Zn} + \frac{RT}{2F} \ln a_{Zn^{2+}}$$

where $a_{Zn^{2+}}$ represents the activity of zinc and corresponds to the product between the molality (concentration in moles per 1000 g of water) and activity coefficient γ as a function of concentration and temperature. With regard to solid materials or metals, it is a normal practice to assign to them an activity of unity, that is, to consider them in their standard state at all temperatures, under atmospheric pressure.

The equilibrium potential of metals corresponds to the formation of a complete double layer. This involves equilibrium between metal/interface/solution. According to the standard scale of comparison of every metal in a solution containing 1 M activity of its ions, it is clear that some metals do not reach this state of equilibrium, for example, the alkali metals such as sodium in water. The state of balance therefore exists completely on the left, while for noble metals such as gold, the balance exists on the other side.

1.5. STANDARD POTENTIALS OF ELECTRODES

1.5.1. Standard States in Solution

The chemical potential of a species A is $\mu_A = \mu^{\circ}(p, T) + RT \ln a_A$, where the activity $a_A = f_A x_A$ is the product between activity coefficient f_A and the molar fraction of the species A, x_A. Note that μ° is the chemical potential in the standard state when $x_A \rightarrow 1, f_A \rightarrow 1$, so that the standard state is also the reference state (the state where the activity coefficient goes to a defined limit, i.e., unity). For many electrolytes, the symmetrical system described above is inconvenient, partly because salts and solvents are often not completely miscible. It is therefore more common to use systems based on molality or concentration. The former has the advantage that it is independent of temperature. Our equation remains valid for the solute but the activity is given by $a_A = \gamma_A(m_A/m^{\circ})$.

Although it is indispensable to note a difference of the potential between a metal and a solution of its ions, there is no absolute means to measure this potential. The convenient measurements give a difference of potential and require a complete circuit with another metal–solution interface (standard is arbitrarily gold) at the same temperature as the hydrogen electrode.

1.5.2. Hydrogen Electrode

By convention, the half-cell potential of the hydrogen reaction under standard conditions is zero at all temperatures: that is, $E^\circ = E^\circ_{H^+/H_2,\,Pt}$ "reduction" $= E^\circ_{Pt,\,H_2,/H^+}$ "oxidation" $= 0$ and the corresponding reaction,

$$2H^+(aq) + 2e^- \rightarrow H_2(gas),\ E^\circ = 0.000\ V$$

When the proton activity of the hydrogen ion (aq) $a_+ = 1$ or the hydrogen fugacity is not equal to unity, the electrode potential is given by the appropriate form of the Nernst equation:

$$E^\circ = -(RT/2F)\ln f_{H_2} + (RT/F)\ln\ a_{H^+}$$

where f is the fugacity of hydrogen gas and a_{H^+} is the proton activity. The effect of nonideal behavior of hydrogen gas may normally be ignored at ambient pressure but becomes significant at high pressure [7].

Applying the Nernst equation, and considering that the ionization constant of water at 298.15 K according to the equation $H_2O = H^+ + OH^-$ is equal to $K = 1.008 \times 10^{-14}$, the potentials of the hydrogen electrode in pure water (10^{-7} M) and in molal hydroxide solution are −0.414 V and −0.828 V, respectively. When the fugacity (atmospheric pressure) is equal to unity, the Nernst equation at 25 °C becomes

$$E = -0.0592\ pH = 0.4144\ V$$

since E° is equal to zero and $pH = -\log a_{H^+}$. The reversible electrode to hydrogen is essentially composed of one electrode of platinized platinum immersed in a solution containing ions of hydrogen and saturated with the hydrogen gas, under atmospheric pressure. The platinized surface is especially prepared with a coating of very finely divided platinum, giving an important geometric surface that is multiplied by a factor of a few hundred. Platinized platinum reacts as a source of electrons for the discharge of hydrogen ions and, at the same time, adsorbs the formed atomic hydrogen. These molecules of hydrogen at the surface are also transformed subsequently into ions.

1.5.3. Positive and Negative Signs of Potentials

The potential sign of a reaction E° denotes the potential at equilibrium under standard conditions, all relative to the standard hydrogen electrode (SHE). The sign or polarity of the electrode (i.e., M, M^{n+}) is determined basically by the difference between the work required to move unit positive charge from infinity to the metal, M, and the work required for transport to the SHE. The electrode requiring the greater amount of work in moving the unit positive charge from infinity will be at a higher potential and is said to be positive relative to the second electrode, which is called the negative electrode. If the electrodes are connected externally through a conductor, conventional positive current, I, will flow from the positive to the negative electrode, although the actual carriers are electrons flowing in the opposite direction.

The International Union of Pure and Applied Chemistry, in 1953, declared that the reduction potential of one electrode is called the potential (E°). It is compliant with the definition of potential by physicists as the necessary work to bring a unit positive charge to the point where the potential is determined ($Zn^{2+} \rightarrow Zn - 2e^-$). It also has the advantage of corresponding in sign to the polarity of a metal in relation to the electrode of hydrogen. For example, zinc has a negative potential of reduction and it is the negative electrode (anode) in a galvanic cell when we consider the standard electrode of hydrogen as the cathode. Practically, the polarity of the electrode whose potential is being measured relative to the SHE is given by the polarity of the terminal of a high-impedance voltmeter or electrometer that must be attached to the electrode to obtain a positive meter reading. Thus if M spontaneously oxidizes to M^{n+} when coupled to the SHE, the M electrode will be negative relative to the SHE, and $E'_{M,M^{n+}}$ will be negative for the half-cell reaction [4], $M = M^{n+} + ne^-$.

It is important to realize that the standard half-cell potential, E°, or the half-cell potential at other than standard conditions, E', is sign invariant with respect to how the equilibrium reaction is written or considered; for example, $E^\circ_{Fe,Fe^{2+}} = -440$ mV(SHE) for both $Fe = Fe^{2+} + 2e$ and $Fe^{2+} + 2e = Fe$ [4]. When applying the Nernst equation we always consider oxidized species (ions) as reactants and reduced agents as products (metals) independent of the stoichiometric presentation of the electrochemical equation [6]. These measurements are usually made with an electrometer ($>10^{14}$ ohms internal resistance). The potentiometer (used frequently) is a variable potential device that is attached to the cell and adjusted until the current flow is zero. At this condition, the potentiometer is applying a potential to the cell that just equals the cell potential, for example, 440 mV for $Fe = Fe^{2+} + 2e$ with the negative terminal of the potentiometer connected to the Fe electrode, that is, $E^\circ_{Fe,Fe^{2+}} = -440$ mV(SHE) [4].

In applying the Nernst equation for unit activities,

$$E = E^\circ - \frac{RT}{nF} \ln \frac{a_{\text{products}}}{a_{\text{reactants}}}$$

for an electrochemical cell such as $Zn/Zn^{2+}||H^+/H_2$,Pt; following the reaction $Zn + 2H^+ \rightarrow Zn^{2+} + H_2$, the emf of this cell is 0.763 V. On the other hand, for the reaction of the cell Pt,$H_2/H^+||Zn^{2+}/Zn$, following the equation $Zn^{2+} + H_2 \rightarrow Zn + 2H^+$, the emf is -0.763 V (|| indicates a bridge between the two electrodes).

1.5.4. Graphical Presentation

It is common practice to present reactions of reduction and oxidation on the internationally accepted scale of reduction, on the vertical or horizontal scales. These scales (vertical or horizontal) are employed often for curves of polarization in corrosion and can show the dominant or more exothermic reactions or less endothermic in every case. Figure 1.6 shows the basics of a frequently used vertical reduction scale in electrochemical studies.

B. ACTIVITY AND CONDUCTANCE OF THE ELECTROLYTE

1.6. ACTIVITY OF THE ELECTROLYTE

1.6.1. Constant and Degree of Dissociation

Let us consider a weak electrolyte AB dissociated reversibly following the equation

$$AB \Leftrightarrow A^+ + B^-$$

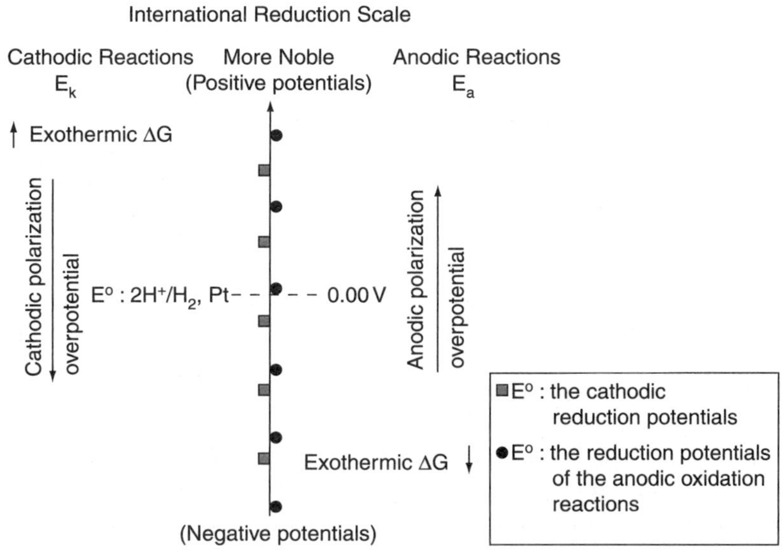

Figure 1.6 Schematic of the potential for reduction and oxidation reactions on the internationally accepted scale of reduction indicating positive and negative potential values.

One has the relation provided by the law of mass action of Guldberg and Waage in considering that the solution is sufficiently diluted:

$$\frac{|A^+||B^-|}{|AB|} = K_{\text{diss}}(T)$$

K_{diss} or K_c (dependent on the temperature) is called the constant of dissociation of the electrolyte. The coefficient or degree of dissociation, α is expressed by

$$\alpha = \frac{|A^+|}{|A^+| + |AB|} = \frac{|B^-|}{|B^-| + |AB|}$$

In the case of a strong electrolyte, the constant of dissociation K_c is therefore infinite and α is equal to 1. K_c is often very weak, on the order of several negative powers of 10; it is advantageous to describe it by

$$pK = -\log K_c$$

1.6.1.1. Variation of α with the Concentration

If we consider that n molecules of AB give αn ions $|A^+|$ and αn ions $|B^-|$, while $n\,(1-\alpha)$ molecules of AB remain nondissociated in solution, one can get the following equation for a determined temperature:

$$\frac{\dfrac{n\alpha}{V} \times \dfrac{n\alpha}{V}}{\dfrac{n(1-\alpha)}{V}} = K_c$$

$$\frac{n\alpha^2}{V(1-\alpha)} = K_c$$

$$\frac{C\alpha^2}{1-\alpha} = K_c$$

This relation between the degree of dissociation of one weak electrolyte and its concentration constitutes the *law of Ostwald dilution*:

$$\alpha \to 0 \quad \text{when } C \to \infty$$
$$\alpha \to 1 \quad \text{when } C \to 0$$

1.6.2. Activity and Concentration

The ionic strength is the electrostatic strength between two ions. The ionic strength between two ions with double charges is four times stronger than the one between two ions with one charge each:

$$\mu = \frac{1}{2}\sum_i m_i z_i^2$$

The ionic strength μ is therefore equal to half of the global sum of the molar concentration of every ion multiplied by the square of the electrovalence (z^2) for every ion (i). The law of mass action, expressed in concentrations, only constitutes a first valid approximation if concentrations remain weak. The mainly electrostatic interactions between the elementary particles of the electrolyte are considered negligible in this situation.

Taking into consideration the interactions between particles, Lewis suggested the term *particle activity* instead of concentration. The law of dissociation can then be expressed as

$$\frac{a_{A^+} a_{B^-}}{a_{AB}} = K(T)\left(\frac{\text{mole}}{\text{dm}^3}\right)$$

a_{A^+} and a_{B^-} are the activities of the ions A^+ and B^- and a_{AB} corresponds to the activity of the nondissociated molecules. The activity of an ion A^+ is expressed as $a = \gamma_\pm C$; $\gamma_\pm$ is generally considered arbitrary, since it is different from γ_+ specially if there is an important difference in mobilities between the cation and the anion. If dilution is sufficiently important (ionic force $\mu < 0.01$), one can apply the Debye–Hückel limiting law:

$$\log \gamma_\pm = -Az_+z_-\mu^{1/2}$$

$$A = \frac{1}{2.303}\frac{1}{2}\frac{e^2}{DkT}\sqrt{\frac{8\pi e^2 N\rho_0}{1000DkT}}$$

where

Dielectric constant of water: $D = 78.54$ (1 for vacuum)

Electron charge: $e = 4.803 \times 10^{-10}$ esu

Avogadro number: $N = 6.02252 \times 10^{23}$ mole^{-1}

Boltzmann constant: $k = 1.38054 \times 10^{-16}$ erg · degree^{-1}

$A = 0.5091$ for water at 25°C ($\rho_0 = 0.997$ in these conditions) [3]

For an electrolyte $A_xB_y \Leftrightarrow xA^{y+} + yB^{x-}$, the average coefficient of activity is

$$\gamma_{\pm} = (\gamma_+^x \gamma_-^y)^{1/(x+y)}$$

For example, the average coefficient of activity of a solution of 1 M of Fe Cl_2 is

$$FeCl_2 \rightarrow Fe^{2+} + 2Cl^-: \; \gamma_{\pm} = (\gamma_+^1 \gamma_-^2)^{1/(1+2)} = (\gamma_+ \gamma_-^2)^{1/3}.$$

Generally, K_c is given when considering concentrations and not activities. However, even for a weak acid (e.g., acetic acid) having $K_c = 1.81 \times 10^{-5}$, where electrostatic attractions between ions are negligible, $K_{\text{thermodynamic}}$ was calculated and found to be slightly different and equal to 1.72×10^{-5} M at 25°C. To calculate $K_{\text{thermodynamic}}$, α can be considered to be 0.04165 from the equation $K_c = C\alpha^2/(1-\alpha)$ and applying the Debye–Hückel limit law to determine the activities of the ions [8].

1.6.3. Theory of More Concentrated Solutions

There is a natural interest to develop a theory that is applicable to solutions that are more concentrated than those for which the Debye–Hückel theory is valid. This is one of the main nonsolved problems of physical chemistry. In the case of a molecule such as HCl in water, there are two fundamental parameters to consider:

1. Speeds with which molecules or the complex split up and reform themselves from ions in solution are very high. The average longevity of a complex or an ion can be on the order of only 10^{-10} second, instead of 1 second as in a gas. In this short lapse of time, few ions become really free and the likeliest phenomenon after separation is nearly an immediate recombination.
2. The dissociation of a molecule to give the hydrated ions requires the separation of ions having opposite charge. The electrostatic attraction between these two ions decreases relatively slowly when they separate, so that they are always associated more or less even though they are separated by a distance of several molecular diameters.

There are then two parameters to add to the Debye–Hückel limiting law to make it apply to more concentrated solutions:

1. The action of the repulsive forces at short distances between charged ions that results from the physical dimensions of the ion, since the charge is distributed on the surface and not concentrated at a simple point. This consideration has the tendency to reduce the electrostatic interactions. In a dilute solution, one can disregard the diameter of an ion in relation to the ionic atmosphere. On the other hand, when the concentration is close to 0.1 mole, the radius of the ionic atmosphere becomes very close to that of an ion, which is on the order of 2×10^{-8} cm. In such solutions, Debye and Hückel stated an improved theory that considers the finite dimensions of the ions. The expression of the activity coefficient is as follows:

$$\log \gamma_{\pm} = -\frac{A_{z_+z_-}\sqrt{\mu}}{1 + Bd\sqrt{\mu}}$$

where B is a constant $= \sqrt{8\pi Ne^2\rho_0/1000DkT}$ and d is the average efficient diameter of the ions (Figure 1.7).

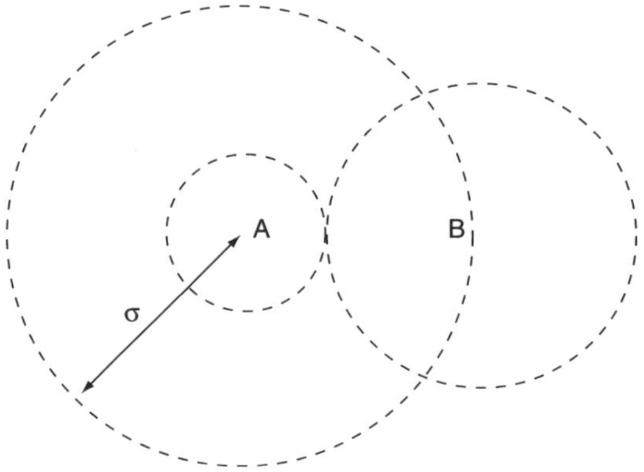

Figure 1.7 Average efficient ionic diameters.

2. The action of ions on molecules of the solvent probably has an even bigger importance. It is admitted that ions are hydrated or solvated. The phenomenon of salting out indicated that the addition of electrolytes reduces the solubility of some nonelectrolytes. For example, the solubility at 25°C of the ethyl oxide in pure water is 0.91 mol · L^{-1} while it is only 0.13 mol · L^{-1} in a solution containing 15% of NaCl. Then, another correction should be added to the Debye–Hückel equation:

$$\log \gamma_{\pm} = -\frac{A_{z_+ z_-}\sqrt{\mu}}{1 + Bd\sqrt{\mu}} + C'\mu$$

C' is called the *salting out constant* and this equation is called the Hückel equation.

Example of Application of Hückel Equation Calculate the average coefficient of activity of 0.005 mol · L^{-1} of zinc chloride at 25°C using (a) the Debye–Hückel limit law and (b) the Hückel equation considering $d = 2.55 \times 10^{-8}$ cm (at 25°C, $B = 0.3286 \times 10^8$) and $A = 0.509$ for H_2O at T = 25°C.

$$\begin{array}{lll} ZnCl_2 & \Rightarrow Zn^{2+} & + 2Cl^- \\ 0.005 & \quad 0.005 & \quad 0.01 \end{array}$$

(a) $\mu = \frac{1}{2}\sum m_i z_i^2 = 0.015$

$\log \gamma_{\pm} = -Az_+ z_- \mu^{1/2}; \; \gamma_{\pm} = 0.7504$

(b) $$\log \gamma_{\pm} = \frac{-0.509 \times 2 \times 1 \times \sqrt{0.015}}{1 + (0.3286 \times 10^8 \times 2.55 \times 10^{-8}\sqrt{0.015})}$$

$\gamma_{\pm} = 0.7708$

It is obvious that even for this dilute solution there is a difference in the activity coefficient in applying the correction of the Debye–Hückel limit law. If the electrolyte is weak enough (e.g., $K < 10^{-2}$), concentrations in free ions are very weak and one can admit that the coefficient of ion activity is always equal to 1. The limit law with and

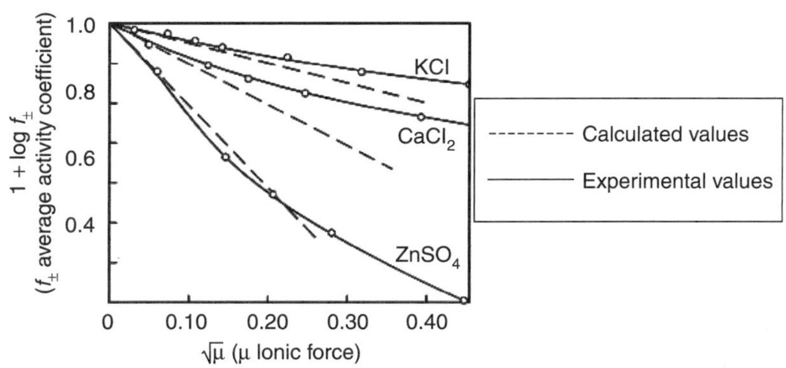

Figure 1.8 Evaluation of Debye–Hückel theory for more concentrated solutions (From Ref. 3).

without the two mentioned corrections for more stronger electrolytes can be applied to achieve the real activity of the ions in solution.

It is important to mention that the theory of ion association developed independently by Bjerrum, Fuoss, and Krauss gives a more lucid physical representation of what can happen in more concentrated solutions. When the concentration of solutions increases, it is likely that definite ion pairs of opposite sign, associated by electrostatic attraction, are formed. This tendency to form ion pairs is much more marked when the dielectric constant of the solvent is lower and the radius of ions is smaller. The degree of association can become very substantial, even in a higher dielectric constant solvent such as water. Bjerrum and co-workers calculated that a normal aqueous solution of one electrolyte to two univalent ions, having a diameter D of 2.82 Å, has 13.8% of associated pairs; for $D = 1.76$ Å there are 28.6% associated ions. For solvents having a lower D than water, the proportion of associated ions will be even more important. This association of ions as pairs must decrease the values of ionic activity coefficients [3]. One can admit then that in the case of concentrated solutions (when the ionic strength is considerably over 0.1), the non-ideal behavior of the solution is not merely electrostatic in nature and it is useless to determine ideal behavior by a simple adjustment to the Debye–Hückel theory [8]. Practical measurements by selective electrodes, chemical cells with and without transport, and concentration cells with and without transport should be used and compared to some electrolytes with standard activities. Figure 1.8 shows the deviation from the theory for some frequently used electrolytes.

1.6.4. Electrolytic Conduction

The term *conductance* expresses the ratio of current to potential difference, while *conductivity* expresses the specific conductance of the electrolyte, k, considering 1 cm^2 of electrode surface and 1 cm^3 of solution between the electrodes [9].

1.7. MOBILITY OF IONS

For simple considerations, the ion is supposed to be spherical, with bigger measurements than that of particles of the solvent. The electric field E is assumed to be constant and the resistance created by the medium obeys Stokes law.

$$\text{Force of resistance}: \; F_r = \sigma\pi\eta r v = Kv$$

where σ is conductivity, η is the coefficient of viscosity, r is the radius of the ion, v is the speed of the ion, and K is a constant for certain conditions.

At the instant where one applies the E field, the ion is going to be displaced by a moving force:

$$F_m = neE$$

where n is the valence of the ion, e is the charge of the electron, and E is the electric field strength.

The force of resistance should be equal to F_m such that the resultant force acting on the moving ion is zero and the speed of the ion is uniform without acceleration:

$$\begin{aligned} F_r &= F_m \\ Kv &= neE \\ v &= \frac{ne}{K}E \end{aligned}$$

The speed limit v settles in a very short time and is considered to be the speed of the ion. v is proportional to the electric field and K is constant. The units for electric field are $V \cdot cm^{-1}$ while $cm \cdot s^{-1}$ are the units for speed. Generally, the mobility u of an ion is defined as the speed of that ion in an electric field E and can be expressed as

$$\begin{aligned} &v = u \; E \\ &v = u \; \text{for} \; E = 1\,V \cdot cm^{-1} \end{aligned}$$

v^- and u^- are used for an anion, and v^+ and u^+ for a cation.

As an example, u values in aqueous solution (0.1 M) are $20 \times 10^{-4}\,cm \cdot s^{-1}$ for OH^-, and $30 \times 10^{-4}\,cm \cdot s^{-1}$ for H^+.

1.7.1. Law of Additivity of Kohlrausch

"The conductance limits of a binary strong electrolyte is the sum of two terms, that of the anion and that of the cation, for a determined solvent and at a certain temperature." The law of additivity is assigned to the ionic conductivity of an equivalent. Applying the law for some electrolytes, we have

$$\begin{aligned} \Lambda_0(KCl) &= \lambda_0(K^+) + \lambda_0(Cl^-) \\ \Lambda_0(CaCl_2) &= \lambda_0(\tfrac{1}{2}Ca^{2+}) + \lambda_0(Cl^-) \\ \Lambda_0(BaSO_4) &= \lambda_0(\tfrac{1}{2}Ba^{2+}) + \lambda_0(\tfrac{1}{2}SO_4{}^{2-}) \end{aligned}$$

In Table 1.1, the values λ_0(cation) and λ_0(anion) of some ions at concentration zero (infinite dilution) in aqueous solution at 25°C are given. These values are valid as long as the degree of dissociation α is equal to one. The equivalent conductance is expressed in $\mho \cdot cm^2 \cdot \text{equivalent}^{-1}$.

From Table 1.1, it is clear that the ions H^+ and OH^- have very high values of λ_0 as compared to that of the average of other ions, which is between 50 and 75. The law of

Table 1.1 Values of λ_0 at 25°C

Cation	λ_0	Anion	λ_0
H^+	349.80	OH^-	198.60
Na^+	50.11	Cl^-	76.34
Ag^+	61.90	Br^-	78.10
$\frac{1}{2}Ba^{2+}$	63.60	$\frac{1}{2}SO_4^{2-}$	80.00
$\frac{1}{2}Ca^{2+}$	56.60	ClO_4^-	67.30

Kohlrausch permits one to calculate Λ_0 or Λ_∞ for the weak electrolytes, such as organic acids, from measured conductivities of their salts, which are strong electrolytes.

$$\Lambda_0(CH_3COOH) = \Lambda_0(CH_3COONa) + \Lambda_0(HCl) - \Lambda_0(NaCl)$$

The equivalent conductance and mobility are

$$\begin{aligned} \Lambda_c &= \alpha Ne(u^+ + u^-) \\ &= \alpha F(u^+ + u^-) \\ &= \alpha(\lambda_+ + \lambda_-) \\ \Lambda_\infty &= (\lambda_+)_\infty + (\lambda_-)_\infty \end{aligned}$$

1.7.2. Ion Transport Number or Index

The contribution of every ion to the transport of the current is a function of its mobility and concentration. The transport number is defined as the fraction of total current, which is carried by an ion.

$$t_+ = \frac{u_+}{u_+ + u_-} \qquad t_- = \frac{u_-}{u_+ + u_-}$$

$$t_+ = \frac{\lambda_+}{\lambda_+ + \lambda_-} \qquad t_- = \frac{\lambda_-}{\lambda_+ + \lambda_-}$$

The transport numbers of positive and negative charge in a cell are evidently between 0 and 1 for the transport of 1 faraday in the cell. The sum of the transport number of positive and negative charges that are carried by the electrolyte should be equal to 1 in order to preserve the neutrality of the electrolyte. This does not mean that half of the current is carried by the positive ions or by the negative ions. To determine the transport number, Hittorf developed a method that is based on the measurement of concentration changes provoked in the neighborhood of electrodes by the passage of a current through the electrolyte, that is, in the anolyte and catholyte solutions [10].

The Hittorf method for the determination of the transport number considers an electrolytic cell divided into three compartments as in Figure 1.9 [3]. The disposition of ions before the passage of a current is represented schematically in Figure 1.9a, where the + or − sign means an equivalent of the corresponding ion. Now, suppose that the mobility of the positive ion is triple that of the negative ion, $U_+ = 3U_-$ and that 4 faradays of electricity is carried across the cell. This means that four equivalent negative ions should be carried at the surface of the anode and four equivalent positive ions at the cathode (Figure 1.9a). Three faradays of electricity is transported by the positive ions across the

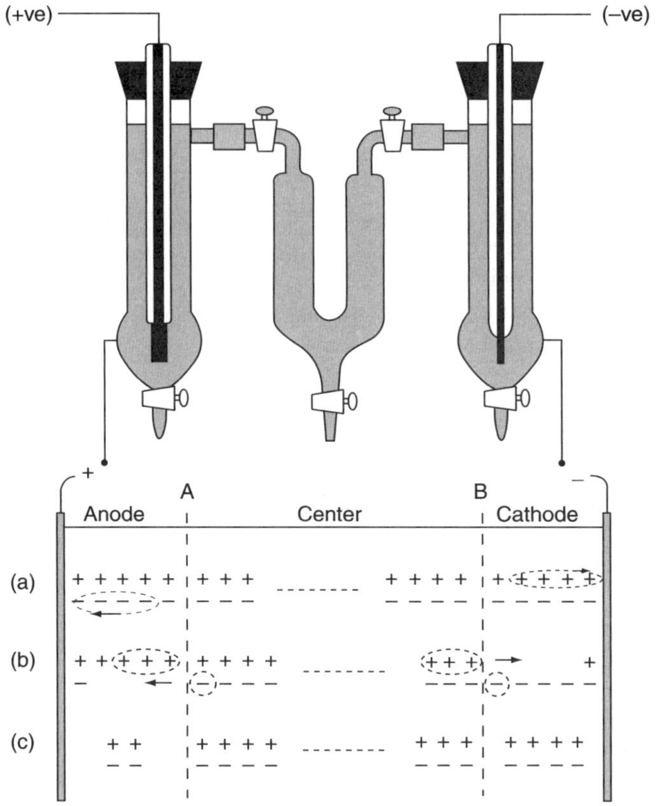

Figure 1.9 Determination of the transport number of an ion [3].

plane P on the right while only 1 faraday is transported on the left by the negative ions (Figure 1.9b). There is a variation of the number of equivalents in the neighborhood of the anode ($\Delta n_a = 5 - 2 = 3$) and in the neighborhood of the cathode ($\Delta n_c = 5 - 4 = 1$) (Figure 1.9c) [10]. The change in concentration in the anodic and cathodic compartments in equivalents can be expressed as follows:

$$\Delta n_a = \frac{q}{F} - t_- \times \frac{q}{F} = 4 - 0.25 \times 4 = 3$$
$$\Delta n_c = 4 - 0.75 \times 4 = 1$$

The report of these concentration variations is inevitably equal to the report of ions mobilities [10]:

$$\frac{\Delta n_a}{\Delta n_c} = \frac{u_+}{u_-} = 3 \quad t_+ = 0.75 \quad t_- = 0.25$$

Then the variation of the concentration at the anodic or cathodic compartments is a function of the mobility of the ions in solution:

$$\Delta n_c = \frac{q}{F} - t_+ \frac{q}{F} = \frac{q}{F}(1 - t_+) = \frac{q}{F} t_-$$

Thus

$$t_- = \frac{\Delta n_c}{q/F} \quad \text{and} \quad t_+ = \frac{\Delta n_a}{q/F}$$

In the last example, the electrodes are inert, while in other cases, electrodes can dissolve in the solution giving positive ions. For example, for a silver anode in a solution of silver nitrate, when an electric current crosses the cell, the number of electrolyte equivalents increases in the neighborhood of the anode to equal the number of incoming silver equivalents in solution in the anodic compartment, less the number of silver equivalents moving to the cathodic compartment.

1.8. CONDUCTANCE

The conductance of one electrolyte is a function of the nature and the number of present ions. The resistance R of a conductor is directly proportional to the length and inversely proportional to the area:

$$R = \rho \frac{l}{A}$$

A conductor's resistance for a section 1 cm^2 and a length of 1 cm is $R = \rho$. The constant ρ is characteristic of every conductor and is called the *specific resistance* or *resistivity.* ρ is measured in units of $\Omega \cdot cm$. The conductance is the inverse of the resistance and is frequently used. For such a system, that means the specific conductance or conductivity is

$$k = \frac{1}{\rho} = \frac{l}{RA} \Leftrightarrow \rho = \frac{RA}{l}$$

k is expressed in $\Omega^{-1} \cdot cm^{-1}$ (mho/cm), and the values of k and ρ are dependent on the concentration of the electrolyte. The electrolytic conductivity is therefore better compared if one considers it in relation to a unit of concentration for all electrolytes, as in $mol \cdot L^{-1}$ or in equivalents $\cdot L^{-1}$. The equivalent conductance is therefore

$$\text{Equivalent conductance} = \frac{\text{Conductivity for a certain concentration}}{\text{Equivalent per cm}^3 \text{ of electrolyte}}$$

This is the most frequently used unit to characterize the conductance of electrolytes.

$$\Lambda_c = \frac{1\,000k}{zC} = \frac{1\,000\,L}{RA\,zC} = \frac{\Omega^{-1} \cdot cm^3 \cdot cm}{cm^2 \cdot \text{equivalent}} = \Omega^{-1} \cdot cm^2 \cdot \text{equivalent}^{-1}$$

The equivalent conductance Λ_c varies to a certain extent with the temperature and the concentration. At constant temperature, the value Λ_c increases when the concentration decreases and in several cases it is possible to define the value of Λ_c for a concentration of zero (Λ_0) or for infinite dilution (Λ_∞).

According to Kohlrausch, this value is determined better if one draws the equivalent conductance against the root of the normality of the electrolyte. It is clear that the determination of Λ_0 is not possible for weak electrolytes by this method.

Arrhenius explained that the growth of the equivalent conductance according to the decrease of the concentration is due to the increase of the degree of dissociation of the electrolyte in ions. One can suppose therefore that, at concentration zero, the dissociation of the electrolyte is complete. The degree of dissociation (α) at C concentration could be determined from the relation Λ_c/Λ_0. This point of view is correct for weak electrolytes but not acceptable for strong electrolytes, especially when concentrations are high. One can apply the Debye–Hückel theory therefore for weak concentrations of strong electrolytes by taking account of the electrostatic attractions to determine the coefficient of activity rather than the coefficient of dissociation. It is interesting to note that generally the equivalent conductance increases 2% by 1°C for temperatures below 40°C, and 3% by 1°C for temperatures above 40°C.

1.9. POTENTIAL OF DECOMPOSITION

To generalize Ohm's law, the $V = E + RI$ relation is introduced, where E is the minimal potential of decomposition necessary for visible electrolysis. In the case of electrolytes, the current is transported by ions and the passage of the current could heat the electrolyte and vary the resistance. In addition, the electrolysis, accompanied by the passage of the current, could result in an impoverishment of ions at the interface or the formation of a nonconductor (passive) film such as aluminum oxide on an aluminum anode.

C. THE DIFFERENT TYPES OF ELECTRODES

1.10. GAS ELECTRODES

A gas electrode consists of bubbling a gas about an inert or noble metal in the form of a wire or a foil immersed in a solution that contains ions for which the gas is reversible. Among gas electrodes are the hydrogen electrode reversible to hydrogen ions, the chlorine electrode reversible to chlorine ions, and the oxygen electrode whose emf depends on the activity of hydroxyl ions. Although hydrogen and chlorine electrodes can be made reversible, no material has been found to establish the state of equilibrium between the oxygen and hydroxyl ions. The precise information on the potential of this electrode is then deduced from free-energy data and not from direct emf measurements.

In the case of the hydrogen electrode ($H^+ + e^- \rightarrow \frac{1}{2}H_2$), the function of the metal, normally platinized platinum foil or wire, is to facilitate the state of equilibrium between hydrogen gas and its ions and to serve as the electric contact. The activity of hydrogen ions in the solution and the pressure of hydrogen around the electrode determine the potential of the hydrogen electrode:

$$E_{H^+/H_2,Pt} = E^{\circ}_{H^+/H_2,Pt} - \frac{RT}{F}\ln\frac{\sqrt{P_{H_2}}}{a_{H^+}}$$

However, $E^{\circ}_{H^+/H_2,Pt} = 0$ when the atmospheric pressure and the activity of ions are equal to 1:

$$
\begin{aligned}
E_{H^+/H_2,Pt} &= -\frac{RT}{F}\ln\frac{\sqrt{P_{H_2}}}{a_{H^+}} \\
&= \frac{RT}{F}\ln a_{H^+} - \frac{RT}{2F}\ln P_{H_2} \\
&= \frac{RT}{F}\ln a_{H^+} = -0.0592\,\text{pH at } 25^\circ\text{C and atmospheric pressure}
\end{aligned}
$$

The chlorine electrode shows the same behavior. The reaction of reduction is

$$\tfrac{1}{2}Cl_2 + e^- \rightarrow Cl^- \quad \text{and} \quad E_{Cl_2/Cl^-} = E^\circ_{Cl_2/Cl^-} - \frac{RT}{F}\ln\frac{a_{Cl^-}}{\sqrt{P_{Cl_2}}}$$

The standard $E^\circ_{Cl_2/Cl^-}$ is 1.3595 at 25°C and atmospheric pressure, and -1.3595 V corresponds to the potential of the oxidation reaction.

1.11. METAL–METAL ION ELECTRODES

These electrodes are considered reversible to their ions, meaning that the potential of each electrode is a function of the activities of its own ions in solution.

$$M^{n+} + ne^- \rightarrow M$$

The $Cu/CuSO_4$ electrode is a robust and economic electrode and often used for field measurements of potentials as in soil for cathodic protection; however, it is less precise than the calomel and silver electrodes. This electrode has a potential of 0.316 V due to the weak activity of copper ions in saturated solution since $E^\circ = 0.337$ V for $Cu^{2+} + 2e \rightarrow Cu$. Solid crystals of copper sulfate are added to keep the solution saturated and the potential steady with a temperature coefficient of $0.7\,\text{mV}\cdot{}^\circ\text{C}^{-1}$.

1.11.1. Alloyed Electrodes

Alloying a noble metal with a more active one can give a new solid solution with an intermediate potential more noble and stable than that of the active one. The amalgam electrodes are a good example. Electrodes of amalgams of metals, more active than mercury, behave essentially like the pure metal with a lower activity because of the dilution with mercury and become more reversible. Currently, for environmental considerations, some laboratories decrease the use of mercury appreciably.

The standard Weston cell has an amalgam electrode Cd(Hg) and a reference electrode. The reaction is $Cd^{2+} + 2e^- + Hg \rightarrow Cd\,(Hg)$ and the potential of this electrode is

$$E_a = E^\circ_{Cd^{2+}/Cd(Hg)} - \frac{RT}{2F}\ln a_{Cd(Hg)} + \frac{RT}{2F}\ln a_{Cd^{2+}}$$

Where E°_a is the standard potential for a given amalgam with a certain composition and equals

$$E^\circ_a = E^\circ_{Cd^{2+}/Cd(Hg)} - \frac{RT}{2F}\ln a_{Cd(Hg)}$$

The E_a° value can be determined by measuring the emf of a cell composed of the amalgam and the pure metal electrodes, immersed in the same solution containing Cd^{2+} ions.

1.12. METAL–INSOLUBLE SALT OR OXIDE ELECTRODES

1.12.1. Metal–Insoluble Salt Electrodes

These are frequently called reference electrodes and can be used for ambient and high-temperature conditions. Reference electrodes such as calomel electrodes, silver–silver chloride electrodes, lead–lead sulfate electrodes, and silver–silver bromide electrodes are examples of metal–insoluble salt electrodes. This kind of electrode is composed of metal in one of its almost insoluble salts and a solution containing the ion present in the salt. The reaction of these electrodes is composed of a chemical dissociation reaction combined with an electrochemical one. The two most used reference electrodes are silver and calomel electrodes in chloride media. The silver electrode should be checked periodically because of a gradual change on aging. Sulfate ions can replace chlorides for both electrodes for metal or system interfaces sensitive to chloride ions.

1.12.1.1. Silver–Silver Chloride Electrodes

The reaction of reduction (Cl^-/AgCl, Ag) is

$$\begin{array}{c} \mathrm{AgCl} \Leftrightarrow \mathrm{Ag^+} + \mathrm{Cl^-} \\ \mathrm{Ag^+} + \mathrm{e^-} \Leftrightarrow \mathrm{Ag} \\ \hline \mathrm{AgCl} + \mathrm{e^-} \Leftrightarrow \mathrm{Ag} + \mathrm{Cl^-} \end{array}$$

The potential of the electrode corresponding to the reduction reaction (Cl^-/AgCl, Ag) or AgCl/Ag is

$$E = E^\circ - \frac{RT}{F} \ln a_{\mathrm{Cl^-}}$$

The inverse reaction is that of oxidation corresponding to $\mathrm{Ag} + \mathrm{Cl^-} \Leftrightarrow \mathrm{AgCl} + \mathrm{e^-}$ and has the negative sign of potential. The value of the standard reduction potential $E^\circ = 0.223$ V from numerous determinations, and for a solution of 0.1 N KCl it is 0.288 V. The temperature coefficient for this concentration is -4.3×10^{-4} V/°C. Knowing that the K_c of AgCl is equal to 1.8×10^{-10}, it is easy to calculate the value of the standard potential E° of the silver–silver chloride electrode in a 1 M KCl solution from the Nernst equation directly:

$$\begin{aligned} E &= E^\circ + 0.0592 \log a_{\mathrm{Ag^+}} \text{ (reaction of reduction)} \\ &= 0.7996 + 0.0592 \log(1.8 \times 10^{-10}), \text{ where } a_{\mathrm{Ag^+}} = \frac{K}{a_{\mathrm{Cl^-}}} \end{aligned}$$

Or we can use the method that considers the sum of the two free energies of the chemical and electrochemical reactions, respectively, at 25°C as follows:

$$\begin{aligned} \Delta G_1 &= -nRTK = -0.0592\,F \log(1.8 \times 10^{-10}) \\ \Delta G_2 &= -nE^\circ F = -0.7996\,F \\ E^\circ &= -\frac{\Delta G}{nF} = 0.223\ \mathrm{V} \end{aligned}$$

Table 1.2 Calomel Electrode Potential at Different Concentrations of KCl

Concentration of KCl	$Hg_2Cl_2/2Hg$, E (volts)	Temperature coefficient (V/°C)
0.1 N	+0.3337	-0.88×10^{-4}
1.0 N	+0.2800	-2.75×10^{-4}
Saturated	+0.2415	-6.60×10^{-4}

The external pressure-balanced reference electrode has been used in high-temperature and high-pressure electrochemistry studies, however, the Ag/AgCl as an internal reference electrode is one of the most accurate and serviceable and has been used for pH measurements for supercritical temperatures. Under every condition, the potential of the Ag/AgCl electrode can be calculated or calibrated against a hydrogen cell that has to be installed separately inside the autoclave [11].

1.12.1.2. Calomel Electrodes

The reaction of reduction (KCl/Hg_2Cl_2, Hg) is

$$\begin{aligned} Hg_2Cl_2 &\Leftrightarrow Hg_2^{2+} + 2Cl^- \\ Hg_2^{2+} + 2e^- &\Leftrightarrow 2Hg \\ \hline Hg_2Cl_2 + 2e^- &\Leftrightarrow 2Hg + 2Cl^- \\ E^\circ_{Hg_2Cl_2/Cl^-} &= 0.268\ \text{V} \end{aligned}$$

Three concentrations of potassium chloride are frequently used, as shown in Table 1.2 [2].

The more convenient solution to prepare is the saturated one, but its response to temperature is somewhat more sluggish [2]. For some corrosion studies, it is preferable to use reference electrodes based on sulfate ions (K_2SO_4/Hg_2SO_4, Hg) instead of chloride ions, which may cause or accelerate localized or pitting corrosion. If K_2SO_4 is saturated instead of potassium chloride, the potential of this electrode is 0.64 V (the reduction potential).

When the measurement is done, for example, against a saturated calomel electrode at 25°C, there are graphical and/or equation methods that can be used to calculate the measured potential with respect to the standard hydrogen electrode scale as explained (Figure 1.10) for a relatively more or less active potential than that of the reference calomel electrode.

1.12.2. Metal–Insoluble Oxide Electrodes

These are similar to the metal–insoluble salt electrodes. Some of them are also used as reference electrodes at high temperatures. The mercury–mercuric oxide electrode Pt, H_2 |KOH(aq)|HgO, Hg has a potential of 0.9256 V and is widely applied as a reference electrode in alkaline medium [11]. The reaction of the electrode is $HgO + H_2O + 2e^- \Leftrightarrow Hg + 2OH^-$ (International Union of Pure and Applied Chemistry (IUPAC) 1985). It is evident that for given water activity and pH, the equilibrium potential depends on the standard potential of the internal reference couple as Hg/HgO following the equation

$$HgO + 2H^+ \Leftrightarrow Hg + H_2O$$

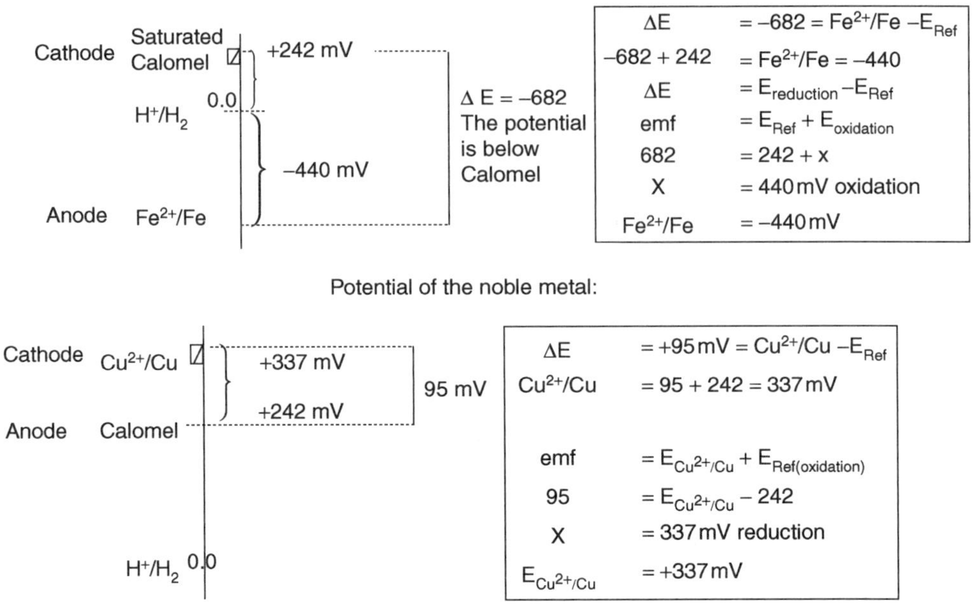

Figure 1.10 Deduction of the electrode potential values to the standard hydrogen scale instead of other reference electrode measurements.

Another example is the antimony–antimony trioxide electrode, $Sb,Sb_2O_3|OH^-$. The antimony rod covered with a thin layer of oxide is dipped in solution containing OH^- ions that is reversible to hydroxide ions according to the equation

$$Sb + 3OH^- \rightarrow \tfrac{1}{2} Sb_2O_3 + \tfrac{3}{2} H_2O + 3e^-$$

Since OH^- and H^+ ions can establish a rapid equilibrium, this electrode is also reversible to H^+ ions. The potential of this electrode for the cubic oxide is 0.152 V. It can be useful in nonaqueous solutions [8,11].

The silver–silver oxide electrode has been investigated for high-temperature measurements and the emf of the cell $Ag,Ag_2O|$ $KOH(aq)|HgO$, Hg was found to be 0.2440 ± 0.005 V at 25°C. The temperature coefficients were obtained between 0°C and 90°C, the value being $-0.000198 \pm 0.000003\ V \cdot deg^{-1}$. This leads to $E = 1.1700$ V at 25°C [11].

1.13. ELECTRODES OF OXIDATION–REDUCTION

This designates a potential created on the surface of an electrode by two forms of ions of a substance in two stages of oxidation [3]. When a platinum wire or an inert metal is immersed in a solution containing, for example ferrous–ferric, cerous–ceric, stannous–stannic, or manganous–permanganate ions, the wire picks up a specific potential for every couple. This corresponds to the tendency of a free-energy decrease of ions in one state to another more stable state. The general reaction is $A^{n_1}(a_1) + ne^- \rightarrow A^{n_2}(a_2)$, where a_1 is the valence in the superior state of oxidation, a_2 is the valence in the inferior state and $n = n_1 - n_2$.

For example, for $Fe^{3+} \rightarrow Fe^{2+}$ or the Fe^{2+}, Fe^{3+}/Pt electrode,

$$Fe^{3+} + e^- \rightarrow Fe^{2+},\ E^\circ = 0.771\ \text{V}$$
$$E = E^\circ - \frac{RT}{nF} \ln \frac{a_{Fe^{2+}}}{a_{Fe^{3+}}}$$

This is an exothermic reaction that can create serious corrosion problems such as when atmospheric oxygen facilitates the cathodic reaction of steel corrosion; this can also initiate microbiological corrosion or can be used as a corrosion inhibitor to deplete oxygen. The oxidation reaction of ferrous to ferric has a potential of −0.771 V and can be designated as $Fe^{2+} \rightarrow Fe^{3+}$ or the Fe^{2+}, Fe^{3+}/Pt electrode.

Another example is the Ce^{+3}, Ce^{+4}/Pt electrode:

$$Ce^{+4} + e^- \rightarrow Ce^{+3},\ E^\circ = 1.61\ \text{V}$$

1.14. SELECTIVE ION ELECTRODES

Some selective electrodes for certain ions are very useful in the field of corrosion science and technology, for example, selective electrodes for hydrogen ions, chloride ions and oxygen.

1.14.1. Glass Electrodes

Some glasses in fine membranes separating two different solutions can achieve a potential difference between the faces in contact with the two different solutions, which depends on the pH difference between the two solutions. How a tension on the glass electrode is established is not known for certain. However, if the surface of glass is in contact with a solution containing some H^+ ions, an exchange occurs between the alkali ions of the glass and the H^+ ions of the solution. This exchange is so weak that it does not alter the composition of the glass. At the interface, there is then no exchange of electrons, but only of ions. Since H^+ ions are more mobile than sodium ions, a double layer is built up on the glass membrane interface in contact with every solution. It is possible to measure the potential of the membrane electrode with the help of an external electrode with known potential. If one of these solutions is standard with known hydrogen activity or pH, the other pH $(-\log a_{H^+})$ can be deduced from measurement of the emf of the cell (Figures 1.11 and 1.12) [12].

It is possible to have electrodes of different shapes. Some permit one to measure the pH of just one drop of solution, while most require 5 mL in general to cover the appreciable part of the electrode completely. The deduced potential of the membrane electrode is

$$\begin{aligned} E &= \frac{RT}{F} \ln \frac{a_{H^+ \text{Soln}C_1}}{a_{H^+ \text{Soln}C_2}} \\ &= 2.3\frac{RT}{F}(\log a_{H^+ C_1} - \log a_{H^+ C_2}) \\ &= 2.3\frac{RT}{F}(\text{pH}_2 - \text{pH}_1) \end{aligned}$$

$$E = \frac{RT}{F} \ln \frac{a_x \cdot \text{Solu } C_1}{a_x \cdot \text{Solu } C_2}$$
$$= 2.3 \frac{RT}{F} (\log a_x \cdot C_1 - \log a_x \cdot C_2)$$
$$= 2.3 \frac{RT}{F} (\text{pH}_2 - \text{pH}_1)$$

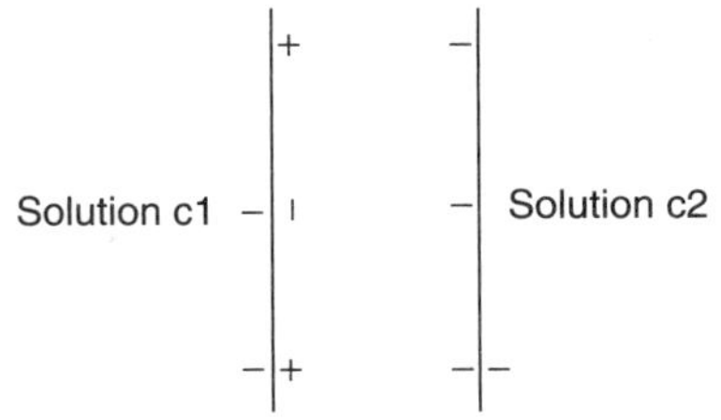

Figure 1.11 Membrane of glass in contact with two different pH solutions.

1.14.2. Copper Ion-Selective Electrodes

Solid-state copper ion-selective electrodes are usually equipped with membranes containing a divalent copper ions with Nernstian slope of 29.6 mV/decade at 25°C in wide concentration ranges down to 10^{-8} M. This can result in good selectivity, short response time, and long lifetime, allowing a variety of successful analytical applications [13]. This extremely sensitive selective electrode can be influenced by the presence of chloride ions in concentrations higher than ~0.1 M [14]. Such electrodes are useful in certain corrosion studies where conventional measurements in corrosion and prevention investigations are difficult to conduct [15–17].

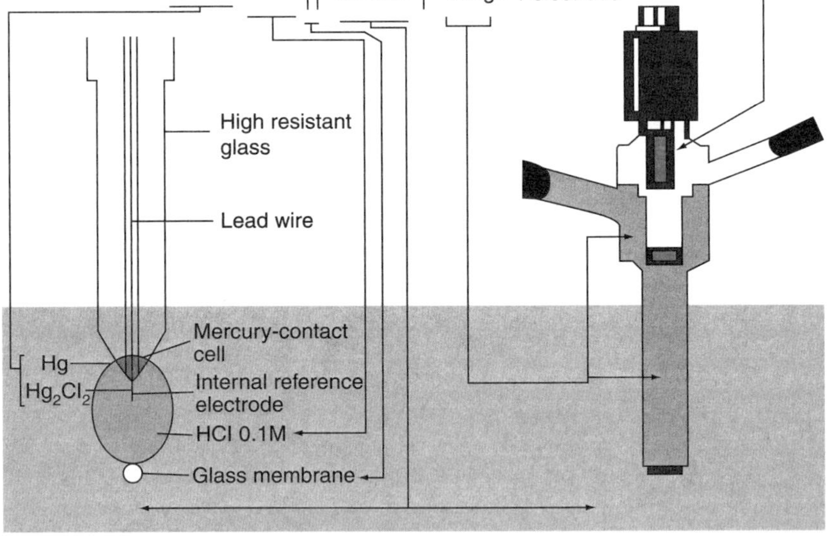

Figure 1.12 Glass electrode and its reference electrode (Adapted from Ref. 12).

D. ELECTROCHEMICAL AND CORROSION CELLS

The electrochemical cells can be divided into chemical and concentration cells. Corrosion cells are treated as electrochemical cells, unless specific features of corrosion mechanisms, such as solvent corrosion cells, are present, in that case they are treated separately. The chemical cell results because of a difference in potential between different electrodes, while the concentration (activity) cell is formed because of the difference in activity of species in the electrodes or in the electrolytes.

1.15. CHEMICAL CELLS

These can be divided into chemical cells with transport and chemical cells without transport.

1.15.1. Chemical Cell with Transport

A current cell of corrosion is composed frequently of two different electrodes with electric contact immersed in a solution containing their dissociated ions such as zinc and steel (iron). This corresponds to a frequently engineered cell in corrosion protection that consists of a zinc coating on a steel rod. The zinc is the sacrificial electrode (anode) and the steel rod (iron) acts as a cathode. In this cell, zinc is corroding, giving zinc ions and electrons. On the surface of the steel, the electrons should be gained to complete the cell, whether through the reduction of zinc ions, which is a highly endothermic reaction, or through the reduction of hydrogen ions (still slightly endothermic reaction, ΔG is positive), or through the reduction of hydrogen ions accelerated (depolarized) by oxygen in atmospheric conditions (exothermic reaction at pH 3–5 depending on the locality and acid rain pH) (Figure 1.13).

A single vertical bar represents a phase boundary between solid, gas, or aqueous solution, while a dashed vertical bar represents the presence of a junction between miscible liquids, and the cell is written as follows:

$$\mathrm{Zn(s)|Zn^{2+}(aq)} \vdots \mathrm{Cd^{2+}(aq)|Cd(s)}$$

The emf of the cell is $E = E_1 + E_2 + \cdots + E_j$. The potential of the junction, E_j, depends on the mobility and concentration of cations to the cathode and anions to the anode.

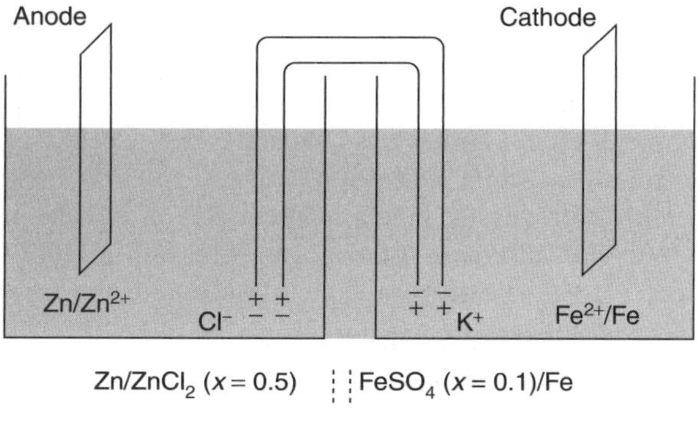

Figure 1.13 Salt bridge in a chemical cell with transport.

Normally, one adds a salt bridge as a saturated solution or 1 N of KCl between the two solutions; the salt bridge is supposed to reduce the junction potential to a minimum. The mobilities of the K^+ and Cl^- ions are nearly equal, and thus there will be two junction potentials with opposite signs at the two ends of the KCl bridge that nearly cancel each other. By convention, a double dashed vertical line represents the presence of a salt bridge at the junction of miscible liquids:

$$Zn(s)|Zn^{2+}(aq) \vdots\vdots Fe^{2+}(aq)|Fe(s)$$

The emf of the cell is

$$E = E^\circ - \frac{RT}{2F}\ln\frac{a_{Zn^{2+}}}{a_{Cd^{2+}}}$$

The certainty of the calculated emf of this cell with respect to the measured one depends on the coefficient of activity of the cations γ_+ (not $\gamma_\pm$ of both cations) and on $E_j \approx 0$.

Many measurements have also been made in cells, which have an explicit liquid junction. For example,

$$Cu|Hg(1)|Hg_2Cl_2(s)|HCl(sat) \vdots CuSO_4(aq)|Cu$$

The Nernst equation for such a cell may be written

$$E = E^\circ - (RT/2F)\ln \geq (1/a_{Cu^{2+}}a^2_{Cl^{-1}}) + E_{LJ}$$

where it must be noted that the activities of Cu^{2+} and Cl^- are in different solutions. Measurements in a cell of this type are usually done by varying the electrolyte concentration in the right-half of the cell while keeping that in the left-half constant. In such measurements $a_{Cu^{2+}}$ and E_{LJ} (the liquid junction potential) vary. Neither of these quantities is independently variable, but when one side of the junction consists of saturated KCl and the other an electrolyte of substantially lower concentration, the theory of liquid junctions suggests that E_{LJ} is small and reasonably constant [7].

The *Daniell battery* is an example of a battery with a liquid junction. This battery consists of a copper electrode in a copper sulfate solution, and zinc electrode in a zinc sulfate solution, containing a liquid junction.

$$Zn/ZnSO_4, H_2O \vdots CuSO_4, H_2O/Cu$$

A porous system, such as fritted glass, separates the two electrolytes. It prevents the mixture of the two solutions, while letting ions migrate from one solution to the other. For most electrochemical batteries, the junction potential is only on the order of a few millivolts. This increases with the difference of concentration and mobility between ions of the two solutions [6].

1.15.1.1. Chemical Corrosion Cells on the Same Metal Surface

In corrosion, the chemical electrode cell between two different electrodes is easy to identify and control, but for electrodes of the same metal it is more difficult to assess. The potential

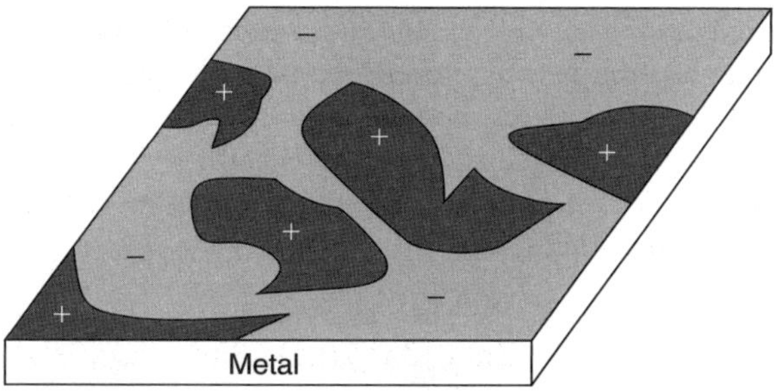

Figure 1.14 Metal surface enlarged, showing schematic arrangement of anodic and cathodic sites of a chemical cell [2].

difference on the same electrode exists due to geometrical, mechanical, and microstructural properties or different phases. Also, any contamination with different conducting particles can act as a different electrode. A deformed metal or cold-worked metal next to the same metal, grains in contact with joint grains, and a monocrystal with an orientation in contact with another crystal with a different orientation are examples of mixed electrodes. A mixed electrode is a metallic surface that possesses some local cells having anodic and cathodic sites (Figure 1.14). The corrosion potential E_{corr} is the open circuit potential, where anodic and cathodic currents are equal: $i_a = -i_c = i_{corr}$. The reaction of the cell can be the dissolution of the metal or alloy and the cathodic reaction can be any other reduction reaction or even an oxidoreduction reaction.

1.15.2. Chemical Cell Without Transport

In order to construct a chemical cell without transport, two electrodes and one electrolyte are chosen so that one electrode is reversible for the cation and the other is reversible for the anion. This is typical of a reference electrode such as

$$\mathrm{Cu(s)|Pt(s)|H_2(g)|HCl(aq)|AgCl(s)|Ag(s)|Cu(s)}$$

The total reaction of the cell is

$$\tfrac{1}{2}\mathrm{H_2}(\mathrm{P_{H_2}}) + \mathrm{AgCl(s)} \leftrightarrow \mathrm{Ag(s)} + \mathrm{H^+}(a_{\mathrm{H^+}}) + \mathrm{Cl^-}$$

It is possible to write the emf of this cell according to the Nernst equation by considering the potentials ($E°$) of the two electrodes and the chemical activities of all reagents:

$$E_{\text{anode}} + E_{\text{cathode}} = E° - \frac{RT}{F} \ln \frac{a_{\mathrm{H^+}} a_{\mathrm{Cl^-}}}{\sqrt{P_{\mathrm{H_2}}}}$$

This cell is useful to determine experimentally the coefficient of activity of the electrolyte because of its precise emf.

1.15.2.1. The Weston Standard Cell

This is a good example of a chemical cell without transport (Figure 1.15). This battery has two stable electrodes: the amalgam electrode and the reference electrode. Its symbol and

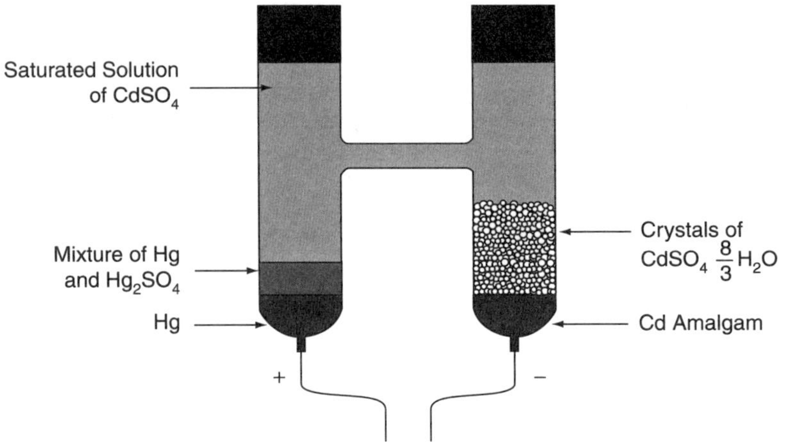

Figure 1.15 Weston standard cell.

reaction are

$$Cd(Hg)|CdSO_4 \cdot \tfrac{8}{3}H_2O|CdSO_4(\text{saturated solution})|Hg_2SO_4,\ Hg,\ 12.5\% \text{ of Cd in Hg}$$

$$\begin{aligned}
&\text{Anodic reaction}: && Cd(Hg) \rightarrow Cd^{2+} + 2e^- + Hg\\
&\text{Cathodic reaction}: && Hg_2SO_4 + 2e^- \rightarrow SO_4^{2-} + 2Hg\\
& && Cd(s) + Hg_2SO_4(s) + \tfrac{8}{3}H_2O \rightarrow CdSO_4 \cdot \tfrac{8}{3}H_2O(s) + 2Hg
\end{aligned}$$

The emf of the cell is

$$E = 1.01485 - 4.05 \times 10^{-5}(T - 20) - 9.5 \times 10^{-7}(T - 20)^2 \text{ V}$$

Then, $E = 1.01485$ V at 20°C and 1.01463 V at 25°C. The weak temperature coefficient of the emf of this battery is a good advantage. The emf can be slightly different from one battery to another and as a function of time, and so these must be verified periodically.

Other examples are

$$Pt, H_2(P_{H_2})|H_2SO_4(a_{H_2SO_4})|Hg_2SO_4, Hg$$

and

$$Cd|CdSO_4(a_{CdSO_4})|Hg_2SO_4(s), Hg$$

1.16. CONCENTRATION CELLS

The change of concentration can occur in either the electrode or the electrolyte. These cells are known as concentration cells, although the correct term should be *activity cells* since it is the activity and not the concentration that is responsible for the emf of the cell.

1.16.1. Concentration Cell with Difference of Activity at the Electrode and Electrolyte

One finds examples of concentration change in gas with different gas pressures or different concentrations of a given metal in an alloy (e.g., amalgam). An example for the gas electrode cell is

$$\mathrm{Pt, H_2(P_{H_2} = P_1)|H^+(a_{H^+})|H_2(P_{H_2} = P_2), Pt\ (P_1 > P_2)}$$

The reaction is $\frac{1}{2}\,\mathrm{H_2}(P_1) \rightarrow \frac{1}{2}\,\mathrm{H_2}(P_2)$.

For a metallic electrode such as

$$\mathrm{Zn(Hg)}(a_{\mathrm{Zn}} = a_1)|\mathrm{Zn^{2+}}(a_{\mathrm{Zn^{2+}}})|\mathrm{Zn(Hg)}(a_{\mathrm{Zn}} = a_2)\ (a_1 > a_2)$$

the emf is

$$E = -\frac{RT}{2F}\ln\frac{P_2}{P_1} \qquad E = -\frac{RT}{2F}\ln\frac{a_2}{a_1}$$

For the two cells, the emf is due to the transference of hydrogen from the electrode having higher pressure or from the amalgam with a higher activity a_1 to the other electrode, and the cell will stop in both cases when the two electrodes in every cell become equal in activity.

Two identical electrodes immersed in the same solution but with different concentrations exist in a concentration cell with a difference in electrolyte concentration. This cell has a liquid junction, and a junction potential therefore, except in the absence of matter transfer between the catholyte and anolyte. The electromotive strength here is not a function of a chemical reaction, but depends on the transfer of one electrode solution from anodic to cathodic compartments or vice versa. A typical example of these cells is $\mathrm{Cu|H_2SO_4}(a_1)\vdots\mathrm{H_2SO_4}(a_2)\mathrm{|Cu}$ (Figure 1.16) [6]. The copper electrode immersed in the less concentrated (active) solution will act as an anode.

The reactions of the cell considering the ion transport are as follows:

$$\begin{array}{ll} \text{Anodic reaction :} & \mathrm{Cu \Rightarrow Cu^{2+} + 2}e^- \\ \text{Cathodic reaction :} & \mathrm{Cu^{2+} + 2e^- \Rightarrow Cu} \\ \hline \text{Total reaction :} & \mathrm{Cu^{2+}_{(C2)} \Rightarrow Cu^{2+}_{(C1)}} \end{array}$$

At the junction of the two liquids, there is the following exchange:

$$t_-\,\mathrm{SO^{2-}_{4(C2)}} \Rightarrow t_-\,\mathrm{SO^{2-}_{4(C1)}}$$
$$t_+\,\mathrm{Cu^{2+}_{(C1)}} \Rightarrow t_+\,\mathrm{Cu^{2+}_{(C2)}}$$

The sum of the reactions at the electrodes and due to the ion transport gives

$$\mathrm{Cu^{2+}_{(C2)}} + t_-\,\mathrm{SO^{2-}_{4(C2)}} + t_+\,\mathrm{Cu^{2+}_{(C1)}} \Rightarrow \mathrm{Cu^{2+}_{(C1)}} + t_-\,\mathrm{SO^{2-}_{4(C1)}} + t_+\,\mathrm{Cu^{2+}_{(C2)}}$$
$$(t_+ + t_-)\mathrm{Cu^{2+}_{(C2)}} + t_-\,\mathrm{SO^{2-}_{4(C2)}} + t_+\,\mathrm{Cu^{2+}_{(C1)}} \Rightarrow (t_+ + t_-)\mathrm{Cu^{2+}_{(C1)}} + t_-\,\mathrm{SO^{2-}_{4(C1)}} + t_+\,\mathrm{Cu^{2+}_{(C2)}}$$
$$t_-\,\mathrm{CuSO_{4(C2)}} \Rightarrow t_-\,\mathrm{CuSO_{4(C1)}}$$

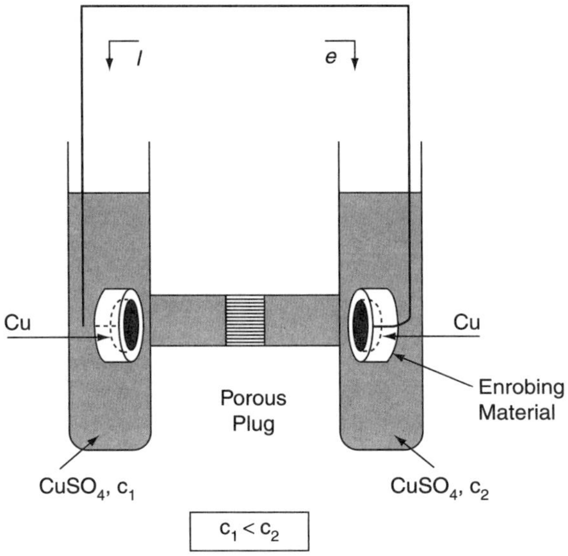

Figure 1.16 Copper concentration cell with transport [6].

The emf of this cell is

$$E = -\frac{RT}{F}\ln\frac{a_1^{t_-}}{a_2^{t_-}}$$

Suggesting that t_- is 0.6, we find the emf of this cell can be on the order of 25 mV if a_1 is 0.1 M and a_2 is 0.5 M and that can initiate pitting corrosion in certain circumstances.

$$\text{FEM} = -t_-\frac{RT}{zF}\ln\frac{[(a_{\text{Cu}^{2+}})(a_{\text{SO}_4^{2-}})]_{\text{C1}}}{[(a_{\text{Cu}^{2+}})(a_{\text{SO}_4^{2-}})]_{\text{C2}}}$$

1.16.1.1. Oxygen Differential Cell

Two different electrolytes at the interface may create a concentration cell. The oxygen differential cell is one of the most important ones in corrosion. This can include two iron electrodes in a dilute solution of sodium chloride (NaCl); the electrolyte around an electrode is well aerated (cathode) while the oxygen around the other solution is expelled by nitrogen bubbling in the solution (anode) (Figure 1.17).

A small quantity of oxygen reaches the metallic surface below the iron oxide (rust), which slows its diffusion (Figure 1.18). The water line corrosion, caused by a differential oxygen cell and observed very frequently on ships in seawater or even boats in river water is illustrated in Figure 1.19. The underground corrosion, caused by a differential oxygen cell, is illustrated in Figure 1.20. The oxygen flux is stronger at the top than that on the bottom. This creates a differential aeration cell with higher oxygen at the surface of the pipe as compared to that of the bottom. The latter with less oxygen plays as anode. Frequently,

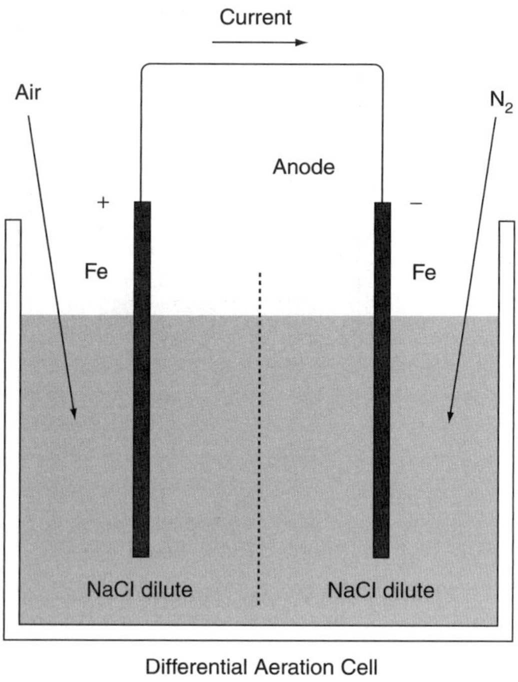

Figure 1.17 Oxygen differential cell.

careful inspection should identify pitting corrosion at the bottom of the pipe rather than on the surface. The main reason explaining the importance of this cell in corrosion is that the cathodic reaction of $2H^+ + \frac{1}{2}O_2 + 2\ e^- \rightarrow H_2O$ is much more exothermic than that of hydrogen $2H^+ + 2\ e^- \rightarrow H_2$, with almost 1200 mV at unity activity of hydrogen ions at 25°C under atmospheric pressure.

1.16.2. Junction Potential

The junction potential (E_j) is the difference between the total emf of the cell and the sum of the reactions at the anode and cathode. A typical example to calculate the junction potential is to consider the reaction at the cell (Figure 1.21), where $a_2 > a_1$:

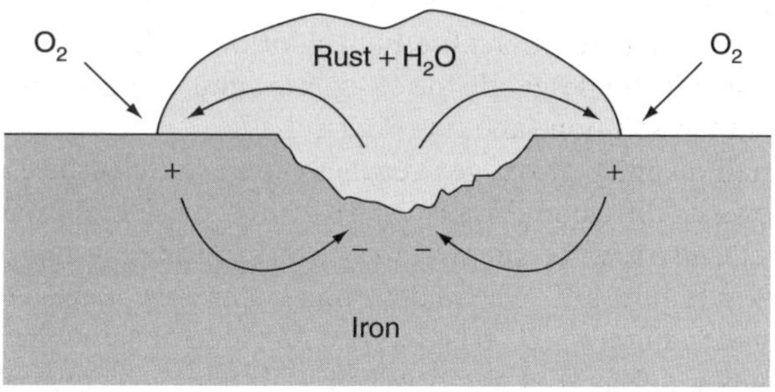

Figure 1.18 Corrosion differential aeration cell formed by rust on iron.

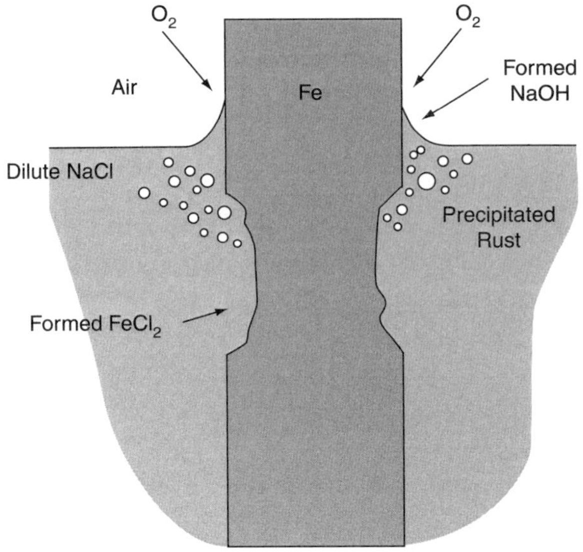

Figure 1.19 Corrosion differential aeration cell at the waterline.

$$\mathrm{Pt,\ H_2(1\ atm)|HCl}(a_1) \vdots \mathrm{HCl}(a_2)\mathrm{|H_2(1\ atm), Pt}$$

$\boxed{\text{anode}}$ $\boxed{\text{cathode}}$

The sum of the reactions at the two electrodes is

$$H^+(a_2) \leftrightarrow H^+(a_1)$$

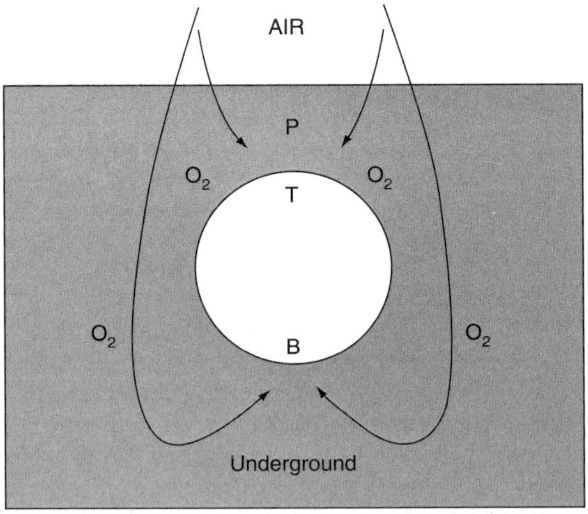

Figure 1.20 Corrosion differential aeration cell of buried pipe [17].

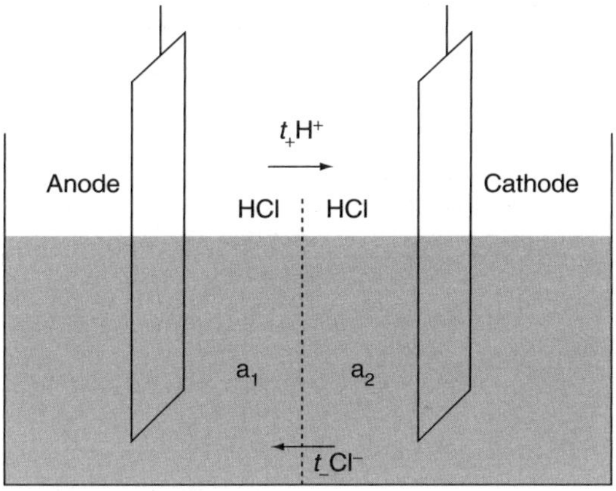

Figure 1.21 Corrosion cell with transport.

The emf due to the reaction at the electrodes is

$$E_1 + E_2 = -\frac{RT}{F}\ln\frac{(a_{H^+})_1}{(a_{H^+})_2}$$

$$E_1 + E_2 = \frac{RT}{F}\ln\frac{(m_{H^+}\gamma_{H^+})_2}{(m_{H^+}\gamma_{H^+})_1}$$

$$E_{total} = \frac{2t_-RT}{F}\ln\frac{m_2\gamma_2}{m_1\gamma_1}$$

$$E_{LJ} = \frac{2t_-RT}{F}\ln\frac{m_2\gamma_2}{m_1\gamma_1} - \frac{RT}{F}\ln\frac{(m_{H^+}\gamma_{H^+})_2}{(m_{H^+}\gamma_{H^+})_1}$$

Considering the two approximations $(m_{H^+}) \approx m_1$, and $(\gamma_{H^+}) = \gamma_1$,

$$E_{LJ} = (t_- - t_+)\frac{RT}{F}\ln\frac{m_2\gamma_2}{m_1\gamma_1}$$

In the case of a cell reversible to the anion, Ag,AgCl(s) |HCl(a_1)¦HCl(a_2) |AgCl(s), Ag, the liquid junction potential is

$$E_{LJ} = (t_+ - t_-)\frac{RT}{F}\ln\frac{m_1\gamma_1}{m_2\gamma_2}$$

It can be observed that in this case $a_1 > a_2$ for a spontaneous exothermic reaction in this direction of the cell.

Since the electromotive strength of these batteries involves the transport number, such batteries can be used to determine the transport number or the coefficient of activity in measuring the emf of the battery and knowing the t or the γ of the studied cation or anion.

The liquid junction potential is the difference between the transport numbers of the cations and anions of the electrolyte. In the case of potassium chloride, the value of $t_+ - t_-$ is equal to 0.02; therefore the junction potential is on the order of 1 mV between two concentrations of KCl, 0.001 N and 0.01 N, at a temperature of 25°C. This is the origin of the current use of a KCl salt bridge since $t_+ = 0.51$ and $t_- = 0.49$. E_j could be ~ zero, positive,

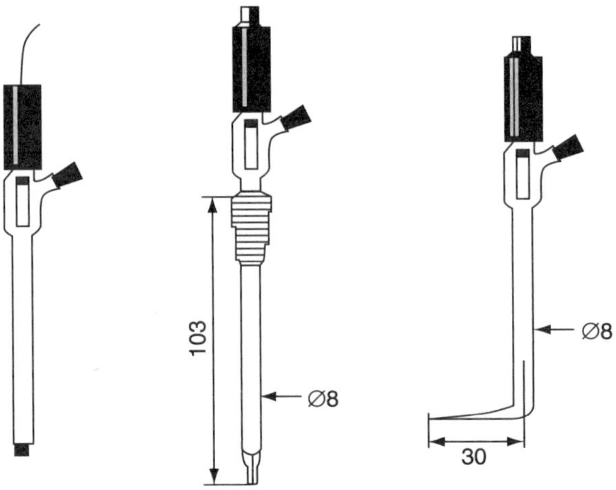

Figure 1.22 Different shapes of reference electrodes, one with a junction bridge. The dimensions are given in millimeters (Tacussel Company electrode configurations, France).

or negative (Figure 1.22). The hydrochloric acid has a very elevated value for $2t_+ - 1 = t_+ - t_-$, roughly equal to 0.65; the junction potential between two concentrations of HCl, 0.001 N and 0.01 N, is on the order of 39 mV. This emf can initiate a local corrosion cell that can lead to pitting corrosion.

When the liquid on each side of the junction is different, the potential is complex and difficult to determine and, in most cases, is function of the geometric features of the junction. For numerous electrochemical batteries, the junction potential is on the order of a few millivolts and increases with the difference of concentration and mobility between ions of the two solutions [6].

In only one general case—a liquid junction between two liquids with the same concentration where all the ions are monovalent, and there is a common ion such as NaCl and KCl solutions—the potential of the liquid junction is independent of the structure of the border and equal to

$$E_\mathrm{j} = \frac{2.303RT}{F} \log \frac{\Lambda_1}{\Lambda_2}$$

where Λ_1 and Λ_2 are the equivalent conductance of the two electrolytes, respectively. It is clear from this equation that when the values of Λ are very close, the potential of the junction becomes very weak. For example, the liquid junction between a solution of 0.1 N of KCl and a solution of NaCl at 25°C is on the order of

$$E_\mathrm{j} = 0.0592 \log \frac{128.96}{106.74} = 0.0049\ \mathrm{V}$$

On the other hand, when the junction is between hydrochloric acid 0.1 N, where the H^+ ions are very mobile, and 0.1 N sodium chloride, the potential of junction is important at 25°C:

$$E_\mathrm{LJ} = 0.0592 \log \frac{391.32}{106.74} = 0.0333\ \mathrm{V}$$

In practice, one could reduce E_j appreciably by means of a bridge composed of a narrow tube full of a solution saturated with potassium chloride or ammonium nitrate.

There is a concentration cell without transport that is in reality a combination of two chemical cells that can monitor the emf of the cell without any interference of the junction potential. This kind of concentration cell is useful to determine the activity coefficient or to deduce the junction potential:

$$\mathrm{Pt, H_2(1\ atm)|HCl}(a_1)\mathrm{|AgCl(s), Ag - Ag, AgCl(s)|HCl}(a_2)\mathrm{|H_2(1\ atm), Pt}$$

and the reaction is

$$\tfrac{1}{2}\mathrm{H_2(1\ atm) + AgCl(s)} - \tfrac{1}{2}\mathrm{H_2(g)(1\ atm) - AgCl(s)} \leftrightarrow \mathrm{Ag(s) + HCl}(a_1) - \mathrm{Ag(s) - HCl}(a_2)$$

$$\mathrm{HCl}(a_2) \leftrightarrow \mathrm{HCl}(a_1)$$

The emf is

$$E = E_1 - E_2$$

1.17. SOLVENT CORROSION CELLS

This type of electrochemical cell depends on the surface properties, atomic structure, and potential level of the electrode, its chemical reactivity with respect to the species in solution (metallic or complex ions), or on some stray electrical current in the electrolyte.

1.17.1. Cathodic Oxidoreduction Reaction

This is a type of cell where its existence depends on an exothermic oxidoreduction electrode reaction. The ferric–ferrous reaction, the most abundant one in this category, has been exploited for chemical machining of highly precise steel instruments or for bacterial leaching of sulfide minerals. In the absence of oxygen, the iron and the divalent ion can coexist at equilibrium states. However, in the presence of oxygen and/or at higher temperatures than ambient, the trivalent ion can be formed and corrosion of iron can be accelerated according to

$$\mathrm{Fe} + 2\mathrm{Fe}^{3+} \rightarrow 3\mathrm{Fe}^{2+}$$

Table 1.3 gives the potential standard of different reactions between dissolved species. It shows that the oxygenated water and the dichromate ions are particularly powerful oxidizers. The oxidizing power of the dissolved oxygen decreases with an increase of the pH [6].

1.17.2. Displacement Cell

The most frequent corrosion reaction of active metals occurs in a liquid environment when ions of a more cathodic metal (e.g., copper) are plated out of solution onto a more anodic one (e.g., aluminum or magnesium alloy). This type of reaction is very serious for aluminum

Table 1.3 Standard Oxidoreduction Potentials

Reaction	$E°$ (V)
$H_2O_2 + 2\,H^+ + 2\,e^- = 2\,H_2O$	1.763
$Ce^{4+} + e^- \rightarrow Ce^{3+}$	1.61
$Cr_2O_7^{2-} + 14\,H^+ + 6\,e^- = 2\,Cr^{3+} + 7\,H_2O$	1.36
$2\,H^+ + 0.5\,O_2 = H_2O$ at 25°C and atmospheric pressure $= 1.23 - 0.0592$ pH	
$Fe^{3+} + e^- = Fe^{2+}$	0.771
$Fe(CN)_6^{3-} + e^- = Fe(CN)_6^{4-}$	0.361
$TiO^{2+} + 2\,H^+ + e^- = Ti^{3+} + H_2O$	0.100
$TiO^{2+} + 2\,H^+ + 2\,e^- = Ti^{2+} + H_2O$	−0.135
$Cr^{3+} + e^- = Cr^{2+}$	−0.424

Source: Reference [6].

and magnesium alloys in the active state, since they are the most active structural materials and so the deposited metallic impurities create active galvanic cells and more frequently pitting corrosion. It is called *deposition corrosion*, which is a combination of pitting and galvanic corrosion.

In the area of hydrometallurgy, a very useful and desired reaction is *cementation*, such as to cement a noble metal like gold or copper onto the surface of an active metal

$$2Au^+ + Fe \rightarrow 2Au + Fe^{+2} \quad \text{or} \quad Zn + Cu^{+2} \rightarrow Zn^{+2} + Cu$$

1.17.3. Complexing Agent Cells

The presence of a complexing agent influences the potential of metals since it replaces the molecules of hydration or solvatation of the dissolved ion. Some powerful complexing ions are chloride, cyanide, and ammonia. The complexing chloride ions for copper and gold, the ammonia ions for copper, and the cyanide ions for gold, iron, and copper are all examples of exothermic corrosion reactions as seen in Table 1.4. They lower the potential standard of reactions of metal dissolution and facilitate corrosion. At the time of the dissolution of a metal, the presence of complexants encourages the formation of ions, the oxidization state of which correlates to the most stable formation of the complex. For example, in the case of copper, monovalent ions $CuCl_2^-$ and $Cu(CN)_2^-$ are formed in the presence of complexing

Table 1.4 Standard Potentials of Electrode Reactions Implying Complexing Agents

Reaction	$E°$ (V)
$Fe(CN)_6^{4-} + 2\,e^- = Fe + 6\,CN^-$	−1.56
$Fe(CN)_6^{3-} + 3\,e^- = Fe + 6\,CN^-$	−0.92
$Au(CN)_2^- + e^- = Au + 2\,CN^-$	−0.595
$Cu(CN)_2^- + e^- = Cu + 2\,CN^-$	−0.44
$Ag(CN)_2^- + e^- = Ag + 2\,CN^-$	−0.31
$CuCl_2^- + e^- = Cu + 2\,Cl^-$	−0.225
$Cu(NH_3)_2^+ + e^- = Cu + 2\,NH_3$	−0.10
$AuCl_4^- + 3\,e^- = Au + 4\,Cl^-$	1.002

Source: Reference [6].

ions Cl^{-1} and CN^{-}, respectively, while its dissolution in a noncomplexing medium produces the bivalent ion Cu^{2+}. These should be identified carefully in every situation since they can initiate general and more severe forms of localized corrosion and even stress corrosion cracking such as in copper in solutions containing ammonia [6].

1.17.4. Stray Current Corrosion Cell

A stray current corrosion cell results from an induced electrical current and is basically independent of the environmental factors such as oxygen concentration or pH that influence other forms of corrosion. A current leaves the intended path because of poor electrical connections within the circuit due to poor insulation around the intended conductive material. It then passes through soil, water, or any other suitable electrolyte to find a low-resistance path, such as a buried metal pipe or some other metal structure, and flows to and from that structure, causing accelerating corrosion. Sources of stray currents include induction from adjacent lines, leakages, variable ground voltages, electric railway systems, grounded electric direct-current power, electric welders, cathodic protection systems, and electroplating plants.

Stray current effects are common on underground cast iron or steel pipelines that are located close to electrical supply lines. Stray currents cause corrosion at the point where they leave the metal (e.g., lead pipe or lead cable sheathing can undergo severe corrosion). Most soils, especially those containing sulfates, will frequently produce graphitic corrosion (a form of dealloying of cast iron) of unprotected gray and nodular cast iron. Tantalum electrically coupled to a less noble metal, such as low-carbon steel, in the presence of stray current may become a cathode and consequently may absorb and become embrittled by atomic hydrogen in the electrolytic galvanic cell [18].

1.18. TEMPERATURE DIFFERENTIAL CELLS

These cells are formed by a difference in the temperature. The constituents of these cells are made of the same metal immersed in one electrolyte with an equal initial composition but each is subjected to a different temperature. The importance theoretical basis of these cells are less obvious than for previous cells. These corrosion cells can be observed in heating elements and furnaces.

In a solution of copper sulfate at an elevated temperature, the copper plays the role of the cathode, while in the same solution at a lower temperature the copper plays the role of the anode. It is important to note that this cell depends on the metal, the medium, and the acceleration of cathodic reactions relative to that of the anodic ones.

1.19. OVERLAPPING OF DIFFERENT CORROSION CELLS

Any classification of corrosion cells cannot be based solely on interdependence of the parameters of these cells. However, it is useful to identify the main initiating reasons of corrosion in case histories.

1. Sometimes, the electrode itself for different concentrations of the same solution could be passive or active and so there is a chemical cell at the beginning that can create two different solutions. This can initiate a concentration cell or accelerate corrosion by acidity, for example.

2. The presence of two identical electrodes immersed in the same solution but with different concentrations can create a cell that depends not only on the charge transport but also on diffusion laws, convection, and temperature and density considerations.
3. Frequently, in the case of corrosion cells, a difference of temperature of the same material, such as copper in contact with the same solution at different temperatures, can give a difference in the electrode potentials as well as a difference in the activities of the corroding solutions.

E. CHEMICAL AND ELECTROCHEMICAL CORROSION

Chemical reactions are governed by the laws of mass action, solubility product, and chemical equilibrium such as those involving oxidation–reduction processes. Electrochemical reactions are defined as those in which free electric charges, or electrons, participate. If a piece of iron is connected to a piece of zinc metal in seawater, zinc dissolves and liberates electrons and becomes an anode while iron can be protected completely by acting as a cathode in accepting these electrons to reduce water in the presence of oxygen on its surface.

1.20. DEFINITION AND DESCRIPTION OF CORROSION

There is a wide range of construction materials—metals and alloys, plastics, rubber, ceramics, composites, wood, and so on—and the selection of an appropriate material for a given application is the responsibility of the designer. Corrosion is not the sole factor of importance in material selection, however; corrosion is frequently considered as the most neglected factor by the design engineer. Corrosion is a major factor in assessing the performance of structural materials. Evans [19] considers that corrosion may be regarded as a branch of chemical thermodynamics or kinetics, such as the outcome of electron affinities of metals and nonmetals, the short-circuited electrochemical cells, or the demolition of the crystal structure of a metal. However, this concerns electrochemical corrosion only. Fontana and Staehle [20] have stated that corrosion includes the reaction of metals, glasses, ionic solids, polymeric solids, and composites with environments that embrace liquid metals, gases, nonaqueous electrolytes, and other non-aqueous solutions.

In the early beginning of corrosion science, corrosion in aqueous solutions was defined as wet corrosion, while corrosion in the absence of water was called dry corrosion. A "wet" reaction of iron in an oxygenated aqueous medium can be represented by the equation

$$Fe + 0.5\,H_2O + 0.75\,O_2 \rightarrow 0.5\,Fe_2O_3 \cdot H_2O$$

A "dry" corrosion of iron where water or aqueous solutions are not involved can be expressed as

$$Fe + 0.75\,O_2 \rightarrow 0.5\,Fe_2O_3$$

If the mechanism of the reaction is considered, our present knowledge of corrosion phenomena shows that wetting of solids by mercury, for example, should be considered a wet reaction in spite of the absence of aqueous solution. Liquid-metal corrosion should then be classified as "wet." On the other hand, the mechanism of corrosion in a wet solution, such as the case of the interior boiler drum corroding in dilute caustic soda at high temperature and high pressure, is best interpreted in terms of a dry corrosion mechanism. Similarly, the reaction of high-temperature water with aluminum and zirconium has been found to show a conventional dry corrosion mechanism [21].

1.21. ELECTROCHEMICAL AND CHEMICAL REACTIONS

If we look profoundly for the mechanism of the majority of wet and dry reactions, we can consider that the interfaces in both reactions involve anodic and cathodic sites. Wet corrosion (Figure 1.23)[21] can be considered as

$$\begin{aligned} \text{Anodic reaction}: &\ M_{\text{lattice}} + H_2O \rightarrow M^{z+}_{\text{aqueous}} + ze \\ \text{Cathodic reaction}: &\ \tfrac{1}{2}O_2 + H_2O + 2\,e \rightarrow 2\,OH^- \\ \text{or} &\ \ \tfrac{1}{2}O_2 + 2\,H^+ + 2\,e \rightarrow H_2O \end{aligned}$$

Similarly, a dry reaction can be divided into two anodic and cathodic reactions (Figure 1.24):

$$\text{Anodic reaction}: \ M \rightarrow M^{z+} \bullet/0 + z\,e \bullet/0$$

where M^{z+} • represents a cation vacancy, e • a positive hole, and/0 the metal–oxide interface. At the gas–oxide interface the gas ionizes.

$$\text{Cathodic reaction}: \ (\tfrac{1}{2}O_2/\text{ads}) + 2(e/X) \rightarrow (O^{2-}/\text{ads})$$

where/X indicates the gas–oxide interface [21].

The formation of a hydrated ion is obligatory in a wet reaction. However, the hydration reaction of most metal ions in a wet reaction is very quick and thus facilitates ionization. The dry reaction involves a direct ionization of oxygen.

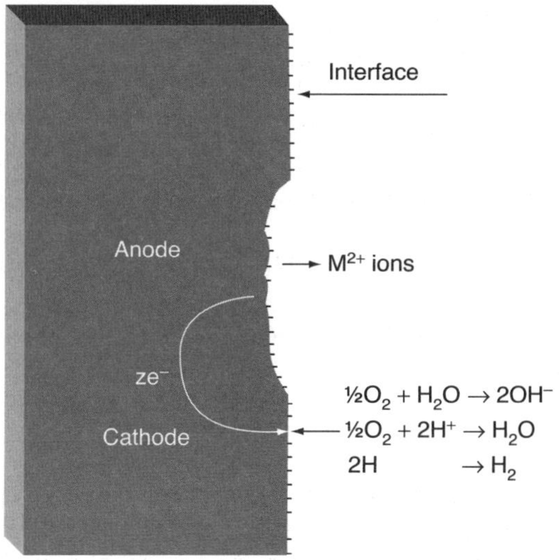

Figure 1.23 Example of electrochemical aqueous corrosion of an active metal in an oxygenated aqueous solution [21].

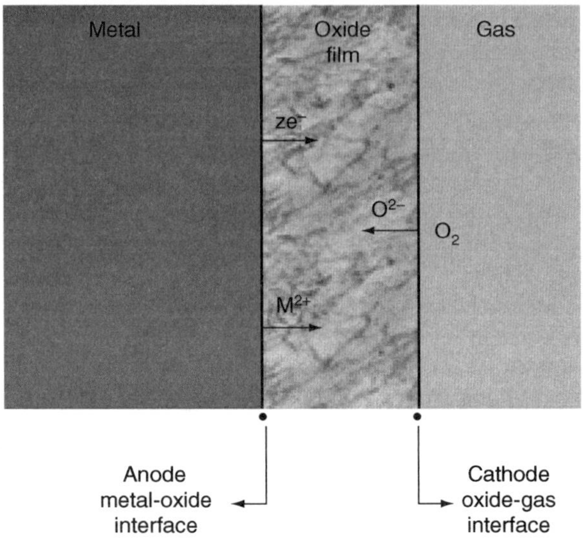

Figure 1.24 An electrochemical reaction at high temperature: corrosion reaction diagram [21].

Shreir [21] suggested that the main types of corrosion reactions could be divided into electrochemical reactions and film-free chemical interactions.

1.21.1. Electrochemical Corrosion

Electrochemical reactions can be divided further into three categories: microelectrochemical cells (inseparable anode/cathode), macroelectrochemical cells (separable anode/cathode areas), and interfacial anode/cathode type.

1.21.1.1. Microelectrochemical Cells (Inseparable Anode/Cathode Areas)

The general uniform dissolution reactions of metals in acid, alkaline, or neutral solutions like zinc in hydrochloric acid or in caustic soda (wet reactions) are typical examples. The anodes and cathodes cannot be distinguished by experimental methods, although their theoretical presence is postulated. It can be admitted that the cathodic and anodic sites are interchangeable. The corrosion current of passive metals, like that of stainless steel in aqueous solutions, is a typical example of mobile anodic sites, although the corrosion rate is very low.

1.21.1.2. Macroelectrochemical Cells (Separable Anode/Cathode Areas)

Certain areas of the metal can be distinguished experimentally as predominantly anodic or cathodic, although the separation distances of these areas may be as small as a fraction of a millimeter. Typical examples could be the general attack of certain regions (localized corrosion), the reaction of iron containing a discontinuous magnetite scale with oxygenated water, crevice corrosion, water-line attack, or "long-line" corrosion of buried iron pipes.

1.21.1.3. Interfacial Anode/Cathode Type

This group includes all metal oxidation reactions in which the charge is transported through a film of reaction product on the metal surface. With parabolic, logarithmic, or asymptotic growth rates, a film can be rate determining, while linear growth rates cannot be rate determining. These reactions are generally considered dry reactions (Figure 1.24). This group also includes the metal–solution reactions, where the uniform formation and growth of a film of reaction product are observed. This includes the reaction of metals with high-temperature water and the reaction of copper with sulfur dissolved in carbon disulfide. Then the reaction of iron with oxygen at room temperature or with oxygen or water at high temperatures is an interfacial electrochemical reaction. For example, in the oxidation of iron at high temperature, the Fe–oxide interface can be considered an anode while the oxide–O_2 interface can be considered a cathode. Reactions with fused salts or nonaqueous solutions (e.g., organic solvent/metal) can be electrochemical in nature. A typical example is the corrosion of copper in a molten salt of ammonium nitrate and bromine in alcohol. In certain cases of uniform general corrosion of metals in acids (e.g., aluminum in hydrochloric acid or iron in reducible acids or alkalis), a thin film of oxide is present on the metal surface. Although the film is not rate determining these reactions cannot be considered film-free ones.

Landolt [6] has characterized the kinetics of corrosion at high temperatures by three parameters:

1. In the absence of an aqueous solution, reactions remain electrochemical in nature (see Figure 1.24).
2. At high temperatures, the volume diffusion and the diffusion at joint grains constitute the basic mechanism of transport in the oxide films. Consequently, the film thickness can be measured in micrometers. In aqueous media, the oxide growth takes place by ionic conduction at high field, limiting the formed film to a thickness of a few nanometers.
3. Since the energy of activation of the electrochemical reaction is superior to that of diffusion phenomenon at high temperatures, an important increase in the corrosion rate occurs with temperature. Equilibrium can be achieved due to the increase in diffusion rate, which becomes sufficiently important. At room temperature, the reaction rate is frequently controlled by the charge transport to the interface.

1.21.2. Film-Free Chemical Interactions

There is a direct chemical reaction of a metal with its environment. The metal remains film-free and there is no transport of charge. The metal–gas reaction forms a volatile oxide or compound, like the reaction of molybdenum with oxygen or the reaction of iron or aluminum with chlorine (dry reactions). The reactions of solid metals with liquid metals like the dissolution of aluminum in mercury and the corrosion of metals in their fused halides (e.g., lead in lead chloride) can be considered chemical reactions. In the same way, the dissolution of metals in nonaqueous solutions (e.g., reaction of aluminum in carbon tetrachloride) can be integrated into this group of chemical reactions.

The first step in the identification of this type of corrosion reaction should then be followed as far as possible by:

1. Corrosion behavior: active or active–passive behavior.
2. Corrosion form, mode, and corrosion type.
3. Material properties.
4. Media description.
5. Corrosion product.
6. Corrosion kinetics and mechanisms.
7. Corrosion testing standards and/or equivalent procedures.

These will be described in detail in the following chapters of this book with special reference to aluminum and magnesium alloys.

REFERENCES

1. G. Milazzo, *Electrochemistry—Theorical Principles and Practical Applications.* Elsevier Publishing, Amsterdam, The Netherlands, 1963.
2. H. H. Uhlig and R. W. Revie, *Corrosion Handbook.* Wiley, Hoboken, NJ, 1985, pp. 8, 35–59.
3. S. H. Maron and J. B. Lando, in *Fundamentals of Physical Chemistry*, edited by C. F. Prutton and S. H. Maron. Macmillan Publishing, New York, 1974, pp. 496–501, 526–529, and 554–623.
4. E. E. Stansbury and R. A. Buchanan, *Fundamentals of Electrochemical Corrosion.* ASM International, Materials Park, OH, 2000, pp. 23–84.
5. D. A. Jones, in *Principles and Prevention of Corrosion*, 2nd edition, edited by W. Stenquist and R. Kernan. Prentice Hall, Upper Saddle River, NJ, 1996, pp. 40–74.
6. D. Landolt, in *Corrosion et chimie de surfaces des métaux*, edited by D. Landolt Presses Polytechniques et Universitaires Romandes, Lausanne, Switzerland, 1993, pp. 13–109.
7. R. Parsons, in *Standard Potentials in Aqueous Solution*, edited by A. J. Bard, R. Parsons, and J. Jordan. Marcel Dekker, New York, 1985, pp. 1–11.
8. W. J. Moore, in *Physical Chemistry*, edited by W. J. Moore. Prentice-Hall, Upper Saddle River, NJ, 1972, p. 531.
9. N. J. Selley, in *Experimental Approach to Electrochemistry*, edited by N. J. Selley. Edward Arnold, London, 1977, pp. 35–60.
10. A. L. Levy, *Journal of Chemical Education* **29**, 384 (1952).
11. D. J. G. Ives and G. J. Janz, *References Electrodes—Theory and Practice.* Academic Press, New York, 1961.
12. H. H. Willard, L. L. Merritt, Jr., and J. A. Dean, *Méthodes physiques de l'analyse chimique.* Dunod, Paris, 1965, p. 505.
13. J. Koryta, *Analytica Chimica Acta* **61**, 329–411 (1972).
14. A. Lewenstam, A. Hulanicki, and E. Ghali, in *Contemporary Electroanalytical Chemistry*, edited by A. Ivaska. Plenum Press, New York, 1990, pp. 213–222.
15. International Union of Pure and Applied Chemistry (IUPAC), *Quantities, Units and Symbols in Physical Chemistry.* Blackwell Scientific Publication, Oxford, UK, 1988.
16. T. S. Licht and A. J. deBéthune, *Journal of Chemical Education* **34**, 433–440 (1957).
17. D. L. Piron, *The Electrochemistry of Corrosion.* National Association of Corrosion Engineers, Houston, TX, 1991, p. 163.
18. *Corrosion—Understanding the Basics.* ASM International, Materials Park, OH, 2000, pp. 100, 162, 214–215, 286, 309, and 513.
19. U. R. Evans, *Corrosion and Oxidation of Metals.* Edward Arnold, London, 1968, pp. 3–11.
20. M. G. Fontana and R. W. Staehle, *Advances in Corrosion Science and Technology.* Plenum Press, New York, 1990.
21. L. L. Shreir, R. A. Jarman, and G. T. Burstein, *Corrosion—Metal/Environment Reactions*, 3rd edition. Butterworth-Heinemann, Oxford, UK, 1995, pp. 1–18.

Chapter 2

Aqueous and High-Temperature Corrosion

Overview

Atmospheric environments—rural, marine, industrial, and a combination of these—are considered. Aqueous corrosion in natural, industrial, and underground media is described. The action of contaminants on corrosion mechanisms is discussed. Organic solvent attack reactions of some materials are summarized. Special consideration is given to the influence of oxygen content on corrosion and formation of scale. Types, properties, and different analytical indexes for the characterization of the activity of water are described.

High-temperature attack, including oxidation (tarnishing, scaling, etc.), sulfidation, carburization, and nitriding, is a form of corrosion observed in the absence of liquid electrolyte. The reaction is generally considered to be electrochemical, especially when the generated oxide film is not volatile. Oxidation resistance could be related to the Pilling–Bedworth ratio (PBR) as well as to the composition, microstructure, and relative properties of the alloy and its oxide. Tensile or compressive stresses, porosity, temperature and temperature cycling, and media could influence the stability of the film greatly. Positive and negative carriers (p and n types) are important features of the metal–gas interface. The corrosion behavior of some alloys at elevated temperatures and hydrogen damage at high temperatures are mentioned. Description of liquid-metal corrosion solid metal induced embrittlement (SMIE), and fused salts corrosion at high temperatures are discussed.

2.1. ATMOSPHERIC MEDIA

2.1.1. Description

Atmospheric corrosion can be defined as the corrosion of materials exposed to air (rather than immersed in a liquid) and their pollutants [1]. By its very nature, atmospheric corrosion has been reported to account for more failures in terms of cost and tonnage than any other factor. The severity of atmospheric corrosion tends to vary significantly among different locations [2]. The atmospheric natural environments (e.g., rural, marine, urban, industrial, and some combination of these) are of first concern. Some contaminated atmospheres such as those containing sulfur dioxide, hydrogen sulfide, and ammonia, should be considered

Corrosion Resistance of Aluminum and Magnesium Alloys: Understanding, Performance, and Testing By Edward Ghali

rigorously [3]. Metals and alloys like stainless steel, titanium, chromium, aluminum, and magnesium develop films, which are protective as far as they are defect-free, nonporous, and self-healing. Weathered steels and copper alloys are good examples of materials that exhibit uniform or near uniform general attack, while passive materials in certain atmospheres show localized corrosion like stainless steel, nickel–chromium alloys, or aluminum alloys. Since increasing humidity or moisture can increase the rate of attack, atmospheric corrosion can be classified as dry, damp, and wet corrosion [2].

Atmospheric variables such as temperature, climatic conditions, and relative humidity, as well as surface shape and surface conditions that affect the time of wetness, are important factors that influence the corrosion rate. The corrosion rate increases rapidly with higher temperatures to the point at which evaporation of the solution takes place. At this temperature, the attack rate will decrease quickly. In an urban or industrial atmosphere, contaminants such as SO_x and NO_x contribute to the corrosion process of metals with moisture, oxygen, and carbon dioxide. Marine atmospheres are usually highly corrosive because of the fine windswept chloride particles that get deposited on surface metal [2]. Industrial atmospheres are more corrosive than rural atmospheres, primarily because of the sulfur compounds produced during the burning of fuels. Sulfur dioxide is selectively absorbed and under humid conditions the metal oxide surfaces catalyze the SO_2 to SO_3, which forms sulfuric acid and accelerates corrosion [4].

Depending on the conditions, rain can either increase or decrease the effects of atmospheric corrosion. Corrosive action is caused by rain when a phase layer of moisture is formed on the metal surface. This activity is increased when the rain washes corrosive promoters such as H^+ and SO_4^{2-} from the air (acid rain). Rain has the ability to decrease corrosive action on the surface of the metal as a result of washing away the pollutants that have been deposited during a preceding dry spell and removing the dust particles that can lead to differential aeration corrosion cells [5].

Microclimate parameters of different parts of the structure are very important since certain areas become wet and retain moisture, which can lead to preferential accelerated attack. Whether rain will increase or decrease the corrosive action is dependent on the ratio of deposition between the dry and wet contaminants. When the dry period deposition of pollutants is greater than the wet period deposition of sulfur compounds, the washing effect of the rain will dominate and the corrosive action will be decreased. In areas where the air is less heavily polluted, the corrosive action of the rain will have much greater importance because it will increase the corrosion rate. High concentrations of sulfate and nitrate and high acidity will be found in areas having an appreciable amount of air pollution. The pH of fog water has been found to be in the range of 2.2–4.0 in highly contaminated areas. This leads to an increase in corrosion [5].

2.1.2. Types of Corrosion

Dry Corrosion This is very slow at ambient temperatures for metals and alloys. The tarnishing of copper and silver in dry air with traces of hydrogen sulfide (H_2S) is an example of a nondesirable film formation at ambient temperature. The sulfide increases the likelihood of defects in the oxide lattice and thus destroys the protective nature of the natural film. Surface moisture is not necessary for tarnishing to occur [4].

Damp Corrosion An invisible thin film of moisture will form on the surface of a metal, providing an electrolyte for electron transfer. The critical relative humidity value is the humidity below which water will not form on a clean metal surface and this depends on the properties of the surface (contaminants, corrosion products, salts, hygroscopic materials, etc.). The critical relative humidity value is between 50% and 70% for iron, copper, nickel, and zinc, depending on the surface properties. Damp corrosion increases with moisture. It has been shown that, for iron, the critical humidity is 60% in an atmosphere free of sulfur dioxide [4].

Wet Corrosion This aqueous corrosion occurs when visible water layers are spread on some areas of a metallic structure. Sea sprays, rain, and drops of dew are the main causes of wet corrosion. Crevices and condensation traps can lead to an almost continuous moistened surface. The corrosion rate increases with soluble corrosion products [4].

Wet–Dry Cycling Conditions Under alternating wet and dry conditions, the formation of an insoluble corrosion product on the surface may increase the corrosion rate during the dry cycle by absorbing moisture and continually wetting the surface of a metal. The rusting of iron and steel and the formation of a patina on copper are examples of wet and/or damp corrosion. Atmospheric relative humidity and its cycling are of major importance for corrosion kinetics and a material's resistance to corrosion. The time of wetness determines the duration of the electrochemical process. The thickness and the chemical composition of the water are both important factors. The nature of the corrosion product can change the time of wetness. Dew formation on metal surfaces can lead to accelerated corrosion because of the tendency of the dew to be acidic as a result of high SO_2 values near the ground. The dew can form on open or sheltered surfaces and can lead to a corrosive attack of galvanized sheet called white rusting [4].

Dew is an important source of atmospheric corrosion, more so than rain, and particularly under sheltered conditions. Dew forms when the temperature of the surface metal falls below the dew point of the atmosphere. This can occur outdoors during the night when the surface temperature of the metal is lowered as a result of radiant heat transfer between the metal and the atmosphere. It is also common for dew to form during the early morning hours when the air temperature rises faster than the metal temperature. Dew may also form when metal products are brought into warm or heated storage after cold shipment. Under sheltered conditions, dew is an important cause of corrosion. Dew is highly corrosive because of the higher concentration of contaminants in dew than in rainwater, which leads to lower pH values. Heavily industrialized areas have reported pH values of dew in the range of 3 and lower. Also, the beneficial washing effect of rain is usually slight or negligible [5].

2.1.3. Atmospheric Contaminants

Chemical and particle contaminants often control the corrosion rate of exposed metallic structures and some of them are more abundant and frequently characterize the corrosion form of some atmospheres. Dust particles, besides their corrosive–erosive action on some materials, can contain aggressive contaminants like chlorides and can absorb water or acids and trap the solution against the surface. Carbon dioxide does not play a significant role for certain metals but could accelerate corrosion of magnesium-based alloys [4].

Dust On a weight basis in many locations, dust is the primary air contaminant. When in contact with metallic surfaces and combined with moisture, dust can promote corrosion by forming galvanic or differential aeration cells that, because of their hygroscopic nature, form an electrolyte on the surface. This is particularly true if the dust contains water-soluble particles on which sulfuric acid is absorbed. Dust-free air therefore is less likely to cause corrosion [5]. Dust particles, which may contain aggressive contaminants like chlorides, can absorb water or acids and trap the solution against the surface [2].

Oxygen From a corrosion standpoint, the most significant contaminant is dissolved oxygen from ambient air. Oxygen is a cathodic depolarizer that reacts with and removes hydrogen from the cathode during electrochemical corrosion, thereby permitting corrosion attack to continue with its exothermic cathodic reaction as compared to that of the endothermic reduction of hydrogen ions in the absence of oxidant [6].

Chlorides Atmospheric salinity distinctly increases atmospheric corrosion rates. Apart from the enhanced surface electrolyte formation by hygroscopic salts such as NaCl and $MgCl_2$, direct participation of chloride ions in electrochemical corrosion reactions is also likely. Steel pillars 25 m from the seacoast will corrode 12 times faster than the same steel pillars 250 m further inland because of the difference in the levels of marine salts in the two locations [2]. Atmospheric contaminants often control the corrosion rate of exposed metallic structures. The corrosive effects of gaseous chlorine and hydrogen chloride present in the atmosphere can intensify atmospheric corrosion damage and tend to be stronger than those of chloride salt anions because of the acidic character of the former species [2, 4].

Sulfur Dioxide Sulfur dioxide, a product of the combustion of sulfur-containing fossil fuels, plays an important role in atmospheric corrosion in urban and industrial atmospheres. It is adsorbed on metal surfaces, has a high solubility in water, and tends to form sulfuric acid in the presence of surface moisture films [2]. Industrial atmospheres are more corrosive than rural atmospheres, primarily because of the sulfur compounds produced during the burning of fuels. Sulfur dioxide is selectively absorbed and under humid conditions the metal oxide surfaces catalyze the oxidation reaction of SO_2 to SO_3, which forms sulfuric acid and accelerates corrosion.

Hydrogen Sulfide This compound can be generated naturally by the decomposition of organic compounds or by sulfate reducing bacteria (SRB) in polluted rivers [4]. Hydrogen sulfide is known to be extremely corrosive to most metals and alloys and could lead to stress corrosion cracking.

Nitrogen Compounds These compounds occur naturally during thunderstorms or are added to the atmosphere by the use of ammonia-based fertilizers [4]. Nitrogen compounds in the form of NO_x also tend to accelerate atmospheric attack. NO_x emissions, largely from combustion processes, have been reported to have increased relative to SO_2 levels.

Ozone Until recently, the effect of ozone (O_3) had been largely neglected in atmospheric corrosion research. It has been reported that the presence of ozone in the atmosphere may lead to an increase in the sulfur dioxide deposition rate. While the accelerating effect of ozone on zinc corrosion appears to be very limited, both aluminum and copper have been noted to undergo distinctly accelerated attack in its presence [2].

Synergistic Effects of Some Contaminants The kinetics of the attack by aggressive contaminants in an atmospheric medium is a function of the exposed material. Weathered steel and galvanized steel were among materials exposed in controlled environment chambers. The direct and synergistic effects of relative humidity, sulfur dioxide, nitrogen dioxide, and ozone in a programmed dew/light cycle on corrosion rate were studied. The important parameters were sulfur dioxide, relative humidity, and interaction between the two for weathered steel, while only the direct effects of sulfur dioxide and relative humidity were relevant to the corrosion of galvanized steel. The SO_2-rich atmosphere and acid rain are important for the quick deterioration of metallic structures and statues exposed to the atmosphere [7].

2.1.4. Corrosion Prevention and Protection

Most grades of carbon steels do not exhibit large differences in atmospheric corrosion rate. However, small additions of copper (0.1%) will increase the resistance of steel to a sulfur-polluted environment by enhancing the formation of a tighter, more protective rust film. Addition of nickel and chromium to carbon steel increases the corrosion resistance by the formation of insoluble sulfates. The combination of minor elements, such as the addition of chromium and nickel with copper and phosphorus, can be very effective for different and severe climates like tropical marine regions [4].

Corrosion prevention with thermal-sprayed zinc and aluminum coatings showed that low-carbon steels can be protected from corrosive effects of rural, industrial, salt air, and salt spray environments for 19 years [8]. Results of 3–7 year exposure periods of galvanized steel specimens in Ontario and Quebec (Canada) highway environments with and without coating are reported. A counting life of 5 years per 25 μm coating can be expected in urban environments and 6–7 years in rural environments. The method of application of zinc appeared to have no important effect [9].

Temporary protection during transport or storage of corrodive materials can be achieved by controlling the electrochemical reacting interface. Lowering the atmospheric humidity by using a desiccant or by heating and using a vapor phase specific inhibitor for the metallic surface is recommended. Temporary and/or permanent indoor heated storage is preferred to avoid excess humidity. A rather permanent protection can include organic, inorganic, and metallic coatings. Proper preparation of the metallic surface is necessary and may require a conversion coating for the surface (e.g., phosphates) in certain cases.

2.2. AQUEOUS ENVIRONMENTS

Natural or industrial waters and extremely dilute inorganic or organic chemicals are included in this category, as well as aqueous corrosion at temperatures above ambient [3].

Fresh Water Fresh water may come from either a surface or ground source and typically contains less than 1% sodium chloride. It may be either "hard" or "soft," that is, either rich in calcium and magnesium salts and thus sometimes forming insoluble curds with ordinary soap or not. Actually, there are gradations of hardness, which can be estimated from Langelier or Ryznar indexes or accurately determined by titration with standardized chelating-agent solutions (e.g., versenates) [6].

Of the dissolved gases occurring in water, oxygen occupies a special position, since it stimulates corrosion reactions. In surface waters, the oxygen concentration approximates saturation, but in the presence of green algae, supersaturation may occur. Underground waters are more variable in oxygen content, and some waters containing ferrous bicarbonate are oxygen-free. Hydrogen sulfide and sulfur dioxide are also usually the result of pollution or of bacterial activity. Both gases may initiate or significantly accelerate corrosion of most metals. Sulfates have an important role in bacterial corrosion under anaerobic conditions [2].

Acid water is more aggressive than neutral or alkaline water. Low pH values can be attributed to a variety of causes, including acidity of the supply water, oxidation of iron sulfides, and bacterial activities. In a low pH environment, it may be necessary for mine operators to take all necessary measures to treat and neutralize mine water that is released into the landscape as a result of mining operations [2].

Corrosion-resistant aluminum alloys are suitable for use with high-purity water at room temperature. The slight reaction with the water that occurs initially ceases almost completely within a few days after development of a protective oxide film of equilibrium thickness. After this conditioning period, the amount of metal dissolved by the water becomes negligible. Corrosion resistance of aluminum alloys in high-purity water is not significantly decreased by dissolved carbon dioxide or oxygen in the water or, in most cases, by the various chemicals added to high-purity water in the steam power industry to provide the required compatibility with steel. These additives include ammonia and neutralizing amines for pH to control carbon dioxide, hydrazine and sodium sulfate to control oxygen, and filming amines (long-chain polar compounds) to produce nonwettable surfaces. Somewhat surprisingly, the effects of alloying elements on corrosion resistance of aluminum alloys in high-purity water at elevated temperatures are opposite to their effects at room temperature: elements (including impurities) that decrease resistance at room temperature improve it at elevated temperatures [10].

Seawater Seawater has a high salt concentration, mainly sodium chloride, a high electrical conductivity, relatively high and constant pH, and a good buffering capacity. The salinity may be weakened in some areas by dilution with fresh water or concentrated in some areas by solar evaporation. Seawater is normally more corrosive than fresh water and the rate of corrosion is controlled by the chloride content, oxygen availability, and the temperature. The 3.5% salt content of seawater produces the most corrosive solution since oxygen solubility is at its maximum and is reduced in more concentrated salt solutions [6]. Seawater as a medium promotes the presence of microorganisms that influence corrosion. The presence of microfouling (e.g., bacteria, slime) and macrofouling (e.g., seaweed, mussels, barnacles) due to corrosion processes are frequently observed. Sulfate reducing bacteria lead to an increase in the corrosion process of many metals [2].

Distilled or Demineralized Water The total mineral content of water can be removed by either distillation or mixed-bed ion exchange. Purity can be described qualitatively in some cases (e.g., triple-distillated water) but is best determined, for both distillated and demineralized water, in terms of specific conductivity. Water also can be demineralized by reverse osmosis or electrodialysis [2].

Steam Condensate Water condensate from industrial steam is called steam condensate. It approaches distilled water in purity, except for contamination (as by dissolved oxygen or carbon dioxide) and the effect of deliberate additives such as neutralizing or filming amines [2].

Mine Waters These waters are not a widely used source of supply but their quality is obviously important in this particular industry. Some of them are very similar to surface waters. They may vary widely according to the strata through which the water drains and may even vary in composition in the same mine. A number of them are very acid and as a result very corrosive [11]. The pH values of mine waters can range from 2.8 to 12.3. The high pH of mine waters [12] may be due to the use of cement in the backfill while the low pH of mine water is due to the oxidation of iron sulfides. The acidic mine water is also produced by bacteria such as *Thiobacillius thio-oxidans* and *Thiobacillius ferro-oxidans*. The rate of sulfuric acid production is four times greater in the presence of thiomic bacteria than in the absence of bacteria. The generated acidity is much greater under aerobic conditions than under anaerobic conditions [13]. Acidic mine water attacks the clay, silicates, and carbonates in the mineral and increases the concentration of silica, aluminum, calcium, magnesium, and manganese in the waters. Neutralization of the acidity by lime, reverse osmosis, and addition of flocculants can be used to control water quality [14].

2.3. ORGANIC SOLVENT PROPERTIES

Process organic solvent media have specific active properties depending on the material used. The nature of corrosion in organic solvents is unpredictable. For example, the corrosion of nickel in different solvents containing 0.05 wt% H_2SO_4 at various temperatures: the corrosion rate in ethanol is far greater than that in aqueous acid whereas in acetone the rate is practically zero. The addition of 0.05% H_2SO_4 to acetic acid decreases the corrosion rate [15, 16].

Heitz [15] classifies corrosion in organic solvents into electrochemical and chemical reactions.

1. ***Electrochemical Reactions.*** The anodic reaction can be the formation of a solvated metal cation M^{z+}_{solv}, a charged or uncharged metal complex MX^-, or a solid compound MX_z, where X is a halogen ion, organic acid anion, and so on. The cathodic reaction can be $H^+_{solv} + e \rightarrow \frac{1}{2}H_2$, or $HA + e \rightarrow \frac{1}{2}H_2 + A^-$, where A^- can be a carboxylic acid anion, alcoholate ion, and so on. Typical examples include reduction of an oxidizing gas like O_2, Cl_2, F_2, Br_2, O_3, and N_2O_4, or reduction of an oxidizing ion such as Cu^{2+}, $MnO_4{}^{2-}$, or $ClO_3{}^-$.
2. ***Chemical Reactions.*** This type of corrosion involves a direct charge transfer between the metal atoms in the lattice of the atom and the oxidizing reagent and can be presented as

$$M + 2\,C_xH_yX_z \rightarrow MX_2 + C_{2x}H_{2y}X_{2z-2}$$

where X is a halogen and M is a divalent metal such as

$$Mg + CH_3Cl \rightarrow CH_3MgCl$$

Metals can also react with organic sulfur:

$$2\,M + 2\,RSH \rightarrow 2\,MS + H_2 + R_2$$

Different forms of corrosion present in aqueous solutions can be observed in organic solvents. Cathodic and anodic protection mechanisms are seriously limited by the resistivity of the solvent and the poor performance of organic coatings.

2.4. UNDERGROUND MEDIA

This category includes underground installations of pipes and vessels and solid structures such as tank bottoms [3]. In general, the classification of soils according to their characteristics is based on their physical and chemical properties rather than on their geologic origin or geographic location, although the soil characteristics may be influenced by both the origin and location. Soil is an aggregate of minerals, organic matter, water, and gases (mostly air). It is formed by the combined weathering action of wind and water and also by organic decay. The proportions of the basic constituents vary greatly in different soil types. For example, humus has very high organic matter content, whereas the organic matter content of beach sand is practically zero. The properties and characteristics of soil obviously vary as a function of depth. The different layers of soil are known as *soil horizons*. The following soil horizons have been classified [2]: surface soil, organic horizon, eluviation horizon, accumulation horizon (rich in metal oxide), and parent material (largely nonweathered bedrock).

Tomashov and Mikhailovsky [17] showed in an exhaustive study conducted by The National Bureau of Standards the different corrosion resistances of steel, iron, copper, lead, and zinc as a function of different soils. It is interesting to note the similarities in high corrosion rates of iron and steel for two soils, while they were almost noncorrosive in the third soil. This reflects in reality a multitude of factors to consider in order to assess the corrosivity of a soil. It is important that long-term corrosion studies for underground media should be developed [17].

Several important variables have already been identified that have an influence on corrosion rates in soil; these include water, degree of aeration, pH, redoxpotential, resistivity, soluble ionic species (chlorides, sulfates), and microbiological activity [2]. Romanoff [18, 19] summarized the factors that could affect corrosion:

1. Aeration factors are those that affect the access of oxygen and moisture to the metal, thereby creating higher conductivities. Oxygen, either from atmospheric sources or from oxidizing salts or compounds, stimulates corrosion by combining with metal ions to form oxides, hydroxides, or salts of metal. These salts could be soluble and could be removed, or they could precipitate to protect the metal, or they could stimulate and localize corrosion through the formation of an oxygen differential corrosion cell.
2. An electrolyte is necessary to carry a current since electrical resistivity is a major factor in controlling corrosion. Also, these dissolved ions control the chemical properties such as acidity, alkalinity, and reactions between corrosion products and the solution at the metal–soil interface.
3. Electrical factors define the size, number, and location of anodic areas and the amount of current that flows from a pipe to the soil, for example.
4. Miscellaneous factors could be a combination of the previous factors or some noncontrolled parameters such as a flowing stray current or backfilling a trench in a different state of compactness after a pipe is laid.

In spite of the multitude of factors and their relative synergies, the electrical conductivity of the medium surrounding the metal is an important factor that emerges frequently to evaluate corrosion. Resistivity measurements may be used to locate areas where corrosion may represent a problem. If the same metallic structure has wide variations of resistivity from point to point along its surface, this could indicate a danger of corrosion damage at low-resistivity areas. Rough indications of soil corrosivity versus resistivity indicated that resistance (in $\Omega \cdot$ cm) lower than 500 is very corrosive while that higher than 10,000 is noncorrosive. Three zones of resistance, 500–1000, 1000–2000, and 2000–10,000 $\Omega \cdot$ cm could correspond to corrosive, moderately corrosive, and highly corrosive soils, respectively [20].

Chloride and Sulfate Ions in Mine Water Ranasooriya et al. [21] have investigated the chemical composition of 24 mine waters in Australia and found the minimum observed Cl^- (1080 ppm) and SO_4^{2-} (300 ppm) for these analyses. Mine air can contain significant amounts of oxides of sulfur and nitrogen. Depending on the composition of mine dust, corrosion rates can reach significant values. Dust particles settling on the surface of a bolt can accelerate corrosion attack as they can absorb harmful gases like sulfur and nitrogen oxides. If a significant deposit of dust results, differential oxygen concentration can initiate vigorous localized corrosion attacks. Different types of bacteria, which can be found in the underground environment, can accelerate the production of acid in mine water. Under these conditions, the corrosion rate can increase considerably. The mentioned values of Cl^- and SO_4^{2-} are considered to be high enough to accelerate steel corrosion especially at negative values of the Langelier saturation index of water [13].

2.5. WATER MEDIA PROPERTIES

Natural waters, atmospheric moisture, and rain as well as prepared solutions are the most frequent media of aqueous corrosion. Water as a solvent, when it has a suitable electrolytic conductivity, can give rise to corrosion of some active materials. The aqueous corrosion of a metal is mostly an electrochemical corrosion. For metal corrosion to occur, an oxidation reaction (generally a metal dissolution and/or an oxide formation) and a cathodic reaction such as proton or oxygen reduction must proceed simultaneously. For example, the corrosion of iron in acid solution can be expressed as

$$\begin{aligned} \text{Oxidation:}\ & \mathrm{Fe} \rightarrow \mathrm{Fe^{2+}} + 2\mathrm{e} \\ \text{Reduction:}\ & 2\mathrm{H^+} + 2\mathrm{e} \rightarrow \mathrm{H_2} \\ \text{and/or}\ & 2\mathrm{H^+} + \tfrac{1}{2}\mathrm{O_2} + 2\mathrm{e} \rightarrow \mathrm{H_2O} \quad \text{(acid solution)} \end{aligned}$$

In most natural waters or waters with a near neutral pH, which contain an appreciable quantity of oxygen, the primary cathodic reaction is

$$\mathrm{H_2O} + \tfrac{1}{2}\mathrm{O_2} + 2\mathrm{e} \rightarrow 2\,\mathrm{OH^-} \quad \text{(alkaline solution)}$$

Several cathodic reactions may simultaneously support the metal corrosion. In the same way, metal corrosion may also be the sum of more than one dissolution process.

$$i_{\text{corrosion}} = \sum i_a = -\sum i_c$$

Although i_a and i_c are equal, the current densities are frequently different depending on the relative surface of the anodic and cathodic sites of the galvanic cell. The cathodic reactions correspond to the consumption of acidic hydrogen ions and/or the production of hydroxide ions at the metal–electrolyte interface, where Fe^{2+} is abundant:

$$Fe^{2+} + 2\,OH^- \rightarrow Fe(OH)_2$$

When the solubility of the $Fe(OH)_2$ is exceeded, it will be rapidly oxidized in presence of oxygen to $Fe(OH)_3$:

$$2\,Fe(OH)_2 + H_2O + \tfrac{1}{2}O_2 \rightarrow 2\,Fe(OH)_3$$

A solid precipitates and can result in the growth of a tubercle on the surface [22].

Generally, corrosion is influenced by the metal or alloy and the environment. Environmental considerations should particularly address the properties of the interface. The important properties of the environment are pH, oxidizing power (potential), temperature (heat transfer), velocity (fluid flow), and concentration of the different ions in solution. The influence of biological organisms on these properties of the environment should be considered. A description of the properties of the mass of the solution and the solid conductor is necessary to describe the metal–solution interface that controls electrochemical corrosion. Since the diffusion of oxygen is frequently the rate-determining factor in aqueous corrosion, large cathode/anode area ratios will frequently result in intense galvanic attack [23].

Water can be corrosive to most metals. Pure water, without dissolved gases (e.g., oxygen, carbon dioxide, and sulfur dioxide), does not cause undue corrosion attack on most metals and alloys at temperatures up to at least the boiling point of water. Even at temperatures of about 450 °C, almost all of the common structural metals, except magnesium and aluminum, possess adequate corrosion resistance to high-purity water and steam.

In summary, the factors influencing the corrosion of materials in water systems are the following: the physical configuration of the system, the flow rate that controls the properties of the interface, and the temperature of the water. The influence of temperature in aqueous solutions on corrosion rate is *very* significant. It can change the state of the solid and the nature of the environment with more or less dissolved gases. It changes the diffusion and reaction kinetics and, most importantly, the physicochemical properties of the interface; solid, chemical, and gas contaminants as well as bacteria are discussed for atmospheric corrosion. The chemical properties of the water—oxidizing power, hardness, salts, chlorides, and dissolved gases—are the most important [6].

2.5.1. Water Composition

Natural or treated waters always contain various amounts of dissolved materials either from the atmosphere or from the ground through which they percolate. However, the water used in industry shows a much wider variation in composition and properties [11].

2.5.1.1. Total Dissolved Salts (TDS)

The main constituent ions in natural waters are positively charged cations, such as Ca^+, Mg^+, Na^+, and H^+, and negatively charged anions, such as Cl^-, SO_4^{2-}, HCO_3^-, CO_3^{2-}, and OH^-. The TDS can be determined directly by evaporating to dryness, the final drying usually being carried out at 180 °C, or can be estimated to a sufficient degree of accuracy from the electrical conductivity of the water. The total dissolved solids (in ppm) is close to 1/15th of the conductivity, in reciprocal ohms (Ω^{-1}) [11].

Chlorides and Sulfates In general, the amount of dissolved chloride is greater than the amount of sulfate and only in certain highly mineralized waters does the sulfate predominant. Large amounts of chloride result mostly from pollution of rivers by sewage and industrial effluents. For example, the 10 and 20 ppm present in unpolluted rivers or streams may be increased in average domestic sewerage to about 100 ppm. Still greater amounts are often found in underground sources, particularly in desert regions or in coastal regions where there may be infiltration of seawater, which has a high chloride concentration of about 35,000 ppm. Very wide variations can occur in river estuaries, particularly those that are tidal such as the Severn. Water high in chloride is often referred to as "brackish" [11].

Carbonate and Bicarbonate These constitute the bulk of the dissolved salts in natural waters. They are closely linked with the carbon dioxide and calcium content of the water [11].

2.5.1.2. Minor Inorganic Constituents

Silica and traces of certain heavy metals are among the minor inorganic constituents present that are indicative of the corrosive nature of the water or its toxicity [11].

Silica Silica concentration in natural waters varies from a trace to over 75 ppm SiO_2, but the amount is usually in the range of 5–30 ppm. Silicates have certain inhibitive properties and are added to soft waters to reduce corrosion and are used as conditioning agents in low-pressure boilers. In high-pressure steam-raising equipment, silica is undesirable since even at small concentrations it forms hard incrustations [11].

Iron Iron is sometimes present in natural waters, often as ferrous carbonate, at concentrations up to 20 ppm. On coming into contact with the air, it is oxidized and rust is precipitated. This "red water" causes unsightly stains and renders the water unsuitable for domestic and many industrial uses. Iron starts to become a nuisance at about 0.2–0.3 ppm [11].

Copper Copper is not normally present in natural waters and when present in tap water it is usually derived from copper pipes and storage cylinders. Very small amounts are capable of stimulating an attack on aluminum, to a less extent on zinc, and to some degree on iron. The most aggressive cations that can accelerate copper corrosion are trivalent iron (oxidizing capacity) and ammonium salts (capability of chelating copper) [24].

Lead Lead is present principally due to the corrosion of lead pipes. It is a cumulative poison and should not be present in an amount greater than 0.1 ppm [11].

2.5.1.3. Dissolved Gases

In its passage through the air, water dissolves nitrogen, oxygen, and carbon dioxide and, in polluted atmospheres, small amounts of hydrogen sulfide, sulfur dioxide, and ammonia. Further amounts of gases derived from decaying vegetation are dissolved during the water's passage through the ground [11].

Oxygen Most public supplies are well oxygenated with an oxygen content of 2–8 ppm at ordinary temperatures. For a given partial pressure of oxygen, the amount dissolved decreases with rise in temperature to just above 100 °C and then increases, the solubility at 200 °C being similar to that at 25 °C [11].

Nitrogen The nitrogen content of water has little direct effect on the corrosion reaction but bubbles of gas can give rise to impingement or cavitation attack [11].

Carbon Dioxide The amount of free carbon dioxide in natural waters is seldom greater than 10 ppm and a part of this is closely connected with the carbonate equilibrium. Carbon dioxide, which is more soluble than oxygen in pure water (1.4 g/L at 85 °F), will convert to carbonic acid, producing a solution having a pH of less than 6, where acid attack can predominate. Besides the potential for increased corrosion, carbon dioxide in natural waters affects the solubility and precipitation of calcium carbonate [24]:

$$CO_2 + H_2O \rightarrow H_2CO_3$$

$$H_2CO_3 \rightarrow H^+ + HCO_3^-$$

$$Ca^{2+} + HCO_3^- + OH^- \rightarrow CaCO_3 + H_2O$$

The solubility of calcium carbonate, and other sparingly soluble inorganic salts, will decrease with an increase in temperature, precipitating and forming a thick deposit at the hottest areas. Existing differential aeration conditions can cause localized corrosion under the deposit [24].

Hydrogen Sulfide Amounts of hydrogen sulfide up to 15 ppm may occasionally be present owing to pollution or, more often, the action of sulfate reducing bacteria (SRB). As little as 0.5% may be detected by its objectionable odor [11].

Chlorine Chlorine, as a dissolved gas, is not naturally found in municipal waters. However, chlorine is added for infection control. The action of chlorine with water produces hypochlorous acid, HClO, which will suppress the pH [11]:

$$Cl_2 + H_2O \rightarrow HClO + HCl$$

2.5.1.4. Organic Matter

The organic matter present in water consists of living organisms and the products of their metabolism or decay [11].

Nonliving Organic Constituents These may be in colloidal or true solution or in suspension and are derived from the decay of vegetable matter in the drainage area, domestic and industrial wastes, and oil contamination [11].

Living Organic Matter There are two main classes of living organisms found in waters: microscopic organisms such as bacteria, slimes, fungi, and algae, and macroscopic marine organisms, such as barnacles. Algae remove carbon dioxide and give off oxygen while other organisms consume oxygen. Hydrogen sulfide may be produced by SRB or by the decay of organic matter. This decay may also give corrosive amino acids, which deposit deleterious sulfide films on copper-alloy condenser tubes [11].

Iron bacteria produce fouling owing to their accumulation of large amounts of ferric hydrate, which may be up to 500 times as great as the volume of bacteria concerned. This leads to blockage, increase in friction owing to the formation of tubercles, and the production of red water [11].

Marine organisms such as barnacles and mollusks have a marked fouling action and attach themselves to metal surfaces, providing the water is stagnant or slow moving [11].

2.5.2. The Oxidizing Power of Solution

Concentrated nitric acid is a highly oxidizing environment, aerated (containing oxygen) acid is mildly oxidizing, and deoxygenated acid is a relatively reducing environment. The oxidizing power of deoxygenated acid is sufficient to corrode both magnesium and iron but is insufficient to corrode copper or gold. An aerated acid, that is, one containing dissolved oxygen, has sufficient oxidizing power to corrode magnesium, iron, and copper. An aerated acid is still insufficient to corrode gold. Concentrated nitric acid has high oxidizing power and corrodes gold, copper, iron, and magnesium. The addition of oxygen dissolved in a solution increases its oxidizing power. Other chemical species, however, also increase the oxidizing power. Ferric ions and cupric ions greatly increase the oxidizing power of a solution. Deoxygenated hydrochloric acid is an example of a highly reducing environment [25]. At certain concentrations and temperatures, highly oxidizing solutions could lead to passivation depending on the active–passive behavior of the metal or the alloy.

The oxygen solubility in water as a function of temperature is shown in Table 2.1 and Figure 2.1. The oxygen solubility decreases as temperature increases from 0 °C (32 °F) through 100 °C (212 °F). At the boiling point, all oxygen is stripped from the water, and the solubility essentially becomes zero. The corrosion rate of an oxygen-saturated solution based on oxygen solubility would be predicted to decrease with increasing temperature. This effect is often offset by increasing reaction kinetics as temperature increases; however, the corrosion rate drops rapidly at the boiling point because of a discontinuous drop in oxygen concentration [25]. The concentration of oxygen corresponding to air saturation at 25 °C is 5.7 mL O_2/L. Deviations from the linear relation between corrosion rate of iron and oxygen concentration occur sooner in distilled water than when chloride ions are present, since the critical concentration of oxygen above which corrosion decreases is about 12 mL O_2/L, which is lower than that in the presence of dissolved salts or at higher temperatures. When corrosion is controlled by diffusion of oxygen, the corrosion rate at a given oxygen concentration doubles for every 30 °C [26].

Table 2.1 Oxygen Solubility in Water

Temperature		Oxygen solubility	
Celsius (°C)	Fahrenheit (°F)	Grams per kilogram water	Parts per million
0	32	0.069	69
20	70	0.043	43
40	105	0.031	31
60	140	0.027	27
80	140	0.014	14
100	212	0.00	<1

Source: Reference 25.

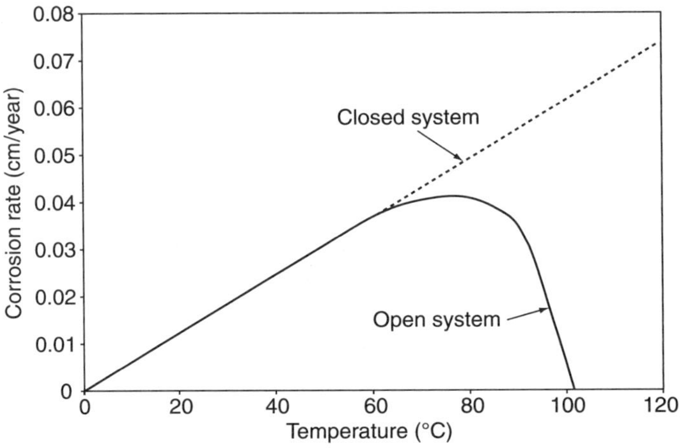

Figure 2.1 Effect of oxygen on the corrosion of steel [27].

An increase in temperature usually brings about an increase in the rate of corrosion. A 10 °C rise in temperature can double the rate of chemical reactions and also increase the rate of diffusion of ions to the interface. In many cases, however, the corrosion rate reaches a maximum at about 80 °C (Figure 2.1). With a further increase in temperature, the corrosion rate will decrease because the solubility of oxygen decreases, as deduced from the corrosion rates in an open system as compared to a closed one.

In supercritical water solutions at high temperature and high pressure corresponding to $T > 374\,^{\circ}C$ and $p > 22.05$ MPa, the oxygen solubility increases with temperature. Consequently, the oxidizing power of the solution increases, and electrochemical processes become more and more important. In the last two decades, these waters have become an interesting medium for many applications: chemical reactions, hydrothermal syntheses, waste oxidation, radioactive waste reduction, biomass conversion, plastic degradation, and synthesis of nanoparticles. The most commonly used reactor materials are stainless steels and nickel-based alloys [28].

2.5.3. Scale Formation and Water Indexes

Encrustation of tubing, boilers, coils, jets, sprinklers, cooling towers, and heat exchangers arises wherever hard water is used. Scale formation can greatly affect heat transfer performance. For example, 1-mm thick scale can add 7.5% to energy costs, whereas 1.5 mm adds 15% and 7 mm can increase costs by more than 70%. Many factors can affect scaling. Scaling, which is basically the deposition of mineral solids on the interior surfaces of water lines and containers, most often occurs when water containing the carbonates or bicarbonates of calcium and magnesium is heated. The saturation level (SL) of water in a mineral phase is a good indicator of the potential for scaling as a result of that specific scalant. SL is a ratio between the ion activity product (IAP) and the thermodynamic solubility product (K_{sp}) of a specific compound in that water. For example, when calcium carbonate ($CaCO_3$) is the scalant, SL is defined as

$$SL = \frac{a_{Ca^{2+}}\ a_{CO_3^{\,2-}}}{K_{sp}}$$

where $a_{Ca^{2+}}\ a_{CO_3^{2-}}$ is the IAP of the two ions involved in the formation of $CaCO_3$, that is, Ca^{2+} and CO_3^{2-} [6].

K_{sp} is a measure of ionic concentration when dissolved ions and undissolved ions are in equilibrium. When a saturated solution of sparingly or slightly soluble salt is in contact with undissolved salt, equilibrium is established between the dissolved ions and undissolved salt. In theory, this equilibrium condition is based on undisturbed water maintained at constant temperature and allowed to remain undisturbed for an infinite period of time. In this example, water is said to be undersaturated ($SL < 1$) if it can still dissolve calcium carbonate. When water is at equilibrium, $SL = 1.0$ by definition. Supersaturated water ($SL > 1$) will precipitate calcium carbonate from water if allowed to rest. As the saturation level increases beyond 1.0, the driving force for the precipitation of calcium carbonate increases [6].

The following sections describe some indexes that have gained wide acceptance in the corrosion community. However, it should be stated that these indexes are designed to indicate the tendency of given waters to deposit scales on metal substrates and not to predict the absolute corrosivity of specific waters. Generally, scales precipitated onto metal surfaces can provide protection of the substrate from general corrosion. If, on the other hand, the scales are defective and contain voids or cracks, they could lead to localized corrosion. The assumption that water below calcium carbonate saturation is corrosive, although occasionally correct, is not reliable [6].

Langelier Saturation Index (LSI) The LSI is frequently used to determine whether or not water has the tendency to deposit scale and is a useful way to quantify water aggressiveness. The total quantity of carbon dioxide can be divided into bicarbonate and free carbon dioxide. The primary reaction in the formation of scale is

$$Ca(HCO_3)_2 \Rightarrow CaCO_3(c) + CO_2(g) + H_2O$$

Bicarbonates act as corrosion inhibitors by precipitating a protective film of calcium carbonate under alkalinity conditions. The free carbon dioxide can be divided into CO_2 equilibrant and aggressive carbon dioxide.

The Langelier Saturation Index ($LSI = pH - pH_s$) is a function of pH, temperature, calcium concentration, alkalinity, and total dissolved solids concentration of the water. The method is based on the calculation of the pH of water at saturation (pH_s) in calcite or calcium carbonate, which is then compared to the measured one. A positive LSI indicates that water is oversaturated with respect to calcium carbonate and has the tendency to form scale. A negative LSI may sometimes (but not always) indicate that water is corrosive, especially if the water contains dissolved oxygen. An LSI of zero indicates that the water is at equilibrium with respect to calcium carbonate and should neither tend to deposit nor dissolve the scale of the calcium carbonic equilibrium [29]:

$$\begin{aligned} pH_s &= C + pCa + pAlc \\ pCa &= -\log(Ca^{2+}) \\ pAlc &= -\log(HCO_3)^- \\ C &= pK_2' - pK_s' \\ pK_2' &= pK_2 - 2\varepsilon \\ pK_2 &= -\log K_2;\ [H^+][CO_3^{2-}] = K_2'[HCO_3] \end{aligned}$$

$$pK_s' = pK_s - 4\varepsilon;\ [Ca^{2+}][CO_3^{2-}] = K'_s$$
$$\varepsilon = \frac{\mu}{1 + 1.4\sqrt{\mu}}$$
$$\mu = \frac{1}{2}\sum c_i z_i^2$$

where μ is ionic force, c_i is ion molar concentration, and z_i is ion valence.

Examples concerning the calculation of LSI for drinking water can be found in References 6 and 22. To calculate the LSI it is necessary to know the alkalinity (mg/L, as $CaCO_3$ or calcite), the calcium hardness (mg/L, Ca^{2+} as $CaCO_3$), the total dissolved solids (mg/L, TDS), the actual pH, and the temperature of the water (°C). If the amount of TDS is unknown but conductivity is known, one can estimate the amount (mg/L) of TDS using a conversion table [6].

Ryznar Stability Index (RSI) The RSI is also sometimes employed for a more meaningful indication. The RSI uses a correlation established between an empirical database of scale thickness observed in municipal water systems and associated water-chemistry data. Similar to LSI, the RSI has its basis in the concept of saturation level. The RSI takes the form $RSI = 2(pH_s) - pH$.

A RSI <3.5 indicates scaling tendency; for RSI <6, the scaling tendency increases as the index decreases; for RSI >7, the calcium carbonate formation probably does not lead to protective corrosion-inhibitor film; RSI >7.5 indicates scale dissolution; and for RSI >8, mild steel corrosion becomes an increasing problem.

Both LSI and RSI may readily be calculated by a computer program operating under Lotus 1-2-3 and available from the American Water Works Association (AWWA) [30].

Puckorius Scaling Index (PSI) The PSI is based on the buffering capacity of the water and the maximum quantity of precipitate that can form in bringing water to equilibrium. This has not been considered in the Langelier and Ryznar indexes. Water high in calcium but low in alkalinity and buffering capacity can have a high calcite-saturation level. The high calcium level increases the ion activity product. Such water might have a high tendency to form scale because of the driving force, but scale formed might be of such a small quantity as to be unobservable. The water has the driving force but not the capacity and ability to maintain pH as precipitate matter forms. The PSI is calculated in a manner similar to the RSI. PSI uses an equilibrium pH rather than the actual system pH to account for the buffering effect [6].

$$PSI = 2(pH_s) - pH_{eq}$$

where pH_s is still the pH at saturation in calcite or calcium carbonate.

$$pH_{eq} = 1.465\ \log_{10}[\text{Alkalinity}] + 4.54$$
$$\text{Alkalinity} = [HCO_3^-] + 2[CO_3^{2-}] + [OH^-]$$

Larson–Skold Index The Larson–Skold Index describes the corrosivity of water toward mild steel. The index is based on evaluation of in situ corrosion of mild steel lines transporting Great Lakes water. The index is the ratio of equivalents per million (epm) of sulfate (SO_4^{2-}) and chloride (Cl^-) to the epm of alkalinity in the form of bicarbonate plus carbonate ($HCO_3^- + CO_3^{2-}$).

$$\text{Larson–Skold Index} = \frac{\text{epm Cl}^- + \text{epm SO}_4^{2-}}{\text{epm HCO}_3^- + \text{epm CO}_3^{2-}}$$

Considering the studied waters for the Larson–Skold Index, extrapolation to other waters, such as those of low alkalinity or extreme alkalinity, goes beyond the range of the original data. The Larson–Skold Index can show the following:

Index <0.8: Chlorides and sulfates probably will not interfere with natural film formation.

0.8< Index <1.2: Chlorides and sulfates may interfere with natural film formation. Higher corrosion rates than desired might be anticipated.

Index >1.2: The tendency toward high corrosion rates of a local type should be expected as the index increases.

Extrapolation to waters other than the Great Lakes, such as those of low alkalinity or extreme alkalinity, goes beyond the range of the original data. The index has proved to be a useful tool in predicting the aggressiveness of once-through cooling waters [6, 31]. Since the index has been developed for steel, aluminum pitting can be admitted as a possibility.

Oddo–Tomson Index The Oddo–Tomson Index accounts for the effect of pressure and partial pressure of carbon dioxide on the pH of water and on the solubility of calcium carbonate. This empirical model also incorporates corrections for the presence of two or three phases (water, gas, and oil). Interpretation of the index is by the same scale as for the LSI and Stiff–Davis Index [6].

Stiff–Davis Index The same concept of saturation level of Langelier index was used to evaluate the tendency of oil field water to deposit calcium carbonate. Considering the total dissolved solids and the common ion effect, Stiff–Davis indices show that water is less scale forming than the LSI calculated for the same water chemistry and conditions. The deviation between the indices increases with ionic strength [6, 32].

2.6. CORROSION AT HIGH TEMPERATURES

2.6.1. Description

High-temperature corrosion is a form of corrosion that does not require the presence of a liquid electrolyte. The reaction of a metal and gas at high temperatures can be referred to as tarnishing, high temperature oxidation, or scaling. In general, the names of the corrosion mechanisms are determined by the most dominant corrosion products. For example, oxides are involved in oxidation, sulfides in sulfidation, and carbides in carburization [32]. The high-temperature reaction is considered generally to be electrochemical in nature, especially when the product (e.g., oxide film) is not volatile. Besides the importance of high-temperature corrosion resistance in a modern society, studies of electrochemical corrosion at high temperatures in a gaseous phase help us to understand the corrosion mechanisms and kinetics in aqueous and organic solutions. A clean reactive metal surface exposed to oxygen follows a sequence, very parallel to that in aqueous solutions: adsorption of oxygen, formation of oxide nuclei, and finally a growth of a continuous oxide film. In certain cases, the corrosion mechanism in dry atmospheres is considered in "wet" or aqueous solutions such as aluminum and zirconium in water at high temperatures and steel in dilute caustic soda at high temperatures and pressures [16].

Oxidation is probably the predominant high-temperature corrosion process encountered in the heat treating industry, owing to the fact that air is the most common heat treating atmosphere for cost reasons. The high-temperature oxidation resistance of stainless steel as well as aluminum- and silicon-bearing alloys relies on the development of tenacious, nonporous, and adherent films. Cr_2O_3, Al_2O_3, and SiO_2 coatings are selected very frequently for their good performance [33, 34].

The high-temperature corrosion processes that are most frequently responsible for the degradation of furnace accessories are oxidation, carburization, decarburization, sulfidation, molten-salt corrosion, and molten-metal corrosion. In each case, the corrosion susceptibility of the components is enhanced due to thermal cycling. Thermal cycling not only promotes the loss of protective surface film but also causes structural distortion. Numerous oxidation tests on commercial alloys have been performed at 980 °C or higher. For example, in one investigation, cyclic oxidation tests were conducted in air and each cycle consists of exposing the samples at 980 °C for 15 min, followed by a 5 min air cooling. Application of high-temperature corrosion and oxidation-resistant coatings has become common for furnace accessories and equipment. These coatings are typically ceramics of thermally dense materials applied using physical vapor deposition, plasma, chemical vapor deposition, and high-velocity oxyfuel spray methods [33, 34].

2.6.2. The Pilling–Bedworth Ratio (PBR)

In one of the earliest scientific studies of oxidation, Pilling and Bedworth [35] proposed that oxidation resistance should be related to the difference in the molar volumes of metal and the oxide scale formed on it whether the molar volume of reaction product is greater or less than the volume of metal from which the product forms. Mathematically, this can be expressed as [35]

$$R = \frac{Wd}{nDw}$$

where W and D are the molecular weight and density, respectively, of the oxide, w and d are the atomic weight and density, respectively, of the metal, and n is the number of metal atoms in the oxide molecule ($n = 2$ for Al_2O_3). The ratio R indicates the volume of oxide formed from a unit volume of metal. Iron can have different PBR values with respect to α-Fe depending on the nature of the oxide: $FeO = 1.68$, $Fe_3O_4 = 2.10$, and $Fe_2O_3 = 2.14$. A Pilling–Bedworth volume ratio of less than 1 produces insufficient oxide to cover the metal and is nonprotective, such as $Li_2O/Li = 0.57$, and the scale has tensile stress [32].

On the other hand, if the oxide volume is very high with a PBR much greater than 1, this tends to introduce large compressive stresses in the oxide, which also causes poor oxidation resistance due to a break in the adhesion between the metal and oxide, cracking, and spalling, such as $SiO_2/Si = 2.15$ and $Cr_2O_3/Cr = 2.07$. Figure 2.2a,b shows the relief of compression constraints in the oxide during curling and, because it has little intrinsic tensile strength, eventually rupturing and spalling. If temperature variations occur, a weakly adherent scale is likely to separate locally from the base metal, a phenomenon sometimes termed blistering (Figure 2.2c). Values of the PBR such as that of $MgO/Mg = 0.79$–0.81 and $Al_2O_3/Al = 1.28$ could be considered protective [25, 35–37].

It has been proved that the PBR alone does not accurately predict oxidation resistance since continuous layers are observed even if PBR < 1; cracks or fissures in oxide layers can

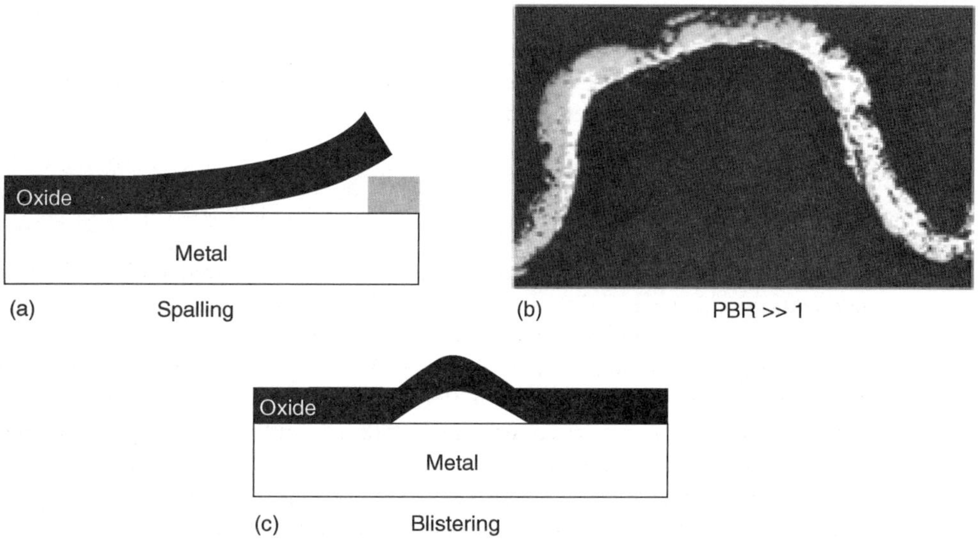

Figure 2.2 (a) Schematic representation of an oxide film spalling [37]; (b) Cr_2O_3 flaking on pure chromium at 1100 °C (PBR = 2.07) [38]; and (c) schematic representation of oxide film blistering [37].

be "self-healing" as oxidation progresses and oxide porosity is not accurately predicted by the PBR parameter. Also, to be protective, an oxide must possess good adherence, a high melting point, a low vapor pressure, good high-temperature plasticity to resist fracture, low electrical conductivity or low diffusion coefficients for metal ions or oxygen, and high-temperature plasticity [25, 31, 36].

The main reason for scale spallation is the stress generation in the oxide during its growth. The stresses start accumulating with the oxide growth process as the weight increases due to oxidation. However, at a certain point, the scale thickness is unable to bear the increased stress and it releases the stress. The release of stress can either be due to cracking in the scale or creep of the substrate metal. Mode of diffusing species, differences in the thermal expansion coefficients between the oxide and the metal, composition of the scale, and geometry of the sample could contribute to increasing stresses [32, 36].

Mechanisms of Formation of the Oxide Film In many cases, the scale formed is very smooth and adherent and does not spall at all, while in other cases, the spallation starts at the very early stages of oxidation and continues with intermediate rebuilding of the scale. The sign and level of the stresses in the scale depend on its growth direction, on the radius of service curvature, and on the Pilling–Bedworth ratio (PBR). Some oxides actually grow at the oxide–air interface as opposed to the metal–oxide interface. Since the lattices of metals and surface oxides are generally dissimilar, nucleation and growth of oxides on metal surfaces may be accompanied by epitaxial stresses. These stresses are expected to be maximum at the metal–oxide interface and decrease to zero at the free surface [26].

The reason for such growth stresses is that the oxide scale does not grow in only one direction: in most cases, it grows by counterdiffusion of the oxide forming species. In this case, part of the oxidation (i.e., oxide formation) takes place in the interior of the oxide scale, thus creating new oxide volume resulting in the buildup of compressive stresses. Intrinsic growth stresses may arise partially from phase changes in the oxide. Buckling and cracking of oxide scales due to intrinsic growth stresses can be suppressed by the addition of active elements to the alloy. There are several possible explanations for the active element effect, which helps to improve adhesion of the scale and which has been proved for a number of different materials [39].

The choice of the mechanism(s) to better control stresses in the oxides depends on the characteristics of each system and these could be summarized as follows [26, 39]:

1. Change of the transport properties in the scale; that is, metal cation diffusion in the outward direction can be suppressed by the presence of certain active elements so that only oxygen-inward transport occurs, leading to scale formation solely at the metal interface. Thus the amount of growth stresses and the scale thickness are reduced.
2. Increase of the number of oxide nucleation sites at the beginning of oxidation, resulting in a more fine-grained oxide and a reduced number of nonprotective oxide nuclei, both leading to a scale that is more tolerant of mechanical stresses.
3. Formation of oxide pegs at the metal grain boundaries leading to a keying effect of the oxide on the metal.
4. Reduction in the number of physical defects like pores in the oxide, thus reducing the susceptibility of the scale to cracking or detachment.
5. Tie-up of sulfur impurities in the metal substrate, thus impeding the segregation of sulfur to the metal substrate interface, which would usually reduce the adhesion strength of the scale.

Positive and Negative Carriers (p and n Types) The corrosion rate increases with temperature. The surface film typically thickens as a result of reaction at the scale–gas or metal–scale interface due to cation or anion transport through the scale, which behaves as a solid electrolyte. High-temperature scales are usually thought of as oxides but may also be sulfides, possibly carbides, or a mixture of these species. Oxides and sulfides are non-stoichiometric compounds and semiconductors. There is a positive carrier (p-type) and a negative carrier (n-type). For diffusion-controlled scaling, the rate of scale growth can be altered by modification of the concentration of the particular defects involved: for example, p-type oxides exhibit increased cationic transport rates at increased oxygen pressures, while transport in n-type oxides is essentially independent of oxygen pressure. Both oxides can be doped by the addition of specific ions to the oxide lattice [38].

On the submolecular level, metal oxides contain defects, in the sense that their composition deviates from their ideal stoichiometric chemical formulas. By nature of the defects found in their ionic lattices, they can be subdivided into three categories [32]:

1. An n-type cation interstitial metal-excess oxide contains interstitial cations, in addition to the cations in the crystal lattice. Charge neutrality is established through an excess of negative conduction electrons, which provide for electrical conductivity (Figure 2.3).
2. An n-type anion vacancy oxide contains oxygen anion vacancies in the crystal lattice. Current is passed by electrons, which are present in excess to establish charge neutrality (Figure 2.4).

The growth of an n-type cation interstitial oxide at the oxide–gas interface is illustrated in Figure 2.3. Interstitial metal cations are liberated at the metal–oxide interface and migrate through the interstices of the oxide to the oxide–gas interface. Conduction band electrons also migrate to the oxide-gas interface, where oxide growth takes place. For the n-type anion vacancy oxide, film growth tends to occur at the metal–oxide interface, as shown in

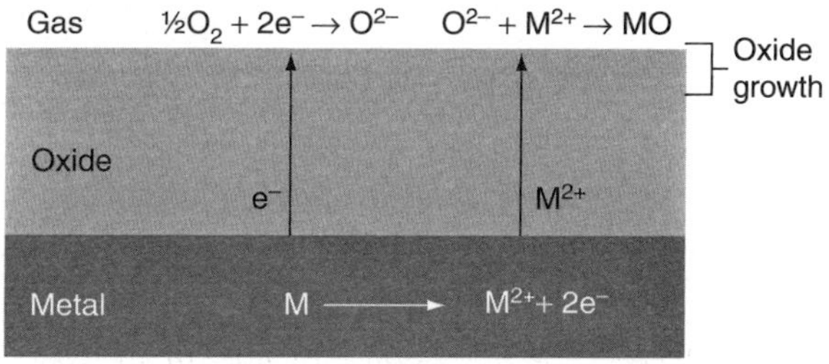

Figure 2.3 Schematic description of the growth of an n-type cation interstitial oxide occurring at an oxide–gas interface [32].

Figure 2.4. Conduction band electrons migrate to the oxide–gas interface, where the cathodic reaction occurs. The oxygen anions produced at this interface migrate through the oxide lattice by exchange with anion vacancies. The metal cations are provided by the anodic reaction at the metal–oxide interface [32].

3. A p-type metal-deficit oxide contains metal cation vacancies. Cations diffuse in the lattice by exchange with these vacancies. Charge neutrality in the lattice is maintained by the presence of electron holes or metal cations of higher than average positive charge. Current is passed by positively charged electron holes (Figure 2.5).

High-temperature electrochemical reactions proceed by a similar process to that in aqueous solution. For example, the reaction $M + \frac{1}{2}O_2 \rightarrow MO$ is composed of $M \rightarrow M^{2+} + 2e^-$ (anodic reaction) and $\frac{1}{2}O_2 + 2e^- \rightarrow O^{2-}$ (cathodic reaction).

In the case of the p-type metal-deficit oxides, metal cations produced by the anodic reaction at the metal–oxide interface migrate to the oxide–gas interface by exchange with

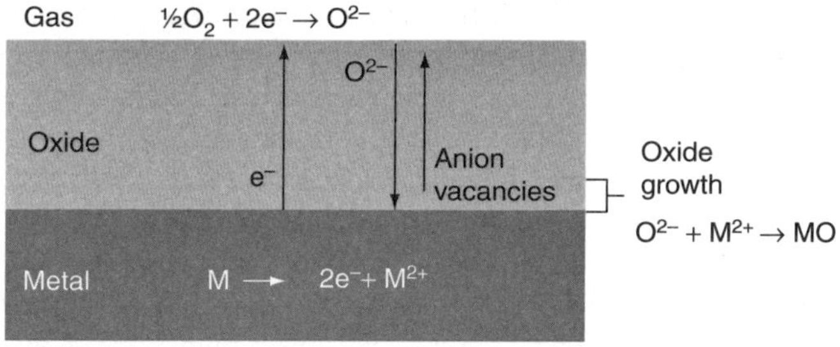

Figure 2.4 Film growth of an n-type anion vacancy oxide occurring at a metal–oxide interface [32].

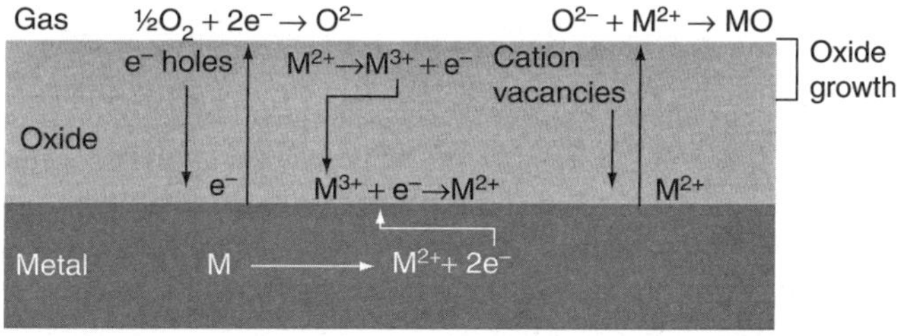

Figure 2.5 Schematic description of a cathodic reaction and oxide growth occurring at oxide–gas interface [32].

cation vacancies. Electron charge is effectively transferred to the oxide–gas interface by the movement of electron holes in the opposite direction (toward the metal–oxide interface). The cathodic reaction and oxide growth thus tend to occur at the oxide–gas interface (Figure 2.5) [32].

The important influence of the diffusion of defects (excess cations, cation vacancies, or anion vacancies) through the oxide film on oxidation rates should be apparent from Figures 2.3–2.5. Doping of the scale is an interesting avenue for metal protection. Conduction electrons (or electron holes) are much more mobile compared to these larger defects and therefore are not important in controlling the reaction rates. For example, if nickel oxide [40] is considered as a p-type metal-deficient oxide, the oxidation rate of nickel depends on the diffusion rate of cation vacancies. If this oxide is doped with Cr^{3+} impurity ions, the number of cation vacancies increases to maintain charge neutrality. A higher oxidation rate is thus to be expected in the presence of these impurities. By this mechanism, a nickel alloy containing a few percentages of chromium does indeed oxidize more rapidly than pure nickel. From these considerations, a clearer picture of requirements for protective oxides has emerged [32]. On the other hand, p-type metal-deficit oxides can be doped by the addition of cations of lower valence than the native cations, resulting in a decrease in the number of cation vacancies and therefore a decrease in the oxidation rate. Sulfides exhibit a greater rate of transport of anions and cations than the oxides of the same metal and so provide scales, but much less protective than oxides [38].

2.6.3. Kinetics of Formation

Initial film growth is usually very rapid on clean surfaces in an oxidizing environment. For a nonporous scale that covers the metal surface completely, the reaction rate decreases with time and the transport of various species through the film becomes rate controlling. The subsequent corrosion rate depends on the details of this transport mechanism. It may be driven by gradients of electrical potential, and/or gradients of concentration or by migration along preferential paths. After the quick formation of a surface film a few thousand angstroms thick the subsequent corrosion rate follows one of the laws given in Figure 2.6 [38].

Linear Rate Law If the scale is porous, nonprotective, or does not cover the surface, a linear rate is expected. The linear rate law is $x = k_1 t$, where k_l is the rate constant. If a vapor species is formed, the reaction can be chemical and the rate is frequently linear. This linear scale is expected for metals having a PBR lower than 1; it can also hold for metals with a PBR higher than 1 when the rate of formation of the outer scale becomes equal to that of the inner scale, such as that of tungsten at 700–1000 °C since it forms an outer porous scale (WO_3) and an inner compact scale of different composition [26].

Metals such as molybdenum, tungsten, osmium, rhenium, and vanadium, associated with volatile oxide formation and having linear oxidation kinetics at a certain temperature, could undergo so-called catastrophic oxidation (also referred to as *breakaway corrosion*) at higher temperatures. In the case of magnesium, ignition of the metal may even occur. The formation of low-melting-point oxidation products (eutectics) on the surface has also been associated with catastrophic oxidation. In this case, a rapid exothermic reaction occurs on the surface, which increases the surface temperature and the reaction rate even further. The presence of vanadium and lead oxide contamination in gases deserves special mention because they pose a risk to inducing extremely high oxidation rates [32, 41].

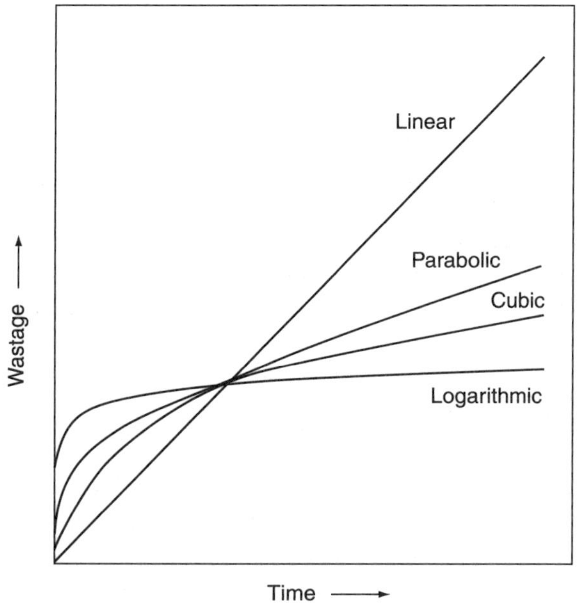

Figure 2.6 The kinetic curve shapes that represent several processes of thermal deterioration [38].

Parabolic Rate Law For continuous nonporous scales, ionic transport through the scale is rate determining. The kinetics usually follow a parabolic law, in which the rate progressively decreases with time. The parabolic rate law assumes that the diffusion of metal cations or oxygen anions is the rate-controlling step and is derived from Fick's first law of diffusion:

$$x^2 = k_p t + x_0$$

where x is the oxide film thickness (or mass gain due to oxidation, which is proportional to oxide film thickness), t is time, k_p is the rate constant (directly proportional to diffusivity of ionic species that is rate controlling), and x_0 is a constant [32].

The concentrations of diffusing species at the oxide–metal and oxide–gas interfaces are assumed to be constant. This implies that the oxide layer has to be uniform, continuous, and of the single-phase type and there is no temperature gradient. However, when the rate-controlling step in the oxidation process is the diffusion of ions through a compact barrier layer of oxide with the chemical potential gradient as the driving force, a parabolic rate law holds well for metals having PBR >1 such as Cu, Ni, Fe, Cr, and Co [32]. However, under certain conditions, the compressive stresses resulting from a Pilling–Bedworth ratio greater than 1 become sufficiently great that the scale or alloy deforms and possibly spalls as a relief mechanism (Figure 2.6) [38]. Tungsten has a PBR $= 3.6$ for WO_3/W and is normally expected to be protective except at high temperatures above 800°C, where it volatilizes. It first oxidizes in accord with the parabolic equation [26].

As the oxide grows thicker, the diffusion distance increases, and the oxidation rate slows down. The rate is inversely proportional to the oxide thickness, that is, $dx/dt = k_p/x$, where k_p is the parabolic rate constant. The equation can also be expressed in logarithmic form [32, 41]:

$$\ln x = \tfrac{1}{2}\ln(2k_p) + \tfrac{1}{2}\ln t$$

Logarithmic Rate Law When metals oxidize initially or at low temperatures to form thin protective films, oxidation is usually observed to follow logarithmic kinetics. The logarithmic rate law is given by $x = k_g \log(bt + 1)$, where k_g and b are constants for a particular set of conditions [41].

Transport processes across the film are rate controlling. This has been found to express the initial oxidation behavior of Al, Fe, Ti, Cu, Ni, Sn, Zn, Pb, Cd, Mn, and Ta but is rarely applicable to high-temperature engineering problems [26, 32]. If, for thin-film behavior, the migration of ions controls the rate and the prevailing electric field within the film is set up by gaseous ion adsorption on the outer surface, the rate of ion migration is an exponential function of the field strength and the inverse logarithmic equation has been reported to hold for Cu and Fe oxidized at low temperatures [26, 42].

Cubic Rate Law Oxidation following the cubic rate law is very probably due to combined mechanisms. The law is expressed as $x^3 = k_c t$ + constant, where k_c is the cubic rate constant. Data obeying this law can also be represented by a two-stage logarithmic equation, where an initial lower rate is followed by a final higher oxidation rate [26].

The oxidation rate for thin or thick films increases with temperatures and obeys the Arrhenius equation:

$$\text{Reaction rate constant} = A\exp(-\Delta E/RT)$$

where E is the activation energy, R is the gas constant, and T is the absolute temperature.

2.6.4. Corrosion Behaviors of Some Alloys at Elevated Temperatures

The influence of temperature on corrosion rate is very significant. It can change the state of the solid and the nature of the environment with more or less dissolved gases. It changes diffusion and reaction kinetics and, most importantly, the physicochemical properties of the interface. The rate of oxidation and the type of oxide scale are functions of the alloy, temperature, and oxygen pressure.

Protective Oxides (e.g., Al_2O_3, Cr_2O_3) The alloys intended for high-temperature application should have a protective oxide; however, the performance of alloys for very high temperatures necessitates a minimum of scale thickness. The oxides that are known to act as protective coatings are Al_2O_3, Cr_2O_3, and SiO_2. The best oxidation resistance is attributed to Al_2O_3 since it has the slowest transport rates for metal and oxygen ions. It has been shown that the parabolic rate constant, the composition, and the thickness of the scale layer vary as a function of the chromium content in a Fe–Cr alloy [38]. It has been found that a percentage of 20% Cr is required to have a protective Cr_2O_3 oxide. If this oxide is lost repeatedly in a certain aggressive environment, the percentage of chromium in the alloy should be increased. The breakdown of the protective oxides is caused mainly through mechanical means: thermal cycling (spallation), abrasion, or impact. Chemical reactions, such as molten species below scales, may also be detrimental to high-temperature corrosion resistance [38].

Sulfidation The partial pressure concentration of sulfur in a gaseous environment can be high enough to form sulfide phases instead of oxide phases. However, in the majority of practical environments, Al_2O_3 or Cr_2O_3 should form in preference to sulfides. The sulfidation attack occurs mainly at sites where the protective oxide has broken down. Once sulfur has entered the alloy, it appears that sulfur ties up the chromium and aluminum as sulfides and interferes with the process of formation and reformation of the protective scale [38]. Once sulfur forms sulfides, discrete sulfide precipitates can be observed beneath the protective oxide. The sulfur can be displaced inward, forming new sulfides, deeper in the alloys in grain boundaries or at preferable sites such as carbides (chromium- or aluminum-rich phases). Since metal sulfides grow much faster and melt more readily than oxides, the quality of protection by sulfides is very poor [38].

Hot Corrosion This type of high-temperature corrosion is typical in gas turbine engines. This can be divided into two types: type I for temperatures of the metal between 850 and 950 °C and type II for temperatures in the range of 650–700 °C.

Type I This sulfidation-based attack on the hot gas path parts involves the formation of condensed salts, which are often molten at the turbine operating temperature. The major components are sodium sulfate (melting point is 884 °C) and/or potassium sulfate. Sulfur comes from the fuel and sodium from the fuel or the ingested air [38].

Very small amounts of sulfur and sodium or potassium in the fuel and air can produce sufficient Na_2SO_4 in the turbine to cause serious corrosion problems because of the concentrating effect of the turbine pressure ratio. It has been suggested that concentrations of sodium $\geq$0.008 ppm by weight are critical to form this type of corrosion. Other fuel or air impurities, such as vanadium, phosphorus, lead, or chloride, may combine with sodium sulfate to form mixed salts having reduced melting temperature, which can accelerate the corrosion rate. Unburned carbon can promote deleterious interactions in the salt deposits. High chromium content in superalloys shows generally good resistance to high-temperature hot corrosion. Some coatings of certain superalloys having alloying elements such as chromium, tungsten, molybdenum, and tantalum, with recommended concentration levels of some of these elements, can be used to resist this type of corrosion [38].

Type II Low-temperature hot corrosion can result from the low melting point of mixed sulfates such as sodium sulfate–cobalt sulfate, which can have an eutectic temperature of 540 °C. Molten salts can cause characteristic pitting of the superalloy. The partial pressure of sulfur trioxide is critical in this type of attack. Cr-rich and cobalt-free nickel-based alloys are recommended. Coatings of these alloys are also advised [38].

Furnace environments get sulfur from fuels, fluxes used for specific operations, and cutting oil left on the parts to be heat treated, among other sources. Sulfur in the furnace environment could greatly reduce the service lives of components through sulfidation attack. It is well known that nickel-based alloys are highly susceptible to catastrophic sulfidation due to the formation of nickel-rich sulfides, which melt at approximately 650 °C [33, 34].

Nitriding Nitriding has been employed for years to provide hard, wear-resistant surfaces on certain low-alloy steels. The nitrogen molecule is relatively stable without important corrosion problems except at very high temperatures over very long periods. However, rapid nitride formation below 540 °C due to active nitrogen, produced by the decomposition of ammonia, is a serious problem. Ammonia is frequently used in chemical processes and is essential in the fertilizer industry. Ni and Cu do not form stable nitrides at elevated temperatures and confer some resistance to metals such as Fe, Al, and Ti that readily form nitrides [37].

Carburization Protective oxides of aluminum and chromium are more stable than sulfides or carbides. However, carburization can occur in many carbon-containing environments. Sufficient high carbon activities can be generated at the alloy surface for carburization by the concentration of carbon monoxide, for example, in the outer porous oxide layer or by the creation of a localized microenvironment. The high solubility of carbon in austenitic steels makes them more vulnerable to carburization than ferritic steels. Iron–chromium alloys containing more than ~20% Cr can absorb considerable amounts of carbon before formation of austenite, giving principally $(CrFe)_{23}C_6$ and ferrite. This leads to pitting attack as has been observed for a reactor made of 310 stainless steel [38].

The environment in the carburizing furnace typically has a carbon activity that is significantly higher than that in the alloy of the furnace component. Therefore carbon is transferred from the environment to the alloy, and the carburized alloy becomes embrittled. A protective atmosphere is maintained to protect against oxidation; that is, the CO/CO_2 ratio in the furnace atmosphere is maintained high to generate a low oxygen partial pressure, $CO + \frac{1}{2}O_2 \rightarrow CO_2$. However, a competing reaction requires that the carbon dioxide activity be high to prevent carbon deposition through the reaction $2CO \rightarrow C + CO_2$. The temperature-dependent competition between these two reactions determines the sensitivity of the heat treatment furnace accessories to oxidation, carburization, and decarburization. Alloys containing strong carbide formers, such as stainless steels, or having high permeability for carbon show poor resistance to carburization, while the alloys with weak carbide formers, such as nickel and aluminum, show better resistance [33, 34].

Metal Dusting Metal dusting problems have also been reported in petrochemical processing. Metal dusting is another frequently encountered mode of corrosion that is associated with carburizing furnaces. Metal dusting tends to occur in a region where the carbonaceous gas atmosphere becomes stagnant. The alloy normally suffers rapid metal wastage. The corrosion products (or wastage) generally consist of carbon attack on the multimetallic alloy. The component can be perforated as a result of metal dusting. Metal dusting has been encountered with straight chromium steels, austenitic stainless steels, and nickel- and cobalt-based alloys. All of these alloys are chromium formers; that is, they form Cr_2O_3 scales when heated to elevated temperatures. No metal dusting has been reported on the alloy systems that form a much more stable oxide scale, such as Al_2O_3, because alumina is a much more thermodynamically stable oxide than chromia. Al_2O_3 scale is much more resistant to carburization attack than Cr_2O_3 scale. Because metal dusting is a form of carburization, it would appear that alumina formers, such as Haynes alloy 214, would also be more resistant to metal dusting [33, 34].

Hydrogen Reactions at High Temperature At high temperatures and pressures, hydrogen can penetrate the metal as atomic hydrogen and react with reducible species. Atomic hydrogen can react with iron carbides to form methane and fissure the metal (decarburization). Chromium carbides are less susceptible to this reaction. Hydrogen can reduce the metallic oxide to form steam and metal. Copper frequently contains small quantities of Cu_2O, which can produce steam within the alloy and result in significant void formation [38].

Liquid-Metal Corrosion at High Temperature The corrosion of metals and alloys by liquid metals at high temperature can be described as alloying. In some special cases, electron transfer processes involving reducible impurities in the liquid metal may modify or over ride the simple dissolution process. Dissolution can be uniform or localized, like the leaching of one component of an alloy or intergranular attack. Temperature gradient mass transfer and dissimilar metal or chemical activity gradient mass transfer can occur. Reducing the thermal-hydraulic performance of heat exchange systems, fouling, and blocking of tubes by deposition and degrading mechanical properties are concrete engineering difficulties [43].

Direct dissolution is the release of atoms of the containment material into the melt in the absence of any impurity effects. If the liquid-metal system is nonisothermal, forced circulation (pumping) of liquid metals used as heat transfer media exacerbates the transport of materials from hotter to cooler parts of the liquid-metal circuit. This leads to irregular attack, which can also be observed as general attack. The presence of impurities and/or the compositional inhomogeneities in the solid are the main factors in this type of corrosion. Which general corrosion is observed, localized corrosion can be caused by the preferential or selective dissolution of some constituents of the molten alloy [44].

The decarburization of steel in lithium and the oxidation of steel in sodium or lead of high oxygen activity are examples of the impurity and interstitial reactions that characterize a net transfer of interstitials or impurities to, from, or across a liquid metal. Alloying between atoms of the liquid metals and those of the constituents of the containment material can lead to the formation of a stable product on the solid [44]. The corrosion of ceramics exposed to liquid metals can be caused by the reduction of the solid by the melt. This concerns the loss of structural integrity by the reduction-induced removal of the nonmetallic element from the solid [44].

Liquid-metal embrittlement is the catastrophic brittle failure of a normally ductile metal when coated with a thin film of a liquid metal and subsequently stressed under tension. The fracture changes from a ductile to a brittle intergranular or brittle transgranular (cleavage) mode; however, there is no change in the yield or flow behavior of the solid metal [45].

Solid Metal-Induced Embrittlement (SMIE) Embrittlement occurs below the melting temperature of the solid in certain liquid-metal environment (LME) couples. The severity of embrittlement increases with temperature, with a sharp and significant increase in severity at the melting point, T_m, of the embrittler. For example, embrittlement of 4140 steel by various liquid metals below their melting point (Pb, Cd, Zn, Sn, and In) increases enormously when the ratio of T/T_m approaches 1, where T is the corrosion temperature and T_m is the melting temperature. Severe embrittlement is observed especially in the region when T/T_m is between 0.8 and 1 [45].

Fused Salts Corrosion The attack of a metal or a material by a salt melt can be classified as direct dissolution without oxidation of the metal, a mechanism similar to attack by liquid metals. If the solubility is accessible, corrosion can occur in some rare cases. Most of the metals of the first and second groups of the periodic table are soluble in their own halides, and in certain cases there is a complete miscibility at high temperatures. If the metal is oxidized to metal ion, this constitutes an electrochemical reaction [46].

Most fused salts are predominantly ionic but contain a proportion of molecular constituents. The relative mobilities of the salt melt and the metal are important. A noble metal in contact with a pure melt of a base-metal cation can react only to a very limited extent provided the anion is not reducible; for example, nickel cannot corrode in molten sodium chloride unless another reducible impurity is present [46].

Predictions of corrosion are difficult or even impossible in engineering systems. The most prevalent molten salts are nitrates and halides and to a lesser extent sulfates, hydroxides, and oxides. Principally, molten or fused salts can cause attack of materials by electrochemical reactions (general, pitting, and selective corrosion), mass transport due to thermal gradients, and reaction of the constituents or the impurities of the molten salt with the container material [47].

REFERENCES

1. A. Aballe, M. Bethencort, F. J. Botana, M. Marcos, and J. M. Sanchez-Amaya, *Corrosion Science* **46**, 1909–1920 (2004).
2. P. Roberge, in *Handbook of Corrosion Engineering*, edited by Robert Esposito. McGraw-Hill, New York, 2000, pp. 55–216, 105–136.
3. C. P. Dillon, *Forms of Corrosion Recognition and Prevention*. International Association of Corrosion Engineers, Houston, TX, 1982.
4. S. L. Pohlman, in *ASM Handbook, Volume 13, Corrosion*, edited by J. R. Davis, ASM International, Materials Parks, OH, 1987, pp. 80–103.
5. P. A. Schweitzer, in *Corrosion Engineering Handbook* edited by Philip A. Schweitzer. Marcel Dekker, New York, 1996. pp. 99–156.
6. P. Roberge, *Corrosion Basics: An Introduction*, 2nd edition. NACE International, Houston, TX, 2006, pp. 125–136.
7. F. H. Haynie, J. W. Spence, and J. B. Upham, in *Atmospheric Factors Affecting the Corrosion of Engineering Metals, ASTM STP 646*, edited by S. K. Coburn. American Society for Testing and Materials (ASTM), Philadelphia, PA, 1978, pp. 30–47.
8. F. N. Longo and G. J. Durmann, in *Atmospheric Factors Affecting the Corrosion of Engineering Metals, ASTM STP 646*, edited by S. K. Coburn. American Society for Testing and Materials (ASTM), Philadelphia, PA, 1978, pp. 97–114.
9. G. German, in *Atmospheric Factors Affecting the Corrosion of Engineering Metals, ASTM STP 646*, edited by S. K. Coburn. American Society for Testing and Materials (ASTM), Philadelphia, PA, 1978, pp. 74–82.
10. J. G. Kaufman, in *ASM Handbook, Volume 13B, Corrosion: Materials*, edited by S. D. Cramer, and B. S. Covino, Jr. ASM International, Materials Park, OH, 2005, 95–124.
11. M. A. Butler and H. C. K. Ison, *Corrosion and Its Prevention in Waters*. Reinhold Publishing, New York, 1966, pp. 18–60.
12. N. S. Rawat, *British Corrosion Journal* **11**, 86–91 (1976).
13. V. S. Sastri, G. R. Hoey, and R. W. Revie, *CIM Bulletin* **87**, 87–99 (1994).
14. J. Hadjigeorgiou, E. Ghali, F. Charette, and M. R. Krishnadev, in *Fracture Analysis of Friction Rock Bolts*, edited by W. B. R. Hammah, J. Curran, and M. Telesnicki. Proceedings of the 5th North American Rock Mechanics Symposium and the 17th Tunnelling Association of Canada Conference: Narms-Tac 2002, Toronto Canada, 2002 (Narms-Tac 2002), pp. 881–886.
15. E. Heitz, in *Advances in Corrosion Science and Technology Volume 4*, edited by M. G. Fontana, and R. W. Staehle, Plenum Press, New York, 1974, p. 149.
16. L. L. Shreir, R. A. Jarman, and G. T. Burstein, *Corrosion—Metal/Environment Reactions*, 3rd edition. Butterworth-Heinemann, Oxford, UK, 1995, p. 1–18.
17. N. Tomashov and Y. Mikhailovsky, Study on soil corrosivity measuring techniques, *Corrosion* **15**, 77t (1959).
18. U.S. Patent, 4,853,035 (August 1989).
19. M. Romanoff, *Underground Corrosion Characteristics of Soils*. NACE International, Houston, TX, 1989, pp. 3–13.
20. C. Li and K. Lindblad,Research Report Tulea, 1995, Lulea University of Technology, 1996.
21. J. Ranasooriya, G. W. Richardson, and L. C. Yap, *Corrosion Behaviour of Friction Rock Stabilities Used in Underground Mines in Western Australia*. Underground Operator's Conference, Kalgloorlie, Western Australia, 1995. Australasian Institute of Mining and Metallurgy, 1995, pp. 9–16.
22. L. D. Benefield, J. F. Judkins, and B. L. We, in *Process Chemistry for Water and Waste Water Treatment*. Prentice Hall, Englewood Cliff, NJ, 1982, pp. 239–266.
23. D. C. Silverman and R. B. Puyear, in *ASM Handbook, Volume 13, Corrosion*, J. R. Davis. ASM International, Materials Parks, OH, 1987, pp. 37–44.
24. B. P. Boffardi and G. W. Schweitzer, *Water quality, corrosion control and monitoring*. Proceedings of the International Congress on Metallic Corrosion, Toronto, 1984, pp. 291–295.
25. ASM International Handbook Committee, in *Corrosion—Understanding the Basics*, edited by J. R. Davis. ASM International, Materials Park, OH, 2000, pp. 21–48.
26. H. H. Uhlig and R. W. Revie, *Uhlig's Corrosion Handbook*. Wiley, Hoboken, NJ, 1985 pp. 8, 35–59.

27. F. N. Speller, *Corrosion*, 3rd edition, McGraw-Hill, New York, 1951, p. 168.
28. P. Kritzer, *Journal of Supercritical Fluids* **29**, 1–29 (2004).
29. L'équilibre calcocarbonique, *Memento Technique de l'eau Tome113.1.1-13.1.2*. Edition du Cinquantenaire, neuvième édition, 1989.
30. *TPC Publication 7*. NACE International, Houston, TX, 1994, pp. 15–16.
31. T. E. Larson, and R. V. Skold, *Journal ofAmerican Water Works Association "AWWA"*, **49**, 1294 (1957).
32. P. R. Roberge, *Handbook of Corrosion Engineering*. McGraw-Hill, New York, 2000, pp. 221–265.
33. G. A. Minick and D. L. Olson, in *ASM Handbook, Volume 13 Corrosion* 9th edition, edited by L. J. Korb, and D. L. Olson, ASM International, Materials Park, OH, 1987, pp. 1293–1298.
34. B. Mishra, in *ASM Handbook Volume 13C*, edited by S. D. Cramer, and B. S. Covino, Jr. ASM International, Materials Park, OH, 2006, pp. 1067–1075.
35. N. Pilling and R. Bedworth, *Journal of the Institute of Metals* **29**, 529–582 (1923).
36. A. S. Khanna, *Introduction to High Temperature Oxidation and Corrosion*. ASM International, Materials Park, OH, 2002, pp. 202–216.
37. P. Roberge, *Corrosion Basics: An Introduction*, 2nd edition. NACE International, Houston, TX, 2006, pp. 217–263.
38. I. G. Wright, in *ASM Handbook, Volume 13 Corrosion*, 9th edition, edited by L. J. Korb, and D. L. Olson, ASM International, Materials Park, OH, 1987, pp. 96–103.
39. M. Schutze, in *Materials Science and Technology*, edited by R. W. Cash, P. Haasen, and E. J. Kramer., Wiley-VCH, Weinheim, Germany, 2000, pp. 94–95.
40. International Union of Pure and Applied Chemistry, *Quantities, Units and Symbols in Physical Chemistry*. Blackwell Scientific Publication, Oxford, UK, 1988.
41. C. Bagnall and W. F. Brehm, ASM Handbook, Volume 13, Corrosion, 9th edition, edited by L. J. Korb, and D. L. Olson. ASM International, Materials Park, OH, 1987, pp. 91–103.
42. D. Gilroy and J. Mayne, *Corrosion Science* **5**, 55–58 (1965).
43. G. Long and A. W. Thorley, *Corrosion, Volume 1*, Shreir L. L. Jarman R. A. and Barstein G. T, Butterworth-Heinemann, Oxford, UK, 1995, pp. 120–129.
44. Tortorelli P. F, in *ASM Handbook, Volume 13 Corrosion*, 9th edition, edited by L. J. Korb and D. L. Olson. ASM International, Materials Park, OH, 1987, pp. 56–60.
45. M. H. Kamdar, in *ASM Handbook, Volume 13, Corrosion*, 9th edition, edited by L. J. Korb and D. L. Olson. ASM International, Materials Park, OH, 1987, pp. 171–187.
46. D. Inman, in *Corrosion Volume 2*, edited by R. A. Jarman, L. L. Shreir, and G. T. Barstein. Butterworth-Heinemann, Oxford, UK, 1995, pp. 130–142.
47. J. W. Koger, in *ASM Handbook Volume 13 Corrosion*, 9th edition, edited by L. J. Korb, and D. L. Olson. ASM International, Materials Park, OH, 1987, pp. 50–55, 88–91.

Chapter 3

Active and Passive Behaviors of Aluminum and Magnesium and Their Alloys

Overview

The E–pH diagrams are based on the reactions between pure metals immersed in pure water and thermodynamic equilibrium states. There are serious issues that arise when we extrapolate the prediction of metal corrosion by this method to corresponding variable states at the metal–solution interface and to alloys that could contain impurities or alloying elements. The term η is generally the sum of four types of overpotentials: transfer overpotential of the electron, diffusion overpotential (η_d), reaction overpotential ($\eta_{d'}$), and resistance overpotential (η_r). The overvoltage of the hydrogen evolution reaction is considered. The depolarized cathodic reaction by oxygen or an oxidant is explained. There are major differences between the thermodynamic calculated values and the open circuit potentials (OCPs). The OCP is a mixed potential of anodic and cathodic reactions that can be influenced by the mentioned overpotent1ials, secondary reactions, Beilby layer, and so on. The galvanic series of some metals and alloys in seawater is given.

Chemical passivity corresponds to the state where the metal surface is stable or substantially unchanged in a solution with which it has a thermodynamic tendency to react. The electrochemical passivation of metal takes place only if its potential exceeds the critical potential of passivation. Some fundamental aspects of the phenomenon of passivation are briefly described. Conventional polarization methods, electrochemical impedance and noise studies, and cathodic reduction of passive films are appropriate to evaluate active and passive behavior.

Active and passive behaviors of aluminum (Al) and magnesium (Mg) are explained by considering the E–pH Pourbaix diagrams and polarization curves, as well as transfer, concentration, and resistance overpotentials. Formation, composition, and protection quality of the passive layers are discussed. Corrosion resistance of passive Al or Mg and different types of pitting corrosion are discussed. Some properties and examples of the performance of passive Al, Mg, and their alloys are given.

Corrosion Resistance of Aluminum and Magnesium Alloys: Understanding, Performance, and Testing By Edward Ghali

3.1. POTENTIAL–pH DIAGRAMS OF ALUMINUM AND MAGNESIUM

3.1.1. Construction of Pourbaix Diagrams

The application of thermodynamics to evaluate the corrosion tendency or trend of metals has been used intensively in E–pH diagrams known as *Pourbaix diagrams*. These diagrams take into consideration the electrochemical potentials and equilibrium as determined from the free enthalpy ΔG and the chemical equilibrium between the different metallic compounds in solution as applied in the Nernst equation [1]. Although these electrochemical potential values correspond initially to some standard conditions and thermodynamic equilibrium, they can be calculated at some operating conditions for predicting the corrosion tendency or trend of metals in certain media.

The Different Types of Reactions For a certain metal, at a certain temperature (usually 25 °C is considered because of available thermodynamic data), the potentials of electrochemical reactions are traced as a function of pH. Chemical reactions should be considered. Generally, there are four types of reactions to consider [1].

1. There are reactions that depend on the electrochemical potential equilibrium between a metal and its ions such as $M^{n+} + ne = M$.
2. There are reactions that depend on both E and pH such as electrochemical equilibrium between a metal and its oxide or electrochemical equilibrium between two oxides with different levels of oxidation:

$$MO_{n/2} + nH^+ + ne = M + \frac{n}{2}H_2O$$
$$MO_{n/2} + nH^+ + ne = MO_{(n-1)/2} + nH_2O$$

3. There are reactions that depend on pH only such as chemical equilibrium in acid or alkaline medium between an oxide and dissolved ions [2]:

$$MO_{n/2} + nH^+ = M^{n+} + \frac{n}{2}H_2O$$
$$MO_{n/2} + 2OH^- = MO^{2-}_{(n+1)/2} + H_2O$$

4. There are some reactions that are independent of E (no exchange of electrons) or pH such as

$$CO_2 + H_2O \leftrightarrow H_2CO_3$$

Equilibrium states of reactions are the same in whichever direction one considers the reactions. However, for E–pH diagrams, the electrochemical reactions are written in the reduced form as gain of electrons on the left according to international convention. It has been assumed that the value of the standard potential $E°$, calculated from the Nernst equation and depending on the activity of metallic ions in solution, is equal to that of the metal in equilibrium with an activity of 1 M of its ions in solution. However, this does not mean that all the metals are in a state of equilibrium at their considered relative potentials; for example, the equilibrium for a metal immersed in an aqueous solution containing 1 mol/L will be closer to the metallic state (to the right) in the case of gold or to the ionic state (to the left) in the case of lithium, magnesium, and aluminum in aqueous solutions at 25 °C.

Considering that all electrochemical reactions are stated in a reduction scale, it is important to deduce and compare the affinity to corrosion or the formation of stable compounds or oxides from these diagrams. If two electrochemical reactions are in competition at a certain pH, the one with the lower value of potential will prevail (consider the *E*–pH diagram where a less positive or more negative reduction potential corresponds to more exothermic or less endothermic oxidation potential).

Since most *E*–pH diagrams are considered for aqueous solutions, the background of the diagram should consider the stability of water, hydrogen, and oxygen evolution.

- ***Hydrogen Evolution.*** (Figure 3.1, line (a))

$$2H^+ + 2e^- \leftrightarrow H_2$$

The reversible potential at 25 °C and atmospheric pressure can be calculated by the Nernst equation:

$$E_{H^+/H_2} = E^{\circ}_{H^+H_2} + \frac{0.0592}{2}\log\frac{(a_{H^+})^2}{p_{H_2}} = -0.0592\,\text{pH}$$

- ***Oxygen Evolution.*** For the same conditions (Figure 3.1, line (b)),

$$\tfrac{1}{2}O_2 + 2H^+ + 2e^- \leftrightarrow H_2O$$

or

$$\tfrac{1}{2}O_2 + H_2O + 2e^- \leftrightarrow 2OH^- \quad \text{(alkaline medium)}$$

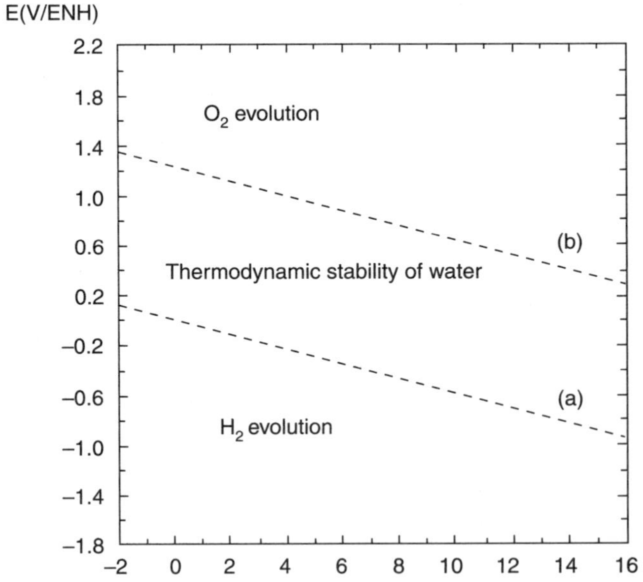

Figure 3.1 Construction of *E*–pH diagram of H_2O at 25 °C [1].

Applying the Nernst equation, we have

$$E = E^\circ + \frac{0.0592}{2}\log(a_{H^+})^2(p_{O_2})^{1/2} = 1.23\text{–}0.0592\,\text{pH}$$

It is clear that both the equilibrium potentials of hydrogen and oxygen evolution depend on pH.

These two relations of balance are represented in Figure 3.1 by the two parallel lines (a) and (b) of slope −0.0592. Between these two lines, there is a field of thermodynamic stability of water under a pressure of 1 atm. For the lower part of line (a), water under an H_2 pressure of 1 atmosphere tends to break up by reduction of the ion H^+. Above line (b), water tends to oxidize which leads to O_2 release. This diagram contains an important consideration for water electrolysis and hydrogen production for storage. Concerning aluminum or magnesium *E*–pH diagrams, the most important reactions are chosen and superposed on the water diagram at atmospheric pressure and 25 °C (3.4 and 3.5, respectively).

3.1.2. Predictions of *E*–pH Diagrams

Metals whose field of immunity is located under line (a) in Figure 3.1 (which indicates balanced H^+/H_2) could be oxidized in the presence of water, even if no oxidizing substance is present, according to whether there is corrosion and/or more or less perfect passivation. Metals whose zone of immunity extends above line (a) in Figure 3.1 cannot be oxidized in the presence of oxidant-free water; but, except for the gold whose field of immunity extends above line (b) (i.e., balanced H_2O/O_2), they could be generally oxidized if water contains oxygen. It is thus logical to bind the degree of nobility of a metal to the extended surface of its immunity region.

Figure 3.2 shows the thermodynamic characterization of Al and Mg together with other chosen metals, Cu, Fe, Zn, Mn, Zr, Al, Ti, and Mg, as examples of frequently used alloying metals or for comparison purposes. The immunity zones are represented by the white zones, the zones of corrosion are hatched by very close lines that slant down from right to left in the case of corrosion by dissolution and from left to right in the less frequent cases of corrosion by gasification by formation of OsO_4, H_2Se, H_2Te, CO_2, CH_4, and AsH_3. Also, the passivation zones by oxides, hydroxides, or hydrides are shown distinctly.

Based on the data of all the elements, two classifications are considered: one is based on the thermodynamic order of nobility for the same elements (classification A) and the other is based on the classification inspired from Figure 3.2 (classification B). This is then based not only on the extent of the field of immunity but also on the whole of the fields of immunity and passivation. This classification by immunity, which is similar to that established by Nernst on the basis of dissolution potential, does not correspond to experimental observations, since several metals such as aluminum, magnesium, tantalum, niobium, titanium, and zirconium resist corrosion better than indicated by classification A. Classification B, on the other hand, is much more close to reality, because it considers that the nobility results not only from one state of immunity but also from the passive state. It should be noted that classification B is limited for a field of pH ranging between 4 and 10, which is met generally in practice.

Classifications A and B are both subject to revision because the diagrams of electrochemical balance on which they are based are themselves only approximations and should be

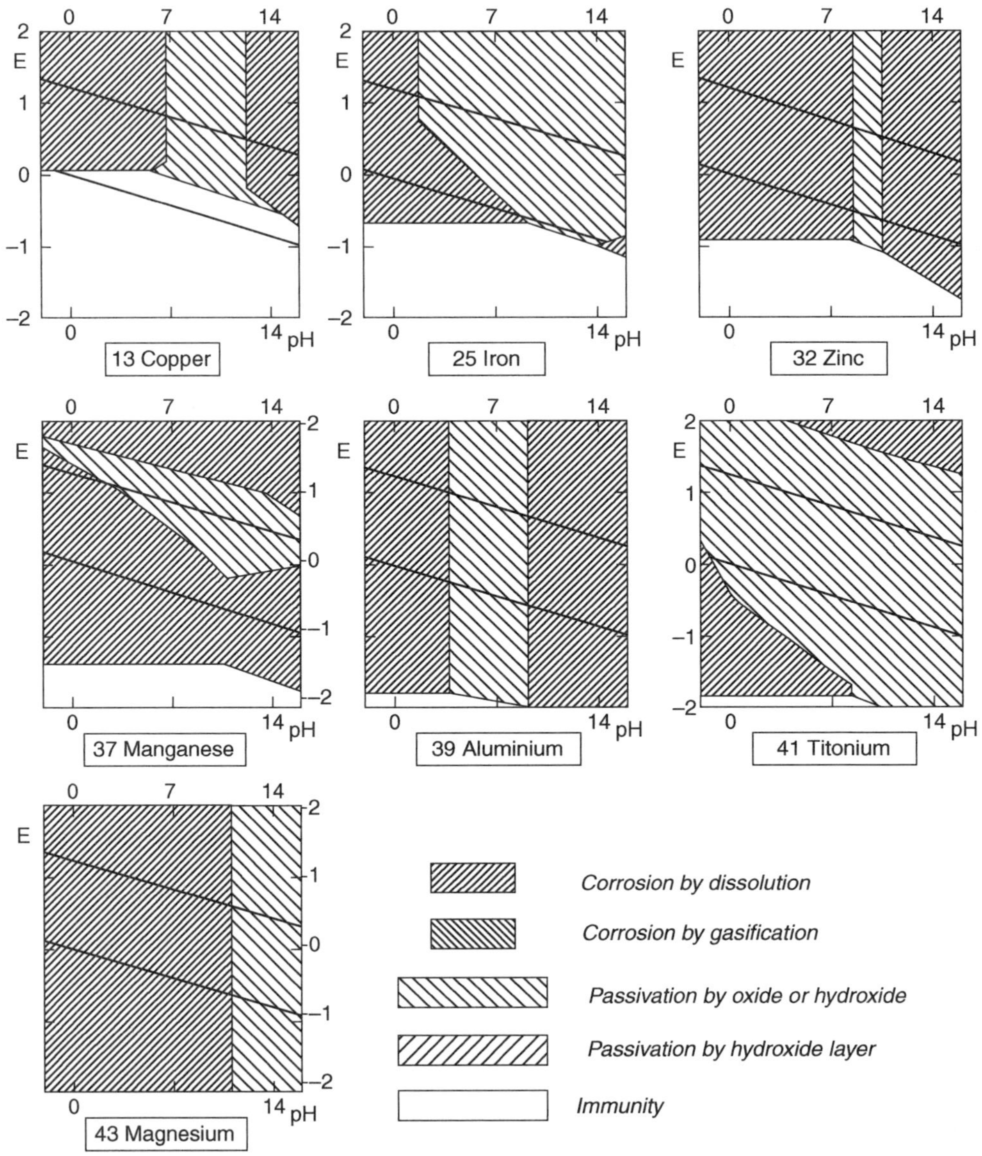

Figure 3.2 Domains of corrosion, immunity, and passivity of Cu, Fe, Zn, Mn, Al, Ti, and Mg [1].

improved. In addition, the reactions of corrosion and/or passivation are sometimes strongly irreversible, such as that concerning carbon. However, in its current form, the comparison indicates the great ennoblement that passivation confers to the following metals: niobium, tantalum, titanium, gallium, zirconium, hafnium, beryllium, aluminum, indium, and chromium. The good corrosion resistance of these metals in many industrial applications is due to their passivation. It should be added that magnesium as calculated is not considered with these ten elements. This could be due to a lack of precise data or to the relatively weak improvement in performance of its passive state as compared to the active state [1].

It is the interesting to note the order of nobility of aluminum and magnesium as compared to other more noble metals. Figure 3.3 gives ΔN, that is, the difference between

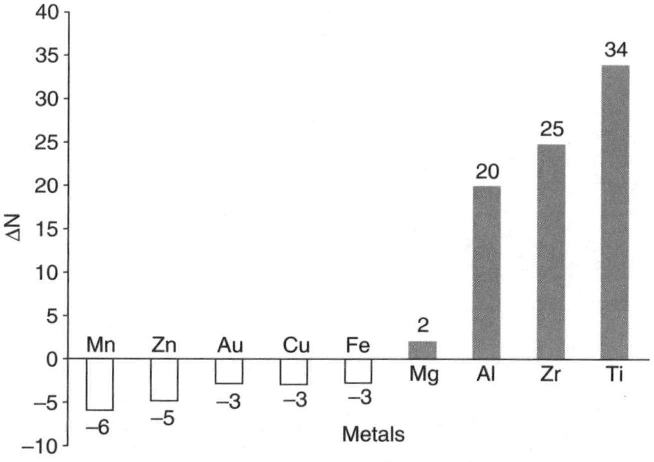

Figure 3.3 Comparison of the thermodynamic and noble practical orders of Al and Mg to seven other elements— Mn, Zn, Au, Cu, Fe, Zr, and Ti. ΔN is the difference between the thermodynamic nobility and the practical nobility as deduced from Pourbaix [1].

the thermodynamic order of nobility and the practical nobility of nine metals inspired from the Pourbaix classification of the nobility order of 43 metals. It can be stated that gold goes to the fourth place and loses 3 points as well as another relatively less noble metal, copper (ΔN). Also, active metals such as iron, zinc, and manganese lose their order of nobility in practice equally. It is interesting to note that aluminum gains 20 points (from 39 in thermodynamic nobility to 19 in practical nobility), influenced largely by its active–passive behavior, and this is also the trend for zirconium and titanium. Magnesium gains and is up two steps from 43 to 41 in the order of nobility but far below that of the other three metals. However, aluminum as well as rare earth metals such as zirconium are used as alloying metals for magnesium.

3.1.3. Utility and Limits of Pourbaix Diagrams

The E–pH diagram is very useful in predicting the reactions that can occur for a system at given conditions. Indeed, it can easily be deduced if a given metal at a certain pH with a precise potential can corrode or not. It is possible then to predict conditions under which corrosion, noncorrosion (immunity), and passivation are possible. It is also possible to predict the types of ions that have promise as oxidizing inhibiting agents. Superposition of the corresponding inhibitor–water pH diagram over that for an aluminium–water system, for example, could predict the region of stability of oxides or possible passive films that coincide with the active regions of aluminum E–pH diagrams [3]. However, predictions based on E–pH diagrams alone have some limitations.

The E–pH diagrams are based on the reactions between pure metals immersed in pure water. In practice, the problems of corrosion are generally due to the presence of salts dissolved in water and the additional reactions that can then occur must be included in the diagrams. Predictions based on the Pourbaix diagram are concerned with equilibrium and do not deal with dynamic, unstable, or transitional states. The properties of the medium at the interface should be considered. In certain cases, the metal is completely immersed in nonagitated or nonhomogeneous electrolytes, creating different successive solutions and potentials at the interface (e.g., galvanic cells at the water line in seawater). Also, in practice, it has already been found, for example, that corrosion can appear in an area corresponding to the immunity field on the Pourbaix diagram because of a drastic change of the corrosive medium localized at the metal–solution interface. Also, in some situations where the metal is expected to corrode, corrosion resistance was achieved because of the establishment of a passive protective film. Thus the reaction product film for passive zones should be tested for its degree and quality of protection in every medium [3].

The polarization of the anodic and cathodic sites and their relative surfaces should also be clarified. The different types of overpotentials on cathodic and anodic reactions could control corrosion rates. It should be underlined also that noncorrosion or immunity could mean hydrogen evolution and that could, in certain circumstances, induce hydrogen embrittlement of certain metals [3].

The impurities included in a metal generate new reactions, which would not occur in the case of a pure metal. There are serious issues that arise when we extrapolate prediction of metal corrosion by this method to corresponding alloys, which could contain additional elements and possible mixed phases, or to a complex aqueous medium. For example, corrosion of silver and gold cannot be predicted from conventional Pourbaix diagrams since pure silver or gold in chloride media could form stable ions involving chloride that are not considered in conventional Pourbaix diagrams. Also, corrosion of iron–nickel alloys could not be predicted simply by superimposing the corresponding pH diagrams for the two metals because of the formation of $NiFe_2O_4$ that is overlooked in both diagrams. Recently, a methodology used by Bale et al. [4], and Thompson et al. [5], called the *Gibbs energy minimization*, is particularly effective in considering all the elements, whether found in the alloy or aqueous medium, that can lead to the formation of significant phases and species [5]. The reaction with the most negative Gibbs energy change is found, thereby identifying the most stable pair of compounds at that particular E_H and pH. However, the alloying elements could give different crystalline phases or amorphous structures with more or less defined composition that lack thermodynamic data.

3.2. ACTIVE BEHAVIOR AND OVERPOTENTIALS

3.2.1. Active Behavior and Polarization

By employing the potentiokinetic method, for example, it is possible to obtain the polarization curve on an electrode in a certain medium by varying the imposed potential of the metal and registering the corresponding current to obtain the polarization curve. One plots the curve of the potential as a function of the current density and generally starts from a cathodic potential with a certain appropriate scan speed versus anodic potentials. Figure 3.4 shows a curve whose form depends on the overpotential of hydrogen evolution as well as that of the anodic reactions. At the $E_{corrosion}$ potential the current is null or the cathodic current is equal to the anodic one. The figure enables us to understand the particular position of the curve of polarization $E = F(I)$ compared to the elementary curves of oxidation and reduction of the two dominant reactions.

3.2.2. Overpotentials

The potential of decomposition of water during electrolysis that corresponds to the evolution of oxygen on the anode and hydrogen on the cathode is much more important than the calculated thermodynamic potential of decomposition (E_r or E_{th}) due to over-

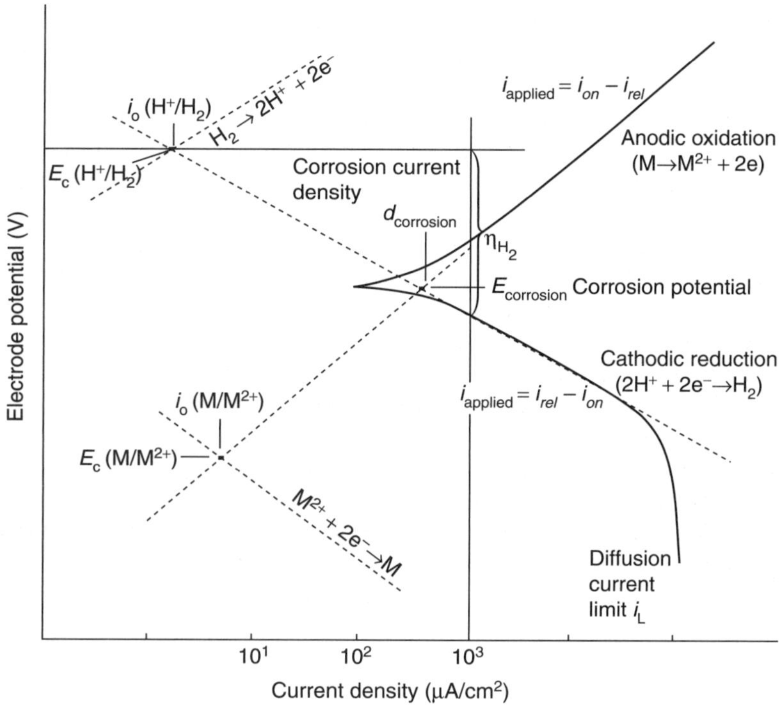

Figure 3.4 Schematic presentation of cathodic and anodic polarization curves of a metal in the active state in a reducing acidic solution.

potentials. In the same manner, the electromotive force of the corrosion cell is less than the thermodynamic calculated one because of the overpotentials η, which can be expressed as

$$\eta = \text{Thermodynamic calculated potential } (E_c - E_a) - \text{Measured potential of the cell } E_{cell}$$

The term η is generally the sum of four types of overpotentials: overpotential of transfer of the electron [6], diffusion overpotential (η_d), reaction overpotential ($\eta_{d'}$), and resistance overpotential (η_r). Very frequently Diffusion and reaction overpotentials are considered in one category as the concentration overpotential since their influence is similar and expressed by the same equation.

3.2.2.1. *Overpotential of Transfer of the Electron*

The transfer overpotential corresponds to the electrochemical reaction of transport of the electrons and can be a loss of electrons (anodic reaction) or a gain of electrons (cathodic reaction) and this corresponds to the primary anodic or cathodic reactions. However, primary reactions can be associated with secondary chemical reactions. The dissolution of the metal, for example, can cause some secondary reactions that lead to the formation of acid, change of the pH, and possible attack of the metal (sometimes localized) such as [2]

$Al \rightarrow Al^{3+} + 3e$	primary electrochemical reaction
$Al^{3+} + 3Cl^- \rightarrow Al\,Cl_3$	secondary chemical reaction
$Al\,Cl_3 + 3H_2O \rightarrow Al(OH)_3 + 3HCl$	secondary chemical reaction

Frequently, the electrochemical reaction is composed not only of one or more electrochemical steps, where there is a gain or loss of electrons, but also of some chemical and physical reactions that take place before or after the primary reaction, such as dehydration, formation of a complex, or dissociation. In this situation the electrochemical step(s) are considered as the primary reaction(s) and the other ones as secondary reactions. Under certain experimental conditions, the secondary reactions can control the corrosion rate because of their relative speed. In the majority of cases, the primary reaction proceeds at a slow rate, which controls the reaction speed.

In 1900 Nernst defined the potential of the electrode for the reaction $Al^{3+} + 3e^- \rightarrow Al$ as

$$E = E^{\circ}_{Al^{3+}/Al} - \frac{RT}{zF} \ln \frac{a_{Al}}{a_{Al^{3+}}}$$

where RT/F is frequently expressed as 0.0592 V, where $R = 8.314$ J/deg · mol is the gas constant, $T = 298.2$ K at 25 °C, and $F = 96{,}500$ C/equiv; z is the number of electrons, and a is the activity of reactants or products in the equation.

In 1905 Tafel showed that there is a linear relation between log I and E for a zone of 150–200 mV located somewhat after the starting area of polarization from the open circuit potential (OCP) (polarization resistance) and before the area controlled by the concentration overpotential of the electrochemical active species. Later, in 1930, Butler and Volmer proved this equation, expressed the overpotentials, and gave more details on the constants of this relation [7]:

$$\eta_c = \frac{RT}{\alpha_c zF} \ln \frac{i_0}{i_c}$$

$$\eta_a = \frac{RT}{\alpha_a zF} \ln \frac{i_a}{i_0}$$

where α_c and α_a are the coefficients of transfer of the cathodic and anodic reactions, respectively. In 1957, Stern and Geary introduced a simple equation [7] to determine the corrosion rate:

$$i_{corrosion} = \frac{i_{applied}}{2.3\eta} \left(\frac{\beta_a \beta_c}{\beta_a + \beta_c} \right)$$

There is a certain approximation in this equation that limits the polarization (±5–30 mV) around the potential of dissolution or open circuit potential. If the exact values of Tafel slopes are well determined, the use of the Stern–Geary reaction has the advantage of expressing the corrosion rate of the metal under conditions close to the OCP without major disturbance.

The Hydrogen Overpotential The cathodic reduction of hydrogen is one of the most examined reactions in electrochemistry because of its importance in controlling corrosion, in hydrogen storage, in electrolysis, and in fuel cells. This includes the overpotential of transfer of the electron [6] that can limit the corrosion rate. Principally, there are two mechanisms that depend on the surface properties of the metal: the mechanism of Volmer–Tafel and that of Volmer–Heyrovsky. The Volmer–Tafel mechanism is composed of two steps: The first electrochemical reaction or step is known as a "Volmer reaction" and forms an adsorbed hydrogen atom on the surface of the electrode. The second is a chemical reaction and consists of the combination of the adsorbed atoms to give molecular hydrogen that evolves (the limiting reaction) [2].

- ***Volmer–Tafel Mechanism***

$$H^+ + e^- \Longleftrightarrow H_{ads}$$
$$H_{ads} + H_{ads} \Longleftrightarrow H_2$$

The Volmer–Heyrovsky mechanism also includes two steps, where the first is a Volmer reaction followed by a "Heyrovsky reaction" that consists of another adsorption reaction in the presence of an adsorbed atom to form molecular hydrogen that can consequently be desorbed [2].

- ***Volmer–Heyrovsky Mechanism***

$$H^+ + e^- \Longleftrightarrow H_{ads}$$
$$H^+ + e^- + H_{ads} \Longleftrightarrow H_2$$

Description of Cathodic Overpotentials of Hydrogen on Metals These overpotentials depend on the properties of the metallic surface, the reaction itself, and the increase with the corrosion rate or the current density (mA/cm^2 or A/m^2). For example, the cathodic reduction of hydrogen ions has different overpotentials and can vary as a function of time due to the formation of corrosion products. It is important to examine the slow step that controls the kinetics of hydrogen evolution for the electrodes of a corroding cell [2]. The determining step of the cathodic reaction is different from metal to metal and three groups can be described with weak, average, and high overpotentials:

1. ***Weak Overpotentials.*** For a platinum electrode, the reactions that proceed are

$$\text{(a)} \quad H^+ + e^- \rightarrow H_{(ads)} \quad \text{(fast step in acid solution)}$$

The second step for hydrogen evolution is the slow determining step of the overpotential:

$$\text{(b)} \quad 2H_{(ads)} \rightarrow H_2$$

2. ***Average Overpotentials.*** After reaction (a), the slow stage is a second electrochemical reaction and this is observed for copper or nickel, steel electrodes [2]:

$$H^+ + H_{(ads)} + e^- \rightarrow H_2$$

3. ***High Overpotentials.*** The discharge of hydrogen ions (or adsorption) is very slow in acidic and alkaline media while the H_2 recombination is much more faster [2]. This is the case, for example, on the surface of mercury or lead.

$$H^+ + e^- \rightarrow H_{(ads)} \qquad \text{(in acidic medium)}$$
$$H_2O + e^- \rightarrow OH^- + \tfrac{1}{2}H_2 \qquad \text{(in alkaline medium)}$$

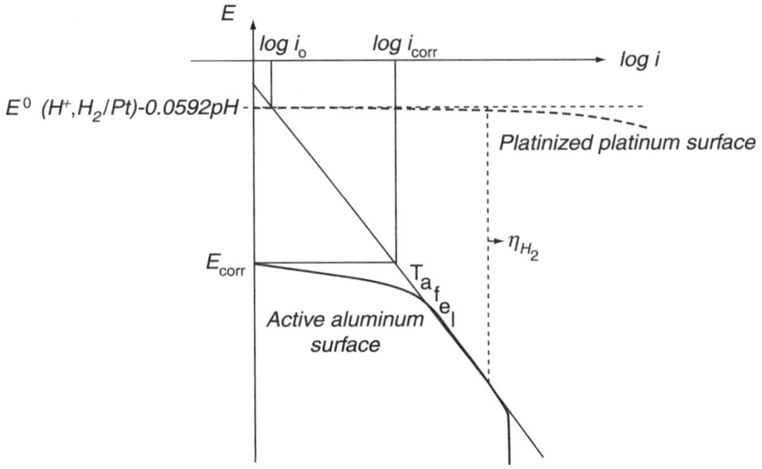

Figure 3.5 Schematic presentation of the hydrogen overpotential on an active metallic surface (strong acidic solution at 25 °C and at 1 atm).

Determination of the Cathodic Overpotential The cathodic overpotential it is the potential difference between hydrogen evolution on the surface of the metal compared to the calculated thermodynamic potential as a function of the pH and temperature at atmospheric pressure.

$$\begin{aligned}\eta &= E_{\text{measured}} - E_{\text{H}_2\,(\text{thermodynamic})} \\ &= E_{\text{measured}} - (-0.0592\,\text{pH}) \\ &= E_{\text{cathodic}} + 0.0592\,\text{pH} \quad (\text{at } 25\,°\text{C})\end{aligned}$$

Presentation of the Cathodic Overpotential of Hydrogen For a certain current density, the determination of the overpotential is easy. It can be seen from Figure 3.5 that a platinized platinum surface has much less overpotential with increasing current density in the same medium and at the same temperature if compared to other metals such as aluminum or iron (Table 3.1).

$$\eta_{\text{H}_2} = \beta \log \frac{i}{i_0} \quad \text{and} \quad \beta = \frac{2.3RT}{\alpha z F}$$

Table 3.1 Examples of Approximate Values of Hydrogen Overpotentials on Metals for Two Current Densities in Acidic Medium

	Overpotential	
Metal	$i = 5 \times 10^{-5}$ A/cm^2	$i = 0.01$ A/cm^2
Platinized platinum	0.005	0.055
Polished platinum	0.090	0.390
Copper	0.230	0.820
Mercury	0.780	1.180

Source: Reference 8.

Since the cathodic overpotential of the hydrogen evolution reaction is negative, β is considered always negative when applying the above Tafel law. In the following, the catalytic activity of the metallic surface for hydrogen ion reduction increases from left to right, corresponding to lower values of overpotentials, respectively.

→ Increasing catalytic activity for hydrogen evolution reaction →

Pb	Sn	Zn	Cu	Ag	Fe	Ni	W	Pd	Pt
1 V					0.3 V				0.0 V

Influence of the Relative Cathodic/Anodic Area Ratios Figure 3.6 explains the influence of the surface area of the cathode to that of the anode on the kinetics of corrosion. The exchange current as well as the corrosion current increase with the relative increase of the cathodic to anodic surface areas ratios. In the case of passive metals or alloys such as that of aluminum or stainless steel, the anodic area can be limited by the porosity of the protective oxide and this leads to highly active pits and perforation. It can be stated from the figure that corrosion potential shifts to more noble values is accompanied by an increase in corrosion current rate. If the anodic surface becomes much more important than that of the cathodic one, the potential shifts to more active values accompanied by an increase in corrosion current.

Hydrogen Reduction Through the Oxygen Depolarized Reaction The depolarized cathodic reaction by the solution saturated with atmospheric oxygen or in the presence of an oxidant or by the presence of a semiconductor oxide on the surface is highly

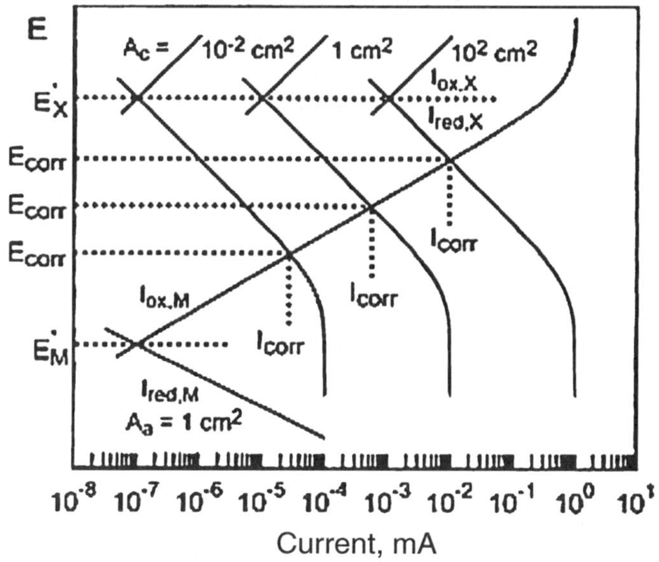

Figure 3.6 Schematic representation of the effect on I_{corr} of different cathodic areas, A_c, and a constant anodic area, A_a [9].

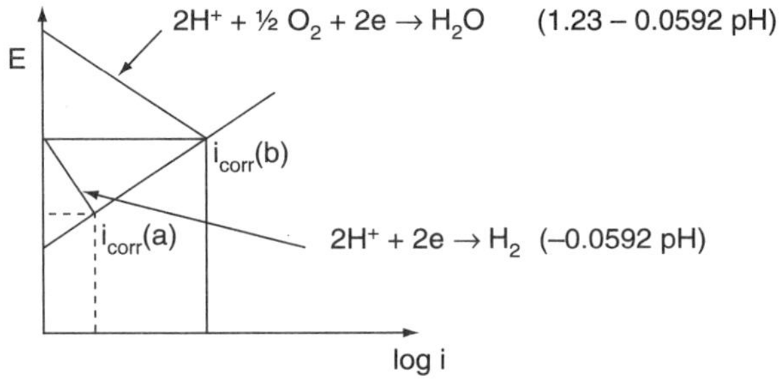

Figure 3.7 Schematic presentation of the influence of depolarizaton on corrosion rates of the cathodic reaction of active metal in a solution saturated with atmospheric oxygen at 25 °C.

exothermic. This accelerating effect acts as a depolarizer for the cathodic reaction that frequently controls the corrosion rate of the metal (Figure 3.7).

$$\tfrac{1}{2}O_2 + 2H^+ + 2e^- \leftrightarrow H_2O \qquad \text{(acidic medium)}$$
$$\tfrac{1}{2}O_2 + H_2O + 2e^- \leftrightarrow 2OH^- \qquad \text{(alkaline medium)}$$
$$E^\circ = 0.401\ \text{V} \quad \text{or} \quad E = 1.23 - 0.0592\text{pH} + 0.0148 \log P_{O_2}$$

The mixed corrosion potential is composed of the anodic and cathodic reactions, and very frequently the hydrogen ion reduction on the active cathodic sites of the metal is the cathodic one. The corrosion rate can increase from (a) to (b) values (Figure 3.7) in the presence of oxygen or an oxidizing agent; but in several cases, because of the partial saturation of the solution with oxygen and also the oxygen diffusion control at the interface, the corrosion rates are intermediate between these two values.

3.2.2.2. Concentration Overpotential

The concentration overpotential can be caused by diffusion or reaction overpotentials or both. The contribution of every phenomenon is deduced and calculated by the same equation; however, the slowest step should control the reaction rate. The diffusion overpotential (η_d) is observed when the exchange of electrons on the metallic surface is slower than the diffusion of the ions to the metal–electrolyte interface. The transport of the mass is controlled by three processes: *convection* that can be caused by heating, density variation, or circulation of the electrolyte; *migration* that is a function of the electrochemical potential, the relative concentration, and the mobility of the ion; and *diffusion*. Diffusion is the gradient of concentration that could control the reaction when the electrochemical active species arrive at the interface of the electrode at a speed slower than that of the exchange of the electrons and will be considered here arbitrarily as the only factor used to derive the values of the concentration electrochemical overpotential.

Figure 3.8a shows a reaction that is almost completely controlled by diffusion at high current densities, showing high diffusion overpotentials. The evolution of the diffusion control at high current densities (transfer of electrons) is shown in Figure 3.8b. The concentration gradient and the layer thickness are functions of the current density $i_1 < i_2 < i_l$ and this simulates the changes of concentration of the ion at the interface at a constant current in the early stages of the process until a stationary state occurs corresponding to the diffusion current limit [10].

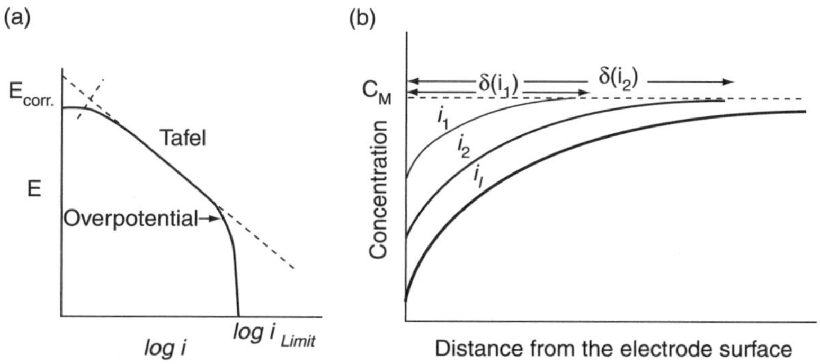

Figure 3.8 (a) Diffusion overpotentials at high current rates and (b) evolution of the thickness of the Nernst diffusion layer δ as a function of current density or time of electrolysis (i_l = stationary diffusion current limit) [10].

The cathodic reaction of hydrogen reduction during electrolysis can be considered ($2H^{2+} + 2e \rightarrow H_2$) as an example, and the application of the Fick rule gives

$$\frac{ds}{dt} = -\frac{AD}{\delta}(C_B - C_x)$$

where ds/dt is the flux of hydrogen ions, A is the area of the exposed electrode surface, D is the diffusion coefficient, δ is the thickness of the Nernst layer, C_B is the concentration (activity) of hydrogen ions at the bulk, and C_x is the concentration (activity) of hydrogen ions at the electrode–electrolyte interface at a distance $x = 0$.

Since the concentration of H^+ ions increases with an increase in distance from the electrode, the equation becomes for a uniform surface

$$\frac{ds}{dt} = \frac{D}{\delta}(C_M - C_x)$$

Also, the migration of hydrogen ions under the influence of an electric field should correspond to the term $t_+ i/zF$ (t_+ is the transport of the hydrogen ion), giving the total rate of hydrogen evolution as

$$\frac{ds}{dt} = \frac{D}{\delta}(C_M - C_{x=0}) + \frac{t_+ i}{zF}$$

The quantity of hydrogen evolved on the uniform cathode surface following Faraday's law is

$$\frac{i}{zF} = \frac{D}{\delta}(C_M - C_x) + \frac{t_+ i}{zF}$$

where i is current density, z is the electrovalence, and F is the Faraday constant.

$$\begin{aligned} i &= \frac{DzF}{(1 - t_+)\delta}(C_B - C_{x=0}) \\ &= \frac{DzF}{t\delta}(C_B - C_{x=0}) \\ &= k(C_B - C_{x=0}) \end{aligned}$$

where the transport number of all the other ions $t = 1 - t_+$.

Since $C_{x=0}$ cannot be determined experimentally, it could be calculated as follows:

$$C_{x=0} = C_B - \frac{i}{k} = \frac{kC_B - i}{k}$$

A concentration cell is created and tends to polarize the potential of the cathode in the positive direction. This concentration cell should be neutralized by an extra potential equal and opposite in sign to that of the created cell. Considering $C_{x=0}$ and C_B as the concentration of hydrogen ions at a distance = 0 and at the bulk of the electrode, respectively, we find

$$E_{\text{anode}} = E^\circ_{\text{oxidation}} - \frac{RT}{zF}\ln C_{x=0}$$

$$E_{\text{cathode}} = E^\circ_{\text{reduction}} - \frac{RT}{zF}\ln\frac{1}{C_B}$$

The emf of the created cell is

$$E_{\text{anode}} + E_{\text{cathode}} = \frac{RT}{zF}\ln\frac{C_B}{C_{x=0}}$$

The diffusion overpotential is then

$$\eta_d = \frac{RT}{zF}\ln\frac{C_{x=0}}{C_B}$$

Replacing $C_{x=0}$ by its calculated value,

$$\eta_d = \frac{RT}{zF}\ln\frac{kC_B - i}{kC_B}$$

where $kC_B = i_{\text{limit}} = i_l$. Then

$$\eta_d = \frac{RT}{zF}\ln\frac{i_l - i}{i_l}$$

Under certain conditions, chemical or physical secondary reactions can control the rate of corrosion as in the case of dissociation of complex ions to give the electrochemical active ion or nonconductive bubble evolution on the surface of the electrode. The values of the reaction overpotential ($\eta_{d'}$) in these cases are expressed by the same equation of the diffusion overpotential.

3.2.2.3. *Electrolytic Resistance or Overpotential*

The values of the ohmic resistance of the electrolyte are generally much higher than the electric resistance in metals. This is important to consider in corrosion studies when the electrolyte is a weak conductor, as in the case of relatively pure water. Some passive nonconducting or weak semiconductor films can increase the ohmic resistance enormously and the overpotential at the metal–solution interface.

3.2.2.4. Comparison of Open Circuit Potential and Thermodynamically Calculated Potentials

The E° standard potential value of a metal is deduced from the thermodynamic equation of the free enthalpy: $\Delta G^\circ = -nFE^\circ$, where n is the number of electrons exchanged in the electrochemical reaction, F is the Faraday constant, and E° is the thermodynamic equilibrium potential. This is frequently considered at standard conditions (temperature of 25 °C and atmospheric pressure) and sometimes compared to open circuit potential (OCP), frequently called the corrosion potential or stationary corrosion potential. The measured OCP can be nobler, equal (rarely), or more active than that calculated from the free enthalpy and the Nernst equation even under standard conditions (in solution activity = 1).

There are major differences between the thermodynamic calculated values and the open circuit potentials. The OCP is a mixed potential of anodic and cathodic reactions that can be influenced by the mentioned overpotentials. Also, every reaction of the mixed potentials can vary as a function of the physical, chemical, and electrochemical conditions of the metallic surface and the solution properties. In the case of active metals, the concentration of the ion can increase at the interface as a function of time, leading to more positive or noble potentials. There is also the possible presence of secondary reactions that control the kinetics of corrosion rates. Generally, it takes a few minutes or 1 hour to reach a relative stable corrosion potential. Initially, one can have dissolution of the surface layer (Beilby) by mechanical polishing and/or atmospheric corrosion. This layer can give generally nobler potentials for short periods at the beginning of the immersion of the metal depending on the aggressiveness of the medium. The corrosion products can also act as a barrier and can lead to resistance overpotential and a passivation can take place. In some situations, stable and equilibrium conditions are not needed for OCP measurements.

3.2.2.5. Galvanic Series in Seawater

The galvanic series is a list of corrosion potentials (OCP), each of which is formed by the polarization of two or more half-cell reactions to a common mixed potential, E_{corr}, measured with respect to a reference electrode such as a calomel electrode. Figure 3.9 shows the galvanic series of some metals and alloys in seawater. The material with the most negative potential has a tendency to corrode when connected to a material with a more positive or noble potential. Some alloys in this medium can have active and passive potentials and sometimes a potential between these two extremes. The galvanic series thus gives qualitative indications of the likelihood of galvanic corrosion in a given medium under certain environmental conditions [11]. The practical change of the potential of every component of a galvanic couple as a function of time is of major importance. If the potential difference between the two metals is sufficient to create a sustained galvanic cell, the potential of every material or electrode can be subjected to certain changes because of the active–passive behavior, the properties of the passive or corrosion barriers, and the change in the ion concentrations.

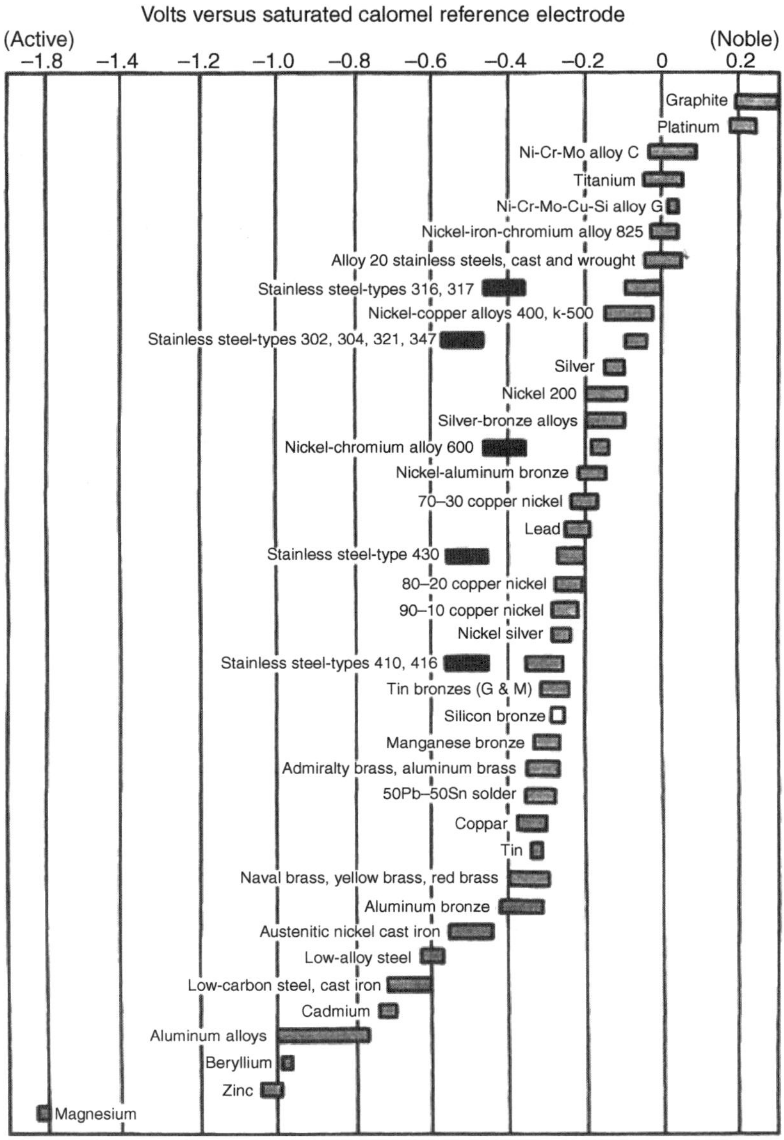

Figure 3.9 Galvanic series in seawater [12].

3.3. PASSIVE BEHAVIOR

3.3.1. The Phenomenon of Passivation

The active state of a metal corresponds to the dissolution or the attack of the metal in a certain environment since this metal is not stable naturally or thermodynamically under its new material–media conditions. There is a marked tendency of the metal to go back to its thermodynamic stable state or to another stable state. Chemical passivity corresponds to the state where the metal surface is stable or substantially unchanged in a solution with which it has a thermodynamic tendency to react. The electrochemical passivation of metal takes place only if its potential exceeds the critical potential of passivation. This could happen in two ways: by anodic polarization of the metal (imposed passivation) or by the presence of oxidant (spontaneous passivation) [2]. Iron, for example, requires anodic polarization or immersion in oxidization inhibiting media such as permanganates or chromates to passivate. This is very frequently the combination of an active attack resulting in an anodic film supported by a cathodic reduction reaction on the surface. Anodic polarization could passivate some metals or alloys at high current densities in an appropriate medium such as sulfuric acid.

Iron immersed in copper sulfate solution dissolves and cements copper on the surface. Once it is passivated by immersion in concentrated nitric acid, copper could not deposit if the passive iron is immersed in the same copper sulfate solution. The early experience of Evans (1925) still shows the profound understanding of this phenomenon. Iron corrodes freely in aerated water (active state), but if we take the same sample of iron prepared in air and carrying air-formed oxide and we keep it moving quickly in aerated water, little corrosion occurs, corresponding to the passive state. Evidently, the rapid rate of oxygen supply to the surface is the factor promoting passivity [10].

The presence of a thick barrier of corrosion products that are relatively protective of the metallic surface could be considered a type of passivation. However, this type of passivity involves any kind of corrosion product film that isolates the metallic surface from the medium rather than the "passive film" mentioned earlier. In this case, the metal is considered passive because it substantially resists corrosion in a given environment despite a marked thermodynamic tendency to react. This corresponds to low corrosion rate and a relatively active potential (e.g., lead in sulfuric acid and iron in an inhibited pickling acid) [13].

The surface of a metal or alloy in aqueous or organic solvent is passivated generally by thin (1–4 nm), compact, and adherent oxide or oxihydroxide film that diminishes the corrosion rates of metals and alloys. The film is invisible to the naked eye and can form at room temperature. In less than a millisecond, atmospheric oxygen attacks the exposed metal and a thin oxide layer begins to form and attains a limiting thickness after a few days. This could be similar in some aspects to high-temperature oxidation, which occurs generally at several hundred degrees Celsius above room temperature [14].

The kinetic rate laws of low-temperature oxidation are direct or inverse logarithmic. At high temperatures, parabolic kinetics is usually observed; however, at intermediate temperatures, logarithmic kinetics can change to parabolic kinetics. At room temperature, the tunneling electrons and an electrochemical galvanic cell could overcome the thermal energy barrier ($\sim$ 1 eV) to transform the adsorbed two-dimensional oxygen layer to a three-dimensional oxide film [15, 16].

The electrochemical behavior of metals can be described by the schematic current potential curve (Figure 3.10). The anodic polarization part of the potentiokinetic curve can be obtained at a suitable scan rate of the potential or potentiostatically. The metallic surface gives a low corrosion rate and is characterized by a more noble potential. It is obvious from this curve that the corrosion current at the level of active potentials (Figure 3.10) is much higher than that of the passive metal since the cathodic curve is shifted to much more positive or noble potentials because of the presence of the oxide film or an appropriate oxidant.

There is a third possibility that can arise from the intersection of the cathodic curve with both active and passive regions of the anodic curve. The polarization curve then locates the

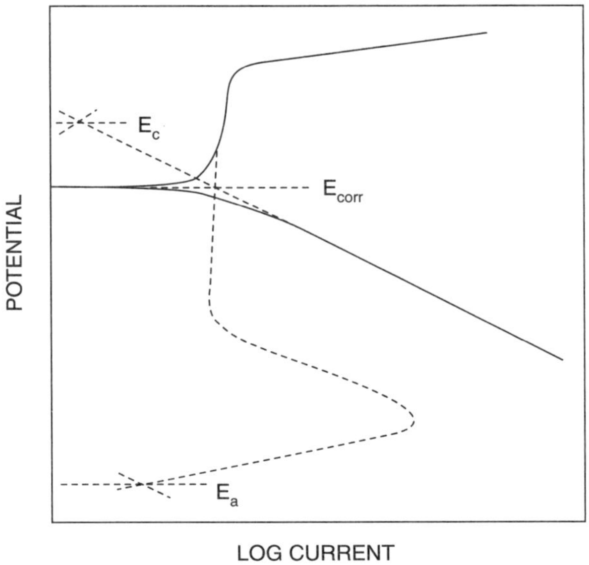

Figure 3.10 Schematic experimental polarization curves: solid curves assume the anodic passive curve and cathodic Tafel slope (dashed curve) in the passive state, while the polarization curves of the metal in the active state are shown in the lower dashed lines [17].

corrosion potential in both regions, where the surface will oscillate between active and passive states, creating unstable conditions. This case can be demonstrated on the laboratory scale when a certain microstructure of steel is immersed in an intermediate concentration of nitric acid.

Aluminum, magnesium, chrome, and stainless steels could passivate upon exposure to natural or certain corrosive media and are used because of their active–passive behavior. Passivity and inhibition are intimately related as corrosion control measures. Inhibition serves to modify the environment by addition of small amounts of inhibitors. Some inhibitors produce films on the anode and hence stifle the corrosion reaction (iron in chromate or nitrite solutions).

Aluminum and other valve metals (titanium, zirconium, hafnium, tantalum, and niobium) show that thicker anodic oxide films can be grown during anodization because oxides are insulators. Thick films of $\sim$100 μm are generally porous, whereas the barrier film is $\lesssim$1 μm. These oxide films formed on metal valves exhibit dielectric breakdown at high voltages. Aluminum can be anodized at a constant current density (typical value is 5 mA/cm^2) in 0.1 M electrolytes. The passive film is composed of an inner Al_2O_3 layer and an outer layer of AlOOH [18].

Normally, good passivity should correspond to a relatively drastic decrease in corrosion rate of the immersed metal and could correspond to a corrosion rate of a few *micro*amperes per square centimeter instead of *milli*amperes/cm^2. The corrosion rate could correspond to minor mobile anodic sites on the metallic surface while the cathodic reaction is generally exothermic and could correspond to the reduction of the oxide film that is continuously generated or by the direct reduction of oxidizing agents such as atmospheric oxygen, hydrogen peroxide, or Fe^{3+} or Cu^{2+} ions available at the cathodic sites of the metal or semiconductor film.

Metal passivation can be obtained by potentiodynamic or galvanodynamic techniques, which correspond to a potential or current scan, respectively. Passivity can be maintained by means of external electromotive force and an auxiliary cathode, as in anodic protection. A current can be applied using a device called a potentiostat, which can set and control the potential at a value greater than the passivating potential E_p or below E_{pit} for environments containing damaging species such as chlorides and bromides [13]. If the anodic sites do not repassivate and exchange sites, localized corrosion occurs, such as pitting, crevice corrosion, stress corrosion, and corrosion fatigue; this is the most serious sequence caused by the passive behavior of materials. The resistance of this anodic oxide film to dissolution is related to the chemical, physical, and structural properties of this film in certain media. The capacity of this film to isolate the metallic substrate from the media is a controlling factor for corrosion inhibition and corrosion rate. Corrosion can proceed through the pores of the passive film (breakdown), but passivation of these pores, followed by creation and corrosion of other pores, avoids localized corrosion. The aggressiveness of the aqueous environment is determined by the pH, temperature, and anion content of the solution.

A good level of understanding of the surface reactions involved in the formation and composition of passive films (passivation/repassivation) is necessary for the creation of highly corrosion-resistant alloys. Good understanding of the metallurgical factors and the rational use of alloyed elements could help to control general and localized corrosion such as pitting. Also, a better comprehension of the mechanism of failure or passivity breakdown of passive films, such as pitting corrosion, is essential for safety purposes, clean environments, and a dynamic modern society [18].

3.3.2. Passive Layers and Their Formation

There are actually pertinent available data on the various physical and chemical aspects of passivity, including the composition, thickness, structure, growth, and properties of passive layers, that should be made more detailed, investigated, and extrapolated to actual practical media, especially with recent physical and chemical techniques of investigation. It is now well accepted that a passive film is not a single layer but rather has a stratified structure. The inner layer plays the role of a barrier layer against corrosion and the outer layer acts as an exchange layer.

Specific oxide thicknesses and properties can be achieved for electronic applications. Materials that grow network-forming oxides are better for these purposes. Metals that grow network-modifying oxides more easily undergo degradation by corrosion. The existence of grain boundaries or other paths of easy ion movement in the oxide allows continued film growth beyond the electron tunneling limit. A partial solution to this problem is to alloy the metal with one that forms a network oxide [14].

Depending on the metal, there are differences in the composition and stoichiometry of the films that influence the stability and growth of oxides. In aqueous solutions, solution anions, halides, and nonhalide types can play a major role in passive film growth and breakdown. Borates, for example, appear to have a beneficial effect. It is necessary to consider the nature of the oxide film, the solution in which the film is formed, and the electrochemical conditions of film formation to evaluate the characteristics of the passive layer. Halide ions such as Cl^- can give rise to severe localized corrosion (e.g., pitting) [19].

The chemical composition is a function of the microstructure of the metal, the pH of the electrolyte, and the level of the anodic potential. Film growth is generally a direct result of the reaction between the metal and the aqueous solution. However, it could also be the result of a dissolution/precipitation (dissolution of metal ions and subsequent precipitation of an oxide, oxihydroxide, or hydroxide) and anodic deposition process that consists of anodic oxidation of metal ions in the solution and deposition on the surface [18].

Passivation kinetics can be composed of three phases: adsorption, nucleation-lateral growth, and growth in thickness.

1. *Adsorption Phase.* It can be agreed on that the first-formed phase is a chemisorbed layer such as oxygen. Physisorption is generally electrostatic (first and reversible) and depends on the electric field of the outer Helmholtz plane of the electric double layer. Water adsorption takes over the adsorption of nonpolar monomers if the potential is shifted far from the potential of zero charge (PZC), as then the surface is highly charged. The adsorption of the monomer prevails in the vicinity of the potential charge. Cations will be adsorbed at potentials negative to the PZC, and anions at potentials positive to the PZC. Chemisorption is slower than physisorption (irreversible and strong) and involves an electron transfer between the substrate and the adsorbed molecule. This can occur between a single pair of electrons of the adsorbed molecule (N, S, P) and empty bands of the solid (orbital overlap). Electron transfer is typical for transition metals having vacant low-energy-electron orbitals [20].

Adsorbed species may act by loosening the metal–metal bond or changing the electric field at the metal–electrolyte interface. They can favor or inhibit the adsorption or the recombination of adsorbed atoms normally involved in the anodic or cathodic reactions. For the formation of bulk compounds, the data of adsorption from gas-phase studies can give the conditions for the presence of a monolayer on a metal surface in contact with an electrolyte. The OH adsorbed groups originating from water dissociation are the precursors of the passive film formation on transition metals [21]. Inhibitors may enhance the formation of passive films on top of the substrate, like benzotriazol on copper or benzoate on iron, or they may form monomolecular adsorption layers and prevent the dissolution of the substrate and the reduction of oxygen by changing the potential drop across the interface and/or the reaction mechanism.

2. *Nucleation-Lateral Growth Phase.* At certain sites on the electrode surface this two-dimensional phase begins to convert to a three-dimensional phase oxide, which spreads across the entire surface. The transition of an adsorbed film to an oxide one can be explained by the nucleation of oxide islands from the adsorbed oxygen that then grow laterally across the surface. Fehlner and Graham [14] proposed a place exchange, requiring a cooperative movement of cations and anions. The "island growth" is the initial stage of three-dimensional oxide film and has been observed for magnesium and barium [15]. The step of transition from island growth to three-dimensional oxide film transforms the linear kinetic growth to a logarithmic one at room temperature [14]. The mechanism of thin oxide growth at room temperature can be explained by the model of Cabrera and Mott [16]. The overall process is quite analogous to anodic oxidation at constant voltage.

3. *Growth in Thickness.* The oxide continues to grow in thickness as long as its rate of formation exceeds its rate of dissolution. Registering current transients during potentiostatic control of the metal [22] in the region of passivation gives the formation rate. It has been found that $i \propto \exp(-d_{ox})$, where i is the current and d_{ox} is the film thickness. A direct logarithmic law, log i–log t plot, where t is the time in seconds, gives a slope of -1. The inverse logarithmic law, which gives the same kinetic results, can also be considered especially for very thin films where the activation barrier is located at the metal–oxide interface [16, 19].

3.3.2.1. Network-Forming Oxide and Modifiers

A network-forming oxide is one in which covalent bonds connect the atoms in a three-dimensional structure. There is short-range order on the atomic scale but no long-range order. For example, oxygen atoms form tetrahedra around ions such as silicon and triangles around aluminum. These continuous random networks can be broken up by the introduction of modifiers. The network formers would be expected to follow inverse logarithmic kinetics. Oxides such as sodium oxide have ionic bonding. When added to a network-forming oxide, they break the covalent bonds in the network, introducing ionic bonds that change the properties of the mixed oxide. It is very probable that direct logarithmic kinetics would be observed for modifying oxides. Ion entry into a growing oxide occurs at the metal–oxide interface for cations and at the oxide–gas interface for anions. One or both species can be mobile in an oxide undergoing anodization [23]. As an example, iron can act as a modifier when it is divalent and can be a network former when it is trivalent. As a trivalent metal network former, iron is useful for the formation of mixed oxide colored glasses [24].

3.3.2.2. Influence of Metal Oxide Structures and Impurities

It has been shown by Rhodin [25] that the (100) face oxidized approximately twice as fast as the (110) and (111) faces for a copper–oxygen system at temperatures from -195 to $50\,^{\circ}\mathrm{C}$. Young et al. [26] showed that impurities could affect the rate of oxidation, especially on the (110) plane. The mentioned single-crystal studies reveal the influence of the crystallographic orientation on oxide growth, but most metals are polycrystalline. The grain boundaries, which randomly separate the oriented grains, must accommodate the discontinuities between the disoriented lattices. The resulting region of disorder serves as a sink for impurities and often a region of fast oxidation. Polycrystalline metals develop very uneven polycrystalline oxides because of the different crystallographic orientations of the grains and the presence of grain boundaries. The advent of glassy metal alloys has overcome some of these problems [27].

The structure of vitreous or glassy oxide and that of a single crystal is expected to be more protective than a polycrystalline would be. Effectively, the grain boundaries in the polycrystalline oxide provide paths for easy ion movement or more rapid oxide growth [28]. Several schemes are used to classify metal oxides as network formers, intermediates, or modifiers [28]. Glass formers tend to have single oxide bond strengths greater than $75\,\mathrm{kcal \cdot mol^{-1}}$. The directional covalent bonds interfere with the crystallization of an oxide while it is being formed on the metal. Intermediates lie between 75 and $\sim 50\,\mathrm{kcal \cdot mol^{-1}}$ and modifiers with ionic bonds lie below $50\,\mathrm{kcal \cdot mol^{-1}}$.

Water is the source of oxygen at high temperatures, but at low temperatures it modifies the structure of the oxides. It may be postulated that water can act as a modifying oxide when added to network-forming oxides and thus can weaken the structure. On the other hand, water incorporation to modifiers may result in polymeric species, which would form a stable protective gel layer [16, 29]. The presence of other impurities such as sodium, chlorine, sulfur dioxide, or nitrogen oxides can change the rate of thin-film formation. In the case of aluminum, alloying elements such as copper in solid solution are beneficial. However, low solubility that leads to the existence of second phases could be detrimental to the resistance of the passive film to breakdown since these could lead to pitting or other types of localized corrosion.

3.3.3. Breakdown of Passivity

Generally, the first step of breakdown of passive films is the formation of unstable or metastable pits that could be transformed to stable ones.

Four conditions have been identified by Hoar [30] to start the process of breakdown:

1. The critical anodic potential should be exceeded E_{bd}.
2. Damaging species such as chlorides or higher atomic weight halides are present.
3. Induction time for initiation necessitates the initiation of the breakdown process and ends with localized conditions that could raise the localized corrosion density.
4. The sites are no longer mobile and are localized, allowing environmental conditions inside the pit to foster propagation [11, 30].

Breakdown of passivity can be complete or partial, leading to the onset of general or localized corrosion. This can be achieved by electrochemical reduction or oxidation, chemical dissolution, undermining attack at film imperfections, or mechanical disruptions (bending, stretching, impact, scratching, etc.). Undermining means that the metal or the substrate, but not the oxide, is attacked by the reagent. Undermining with disruption of the film can occur at any fortuitous breaks in the film. Mercury does not wet passive metal, although it quickly amalgamates with metal surfaces freed from oxide film by acid pickling or by reduction in hydrogen. An extreme example is the breakdown of most oxide films on aluminum in the presence of mercury, which dissolves any exposed metal—a hazard in the use of aluminum thermometers in aluminum vessels, and a useful technique in the production of oxide-film replicas in the electron microscopy of aluminum alloys [10].

A schematic presentation is given in Figure 3.11: the intersection of cathodic curve 1 with the anodic curve of a system exhibiting passivity results in the breakdown of passivity and leads to pitting, while the intersection of cathodic curve 2 at a potential below E_{bd} and in the passive region results in no breakdown [13].

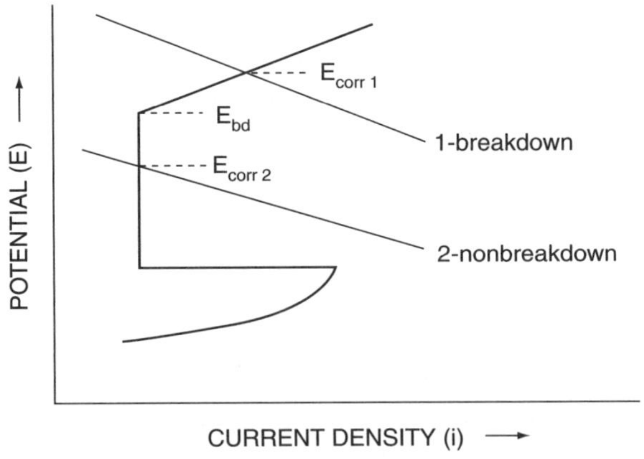

Figure 3.11 An anodic polarization curve for a system capable of exhibiting passivity but subject to breakdown at potentials above the breakdown potential E_{bd}, where pitting is initiated [13].

There are two forms of pitting that follow the breakdown of a passive metal or alloy surface—pitting at low and high potentials. The one at low potential is influenced by cathodic or self-activation and leads to merging etch pits that can eventually extend to general corrosion with etching. High-potential pitting takes the form of hemispherical pits corresponding to anodic dissolution in the electrobrightening mode. This requires a random dissolution due to the presence of film-breakdown factors and is independent of the metal crystal structure. This takes place through a randomly defective solid film that is nonetheless a very good ion conductor. Pits of either kind lead to occluded corrosion [10].

Breakdown of passivity is the first stage of pitting corrosion. Pit growth and repassivation phenomena are characteristic of every corrosion passivation system. Metals show different patterns of passivation. Al and Cu are not passive in strongly acidic electrolytes, while Fe, Ni, and steels are passive even in strongly acidic electrolytes, in disagreement with the predictions of Pourbaix diagrams. Localized acidification by the hydrolysis of corrosion products may serve as a stabilizing factor for pitting in certain metals. The tendency of halides to complex with metal cations is very important in understanding the stabilization of a corrosion pit by prevention of the repassivation of a defect site within the passive layer. Enhancement of the transfer of metal cations from the oxide to the electrolyte by halides, especially the strongly complexing fluoride, holds for many metals. The slow dissolution kinetics of the Cr(III) salts can explain the resistance of chromium to localized corrosion [31]. Detrimental effects of sulfur species have been encountered in a large number of service conditions; however, in the area of passivity, the effects of chloride ions have been investigated more thoroughly. Recent data show a direct link between atomic-scale surface reactions of sulfur and macroscopic investigations like enhanced dissolution, passivating blocking or retarding, and passivity breakdown [32].

3.3.4. Electrochemical and Physical Techniques for Passive Film Studies

Cyclic voltammetry, potentiodynamic, potentiostatic, and galvanostatic techniques are necessary for evaluation of the passive region, aptitude to passivation, stability, and quality of passivation (corrosion rate). Impedance measurements and noise electrochemistry are excellent techniques for detailed studies and monitoring. Cathodic reduction of passive films formed on different substrates is also used as a function of time of passivation. Noise electrochemistry may clarify the initial conditions for pit initiation. Pitting can be random and amenable to stochastic (statistical) theory, and can be considered as deterministic but very sensitive to experimental parameters, such that reproducibility of induction time and electrochemical properties are not achieved [19]. Pitting of a passive metal is associated with a particular combination of film thickness and halide concentration. It has been shown that Q_A(pitting) depends on halide concentration. A smaller value of Q_A indicates more susceptibility to pitting (Cl^- is more aggressive than Br^-, IM of acetate inhibits pitting compared with 0.4 borate). It is generally agreed that well-developed pits should have high chloride ion concentration and low pH values in the pit.

Passive films could be studied by in situ techniques (e.g., Mössbauer spectroscopy) and ex situ techniques (e.g., SIMS) have been used to show the profile and the composition of the passive film [19]. It has been shown that the ex situ and in situ films, formed for iron in nitrite and chromate, have major structural differences by EXAFS (extended X-ray adsorption fine structure) [33]. Great care has to be taken when employing ex situ techniques, such as the ultrahigh vacuum (UHV) spectroscopies or reflection high-energy electron diffraction

(RHEED), since some films formed during immersion may change structurally during drying. Some tests, using ^{18}O/SIMS to identify changes of the passive film, were conducted on a previously passivated metal—nickel in a borate solution containing ^{18}O, exposed to a pH sodium sulfate solution free of ^{18}O. The passive film became completely free of ^{18}O, indicating that breakdown and repair events of the passive film are continuous events [19].

Several methods should be performed in order to get a comprehensive picture of metal oxide system behavior. Kinetics of oxide growth could be monitored in a controlled atmosphere furnace by manometric techniques, supplemented by recording mircrobalances for weight gain or metal loss. Resistivity measurements can supplement the weight gain measurements. The initial stages of a three-dimensional oxide film can be studied using X-ray photoelectron spectroscopy (XPS) and Auger electron spectroscopy. The valence state can be determined from energy shifts of the characteristic peaks. The escape depth of excited electrons limits application to 1–2 nm thickness. Low-energy diffraction [34] and reflection high-energy electron diffraction (RHEED) can reveal the structure of the adsorbed layer and the three-dimensional layer [14].

Depth profiles, performed by XPS, Auger, or secondary ion mass spectrometry (SIMS), and Rutherford backscattering spectroscopy (RBS) on thick oxide films can give essential information on the homogeneity of the oxide. In SIMS, an ion beam is used to bore a hole through the oxide while RBS is a nondestructive test. The energy distribution of back-scattered ions such as helium is analyzed to reveal atomic mass as well as depth information. Transmission electron microscopy (TEM) and transmission electron diffraction (TED) are good techniques to study the plane and cross section of the film and its crystallography. The atomic arrangement on the surface can be monitored by tunneling microscopy (STM) and atomic force microscopy [14].

3.4. ACTIVE AND PASSIVE BEHAVIORS OF ALUMINUM AND ITS ALLOYS

3.4.1. The *E*–pH Diagram of Aluminum

By using the free enthalpy of the chemical compounds in Table 3.2, it is possible to calculate the E° of the equations used for the simplified scheme of Al/H_2O [2].

Table 3.2 The Free Enthalpy of Some Key Compounds for an Aluminum–Water System

Chemical compound	$\Delta G^\circ (kJ \cdot mol^{-1})$
$Al^{3+}_{(aq)}$	−485
$Al(OH)^-_{4(aq)}$	−1297.8
$Al(OH)^{2+}_{(aq)}$	−694.1
$OH^-_{(aq)}$	−694.1
H_2O	−237.2
$Al(OH)_3$ amporphous($Al_2O_3 \cdot 3H_2O$)	−1137.6
$Al(OH)_3$ Gibbsite($Al_2O_3 \cdot 3H_2O$)	−1154.9
$Al_2O_3 \cdot$ H2O Boehmite	−1825.4

Source: Reference 2.

The following six equations for Al and its ions could present some important characteristics of the diagram of aluminum in aqueous solution when considering one activity (concentration) of the ion 10^{-6} mol/L for simplification [2].

$$\text{①}\ Al^{3+} + 3e^- \leftrightarrow Al \qquad E_{rev} = -1.676 + \frac{0.059}{3}\log a_{Al^{3+}}$$

$$a_{Al^3} = 10^{-6} \rightarrow E = -1.794\ V$$

$$\text{②}\ Al(OH)_3 + 3H^+ + 3e = Al + 3H_2O \quad E_{rev} = -1.563 - 0.059\,pH$$

$$\text{③}\ Al(OH)_4^- + 3e = Al + 4OH^- \qquad E_{rev} = -1.20 + \frac{0.059}{3}\log a_{Al(OH)_4^-} - 0.079\,pH$$

$$a_{Al(OH)_4^-} = 10^{-6} \rightarrow E = -1.32 - 0.079\,pH$$

$$\text{④}\ Al(OH)_3 + 3H^+ = Al^{3+} + 3H_2O \qquad pH = 2.44 - \tfrac{1}{3}\log a_{Al^{3+}}$$

$$a_{Al^{3+}} = 10^{-6} \rightarrow pH = 4.44$$

$$\text{⑤}\ Al(OH)_3 + OH^- = Al(OH)_4^- \qquad pH = 16.53 + \log a_{Al(OH)_4^-}$$

$$a_{Al(OH)_4^-} = 10^{-6} \rightarrow pH = 10.53$$

The relative stability of two dissolved substances Al^{3+} and AlO_2^- is [35]

$$\text{⑥}\ Al^{3+} + 2H_2O \rightarrow AlO_2^- + 4H^+ \quad \log(AlO_2^-/Al^{3+}) = -20.30 + 4\,pH$$

The limit of the domains of relative predominance of these two ions are: Al^{3+}/AlO_2^-, $pH = 5.07$.

Figure 3.12 shows the E–pH diagram of aluminum considering water stability at $P_{H_2} = 1$ and $P_{O_2} = 1$ atm); $Al(OH)_3$ corresponds to gibbsite (also called hydrargillite) and the concentrations of the different dissolved species (Al^{3+} and $Al(OH)_4^-$) are in equilibrium and equal to 10^{-6} mol/L. This diagram permits one to identify the regions of stability of Al with its oxide and the ions Al^{3+} and $Al(OH)_4^-$. Other oxides or hydroxides of aluminum are not considered. The concentration is assumed to be equal to the activity or, in other terms, the coefficient of activity is equal to 1.

In aqueous solutions, the potential–pH diagram according to Pourbaix [1] in Figure 3.12 expresses the thermodynamic conditions under which the film develops. The line representing reaction ① indicates the potential of protection for Al. At lower potentials, the corrosion rate of aluminum is negligible. In practice, potentials of cathodic protection that correspond to complete protection of aluminum should be sufficiently lower than this one. Aluminum is amphoteric in nature (line corresponding to reaction ⑥) and corrodes under both acidic (to yield Al^{3+} ions) and alkaline conditions to yield AlO_2^- (aluminate ions). It is clear from Figure 3.12 that the potential of hydrogen evolution (line (a)) is higher than that of reactions ① or ③, describing the equilibrium between the aluminum and the dissolved species Al^{3+} and $Al(OH)_4^-$ respectively. For both conditions, the metal can corrode and the evolution of hydrogen is thermodynamically possible at the cathodic sites of the local galvanic cells of the metal. Then, outside the limits of the region of passivity, aluminum can

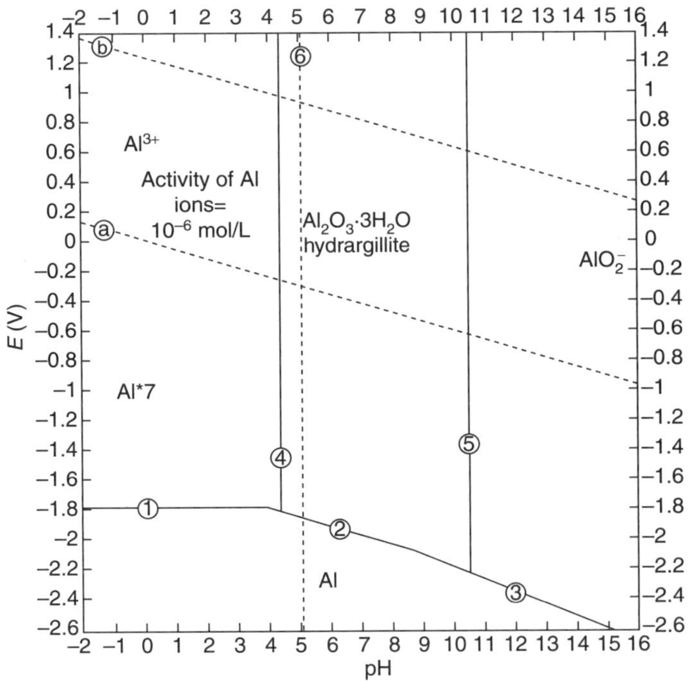

Figure 3.12 Potential versus pH diagram for Al/H_2O system at 25 °C. The concentrations of dissolved ionic species are equivalent to $C_{Al^{3+}}$ and $c_{Al(OH)_4^-}$ and 10^{-6} mol/L [1].

corrode according to the following two equations [2]:

$$\text{In acidic medium:} \quad 2Al + 6H^+ = 2Al^{3+} + 3H_2$$

$$\text{In alkaline medium:} \quad 2Al + 6H_2O + 2OH^- = 2Al(OH)_4^- + 3H_2$$

Figure 3.12 also shows that aluminum is passive only in the pH range from about 4 to 9, which corresponds to a stable oxide film. In this pH region, aluminum and its alloys normally undergo localized corrosion rather than general uniform corrosion. The various forms of aluminum oxide exhibit minimum solubility at about pH 5 at 25 °C. The limits of passivity depend on the temperature, the form of oxide present, and the low dissolution of aluminum. Stoichiometric and thermodynamic information, given in Pourbaix diagrams, concern bulky thick oxides, while passive films are frequently very thin and on the order of 1 nm. Strehblow [31] mentions that Fe, Ni, and steels are passive, even in strongly acidic solution, in disagreement with the predictions of Pourbaix diagrams.

Figure 3.13 shows the rate of corrosion of aluminum as a function of pH. As an example, the pH of acid rain ranges typically from 4 to 5.5 and rarely is less than 3.5. As such, acidic precipitation does not cause severe damage to aluminum and its alloys from the standpoint of structural integrity. However, acid rain can cause cosmetic problems, such as dark brown to black stains [36, 37].

Notable exceptions have been found concerning the stability of the film, either where the oxide film is not soluble in specific acidic or alkaline solution, or where it is maintained by the oxidizing nature of the solution: for example, nitric acid above 80% concentration by weight, sulfuric acid of 98–100% concentration, and glacial acetic acid at pH 1 and lower, or ammonium hydroxide above 30% concentration by weight and sodium disilicate at pH 11–13 [39].

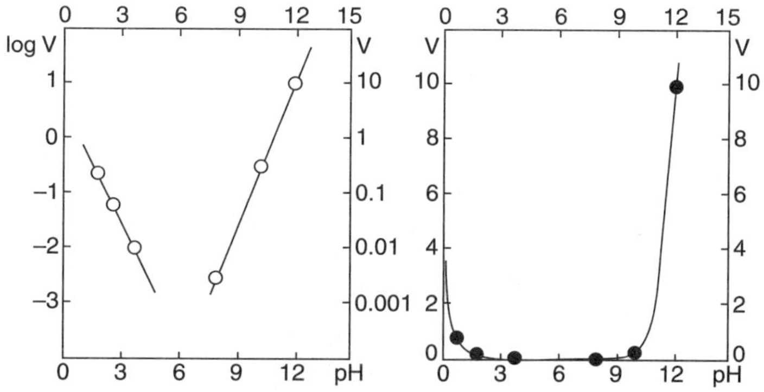

Figure 3.13 Influence of pH on the corrosion rate of aluminum [37, 38].

3.4.2. Active and Passive Behaviors

Aluminum is a highly reactive metal that has a high resistance to corrosion in many environments because of the presence of a thin, highly adherent film of aluminum oxide. When a fresh surface of aluminum is exposed to air or water, a surface film of aluminum oxide immediately begins to form and grow rapidly. In the passive region of pH, aluminum is protected by its oxides and hydroxides. In contact with wet environments, the external side of the oxide film hydrolyzes to produce hydrated oxides such as bayerite ($Al_2O_3 \cdot 3H_2O$) formed below 70 °C and boehmite ($Al_2O_3 \cdot H_2O$ or A1OOH) formed above 100 °C [36, 37, 40]. At lower temperatures, the predominant forms produced by corrosion are bayerite, aluminum trihydroxide $Al(OH)_3$, while at higher temperatures, it is boehmite $Al_2O_3 \cdot H_2O$.

The aluminum hydroxide gel is not stable, but crystallizes with time to give, first, the rhombohedral monohydrate $Al_2O_3 \cdot H_2O$ or boehmite, then the monoclinic trihydrate $Al_2O_3 \cdot 3H_2O$ or bayerite, and finally another monoclinic trihydrate, hydrargillite (or gibbsite), especially if ions of alkali metals are present. This development of aluminum hydroxide is known as *aging* [37, 38]. At higher temperatures, thicker films are formed; these may consist of a thin amorphous barrier layer next to the aluminum and a thicker crystalline layer next to the barrier layer. Relatively thick, highly protective films of boehmite, aluminum oxide hydroxide AlOOH, are formed in water near its boiling point, especially if it is made slightly alkaline, and thicker, more protective films are formed in water or steam at still higher temperatures.

The protective oxide film formed in water and atmospheres at ambient temperature is only a few nanometers thick (2–4 nm) and amorphous (Figure 3.14) [41]. During atmospheric corrosion, aluminum initially forms a layer of aluminum oxide, γ-Al_2O_3, a few nanometers thick, which after prolonged exposure in humidified air is covered by aluminum oxyhydroxide, γ-AlOOH, and subsequently by various hydrated aluminum oxides and aluminum hydroxides. The stability of the compounds decreases with acidity and their dissolution gives Al^{3+}. Aluminum has the ability to form oxygen-containing corrosion products such as basic or hydrated aluminum sulfates, but no detection of aluminum sulfide has been observed. The sulfates found on aluminum are poorly soluble, amorphous, and highly protective. Rates of uniform corrosion are relatively lower than for most other structural materials and on the order of 0.0–0.1(rural), 0.4–0.6 (marine), and ~1 (urban) μm/year. Atmospheric corrosion rates are influenced by chloride ions and cause localized attack more than general uniform attack [42].

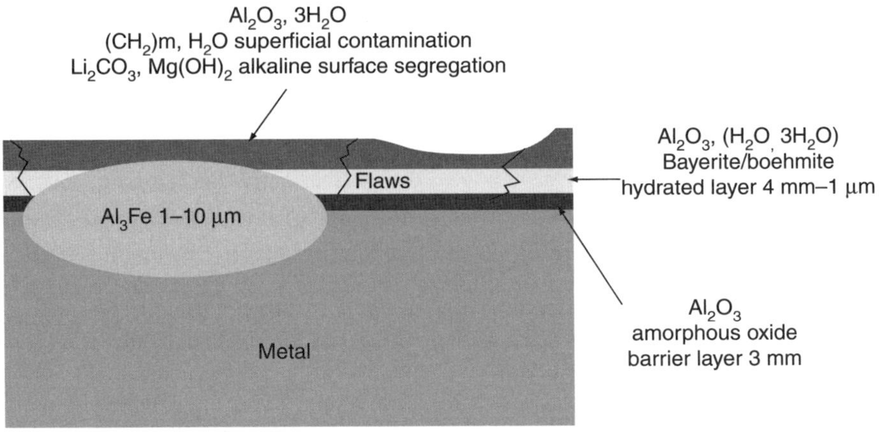

Figure 3.14 Schematic view of aluminum oxide film on rolled products [41].

It is interesting to note that the Pilling–Bedworth ratio (PBR) for Al_2O_3/Al is 1.28, generally satisfactory since the oxide covers the substrate without excessive compressive stresses. However, this can change as a function of the composition and thickness of the scale (oxide and/or hydroxide that can be formed). Effectively, the stresses start accumulating with the formation and growth of the oxide film on the substrate. These can lead to the accumulation of compressive stresses in the growing oxide scale as a function of time (thickness) and temperature [43]. At a certain time, the scale thickness is unable to bear the increasing stresses (point t_c, Figure 3.15) and there is a release of the stress; this can occur through cracking or creep of the passive layer.

In practice, it has frequently been found that PBRs are poor predictors of the actual protective properties of scales since there are many factors that act simultaneously and can lead to blistering or buckling, shear cracking, flaking, and even failure of the corrosion product. Besides the buildup of different corrosion products, epitaxial stresses, the mode of diffusion of the species, the difference in the thermal expansion coefficients between the oxide and the metal, and the geometry of the sample are also important factors. Also, the oxide porosity is not considered in the PBR and it plays a major role in pitting corrosion of aluminum alloys.

Aluminum may corrode because of defects in its protective oxide film. Resistance to corrosion improves considerably as purity is increased, but the oxide film on even the purest aluminum contains a few defects where minute corrosion can develop. In less pure

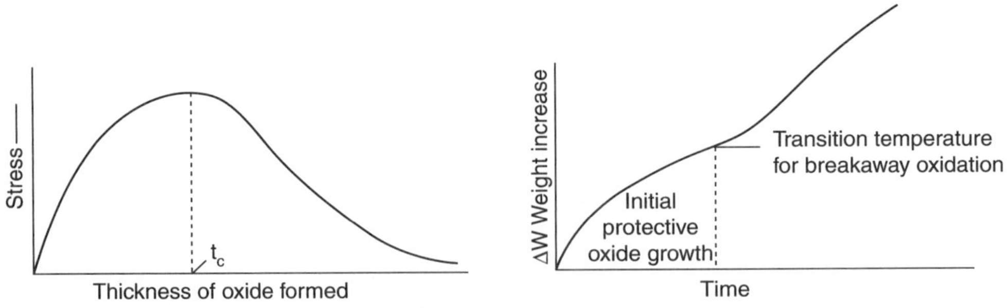

Figure 3.15 Stress accumulation in the growing oxide scale with time and temperature [43].

aluminum of the 1XXX series and in aluminum alloys, the presence of second phases is an important factor. These phases are present as insoluble intermetallic compounds produced primarily from iron, silicon, and other impurities, and, to a lesser extent, precipitate of compounds produced primarily from soluble alloying elements. Most of the phases are cathodic to aluminum, but a few are anodic. In either case, they produce galvanic cells because of the potential difference between them and the aluminum matrix [37, 39].

Alloying elements such as Li, Mg, and Be, which are more active (less noble) than aluminum, oxidize first, forming poorly protective oxides at the extreme surface. On the other hand, alloying elements nobler than aluminum, present in solid solution or in the form of small coherent precipitates (size 0.5–50 nm), produce a mixed oxide film. In contrast, the largest precipitates (size approximately 1–10 μm) formed from these elements are not coherent with the matrix and often remain unoxidized [44]. Aluminum oxide film forms on rolled products. Some oxides, such as those of Mn and Mg, can improve the resistance to general corrosion when they are partially introduced into the oxide film, whereas other oxides can deteriorate the protective quality of this film [45].

It is important to examine the domain of passivity and passivation quality as predicted from the *E*–pH Pourbaix diagram by different polarization techniques. Figure 3.16a shows an anodic polarization curve obtained by potentiodynamic techniques at an appropriate scan rate of the potential for an active–passive behavior of a metal in aqueous solution. The first observed region corresponds to the active state of the metal, followed by the passive one, and then finally by the transpassive region. Figure 3.16b shows the possibility of detecting the metastable and stable pitting of the passive metal in chloride medium as a function of anodic polarization. The potentiodynamic polarization method can also determine the pitting potential of the metal at a certain pH for a certain chloride concentration (Figure 3.16) (*see* Chapter 5). It should be underlined that these simple tests describe the aptitude, quality, and corrosion resistance of the passive metal [18].

A representative anodic polarization curve for 99.99% Al in deoxygenated 0.1 M NaCl shows the initial corrosion potential is about −750 mV standard hydrogen electrode (SHE), from which the potential and current density continuously increase characteristic of a preexisting passive film. The sharp increase in current density at −450 mV (SHE) corresponds to the onset of pitting and identifies the critical pitting potential, $E_{b,pit}$, for this chloride concentration. The dependence of the pitting potential on chloride ion

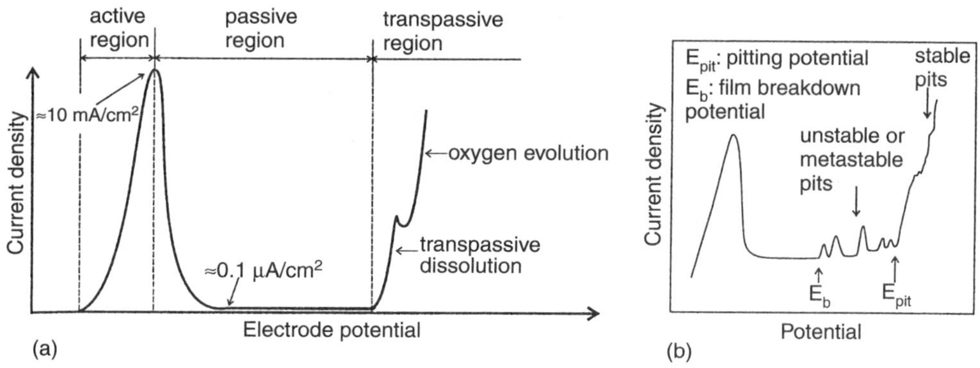

Figure 3.16 (a) Anodic polarization curve of a metal showing active–passive behavior in aqueous solution. (b) Metastable and stable pitting of passive metal in chloride ion-containing solution develops progressively as a function of chloride ion concentration and the level of anodic potential [18].

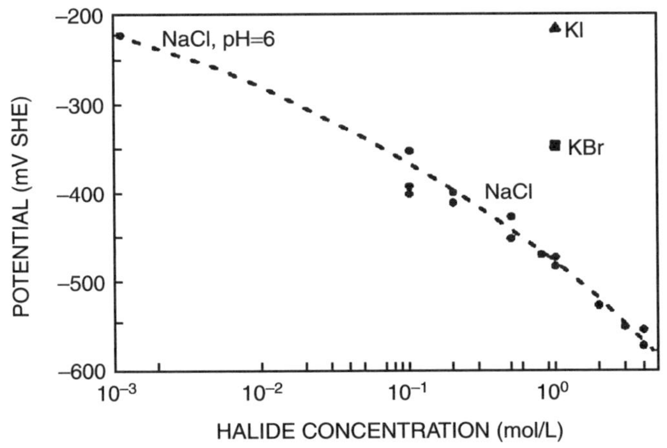

Figure 3.17 Pitting potentials of 99.99 wt % aluminum exist in all halide environments of pH = 11 except as indicated for pH = 6 [9, 17, 46].

concentration is shown in Figure 3.17. Although the curve implies a limiting $E_{b,pit} = -220$ mV (SHE) in the chloride-free medium, pitting occurs only at very high potentials in the absence of pit-inducing anions [17, 46].

3.4.3. Pitting Corrosion of Aluminum Alloy 5086

Gimenez et al. [47] studied the corrosion behavior of the alloy AA5086 in sodium chloride solution (30 g/L) and conceived an experimental *E*–pH diagram predicting the zone of different forms and types of corrosion of aluminum. In this diagram, the following potentials are considered in deaearated solutions with respect to saturated calomel electrode (Figure 3.18): E_0 is the corrosion potential, which is a mixed potential of the metal immersed in water at 25 °C; E_c is the pitting potential; E_p is the protection potential that corresponds to the value below which there is no pitting; E_{ga} is the potential of uniform anodic attack at which pitting corrosion starts spreading over the entire surface, thus giving rise to a uniform attack; E_{cc} corresponds to the cathodic pitting corrosion potential; and E_{gc} shows a general cathodic attack that is expected to appear much more quickly with very high pH values.

The Corrosion Potential and Anodic Pitting Considering AA 5086, the corrosion potential in aerated 3% NaCl solution was approximately −740 to −755 mV saturated calomel electrode (SCE) at pH 4–9. In deaerated 3% NaCl solution, the potential was approximately −890 to −1120 mV (SCE) or irreproducible. Gimenez et al. [47] were able to define the zone of formation of crystallographic pits on the order of 26 mV at pH 8.2, while there is another more negative zone on the order of 100 mV, where the existing pits can grow only [47]. Figure 3.19 shows the typical morphology of the crystallographic anodic pits.

Cathodic Corrosion Forms of Aluminum There are two corrosion forms, cathodic pitting and cathodic general uniform attack. Cathodic pitting is produced in localized alkaline medium. This occurs since the cathodic reaction corresponds to the consumption of hydrogen ions to form hydrogen gas in the beginning at acidic pH, and then in a basic environment, water is reduced to evolve hydrogen and form OH^- ions, causing under certain conditions localized alkalinization. The natural protective layer dissolves in such an

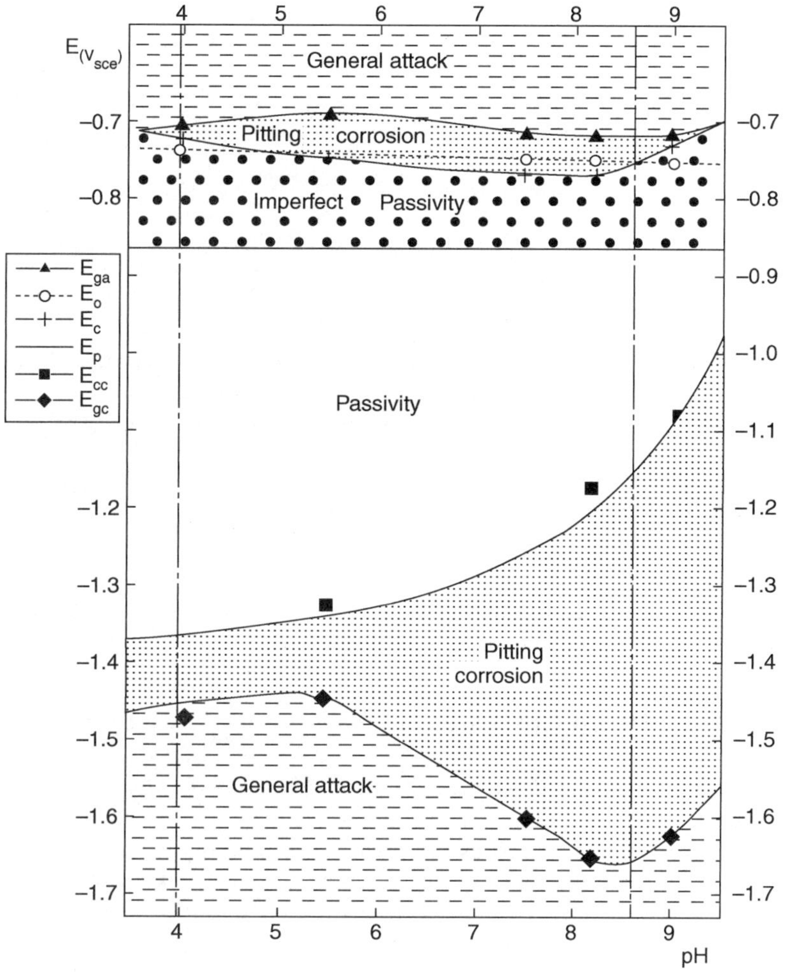

Figure 3.18 Experimental *E*–pH diagram of the aluminum alloy 5086 in deaerated 3% chloride solution. Thermodynamic *E*–pH graph done in mixed lines shown by key at left [47].

alkaline medium. Uniform cathodic attack is a catastrophic corrosion that may dissolve up to 10 μm/h under cathodic polarization [45]. Uniform corrosion of aluminum is caused because aluminum is very active and has no immunity region, and water is unstable under conditions of cathodic polarization of aluminum and aluminum alloys. In a lime-saturated solution, AA5086 fluctuates between −1200 and −1400 mV (SCE). The corrosion observed is identical to uniform cathodic attack; however, this type of attack proceeds in the form of hemispherical shaped pits (Figure 3.20) [47].

The following zones are deduced from the experimental *E*–pH diagram of AA5086 in deoxygenated sodium chloride solution at pH 8.6 similar to that observed in seawater (Figure 3.18) [47]. The passivation zone starts at −863 mV (SCE) and finishes at −1175 mV (SCE). The imperfect passivity area is between −863 and −764 mV (SCE) and is characterized by the possibility of growing the existing pits but not creating or initiating new ones. The anodic pitting corrosion area starts at −764 to −720 mV (SCE), where crystallographic pits are formed. A uniform general anodic attack zone occurs for higher anodic potentials above −720 mV (SCE). A cathodic corrosion area can be found in the

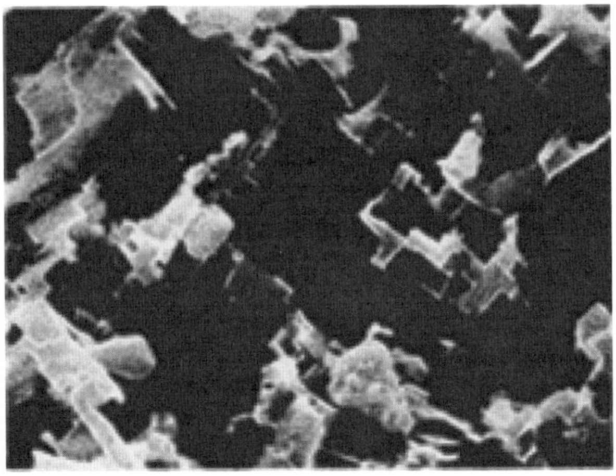

Figure 3.19 Anodic pitting obtained by a potentiodynamic test in 3% NaCl solution at pH 7 [47].

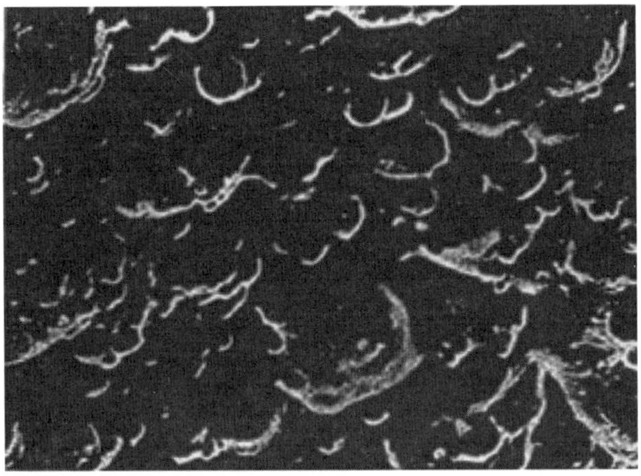

Figure 3.20 Scanning electron microscope photograph showing cathodic pitting obtained by immersion of alloy AA 5086 in a lime-saturated solution (NaCl 30 g/L and at pH = 11.6) [47].

range below −1175 mV (SCE). First, semispherical pits are formed, then a uniform attack is produced with higher cathodic polarization [47].

3.5. ACTIVE AND PASSIVE BEHAVIORS OF MAGNESIUM AND ITS ALLOYS

3.5.1. *E*–pH Diagram of Magnesium

The probable primary overall corrosion reaction for magnesium in aqueous solutions is

$$Mg(s) + 2H_2O(\ell) \rightarrow Mg(OH)_2(s) + H_2(g)$$

This overall reaction can be described in terms of anodic and cathodic reactions as follows.

Anodic reaction—dissolution of Mg:

$$Mg \rightarrow Mg^{2+} + 2e^- \quad \text{and/or} \quad Mg(s) + 2(OH)^- \rightarrow Mg(OH)_2(s) + 2e^-$$

Cathodic reaction—evolution of hydrogen gas:

$$2H^+ + 2e^- \rightarrow H_2(g) \quad \text{and/or} \quad 2H_2O + 2e^- \rightarrow H_2(g) + 2(OH)^-$$

Construction of Mg–pH Diagram The following reactions [48, 49] are considered in the Pourbaix (potential–pH) diagram at 25 °C and atmospheric pressure (Figure 3.21a):

$$\text{(a)}\ \ 2H^+ + 2e^- \rightarrow H_2, \ E = -0.0592\,\text{pH}$$

$$\text{(10)}\ \ MgH_2 \rightarrow Mg^{2+} + H_2 + 2e^-, \ E^\circ = -2.186\ \text{V (NHE) (normal hydrogen electrode)}$$

$$\text{(11)}\ \ MgH_2 + 2OH^- \rightarrow Mg(OH)_2 + H_2 + 2e^-, \ E^\circ = -2.512\ \text{V(NHE)}$$

$$\text{(14)}\ \ Mg^{2+} + 2OH^- \rightarrow Mg(OH)_2, \quad \log(Mg^{2+})^* = 16.95 - 2\,\text{pH}$$

$$\text{(25)}\ \ Mg^+ \rightarrow Mg^{2+} + e^-, \ E^\circ = -2.067\ \text{V(NHE)}$$

$$\text{(27)}\ \ Mg^+ + 2OH^- \rightarrow Mg(OH)_2 + e^-, \ E^\circ = -2.720\ \text{V(NHE)}$$

$$Mg^+ + 2H_2O \rightarrow Mg(OH)_2 + 2H^+ + e^-, \ E^\circ = -1.065\ \text{V(NHE)}$$

$$\text{(48)}\ \ MgH_2 \rightarrow Mg^+ + H_2 + e^-, \ E^\circ = -2.304\ \text{V(NHE)}$$

Perrault [50] considered the formation of MgH_2 and Mg^+ and assumed that thermodynamic equilibrium cannot exist for a magnesium electrode in contact with aqueous

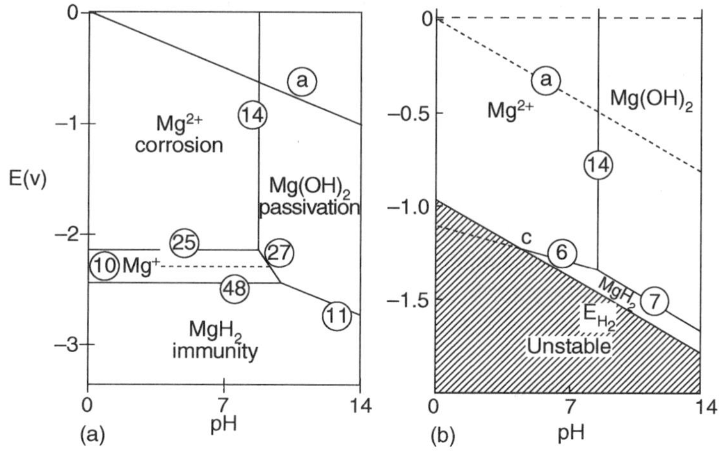

Figure 3.21 (a) Equilibrium of Mg–H_2O system in the presence of H_2 molecules at 25 °C. (b) Stability domains of the magnesium compounds in aqueous solutions with hydrogen overvoltage of 1 V at 25 °C [50].

solutions. Such equilibrium however, is, possible if the hydrogen overpotential is about 1 V and the pH is greater than 5 (Figure 3.21b). Although magnesium has a standard electrode potential at 25 °C of −2.37 V, the corrosion potential of Mg is slightly more negative than −1.5 V in dilute chloride solution or a neutral solution with respect to the standard hydrogen electrode due to the polarization of the formed $Mg(OH)_2$ film. This indicates that the metal corrodes with an accompanying fairly stable film of rather low conductivity even in acidic solutions. Perrault [50] considered the hydride–ion equilibrium (equation ⑥) and the hydride–hydroxide equilibrium (reaction ⑦) [35]:

$$⑥\ MgH_2 \rightarrow Mg^{2+} + 2H^+ + 4e^-, \quad E^\circ = -1.114\ V(NHE)$$

$$⑦\ MgH_2 + 2OH^- \rightarrow Mg(OH)_2 + 2H^+ + 4e^-, \quad E^\circ = -1.256\ V(NHE)$$

Electrochemical dissolution of magnesium can be carried out through two successive steps (the electrochemical formation of Mg^+ and its oxidation by a chemical reaction to divalent one), or through the electrochemical formation of divalent ion directly, or through both. Dissolution could occur through the pores of the passive hydroxide film (if present) or the dissolution of the film itself. Also, the mechanism of dissolution through the stability of the magnesium dihydride as a function of pH at the interface could be deduced from the E–pH diagram (Figure 3.21). However, there are no equilibrium or kinetic considerations that depend on the metal and the medium at the metal–solution interface such as the level of open circuit corrosion potential, the presence of an oxidant as hydrogen peroxide, or cathodic or anodic polarization [51].

The Pourbaix potential–pH diagram shows possible protection of magnesium at high pH values starting at 8.5 when the activity of magnesium ions is equal to 1 mol at 25 °C under atmospheric pressure, while aluminum oxide film is stable in the pH range of 4.0–9.0 [1, 50]. At acidic and neutral pH, the barrier layer on magnesium is difficult to detect; however, at pH 9, a thick white precipitate of magnesium hydroxide begins to form on the outside of the inner film. This surface film protects magnesium in alkaline environments and poorly buffered environments, where the surface pH can increase. The relatively high pH of magnesium hydroxide (10.4) allows magnesium to resist strong bases well. The pH values between 8.5 and 11.5 correspond to a relatively protective oxide or hydroxide film; however, above 11.5 a passive magnesium hydroxide layer dominates the electrochemical behavior of Mg [52].

Magnesium is shown to dissolve over a wide range of pH and potential as Mg^+ or Mg^{2+} in the absence of substances that can form soluble complexes, such as tartrate and metaphosphate, or insoluble salts, such as oxalate, carbonate, phosphate, and fluoride. Alloying affects the nature of this film, but the effects are poorly understood. The corrosion of magnesium and its alloys is strongly dependent on the absence of impurity elements, some of which have well-defined tolerance levels above which corrosion resistance drops dramatically. For conventional magnesium alloys and recycling processes, these tolerance limits must be observed even if extensive surface treatments are applied [53].

At high pH, the corrosion product film of magnesium hydroxide (brucite) that forms on the surface is only semiprotective. The solubility of the metal as related to the concentration of log Mg^{2+} in solution decreases linearly with pH starting at approximately pH 8.5.

The *negative difference effect* (NDE) phenomenon is observed for Mg since the rate of the cathodic reaction (hydrogen evolution) can increase even when the driving force for reduction decreases upon application of potentials anodic to the open circuit potential [52]. Song [54] stated that the reason for the NDE is the anodic dissolution of magnesium in the surface-film broken areas, giving the metastable Mg^+ that is oxidized chemically as follows:

$$Mg^+ + H_2O \rightarrow Mg^{2+} + OH^- + \tfrac{1}{2}H_2$$

This does not exclude the direct dissolution of magnesium as a bivalent ion (see Figure 3.21). Hawke et al. [52] mentioned that exposing active metal by mechanical and chemical attacks on the protective film, the formation of magnesium hydride, and the loss of metal by disintegration (chunk effect) are also possible causes [52]. MgH_2 is also considered as an intermediate of the anodic dissolution process.

3.5.2. Passive Mg Layers (Films)

Passivity of magnesium is destroyed by several anions, including chloride, sulfate, and nitrate. Chlorides, even in small amounts, usually break down the protective film on magnesium. Fluorides form insoluble magnesium fluoride and consequently tend to passivate. The presence of oxidants such as chromate, vanadates, and phosphates, which promote the formation of a protective layer, tend to retard corrosion [48, 52].

Properties and Formation of the Barrier Film Magnesium exposed to air is recovered by a gray oxide film, which can offer considerable protection to magnesium exposed to atmospheric corrosion in rural, industrial, and marine environments. In aqueous media, magnesium may form a surface film, which protects it in alkaline environments and poorly buffered environments. This may result from $Mg(OH)_2$ formation during the corrosion reaction and can increase the pH. In aqueous solutions, magnesium dissociates by electrochemical reaction with water to produce a crystalline film of magnesium hydroxide, $Mg(OH)_2$, and hydrogen gas, a mechanism that is highly insensitive to the oxygen concentration, generally in the absence of oxidizing agents. Sites of hydrogen discharge can then control the corrosion rate according to the total reaction:

$$Mg(s) + 2H_2O(\ell) \rightarrow Mg(OH)_2(s) + H_2(g)$$

The oxide layer is composed of $MgO \cdot H_2O$. The film formed in air immediately after scratching the metal surface is initially thin, dense, amorphous, and relatively dehydrated. The oxide thickness on pure magnesium after exposure for only ~10 seconds to ambient air is 2.2 ± 0.3 nm (~ seven monolayers of MgO) and increases slowly and linearly with the logarithm of exposure as determined during a 10 month test period [55]. Continuing exposure to humid air or water leads to the formation of a thicker hydrated film adjacent to the metal. Exposure to ambient air for a period of 15–60 minutes gives a film thickness of 20–50 nm, while exposure to humid air with ~65% relative humidity during 4 days gives a thickness of 100–150 nm [56]. In the case of aluminum, the air-formed oxide film on the surface is an amorphous aluminum oxide, 2–4 nm thick at room temperature, and this oxide appears to reach a terminal thickness after 1 hour exposure [55].

The oxide films on magnesium, formed by immersion in distilled water after 48 hours, have a three-layer structure, consisting of an inner cellular structure (0.4–0.6 μm), a dense intermediate region (20–40 nm), and an outer layer with a platelet-like morphology (around 2 μm); see Figure 3.22. The hydrated inner and the intermediate layers are similar in structure to the air-formed film [56].

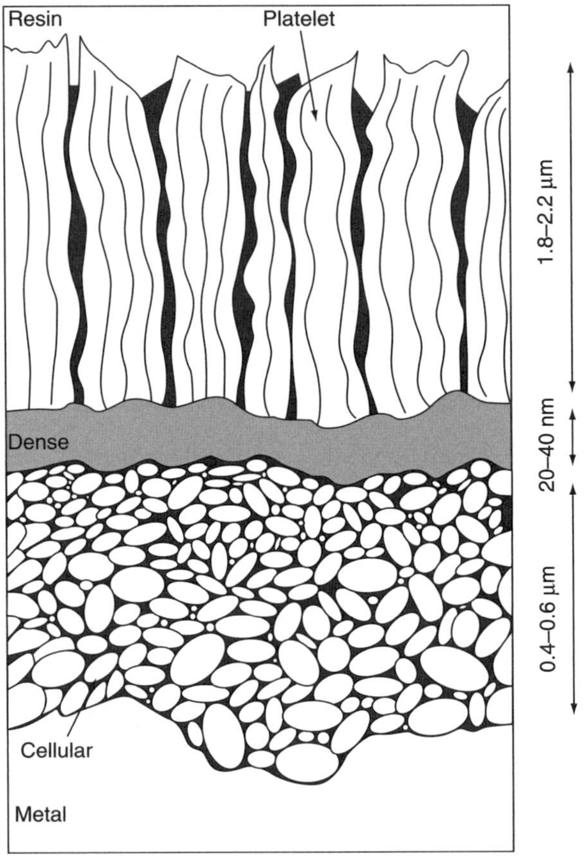

Figure 3.22 Schematic presentation of the three-layer structure of the oxide films on magnesium [56].

The Pilling–Bedworth ratio (PBR) in the case of MgO/Mg is 0.81, meaning that the scale is formed in tension and tends to be nonprotective. However, the PBR is not a unique factor to consider for predicting the quality of the passive layer (see Chapter 2). Prolonged exposure to humid air or water leads to the formation of a hydrated oxide (or hydroxide) layer under the initial dense layer that becomes separated from the substrate. The hydroxide film, brucite, has a hexagonal crystalline structure that is layered, alternating between Mg and hydroxide ions, facilitating easy basal cleavage. Cracking and curling of the film have been noted, although it is not clear whether it is from the properties of the film or the evolution of hydrogen gas. The PBR for $Mg(OH)_2$ is 1.77, which indicates a resistant film in compression. A combination of internal stresses and easy basal cleavage may account for a portion of the cracking and curling of the film. Thus the structure of the corrosion product directly influences the corrosion behavior of the base metal [37].

3.5.3. Passive Properties and Stability

The quasipassive hydroxide film on magnesium is much less stable than the passive films formed on aluminum or stainless steels, for example [57]. Generally, the corrosion rate of magnesium alloys lies between that of aluminum and that of mild steel. Amorphous oxides are in general regarded to have better passive properties than the crystalline ones. Aluminum oxide is a stable amorphous one while magnesium oxide is more prone to crystallization in the form of MgO, as a result of dehydration [56]. There are two main processes of attack of the passive magnesium surface: conversion of the protective surface film to soluble bicarbonates, sulfites, and sulfates, which are washed away by rain, and/or stimulation of local cell action by chloride ions [58].

The film is amorphous and has an oxidation rate less than 0.01 μm/yr. In general, the magnesium corrosion products resulting from the anodic reaction depend on the environment and may include carbonate, hydroxide, sulfite, and/or sulfate compounds. Carbon dioxide and sulfur dioxide play an important role in the stability and composition of the film. A mixture of crystalline hydroxycarbonates of magnesium hydromagnesite $MgCO_3 \cdot Mg(OH)_2 \cdot 9H_2O$, nesquehonite $MgCO_3 \cdot 3H_2O$, and lansfordite $MgCO_3 \cdot 5H_2O$ is reported to be an oxidation product on magnesium; hydromagnesite and hydrotalcite $Mg_6Al_2(OH)_{16}CO_3 \cdot 4H_2O$ are formed on AZ31B. In an industrial atmosphere with high SO_2 content, traces of $MgSO_4 \cdot 6H_2O$ and $MgSO_3 \cdot 6H_2O$ were detected in addition to the hydroxycarbonate products for unalloyed ingot [59].

Influence of Oxygen and Some Active Ions The influence of oxygen concentration in aqueous media on magnesium corrosion is not completely agreed upon. Dissolved oxygen plays no major role in the corrosion of magnesium in either fresh water or saline solutions [60]. Magnesium dissolution in aqueous environments generally proceeds by an electrochemical reaction with water to produce magnesium hydroxide and hydrogen gas, so that magnesium corrosion is relatively insensitive to the oxygen concentration. Some schools perform experiments without deaeration or stirring to simulate practical conditions, and since it has been mentioned that the percentage of oxygen does not influence corrosion rates, at least under certain circumstances [61].

However, the presence of oxygen is an important factor in atmospheric corrosion [53]. The most positive potentials are observed in pure water and alkaline solutions containing subcritical amounts of certain anions. These potentials are usually near the hydrogen electrode reversible potential or readily rose thereto by application of a very small anodic current. Only in environments of this type, and then under good aeration, does oxygen reduction play a significant role [58]. As the potential is lowered due to the presence of anions, oxygen reduction becomes negligible relative to the hydrogen evolution.

The solubility of air and oxygen in saline solutions decreases with increasing concentration of the salt, but salt increases the solution conductivity. The two effects combine in oxygen reduction cathodic systems to produce increasing corrosion rates in up to about 3.5 wt. % sodium chloride solutions and decreasing corrosion rates above that [62]. It has been shown that oxygen plays a major role in the initiation of pitting of AZ91, HK31, and some Mg–Zn alloys in 5 wt % sodium chloride solution at room temperature at relatively high corrosion potentials. This concern of initiation of pitting by the oxygen reduction reaction can be extrapolated to other forms of localized corrosion. In acidic solution, and at more negative potentials, it seems that hydrogen reduction is the main cathodic reaction [63]. Hanawalt et al. [64] stated that the high hydrogen overvoltage on pure magnesium is greatly lowered by iron and that there is no correlation of corrosion effect with hydrogen overvoltage.

The ability of an anion to reduce the magnesium potential appears to depend on the solubility of its magnesium salt. It has been suggested that anions are carried by electrochemical transport to anodic sites on the metal surface, where they form magnesium salts that are acidic to the magnesium hydroxide film. The rapid uniform corrosion rate observed in 3 M $MgCl_2$ at a lower electrode potential supports this mechanism [65]. Examples of these activating anions are Cl^-, Br^-, SO_2^{2-}, and ClO_4^-. In the presence of salt solutions of these anions, magnesium becomes several tenths of a volt active to the hydrogen electrode potential. Hydrogen discharge becomes the controlling factor on effective sites as an element of low hydrogen overvoltage other than that of the hydroxide film itself because of its poor electronic conduction [58].

Baril and Pebere [66] studied the corrosion behavior of pure magnesium in aerated and deaerated solutions (0.01 and 0.1 M) by steady-state current–voltage and electrochemical impedance measurements. It was shown that the anodic current densities were lower and the resistance values higher in deaerated media. They have stated that the presence of oxygen does not influence the cathodic reaction and so oxygen has no effect on magnesium corrosion; however, the shift of the potential in the cathodic direction in aerated solutions and higher anodic corrosion current densities can be explained by the presence of bicarbonate ion in natural conditions (40 mg HCO_3^-/L). Around the corrosion potential, on the anodic side, the current was dependent on the rotation rate and so the process should be partially controlled by diffusion [66]. They showed that the corrosion rate is dependent on HCO_3^- concentration and as a consequence on the presence of CO_2 in solution. The impedance measurements showed that the corrosion rate of magnesium rapidly reached a plateau during immersion in sodium sulfate and a porous layer on the magnesium surface is formed. They stated that the HCO_3^- increased the rate of dissolution by formation of soluble salts.

Nakatsugawa [67] developed a method to create a hydrogen-rich layer onto AZ91D by way of cathodic charging. MgH_2 is a reductant and decomposes gradually to form the hydroxide $Mg(OH)_2$ in an aqueous environment. The treated Mg or Mg–Al alloys show a pseudopassive behavior in the anodic region in 5% NaCl solution and an increase of the Tafel slope in the cathodic region. The corrosion resistance of this coating is superior to a Cr^{6+}-based conversion coating and has a fairly good adhesion to paint [68].

Influence of Agitation Agitation or any other means of destroying or preventing the formation of a protective film leads to corrosion. When magnesium is immersed in a small volume of stagnant water, its corrosion rate is negligible. When the water is completely replenished, the solubility limit of $Mg(OH)_2$ is never reached and the corrosion rate may increase. In stagnant distilled water at room temperature, magnesium alloys rapidly form a protective film that prevents further corrosion. Small amounts of dissolved salts in water, particularly chlorides or heavy metal salts, will break the protective film locally, which usually leads to pitting [60].

3.5.4. Temperature Influence in Aqueous Media

Pure magnesium (99.5% + % purity < 10 ppm (Fe + Ni + Cu)) immersed in distilled water, from which acid atmospheric gases have been excluded, is also highly protected. However, this good resistance to corrosion in water at room temperature decreases with increasing temperature, corrosion becoming particularly severe above 100 °C. The corrosion of magnesium alloys by pure water increases substantially with temperature. At 100 °C, the AZ alloys corrode typically at 0.25–0.50 mm/yr. Pure magnesium and the alloy ZK60A corrode excessively at 100 °C with rates up to 25 mm/yr. At 150 °C, all alloys corrode excessively. Another example, the alloy AZ31B has a corrosion rate of 0.43 mm/yr (17 mpy) at 100 °C, but 30.5 mm/yr at 150 °C [58]. Water vapor in air or in oxygen sharply increases the rates of oxidation of magnesium and its alloys above 100 °C, but boron trifluoride (BF_3), SO_2, and BF_6 are effective in reducing the oxidation rates [60].

The increasing rate of corrosion, with increase in temperature, of ternary alloys is higher than that of pure magnesium and may be due to the activation of some impurities in the ternary alloy at higher temperatures. It appears that the onset of pitting in a given alloy and in certain media depends on a critical pitting temperature, below which only uniform corrosion is encountered. The corrosion of AZ31 in magnesium perchlorate with increasing temperature revealed only a gradual increase without pitting [57]. Increasing temperature sometimes precipitates protective salts, such as calcium carbonate, which decrease corrosion rates in normal-to-hard waters. Temperature differentials between points in a flow system can produce accelerated attack due to differences in ionic activity. The hot zones are generally anodic at the start; however, protective scales occasionally precipitate on the high-temperature metal surface zone and attack proceeds at the cooler sites [62]. Carbonate scaling and its measurement through the Langelier index should be determined. Temperature increase generally decreases solubility of gases in open aqueous solutions, particularly oxygen. This reduces the cathodic action or, more exactly, that portion due to oxygen reduction; this has not yet been determined precisely for magnesium and magnesium alloys and so the decrease of the amount of anodic reaction is not guaranteed.

The diffusivity (D) of Mg within the MgO lattice at 400 °C is as low as 2.24×10^{-18} m^2/s, justifying negligible weight gains. Due to the large difference in densities between the oxide and metal, expressed by the MgO to Mg volume ratio of 0.81, the scale should not form a compact layer. However, systematic spallation of the oxide has been shown in the thin-film range where it exhibits highly protective behavior. There is an inherent strength of the thin MgO film in which stress is operating in an essentially two-dimensional system, and the oxide can withstand the tensile stress necessary to adapt to the dimensions of the metal. Rupturing occurs only after the film exceeds a critical thickness as a function of longer times and higher temperatures. Accelerated, nonprotective oxidation is expressed by the growth and coalescence of oxide nodules, subsequently transformed to scales with a loose structure [69].

3.5.5. Atmospheric and High-Temperature Oxidation

At room temperature, there is a significant difference in corrosion resistance between alloy constituents; and in an NaCl aqueous environment, the $Mg_{17}Al_{12}$ phase exhibits a better resistance by nearly one order of magnitude compared to α-Mg. Although the majority of present applications of magnesium alloys cover room temperature environments, the alloys are subjected to high temperatures and detrimental contact with an oxidizing medium at various stages of manufacturing, including heat treatment, welding, casting, various routes of semisolid processing, or future automobile applications. Removal of the skin layer, degraded by reheating prior to thixocasting and thixoforming, not only causes loss of material but also contributes to a reduction in the billet temperature, important for the further stage of die-cavity filling [69].

Figure 3.23 shows that the commercial AZ91D magnesium alloy tested at a temperature of 197 °C did not increase its weight for time periods as long as 10 h. The measurement conducted at 437 °C revealed an accelerated weight gain after approximately 30 min of the

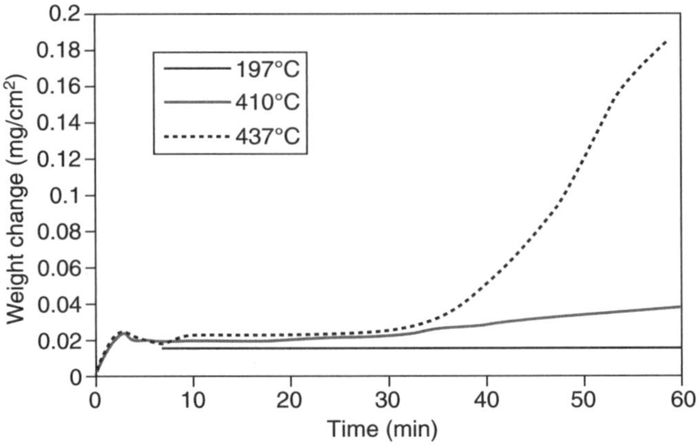

Figure 3.23 Weight increase of AZ91D magnesium alloy during 1 h at a temperature of 197 °C [69].

reaction and this indicates that the capacity of protection of the oxide barrier disappears progressively at a critical temperature zone [69].

REFERENCES

1. M. Pourbaix, *Atlas of Electrochemical Equilibria in Aqueous Solutions*. NACE International and CEBELCOR (Centre Belge d'Étude de la Corrosion), 1974, pp. 100–145 and 168–175.
2. D. Landolt, in *Corrosion et chimie de surfaces des métaux*. Presses Polytechniques et Universitaires Romandes, Lausanne, Switzerland, 1993, pp. 13–109.
3. E. D. Verink, Jr., in *Uhlig's Corrosion Handbook*, 2nd edition, edited by R. W. Revie. Wiley, Hoboken, NJ, 2000, pp. 111–124.
4. C. W. Bale, A. D. Pelton, and W. T. Thompson, *Facility for the Analysis of Chemical Thermodynamics—User Manual 2.1*. Ecole Polytechnique de Montreal/McGill University, Montreal 1996.
5. W. T. Thompson, M. H. Kaye, C. W. Bale, and A. D. Pelton, in *Uhlig's Corrosion Handbook*, 2nd Edition, edited by R. W. Revie. Wiley, Hoboken, NJ, 2000, pp. 125–136.
6. H. H. Uhlig, J. Vœltzel (translator), *Corrosion et protection*. Dunod, Paris, 1970 pp. 98–108, 136–143, and 148–157.
7. L. L. Shreir, R. A. Jarman, and G. T. Burstein. *Corrosion—Metal/Environment Reactions*, 3rd edition. Butterworth-Heinemann, Oxford, UK, 1995, pp. 1–18.
8. H. H. Uhlig and R. W. Revie, Polarization and corrosion rates, Chapter 4, in *Corrosion and Corrosion Control*, Wiley, Hoboken, NJ, 1985, pp. 8, 35–59.
9. E. E. Stansbury and R. A. Buchanan, *Fundamentals of Electrochemical Corrosion*. ASM International, Materials Park, OH, 2000, p. 149.
10. L. L. Shreir, *Corrosion*, Volume 1. Newnes–Butterworths, London, 1976, pp. 126 and 114–129.
11. V. S. Sastri, E. Ghali, and M. Elboujdaini, *Corrosion Prevention and Protection—Practical Solutions*. Wiley, Chichester, UK, 2007, pp. 331–459.
12. E. D. Verink, Jr. *Uhlig's Corrosion Handbook*, 2nd edition, edited by R. W. Revie. Wiley, Hoboken, NJ, 2000, pp. 97–110.
13. J. Kruger, in *Uhlig's Corrosion Handbook*, edited by R. W. Revie. Wiley, Hoboken, NJ, 2000, pp. 165–170.
14. F. P. Fehlner and M. J. Graham, in *Corrosion Mechanisms in Theory and Practice*, edited by P. Marcus and J. Oudar. Marcel Dekker, New York, 1995, pp. 123–141.
15. M. Keddam, in *Corrosion Mechanisms in Theory and Practice*, edited by P. Marcus and J. Oudar. Marcel Dekker, New York, 1995, pp. 55–122.
16. N. Cabrera and N. F. Mott, *Reports on Progress in Physics* **12** 163 (1949).
17. E. E. Stansbury and R. A. Buchanan, *Fundamentals of Electrochemical Corrosion*. ASM International, Materials Park, OH, 2000, p. 237.
18. P. Marcus and V. Maurice, in *Corrosion and Environmental Degradation*, edited by M. Schutze. Wiley-VCH, Weinheim, Germany, 2000, pp. 131–169.
19. B. MacDougall and M. J. Graham, in *Corrosion Mechanisms in Theory and Practice*, edited by P. Marcus and J. Oudar. Marcel Dekker, New York, 1995, pp. 143–171.
20. M. Stratmann, W. Furbeth, G. Grundmeier, R. Losch, and C. R. Reinartz, in *Corrosion Mechanisms in Theory and*

Practice, edited by P. Marcus and J. Oudar. Marcel Dekker, New York, 1995, pp. 373–419.

21. J. Oudar, in *Corrosion Mechanisms in Theory and Practice*, edited by P. Marcus and J. Oudar. Marcel Dekker, New York, 1995, pp. 143–173.
22. D. L. Piron, in *The Electrochemistry of Corrosion*. National Association of Corrosion Engineers, Houston,TX, 1991, p. 163.
23. J. A. Davies, B. Domeij, J. P. S. Pringle, and F. Brown, *Journal of the Electrochemical Society* **112**, 675 (1965).
24. W. A. Weyl, *Coloured Glasses*. Dawson's of Pall Mall, London, 1959.
25. J. N. Rhodin, Jr. *Journal of the American Chemical Society* **72**, 5102 (1950).
26. F. W. Young, Jr., J. V. Cathcart, and A. T. Gwathmey, *Acta Metallurgica* **4**, 145 (1956).
27. D. J. Siconolfi and R. P. Frankenthal, *Journal of the Electrochemical Society* **136**, 2475 (1989).
28. F. P. Fehlner and N. F. Mott, *Oxidation of Metals* **52**, 59 (1970).
29. C. S. G. Philipps and R. J. P. Williams, *Inorganic Chemistry, Volume 1*. Oxford University Press, New York, (1965).
30. T. P. Hoar, *Corrosion Science* **7**, 355 (1967).
31. H. H. Strehblow, in *Corrosion Mechanisms in Theory and Practice*, P. Marcus and J. Oudar. Marcel Dekker, New York, 1995, pp. 201–237.
32. P. Marcus, in *Corrosion Mechanisms in Theory and Practice*, edited by P. Marcus and J. Oudar. Marcel Dekker, New York, 1995, pp. 239–263.
33. G. G. Long, J. Kruger, D. R. Black, and M. Kuriyama, *Journal of Electroanalytical Chemistry* **150**, 603 (1983).
34. R. P. George and P. Muraleedharan, *Corrosion of Austenitic Stainless Steels: Mechanism, Mitigation and Khatak Monitoring*, H. S. Khatak and B. Raj. Narosa Publishing House, New Delhi, 2002, pp. 218–240.
35. ASTM B296 (reapproved 1990), in *Annual Book of ASTM Standards, Volume 02.02, Aluminum and Magnesium Alloys*. ASTM, Philadelphia, PA, 1994, pp. 288–289.
36. B. W. Lifka, in *Corrosion Tests and Standards Application and Interpretation*, 2nd edition, edited by R. Baboian. ASM International, Materials Park, OH, 2005, pp. 547–557.
37. E. Ghali, in *Uhlig's Corrosion Handbook*, 2nd edition, edited by R. Winston Revie Wiley, Hoboken, NJ, 2000, pp. 677–715.
38. R. B. Mears, in *Corrosion Handbook*, edited by H. H. Uhlig. Wiley, Hoboken, NJ, 1976, pp. 39–55.
39. E. H. Hollingsworth and H. Y. Hunsicker, in *Corrosion and Corrosion Protection Handbook*, edited by P. A. Schweitzer. Marcel Dekker, New York, 1983, 111–145.
40. E. H. Hollingsworth and H. Y. Hunsicker, in *ASM Metals Handbook, Volume 13, Corrosion*, 9th edition, edited by J. R. Davis, Senior Editor. ASM International, Materials Park, OH, 1987, 583–609.
41. H. Reboul and R. Canon, Corrosion galvanigue de l'aluminium. Mesures de protection, *Revue de l'aluminium*, 403–426 (1977).
42. C. Leygraf, in *Corrosion Mechanisms in Theory and Practice*, edited by P. Marcus and J. Oudar. Marcel Dekker, New York, 1995, 457–500.
43. A. S. Khanna, *Introduction to High Temperature Oxidation and Corrosion*. ASM International, Materials Park, OH, 2002, 202–216.
44. P. Delahay, M. Pourbaix, and P. Van Rysselberghe. *Diagramme d'équilibres Potentiel-pH de quelques éléments*. C. R. 3e réunion du CITCE, Berne, 1951.
45. C. Vargel, *Corrosion of Aluminium*. Elsevier, Boston, 2004, pp. 75–85 and 181–195.
46. H. Kaesche, in *Pitting Corrosion of Aluminum and Intergranular Corrosion of Aluminum Alloys* edited by R. Staehle, B. F. Brown, J. Kruger, and A. Agrawal. NACE International, Houston, TX, 1974, pp. 516–525.
47. P. Gimenez, J. J. Rameau, and M. C. Reboul, *Corrosion* **37**, 673–682 (1981).
48. E. Ghali, in *Some Aspects of Corrosion Resistance of Magnesium Alloys*, edited by M. O. Pekguleryuz and L. W. F. Mackenzie. International Symposium on Magnesium Technology in the Global Age: Magnesium in the Global Age. Canadian Institute of Mining, Metallurgy and Petroleum, Montreal, 2006, pp. 271–293.
49. Topic 14107, 1999, Water Staining, Available at http://www.oclu-info.dk.
50. G. G. Perrault, *Electroanalytical Chemistry and Interfacial Electrochemistry* **51**, 107–119 (1974).
51. G. G. Perrault, in *Encyclopedia of Electrochemistry of the Elements*, Volume 8, edited by A. J. Bard. Marcel Dekker, New York, 1978, pp. 263–319.
52. D. L. Hawke, J. Hillis, M. Pekguleryuz, and I. Nakatsugawa, in *ASM Specialty Handbook: Magnesium and Magnesium Alloys*, edited by M. M. Avedesian and H. Baker ASM International, Materials Park, OH, 1999, pp. 194–210.
53. G. L. Makar and J. Kruger, *Journal of the Electrochemical Society* **13**, 414–421 (1990).
54. G. L. Song, *Advanced Engineering Materials* **7**, 308–317, 563–586. (2005).
55. N. S. McIntyre and C. Chen. *Corrosion Science* **40**, 1697–1709 (1998).
56. J. H. Nordlien, S. Ono, N. Masuko, and K. Nisancioglu. *Journal of the Electrochemical Society* **142**, 3320–3322 (1995).
57. G. L. Song and A. Atrens, *Advanced Engineering Materials* **1**, 11–33 (1999).
58. W. A. Ferrando, *Journal of Engineering Materials* **11**, 299–313 (1989).

59. J. E. Hillis, in *ASTM Manual Series: MNL 20, Corrosion Testing and Standards: Application and Interpretation*, edited by R. Baboian. ASTM, Philadelphia, PA, 1995, pp. 438–446.
60. A. F. Froats, T. Kr. Aune, D. Hawke, W. Unsworth, and J. Hillis, in *Corrosion, Volume 13*, edited by L. J. Korb D. L. Olson, and J. R. Davis. ASM International, Materials Park, OH, 1987, pp. 740–754.
61. S. Mathieu, C. Rapin, J. Hazan, and P. Steinmetz, *Corrosion Science* **44**, 2737–2756 (2002).
62. C. H. Baloun, in *ASM Metals Handbook, Volume 13, Corrosion*, 9th edition, ASM International, Materials Park, OH, 1987, pp. 207–208.
63. B. Y. Hur and K. W. Kim, *Corrosion Reviews* **16**, 85–94 1998.
64. J. D. Hanawalt, C. E. Nelson, and J. A. Peloubet, *Transactions of American Society of Mining and Metallurgical Engineering* **147**, 273–299 (1942).
65. J. L. Robinson and P. F. King, *Journal of the Electrochemical Society* **108**, 36–41 (1961).
66. G. Baril and N. Pebere, *Corrosion Science* **43**, 471–484 (2001).
67. I. Nakatsugawa, *Cathodic Protection Coating on Magnesium or its Alloys Methods of Production,* USA Patent 6, 291, 076 and Canada Patent 218, 983(2001).
68. I. Nakatsugawa, J. Renaud, and E. Ghali, in *Environmental Degradation of Materials and Corrosion Control of Metals*, edited by M. Elboujdaini and E. Ghali. METSOC, 38th Annual Meeting of Metallurgists of CIM, 1999.
69. F. Czerwinski, in *The Oxidation of Magnesium Alloys in Solid and Semisolid State*, edited by H. I. Kaplan. Magnesium Technology 2003. The Minerals, Metals and Materials Society, Warrendale, PA, 2003, 39–42.

Part Two

Performance and Corrosion Forms of Aluminum and Its Alloys

Chapter 4

Properties, Use, and Performance of Aluminum and Its Alloys

Overview

The physical and general properties of aluminum and its alloys are given. Properties of cast and wrought aluminum alloy series are mentioned with special attention given to some key electrochemical properties. Some examples of conventional and advanced powder metallurg (P/M) prepared alloys as well as metal matrix composites (MMCs) and their properties are described.

Conventional uses in the aerospace, automotive, shipping, and building industries, packaging, and electrical conductors are briefly explained. Some promising uses of Al and its alloys are identified, such as pure Al for Al/air batteries. General and special uses of cast aluminum alloys are discussed. Pure Al rotor, piston, elevated temperature, and Al–Sn bearing alloys are mentioned. The use of wrought alloys in automotive sheetmetal, structural alloy, or electrical conductor alloy as well as in the area of shipping, building, construction, and packaging is described.

Resistance of Al alloys to natural and industrial atmospheres and factors affecting atmospheric corrosion (O_2, N_2, CO_2, O_3, SO_2, and SO_3) are discussed. Contact with inorganic material (e.g., wool or cloth) in exterior or interior atmospheres can cause poultice corrosion. Indoor atmospheric conditions differ greatly and performance depends on constitutent gases and ambient temperature. Aluminum performance in aqueous solutions at different pH values and containing salts or organic compounds is explained. The resistance of pure Al to attack by most acids and neutral solutions is higher than that of aluminum of lower purity or of most of the aluminum-based alloys. Corrosion resistance of Al alloys in seawater and soil is discussed. Performance of the alloys in dry and aqueous organic compounds, acids, alkalis, gases, and mercury is reviewed. Critical points concerning the performance of the different series of cast and wrought alloys in different media are given. Finally, high-temperature corrosion resistance of aluminum alloys is discussed. Al–Ni–Fe alloys show good elevated-temperature resistance to high-purity water.

Corrosion Resistance of Aluminum and Magnesium Alloys: Understanding, Performance, and Testing By Edward Ghali

A. PROPERTIES OF ALUMINUM

4.1. PHYSICAL AND GENERAL PROPERTIES OF ALUMINUM

Aluminum is the third most abundant (7.5–8.1%) metal in the crust of the earth, almost twice as plentiful as iron. Aluminum is second only to iron as the most important metal used in industry and commerce. Aluminum is very rare in its free form and is found primarily in the ore bauxite mainly as silicates ($3Na_2O \cdot k_2O \cdot 4Al_2O_3 \cdot 9SiO_2$). Hydrated alumina occurs in bauxite as gibbsite (hydrargillite), $Al_2O_3 \cdot 3H_2O$, and as bohmite and diaspore, two forms of monohydrate $Al_2O_3 \cdot H_2O$. The name aluminum is derived from the ancient name for alum (potassium aluminum sulfate), which was *alumen* (Latin, meaning bitter salt) [1].

Several gemstones are made of the clear crystal form of aluminum oxide known as corundum. The presence of traces of other metals creates various colors: cobalt creates blues sapphires, and chromium makes red rubies. Both of these are now easy and inexpensive to manufacture artificially. Topaz is aluminum silicate colored yellow by traces of iron. Recovery of this metal from scrap (via recycling) has become an important component of the aluminum industry. Industrial production worldwide of new metal is around 20 million tons per year, and a similar amount is recycled. Known reserves of ores are 6 billion tons [1].

Aluminum is a silvery and ductile metal. Aluminum has an atomic number of 13 while that of magnesium is 12. Both are naturally occurring and their atomic weights are 26.982 and 24.305 g, respectively. Aluminum is below boron in the periodic table, a metalloid with which it has little in common, and Mg is positioned with the alkali metals. Habashi [2] states that Al seems to be misplaced in the periodic table since Al should be transferred to be above scandium and to the right of or beside Mg, together with the alkali metals and the alkaline earth metals, since these metals represent the typical metals with related properties and electronic configurations.

Pure aluminum has a relatively low strength. The density of all alloys (99.65–99.99%) is on the order of 2.7 g/mL, one-third that of steel. After magnesium, aluminum is the lightest of the common metals. Atomic volume is the volume occupied by 1 gram atomic weight of the element in the solid state (atomic weight/density). The atomic volume of aluminum is 10 mL as compared to magnesium at 14.2 and to iron at 7.5 [2]. In addition to recycling and new smelting processes, aluminum and its alloys provide a high ratio of strength to weight. Salts of aluminum do not damage the environment or ecosystems and are nontoxic. Aluminum and its alloys are nonmagnetic and have high electrical conductivity, high thermal conductivity, high reflectivity, and noncatalytic action [3, 4].

Aluminum has a relatively low cost and its properties give it one of the best performance/weight $\times$ price ratios and so it is used extensively in airframe construction. With the appropriate choice of alloy and design, Al is easy to cast. Its relatively low melting temperature makes it possible to use permanent metallic dies. In the end, its cost is competitive with respect to other materials that are a priori less expensive [5].

Magnesium can be cast as readily as aluminum, but its mechanical performance is generally poorer when hot (from 100 °C), especially in creep, and it is more sensitive to corrosion generally. The global weight saving in using light metals includes the weight saved in the casting itself and all induced weight savings. For example, making a wheel lighter makes it possible in turn to reduce the weight of the brakes, of suspension parts, and so on. Plastics are less expensive than aluminum and allow the same freedom of shape but have the following drawbacks: poor heat resistance, brittle with aging, and recycling problems. Composites of aluminum and magnesium (in development) with plastics can be

Table 4.1 Physical Data and Properties of Aluminum

Property	Aluminum	Reference
Atomic mass	26.9815386 g	5, 18
Tensile strength	77 MN/m^2	5, 18
Maximum oxidation number	3+	7 (Chapter 3)
Manimum oxidation number	0	1
Crystal structure	Face centered	7 (Chapter 3)
Spectroscopy	$(1s)^2,(2s)^2,(2p)^6,(3s)^2,(3p)^1$	7 (Chapter 3)
Ionization potential	I(5.97 eV), II(18.8 eV)	1
Potential standard	−1.663 V	9, 13
Pauling electronegativity	1.5	1
Density at 20 °C	2.7 g/cm^3	7 (Chapter 3)
Melting point	659.7–660.1 °C	7 (Chapter 3)
Boiling point	2400–2450 °C	7 (Chapter 3)
Specific heat at 20 °C	0.891 kJ/kg °C	14
Thermal conductivity at 20 °C	272 W/m · K	9, 14
Van der Waals radius	0.143 nm	1
Ionic radius	0.05 nm	1
Number of common isotopes	1 (isotope 27)	

made that have excellent mechanical properties; however, their cost is prohibitive for many applications [5]. Table 4.1 shows the physical data and the properties of aluminum.

4.2. CAST ALUMINUM ALLOYS

Aluminum alloys are produced by squeeze-casting, rheocasting, thixocasting, or thixoforming. Pressure casting or spray-forming are used in repetitive series production [6]. The ratio of cast to wrought aluminum alloy products is increasing primarily because of the larger amounts of castings being used for automotive applications. This ratio varies from country to country and in 2004 it was approximately 1:2 in North America [7].

A wide range of cast aluminum alloys are available for commercial use and, for example, nearly 300 compositions were registered with the United States Aluminum Association in 2005. The most widely used are those based on the Al–Si, Al–Si–Mg, and Al–Si–Cu systems. In general, alloys are classed as "primary" if prepared from new metals and "secondary" if recycled materials are used. Secondary alloys usually contain more undesirable impurity elements that complicate their metallurgy and often lead to properties inferior to those of the equivalent primary alloys. In all areas, except creep, castings normally have mechanical properties that are inferior to wrought products [7].

Cast aluminum products are produced by sand casting, die casting, permanent mold (gravity die) casting, and cold chamber and hot chamber pressure die casting methods. Their selection involves consideration of casting properties as well as of physical properties. However, die, permanent mold, and sand casting account for the greatest proportion. Many alloys respond to thermal treatment based on phase solubility. These treatments include solution heat treatment, quenching, and precipitation, or age hardening. For either cast or wrought alloys, such alloys are described as heat treatable. Heat treatable cast alloys are strengthened by dissolution of soluble alloying elements and their subsequent precipitation. Non-heat-treatable alloys are strengthened by intermetallic compounds formed of insoluble or undissolved alloying elements.

Commercial ingots are cast by continual processes as rounds or rectangles; hence they can be of various lengths with as much as a 915 mm diameter or 619 mm thick. The most basic form of cast aluminum product is a large ingot produced for subsequent fabrication into a wrought product. Although the homogenization (preheating) and hot and cold working processes performed on the ingot will affect the metallurgical structure and corrosion resistance of the final product, corrosion tests are rarely made for products still in the ingot stage. It is therefore important to establish, whether changes in ingot casting or processing affect the corrosion performance of the finished product [3, 8, 9].

4.2.1. Designation of Cast Aluminum Alloys and Ingots

The system for designating aluminum and aluminum alloys that incorporate the product form (wrought, cast, or foundry ingot), and its respective temper (with the exception of foundry ingots, which have no temper classification) are covered by American National Standards Institute (ANSI) Standard H35.1. The Aluminum Association is the registrar under ANSI H35.1 with respect to the designation and composition of aluminum alloys and tempers registered in the United States. No internationally accepted system of nomenclature has so far been adopted for identifying cast aluminum alloys and foundry ingot. However, the Aluminum Association (AA) of the United States has introduced a revised system, which has some similarity to that adapted to wrought alloys [10].

The AA system consists of four numeric digits, with a period between the third and the fourth. The first digit indicates the principal alloying group or constituent(s). For the 1xx.x group, the second and third numbers indicate the purity of the alloy: they are the decimal of 99.xx%, for example, 182.0 contains 99.82% Al. For the 2xx.x to 9xx.x series, the second and third digits identify individual alloys and have no numerical significance. The last digit indicates the product form: 1xx.0 indicates castings, and 1xx.1 indicates ingot or ingot having composition ranges narrower than but within those of standard ingot by 2 [7, 11]. When variations in the composition limits are too small to require a change in numeric designation, a preceding serial letter (A, B, C, etc.) is added, omitting I, O, Q, and X, the X being reserved for experimental alloys. Example, the original alloy is 346.0, no letter prefix; for the first variation, an A is added, A346.0; for the second variation B is added, B346.0, and so on.

For 2xx.x through 9xx.x (excluding 6xx.x alloys), the alloy group is determined by the alloying element present in the greatest mean percentage, except in cases in which the composition being registered qualifies as a modification of a previously registered alloy. If the greatest mean percentage is common to more than one alloying element, the alloy group is determined by the element that comes first in the sequence [10]. After the first pure aluminum series, aluminum alloys are then grouped as a function of major alloying element(s) as follows [11]:

- 1xx.x, pure aluminum ($\geq$99.00%)
- 2xx.x, copper alloys
- 3xx.x, silicon with added copper and/or magnesium
- 4xx.x, silicon
- 5xx.x, magnesium
- 6xx.x, unused series

- 7xx.x, zinc
- 8xx.x, tin
- 9xx.x, other elements

In addition, cast alloys are separated in two types: non-heat-treatable, designated by an "F" for which strengthening is produced primarily by intermetallic compounds; and heat-treatable, designated by a "T," corresponding to the same type of wrought alloys where strengthening is produced by dissolution of soluble alloying elements and their subsequent precipitation. Alloys of the heat-treatable type are usually thermally treated subsequent to casting, but in a few cases, where a significant amount of alloying elements are retained in solution during casting, they may not be given a solution heat treatment after casting; thus they may be used in both the "F" and fully strengthened "T" tempers (Tables 4.2 and 4.3) [4].

Most alloys are covered by the British Standard 1490 and compositions for ingots and castings are numbered in no special sequence and have the prefix LM. The condition of castings is indicated by the following suffixes [7]:

M	As-cast
TB	Solution treated and naturally aged (formerly designated W)
TB7	Solution treated and stabilized
TE	Artificially aged after casting (formerly P)
TF	Solution treated and artificially aged (formerly WP)
TF7	Solution treated, artificially aged and stabilized (formerly WP-special)
TS	Thermally stress-relieved

The absence of a suffix indicates that the alloy is in ingot form.

Table 4.2 Nominal Chemical Compositions of Representative Aluminum Casting Alloys

	Percentage of alloying elements				
Alloy	Si	Cu	Mg	Ni	Zn
	Alloys not normally heat treated				
360.0	9.5		0.5		
380.0	8.5	3.5			
443.0	5.3				
514.0			4.0		
710.0		0.5	0.7		6.5
	Alloys normally heat treated				
295.0	0.8	4.5			
336.0	12.0	1.0	1.0	2.5	
355.0	5.0	1.3	0.5		
356.0	7.0		0.3		
357.0	7.0		0.5		

Source: Reference 3.

Table 4.3 Typical Tensile Properties of Representative Cast Aluminum Alloys in Various Tempers[a]

Alloy and temper		Type casting	Strength (MPa) Ultimate	Yield	Percent elongation In 50 mm[b]
295.O	–T6	Sand	250	165	5
336.O	–T5	Permanent mold	250	195	1
355.O	–T6	Sand	240	170	3
	–T6	Permanent mold	375	240	4
	–T61	Sand	280	250	3
	–T62	Permanent mold	400	360	1.5
356.O	–T6	Sand	230	165	3.5
	–T6	Permanent mold	255	185	5
	–T7	Sand	235	205	2
	–T7	Permanent mold	220	165	6
357.O	–T6	Sand	345	295	2
	–T6	Permanent mold	360	295	5
	–T7	Sand	275	235	3
	–T7	Permanent mold	260	205	5
360.O	–F	Pressure die	325	170	3
380.O	–F	Pressure die	330	165	3
443.O	–F	Pressure mold	160	60	10
514.O	–F	Sand	170	85	9
710.O	–F	Sand	240	170	5

[a]Averages for separate cast test bars; not to be specified as engineering requirements or used for design purposes.
[b]A 1.60 mm thick specimen.
Source: Reference 3.

4.2.2. Alloying Elements

Major alloying elements define the ranges of elements that control castability and property development. Minor alloying elements control solidification behavior, modify eutectic structure, refine primary phases, refine grain size and form, promote or suppress phase formation, and reduce oxidation. Impurity elements influence castability and the form of insoluble phases that at times limit or promote desired properties [11]. Cast alloys contain a greater amount of alloying additions than those used for wrought products and are added to improve castability or to strengthen wrought alloys. This results in a largely heterogeneous cast structure with a substantial volume of second phases. Any coarse, sharp, and brittle phase can create harmful internal notches and can nucleate cracks in service, which may lead to fatigue, corrosion fatigue, and stress corrosion cracking [12].

The traditional additions are chromium, manganese (these two elements improve weldability), nickel (added to certain 3xxx cast alloys used for pistons, cylinder blocks, and other engine parts to improve resistance at high temperatures), titanium (refining the as-cast structure), beryllium, zirconium, and lead (free machining alloys). Cast aluminum-bearing alloys may contain tin [9]. An alloy can contain more than one additive, and their concentrations may exceed 1% in certain cases. All the alloying elements can also be additives in another series of alloys [13].

Iron and silicon are the two main impurities of unalloyed aluminum in the 1000 series; their total concentration determines the purity of the metal. The iron/silicon ratio is close to 2, for most grades, unless it is deliberately modified, as with the 8000 series. The concentration of impurities can vary, depending on the alloy, from a few parts per million (ppm) in refined aluminum (1199) up to 1000–2000 ppm in most wrought alloys. The impurity level of cast alloys can be higher for alloys based on secondary aluminum [13].

4.2.3. Cast Alloys Series

Alloys belonging to the same series exhibit a set of common properties such as castability, mechanical properties, extrudability, and corrosion resistance. These properties can vary considerably from one series to another and as a function of metallurgical tempers [13]. Aluminum castings are widely used in automobile, electronic, aviation, and aerospace industries. More than 300 alloys are in international use. Properties displayed by these alloys are shown in Table 4.4 [11].

Cast aluminum products normally have an equiaxed grain structure. Special processing routes can be taken to produce fine, equiaxed grains in thin rolled sheets and certain extruded shapes (ASTM G34, Test Method for Exfoliation Corrosion Susceptibility in 2xxx and 7xxx Series Aluminum Alloys) [14]. Unlike wrought alloys, their selection involves consideration of casting characteristics as well as of properties [4].

As with wrought alloys, copper is the alloying element most deleterious to general corrosion. Alloys such as 356.0, A356.0, B443.0, 513.0, and 514.0 that do not contain copper as an alloying element have a high resistance to general corrosion comparable to that of non-heat-treatable wrought alloys. In other alloys, corrosion resistance becomes progressively less the greater the copper content. More so than with wrought alloys, a lower resistance is compensated by the use of thicker sections usually necessitated by requirements of the casting process [3, 4].

Cast aluminum and magnesium alloys have many competitors such as cast iron, zinc, copper, plastic, wrought steel, and machine-welded steel. Some comparative parameters are given in Table 4.5 [5].

Table 4.4 General Properties of Cast Aluminum Alloys

Property	Value
Tensile strength, MPa (ksi)	70–505 (10–72)
Yield strength, MPa (ksi)	20–455 (3–65)
Elongation, %	$<$1–30
Hardness, HB	30–150
Electrical conductivity, %IACS	18–60
Thermal conductivity, W/m · K at 25 °C (Btu · in./h · ft^2 °F at 77 °F)	85–175 (660–1155)
Fatigue limit, MPa (ksi)	55–145 (8–21)
Coefficient of linear thermal expansion at 20–100 °C (68–212 °F)	(17.6–24.7) $\times$ 10^{-6}/°C ((9.8–13.7) $\times$ 10^6/°F)
Shear strength, MPa (ksi)	42–325 (6–46)
Modulus of elacticity, GPa (10^6 psi)	65–80 (9.5–11.2)
Specific gravity	2.57–2.95

Source: Reference 11.

Table 4.5 Comparative Classification of Different Cast Materials

Property	Aluminum	Magnesium	Cast iron	Zinc	Copper	Plastic
Density	3	2	5	4	6	1
Specfic mechanical properties	3	4	1	5	2	6
	1	1	1	1	1	1
Gravity pressure die	2	3	1	4	2	—
	3	4		2	5	1
Electrical conductivity	2	3	3	3	1	4
Surface condition (as-cast)	3	5	6	1	6	2
Corrosion	2	5	3	4	1	1
Heat resistance	3	3	1	4	2	5
Cost	3	4	1	2	5	1

Source: Reference 5.

4.3. WROUGHT ALUMINUM ALLOYS

4.3.1. Designation of Wrought Aluminum Alloys

A four-digit numerical designation system is used to identify wrought aluminum and aluminum alloys. As shown below, the first digit of the four-digit designation indicates the group [10]:

Aluminum, $\geq$99.00%	1xxx

The following designations are for aluminum alloys grouped by major alloying element(s):

Copper	2xxx
Manganese	3xxx
Silicon	4xxx
Magnesium	5xxx
Magnesium and silicon	6xxx
Zinc	7xxx
Other elements	8xxx
Unused series	9xxx

For the 2xxx through 7xxx series, the alloy group is determined by the alloying element present in the greatest mean percentage. An exception is the 6xxx series alloys in which the proportions of magnesium and silicon available to form magnesium silicide (Mg_2Si) are predominant. Another exception is made in those cases in which the alloy qualifies as a modification of a previously registered alloy. If the greatest mean percentage is the same for more than one element, the choice of group is in order of group sequence: copper, manganese, silicon, magnesium, magnesium silicide, zinc, or others [10].

In the 1xxx group, the series 10xx is used to designate unalloyed compositions that have natural impurity limits. The last two of the four digits in the designation indicate the minimum aluminum percentage. These digits are the same as the two digits to the right of the decimal point in the minimum aluminum percentage when expressed to the nearest 0.01%. A designation having second digits other than zero (integers 1 through 9, assigned consecutively as needed) indicate special control of one or more individual impurities [10].

In the 2xxx through 8xxx alloy groups, the second digit in the designation indicates alloy modification. If the second digit is zero, it indicates the original alloy; integers 1 through 9, assigned consecutively, indicate modifications of the original alloy. Explicit rules have been established for determining whether a proposed composition is merely a modification of a previously registered alloy or if it is an entirely new alloy. The last two of the four digits in the 2xxx through 8xxx groups have no special significance, but serve only to identify the different aluminum alloys in the group [10].

Wrought aluminum products are produced by all of the standard hot and cold working processes. In general, the commercial alloys and tempers cross product lines. As such, the main effect of various products on corrosion is that attributable to variations in grain structure. Wrought alloys are of two types: non-heat-treatable, of the 1xxx, 3xxx, 4xxx, and 5xxx series, and heat treatable, of the 2xxx, 6xxx, and 7xxx series. An 8xxx series is reserved for miscellaneous alloys not covered by the previous groups [9].

Strengthening is produced by strain hardening, which can be increased by solid solution and dispersion hardening for the non-heat-treatable alloys. In the heat-treatable type, strengthening is produced by (1) solution heat treatment at 460–565 °C (860–1050 °F) to dissolve soluble alloying elements; (2) quenching to retain them in solid solution; (3) a precipitation or aging treatment, either naturally at ambient temperature or, more commonly, artificially at 115–195 °C (240–380 °F), to precipitate these elements in an optimum size and distribution; (4) solution heat treatment and natural aging; (5) air-quenching and aging; (6) solution heat treatment and annealing; (7) like 6, but overaged; (8) like 3, but with accelerated aging; and (9) like 6 but followed by strain hardening (cold working) [4].

The basic temper designations are presented in Table 4.6. Strengthened tempers of non-heat-treatable alloys are designated by an "H" following the alloy designation, while for heat-treatable alloys, tempers are designated by a "T"; suffix digits designate the specific treatment (e.g., 1100-H14 and 7075-T651). In both cases, the annealed temper, a condition of maximum softness, is designated by an "O" [3]. The temper designation system is used for all forms of wrought and cast aluminum and aluminum alloys except ingot cast materials. Basic temper designations consist of letters; subdivisions of the basic tempers, where required, are indicated by one or more digits following the letter [15].

4.3.2. Alloying Elements

Alloying elements are added to wrought alloys in quantities ranging from 1% to 7% (in mass percent), and in higher quantities, up to 20% silicon, to cast alloys.

The metallurgy of industrial aluminum alloys is therefore based on six systems:

- Aluminum–copper
- Aluminum–manganese
- Aluminum–silicon (with or without magnesium)
- Aluminum–magnesium
- Aluminum–magnesium–silicon
- Aluminum–zinc (with or without copper)

Table 4.6 Basic Temper Designations

	Basic temper designations
F	As fabricated. Applies to products in which no thermal treatments or strain-hardening methods are used to shape the product.
O	Annealed. Applies to wrought alloys that are annealed to obtain the softest temper, and to cast alloys that are annealed to improve ductility and dimensional stability.
W	Solution heat treated. An unstable temper applying to certain of the (7xxx) heat-treatable alloys that, after heat treatment, spontaneously age harden at room temperature. Only when the period of natural aging is indicated (e.g., W 1 hour) is this a specific and complete designation.
H	Strain hardened. Applies to those wrought products that have had an increase in strength by reduction through strain-hardening or cold working operations.
T	Thermally treated to produce tempers other than F, O, or H. Applies to those products that have had an increase in strength due to thermal treatments, with or without supplementary strain-hardening operations.
	Subdivisions of "T" temper heat-treatable alloys
T1	Cooled from an elevated temperature shaping process and naturally aged to a sub stantially stable condition. Usually associated with extruded products and limited to the 6xxx series alloys.
T2	Annealed. Cold worked to improve strength after cooling from an elevated temperature shaping process, or cold worked in flattening or straightening has a significant effect in mechanical property limits. Usually associated with cast products.
T3	Solution heat treated, cold worked, and naturally aged to a substantially stable condition (T4 + cold work).
T4	Solution heat treated, and naturally aged to a substantially stable condition. T42 indicates material is solution heat treated from the O or F temper to demonstrate response to heat treatment, and naturally aged to a substantially stable condition.
T5	Cooled from an elevated temperature shaping process and artificially aged. Usually associated with extruded products in the 6XXX series alloys (T1 + artificial age).
T6	Solution heat treated, and artificially aged (T4 + artificial age). T62 indicates material is solution heat treated from the O or F temper to demonstrate response to heat treatment, and artificially aged.
T7	Solution heat treated, and overaged/stabilized. Applies to products that are stabilized after solution heat treatment to carry them beyond the point of maximum strength to provide control of some special property.
T8	Solution heat treated, cold worked, and artificially aged (T3 + artificial age).
T9	Solution heat treated, artificially aged and cold worked (T6 + artificial age).
T10	Cooled from an elevated temperature shaping process, cold worked, and artificially aged. Usually associated with cast products (T2 + artificial age).
	The following specific digits have been assigned for stress-relieved tempers of wrought products
T-51	Applies to cold finished rod or bar when stress-relieved by stretching 1–3% permanent set. Stretching is performed after solution heat treatment or after cooling from an elevated temperature shaping process. No straightening takes place after stretching.
T-510	Applies to extruded products and to drawn tube when stress-relived by stretching 1–3% permanent set. Stretching is performed after solution heat treatment or after cooling from an elevated temperature shaping process. No straightening takes place after stretching 1–3% permanent set.

Table 4.6 (*Continued*)

T-511	Applies to extruded products and to drawn tube when stress-relieved by stretching 1–3% permanent set. Stretching is performed after solution heat treatment or after cooling from an elevated temperature shaping process. These products may receive minor straightening after stretching to comply with standard tolerance.
	Subdivisions of "H" temper non-heat-treatable alloys
H1	Strain hardened only. Applies to products that are strain hardened or cold worked to obtain the desired strength level without supplementary thermal treatments.
H2	Strain hardened and partially annealed. Applies to products strain hardened or cold worked more than the desired level by partial annealing. The number following this designation indicates the degree of strain hardening remaining after the partial annealing process.
H3	Strain hardened and stabilized. Applies to products in the magnesium–aluminum class, which will age-soften at room temperature after strain hardening. These products are strain hardened to the desired amount and then subjected to a low-temperature thermal operation that results in a improved ductility. The number following this designation indicates the degree of strain hardening remaining after the stabilization treatment.
H1x, H2x, H3x	The second digit following the designations HI, H2, H3 indicates the final degree of strain hardening. The number 8 has been assigned to tempers having a final degree of strain hardening equivalent to that resulting from approximately 75% reduction in area. Tempers between that of the 0 temper.
Hxxx	The third digit indicates a variation of the two digit H temper. It is used when the degree of temper is close to the 2 digit H temper.

Sources: References 16 and 17.

Silicon and magnesium are added for cast alloys of the series 4000 [13]. Lead and bismuth are added to alloys 2011 and 6262 to improve chip breakage and other machining characteristics. Nickel is added to wrought alloys 2018, 2218, and 2618, which were developed for elevated-temperature service [4].

The concentrations of alloying elements and additives are given in Table 4.7 for strain-hardenable alloys (non-heat-treatable) and age-hardenable wrought aluminum alloys [13].

The 4xxx series (Si 0.8–1.7%) are considered heat-treatable alloys [13] since they are used for welding electrodes (e.g., 4043) and as brazing rods (e.g., 4343) [7].

4.3.3. Wrought Aluminum Alloys Series

4.3.3.1. Non-Heat-Treatable Alloys

The non-heat-treatable alloys can be the 1xxx (almost pure aluminum), the 3xxx (manganese-containing alloys), the 5xxx (magnesium-containing alloys), and the 8000 series. Strengthening is produced by strain hardening, which can be increased by solid solution and dispersion hardening for the non-heat-treatable alloys [9]. All non-heat-treatable alloys have a high resistance to general corrosion. Aluminum alloys of the 1xxx series representing unalloyed aluminum have a relatively low strength [3, 4, 13]. Strain hardening involves a modification of the structure due to plastic deformation. It occurs not only during the manufacturing of semiproducts in the course of rolling, stretching, and drawing, but also during subsequent manufacturing steps such as forming, bending, or fabricating operations. Strain hardening increases the mechanical resistance and hardness, but decreases ductility [13].

Table 4.7 Nominal Chemical Compositions of Representative Wrought Aluminum Alloys

	Percentage of alloying elements								
Alloy	Si	Cu	Mn	Mg	Cr	Zn	Ti	V	Zr
				Non-heat-treatable alloys					
1060	99.60%	Minimum Al							
1100	99.00%	Minimum Al							
1350	99.50%	Minimum Al							
3003		0.12	1.20						
3004			1.20	1.0					
5052				2.5	0.25				
5454			0.80	2.7	0.12				
5456			0.80	5.1	0.12				
5083			0.70	4.4	0.15				
5086			0.45	4.0	0.15				
7072[a]						1.0			
				Heat-treatable alloys					
2014	0.8	4.400	0.80	0.5					
2219		6.30	0.30				0.06	0.10	0.18
2024		4.40	0.60	1.5					
6061	0.6	0.28		1.0	0.20				
6063	0.4			0.7					
7005			0.45	1.4	0.13	4.5	0.04		0.14
7050		2.30		2.2		6.2			
7075		1.60		2.5	0.23	5.6			

[a]Cladding for Alclad products.

Sources: References 3 and 4.

Properties of representative non-heat-treatable wrought aluminum alloys are given in Table 4.8.

4.3.3.2. Heat-Treatable Alloys

Heat-treatable alloys are mainly of the 2xxx (copper-containing alloys), 6xxx (silicon- and magnesium-containing alloys), and 7xxx (zinc-containing alloys) series. In the heat-treatable type, strengthening is produced generally by (1) a solution heat treatment at 460–565 °C to dissolve soluble alloying elements, (2) quenching to retain them in solid solution, and (3) aging treatment. Natural or artificial aging (115–195 °C) can precipitate these elements in an optimum size and distribution. Annealing, overaging, accelerated aging, and cold working are used to obtain certain properties of these alloys. Alloys in the 2xxx, 6xxx, and 7xxx series can be strengthened by heating and then quenching, or rapid

Table 4.8 Typical Tensile Properties of Representative Non-Heat-Treatable Wrought Aluminum Alloys in Various Tempers[a]

Alloy and temper		Strength (MPa)		Percentage	
		Ultimate	Yield	In 50 mm[b]	In 5D[c]
1060	-O	70	30	43	
	-H12	85	75	16	
	-H14	100	90	12	
	-H16	115	105	8	
	-H18	130	125	6	
1100	-O	90	35	35	42
	-H14	125	125	9	18
	-H18	165	150	5	13
3003	-O	110	40	30	37
	-H14	150	145	8	14
	-H18	200	185	4	9
3004	-O	180	70	20	22
	-H34	240	200	9	10
	-H38	285	250	5	5
5052	-O	195	90	25	27
	-H34	260	215	10	12
	-H38	290	255	7	7
5454	-O	250	115	22	
	-H32	275	205	10	
	-H34	305	240	10	
	-H111	260	180	14	
	-H112	250	125	18	
5456	-O	310	160		22
	-H111	325	230		16
	-H112	310	165		20
	-H116, H321	350	255		14
5083	-O	290	145		20
	-H116, H321	315	230		14
5086	-O	260	115	22	
	-H116, H32	290	205	12	
	-H34	325	255	10	
	-H112	270	130	14	

[a]Averages for various sizes, product forms, and methods of manufacture; not to be specified as engineering requirements or used for design purposes.

[b]For 1.60 mm thick specimen.

[c]For 12.5 mm diameter specimen.

Source: Reference 3.

cooling and then further strengthened by cold working (deformation at room temperature). For example, heat treatment and cold working can increase the ultimate yield strength of the aluminum alloy 2024 in the fully annealed O-temper by $2\frac{1}{2}$ times [9, 18].

Properties of representative heat-treatable wrought aluminum alloys are given in Table 4.9.

Table 4.9 Typical Tensile Properties of Representative Heat-Treatable Wrought Aluminum Alloys in Various Tempers[a]

Alloy and temper		Strength (MPa)		Percentage of Elongation	
		Ultimate	Yield	In 50 mm[b]	In 5D[c]
2014	-O	185	95		16
	-T4, T451	425	290		19
	-T6, T651	485	415		11
2219	-O	170	75	18	
	-T37	395	315	11	
	T-87	475	395	10	
2024	-O	185	75	20	20
	-T4, T351	470	325	20	17
	-T851	480	450	6	
	T-86	515	490	6	7
6061	-O	125	55	25	27
	-T4, T451	240	145	22	22
	-T6, T651	310	275	12	15
6063	-O	195	90	25	27
	-H34	260	215	10	12
	-H38	290	255	7	7
7005	-O	195	85		20
	-T63, T6351	370	315		10
7050	-T76, T7651	540	485		10
	-T736, T73651	510	455		10
7075	-O	230	105	17	14
	-T6, T651	570	505	11	9
	-T76, T7651	535	470		10
	-T736, T7351	500	435		11

[a]Averages for various sizes, product forms, and methods of manufacture; not to be specified as engineering requirements or used for design purposes.

[b]For 1.60 mm thick specimen.

[c]For 12.5 mm diameter specimen.

Source: References 3 and 4.

4.3.4. Description of the Wrought Alloys Series

4.3.4.1. The 1xxx Series

The 1000 series alloys have 99% pure aluminum or higher. This series has excellent resistance to general, localized corrosion and high electrical and thermal conductivities, but poor mechanical properties [9]. Wrought aluminum alloys of the 1xxx series conform to composition specifications that set maximum individual, combined, and total contents for several elements present as natural impurities in the smelter-grade or refined aluminum used to produce these products. Alloys 1100, 1120, and 1150 differ somewhat from the others in this series by having minimum and maximum specified copper contents. Under many conditions, it decreases slightly with increasing alloy content [19]. The refined alloys (1199, 1198) have a degree of purity between 99.90% and 99.999%. Depending on their purity, they are used in the manufacture of electrolytic condensers and lighting devices and for decorative applications in the building sector and luxury packaging (cosmetics, perfumes). The metal is usually anodized. The 1050A alloy is more than 99.50% pure and is one of the most widely used grades. It has a wide range of applications: for example, packaging, buildings, sheet metal working, fins and tubes for heat exchangers, and electrical conductors. The 1200 alloy is between 99% and 99.5% pure and replaces 1050A whenever its plastic formability is adequate (packaging, circles for kitchen utensils) [20].

4.3.4.2. The 2xxx Series

The 2xxx wrought alloys and 2xx.x cast alloys, in which copper is the major alloying element, are less resistant to corrosion than alloys of other series, which contain much lower amounts of copper. Alloys of this type are used in structural applications, particularly in aircraft and aerospace applications [9].

2xxx Wrought Alloys Containing Lithium The 2xxx series alloys have strong electrochemical effects due to the presence of copper. Variations of copper concentration (heterogeneous segregation) in solid solution lead to variations in electrode potentials and thus local galvanic corrosion cells. Copper also has the tendency during electrochemical corrosion reactions to replate on aluminum surfaces and form small cathodic sites, causing additional local galvanic cells. These alloys are susceptible to pitting and stress corrosion cracking [21]. Lithium additions decrease the density and increase the elastic modulus of aluminum alloys, making aluminum–lithium alloys good candidates for replacing the existing high-strength alloys, primarily in aerospace applications. Although lithium is highly reactive, addition of up to 3% Li to aluminum shifts the pitting potential of the solid solution only slightly in the active direction in 3.5% NaCl solution [19].

4.3.4.3. The 3xxx Series

One of the most widely used alloys, 3003, has moderate strength, good workability, and very high resistance to corrosion and can be inhibited in certain media [9]. Alloys of the 3xxx series (Al–Mn, Al–Mn–Mg) have the same desirable characteristics as those of the 1xxx series, but somewhat higher strength. Almost all the manganese in these alloys is precipitated as finely divided phases (intermetallic compounds), but corrosion resistance is not impaired because of the negligible difference in electrode potential between the phases and the aluminum matrix in most environments does not create a galvanic cell [3, 4]. The manganese is present in the aluminum solid solution, in submicroscopic particles of precipitate, and in larger particles of $Al_6(Mn,Fe)$ or $Al_{12}(Mn,Fe)_3Si$ phases, both of which have solution potentials almost the same as that of the solid-solution matrix. Such alloys are widely used for cooking and food-processing equipment, chemical equipment, and various architectural products requiring high resistance to corrosion [19]. Magnesium ($\sim$1.2%) significantly enhances the mechanical properties of aluminum and adds between 40 and 50 MPa to the minimum guaranteed tensile strength values, while retaining good formability. Adding up to 0.20% copper provides a further increase in mechanical resistance, and adding up to 0.7% copper makes it possible to obtain a fine-grained structure [20]. Moreover, Mg added to some alloys in this series provides additional strength through solid-solution hardening, but the amount is low enough that the alloys behave more like those with manganese alone than like the stronger Al–Mg alloys of the 5xxx series [3, 4].

The main applications of 3003 are in the building sector (cladding panels, roofing sheet), fabrication, sheet metal work, heat exchanger tubing, and circles for kitchen utensils. The 3004 alloy, with roughly 1% magnesium added, offers slightly better mechanical properties, while retaining the overall properties of 3003. It is used chiefly for cans (food cans), for circles for kitchen utensils, and in buildings (coil-coated sheets) [20].

4.3.4.4. The 4xxx Series

These alloys are in demand for architectural uses because of the color effects that can be obtained when anodic coatings are applied. Alloys in this series have good corrosion resistance and can be inhibited. Alloys of the 4xxx series (Al–Si) are low-strength alloys used mainly for brazing and welding products (because of their lower melting points) and for cladding in architectural products. These alloys develop a gray appearance upon anodizing [3, 4, 9]. Elemental silicon is present as second-phase constituent particles in wrought alloys of the 4xxx series, in brazing and welding alloys, and in cast alloys of the 3xx.x and 4xx.x series. Silicon is cathodic to the aluminum solid-solution matrix by several hundred millivolts and accounts for a considerable volume fraction of most of the silicon-containing alloys. However, the effects of silicon on the corrosion resistance of these alloys are minimal because of low corrosion current density resulting from the fact that the silicon particles are highly polarized [11].

4.3.4.5. The 5xxx Series

Alloys of the 5xxx series (Al–Mg) are the strongest non-heat-treatable aluminum alloys, and in most products, they are more economical than alloys of the 1xxx and 3xxx series in terms of strength per unit cost. Magnesium is one of the most soluble elements in aluminum, and when dissolved at an elevated temperature, it is largely retained in solution at lower temperatures, even though its equilibrium solubility is greatly exceeded. It produces considerable solid-solution hardening, and additional strength is produced by strain hardening. Alloys of the 5xxx series not only have the same high resistance to general corrosion as other non-heat-treatable alloys in most environments, but in slightly alkaline environments they have a better resistance than any other aluminum alloy. They are widely used because of their high as-welded strength when welded with a compatible 5xxx series filler wire, reflecting the retention of magnesium in solid solution [3, 4].

Wrought alloys of the 5xxx series (Al–Mg–Mn, Al–Mg–Cr, and Al–Mg–Mn–Cr) and cast alloys of the 5xx.x series (Al–Mg) have high resistance to corrosion, and this accounts in part for their use in a wide variety of building products and chemical-processing and food-handling equipment, as well as marine applications involving exposure to seawater [19]. Prolonged holding at a high temperature leads to the precipitation of the intermetallic compound $A1_3Mg_2$ at the grain boundaries. If required by the application, a stabilization heat treatment can be carried out on alloys containing 3% magnesium or more (H321 and H116 tempers). Other possible additions are manganese, chromium, and titanium, which provide a further increase in tensile strength and/or certain properties such as corrosion resistance and weldability [20].

Surface treatments such as brightening or anodizing can give these alloys a very attractive surface appearance, especially when the alloy is derived from base metal that is low in iron and silicon; this is the case of alloy 5657 (base metal 1080). Alloy 5052, with 2.5% magnesium and added chromium, is a good compromise between mechanical resistance, formability, fatigue resistance, and corrosion resistance. It is widely used in the H28 temper for food cans and in a large number of applications in fabricating, commercial vehicle bodies, and road signs. Alloy 5049 is a variant of 5052 containing manganese but no chromium. Coil in 5049 is widely used for thermal insulation and for sheet metal forming [20].

4.3.4.6. The 6xxx Series

The 6xxx series alloys, silicon- and magnesium-containing alloys, are present in the ratio required to form magnesium silicide. These alloys have good corrosion resistance and may be inhibited effectively [9]. Among heat-treatable alloys, those of the 6xxx series, which are moderate-strength alloys based on the quasi-binary Al–Mg_2Si (magnesium silicide) system, provide a high resistance to general corrosion equal to or approaching that of non-heat-treatable alloys [19]. Considerable industrial interest exists for the 6xxx series alloys because of their attractive combination of properties such as medium strength, good corrosion resistance, formability, weldability, and low cost. The Al–Mg–Si alloy 6061, having a balanced ratio of 1% magnesium and 0.6% silicon to form Mg_2Si, has set up as the standard for light-weight, economical material for general purpose structural use. It contains an addition of 0.3% copper to achieve a higher strength in the T6 temper compared to copper-free alloys with balanced composition in Mg and Si. The lower fuselage of the high capacity aircraft Airbus 380 was built up with welded panels of the alloys 6013 and 6056, which contains a high amount of copper in the range from 0.6% to 1.1%. Addition of copper to Al–Mg–Si alloys refined the precipitated structure, induced the formation of the quaternary strengthening phase Q', and increased the hardness. However, a high alloying amount of copper deteriorates the good corrosion resistance of Al–Mg–Si alloys, inducing sensitivity to localized corrosion [22].

4.3.4.7. The 7xxx Series

Heat-treatable alloys of the 7xxx series (Al–Zn–Mg) that do not contain copper as an alloying addition also provide a high resistance to general corrosion [3, 4]. Moderately high strength and very good resistance to corrosion make the heat-treatable wrought alloys of the 6xxx series (Al–Mg–Si) highly suitable in various structural, building, marine, machinery, and process-equipment applications [11]. The zinc-containing alloys, the 7000 series alloys, may also contain smaller percentages of magnesium, copper, and chromium. These alloys can have very high strengths, for example, 7075, which is one of the highest strength aluminum alloys. Inhibitors may be used with the 7000 series [9]. The 7xxx wrought alloys and the 7xx.x cast alloys contain major additions of zinc, along with magnesium or magnesium plus copper in combinations that develop various levels of strength. Those containing copper have the highest strengths and have been used as construction materials, primarily in aircraft applications, for more than 40 years [11].

The 7xxx series alloys contain zinc and magnesium and include a great number of alloys that also contain copper (e.g., Al 7075). These alloys are strengthened by precipitation of solute-rich zones and are among the highest strength materials available on the basis of strength-to-weight ratios. The 7xxx series alloys are also more resistant to general corrosion than the 2xxx series alloys. However, they are susceptible to stress-corrosion cracking and exfoliation corrosion. The 7xxx series alloys with higher copper content allow for higher aging temperatures without excessive loss of strength. The T7 heat treatment has been developed to improve resistance to exfoliation and stress-corrosion cracking [21].

4.3.4.8. The 8xxx Series

Alloys in the 8xxx series encompass a wide range of compositions. The simultaneous addition of iron (which yields a fine-grained structure) and silicon improves the mechanical properties of aluminum. With their fine-grained structure and good isotropy, these alloys have good formability under difficult conditions, even as foil (between 50 and 200 m thick). This explains their increasing use as fins for heat exchangers, spiral tubes, dishes, and thin foil [20]. Alloy 8011 (Al–0.75Fe–0.7Si) is used for bottle caps because of its good deep drawing qualities and several other dilute compositions as electrical conductor materials.

These alloys and other dilute compositions containing transition metal elements, such as 8006 (Al–1.6Fe–0.65Mn), are used for producing foil and finstock for heat exchangers. This series contains several dilute alloys, for example, 8001 (Al–1.1Ni–0.6Fe), which is used in nuclear energy installations where resistance to corrosive attack by water at high temperatures and pressures is the desired characteristic. Its mechanical properties resemble 3003. Alloys such as 8280 and 8081 serve an important role as bearing alloys based on the Al–Sn system but are not widely used in motor cars and trucks, particularly where diesel engines are involved. Some new, lithium-containing alloys, designated 8090 (Al–2.4Li–1.3Cu–0.9Mg–0.1Zr) and 8091 (Al–2.6Li–1.9Cu–0.9Mg–0.12Zr), have been developed in Britain and France [7].

4.4. ALUMINUM POWDERS AND ALUMINUM MATRIX COMPOSITES

4.4.1. Aluminum Powders

The lack of crystalline atomic periodicity is the primary distinguishing feature of an amorphous metal. The chemical homogeneity and lack of grain boundaries and line defects, such as dislocations, inherent in amorphous metals suggest that superior corrosion resistance might be achievable. Their already highly disordered structures would appear to be reasonably resistant to radiation damage, suggesting their utilization where conductivity or mechanical properties must remain constant under irradiation [23]. For example, very slow corrosion rates have been observed for Cr-containing metallic glasses exposed to standard test solutions compared to those of ordinary stainless steels since, for example, the amorphous alloys are resistant to pitting attack in sulfuric acid solutions containing chloride ions. ESCA studies indicate that the passive film formed on the amorphous alloy is similar in composition to that found on stainless steels. The superior long-time performance of this film must therefore be due to its microscopic homogeneity [24].

Aluminum powders have an impressive variety of applications because of the following properties:

- Exceptional mechanical properties
- Exceptional fatigue properties
- Low density
- Good ductility
- Nonmagnetic properties
- Corrosion resistance
- High thermal and electrical conductivity
- Excellent machinability

- Good response to a variety of finishing processes
- Competitive cost per unit volume basic

The primary driver to use powder metallurgy (P/M) for aluminum is the ability to produce complex net or near net shape. Mechanical properties can vary from 110 to 345 MPa (16–50 ksi) depending on composition, density, sintering practices, and thermal treatments. The two basic classes of commercial press and sinter alloys are 601AB (Al–0.25Cu–0.6-Si–1Mg) and 201AB (Al–4.4Cu–0.8Si–0.5Mg). The 601AB alloy displays moderate strength with excellent corrosion resistance while 201AB alloy has high mechanical properties in both the as-sintered and heat-treated condition. Aluminum powder is used in blasting agents and in solid-fuel rockets used for national defense and to launch space probes. The combustion of the powder releases concentrated energy from high heat. This characteristic is also used for a heat source, a reducing agent, hot topping compounds, stress relief, exothermic welding, and powder lancing in metallurgical industries. Rapid solidification technology produces aluminum powder prealloyed with strength, toughness, fatigue and corrosion resistance, and elevated-temperature performance not achievable with conventional wrought alloys [25]. Powder compounds derived from Al P/M also have a wide variety of uses in the chemical and plastics industries. Aluminum nitride P/M ceramics find applications in the electronic area, like multichip modules because of their thermal expansion and heat transfer characteristics. Aluminum powder is an interesting way to produce Al foam components [26].

The compositions of some aluminum P/M alloys appear in Table 4.10.

Conventional P/M Conventional alloys consist of blends of elemental powders, often containing lubricants, which are consolidated by press and sinter processing [28]. Carbowax and stearic acid are suitable lubricants for the die wall to prevent wear [27]. Powder metallurgy (P/M) technology provides a useful means of fabricating net-shape components, enabling machining to be minimized and thereby reducing costs. Aluminum P/M alloys can therefore compete with conventional aluminum wrought and cast alloys, as well as with other materials, for cost-critical applications [29]. In addition, P/M allows the development of alloys and microstructures showing unique properties as well as a great latitude in these properties. Mechanical properties of these alloys are shown in Table 4.11.

Table 4.10 Composition of Some Aluminum P/M Alloys

Alloy designation	Composition
201AB (conventional alloy)	Al–4.4Cu–0.8Ci–0.5Mg–1.5other
202AB (conventional alloy)	Al–4.0Cu–1.5other
601AB (conventional alloy)	Al–0.25Cu–0.6Si–1.0Mg–1.5other
602AB (conventional alloy)	Al–0.6Mg–0.4Si–1.5other
7090 (advanced alloy)	Al–8.0Zn–2.5Mg–1.0Cu–1.5Co–0.35O
7091 (advanced alloy)	Al–6.5Zn–2.5Mg–1.5Cu–0.4Co
X7090 (advanced alloy)	Al–0.12Si–0.15Fe–(0.6–1.3)Cu–(2–3)Mg–(7.3–7) Zn–(1.0–1.9)Co
X7091 (advanced alloy)	Al–0.12Si–0.15Fe–(1.1–1.8)Cu–(1–3)Mg–(5.8–7.1) Zn–(0.2–0.6)Co
X7093 (advanced alloy)	Al–9.0Zn–2.2Mg–1.5Cu–0.14Zr–0.1Ni
8009 (advanced alloy)	Al–8.5Fe–2.4Si–1.3V
X8019 (advanced alloy)	Al–8.3Fe–4.0Ce

Source: Reference 27 (p. 840).

Table 4.11 Mechanical Properties of Conventional P/M Alloys

Alloy	Compacting pressure (MPa)	Tensile strength (MPa)	Yield strength MPa	Elongation (%)	Hardness (HRE)
601AB-T6	96	183	176	1	70–75
	165	232	224	2	75–80
	345	238	230	2	80–85
602AB-T6	165	179	169	2	55–60
	345	186	172	3	65–70
201AB-T6	110	248	248	0	80–85
	180	323	322	0.5	85–90
	413	332	327	2	90–95
202AB Compacts- T6	180	227	147	7.3	45–50
202AB Cold-formed 19% T6	180	274	173	8.7	85

Sources: References 10 and 28 (p. 1269).

Aluminum powder production can be divided into four main routes, which all give different final properties: commercial atomization, rapid solidification processing (RSP), mechanical alloying and processing [30], and reaction milling. Moreover, aluminum P/M alloys fall into two major groups—conventional alloys and advanced alloys. The conventional alloys are mainly obtained by commercial atomization techniques (inert gas, water and air atomization), while advanced alloys are obtained via the three other production routes [28].

4.4.2. Rapid Solidification Processing

Recently developed aluminum alloys can provide nearly custom-engineered strength, fracture toughness, fatigue resistance, and corrosion resistance for aircraft forgings and other critical components. Rapid solidification processing (RSP) is the basis for these new alloy systems, called wrought P/M alloys [31]. The term wrought P/M is used to distinguish the technology from the conventional press-and-sinter P/M technology. Grades 7090 and 7091 are the first commercially available wrought P/M aluminum alloys. These alloys can be handled like conventional aluminum alloys on existing aluminum-fabrication facilities [31].

The RSP is used to develop new alloys that fall into four basic groups: (1) high-strength corrosion-resistant alloys based on traditional 7000 series aluminum, (2) lower density Al–Li alloys having higher Li levels than possible by conventional means, (3) high-temperature alloys containing normally low solubility elements such as Fe, Mo, Ni, and rare earth elements, and (4) Al–Si alloys with improved wear and modulus and decreased thermal expansion coefficients. The properties of some alloys that fall into these groups can be found in Tables 4.11 and 4.12 [28].

Table 4.12 shows the properties of mechanical alloying and processing (MAP) and RSP aluminum alloys and composites. Mechanical alloying is used for fabricating oxide-dispersion-strengthened alloys and discontinuously reinforced composites [27].

4.4.3. Aluminum Matrix Composites and P/M- MMCs

The recent worldwide interest shown in the metal matrix composite (MMC) materials has been fueled by the fact that mechanical properties of light alloys can be enhanced by incorporating reinforcing fibers (usually ceramic). Several manufacturers are marketing a

Table 4.12 Properties of Some MAP and RSP Aluminum Alloys and Composites

Material	0.2% Yield strength (MPa)	Ultimate tensile strength (MPa)	Elongation (%)	Young's modulus (GPa)
MAP AA2014 as extruded		450		
MAP AA2014-T6		563		
MAP 7010, at room temperature	412	498	1.85	74.2
RSP 7090	595	637	10	
RSP X7090-T6E192	641	676	10	73.8
RSP X7091-T6E192	558	614	11	72.4

Source: Reference 27 (p. 846).

range of particulate reinforced MMC products with different compositions, for example, 12% alumina, 9% carbon fiber, reinforced Al–12% SiC, and particulate SiC/Al ingots. The major reinforcements used in aluminum-based MMCs are boron, graphite, silicon carbide, and alumina [4].

Generally, long-term tests have shown that the introduction of a reinforcement phase reduces the resistance to corrosion. The extent of this reduction largely depends on the reinforcement species and form. As with conventional aluminum alloys, fabrication method and heat treatment influence the corrosion resistance of MMCs and must be carefully controlled. As surface protection is advisable in certain applications, it is encouraging to see a variety of standard techniques showing promise for MMCs. From the studies performed on the corrosion fatigue of MMCs in saline environments, it appears that they are marginally inferior to their matrix alloys [32].

Composites Grouped According to the Reinforcement Used Aluminum matrix composites (AMCs) can be found in an ever-expanding number of domains. Even though aerospace is the main field of expertise for these composites, important applications also exist in the automotive and electrical industries. The reinforcements used are extremely varied and the properties of some of them are summarized in Table 4.13 [32].

Reaction milling is used to produce composites of aluminum alloys and aluminum nitride. Once incorporated in the structure, the AlN stabilizes the grain size and, it was found that powders could be sintered to 99.6% density while retaining most of its refined grain size during either extrusion or hot isostatic pressing [28]. Aluminum alloys reinforced with silicon carbide, graphite, alumina, boron, or mica show promise as MMCs with increased modulus and strength and are potentially well suited to lightweight structural applications, including aerospace and military needs. The structures of continuous fiber MMCs are equivalent to those in polymer matrix composites.

Industrial applications have emerged recently, for example, reinforced pistons for assembly in light diesel engines, 12% alumina, with 9% carbon fiber reinforced A1–12.7% Si MMC cylinder liner [4, 19, 33]. Manufactured powder metallurgy metal matrix composites (P/M-MMCs) offer economical solutions for the production of high-performance materials. With P/M-MMCs, many disadvantages associated with the fusion metallurgical production of composite materials with a metallic matrix can be avoided [34].

However, fusion metallurgical procedures are used primarily to produce composite materials. Nevertheless, a set of composite materials has also been manufactured by powder metallurgy methods. Thus it has to be pointed out that the powder metallurgy manufacturing route of MMCs is ideally suitable for inserting a high volume percentage of reinforcement components into the material. Values up to approximately 50 vol % are obtainable. This

Table 4.13 Properties of Some of the Reinforcements Used in AMCs

Particular type	Density (g/cm^3)	Melting point (°C)	Modulus (GPa)	Thermal expansion ($K^{-1} \times 10^{-6}$)	Size range (μm)
Borides					
CrB_2	5.10	2100	n/a	7.5	—
MoB	8.65	2180	n/a	n/a	—
TiB_2	4.50	2800	515–574	4.6	> 40
ZrB_2	6.20	3200	n/a	5.9	—
Carbides					
B_4C	2.51	2350	450	4.5	—
CrC	7.00	3660	370	11.0	—
HfC	12.70	3890	352	6.3	—
SiC	3.22	2300	450	4.5	1–7 or 10–15
TiC	4.95	3000	460	7.6	—
WC	15.50	2800	700	4.9	—
ZrC	6.75	3500	350	6.6	—
Nitrides					
AlN	1.30	2200	320	5.5	> 40
BN	3.48	2500	195	7.5	3–40
HfN	14.00	3300	n/a	6.9	—
Si_3N_4	3.60	1750	300	3.7	—
TiN	550	2900	n/a	9.4	—
ZrN	7.30	3000	n/a	7.0	—
Oxides					
Al_2O_3	3.97	2015	380	8.0	1–80
BeO	3.06	2500	380	10.3	—
HfO_2	9.68	2760	n/a	5.8	—
MgO	3.75	2620	275	13.0	—
SiO_2	2.65	1610	110	0.55	—
ThO_2	9.9	3200	240	10.4	—
TiO_2	4.26	1800	88	6.8	—
Y_2O_3	5.01	2375	n/a	9.3	—
ZrO_2	6.27	2500	185	8.0	—
Whisker/short fiber reinforcements					
Al_2O_3 + 4% SiO_2	2.1	—	280	5–9	3 diameter, 500 length
β-SiC	3.2	—	410	~4.3	0.05–1.5 diameter, 5–2000 length
α-Si_3N_4	3.18	—	358	~3.2	0.1–0.6 diameter, 5–200 length
β-Si_3N_4	3.2	—	379	~3.2	0.1 diameter 15 length
C	2.25	—	700–800	—	0.2–1 diameter 10–200 length
Continuous fibers					
Al_2O_3	3.9	—	385	5.7	20
C(PAN)	1.76–1.81	—	230–390	−0.1 to −1.2 longitudinal and 0.7–1.2 perpendicular	6.5–7

Table 4.13 (*Continued*)

Particular type	Density (g/cm^3)	Melting point (°C)	Modulus (GPa)	Thermal expansion ($K^{-1} \times 10^{-6}$)	Size range (μm)
C (pitch)	1.9–2.08	—	160–520	−0.1 to −1.2 longitudinal 0.7–1.2 perpendicular	11
Si–C–O (Nicalon)	2.55	—	190–616	3.1	15
δ-$Al_2O_3 \cdot SiO_2$	3.2	—	173–300	—	3
Monofilaments					
SiC (CVD on W)	3.1	—	400	3.1	100
B (CVD on W)	2.6	—	410	—	140

Source: Reference 32 (p. 7).

differentiates such systems from particle-strengthened MMCs, which can be manufactured using fusion metallurgy, where typically about 20 vol% of reinforcement components can be brought into the matrix.

A further advantage of the powder metallurgy manufacturing of MMCs is the homogeneous structure. Besides mixing source materials with subsequent consolidation, mechanical alloying is used or in situ reaction for aluminum–based P/M-MMCs. Examples of mechanically alloyed powders are the systems for dispersion-solidified aluminum (Al/Al_2O_3, Al/Al_4C_3). Other materials manufactured by mechanical alloying are aluminum alloys, where carbon is ground in an oxygen-containing ball mill atmosphere, which leads to the formation of Al_2O_3 and Al_4C_3. P/M-MMCs show a large variation in material systems and powder manufacturing processes, which gives a variety of custom-made powders suitable for each application. Accordingly, the application possibilities are huge [34].

The P/M process usually involves mixing of powders of the matrix alloy with the reinforcing particles, followed by compacting and solid-state sintering. This means using lower temperatures than alternative processing methods, with less interaction between the matrix and the reinforcement. It is very important that all particles are homogeneously distributed in the mixture in order to obtain a good microstructure. When whiskers are used as reinforcement, smaller particles for the matrix alloys are required for the improvement of the packing effect and to obtain a good dispersion of the fibers in the matrix. All the properties of the MMCs obtained by P/M (some are given in Table 4.14) can be improved through liquid-phase sintering with or without extra pressing, and usually through final steps such as extrusion, forging, or rolling [35].

Many aluminum P/M alloys just like the wrought alloys can be age treated. However, the temper designation for P/M parts is a little different from those used for wrought alloys. The following designations are often used for conventional P/M aluminum alloys [28]:

T1	As-sintered
T2	As cold formed (after sintering)
T4	Solution heat treated and at least 4 days at room temperature
T6	Solution heat treated and artificially aged

Other designations exist and mean other processing steps where applied to the part. For example, these may be repressing, cold deformation, or overaging steps similar to the T7 and T8 designations for wrought alloys [28].

Table 4.14 Mechanical Properties for Different Aluminum MMCs

Materials	Manufacturing	Young's modulus (GPa)	$\sigma_{0.2}$ (MPa)	σ_{max} (MPa)	Elongation (%)	Fracture toughness $(MPa \cdot \sqrt{F})$
Al–Cu	Squeeze cast	70.5	174	261	14.0	—
Al–Cu + Al_2O_3 (Vf = 0.2 fiber)	Squeeze cast	95.4	238	374	2.2	—
Al–Cu–Mg (T6), 2014	Spray (sheet)	73.8	432	482	10.2	—
Al–Cu–Mg + SiC (T6), Vf = 0.1–10m part	Spray (sheet)	93.8	437	484	6.9	—
Al–Cu–Mg(T4),2124	Powder rolling (sheet)	72.4	360	525	11.0	—
Al–Cu–Mg + SiC (T4), Vf = 0.17–3 μm part	Powder rolling (sheet)	99.3	420	610	8.0	18
Al–Cu–Mg (T6), 2124	Powder rolling (sheet)	73.1	425	474	8.0	26
Al–Cu–Mg + SiC (T6), Vf = 0.17–3 μm part	Powder rolling (sheet)	99.6	510	590	4.0	17
Al–Si–Mg (T6), 6061	Spray + rolling (sheet)	69.0	240	264	12.3	—
Al–Si–Mg + SiC (T6), Vf = 0.1–10 μm part	Spray + rolling (sheet)	91.9	321	343	3.8	—
Al–Zn–Mg–Cu (T6), 7075	Spray + extrusion	71.1	617	659	11.3	—
Al–Zn–Mg–Cu + SiC (T6), Vf = 0.12–10 μm part	Spray + extrusion	92.2	597	646	2.6	—
Al–Li–Cu–Mg (T6), 8090	Spray (sheet)	70.5	420	505	6.5	38
Al–Li–Cu–Mg + SiC (T6), Vf = 0.17–3 μm part	Spray (sheet)	104.5	510	550	2.0	—

Source: Reference 35.

In the group of MMC materials the infiltration of a porous preform, using ceramic reinforcement components in the form of short fibers and/or particles, with molten light metal alloy surely represents one of the most promising technologies with regard to the range of the attainable properties of the final composite material. Some applications (e.g., fiber-reinforced aluminum diesel pistons for trucks) have been in production for over 10 years with up to several hundred thousand pieces produced annually, and it has been proved that this technology is controllable for series quantities also. Nevertheless, the widespread use of MMC materials has not yet taken place. The reason for this is the manufacturing cost, since among other things special pressure casting processes such as squeeze casting are necessary. The samples with a perform porosity of 70% showed after 6000 cycles an average total crack length of only 12 mm and no fracture. This is in the same range as a 20 vol.% Al_2O_3 fiber-reinforced alloy, such as an aluminum piston with fiber-reinforced combustion bowl. Light metal alloy composite materials can help to reduce masses within the vehicle due to their high specific properties [36].

4.4.4. Al MMC Particles and Formation

4.4.4.1. Particle Reinforcing Aluminum Alloys Matrix

SiC Depending on the intended use, the reinforcement is either a whisker, a particle, or in a few cases monofilaments. SiC whiskers are discontinuous, rod- or needle-shaped fibers in the size range of 0.1–1 μm in diameter and 5–100 μm in length. Because they are nearly single crystals, the whiskers typically have very high tensile strengths (up to 7 GPa) and elastic modulus (up to 550 GPa). SiC particles have a lower cost, and since they have an irregular shape the composites produced show isotropic properties. Duralcan™ is an example of a readily available Al/SiC_p (p for particle) composite; examples of applications include brake disks, drums, calipers, and backplate, stabilizer bars, train brake rotors, and bike and golf components. SiC particles are added to Al–Si casting alloys where the Si in the alloy inhibits the formation of Al_4C_3 [32]. The high hardness of silicon carbide makes the composite resistant to wear. The typical product consists of AA359 or AA360 aluminum matrix reinforced with 10–20% SiC particles [37, 38]. Table 4.15 compares the properties of as-cast AA359 and AA360-T6 alloys with those of the composites.

The AA359 composite is used with gravity casting techniques and segregation of the reinforcement may occur during solidification. On the other hand, the AA360 composite is intended for high-pressure die casting and the high cooling rates associated with this

Table 4.15 Comparison of AA359 and AA360 Alloy Properties with Same Alloys Reinforced with 20% SiC Particles

Material	UTS (MPa)	Yield (MPa)	Elastic modulus (GPa)	Elongation (%)	Coefficient of thermal expansion, CTE
AA360—F	300	170	71	2.5	20.9
AA360 + 20% SiC—F	303	248	108	0.5	16.6
AA359—T6	310	234	72.4	4	20.9
AA359 + 20% SiC—T6	359	338	98.4	0.4	17.5

Source: Reference 37.

technique tend to minimize segregation and grain size while maximizing mechanical properties [37].

A similar type of discontinuously reinforced aluminum (DRA) composite may also be obtained via powder metallurgy techniques, which results in a good distribution of the reinforcement. Aluminum powders (mainly 2009 and 6092 alloys) are blended with SiC particles or whiskers and vacuum-hot-pressed to produce billets, which are then extruded and heat treated to make the desired composite. The composites show excellent specific strength and elastic modulus, good high-temperature mechanical properties, as well as outstanding fatigue and creep resistance. Table 4.16 presents the mechanical properties of two particulate-reinforced composites and one whisker-reinforced composite commonly encountered [39, 40]. Examples of use include F-16 fuel access door covers, ventral fins and fan exit guide vanes, Eurocopter blade sleeves, and rollercoaster brakes.

Another possibility with SiC particles is to use the Osprey process to produce functionally graded materials (FGMs). The idea is to atomize an aluminum alloy, which is mixed with reinforcing particle onto a substrate to be covered. The mix solidifies on the substrate, making it possible to optimize the properties of the material where it is needed [32].

Because of the difficulty in incorporating SiC monofilaments in an aluminum matrix, the work that exists on the subject is neither extensive nor very successful in application [41]. According to the literature, Textron Corporation produces Al/SiC_f composites commercially.

Al_2O_3 Grades of Duralcan containing Al_2O_3 particles instead of SiC are available when wrought alloys, which contain lower amounts of Si, are needed, but the most recent innovation is the production of composite electrical conductor cables using continuous alumina fibers. 3M's Nextel™ cable was developed with the intent of allowing higher current transport, high strength, and stiffness and creep resistance while having a lower cable density than the original steel core cable. The only drawback is the price of the cable since the cost of the alumina fibers alone is approximately three times the cost of the whole steel cable. The wires consist of oriented continuous alumina fibers reinforcing either a pure or a 2% Cu aluminum matrix. Table 4.17 compares the properties of the conventional and composite wires [41].

Another application for the Nextel 610 Al–2%Cu composite is for rotors used in aerospace applications where thermal and environmental stability are needed [41].

Intermetallics Recently, a new family of particle reinforcement has been used with promising results: intermetallics. The most used systems are Ni–Al (probably the most promising) and Fe–Al, but other systems such as Al–Nb can suggest multiple improvements in the composite properties. As a general rule, intermetallics offer an increase in wear and

Table 4.16 Typical Properties of DRA Composites Obtained Via Powder Metallurgy

Material	Ultimate strength (MPa)	Yield strength (MPa)	Young's modulus (GPa)	Strain to failure (%)
2009/SiC/30p	561	488	123	1.3
6092/SiC/25p	478	399	112	2.1
2009/SiC/15w	635	355	n.a.	3.9

Source: References 39 and 40.

Table 4.17 Conventional Versus Composite Wires

Property	Al/Al_2O_3 core	Steel core
Density (g/cm^3)	2.8	7
Tensile strength (MPa)	1600	1600
Electrical resistance (copper = 100%)	32%	8%
CTE (K^{-1})	7×10^{-6}	7×10^{-6}
Thermal conductivity (W/m/K)	100	20

Source: Reference 41.

corrosion behavior as well as an improvement in mechanical properties. One of the problems that must be controlled in reinforcing with intermetallics is their higher reactivity with the matrix, which can reduce the age hardenability of the matrix alloy [35].

Aluminum Carbon and Boron Fibers Because of the exceptionally high specific stiffness and close to zero coefficient of thermal expansion (CTE) of these AMCs, carbon fiber Al/C_f reinforced aluminum composites are mainly used for structural components in aerospace applications. For example, the high-gain antenna of the Hubble space telescope is manufactured by diffusion bonding of laid-up, melt-infiltrated Al/C_f wires and serves two functions—to support the communications antenna away from the telescope and to carry radio signals between the craft and the antenna [41]. Table 4.18 summarizes the properties of the Al/C_f composite antenna.

The substrate for the Standard Electronic Module Series-E is another less known but notable application. The SEM-E modules are used in U.S. naval and aerospace applications to carry microelectronic devices.

Research is being done in the European NACE project on Al/C_f composites for overhead electrical wires. The idea is to make a product with similar properties to 3M's Al/Al_2O_3 wire but with a much lower cost (due to the use of the cheaper carbon fibers). Unfortunately, major problems are associated with the use of carbon fibers rather than alumina fibers. First, since aluminum does not naturally wet carbon, high pressures are necessary to create a good interface between the reinforcement and the matrix. Second, the

Table 4.18 Properties of 6061 Al/40 vol% P-100C_f Composite Tubes

Longitudinal properties	
Thickness (mm)	0.058
Density (g/cm^3)	2.48
Modulus (GPa)	339
UTS (MPa)	680
Strain-to-failure (%)	0.201
CTE (ppm/°C)	0.29
Transverse properties	
Poisson's ratio	0.297
Modulus (GPa)	27.9
UTS (MPa)	20.9
Strain-to-failure (%)	0.099

Source: Reference 41.

temperature of the melt has to be kept to a minimum in order to reduce Al_4C_3 formation. Finally, unlike Al_2O_3 and SiC fibers, graphite fibers can induce galvanic corrosion [42, 43].

The first use of boron/aluminum MMCs has been tubular struts in the mid-fuselage structure of the space shuttle. The space shuttle struts were made by General Dynamics/Convair and Amercom, Inc. in 1975 and are still in service on the shuttle fleet. The strut tubes consist of 6061 Al, unidirectionally reinforced with approximately 50 vol% of boron fibers and resulted in a 44% weight savings as compared to the originally specified aluminum extrusions (Table 4.19) [44].

Superplastic Superplastic forming of metal, a process similar to vacuum forming of plastic sheet, has been used to form low-strength aluminum into nonstructural parts such as cash register housings, luggage compartments for passenger trains, and nonload-bearing aircraft components. New in this area of technology is a superplastic-formable high-strength aluminum alloy, now available for structural applications and designated 7475-02. The strength of alloy 7475 is in the range of aerospace alloy 7075, which requires conventional forming operations. Although the initial cost of 7475 is higher, the finished part cost is usually lower than that of 7075 because of the savings involved in the simplified design/assembly [31].

4.4.4.2. Aluminum Matrix Composite Formation

In Situ The commonly known example of in situ processing is unidirectional eutectic solidification. However, newly developing processes are based on two principles: (1) controlled reaction between a molten alloy and a gas and the subsequent forming of reinforcement in the molten metal, and/or (2) endothermic reactions between the components in order to produce the reinforcement. The latter process is known as self-autopropagating high-temperature synthesis (SHS). One example of controlled reactions in a liquid is the in situ oxidizing process, called the lanxide process. In this process, molten Al oxidizes to produce a mixture of Al and Al_2O_3 [35].

Spray Forming One of the P/M processes to obtain MMCs is spray forming. This is based on powder gas atomizing (which consists of a melt of metal that is atomized by a gas at high pressure). In the case of spray forming, the atomized beam strikes an intermediate preform, which is the matrix of the composite with the desired shape. MMCs manufactured

Table 4.19 Properties of Al/B_f 50 vol% Composite (0 °C)

Density (g/cm^3)	0.7
Poisson ratio (n_{xy})	0.23
Longitudinal properties	
Young's modulus (GPa)	235
UTS (MPa)	1100
CTE(10^{-6}/K)	5.8
Transverse properties	
Young's modulus (GPa)	138
UTS (MPa)	110

Source: Reference 44.

by this method are made by the introduction of reinforced particles inside the atomizing beam for incorporation into the solidified alloy. The contact time between the liquid metal and the reinforcing particles is short. This fact and the high cooling rate of the molten particles reduce the interfacial reaction possibilities. In this way, the formation of brittle and undesired interfacial compounds is minimized. The atomizing melting rate is close to 5 kg/min, and the obtained perform has a density of 95% of the theoretical value. After that, a finishing operation must be done (such as forging, extrusion, or rolling) in order to obtain the full density [35].

This processing method gives the obtained parts a fine microstructure with a very homogeneous distribution of the reinforcing material, and they can retain a high amount of alloying elements in solution [35].

B. USE OF ALUMINUM AND ALUMINUM ALLOYS

Structural components made from aluminum are vital to the aerospace industry and very important in other areas of transportation and building in which light weight, durability, and strength are needed. The use of aluminum exceeds that of any other metal except iron. Pure aluminum easily forms alloys with many elements such as copper, zinc, magnesium, manganese, and silicon. Nearly all modern mirrors are made using a thin reflective coating of aluminum on the back surface of a sheet of float glass. Telescope mirrors are also coated with a thin layer of aluminum. Other applications are electrical transmission lines and packaging (cans, foil, etc.). Because of its high conductivity and relatively low price compared to copper, aluminum was introduced for household electrical wiring to a large degree in the United States in the 1960s. Unfortunately, function problems were caused by its greater coefficient of thermal expansion and its tendency to creep under steady sustained pressure, both eventually causing loosening of connections and galvanic corrosion, increasing the electrical resistance. The most recent development in aluminum technology is the production of aluminum foam by adding to the molten metal a compound (a metal hybrid) that releases hydrogen gas. The molten aluminum has to be thickened before this is done and this is achieved by adding aluminum oxide or silicon carbide fibers. The result is a solid foam that is used in traffic tunnels and in the space shuttle [1].

The widespread use of aluminum in processing, handling, and packaging of foods, beverages, and pharmaceutical and chemical products is based on economic factors and the excellent compatibility of aluminum with many of these products. In addition to high corrosion resistance in contact with such products, many of these applications depend on the nontoxicity of aluminum and its salts, as well as its freedom from catalytic effects that cause product discoloration. Aluminum for packaging foods, beverages, and pharmaceutical products accounts for approximately 20% of the aluminum marketed in the United States. Large quantities of aluminum foil, either uncoated or with plastic coatings, are used in flexible packages. Packaging foils are produced from unalloyed aluminum corresponding to composition limits for aluminum 1230. Sheet for beverage can bodies is generally made from alloy 3004, 5352, or 5050 and can ends are made from alloy 5182. These alloys have high corrosion resistance and are not normally subject to corrosion problems in such applications. Aluminum alloy household cooking utensils, usually made of alloy 1100 or 3003, have been used for many years. These utensils, as well as commercial food processing equipment, do not require protective coatings. Alloys used in commercial food processing include alloy 3003, 5xxx alloys, and cast alloys 444.0 and 514.0 [19].

Aluminum–lithium alloys are used in cryogenic applications and in the production of aircraft parts and in space applications [18]. Weldalite alloy (5.4%Cu, 1.3%Li, 0.4%Mg, 0.14%Zr, 0.4%Ag 0.0%Zn, 0.0%Mn, 0.0%Cr, Bal.% Al (wt%)), AA2195, has received increasing research interest in recent years due to its commercial potential [45]. Al–Li alloy AA2195 is a good candidate material for the next generation of space shuttle vehicles. Its high specific strength and stiffness will improve lift efficiency, fuel economy, and performance, and would increase payload capabilities of aircrafts and spacecrafts [46, 47].

The Al/Air Battery (Potential Future Application for High-Purity Aluminum)
The conceptual aluminum/air battery is composed of pure Al as the anode and an air electrode as the cathode. The electrolyte may be either a neutral chloride (e.g., NaCl) or an alkali (e.g., NaOH) and the net reaction is $4Al + 6H_2O + 3O_2 \rightarrow 4Al(OH)_3$. Refueling is also conceptually easy and quick, since it involves only the regular addition of water and removal of the solid reaction products, together with the occasional replacement of the aluminum anodes. An Al–Sn–Mg anode exhibiting the desired characteristics was patented by Alcan in 1988 and development since then has concentrated on strategies that would allow the use of lower-purity aluminum. Essentially this has involved coping with iron impurities and up to 40 ppm can now be tolerated, which has more than halved the cost of the aluminum. Individual cells generate approximately 1.4 V and are connected in series to give the desired power output.

Interest in the aluminum/air battery was first stimulated by its potential as a power unit for electric vehicles. Using an alkaline electrolyte, this battery was seen as a viable alternative to the internal combustion engine as far as acceleration, refueling time, and range were concerned. Specific energy yields of around $4\,W \cdot h \cdot g^{-1}$ were obtained and it has been claimed that such a battery would be capable of providing the power needed to drive a conventional sized motor car some 400 km between stops for water to replenish the alkaline electrolyte, and 2000 km before more aluminum was needed. To date, however, the system has not proved to be competitive with engines powered by petroleum fuels. One limited application has been the use of aluminum/air batteries as reliable and compact reserve units to back up dc electrical systems. In this regard, they provide a quiet and clean alternative to more costly diesel generators [7].

4.5. USE OF CAST ALUMINUM ALLOYS

Cast aluminum alloys are often grouped into categories as a function of the metal, the alloy, the quality of the casting processes, and the direct use.

4.5.1. Standard General Purpose Aluminum Alloys

These alloys containing silicon as the major constituent are by far the most important commercial cast alloys mainly due to superior casting characteristics. Binary Al–Si alloys show high corrosion resistance, good weldability, and low specific gravity; however, they are difficult to machine. Si is on the order of 7%, while 12% Si, close to eutectic composition, is characterized by its high fluidity. Al–Si–Cu alloys with silicon (range 3–10.5%) and copper (range 2–4.5%) are chosen for higher strength and improved machinability, while at the same time Cu leads to reduced ductility and lower corrosion resistance. General purpose alloys are used in the F temper, while the T5 temper is considered for some of these alloys with improved hardness and machinability [18].

Al–Si–Mg alloys are selected for their excellent casting characteristics and resistance to corrosion (7% Si and 0.3% Mg). This justifies its use in large quantities for sand and permanent mold-castings. With the help of several heat treatments, required tensile and physical properties are achieved. The excellent properties of these alloys are achieved for certain aerospace and military applications through T6 heat treatment. Al–Si–Mg–Cu alloys have higher strengths because of their greater response to heat treatment due to copper addition, with some sacrifice in ductility and corrosion resistance [18]. A low iron version of these alloys gives higher tensile properties and premium quality for sand and permanent mold castings [18].

Magnesium content is usually minimized to control oxidation during the casting process. Iron content on the order of 0.7% or greater is preferred in most processes to maximize the casting process; however, a reduced Fe as low as 0.25% is recommended for improved ductility. Zinc additions are sometimes considered for enhanced fluidity of the 380.0 type, for example [18].

Premium Cast Alloys These reflect higher levels of quality and reliability than that found in conventionally produced parts especially in better mechanical properties, extreme soundness, dimensional accuracy, and better finish. This can reflect fine dendrite arm spacing, and well-refined grain structure in the microstructure through optimum concentrations in hardening elements and restricted impurities. In aluminum silicon alloys, for example, iron should be controlled at or below 0.01% with measurable advantages to the range of 0.03–0.05%, the practical limit of commercial smelting capacity. Beryllium is present in A357 and A158 cast alloys to alter the form of the insoluble phase to a more nodular less detrimental form to ductility and to inhibit oxidation as a corollary benefit [18].

4.5.2. Some Specific Uses

Rotor Pure Al Alloys Pure alloys (99.0–99.7% Al) with controlled impurities are used to minimize variations in electrical conductivity as a parameter and to minimize micro-shrinkage and cracks during casting.

Heat-Treatable Duralumin Alloys These commercial alloys were the first heat-treatable alloys and have been used extensively as cast or wrought alloys where high strength and toughness are required. Al–Cu–Mg and Al–Cu–Si alloys were developed after World War I in Europe and the United States, respectively. More recently, unusual strength and toughness have been achieved for Al–Cu–Mg alloys through the solving of castability problems by modern foundry equipment and control techniques [18].

Piston and Elevated-Temperature Alloys The most used alloy for passenger car pistons is 332.0-T5 which has a good combination of foundry, mechanical, and physical characteristics, including low thermal expansion. Heat treatment improves hardness for improved machinability, and eliminates any permanent changes in dimension due to aging at operating temperatures. For other applications, such as for airplanes and motorcycles, 10% Cu alloy 222.0-T61 is replaced by 242.0 and 243.0 compositions because of their better properties at elevated temperatures [18].

Aluminum–Tin Bearing Alloys These alloys contain 6% Sn and small amounts of Cu and Ni for strengthening. They are used for cast bearings because of the excellent lubricity imparted by tin. Al–Sn bearing alloys are superior overall to bearings made using most other materials. They are applied in certain applications where load-carrying capacity, fatigue strength, and resistance to corrosion by internal combustion lubricating oil are important criteria (e.g., connecting rods and crankcase bearings for diesel engines) [18].

4.6. USE OF WROUGHT ALUMINUM ALLOYS

4.6.1. Aerospace Applications

Aircraft designers require materials that will allow them to produce lightweight, cost-effective structures that are durable and damage tolerant at ambient, subzero, and occasionally elevated temperatures. Strong aluminum alloys date from the accidental discovery of the phenomenon of age hardening by Alfred Wilm in Berlin in 1906. His work led to the development of the wrought alloy known as Duralumin (Al–3.5Cu–0.5Mg–0.5Mn), which was quickly adopted in Germany for structural sections of Zeppelin airships, and for Junkers F-13 aircraft that first flew in 1919. Since that time, wrought aluminum alloys have been the major materials for aircraft construction, which, in turn, has provided much stimulus for alloy development. Duralumin was the forerunner of a number of 2xxx series alloys including 2014 and 2024 that are still used today. The other major aircraft group of alloys is the 7xxx series [7].

Novel dispersoids and dispersoid combinations have enabled further improvements in the performance of existing alloy families. For example, appropriate Sc and Zr additions have a significant impact on the grain structure of 2xxx alloys and thus on performance. Another high potential approach for alloy performance improvements is the optimization of Al–Cu–Li–(Mg–Ag–Zn) alloys. These "third generation Al–Li alloys" were principally developed for military and space applications; in order to meet the demands of future commercial airframes, more damage-tolerant variants are being developed [48]. The current choices in aluminum alloys for applications requiring higher strength with improved stress-corrosion and exfoliation resistance are overaged 7050, 7150, and 7055 plates and extrusions for upper wing skin, and 2324-T39 or 2024-T39 plates for the lower wing skins.

Clad 2024-T3 or 2524-T3 sheets or 2024-T351/2524-T351 plates are used on the pressurized fuselage. Cladding on the exterior skin of the fuselage with 2xxx and 7xxx series alloys, consisting of 1230 and 7032, respectively, is provided for increased corrosion resistance. 2xxc-T3x alloys have relatively low exfoliation and stress-corrosion cracking resistance and therefore must be very well protected. The chemical-milled or machined surfaces of 2024-T351 clad aluminum plate used for bilge skin has proved to be a corrosion concern, regardless of whether the milled cladding is internal, as in the Boeing 747–400, or external, as in the Boeing 777 [49].

A new process called creep age forming (CAF) has been developed, which offers substantial cost benefits for the production of curved aluminum alloy components, such as large wing panels for aircraft. CAF has been applied successfully to alloys of the 2xxx and 7xxx series, as well as to the lithium-containing alloy 8090. It is being used for the production of upper wing panels for several civil and military aircraft, including the new Airbus 380 [7].

4.6.2. Automotive Sheet and Structural Alloys

Serious attention to weight savings in motor vehicles first arose in the 1970s following steep increases in oil prices imposed by Middle Eastern countries. More recently, impetus has come from legislation in some countries to reduce levels of exhaust emissions through improved fuel economies. In this regard, each 10% reduction in weight is said to correspond to a decrease of 5.5% in fuel consumption. Moreover, each kilogram of weight saved is estimated to lower CO_2 emissions by some 20 kg for a vehicle covering 170,000 km. Particular attention has focused on the replacement of steel and cast iron by aluminum alloys, which usually results in weight savings of 40–50% [7].

The first alloys selected for automotive sheet were 3004, 5052, and 6061. However, the low strength of 3004, problems with Lüders band formation drawing of some 5xxx series alloys, and the limited formability of 6061 led to the development of new compositions such as the copper-containing alloys 2008 and 2036, and other 6xxx series alloys including 6009 and 6010. Now focus for producing the "body in white" vehicle is on the use of the non-heat-treatable Al–Mg alloys or several Al–Mg–Si alloys of the 6xxx series that respond to age hardening [7]. Caceres [50] has made a cost analysis showing that direct equal-volume, Al alloy substitutions of cast iron and steel are the most feasible in terms of the Corporate Average Fuel Economy (CAFE) liability. The current higher recycling efficiency of cast Al alloys confers on Al a significant advantage over Mg alloys used for automotive applications [50]. The use of composite materials offers advantages when the characteristic profile of a standard material for an application is no longer sufficient. In their use in combustion engines, the following objectives are of importance [51]:

- Increase in mechanical strength (in particular, at higher temperatures)
- Increase in thermal shock stability
- Increase in stiffness (Young's modulus)
- Improvement in wear resistance and tribological characteristics
- Reduction of thermal expansion

A cylinder surface technology, by which the cylinder surfaces in an aluminum cylinder crankcase are enriched by perform infiltration with silicon, is presented. This method develops a local metal matrix composite material. This technology went into mass production for the first time in 1996 with the cylinder crankcase of the Porsche Boxster. Monolithic and quasi-monolithic concepts are to be preferred to heterogeneous solutions due to their technical advantages [51].

The proven LOKASIL® concept, which links the advantages of the monolithic cylinder crankcases with highly productive die casting processes, is surely the best solution for the advancement of perform production and composition, with only small extra costs compared to gray cast iron liners. Plasma-coated cylinder surfaces find straightforward introduction into series application. Their operability is beyond doubt [51].

Aluminum alloys have overcome various difficulties to successfully realize application throughout the vehicles, from engine parts to body and chassis components. However, there is still room to improve technologies to assure part reliability, with no defects, and casting designs for part integration. Also, there are still many issues to address with respect to the cost competitiveness of aluminum. It is necessary to reduce the cost of materials and processes, such as forging and stamping. Moreover, automakers' globalization of manufacturing plants makes it necessary to assure the availability of high-quality materials worldwide. At the same time, recycling systems are becoming more important to improve Life Cycle Assessments of all materials, including aluminum. These concerns cannot be addressed by automakers alone, so we would like to promote cooperation with aluminum manufacturers to work together for the resolution of these and other concerns [52].

4.6.3. Shipping

For passenger vessels, the use of aluminum alloys makes possible an increase in the volume and height of the superstructure without loss of stability, which, in turn, allows for the inclusion of more passenger decks than is possible with an equivalent design built in steel [7]. The most commonly used alloys for plate are the 5xxx series based on the composition 5083. A more recent alloy is 5383, for which the nominal levels of magnesium and manganese have been slightly increased [7].

4.6.4. Building and Construction

Significant use of aluminum and its alloys for building materials commenced some 65 years ago after the end of World War II. Advantages of aluminum include its good decorative appearance, high corrosion resistance in most environments, light weight, ease of fabrication, and the fact that extruded sections can easily be prepared for the provision of double glazing or the insertion of insulation and blinds. Applications include facades, roofing, gutters, window frames, sun shades, curtain walls, and balustrades. Aluminum alloys are also often used as external cladding to retain spalled fragments and disguise discoloration in old stone and concrete buildings. More limited use is being made to construct small bridges. The alloys in common use for rolled products are those based on the 5xxx series (e.g., 5083) and 3xxx series (e.g., 3003), whereas extrusions are usually made from the 6xxx series (e.g., 6063) [7].

4.6.5. Packaging

During the last two decades, the use of aluminum in packaging has increased to an extent that, at times, it has been the largest market for this metal in some countries including the United States. In 2004, packaging placed second in the United States and also in Japan, which is one of the other two leading consumers of aluminum. This situation has arisen because aluminum is an attractive container for food and beverages since it has generally high corrosion resistance, offers good thermal conditions, and is impenetrable by light, oxygen, moisture, and microorganisms [7].

4.6.6. Electrical Conductor Alloys

The use of aluminum and its alloys as electrical conductors has increased significantly in recent decades, due mainly to fluctuations in the price and supply of copper. The conductivity of electrical conductor (EC) grades of aluminum and its alloys average about 62% that of the International Annealed Copper Standard (IACS). However, aluminum conducts more than twice as much electricity, because of its lower density, compared to an equivalent weight of copper. As a consequence, aluminum is now the least expensive metal with a conductivity high enough for use as an electrical conductor and this situation is unlikely to change in the future. Aluminum is also widely used for insulated power cable, especially in underground systems [7].

The aim of MMC developments is the substitution of aluminum for heavy materials such as gray cast iron, and the design of more filigree aluminum components, so that they can be made more compact. Magnesium alloys are also attaining increasing importance because of their low specific gravity. Promising results were obtained in work with carbon short fiber preforms and carbon fiber hybrid preforms [6, 36].

Above all, hybrid preforms offer a tool to influence light metal alloy characteristics such as Young's modulus, tensile strength, fatigue behavior, creep stability, hardness, wear resistance, thermal expansion, and thermal conductivity and to tailor properties within certain limits. This can be further extended by the development of multiphase hybrid preforms and the adjustment of controlled property gradients in the direction of local material engineering in both the MMC component and the preform itself [36].

It is well known that MMC materials are significantly superior to most single-phase material systems in their oscillation and absorption behavior. A further application potential exists in the range of noise reduction—noise vibration harshness (NVH) [36].

C. ALUMINUM PERFORMANCE

4.7. RESISTANCE OF ALUMINUM ALLOYS TO ATMOSPHERIC CORROSION

It should be mentioned that the rate of attack greatly decreases with increasing time of exposure. In atmospheric corrosion, slightly elevated temperatures can be beneficial by reducing the time of wetness. An example is electrical conductors that operate slightly above ambient temperatures and which usually incur little corrosion because the elevated operating temperature keeps them dry [53].

The aluminum-based alloys as a class are highly resistant to normal outdoor exposure conditions. The alloys containing copper as a major alloying constituent (over about 1%) are somewhat less resistant than the other aluminum-based alloys such as 1100, 3300, 5052, 6053, Alclad 3300, Alclad 1017-T, and Alclad 2024-T. They will all discolor or darken appreciably under most outdoor exposures (particularly that caused by SO_2), but will suffer no structurally appreciable changes in properties unless exposed in relatively thin sections below 0.076 mm (0.03 in.) thick [53, 54].

Results of typical outdoor exposure tests are based on exposure of machined tensile specimens 103.1 mm (4.06 in.) thick. Loss in tensile strength is generally on the order of 1–2% for the first year depending on the alloy and the atmosphere. An alloy such as 2017-T can lose up to 17% in tensile strength during the first year. If the specimens had been thinner, obviously the losses would have been relatively greater; whereas if they had been thicker, the losses would have been smaller. This effect of thickness is especially pronounced in the case of aluminum-based alloys, since the rate of attack greatly decreases with increasing time of exposure [8].

Specimens were freely exposed to outdoor locations. If they had been partially sheltered, the rate of attack was somewhat greater; if they had been largely sheltered, very little attack occurred. Apparently, in the case of aluminum-based alloys, periodic exposure to rain is beneficial, probably because the rain washes off corrosive products that settle from the air. Evidently, free exposure to rain is not harmful but, on the contrary, is beneficial [4]. Nevertheless, there is appreciable public concern over the effect of acidic precipitation ("acid rain") on all construction materials. Typically, the pH ranges from 4 to 5.5 and rarely is less than 3.5. As such, acidic precipitation does not cause severe damage to aluminum and its alloys from the standpoint of structural integrity. However, acid rain can cause cosmetic problems, such as dark brown to black stains [53].

Table 4.20 Results of Atmospheric Exposure of Different Aluminum Materials in a Wide Variety of Testing Sites Around the World

Alloy	Atmosphere	Location	Exposure (years)	Rate (μm/yr)
Alclad 2017-T3	Industrial	New York	20.55	20.3
3003-H14	Industrial	New York	20.55	19.3
6051-T4	Industrial	New York	20.55	18.3
1100-H14	Industrial	New York	20.55	15
7075-T6	Marine	Aruba, Dutch Antilles	7	10.2
1100	Marine	Panama	16	17.3
6061-T	Marine	Panama	16	17.3
2014-T3	Marine	Aruba, Dutch Antilles	7	17.8
2017-T3	Marine	La Jolla, California	18.15	45.2

Source: Reference 55 (p. 603).

The gases ordinarily found in industrial atmospheres have little effect in accelerating the corrosion of aluminum-based alloys. Carbon particles from the atmosphere may accelerate corrosion by galvanic action. Under outdoor atmospheric exposure conditions, this factor is of secondary importance even in intensely industrial regions. Sulfur compounds, such as H_2S, have no specific effect in accelerating the tarnishing or corrosion of aluminum alloys. However, the highly acidic nature of water containing dissolved SO_2 or SO_3 causes it to become somewhat corrosive [8].

Some results of atmospheric exposure are given in Table 4.20. The industrial atmosphere shows a corrosion rate of the concerned alloys between 15 and 20 μm/year. In the case of a marine atmosphere, the corrosion rate values were mostly between 10 and 18 μm/year. However, the increased agressivity shown in the case of La Jolla, California, could be due not only to the alloy composition and microstructure but also to the distance of the exposed samples from the shore [55].

4.8. FACTORS AFFECTING ATMOSPHERIC CORROSION OF ALUMINUM ALLOYS

Effect of O_2 Oxygen does influence the corrosion of aluminum. The corrosion of aluminum is very slow in deoxygenated solutions. In the presence of atmospheric dissolved O_2, corrosion is accelerated. In general, high concentrations of dissolved oxygen tend to stimulate attack, especially in acid solutions, although this effect is less pronounced than for most of the other common metals [8].

Effect of Hydrogen and Nitrogen Hydrogen and nitrogen have no effect, except as they influence the oxygen content in a solution, but aqueous solutions of hydrogen chloride are strongly corrosive to aluminum [8].

Effect of CO_2 Carbon dioxide and hydrogen sulfide, even in high concentrations, appear to have a slight inhibiting action on the effect of aqueous solutions on aluminum alloys [4]. Carbon dioxide strongly inhibits aluminum corrosion in the presence of $AlCl_3 \cdot 6H_2O$ and especially NaCl, but shows more profound localized corrosion. The inhibitive effect of CO_2 in the case of NaCl is attributed to its acidity. Carbon dioxide neutralizes the alkaline solution formed in the cathodic areas and forms solid carbonates. CO_2 decreases pH in the surface electrolyte, resulting in a positively charged alumina film. Chloride adsorption on the passive film causes local depassivation, explaining the predominance of pitting corrosion in the presence of CO_2. The slowing down of aluminum chloride-induced corrosion of aluminum by CO_2 may be connected to the formation of amorphous precipitates, aluminum hydroxy carbonates. Carbon dioxide is slightly corrosive in the presence of $MgCl_2 \cdot 6H_2O$. It is suggested that CO_2 accelerates the magnesium chloride-induced corrosion of aluminum because it acidifies the electrolyte, keeping Mg^{2+} in solution [56].

Effect of Ozone Ozone plays a significant role in the corrosion of aluminum and can be a corrosion accelerator. In the past, this role was mainly attributed to oxidizing H_2S, S(IV), and nitrogen species. Oesch and Faller [57] have shown that ozone can enhance the aluminum corrosion processes substantially on its own. The high corrosion rate observed can be attributed to the electrochemical reduction reactions of ozone:

$$O_3 + 2H^+ + 2e^- \rightarrow H_2O + O_2, \quad E^o = 2.08 - 0.06\,\text{pH} + 0.03\log(p_{O3}/p_{O2})$$

$$O_3 + 6H^+ + 6e^- \rightarrow 3H_2O, \quad E^o = 1.5 - 0.06\,\text{pH} + 0.098\log(p_{O3})$$

or one of its reaction products (i.e., the hydroxy radical), which is balanced by the metal dissolution. Exposure to ozone leads to locally enhanced attack and to the formation of aluminum oxides or hydroxides.

Effect of Sulfur Dioxide Oesch and Faller [57] have shown that exposure to sulfur dioxide results in a thin surface layer consisting, according to XRD and EDX measurements, of $Al_3(SO_4)_2(OH)_5 \cdot 9H_2O$ and aluminum oxide or hydroxide and leads to localized corrosion. This layer flakes off the base metal due to an increase in volume during the transformation from aluminum oxides or hydroxides to the mentioned sulfate. Its effect on corrosion processes is greater than nitrogen dioxide and smaller than ozone.

The corrosion rate of the aluminum alloy AA3003 by 0.01% or up to 1% SO_2 (285 to 28,500 mg/m^3, respectively) had no influence on the corrosion rate of the alloy 3003 at relative humidity (RH) 66% (0.1 mg/m^3). At RH 98% the corrosion rate increased to 0.15 for 0.01% SO_2, while for the strong concentration of 1% SO_2, the corrosion rate increased drastically to 1.8 mg/cm^2 [13].

Effect of Sulfur Trioxide Dry SO_3 has no effect on aluminum corrosion processes but it reacts with water to form sulfuric acid. This acid is quite aggressive and strongly attacks aluminum [13].

Effect of Nitrogen Dioxide or Nitrogen Tetraoxide Nitrogen dioxide leads to localized corrosion and to the formation of aluminum oxides and hydroxides [57]. N_2O_4 corrodes aluminum alloy 5086 at a rate of 1.25 mm/year at 0.2% humidity [13].

Contact with Nonmetallics The weather resistance of aluminum can seriously be affected when aluminum is used in contact with nonmetallic substances that either become saturated by moisture or are hygroscopic. Moist wood, insulation, or masonry in contact with aluminum can stimulate accelerated corrosion simply by keeping the aluminum wet for prolonged periods. These moist materials can also create a poultice corrosion, which establishes corrosion-conductive differently aerated cells [58].

Indoor Exposures The effects of indoor exposure differ greatly, depending on the exposure conditions. Exposure indoors in homes or offices ordinarily causes, at most, only a mild surface dulling of aluminum-based alloys even after prolonged periods of exposure. In damp locations, especially where there is contact with moist insulating materials, such as wood, cloth, and paper insulation, attack may be more appreciable (e.g., poultice corrosion). In factories or chemical plants, fumes or vapors incident to the operations being conducted may cause a definite surface attack. However, in most indoor atmospheres where pools of contaminated water do not remain in prolonged contact with aluminum alloys, or where extended contact with moist, porous materials is avoided, no appreciable loss of mechanical properties through corrosion will occur. In particular, aluminum alloys are highly resistant to warm, humid conditions where there is appreciable moisture condensation as long as contact with porous materials is avoided. Bare aluminum alloy panels have been used in constructing humidity cabinets that operate just above the dew point at 50 °C. After 5 years of use, there was no corrosion other than minor surface staining [8].

4.9. WATER CORROSION

Aluminum-based alloys are not appreciably corroded by distilled water even at elevated temperatures (up to 180 °C [350 °F] at least) [53]. Somewhat surprisingly, the effects of alloying elements on corrosion resistance of aluminum alloys in high-purity water at elevated temperatures are opposite to their effects at room temperature; elements (including impurities) that decrease resistance at room temperature improve it at high temperature [58]. Aluminum alloys of the 1xxx, 3xxx, 5xxx, and 6xxx series are resistant to corrosion by many natural waters. The more important factors controlling the corrosiveness of fresh waters on aluminum include water temperature, pH, and conductivity; availability of cathodic reactant; presence or absence of heavy metals; and the corrosion potentials of the specific alloys [19].

Corrosion of aluminum requires the presence of moisture and oxygen. Aeration and oxygenating conditions will accelerate corrosion. Conversely, deoxygenation will retard corrosion. The amount of water may be minuscule and present as isolated droplets or a continuous film. Soft waters tend to be less corrosive than hard waters. At ambient temperatures, aluminum initially reacts with high-purity water, but this ceases after a few days as a result of the development of a more protective oxide film. A small amount of water can drastically affect resistance to certain anhydrous organic solutions, particularly halogenated hydrocarbons [59].

Certain waters may cause severe localized attack or pitting. The Alclad products are much more resistant to perforation by pitting than are the other aluminum alloys. Therefore, wherever the characteristics of specific water are not known in advance, it is safer to employ aluminum alloys such as Alclad 3003. Pitting is of most importance where the metal section thickness is small, since the rate of attack at the pits generally falls off with increasing time of exposure. In general, the time necessary to perforate an aluminum alloy sheet 0.10 cm thick or greater is prolonged, as attested to by the wide and successful use of aluminum tea kettles [4]. Water staining, a type of crevice corrosion of aluminum, can occur. Water vapor in the air is sufficient to cause staining upon condensation and to support stress-corrosion cracking (SCC).

Steam Condensate Condensates from steam boilers, if free from carry-over of water from the boiler, are similarly inert to aluminum-based alloys. Thus either wrought or cast aluminum alloys are used successfully for steam radiators or unit heaters. Where aluminum alloys are used, it is desirable to install suitable traps in the steam lines, since entrapped boiler water, especially if alkaline water-treating compounds are employed, may be corrosive [4]. Unlike steel, aluminum is resistant to steam, and steam or boiling water treatments can actually increase the protective oxide [59].

Chloride Ions Halides, particularly the chloride ion, are corrosive to aluminum. Various mechanisms have been proposed, the most probable being localized breakdown and penetration of the oxide film. The corrosiveness of the chloride ion is a major concern because the ion is ubiquitous. Potable water and well water often are used to make up processing solutions and these normally contain small amounts of chlorides p59].

Heavy Metal Ions Small concentrations of heavy metals in solution (particularly copper, iron, lead, and even trace amounts of mercury) can plate out on the aluminum surface and can cause rapid localized pitting. Interactive effects can also occur, for example, the rate of pitting as a function of chloride content is highly accelerated by the addition of a few parts per million (ppm) of copper ions. Consequently, the composition of etchants and other process solutions should be monitored and reviewed on a regular basis. The depth of pitting occurring in such operations may be shallow, but can be adverse for cosmetic reasons, or can become a site for subsequent corrosion [59].

Inhibitors Some ions are inhibitors and will reduce either the anodic or cathodic reactions that occur during the corrosion of aluminum. Examples are chromates (an anodic inhibitor) and phosphates (a cathodic inhibitor). Inhibition of circulating water systems is complex and professional consultation is recommended for the design of water treatment systems [59].

Corrosion rates tend to decrease as the electrolyte becomes spent and saturated with aluminum ions. Consequently, a greater amount of corrosion can be expected for conditions that prevent this, such as high flow rates, and a low ratio of area of metal surface to volume of solution. Contamination of a pure solution can increase, or decrease, the corrosion rate. Therefore predictions of the corrosion performance should be obtained from published data or by experimentation [59].

4.10. SEAWATER

Corrosion of aluminum alloys in seawater is mainly of the pitting type, as would be expected from its salinity and enough dissolved oxygen to act as a cathodic reactant to polarize the alloys to their pitting potentials. Rates of pitting usually range from 3 to 6 μm per year during the first year and from 0.8 to 1.5 μm per year averaged over a 10 year period; the lower rate for the longer period reflects the tendency for older pits to become inactive. The corrosion behavior of aluminum alloys in deep seawater, judging from tests at 1.6 km, is generally the same as at the surface except that the effect of crevices is greater [60]. The 5xxx series of the wrought alloys have the highest resistance to seawater and are widely used for these media. A low-carbon steel corrodes 100 times more rapidly than aluminum alloys of these series in seawater [19]. The 356.0 and 514.0 cast alloys are used extensively for marine applications. For testing, measurement of change in tensile strength is the most commonly used criterion [3].

4.11. SOIL CORROSION

The corrosion rate of the copper-containing 2xxx and 7xxx series alloys in moist low-resistivity soils is several times greater than the corrosion rate of the more resistant 1xxx, 3xxx, 5xxx, and 6xxx series alloys. Aluminum alloys 3003, 6061, and 6063 are most frequently used for surface and underground pipelines for irrigation, petroleum, and mining applications. Soil resistivity provides a useful guideline to soil corrosiveness; corrosion problems are usually limited to soils having resistivity less than 1500 $\Omega \cdot$ cm [19]. The extent of attack that occurs on aluminum alloys buried underground varies greatly, depending on the soil composition and climatic conditions. In dry, sandy soil, corrosion is negligible. In wet, acid or alkaline soils, attack may be severe. Generally, in well-drained soil, attack on several aluminum-based alloys, except 2017-T, is mild after 5 years of testing. Chemical dip and sulfuric acid anodic coatings are generally protective for 6053-T and presumably for the other aluminum alloys [59].

Results of soil corrosion tests in two locations are summarized in Table 4.21 [8]. In both these locations, panels of the various alloys were buried in clay soil of the Aluminum Research Laboratories' properties in New Kensington, Pennsylvania. One location was in relatively well-drained soil and the other was in a marshy area less than 100 ft away. In the well-drained soil, attack on all the aluminum-based alloys, except 2017, was mild after 5 years. The alloy 2017-T was severely attacked although not as much as the steel [8].

In the marshy soil, maximum depths of attack on all the uncoated aluminum-based alloys, except Alclad 2024-T, were appreciable and of the same order of magnitude as on steel, although the relative loss in tensile strength was definitely less for most of the aluminum-based alloys than for steel. In the case of the Alclad 2024-T, the attack that occurred was all confined to the coating, as would be expected. Chemical dip and sulfuric acid anodic coatings were definitely protective for 6053-T and presumably for the other aluminum alloys [4].

Aluminum alloys are used relatively rarely in buried applications, although some pipelines and underground tanks have been constructed from these alloys. Aluminum alloys tend to undergo localized corrosion damage in chloride-contaminated soils [61].

4.12. SOME AGGRESSIVE MEDIA: ACID AND ALKALINE SOLUTIONS

There is no general relationship between pH and rate of attack because the specific ions present largely influence the behavior. Thus most aluminum alloys are inert to strong nitric or acetic acid solutions, but are readily attacked in dilute nitric, sulfuric, or hydrochloric acid solutions. Similarly, solutions with a pH as high as 11.7 may not attack aluminum alloys, provided silicate inhibitors are present, but, in the absence of silicates, attack may be appreciable at a pH as low as 9.0. In chloride-containing solutions, generally less corrosion occurs in the near-neutral pH range, say, 5.5–8.5, than in either distinctly acid or distinctly alkaline solutions. However, the results obtained vary somewhat, depending on the specific aluminum alloy under consideration [3].

The aluminum oxide film generally is stable in the pH range of 4–9, but is readily dissolved in strong acids and alkalis, with some notable exceptions as mentioned in Section 3.4.2 ("Active and Passive Behaviors"). The rate of corrosion cannot be predicted solely by the pH, but depends on the specific ions present, their concentration, and the temperature. For example, the dissolution rate of aluminum in sulfuric acid becomes

Table 4.21 Soil Burial Tests of Five Years' Duration with Aluminum Alloy Specimens[a]

	Well-drained soil			Marshy soil		
Alloy	Maximum depth of attack[b] (inches)	Percentage change in tensile strength[c]	Remarks	Maximum depth of attack (inches)	Percentage change in tensile strength	Remarks
1100-$^1/_2$H	0.0017	−1	Mild general etching	0.0280	−7	Pitted
5052-$^1/_2$H	0.0007	1	Mild general etching	0.0140	0	Pitted
6053-T	0.0007	0	Mild general etching	0.0150	0	Pitted
6053-T, Alrok #13 coated	0.0006	0	Mild general etching	0.0006	0	Mild general etching
6053-T, aluminite #204 coated	0.0003	0	Mild general etching	0.0002	2	Mild general etching
2017-T	0.0380	−20	Severe pitting	0.0310	−41	Severely pitted
Alclad 2024-T	0.0013	0	Mild general etching	0.0028	−1	Generally etched
Steel	0.0640	−27	Completely perforated at three spots	0.0190	−17	Pitted

[a]Specimens in the form of panels 3 in. × 9 in. × 0.064 in. (7.6 cm × 22.9 cm × 0.16 cm) thick were buried to a depth of 61 cm in soil at the property of the Aluminum Research Laboratories in New Kensington, Pennsylvania.

[b]Depth of attack determined by microscopic examination of cross sections.

[c]Change in tensile strength determined by machining tensile specimens from the panels after exposure and comparing their strength with that of unexposed tensile specimens of the same materials.

Source: Reference 8.

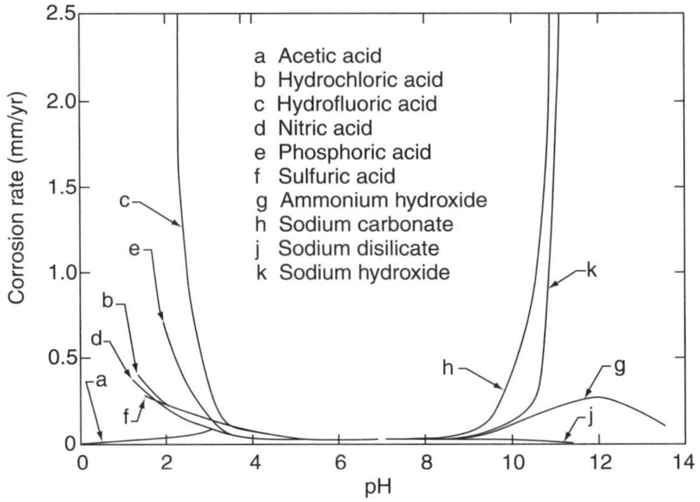

Figure 4.1 Relation to pH of the corrosiveness toward 1100-H14 alloy sheet of various chemical solutions [3, 58].

appreciable between 50% and 100% concentration, with the maximum rate occurring at 70–90% concentration. Furthermore, the corrosion rate at 50 °C can be as much as four times more rapid than at 25 °C [59].

Figure 4.1 shows the corrosion performance of pure aluminum in various acidic and alkaline media. Aluminum 1100 is pH sensitive because of its active–passive behavior and is especially extremely dependent on the nature of the anion such as that of HF and H_3PO_4 acids and Na_2CO_3 and NaOH alkalis.

A saturated solution of aluminum chloride ($AlCl_3$) has a pH of 3.0–3.5, just into the range where the natural oxide is unstable. A saturated concentration can occur when aluminum corrodes in a chloride-containing solution under conditions where the electrolyte cannot be readily replenished, such as crevices, deep pits, and cracks. It also occurs in cyclic wet–dry environments, where the solution gradually evaporates and becomes concentrated. This tends to keep corrosion active at such a site, as opposed to the usual self-limiting effect [59].

Neutral or nearly neutral (pH from about 5 to 8.5) solutions of most inorganic salts cause negligible or minor corrosion of aluminum-based alloys at room temperature. This is true for both oxidizing and nonoxidizing solutions. Any attack that does occur in such solutions is likely to be highly localized (pitting) with little or no general corrosion. Solutions containing chlorides are likely to be more active than other solutions. The simultaneous presence of salts of the heavy metals, especially copper, and chlorides may be very detrimental. Distinctly acid or distinctly alkaline salt solutions are generally somewhat corrosive. The rate of attack depends on the specific ions present. In acid solutions, chlorides, in general, greatly stimulate attack. In alkaline solutions, silicates, for example, greatly retard attack [59].

4.12.1. Acids

Acid mine waters are corrosive to aluminum-based alloys. The extent of attack depends on the specific composition of the water. Some use of aluminum pipe has been made in soft coal mines for handling acid mine waters. It has been found that pipe of aluminum alloy 3003

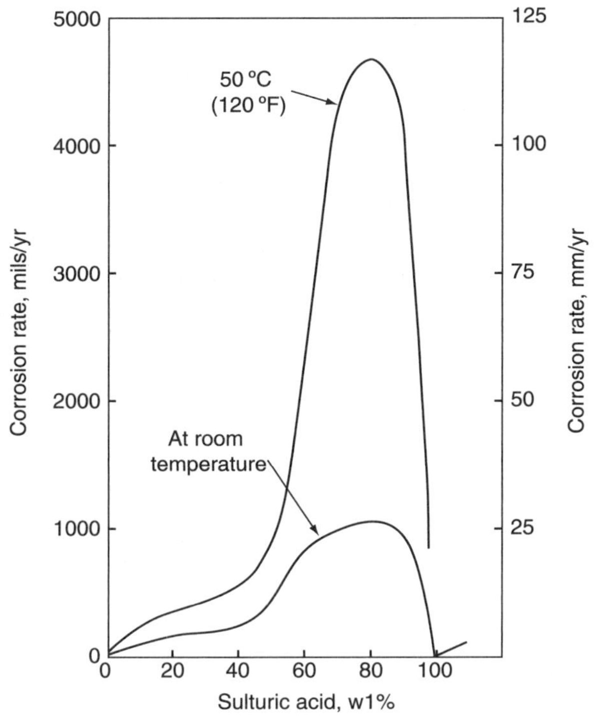

Figure 4.2 Corrosion of the alloy AA1100 as a function of the concentration of sulfuric acid at room and 50 °C [58].

greatly outlasts bare or galvanized steel pipe in this application. Many aluminum-based alloys are highly resistant to nitric acid in concentrations of about 80–99%. Alloys such as 1100, 3003, and 6061 have received the widest use for handling nitric acid at these concentrations. Nitric acid of lower concentrations is less corrosive [4, 8].

Dilute sulfuric acid solutions, up to about 10% in concentration, cause some attack on aluminum-based alloys, but the action is not sufficiently rapid at room temperature to prevent their use in special applications. In the concentration range of about 40–95%, rather rapid attack occurs. In extremely concentrated or fuming acid, the rate of attack drops again to a very low value [4, 8]. The curve of attack of aluminum in sulfuric acid as a function of the concentration shows a maximum around 80 wt % (Figure 4.2). This shows that the protective film loses its efficiency at this concentration. This phenomenon can be observed from other reagents at certain concentrations [58].

The action on aluminum (1100) of solutions containing sulfuric acid, nitric acid, and water is illustrated in Figure 4.3. Aluminum is most resistant to solutions dilute in both acids or high in nitric acid concentration (above 82%), or in 100% sulfuric acid. Hydrofluoric, hydrochloric, and hydrobromic acid solutions, except at concentrations below about 0.1%, are definitely corrosive to aluminum alloys. The rate of attack is greatly influenced by temperature (Figures 4.3 and 4.4) [4, 58].

Both perchloric and phosphoric acid solutions in intermediate concentrations definitely attack aluminum. Dilute (below 1%) phosphoric acid solutions have a relatively mild, uniform etching action that makes them useful for cleaning aluminum surfaces. Boric acid solutions in all concentrations up to saturation have negligible action on aluminum alloys. Chromic acid solutions in concentrations up to 10% have a mild, uniform etching action.

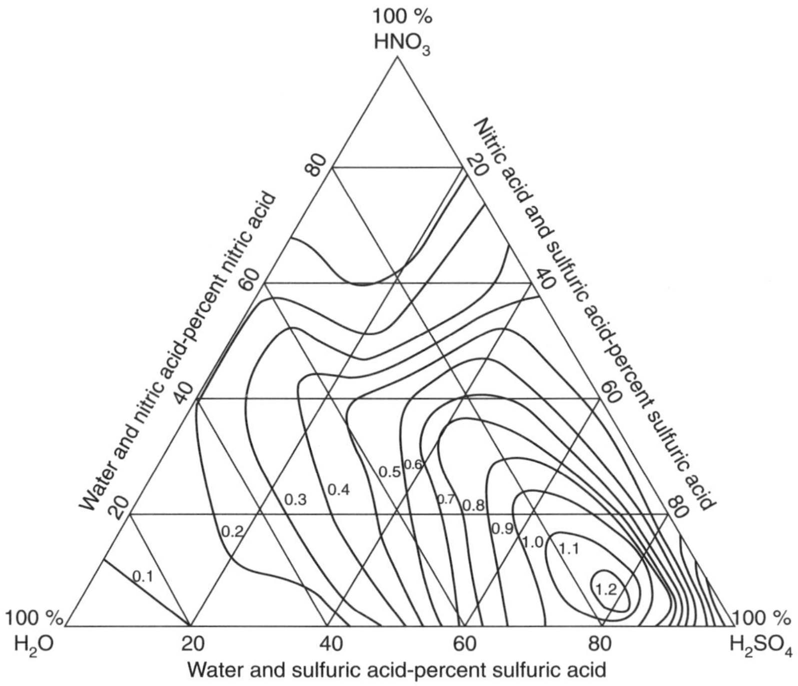

Figure 4.3 Action of mixtures of nitric and sulfuric acids on aluminum alloy 1100: 24 hour tests at room temperature; contours labeled in inches per year (25.4 mm) [8].

Mixtures of chromic acid and phosphoric acid have practically no action on a wide variety of aluminum alloys, even at elevated temperatures. Such mixtures are used for quantitatively removing corrosion products or oxide coatings from aluminum alloys. The resistance of pure aluminum to attack by most acids and many neutral solutions is higher than that of aluminum of lower purity or of most of the aluminum-based alloys [8].

4.12.2. Alkalis

Solutions of sodium hydroxide or potassium hydroxide in all but the lowest concentrations (less than 0.01%) rapidly attack aluminum and its alloys. Attack by the very dilute caustic

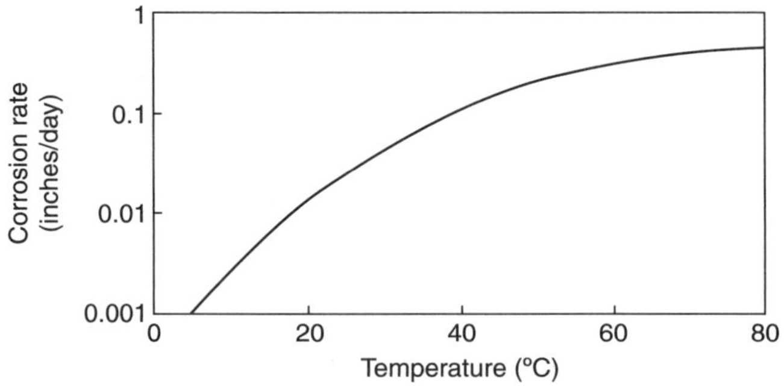

Figure 4.4 Effect of temperature on corrosion rate of aluminum 6053-T in 10% HCl (1 inch/day = 25.4 mm/day) [8].

solutions can be inhibited by corrosion inhibitors, such as silicates, but in more concentrated solutions none of the usual inhibitors are very effective. The aluminum alloys containing more than about 4% magnesium are somewhat more resistant to attack by alkalis than are the other aluminum-based alloys and this can be supported by Pourbaix diagrams of magnesium. Lime or calcium hydroxide solutions are also corrosive, but the maximum rate of attack is limited due to their low solubility [59]. Solutions of lithium hydroxide strongly attack aluminum alloys. Any application with LiOH such as a lithium battery must be avoided. Solutions containing trisodium phosphate are highly alkaline (10% of Na_3PO_4 gives a pH around 13) and strongly attack aluminum alloys [20].

The aluminum-based alloys are highly resistant to ammonia and ammonium hydroxide. The alloys that contain appreciable magnesium tend to be even less affected by ammonium hydroxide solutions than the other aluminum alloys [59]. The amines generally have little or no action on aluminum alloys after a given concentration. For over 30% for methylamine and 70% for butylamine no corrosion is reported on aluminum 1100, although the pH of such solutions is as high as 13 or 14. However, below these concentrations the corrosion rate increases rapidly to reach a critical point. At 10% methylamine, the corrosion rate is ~53 mm/year, and at 20% butylamine, the corrosion rate is ~26 mm/year [Ref. 20, pp 404–405]. A similar situation is observed with monoethanolethylenediamine (MEEDA). Over 60%, negligible corrosion rates on aluminum 3103 are reported. Below this concentration, the corrosion rate increases drastically to reach 9.9 mm/year at ambient temperature and over 50 mm/year at 50 °C [20].

4.13. DRY AND AQUEOUS ORGANIC COMPOUNDS

Dry Organic Compounds At elevated temperatures, some organic compounds, such as methyl alcohol and phenol, definitely become corrosive, especially when they are completely anhydrous. A small amount of water can drastically affect resistance to certain anhydrous organic solutions, particularly halogenated hydrocarbons [59]. Moreover, under most conditions, particularly at room temperature, aluminum alloys resist halogenated organic compounds, but under some conditions, they can react rapidly or violently with some of these chemicals. If water is present, these chemicals can hydrolyze to yield mineral acids that destroy the protective oxide film of aluminum. Reactivity of aluminum alloys with halogenated organic chemicals is inversely related to the chemical stability of these reagents. Thus they are most resistant to chemicals containing fluorine and are decreasingly resistant to those containing chlorine, bromine, and iodine. Aluminum alloys resist highly polymerized chemicals, reflecting the high degree of stability of these chemicals [58]. Phenols and carbon tetrachloride nearly dry or near their boiling points are very corrosive to aluminum alloys. This behavior can be prevented by the presence of trace water [59]. The corrosion rate of aluminum in phenol reaches more than 50 mils/year (1.27 mm/year) at phenol's boiling point [62].

Aqueous Organic Compounds and Acids Aqueous solutions of organic chemicals having a substantially neutral reaction are generally not corrosive to aluminum-based alloys, unless these solutions are contaminated with other substances, particularly chlorides and heavy metal salts. At room temperature or slightly above, most organic compounds, such as organic sulfur compounds, in the absence of water are completely inert to aluminum-based alloys [59].

Most organic acids are well resisted by aluminum alloys at room temperature. In general, rates of attack are highest for solutions containing about 1% or 2% of the acid. Formic acid, oxalic acid, and some organic acids containing chlorine (such as trichloroacetic acid) are exceptions and are definitely corrosive. Equipment made of aluminum alloys, such as 1100 or 3003, is widely and successfully used for handling acetic, butyric, citric, gluconic, malic, propionic, and tartaric acid solutions [4]. Aluminum alloys also have a high resistance to the action of uncontaminated natural fruit acids. Contamination of these substances by heavy metal compounds may cause them to become corrosive. In contrast, the addition of sugar to fruit acids causes them to become even less corrosive [8].

4.14. GASES

Most gases, in the absence of water and at or near room temperature, have little or no action on aluminum-based alloys. In the presence of water, the acid gases, such as HCl and HF, are corrosive, and wet SO_2 causes corrosion (Table 4.22). Hydrogen sulfide or ammonia, either in the presence or absence of water and at room temperature or slightly above, has negligible action on aluminium-based alloys. Halogenated hydrocarbons, such as dichlorodifluoromethane, dichlorotetrafluoromethane, and monochlorodifluoromethane, are almost completely inert to aluminum. However, methyl chloride and methyl bromide are corrosive and should not be used in contact with aluminum-based alloys [8].

4.15. MERCURY

When dry, metallic mercury reacts only with difficulty because of the oxide film on the aluminum surface and it should penetrate the natural oxide film to attack the metal. The presence of traces of acidity or halides on the surface causes rapid attack. Solutions containing mercury ions tend to cause rapid pitting of aluminum alloys because mercury

Table 4.22 Resistance of Aluminum to Aqueous Solutions of Several Gases

	Carbon dioxide[a] and water		Sulfur dioxide[b], air, and water		Hydrogen sulfide[c] and water	
Metal	Average weight loss (grams)	Average[d] inches per year	Average weight loss (grams)	Average[d] inches per year	Average weight loss (grams)	Average[d] inches per year
Aluminum 1100	0.0003	0.00004	0.150	0.0498	0.002	0.00028
Copper	—	—	0.681	0.0701	0.237	0.01030
Steel	0.2153	0.00977	8.583[e]	1.02[e]	1.366	0.06800

[a]Metal specimens 1 in. × 4 in. × $\frac{1}{16}$ in. (2.5 cm × 10.2 cm × 0.16 cm) were partially immersed (to a depth of 51 mm).

[b]Metal specimens 2.5 cm × 10.2 cm × 0.16 cm thick were partially immersed (to a depth of 2 in.) in distilled water through which air and sulfur dioxide were bubbled. The total period of exposure was 135 hours at room temperature.

[c]Metal specimens 1 in. × 4 in. × $\frac{1}{16}$ in. thick were partially immersed (to a depth of 51 mm) in distilled water through which hydrogen sulfide was bubbled. The total period of exposure was 320 hours at room temperature.

[d]This calculation was based on the assumption that all corrosion was confined to the immersed areas of the specimens.

[e]Steel specimen corroded completely through at the water line.

Source: Reference 8.

plates out in localized areas. The mercury penetration tends to proceed along grain boundaries, and if tensile stresses are present in the metal, drastic splitting and the exposure of further film-free metal occurs. For example, attack by mercury and zinc amalgam combined with residual stresses from welding causes cracking of the weldment [59].

Mercury tends to amalgamate readily with aluminum at room temperature to produce an extraordinary corrosion rate in the presence of moisture with the production of voluminous columnar corrosion products mainly aluminum oxide. When that reaction is started, the rate of corrosion depends on relative humidity. Amalgamation of aluminum, once initiated, is an autocatalytic reaction. Mercury can plate out of aqueous solutions to produce this effect. A mercury content of greater than 0.01 ppm is cause for concern. Detection of even lesser amounts of mercury may indicate a problem, since mercury tends to evaporate and low levels are difficult to analyze. The effect can be severe when stress is present. The corrosive action of mercury can be attributed to the galvanic cell, and especially to the prevention of the formation of aluminum oxide. The corrosion rate can be extremely high, up to 1270 mm/year [4, 59].

Common sources of mercury are broken thermometers and mercury vapor bulbs, or mercury manometers that have been overpressurized. Mercury can be removed from aluminum surfaces by treatment with 70% nitric acid. Mercury can be distilled away from an aluminum surface by treatment with steam or hot air [59].

4.16. CORROSION PERFORMANCE OF ALLOYS

Resistance to corrosion is greatly dependent on alloy content and its effect on the oxide film. Principally, every alloying element can influence the potential of the alloy through solid solution, through precipitate formation (new phases), and through the change in the microstructure and castability. This can influence corrosion rate, passivation, and type of corrosion. Also, there is a synergetic influence between certain alloying elements which can modify the properties and consequently the corrosion resistance of the alloy [63].

Some added minor alloying elements can be useful or harmful for corrosion resistance. Antimony enhances corrosion resistance in salt water by forming a protective film of antimony oxychloride. Oxidation and discoloration of wrought aluminum–magnesium products are greatly reduced by small amounts of beryllium Up to 0.3% cadmium increases strength and increases the corrosion resistance of some alloys. Chromium addition ($\leq$0.35%) to Al–Mg, Al–Mg–Si, and Al–Mg–Zn groups increases corrosion resistance. Chromium develops a fibrous structure that reduces stress corrosion cracking. Nickel additions with Fe improve corrosion resistance to high-pressure steam. Silver in small additions (0.1–0.6%) is effective in improving strength and stress corrosion cracking of Al–Zn–Mg alloys [63].

Nickel promotes pitting corrosion in alloys such as 1100. Carbon is an infrequent impurity in the form of oxycarbides and carbides, of which the most common is Al_4C_3. Al_4C_3 decomposes in the presence of water and water vapor, and this may lead to surface pitting. Very small amounts of calcium (10 ppm) increase the tendency of molten aluminum alloys to pick up hydrogen [63].

4.16.1. Performance of the Cast Series

1xx.x 99.0% Minimum Al Controls unalloyed (pure) compositions, especially for rotor manufacture. The alloying elements and the presence of different phases decrease the corrosion resistance of aluminum [63]. The resistance of pure aluminum such as alloy 1100 (2S) to attack by most acids and many neutral solutions is higher than that of aluminum of lower purity or of most of the aluminum-based alloys. Very-high-purity aluminum, 99.99% or purer, is highly resistant to pitting. Any alloying addition will reduce resistance to pitting corrosion [3].

2xx.x Al–Cu Alloys Copper is the principal alloying element, but other alloying elements may be specified. These alloys are used in heat-treatable, sand, and permanent mold castings [63]. Copper often reduces resistance to general corrosion and, in specific compositions and certain conditions, stress-corrosion susceptibility. For cast and wrought alloys, copper is the alloying element most subject to general corrosion. Al–Cu alloys have poor corrosion resistance and cannot be exposed to natural atmospheres without appropriate protection and should be protected. Most of the 2xx.0 alloys are inherently susceptible to stress-corrosion cracking in the peak hardness T6 temper, while it has a much better stress-corrosion resistance in the T7 overaged temper [64]. Thus use of this kind of alloy in marine atmospheres, in aqueous media, and in industrial media (chemicals) is not recommended [54].

3xx.x Al–Si + Mg or Cu or Mg + Cu Alloys The 3xx.x series comprises nearly 90% of all shaped castings produced. These alloys are used in heat-treatable, sand, permanent mold, and die castings. These alloys show good corrosion resistance in most natural fresh waters and in chemical media [54]. Silicon improves corrosion resistance. Al–Si alloys containing copper (Al–Si–Cu) are applied in the automotive industry and do not have good corrosion resistance. For better corrosion resistance, alloys lower in copper should be chosen [20, 65].

4xx.x Al–Si Alloys These alloys are used in non-heat-treatable, sand, permanent mold, and die castings. Silicon is the principal alloying element and improves the overall corrosion resistance of these alloys [63].

5xx.x Al–Mg Alloys Magnesium is the principal alloying element. These alloys are used in non-heat-treatable, sand, permanent mold, and die castings. These alloys show good corrosion resistance in most natural fresh waters and in chemical media [54]. Al–Mg alloys are suitable for welding, have good machinability, and have an attractive appearance when anodized. The primary advantage of cast Al–Mg alloys is the high corrosion resistance, especially to seawater and marine atmospheres. Solid and gaseous impurities should be kept as low as possible during casting and handling for better corrosion resistance. Magnesium addition in the 5xx.0 series is usually minimized to control the generation of oxides in the casting process. Alloys containing magnesium in solid solution, or in a separate phase such as Al_8Mg_5 particles dispersed uniformly through the matrix, are generally as corrosion resistant as commercially pure aluminum. Almost intergranular continuous Al_8Mg_5 precipitation can lead to stress-corrosion cracking (SCC). It should be noted that the alloy 518.0 is occasionally specified when the highest corrosion resistance is required. The alloy 535.0 (7% Mg) has the highest resistance to corrosion of any of the common cast alloys [65, 67]. High copper or nickel content greatly decreases resistance to corrosion. High iron, silicon, or manganese content adversely affects mechanical properties. The alloy 520.0 (10% Mg) is a relatively high-strength, heat-treatable, sand casting alloy with excellent resistance to corrosion but it is highly susceptible to stress-corrosion cracking and has causes many stress-corrosion service failures in 520.0-T4 aluminum alloy aircraft parts. To minimize this susceptibility, a "slow" quench in hot oil after solution heat treatment is somewhat effective [63, 66].

7xx.x Al–Zn Alloys Zinc is the principal alloy element, but other alloying elements such as copper and magnesium may be specified. These alloys are used in heat-treatable, sand, and permanent mold castings. The 7xx.0 series generally also contain small amounts of magnesium. The Al–Zn–Mg alloys have good machinability and good resistance to general corrosion; however, susceptibility to stress-corrosion cracking is observed. Their corrosion resistance is not as good as that of the 5xx.x group. The tensile properties of these alloys in the as-cast condition increases rapidly during the first weeks of room temperature aging, due to precipitation hardening. At the same time, stress-corrosion resistance deteriorates steadily. Copper accelerates general corrosion but reduces SCC susceptibility. It allows the overaging temper such as T73, which couples high strength to excellent resistance to SCC. Depending on the alloy, stress corrosion can be controlled by certain heat treatments (overaging, cooling rate after solution treatment), by alloying additions (Cu or Cr), and by adjusting the Zn/Mg ratio close to 3 [65]. Thus use of this kind of alloy in marine atmospheres, in aqueous media, and in industrial media (chemicals) is not recommended [54]. Minor alloy additions such as chromium and zirconium have a marked influence on the mechanical properties and corrosion resistance. Zirconium can be used to achieve a nonrecrystallized structure [63, 65].

8xx.x Alloys Tin the principal alloying element. These alloys are used in heat-treatable, sand, and permanent mold castings. Al–Sn has good load-carrying capacity, fatigue strength, and resistance to corrosion by internal-combustion lubricating oil. Cast aluminum–tin alloys are applied principally for connecting rods and crank case bearings for diesel engines [63, 65].

4.16.2. Performance of the Wrought Series

The 5xxx Al–Mg alloys and the 3xxx Al–Mn alloys resist pitting corrosion almost equally well. The pure metal and the 3xxx, 5xxx, and 6xxx series alloys are resistant to the more damaging forms of localized corrosion, exfoliation, and stress-corrosion cracking (SCC). However, cold-worked 5XXX alloys containing magnesium in excess of the solid solubility limit (above 3% magnesium) can become susceptible to exfoliation and SCC when heated for long times at temperatures of about 80–175 °C [67]. Most wrought products (rolled, forged, drawn, or extruded products) normally have a highly directional, anisotropic grain structure. Rectangular products have a three-dimensional (3D) grain structure. These directional structures markedly affect resistance to SCC and to exfoliation of high-strength alloy products [59].

The solution heat-treated tempers are usually more corrosion resistant and more amenable to corrosion inhibition than are the hardened alloys. The strain or work hardened alloys are somewhat more readily inhibited than are the alloys hardened by aging treatments [4]. Several major test programs have been conducted under the supervision of ASTM International to investigate the weathering of aluminum alloy sheet. The first program, started in 1931, was limited in the variety of alloys tested but included desert, rural, seacoast, and industrial exposures. Corrosion rates were calculated from cumulative weight loss after 20 years, and average and maximum depths of attack were measured microscopically. In aggressive (seacoast and industrial) environments, the bare (non-

Alclad) heat-treated alloys—2017-T3 and, to a lesser extent, 6051-T4—exhibited more severe corrosion and greater resulting loss in tensile strength than the non-heat-treatable alloys. Alclad alloys are duplex wrought products, supplied in the form of sheet, tubing, and wire, which have a core of one aluminum alloy and a coating on one or both sides, of aluminum or another aluminum alloy. Generally, the core comprises 90% of the total thickness with a coating comprising about 5% of the thickness on each side. The coating is metallurgically bonded to the core over the entire area of contact. The coating is usually selected to be anodic to the core alloy in most natural environments and will protect the core (galvanic cell) where it is exposed at cut edges, rivet holes, or scratches. Such Alclad alloys are usually more resistant to penetration by neutral solutions than are any of the other aluminum-based alloys [4, 8]. Alclad 2017-T3, although as severely corroded as the non-heat-treatable materials, did not show measurable loss in strength: in fact, some specimens of this alloy were 2–3% higher in strength after 20 years because of long-term natural aging [19]. Data from weathering programs demonstrate that differences in resistance to weathering among non-heat-treatable alloys are not great, that Alclad products retain their strength well because corrosion penetration is confined to the cladding layer, and that corrosion and resulting strength loss tend to be greater for bare (non-Alclad) heat-treatable 2xxx and 7xxx series alloys [19].

4.17. ALUMINUM HIGH-TEMPERATURE CORROSION

At low temperatures (4 °C [40 °F] or below), the action of most aqueous solutions is much slower than at room temperature. However, in many solutions, increasing temperatures above about 80 °C (180 °F) results in a decrease in the rate of attack. Thus a temperature of 70–80 °C (160–180 °F) is likely to result in more severe corrosion than temperatures of 20 °C (70 °F) or 100 °C (212 °F) [4]. At 200 °C, high-purity aluminum of sheet thickness disintegrates completely within a few days of reaction with high-purity water to form aluminum oxide. In contrast, Al–Ni–Fe alloys have the best elevated-temperature resistance to high-purity water of all aluminum metals, for example, alloy X8001 (1.0Ni–0.5Fe) has good resistance at temperatures as high as 315 °C [19]. Above a temperature of about 230 °C (445 °F), a protective film no longer develops in water or steam, and the reaction progresses rapidly until eventually all the aluminum exposed in these media is converted into oxide [68].

In addition to affecting the corrosion reactions, elevated temperatures can cause additional precipitation and changes in the metallurgical structure of the metal. The 2xxx and 7xxx alloys are affected beginning at temperatures of about 120 °C. This effect initially is adverse, but with continued heating past peak strength conditions, a reversal and improved resistance can occur. The 5xxx series alloys containing more than 3% magnesium are sensitized by exposure to temperatures on the order of 80–175 °C. Exposure of these 5xxx alloys to even higher temperatures results in a coarsening of the precipitate, producing discontinuous grain boundary precipitation, which reduces or eliminates the sensitization effect [69].

Dry gaseous Cl_2 at ambient temperature has no effect on aluminum corrosion. Nevertheless, at high temperature (above 130 °C) the corrosion reaction is violent and strong [20]. The corrosion rate of high-purity aluminum alloy 1100 in the presence of dry chlorine gas at 150 °C is 1.5–3 mm/year and increases abruptly, being 10 times more severe at 180 °C [70].

In general, corrosion rates increase with increasing temperatures. In some cases this effect can be very marked, as noted previously for sulfuric acid [59]. Certain Al–Cu alloys such as 2219 and 2618, used in aerospace applications, and castings of some Al–Si–Cu alloys such as 390.0, used in internal combustion engines, can withstand higher working temperatures between 150 and 300 °C for long periods. Among the most resistant cast alloys are the piston alloys, which contain up to 20% silicon (highly hypereutectic composition). Above 300 °C, alloys made by thermomechanical alloying of metal and ceramic particles are very strong and fairly conductive at temperatures up to 500 °C. At room temperature, aluminum reacts with oxygen (in the absence of water) to form a very thin oxide layer (0.01 μm thickness). At elevated temperatures above 500 °C, aluminum reacts with oxygen to form thicker oxide layers, and at ~600 °C the oxide layer attains 0.1–1 μm thickness [71].

Sometimes high-temperature oxidation is a misnamed condition of hydrogen diffusion that affects surface layers during elevated-temperature cases. This condition can result from moisture contamination in the furnace atmosphere and is sometimes aggravated by sulfur or other furnace refractory contamination [65]. The 242.0 and 243.0 alloys have better properties at elevated temperatures than the 10% Cu alloy 222.0-T61 and are used for air-cooled cylinder heads for airplanes and motorcycles [63].

REFERENCES

1. Lenntech Periodic Table. Available at http://www.lenntech.com/Periodic-chart-elements/.
2. F. Habashi, *Science Reviews* **22**(1), 53–60 (1997).
3. E. H. Hollingsworth and H. Y. Hunsicker, in *Corrosion and Corrosion Protection Handbook*, edited by P. A. Schweitzer. Marcel Dekker, New York, 1983, pp. 111–145.
4. E. Ghali, in *Uhlig's Corrosion Handbook*, 2nd edition, edited by R. W. Revie. Wiley, Hoboken, NJ, 2000, pp. 677–715.
5. *Aluminum Cast Alloys* [in English and French]. Centre technique des industries de la fonderie and Chicoutimi, Canada, and Centre Québécois de Recherche et de Développement de l'Aluminium, Cedex, France, 2002.
6. K. U. Kainer, *Magnesium Alloys and Technology*, Wiley-VCH, Weinheim, Germany, 2003, pp. 1–22.
7. I. J. Polmear, *Light Alloys from Traditional Alloys to Nanocrystals*. Elsevier, Sydney, Australia, 2006.
8. R. B. Mears, in *Corrosion Handbook*, edited by H. H. Uhlig. Wiley, Hoboken, NJ, 1976, pp. 39–55.
9. B. W. Lifka, in *Corrosion Engineering Handbook*, edited by P. A. Schweitzer. Marcel Dekker, New York, 1996, pp. 99–106.
10. ASM International Handbook Committee I, in *Metals Handbook*, 2nd edition, edited by J. R. Davis. ASM International, Materials Park, OH, 1998, pp. 423–426.
11. J. G. Kaufman and E. L. Rooy, in *Corrosion Test and Standards, Application and Interpretation*, 2nd edition, edited by R. Baboian. ASM International, Materials Park, OH, 2005, pp. 1–8.
12. Aluminum Association Committee Aluminum Powder Metallurgy. PDF file, available at www.aluminum.org/Content/NavigationMenu/The_Industry/Powder_and_Paste/PowderMetallurgy.PDF.
13. C. Vargel, in *Corrosion de l'aluminium*, edited by C. Vargel. Dunod, Paris, 1999, pp. 199–217.
14. S. Lee and B. W. Lifka, *New Methods for Corrosion Testing of Aluminum Alloys, ASTM STP 1134*. American Society for Testing and Materials, Philadelphia, 1992.
15. *Aluminum Standard*. Aluminum Association, Inc., Washington, DC, 1984, p. 12.
16. T. Martin, Aluminium Alloys, Tempers and Terminology. Available at http://shopswarf.orcon.net.nz/alalloy.html.
17. eFunda Temper Designations of Aluminium Alloys. Available at http://www.efunda.com/Materials/alloys/aluminum/temper.cfm.
18. ASM Specialty Handbook Committee, in *Aluminum and Aluminum Alloys*, edited by J.R. Davis. ASM International, Materials Park, OH, 1993, pp. 3–55, 88–120, 579–731. in 1st, 2nd and 5th parts
19. J. G. Kaufman, in *ASM Handbook, Volume 13B, Corrosion: Materials*, edited by S. D. Cramer and B. S. Covino Jr. ASM International, Materials Park, OH, 2005, pp. 95–124.
20. C. Vargel, in *Corrosion de l'aluminium*, edited by C. Vargel. Dunod, Paris, 1999, pp. 50–58, 389–415.
21. J. E. Benfer, in *ASM Handbook, Volume 13C, Corrosion: Environments and Industries*, edited by S. D. Cramer and B. S. CovinoJr. ASM International, Materials Park, OH, 2006, pp. 184–194.

22. R. Braun, *Materials Science Forum* **519–521**, 735–740 (2006).
23. E. P. Polk and B. C. Giessen, *Metallic Glasses*. ASM International, Materials Park, OH, 1976, pp. 1–35.
24. K. Hashimoto and T. Masumoto, *Materials Science Engineering* **23**, 285(1976).
25. A. Ünal, D. D. Leon, T. B. Gurganus, and G. J. Hildeman, in *ASM Handbook, Volume 7, Powder Metal Technologies and Applications*, edited by S. R. Lampman. ASM International, Materials Park, OH, 1998, pp. 148–159.
26. A. R. Kennedy and S. Asavavisitchai, *Scripta Materialia* **50**(1), 115–119 (2004).
27. R. B. Bhagat, in *ASM Handbook, Volume 7 Powder Metal Technologies and Applications*, edited by S. R. Lampman. ASM International, Materials Park, OH, 1998, pp. 840–848.
28. J. W. Newkirk, in *Handbook of Aluminum, Volume 1, Physical Metallurgy and Processes*, edited by G. E. Totten and D. S. MacKenzie. Marcel Dekker, New York, 2003, pp. 1251–1277.
29. Key to Metals Task Force & INI International High-Strength Aluminium P/M Alloys. Available at http://www.key-to-metals.com/Article62.htm.
30. G. C. Wood, J. A. Richardson, M. F. Abd Rabbo, L. B. Mapa, and W. H. Sutton, in *Passivity of Metals*, edited by R. P. a. J. K. Frankenthal. Electrochemical Society, Princeton, NJ, 1978.
31. C. Vargel, *Corrosion of Aluminum*, translated by M. P. Schmidt. Elsevier, Oxford, UK (2004) 626 pages.
32. K. A. Lucas and H. Clarke, *Corrosion of Aluminium-based Metal Matrix Composites*. Research Studies Press Ltd, Taunton, Somerset, UK 1993.
33. A. Walker, *Materials Edge* **34**(13) x-x (1992).
34. N. Hort and K. U. Kainer, in *Metal Matrix Composites, Custom-made Materials for Automotive and Aerospace Enginnering*, edited by K. U. Kainer. Wiley-VCH, Weinheim, Germany, 2006, pp. 243–276.
35. J. M. Torralba, C. E. da Costa, and F. Velasco, *Journal of Materials Processing Technology* **133**, 203–206 (2003).
36. R. Buschmann, in *Metal Matrix Composites, Custom-made Materials for Automotive and Aerospace Engineering*, edited by K. U. Kainer. Wiley-VCH, Weinheim, Germany, 2006.
37. P. Degisher, Assessment of Metal Matrix Composites for Innovations. Available at http://mmc-assess.tuwien.ac.at.
38. Alcan Engineered Cast Products:Duralcan MMC. PDF file. Available at http://www.temponik.com/files/Guidelines%20Mar%2004%203p%20sec%201.pdf.
39. A. L. Geiger and P. Welch, *Journal of Materials Science* **32**, 2611–2616 (1997).
40. K. Shin, S. Lee, S. J. Kim, and K. Cho, *Advanced Performance Materials* **5**, 307–318 (1998).
41. T. W. Clyne, in *Comprehensive Composite Materials*, Vol. **3**, edited by A. Kelly and C. Zweben. Elsevier, Amsterdam, 2000, pp. 1–26.
42. D. A. Jones, in *Principles and Prevention of Corrosion*, 2nd edition, edited by W. Stenquist and R. Kernan. Prentice Hall, Upper Saddle River, NJ, 1996, pp. 40–74.
43. C. Margueritat-Regenet, *Elaboration et caractérisation de fils composites C/Al: infiltration spontanée et continue par activation chimique du mouillage*. ENSMP, Paris, 2002.
44. S. Rawal, *Journal of the Minerals, Metals & Materials Society* **53** (4), 14–17 (2001).
45. G. T. Kridli, A. S. El-Gizawy, and R. Lederich, *Materials Science and Engineering A* **A244** (2), 224–232 (1998).
46. D. Furrer and R. Noel, *Advanced Materials & Processes* **151**, 59–60 (1997).
47. L. P. a. J. A. W. Troeger, *Journal of Materials Processing and Manufacturing Science* **9** (3), 205–222 (2001).
48. T. Warner, *Materials Science Forum* **519–521**, 1271–1278 (2006).
49. A. Adjorlolo, in *ASM Handbook, Volume 13C, Corrosion: Environments and Industries*, edited by S. D. Cramer and B. S. CovinoJr. ASM International, Materials Park, OH, 2006, pp. 598–612.
50. C. H. Caceres, *Materials Science Forum* **519–521**, 1801–1808 (2006).
51. E. Köhler and J. Niehues, in *Metal Matrix Composites, Custom-made Materials for Automotive and Aerospace Enginnering*, edited by K. U. Kainer, Wiley-VCH, Weinheim, Germany, 2006, pp. 95–109.
52. M. Suzuki, *Materials Science Forum* **519–521**, 11–14 (2006).
53. B. W. Lifka, in *Corrosion Testing and Standards: Application and Interpretation*, edited by R. Baboian. American Society for Testing and Materials, Philadelphia, PA, 1995, pp. 447–457.
54. P. A. Schweitzer, in *Corrosion Engineering Handbook*, edited by P. A. Schweitzer. Marcel Dekker, New York, 1996, pp. 99–156.
55. P. Roberge, in *Handbook of Corrosion Engineering*, edited by R. Esposito. McGraw Hill, New York, 2000, pp. 584–611.
56. D. Bengtsson Blücher, J.-E. Svensson, and L.-G. Johansson, *Corrosion Science* **48**, 1848–1866 (2006).
57. S. Oesch and M. Faller, *Corrosion Science* **39** (9), 1505–1530 (1997).
58. ASM Specialty Handbook Committee, *Corrosion of Aluminum and Aluminum Alloys*. ASM International, Materials Park, OH, 1999, p. 313.
59. B. W. Lifka, in *Corrosion Tests and Standards, Application and Interpretation*, 2nd edition, edited by R. Baboian. ASM International, Materials Park, OH, 2005, pp. 547–557.

60. F. M. Reinhart, *Corrosion of Metals and Alloys in the Deep Ocean*. U.S. Naval Engineering Laboratory, Port Heceneme, California, 1976.

61. P. Roberge, in *Handbook of Corrosion Engineering*, edited by R. Esposito. McGraw Hill, New York, 2000, pp. 55–216.

62. P. A. Schweitzer, in *Corrosion Resistance Tables, Metals, Plastics, Nonmetallics and Rubbers*, 2nd edition, edited by P. A. Schweitzer. Marcel Dekker, New York, 1986, p. 830.

63. E. Ghali, in *Aluminum Cast Alloys* [in English and French]. Centre techniques des industries de la fonderie, Cedex, France, 2002, p. B0430.

64. M. O. Speidel, in *NATO Advanced Study Institute on Stress Corrosion Cracking*, edited by J. C. Scully. NATO Advanced Study Institute, Copenhagen, Denmark, 1971, pp. 345–354.

65. *ASM Specialty Handbook*, edited by J. R. Davis. ASM International, Materials Park, OH, 1991, pp. 3–55, 135–160, 579–622.

66. A. Kearney and E. L. Rooy, *Nonferrous Alloys and Special-Purpose Materials* **2**, 121–151 (1990).

67. E. H. Hollingsworth and H. Y. Hunsicker, in *Metals Handbook, Volume 13, Corrosion*, 9th ed., edited by L. J. Korband and D. L. Olson. ASM International, Materials Park, OH, 1987 pp. 583–609.

68. M. H. Brown, R. H. Brown, and W. W. Binger, in *High Purity Water Corrosion of Metals*, Vol. **82**, edited by N. E. Hammer. National Association of Corrosion Engineers, Houston, TX, 1960.

69. E. H. Dix, W. A. Anderson, and M. B. Shumaker, *Technical Paper 14*. Aluminum Company of America, Pittsburgh, PA, 1958.

70. ASM International Handbook Committee, *Handbook of Corrosion Data*, 2nd edition. ASM International, Materials Park, OH, 1995.

71. D. G. Altenpohl, *Aluminum: Technology, Applications, and Environment, A Profile of a Modern Metal*, 6th edition. The Minerals, Metals & Materials Society, Warrendale, PA, 1998.

Chapter 5

General, Galvanic, and Localized Corrosion of Aluminum and Its Alloys

Overview

Aluminum oxide is amphoteric since the oxide is stable at pH ~4–8 only. General corrosion can be uniform (even), quasi-uniform (near-uniform corrosion), or uneven. Underground corrosion is frequently observed as localized corrosion. Oxidation of aluminum or tarnishing in air gives a smooth surface with thin, tightly adherent protective films. High temperature conditions result in uniform attack normally; however, subsurface corrosion films within the matrix of the alloy can be observed in several alloys.

Galvanic action is much more pronounced in marine or seacoast atmospheres than in rural or industrial locations. Contact between stainless steel and aluminum in seawater or other saline solutions usually results in less galvanic action on the aluminum than that of aluminum with steel. Aluminum alloys with high magnesium or zinc contents will be more anodic (negative) by as much as 260 mV, while high copper content alloys will be more cathodic up to about 140 mV. Care must therefore be taken that all alloys and tempers are compatible, even in the same aluminum structure. Any concentration in a solution of more than a few parts per billion of mercury can be detrimental. The presence of acidity on the surface often provides the clue that reveals unexpected stray current activity. Deposition galvanic corrosion is due to the reduction of heavy metal ions, which leads to deposition of the more noble metal on the aluminum surface.

Corrosion starts at the original defects of the protective oxide film or at the break of the passive film or the coating and continues to undercut the coating, forming a rather heavy tubercle of hard rust or scale with the pit in the original metal underneath. It has been shown that the pit density of aluminum increases with temperature in water and that the depth of pits decreases with temperature. Three stages of pitting could be identified: nucleation, metastable pit formation, and stable pitting. Potential drops may stabilize localized corrosion for large pits but never for small pits during the initial stages.

Water staining is the most common case of aluminum crevice corrosion and occurs by the entrapment of moisture between the adjacent surfaces of closely packed material during transport or storage. Poultice corrosion takes the form of pitting when absorptive debris such

Corrosion Resistance of Aluminum and Magnesium Alloys: Understanding, Performance, and Testing By Edward Ghali

as paper, wood, asbestos, sacking, or cloth is in contact with a metal surface that becomes wetted periodically. Deformation caused by corrosion in lap joints of commercial airlines is accompanied by a bulging ("pillowing").

Filiform corrosion is a special type of crevice corrosion that can occur on an aluminum surface under a thin organic coating (typically 0.1 mm or 4 mils thick). Two-coat polyurethane paint systems give effective protection. Filiform corrosion rarely occurs when bare aluminum is chromic acid anodized or primed with chromate or chromate–phosphate conversion coatings.

A. GENERAL CORROSION

5.1. GENERAL CONSIDERATIONS

The first and most common form of corrosion is general corrosion: this can be uniform (even), or quasi-uniform (near-uniform corrosion), or uneven. General corrosion accounts for the greatest loss of metal or material. However, it is predictable and the designer can avoid catastrophic accidental corrosion problems. Frequently, electrochemical general corrosion in aqueous media can include galvanic or bimetallic corrosion, atmospheric corrosion, stray-current dissolution (treated in galvanic corrosion), and biological corrosion. Dissolution of steel or zinc in sulfuric or hydrochloric acid is a typical example of uniform electrochemical attack. Steel and copper alloys are more vulnerable to general corrosion than other alloys. Uniform corrosion often results from atmospheric exposure (polluted industrial environments); from exposure in fresh, brackish, and salt waters; or from exposure in soils and chemicals. The rusting of steel, the green patina on copper, the tarnishing of silver, and the white rust on zinc upon atmospheric exposure are due to uniform corrosion [1].

Underground corrosion is frequently observed as localized corrosion. Oxidation, sulfidation, carburization, hydrogen effects, and hot corrosion can be considered frequently as types of general corrosion. Liquid metals and molten salts at high temperature lead very frequently to general corrosion [2].

5.2. DESCRIPTION

General uniform corrosion of aluminum is rare, except in special, highly acidic, or alkaline corrosive reagents as can be deduced from Pourbaix diagrams; however, we frequently observe the oxidation of aluminum in air. In specific industrial atmospheres, salt waters, or chemicals, where aluminum passivation is not attained, uniform and nonuniform corrosion can be observed. For example, aluminum is attacked in a general uniform corrosion in 5% by weight sodium hydroxide at 80 °C. Under controlled anodic polarization conditions and/or nonpassivating media (e.g., containing chloride ions), the rate of dissolution is quite uniform and linear, depending primarily on the chemical concentration and temperature of the electrolyte. Several industrial processes use this principle to electrochemically machine or size parts. For linear, steady dissolution rate applications, thickness measuring devices or gravimetric procedures can be used [3].

General electrochemical corrosion is a function of material, the environment, and the interface. The nature of the metal or the alloy, the solid solution, phases, inclusions and precipitates, and the homogeneity of the microstructure contribute to the relative stability of an active material. The environment (such as oxygen) and its uniformity, temperature, and the activity of active species, and diffusion, convection, and movement of the solution are the essential parameters. Oxidation at high temperature frequently reflects the same electrochemical attack in aqueous solutions and causes uniform attack of the surface [4]. The following factors should be considered for appropriate general form studies:

1. The agitation of the medium or the level of agitation of the electrolyte has a great influence on the corrosion performance of most metallic alloys since agitation causes acceleration of diffusion of aggressive species or mechanical destruction of the passive layer.

2. Acid pH accelerates corrosion for most of the alloys, since for an active metal such as iron or zinc, the cathodic reaction controls the rate of the reaction according to $E = E^{\circ} - 0.0592$ pH. The Evans diagrams can be obtained by extrapolation of the Tafel slopes for the cathodic and anodic polarization curves. In general, cathodic Tafel slopes are more reproducible and more reliable to evaluate corrosion rates since they represent the almost noncorroded or original state of the surface. It has been observed from the diagrams that there is a marked increase in corrosion current for more acidic solutions. It should be noted that the influence of pH also depends on the composition of the alloy. In the case of zinc, amalgamation with a more noble metal such as mercury decreases the corrosion rate in addition to the slower hydrogen evolution reaction, which requires high overpotentials. Platinum gives high corrosion rates because it provides effective cathodic sites for hydrogen evolution. Another factor is the stability of the passive film of the metal–solution system in acid, neutral, or alkaline pH. Magnesium fluoride, for example, is protective for magnesium in alkaline medium. Aluminum oxide is amphoteric and is stable at pH ~4–8 [5] (see Chapters 3 and 17).

3. A difference in temperature in the case of copper tubing can create a corrosion cell. Generally, the increase in temperature accelerates corrosion. For temperatures between 15 and 70 °C, the rate of corrosion of steel in dilute acidic solutions can be doubled for every increase of 10 °C. Above this range of temperatures, the solubility of oxygen in water is low and the rate of corrosion cannot be doubled as before since oxygen plays an accelerating effect for the cathodic reaction.

4. Protective passive films similar to that of stainless steel or aluminum result in uniform corrosion because of the mobility of active sites that passivate readily. Corrosion products and/or passive films are characteristic of electrochemical corrosion of alloys. A film is protective depending on coverage capacity, conductivity, partial pressure, porosity, toughness, hardness, and resistance to chemicals and gases. Rust (Fe) and white rust (Zn) are generally not protective, while patina (Cu), Al_2O_3, MgO, and Cr_2O_3 are protective in certain environments. Corrosion is generally controlled by diffusion of active species through the film. The film-free reaction is generally considered a chemical reaction [1].

The most common expression of the corrosion rate in practice is in milligrams per square decimeter per day (mdd) that can be converted to volume (millimeters per year, $mm \cdot yr^{-1}$):

$$\text{mdd}\frac{0.0365}{\rho} = \text{mm} \cdot \text{yr}^{-1}$$

where ρ is the density of the metal in $g \cdot cm^{-3}$ [2].

5.3. MECHANISMS

Generally, the galvanic cell in corrosion is complex and corresponds to the dissolution of the active metal and oxygen reduction or hydrogen evolution on the cathode surface. Microelectrochemical cells result in uniform general corrosion. Dissolution of metals in acids is due to the presence of indistinguishable anodic and cathodic sites. Uniform general corrosion can be observed during chemical reaction, electrochemical polishing, and passivity, where anodic and cathodic sites are physically inseparable. A polished surface of a pure active metal immersed in a natural medium (atmosphere) can suffer from galvanic cells. Most of the time, the metallic asperities act as anodes and that of the cavities are considered as cathodes. If these anodic and cathodic sites are mobile and change in a continuous dynamic way, uniform or quasi-uniform corrosion is observed. If some anodic sites persist and are not covered by protective corrosion products or do not passivate, localized corrosion is observed [2]. Some macroelectrochemical cells can cause a uniform or near-uniform general attack of certain regions. General uneven corrosion or quasi-uniform corrosion is observed in natural environments and is much more common.

In reactions such as oxidation of aluminum or tarnishing of silver in air, thin, tightly adherent protective films are formed, and the metal surface remains smooth. For some metals or alloys, uniform corrosion produces a somewhat rough surface by removal of a substantial amount of metal, which either dissolves in the environment or reacts with it to produce a loosely adherent, porous coating of corrosion products. As an example, following a careful removal of the rust after general atmospheric corrosion of steel, the surface reveals an undulated surface indicating nonuniform attack of different areas [2]. In natural atmospheres, the general corrosion of metals can be localized. The conductivity, ionic species, temperature of the electrolyte, alloy composition, phases and homogeneity in the microstructure of the alloy, and differential oxygenation cells can influence the corrosion morphology. High temperature shows a uniform attack normally. However, subsurface corrosion films within the matrix of the alloy can be observed by microscopic examination due to film formation at the interface of certain microstrucrures in several alloys at high temperature [6].

5.4. PREVENTION

5.4.1. Design Considerations

In design, a metal or alloy that forms a stable passive film is recommended. In establishing the design parameters, one must consider the predicted general corrosion penetration for the expected life of the structure and double this estimation for safety considerations. Taking the example of a steel tank: the wall thickness for mechanical considerations should be considered for the integrity of the tank, and add to that two times the uniform corrosion thinning for the use period [7].

5.4.2. Surface Pretreatment

A surface pretreatment in oxidized solutions has been adopted for stainless steels and is recommended in many circumstances. The environment can be modified in the bulk and should be effective at the interface in adding oxidizing agents, such as nitrite or strong nitric acid, that maintain the passive state on some metals and alloys [8].

5.4.3. Corrosion Control

Generally, corrosion inhibitors, cathodic protection, anodic protection, and coatings, or a combination of them, are used for corrosion control. However, cathodic protection is the only method that avoids corrosion completely if the system is not sensitive to hydrogen embrittlement or to alkaline medium corrosion. Anodic protection is a viable approach when the metal can be passivated in the corrosive solution. In this technique, a current can be applied using a potentiostat, which can set passivation and control the potential at a value greater than the passive potential E_p or below the pitting potential E_{pit} for environments containing corrosive species such as chlorides or bromides.

5.4.4. Aluminum Alloys and Resistance to General Corrosion

All non-heat-treatable alloys have a high resistance to general corrosion. Alloys of the 1xxx series have the highest resistance to general corrosion. Almost all the manganese in the alloys of the 3xxx series is precipitated as finely divided phases (intermetallic compounds) with negligible difference in electrode potential between the phases and the aluminum matrix. Magnesium, if present, provides additional strength through solid solution hardening, but the amount is low enough that the alloys behave more like those with manganese alone. Their resistance to general corrosion is comparable to that of the 1xxx series. Alloys of the 4xxx series (Al–Si) are low-strength alloys used for brazing and welding products and for cladding in architectural products. These alloys develop a gray appearance upon anodizing. The silicon, most of which is present in elemental form as a second-phase constituent, has little effect on corrosion. Alloys of the 5xxx series have not only the same high resistance to general corrosion as other non-heat-treatable alloys in most environments, but, in slightly alkaline ones, they have a better resistance than any other aluminum alloy. They are widely used because of their high as-welded strength when welded with a compatible 5xxx series filler wire because of the retention of magnesium in solid solution [9].

Heat-treatable wrought alloys have a significantly lower resistance to general corrosion. These include all alloys of the 2xxx series (Al–Cu, Al–Cu–Mg, Al–Cu–Si–Mg) and those of the 7xxx series (Al–Zn–Mg–Cu) that contain copper as a major alloying element. The lower resistance is caused by the presence of copper in these alloys and protective measures are recommended [1]. However, those of the 6xxx series, which are moderate-strength alloys based on the quasi-binary Al–Mg_2Si system, provide a high resistance to general corrosion equal to or approaching that of non-heat-treatable alloys. Also, heat-treatable alloys of the 7xxx series (Al–Zn–Mg) that do not contain copper as an alloying addition also provide a high resistance to general corrosion [9].

Aluminum alloys containing copper (2000 series) and zinc (7000 series) as major alloying elements are generally less corrosion resistant than those without these elements. For this reason, corrosion of aluminum alloys in these two series is usually difficult to inhibit. Alloys in both series are high strength and widely used. The solution heat-treated tempers are usually more corrosion resistant and more amenable to corrosion inhibition than are the hardened alloys. The strain or work hardened alloys are somewhat more readily inhibited than are the alloys hardened by aging treatments [9].

B. GALVANIC CORROSION

5.5. GENERAL CONSIDERATIONS

Galvanic corrosion is a term generally used to consider a galvanic cell between two different metals (bimetallic or corrosion). However, the same metal or alloy could have a microstructure containing a solid solution and different phases or precipitates with different potentials that create galvanic local cells. These types of local galvanic cells due to the microstructure of the alloy are related closely to metallurgically influenced corrosion form and will be treated in more detail later.

The rate of the attack in a galvanic cell in the presence of a corrosive medium is controlled by:

- The difference of potential between the two structural alloys or metals in the corrosive medium.
- The electric resistance between the two conductors, which is frequently low. The electrolyte (e.g., seawater) with a low resistivity (a few ohms/cm^2) is particularly aggressive.
- The formation of passive, semiconductive, or, nonconductive films very frequently increases the resistance at the anode interface and the reaction becomes controlled by diffusion through the pores of recovering films.
- The surface area of the anode as compared to the cathode [9, 10].

Galvanic corrosion can then give rise to a uniform attack if the conductivity of the solution at the interface is considerable and if the electrolyte is renewed constantly and in close contact with numerous microgalvanic cells in a metal with a homogeneous microstructure, and is the case of dissolution of zinc, for example, in a well-agitated acidic solution. This excludes the formation of corrosion products that form a barrier at the metal–solution interface. However, in numerous situations, localized galvanic corrosion occurs as shown in Figure 5.1.

5.6. GALVANIC SERIES OF ALUMINUM ALLOYS

If aluminum-based articles are exposed outdoors or in moist locations in contact with parts made of other metals, galvanic attack of the aluminum surfaces adjacent to the dissimilar

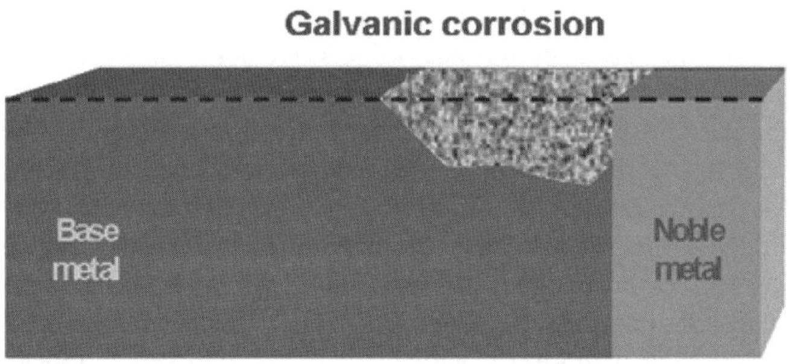

Figure 5.1 Galvanic localized corrosion of Al in the active state in contact with a structural metal having a more positive potential (e.g., Cu or Fe).

metal is likely to occur since aluminum is anodic to the majority of structural metals with the exception of magnesium and zinc. Galvanic action is much more pronounced in marine or seacoast atmospheres than in rural or industrial locations [9, 10]. It is generally recognized that protection must be provided if aluminum and aluminum alloys contact copper, copper alloys, iron, or steel. Aluminum alloys are quite anodic to these metals and an alloy change will have little effect. An exception is 2xxx series aluminum alloys containing 2% or more copper in the solution heat-treated T3 or T4 tempers. These alloys can approach the corrosion potential of mild steel [3].

For natural atmospheric corrosion, aluminum can be compared to other passive alloys in this atmosphere, such as stainless steels. Contact with copper or copper-based alloys causes more pronounced galvanic attack than does contact with most other metals like stainless steels or steel. In rural or industrial locations, contact with steel does not generally cause a very pronounced acceleration in rate of attack of aluminum-based alloys (especially the Al–Cu alloys such as 2017 and 2024). In seacoast locations, attack may be appreciably accelerated. In certain solutions and in some natural waters, this action may be reversed, so that attack of the steel is accelerated and the aluminum is protected [9].

Contact between stainless steel and aluminum in seawater or other saline solutions usually results in less galvanic action on the aluminum than does contact of aluminum with steel. However, no cases of reversal of the stainless steel–aluminum couple, such as occurs with steel, are known. Cadmium has about the same potential as aluminum, and therefore contact with cadmium usually results in negligible galvanic action. Zinc is anodic to aluminum in most neutral or acid solutions; hence, in such solutions, contact with zinc results in protection of the aluminum. In alkaline solutions, the potentials reverse so that, in these media, contact with zinc can cause accelerated attack of aluminium [9].

Designers often use aluminum in contact with stainless steel and titanium. Both these materials are quite cathodic to aluminum, but the effect of the couple depends to a large extent on whether or not the stainless steel or titanium passivates or stays active. If these metals passivate, such a couple often is tolerable, but one cannot assume passivity. If, for example, one of them did not passivate in a less oxygenated medium (crevice) and/or acid chloride-rich electrolyte, coating should be prescribed in this situation. The designer really needs to determine the expected environmental conditions and establish the performance of the stainless steel and titanium. Conditions such as crevices, acidity, and oxygen availability are important [3].

Titanium and aluminum alloys are widely used in the aeronautical industry because of their high ratio of strength to weight as well as good corrosion resistance. The average galvanic current of titanium alloy coupled to LC4 (copper) was higher than LC12 (zinc) in 3.5% NaCl. There is a poor correlation between potential difference and galvanic current density in alloys in aqueous 3.5% NaCl. Anodizing according to HB/Z 5076-78 (a Chinese standard method for galvanic corrosion with film thickness of 10–15 μm) slows corrosion at the very beginning but galvanic corrosion produces cracks in the oxide film and galvanic current density was similar to that of uncoated aluminum. Anodizing is then not beneficial for prevention of galvanic corrosion of these aluminum alloys; however, details about sealing treatment after anodizing were not described [11].

When exposed to a given environment, the potential of a metallic material is determined by many factors, such as temperature, liquid flow rate, and level of aeration. However, the relative ranking of alloys in the galvanic series remains largely unaffected by such changes, even though their corrosion potential may shift by several hundred millivolts. The difference between the most electropositive and the most electronegative metals is nearly 2 V, and the coupling of these materials could generate serious currents and hence high corrosion rates on the most anodic material [12]. However, the magnitude of the potential difference alone will not necessarily indicate the amount of galvanic corrosion. For instance, metals with a potential difference of only 50 mV have shown severe galvanic-corrosion problems, whereas metals with a potential difference of 800 mV have been successfully coupled together. This is because the potential difference gives no information about the kinetics of galvanic corrosion, which depends on the current flowing between the two metals in the couple [12].

Because more observations of potentials and galvanic behavior have been made in seawater than in any other single environment, an arrangement of metals in a galvanic series based on observations in seawater is frequently used as a first approximation of the probable direction of the galvanic effects in other environments (Section 3.2.2.5 in Chapter 3) [9]. For the sake of comparison, Table 5.1 shows the potentials of metals exposed to neutral soils and water. Frequently, potentials in neutral soils and water are measured versus the practical saturated reference electrode $Cu/CuSO_4$ [12].

Anode/Cathode Surface Area Ratios The larger the cathode compared with the anode, the more oxygen reduction, or other cathodic reaction, can occur, and hence the greater the galvanic current. Under static or slow-flow conditions, where the galvanic-corrosion current is often dependent on the rate of diffusion of dissolved oxygen to the cathode, the total amount of galvanic corrosion is independent of the size of the anode and proportional to the area of the cathodic metal surface. This is sometimes known as the catchment area principle and has important implications in designing to minimize the risk of galvanic corrosion. Thus, for a constant area of cathode metal, the total amount of corrosion of the anode is constant, but the corrosion per unit area increases as the area of the anode is decreased [12].

If the surface area of the anode (aluminum or its alloys) is very low with respect to the cathodic surface, the rate of general corrosion will be very high at the limited anodic surfaces. In some cases, severe localized corrosion occurs, which can lead to perforation, especially if the resistance of the corrosive medium is high; for example, seawater, with a low resistivity (a few ohms/cm^2), is particularly aggressive and the solution is stagnant. The rate of attack is controlled by the difference of potential between the two structural alloys or metals in the corrosive medium. The electric resistance between the two conductors

Table 5.1 Practical Galvanic Series for Metals in Neutral Soils and Water

Metal	Potential (SCE)
Commercially pure magnesium	−1.75 V
Magnesium alloy (6% Al, 3% Zn, 0.15% Mn)	−1.6 V
Zinc	−1.1 V
Aluminum alloy (5% zinc)	−1.05 V
Commercially pure aluminum	−0.8 V
Mild steel (clean and shiny)	−0.5 to −0.8 V
Lead	−0.5 V
Copper, brass, bronze	−0.2 V
Platinum	0 to −0.1 V
Carbon, graphite, coke	+0.3 V

Source: Reference 12.

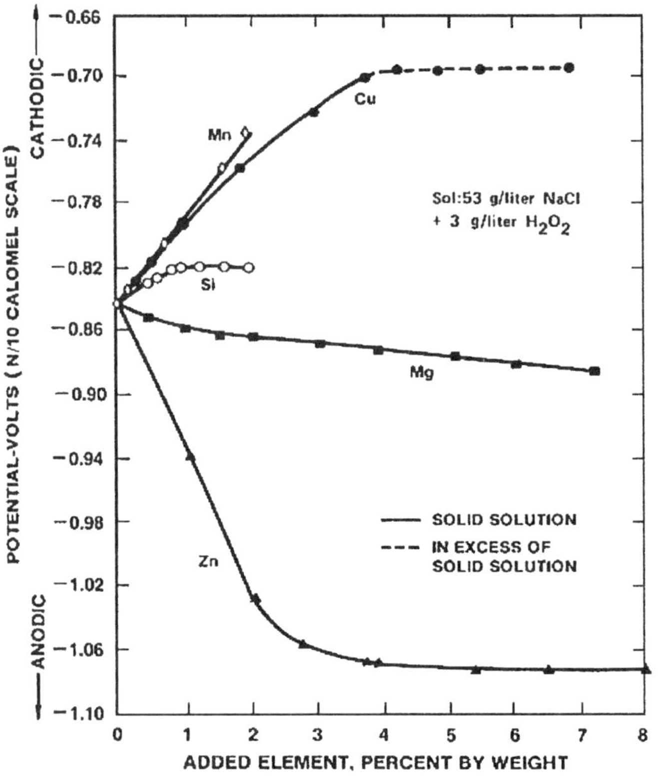

Figure 5.2 Effect of alloying elements on the electrode potential of aluminum [9, 14].

frequently becomes low because of oxide formation accompanied by diffusion control and polarization of the two electrodes [9, 10].

Figure 5.2 concerns the effect of alloying elements in determining the position of aluminum alloys in the series; these elements, primarily copper and zinc, affect electrode potential only when they are in solid solution [9].

In the presence of a good electrolyte, as little as 15 mV difference in corrosion potential of the two metals can have an effect, and if the difference is 30 mV or greater the anodic material will definitely corrode sacrificially to protect the contacting less active cathodic metal [13]. Many users do not recognize that aluminum alloys themselves span a range of about 400 mV in their respective corrosion potentials. In aerated sodium chloride solutions, pure aluminum, 3xxx alloys, and many other alloys have a potential of about −740 mV (SCE). Aluminum alloys with high magnesium or zinc contents will be more anodic (negative) by as much as 260 mV, while high copper content alloys will be more cathodic up to about 140 mV. Care must therefore, be taken that all alloys and tempers are compatible, even in the same aluminum structure [3].

Table 5.2 shows a galvanic series of aluminum alloys and other metals representative of their electrochemical behavior in seawater and in most natural waters and atmospheres. Aluminum (and its alloys) becomes the anode in galvanic cells with most metals, protecting them by corroding sacrificially. Only magnesium and zinc are more anodic and corrode to protect aluminum. This type of dissimilar metal corrosion can be found in strong acidic or strong basic solutions. The rate of corrosion can vary from several micrometers per year to several micrometers per hour.

Table 5.2 Electrode Potentials of Representative Aluminum Alloys and Other Metals[a]

Aluminum alloy or other metal[b]	Potential (V)
Chromium	+0.18 to −0.40
Nickel	−0.07
Silver	−0.08
Stainless steel (300 series)	−0.09
Copper	−0.20
Tin	−0.49
Lead	−0.55
Mild carbon steel	−0.58
2219-T3, T4	−0.64[c]
2024-T3, T4	−0.69[c]
295.O-T4 (SC or PM)	−0.70
295.O-T6 (SC or PM)	−0.71
2014-T6, 355.O-T4 (SC or PM)	−0.78
355.O-T6 (SC or PM)	−0.79
2219-T6, 6061-T4	−0.80
2024-T6	−0.81
2219-T8, 2024-T8, 356.O-T6 (SC or PM), 443.O-F (PM), cadmium	−0.82
1100, 3003, 6061 T-6, 6063-T6, 7075-T6, 443.O-F (SC)	−0.83
1060, 1350, 3004, 7050-T73, 7075-T73[c]	−0.84
5052, 5086	−0.85
5454	−0.86
5456, 5083	−0.87
7072	−0.96
Zinc	−1.10
Magnesium	−1.73

[a]Measured in an aqueous solution of 53 g of NaCl and 3 g of H_2O_2 per liter at 25 °C versus 0.1 N calomel reference electrode.

[b]The potential of an aluminum alloy is the same in all tempers wherever the temper is not designated.

[c]The potential varies ±0.01 to 0.02 V with quenching rate.

Source: Reference 14.

5.7. MECHANISMS

5.7.1. Cu–Al Galvanic Cell

Galvanic coupling between different α and θ phase-containing model Al–Cu alloys, deposited by magnetron sputtering, has revealed that the anodic α phase did not suffer corrosion and remained in the passive state in sulfate solution. Conversely, sulfate ions induced pitting of the cathodic θ phase. Pitting susceptibility of the cathode increased when the difference between the copper content of the anode and cathode increased. Similar observations were made for all the galvanic couples; furthermore, the higher the copper content of a phase, the greater its susceptibility to pitting [15].

5.7.2. Mg–Al Galvanic Cell

Aluminum can be attacked by the strong alkali generated at the cathode when magnesium corrodes sacrificially in static NaCl solutions. Such attack destroys compatibility in alloyscontaining significant iron contamination, apparently by exposing fresh, cathodicactive sites with low overvoltage. Magnesium and its alloys are definitely anodicto the aluminum alloys, and thus contact with aluminum increases the corrosionrate of magnesium. However, such contact is also likely to be harmful to aluminum, sincemagnesium may send sufficient current to the aluminum to cause cathodic corrosion in alkaline medium. Such damage is frequently encountered in seacoast locations [9, 10].

Cathodic corrosion of aluminum is much less severe in seawater than in NaCl solution, because the buffering effect of magnesium ions reduces the equilibrium pH from 10.5 to about 8.8. The compatibility of aluminum with magnesium is, accordingly, better in seawater and is less sensitive to iron content [16].

Aluminum oxide is amphoteric, that is, soluble in alkaline as well as acid solution. The standard potentials of these two half-reactions are

$$\text{Acid:}\quad Al^{3+} + 3e^- = Al(-1.66\ V) \tag{5.1}$$

$$\text{Alkaline:}\quad H_2AlO_3^- + H_2O + 3e^- = Al + 4OH^-\ (-2.35\ V) \tag{5.2}$$

Half-reaction (5.1) has nearly the same standard potential as that for acidic dissolution of magnesium: $Mg^{2+} + 2e^- = Mg\ (-2.37\ V)$.

Commercial aluminum alloys contain several thousand parts per million (ppm) of iron in the form of the intermetallic $FeAl_3$. The mutually destructive galvanic action between magnesium and commercial aluminum alloys in salt water proceeds as follows:

1. Rise in the pH of the liquid in contact with the aluminum member. This is most likely the result of galvanic current flow between the magnesium and the initially passive aluminum.
2. Shift of the aluminum potential in the active direction in accordance with the half-reaction (5.2).
3. Exposure of iron aluminum intermetallic particles (e.g., $FeAl_3$), which then engage in separate galvanic activity with the magnesium. This galvanic current flow accounts for the severe sacrificial corrosion of the magnesium, and the alkali generated at the cathode ensures continued corrosion of the aluminum in accordance with half reaction (5.2) [16].

Certain aluminum-based alloys (such as 5056) are less affected by contact with magnesium than are other aluminum alloys. For this reason, 5056 rivets have been employed extensively in assembling magnesium alloy structures. A 5052 alloy would meet the essential requirement for a fully compatible aluminum alloy with a maximum of 200 ppm Fe, or a 5056 alloy with a maximum of 1000 ppm Fe. In designing outdoor structures, it is often necessary to combine dissimilar metals in the structure. Suitable protective methods are available that, if adopted, will greatly reduce the risk of galvanic corrosion [9, 10].

5.7.3. Galvanic Effect of a Coating

Generally, the main reason is the presence of a relatively small anode and a many-times-larger cathode. The galvanic cell could be created efficiently where protective coatings are applied over metal and where there is a break in the coating so that the large coated area acts as a cathode, even though a very weak one. The coating breakdown occurs because of moving of the electrolyte, gases (oxygen), and moisture through the film to the pinholes and micropores. Water and gases pass through the coating film and cause osmotic pressure. Water diffusion and visual blistering can be observed. Permeation is at its best at 65–96 °C, accelerated by osmotic pressure, electroendosmotic pressure, thermal agitation, and vibration of the coating film molecules. The electroendosmotic gradient is created between the corroding area and the protected areas in electrical contact. Resistance to filiform corrosion depends more on factors such as the environment (e.g., humidity and chloride ions), the type and thickness of coating, metal surface preparation, and coating application procedures than on the metal itself. However, there is evidence that higher copper content aluminum alloys are more susceptible [3].

5.8. DEPOSITION CORROSION

Deposition corrosion is a particular case of galvanic corrosion that leads to pitting. Aluminum reduces ions of many metals, of which copper, cobalt, lead, mercury, nickel, and tin are the ones encountered most commonly. Reduction of these heavy metal ions leads to corrosion of aluminum and deposition of the more noble metal on the surface of the aluminum. Once this metal is deposited, it leads to serious attack of the alloy due to the galvanic cell that is established. Reducible metallic ions are of most concern in acidic solutions since their solubilities are greatly reduced in alkaline solution. Copper is the heavy metal most commonly encountered in applications of aluminum. A copper-ion concentration of 0.02–0.05 ppm in neutral or acidic solutions is generally considered to be the threshold value for initiation of pitting on aluminum. Ferric ions, Fe^{3+}, can be reduced by aluminum but do not usually form a metallic deposit. At room temperature, the most anodic aluminum alloys (those with a corrosion potential approaching -1.0 V versus the SCE) can reduce ferrous ions, Fe^{2+}, to metallic iron and produce a metallic deposit on the surface of the aluminum. The presence of Fe^{2+} ions also tends to be rare in service; it exists only in deaerated solutions or in other solutions free of oxidizing agents [17].

Galvanic corrosion by copper-bearing waters can often be the cause of localized attack on a number of metals. The critical amount of copper, necessary in the natural water or dissolved from copper or copper-alloy parts in the system, decreases in the order iron, zinc, and aluminum, the latter being sensitive to a concentration as low as 0.02 ppm. Provided there is also dissolved oxygen, calcium bicarbonate, and chloride present—which is usually the case—this and larger amounts of copper will induce localized pitting of aluminum. It may usually be recognized by the existence of a bluish ting or flecks in the corrosion product. It is relatively easy to check for copper if it is suspected as the pit initiator. An example is a galvanized cooling-water tank on which intense localized attack had resulted in deep pits containing copper compounds overlaid with nodular corrosion product [5].

Deposition Corrosion and Mercury In the case of mercury, any concentration in a solution of more than a few parts per billion can be detrimental. No amount of metallic mercury should be allowed to come into contact with aluminum. Aluminum in contact with a solution of a mercury salt forms metallic mercury, which then readily amalgamates the aluminum. Of all the heavy metals, mercury can cause the most corrosion damage to aluminum. The effect can be severe when stress is present. An extreme example of breakdown by undermining is the breakdown of most oxide films on aluminum in the presence of mercury, which dissolves any exposed metal—a hazard in the use of aluminum thermometers in aluminum vessels, and a useful technique in the production of oxide-film replicas in the electron microscopy of aluminum alloys [9, 14]. Also, the attack by mercury and zinc amalgam combined with residual stresses from welding caused cracking of the weldment. The corrosive action of mercury can be serious with or without stress because amalgamation, once initiated, continues to propagate unless the mercury can be removed. The corrosive action of mercury was attributed not only to the galvanic cell, but also to the destruction and prevention of formation of aluminum oxide. The corrosion rate can be extremely high, up to 1270 mm/year [9].

5.9. STRAY CURRENT CORROSION

Stray current corrosion is created from direct flow through paths other than the intended circuit, for example, by an extraneous current in the earth. Since it is a general attack of the anodic part of the conductor, it is better to classify this phenomenon under galvanic corrosion. Whenever an electric current (ac or dc) leaves an aluminum surface to enter an environment, such as water, soil, or concrete, aluminum is corroded at the area of current passage in proportion to the amount of current passed. This is known as *stray current corrosion* or electrolysis (a poorly chosen, ambiguous term, but one firmly entrenched in pipeline and shipping technology). Examples of stray current corrosion of aluminum have been reported in concrete (electrical conduit), in seawater (boat hulls), and in soils (pipelines and drainage systems). At low current densities, corrosion may take the form of pitting, whereas at higher current densities considerable destruction of the metal can occur. The corrosion rate does not diminish with time. Stray currents encountered in practice are usually direct current (e.g., from a welding generator), but may also be alternating current. For most metals, ac corrosion is negligible, but with aluminum it can be appreciable. Below a critical small ac current density, no corrosion of aluminum occurs. Since the aluminum surface from which the current leaves functions as an anode, oxidation (corrosion) occurs, and the area becomes acidic. The presence of acidity on the surface often provides the clue that reveals unexpected stray current activity. Local acidity can develop even in an alkaline environment such as concrete [9].

5.10. PREVENTION

1. For natural atmospheric corrosion, aluminum can be compared to other passive alloys in this atmosphere, such as stainless steels. Direct assembly between aluminum and copper should be avoided. This has not been observed for all applications since some electric industries have introduced a bivalent sheet of aluminum and copper in direct contact with the rest of the structure in aluminum and copper, respectively [9, 18].

2. Galvanic corrosion can be prevented by breaking electric contact between the two metals, using, for example, gum, paint, rubber, nonconducting polymers, or washers with sufficient thickness. Isolation of the cathodic metal of the galvanic cell (the cathodic alloy in contact with the structural aluminum) by a resistant paint is common practice.

For example, the steel in an assembly of aluminium–steel in seawater is generally coated with zinc (metallization) and by an adherent resistant paint [9].

3. The corrosion rate of aluminum when coupled to a more cathodic metal depends on the extent to which it is polarized in the galvanic cell. It is especially important to avoid contact with a more cathodic metal where aluminum is polarized to its pitting potential because (see Section 5.12) a small increase in potential produces a large increase in corrosion current [9].

4. By removing the cathodic reactant, galvanic corrosion is reduced because the aluminum is less likely to be polarized to its pitting potential. Thus the corrosion rate of aluminum coupled to copper in 3.5% NaCl solution is greatly reduced when the solution is deaerated. In closed multimetallic systems, the corrosion rate, even though it may be high initially, decreases to a low value whenever the cathodic reactant is depleted. Galvanic corrosion is also low where the electrical resistivity is low, as in high-purity water. Some semiconductors, such as graphite and magnetite, are cathodic to aluminum, and in contact with them, aluminum corrodes sacrificially. Galvanic corrosion of aluminum by more cathodic metals in solutions of nonhalide salts is usually less than in solutions of halide ones [9].

5. To minimize corrosion of aluminum in contact with other metals, the ratio of the exposed area of aluminum to that of the more cathodic metal should be kept as high as possible (since such a ratio reduces the current density on the aluminum). Paints and other coatings for this purpose may be applied to both the aluminum and the cathodic metal, or to the cathodic metal alone, but they should never be applied to the aluminum only, because of the difficulty in applying and maintaining them free of defects [19].

5.11. BASIC STUDY OF Al–Cu GALVANIC CORROSION CELL

Arrays of engineered copper islands on an aluminum thin-film matrix have been employed to investigate the role of copper in galvanic corrosion of Al–Cu alloys. When exposed to dilute NaCl solutions, the engineered samples corrode, showing morphology similar to that observed in second-phase particles in real alloys. In situ fluorescence microscopy allows the observation of oxygen reduction at copper islands during corrosion of the underlying aluminum thin-film matrix [19]. In situ fluorescence microscopy was used to image local OH^- production during open-circuit exposure of the samples to a 0.05 M NaCl solution, at pH $\approx$6.5.

An Al–Cu galvanic couple immersed in aerated chloride solution is expected to support two types of electrochemical reactions leading to the corrosion of aluminum. The first is oxygen reduction on the Cu:

$$O_2 + 2H_2O + 4e^- \Rightarrow 4OH^-$$

The OH^- produced by this reaction may locally increase the pH, making the solution more alkaline in the vicinity of the copper. This oxygen-reduction reaction is balanced by the oxidation of Al:

$$Al \Rightarrow Al^{3+} + 3e^-$$

followed by aluminum hydrolysis:

$$Al^{3+} + nH_2O \Rightarrow Al(OH)_n + nH^+$$

The H^+ produced in the region where aluminum is corroding may locally decrease the solution pH, creating acidic regions. Aluminum complex ions are expected to be soluble within the pit, diffusing out to the solution region above the copper island, where they remain soluble in the alkaline environment. As they diffuse further toward the bulk solution and into a more neutral environment, precipitation of aluminium hydroxides and oxyhydroxides can occur, resulting in the corrosion product halos [19].

However, when the spacing between islands is reduced to 10 μm, the corrosion product halo precipitates outside the entire 100-element array, suggesting that the elevated pH regions generated at individual copper islands overlap. Energy dispersive spectroscopy analysis showed that this overlap between alkaline regions leads to a higher corrosion rate of the aluminum matrix between, relative to beneath, the copper islands. Such an influence on matrix corrosion rate due to the spatial distribution between noble particles has not been previously identified and observed. This result shows that the noble particle distribution may have a strong effect on time to failure in Al–Cu alloys [5, 20].

C. LOCALIZED CORROSION

All forms or types of general corrosion that are not distributed uniformly on the surface could be considered as localized corrosion. Localized corrosion is the most insidious corrosion because it is far less predictable than general corrosion and can have serious consequences, such as putting some or all of the equipment out of service, causing fatal accidents in some circumstances. The three types closely identified with localized corrosion treated here are pitting corrosion, crevice corrosion, and filiform corrosion. Fundamentally, the kinetics of their propagation is autocatalytic, based on the same mechanism, but initiation of corrosion is different especially between pitting and that of crevice and filiform corrosion [21, 22]. Considering the major reasons behind localized corrosion failures, some of them could be described in the context of other forms of corrosion, such as metallurgically influenced corrosion, mechanically assisted corrosion, or environmentally induced cracking. Fatigue corrosion and stress-corrosion cracking are evidently dangerous and could start from, give, or result from pitting.

In all types of localized corrosion, active and passive surface states are simultaneously stable on the same metal surface over an extended period of time, so that local pits can grow to macroscopic size [21, 22]. This type of attack results frequently from a concentration cell formed between the electrolyte within the pit or crevice, which is oxygen starved, and the electrolyte outside the crevice, where oxygen is more plentiful (the oxygen differentiation cell). The material within the crevice acts as the anode, and the exterior material becomes the cathode. A crevice may be produced by design or by accident. Crevices caused by design occur at gaskets, flanges, rubber O-rings, washers, bolt holes, rolled tube ends, threaded joints, riveted seams, overlapping screen wires, lap joints, beneath coatings (filiform corrosion) or insulation (poultice corrosion), and anywhere close-fitting surfaces are present [20]. The crevice cell could be formed due to different concentrations of the metallic ions inside and outside, but no published work shows that this is the case for crevice corrosion of aluminum alloys [20–22].

5.12. PITTING CORROSION

5.12.1. Occurrence and Morphology

Although in appearance pitting corrosion doesn't seem important, the depth of the pit and pit propagation speed can become extremely dangerous: pitting is one of the most serious types of localized corrosion. Pitting is often a concern in applications involving passive metals and alloys in aggressive environments. Pitting can also take place under atmospheric conditions. In practice, pitting corrosion of passive metals is commonly observed in the presence of chlorides or other halides; however, it can also occur in nonpassivating alloys with protective coatings or in certain heterogeneous corrosive media such as aluminum in nitrate solutions at high potentials or carbon steel in high-purity water at elevated temperature [22, 23].

Corrosion starts at the original defects of the protective oxide film or at the break of the passive film or the coating and continues to undercut the coating, forming a rather heavy tubercle of hard rust or scale with the pit in the original metal underneath. These are common in the marine area as well as various industries, where strong corrosive conditions exist [9, 24]. Resistance to corrosion improves considerably as the purity is increased, but the oxide film on even the purest aluminum contains a few defects where minute corrosion can develop. Metallurgical microstructure has little effect on the pitting potential of aluminum, nor do second phases in the amounts present in its alloy have significant effect. Severe cold work makes the potential more anodic by a few millivolts, and this change, although small, is sufficient to affect the extent to which pitting develops (e.g., more pitting on machined or sheared edges) [21].

In all types of localized corrosion, active and passive surface states are simultaneously stable on the same metal surface over an extended period of time, so that local pits can grow to macroscopic size. When the active state within pits and crevices and under a deposit (film, corrosion product, conversion coating, or paint) is maintained over an extended period of time, rapid metal dissolution usually occurs. The resulting pit and crevice geometries as well as the surface state within the pits vary markedly from open and polished hemispherical pits on free surfaces to etched crack-like shapes within crevices, depending largely on the type of rate-controlling reactions during the growth stage and the level of potential values of the anodic sites. Figure 5.3 shows the morphology of the most abundant shapes of pits [22, 25].

5.12.2. Kinetics

The pitting density can be obtained for a surface area of about 1 dm^2. The kinetics of perforation is obtained for different periods of immersion—1, 2, 3, 6, and 12 months for aluminum. Aziz [27] expressed the pitting kinetics by the following relation: $P = K + \Gamma/3$, where P is the depth of the pit, K is a constant, and Γ is the corrosion time in years. In the case of aluminum, the rate of perforation decreases with time. Figure 5.4 gives an expression of the depth of the pit as a function of time of immersion. It has been suggested that extended contact of the metal with water favors the formation of a protective oxide film. The average and maximum perforation should be considered. The maximum perforation rate should be used for design and prevention methods, since doubling the sheet thickness can multiply the service life by a factor of 8 in several media [5].

It has been shown that the pit density of aluminum increases with temperature in water and that the depth of pits decreases with temperature (Figure 5.4). Ordinary chemical

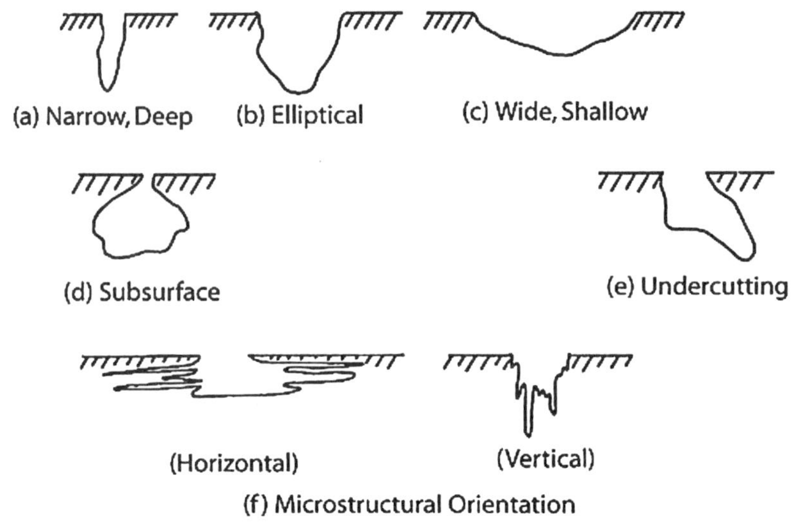

Figure 5.3 Sections of the most abundant types or shapes of pits (ASTM G46) [9, 22] (From Reference 25.)

dissolution of passive oxide films is frequently an exothermic reaction, even in the same solution where passivation is produced. Increase in temperature accelerates dissolution rate of oxides such as Fe_2O_3, Cr_2O_3, and Al_2O_3. Higher valent oxides are usually the best passive films because of their slow rate of dissolution [3, 26]. However, it has been suggested that extended contact of aluminum with water favors the formation of a protective oxide film [9, 27].

Metal oxides such as those of aluminum and zinc are not cathodically reducible at the potentials obtainable in aqueous media; hydrogen is reduced instead. For aluminum, the very feebly electron-conducting aluminum oxide can produce hydrogen gas that can push the film off and can be followed even by protons entering the film without discharge and

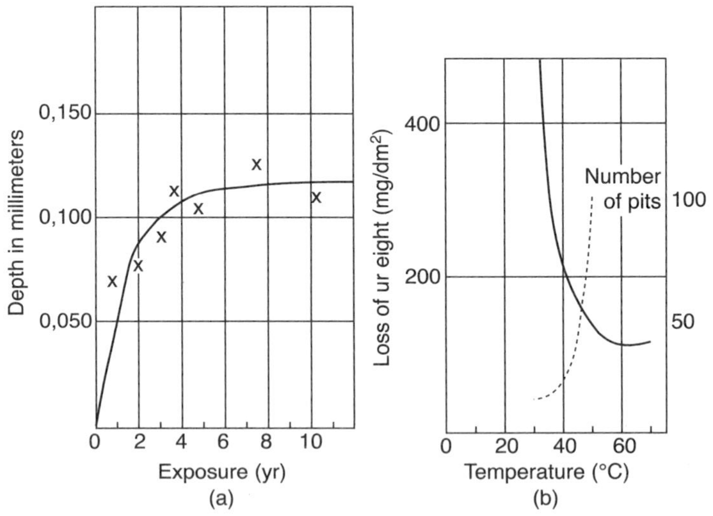

Figure 5.4 Pit depth as a function of (a) exposure period [9, 27] and (b) influence of water temperature [9, 18].

rendering it conducting. Under high field conditions, aluminum ions may even transfer from film to metal since the metal–film interface is nonaqueous. In the case of Zn, the vigorous evolution of hydrogen gas assisted by the zinc oxide electron conduction can accelerate the breakdown of passivity [28].

Al and Cu are not passive in strongly acidic electrolytes and localized acidification by the hydrolysis of corrosion products may serve as a stabilizing factor for their pitting. However, this factor could not have the same influence in the case of Fe, Ni, and steels, which are passive even in strongly acidic electrolytes, in disagreement with the predictions of Pourbaix diagrams. Electrochemical breakdown of oxides of some metals is possible, such as copper, tin, and lead, since they are readily reduced cathodically to the metal in many solutions while ferric oxide is reduced to ferrous ions in aqueous solutions [29].

In aqueous solutions, solution anions, halide and nonhalide types, can play a major role in passive film growth and breakdown. Borates, for example, appear to have a beneficial effect. Halide ions such as Cl^- can give rise to severe localized corrosion (e.g., pitting). Pitting is associated with a particular combination of film thickness and halide concentration. Cl^- is more aggressive than Br^-. It is agreed that well-developed pits have high $[Cl^-]$ and low pH. It is necessary to consider the nature of the oxide film, the solution in which the film is formed, and the electrochemical conditions of formation of the film to evaluate the characteristics of the passive layer [30].

The potency of halides to complex with metal cations is very important in understanding the stabilization of a corrosion pit by prevention of the repassivation of a defect site within the passive layer. Enhancement of the transfer of metal cations from the oxide to the electrolyte by halides, especially the strongly complexing fluoride ions, holds for many metals. The detrimental effects of sulfur species have been encountered in a large number of service conditions; however, in the area of passivity, the effects of chloride ions have received more attention. Recent data show a direct link between atomic-scale surface reactions of sulfur and macroscopic changes, like enhanced dissolution, passivating blocking or retarding, and passivity breakdown [9].

5.12.3. The Pitting Potential

The pitting-potential (E_p) principle establishes the conditions under which metals in the passive state are subject to corrosion by pitting. E_p is that potential of a certain alloy in a particular solution above which pits will initiate and propagate and below which they may form but will not propagate. The most widely used method to determine E_p is the controlled potential, in which the potential of a specimen usually immersed in a deaerated electrolyte is made more positive and the resulting current density is measured. E_p corresponds to the beginning of a sharp increase in current density in the passive region of the alloy and is frequently called the *breakdown potential of the oxide film*.

Corrosion of aluminum is controlled by an anodic reaction (oxidation), which leads to metallic dissolution, and a cathodic 1reaction (reduction) of environmental species. The relation where the anodic reaction occurs on aluminum, and thus leads to its corrosion, is shown in Figure 5.5. The anodic polarization curve shown is typical for aluminum and its alloys when they are polarized anodically in an electrolyte free of a readily available cathodic reactant (e.g., in a deoxygenated electrolyte), whereas the polarization curves for the cathodic reactions are schematic only. The corrosion current developed by the two reactions (which determines the rate of corrosion of the aluminum) is indicated by the

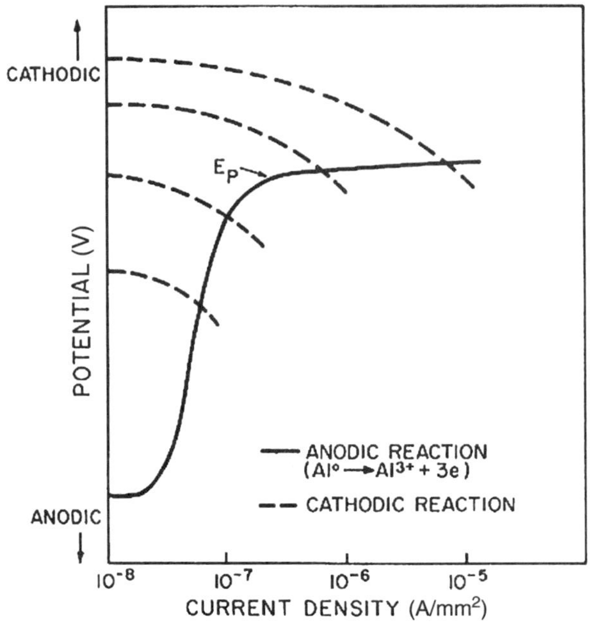

Figure 5.5 Typical anodic polarization curve (solid line) and schematic cathodic polarization curves for an aluminum alloy in a chloride electrolyte free of oxygen. E_p is the pitting potential of the alloy [9, 31, 32].

intersection of the anodic polarization curve for aluminum with one of the cathodic polarization curves [9, 31, 32].

Figure 5.5 shows a typical anodic polarization curve (solid line) for an aluminum alloy in an electrolyte free of a readily available cathodic reactant (commonly oxygen); E_p is the pitting potential of the alloy. The intersection of this curve with one of the cathodic polarization curves (schematic) determines the corrosion current of the alloy. Hollingsworth and Hunsicker [31] mentioned that the values of E_p for 5xxx and 6xxx alloys at 25 °C is on the order of -0.4 to -0.7 V (SCE) in a deoxygenated solution containing 3–0.3% Cl^- at ambient temperature [9, 32].

5.12.4. Mechanisms

Pitting corrosion is observed when aluminum and its alloys are in the pH range where it is passive. On increasing acidity or alkalinity beyond the passive range of pH, corrosion attack becomes more nearly uniform and polarization of potential reaches at least the pitting potential. In aerated solutions, the cathodic reaction is oxygen reduction, while the anodic reaction is accelerated by halide ions, of which chloride is the one most frequently encountered in service. Pitting is observed in aerated solutions of halides in the passive region of pH [9, 33].

Localized corrosion can be initiated by the same local aggressive solution. For example, a saturated solution of aluminum chloride ($AlCl_3$) has a pH of 3.0–3.5, just into the range where the natural oxide is unstable. A saturated concentration can be achieved when aluminum corrodes in a chloride-containing solution under conditions where the electrolyte cannot readily be replenished, such as crevices, deep pits, and cracks. It also occurs in cyclic wet–dry environments, where the solution gradually evaporates and becomes concentrated and aggressive at certain sites of the surface. Generally, aluminum does not pit in aerated solutions of nonhalide salts, because the pitting potential is considerably nobler than it is in halide solutions, and aluminum is not polarized to this level of potential in normal service. Pitting corrosion initiates at weak points of the oxide or hydroxide passivating film of the alloy [28].

Sometimes a very localized attack on a specific area starts with a spot. The anodic current is distributed on a small anodic surface that is extremely limited compared to the surrounding cathodic areas and this leads to deep perforation. Pits usually, but not always, progress in the gravitational sense. Once the process of dissolution starts, the dissolution doesn't need to be stimulated anymore because the process is generally autocatalytic. In certain cases, propagation can be blocked, temporarily or permanently, if some impervious products precipitate on the active sites or there is slow kinetics of pit growth or a difference in the microstructure of the alloy [34].

The positive charge of dissolved cations attracts chloride ions inside the pit. These ions facilitate the anodic reaction and form aluminum chloride, which gives hydroxides and acids by hydrolysis. This helps to shift the pH to acidic values at the anodic sites by hydrolysis:

$$AlCl_3 + 3H_2O \rightarrow Al(OH)_3 + 3HCl$$

Possible reactions at the cathode are

$$3H^+ + 3e^- \rightarrow \tfrac{3}{2}H_2 \quad \text{or} \quad \tfrac{1}{2}O_2 + H_2O + 2e^- \rightarrow 2OH^-$$

The cathodic sites are frequently more alkaline because of the local formation of hydroxides and if the metal hydroxide precipitates at the mouth or the sides of the pit, this can help the autocatalytic nature of penetration of the pit. The presence of oxygen and/or another oxidant is essential for pitting. Considering a neutral solution, the consumption of hydroxide ions at the anodic sites can change the pH to a more acidic value, to the level of 3–4. In natural seawater, the pH within a pit solution decreases from 5 to 2.5 after 100 h of exposure and chloride anions increase. The pit bottom indicates a selective dissolution along certain crystallographic planes. Using a freezing method, the pH of the solution measured in pure Al was found to be between 3 and 4 when the pH of the bulk solution was 11. Kaesche [34] found a pH of 2 in pits and Hoch [35] measured a pH of 1 at the active heads of filiform corrosion [9, 29, 36, 37].

Pitting can be random and amenable to stochastic (statistical) theory, and can be considered as deterministic but very sensitive to experimental parameters such that reproducibility of induction time and electrochemical properties is not achieved. Noise electrochemistry may clarify the initial conditions for pit initiation [29]. Modeling passivation processes can provide new insights into alloy performance and new alloy design concepts [37, 38]. A Gaussian distribution was examined but the Poisson distribution was found to be a better approach for pit generation. The results indicate that different pit generation rates can be observed as a function of time. Two model groups considering either pit generation events alone or assuming pit generation and subsequent repassivation processes were proposed.

5.12.5. Possible Stages of Pitting

Although pitting is self-initiating and self-propagating and it is difficult to determine borders for every stage, generally three stages of pitting could be identified: nucleation, metastable pit formation, and finally stable pitting.

5.12.5.1. Pit Nucleation Stage

The nucleation stage is tiny and very fast and could lead to the formation of pit precursors or metastable pits. It is believed that a critical size of pit embryo for pit initiation is on the order of 10–20 μm [21]. There are three main considerations for pit initiation and theoretical models describe the initiation process leading to passive film breakdown in three classes: (1) mechanical film breakdown theories [5, 20], (2) adsorption and adsorption-induced mechanisms, where the adsorption of aggressive ions like Cl^- is of major importance, and (3) ion migration and penetration models. A mechanical breakdown of the film can be achieved by bending, scratching, stretching, impact, and other mechanical forces depending on the volume of the oxide film with respect to that of the basic metal. Self-healing by chemical or electrochemical processes is favored in some aqueous media other than natural atmospheres: for example, stainless steel passivated in nitric acid is better than the natural air-formed film [28]. The film breaking mechanism is probably important for nonsteady states of the passive layer; sudden changes of the electrode potential cause stresses in the film, due to changes in the chemical composition or electrostriction. Rupture of the film is the most probable mechanism kinetically in the case of nonstationary conditions due to abrupt change of potential.

The second consideration is the adsorption of anions on the passive film that influence the passive current density; for example, SO_4^{2-} or ClO_4^- and halides have different accelerating effects [39]. Considering stationary conditions of a passive film, it seems that the adsorption mechanism should be the dominant one depending on the metal and the corroding solution at the interface. The third consideration is the penetration mechanism, which requires transfer of aggressive anions through the passive layer to the metal–oxide interface due possibly to the high electric field and a high defect concentration and disorder in the passive film [40].

Other mechanisms cannot be neglected under certain conditions in the initiation of pits such as the dissolution of inclusions or the presence of impurities, a discontinuity in the film leading to pores, or a difference in the phases of the microstructure. It has been shown that, irrespective of any compositional changes induced in preexisting air-formed or anodic films on aluminum upon exposure to halide solutions, breakdown of passivity and initiation of pits occur at preexisting flaws in the surface film [41]. The local breakdown of passivity of commercially available engineering materials, such as stainless steels, nickel, or aluminum, occurs preferentially at sites of local heterogeneities, such as inclusions, second-phase precipitates, dislocations, flaws, or sites of mechanical damage. The size, shape, distribution, and chemical or electrochemical dissolution behavior (active or inactive) of these heterogeneities, in a given environment, determine to a large extent whether pit initiation is followed by repassivation (metastable pitting) or stable pit growth. Pit initiation in an oxygenated chloride solution is generally controlled by the cathodic reaction kinetics.

The potential drop within the electrolyte for an open hemispherical pit can be estimated by $\Delta U = a i_{c,p} r / K$, where a is a geometric factor, $i_{c,p}$ is the local current density, r is radius, and K is the specific conductivity. It depends on the specific situation where ΔU is large enough to shift the potential below the flade potential, E_F, that is, in the active range of the polarization curve (Figure 5.6). Potential drops may stabilize localized corrosion for large pits but never for small pits during the initial stages. The conductivity of the electrolyte is another factor that should be taken into consideration. Evidence of potential control of pit growth is given for Al by Hunkler and Böhni [41] since they found a linear dependence of the product $i_{c,p} r$ with the voltage drop ΔU.

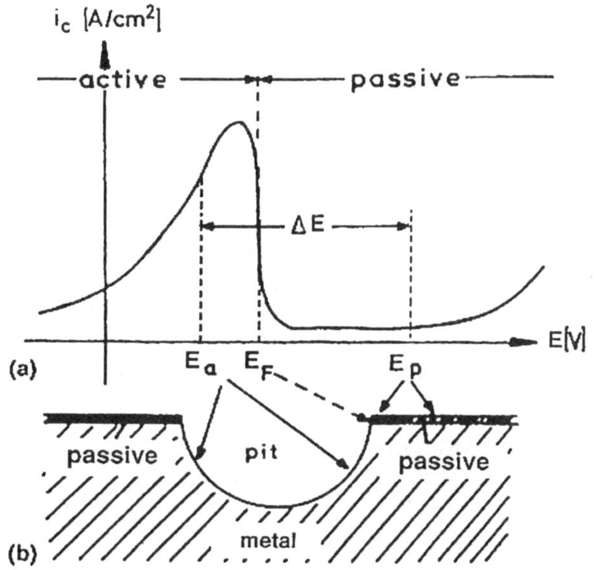

Figure 5.6 (a) Typical current density–potential curve of a passive metal. (b) Hemispherical pit with potentials E_a inside and E_p outside the pit according to the theory frequently used for small pits [28].

Recently, a new microelectrochemical technique applying microcapillaries as electrochemical cells has been developed. Only small surface areas of a few micrometers or even nanometers in diameter are exposed to the electrolyte. This leads to a strongly enhanced current resolution, down to picoamperes. Microelectrochemical techniques, combined with statistical evaluation of the experimental results, allow one to gain more insight into the mechanism of these processes [20, 42–45]. The resulting microstructure is either nanocrystalline or amorphous. It was recently shown that sputter-deposited aluminum alloys containing only a few atomic percent of metal solute, such as Cr, Ta, Nb, W, Mo, or Ti, exhibit a strong increase of E_p of 0.2–1 V. The increase in pitting resistance was explained by the reduced pit initiation tendency as well as by a more protective passive film, favoring rapid repassivation [46].

5.12.5.2. Metastable Pit Formation Stage

In this stage, propagation of the pit is not yet fully established but growth is sustained by the surrounding and covering geometry of the passive surface from which the pit began. Several authors assume that the pit precursors cannot grow until E_p is reached. This indicates that film breakdown is a necessary but not sufficient condition for pit growth to occur. It has been deduced that the lifetime of each detected metastable pit as determined by imaging was less than 6 seconds for austenitic stainless steel in situ at the open-circuit corrosion potential [47]. Electrochemical noise measurements (ENMs) of pure aluminum in the presence or absence of chloride, obtained by analyzing the corrosion potential (or current) fluctuation in time, have provided a helpful approach for studying corrosion processes. The source of the chemical noise is assumed to be metal dissolution and repassivation transients resulting from the exposure of a fresh metal surface, following the rupture of the passive film. Rozenfeld et al. [47] reported that each minimum and maximum on the charging curve corresponds to the formation of an active center and to the beginning of passivation, respectively. The metastable pits form in high numbers at potential less negative than the pitting potential but more positive than the repassivation potential and usually they form with increasing intensity close to the stable pitting potential [48].

Davis et al. [48] studied the influence of the crystallographic orientation $\langle 111 \rangle$, $\langle 100 \rangle$, and $\langle 110 \rangle$ on the metastable pitting of Al in chloride solution [37, 49]. The $\langle 111 \rangle$ surface had the highest atomic planar density and exhibited the highest number of events at any potential and the lowest pitting potential. The $\langle 100 \rangle$ surface with the next highest atomic planar density exhibited the second highest number of events and the next lowest potential. The pitting potential of 99.999% aluminum single crystals in 0.1 M NaCl increased in the order $E_{p\langle 111 \rangle} < E_{p\langle 100 \rangle} < E_{p\langle 110 \rangle}$. The XPS indicated the highest ratio of Cl^-_{oxide} : Al^{3+} on the $\langle 110 \rangle$ surface and the lowest ratio on the $\langle 111 \rangle$ surface [28].

5.12.5.3. Pit Stabilization Stage

This is the most critical stage since a large number of metastable pits die. Generally, the accumulation of aggressive anions is the most important factor in stabilizing pit growth and the Cl^- causes the formation of a concentrated $AlCl_3$ solution within active pits. At the same time, the kinetics of the repassivation of small pits in the early stage depends on the transport of the aggressive anions out from the pit region to the bulk solution. If this transport is the rate-determining step, one expects the repassivation time to increase with the depth of the pit during its lifetime. The dependences expected have been confirmed by potentiostatic pulse measurements for stainless steels [37].

Very concentrated chlorides of dissolved metal cations, very low pH, a salt film at the bottom of the pit, and the usual presence of some metal and corrosion products as a cover influence pit development or growth and play an important part in stabilizing metastable pits [50–52]. The pit cover acts as a physical barrier against the current flow and diffusion that helps to maintain a concentrated aggressive environment inside the pit, while metal dissolution occurs through perforation of this cover [37]. If the remnant of the film is strongly disrupted and a bulk solution enters the pit, diluting the pit solution, repassivation of the pit may occur. A stabilizing effect of the pit remnant depends on the passive film chemical composition, porosity, and its strength. A strong, resistant cover facilitates pit stability, while a weak or stressed cover hinders it [53].

Within Al pits a salt layer exists during pit nucleation or at the early stage of pit growth. For aluminum, there are two possible pit salts: aluminum chloride, $AlCl_3$, and aluminum oxychlorides, $Al(OH)_2Cl$ and $Al(OH)Cl_2$. Depending on the kind of salt, a difference of pH of the solution can be expected. In the case of the appearance of $AlCl_3$, the pH should be as low as 1, because the pH of saturated $AlCl_3$ is -0.3 [54]. According to Hagyar and Williams [54], the following sequence of reactions occur in a pit: ionization of the bare surface of Al occurs rapidly and Al^{3+} undergoes hydrolysis rapidly; also, aluminum hydroxide reacts with chloride producing $Al(OH)Cl^+$ and then with water producing acidic conditions [56]:

$$Al(OH)Cl^+ + H_2O \rightarrow Al(OH)_2Cl + H^+$$

Pit initiation of passive aluminum alloy necessitates, for example, oxygenated chloride solution. The oxidizing agent facilitates the cathodic reaction, which controls the corrosion kinetics and the sufficient chloride ion concentration to cause the breakthrough of the passive film and the penetration of the ion. The corrosion potential exhibits time

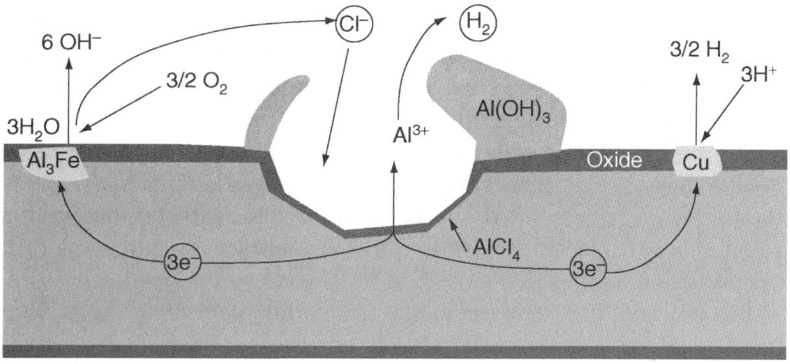

Figure 5.7 A proposed scheme for the propagation of the pit in aluminum alloys [9, 56].

fluctuations, corresponding to the elementary depassivation–repassivation events. Reboul and Canon [56] proposed a ten-step mechanism for the pitting initiation and propagation of aluminum in the presence of chloride ions (Figure 5.7):

1. Cl^- adsorption in microflaws of the oxide film, assisted by the high electric field (10^7 V/cm) through the barrier oxide film, resulting from the Al–air corrosion cell (emf 2.9 V).
2. Slow oxygen reduction on the cathodic area, charging the double-layer capacitance (50 $\mu F/cm^2$).
3. Dielectric breakdown of the oxide film at weak points corresponding to the microflaws.
4. Fast aluminum oxidation of bare aluminum, producing soluble chloride and oxychloride complexes at the bottom of flaws.
5. Dissolution of chloride complexes and repassivation of pits.

   (These first five steps produce $10^6/cm^2$ micropits of size 0.1–1 μm.)
6. Exceptionally, and for some different (often unexplained) reasons, a few micropits propagate. This propagation requires the stabilization of a chloride/oxychloride layer at the active bottom of pits. This layer should be renewed faster than it dissolves, which implies a large enough cathodic area, resulting from the repassivation of the surrounding competitive pits, formed during step 4.
7. Hydrolysis of soluble chlorides/oxychlorides, resulting in the acidification (to pH = 3) of the solution within pits.
8. Hydroxide dissolution inside pits and precipitation of aluminum hydroxide outside pits, resulting in the formation of cone-shaped accumulations of corrosion products at the mouths of pits.
9. Aluminum corrosion inside the pits propagates due to the aggressive hydrochloric acid.
10. Repassivation and pit death occur when I_{pit}/r_{pit} (r is the radius of the pit) decreases to 10^{-2} A/cm. The chloride/oxychloride film is dissolved and replaced by a passive oxide film. The solution within the pit reverts to the composition of the bulk solution [56].

Ito et al. [57] studied the pitting of Al in water with added Na_2HPO_4, NaCl, and chlorine. They reported that pits formed in this aggressive solution have a crystallographic and tunnel-like morphology preceded by {100} faceting dissolution. Previously, Edeleanu [58] showed that tunneling occurred in neutral NaCl along the crystallographic {100} planes. Ito et al. [57] noted that pits exhibited a crystallographic attack even at fairly high anodic potentials. They therefore assumed that the transformation from crystallographic to hemispherical pit morphology could not be attributed to the variation of potential to more noble values during corrosion. It was concluded that corrosion of Al in fresh water occur in the following stages: initiation of crystallographic pits, formation of concentrated $AlCl_3$ solution in pits, and anodic dissolution of the edges of crystallographic pits and formation of hemispheric pits [58].

Scully and Rudd [59] found that the morphology of pits on Al depends on the kind of inhibitor added to a chlorine solution. Pits formed in solutions containing citrate, tartrate, or acetate ions had crystallographic features. Pits formed in citrate-containing solutions were deep and widely spaced, while those formed in tartrate-containing solutions were flat with islands of residual oxide in the center of the pitted region. In acetate solutions, small irregular pits were formed. In phosphate solutions, pits were covered with corrosion products [60].

Pitting on titanium and aluminum at free open circuit or corrosion potentials occurs generally at high ohmic-limited current densities. Generation of large amounts of hydrogen bubbles within the pit strongly increases the mass transport rate. Therefore the fluid flow of the bulk electrolyte has little effect on pit growth under such conditions. More detailed support for ohmic-controlled pit growth on aluminum was obtained by Hunkeler and Böhni [21, 61]. For small Tafel constants, as in the case of aluminum, and sufficiently large pits ($> 10\ \mu m$), contributions from charge transfer as well as ohmic transport outside the pit may be neglected and a simple parabolic rate law can be derived, in which the preexponential factor depends directly on the electrolytic conductivity of the bulk electrolyte. Due to the generation of hydrogen bubbles during pitting of aluminum, no significant change in the composition of the electrolyte within the pit takes place, in contrast to situations in which diffusion processes control pit growth. These findings are in excellent agreement with the evaluation of long-term pit growth measurement under open-circuit conditions on aluminum in tap water of known conductivity (Figure 5.8) [21].

A pit is an actively corroding spot of free metal surface surrounded by the passive layer, which can act as a cathode [28]. Extrapolation of the dissolution rates of the free metal

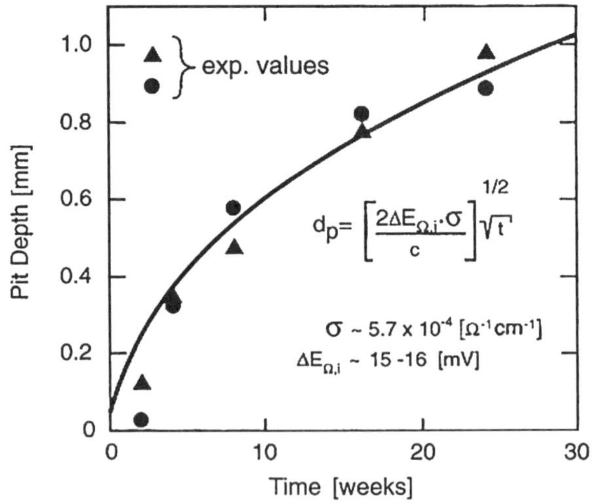

Figure 5.8 Pit growth on aluminum in tap water at open-circuit conditions [61].

corrosion to the passive range would lead to extremely high local current densities of 10^3–10^6 A/cm^2, and these large current densities would cause precipitation of a salt film within 10^{-4}–10^{-8} s. Kaesche [34] found that the current density of aluminum dissolution in pits is potential independent. At the breakdown potential, the current density in the pit is independent of the Cl^- concentrations; however, it does not drop to zero. At high Cl^- concentrations and potentials greater than the breakdown potential, the current density in the pits can reach very high values. These results suggest that, at the breakdown potential, conditions inside the pits become independent of the composition of the bulk solution. If the current density drops below 0.3 A/cm^2, the pit does not grow [62].

Galvele [62] proposed the condition for pit stabilization based on his model of the pit growth. He suggested that for each metal and alloy a critical acidification in the pit environment is required for pit stability. He developed a pit model under the assumption that metal ions hydrolyze inside the flaws or micropits already existing on the passive film and that pit growth is controlled by active dissolution in the pit environment. Hydrogen ions are produced from hydrolysis and a corresponding high concentration of Cl^- from the bulk solution. When a pit develops in a supersaturated pit solution, a salt film may form. Galvele suggested that the product of current density and depth of a one-dimensional pit, $i \times r$ (current density times pit radius), must be greater than a critical value in order for the pH at the pit surface to be sufficiently low to maintain active conditions for pit growth: $i \times r > z\text{FD}\,\Delta C/\pi$ (where z is the average charge of the metal ions, F is the Faraday number, D is the diffusion coefficient, and ΔC is the difference in metal ion concentration from the corroding pit surface to the bulk solution). The pit stability product $i \times r$ has been found close to 10^{-2} for aluminum and austenitic stainless steel [21].

The stability of the passive film is decisive for the corrosion resistance of passive metals and alloys. Fast and effective repassivation, necessary for highly corrosion resistant alloys, may only occur if highly stable films are formed during repassivation. Therefore further investigations should be focused on the initiation of localized corrosion. The stability of passive films is often reflected by the semiconductive properties of these films. Thus electrochemical impedance spectroscopy, photoelectrochemical methods, and in situ analytical techniques are valuable tools to study the chemical and electrochemical behavior of these passivating oxide films [63].

Burstein and Mattin [63] have observed anodic current transients on type 304 stainless steel and established that the product of the pit depth and current density must exceed a minimum value for maintaining a sufficiently aggressive solution at the dissolving surface so that the pit does not repassivate and grow. However, this minimum is generally not achieved in the first stages of pitting, and the pit growth requires the presence of a barrier of diffusion at the pit mouth, which is thought to be either the remnant of the passive film or a trace of the outer surface of the metal itself. The rupture of this "pit cap" leads to repassivation of the metastable pit. The higher current density inside the metastable pit leads to the greater probability of the onset of stable pitting. It was found that the current density in each pit to be independent of the potential, since the growth is diffusion controlled [64].

5.12.6. Prevention of Pitting Corrosion

5.12.6.1. Conception and Design

Selection of materials and alloys for a certain medium and usage is the prerequisite for good corrosion resistance. This includes the kinetics and quality of passivation, composition of the passive layer, and resistance of the microstructure to initiation and propagation of pitting. A careful design to prevent the stagnant electrolyte or the harmful crevices or the nonprotective coatings is the best approach to avoid pitting and crevice corrosion.

5.12.6.2. Surface Preparation, Conversion, and Coatings

Anodizing and sealing are effective methods for natural media, accompanied by appropriate painting for more aggressive ones. Phosphating before painting or using corrosion inhibitors as a pigment in paint (chromate, phosphate, molybdate, etc.) or zinc-rich primers (70–75 μm in thickness) are good options (see Chapter 14). Sacrificial coatings such as zinc (galvanization or plating) for Al are used. A thin organic coating for sacrificial zinc can multiply the useful life of zinc by more than 10. A compatible organic coating is best with accompanying cathodic protection, with special care given to cathodic corrosion of aluminum.

5.12.6.3. Avoiding Oxidizing Agents

The development of pitting can be prevented by removal of the reducible species required for a cathodic reaction. In neutral solutions, this species is usually oxygen. Thus its removal by deaeration prevents the development of pitting in aluminum even in most halide solutions because, in its absence, the cathodic reactions are not sufficient to polarize aluminum to its pitting potential. To minimize corrosion of aluminum in contact with other metals, the ratio of the exposed area of aluminum to that of the more cathodic metal should be kept as high as possible (since such a ratio reduces the current density on the aluminum).

5.12.7. Corrosion Resistance of Aluminum Cathodes

Electrolytic zinc extraction using the "electrowinning method" accounts for about 75% of zinc production worldwide. The corrosion and failure of aluminum cathodes have been a major problem of the zinc electrowinning operation. The useful life of the aluminum cathode plate is short, generally 18–24 months for a 6–6.5 mm thick plate. This case includes other applications using cathodically protected systems and integrated circuits [65]. Cathodes are made from an alloy of the set 1xxx containing at least 99.0% of aluminum. After casting, the aluminum ingots are preheated and laminated in order to get the required thickness as well as the wanted mechanical properties [65].

Aluminum cathode failure often results from intense localized pitting corrosion just above the electrolyte–air interface due to a formed thin film. Zn is electroextracted from an acidified zinc bath. An applied cathodic current density of $\sim 400\ A \cdot m^{-2}$ deposits high-purity Zn onto the Al cathode and generates oxygen evolution and H^+ at the lead anode. After plating for 16–72 hours, the Al cathode plates are removed from the cells and the deposited zinc is stripped and recovered. The zone situated between 0 and 40 mm above the interface shows the greatest corrosion damage. The maximum damage occurred at 30 mm above the electrolyte surface and extended across the width of the plate [65]. Some plates exhibited severe thinning and perforation, which often occurred near the electrical contact edge. Plates usually fail in service by fracture because of intense thinning, which cannot withstand stresses generated during the zinc stripping operation. The reduction of the surface was about 80% [65].

Corrosion under a thin film could be due to the concentration of Cl^- being more than the electrolyte because of the reduction of dissolved chlorine, the reduction of abundant oxygen, and/or the presence of a nonprotective or porous passive film. Surface conditioning, phosphating, and anodization can improve the performance of the cathode. A hard and resilient corrosion inhibited paint could also be required. In general, coatings will be severely challenged in this application simply by the harshness of the stripping process and by the presence of Cl^-, F^-, and acidic electrolyte. A corrosion inhibition tape (CIT) would have to be resistant to acid electrolyte and acid mist. A sprayed Al–5Mg alloy could also be tried since magnesium fluoride and oxide readily form high corrosion films that could improve the thin-film corrosion situation. Spraying a more noble aluminum alloy than the substrate is not excluded as an approach [20, 66, 67].

5.13. CREVICE CORROSION

5.13.1. General Considerations and Description

This type of corrosion is due to the presence of a corrosive solution quantity that is stagnant to the neighborhood of a hole, under a deposit, or any geometric shape that can form a crevice. It is called *cavernous corrosion* or *corrosion under deposit.* Oxygen differential cells could be established between the oxygenated seawater outside or at the opening of the crevice surfaces, for example, and inside crevice anodic areas (Figure 5.9). This crevice must be sufficiently large to permit the entry of the corrosive solution but narrow enough to form a stagnant state and hold a solution with required characteristics. The opening is generally on the order of 50–200 μm. The space between the two materials is less aired, has a weak surface, and contains a solution often rich in salt. Hydrolysis reactions within the crevices could produce changes in pH and chloride concentration in the crevice environment [37].

The most serious crevice corrosion of aluminum takes place in the aerospace industry and in the beverage can industry [13]. If an electrolyte is present in a crevice formed between two faying aluminum surfaces, or between an aluminum surface and a nonmetallic material, such as a gasket, localized corrosion in the form of pits or etch patches may occur. Corrosion rates were low for crevice openings greater than 254 μm. Localized corrosion occurs in the oxygen-depleted zone (anode) immediately adjacent to the oxygen-rich zone (cathode) near the mouth of the crevice. Once crevice attack has been initiated, the mechanism of

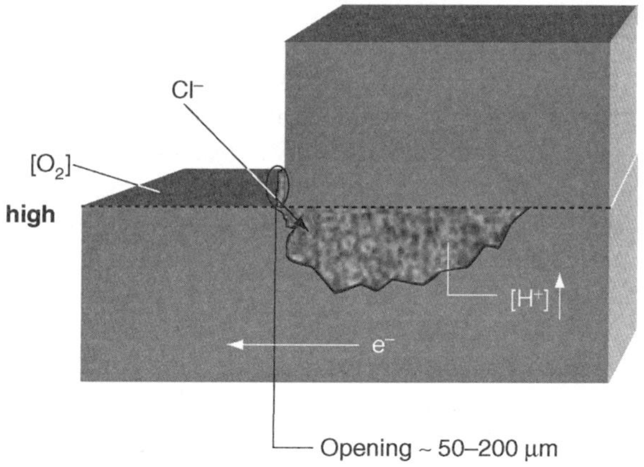

Figure 5.9 Schematic of crevice corrosion mechanism.

propagation is autocatalytic, the anodic area becomes more acidic, and the cathodic area becomes more alkaline [13].

The amount of aluminum consumed by crevice corrosion is small and is of practical importance only when the metal is of thin cross section or in cases where surface appearance is important. The expansion force of corrosion products produced in a confined space can be more serious. These corrosion products are about five times the volume as the metal from which they were produced and about twice the volume as rust on steel, and can distort even heavy sections of metal. Aluminum–copper and aluminium–zinc–magnesium–copper alloys corroded many times faster than 1100, 3xxx, or 5xxx alloys. Crevice corrosion is generally critical for atmospheric corrosion when the thickness of the sheet is less than about 1 mm and the required service life is more than about 5 years [13].

In most fresh waters, crevice corrosion of aluminum is negligible. In seawater, crevice corrosion takes the form of pitting, and the rate is low. Resistance to crevice corrosion has been found to parallel resistance to pitting corrosion in seawater and is higher for aluminium–magnesium alloys than for aluminium–magnesium–silicon alloys [68, 69].

Tarnished Appearance of the Oxide Film It is necessary first to distinguish the corrosion by the water stain, the oil stain, and the dull or lusterless appearance of the oxide. Naked aluminum or its alloys, when not protected, can lose their brilliant aspect progressively when exposed to pollutants. The outside surface becomes drab after a prolonged period. This results from a modification of the optic properties of the natural oxide layer because of its evolution under atmospheric conditions. Anodizing is a sure means to avoid the tarnished appearance [3].

Oil Stains The literature discusses the possibility of oil stains. The brilliance of aluminum surfaces can be dulled by oil or by products of oil reaction with the surface of the aluminum or with the atmospheric pollutants. The latter could be avoided by application of chemicals that prevent both oil and water stains.

Water Stains Water stains are the most common case of aluminum crevice corrosion occuring by the entrapment of moisture between the adjacent surfaces of closely packed material during transport or storage in packages of sheets or wraps of coil or foil. Bright aluminum surfaces incur staining in certain aqueous solutions (Figure 5.10). Frequently, stains result from a roughening and thickening of the aluminum oxide, which causes a change in the refraction of incident light. This is important in applications such as bright trim and reflector sheet. The appearance of water staining varies from iridescent in mild cases to white, gray, or black in more severe instances, depending on the alloy or degree of oxidation. In some cases, the stain pattern shows a series of irregular rings, like the lines on a contour map. These may indicate the outlines of a receding water pool at various stages of evaporation. Assessment of damage is by visual inspection, photographs, and measurement of reflectivity or image clarity [70].

In severe cases, the corrosion product cements the two surfaces together and makes separation difficult [68, 71]. Water stains are the result of inadequate protection from rain and condensation within the crevice when the metal surface temperature falls below the dew point. The stained areas are not more susceptible to subsequent corrosion; on the contrary, they are more resistant because they are covered with a thickened oxide film. Alloys containing magnesium (series 5000 and 6000) would be more susceptible to this phenomenon because of the presence of magnesium oxide on the surface [72].

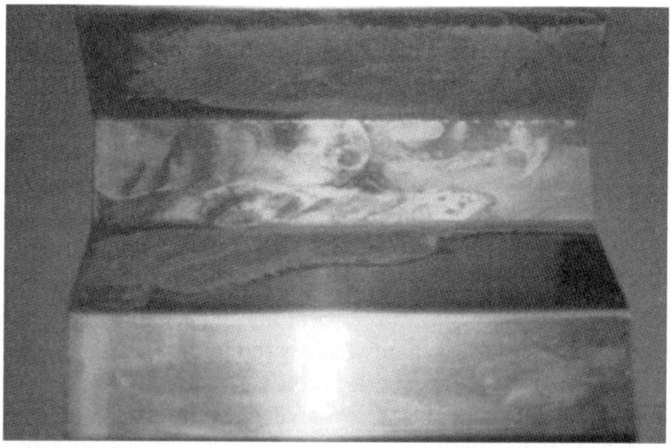

Figure 5.10 Water stain on aluminum sheet of the series of AA3105 after 2 months of exposure to a hot humid atmosphere. The dimension of every rectangle is 14.5 cm × 5.5 cm [3].

5.13.2. Poultice Corrosion

This is a special case of localized corrosion due to differential aeration, which usually takes the form of pitting. It occurs when an absorptive material such as paper, wood, asbestos, sacking, or cloth is in contact with a metallic surface that becomes wetted periodically. During the drying periods, adjacent wet and dry regions develop. Near the edges of wet zones and because of limited quantities of dissolved oxygen, differential aeration cells develop and this leads to pitting. An example is the extensive damage observed for aluminum surfaces of fuel tanks in aircraft because of the accumulation of organic materials inside the tanks due to bacterial and fungal growths in jet aviation fuel. Poultice corrosion can be prevented by avoiding contact of absorptive materials with a metallic surface, by design, or by painting [73].

Poultice corrosion is a form of crevice corrosion that occurs beneath hygroscopic attachment or insert [74]. This could be a lamination of paper, cloth, or wood to a single layer of aluminium with chemicals that are corrosive to aluminum. For example, depending on the species, freshly cut wood contains over 50% moisture and organic acids that can be quite corrosive. Wood treated against disease and insects can contain chemicals that can leech out and be corrosive. Design prevention measures would be to use laminated material that does not absorb moisture and to seal edges. Periodic cleaning and drying are also good preventative measures [75, 76]. Poultice corrosion also occurs under deposits of road debris, such as mud that is deposited on the underside of automobile fenders and at other locations. These deposits hold corrosive substances such as road salt and abraded metal particles (e.g., brake dust) in contact with the body material and retard or prevent runoff. Areas on automobiles where poultice corrosion has occurred, for example, include the hem flange within doors, wheel wells, and inside frames. These areas remain wet almost continuously with a highly corrosive liquid because of the moisture-entrapment effect of the poultice. The aggravation caused by deicing salts can be quite serious in these areas because of wet–dry cycling and accumulation that can reach saturation. The stationary electrolyte can become increasingly acidic in this particular form of crevice corrosion [12]. Electrolyte composition gradients are thought to be the most common cause of this form of poultice corrosion. Limited studies to date indicate the aluminum is much more sensitive to the effects of road debris than steel in auto body parts—"body-in-white" (structural shell/skin) [12].

A detailed analysis of poultices collected on 50 cars driven in four major North American cities has revealed the presence of large quantities of ions such as sodium, calcium, sulfate, and chloride. In these cities, sodium, calcium, and chloride deposit in poultices because of road deicing and dust-control practices. Sulfate can be attributed to acid deposition commonly associated with pollution [12]. Deformation caused by corrosion in lap joints of commercial airlines is accompanied by a bulging ("pillowing") between rivets, resulting from the increased volume of the corrosion product over the original material. Corrosion processes and the subsequent buildup of voluminous corrosion products inside the lap joints leads to pillowing, whereby the faying surfaces are separated. The buildup of voluminous corrosion products, especially for aluminum oxides and hydroxides as compared to the metal volume, also leads to an undesirable increase in stress levels near critical fastener holes. Rivets have been known to fracture because of the high tensile stresses resulting from pillowing [12].

5.13.3. Mechanisms

Siitari and Alkire [77] noticed a correlation between the pH behavior and current distribution inside the crevice corrosion of aluminum. In the first stage of crevice initiation, a sharp peak in the anodic currents coincides with a similar rise and fall of the pH. The initial rise in pH in the crevice is explained by the consumption of hydrogen ions by the oxygen reduction reaction. Immediately after formation of the crevice, the pH reached a maximum of $\sim$7 and then gradually decreased to a value of $\sim$4 at the breakdown. Hebert and Alkire [78] stated that a mechanism for initiation of crevice corrosion based on acidification only has to be excluded. Their model predicts a rapid depletion of O_2 and a rapid acidification of the crevice corrosion, followed by the gradual buildup of dissolved metal species, which trigger breakdown. They stated that there is a necessary critical concentration of Al^{3+} ions to activate the metal [78]. Alavi and Cottis [79] revealed the complex nature of aluminum crevice corrosion. They found a different pH along the crevice from mildly acidic to alkaline in the deeper part of the crevice. Connolly et al. [80] demonstrated that in crevices of AA3104 alloy and 99.99% Al, anodic and cathodic processes are separated. After the experiments, the analyzed crevice solutions show anodic sites with $pH \approx 3.6$ and cathodic sites with $pH \approx 10$. Stable pitting was observed in the acidic region. Crevice corrosion was found only in aerated chloride solution. It is concluded that pitting is a prerequisite for crevice corrosion. They proposed that the gap differential associated with the subcrevicing phenomenon leads to a localized separation of the anodic and cathodic zones. The coalescence of pits led to the crevice corrosion in the subcreviced region [81, 82].

5.13.4. Water Stains on AA3xxx

The first consignment of the sheet metal 3105, with a total quantity of 200,000 square feet, was made in the United States and received in a northern city in Canada during the month of October 1998 (approximately 1600 km). Stain formation was not noticed on the material of this first consignment. A second similar consignment of sheet metal manufactured by another American company that specializes in recycling of aluminum alloys was received in Canada in July 1999. Stains were observed on the surface of the sheet metal of the second consignment, in violation of the contract.

The transportation had been done in nonheated trucks and the products were covered only with a canvas. Thereafter shipments arrived from the United States in Canada within 3 days of shipment. Out of a packet of 45 metal sheets, 10 sheets showed definite staining. A protective oil was transparent but not evenly distributed on the surface. One month later, numerous water stains on the surface of the sheet metal were observed.

The degree of staining may be judged by the relative roughness of the stained area. If the surface is reasonably smooth, its appearance can be improved by mechanical and some chemical treatments. Water staining is removed by grinding using steel wool and oil. A chemical dip without undue etching is preferred using an aqueous solution containing 10% volume sulfuric acid and 3% per weight of chromic acid at about 10 °C [81].

5.13.4.1. Suggested Causes

The crevice corrosion of aluminum alloys requires the presence of water and a crevasse of 5–25 μm, somewhat below the current width of conventional crevices. The very small crevice is able to attract water by capillarity and because it is sufficiently narrow a differential oxygen concentration cell can form. Once a differential oxygen cell is initiated, the formation of the water stain is imminent. A medium containing salt or atmospheric pollutants can accelerate the kinetics of stain formation. The mechanism of propagation is autocatalytic reaction.

The second consignment, received in July 1999, could have experienced water accumulation in crevasses between the sheets, which did not happen with the first consignment received in October 1998. It is the difference of temperature between day and night that is a determining factor in the accumulation of water. It is necessary to underline that corrosion rates are greater in July than in October for the weather of northern Quebec since the rate of corrosion approximately doubles for every increase of 10 °C.

Alloy AA3105 of the second consignment was a green alloy, recycled and relatively new, and its trace elements such as lead or copper could have initiated galvanic cells. Also, it is possible that the even application of the protective oil was different for the two consignments.

5.13.4.2. Prevention of Water Stains

The following considerations could be used separately or jointly as it applies to every case:

1. The purer aluminum alloys are more resistant to water stains, while stains are most pronounced on those alloys having high magnesium content. Pure aluminum is less susceptible to corrosion than alloy AA3003, and alloy AA3003 is less susceptible than AA5000 [71]. The manganese present in the 3xxx series of alloys is recognized to give good corrosion resistance [68, 74, 76]. AA5000 is susceptible to the formation of water stains because of its Mg content. The magnesium in AA3105 (0.2–0.8) could form magnesium oxide, which accelerates the formation of water stains [71].

2. Water staining can be prevented by avoiding exposure to rain and condensation conditions. The metal temperature must be maintained above the dew point, either by providing a low relative humidity or preventing abrupt cooling of the metal [83]. The dew point should be considered in order to minimize the condensation of water in crevasses during transport and storage.

3. Use of embossed metal sheet guarantees the circulation of air in crevasses and prevents water stains. The metal should be protected by galvanized steel on every side of the sheet packet during transport and storage. Steel strapping is used to reinforce skids and boxes and to bind wrapped bundles. Avoid long transportation distances as far as possible. However, it is most important to avoid a lowering of temperature during the transportation of the aluminum between two places. If the difference is more than 11 °C, the transported material must be used immediately. Also, it is bad practice to introduce cold transported sheets into heated environments directly. If required, any crevices should be filled with a joint sealing compound to prevent the ingress of water; if this is not done properly it may create a narrower and more serious crevice [83, 84].

4. An oil finish could be used as a lubricant and to penetrate the metal surface. The resulting films should be oily, transparent, uniform, and nonstaining and should act as a water-displacing, nonstaining corrosion preventative, which will rapidly separate water displaced from metal surfaces after machining operations or after alkali cleaning. These ultrathin, highly protective films are effective for atmospheric corrosion protection even under conditions of high humidity. Best results are obtained when the film is sprayed above 16 °C from a distance of 30–40 cm in light even coats. Corrosion inhibitor could be used as an additive for protecting lubricated metal surfaces against chemical attack by water or other contaminants. There are several types of corrosion inhibitors. Polar compounds wet the metal surface preferentially, protecting it with a film of oil. Other compounds may absorb water by incorporating it in a water-in-oil emulsion so that only the oil touches the metal surface. Another type of corrosion inhibitor combines chemically with the metal to present a nonreactive surface [83, 84].

5. Structures are additionally interleaved with special paper with weak acidity to protect the surfaces against abrasion (most of the time from AA2024, AA6061, and AA7075); these interleaving papers contain an additive to inhibit water staining. Also sheets of aluminum are isolated from one another by an adherent plastic film, but this is less common [67].

5.14. FILIFORM CORROSION

5.14.1. General Considerations

Filiform or underfilm corrosion is selective worm-track pitting corrosion of the surface of the metal beneath a pliable film and is a special type of crevice corrosion. It appears as a blister under the paint. The underfilm filament propagation can be reflected, can be split, or can join together. Since the tracks propagate in direct lines, some of them reflect because of obstacles, such as adhesive parts of the film, to the substrate and become trapped in a very narrow place (death trap) [20, 85].

The discontinuity of an organic film permits air and water to penetrate through the coating and reach the underlying metal. This adjacent humid layer becomes saturated or rich in corrosive ions from soluble salts. Metals, in the presence of salts, pass into solution. This forms a zone called an *active head*. The dissolution of the metal decreases as the solubility of oxygen in solution increases. The metallic ions oxidize and form compounds or corrosion products. These zones are called *tails*. Morphology and directionality of filaments are determined by the material microstructure such as compositional and crystallographic factors [12].

Filiform corrosion is a special form of oxygen-cell corrosion beneath organic or metallic coatings on steel, zinc, aluminum, or magnesium. Filiform corrosion normally starts at small, sometimes microscopic, defects in the coating. Lacquers and "quick-dry" paints are most susceptible to the problem. The attack results in a fine network of random "threads" of corrosion product developed beneath the coating material with a shallow grooving of the metal surface. Such attack develops beneath semipermeable films in a high-humidity environment on such items as coated cans, office furniture, cameras, aircraft structures, and auto interiors and exteriors [12].

The head of the advancing filament (about 0.1 mm wide) becomes anodic, with a low pH and a lack of oxygen as compared with the cathodic area immediately behind the head, where oxygen is available through the semipermeable film. The water and oxygen present in the cathodic area convert the anodic products to the usual oxides of the metal [9].

Filiform corrosion takes the form of randomly distributed thread-like filaments on an aluminum surface under an organic coating and is sometimes called vermiform or worm-track corrosion. The corrosion products cause a bulge in the surface coating much like molehills in a lawn. When dry, the filaments may take on an iridescent or clear appearance because of internal light reflection. The tracks proceed from one or several points where the coating is breached. The surface film itself is not involved in the process, except in the role of providing inadequate zones of poor adhesion that form the crevices in which corrosion occurs upon exposure to moisture with restricted access of oxygen [76].

5.14.2. Aluminum Alloys and Filiform Corrosion

Filiform corrosion is a special type of crevice corrosion that can occur on an aluminum surface under a thin organic coating (typically 0.1 mm, or 4 mils thick). It is commonly observed on aluminium sheet, plate, and foil. The corrosion products are gelatinous and milky in color. The pattern of attack is characterized by the appearance of fine filaments emanating from one or more sources in semirandom directions. The filaments are fine tunnels composed of corrosion products underneath the bulged and cracked coating. Filiforms are visible at arm's length as small blemishes. Upon closer examination, they appear as fine striations shaped like tentacles or cobweb-like traces [9].

The filiform cell consists of an active head and a tail that receives oxygen and condensed water vapor through cracks and splits in the applied coating. The cell is driven by a difference of potential between the head and the tail on the order of 0.1–0.2 V. In aluminum, the head is filled with flowing flocks of opalescent alumina gel moving toward the tail. Gas bubbles may be present if the head is very acidic. In aluminum, filiform tails are whitish in appearance. The corrosion products are hydroxides and oxides of aluminum. Anodic reactions produce Al^{3+} ions, which react to form insoluble precipitates with the hydroxyl ions produced in the oxygen reduction reaction occurring predominately in the tail [9].

Aluminum is susceptible to filiform corrosion in the relative humidity range of 75–95%, with temperatures between 20 and 40 °C (70 and 105 °F). Relative humidity as low as 30% in hydrochloric acid (HCl) vapors has been reported to cause filiform corrosion. The source of initiation is usually a defect or mechanical scratch in the coating. This type of attack is rare on aluminum below about 55% relative humidity or above 95%. In natural atmospheres, it occurs most readily on aluminum at relative humidity between 85% and 95% [9]. Filiform corrosion has occurred on lacquered aluminum surfaces in aircraft exposed to marine and other high-humidity environments. Filiform attack is particularly severe in warm coastal and tropical regions that experience salt fall or in heavily polluted industrial areas. Rougher surfaces also experience a greater severity of filiform corrosion [9]. Typical filament growth rates average about 0.1 mm/day (4 mils/day). Filament width varies with increasing relative humidity from 0.3 to 3 mm (12 to 120 mils). The depth of penetration in aluminum can be as much as 15 μm (0.6 mil). Numerous coating systems used on aluminum are susceptible to filiform corrosion, including epoxy, polyurethane, alkyd, phenoxy, and vinyl coatings. Condensates containing chloride, bromide, sulfate, carbonate, and nitrate ions have stimulated filiform growth on coated aluminum alloys [3]. This type of corrosion usually penetrates only a few hundredths of a millimeter; hence the concern generally is aesthetic appearance, except in the case of thin foils where perforation can occur, or where the packaged contents can be contaminated [76].

5.14.3. Kinetics, Mechanism, and Prevention

The mechanisms of initiation and propagation of filiform corrosion in aluminum are the same as for coated iron and steel, as shown in Figure 5.11. The acidified head is a moving pool of electrolyte, but the tail is a region in which aluminum transport and reaction with hydroxyl ions take place. The final corrosion products are partially hydrated and fully expanded in the porous tail. The head and middle sections of the tail are locations for the various initial reactant ions and the intermediate products of corroding aluminum in aqueous media [9, 86].

In contrast to steel, aluminum has shown a greater tendency to form blisters in acidic media, with hydrogen gas evolved in cathodic reactions in the head region. The corrosion product in the tail is aluminum trihydroxide, $Al(OH)_3$, a whitish gelatinous precipitate. If filiform corrosion is neglected, more serious structural damage caused by other forms of corrosion may develop [9, 87].

In aircraft, filiform corrosion was observed on 2024 and 7000 series aluminum alloys coated with polyurethane and other coatings. Two-coat polyurethane paint systems experienced far fewer incidences of filiform corrosion than single-coat systems did. Filiform corrosion rarely occurred when bare aluminum was chromic acid anodized or primed with chromate or chromate-phosphate conversion coatings [9, 86].

Reducing relative humidity below 60%, especially for long-term storage, can prevent filiform corrosion. Also, the use of zinc and zinc primers on steel, chromic acid anodizing, and chromate or chromate-phosphate coatings have provided some relief from filiform corrosion. Multiple-coat systems resist penetration by mechanical abrasion and have a more uniform surface [12].

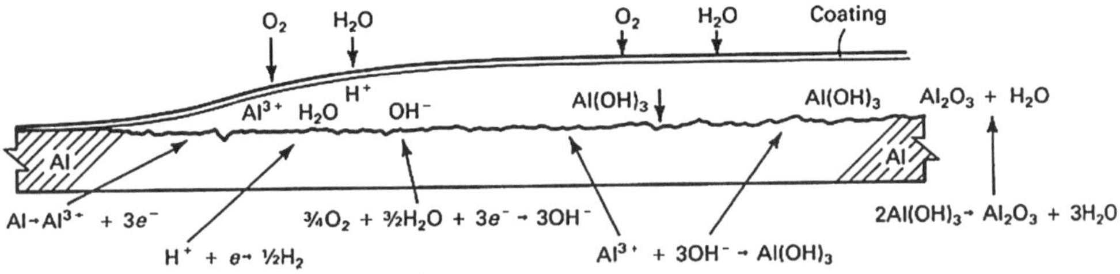

Figure 5.11 Filiform corrosion of aluminum [9, 86].

5.14.4. Filiform Occurrence

Filiform Corrosion on the Skin of an Aircraft For aluminum, an electrochemical potential at the front of the head of −0.73 V (SHE) has been reported, together with a 0.09 V difference between the front and the back of the head. Acidic pH values close to 1 at the head have been reported, with higher fluctuating values in excess of 3.5 associated with the tail (courtesy of Kingston Technical Software) [76].

Lower Wing Skin When the Boeing 747 aircraft was first placed into service, filiform corrosion was detected on the lower wing skins of one of these aircraft. This corrosion developed from intergranular corrosion around titanium fasteners that were inserted into the airframe structure [76].

Pylon Tank Filiform corrosion caused the perforation of one area of an aluminum alloy 6061-T6 pylon tank. Pitting and intergranular corrosion were also detected on the pylon tank during investigation of this problem [76].

Auto Body Present experience indicates that the potential for filiform corrosion of aluminum auto body sheet is increased with the following [76]:

- Certain constituents of the alloy, especially copper
- Mechanical surface treatment (sandblasting) of the sheet metal
- Lack of a conversion coating or an unsuitable conversion coating

Surface sandblasting of the 2008 alloy results in much more severe filiform corrosion than with the 6009 alloy. In the light of other test results, 6016 can be regarded as equivalent to 6009. All three alloys give equally good results with nonsanded surfaces, showing little susceptibility to filiform corrosion.

Test panels in T6 temper condition have a slightly better resistance to filiform corrosion than panels in the T4 temper condition. It is recommended that sheet panels are first conversion coated by phosphating and then painted with a three-coat paint system consisting of a cathodic electrocoat, a primer/surface coat, and a top coat.

Thin Foil If the aluminum foil is consumed by filiform corrosion, the product can be contaminated, lost, or dried out because of breaks in the vapor barrier. Typical coatings on aluminum foil are nitrocellulose and polyvinyl chloride (PVC), which provide a good intermediate layer for colorful printing inks. Degradation of foil-laminated paperboard can occur during its production or during its subsequent storage in a moist or humid environment. Coatings with water-reactive solvents, such as polyvinyl acetate, should not be used [9]. Aluminum is widely used for cans and other types of packaging. Aluminum foil is routinely laminated to paperboard to form a moisture or vapor barrier [9].

Degradation of the foil-laminated paperboard may occur during its production or its subsequent storage in a moist or humid environment. During the production of foil-laminated paperboard, moisture from the paperboard is released after heating in a continuous-curing oven. Heat curing dries the lacquer on the foil. Filiform corrosion can result as the heated laminate is cut into sheets and stacked on skids while the board is still releasing stored moisture [9].

As shown in Figure 5.12, the hygroscopic paperboard is a good storage area for moisture. Packages later exposed to humidity above 75% in warm areas can also experience filiform attack. Any solvents entrapped in the coating can weaken the coating, induce pores,

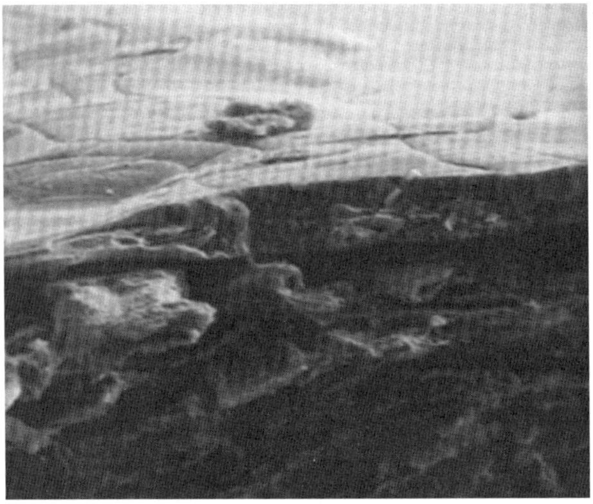

Figure 5.12 Cross section (SEM, 650×) of aluminum foil laminated on paperboard showing the expansion of the PVC coating by the corrosion products of filiform corrosion. Note the void spaces between the paperboard fibers that can entrap water [9, 86].

or provide an acidic medium for further filament propagation. Harsh curing environments can also result in the formation of flaws in the coating due to uneven shrinkage or rapid volatilization of the solvent. Rough handling can induce mechanical rips and tears [9, 86].

REFERENCES

1. S. L. Pohlman, in *ASM Handbook, Volume 13, Corrosion*, edited by J. R. Davis. ASM International, Materials Parks, OH, 1987, pp. 80–103.
2. L. L. Shreir, R. A. Jarman, and G. T. Burstein, *Corrosion—Metal/Environment Reactions*, 3rd edition. Butterworth-Heinemann, Oxford, UK, 1995, pp. 1–18.
3. B. W. Lifka, in *Corrosion Tests and Standards, Application and Interpretation*, 2nd edition, edited by R. Baboian. ASM International, Materials Park, OH, 2005, pp. 547–557.
4. ASM International Handbook Committee, *Corrosion—Understanding the Basics*. ASM International, Materials Park, OH, 2000, pp. 21–48, 100, 162, 214–215, 276, 286, 309, 513.
5. L. L. Shreir, *Corrosion*, Vol. **1**. Newnes-Butterworths, London, 1976, pp. 114–129.
6. C. P. Dillon, *Forms of Corrosion Recognition and Prevention*. International Association of Corrosion Engineers, Houston, TX, 1982.
7. R. J. Landrum, *Fundamentals of Designing for Corrosion Control*. NACE International, Houston, TX, 1992, pp. 1–24.
8. J. Kruger, in *Uhlig's Corrosion Handbook*, edited by R. W. Revie. Wiley, Hoboken, NJ, 2000, pp. 165–170.
9. E. Ghali, in *Uhlig's Corrosion Handbook*, 2nd edition, edited by R. W. Revie. Wiley, Hoboken, NJ, 2000, pp. 677–715.
10. R. B. Mears, in *Corrosion Handbook*, edited by H. H. Uhlig. Wiley, Hoboken, NJ, 1976, pp. 39–55.
11. J. Liu, H. Wu, and Z. Sun, in *Galvanic Corrosion Between Titanium Alloy and Aluminium Alloys in 3.5% NaCl*, Second International Conference on Environment Sensitive Cracking and Corrosion Damage. ESCCD, Qingdao, China, 2001, pp. 46–48.
12. P. Roberge, *Corrosion Basics: An Introduction*, 2nd edition. NACE International, Houston, TX, 2006, pp. 125–136.
13. B. W. Lifka, in *Corrosion Testing and Standards: Application and Interpretation*, edited by R. Baboian. American Society for Testing and Materials, Philadelphia, PA, 1995, pp. 447–457.
14. E. H. Hollingsworth and H. Y. Hunsicker, in *Corrosion and Corrosion Protection Handbook*, edited by P. A. Schweitzer. Marcel Dekker, New York, 1983, pp. 111–145.
15. J. Idrac, G. Mankowski, G. Thompson, P. Skeldon, Y. Kihn, and C. Blanc, *Electrochimica Acta*, **52**(27), 7626–7633 (2007).

16. B. Arsenault and E. Ghali, *Corrosion and Mechanical Properties of Coated Structural Aluminum Alloys as Function of Various Surface Preparation Techniques*, International Thermal Spray Conference and Exposition ITSC and Aéronat, Seattle, Washington, USA, May 15–17, 2006.
17. P. Delahay, M. Pourbaix, and Van Russelberghe, *Diagramme d'équilibres Potentiel–pH de quelques éléments*. CITCE, Berne, 1951.
18. C. Vargel, *Le Comportement de l'aluminium et de ses alliages*. Dunod, Paris, 1976.
19. N. Missert, J. C. Barbour, R. G. Copeland, and J. E. Mikkalson, *JOM*, **53**(7), 34–36 (2001).
20. V. S. Sastri, E. Ghali, and M. Elboujdaini, *Corrosion Prevention and Protection—Practical Solutions*. Wiley, Chichester, West Sussex, UK, 2007, pp. 331–459.
21. H. Böhni, in *Uhlig's Corrosion Handbook*, edited by R. W. Revie. Wiley, Hoboken, NJ, 2000, pp. 173–190.
22. D. O. Sprowls, in *ASM Handbook, Volume 13, Corrosion*, edited by J. R. Davis. ASM International, Materials Park, OH, 1987, pp. 231–233.
23. W. F. Bogaerts, *K. S. Agena Active Library on Corrosion*. Elsevier, Amsterdam, in conjunction with NACE, Houston, TX, 1996.
24. R. B. Mears and R. H. Brown, *Industrial and Engineering Chemistry* **33**, 1001 (1941).
25. K. G. Compton, A. Mendizza, and W. W. Bradley, *Corrosion*, **11**, 383 (1955).
26. D. O. Sprowls and R. H. Brown, *Metals Progress*, April, pp. 79–85 and May, pp. 77–83 (1962).
27. P. M. Aziz, *Corrosion* **12**, 35–46 (1956).
28. H. H. Strehblow, *Corrosion Mechanisms in Theory and Practice*. Marcel Dekker, New York, 1995, pp. 201–237.
29. B. MacDougall and M. J. Graham, in *Corrosion Mechanisms in Theory and Practice*, edited by P. Marcus and J. Oudar. Marcel Dekker, New York, 1995, pp. 143–174.
30. P. Marcus, in *Corrosion Mechanisms in Theory and Practice*, edited by P. Marcus and J. Oudar. Marcel Dekker, New York, 1995, pp. 239–263.
31. E. H. Hollingsworth and H. Y. Hunsicker, in *Metals Handbook*, 9th edition, Vol. **2**, edited by D. Benjamin. ASM International, Materials Park, OH, 1979, p. 204.
32. M. Elboujdaini, E. Ghali, R. G. Barradas, and M. Girgis, *Corrosion Science* **30**(8–9), 855–867 (1990).
33. E. H. Hollingsworth and H. Y. Hunsicker, in *Metals Handbook, Volume 13, Corrosion* 9th edition, edited by L. J. Korband and D. L. Olson. ASM International, Materials Park, OH, 1987, pp. 583–609.
34. H. Kaesche, in *Pitting Corrosion of Aluminum and Intergranular Corrosion of Aluminum Alloys*, edited by R. Staehle, B. F. Brown, J. Kruger, and A. Agrawal. NACE International, Houston, TX, 1974, pp. 516–525.
35. G. M. Hoch, *Journal of the Electrochemical Society*, **137**, 134 (1974).
36. K. P. Wong and R. C. Alkire, *Journal of the Electrochemical Society*, **137**, 3010 (1990).
37. Z. Szklarska-Smialowska, *Pitting and Crevice Corrosion*. NACE International, Houston, TX, 2005, pp. 327–329.
38. C. R. Clayton and I. Olefjord, in *Corrosion Mechanisms in Theory and Practice*, edited by J. O. P. Marcus. Marcel Dekker, New York, 1995, pp. 175–199.
39. G. Mankowski, C. Lemaitre, and H. H. Strehblow, in *Corrosion Localisée*, edited by F. Dabosi, G. Beranger, and B. Baroux. Les Editions de la Physique, Les Ulis Cedex A, France, 1994, pp. 173–239.
40. G. C. Wood, J. A. Richardson, M. F. Abd Rabbo, L. B. Mapa, and W. H. Sutton, in *Passivity of Metals*, edited by R. P. a. J. K. Frankenthal. Electrochemical Society Incorporated, Princeton, NJ, 1978.
41. F. H. Hunkler and H. Böhni, *Werkstoffe und Korrosion J*, **34**, 593 (1983).
42. L. Muller and J. R. Galvele, *Corrosion Science* **17**(3), 179–193 (1977).
43. H. Böhni, T. Suter, M. Büchler, P. Schmuki, and S. Virtanen, *Metallurgical Foundry Engineering* **23**, 139 (1997).
44. W. C. Moshier, G. D. Davis, T. L. Fritz, and G. O. Cote, *Journal of the Electrochemical Society* **137**, 1317 (1990).
45. G. S. Frankel, M. S. Russak, C. V. Jahnes, and V. A. Brusic, *Journal of the Electrochemical Society* **136**, 1243 (1989).
46. Y. Golzalez-Garcia, G. T. Burstein, S. Gonzalez, and R. M. Souto, *Electrochemistry Communications* **7**, 637–642 (2004).
47. L. Rozenfeld and I. S. Danilov, *Zashchita Metallov J*, **6** (7), 4–20 (1970).
48. B. W. Davis, P. J. Moran, and P. M. Natishan, in *Proceedings 98-17*, edited by R. G. Kelly, P. M. Natishan, G. S. Frankel, and R. C. Newman. Electrochemical Society, Pennington, NJ, 1999, pp. 215–222.
49. S. Y. Yu and P. M. Natishan, in *Proceedings 98-1*, edited by R. G. Kelly, P. Natishan, G. S. Frankel, and R. C. Newman. Electrochemical Society, Pennington, NJ, 1998, p. 256.
50. H. S. Isaacs, *Corrosion Science* **29**, 313 (1989).
51. P. C. Pistorius and G. T. Phil Burstein, *Transactions of the Royal Society of London Series A* **341**, 531 (1992).
52. L. F. Garfias-Mesias and J. M. Sykes, *Corrosion Science* **41**, 959 (1999).
53. D. A. Vermilyea, *Journal of the Electrochemical Society* **118**, 529 (1971).
54. T. Hagyar and J. Williams, *Transactions of the Faraday Society* **57**, 2288 (1961).
55. T. R. Beck, *Electrochimica Acta* **29**, 18–50 (1984).

56. H. Reboul and R. Canon, *Revue de l'aluminium* 403–426 (1984).
57. G. Ito, K. Goto, and Y. Shimizu, *International Congress on Metal Corrosion*. Australian Corrosion Association, Victoria, Australia, 1975, p. 1192.
58. C. Edeleanu, *Journal of the Institute of Metals* **89**, 90 (1960).
59. J. C. Scully and W. J. Rudd, *Corrosion* **79**, Reprint No. 160 (1979).
60. F. Hunkeler, *On the Pitting Mechanims of Aluminum with Special Emphasis on Pit Growth Kinetics*. ETH, Zürich, 1980.
61. F. Hunkeler and H. Böhni, *Corrosion* **40**, 10 (1984).
62. J. R. Galvele, *Journal of the Electrochemical Society* **123**, 123 (1976).
63. G. T. Burstein and S. P. Mattin, in *Proceedings 95-15*, edited by G. S. Frankel, R. C. Newman, and R. G. Kelly. Electrochemical Society, Pennington, NJ, 1996, p. 1.
64. R. Gundersen and K. Nisancioglu, *Corrosion* **46**, 279 (1990).
65. M. Buarzaiga, *An Investigation of the Failure Mechanisms of Aluminum Cathodes in Zinc Electrowinning Cells*. University of British Columbia, Vancouver, 1999.
66. R. M. Kain, *Corrosion* **40**(6), 313–321 (1984).
67. M. G. Fontana and N. D. Greene, *Corrosion Engineering*. McGraw-Hill, New York, 1978, pp. 7–27.
68. C. Vargel, in *Corrosion de l'aluminium*, edited by C. Vargel. Dunod, Paris, 1999.
69. C. Vargel, in *Corrosion de l'aluminium*, edited by C. Vargel. Dunod, Paris, 1999, pp. 199–217.
70. E. Ghali, *Les tâches de l'alliage 3105*. Consultants, experts et conseils, Ghali Engineering, Quebec, 2000, p. 35.
71. J. C. W. Hinchiffle, *Australian Corrosion Engineering* 7–15 (1972).
72. H. P. Godard, *NACE Basic Corrosion Course*. NACE, Houston, TX, 1970, pp. 8:1–15.
73. B. W. Lifka, in *Corrosion Engineering Handbook*, edited by P. A. Schweitzer. Marcel Dekker, New York, 1996, pp. 99–106.
74. ASM International Handbook Committee, in *Corrosion of Aluminum and Aluminum Alloy*, edited by J. R. Davis. ASM International, Materials Park, OH, 1999, pp. 63–74, 161–178.
75. R. Dietz, *Corrosion Protection Measures on an All-Aluminum Body*, Sixth Automotive Corrosion and Prevention Conference. Society of Automotive Engineers, Dearborn, Michigan, USA, 1993, pp. 355–361.
76. ASM Specialty Handbook Committee, *Corrosion of Aluminum and Aluminum Alloys*. ASM International, Materials Park, OH, 1999, p. 313.
77. D. Siitari and R. Alkire *Journal of the Electrochemical Society* **129**, 481 and 488 (1982).
78. K. Hebert and R. Alkire, *Journal of the Electrochemical Society* **130**, 1001 and 1007 (1983).
79. A. Alavi and R. Cottis, *Corrosion Science* **27**, 443 (1987).
80. B. J. Connolly, J. R. Scully, and R. S. Lillard, in *Proceedings 98-17*, edited by R. G. Kelly, P. M. Natishan, G. S. Frankel, and R. C. Newman. Electrochemical Society, Pennington, NJ, 1999, pp. 409–420.
81. Ullrich Aluminium Company Limited, New Zealand, 1998. http://ulrich-aluminium.co.nz/index.htm, A:\Ulllrich Aluminium (Rolled and Extruded Products) Handling and storing.htm.
82. Topic 14107 Water Staining Available at http://www.oclu-info.dk.
83. *Guidelines for Minimizing Water Staining of Aluminum*. Aluminum Association, 1999.
84. Topic 11140 Water Staining. Available at www.oclu-info.uk.
85. O. Lunder, T. Kr. Aune, and K. Nisancioglu, *Corrosion* **43**, 291 (1987).
86. S. C. Dexter, in *Metals Handbook, Volume 13, Corrosion*, 9th edition, edited by L. J. Korband and D. L. Olson. ASM International, Materials Park, OH, 1987, pp. 103–122.
87. N. E. Ryan, *Society for the Advancement of Material and Process Engineering* 1 (May), 638–648 (1979).

Chapter 6

Metallurgically and Microbiologically Influenced Corrosion of Aluminum and Its Alloys

Overview

Dealloying, intergranular, and exfoliation types of corrosion are discussed. The addition of alloying elements to improve the mechanical properties or resistance to general uniform corrosion can lead to increased susceptibility to localized corrosion. If the β-additions (Cu or Mg) are in considerable quantities to aluminum beyond the solubility limit, two-phase structure results and this can create galvanic corrosion cells. Aluminum-based alloys can be susceptible to intergranular attack due to direct corrosion of the precipitate, which is more active than the matrix, or to corrosion of a denuded zone adjacent to a nobler phase.

Exfoliation corrosion is a form of severe intergranular corrosion that occurs at the boundaries of grains elongated in the rolling direction. The corrosion begins as lateral intergranular corrosion on subsurface grain boundaries parallel to the metal surface, but entrapped corrosion products create internal stresses that tend to lift off the overlying metal. This spalling off of the metal creates fresh metal surfaces for continued corrosion. Some authorities regard it as a form of stress-corrosion cracking. Exfoliation corrosion may occur on material that has a marked fibrous structure caused by rolling or extrusion.

The influence of different types of welding on corrosion forms is discussed. The resistance to corrosion of weldments of aluminum alloys is determined by the basic alloy, the filler alloy, and the welding process. The formed galvanic corrosion cells are due to potential differences of the different phases of the microstructure. Incomplete removal of fluxes may also cause corrosion. Electron beam welds of aluminum alloy 2219 offer much higher strength compared to gas tungsten arc welds of the same alloy and the reasons for this are explored. Fusion solid welds have superior mechanical properties to fusion welds owing to the lower heat input and greater microstructural homogeneity of the three regions: the nugget, the thermomechanically affected zone, and the heat affected zone.

Light metals like aluminum undergo microbiologically influenced corrosion (MIC). MIC occurs everywhere in the biosphere and even in oxygen-free media, where microorganisms are present. Aluminum MIC in space, tropical atmosphere, polluted fresh water, polluted seawater, industrial seawater, and kerosene are discussed. MIC mechanisms of acceleration and inhibition are given. Microorganisms could produce polymer or slime, which could have an accelerating effect on localized corrosion and sometimes a beneficial inhibiting effect. They could produce organic acids, oxidize sulfur to SO_4^{2-}, oxidize nitrate to NO_2, and reduce sulfate to H_2S or NO_3^- to NH_3. They could break down passive films and/or cause hydrogen embrittlement.

A. METALLURGICALLY INFLUENCED CORROSION (METIC)

6.1. FUNDAMENTALS OF METIC

Since corrosion is mostly an electrochemical phenomenon, it might be expected that alloys composed of one homogeneous phase or of two or more phases, all of which have very similar electrochemical (galvanic) potentials, would be more resistant to corrosion than alloys composed of two or more phases with widely different potentials. This expectation is generally correct. Thus pure aluminum or single-phase alloys of aluminum and copper, aluminum and magnesium, or aluminum and silicon are all relatively resistant to corrosion. The solid solution of Al–Cu has a new electrochemical identity and its potential in an aggressive solution should be intermediate between that of aluminum and that of copper. More noble or positive potentials for the solid solution have more copper or more silicon.

The addition of alloying elements to improve the mechanical properties or resistance to general or uniform corrosion can cause increased susceptibility to localized corrosion processes, such as pitting or intergranular corrosion. If the β-additions (Cu or Mg) to aluminum are in considerable quantities beyond the solubility limit, two-phase structure results. Al–Cu alloys heat-treated and quenched to retain the copper in solid solution are much more resistant to corrosion than are similar alloys treated so that the copper precipitates out of solution as a constituent, $CuAl_2$, which differs in solution potential from the matrix solid solution and may cause intergranular corrosion. Al–Mn alloys (such as 3003) are highly resistant to corrosion because the manganese constituent that is present as a separate phase has a potential very similar to that of the matrix [1].

In metallographic examinations, the specimen surface is polished to a mirror finish and then exposed to chemical etchants. The chemical solution preferentially attacks particular constituents of the alloy, and thus the microstructure of the alloy is revealed. The microstructure is made up of small islands of beta (β) phase ($CuAl_2$ or Mg_2Al_3) distributed throughout a continuous matrix of alpha (α) phase [2].

Other second-phase particles can be undesirable. Undesirable precipitates such as oxides and sulfides precipitate in the metal from dissolved oxygen and sulfur in the metal-producing process. This results in a distribution of inclusions (small particles of oxide, sulfide, etc.) throughout the alloy [2].

When these inclusions are exposed at the metal surface to a corrosive environment, they can affect corrosion behavior. If the inclusion is active, that is, less corrosion resistant than the matrix, then the inclusion dissolves, leaving a hole or pit in the metal surface. If only portions of the inclusion are active, then the exposed portions are attacked, leaving the other portions intact. If the inclusion is noble (more corrosion resistant than the matrix), then accelerated attack of the matrix adjacent to the noble inclusion can be observed. In other cases where the inclusion is inert to attack, accelerated corrosion adjacent to the inclusion can still occur because of a crevice generated between the inclusion and the matrix [2].

For many aluminum-based alloys, metallurgical factors have relatively little effect on resistance to corrosion. Alloys such as 1100, 3003, 5052, 6053, and 6061 are relatively insensitive in this respect. However, higher strength alloys are produced by the addition of magnesium and silicon (6xxx series); and by the addition of copper plus silicon, or copper plus magnesium and silicon (2xxx series). The highest strength alloys are produced by the addition of zinc plus magnesium, or magnesium and copper. The additions of these elements change the electrochemical potential of the alloy, which affects corrosion resistance even when the elements are in solid solution. Zinc and magnesium tend to shift the potential markedly in the anodic direction, while silicon has a minor anodic effect. Copper additions cause marked cathodic shifts. This results in local anodic and cathodic sites in the metal that affect the type and rate of corrosion [1].

6.1.1. Influence of Metallurgical and Mechanical Treatments

Metallurgical and mechanical treatments often are interactive with regard to producing chemical microstructural features in aluminum alloys, such as dislocations and precipitates, and the microstructural morphology (grain size and shape). However, these two treatments are examined separately [1].

Effect of Metallurgical Treatments Variations in thermal treatments such as solution heat treatment, quenching, and precipitation heat treatment (aging) can have marked effects on the local chemistry and hence the local corrosion resistance of high-strength, heat-treatable aluminum alloys. Ideally, all the alloying elements should be fully dissolved and the quench cooling rate should be rapid enough to keep them in solid solution. The first objective usually is achieved, except when alloying elements exceed the solid solubility limit (e.g., alloy 2219); but a sufficiently rapid quench often is not obtained, either because of the physical cooling limitations, or the need for slower quenching to reduce residual stresses and distortion. Generally, practices that result in a nonuniform microstructure will lower corrosion resistance, especially if the microstructural effect is localized [1].

Precipitation treatment (aging) is conducted primarily to increase strength. Some precipitation treatments go beyond the maximum strength condition (T6 temper) to markedly improve resistance to intergranular corrosion (IGC), exfoliation, and stress-corrosion cracking (SCC) through the formation of randomly distributed, incoherent precipitates (T7 tempers). This diminishes the adverse effect of highly localized precipitation at grain boundaries resulting from slow quenching, underaging, or aging to peak strengths [1].

Effect of Mechanical Treatments Mechanical working influences the grain morphology and the distribution of alloy constituent particles. Both of these factors can affect the type and rate of localized corrosion. Mechanical deformation could generate active slip steps, which can weaken the protective film on metals and alloys and lead to localized corrosion [1]. In many types of exposure, cold work does not appreciably affect the resistance to corrosion of a wide variety of aluminum-based alloys. In solutions of nonoxidizing acids, however, cold work stimulates corrosion to some extent and also indirectly stimulates corrosion of aluminum alloys containing over about 5% magnesium. With the latter alloys, severe cold work increases the tendency for a magnesium–aluminum constituent to precipitate from solid solution. Upon exposure to certain media, selective attack of this constituent then occurs [1].

6.2. TYPES OF METALLURGICALLY INFLUENCED CORROSION

6.2.1. Dealloying (Dealuminification)

Dealloying is a corrosion process in which one or more elements are selectively dissolved, leaving behind a porous residue of the remaining element(s). For example, in the silver–gold system, silver can be almost 100% removed in various acid electrolytes, leaving behind porous gold. This bicontinuous metal–void structure is highly brittle in nature and has been linked to stress-corrosion cracking in many alloy systems. The seasonal cracking of brass is perhaps the best recognized. It is now known that dealloying can occur in nearly any system in which a large difference in equilibrium potential exists between the alloying components and the fraction of the less noble constituent(s) is significantly high [3].

Recent investigations have shown the importance of the dealloying of S-phase (Al_2CuMg) particles on the corrosion of aluminum aircraft alloys, specifically aluminum alloy 2024-T3. In 2024-T3, the S-phase particles represent approximately 60% of the particle population. These particles are on the order of 1 μm in diameter, with a separation on the order of 5 μm representing an area surface fraction of 3%. The selective removal of aluminum and magnesium from Al_2CuMg particles leaves behind a porous copper particle that becomes the preferential site for oxygen reduction [3]. There are three models to explain this phenomenon of selective dissolution:

1. ***Volume Diffusion Model for Selective Dissolution.*** In this mechanism, the less noble element in the alloy undergoes selective dissolution. The dissolution process is maintained beyond the first few monolayers by volume diffusion of both elements in the solid phase. The inherent problem with this mechanism is that, at room temperature, the rate of transport of the less noble element to the surface is not sufficient to support the dealloying current densities greater than 10 mA/cm^2 observed experimentally [3].

2. ***Surface Diffusion/Structural Rearrangement Model for Selective Dissolution.*** This model proposes that the less noble element is preferentially dissolved. The remaining more noble element is now in a highly disordered state and begins to reorder by surface diffusion and nucleation of islands of almost pure noble metal. The coalescence of these islands continues to expose fresh alloy surface, where further dissolution will occur, leading to the formation of tunnels and pits [3].

3. ***Percolation Model for Selective Dissolution.*** In many alloy systems, a sharp critical composition of the less noble element exists, below which dealloying does not occur. This model extends the surface diffusion model to include the importance of the atomic placement of atoms in the randomly packed alloy. The model predicts that, as a minimum requirement, a continuous connected cluster of the less noble atoms must exist in order for the selective dissolution process to be maintained for more than just a few monolayers of the alloy. This percolating cluster of atoms provides a continuous active pathway for the corrosion process as well as a pathway for the electrolyte to penetrate the solid [3].

6.2.2. Intergranular Corrosion

Intergranular corrosion is the selective attack of the grain boundary zone, with no appreciable attack of the grain body or matrix. Electrochemical cells are formed between second-phase microconstituents and the depleted aluminum solid solution from which these microconstituents formed [1]. Aluminum-based alloys can be susceptible to intergranular attack. The likelihood and severity of attack depend on the composition and structure of the alloy and the corrosiveness of the environment [2].

Intergranular corrosion of aluminum alloys can be caused by direct dissolution of a precipitate that is less corrosion resistant (more active) than the matrix or by corrosion of a denuded zone adjacent to a noble phase. Although the aluminum alloys are more resistant to intergranular corrosion in the solution-treated condition, avoiding precipitates is not a practical means of avoiding intergranular corrosion in these systems. The precipitates are important to the strengthening of the alloys and are necessary for their performance. Whether or not the alloy will be subject to intergranular corrosion in a particular environment is an important part of the alloy selection process. Figure 6.1 shows the selective attack of the grain boundary zone in cast, unrecrystallized, and recrystallized wrought aluminum microstructures. Since intergranular corrosion is involved in stress-corrosion cracking and exfoliation of aluminum alloys, it is often presumed to be more deleterious than pitting or uniform corrosion [1].

The microconstituents have a different corrosion potential than the adjacent depleted solid solution. In some alloys, such as the aluminum–magnesium and aluminum–zinc–magnesium–copper families, the precipitates Mg_2Al_3, $MgZn_2$, and $Al_x Zn_x Mg$ are anodic to the adjacent solid solution. In other alloys, such as aluminum–copper, the precipitates ($CuAl_2$ and Al_xCu_xMg) are cathodic to the depleted solid solution. In either case, selective attack of the grain boundary region occurs [1].

The degree of intergranular susceptibility is controlled by fabrication practices that can affect the quantity, size, and distribution of second-phase intermetallic precipitates. Resistance to intergranular corrosion is obtained by heat treatments that cause precipitation to be more general throughout the grain structure, or by restricting the amount of alloying elements that cause the problem. Alloys that do not form second-phase microconstituents at grain boundaries or form phases having similar corrosion potentials to the matrix (e.g., $MnAl_6$) are not susceptible to intergranular corrosion. Examples of alloys of this type are 1100, 3003, and 3004 [1].

Intergranular (intercrystalline) corrosion (IGC) also occurs randomly, over the entire surface, but corrosion is limited to the immediate grain boundary region and often is not apparent visually. This localization results from compositional differences between precipitates on the boundaries, the solute-depleted grain margins, and the higher (normal) solute grain interior. IGC penetrates more quickly than does pitting corrosion, but it too reaches a self-limiting depth. This is due to limiting transport of oxygen and the corroding agent down the narrow corrosion path. When the limiting depth has been reached,

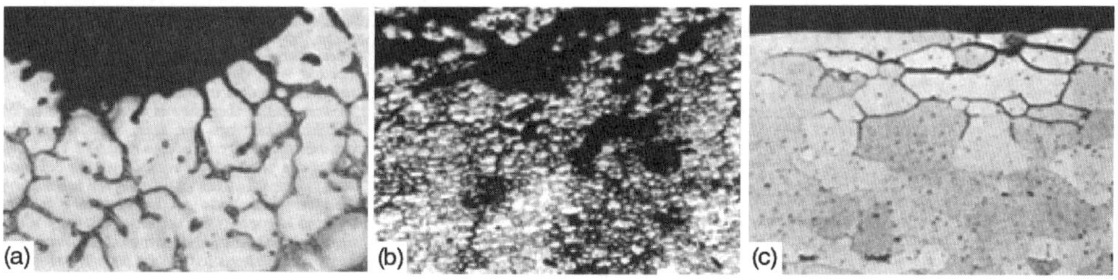

Figure 6.1 Various types of intergranular corrosion: (a) interdendritic morphology in cast microstructure, (b) interfragmentary morphology in unrecrystallized wrought microstructure, and (c) intergranular morphology in recrystallized wrought structure (500×) [4].

intergranular attack spreads laterally. If IGC results in splitting or exfoliation, the corrosion will not be self-limiting. IGC has much sharper tips than pitting corrosion; hence it is a more drastic stress riser and has a more damaging contribution to corrosion fatigue [1].

The Anodic Path The location of the anodic path varies with the different alloy systems. In 2xxx series alloys, the location of the anodic path is a narrow band on either side of the grain boundary that is depleted of copper. As an example, in the 2024 alloy, $CuAl_2$ precipitates are nobler than the matrix and act as cathodes, accelerating the corrosion of a depleted zone adjacent to the grain boundary. A similar phenomenon is observed for alloy 7075 [1]. In the 5xxx series alloys, it is the anodic constituent Mg_2Al_3 that forms a continuous path along the grain boundary. In copper-free 7xxx series alloys, the anodic path is generally along the zinc-bearing and magnesium-bearing constituents on the grain boundary. In the copper-bearing 7xxx alloys, the anodic path appears to be the copper-depleted band along the grain boundaries. The 6xxx series alloys generally resist this type of corrosion, although slight intergranular attack has been observed in aggressive environments [2].

Slow and Rapid Quenching When the precipitate particles are in a very finely divided state, the alloys are relatively resistant to corrosion. It is in this quenched and room-temperature-aged condition that they are generally used and, as such, are susceptible only to pitting corrosion with no selective attack at grain boundaries. If the alloys are quenched more slowly from the heat-treating temperature, that is, quenched in boiling water instead of cold water, they become susceptible to selective grain boundary attack (intergranular corrosion).

Pure Aluminum In the less pure aluminum 1xxx series and in aluminum alloys, the presence of second phases is an important factor. These phases are present as insoluble intermetallic compounds produced primarily from iron, silicon, and other impurities and, to a lesser extent, precipitates of compounds produced primarily from soluble alloying elements. Most of the phases are cathodic to aluminum, but a few are anodic. In either case, they produce galvanic cells because of the potential difference between them and the aluminum matrix [1].

Binary Alloys Ramgopal and Frankel [5] measured the dissolution kinetics of aluminium binary alloys (Al–Zn, Al–Cu, and Al–Mg) in an artificial crevice and a repassivation potential. All experiments were carried out in 0.5 M NaCl solution. The addition of Cu from 0.2% to 3.9% and Zn from 0.2% to 5.6% increases and decreases the overpotential of the anodic reaction, respectively. The addition of Mg does not affect the dissolution process. The overpotential was found to be dependent on the exchange current density and the Tafel slope. However, the crevice solution was not analyzed. If compared to aluminum, the repassivation potentials of Al–Cu and Al–Zn were more positive and more negative, respectively, and exhibited a very smooth dependence on the concentration of alloying additions [6].

Al–Cu Alloys In aluminum–copper–magnesium alloys (2xxx series), thermal treatments that cause selective grain boundary precipitation lead to intergranular corrosion susceptibility. In this respect, the Al–Cu alloys of the Duralumin type have been well studied. Such alloys contain about 4% copper (alloys 2017-T and 2024-T). This amount of copper is soluble in solid aluminum at elevated temperatures (above 480 °C) but is not entirely soluble at room temperature. After fabrication, such alloys are commonly heat treated at about 490 °C in order to dissolve the copper in the aluminum. They are then immediately quenched in cold water to retain the copper in solution. During aging at room temperature, the hardness and strength of the alloys increase, approaching maximum values after about 4 days. It is generally assumed that this age hardening is caused by the precipitation of a $CuAl_2$ constituent from the Al–Cu solid solution [1].

Consequently, the depleted zone corrodes, giving an intergranular form of corrosion. Somewhat similar results occur if the rapidly quenched Al–Cu alloy is heated (artificially aged) to a somewhat elevated temperature (above 120 °C) for a critical period of time. This heating also causes the alloy to become susceptible to intergranular corrosion. However, if the heating is carried out for a sufficiently extended period of time, the susceptibility to intergranular corrosion again disappears, probably because substantially all the copper has precipitated out of solid solution and thus the zones adjacent to the grain boundaries are no more depleted in copper than are the other areas in the grain boundaries [1].

Al–Mg Alloys Aluminum–magnesium alloys (5xxx series) containing less than 3% magnesium are resistant to intergranular corrosion. Aluminum–magnesium alloys that contain more than 3% Mg (e.g., 5083) may become susceptible to intergranular corrosion because of preferential attack of Mg_2Al_3 (anodic constituent). Intergranular corrosion does not occur when these alloys are correctly fabricated and used at ambient temperatures. These alloys can become susceptible to intergranular corrosion, however, after prolonged exposure to temperatures above 27 °C (sensitization). Susceptibility increases with magnesium content, time, temperature, and amount of cold work [1].

Al–Mg–Si Alloys Aluminum–magnesium–silicon wrought alloys (6xxx series) usually show some susceptibility to intergranular corrosion. With a balanced magnesium–silicon composition that results in the formation of Mg_2Si constituent, intergranular attack is minor, and less than that observed with aluminum–copper (2xxx) and aluminum–zinc–magnesium–copper (7xxx series) alloys. When the 6xxx alloy contains an excessive amount of silicon (more than that needed to form Mg_2Si), intergranular corrosion increases because of the strong cathodic nature of the insoluble silicon [1].

The microstructure and corrosion behavior of 6061 and 6013 sheet material were investigated in the naturally aged and peak-aged heat treatment conditions. Transmission electron microscopy did not reveal strengthening phases in the naturally aged sheet. In the peak-aged temper, β' precipitates were observed in alloy 6061, whereas both β' and Q' phases were present in 6013-T6 sheet. Marked grain boundary precipitation was not found. Corrosion potentials of the alloys 6061 and 6013 shifted to more active values with increasing aging. For the copper-containing 6013 sheet, the potential difference between the tempers T4 and T6, was more pronounced. When immersed in an aqueous chloride-peroxide solution, alloy 6061 suffered predominantly intergranular corrosion and pitting in the tempers T4 and T6, respectively. On the contrary, 6013 sheet was sensitive to pitting in the naturally aged condition, and intergranular corrosion was the prevailing attack in the peak-aged material. Both alloys 6061 and 6013 were resistant to stress-corrosion cracking in the tempers T4 and T6 [7].

Al–Mg–Zn Alloys In aluminum–magnesium–zinc alloys such as 7030, the compound $MgZn_2$ is attacked. Intergranular corrosion in aluminum–zinc–magnesium–copper (7xxx series) alloys can be affected by thermal treatments. Heat treatment, sometimes in combination with strain hardening, is used to provide good resistance to intergranular corrosion [1].

6.2.2.1. Pitting Potential (E_p) of Different Aluminum Alloys

In solution-treated alloys of aluminum with copper, the E_p was found to increase with an increasing Cu content [8] (from −0.5 without Cu to −0.32 V for 5% Cu/SHE). The highest E_p was limited by the solubility of Cu in Al. In an Al-4Cu alloy, Ep depends on heat treatment and showed a significant drop to more negative values when the alloy came close to peak hardness [9]. Obviously, this was associated with changes in the microstructure of the alloy during aging. In the over aged condition, the $CuAl_2$ phase is in equilibrium with the matrix Al–0.25Cu. The pitting potential of an overaged alloy is equal to that of the matrix. For intermediate levels of aging, the matrix-surrounded $CuAl_2$ particles are depleted of Cu, and these zones are locally attacked. When the grain boundaries are impoverished in Cu, the alloy also becomes susceptible to intergranular corrosion.

The behavior of the 2024 alloy is influenced by the presence of precipitates enriched with copper. For low potentials, these particles can dissolve and copper becomes deposited around them. When chloride ions are added, pits form in the zones enriched with copper because of the aggressiveness of both chloride and sulfate toward copper. So the 2024 alloy does not have the same behavior as pure aluminum: its behavior is related to that of copper. For higher potentials, the 2024 alloy has the same behavior as pure aluminum: there are no copper deposits and pits form in the matrix because of the aggressiveness of the chloride ions only. However, the 2024 alloy is more susceptible to chloride ions than pure aluminum because the passive film that develops on the alloy is more heterogeneous. The behavior of the 6065 alloy, which contains a lower quantity of copper, is similar to that of pure aluminum. Therefore the 6056 alloy appears to be more suitable than the 2024 alloy in conditions where pits can appear [10].

The addition of magnesium (0.95%, 2.7%, and 4.5%), manganese (0.8%), or silicon (0.83%) to aluminum does not significantly affect the pitting potential of the alloy when it is submersed in synthetic seawater. Holtan and Sigurdsson [11] studied the influence of small amounts of chromium (0–0.47%), manganese (0–0.94%), or antimony (0–0.4%) on the pitting of aluminum alloyed with 3–5% magnesium in 3% NaCl. The effect of these conditions was negligible. Molybdenum implantation was found to increase the E_p of Al and several aluminum alloys. The presence of tin in pure Al decreases E_p in a NaCl solution. Zinc added to Al at a concentration higher than 1% shifts pitting potential in chloride solution to more negative values [12].

Zn decreases the protection potential E_r as well. A decrease of the resistance to pitting of Al–Zn alloy in comparison to Al [13] is explained by the enhancement of the dissolution kinetics within the pits. Sato and Newman [13] assume that the activation is caused by interference of single-activator atoms with the connectivity of a surface oxide monolayer on the Al surface, which catalyzes the dissolution of periphery Al atom [6]. McCafferty [14] showed that implanted silicon, molybdenum, tantalum, niobium, zirconium, or chromium all increased the pitting potential of aluminum in a chloride solution [6].

6.2.2.2. Modeling of Intergranular Corrosion

Ruan et al. [15, 16] proposed a brick wall model to describe the relationship between the microstructure and the IGC growth rate of AA2024-T3. The short transverse direction was the only one calculated for IGC growth while that in the longitudinal and transverse directions is not considered. Also, when the corrosion path met an intersection, it was assumed to be a four-way intersection in the simulation, even though by nature of the aluminum alloy, three-way intersections are more common.

A modified generalized brick wall model is suggested by Zhao et al. [17] to describe intergranular corrosion in equiaxed AA7178-T6 and wing skin AA7178-T6 aluminum alloys. The intergranular corrosion rate is highly related to grain size and shape. High-strength aluminum alloys are often elongated and anisotropic, with the fastest nominal IGC growth rate in the longitudinal direction (L) or long transverse direction (T) and the slowest in the short transverse direction (S). The corrosion growth kinetics for the three directions is considered. A three-way intersection model and its use are proposed to simulate the corrosion kinetics for each direction. Intersections are divided into two three-way types ("$\dagger$" and "$\perp$") and, in simulation, a random mechanism to decide whether a given intersection is upward ("$\dagger$") or downward ("$\perp$") was used. If the corrosion path turns upward and arrives at the surface or the corrosion path turns downward and reaches the bottom, it is assumed that it is terminated in both cases. However, if it does not turn upward or downward, it continues to propagate along the horizontal grain boundary, which assumes that an intersection is skipped. With a proper combination of model parameters, the generalized IGC model provides a good fit to experimental data developed by the foil penetration technique [17].

From the study by Huang and Frankel [18], the grain sizes in all directions have distributions that are skewed to the right, so that gamma distributions are appropriate for modeling the grain size distributions for all three directions. The method of moments to estimate the parameters of these gamma distributions was used. In the experiment, we obtain the sample means, M, and sample standard deviations, S, of the grain sizes for each of the three directions. Method of moment estimates of the gamma distribution parameters α and β can then be calculated from the following two equations:

$$\alpha\beta = M \quad \text{and} \quad \alpha\beta^2 = S^2$$

The brick wall model was then used [15, 16] to describe the influence of the grain size on the minimum IGC path length. Let p_{up} represent the probability that a corrosion path turns upward at a "$\perp$" type intersection and let p_{sup} represent the probability that a corrosion path skips a "$\perp$" type intersection. Similarly, let p_{down} denote the probability that a corrosion path turns downward at a "$\dagger$" type intersection and let p_{sdown} denote the probability that a corrosion path skips a "$\dagger$" type intersection. Then we have [17]

$$p_{\text{up}} + p_{\text{sup}} = 1 \quad \text{and} \quad p_{\text{down}} + p_{\text{sdown}} = 1$$

In addition, let p_{split} represent the probability that a corrosion path splits into two branches. Suppose there are m initial corrosion paths at the surface of the aluminum alloy. Let $u \geq 0$ be the number of additional paths resulting from splitting, and let $v \geq 0$ be the number of branches terminated at the top surface. Then there are a total of $m + u - v$ corrosion paths traveling through the alloy from the top surface to the bottom. Let $W_{\min,D}$ denote the minimum IGC path length for thickness D. Then [17].

$$W_{\min,D} = \min W_{iD}, \quad I = 1, \ldots, m+u-v$$

Since the fastest IGC rate is the matter of concern, the minimum IGC length for each of the three directions (L, T, and S) was examined in this simulation and the programming language R was used to simulate the minimum IGC length in each of these three directions. For simplicity, it is assumed that the probability of skipping an intersection is the same for both types ("†" and "⊥") intersections; that is, $p_{sup} = p_{sdown} = p_{skip}$. Let $bj, j = 1, \ldots, k$, be the thickness of the jth layer, which is generated from a gamma distribution with parameters obtained from $\alpha\beta = M$ and $\alpha\beta^2 = S^2$ [17].

From the studies by Ruan and Wolfe [15, 16], grain sizes of equiaxed AA7178-T6 and wing skin AA7178-T6 are given [17]. Gamma distributions to model grain sizes for the three directions in both alloys were used. The simulation results suggest that p_{up} does not have an obvious effect on the ratio when both p_{skip} and p_{split} are small. On the other hand, p_{skip} has a clear effect on the ratio when p_{split} is small. This can be explained from the simulation algorithm. It is assumed that after the vertical step, R_a and R_b were compared to decide if the first intersection in the horizontal direction is "†" or "⊥". Note that R_b can be represented by rG, a random value generated from the appropriate estimated gamma distribution, while $R_a = rU(1) \cdot rG$, where $rU(1)$ is a random number generated from the uniform (0, 1) distribution. Thus $R_b > R_a$ in most cases means that the first intersection is usually "†". Thus p_{up} will not have much effect on the ratio when p_{skip} is small, since the corrosion path will not often have an opportunity to meet a "⊥" intersection and turn upward [17]. (See Chapter 8 for intergranular stress-corrosion cracking (SCC).

6.2.3. Exfoliation

Exfoliation corrosion (layer corrosion or lamellar corrosion) is a form of severe intergranular corrosion (IGC) that occurs at the boundaries of grains elongated in the rolling direction. The corrosion begins as lateral intergranular corrosion on subsurface grain boundaries parallel to the metal surface, but the entrapped corrosion product that forms has a greater volume than the volume of the parent metal, and the increased volume forces the layers apart, causing strips of metal to exfoliate (delaminate). Exfoliation is characterized by leafing, or alternate layers of thin, relatively uncorroded metal and thicker layers of corrosion product of lager volume than the original metal. The layers of corrosion products cause the metal to swell. This spalling off of the metal creates fresh metal surfaces for continued corrosion. Consequently, exfoliation corrosion does not become self-limiting. In an extreme case, an ~1.3 mm (0.050 in.) thick sheet was observed to swell to a thickness of 25 mm (1 in.) [1].

Exfoliation has also been observed along striations of insoluble constituents that have strung out in parallel planes in the direction of working. Exfoliation occurs predominantly in relatively thin products with highly cold-worked, elongated grain structures. The intensity of exfoliation increases in slightly acidic environments and when the aluminum is coupled to a cathodic dissimilar metal. Some evidence also shows that pitting can develop exfoliation. Exfoliation usually proceeds inwards laterally from a sheared edge, rather than inward from a rolled or extruded surface. In mild cases, it takes the form of blisters that resemble volcanoes, with corrosion products swelling up in the center. In this case, pits occur first and proceed inward until the susceptible layer is encountered. The attack then changes to lateral penetration with generation of less dense corrosion products that cause the blisters to develop [1].

Since exfoliation corrosion may occur on material that has a marked fibrous structure caused by rolling or extrusion, some authorities regard it as a form of stress corrosion, the stress being either inherent in the metal or produced through the pressure of the larger volume of the corrosion product. It is rare, occurring mainly in copper-bearing aluminum alloys, but can occur in a number of environments, including some regarded as only mildly corrosive. Suitable adjustments of aging treatments and copper content may largely overcome the effect in the higher-strength Al–Cu type alloys [1].

There are two ASTM standards that cover exfoliation: the EXCO test for high-strength 2xxx and 7xxx series (G34) and the ASSET test (Assessment of Exfoliation Corrosion Susceptibility) of the 5xxx series aluminum alloys (G66). Highly cold-worked tempers of certain 3xxx and 5xxx alloys can incur a less aggressive form of exfoliation that proceeds in a transgranular mode, following selective precipitation along slip bands [1].

6.2.3.1. Initiation and Propagation of Exfoliation

In aggressive environments, the volume of the nonsoluble corrosion product formed is approximately three times that of the aluminum from which it forms; this results in wedging stresses that lift the surface grains. The wedging effect of corrosion products can be a simple consequence of IGC and can be essential to continue propagation by a stress-corrosion mechanism. Wrought products of aluminum alloys in certain age-hardened heat treatment conditions are subject to corrosion by exfoliation [19].

The effect of quench rate on corrosion has been addressed. The sensitivity to exfoliation corrosion of AA7449 in relation to the intergranular and stress-corrosion cracking sensitivity has been addressed in a program of controlled quenches followed by thermal treatments. It has been demonstrated that the quench rate has a strong effect on intergranular corrosion and exfoliation corrosion sensitivity and to a lesser extent on stress-corrosion cracking. In the first moments of the EXCO test, the initiation of corrosion follows the same trends as those revealed by the ASTM G110 test concerning intergranular corrosion in $NaCl{\cdot}H_2O_2$ solution. Intergranular initiation has been observed for the slow quench rate ($\sim$5 °C/s) and pitting initiation for samples quenched between 50 and 500 °C/s. On the contrary, the final EXCO corrosion quotations do not seem to correlate with the intergranular resistance but rather with SCC resistance [19].

In peak-aged temper, exfoliation susceptibility decreases by increasing the quench rate, whereas high quench rate tends to delay SCC failure. EXCO performance in the peak-aged temper is better for a product that has undergone a faster quench. Whatever the quench rate, SCC failures were observed before 40 days at 50% of the yield strength. IGC resistance in the peak-aged temper is better for a product that has undergone a faster quench. IGC sensitivity is not a prerequisite for exfoliation corrosion. The first moments of EXCO kinetics are correlated with IGC sensitivity: worse EXCO results correspond to a product that is sensitive to IGC (7449-T6 slow quench, 7449-T7X intermediate quench rate). Final EXCO corrosion quotations are correlated with SCC resistance [19].

Apart from the blister size and their quantity, few differences on the propagation mode, whatever the EXCO rating, are observed: grain boundary dissolution, followed by the attack of adjacent grains and growing of oxides. Two sorts of initiation exfoliation corrosion can be considered depending on the IGC sensitivity of the alloy: intergranular corrosion for IGC-sensitive materials or exfoliation corrosion for IGC-insensitive materials. Propagation proceeds by the induced wedging effect and thus a SCC-like phenomenon at grain boundaries, and then formation of blisters [19].

6.2.3.2. The Aluminum Alloy Series and Exfoliation

Metallographic examination, visual rating, and weight loss measurements after exposure to corrosive environments (solutions and sprays) at ambient and elevated temperatures can be used to test for exfoliation corrosion susceptibility. The commercial-purity aluminum (1xxx) and aluminum–manganese (3xxx) alloys are quite resistant to exfoliation corrosion in all tempers. Exfoliation has been encountered in some highly cold-worked aluminum–magnesium (5xxx) materials especially in seawater media. However, cold-worked 5xxx alloys containing magnesium in excess of the solid solubility limit (above 3% magnesium) can become susceptible to exfoliation and SCC when heated for long times at temperatures of about 80–175 °C [1].

In the heat-treatable aluminum–copper–magnesium (2xxx) and aluminum–zinc–magnesium–copper (7xxx) alloys, exfoliation corrosion has usually been confined to relatively thin sections of highly worked products with an elongated grain structure. In 2124-T351 plate, for example, 13 mm plate was quite susceptible in laboratory and atmospheric tests, while 50 mm and 100 mm plate, with less directional microstructures, did not exfoliate. In extrusions, the surface is often quite resistant to exfoliation because of the recrystallized grain structure. Subsurface grains are unrecrystallized, elongated, and vulnerable to exfoliation [1].

In aluminum–zinc–magnesium alloys containing copper, such as 7075, resistance to exfoliation can be improved markedly by averaging designated by the temper designations of T7xxx for wrought products. While a 5–10% loss in strength occurs, improved resistance to exfoliation is provided. In copper-free or low-copper 7xxx alloys, exfoliation corrosion can be controlled by over aging or by recrystallizing heat treatments and can also be controlled to some extent by changes in alloying elements. In aluminum–copper–magnesium (2xxx) alloys, artificial aging to the T6 or T8 condition provides improved resistance [1].

6.2.3.3. Exfoliation of the High-Strength 7000 Series with Different Zn Percentage

It is generally recognized that susceptibility to exfoliation corrosion of copper-rich 7000 alloys results from sensitivity to intergranular corrosion, due to a more anodic potential of the grain boundarys precipitates η as compared to the matrix. Overaging beyond maximum hardness (i.e., to T7 temper) reduces the susceptibility. The incorporation of solute (Zn, Cu) in the hardening precipitate and in the grain boundary precipitate reduces the difference in potential between the precipitate grain boundaries and the matrix. It seems that PA766, which has a higher Zn and lower Cu content, has a smaller susceptibility to exfoliation corrosion than the two other alloys for a T6 temper [20].

Alloy development in the high-strength 7000 series alloys aims at increasing particularly the yield strength. However, it is known that for a given alloy the yield strength has a negative correlation with corrosion resistance, and particularly exfoliating corrosion, as measured, for instance, by the EXCO test. A usual way to reduce the corrosion susceptibility is to overage the material, to the level of a 15% decrease of strength. Exfoliation corrosion susceptibility of each material was assessed using the standard EXCO test (ASTM G34) on a 50 mm × 100 mm surface of the plate [20].

The microstructural evolution has been investigated in three alloys of the 7000 series possessing increasing zinc contents by combining small-angle X-ray scattering, differential scanning calorimetry, and transmission electron microscopy, in order to gain understanding on the evolution of the compromise between yield strength and corrosive resistance. The three chosen materials show qualitatively identical precipitation sequences; however, the precipitated volume fraction is shown to increase in parallel to the Zn content. Moreover, the precipitate size evolution is faster in the high Zn content alloy.

Exfoliation corrosion sensitivity and structural properties depend directly on microstructure. The microstructure evolution has been investigated, during heat treatment, for the three alloy compositions. The results elucidate the following influence of solute [20]:

- The composition of the hardening precipitate seems to depend on the initial matrix composition.
- The increase of hardness with higher Zn content is related to a higher precipitate volume fraction.
- Higher Cu content seems to limit the precipitate coarsening during the aging treatment.
- Higher Zn content and lower Cu content do not result in a degradation of EXCO rating.

6.2.3.4. Modeling of Exfoliation Corrosion

In aircraft materials, exfoliation corrosion is most common in the heat-treatable Al–Zn–Mg–Cu (7000 series), Al–Cu–Mg (2000 series), and Al–Mg alloys, but it has also been observed in Al–Mg–Si alloys. Generally, exfoliation occurs when there is a combination of three factors: a highly directional microstructure, a preferential anodic path, and a specific type of corrosive environment. The current costly "find-it–fix-it" approach to corrosion maintenance requires that even the smallest of corrosion damage be removed by grind-out. The grinding of corroded material can be carried out until the allowable limit is reached, at which stage the skin may require repair or replacement at a significant cost to the operator. A better and appropriate corrosion management is to anticipate, plan, and manage corrosion. To implement this approach, analytical models need to be developed to evaluate the impact that exfoliation has on structural integrity. Corrosion evaluation of different forms and modeling studies on the remaining fatigue life of aircraft wing skins containing natural exfoliation corrosion are then carried out [21].

Specimens were taken from 7075-T6511 upper wing panels containing natural exfoliation. The maximum depth of the exfoliation damage was determined by an ultrasonic nondestructive inspection (NDI). Fatigue tests were carried out under fully reversed constant amplitude loading ($R = -1.0$), and fractographic analyses were performed to examine the cracking mechanisms in the exfoliation region [21]. The tests were carried out in both dry air with relative humidity (RH) below 20% and saturated air with 90% RH.

Static tests carried out on naturally preexfoliated specimens showed that natural exfoliation may not have a detrimental effect on the residual strength in the elastic and near plastic regimes, such as the compressive yield and bearing strength; however, it may have an effect on the strength in the large plastic region, such as the compression stress at 4% compressive deformation. Concerning fatigue, some studies indicated that natural prior exfoliation reduced the fatigue life of aircraft structures by 40–60% under constant amplitude loading or low–high block loading. Another test showed that fatigue crack growth rate was enhanced by prior exfoliation from service [21].

This work was planned to address the effects of natural exfoliation on the fatigue properties of aircraft material and structures, since several fatigue models had already been developed for estimating the fatigue life of artificially exfoliated specimens. Fatigue tests on naturally exfoliated 7075-T6511 specimens are briefly reported. The results from the tests are then used for examining an existing fatigue model developed from artificial exfoliation tests. Finally, a practical fatigue model for estimating the remaining fatigue life of the naturally exfoliated specimen is examined [21].

The materials for the test specimens were cut from various 7075-T6511 upper wing skin panels, which were manufactured for the aircraft. These panels, which became corroded in storage and therefore were never used in service, contained various levels of exfoliation ranging from barely visible to extensive. Extensive damage characterization was performed by Bellinger et al. [22] on sections taken from the skins, and progressive polishing showed that the exfoliation formed from IGC that originated at corrosion pits. Ultrasonic (UT) and nondestructive inspection (NDI) techniques were used to examine the wing skins from the back (noncorroded) side [21].

Among all specimens, 13 out of 27 failed from the noncorroded corner at the end of a fillet, even though these specimens had exfoliation damage located in their central sections. The cracks nucleated from discontinuities such as intrinsic particles and manufacturing marks like indents and scratches. In other words, when the exfoliation damage is not very severe (8% in this case), the discontinuities present at the end of a fillet, where the stress concentration K_t was 1.23, could override the exfoliation damage and become the primary crack origin.

The other 14 specimens (52%) failed from exfoliation damage ranging from 152 to 838 μm (0.006–0.033 in.) in depth. Eight of these specimens cracked from corrosion pits at or near the corner at the end of a fillet, and six specimens cracked from the exfoliation damage located in the center of the specimen. On most specimens, the cracking origins were not at the locations where the maximum exfoliation damages were determined on that specimen by ultrasonic NDI. Three cracking mechanisms in the exfoliation region were identified: (1) from pits on the bottom of the exfoliated surface (three specimens), (2) from a tip of IGC (two specimens), and (3) from a particle or grain denuded by IGC/exfoliation (two specimens). Examples of (1) and (2) are shown in Figure 6.2 [21].

In summary, exfoliation above a "critical" level could significantly decrease the fatigue life of the naturally exfoliated specimens. Within a certain level (below the critical level), the stress concentration site at the fillet ($K_t = 1.23$) overrode the exfoliation damage and became the primary cracking origin. Above the critical level, the cracks primarily nucleated from corrosion pits and IGC-related features in the exfoliation region. However, to determine this critical level, more tests with different levels of exfoliation are needed. On the other hand, a reliable analytical model would be helpful to save some testing [21].

Based on the test findings, a simplified fatigue model was developed to estimate the remaining fatigue life of the corroded exfoliated specimens, where the cracks nucleated from the exfoliation damage. In this model, the exfoliation damage was assumed to be a surface crack with a depth that was presumably available from nondestructive inspection or grind-out database. The model is based on a fracture mechanics approach, which assumed that the total fatigue life is the crack growth life, including short/small crack growth life. The fatigue analysis tool and material model were verified with noncorroded test results, and then used as baselines for developing a model for exfoliation fatigue. The comparison indicates that the simplified model gave a good estimation for the remaining fatigue life of the naturally exfoliated specimens [21].

In reality, the majority of exfoliation on the upper wing skins was found around the fastener holes. The crack nucleated from either corrosion pits or IGC in the alloy 7178-T651 of the upper wing skin. Although the fatigue models were mainly used for smooth specimens

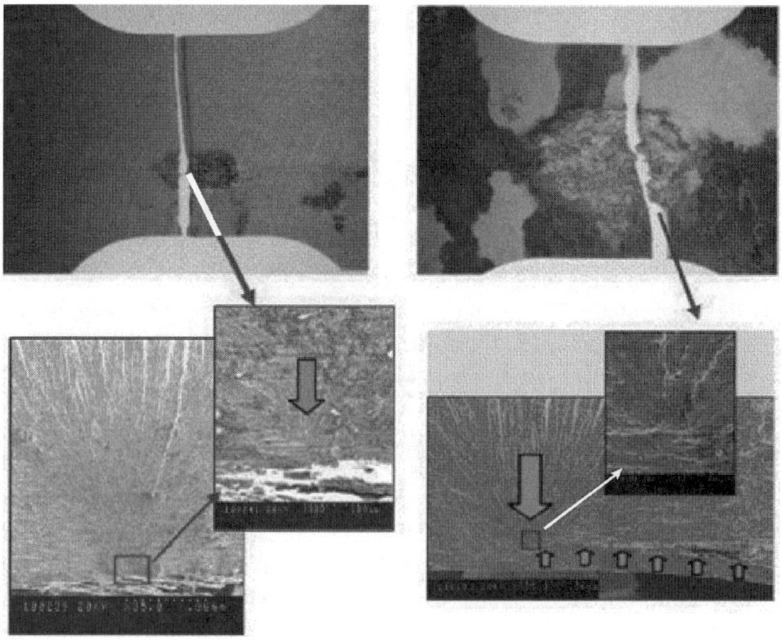

(a) from corrosion pits on bottom exfoliated surface

(b) from a tip of IGC

Figure 6.2 Fatigue cracking mechanisms from exfoliation damage started from (a) a pit or (b) intergranular corrosion [21].

with prior exfoliation, the cracking mechanisms found on the smooth specimens are very similar to the findings on exfoliated fastener specimens. For the exfoliated fastener holes, the corrosion pits were located on either the countersink or bore of the hole, or the faying surface, and IGC was found to extend into either the countersink or bore of the hole. In these scenarios, the complex stress distribution needs to be determined for the fastener hole, interference fit, and exfoliation surrounding the hole, and then included in the fatigue model [21].

6.2.3.5. Exfoliation Corrosion of Welded AA7004 Floor Plate

AA7004 has been chosen for 20 locomotives and 50 passenger train cars as a support plate for the floor. A lead plate was introduced into the structure of the floor to insulate it from the noise of the moving structures and was occasionally in contact with the aluminum supporting plate. This Al–Zn–Mg alloy has an excellent mechanical resistance and good welding properties but is susceptible to exfoliation and stress-corrosion cracking under these conditions of service. After 2 years of service in a northern country, the corrosion resistance of the supporting aluminum floor plate was exhibiting inferior performance and exfoliation corrosion was evident [23].

Some plates, having a T form in certain zones, were welded and placed in different zones of the structure. These T form AA7004 plates in the lavatory were the ones that showed severe corrosion attack very early, as compared to those plates in the kitchen or those very close to the entry stairs. Because of welding, the thermally affected zone is very susceptible to galvanic corrosion. The localized corrosion in the form of crevices in certain regions led to almost complete perforation. The corrosion products included components of

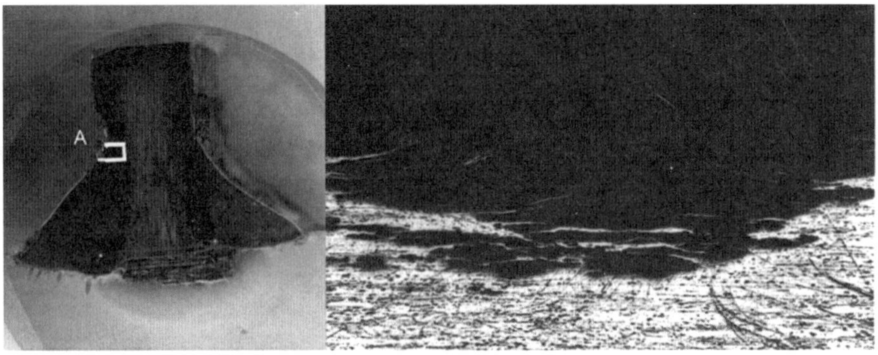

Figure 6.3 Typical exfoliation of the AA7004 plate at the extremities of the superior part of the T welded structure in the toilet supporting floor of a passenger train car (on the left) and evidence of the disintegrated layers in the zone (A) on the right (50×) [23].

the aluminum plate: lead was an important quantity but much less than other heavy metals and mercury. Different parts of the weld were polished and chemically attacked to examine the microstructure. The corrosion attack was very evident and very close to the welded regions. The superior part of the weld is shown in Figure 6.3 and one can observe at the extreme down end of this piece the successive layers of corroded metal as evidence of typical exfoliation. The magnified micrograph of this figure shows the disintegrated metal layers and the effect of severe corrosion [23].

The plate shows different forms of corrosion: galvanic corrosion is deduced, because of the presence of severe corrosion beside the welded sections, and exfoliation as a dominating type of corrosion. Certain evidence of stress-corrosion cracking has also been identified. Nonuniform general corrosion was monitored by the thinning of the plate.

Close to the weld nugget and heat-affected zone, exfoliation corrosion was found on all sides of the welding nugget due to the localized galvanic cell between the welded regions themselves and that of the matrix. In this region, the precipitated Al–Mg–Si particles were anodic with respect to that of the matrix. Also, the car was not well isolated and there was important infiltration of different atmospheric pollutants and some rainwater beside the chloride ions on the floor coming from the distributed calcium chloride salts on the icy roads during winter in northern regions. The presence of mineral wool and other similar water-absorbing materials retained the corrosive solution in contact with the supporting aluminum alloy plate [24].

Certain metallic alloy coatings of AA7004 have been tried. The alloys Al–Ni (Bondarc), Al–Zn, and Al–Mg were considered as possible surface coatings of the Al plate 7004 for corrosion prevention as sacrificial anodes. Only the magnesium coating was efficient after a severe corrosion test in a solution of 3% NaCl and 1% NaOCl at pH 12 during 2 hours of immersion. However, the Al–Ni and Al–Zn alloys gave partial protection. Sacrificial metallic coatings of the plate also need nonconducting inorganic materials or organic paint as a cover for corrosion control because of the aggressive salt corrosion media.

The welded zones showed evident galvanic corrosion because of different microstructures that lead to exfoliation and SCC and these should be coated. Also, the presence of lead compounds in the corrosion product indicated that lead sheet was the subject of corrosion attack due to local galvanic cells in the presence of chlorides. Mercury was also identified in traces and it is known that Hg (ppm) is very aggressive in the pitting of aluminum alloys and should be avoided completely. Contact of an aluminum surface with lead ions and other similar heavy metal ions should be avoided because of metal deposition on the aluminum surface and creation of galvanic cells [25].

A complete system of corrosion prevention is suggested for the aluminum supporting plate containing a phosphate conversion coating or anodized coating, followed by a primer and intermediate and finishing organic coating layers. The lead sheet that created galvanic cells because of its contact with the aluminum plate and that was acting as a cathode should be isolated completely from the AA7004; the aluminum plate (anode) should also be coated as explained previously [25].

6.3. JOINING AND WELDING

6.3.1. Corrosion Resistance of Brazed, Soldered, and Bonded Joints

6.3.1.1. Corrosion of Brazed Joints

Brazing is a high-temperature joining process usually above 450 °C. It is a widely used manufacturing technique because of the adaptability to dissimilar metals and the strength levels meeting or exceeding the base materials. Brazed joints in aluminum alloys have good resistance to corrosion. Corrosion resistance of aluminum alloys generally is unimpaired by brazing if a fluxless brazing process is used. However, brazing of aluminum generally requires the use of fluxes.

Brazing performed in air or other oxygen-containing atmospheres requires the use of a chemical flux to promote wetting and flow of the filler metal. The flux contains chlorides and/or fluorides; therefore they must be completely removed after joining. Aluminum alloys best suited for brazing are also among those most resistant to corrosion. Observed excessive corrosion is usually caused by fluxes that are not removed completely, or that are removed by a treatment that, together with the fluxes, may cause corrosion. The presence of moisture in the flux can lead to interdendritic attack on the filler metal at joint faces and to intergranular attack of the base metal [1, 26, 27].

Filler metals of the aluminum–silicon type have high corrosion resistance comparable to the base metals usually brazed. Filler metals containing zinc or copper are less corrosion resistant but are usually suitable, except for service in severe environments. The potential of joints brazed with filler metal $BAlSi_3$ (contains ~3.5% Cu and ~10% Si) depends on the cooling rate after brazing. For slow cooling, the joint has a potential of −0.82 V. If the cooling is rapid enough to retain a certain amount of copper in solid solution, the potential is about −0.73 V [26, p. 421].

Galvanic corrosion may be observed since the brazed joint consists of a bond between the dissimilar base metal and a filler metal. Exposure to salt water or some other electrolyte can result in attack on the more anodic alloy. This condition is aggravated if the anodic part is relatively small compared with the other piece; the anodic alloy should be the larger of the two brazed components [26].

The 1xxx, 3xxx, and low magnesium content 5xxx aluminum alloy series are the most successfully brazed. The commonly brazed heat-treatable wrought alloys are the 6xxx series [26].

Filler metal alloys used in flux brazing usually contain between 7% and 12% silicon-balanced aluminum. Filler metal alloys used in the fluxless brazing method use a higher percentage of silicon (>9%) and have varying additions of magnesium to enhance oxide film modification to promote wetting [26].

A thorough water rinse followed by a chemical treatment is the most effective means of complete flux removal. Immersing the part in an overflowing bath of boiling water just after the filler metal has solidified removes much of the flux. Any of several acid solutions can remove flux that remains after washing [26].

6.3.1.2. Corrosion of Soldered Joints

Soldering is a widely used low-temperature joining technique carried out generally at temperatures below 450 °C. Generally, it is applied in air without protective atmosphere. There are three categories of soldered joints: low–temperature lead-tin, intermediate temperature zinc–cadmium or zinc–tin, and high-temperature zinc or zinc–aluminum. The tenacious, refractory oxide film of aluminum requires active fluxes. All soldered aluminum joints have a lower resistance to corrosion than joints that are welded or brazed. Prior electroplating of aluminum improves the corrosion resistance of low-temperature soldered joints. Plating of copper, iron, or nickel prevents the formation of a high-potential interface between the solder and the aluminum [26].

Corrosion resistance of soldered joints in aluminum alloys depends on alloy composition, flux composition, joint design, protective coating, and environment. The base alloy and temper have a small effect on corrosion resistance. Unprotected low-temperature soldered joints can provide excellent service in dry atmospheres but they may fail in a short time in humid or marine atmospheres. Soldered joints have a satisfactory resistance to corrosion for applications in milder environments, but not for those in more aggressive ones. Environment is less critical for unprotected zinc-soldered joints but they still require protection in the most corrosive environments [26].

In the presence of an electrolyte, electrochemical corrosion can occur because of galvanic cells created between the base metal, the solder phases, and the diffusion layer formed at the aluminum–solder interface. When the galvanic cells are established, the material with the highest negative electrode protects the remainder of the assembly and therefore corrodes preferentially [26].

In a low-temperature soldered joint, the interface corrodes preferentially to protect the aluminum and the solder. The cross section and total amount of interfacial layer are relatively small compared to the rest of the assembly; therefore this area can corrode rapidly. In zinc-soldered joints, the solder is the most anodic part of the joint, so it corrodes preferentially to protect the interfacial layer and the aluminum. Because there is a greater volume of solder than interfacial layer, the corrosion resistance of the zinc-soldered joint is better than the low-temperature soldered joint. Flux composition can affect corrosion resistance if residues are not completely removed. Flux containing chlorides can cause severe corrosion when trapped into the assembly. Chloride-free flux generally causes little or no corrosion [26, 27].

Most of the corrosion associated with the soldered joints is due to the use of improper cleaning methods to remove the fluxes. Aluminum has a good corrosion resistance due to the formation of the oxide layer. If the flux is not completely removed, it may continue to react with the oxide layer, exposing the base metal to the atmosphere and the flux and this results in severe localized corrosion over time.

6.3.1.3. Corrosion Resistance of Adhesive Bonded Joints

Although adhesive bonding eliminates many of the corrosion problems associated with welding, brazing, and soldering, environmental susceptibility of bonded structures is of concern. Stable oxide preparation is an essential part of the bond foundation [26].

When exposed to a wet environment, water molecules will migrate and be preferentially adsorbed onto the interface region. This is because joint substrates, such as metal or metal oxides, have very high surface energies and water permeates through all organic adhesives. In a typical joint such as epoxy–aluminum, water generally enters a joint system by diffusion through the epoxy rather than by passage along the interface. Hydration of the metal oxide layer at the interface can degrade the strength of adhesive joints. The resulting metal hydrates become gelatinous, and they act as a weak boundary layer because they exhibit very weak bonding to their base metals [26, 27].

Phosphoric acid anodization (PAA) produces an oxide surface that outperforms some adhesive bonded joints. Performance of the PAA surface is attributed to the oxide morphology, which contains a thicker hexagonal cell structure with longer whisker-like protrusions. This provides a polymer–oxide film interface similar to the fiber-reinforced structure with more effective mechanical interlocking. A primer solution can be applied to the aluminum surface prior to bonding with the adhesive to improve wet strength [26].

6.3.2. Welding Fundamentals

Primary welding methods used to weld aluminum are gas-shielded arc welding processes such as GTAW and GMAW. Other welding methods commonly used include oxyfuel gas welding processes, high energy density welding processes, electron beam and laser beam welding, and resistance and friction welding processes, such as friction stir welding [26]. The following are the most common welding processes and metallurgical zone abbreviations encountered in aluminum welding:

GTAW	Gas tungsten arc welding
GMAW	Gas metal arc welding
FSW	Friction stir welding
EBW	Electron beam welding
LBW	Laser beam welding
SMAW	Shielded metal arc welding
OFW	Oxyfuel gas welding
RW	Resistance welding
SW	Stud welding
BM	Base material (unaffected metal away from the weld)
FZ	Fusion zone
HAZ	Heat-affected zone
TMAZ	Thermomechanically affected zone (specific to FSW)
DXZ	Dynamically recrystallized zone or weld nugget (specific to FSW)
ac	Alternating current
dc	Direct current

Aluminum and aluminum alloys can be joined by many joining methods. If filler alloy is needed, the following factors should be considered for its selection: ease of welding and absence of cracking, tensile or shear strength of the weld, weld ductility, service temperature, corrosion resistance, and color match between the weld and base metal after anodizing [26].

6.3.2.1. Gas-Shielded Arc and Arc Welding Processes

Gas Tungsten Arc Welding (GTAW) Process GTAW has been used to weld thicknesses from 0.25 to 150 mm in multipass welding. It is a relatively slow welding process from 3 to 5 mm/s, easily maneuverable for circular welds and variable shapes. The GTAW process permits good penetration control. The ac-GTAW process provides an arc cleaning action to remove surface oxide during the positive electrode half-cycle and a penetrating arc when the electrode is operated at negative polarity. It is used to weld sections up to 6.3 mm thick. The ac-GTAW process is particularly useful for aluminum pipe welding. An integral backing is suitable for structural and electrical bus applications; however, for fluid flow of water, gas, oil, and chemicals, crevice corrosion can result at the backing interface within the pipe [26].

The dc-GTAW process provides a deep narrow penetration profile suitable for welding a square groove joint of thick aluminum sections up to 13 mm. The narrow penetration profile allows welding of heat-treatable alloys with a lower heat input than ac-GTAW. There is no arc cleaning, therefore the surface oxide must be minimized to ensure a sound weld. A chemical etching followed by a mechanical scraping of the joint surface is necessary [26].

Gas Metal Arc Welding (GMAW) Process This process is the major high-speed production process for arc welding of aluminum. It uses positive electrode dc power, which gives it a continuous cleaning action and concentrates the arc to produce rapid melting. Thicknesses of 1 mm and higher can be welded by GMAW. The process is best used in lap, fillet, or groove joints with integral or temporary backing. Heat input needs to be constant for uniform penetration. Argon gas shielding is most often used. Due to a higher welding speed (up to *42 mm/s* for mechanized welding) and less energy applied to the aluminum parts to be welded than with GTAW, welds done with GMAW are less susceptible to corrosion [26, 27].

6.3.2.2. Shielded Metal Arc Welding

Shielded metal arc welding (SMAW) with flux coated rods has been replaced largely by the GMAW process. SMAW can be effective on 9.5 mm and thicker aluminum where high heat inputs are necessary. The flux is highly corrosive and should be removed after welding to avoid corrosion [26].

6.3.2.3. Stud Arc Welding Process (SW2)

Stud arc welding and capacitor discharge stud welding processes can be used to join aluminum alloys. For stud arc welding, the 1xxx, 3xxx, and 5xxx series alloys are considered best, while the 2xxx and 7xxx series alloys are considered poor. For capacitor discharge stud welding, the 1xxx, 3xxx, 5xxx, and 6xxx series alloys are considered excellent while the 2xxx and 7xxx series alloys are passable under certain conditions. Studs are made of aluminum–magnesium alloys (5086, 5356, and 5456) [26].

6.3.2.4. Oxyfuel Welding Process

Oxyfuel gas welding (OFW) processes use a flux and either an oxyacetylene or oxyhydrogen gas flame. The best visibility, control, and weld speed are obtained when an oxyhydrogen flame is used with aluminum alloys. The flux (composed of chlorides and fluorides) must be removed after welding to avoid corrosion. The use of a flux limits the alloys for which it is suitable and produces the greatest heat input [26]. Higher heat inputs can promote corrosion of certain aluminum alloys.

Gas metal arc welding (GMAW) and gas tungsten arc welding (GTAW) are the primary used welding methods. These methods eliminate the potential hazard of flux removal inherent with oxyfuel gas welding and shielded metal arc welding (SMAW). Flux residues, of course, are corrosive. If the welding method requires flux, the joint must permit thorough flux removal [27].

6.3.2.5. High Energy Density Welding Processes

Electron Beam Welding EBW Process The EBW process in a high-vacuum chamber produces a very deep, narrow penetration at high welding speeds. The low overall heat input produces the highest as-welded strengths in the heat-treatable alloys. The high thermal gradient from the weld into the base metal creates very limited metallurgical modifications and is least likely to cause intergranular corrosion in butt joints when no filler is added. Alloys are more prone to have porosity when welded in vacuum. Fixturing to provide transverse compressive loading on the joint can be very helpful in avoiding intergranular cracking in the fusion zones or heat-affected zones [26].

Laser Beam Welding (LBW) Process The LBW process is now a viable fusion joining process for aluminum since stable high-power laser systems are commercially available. Because of the aluminum high reflectivity, high power density is necessary. With power densities on the order of 10^6 W/cm^2, laser welds of aluminum can be produced with minimal distortion at a high processing speed. Inert gas shielding is necessary and a filler metal must be used when welding heat-treatable aluminum alloys [26].

6.3.2.6. Hybrid GMAW–LBW Process (HLBW)

This process combines the gap-bridging ability of the GMAW process and the deep narrow penetration of the LBW process. Higher energy density input is achieved with this process, which operates at a higher welding speed than GMAW and sometimes even higher than LBW. Therefore less metallurgical transformations occur with this process than with the GMAW process; less distortion and residual stresses are also possible. This hybrid welding process is generally referred to as the GMAW–LBW process, but other hybrid welding processes also exist, such as the GTAW–LBW or PLASMA–MIG welding process [26].

6.3.2.7. Resistance Welding (RW) Processes

Non-heat-treatable and heat-treatable alloys can be resistance welded. Resistance spot and seam welding are used in many manufacturing processes, from utensils to automotive components. Selective corrosion attack of resistance spot welded joints can develop in service. Crevice corrosion can occur in spot-welded assemblies. The weld bond technique is used to solve this problem [26].

6.3.2.8. Friction Welding Processes

Aluminum alloys are friction welded to similar or dissimilar aluminum alloys, to copper alloys, and to steels. Most applications involve joining aluminum to steel. High thermal conductivity, large differences in forging temperatures, and the formation of brittle intermetallic compounds are the primary problems encountered. Usually, the friction weld of aluminum alloys may develop a joint strength of only 60–70% that of the weaker base metal. These joints are useful for pressure sealing and for joints that require good electrical and thermal conductivity rather than high strength [26].

Friction Stir Welding (FSW) Process Friction stir welding is a solid-state joining process in which a rotating pin plastically deforms metal along the joining face of two proximal metallic components to form an integral metallurgical bond. During the joining process, the metal is heated but not melted. Joining is accomplished by superplastic deformation and mixing of material from the two components. This process enables joining of alloys that are normally difficult to weld by traditional methods including Cu-bearing 7xxx Al alloys [28]. FSW is highly suited to aluminum alloys and particularly to the so-called difficult-to-weld heat-treatable alloys [29].

FSW can be characterized as a forging and extruding metal-forming process. In the initial stage, a cylindrical tool consisting of a probe and a shoulder is rotated and slowly plunged into the joint line of the materials to be joined. The weld tool is nonconsumable and it generates heat from friction and plastic strain energy release during the mechanical deformation of the assembly, softening the material. As the tool is moved along the joint line, material is extruded around the tool probe and is simultaneously forged into a consolidated joint by the pressure applied by the weld tool shoulder. The joints produced have higher strength than riveted joints and much lower residual stresses than typical fusion weld joints [29].

The advancing side of the weld is where the rotational velocity of the tool has the same direction as its travel velocity, whereas on the retreating side of the weld, the two velocity components have opposite directions [30]. The welds produced by FSW are typically characterized by three primary zones: the heat-affected zone (HAZ), the thermomechanically affected zone (TMAZ), and the dynamically recrystallized zone (DXZ) or weld nugget. Tensile failure occurs in the HAZ or in the section between the HAZ and the TMAZ. Elevated temperatures are generated during FSW in this region, dissolving Guinier-Preston (G-P) zones and coarsening precipitates, creating local strength minima. This region is therefore generally susceptible to corrosion [26].

6.3.3. Welding Influence on Behavior of Aluminum Alloys

The corrosion resistance of welds may be inferior to that of the properly annealed base metal because of microsegregation, precipitation of secondary phases, formation of unmixed zones, recrystallization and grain growth in the weld heat-affected zone (HAZ), volatilization of alloying elements from the molten weld pool, and contamination of the solidifying weld pool. There are numerous factors that can be considered for failure prevention of a welded assembly, such as weldment design, fabrication technique, welding practice, welding sequence, moisture contamination, organic or inorganic chemical species, oxide film and scale, weld slag and spatter, incomplete weld penetration or fusion, porosity, cracks (crevices), high residual stresses, improper choice of filler metal, and final surface finish [27, 31].

In general, the resistance to corrosion of weldments of aluminum alloys is determined, in part, by the alloy welded, by the filler alloy, and by the welding process. Galvanic cells that cause corrosion may be created because of potential differences among the parent alloy, the filler alloy, and the heat-affected zones, where microstructural changes occur. Incomplete removal of fluxes after welding may also cause corrosion [1].

Heat-Treatable and Non-Heat-Treatable Alloys Weldments in non-heat-treatable alloys generally have good resistance to corrosion. Microstructural changes in the heat-affected region in these alloys have little effect on potential, and the recommended filler alloys have potentials close to those of the parent alloys. In some heat-treatable alloys, however, the effect on potential of microstructural changes may be large enough to cause appreciable corrosion in more aggressive environments; the corrosion is selective, either in the weld bead or in a restricted portion of the heat-affected zone. To a considerable degree, the effect of microstructural changes on corrosion in the heat-affected zone can be eliminated by post-weld heat treatment. Stress-corrosion cracking in weldments is caused by residual stresses introduced during welding, but its occurrence is rare [1] (see Chapter 8).

Considering the corrosion resistance of wrought aluminum alloys of the different aluminum series, the following observations can be made [26].

AA1xxx and AA3xxx

AA1050, AA1100, AA3003 Generally, no weld corrosion has been observed for these alloys under current conditions. The presence of specific chemical reactants can impair the corrosion resistance for some alloys. Welded alloys AA1100 and AA3003 are recommended for nitric acid processing plants and can be used in hydrogen peroxide fabrication plants.

Alclad AA3003 The gas metal arc welding (GMAW) process is recommended for this alloy since the accelerated welding speed reduces the assembly (core and cladding) exposure to heat. Coating by hot metal projection of the HAZ and the weld bead is a good measure of protection. Filler metal must be compatible with the base metal to reduce cracking [16].

AA2xxx

AA2014, AA2017, AA2024 Heat generated by welding affects the microstructure of these alloys, and generally weld beads and areas close to the weld bead have low corrosion resistance. These alloys must be welded with great precaution because of their Cu composition, which ranges from 3.5% to 5%. Alloy AA2219, for example, containing 5.8–6.8% Cu is more sensitive to cracking [26].

AA5xxx

AA5052 AA5356 filler metal is preferably used with this alloy, but AA4043 filler metal can be used without noticeable problems of corrosion, except in the presence of some chemical reactants.

AA5454 This alloy, containing a maximum of 3% magnesium, is recommended for chemical storage vessels and pressurized vessels at temperatures above 65 °C. Filler metal AA5554 is recommended for this alloy.

AA5652, AA5254 These alloys have a low percentage of manganese; they are often used for hydrogen peroxide storage. These alloys must be welded with AA5654 filler metal. Alloys with manganese as the alloying element have a tendency to corrode preferentially in the manganese-rich regions.

AA5083, AA5086 These alloys are usually welded with filler metals AA5356, AA5556, or AA5183. Parts welded with these filler metals have good corrosion resistance in most natural environments. AA5083 is sensitive to exfoliation corrosion in the HAZ if the alloy is not carefully elaborated. Alloys with magnesium content above 3% should not be used in service above 65 °C. AA5083 has good corrosion resistance but can be sensitive to stress-corrosion cracking if the part is preheated before welding. AA5086 has been used with success in the fabrication of welded train cars for the transport of chemicals such as sodium chlorate but control of heat input is obligatory [26, 32].

AA6xxx

AA6063 Corrosion resistance of AA6063 is better than AA6061 because it contains no Cu and less Mg_2Si precipitate. This condition is true for the base metal as well as the HAZ. Filler metal AA5356 is currently used for good corrosion resistance and color coordination after anodizing. Filler metal AA4043 can be used if the part is not anodized [26].

AA7xxx

AA7004, AA7005, AA7020, AA-7039 These aluminum–magnesium–zinc alloys are weldable and have the property to naturally age at room temperature after welding by recovering most of their mechanical properties lost in the welding process. These alloys contain zirconium (Zr), which has the properties of stopping grain growth and favoring a lamellar or fibrous grain structure. This grain structure has the tendency to lower the risk of microcracking in the HAZ. However, this created microstructure is susceptible to layer or exfoliation corrosion in the HAZ. Alloys in the T6 temper have good exfoliation corrosion resistance but heat generated in welding destroys the T6 temper [26]. As a general rule, filler metal AA5356 is used in welding because, after natural aging at room temperature, the weld bead must have good mechanical resistance, better than the base metal [26, 33].

AA7017, AA7075 AA7075 is not weldable because of its high Cu content, which can lead to cracking of sensitive microstructure and is highly susceptible to SCC. AA7017 is used in military applications and is considered weldable because it does not contain copper. A humid environment promotes SCC. With certain alloys, particularly those of the heat-treatable 7xxx series, thermal treatment after welding is sometimes used to obtain maximum corrosion resistance [26, 27, 33, 34].

6.3.4. Frequent Corrosion Types of Welded Aluminum Alloys

6.3.4.1. Galvanic Corrosion

Galvanic cells that cause corrosion can be created because of corrosion potential differences among the base (parent) metal, the filler metal, and the heat-affected regions where microstructural changes have been produced [27]. Aluminum alloys can be welded autogenously but the use of a filler metal is preferred to avoid cracking during welding and to optimize corrosion resistance. With certain alloys, particularly those of the heat-treatable 7xxx series, thermal treatment after welding is sometimes used to obtain maximum corrosion resistance [27].

Corrosion resistance of the non-heat-treatable alloys is not altered significantly by the heat of welding. The 2xxx and 7xxx series heat-treatable alloys, which contain substantial amounts of copper and zinc, respectively, can have their resistance to corrosion altered by the heat of welding. In aluminum–copper alloys, the HAZ becomes cathodic; in aluminum–zinc alloys, the HAZ becomes anodic. The differences in corrosion or solution potentials can lead to localized corrosion. In general, the welding procedure that puts the least amount of heat into the metal has the least influence on microstructure and the least chance of reducing the corrosion-resistant behavior of aluminum weldments. Selective corrosion can result in immersed service, where the base alloy and the weld metal possess significant differences in potential [27].

The alloy with the more negative potential in the weldment will attempt to protect the other part. Thus if the weld metal is anodic to the base metal (as is a AA5356 weld in AA6061-T6), the small weld can be attacked preferentially to protect the larger surface area of the base metal. The greater the area to be protected and the greater the difference in electrode potential, the more rapidly will corrosion action occur. The solution potential of the base alloy and the filler alloy should be the same to guarantee maximum protection. If it is not practical, then a preferred arrangement is to have the larger base alloy surface area be anodic to the weld metal, such as AA7005-T6 welded with AA5356 filler. Fabrications in the 7xxx alloys are usually painted to avoid galvanic corrosion. As an additional safety precaution in some cases, the weld area is metalized with another aluminum alloy to prevent galvanic corrosion if a void occurs in the paint coating [27].

In some cases, an alloy constituent can be formed by alloying components of the base and filler alloys to produce an anodic zone at the transition of the weld and base metal. If a 5xxx series alloy is welded with aluminum–silicon filler, then a magnesium silicide constituent can be formed. The magnesium silicide can be highly anodic to all other parts of the weldment. A very selective knife-line corrosive attack results from this immersed service. In aluminum–lithium alloys, two experimental alloys with high lithium content (2.9 wt % Li and 3.0 wt % Cu), welded with either AA2319 or AA4043 fillers, displayed a narrow region within the HAZ that was highly anodic to both base metal and weld metal. In contrast, a 2090-type alloy showed a continuously increasing (cathodic) potential when going from base metal to weld metal and was resistant to pitting attack [27].

6.3.4.2. Pitting and Crevice Corrosion

AA6061 is considered to have good corrosion resistance, but it's relatively high levels of Cu (between 0.15% and 0.40%) have impaired its reputation. Therefore AA6082 is used in

Europe instead of AA6061 because it contains less Cu (nominal composition of 0.1%) [26]. Under certain conditions, possible galvanic corrosion cells between the HAZ, the filler alloy, and the bead lead to pitting corrosion of the more anodic, less protected microstructure.

In chloride solutions, for example, AA6061, which contains 0.15–0.4% Cu, is still sensitive to galvanic corrosion in laboratory testing. Filler metals AA4043 and AA5356 are used for welding of AA6061. Corrosion appears in two different ways depending on the filler metal used for low dilution welds. For AA5356, pitting corrosion appears in the weld bead while there is almost no corrosion in the HAZ. For AA4043, there is almost no corrosion in the weld bead while there is pitting corrosion in the HAZ. Galvanic corrosion created by the welding process explains this phenomenon. In the AA5356 weld, the weld bead is anodic to the HAZ, protecting it. In the AA4043, the HAZ of AA6061 containing magnesium is anodic and protects the weld bead containing lower levels of magnesium[26].

Crevice corrosion is critical generally for welded assemblies because of frequent formation of cracks with critical dimensions leading to localized corrosion with the help of corrosion product accumulation (see Section 6.3.5.2). Strong 99% HNO_3 is particularly aggressive toward weldments that are not made with full weld penetration [27].

Resistance Spot Welding In the case of high-strength 2xxx and 7xxx alloys, selective corrosion attack of the welds can develop in service. Crevice corrosion can occur in spot-welded assemblies. The weld bond technique is used to solve this problem. The pieces to be joined are first bonded by adhesives that seal the crevices, followed by resistance spot welding. A recent development involves joining aluminum to dissimilar metals by the use of transition joints. In this case, aluminum is first spot welded to a compatible metal that in turn is joined to the dissimilar metal. This procedure improves resistance to galvanic corrosion by minimizing dissimilar metal contact and also eliminates brittle intermetallic compounds that form at the joint interface [27].

6.3.4.3. Cavitation Damage

Cavitation damage usually occurs on propellers, hydraulic turbine blades, vanes, ultrasonic devices, and pipelines. Copper–manganese–aluminum (CMA) alloy is reported to have an excellent combination of mechanical properties and possesses erosion and corrosion resistance to high-velocity seawater. Its foundry and welding characteristics are better than conventional aluminum bronzes. Welding is a common method for repairing damaged ship propellers, especially by cavitation erosion. CMA weldment was prepared by tungsten inert gas (TIG) welding, and its cavitation erosion behavior and corrosion behavior in 3.5% NaCl aqueous solution were studied, respectively, with a magnetostrictive vibratory device and an electrochemical device. Results show that the weld zone (WZ) of the weldment exhibits better cavitation erosion and corrosion resistance than the heat-affected zone (HAZ) and the base metal. The cumulative mass loss of the WZ is only one-quarter that of the base metal. SEM analysis of eroded specimens reveals that the base metal is attacked most severely; the HAZ less and the WZ least. The microcracks causing cavitation damage initiate at the phase boundaries. Among the three zones of the weldment, the WZ is the noblest with a corrosion potential of -266 mV, while the corrosion potential of the HAZ is -284 mV and that of the base metal is -279 mV, when exposed for about 60 h to 3.5% NaCl aqueous solution. The WZ corrosion current density is the lowest, about 0.035 $A.m^{-2}$, while the corrosion current density for the HAZ is 0.078 $A.m^{-2}$ and that for the base metal is ± 0.79 $A.m^{-2}$ [35].

6.3.4.4. *Corrosion Fatigue, SCC, and Knife-Line Attack*

Fatigue crack propagation rates in the white zone of MIG AA7017 welds in an aqueous salt chromate environment showed a pronounced enhancement compared with tests performed in air. Intergranular corrosion is initiated by galvanic cells and can lead to intergranular stress-corrosion cracking. Also, knife-line attack, a sort of SCC, is observed in a very thin region of corrosion for most aluminum alloys adjacent to the weld above 50 °C and the depth of attack increases markedly with temperature [27, 31].

An early investigation of the SCC of the 7xxx series stated that when the sum of Mg and Zn contents is higher than 6%, the alloy is susceptible to SCC [36]. Under certain conditions, AA7004 and AA7005 are susceptible to SCC, for example. Welding of SCC-sensitive base metals creates SCC-sensitive welds under certain conditions because fusion welding creates stresses in the structure. These alloys cannot be strain hardened after welding, since they become sensitive to SCC. When these alloys are damaged in service, a stress-corrosion crack is induced in the damaged region under certain ambient conditions [26, 33, 34]. (See Chapter 8.)

6.3.5. Corrosion Resistance of Wrought and Cast Al Alloys

In order to reduce corrosion susceptibility of aluminum welded joints, low heat input density processes, higher welding speeds, and narrower penetration profiles should be used. For heat-treatable alloys, a lower energy input minimizes the width of the HAZ and therefore the corrosion occurs closer to the weld bead. Also, lower energy input reduces the possibility of dissolution and the segregation of precipitate elements at the grain boundaries. When segregation occurs, the electrode potential difference between the grain boundary and the adjacent grain produces a selective corrosion of the grain boundary. This is one of the main reasons for intergranular corrosion. Exfoliation corrosion is expected also in this type of welding [26].

Post-weld heat treatment has a beneficial effect on corrosion behavior. Post-weld aging creates a uniform microstructure and electrode potential across different weld regions. The corrosion potential differential between the weld regions BM, HAZ, and FZ is lowered, and therefore corrosion susceptibility is lowered too [26]. It has been observed that the determination of the potential of different microstructures created after welding can reflect the importance of the corrosion galvanic cells that lead to localized corrosion (selective leaching). Figure 6.4 gives the variation of potential across the welded assembly of an alloy and filler from the same series in aqueous chloride oxidizing solution. Generally, corrosion occurs in the HAZ for the as-welded condition since electrochemical potential in the HAZ is less noble than that in the base metal and in the weld bead. If electrochemical potential is the same in the HAZ as in the base metal, it is predicted that corrosion should be less important [27].

6.3.5.1. *Corrosion Resistance of Electron Beam Welds*

Electron beam welds of AA2219 offer much higher strength compared to gas tungsten arc welds of the same alloy and the reasons for this have been explored. AA2219 contains 6.5% Cu and 0.3% Mn as the major alloying elements. Weld joints were made on 8.5 mm thick AA2219-T87 plates using both GTAW and EBW. In both cases closed square butt joint configuration and single-pass welding were used. In the case of gas tungsten arc welds

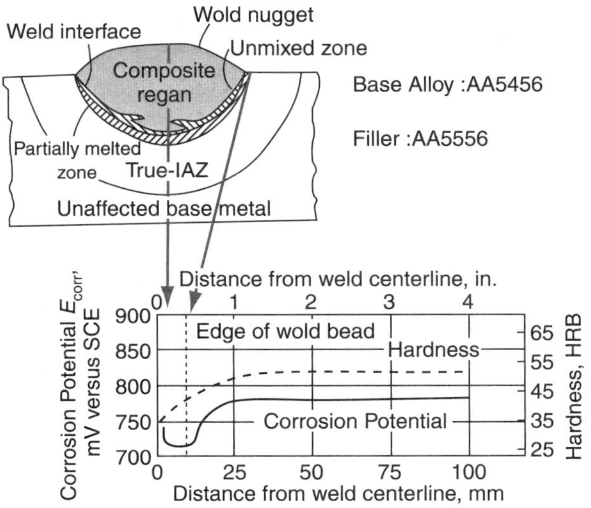

Figure 6.4 Schematic of the welded assembly showing the effect of welding heat on microstructure hardness and corrosion potential of the welded assembly of AA5456-H321, base metal, and AA5556 filler (three pass metal inert gas welding) in 53 g/L NaCl and 3 g H_2O_2/L solution [37].

AA2319 filler material was used. Vickers hardness values in the weld metal and HAZ regions, on the weld metal surface, and in the through thickness direction were measured. Tensile specimens were cut from the joints, transverse to the weld direction, according to the ASTM E8 standard. Tensile tests were conducted on a computer controlled Instron universal testing machine [38].

Active–passive corrosion behavior studies were conducted according to the ASTM G3 standard in 3.5% NaCl solutions with pH adjusted to 10. The potential scan of 1 cm^2 was carried out at a 0.166 mV/s scanning rate, with initial potential of −0.25 V to open circuit potential and then to final potential of pitting. The potential at which current increases drastically is treated as the critical pitting potential (E_{pit} or E_p). The specimens for this purpose were cut from weld metal and the HAZ using a diamond saw cutter so that the specimen consists of only weld metal or only HAZ. Dynamic polarization curves show that the pitting potential values of weld metals were found to be -485 mV/SCE for EBW (higher resistance) and −606 mV for GTAW. Copper segregation at grain boundaries has significant effect on the pitting corrosion behavior of weld metals [38]. (See Chapter 17 for more details.)

Large columnar grains form in the gas tungsten arc weld metal. An SEM-EDX analysis of these grain boundary phases shows that copper content corresponds to the Al–Cu eutectic composition. In gas tungsten arc welds, segregation of copper at grain and sub–grain boundaries results in a matrix depleted of copper, affecting the weld strength. Transmission electron micrographs of the heat-affected zones revealed the precipitate disintegration and overaging in gas tungsten arc welds, resulting in depletion of copper inside the grains [38].

In the case of electron beam weld metal, there was no evidence of any dendritic solidification. This is mainly due to the very high solidification rates associated with EBW. Extremely fine equiaxed grains form in electron beam weld metal; copper distribution is very fine and spread more evenly in the matrix. It has been shown that electron beam welds exhibit superior tensile strength, finer microporosity, and better corrosion properties as compared to gas tungsten arc weld metal. Fine rounded precipitates were present in electron beam welds, which could be the product of the original $CuAl_2$ precipitates of the base material [38].

However, study of the heat input density of AA7003 shows different effects on corrosion depending on the welding process used and the severity of corrosion testing. A gas metal arc weld of this alloy was submitted to a strong acid pulverization corrosion test. Strong corrosion was found in the HAZ following a line near the 275 °C isotherm along the weld. The same alloy was welded with a high-density electron beam welding (EBW) process and then submitted to the same test. Strong corrosion was observed in the HAZ along the weld near the 275 °C isotherm but closer to the weld. In the EBW process, the corrosion line was thinner and deeper than in the GMAW process. A high energy density welding process does not prevent corrosion but moves it closer to the weld. Corrosion is deeper and therefore more critical than in a lower energy density process [26].

6.3.5.2. Corrosion Resistance of Friction Stir Welded Aluminum Alloys

Welds achieved by FSW are typically characterized by three primary zones: HAZ, TMAZ, and DXZ. Tensile failure occurs in the HAZ or in the section between the HAZ and the TMAZ since elevated temperatures are generated during FSW in this region, dissolving Guinier–Preston (GP) zones and coarsening precipitates, creating local strength minima. This region is therefore generally susceptible to corrosion. When joining higher strength 7000 series aluminum alloys, post-weld artificial aging (PWAA) is necessary to stabilize the microstructure and improve corrosion resistance. As an example, the pitting response observed in AA7075-T6 was absent from AA7075-T73 samples with PWAA [29].

FSW is a solid-state process performed by plunging a spinning tool piece into the junction between two pieces of metal; local heating causes plastic flow of the metal causing mixing. The process produces a weld with three microstructural regions: the nugget, where the tool piece pin has caused a high level of deformation, usually leading to a recrystallized structure; the thermomechanically affected zone (TMAZ), where the original grains have been deformed by plastic flow; and the heat-affected zone (HAZ), where the microstructure has been affected by heat alone, rather than plastic deformation. FSW gives superior mechanical properties over fusion welding owing to the lower heat input and greater microstructural homogeneity. However, concerns exist over the welds' resistance to localized corrosion [39].

Several researchers investigated the corrosion susceptibility of welds created by FSW in aluminum alloys and they found corrosion in the HAZ and/or nugget region. This varies considerably between different alloys and the use of different welding conditions. FSW is widely used for joining a range of alloys, particularly aluminum alloys in aerospace and marine applications. The effect of processing parameters on the corrosion susceptibility of welds in AA2024 has been explored, together with the use of laser surface melting to protect welds from corrosion [39].

The corrosion susceptibility of friction stir welds in AA2024-T352 was found to vary with the weld processing parameters. Corrosion attack was investigated with in situ X-ray tomography, which showed how the penetration of corrosion into the interior of the structure varied with weld microstructure. The susceptibility to corrosion was related to the degree of over aging by comparing the corrosion behavior to samples of the base alloy that had been aged at different temperatures. A systematic increase first in the anodic reactivity and then the cathodic reactivity of the overaged structures with temperature can be used to predict the location of the region of the weld with the highest susceptibility to corrosion [39].

The corrosion susceptibility of FSW in AA2024 is affected by processing parameters, particularly tool piece rotation speed: a higher heat input from a higher rotation speed leads to high cathodic reactivity in the nugget owing to precipitation of S-phase particles and high anodic reactivity in the heat-affected zone as a result of sensitization of grain boundaries. Joining of AA2024 to AA7010 by FSW leads to galvanic corrosion because the AA7010 nugget has elevated anodic reactivity whereas the AA2024 nugget has elevated cathodic reactivity. Laser surface melting leads to a highly homogeneous surface layer that is 5–10 μm thick, with low anodic and cathodic reactivity. This melted and rapidly solidified layer over the weld surface is highly effective in protecting AA2024 from corrosion in FSW [39].

Friction stir welded AA2024-T351 is susceptible to corrosion in both the nugget region and HAZ (grain boundary sensitization). Intergranular corrosion attack of Al–Cu alloys is due to precipitation of Cu-rich intermetallic particles at grain boundaries, which results in the formation of anodically active Cu-depleted regions adjacent to the boundaries. The enhanced cathodic reactivity observed in the nugget region is due to the precipitation of coarse S-phase particles of Al_2CuMg. The rotation speed is the main processing factor in determining the location of corrosion for welds manufactured with the different parameters in this study. For low rotation speeds, the intergranular attack in sodium chloride and hydrogen peroxide solution (ASTM G110) was in the nugget region due to the significant increase in anodic reactivity in this region. For higher rotation speeds, the corrosion attack was in the HAZ owing to the presence of sensitized grain boundaries in this region; the nugget region is less anodically active than the HAZ so the nugget is polarized cathodically (partially protected), supporting the high anodic reactivity in the HAZ. The reason for the different regions of attack is the balance between anodic and cathodic reactivity in the weld regions [40].

Heat treatment of AA2024 alters the microstructure by changing the distribution of submicron precipitate particles, but has little or no effect on the micron-sized constituent particles. Three zones of temperatures corresponding to three distinct features were examined. Zone I covers the untreated T351 temper to aging treatments at temperatures below 200 °C. In this regime, as the aging temperature is increased, more precipitation and growth of S-phase particles starts to occur predominantly at the grain boundaries due to the higher nucleation rate at the grain boundaries. This leads to increasing anodic reactivity that is associated with intergranular attack, owing to copper depletion at the grain boundaries [40].

In Zone II (200–350 °C), the growth of S-phase particles occurs not only in the grain boundaries but also in the matrix. The anodic reactivity shows a maximum value at 250–300 °C associated with intergranular corrosion. Intergranular corrosion is found in both Zones I and II and anodic reactivity reaches a maximum in Zone II. This is probably associated with increasing copper depletion along grain boundaries due to growth of grain boundary precipitates. For example, an increase in the length of Cu-rich particles at the grain boundaries was found with increasing aging time at the aging temperature of 190 °C for the Al–4Cu system. In addition, the copper levels in the Cu-depleted zones were found to decrease dramatically with increasing the aging time of Al–Cu at 240 °C from 15 to 144 min.

It should be noted that intergranular corrosion is not observed at the aging temperature of 350 °C. The increase in formation of precipitate particles in the matrix leads to a decrease in Cu content in the matrix, causing a decrease in breakdown potential in the matrix and reducing the difference in pitting potential between matrix and Cu-depleted zones, making intergranular attack less favorable. In Zone III (400–490°C), the S-phase precipitates dissolve back into the matrix, causing an increase in the copper content in the matrix. It is

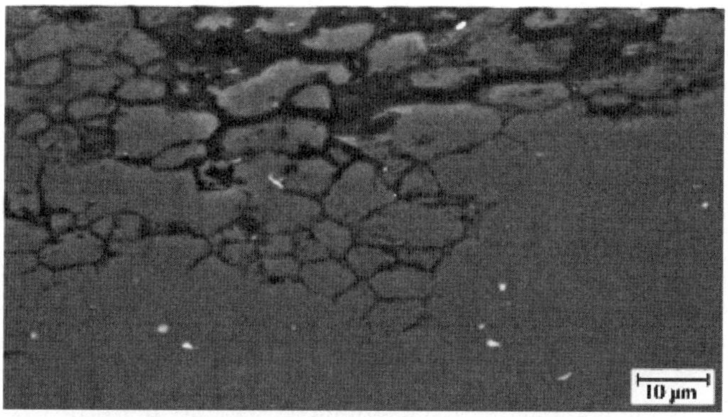

Figure 6.5 SEM reveals intergranular corrosion of the cross section of the stir welded AA7108-T79 after 72 h immersion in the modified EXCO solution [30].

notable that intergranular corrosion is not observed in this regime as a result of more homogeneous distribution of copper in the matrix [40].

The corrosion behavior of extruded sections of welded commercial alloy AA7108-T79 that is 3 mm thick was studied. In the T79 condition, friction stir welding was carried out at a steady welding travel speed of about 1 m/min. Following welding, AA7108 exhibited natural aging and, after 30 days, the heat-affected zone (HAZ) recovered its strength to about 90% of the parent material. The welded alloy showed the expected zones associated with friction stir welding, namely, nugget, thermomechanically affected zone, and heat-affected zone. Samples 10 mm length in the welding direction and 60 mm length in the transverse direction with the weld at the center, were exposed to the test solution for 72 h. A modified ASTM G34, EXCO test employing 15 vol% dilution of a solution of 4.0 M NaCl, 0.5 M KNO_3, and 0.1 M HNO_3 was used. Intergranular corrosion appeared within the thermomechanically affected zone (TMAZ) and extended into the HAZ (Figure 6.5) [30].

Open circuit potential measurements were carried out at various locations along the welded sample. Both the cross section and the top surface of the weld were examined. A commercial Ag, AgCl/KCl saturated reference electrode was used. Figure 6.6 shows the variation of the open circuit potential profiles at various locations across the weld, measured

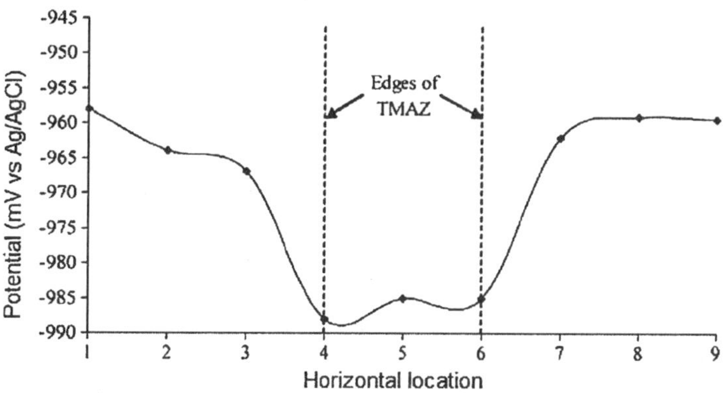

Figure 6.6 Potential profile across the top surface of the stir welded AA7108-T79 [30].

over the cross section and top surfaces. For the top surface, the potential increases from the edges of the TMAZ to the parent regions; the highest potential, approximately −955 mV, is recorded within the parent region, and the lowest potential, approximately −988 mV, is evident within the TMAZ. The potential profile measured on the cross section shows similar behavior, with the lowest potential, approximately −980 mV, recorded at the edges of the TMAZ, and the highest less active potential, approximately −945 mV, evident within the parent region [30].

These suggest that such regions in the weld have been sensitized by the thermal transient caused by heat generated during the welding process and are most susceptible to corrosion. Transmission electron microscopy of tested samples confirmed that corrosion proceeded intergranularly within the TMAZ. As revealed by electron microscopy, precipitates in the parent alloy are distributed relatively uniformly across the grains, with the potential across the grains also being relatively uniform. However, within the TMAZ, the $MgZn_2$ phase, precipitated at grain boundaries, is associated with relatively negative potentials. Thus the grain boundaries represent favorable sites for anodic activity compared with the grain matrix. As a result, intergranular localized corrosion and nonuniform penetration of the sensitized alloy occurred due to the nonuniform distribution of *n*/*n*′ ($MgZn_2$) precipitates within the thermomechanically affected zone. Further more, the relatively open morphology of the corrosion product allows continued access of the environment, which assists propagation along preferred grain boundaries [30].

FSW of Cast Aluminum 7050 Ingots with Scandium Additions Microstructure, corrosion, and environmental cracking behavior of friction stir welded cast aluminum 7050 ingots alloyed with scandium additions were investigated. An overaged (T7451) and a homogenization (24 h/475 °C) temper were applied to the as-cast plates before friction stir welding, while a post-weld heat treatment was applied to all welds to verify the changes in the corrosion behavior. The scandium does not significantly dissolve in any of the phases present in the AA7050 plates and remains homogeneously distributed within the matrix. Zinc is also homogeneously distributed across the grains, so that the grain boundary phases are mainly enriched in Cu and slightly in Mg but not in Sc [41].

The as-cast ScAl FSW microstructure exhibits coarse grain boundary phases away from the nugget region and wide precipitate-free zones and coarse intragranular precipitates. The tensile strength of this weld can only be increased with a homogenization post-weld heat treatment (1 h at 480 °C, 1 h in boiling water and quench). The corrosion tests of the as-cast Sc weld immersed in a 3.5 wt% NaCl solution indicate that the main formation of corrosion products takes place outside the weld nugget region from the thermomechanically affected zone through the parent metal. Less corrosion products are observed within the parent metals of the Sc (T7451) and the Sc (24 h/470 °C H_2O Q) welds, while the main corrosion is localized on the thermomechanically affected and the heat-affected zones. The post-weld heat treatment (1 h/480 °C–1 h/120 °C boiling H_2O quench) does not significantly reduce the formation of corrosion products for all the plate groups, but does change their localization. This treatment increased the tensile strength of the as-cast weld, but decreased the strength of the heat-treated welds. The heat treatment of the as-cast samples to an over aging (T7451) and homogenization (24 h/475 °C) temper increased the general corrosion susceptibility of the friction stir welds [41].

There is an interest in Sc as an alloying addition to Al because of its ability to impart an attractive combination of strength, ductility, and enhanced crack growth resistance. Low-level scandium additions to Al lead to the formation of Al_3Sc dispersoids (L12-type phase), which have a significant effect on grain refinement during thermomechanical processing. A microelectrochemical capillary cell was used to examine the influence of Al_3Sc particles on the initiation and propagation of pitting corrosion in aluminum alloys exposed to dilute chloride solutions. Al_3Sc is found to be spontaneously passive with a low self-dissolution rate [42]. In dilute chloride solutions Al_3Sc is nobler than pure Al and possesses a low self–dissolution rate and a good resistance to breakdown. Al_3Sc particles are comparatively weak local cathodes since the oxygen reduction reaction is slower on Al_3Sc than on some other related dispersoid intermetallic compounds [42].

6.4. METAL MATRIX COMPOSITES FOR NUCLEAR DRY WASTE STORAGE

The prospective use of aluminum metal matrix composites based on boron for nuclear dry waste storage, Al-MMC/B_4C, is discussed as a corrosion prevention case history. Development of metal matrix composites is a major practical challenge to extend the use of aluminum alloys to new areas, necessitating better mechanical properties and higher corrosion resistance for nuclear applications. The incorporation of a second phase into the alloy matrix may change the corrosion behavior of the material along with significant changes in physical and mechanical properties. The modification of the microstructure of the matrix during manufacture of the metal matrix composite (MMC) alters the size and distribution of intermetallic phases or introduces residual stresses between reinforcement and matrix. Corrosion may initiate at the barrier anodic film itself or at the interfaces between reinforcement and matrix in composite materials.

In the case of Al matrix composites in contact with high-temperature water containing few aggressive ions during nuclear applications, the pH at the pit shifts to acidic values because of the autocatalytic nature of the localized attack. For example, pitting of aluminum matrix composites 1050 and 2124, each reinforced with silicon carbide particles (SiCp) has been studied in 1 N NaCl solution. Pores and crevices at SiCp–matrix interfaces strongly influence pit initiation, which is further aided by the cracking of large SiC particles during processing. The presence of $CuAl_2$ and $CuMgAl_2$ precipitates in the 2124-SiCp composite also promotes pitting attack at the SiCp–matrix and intermetallic–matrix interfaces. It is evident that galvanic corrosion of the Al-MMC occurs since SiCp as well as B_4C are active cathodic sites in the matrix [43–45].

Satisfactory tests on the neutron absorbing capacity and mechanical properties have been done on Al B_4C materials. In the nuclear industry, aluminum-based material reinforced with B_4C is being used. "Boral" is based on pure aluminum while Metamic is fabricated by powder metallurgy (AA6061/21%B_4C) and presents corrosion and homogeneity problems due to the forming method and the composition of AA6061. Other alloys based on AA6063/10%B_4C or AA6351/16%B_4C (Al–1%Si–6%Mg–0.6%Mn) have recently been developed. The 6000 series of alloys are heat treatable with good strength and elevated temperature resistance, and they can be anodized and hard coated to improve corrosion resistance, abrasion resistance, and emissivity, all of which are significant in dry cask nuclear storage applications [46–48].

Dry storage is much more important than wet storage because of increasing waste material, space considerations, and safety. It is a very sensitive area since materials can be subjected to frequent humid and oxidizing atmospheres; their possible oxidation at high temperatures up to 200 °C must also be considered. The development of alloys for commercial use in dry fuel storage requires additional properties such as corrosion performance. In recent years, aluminum metal matrix composites (Al-MMCs) containing reinforcing particles of B_4C have been increasingly used in the nuclear industry. Boron and B_4C are employed to control the activity of highly radioactive nuclear waste in nuclear reactors, being materials with high stability under neutron irradiation.

These materials are commonly produced in the form of plates, sheets, rods, and liners and are used to fabricate the inside baskets of storage and transport containers in industries associated with nuclear energy production. Dry casks are used to store spent nuclear reactor fuel prior to long-term disposal. Further more, because of their high strength, the 6000 series of alloys are capable of being used as structural elements. The materials are placed between nuclear fuel compartments and one common usage is to ensure subcriticality during normal and off-normal/accident service conditions inside storage casks and transportation packages. They serve the following basic functions: nuclear criticality safety, structural support of the fuel assemblies, and heat removal [49–51].

During dry storage, there is little possibility of corrosion because of the helium atmosphere, although the condition of a basket is expected to be approximately 200 °C. Moreover, during loading of a cask, absorbers will temporarily be in contact with reactor pool water and, for wet storage applications, this contact is continuous. In both cases, corrosion and contact corrosion may occur. In addition, there is a possibility of contact with water that contains boric acid, especially in the case of a pressurized water reactor (PWR) plant. Considering the dry and wet conditions of the corrosion media during use of these composites, the most common type of corrosion to be considered for this study is pitting, for bare MMCs. Pitting corrosion is observed when aluminum and its alloys are in the pH range where it is passive. At high-temperature conditions (200 °C), corrosion shows uniform oxidation of aluminum alloys that is closely related to electrochemical uniform and nonuniform types of corrosion at ambient temperatures [50].

Galvanic corrosion, chemical degradation, microstructure influenced corrosion, and processing-induced corrosion are the types of corrosion experienced by MMCs. Generally, the galvanic action is governed by the potential difference between the aluminum matrix and the reinforcing material (cathodic to aluminum) in an aggressive medium. Galvanic local cells can shift general uniform corrosion to nonuniform and even localized corrosion especially in nonagitated solutions and at ambient temperatures. The composite's susceptibility to pit initiation in the range of pH between 4 and 8 is similar to unreinforced alloy but the rate of pit propagation is higher for composites [1].

The functional performance of these composite materials can largely be improved by coatings. Electrochemical anodization or chemical conversion coating can be followed by a chosen organic coating depending on the medium, the temperature, and the operating conditions. Crevice and filiform corrosion for coated MMCs should be of concern. Chloride ions and high humidity are required for initiation and propagation of crevice and filiform corrosion. Recommended laboratory tests that have been used are the 3.5% NaCl alternate immersion test, ASTM G44, or exposure to hydrochloric acid vapors for 24 h followed by prolonged exposure to high humidity at slightly elevated temperatures of about 50–65 °C.

Upon increasing acidity, due to the autocatalytic localized mechanism of attack, corrosion attack becomes more nearly uniform at first and then polarization up to its pitting potential can occur (See Chapter 3). Localized galvanic attack may lead to intergranular corrosion of the matrix, stress-corrosion cracking, or fatigue cracking (see Chapter 8) [47].

B. MICROBIOLOGICALLY INFLUENCED CORROSION: THE BASICS

6.5. MICROORGANISMS

Microbiologically influenced corrosion corresponds to the acceleration of material deterioration by various microorganisms. Biodeterioration then proceeds by the processes of staining, patina formation, pitting, etching, and disaggregation, and exfoliation biological attack frequently leads to the formation of a tubercle that covers a deep pit. When bacteria are present, the tubercle structure is usually less brittle and less easily removed from the metal surface than when they are absent. The organisms grow either in continuous mats or sludge or in volcano-like tubercles with gas bubbling from the center [12]. Many cases have been documented for the biodeterioration by bacteria and/or fungi of architectural building materials, stoneworks, fiber-reinforced composites, polymeric coatings, and concrete [53]. The following four groups organisms could influence corrosion in similar or different ways: bacteria, fungi (including yeast), algae, and lichens.

6.5.1. Bacteria (Prokaryotes)

The smallest organisms that live on their own are bacteria and cyanobacteria (blue-green algae). The lack of a nucleus led to the name of this group (i.e., prokaryotes), in contrast to the eukaryotes. Usually, the enzymes involved in the protein synthesis of prokaryotes have a different structure from those for eukaryotes. That is why antibiotics against bacteria are inactive against eukaryotes (and vice versa). Under favorable conditions, some bacteria can double in number every 20 min or less. The bacteria as a group can survive from $-10\ ^{\circ}C$ to $>100\ ^{\circ}C$, at pH $\approx$ 0–10.5, in dissolved oxygen (0 to saturation), under pressure (vacuum to >31 MPa), and in saline conditions (parts per billion to about 30%). Most bacteria that have been implicated in corrosion grow best at temperatures of 15–45 °C and at a pH of 6–8 [54].

6.5.2. Fungi and Yeast (Eukaryotes)

Contrary to plants or animals, fungi digest food externally and absorb the nutrient molecules into their cells. Mushrooms and yeasts are members of the fungi kingdom and belong to the eukaryotes. A yeast cell (class of unicellular fungi) has a nucleus containing genetic information, ribosomes that are larger than those of prokaryotes, and enzymes for protein synthesis. In contrast to bacteria, fungi and yeasts can live and grow under severe water limitation. *Aspergillus* and *Penicillium* species are also able to tolerate a variation of hydrogen ions in their environment (pH above 12 and below 2)[55, 56]. Alekhova et al. [57] have shown that fungi-induced MIC of Al alloys on the Mir Space Station.

6.5.3. Algae (Eukaryotes)

Algae are similar to fungi but distinguished by their ability to grow with light and to build up their cell organs from CO_2 and air, just like green plants [55]. Algae can survive in aqueous environments (salted or not), terrestrial environments, and air environments. Under photosynthetic metabolism, algae produce oxygen when exposed to light and products such as organic acids that initiate corrosion. Carbon dioxide is also excreted from algae as a metabolite in the absence of light [55, 58]. Fouling and the resulting corrosion damage have been linked to algae. Corrosive by-products, such as organic acids, are also associated with these organisms. Moreover, they produce nutrients that support bacteria and fungi [59].

6.5.4. Lichens

Lichens live in association with (biocenosis) algae and fungi. Biocenosis is the interaction of two different types of organisms living in the same area. The algae produce the nutrients via their photosynthesis, while the fungi provide water and mineral salts. Cyanobacteria can also form lichens with fungi [55]. Lichens are more involved in stone material degradation than in metal corrosion [60]. Nevertheless, a type of lichen (oral lichen planus) seems to be involved in dental metallic material corrosion [56, 61].

6.6. NATURAL AND ARTIFICIAL MEDIA

6.6.1. Air Media

Biofilms are not confined to solid–liquid interfaces; they can also be found at solid–air interfaces. Airborne deteriogens (microorganisms involved in corrosion processes) have been shown to be important factors in the biodeterioration of surface coatings [60]. Microorganisms such as algae and fungi (rather than bacteria) often play the major roles [56, 62].

6.6.2. Aqueous Media

A great variety of microscopic organisms (microorganisms) are present in virtually all natural seawater environments, such as bays, estuaries, harbors, coastal and open ocean seawaters, as well as rivers, streams, lakes, ponds, aqueous industrial fluids, and waste waters. In freshwater environments, bacteria, algae, yeasts, and molds exist. Molds are macroorganisms and not microorganisms. Untreated fresh waters may contain microorganisms, such as gallionella, which cause corrosion. Marine environments show heavy fouling frequently. Larger, macroscopic organisms, such as the well-known barnacles and mussels, are also present in many aqueous environments. In natural conditions, sulfate reducing bacteria (SRB) grow in association with other microorganisms and use a range of carboxylic acids and fatty acids, which are common by-products of other microorganisms. Common bacteria in aqueous media (e.g., *Pseudomonas* and *Flavobacterium*) can secrete large amounts of organic material under both aerobic and oxygen-free (anaerobic) conditions. Biological slimes are commonly found in the water phases of industrial process plants, such as in chemical processing, energy generation, pulp and paper production, hydraulic systems, fire protection systems, water treatment, sewage handling and treatment, highway maintenance, building and stonework fabrication, aviation, underground pipelines, and onshore and offshore oil and gas equipment. The problems depend on the material and the characteristics of the medium in every industry [52, 54].

6.6.3. Soils

Many microorganisms are present in soils, such as SRB, which are found where sulfates are abundant [63].

6.7. ANAEROBIC AND AEROBIC BACTERIA IN ACTION

Considering their related oxygen dependence, four categories of microorganisms can be distinguished:

1. Obligate aerobes require oxygen for growth. These organisms depend on oxygen to survive at a concentration normally found in fresh water or in the atmosphere (21% oxygen).
2. Microaerophilic species are capable of oxygen-dependent growth but cannot grow in the presence of a level of oxygen equivalent to that present in an air atmosphere (21% oxygen). Oxygen-dependent growth occurs only at low oxygen levels.
3. Facultative anaerobes grow with or without oxygen. In the absence of oxygen the microorganism is able to switch its metabolism to fermentation. The yeasts are good examples of facultative anaerobes.
4. Anaerobic species are obligate anaerobes that grow only in the absence of oxygen. Many of these bacteria are known to be sensitive to oxygen, and the presence of even low levels (i.e., 0.1 mg O_2/L) will kill the cells [64]. SRB are a good example of this type.

Some bacteria are involved directly in the oxidation or reduction of metal ions, particularly iron and manganese. Some microbes can produce organic acids, such as formic and succinic, or mineral acids such as sulfuric acid. Some bacteria can oxidize sulfur or sulfide to sulfate or reduce sulfates, very often to hydrogen sulfide as the end product [54]. Hydrogen embrittlement of metals and alloys could be accelerated by microorganisms through the production of molecular hydrogen, atomic hydrogen, and hydrogen sulfide and local attack of a protecting oxide film [64].

6.7.1. Anaerobic Bacteria

Desulfovibrio, Desulfotomaculum, and *Desulfomonas* in anaerobic microenvironments can exist under biodeposits of aerobic organisms, in crevices built into the structure, and at flaws in various types of coating systems. The most commonly encountered type of SRB is known as *Desulfovibrio*. The most corrosive environments are often those in which alternating aerobic–anaerobic conditions exist because of the action of variable flow hydrodynamics or periodic mechanical action. Conditions at the base of even thin slimes (biofilms) can be ideal for the growth of SRB: a high organic nutrient status, lack of oxygen, a low redox potential, and protection from biocidal agents. SRB-induced corrosion is frequently encountered in the oil and gas industry [52, 55, 63].

Sulfate Reducing Bacteria (SRB) SRB are found in fresh water, salted water, and soils, where sulfate is abundant [63]. Some SRB species are thermophilic, such as *Desulfotomaculum*, which can facilitate grow between 30 and 65 °C [65]. They are anaerobic bacteria that obtain their required carbon from organic nutrients and their energy from the reduction of sulfate ions to sulfide. Sulfide appears as H_2S (dissolved or gaseous), HS^- ions, S^{2-} ions, metal sulfides, or a combination of these, according to the conditions. Sulfides are highly corrosive [63]. SRB facilitate the cathodic reaction that controls the corrosion rate in these media.

In anaerobic conditions, SRB cause anodic stimulation by precipitation of metallic ions produced at the anode with biological sulfur. Acceleration of anodic half-reactions occurs by tubercle formation, under-deposit corrosion, acid production, and breakdown of protective films and coatings by the metabolic activities of microorganisms. Moreover, SRB are able to take up molecular hydrogen while reducing sulfate for the synthesis of sulfur-containing substances. Thus in a medium containing high metallic ion concentrations, sulfides produced by SRB may precipitate as metallic sulfide and can form a galvanic couple with the naked metal, accelerating the cathodic reaction [66].

6.7.2. Aerobic Bacteria

Ammonia and amines are produced by microbial decomposition of organic matter under both aerobic and anaerobic conditions (ammonification). These compounds are oxidized to nitrite by aerobic bacteria such as *Nitrosomonas* or *Nitrobacter* species. *Nitrobacter* is very efficient at destroying the corrosion-inhibition properties of nitrate-based corrosion inhibitors by oxidation, unless a biocidal agent is included in the formulation. The release of ammonia at the surfaces of heat-exchanger tubes has a detrimental effect [63].

The bacteria of the genus *Thiobacillus* obtain energy not by oxidation of organic compounds but by oxidation of inorganic sulfur compounds (including sulfides) to sulfuric acid according to the following reaction

$$4FeS_2 + 15O_2 + 2H_2O \rightarrow 2Fe_2(SO_4)_3 + 2H_2SO_4$$

6.7.3. Co-action of Anaerobic and Aerobic Bacteria

Most SRB are obligate anaerobes, yet they are known to accelerate corrosion in aerated environments. This is possible when aerobic organisms form a film or colony and then, through their metabolism, create a microenvironment favorable for anaerobic bacteria. Aerobic organisms near the outer surface of the film consume oxygen and create a suitable habitat for the SRB at the metal surface. The accompanying flora delivers the nutrients that SRB need (e.g., acetic acid and butyric acid) and consumes the oxygen that is toxic for the SRB [52, 54, 55, 67].

Some organisms have a fermentative type of metabolism that produces carbon dioxide (CO_2) and hydrogen (H_2); while other microbes can use CO_2 and H_2 as sources of carbon and energy, respectively. Numerous species of bacteria and algae either produce or use oxygen (Figure 6.7). One series of bacteria can reduce nitrates to nitrogen gas, others can convert nitrates to nitrogen dioxide, or vice versa, or they can break it to ammonia. Some of these gases can cause corrosion [52].

The Tubercle Bacterial attack frequently leads to the formation of a tubercle that covers a deep pit. When bacteria are present, the tubercle structure is usually less brittle and less easily removed from the metal surface than when they are absent. The organisms grow either in continuous mats or sludge or in volcano-like tubercles with gas bubbling from the center, as shown schematically in Figure 6.8 [54]. The biofilm generally creates at the metal interface an aggressive local medium that has different properties than the principal solution [64].

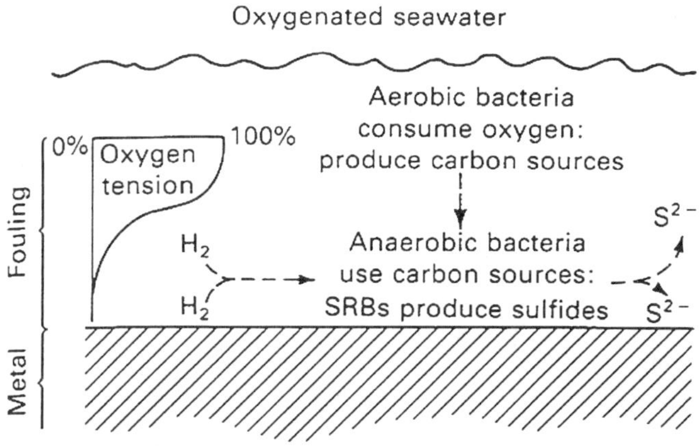

Figure 6.7 Attacks by aerobic and anaerobic bacteria [54].

Consider a buried aluminum specimen in a soil containing a near-neutral pH solution. The presence of sulfate reducing bacteria accelerates the electrochemical reaction of corrosion according to the following equations [54]:

$$8Al \rightarrow 8Al^{3+} + 24e^{-} \qquad \text{(anodic)}$$

$$24H^{+} + 24e^{-} \rightarrow 24\dot{H} \qquad \text{(cathodic)}$$

$$3SO_4^{2-} + 24H \xrightarrow{SRB} 3\,S^{2-} + 12H_2O$$

$$Al^{3+} + 3(OH)^{-} \rightarrow Al(OH)_3 \qquad \text{(corrosion product)}$$

$$2H^{+} + S^{2-} \rightarrow H_2S \qquad \text{(possible gas product)}$$

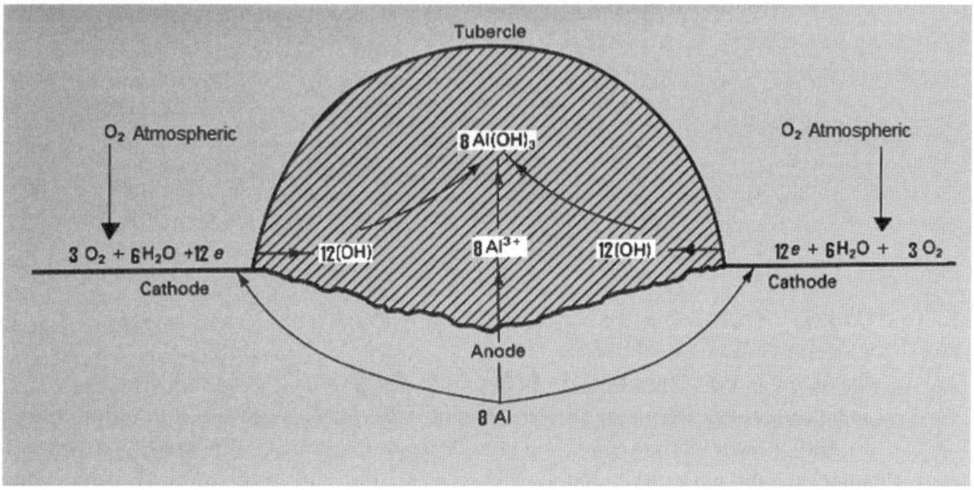

Figure 6.8 Schematic of tubercle formed by bacteria on an aluminum alloy surface [54]

Hydrogen sulfide can also be produced. The SRB contribute to the corrosion of aluminum alloys [54, 68]. Aluminum sulfate or sulfide is expected to be formed under these conditions.

6.8. MIC OF ALUMINUM AND ALUMINUM ALLOYS

In seawater, for example, pure aluminum and aluminum alloys are often damaged by localized corrosion, which includes pitting corrosion, crevice corrosion, stress-corrosion cracking, and exfoliation corrosion. The localized corrosion is the synergetic result of all sorts of ions and all kinds of microorganisms in the seawater. The activity of microorganisms changes the local conditions near the surface of a metal substrate, and corrosion is accelerated. Microorganisms attach themselves to the surface of materials, colonize, proliferate, and produce a biofilm. Gradients of pH, dissolved oxygen, chloride, and sulfate exist in the biofilm, and localized corrosion conditions are created. There are two sorts of microorganisms: aerobic bacteria, such as sulfide oxidizing bacteria and iron bacteria, and anaerobic bacteria, such as sulfate reducing bacteria (SRB), which exist widely in sea bottom soil, seawater, and underground pipelines. Today, SRB are the most widely recognized and studied bacteria as the cause of microbiologically influenced corrosion (MIC) [69]. The actions of the most common microorganisms are summarized next.

6.8.1. Fungi and Bacteria (Space)

Fungi and bacteria isolated from the surfaces of structural materials from the Mir Space Station induced microbiologically influenced corrosion of aluminum alloy AMG-6 used for space applications. The fungal species identified are known to produce active organic acids. Microorganisms inhabiting space stations are of interest for testing the corrosion resistance of various materials used in space products (Figure 6.9) [57].

6.8.2. *Geotrichum* (Tropical Atmosphere)

In the hot and sticky Central American climate, a CD had stopped working and had developed an odd discoloration that left parts of it virtually transparent. Dr. Cardenes and co-workers discovered a fungus was steadily eating through the supposedly indestructible disk. The fungus had burrowed into the CD from the outer edge and had devoured the thin aluminum layer and some of the data-storing polycarbonate resin. Biologists had never seen this fungus, but concluded that it belonged to a common genus called *Geotrichum* [71].

6.8.3. Cyanobacteria and Algae (Polluted Freshwater)

Mariners at Vaal Dam Reservoir (freshwater) in South Africa experienced a high level of organic activity that led to the rapid corrosion of submerged aluminum components. In general, however, the type of die cast alloy used in this service is a low copper Al–Si alloy such as 360.0 or A360.0. Outboard motors are normally supplied with a corrosion-resistant

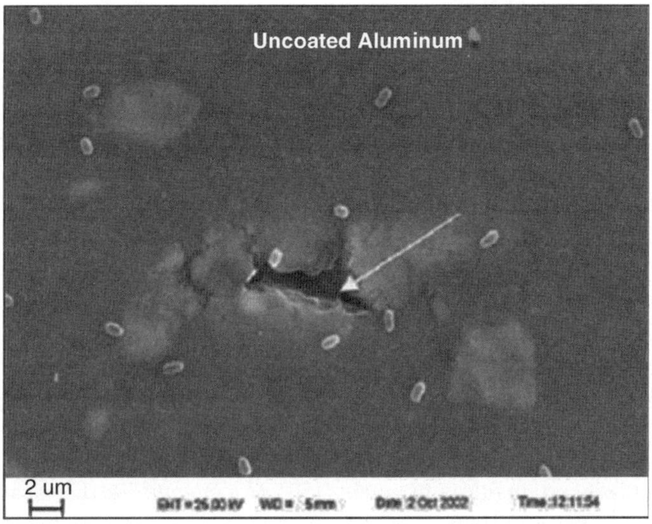

Figure 6.9 SEM image of endospores of *Bacillus. subtilis* dispersed onto uncoated Al spacecraft materials [70].

coating whose formulation is also protected for commercial reasons. The freshwater has the following characteristics (from a typical analysis): pH 7.9, hardness 109 ppm, alkalinity 89 ppm, chlorides 16 ppm, conductivity 32 uS/cm. This freshwater location contains an important population of bacteria and algae that might be involved in aluminum degradation [72].

6.8.4. Rod-Shaped Bacteria and Algae (Polluted Seawater)

Corrosion behavior of pure aluminum (99.7%) in contact with polluted harbor seawater during the early stages of microfouling formation (principally composed of rod-shaped bacteria and algae) shows sparsely distributed areas of pitting. An increase in the pollutant content of the seawater facilitates microbial settlement, shortening the period of colonization of the organisms [59, 73].

6.8.5. SRB (Industrial and Seawater)

Sulfate reducing bacteria (SRB) such as *Desulfovibrio*, *Desulfotomaculum*, and *Desulfomonas* produce tubercules on aluminum surfaces and induce pitting corrosion. Pitting corrosion was more serious on AA7075 than on AA2024. In oxygen-free media, it was shown that SRB generate tubercles on aluminum surfaces and induce pitting corrosion [68]. Many other bacteria can also induce corrosion on aluminum surfaces, such as *Pseudomonas aeruginosa* [74]. Finally, a bacteria called *Bacillus subtilis* seems to be involved in a corrosion-inhibition process in marine environments [74–76].

These recent case histories emphasize that engineers commissioning new stainless steel plants must be aware of potential MIC problems arising from stagnant water lying in the plant and hydrotesting with contaminated water. The bacteria that can cause rapid MIC problems are always lurking, waiting to strike. This argument can be extended to aluminum alloys.

Correct design, appropriate fabrication and construction, careful planning of the hydrotesting and commissioning procedures, observant operation of the plant, and good maintenance will avoid MIC problems. However, MIC can strike unexpectedly if any one of these aspects is neglected.

6.8.6. *Hormoconis resinae* (Kerosene)

The principal type of microorganism involved in aluminum corrosion is a fungus called *Hormoconis resinae* (formerly classified as *Cladosporium resinae*). Brown, slimy mats of *Hormoconis resinae* may cover large areas of aluminum alloy, causing pitting, exfoliation, and intergranular attack due to organic acids produced by the microbes and the differential aeration cells [63].This fungus produces a variety of organic acids (pH 3–4 or lower) and metabolizes certain fuel constituents [52]. *Hormoconis resinae* has the ability to thrive in the presence of kerosene and other hydrocarbons, which it uses as a carbon source for oxidation. It also produces spores that can survive extremes of temperature, only to "germinate" when more moderate conditions prevail. This fungus is a continuing problem in fuel storage tanks and in aluminum integral fuel tanks of aircraft. The problem of fungal growth in the aluminum fuel tanks of jet aircraft has diminished in recent years as the design of fuel tanks has improved to facilitate better drainage of condensed water, and as biocides such as ethylene glycol monoethyl ether and organoboranes have gained acceptance as fuel additives [63, 77].

6.9. MECHANISMS OF MIC AND INHIBITION

6.9.1. Corrosion Mechanisms

Microbiological organisms can initiate either general or localized corrosion. This derives from the ability of the organisms to change variables such as pH, oxidizing power, velocity of flow, and concentration of chemical species at the metal–solution interface [52]. Several investigators have reported that microorganisms produce different oxidizing agents, which lead to corrosion [76]. Biofilm formation frequently supports alterations at the metal–solution interface.

Biofilm (Slime) Microorganisms take up substances that are dissolved in water (nutrients) and produce cell material. Some microorganisms can secrete metabolic products or extracellular polymeric materials (glycocalyx). The organisms attach themselves to and grow on the surface of structural materials, resulting in the formation of a biofilm. The film itself can range from a microbiological slime film (poultice) on freshwater heat transfer surfaces to a heavy encrustation of hard-shelled fouling organisms on structures in coastal seawater. The slime helps glue the organisms to the surface, helps to trap and concentrate nutrients as food for microbes, and shields the organisms from biocides. This slime can change the pH and concentrations of different elements at the electrochemical interface by acting as a diffusion barrier. Its discontinuity, defects, or porosity can create the oxygen differential electrochemical cell [54, 60]. Microbial films will affect the general corrosion rate only when the film is continuous. However, this is not frequently the case since microorganisms form in discrete deposits or colonies, and the resulting corrosion is likely to be localized.

The biofilm can cause different oxidation–reduction conditions at the metal–solution interface, can alter the structure of inorganic passive layers, and can increase their dissolution and removal from the metal surface. It can facilitate the mechanical removal of protective films when the biofilm detaches [78].

Heavy Fouling In marine environments, microbial biofilms may contribute to the attachment of macroorganisms involved in heavy fouling. A heavy fouling of macroorganisms (barnacles, mussels, shellfish, etc.) decreases the amount of dissolved oxygen at the interface and acts as a barrier on structural steel at the splash zone, thus shielding the metal from the damaging effects of wave action. Also, a continuous film of bacteria, algae, and slime (microorganisms) can have the same beneficial effect as that of the macroorganisms. However, in most cases, these films are not continuous and an oxygen preferential cell is created. Microbial films are suspected of being capable of inducing pit initiation on aluminum, stainless steels, and copper alloys in marine and aqueous environments. Natural seawater is more corrosive than artificial seawater because of the living organisms [54, 76].

Corrosion can be influenced principally by the mechanisms discussed next.

6.9.1.1. Production of Differential Aeration Cells

A scatter of individual barnacles on a stainless steel surface creates oxygen concentration cells. The formation of a biofilm generates several critical conditions for corrosion initiation. Uncovered areas will have free access to oxygen and act as cathodes, while the covered zones act as anodes. Under-deposit corrosion (crevice corrosion) or pitting can occur. Insoluble corrosion products such as $Fe(OH)_2$ can help bacterial film to control the diffusion of oxygen to the anodic sites in the pit. Depending on the oxidizing capacity of the bacteria and the chloride ion concentration, the corrosion rate can be accelerated) [52].

6.9.1.2. Production and Consumption of Chemicals

This may involve the production of FeS and $Fe(OH)_2$ and an aggressive chemical agent such as hydrogen sulfide (H_2S) or acidity. Microorganisms may also consume chemical species that are important in corrosion reactions (e.g., oxygen or nitrite inhibitors). They may also break down the desirable physical properties of lubricating oils or protective coatings. The sulfur oxidizing bacteria can produce up to about 10% H_2SO_4. Other bacteria can produce organic acids such as formic and succinic acids [52].

It is important to note that the presence of a biofilm does not necessarily mean that there will always be a significant effect on corrosion. A uniform slime film formation on the piping of potable water handling systems and on the heat-transfer surfaces of low-temperature heat exchangers is inconsequential unless it leads to obstruction of the flow, a health hazard due to growth of the organisms, or localized corrosion [54].

6.9.1.3. Influence of Metallurgical Variables on MIC

Microstructure seems to play an important role in MIC. As an example, the two-phase weld metal appears to be the most susceptible area for an MIC attack. The surface and the microstructural inhomogeneities of a weld make it a high energy area and consequently it forms the anode of the corrosion cell even during non-MIC. Also, the surface condition, altered by the welding process, seems to be influential in the initiation of the preferential MIC attack [66].

6.9.2. Influence of Biofilms on Passive Behavior of Aluminum

Production of organic acids, mainly excreted by the biofilm fungi. These acids can induce or accelerate corrosion [63].

- Hindering of the transport of chemical species necessary for passivation of the metal surface [73]. Some bacteria and fungi, including *Hormoconis resinae* and *Pseudomonas aeruginosa*, consume corrosion-inhibiting ions such as NH_4^+ and NO_3^- [74].
- Facilitation of the removal of passive layers as biofilm detachment occurs.
- Formation of differential aeration cells as a result of a patchy distribution of the biofilm.
- Alteration of oxygen concentration gradients by acting as a diffusional barrier or through the direct use of oxygen in the microorganism's respiration [73].

6.9.3. Corrosion Inhibition by Microorganisms

Stabilization of protective films on the metal (e.g., biofilm exopolymers with metal-binding capacity) [78].

- Decrease in the corrosiveness of the electrolyte in restricted areas of the metal–solution interface by diminishing the dissolved oxygen concentration by respiratory activity or by neutralizing acidity. Altering or neutralizing corrosive substances (such as by catalysis), an enzyme secreted by many bacteria can induce the decomposition of H_2O_2 [78, 79].
- Decrease in hydrogen embrittlement and cracking by hindering the dissolution, dissociation, and absorption of hydrogen by organic compounds related to biofilm formation [55].

MIC Inhibition of Aluminum Bacteria can influence corrosion reactions in a beneficial way by causing microbiologically induced corrosion inhibition (MICI). A protective biofilm contains bacteria involved in the production of antimicrobial proteins active against SRB or other deleterious bacteria. These phenomena led to a new approach called *corrosion control using regenerative biofilms* (CCURB). EIS has been used to follow the pitting process of AA2024 during exposure to sterile artificial seawater (AS) for 30 days. In the presence of a bacterial biofilm produced by *Bacillus subtilis*, pitting was also observed during the first 2 days; however, for the remainder of the exposure period the Al alloy was passive. When the biofilm was genetically engineered to secrete polyglutamate or polyaspartate, an additional small increase in corrosion inhibition occurred. CCURB on AA2024 in AS cannot be solely due to a reduction of the oxygen concentration at the metal surface since the experimental value of the corrosion potential E_{corr} became more noble in the presence of bacteria, suggesting that production of an inhibiting species retained in the biofilm contributes to CCURB [75].

6.10. MIC PREVENTION AND CONTROL

The general approaches to maintaining a system free of biocorrosion problems vary with the materials of construction, environment, economics, and duty cycle of the equipment. The most common approaches involve the use of sterilization to keep the system clean, coatings, cathodic protection, and appropriate selection of materials. The most important step in prevention is to start with a clean system and to keep it clean. The three main ways that biocides work are as enzyme poisons or protein denaturants, oxidizing agents, and surface-active agents.

Sterilization by Physical Methods Flushing is of limited efficacy alone, but it can be supported by cleaners or jointly with chemical agents that induce biofilm detachment. Abrasive sponge balls can damage protective passive films, and nonabrasive sponge balls are not very effective with thick biofilms. Sterilization by physical methods such as irradiation (gamma or UV) for disinfection of materials and environments is also considered. Filtration can be used, for example, for the elimination of cells or spores from solutions, or for the sterilization of air for sterile rooms. Dry sterilization is based on the fact that microorganisms need a minimal water content for growth [55].

Sterilization by Chemical Methods Biocidal action has widely been used for many years to control biofilm formation in closed systems, such as heat exchangers, cooling towers, and storage tanks [54, 55]. These can be either oxidizing or nonoxidizing toxicants. Chlorine, ozone, and bromine are three typical oxidizing agents of industrial use. Nonoxidizing biocides are reported to be more effective than oxidizing biocides for overall control of algae, fungi, and bacteria, as they are more persistent, and many of them are pHindependent. Combinations of oxidizing and nonoxidizing biocides or of two nonoxidizing biocides are often used to optimize the microbiological control of industrial water systems. Typical biocides of the second type are formaldehyde, glutaraldehyde, isothiazolones, and quaternary ammonia compounds [60]. THPS (tetrakis-hydroxymethyl phosphonium sulfate) is a new promising compound with wide-spectrum efficiency on bacteria, fungi, and algae and low environmental toxicity. It is being widely used in the oil industry due to its ability to dissolve ferrous sulfide [78].

For situations where chemical treatments of the environment by biocides are not possible, the following options could be considered:

- Provision of nonaggressive surroundings, such as a backfill of sand or chalk around the material to ensure good drainage and aeration.
- Use of cathodic protection by sacrificial anodes or impressed voltage, sometimes in conjuction with a coating. However, in the case of aluminum alloys, cathodic protection could destroy the passive film and create a high aggressive pH.
- Use of protective coatings. In this respect the coating must be resistant to biodegradation and to chemical attack by the specific metabolic products of microbial activity, principally hydrogen sulfide and sulfuric acid [77].

Van Ooij et al. [80] used a silane pretreatment consisting of a mixture of hydrolyzed bis-amino silane and vinyltriacetoxy silane to prevent the formation of biofilms on Al surfaces. Silver nitrate, a mild biocide, was incorporated into the mixture at a 500 ppm level. Treatment of the metal was done by dip coating. Curing was done at 100 °C for 10 minutes. Pure cultures of bacteria (*Escherichia coli, Pseudomonas aeruginosa, Acinetobacter calcoaceticus,* etc.) strains were maintained under aerobic conditions. Aluminum coupons (AA2024-T3) were incubated at room temperature under aerobic conditions in a stationary position for 96 hours. The results confirmed that bacteria biofilms are inhibited from growing on aluminum surfaces by this treatment [80]. Norouzi et al. [76] used an organic dye to reduce the intensity of biocorrosion caused by *Pseudomonas aeruginosa.* They concluded that coloring aluminum with organic dye such as Quinizarin may protect the aluminum surfaces from MIC.

REFERENCES

1. E. Ghali,in *Uhlig's Corrosion Handbook*, 2nd edition, edited by R. W. Revie. Wiley, Hoboken. NJ, 2000, pp. 677–715.
2. ASM International Handbook Committee, in *Corrosion—Understanding the Basics*, edited by J. R. Davis. ASM International, Materials Park, OH, 2000, pp. 21–48.
3. S. G. Corcoran, in *ASM Handbook, Volume 13A, Corrosion: Fundamentals, Testing and Protection*, edited by S. D. Cramer and B. S. Covino Jr. ASM International, Materials Park, OH, 2003, pp. 287–293.
4. J. E. Hatch, *Aluminum: Properties and Physical Metallurgy*. ASM International, Materials Park, OH, 1984.
5. T. Ramgopal and G. S. Frankel, *Corrosion* **57**, 72 (2001).
6. Z. Szklarska-Smialowska, *Pitting and Crevice Corrosion*. NACE International, Houston, TX, 2005, pp. 327–329.
7. R. Braun, *Materials Science Forum* **519–521**, 735–740 (2006).
8. L. Muller and J. R. Galvele, *Corrosion Science* **17**, 179–193 (1977).
9. J. R. Galvele, S.W. De Micheli, I. L. de Muller, S. B. de Wexler, and I. L. Alanis, in *Localized Corrosion*, edited by R. Staehle, B. Brown, J. Kruger, and A. Agrawal. NACE-International, Houston, TX, 1984, pp. 580–599.
10. C. Blanc, B. Lavelle, and G. Mankowski, *Susceptibility to Pitting Corrosion of Pure Aluminum, 2024 Alloy and 6056 Alloy in Chloride-Containing Sulphate Solutions. Aluminum Alloys: Their Physical and Mechanical Properties*, Transtec Publications, Paris, 1996 pp. 1553–1558.
11. H. Holtan and H. Sigurdsson, *Korrosion* **28**, 475 (1977).
12. J. B. Bessone, R. A. Suarez Baldo, and S. M. De Micheli, *Corrosion* **37**, 533 (1981).
13. F. Sato and R. C. Newman, *Corrosion* **55**, 3–9 (1999).
14. E. McCafferty, *Corrosion* **57**, 1011 (2001).
15. S. Ruan, D. A. Wolfe, W. Zhang, and G. S. Frankel, *Technometrics* **46**, 69–75 (2004).
16. S. Ruan, D. A. Wolfe, and G. S. Frankel, *Journal of Statistical Planning and Inference* **126**(2), 553–568 (2004).
17. S. Zhao, D. A. Wolfe, T.-S. Huang, and G. S. Frankel, *Journal of Statistical Planning and Inference* **137**, 2405–2412 (2007).
18. T. Huang and G. S. Frankel, in *Proceedings of the 206th Meeting of The Electrochemical Society*, Honolulu, Hawaii. Electrochemical Society, Pennington, NJ, 2004, pp. 434–446.
19. C. Henon, G. Pouget, and T. Warner, *Materials Science Forum* **519–521**, 693–698 (2006).
20. T. Marlaud, B. Baroux, A. Deschamps, J. L. Chemin, and C. Hénon, *Materials Science Forum* **519–521**, 455–460 (2006).
21. M. Liao, G. Renaud, and N. C. Bellinger, *International Journal of Fatigue* **29**, 677–686 (2007).
22. M. A. Bellinger, M. Harrison, and T. Reeb, in *Proceedings of the USAF 2003 Aircraft Structural Integrity Program Conference*, Savannah, GA. DoD/FAA/NASA. New Orleans, LA, 2003.
23. E. Ghali, R. Potvin, and A. Galibois, *Opinion on the Corrosion of the Supporting Plate in Aluminum Alloy AA7004 of a Passenger Car*. André Galibois and Associates, Inc., Quebec, Canada, 1986.
24. R. Potvin, E. Ghali, and A. Galibois, *Opinion on the Corrosion of the Supporting Plate in Aluminum Alloy AA7004 of a Passenger Car*. André Galibois and Associates, Inc., Quebec, Canada, 1986.
25. E. Ghali, M. Elboujdaini, H. H. Le, and C. Lincourt, Protection de l'alliage d'aluminium AA7004 et de sa structure soudée par l'alliage CA 420, Report for Bombardier Inc., Montreal, Department of Mining, Metallurgy and Materials Engineering, Laval University, Quebec, Canada, 1988, p. 59.
26. ASM International Handbook Committee, in *ASM Specialty Handbook, Aluminum and Aluminum Alloys*, edited by J. R. Davis. ASM International, Materials Park, OH, 1998, pp. 199–484, 579–731.
27. ASM International Handbook Committee, in *Corrosion of Aluminum and Aluminum Alloy*, edited by J. R. Davis. ASM International, Materials Park, OH, 1999, pp. 63–74, 161–178.
28. R. G. Buchheit and C. S. Paglia, in *Proceedings of the 2003 Meeting of the Electrochemical Society*, edited by R. G. Buchheit, R. G. Kelly, N. A. Missert and B. A. Shaw,

Electrochemical Society, Pennington, NJ, 2004, pp. 94–103.

29. D. Burford, C. Widener, and B. Tweedy, *Airframer* **November**, 3–7 (2006).
30. D. A. Wadeson, X. Zhou, G. E. Thompson, P. Skeldon, L. Djapic Oosterkamp, and G. Scamans, *Corrosion Science* **48**, 887–897 (2006).
31. K. F. Krysiak and ASM Committee on Corrosion of Weldments, in *Metals Handbook, Volume 13, Corrosion*, 9th edition, edited by J. R. Davis. ASM International, Materials Park, OH, 1987, pp. 344–368.
32. E. Bilgram and P. J. Winkler, in *Proceedings of the Third International Conference on Aluminium Weldments*, Aluminium-Verlag, Duesseldorf, West Germany, 1985, pp. V. 7.1–V. 7.14.
33. Aluminum Design Association, Washington, D C (2000).
34. N. J. H. Holroyd, W. Hepples, and G. M. Scamans, in *Proceedings of Corrosion Cracking Program*, ASM, Materials Park, OH, 1986, pp. 291–302.
35. X. Y. Li, Y. G. Yan, L. Ma, Z. M. Xu, and J. G. Li, *Materials Science and Engineering A* **382**, 82–89 (2004).
36. K. G. Kent, *Journal of the Australian Institute of Metals*, **15**(3), 171–178 (1970).
37. *Corrosion of Nonferrous Alloy Weldments*, Vol. 13A. ASM International, Materials Park, OH, 2003.
38. S. R. Koteswara Rao, G. Madhusudhan Reddy, K. Srinivasa Rao, M. Kamaraj, and K. Prasad Rao, *Materials Characterization* **55**, 345–354 (2005).
39. A. J. Davenport, M. Jariyaboon, C. Padovani, N. Tareelap, B. J. Connolly, S. Williams, and E. Siggs, *Materials Science Forum* **519–521**, 699–704 (2006).
40. M. Jariyaboon, A. J. Davenport, R. Ambat, B. J. Connolly, S. W. Williams, and D. A. Price, *Corrosion Science* **49**, 877–909 (2007).
41. C. S. Paglia, K. V. Jata, and R. G. Buchheit, *Materials Science and Engineering A* **424**, 196–204 (2006).
42. M. K. Cavanaugh, N. Birbilis, R. G. Buchheit, and F. Bovard, *Scripta Materialia* **56**, 995–998 (2007).
43. J. Datta, S. Datta, M. K. Banerjee, and S. Bandyopadhyay, *Composites* **A35**, 1003–1008 (2004).
44. L. H. Hihara, in *Metal Matrix Composites*, 2nd edition, edited by R. Baboian. ASTM International, West Conshohocken, PA, 2005. pp. 637–655.
45. D. M. Aylor and R.M. Kain, in *Assessing the Corrosion Resistance of Metal Matrix Composite Materials in Marine Environments: United States and Japan*, edited by J. R. Vinson and M. Taya. American Society for Testing and Materials, Philadelphia, 1983, pp. 632–647.
46. M. C. Chaturvedi and D. L. Chen, *Materials Science and Engineering A* **387–389**, 465–469 (2004).
47. X.-G. Chen, *Interface Reaction of Boron Carbide in Aluminum Matrix Composites and Its Control.* EPD Congress 2005, edited by M. E. Schlesinger, The Minerals, Metals and Materials, TMS Annual Meeting at San Francisco, TMS, Warrendale, PA, 2005, pp. 101–106.
48. H. Chen, J. Liu, and W. Huang, *Materials Science and Engineering A* **415**, 291–296 (2006).
49. J. Abenojar, M. A. Martinez, and F. Velasco, *Journal of Alloys and Compounds* **422**, 67–72 (2006).
50. R. Emmerich, W. Ensinger, and B. Enders, *14th International Symposium on the Packaging and Transportation of Radioactive Materials*, PARTAM 2004, Paper 038, Ramtrans Publishing, UK, Vol. 16(2), 2005, pp. 151–155.
51. K. Hayashi, K. Iwasa, K. Araki, and R. Asano, *14th International Symposium on the Packaging and Transportation of Radioactive Materials*, PARTAM 2004, Paper 100, Ramtrans Publishing, UK, Vol. 16(4), 2005, pp. 295–300.
52. S. C. Dexter, in *Metals Handbook, Volume 13A, Corrosion*, edited by S. D. Cramer and B. S. Covino, Jr. ASM International, Materials Park, OH, 2003, pp. 398–416.
53. C. J. McNamara, T. D. Perry, M. Zinn, M. Breuker, and R. Mitchell, Biodeterioration of concrete and stone, in *Proceedings of The Corrosion Research Tropical Symposium, Microbiologically Influenced Corrosion*, edited by B. Little, NACE International, Houston, 2002, pp. 113–121.
54. S. C. Dexter, in *Metals Handbook, Volume 13, Corrosion*, 9th edition, edited by L. J Korb and D. L. Olson. ASM International, Materials Park, OH, 1987, pp. 41–44, 87–88.
55. W. Sand, in *Corrosion and Environmental Degradation*, edited by M. Schutze. Wiley-VCH, Weinheim, Germany, 2000, pp. 172–202.
56. J.-P. Gravel, Training Report in Chemical Engineering—MIC of Aluminum and Aluminum Alloys and Rational Biocorrosion of Magnesium and Magnesium Alloys, Laval University, Quebec, Canada, 2007.
57. T. A. Alekhova, A. A. Aleksandrova, T. Yu. Novozhilova, L. V. Lysak, N. A. Zagustina, and A. M. Bezborodov, *Applied Biochemistry and Microbiology* **41**, 382–389 (2005).
58. N. Häubner, R. Schumann, and U. Karsten, *Microbial Ecology* **31**, 285–293 (2006).
59. P. Roberge, in *Handbook of Corrosion Engineering*, edited by R. Esposito. McGraw-Hill, New York, 2000, pp. 55–216.
60. L. H. G. Morton and S. B. Surman, *Biodeterioration & Biodegradation*, 203–221 (1994).
61. M. D. Martin, S. Broughton, and M. Drangsholt, *Contact Dermatitis* **48**, 331–336 (2003).
62. G. R. Lloyd, in *Biodeterioration of Constructional Materials: Proceedings of the Summer Meeting of the*

Biodeterioration Society, edited by L. H. G. Morton. Biodeterioration Society, Delft, The Netherlands, 1987, p. 101.

63. J. F. D. Stott, in *ASM Handbook, Volume 13A, Corrosion: Fundamentals, Testing, and Protection*, edited by S. D. Cramer and B. S. Covino, Jr. ASM International, Materials Park, OH, 2003, pp. 644–649.
64. B. J. Little, P. A. Wagner, and F. Mansfeld, *Microbiologically Influenced Corrosion, Corrosion Testing Made Easy*, Vol. **5**. NACE International, Houston, TX, 1997.
65. H. a. H. C. Sass, *Systematic and Applied Microbiology* **27**, 541–548 (2004).
66. R. P. George and P. Muraleedharan, in *Corrosion of Austenitic Stainless Steels; Mechanism, Mitigation and Monitoring*, edited by H. S. Khatak and B. Raj. Narosa Publishing House, New Delhi, 2002, pp. 218–240.
67. W. A. Hamilton and S. Maxwell, in *Natural Biofilms, Proceedings of Biologically Induced Corrosion*, edited by S. C. Dexter. NACE International, Houston, TX, 1986, pp. 131–136.
68. X. L. Wu, J. H. Liu, and X. Y. Chen, *Acta Metallurgica Sinica* **12**, 867–873 (1999).
69. J. Liu, H. Wu, and S. Li, in *Proceedings of the Second International Conference on Environment Sensitive Cracking and Corrosion Damage*, edited by H. N. M. Matsumura, K. Nakasa, and Y. Isomoto. ESCCD, Hiroshima, Japan, 2001, pp. 109–112.
70. A. C. Schuerger, J. T. Richards, P. E. Hintze, and R. G. Kern, *Astrobiology* **5**, 545–549 (2005).
71. J. Garcia-Guinea, V. Cárdenes, A. T. Martínez, and M. J. Martínez, *Naturwissenschaften*, **88**, 351–354 (2001).
72. P. J. B. Scott and M. Davies, *Aluminium Today* **12**, 49–50 (2000).
73. H. A. Videla, M. F. L. de Mele, and G. Brankevich, *Corrosion* **44**, 423–426 (1988).
74. E. S. Ayllon and B. M. Rosales, *Corrosion* **44**, 638–644 (1988).
75. D. Örnek, A. Jayaraman, T. K. Wood, Z. Sun, C. H. Hsu, and F. Mansfeld, *Corrosion Science* **43**, 2121–2133 (2001).
76. P. Norouzi, J. Hamedi, M. R. Ganjali, and A. Ahmadi, *Materials and Corrosion* **57**, 538–542 (2006).
77. J. F. D. Stott, *Metal and Materials* **4**, 224–229 (1988).
78. H. A. Videla and L. K. Herrera, *International Microbiology* **8**, 169–180 (2005).
79. J. P. Busalmen, M. Vasquez, and S. R. de Sanchez, *Electrochimica Acta* **47**, 1857–1865 (2002).
80. W. J. van Ooij, A. Seth, T. Mugada, K. Suryanarayanan, and D. B. Oerther, *Protection of Aluminium Alloys against Corrosion Fatigue Cracking, Erosion Corrosion and Microbiologically-Induced Corrosion, Report*, Department of Chemical and Metallurgical Engineering, University of Cincinnati, OH, 2004, 3 pp.

Chapter 7

Mechanically Assisted Corrosion of Aluminum and Its Alloys

Overview

Mechanically assisted corrosion can be divided into two categories that can interact: *erosion* and *fatigue*. Mechanical parameters interacting with corrosion kinetics are the cause of the premature wear of aluminum. Environmental conditions, chloride concentration, and suspended particles largely influence the progression of mechanically assisted corrosion of aluminum alloys. The effects of various parameters that influence erosion corrosion such as water drop impingement, suspended particles, cavitation, and fretting are examined. The parameters, kinetics, and mechanisms of corrosion fatigue, as observed mainly for aluminum alloys, are given. We give some suggestions on how to reduce or control those types of corrosion.

The aluminum alloys have a relatively low resistance against corrosion fatigue. For low-stress, high-cycle fatigue, crack initiation spans a large portion of the total lifetime. In aluminum alloys exposed to aqueous chloride solutions, localized corrosion, such as pitting or intergranular corrosion, provides stress concentrations, greatly lowers fatigue life, and initiates cracks. Initial crack propagation is normal to the axis of principal stress. Corrosion fatigue failures of aluminum alloys are characteristically transgranular and thus differ from SCC failures that are normally intergranular.

Key parameters concerning cyclic stresses and the environment are discussed. The two main mechanisms of corrosion fatigue are *anodic slip dissolution* and *hydrogen embrittlement*. Corrosion fatigue is influenced by the phases that characterize the microstructure of aluminum alloys after different heat treatments.

Case histories and laboratory research studies of corrosion fatigue of high-strength aluminum alloys and some susceptible Al–Mg–Si alloys are detailed. Examples of modeling of the propagation of fatigue cracks in aluminum alloys are given. One or a combination of the following procedures is recommended for corrosion fatigue prevention: good design, appropriate selection of the alloy, and surface treatment and preparation such as peening of alloys and welded assemblies. Inhibitors, organic coatings, and cathodic protection are also recommended with some precautions to consider.

Corrosion Resistance of Aluminum and Magnesium Alloys: Understanding, Performance, and Testing By Edward Ghali

A. EROSION CORROSION

The effects of mechanical stresses (residual, structural, and cyclic) can lead to one or more forms of corrosion, such as uniform corrosion or the initiation and propagation of localized corrosion (e.g., pitting), or to failure by corrosion fatigue and/or stress-corrosion cracking. The effects of wear, assisted by the processes of corrosion, reduce the useful lifetime of aluminum alloys. Frequent corrosion failures are caused by erosion corrosion, fretting corrosion, fretting fatigue corrosion, and corrosion fatigue. It is important to consider and understand this complex phenomenon, especially for alloys with low or average mechanical properties. The interactions between the mechanical phenomena of destruction and corrosion are still poorly understand, especially for aluminum alloys, even though detailed studies are in progress.

Aluminum alloy designs are closely related to the development of aircraft, which are generally exposed to erosion corrosion. As an example of the importance of this form of corrosion, the Boeing 747, built during the 1960s, contains up to 80% aluminum whereas the newer model Boeing 787 contains only 20% of aluminum alloys [1, 2]. In noncorrosive environments, such as high-purity water, the stronger aluminum alloys have the greatest resistance to erosion corrosion because resistance is controlled almost entirely by the mechanical components of the system. In a corrosive environment, such as seawater, the corrosion component becomes the controlling factor: thus resistance may be greater for the more corrosion-resistant alloys even though they are lower in strength. In the case of neutral solutions, the velocity of the solution, up to about 6 m/s, has little effect on the rate of attack. In some cases, increased movement of the liquid may actually reduce attack by assuring greater uniformity of the environment. However, increases in velocity decrease the variation in pH that can be tolerated without erosive attack occurring [3].

The current trend in industrial processes is to maximize productivity through higher flow rates. Components subjected to liquid flows containing solid particles show higher wear rates. This type of wear is considered to be erosion corrosion and is caused by the synergistic effect of physical abrasion and electrochemical corrosion. Tribocorrosion is a function of two-body or three-body erosion corrosion and can be investigated through different experimental methods since the tribological contact has various natures: impingement, rolling, sliding, and fretting [2, 4].

7.1. IMPINGEMENT WITH LIQUID-CONTAINING SOLID PARTICLES

The phenomenon of corrosion by destruction of the passive layer can be started by a projection of solid particles. At high altitude (above 25,000 feet), particles of great hardness, like silica dust, can strike the exposed surfaces of an airplane. That will involve a destruction of the passive layer of aluminum and this accelerates corrosion. The phenomenon can also appear on the turbine blades of the engine (Figure 7.1).

Smith et al. [4] used highly sensitive measurements and analog–digital conversion devices to realize single-particle impact and detect the repassivation transients after particle impingement on pure aluminum (99.99% Good fellow) at different angles. Simultaneous use of newly constructed slurry-jet and microelectrodes as targets allowed the highly reproducible single-particle impacts. The specimens were polished in a mixture of perchloric and acetic acids (22% to 78%) and the jet was turned on for 60 s at 20 V and room temperature. A high-speed pump was used to circulate the electrolyte (0.1 M acetate buffer, pH 6.0). The samples were polarized to a fixed potential (2 V/hydrogen electrode same

Figure 7.1 Appearance of corrosion that destroys turbine blades. Note the folded and ruptured appearance [5].

solution (HESS)) to ensure well-defined oxide conditions. This in turn allowed correlation between the charge consumed during repassivation in a potentiostatic experiment and the resulting damage.

Smith et al. [4] examined the effects of the angle of impact of projection on erosion corrosion (tribocorrosion) and the possibility of detecting the repassivation after projection of the particle on the aluminum surface. A variation of the jetting angle (30°, 45°, 60°, and 90°) can have a correlation with the static potential, which causes the surface damage, as can be deduced from the crater morphology. Figure 7.2 shows two electronic photographs of these same craters that correspond to the experimental angles of 90° and 30°. The surface areas of the craters were between 200 and 300 μm^2.

The impact crater formed under perpendicular impingement is nearly perfectly round, whereas that at lower angles is elongated and shows scratch marks. They are not completely activated after the impact since the oxide layers are cracked or altered by the shock of the particles. The angle of attack of these particles also affects the level of deterioration of the passive layer. One can note that a perpendicular impact angle causes less damage than that at 30°. This can be the result of the malleability of aluminum as compared to the nano-roughness of abrasive zirconium particles with a nominal 125 μm diameter. As a result of the nano-roughness of the impacting zirconium particles used, it was possible to ascribe the difference in the two mechanisms of erosion as *indentation* and *scratching*. A model that links the normal and lateral components of the kinetic energy to these two wear mechanisms has been developed [2, 4].

The effects of these impacts are divided into two components: a *normal moment* and the *side impact movement* that removes material. The suggested equation uses the kinetic

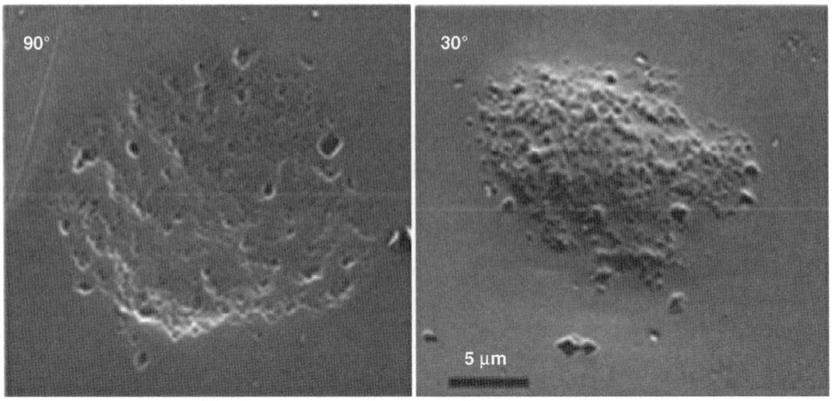

Figure 7.2 Two photographs of the craters as a function of the examined angles of the jet [2, 4].

energy E_{Kin} due to the impact of the particle and can be compared to the impulse $p = mv$ drawn from the mass of the particle:

$$E_{Kin} = p^2/2m$$

One must mention that this formula is not based strictly on a kinetic model, but was obtained by comparison of the various behaviors. Apparently, the kinetic energy of the particle impact activates normal and lateral components that correspond to two different mechanisms of erosion—indentation and scratching of the passive layer.

Figure 7.3 shows the transients for the two extreme jetting angles among the four studied: 90°, 60°, 45°, and 30°. The recorded transients are subsequently analyzed in order to determine the following values: the peak current (I_{max}), the time of the peak current (t_o), the uninterrupted current (I_{rise}), repassivation time (Δt), the consumed load (Q), and the recovery current (I_{back}). Based on these data, the calculated depassivated surface area for the four jetting angles after impingement is between 2 and 7 μm^2 (Table 7.1). For the sake of comparison, the value of the depassivated area for the jetting angle of 30° was more than double that of 90° because of the existence of the two possible mechanisms of erosion—indentation and maximum scratching of the passive layer at 30°.

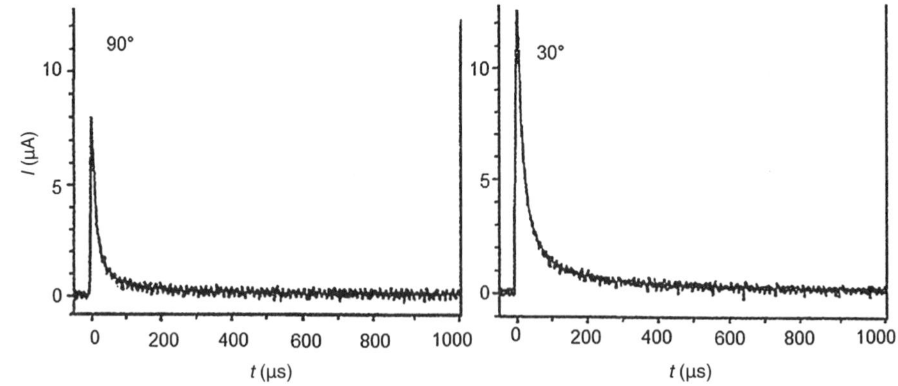

Figure 7.3 Mechanism of repassivation as a function of two extreme experimental angles of impact [2, 4].

Table 7.1 Calculated Depassivated Areas of the Four Jetting Angles

Angle	Q (nC)	n (Al_2O_3) (amol)	V (Al_2O_3) ($\times 10^{-20}\,m^3$)	A_{ox} (Al_2O_3) (μm^2)
90°	0.34	0.58	1.51	2.89
60°	0.30	0.51	1.33	2.55
45°	0.55	0.95	2.44	4.68
30°	0.80	1.38	3.55	6.81

Obviously, the crater is not completely depassivated on impact; rather, the oxide is cracked or/and pierced by the impacting particle. At lower angles, scratching contributes an additional component to the total amount of wear. The acquired transients show that the current recedes back to the initial background level in well under 1 s. An additional determination of the charge can be performed on the linear part of the double logarithmic plot of the transient by extrapolating this current decay to 100 s. Subsequent integration of this extrapolated transient yields a better measure of the true repassivation current.

Li et al. [6] examined the influence of the composition of the environment and the electrochemical potential on the erosion corrosion (EC) of commercially pure aluminum alloy AA1100: Cu $\leq$ 0.10 wt %; Si $\leq$ 0.50 wt %; Fe $\leq$ 0.70 wt %; Mn $\leq$ 0.10 wt %; Zn $\leq$ 0.10 wt %; Al $\geq$ 99 wt %. The electrolyte was composed of aqueous silica slurries containing different solutions of 0.5 M sodium chloride, phosphate buffer, 0.1 M acetic acid + 0.05 M sodium acetate solution, and 0.1 M sodium carbonate. The aqueous phase is driven by a pump into the ejector, creating a pressure differential, which sucks up a concentrated slurry from the reservoir. A broad potential range of voltage from −2.7 to 2.7 V/SCE was examined and the scan rate was 10 mV/s.

Three regimes of EC rates can be identified in the passive film-forming slurries as a function of the imposed potential. Regime I corresponds to an applied potential up to −0.2 V and depends on the pH of the slurry. In this regime, there is a type of cathodic protection and the EC rate expressed as the mass loss per unit mass of impinging particles is approximately 0.15 mg/kg, independent of applied potential and slurry composition. The damage could be mechanical; however, the influence of hydrogen ion reduction at negative or active potentials can create certain damage and possible embrittlement of the oxide coating.

In regime II, the EC rate increases almost linearly with the applied potential and is different in different slurries. As the applied potential increases further, the EC rate increases up to a certain potential, E_s, above which the effect of applied potential is small (regime III). E_s varies for different slurries (phosphate buffer and acidic slurries). In the case of sodium chloride solution, shown as an example in Figure 7.4, scanning electron microscope (SEM) studies showed that regime I gave typical ductile erosion behavior. In regime II, the morphology of the eroded surface changed little, while in regime III, above pitting potential, extensive pitting corrosion was evident. The pitting potential E_p was about −75 V/SCE and the EC rate increased sharply for more positive or higher applied potentials.

In sodium carbonate slurry, the dependence of erosion corrosion on the applied potential is totally different from that observed in the passive film-forming slurries. At potentials below −2.3 V/SCE, the EC rate is independent on the imposed potential. The EC rate then decreased with increasing potential in the range from −2.3 to −0.5 V/SCE. The EC rate increased again as the applied potential was increased further [6].

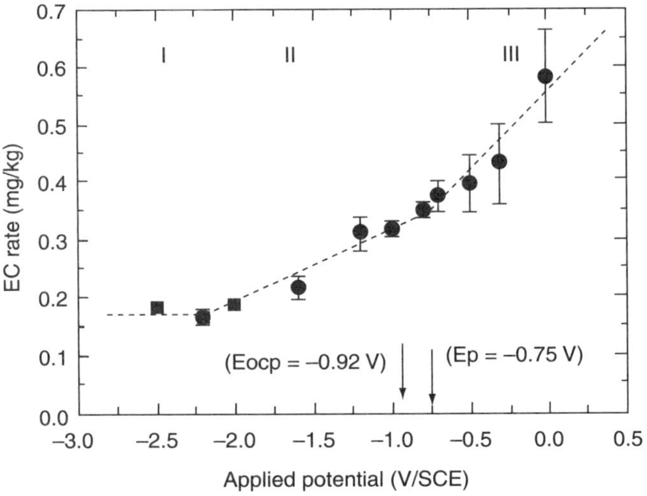

Figure 7.4 Erosion corrosion rate versus applied potential for aluminum in 0.5 M NaCl slurry under un impact velocity of 3.58 m/s, particle concentration of 2%, and impinging angle of 50° for 0.5 h [6].

7.2. CORROSION BY CAVITATION

Cavitation occurs when gas bubbles are in suspension in a fluid. When bubbles that are in contact with or very close to a solid surface collapse, they collapse asymmetrically. Consider a spherical bubble impacting a plane solid surface. The bubble becomes elongated with a tail and then collapses. The jet from the bubbles is believed to cause the cavitation erosion on a solid wall. The gas bubbles go through an implosion when they strike a surface, as shown in Figure 7.5. Implosion of vapor bubble creates a microjet of liquid—a microscopic "torpedo" of water that is ejected from the collapsing bubble at velocities that may range from 100 to

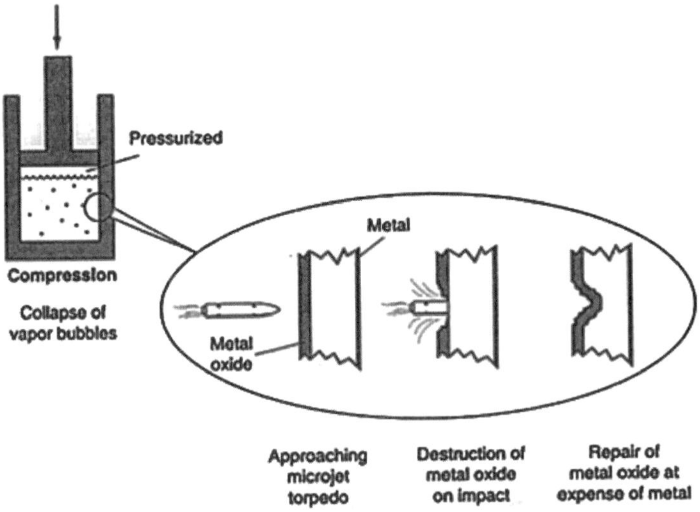

Figure 7.5 Schematic representation of the destruction of the passive layer produced by the impact of a microjet [7].

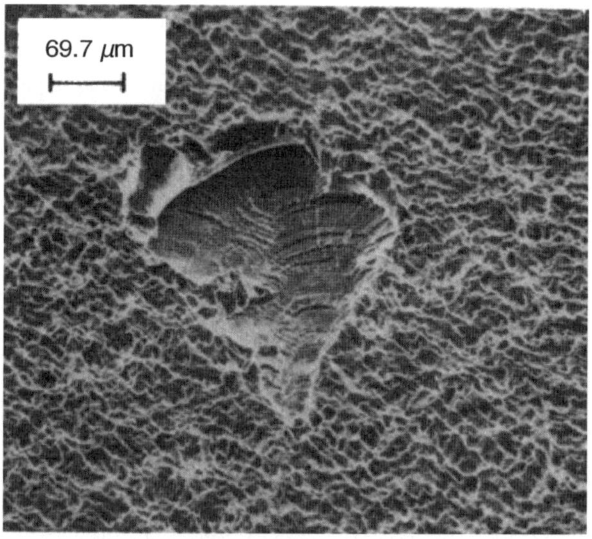

Figure 7.6 Erosion pit in as-quenched Al–4Cu after exposure to cavitation for 17.5 min [7, 8].

500 m/s. When the torpedo impacts on the metal surface, it dislodges protective surface films and/or locally deforms the metal itself [7].

Face-centered cubic (fcc) metals like aluminum are less sensitive to strain rate than are body-centered cubic (bcc) and hexagonal close packed (hcp) metals. Consequently, their response to cavitation is similar to their quasistatic mechanical behavior in that they are highly ductile and fail by a void growth and coalescence mechanism or by a ductile rupture. Studies of the multiphase Al–Mg, Al–Cu, and Al–Zn–Mg–Cu alloys have shown that the size and dispersion of the second phases are the determining factors for cavitation corrosion. Al–Mg alloys exhibit generally better cavitation corrosion resistance than Al–Cu alloys because of the greater propensity for strain aging in Al–Mg alloys. Depending on the increase of solute content or other phases and the degree of hardening, the mode of failure changes from ductile rupture characteristic of fcc metals to the development of flat-bottomed pits that grow parallel to the surface and exhibit striated surfaces reminiscent of fatigue fracture surfaces (Figure 7.6) [7, 8].

A case history of erosion cavitation of a water-cooled aluminum alloy, 6061-T6, shows also the importance of proper heat treatment and the microstructure of the alloy. It is recommended that the pressure in the coolant be raised in order to suppress cavitation bubbles [8].

7.3. WATER DROP IMPINGEMENT CORROSION

The general term could be liquid impingement corrosion or impact by liquid drops or jets. A common type of corrosion seen on the leading edges of helicopter blades and the wings of airplanes is erosion corrosion caused by the impact of water drops (water drop impingement). It is similar to cavitation in that it causes pitting of surfaces and may involve a cavitation mechanism; however, propagation of the water drop impingement rupture in ductile materials shows a directionality that is related to the angle of attack of the drops [9]. Two areas are most notable for this type of corrosion: steam turbines and helicopter rotor blades. In turbines, condensation of steam produces droplets that are carried into the rotor blades, where they can cause surface damage. Raindrop erosion on helicopter blades is the result of elastic compression waves produced by multiple impacts and their interaction. This action generates tensile stresses just below the surface and causes cracking. The raindrops strike the helicopter blades and create a wave of elastic compressions on the attacked surfaces that can exceed the yield stress of aluminum (from 120 to 390 MPa). This creates a synergy that accelerates the process of degradation, mainly on the leading edges of the wings and the control surfaces. Corrosion influence can be amplified with acid rain and pollutants such as chloride ions or deposited particles can lead to localized attack and initiate pitting. Activation of the aluminum alloy is a possibility if the pH of the drop is out of the passive zone (E–pH diagram) of the considered aluminum alloy [2].

As an example, in 2000 a helicopter blade split, which led to the destruction of the apparatus as well as the death of the pilot. Laboratory analysis showed that the failure of the blade, which was manufactured out of an aluminum alloy, occurred by a ductile fracture. This fracture was caused by a loss in the thickness of the transverse wall of the blade, the aluminum being degraded by water drop impingement corrosion. A thin layer of polymer fixed on the external surface of the blade, which was expected to protect the aluminum from corrosion, fell apart partially by water drop impingement, leaving the surface without protection from raindrops. A complete coating system consisting of adherent layers of inorganic or organic compounds, applied as primary, secondary, and finishing coats, should adequately protect the product. This should be coupled with maintenance, regular inspection, improved design, and proper alloy choice as recommended by the Safety Board [10].

Two-Phase Flow In the valves and elbows of pipes, fluid flow can quickly change direction because of the change in the geometry of the course. One often notes important degradation in these types of parts. This damage is caused by erosion corrosion created not only by the dynamic pressure of the fluid, but also by the impact of droplets of the following fluid. This phenomenon has also been observed on the airfoils of planes. This study is important for the aeronautical industry, since several types of planes have wing surfaces, control surfaces, and fuselages in a nonprotected state [11].

The damage created by corrosion and by two-phase movement on AA5056BD at high speed (177 m/s) and high temperature (392 K) was examined. The projection of liquid and vapor phases was produced by an injector; the two-phase flow was set up, and their relative speeds were held constant. The mass loss of aluminum was evaluated for the impact of the fluid drops and vapor at the metal surface and assisted by corrosion.

The loss of metal mass due to pits is generally minimal since perforation or penetration is the major issue in pitting. Uniform corrosion causes more metal loss and degrades more surface than localized corrosion. Effectively, when droplets frequently impact a protective film, fracture and recovery of the film could occur repeatedly and the specimen is uniformly attacked. Finally, it seems that, within a certain range, the mass loss at the metal surface was independent of the length of the droplet [2].

Metal mass loss by corrosion only of the alloy is negligible in a solution containing a static liquid. Erosion is an important mechanism when combined with corrosion produced by a jet of fluid in liquid and gaseous states. Two-phase flow of fluids, in the liquid and vapor states, is often observed in production pipelines. In the case of the moving phases, the inclusions in AA5056 are mainly made up of iron and silicates and multiple pits are often formed on the uniformly damaged surface. The production of pits and pitting growth are related to inclusions that were observed in the SEM cross section of the pit as well as to cracks generated in the passive layer of the alloy [11].

If drops of the flowing fluid frequently strike the pacified layer, pits have a tendency to form a uniformly degraded surface. Instantaneous mechanical damage occurs due to the impact of the drop. The attacked surface corrodes, quickly, assisted by the electrochemical cell between the inclusion and the base metal. The aluminum around silicon and iron inclusions in the microstructure corrodes quickly by galvanic effect. Finally, the inclusion is forced out of the pit by subsequent droplet bombardments and a deeper pit appears at the bottom of the existing pit.

7.4. FRETTING CORROSION

Fretting is the abrasive wear of two touching surfaces subject to cyclic relative motions of extremely small amplitude, that is, not more than 0.1 mm. Fretting occurs by contacting asperities on the mating surfaces continually welding together and then breaking. This leads to surface pitting and the transfer of the metal from one surface to another. In addition, the small fragments of broken metal oxidize, forming oxide particles that are harder than the metal itself for most of engineering metals. Soft aluminum alloys exhibit higher susceptibility to fretting than harder ones of similar type. Fretting is more serious in the presence of oxygen and is greatest under perfectly dry conditions. Fretting damage increases with contact load, slip amplitude, and number of oscillations.

The production of oxide debris in normal atmospheric conditions has led to the term *fretting wear* or *fretting corrosion.* Fretting corrosion is an increased degree of deterioration that occurs because of repeated corrosion or oxidation of the freshly abraded surface and the accumulation of abrasive corrosion products between these surfaces. Although fretting is often limited to small localized patches of wear, it can provide a path for leakage (e.g., valve seats) or an initiation site for fatigue. Hard abrasive particles break off from oxide corrosion products and abrade the metal, maintaining fresh active surface for further corrosion.

Fretting is often found on fuel tank access doors. Fretting corrosion can be controlled by lubrication of the faying surfaces, by restricting the degree of movement, or by the selection of materials and combinations that are less susceptible to fretting. Couples, such as aluminum on aluminum, aluminum on steel, and zinc-plated steel on aluminum, show low resistance to fretting corrosion. Zinc, copper plate, nickel plate, and iron plate on aluminum show moderate resistance to fretting corrosion, whereas silver plate on aluminum plate shows high resistance to fretting corrosion [3].

7.5. FRETTING FATIGUE CORROSION

Fretting fatigue corrosion or *contact fatigue* [1] has a more damaging effect than fretting corrosion when one of the contacting surfaces is subjected to a cyclic stress. The stresses can be moderate to high bearing stresses and can be seen in practice in rolling surfaces of flap trucks, flap carriage component journals, pins, and bearings. Fatigue strength or endurance limits can be reduced by as much as 50–70%, well below that of nonfretted specimens. During fretting fatigue, cracks can be produced at very low stresses, while in fatigue without fretting, the initiation of small cracks can represent 90% of the total component life.

Propagation and Morphology Under cyclic loading, fretting contact between the mating crack faces, pumping of the aqueous environments to the crack tip by the crack walls, and continual blunting and resharpening of the crack tip by the reversing load influence the rate of dissolution [12]. The fretting fatigue or the combined action of fretting and reversing

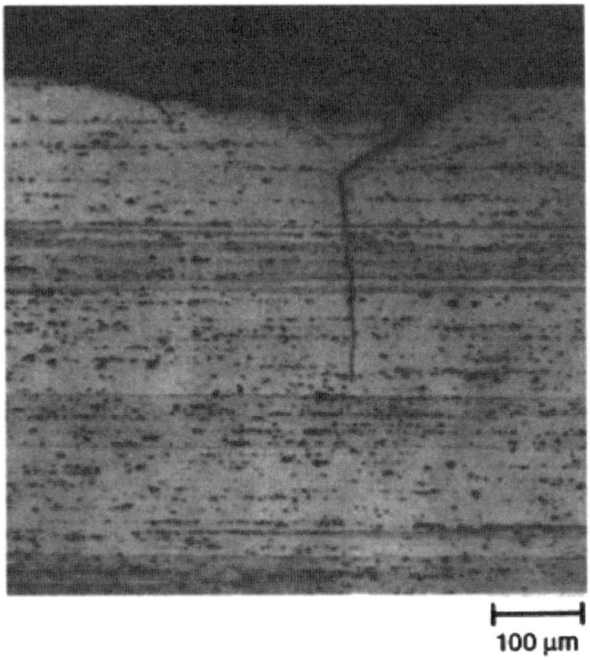

Figure 7.7 Section through a bar of aged Al–4Cu alloy showing a crack initiated by fretting fatigue corrosion or contact corrosion [9, 13].

bending stress accelerates the crack initiation and increases the rate of crack propagation, leading to reduced fatigue strength under fretting.

A fretting fatigue crack initiates in the fretting scar zone and is located at the boundary of a fretted zone. As the crack opens, fretting debris enters and this increases the propagation force. It propagates into the surface at an angle to the surface, as shown in Figure 7.7 for aged Al–4 Cu alloy [9]. Once the crack propagation reaches a depth where it is no longer influenced by surface contact stress, it progresses as a fatigue crack normal to the surface [13].

7.6. PREVENTION OF EROSION CORROSION

Corrosion inhibitors and cathodic protection have been used to minimize erosion corrosion, impingement, and cavitation on aluminum alloys. Lubricants that reduce the frictional force between the contacting surfaces, particularly when used in conjunction with a surface treatment such as phosphating, reduce fretting damage. Molybdenum disulfide is particularly effective.

B. CORROSION FATIGUE

7.7. GENERAL CONSIDERATIONS AND MORPHOLOGY

Corrosion fatigue (CF) causes the sequential stages of metal damage that evolve with accumulated load cycling in an aggressive medium and result from the interaction of irresistible cyclic plastic deformation with localized chemical or electrochemical reaction. Corrosion fatigue has no minimum safe cyclic stress amplitude limit in air, while fatigue in current atmospheres could have one and this obviously reflects our interest to follow up the published fatigue studies. Damage due to corrosion fatigue corresponds to the sum of the damage by corrosion and fatigue acting separately and the synergetic effect of these two parameters. As an example, the shaft of a ship's propeller, slightly above the water line, can normally function until a leak occurs, allowing the water to impinge on the shaft in the area of maximum alternating stress. Also, pipes carrying steam or hot liquids of variable temperature may fail because of periodic expansion and contraction (thermal cycling) [14,15].

Corrosion fatigue can be initiated through pitting and could be influenced by the phases that characterize the microstructure of these alloys after different heat treatments, and these studies should help to draw some key conclusions on the corrosion fatigue of these microstructures. Localized corrosion, such as pitting or intergranular corrosion, provides stress concentrations and greatly lowers fatigue life [3].

Morphology Corrosion fatigue produces fine-to-broad cracks with little or no branching and this is different from SCC, which often exhibits considerable branching. The cracks may occur singly but commonly appear as families or parallel cracks. They are typically filled with dense corrosion product. They are frequently associated with pits, grooves, or some other form of stress concentrator. Transgranular fracture paths, frequently ramified or branched, are more common than intergranular fractures (exception: lead and tin) and some systems show a combination of both paths [14].

In aluminum alloys exposed to aqueous chloride solutions, corrosion fatigue cracks originate frequently at sites of pitting or intergranular corrosion. Initial crack propagation is normal to the axis of principal stress. This is contrary to the behavior of fatigue cracks initiated in dry air, where initial growth follows crystallographic planes. Initial corrosion fatigue cracking normal to the principal stress axis also occurs in aluminum alloys exposed to humid air, but pitting is not requisite for crack initiation [9].

7.8. PARAMETERS

Environmental considerations, stress descriptions, and material properties must be considered in corrosion fatigue investigations.

7.8.1. Environmental Considerations

The environment may affect the probability of fatigue crack initiation, the fatigue crack growth rate, or both. Growth rates are affected by environmental chemical variables, for example, temperature, gas pressure, impurity content, electrolyte pH, potential, conductivity, and halogen or sulfide ion content. Tests for fatigue consist in subjecting a metal to alternate cyclic stresses of compression–tension of different values and measuring the time (number of cycles, N) before rupture. A short characteristic of the fatigue test is the C–N curve, giving the number of cycles N to rupture. The value of the maximal stress for which an infinite number of cycles can be supported without rupture is called the *endurance limit* or *fatigue limit* and is roughly equal to the half of the tensile strength. Nonferrous metals such as aluminum, magnesium, and alloys of copper do not possess a fatigue limit. In these cases, one refers to fatigue strength or resistance to a certain arbitrary number of cycles, such as 10^8 cycles [14].

The fatigue strengths of aluminum alloys in demineralized water, hard tap water, or brine are almost equal and are relatively one-half the fatigue strength in air and one-quarter of the original ultimate strength of the material. The corrosion fatigue strength of an alloy is not greatly affected by variations in heat treatment. Localized corrosion of an aluminum surface, such as pitting or intergranular corrosion, provides stress concentrations and greatly lowers fatigue life [3].

Aqueous and Gas Environments Environmental effects can usually be identified by the presence of corrosion damage or corrosion products on fracture surfaces or within growing cracks. Corrosion products, however, may not always be present. Increasing the chemical activity of the environment—for example, by lowering the pH of a solution, by increasing the concentration of the corrosion species, or by increasing the pressure of a gaseous environment—generally decreases the resistance of a material to corrosion fatigue [9].

Decreasing the chemical activity of the environment improves resistance to corrosion fatigue. In aluminum alloys and high-strength steels, for example, corrosion fatigue behavior is related to the relative humidity or partial pressure of water vapor in air. Corrosion fatigue crack growth rates for these materials generally increase with increasing water vapor pressure until a saturation condition is reached, as has been demonstrated by the appearance of fatigue fracture surfaces of an aluminum alloy tested in argon and in air with water vapor present. Temperature can have a significant effect on corrosion fatigue. The effect is complex and depends on temperature range and the particular material–environment combination in question, among other factors. The general tendency, however, is for fatigue crack growth rates to increase with increasing temperature [9].

Electrode Potential The electrode potential of an aluminum alloy is influenced by certain factors such as dissolved oxygen, flow rate, ion concentration, alloy composition, and microstructure. The level of potential strongly influences corrosion fatigue crack propagation rates in aqueous environments. Controlled changes in the potential of a specimen can result in either the complete elimination or the dramatic enhancement of brittle fatigue cracking. The observed influence depends on the magnitude of anodic or cathodic potential in the examined medium (Figure 7.8). AA7079-T651 is degraded by corrosion fatigue in halide solutions at the free corrosion potential (e.g., sodium iodide 25%) by more than an order of magnitude if compared to dry argon atmosphere. The corrosion fatigue crack (CFC) growth rate was further increased by anodic polarization above about −0.6 V versus standard hydrogen electrode but the rate is suppressed by cathodic polarization. The horizontal arrow indicates that the failure condition has not been attained. Moreover, an increase in the chemical activity of the medium, which can result, for example, in a fall of pH or increase in the concentration of the corrosive species, generally makes materials less resistant to the corrosion fatigue [13].

7.8.2. Cyclic Stresses

Corrosion fatigue cracks are always initiated at the surface, unless there are near-surface defects that act as stress concentration sites and facilitate subsurface crack initiation. Indeed, the amplitude of the load applied, the cyclic frequency of load, the R ratio of stresses (R = load minimum/load maximum), the potential of the electrode in the aqueous medium (Figure 7.8), and the composition of the medium are important factors to be considered for corrosion fatigue studies [9].

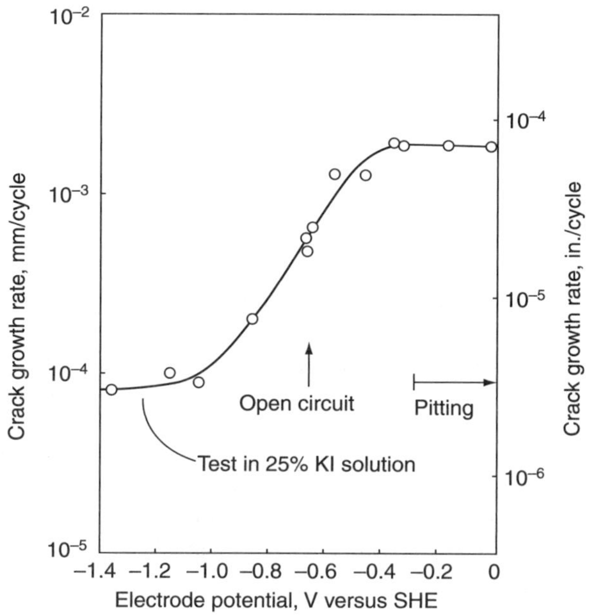

Figure 7.8 Effect of the electrode potential at $\Delta K = 6.7$ MPa $\cdot \sqrt{\text{m}}$ on corrosion fatigue behavior of AA7079-T651 plate in 25% KI solution at 23 °C (S-L orientation, 4 cycles/s, stress ratio $R = 0$) (MPa: mega Pascal, m: meter) [13].

Stresses The main mechanical properties to consider are maximum stress or stress-intensity factor, σ_{max} or K_{max}, cyclic stress or stress-intensity range, ΔK or $\Delta\sigma$ ($K_{max} - K_{min}$), stress ratio *R,* cyclic loading frequency, cyclic load waveform (constant-amplitude loading), load interactions in variable-amplitude loading, state of stress, residual stress, and crack size and shape and their relation to component size geometry [7].

Expressing the crack growth rate *da*/*dN* (where *a* is crack length and *N* is number of cycles) as a function of ΔK provides results that are independent of specimen geometry, and this enables the exchange and comparison of data obtained from a variety of specimen configurations and loading conditions. The growth or extension of a fatigue crack under cyclic loading is principally controlled by maximum load and stress ratio (minimum/maximum stress). However, as in crack initiation, there are a number of additional factors that may exert a strong influence, especially with the presence of an aggressive environment [16]. Most corrosion fatigue crack growth-rate investigations attempt to follow the general provisions of standard test method ASTM E647.

Stress-Intensity Range For embrittling environments, crack growth generally increases with increasing stress intensity (ΔK); the precise dependence, however, varies markedly.

Cyclic Load Frequency This is the most important factor that influences corrosion fatigue for most material environment and stress-intensity conditions. The dominance of frequency is related directly to the time dependence of the mass transport and chemical reaction steps involved for brittle cracking. The corrosion influence in corrosion fatigue is more pronounced at weak frequencies because contact between the corrosive species and metal is longer.

Stress Ratio Rates of corrosion fatigue crack propagation generally are enhanced by increased stress ratio, R, which is the ratio of the minimum stress to the maximum stress.

Variable Amplitude Load Spectrum The corrosion fatigue failure of AA7075-T651 subjected to periodic overloads was examined. Axial fatigue specimens were subjected to a loading spectrum that consisted of a fully reversed periodic overload of near-yield magnitude followed by 200 smaller cycles at high R ratio. This simulates the use of this alloy as a structural material for the wing spars of aircraft. Many components are also subject to corrosion while subjected to a variable-amplitude load spectrum. The specimens were fatigue tested while they were fully immersed in an aerated and recirculated 3.5% NaCl solution and compared to reference experiences exposed to laboratory air [17].

Figure 7.9 shows the results of fatigue testing of AA7075-T651 with periodic overloads in lab air as well as simulated seawater as a function of the stress range for small cycles, ΔS_{sc}, and the total number of cycles to failure, N_f (Figure 7.9). For comparison, the results have been plotted in conjunction with those for constant-amplitude fully reversed ($R = -1$) in air, constant-amplitude fully reversed loading ($R = -1$) in simulated seawater, and periodic reversed overloads in air. The reduction in fatigue life due to corrosion was especially evident in the constant-amplitude corrosion data, which displayed a distinct decrease in life compared to the constant-amplitude cycling in lab air. There was a significant reduction in the fatigue strength of the 7075-T651 alloy when the alloy was subjected to overloads in a corrosive medium, particularly at low stress ranges [17].

It is postulated that anodic slip dissolution would have a greater effect at higher stress levels while pitting should have a greater effect at lower stress levels, since at higher stress levels, corrosion pits do not have enough time to form and initiate a crack. The reduced fatigue life of a susceptible material can be due primarily to premature crack initiation through the formation of corrosion pits and a combination of anodic dissolution at the crack tip and hydrogen embrittlement.

Crack Propagation Three types of behavior have been reported. Figure 7.10a illustrates schematically the sigmoidal variation of fatigue crack growth as a function of stress-intensity factor range on a log–log scale under purely mechanical loading conditions.

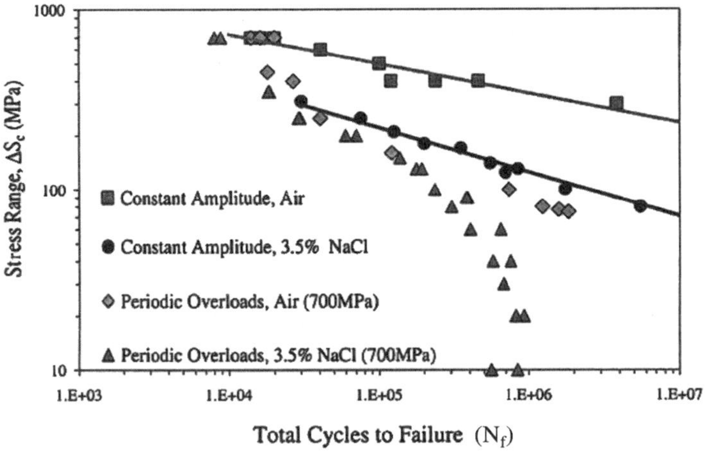

Figure 7.9 Fatigue life data for AA7075–T651 in air and simulated seawater [17].

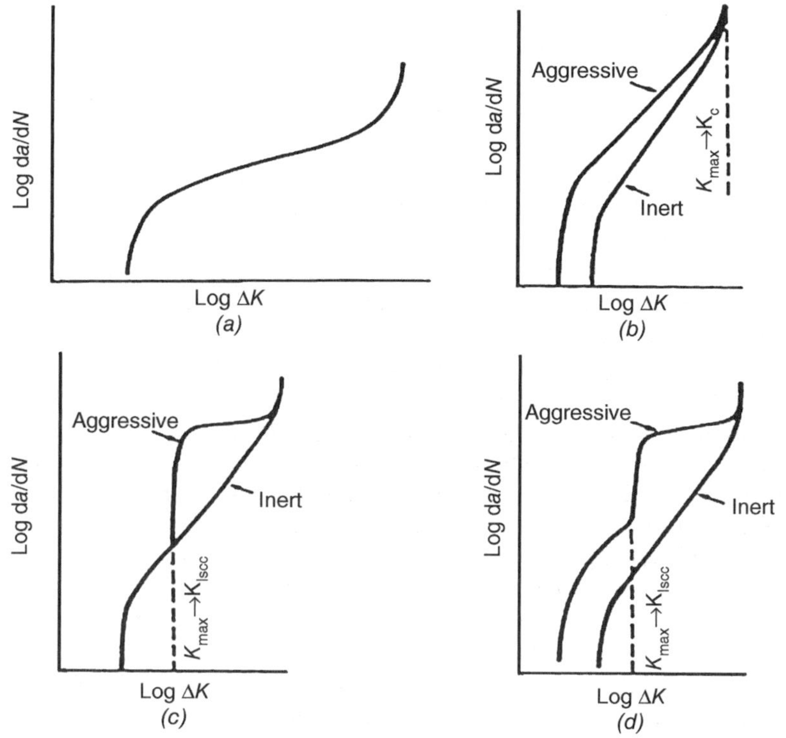

Figure 7.10 Schematic representations of the three types of corrosion fatigue Crack growth behavior: (a) fatigue crack growth in inert atmosphere, (b) true corrosion fatigue, (c) SCC + superposition of mechanical fatigue, and (d) combination of (b) + (c) modes [12,19].

Figure 7.10b illustrates type 1 true corrosion fatigue crack growth behavior. The crack growth rate obeys a Paris law with an increase crack growth rate and a decreased fatigue threshold compared to the behavior in air [18].

Figure 7.10c shows the stress-corrosion fatigue process, type 2, and occurs only when $K_{max} > K_{ISCC}$. In this model, the cyclic character of loading is not important. This behavior is characterized by a plateau region, which prevails above a definite threshold K_{th} of SCC. The combination of true corrosion fatigue and stress-corrosion fatigue results in type 3, the most general form of corrosion fatigue crack propagation behavior (Figure 7.10d) [12,19].

7.8.3. Material Factors

The main metallurgical properties that can influence initiation and growth rates are alloy composition, distribution of alloying elements and impurities, microstructure and crystal structure, heat treatment, mechanical working, preferred orientation of grains and grain boundaries (texture), and mechanical properties (strength, fracture toughness, etc.) [7].

7.9. MECHANISMS OF CORROSION FATIGUE

The two main mechanisms of corrosion fatigue are anodic slip dissolution and hydrogen embrittlement.

Anodic Slip Dissolution Model The cracks grow by slip dissolution due to diffusion of active water molecules, halide ions, and so on to the crack tip followed by a rupture of the protective oxide film by strain concentration and fretting contact between the crack faces. This is followed by dissolution of the fresh exposed surface and growth of the oxide on the bare surface. In the case of metals where the surface is in the active dissolution state, emerging persistent slip bands (PSBs) are preferentially attacked by dissolution. This preferential attack leads to mechanical instability of the free surface and the generation of new and larger PSBs, followed by localized corrosion attack resulting in crack initiation. Under passive conditions, the relative rates of periodic rupture and reformation of the passive film control the extent to which corrosion reduces fatigue resistance. When bulk oxide films are present on the surface, rupture of the films by PSBs leads to preferential dissolution of the fresh metal that is produced [20, 21].

Hydrogen Embrittlement Model In aqueous media, the critical steps involve the diffusion of water molecules or hydrogen ions to the crack tip, reduction to hydrogen atoms at the crack tip, surface diffusion of adsorbed atoms to preferential surface locations, and absorption and diffusion to critical locations in the microstructure (e.g., grain boundaries, ahead of a crack tip, or void) [12].

The presence of pitting corrosion has been reported to reduce the fatigue life of aluminum alloys by a factor of 2 all the way up to an order of magnitude. In the current era of aging aircraft, it is therefore of great importance to understand the mechanism of degradation. Van der Walde and Hillberry [22] carried out interrupted fatigue testing of precorroded bare AA2024-T3 sheet specimens. Several specimen orientations and stress levels were considered. Cycle counts at which the specimens were interrupted (overloaded) to expose crack development were set at percentages of total expected life as determined from previous fatigue-to-failure testing done under the same conditions.

The interrupted testing provided a great deal of information concerning the process of corrosion-nucleated fatigue crack growth. Application of cyclic loading to a premature aluminum specimen shows an early stage of crack growth at multiple points, producing widespread microcracking. It was conclusively deduced that crack initiation is essentially immediate upon the application of cyclic loading, including those tested for just 10% of their total expected life. This can be due primarily to the stress concentration by pitting. The microstructural features in the region of the pit also contribute to crack nucleation. The initial cracks were found to carry resemblance to the grain structure in terms of size, shape, and orientation, and were noted to originate from pits. It was additionally stated that surface breaking states did not consistently appear until 25% of the expected life had been consumed [22].

7.10. CORROSION FATIGUE OF ALUMINUM ALLOYS

The aluminum alloys have a relatively low resistance versus corrosion fatigue. The localized corrosion of an aluminum surface increases the stress and decreases the fatigue strength. For low-stress, high-cycle fatigue, crack initiation spans a large portion of the total lifetime [12,13].

Corrosion fatigue is not appreciably affected by stress orientation, and corrosion fatigue failures can be recognized by a characteristic oyster shell pattern on the fractured surfaces. Corrosion fatigue failures of aluminum alloys are characteristically transgranular, and thus differ from SCC failures that are normally intergranular [9, 23].

Corrosion environments produce smaller reductions in fatigue strength in the more corrosion-resistant alloys. Corrosion fatigue cracks in copper and various copper alloys

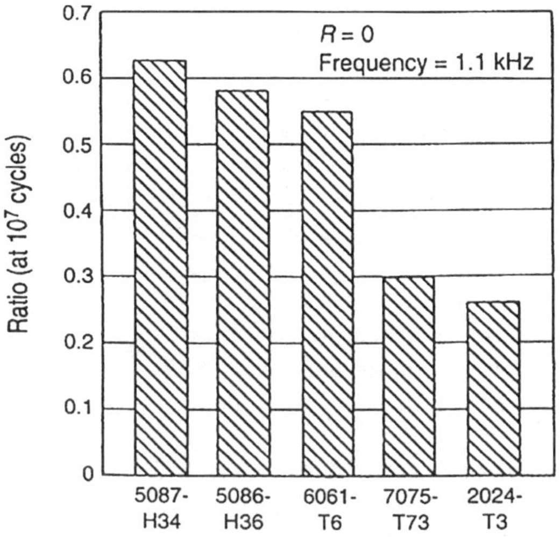

Figure 7.11 Ratio of axial-stress fatigue strength of aluminum alloy sheet in 3% NaCl solution to that in air. Specimens were 1.6 mm (0.064 in.) thick [25].

show intergranular initiation and propagation. Copper–zinc and copper–aluminum alloys, however, exhibit a marked reduction in fatigue resistance, particularly in aqueous chloride solutions [3].

Fatigue strengths of aluminum alloys are lower in such corrosive environments as seawater and other salt solutions than in air, especially when evaluated by low-stress, long-duration tests. As shown in Figure 7.11, such corrosive environments produce smaller reduction in fatigue strength in the more corrosion-resistant alloys, such as the 5xxx and 6xxx series, than in the less resistant alloys, such as the 2xxx and 7xxx [24, 25].

7.10.1. Corrosion Fatigue of AA7017-T651

The effect of loading frequency and stress-intensity range on the corrosion fatigue crack growth rate and the cracking morphology of Al–Zn–Mg 7017 alloy in natural seawater are critical to consider. The actual crack velocity (per second) increases with increasing frequency (Figure 7.12). However, if the crack growth is considered per cycle, the enhancement is more marked at the lower test frequencies. This is evident because of the more synergistic effect of cyclic stresses and electrochemical corrosion that is slower than the effect of the mechanical stress [26].

The log plot of crack growth rate per cycle against test frequency for specific values of ΔK shows a complex dependence of crack growth rate upon frequency (Figure 7.13) [26].

At high frequencies, where little time is available for corrosion to take place, crack growth rates are insensitive to loading frequency but are sensitive to environment: that is, crack growth rates are higher in seawater than in dry air. As the frequency is reduced and more time becomes available for corrosion reactions, flat transgranular and then intergranular modes are introduced, particularly at low ΔK (where propagation rates are inevitably slower). As a result (Figure 7.14), the ΔK value at which transition would occur from ductile to flat transgranular (transition A) and from flat transgranular to intergranular (transition B) increases as the frequency decreases [26].

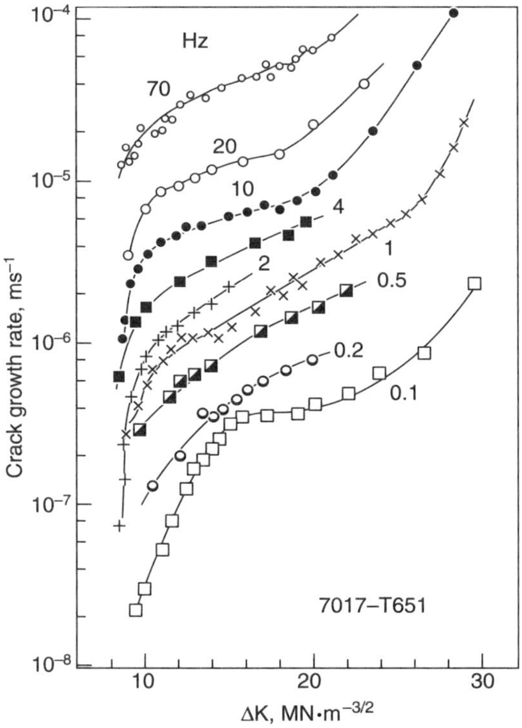

Figure 7.12 Corrosion fatigue of Al–Zn–Mg alloy: crack velocities during fatigue of AA7017-T651 in natural seawater at cyclic loading frequencies from 0.1 to 70 Hz, as a function of ΔK [26].

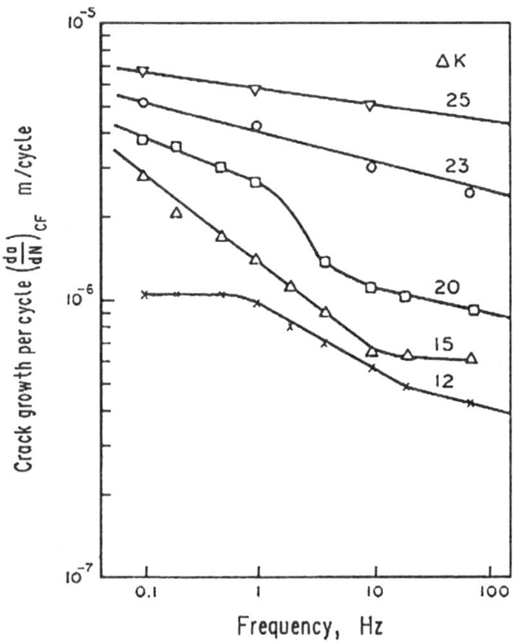

Figure 7.13 Variation of the crack growth rate with frequency during fatigue of AA7017-T651 as a function of ΔK in seawater [26].

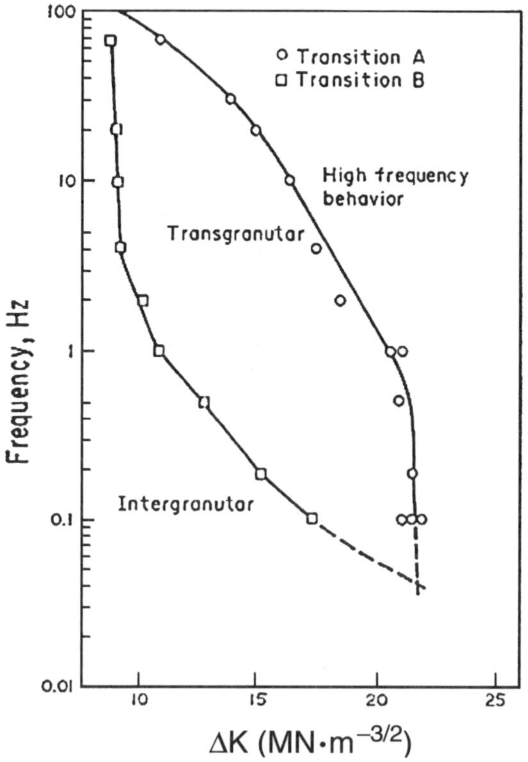

Figure 7.14 Combinations of loading frequency and ΔK responsible for various modes of failure during the corrosion fatigue of AA7017-T651 in seawater [26].

7.10.2. Corrosion Fatigue of AA7075-T6

Figure 7.15 shows an example of a high-strength aluminum alloy with a high resistance to SCC. This alloy had a corrosion fatigue crack growth rate ranging up to 1 order of magnitude higher in 3.5% sodium chloride (NaCl) solution compared to that in dry air. These data also illustrate that corrosion fatigue behavior is similar when tested in the same environment with either a *K*-increasing (remote load) or a *K*-decreasing (wedge force) loading method [27].

7.10.3. Corrosion Fatigue of Al–Mg–Si Compared to Al–Mg Alloys

High-strength aluminum alloys (HSAAs) are widely used in aircraft and other heavy stress-bearing engineering structures. The Al–Mg–Si alloys constitute a large class of materials and are produced mainly by the semicontinuous vertical dc-casting route. Unfortunately, these alloys are susceptible to various forms of corrosion, particularly in the presence of chloride-containing media. The problem of corrosion fatigue in HSAAs has received considerable attention for more than 50 years. The combined effects of cyclic loading and corrosion on a structure can dramatically decrease its service life. The particular problem of pitting-corrosion fatigue is one of the main topics of concern. This is due to the high susceptibility of HSAAs to pitting corrosion. The mechanisms involved in the initiation and growth of corrosion-initiated fatigue cracks have been studied extensively [28].

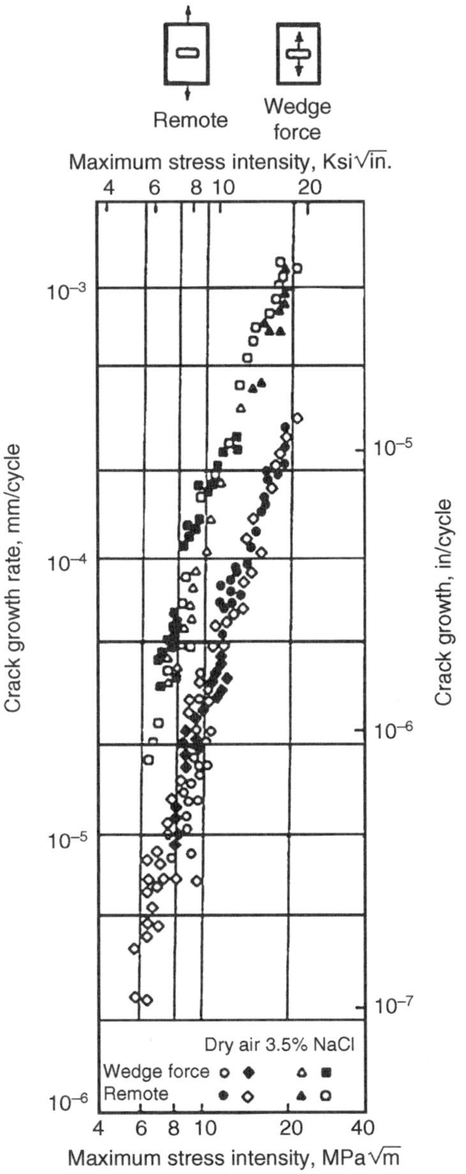

Figure 7.15 Crack tip stress-intensity control of fatigue crack propagation in AA7075-T6 sheet with long transverse loading: remote and wedge force methods of loading specimens in aqueous 3.5% sodium chloride environment and benign dry air environment [27].

Generally, the initiation of corrosion cracks in HSAAs is due to corrosion pits initiated by the galvanic electrochemical cell between the constituent particles and the surrounding matrix. The presence of the preexisting corrosion pits significantly reduces the crack initiation life and the fatigue initiation threshold by as much as 50%. It has been stated that AA6013 has slightly lower corrosion fatigue life compared to AA2024 bare surface and much lower than Alclad-coated 2024 alloy. AA6013 is now used in aerospace applications and in the transportation industry to replace the traditional AA2024-T3 [28].

Laser surface melting, using an excimer laser, was employed to improve the resistance to fatigue cracking of AA6013 induced by pitting corrosion. The results of the corrosion fatigue tests showed that the total fatigue life of the alloy increased noticeably after the laser surface treatment. Effectively, the corrosion fatigue lifetimes of the laser air-treated and the laser nitrogen-treated specimens were 2–4 times longer than that of the untreated specimens. However, the overall fatigue crack propagation rate of the laser-treated specimens was higher than that of the untreated specimens, very probably due to the fewer created corrosion fatigue cracks leading to a certain concentration of cyclic stresses [28].

Comparison of Corrosion Fatigue of Al–Mg AA5083-H3 and Al–Mg–Si AA6061-T6 Al–Mg AA5083-H3 dealt mainly with the influence of pitting on corrosion fatigue and stress-corrosion phenomena of these chosen representative two alloys. Basically, general uniform corrosion is observed on these alloys when the alloy is not in the passive state. More dangerous localized corrosion such as pitting is observed when the alloy is in the passive state, especially in the presence of chloride ions' which leads to the breakdown of the passive film. Under loading, this pitting leads to corrosion fatigue and stress-corrosion cracking [29].

Al–Mg alloys have coarse intermetallic compounds such as $Al_6(Mn,Fe)$ constituent particles that act as cathodic sites. Also, a small discretely distributed β phase (e.g., Al_3Mg_2 or Al_8Mg_5) is strongly anodic to the Al–Mg solid solution matrix. The Al–Mg–Si AA6061-T6 microstructure is characterized by the strengthening particle β-Mg_2Si, which forms during aging. Coarse intermetallics of Al_3Fe, Al_6Fe, Al_8Fe_2Si, Al–Si–Mn–Fe, and Al–Mg–Si are observed. The corrosion fatigue tests were done for two alloys, 5083 and 6061, in 3% NaCl under different stresses in the elastic domain at the free corrosion potential.

The open circuit potential values during corrosion fatigue testing for both aluminum alloys are shown in Figure 7.16. The evolution of corrosion potential of the two alloys during corrosion fatigue testing (150 h) shows some interesting common characteristics. There is a shift to more positive or less negative potential values (20–40 mV), indicating very possibly a certain type of mild passivation or formation of corrosion products in this solution, and this is followed by an important shift to active values (150 mV) during the period between 10 and 70 h and this can correspond to the detection of a more active and negative potential at an acidic pH, very probably due to pitting. The last period for this aggressive pitting medium shows a stable level of potential and we can see that the potential of 6061 is more negative than that of 5083 [30].

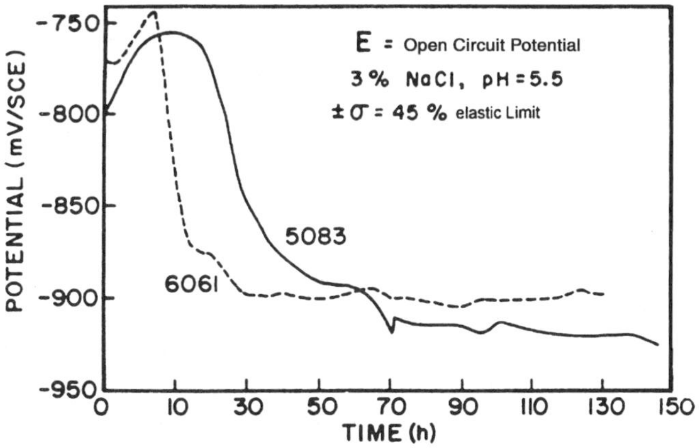

Figure 7.16 Open circuit potential during corrosion fatigue testing for both aluminum alloys [30].

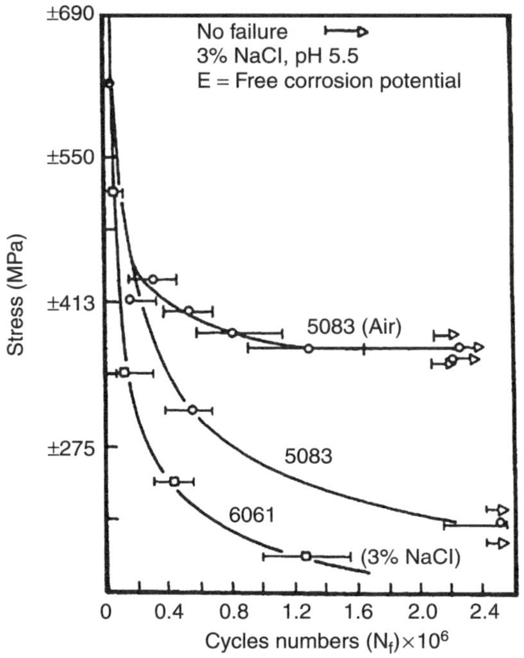

Figure 7.17 Relationship between applied stress and time to failure for AA5083 in 3% NaCl solution as compared to AA5083 in air and AA6061 in chloride medium [30].

Figure 7.17 shows the endurance curve result for materials subjected to alternating flexion ($R = -1$). At high stresses both alloys 5083 and 6061 have generally identical lifetimes. The difference becomes increasingly evident at lower stresses, where test times are longer, allowing more time for electrochemical reactions to occur. At lower stress levels, there is a drop in lifetime of AA6061 compared to AA5083, which can be attributed to the morphology and localization of pitting. The pitting in AA5083 is numerous but less profound, whereas in AA6061 the pits are deeper and more profound. One can note also the influence of the corrosive aqueous chloride medium as compared to that of air, where the synergistic effect of corrosion and fatigue cancels the fatigue limit [30].

Almost all fatigue cracks initiated from constituent particles were on the surface. As an example, a fatigue crack of a 6061-T6 alloy stressed at 72% was initiated from constituent particles in 3% NaCl solution at the free corrosion potential (Figure 7.18). These particles acted as effective stress concentrators to raise the local stresses in addition to their role in creating a galvanic cell and facilitating fatigue crack initiation and propagation [29].

Also, corrosion fatigue of these two alloys was studied as a function of applied potential, for aluminum alloys 6061-T6 and Al–Mg 5083–H321 in aqueous solution, to examine the influence of pitting potential on the initiation and propagation morphology of fracture in 3% chloride solution. Argon and silicon oil were used as reference for an inert medium [29]. Figure 7.19a,b shows that the pit initiation is the trigger for corrosion fatigue cracking. It is known that this form of pitting can accentuate more or less the stress concentration and accelerate crack initiation for both corrosion fatigue and stress-corrosion cracks. For both alloys, the kinetics of fracture is accelerated by the critical sharp pit shape and the electrochemical conditions at the pit tip (Figure 7.19c,d). In both alloys, the fracture exhibited a pronounced intergranular morphology, which is related to the corrosion fatigue mechanism of propagation. It should be underlined that, under the same conditions and for

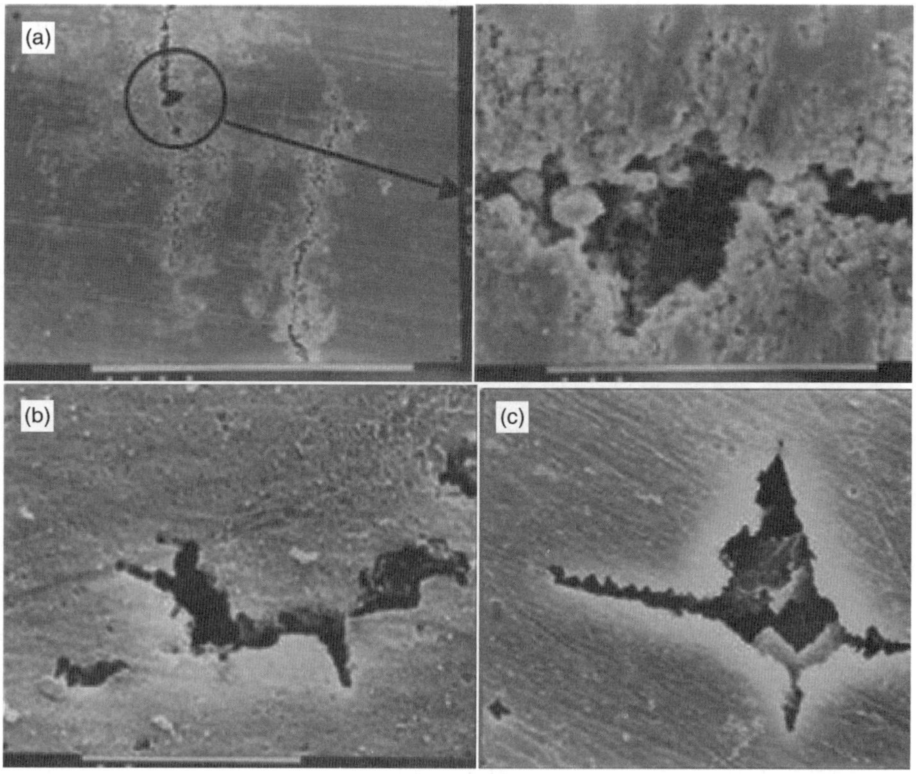

Figure 7.18 Microscopic examination of lateral side surfaces in corrosion fatigue samples cycled in 3% NaCl: (a) crack nucleates and propagates at pit; (b, c) ramified crack from local attack [29].

both alloys, the main morphology of the fracture propagation at the open circuit potential was transgranular. Finally, at an imposed pitting potential the low local pH and the high level of chloride ion at the mouth of the pit caused a considerable drop in the resistance to corrosion fatigue as well as to stress-corrosion cracking that has been further proved.

7.10.4. Modeling of the Propagation of Fatigue Cracks in Aluminum Alloys

Modeling of corrosion fatigue cracking is an important parameter for security considerations, for example, in aerospace applications. Bergner and Zouhar [31] examined the long fatigue cracks in commercial heat-treatable thin-sheet aluminum alloys with focus on the midregion of the cyclic stress-intensity factor. The experimental observations support the distinction of two groups of alloy conditions according to the degree of coherency of the strength-controlling precipitates, based on the fatigue crack growth behavior observed at stress ratios $R = +0.1$ and 0.5.

A simple model is proposed for the first group that is characterized predominantly by incoherent precipitates. The Paris lines are strongly focused at a ΔK position, which scales with the shear modulus. The limits of the observed slope are $m \sim 2$ (e.g., AA6013-T6) and $m \sim 4$ (e.g., AA2024-T81). For the second group, a retardation relative to the first group was observed and caused by a high degree of coherency of the strength-controlling precipitates, localized

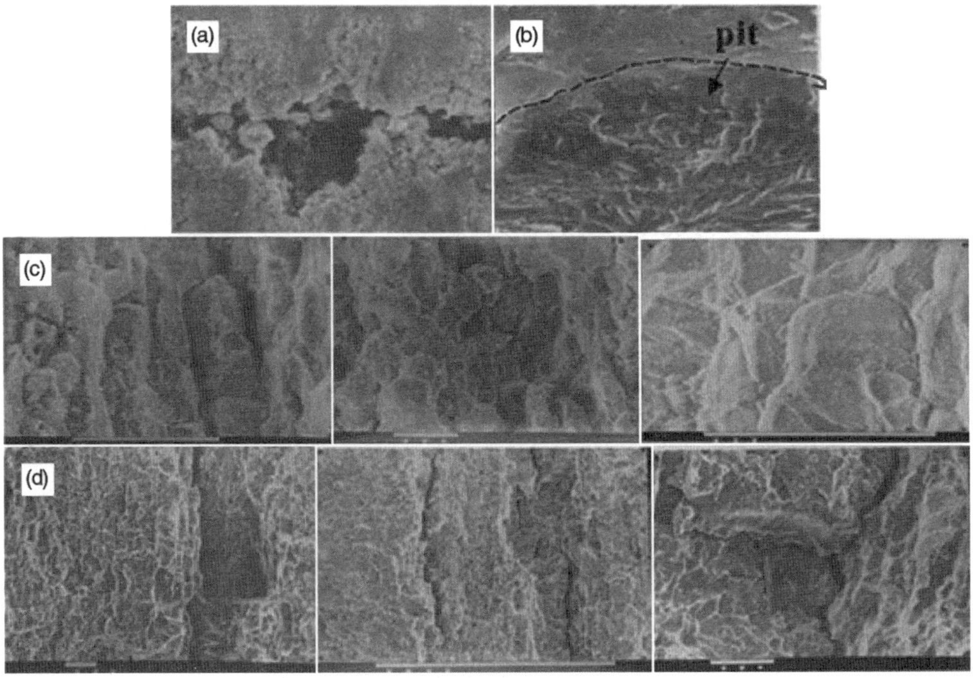

Figure 7.19 Fracture surface of sample tested at pitting potential (E_{pit}). (a) fatigue cracks emanating from pits, (b) growing crack from pit, (c) intergranular crack in AA6061-T6, and (d) secondary cracks along grain boundaries in AA5083-H321 [29].

planar slip, crack deflection combined with a nonproportional mode-II component of the crack opening displacement, and finally by the roughness-induced crack closure [31].

The experimental and frequency effects on fatigue crack growth in aluminum alloys are studied theoretically and experimentally for two alloys of the 2xxx and 7xxx series. The two tested alloys were 2024-T351 and 7075-T651. They were carried out in humid air, purified nitrogen, and vacuum. A crack propagation of da/dN versus ΔK_{eff} does not reflect the plateau-like region. ΔK_{eff} describes the effective stress-intensity factor and corresponds to $\Delta K_{eff} = K_{max}\ (1 - R_{eff})$, where $R_{eff} = 0.45 + 0.2R + 0.25R^2 + 0.1R^3$.

A new model of crack growth is presented which considers the formation and subsequent fracture of a crack tip oxide layer that has been observed experimentally by X-ray photon electron microscopy. At higher loads, other mechanisms are understood to be active. The model parameters are determined from constant-amplitude stress and are valid for a given material and environment [32].

Under constant-amplitude loading, the levels of ΔK_{eff} and K_{max} in a given cycle define which mechanisms are active. As long as ΔK_{eff} is lower than the threshold value, ΔK_{th} elastic deformations occur at the crack tip, and the crack does not propagate (regime 1) (Figure 7.20). In fatigue cycles where ΔK_{eff} is small but greater than ΔK_{th}, crack growth is due either to a cyclic slip mechanism in noncorrosive environments or to oxide layer fracture in corrosive environments (regime 2). Upon a further increase in load for a higher ΔK_{eff}, crack tip blunting occurs following either cyclic slip or oxide layer fracture (regime 3). If the load is increased further, and the K_{max} in the cycle approaches a critical value, K_c, macroscopic plastic deformations occur following crack tip blunting (regime 4). A catastrophic failure occurs once K_{max} reaches K_c (regime 5) [32].

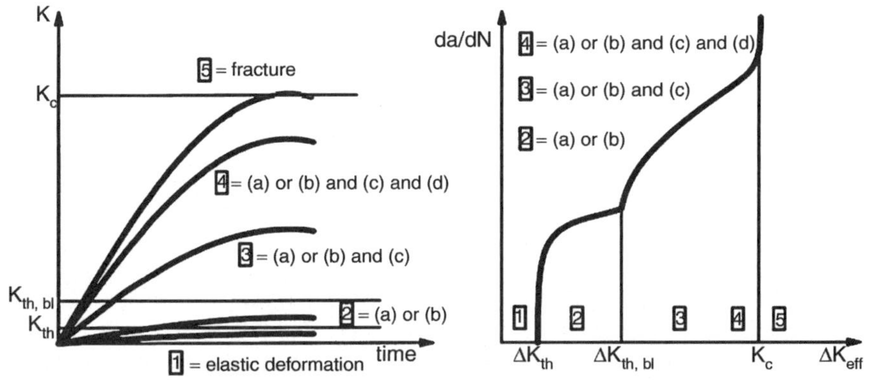

Figure 7.20 Schematic illustration of the crack growth mechanism in a given loading regime (a–d). As the load increases during a fatigue cycle, a different mechanism is active. A given ΔK regime on a *da*/*dN* versus ΔK_{eff} plot (on the right) takes place depending on the level of the peak load during the fatigue cycle [32].

7.11. PREVENTION OF CORROSION FATIGUE

1. It is often more safe and more cost effective to reduce the magnitude of the stress fluctuation through redesigning than reducing the maximum stress level [12].
2. Select a material or heat treatment with higher corrosion fatigue strengths. Corrosion environments produce smaller reductions in fatigue strength in the more corrosion-resistant alloys, such as the 5xxx and 6xxx series, than in the less resistant alloys, such as the 2xxx and 7xxx series.
3. Improving surface conditions is very useful. Application of surface treatments, such as shot peening or sandblasting of the metal surface, that produce constraints of compression are beneficial [33]. Care must be taken not to overpeen the surface to the extent where excessive plastic deformation may cause susceptibility to exfoliation or SCC and to avoid embedding metallic particles into the relatively soft aluminum alloy.
4. Welding lowers both fatigue and corrosion fatigue life, but peening after welding increases the corrosion fatigue life. A recognized process to increase the fatigue strength of the welded joints is by hammering the surface, leaving compressive residual stresses, although it may decrease the fatigue strength of the surface.
5. Use of corrosion inhibitors, reduction of oxidizers, or pH increase can delay the initiation of corrosion fatigue cracks depending on the system and the environment.
6. Organic coatings can successfully impede corrosion fatigue if they contain some inhibitory pigments in the primary layer, and the highest corrosion fatigue life for welded specimens is achieved by peening followed by coating [3].
7. Cathodic protection suppresses the dissolution rate and prevents pit formation, but hydrogen effects can increase crack growth rates of well-defined cracks. It should be added that cathodic protection of aluminum alloys in stagnant solutions can lead to a dangerous alkaline type of pitting of aluminum alloys and this should be observed and avoided.

REFERENCES

1. A. Adjorlolo, in *ASM Handbook, Volume 13C, Corrosion: Environments and Industries*, edited by S. D. Cramer and B. S. Covino, Jr. ASM International, Materials Park, OH, 2006, pp. 598–612.
2. J. Desbiens, *Corrosion des alliages d'aluminium assistée mécaniquement*. Departement of Mining, Metallurgical and Materials Engineering, Université Laval, Quebec City, 2007, pp. 1–24.
3. B. W. Lifka, in *Corrosion Tests and Standards, Application and Interpretation*, 2nd edition, edited by R. Baboian. ASM Internantional, Materials Park, OH, 2005, pp. 547–557.
4. A. J. Smith, M. Stratmann, and A. W. Hassel, *Electrochimica Acta* **51**, 6521–6526 (2006).
5. Academie de Nancy METZ, Physique-chimie. Available at http://www.Ac-nancy-metz.fr.
6. Y. Li, G. T. Burstein, and I. M. Hutchings, *Wear* **181–183**, 70–79 (1995).
7. *Corrosion of Aluminum and Aluminum Alloys*, edited by J. R. Davis, ASM International, Materials Park, OH, 2001, pp. 1–81.
8. S. Vaidya and C. M. Preece, *Metallurgical Transactions A* **9**, 299–307 (1978).
9. W. Glaesar and I. G. Wright, *Forms of Mechanically Assisted Degradation*, Vol. 13A. ASM International, Materials Park, OH, 2003.
10. B. Safety, *Rupture d'une pale du rotor principal*. Canadian Board, 2000, p. 5.
11. H. Nanjo, Y. Kurata, N. Sanada, K. Miyauchi, R. Ohshima, and K. Koike, *Wear* **186–187**, 573–578 (1995).
12. Y.-Z. Wang, in *Uhlig's Corrosion Handbook*, 2nd edition, edited by R.W. Revie. Wiley-Interscience, Hoboken, NJ, 2000, pp. 221–232.
13. ASM International Handbook Committee, in *Corrosion of Aluminum and Aluminum Alloys*, edited by J. R. Davis. ASM International, Materials Park, OH, 1999, pp. 63–74, 135–160.
14. H. H. Uhlig and R. W. Revie, *Uhlig's Corrosion Handbook*. Wiley, Hoboken, NJ, 1985, pp. 8, 35–59.
15. V. S. Sastri, E. Ghali, and M. Elboujdaini, *Corrosion Prevention and Protection—Practical Solutions*. Wiley, Chichester, West Sussex, UK 2007, pp. 331–459.
16. S. L. Kerr and K. Rosenberg, *Transactions of the ASME* **80** (6), 1308–1314 (1958).
17. R. M. Chlistovsky, *International Journal of Fatigue* **29** (9-11), 1941–1949 (2007).
18. T. Magnin and P. Combrade, in *Materials Science and Technology Series*, Vol. 1, edited by R. W. Cahn, P. Haasen, and E. J. Kramer. Wiley-VCH, Weinheim, Germany, 2000, pp. 216–318, 537.
19. A. J. Mc Evily and R. P. Wei, in *Corrosion Fatigue: Chemistry, Mechanics and Microstructure*, edited by O. Devereux, A. J. McEvily, and R. W. Staehle. NACE, Houston, TX, 1972, pp. 381–395.
20. D. J. Duquette, in *Environment-Induced Cracking of Metals*, edited by R. P. Gangloff and M. B. Yves. NACE, Houston, TX, 1990, p. 45.
21. V. S. Sastri, E. Ghali, and M. Elboujdaini, *Corrosion Prevention and Protection—Practical Solutions*. Wiley, Chichester, UK, 2007, pp. 109–176.
22. K. Van der Walde and B. M. Hillberrya, *International Journal of Fatigue* **29**(7), 1269–1281 (2007).
23. E. Ghali, in *Uhlig's Corrosion Handbook*, 2nd edition, edited by R. W. Revie. Wiley, Hoboken, NJ, 2000, pp. 677–715.
24. J. G. Kaufman and E. L. Rooy, in *ASM Handbook, Volume 13B, Corrosion: Materials*, edited by S. D. Cramer and B. S. Covino, Jr. ASM International, Materials Park, OH, 2005, pp. 1–8.
25. J. G. Kaufman, in *ASM Handbook, Volume 13B, Corrosion: Materials*, edited by S. D. Cramer and B. S. Covino, Jr. ASM International, Materials Park, OH, 2005, pp. 95–124.
26. H. D. Holroyd and N. J. Hardie, *Corrosion Science* **529**, 533–535 (1983).
27. B. Phull, in *ASM Handbook, Volume 13A, Corrosion* edited by S.D. Cramer and B. S. Covino, Jr. ASM International, Materials Park, OH, 2003, pp. 575–616.
28. W. L. Xu, T. M. Yue, H. C. Man, and C. P. Chan, *Surface & Coating Technology* **200** (16–17), 5077–5086 (2006).
29. M. Elboujdaini, E. Ghali, and M. T. Shehata, in *Effect of Applied Stress and Potential on SCC and Corrosion Fatigue Behaviour of Aluminum in Chloride Media*, 23rd Annual Conference of Egyptian Corrosion Society, Nozha Beach Resort, Red Sea, Egypt, December 6–10, 2004, p. 14.
30. M. Elboujdaini and E. Ghali, *Materials Science Forum* **44–45**, 153–168 (1989).
31. F. Bergner and G. Zouhar, *International Journal of Fatigue* **25** (9-11), 885–889 (2003).
32. S. A. Michel, R. Kieselbach, and M. Figliolino, *Fatigue Fracture and Engineering Material Structure*, **28**, 205–219 (2005).
33. L. L. Shreir, R. A. Jarman, and G. T. Burnstein, *Corrosion*, 3rd edition. Butterworth-Heinemann, Woburn, MA, 1994, Vol. **1**, section 8, pp. 3–242.

Chapter 8

Environmentally Induced Cracking of Aluminum and Its Alloys

Overview

Stress-corrosion cracking (SCC) or environmentally induced cracking (EIC) generally concerns a certain interaction among the electrochemical dissolution of the metal, hydrogen absorption, and the mechanical loading conditions (stress, strain, and strain rate). SCC susceptibility under certain operating conditions is a type of "allergy" between a material with a certain chemical composition and microstructure and an environment. The stresses can be externally applied, but residual stresses often cause SCC failures. Stresses can be created by lamination, bending, machining, rectification, drawing, drift, riveting, thermal treatments, adherent corrosion products, and welding. Thermal treatments cause changes in the microstructure that can induce dilation and contraction of metal. Welded metals contain residual stresses near the yield point and stress-relief annealing is recommended.

Critical potentials for SCC of a metal–solution system can be related to its *E*–pH diagram. The zone of pitting leading possibly to SCC in the *E*–pH diagram is explained. Halide ions have the greatest effects in accelerating attack. Water or water vapor is the key environmental factor required for producing SCC in aluminum alloys. Hydrogen embrittlement of aluminum alloys is also discussed.

The relation and the influence of corrosion forms, such as corrosion fatigue, intergranular corrosion, and pitting, on SCC of different aluminum alloys of different series are discussed. This is detailed especially for the high-strength aluminum 2xxx and 7xxx series and for welded alloys. Corrosion resistance of some welding processes, such as laser beam and fusion stir welding, is explained using laboratory research studies and concrete case histories. Special care is given to update the possible methods of SCC prevention.

8.1. INTRODUCTION AND DEFINITION OF SCC

Environmentally induced cracking (EIC) or stress-corrosion cracking (SCC) in an aqueous medium (or organic solvent), containing a certain ion at a certain pH, concentration, and

Corrosion Resistance of Aluminum and Magnesium Alloys: Understanding, Performance, and Testing By Edward Ghali

temperature, can cause a microscopically brittle fracture of materials under levels of mechanical stress that may be far below those required for general yielding or those that could lead to significant damage in the absence of an environment. SCC or EIC susceptibility in certain environments and under certain conditions is controlled by the chemical composition and microstructure of the alloy. It is a type of "allergy" between a material and environment [1]. Most of the time, this form of corrosion necessitates a certain interaction among the electrochemical dissolution of the metal, the hydrogen absorption, and the mechanical loading conditions (stress, strain, and strain rate) [2]. Fracture modes included in this category are stress-induced failures (tension, compression, flexure, and shear), overloading, deformation, delamination, and time-dependent modes, such as fatigue, creep, SCC, and embrittlement.

Environmentally assisted cracking (EAC) is a general term that can be divided into mechanically assisted cracking (MAC) and environmentally induced cracking. Mechanically assisted cracking includes wear corrosion, corrosion fatigue, and fretting fatigue corrosion leading to fracture modes (see Chapter 7). Corrosion fatigue cracking occurs only under cyclic or fluctuating operating loads, while SCC and hydrogen embrittlement (HE) occur under static or slowly rising loads. However, in certain situations, a combination of two out of three or all three phenomena are possible.

The EAC term covers a whole spectrum of phenomena from full (brittle) fracture to complete electrochemical corrosion processes and several related mechanisms and models are sometimes involved. EAC is not limited to metals, and it also occurs in glasses (Plexiglas), ceramics, and polymers. Structural failures due to EAC are often sudden and unpredictable, occurring after a few hours of exposure, or after months or even years of satisfactory service. At this time, the costs of EAC of materials annually exceed billions of dollars and are escalating throughout the world [1, 3, 4].

Morphology Failed specimens look macroscopically brittle and exhibit highly branched cracks that propagate transgranularly and/or intergranularly, depending on the metal–environment combination. Transgranular stress-corrosion crack propagation is often discontinuous on the microscopic scale and occurs by periodic jumps on the order of a micrometer, while intergranular cracks are believed to propagate continuously or discontinuously, depending on the system [2]. Transitions in crack modes from intergranular to transgranular cracking are observed and often occur simultaneously in the same alloy.

SCC ruptures are fragile and are sometimes characterized by the presence of cleavages, notably in the case of hydrogen embrittlement. Rupture by cleavage is a fragile rupture occurring along the crystallographic planes. Cleavages possess a definite orientation and can be identified by optical microscopic observation since they present brilliant and plane facets with dimensions related to the size of the grains of the studied material. Using a scanning electron microscope to examine the rupture faces, one can often distinguish the apparent plane zones and the presence of microreliefs, which have the aspect of a hydrographic network corresponding to junction steps between parallel planes on which the crack propagates [1, 5].

Considering aluminum and its alloys, the susceptibility to SCC is affected by the chemical composition, the preferential orientation of grains, the composition and distribution of precipitates (particularly intergranular), the interaction of dislocations, the progression of phase transformations, and cold work processes [5]. However, two key parameters should be addressed first: the stresses and the specific environment for SCC.

8.2. KEY PARAMETERS

8.2.1. Stress

The stresses applied to a metal are nominally static or slowly increasing tensile stresses. The stresses can be applied externally, but residual stresses often cause SCC failures. Stresses can be introduced by cold work processes, such as lamination, bending, machining, rectification, drawing, drift, riveting, thermal treatments, and welding, as well as by adherent corrosion products. Thermal treatments cause changes in the microstructure that can induce dilation and contraction of metal. Welded metals contain residual stresses near the yield point and stress-relief annealing is recommended. Corrosion products have been shown to be another source of stress and can exert a wedging action.

Figure 8.1a concerns the growth of cracks in metallic materials under sustained loading in the presence of an environment and shows the presence of three regions depending on the stress-intensity level [6]. The environment has no effect on fracture behavior below a static intensity factor K_{ISCC}, the threshold stress-intensity factor for the growth of stress-corrosion cracks in tensile opening mode. Above K_{ISCC}, the crack velocity increases precipitously with increasing stress-intensity factor K (region 1). The threshold value is specific for every metal–environment system and lies typically between 10 and 25 MPa $\cdot$ m$^{1/2}$ and σ_1 is usually on the order of 60–100% of the yield stress, but much lower values could be observed for certain conditions as weak as 10% of the elastic limit. At intermediate stress-intensity levels (region 2), the crack propagation rate shows a plateau velocity $V_{plateau}$ that is virtually independent of the mechanical stress. Region 3 corresponds to the critical stress-intensity level for mechanical fracture in an inert environment [2, 5, 7, 8].

Stress corrosion of precracked specimens of the same material (Al–Zn–Mg–Cu) in seawater at room temperature (Figure 8.1b) produces a similar crack velocity v–stress-intensity factor (K) relationship to that obtained for other Al–Zn–Mg–Cu alloys by other workers, and comparison of the fractures with those produced in fatigue is very fruitful.

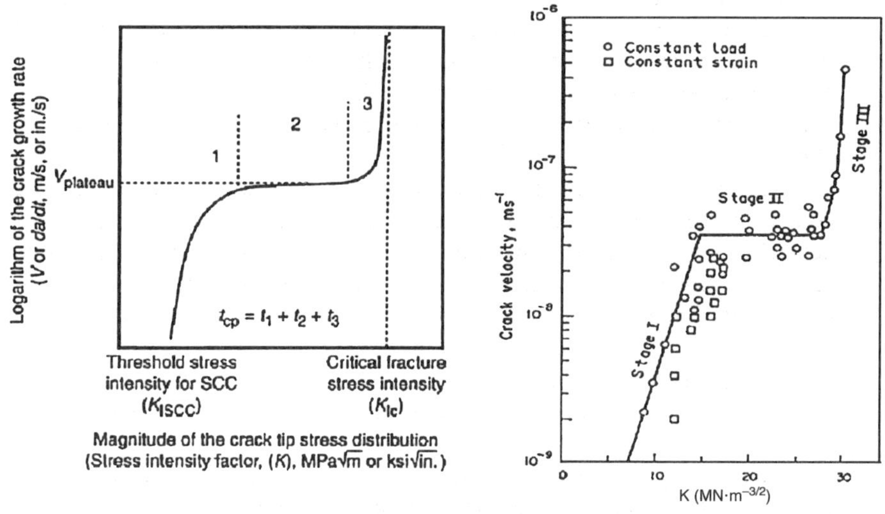

Figure 8.1 (a) Schematic of the regions of crack propagation as a function of the crack tip stress-intensity magnitude factor K expressed in MPa. $\sqrt{}$m [6] (b) SCC relationship between crack velocity and stress-intensity factor for stress corrosion of 7017-T651 in seawater at room temperature [8].

SCC produced in the Stages I and II of the v–K curve was intergranular and fractographically indistinguishable from intergranular corrosion fatigue, whereas SCC produced in Stage III was transgranular and closely resembles the flat transgranular mode in corrosion fatigue [8].

8.2.2. Environment

The E–pH diagram of the cracking metal–solution interface is a basic tool to evaluate and understand the mechanism of SCC; however, there are certain difficulties for precise evaluation since these diagrams only describe the general trend for standard conditions at which film formation and metal corrosion or even pitting occur (see Chapter 5). Critical potentials for SCC of a metal–solution system can be related to its E–pH diagram. However, the regions of intergranular corrosion, oxygen content, and temperature are of major importance.

The crack tip chemistry can differ from the bulk solution depending on the material and the examined solution. Shifts of the potential of the crack tip can generally be on the order of 300 mV in the anodic direction, and this shift is largely reduced under steady-state conditions when the walls of the crack are passivated. However, exact measurement of the potential of the crack tip is difficult since the size of the crack tip is on the order of 1 μm and some authors assume that the local environment could be composed of hydrated salts and oxyhydroxides rather than a liquid solution. This can apply to all types of EAC, including corrosion fatigue, in spite of the fact that cyclic loading may limit the variation of the crack tip chemistry [9–12].

Some of the environment–alloy combinations known to result in SCC are high-purity hot water, aqueous chloride, and cyanide solutions [1, 13].

8.2.2.1. Hydrogen Damage

The interaction between hydrogen and metals can result in the formation of solid solutions of hydrogen in metals, molecular hydrogen, gaseous products that are formed by reactions between hydrogen and elements constituting the alloy, and hydrides. Some hydrogen damage scenarios are described next [13, 14].

Environmental Hydrogen Embrittlement Environmentally assisted cracking (EAC) has been considered by some researchers to have two forms of corrosion: stress-corrosion cracking and environmental hydrogen embrittlement (EHE) [15]. This occurs during the plastic deformation of alloys in contact with hydrogen-bearing gases or during a corrosion reaction and is therefore strain rate dependent. The cracking mechanism with hydrogen depends on the hydrogen fugacity, the strength level of the material, the heat treatment/microstructure, the applied stress, and the temperature. Hydrogen absorption can favor local plasticity, very near the crack tip region, due to enhanced dislocation velocities with hydrogen.

Hydrogen penetration can be accelerated very near the crack tip region by stress-assisted diffusion and dislocation transport. For example, EHE is proposed as the dominant mechanism for corrosion fatigue (CF) crack propagation for aluminum alloys in water vapor as example) [16, 17]. Hydrogen stress cracking is characterized by a brittle fracture under sustained load in the presence of hydrogen. Generally, there is a critical minimum stress, below which delayed cracking will not take place at any time. The critical stress decreases with increase in hydrogen concentration. Hydrogen stress cracking usually produces sharp, singular cracks in contrast to the extensive branching observed for SCC. Sulfide stress cracking is considered a special case of hydrogen stress cracking [17–19].

Hydrogen-Induced Blistering Hydrogen-induced blistering is a cracking process caused by absorbed hydrogen atoms, commonly referred to as hydrogen-induced cracking (HIC) [13]. If some atoms of hydrogen diffuse into a void, they combine into molecular hydrogen. Molecules of hydrogen don't diffuse and the pressure of the hydrogen gas within the void increases. The pressure of the molecular hydrogen in contact with the atomic hydrogen is several hundred thousand atmospheres—sufficient to cause the rupture of any known engineering material [5]!

Formation of Metallic Hydrides Precipitation of a brittle metal hydride at the crack tip causes significant loss in strength and large losses in ductility and toughness of some metals such as magnesium, tantalum, niobium, vanadium, thorium, uranium, zirconium, titanium, and their alloys in hydrogen environments [13]. Alloy systems that form hydrides are generally ductile between 100 and 300 K and ductile fractures occur at these temperatures. Some evidence exists that nickel and aluminum alloys may also form a highly unstable hydride that contributes to hydrogen damage of these alloys; however, some of these alloys are susceptible to failure in hydrogen through other mechanisms [17].

8.2.2.2. Liquid- and Solid-Metal-Induced Embrittlement

Liquid-Metal-Induced Embrittlement (LMIE) This is the catastrophic brittle failure of a normally ductile metal when coated with a thin film of a liquid metal and subsequently stressed in tension. The fracture mode typically changes from a ductile to a brittle intergranular or brittle transgranular (cleavage) mode; however, there is no change in the yield or flow behavior of the solid metal. The velocity of crack propagation can be as large as 10–100 cm/s. In fact, one monolayer is sufficient for this type of fracture. The presence of liquid at the tip of the crack seems necessary for fast fracture; however, a crack initiated in the presence of liquid can propagate in the absence of liquid. It has been suggested that embrittlement is associated with liquid-metal adsorption-induced localized reduction in the strength of the atomic bonds at the crack tip or at the surface of the solid metal at sites of stress concentrations [20].

Solid-Metal-Induced Embrittlement (SMIE) Embrittlement occurs below the melting temperature of the solid in certain solid- and liquid-metal environment couples. The severity of embrittlement increases with temperature, with a sharp and significant increase in severity at the melting point, T_m, of the embrittler. Severe embrittlement is observed especially in the region when T/T_m is between 0.8 and 1 [21].

LMIE and SMIE of Aluminum Alloys LMIE is specific to certain solid-metal–liquid-metal systems, for example, liquid gallium embrittles aluminum but not magnesium and liquid mercury embrittles zinc but not cadmium. The equilibrium phase diagrams for many embrittled couples show that the two metals form binary systems with little or no solid solubility, immiscible in the liquid phase, and they do not form intermetallics. Although far less frequent than brittle fracture, embrittlement can also occur by a ductile dimpled rupture mode in certain steels, copper alloys, and aluminum alloys. SMIE has been observed only in those couples in which LMIE is possible. However, SMIE may occur in the absence of

LMIE if a brittle crack cannot be initiated at the melting point of the embrittling agent. SMIE and LMIE are of significant scientific interest in understanding the mechanisms of embrittlement [20].

8.3. SCC PARAMETERS OF ALUMINUM ALLOYS

Only aluminum alloys that contain appreciable amounts of soluble alloying elements, primarily copper, magnesium, silicon, and zinc, are susceptible to SCC. For most commercial alloys, tempers have been developed that provide a high degree of immunity to SCC in most environments [22].

The complex interactions among the factors that lead to SCC of aluminum alloys are not yet fully understood [23]. Since susceptibility to SCC necessitates soluble alloying elements and the morphology of the rupture is mostly intergranular, in many circumstances galvanic cells and intergranular corrosion are associated with SCC of aluminum alloys. The electrochemical properties associated with the microstructure of different aluminum alloys continue to be the basis for developing aluminum alloys and tempers resistant to SCC [24]. Thus susceptibility to intergranular corrosion is a prerequisite for susceptibility to SCC, and treatment of aluminum alloys to improve resistance to SCC also improves their resistance to intergranular corrosion. For most alloys, however, optimum levels of resistance to these two types of corrosion require different treatments, and resistance to intergranular corrosion is not a reliable indicator of resistance to SCC.

8.3.1. Influence of Stress

SCC in a susceptible aluminum alloy depends on both magnitude and duration of tensile stress acting at the surface. The effects of stress have been established by accelerated laboratory tests, and the results of one set of such tests are shown in the shaded bands in Figure 8.2. Despite the introduction of fracture mechanics techniques for determining crack growth rates, such tests continue to be the basic tools used in evaluating resistance of aluminum alloys to SCC. These tests suggest a minimum (threshold) stress that is required for cracking to develop.

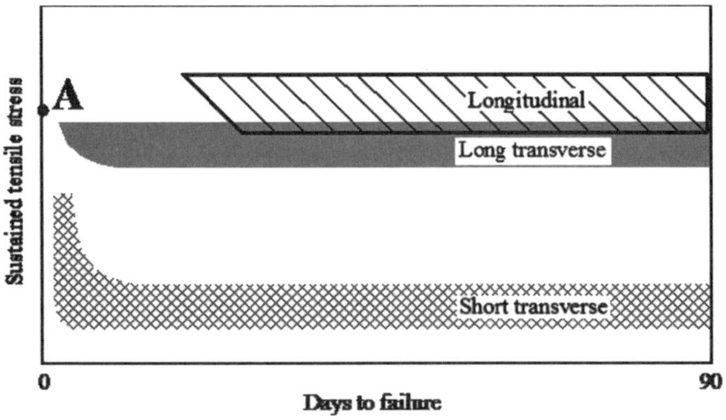

Figure 8.2 Shaded bands indicate combinations of stress and time known to produce SCC in specimens of the alloy 7075-T651 plate intermittently immersed in 3.5% NaCl solution. Point A is minimum yield strength in the long transverse direction for a 75 mm (3 in.) thick plate [22, 25, 26].

For some alloy–temper combinations, results of accelerated laboratory tests reliably predict stress-corrosion performance in service; for example, results of an 84 day alternate immersion test of alloy 7075 and alloy 7178 products correlated well with performance of these products in a seacoast environment [25].

Effects of Grain Structure and Stress Orientation Many wrought aluminum alloy products have highly directional grain structures and are highly anisotropic with respect to resistance to SCC (Figure 8.2). Resistance, measured by magnitude of tensile stress required to cause cracking, is highest when the stress is applied in the longitudinal direction, is lowest in the short-transverse direction, and is intermediate in other directions. These differences are most noticeable in the more susceptible tempers, but are usually much lower in tempers produced by extended precipitation treatments, such as T6 and T8 tempers for 2xxx alloys and T73, T736, and T76 tempers for 7xxx alloys [26].

One of the most serious practices associated with SCC problems is machining, which leads to high tensile stress areas in the material. If the exposed tensile stresses are in a transverse direction or have a transverse component, and if a susceptible alloy or temper is involved, the probability of SCC is present [27].

8.3.2. Role of Environment

Research indicates that water or water vapor is the key environmental factor required for producing SCC in aluminum alloys. Halide ions have the greatest effects in accelerating attack. Chloride is the most important halide ion because it is a natural constituent of marine environments and is present in other environments as a contaminant. Because it accelerates SCC, Cl^- is the principal component of environments used in laboratory tests to determine susceptibility of aluminum alloys to this type of attack. In general, susceptibility is greater in neutral solutions than in alkaline solutions and is greater still in acidic solutions [25].

Testing in specific hydrogen environments has revealed the susceptibility of aluminum to hydrogen damage. Hydrogen damage in aluminum alloys may take the form of intergranular or transgranular cracking or blistering. Blistering is most often associated with the melting or heat treatment of aluminum, where reaction with water vapor produces hydrogen. Blistering due to hydrogen is frequently associated with grain boundary precipitates or the formation of small voids. Blister formation in aluminum is different from that in ferrous alloys in that it is more common to form a multitude of near-surface voids that coalesce to produce a large blister [28].

Hydrogen diffuses into the aluminum lattice and collects at internal defects, most frequently during annealing or solution treatment in air furnaces prior to age hardening. Dry hydrogen gas is not detrimental to aluminum alloys; however, with the addition of water vapor, subcritical crack growth increases dramatically (Figure 8.3) [22]. The threshold stress intensity for cracking of aluminum also decreases significantly in the presence of humid hydrogen gas at ambient temperature [23].

Temperature Stress-corrosion cracking could occur above a certain temperature. Also, increasing the temperature generally lowers the threshold for cracking (σ_1 and K_{ISCC}) and increases the growth rate of propagation. Hydrogen permeation and the crack growth rate are functions of potential, increasing with more negative potentials, as expected for hydrogen embrittlement behavior. The ductility of aluminum alloys in hydrogen is temperature dependent, displaying a minimum in the area below 0 °C; this behavior is similar to that of other face centered cubic (fcc) alloys [23]. Some evidence for a metastable aluminum

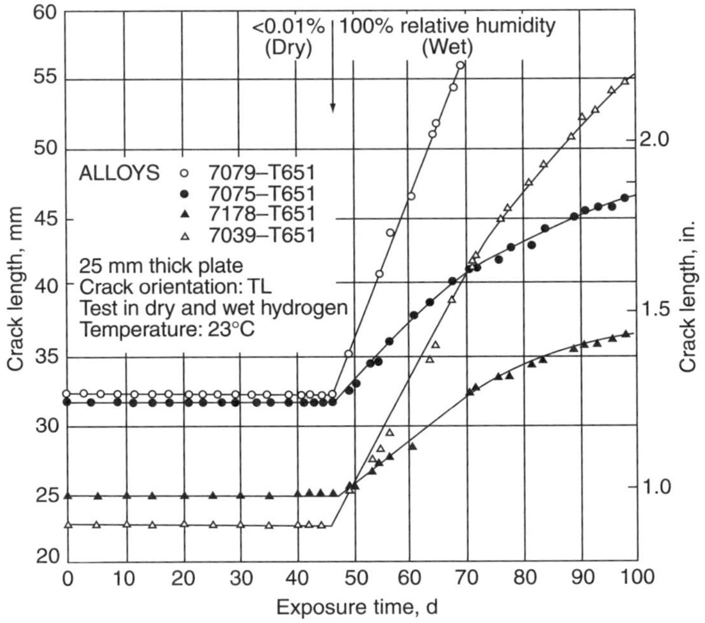

Figure 8.3 Effect of humidity on subcritical crack growth of high-strength aluminum alloys in hydrogen gas (TL = transgranular) [22, 23].

hydride has been found that would explain the brittle intergranular fracture of aluminum–zinc– magnesium alloys (the 7xxx series) in water vapor. However, the instability of the hydride is such that it has been difficult to evaluate. Another explanation for intergranular fracture of these alloys is preferential decohesion of grain boundaries containing segregated magnesium. Overaging of these alloys increases resistance to hydrogen embrittlement in much the same way as for highly tempered martensitic steels [23].

Intergranular, Pitting, and SCC Potential Domains Anodic polarization studies (ASTM G5) have shown that the corrosion of a susceptible microstructure in aqueous solution concerning an alloy of the series 7xxx is exclusively intergranular for a limited range of potentials between the first and the second breakdown potentials (E_{BR1} and E_{BR2}) determined from the polarization curves (Figure 8.4) [18]. E_{BR1} approximates the critical pitting potential of the active corrosion path at the grain boundaries and E_{BR2} could correspond to the pitting potential of the grain bodies. Above E_{BR2}, there is the region of pitting that can lead to SCC [18].

Stress-corrosion cracking in aluminum alloys is characteristically intergranular. This type of corrosion requires a condition along grain boundaries that makes them anodic to the rest of the microstructure so that corrosion propagates selectively along them. Such a condition is produced by localized decomposition of solid solution, with a high degree of continuity of decomposition products along the grain boundaries. The most anodic regions may be either the boundaries themselves (most commonly, the precipitate formed in them) or regions adjoining the boundaries that have been depleted of solute.

As an example, intergranular corrosion of the 5xxx series is the leading form in SCC. Effectively, alloys in this series having more than approximately 3% Mg are rendered susceptible to intergranular corrosion (sensitization) by certain manufacturing conditions

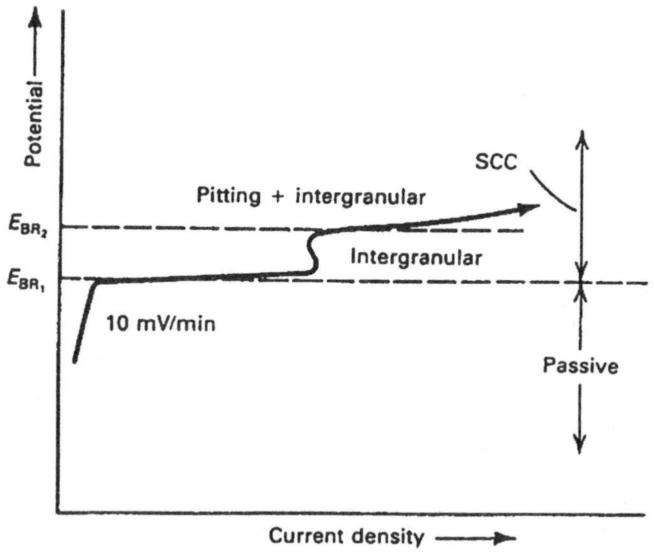

Figure 8.4 Anodic polarization curve of aluminum alloy 7075-T651 in deaerated 3.5% sodium chloride solution showing the domain predicted for pitting, intergranular, and SCC forms of corrosion [18].

or by being subjected to elevated temperatures up to 175 °C (strain hardening). This is the result of continuous grain boundary precipitation of the highly anodic Mg_2Al_3 phase, which corrodes preferentially in most corrosive environments [18]. In 2xxx alloys, the solute-depleted regions are the most anodic, while the most anodic grain boundary regions in other alloys have not been identified with certainty [25].

8.4. SCC MECHANISMS

8.4.1. Overlapping of Cracking Phenomena

It is generally accepted that in many circumstances and service conditions SCC, HE, and CF of different alloys can coexist. For example, CF can help initiation of the fracture while SCC and/or HE could assist (more or less intensively) the crack propagation. In many circumstances, pitting fatigue corrosion is accompanied later by SCC or in other circumstances excessive residual stresses can be joined by cyclic stresses. One or the other can dominate the morphology and mechanism of fracture propagation. However, the distinction between HE and SCC or CF is often very difficult, since it is likely that hydrogen–metal interaction near the crack tip is the controlling process of SCC or CF crack propagation in several circumstances. The most important practical situations are when all three phenomena interact, which is probably representative of many situations involving ductile alloy–aqueous environment systems [2, 18].

Both the decrease of K_{ISCC} and the acceleration of crack growth rates due to the increase of the metal strength become more apparent under cyclic loading, which corresponds more to the actual loading modes of practically all components, structures, and vehicles under in-service conditions. K_{ISCC} is always much higher than K_{th}, the threshold stress-intensity range at the corrosion fatigue crack growth rate $\leq 10^{-10}$ m/cycle. *da*/*dN* in aqueous solutions is much higher (by about a dozen times) than that in ambient air [3, 29].

For materials that exhibit classical active–passive behavior, passivation is more conducive under static rather than dynamic conditions. For the latter, the frequency of cyclic loading is often one of the critical factors that influences CF in corrosive environments. Cathodic protection generally mitigates CF and SCC but increases the probability of HE of susceptible materials. Electrode potential and pH at an active crack tip may be significantly different from those on exposed surfaces of a material. Low-pH conditions can lead to local dissolution of metal and crack tip blunting, which reduces stress concentration effects. In contrast, low-pH conditions favor hydrogen generation and consequently increase risk of HE. Reduction in local ductility associated with HE is more likely to produce sharp crack tips, which, in turn, can exacerbate stress concentration effects for any synergistic SCC or CF [18].

Relation of SCC and CF Cracking of Al–Zn–Mg Alloy Magnin and Rieux [30] examined the SCC and the corrosion fatigue (CF) of smooth specimens of a weldable Al–Zn–Mg (AA7020) alloy deformed at an imposed strain rate, paying particular attention to the microcracking processes at the surface as a function of the heat treatment and the electrochemical conditions. Specimens were machined from rolled plates (fibrous structure) with the tensile axis parallel to the rolling direction.

The specimens were solution treated at 470 °C for 1 hour and then water quenched. Two heat treatments, T4 and T6 DR, were considered. The T4 treatment corresponds to aging at room temperature for 8 days (underaging) and the T6 DR corresponds to T4 + 24 h at 120 °C + 24 h at 140 °C (peak aging). Corrosion fatigue tests were conducted at constant strain rate while SCC tests were carried out using the slow strain rate method. The sensitivity to SCC and CF is quantitatively defined by the ratios $A(\text{NaCl})A(\text{air})$ and $N_i(\text{NaCl})/N_i$ (air), respectively, where A is the elongation to fracture and N_i is the number of cycles to crack initiation, respectively (Figure 8.5).

At the free corrosion potential for both anodic dissolution and hydrogen embrittlement, a marked reduction of the fatigue life is obtained, whatever the strain rate (10^{-8} to $10^{-2}\,s^{1}$), while SCC takes place only at slow strain rates below $2 \times 10^{-5}\,s^{-1}$ for the 7020-T4 and below $8 \times 10^{-7}\,s^{-1}$ for the 7020-T6 DR alloy. At slow strain rates, where SCC is not

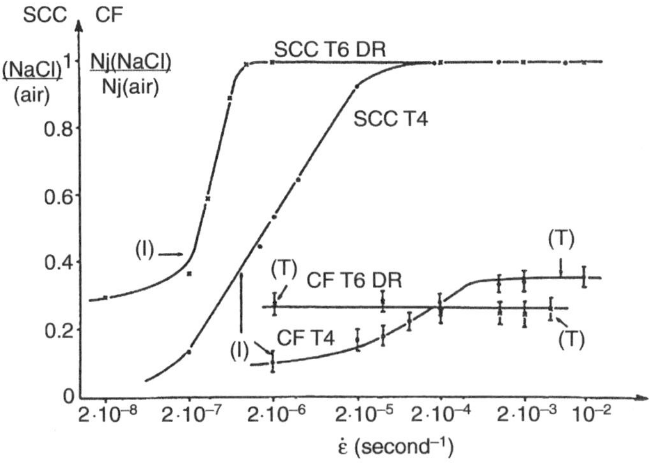

Figure 8.5 SCC versus CF for 7020 Al–Zn–Mg alloys in 3.5% NaCl solution at free corrosion potential as a function of the strain rate (crack initiation can be transgranular (T) or intergranular (I) [2, 30].

possible, transgranular CF cracks were formed by a combined effect of the plastic strain localization due to fatigue and the breakdown of the passive film (Figure 8.5). Both anodic dissolution and hydrogen embrittlement are effective for CF propagation. In the case of SCC, intergranular microcracks were formed. Intergranular SCC, which was observed only at low strain rates, also decreases the fatigue resistance. Magnin and Rieux [30] proved that CF is generally more deleterious than SCC for 7020 Al–Zn–Mg alloys in 3.5% NaCl solution at the free corrosion potential [2].

8.4.2. Significance of the Magnitude of Strain Rates

The most significant variable in slow strain rate testing is the *magnitude* of the strain rate. Relatively low strain rates on the order of 10^{-5} to $10^{-7}\,s^{-1}$ are appropriate since too high rates give rise to ductile fracture while too low strain rates may lead to passivation of the aluminum alloy in the concerned medium and this depends on the alloy, the microstructure, and the environment. The repassivation reaction that is observed at very low strain rates and that prevents the formation of anodic SCC does not occur when cracking is the result of embrittlement by hydrogen produced by the electrochemical corrosion cell. This difference in corrosion mechanism is advantageously used to distinguish between anodic SCC (active path corrosion) and cathodic SCC (or hydrogen- induced cracking) (Figure 8.6) [18].

The most relevant strain rates for different aluminum alloys are given in Figure 8.7 [18]. These trends show that the starting strain rate level should correspond to the considered alloy in a specific medium [18]. Generally, lower stress corrosion crack propagation requires slower strain rate testing. The most severe strain rate should be determined for every case. Rates of 10^{-4} to 10^{-7} are recommended for Al in chloride solutions. For the sake of comparison, the rate of 10^{-5} is frequently chosen for Mg and its alloys in chromate–chloride solutions.

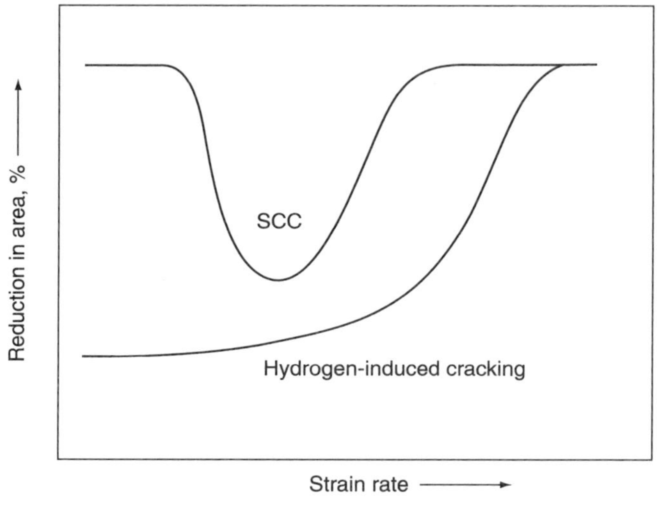

Figure 8.6 Effect of strain rate on SCC and hydrogen-induced cracking [18].

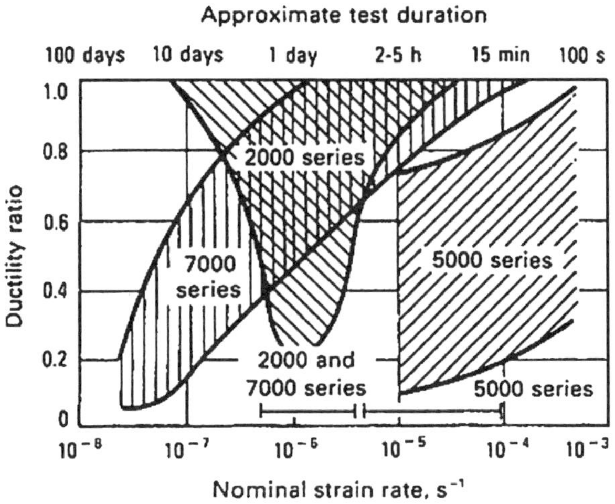

Figure 8.7 Recommended strain rate regimes for different aluminum alloys in 3% NaCl plus 0.3% H_2O_2 [18].

8.4.3. Cracking Initiation and Propagation

The cracking initiation mechanism has been difficult to define or limit and understand until now. Effectively, most of the SCC systems exhibit short initiation times ranging from minutes to weeks and cracking occurs often due to the change in the environment rather than to a very long initiation time. Stress-corrosion crack growth rates are usually 10^{-11} and 10^{-6} m.s^{-1}. Surface films appear to play a major role in the initiation of SCC and may also contribute to hydrogen embrittlement effects. It is assumed that the main role of the surface film is to localize the damage inflicted on the material by the environment. This can be caused by mechanical breakdown of the protective film by a slip dissolution step or electromechanical breakdown of the passive film [2].

If the environment provides hydrogen species that are adsorbed at the crack tip to reduce the effective bond strength, then the surface energy is effectively lowered; alternatively, hydrogen atoms may diffuse into the metal. Decohesion of atoms can result from hydrogen influx to the dilated lattice; certain interactions may occur in advance of the crack tip where the stress and/or the strain conditions are particularly appropriate for nucleation of a crack. This can be followed by the formation of a brittle phase, such as a metal hydride. On the other hand, the recombination of two atoms of hydrogen can form molecular hydrogen, which can cause pressure inside the metallic network and cause inflation; this is a different mechanism than the one for the formation of a blister [31].

Possible mechanisms of hydrogen embrittlement are (1) chemical adsorption of hydrogen, (2) absorption of atomic hydrogen, (3) decohesion of atoms, and (4) possible brittle hydride particles at the tip. Hydrogen-induced crack growth as the dominant SCC mechanism has been suggested for ferritic steels, nickel-based alloys, titanium alloys, and aluminum alloys. A review of the combined effects of impurity segregation and hydrogen embrittlement concluded that grain boundary impurities behave in the same way with cathodic and gaseous hydrogen in that they enhance crack growth by a combined mechanism of grain boundary embrittlement, but not by enhanced hydrogen uptake [6].

Localized corrosion and stress-corrosion cracking are quite often correlated. Stress-corrosion cracks invariably initiate at pits in numerous systems. The role of pitting is to disrupt films that otherwise prevent the ingress of hydrogen and, although there is some stress concentration associated with the pits, this is of lesser importance than the environmental implications of the presence of pits [32, 33].

Propagation Models Usually stress-corrosion cracking occurs at stresses below general yield and propagate in an essentially elastic body, even though local plasticity may be necessary for the cracking process. Thus linear elastic fracture mechanics (LEFM) is used for studying SCC. Significant progress has been made in the conception and development of different models of crack propagation. Stress-corrosion crack propagation mechanisms can be divided into dissolution and mechanical fracture models and the latter can be divided into ductile and brittle crack extension processes. Dissolution models include film rupture and active path processes. Ductile mechanical models include corrosion tunneling and adsorption-enhanced plasticity models. Brittle mechanical models include the tarnish rupture and film-induced cleavage models [1].

Propagation of the crack is very frequently due to periodic microruptures, although anodic dissolution is an important or controlling process. The local environment and water transport to the crack tip could also be important and controlling processes. Except for the slip dissolution model, the EAC models have not provided quantitative prediction of the crack propagation rate [2].

8.5. SCC OF ALUMINUM ALLOYS

Stress-corrosion cracking (SCC) in aluminum alloys is characteristically intergranular. There is a condition along grain boundaries that makes them anodic to the rest of the microstructure, so that corrosion propagates selectively along them. This condition is produced by localized decomposition of solid solution, with a high degree of continuity of decomposition along the grain boundaries. In 2xxx alloys, solute-depleted regions along the grains are the most anodic regions, while for the 5xxx alloys, the Mg_2Al_3 precipitate along the grain boundaries is the most anodic.

Since SCC in aluminum alloys characteristically is intergranular, susceptible alloys and tempers are most prone to SCC when the tensile stress acts in the short-transverse (or thickness) direction, so that the crack propagates along the aligned grain structure. The same material (e.g., 7075-T651 plate) will show a much higher resistance to stress acting in the longitudinal direction, parallel to the principal grain flow. In this case, the intergranular crack must follow a very meandering path and usually does not propagate to any major extent. Special aging processes to various, highly resistant T7 tempers have been developed to counteract this adverse effect of directional grain structure. Various artificially aged tempers are available for both 2xxx and 7xxx alloys that provide a range of choices between maximum strength and maximum resistance to exfoliation and SCC [22]. (Intergranular corrosion and corrosion fatigue propagation are detailed in Chapters 6 and 7, respectively.)

Exfoliation and Intergranular SCC Exfoliation or layer corrosion and stress-corrosion cracking of aluminum alloys can be considered as forms of intergranular corrosion. On Figure 8.8a, the cross section shows exfoliation corrosion in an alloy 7178-T651 plate exposed to a seacoast environment. Exfoliation develops by corrosion along grain boundaries of thin elongated grains. Wedge stresses of exfoliation are created due to the expansion of the products of corrosion and SCC follows the grain boundaries. This can be distinguished in morphology from that of stress intergranular SCC propagation of high-strength aluminum alloy as seen on Figure 8.8b.

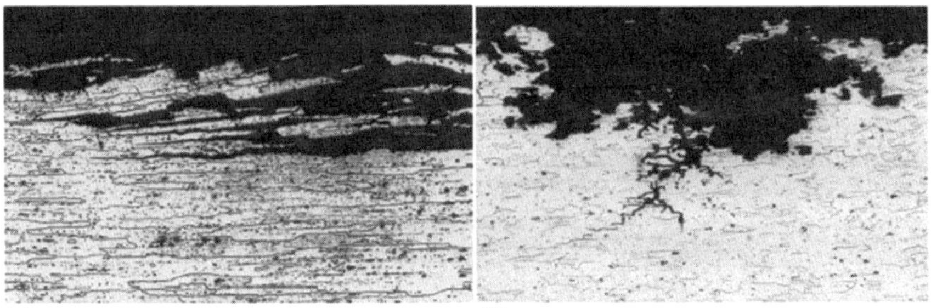

Figure 8.8 A cross section showing (a) exfoliation corrosion of alloy 7178-T651 plate exposed to a seacoast environment [34] and (b) intergranular stress-corrosion cracking of high-strength aluminum alloy.

Intergranular attack creates tension stresses for crack initiation at the grain boundaries because of the electrochemical galvanic cells, such as can be formed at the pit edge in aluminum alloys. SCC propagation then follows the intergranular passage [34].

Figure 8.9 shows different microstructures of an alloy containing 5% Mg after different heat treatments. These microstructures have low to high resistance to SCC depending on the heat treatment. High resistance to SCC treatments produces microstructures free of precipitates along grain boundaries (Figure 8.9a) or with uniformly distributed precipitates within grains (Figure 8.9d). Uniform precipitate distribution increases the ratio of the anodic to cathodic region throughout the microstructure, reducing the corrosion current on anodic regions [35].

Threshold Stress Relief SCC development in a susceptible alloy depends on the magnitude and duration of tensile stress at the surface. Accelerated laboratory tests are the basic tools used in evaluating the SCC resistance of aluminum alloys. A minimum threshold is necessary to develop SCC. Although empirical, these tests provide a valid measure of relative susceptibilities for different aluminum alloys in specific environments. Solution heat treating and quenching creates residual stresses in aluminum alloys. The quenching process places the surfaces in compression and the center in tension. Compressive surface stresses enhance resistance to SCC. Machining into the residual high tensile stress areas of material that has not been stress-relieved promotes SCC. Mechanical stretching is a relatively economical method of stress relief for constant cross section aluminum products. Stretching must be done after quenching and for some alloys before artificial aging [35].

8.5.1. SCC Resistance of Aluminum Alloys

Cast aluminum products normally have an equiaxial grain structure. Special processing routes can be taken to produce fine, equiaxial grains in thin rolled sheet and certain extruded shapes, but most wrought products (rolled, forged, drawn, or extruded products) normally have a highly directional, anisotropic grain structure. Rectangular products have a three-dimensional (3D) grain structure. Rolled plate has a three-directional grain structure—longitudinal (principal working direction), long-transverse, and short-transverse—while rolled rod has a two-directional grain structure. These directional structures markedly affect resistance to SCC and to exfoliation of high-strength alloy products for the 2xxx and 7xxx series aluminum alloys [22, 36].

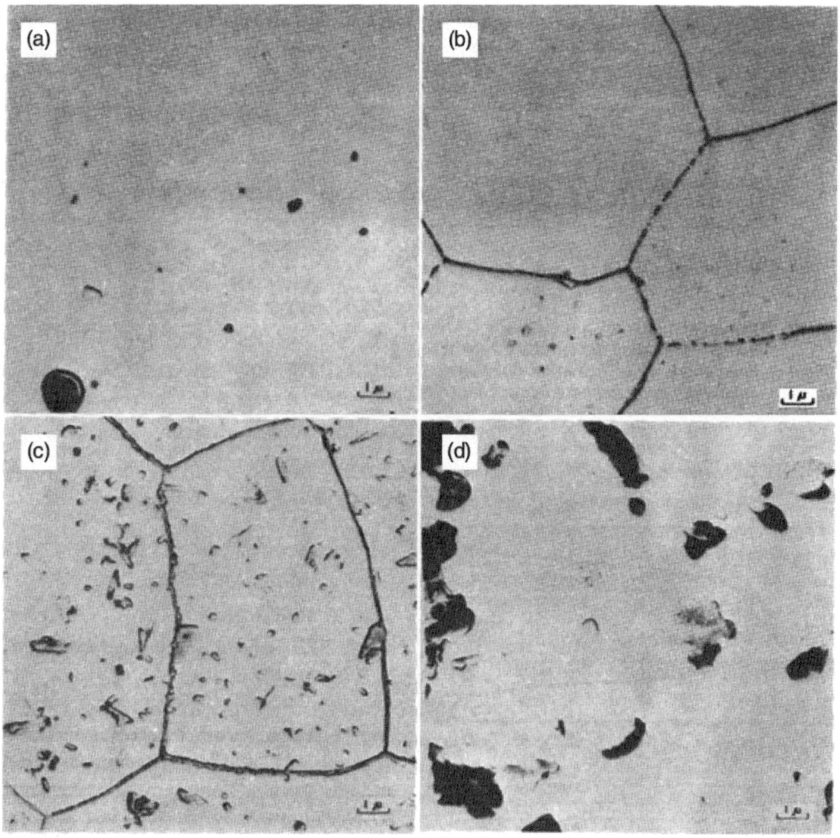

Figure 8.9 Heat treatments of cold-rolled (20%) alloy 5356-H12 to produce varying degrees of susceptibility to SCC: (a) cold-rolled as reference, highly resistant; (b) heat treated for 1 year at 100 °C, highly susceptible; (c) heated for 1 year at 150 °C, slightly susceptible; and (d) heated for 1 year at 205 °C, highly resistant [35].

Most die forgings and many extrusions with irregular, complex cross sections have a metal flow that varies with the product contour. Evaluation of such products requires knowledge of the metal flow pattern through either prior experience or macroetching. The grain structure and resultant corrosion behavior also vary from surface to center in products with appreciable thickness. This factor begins to be important at a thickness of about 12 mm. Almost all forms of corrosion, even pitting, are affected to some degree by this grain directionality [22].

Many wrought aluminum products have a highly directional grain microstructure. This is most noticeable in the susceptible tempers but lower in the tempers produced by extended precipitation treatments. For thin-section applications, where low short-transverse stresses are present, a medium resistant temper may suffice. In thick-section applications, where high tensile stresses are present in the short-transverse direction, more resistant tempers are preferred [35].

Wrought aluminum alloys of the 2xxx, 5xxx (high magnesium content), and 7xxx series are susceptible to SCC. For the 7xxx alloys, the magnesium–zinc grain boundary precipitate is susceptible to SCC. For the 5xxx alloys, the beta phase Mg_2Al_3 is the most susceptible to SCC [35].

2xxx Alloys Second-phase precipitation sites depend on the cooling rate, which in turns depends on the material thickness. Thick sections of 2xxx alloys naturally aged in the T3 or T4 temper show poor resistance to SCC in the short-transverse direction. Grain boundary precipitation doesn't occur in thin sections because of a higher cooling rate. Cast alloys 240.0F, 242.0-T21, 242.0-T77, 242.0-T571, 242.0-T75, and 295.0 in the T2, T6, and T62 tempers exhibit high susceptibility to SCC. These alloys have high copper contents and have a high susceptibility because of the different cooling rates of the cast parts [35].

5xxx Alloys Beta-phase particle Mg_2Al_3 precipitates at the grain boundaries create favorable conditions for SCC. SCC in this case occurs in alloys with magnesium contents higher than 3% and the magnesium content in the solute solution is high enough to precipitate at high temperature and to form a continuous beta phase along the grain boundaries. High temperature and a certain amount of strain are necessary to form precipitates along the grain boundaries [35].

6xxx Alloys There is no evidence that SCC affects these alloys [35].

7xxx Alloys In the case of the 7075 alloy, two-step special heat treatments are available to promote SCC resistance. It has been demonstrated that small copper additions reduced the amount of precipitation by reducing the zinc concentration within the grain boundaries. Cast alloy 712.0-T5 has a low resistance to SCC, while alloys 710.0-T5 and 713-T5 have a higher resistance. These alloys have high zinc and magnesium contents. Alloys 710-T5 and 713-T5 have a small amount of copper; these alloys have a higher resistance to SCC because of the reduced presence of zinc-rich constituents within the grain boundaries [35].

In 7xxx alloys, a small zirconium addition promotes the precipitation of the magnesium–zinc precipitates within the grains. The sum of the magnesium–zinc weight percent should be kept below 6% to enhance SCC resistance. On the other hand, a slight 1% magnesium + zinc composition increase results in 25 MPa higher yield strength. These alloys are mostly used close to the maximum susceptibility limit; therefore SCC failure is of prime concern. In SCC of aluminum alloys, there is no corrosion product visible when the crack propagates at the surface; SCC failure occurs spontaneously without warning. SCC crack detection is possible with nondestructive techniques (NDTs) [35].

8.5.2. Influence of Heat Treatments on Corrosion Forms

Heat treatments can lead to microstructures that are free of precipitates along grain boundaries or to uniform distribution of precipitates within grains. Some residual stresses are induced in aluminum alloy products when they are solution heat treated and quenched and stress relief is frequently recommended.

Impact of the Heat Treatment of the 7000 Series Contrary to AA2024, the alloys in the 7xxx series have shown good corrosion resistance as well as offering the required mechanical properties (Figure 8.10). However, aluminum alloys such as 7079-T6 in all forms and 7075-T6 plate, forgings, or extrusions should be avoided because of their inferior resistance to SCC. AA7079 does not respond to the usual T76 and T73 heat treatments for better corrosion resistance, but AA7075 does. Therefore the effectiveness of the usual heat treatments of this type must be established for the introduction of new 7xxx alloys (Table 8.1). Alloys such as AA7178 and AA2020 are generally not recommended in aircraft structures by the International Air Transport Association [37, 38].

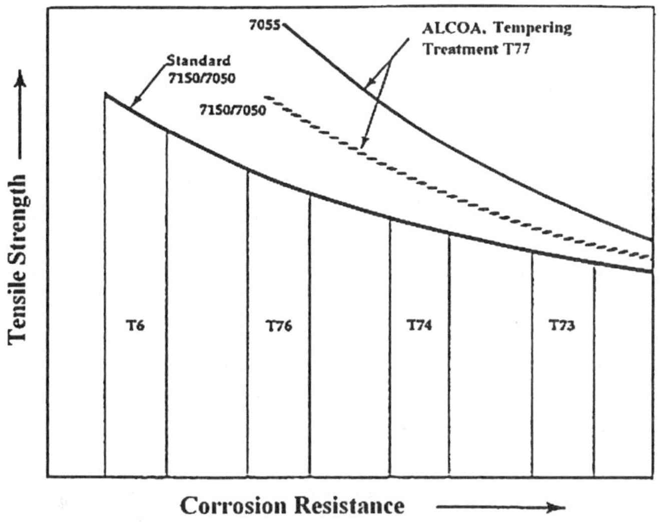

Figure 8.10 Compromise between corrosion resistance and tensile strength [38].

Table 8.1 Stress-Corrosion Cracking, Exfoliation, and Intergranular Corrosion Forms as Related to Some Aluminum Alloys and Their Heat Treatments

Alloy	SCC rating[a]	Susceptible to exfoliation	Probable susceptibility to intergranular corrosion (MIL -H-6088E)
2011-T3	D	Yes	Yes
2011-T8	B	No	Yes
2021-T8	B	No	Yes
2024-T6	C	Yes	Yes
2024-T72	B	No	Yes
2024-T8	B	No	Yes
2219-T8	B	No	Yes
6061-T6	A	No	Yes
7075-T73	B	No	Yes
7075-T76	C	No	Yes
7175-T36	C	No	Yes
7178-T76	C	No	Yes
2024-T3,T4	D	Yes	Yes
2219-T3	D	Yes	Yes
7075-T6	D	Yes	Yes
7175-T66	D	Yes	Yes
7079-T6	D	Yes	Yes
7178-T6	D	Yes	Yes

[a]SCC ratings are based on service experience excluding special chemicals or elevated temperatures and on 4 year exposures to seacoast or industrial atmospheres or 84 days to 3.5% NaCl alternate immersion. Ratings are only for thin sections that can be heat treated and quenched to achieve a high cooling rate: A, no SCC failure observed in service or laboratory tests; B, no SCC failure observed in service but SCC can occur in extreme laboratory tests of short-transverse experiments; C, SCC not anticipated in service or in laboratory tests at sustained tension below typical design stresses or residual stresses resulting from heat treatment, welding, or controlled assembly stresses; D, instances of SCC in service unlikely with metal stressed parallel to direction of grain flow but occasionally experienced with sustained tension in the transverse or short-transverse directions [40].

Sources: References 39 and 40.

High-strength heat-treatable aluminum alloys (2xxx and 7xxx series) have shown corrosion failures due to exfoliation and SCC. Although both of these are generally intergranular phenomena, it has been observed that the serviceability of aluminum alloy products cannot be related in a general way to the susceptibility to IGC or to any other single characteristic such as microstructure, tensile strength, or fracture toughness. For example, artificially aged products of alloys such as 2011, 2021, 2024, and 2219 will provide maximum resistance to SCC even though they can be susceptible to IGC. The same applies for alloy 6061-T6, which is susceptible to IGC but virtually immune to SCC (Table 8.1).

It can generally be stated that alloys that are susceptible to intergranular corrosion (IGC) are more prone to exfoliation (Table 8.1) than corresponding alloys with susceptibility only to pitting corrosion. However, some alloys that are susceptible only to pitting corrosion when exposed in the absence of stress can still be susceptible to SCC, particularly when stressed in the short-transverse or transverse direction. For example, high resistance to SCC in the transverse direction of 64 mm diameter cold finished rolled rod was provided only by 7075-T351 and 2219-T87 even though all five alloys from these series exhibited pitting corrosion after 7 years of seacoast atmospheric exposure in Texas (USA) [39].

Finally, it should be underlined that the type of attack produced in an accelerated laboratory test is of limited usefulness and must be interpreted in the light of the metallurgical history of the metal [39].

8.6. SCC OF WELDED ALUMINUM ALLOYS

The growth of precipitate particles such as $CuAl_2$, $MgZn_2$, and Mg_2Al_3 is promoted by heat treatments and can lead to galvanic corrosion effects [41]. The heat treatments that are beneficial mechanically, because they improve hardness by causing precipitate formation, are at the same time detrimental to corrosion resistance because the growth of different phases within the material favors the possibility of galvanic corrosion. The severity of subsequent attack depends on the characteristics of the heat treatments to the extent that they control the size and distribution of intergranular precipitates. Generally, alloys that are heat treated to peak hardness are most severely attacked by corrosion, while those that undergo overaging are somewhat less susceptible [42]. With certain alloys, particularly those of the heat-treatable 7xxx series, thermal treatment after welding is sometimes used to obtain maximum corrosion resistance [35, 43–46].

8.6.1. Galvanic Corrosion and SCC of Welded Assemblies

Although some aluminum alloys can be autogenously welded, the use of a filler metal is preferred to avoid cracking during welding and to optimize corrosion resistance in corrosive media. The variation of corrosion potential across three welds was examined for alloys 5456, 2219, and 7039 with different fillers 5556, 2319, and 5183, respectively. Figure 8.11 (on the left) shows the evolution of potential for the alloy 7039-T651 based metal with the 5183 alloy filler (two-pass tungsten inert gas weld). The corrosion potential shows an important difference between the edge of the weld bead and that at a distance of 25 mm from the weld. There is an important galvanic corrosion cell that can accelerate different types of localized corrosion and can lead to SCC. The choice of the matching filler for every alloy is of major importance and could be a function of the corrosive medium. The other two welded assemblies of 5456 and 2219 in chloride solutions did not show the same trend of

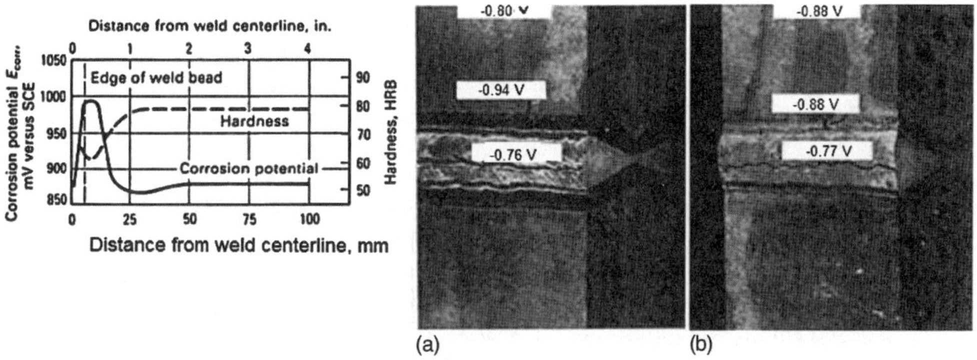

Figure 8.11 On the left, the open circuit potential determination and hardness of welded assembly of 7039–T651, 5183 filler (two-pass tungsten inert gas weld). Part (a) shows inferior corrosion resistance to localized corrosion of 7005 with 5356 filler after a 1 year exposure to seawater as compared to that after post-weld aging (b) [47].

potential evolution, while the trend for hardness evolution was almost similar for the three considered weld assemblies [47].

Principally, the minimum amount of heat during welding corresponds to the least influence on microstructure and lead to a better corrosion resistance. This should be accompanied by a good matching choice for the filler alloy. Post-weld aging is recommended to avoid localized corrosion and SCC for heat-treatable alloys. In particular, for the heat-treatable 7xxx series, thermal treatment after welding is currently used to obtain maximum corrosion resistance. In Figure 8.11, one can observe the beneficial effect of post-weld aging following alloy welding after a 1 year exposure to seawater. Severe corrosion was observed in the HAZ of the welded structure of alloy 7005 (Figure 8.11a) while Figure 8.11b shows the beneficial effects of post-weld aging. The corrosion potentials of the different areas were performed in 53 g/L NaCl plus 3 g/L H_2O_2 calculated to saturated calomel electrode (SCE); smaller galvanic cells were created in the post-weld aged material compared to that of the as-welded material [47].

Preheating affects SCC resistance for aluminum–magnesium alloys because it exposes the material for a short period of time to temperatures between 121 and 204 °C. Studies showed that filler metal AA5356 is sensitive to SCC when exposed to high temperatures [48]. When aluminum–magnesium alloys are heated, Mg_2Al_3 (beta phase) precipitates at grain boundaries. This beta phase is anodic to the rest of the microstructure and therefore, in a corrosive service, creates microscopic galvanic cells and the beta phase corrodes [35, 49]. AA5086 is successfully used for transport containers for sodium chlorate, but heat input must be controlled. The SCC tendency can be reduced if the total composition of Mg + Zn in these alloys is lower than 6% [50]. The SCC tendency in arc welding can be avoided if an anodic (to the base metal) aluminum alloy is projected on the weld bead [35].

8.6.2. SCC Knife-Line Attack

Although the dangerous knife-line attack type of SCC has not been observed for any commercial aluminum alloy or weldment even at 70 °C, it is wise to consider it. Both welded and unwelded aluminum alloys have good resistance to knife-line attack at temperatures $\leq 50\,^{\circ}C$. One exception was found in the case of a fusion-welded 1060 alloy, where no knife-line attack was observed even at 70 °C. Inhibited fuming nitric acid with at least 0.1% HF did not cause knife-line attack in laboratory experiments up to 70 °C [47].

8.6.3. Localized Corrosion and SCC of LBW AA6013

Aluminum alloy 6013 sheet was butt welded using a 3 kW Nd:YAG laser and different filler powders (laser beam welding). Two kinds of filler metals were used: gas atomized powders of the aluminum alloys $AlMg_5$, $AlSi_{12}$, $AlSi_{12}Mg_5$, and $AlSi_{10}Mg$, as well as mixtures of powders of the elements Al and Si and the binary alloys Al–5Mg, Al–5Zr, Al–5Cr, and Al–10Mn. Microstructure, hardness, tensile properties, and corrosion behavior of the welds were investigated in the as-welded T4 and T6 conditions and after a post-weld heat treatment to the T6 temper [51].

The material used was a 1.6 mm thick sheet of the alloy 6013 received in the naturally aged T4 condition. To achieve peak strength, part of the sheet was artificially aged at 191 °C for 4 h. Butt welds of alloy 6013 in the tempers T4 and T6 were produced in a shielded atmosphere using a 3 kW continuous wave Nd:YAG laser. The shielding gas was helium at a flow rate of 10 L/min. The focal length of the employed lens was 150 mm. The laser beam was normal to the sheet and focused 1.0–1.5 mm above the sheet surface. The specimens being 400 mm × 100 mm in size were tightly fastened. Laser power and welding speed ranged from 2300 to 2600 W and from 3 to 4 m/min, respectively. The weld direction was parallel to the rolling direction of the sheet. Filler materials used were gas atomized powders of the aluminum alloys $AlMg_5$ (4.5–5.5% Mg), $AlSi_{12}$ (10.5–13.5% Si), $AlSi_{12}Mg_5$ (10.5–13.5% Si, 4.5–5.5% Mg), and $AlSi_{10}Mg$ (9.0–11.0% Si, 0.2–0.5% Mg).

The corrosion behavior was studied using modified salt spray tests according to ASTM standard G85. Panels of 50 mm × 100 mm in size were exposed to an intermittent acidified salt spray fog for 2 weeks. The stress-corrosion cracking (SCC) behavior was investigated by performing constant-load tests under permanent immersion conditions in an aerated aqueous solution of 0.6 M NaCl + 0.06 M $NaHCO_3$. The fusion zone in the center of the flat tensile specimens (6.5 mm × 15 mm) was machined, reducing the thickness to that of the base sheet. The loading axis was in the transverse direction and the maximum exposure time was 30 days. The failure criterion was fracture.

The weld region exhibited a dendritic cellular structure in the fusion zone and a partially melted zone adjacent to the fusion boundaries. The hardness of the fusion zone depended on the filler metals used. Post-weld artificial aging to the T6 temper improved the hardness, being associated with precipitation of strengthening phases. Joint efficiencies achieved for post-weld heat-treated joints ranged from 80% to 90%. Strengths of welds in the as-welded T6 condition were lower due to the softened fusion zone and heat-affected zone, being between 70% and 75% of the ultimate tensile strength of the base alloy 6013–T6. Optimum tensile properties were obtained with joints made with the filler powder $AlSi_{12}$ and were better than that with mixtures of elemental and binary alloy powders. The dispersoid forming elements Zr, Cr, and Mn added to a mixed Al–7Si powder did not have beneficial effects on weld quality [51].

Corrosion Behavior of 6013 Joints According to visual inspection, base alloy 6013 was susceptible to pitting in both tempers T4 and T6. For the peak-aged sheet, a more

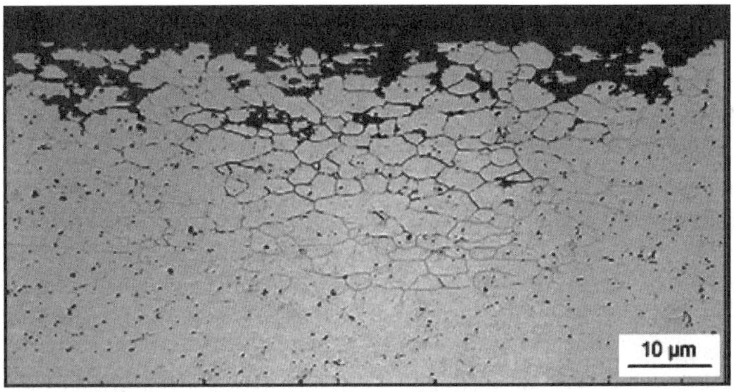

Figure 8.12 Evidence of intergranular and pitting corrosion in a section of 6013 alloy joint, made with the filler powder $AlSi_{12}Mg_5$ in the post-weld heat-treated T6 temper after exposure to an intermittent acidified salt spray fog for 2 weeks [52].

uniform corrosion attack was observed. The surface appearance of panels of joints made with the filler powder $AlSi_{10}Mg$ in the as-welded T4 and post-weld heat-treated T6 conditions was similar to that of the base material in the different tempers. Metallographic sections of base alloy panels revealed pitting for the naturally aged sheet. The maximum depth of the predominantly horizontal pits was 150 μm. In the T6 temper, the base alloy was susceptible to intergranular corrosion and pitting with a maximum depth of attack of 235 μm. Panels of as-welded 6013–T4 joints after 2 weeks of cyclic salt spray testing showed that the prevailing corrosion attack was pitting. In the heat-affected zone, intergranular corrosion was also observed at a distance of 650–1000 μm from the fusion boundary [52].

Similar to base alloy 6013–T6, panels of post-weld joints heat treated to the T6 temper exhibited intergranular corrosion and pitting (Figure 8.12). Weld beads suffered pitting. An aggravation of the corrosion attack in the weld region was not observed. The microstructure of the heat-affected zone of joints in the as-welded T6 condition corresponded mainly to a naturally aged alloy, as has been indicated by hardness profiles. Owing to the difference in corrosion potential, the region adjacent to the fusion boundary was probably cathodically protected by the adjoining peak-aged parent material exhibiting a more active corrosion potential [52].

However, corrosion-free zones were not observed with as-welded 6013–T6 coupons, which were alternately immersed in 3.5% NaCl solution. Cathodic protection was not efficient when the galvanic coupling was interrupted during specimen drying. No significant effect of the filler metal on the corrosion performance of alloy 6013 welds was observed. Cyclic acidified salt spray testing indicated similar corrosion behavior for panels of welded joints using different aluminum alloy powders and powder mixtures [52].

Time-to-failure data of 6013 joints are obtained from constant-load tests using an aqueous chloride–bicarbonate solution. Flat tensile specimens of joints in the as-welded T4 condition failed within 10 days of exposure at an applied stress of 100 MPa in the transverse direction. Failure occurred in the heat-affected zone ~1 mm away from the fusion boundary, regardless of the filler powder used. Fractographic examination revealed an intergranular fracture. No failure was observed for joints in the post-weld heat-treated T6 and as-welded T6 conditions at applied stresses of 200 and 170 MPa, respectively [52].

In summary, when exposed to an intermittent acidified salt spray fog, joints in the as-welded T4 and post-weld heat-treated T6 conditions exhibited corrosion behavior similar to that of the base sheet in the T4 and T6 tempers. As-welded 6013–T4 joints were susceptible to stress-corrosion cracking when immersed in an aqueous solution of 0.6 M NaCl + 0.06 M $NaHCO_3$. Sensitivity to environmentally assisted cracking was associated with grain boundary precipitates in the heat-affected zone [52]. EDX analysis in the transmission electron microscope confirmed that the precipitates were enriched with silicon and magnesium, probably being the Mg_2Si phase. These grain boundary particles were not observed in naturally aged parent material. Their formation was induced by the heat input during laser beam welding. The precipitates promoted intergranular stress-corrosion cracking.

When panels of welds in the as-welded T4 condition were exposed to an acidified salt spray fog, the heat-affected zone was sensitive to intergranular corrosion, whereas naturally aged parent sheet exhibited only pitting. Thus SCC failure occurring in the heat-affected zone of as-welded 6013–T4 joints might be caused by stress-assisted intergranular corrosion. The SCC sensitivity was eliminated in the artificially aged microstructure, probably due to coarsening of the precipitates during post-weld heat treating to the T6 temper [52].

8.6.4. Mechanically Influenced Corrosion and SCC of Welds

Wrought aluminum alloys usually have greater resistance to SCC in the longitudinal (direction of working) than in the transverse orientation or in the short-transverse orientation (thickness). Welding of the 7xxx series alloys near a base metal edge can result in a tensile stress in the short-transverse direction sufficient to cause SCC in the exposed edge. "Buttering" the edge with weld metal provides compressive stress at the edge and overcomes the SCC problem [46].

In some welding processes, the welding operation creates residual stresses. Tensile stress at the surface of the welded region promotes SCC. Solution heat treating and quenching reduce the susceptibility for SCC by changing the state of stress at the surface of the weld. Compressive residual stress at the weld surface is one of the ways to prevent SCC. Shot peening also creates compressive residual stresses at the surface [35].

Generally, special precautions must be considered to reduce the SCC tendency in SCC-susceptible alloys in the short-transverse direction. For example, the 7xxx series alloys must not be punched or sheared when they are fabricated. These operations create tension stress that, coupled with a corrosive environment, forms SCC. In military applications, arc-welding cladding of pure aluminum is used on machining edges to reduce SCC in the short-transverse direction. Corrosion protection of the 7xxx series alloys is achieved when a protective cladding is added on the assembly [35].

8.6.5. Corrosion Fatigue of Friction Stir Welding White Zone

Fusion welding is the dominant method of joining structural members in the engineering construction industry and weldable aluminum alloys appear highly promising for structural applications where weight is a major consideration. In particular, the medium-strength (7xxx series) Al–Zn–Mg alloys have the unique advantage that the heat-affected zone (HAZ) of weldments, rather than showing as annealed properties, recover up to 80% of the parent material strength through precipitation hardening at ambient temperatures.

Environment-sensitive fracture cracking in service usually occurs in a narrow recrystallized band, known as the "white zone" (WZ), at the interface between the weld bead and the HAZ. Although the environment-sensitive fracture behavior of Al–Zn–Mg wrought products has been studied in detail, the sensitivity of the welding zone to environment-induced fracture is difficult to quantify, owing mainly to the small amount of material available in a typical weld [53]. The environment-sensitive fracture under cyclic loading of the white zone of 7017-T651 aluminum alloy has been investigated.

The material used in this work was the medium-strength Al–Zn–Mg alloy AA7017 in the T651 (solution treated, controlled stretched, and artificially aged) condition, supplied as rolled plate of 22 mm thickness. A 15 mm deep "V" groove was machined on the plate in the transverse direction and a single pass bead-on-plate MIG weld was produced by an automated welding plant using 5556A (Al–Mg) filler wire, of typical composition by weight and 1.8 mm in diameter. A current of 365 A and a voltage of 29.4 V were applied during welding. The tested solution was 2.5% NaCl + 0.5% Na_2CrO_4 solution, acidified to pH 3 with HCl. The specimens were fatigue precracked at 20 Hz ($\Delta K = 8.5$ MN m$^{3/2}$ $s_{min}/s_{max} = R = 0.1$). All tests were carried out under load control at room temperature over a range of frequencies (0.01 ± 10 Hz) $R = 0.1$, using a sinusoidal load waveform. Reference tests were carried out in laboratory air at a loading frequency of 4 Hz. The effect of loading strain rate on the corrosion fatigue cracking behavior of the WZ was investigated by performing tests using a triangular load waveform at frequencies of 0.1, 0.5, and 1 Hz. All fatigue testing was carried out in a DARTEC 60 kN capacity servohydraulic testing machine. The ac potential drop (ACPD) technique was employed to monitor crack growth [53].

Fatigue crack propagation rates in the WZ of AA7017 welds in an aqueous salt chromate environment show a pronounced enhancement compared with tests performed in air. The crack growth rate enhancement is more pronounced at low cyclic frequencies. The crack growth rate at which the transition from intergranular (IG) to transgranular (TG) cracking takes place is frequency dependent [53].

For frequencies of 0.1 Hz and above, fatigue crack growth is dependent on the square root of the time available during each cycle, behavior indicative of environment-enhanced fatigue crack growth involving hydrogen diffusion ahead of the crack tip during each cycle. For cyclic loading frequencies below 0.1 Hz, corrosion fatigue crack growth is considered to be a time-dependent rather than a cycle-dependent phenomenon, suggesting that crack growth is a kind of stress corrosion under cyclic loading. Striations observed on the fatigue fracture surface indicate a discontinuous crack propagation mechanism. A model relating the critical stress-intensity range for IG/TG fracture mode transition and fatigue frequency, based on the assumption that discontinuous crack advance takes place by hydrogen accumulating at a region of high stress ahead of the crack tip initiating cracking there, has been demonstrated [53].

The strain dependence of crack growth rates in the WZ is indicative of a hydrogen diffusion-controlled cracking mechanism. A steady-state hydrogen diffusion model applied to crack growth in the WZ suggests that for intergranular cracking to take place a higher concentration of hydrogen is needed than for transgranular fracture [53].

8.6.6. SCC of Friction Stir Welded 7075 and 6056 Alloys

It has been reported that the development of pits has not only shortened the fatigue life of AA7075 by a factor of 3, but has also decreased the fatigue crack initiation threshold by about 50% [54]. Najjar et al. [55] stated that in the SCC of AA7050 the main role of anodic dissolution is to produce critical defects, which not only promote a localization of plastic deformation but also promote hydrogen discharge localization, entry, and subsequent embrittlement [56].

Two aluminum alloys, AA7075 and AA6056, were friction stir welded, with the AA7075 specimen placed on the advancing side of the welding tool. Microstructural observations revealed the development of a recrystallized fine-grained weld nugget, with two different grain sizes, resulting from the two different base materials. Slow strain rate tensile (SSRT) tests in air have shown that the weld nugget is marginally overmatched in the weldment, and the fracture occurred in the relatively weaker thermo mechanically affected zone/heat-affected zone (TMAZ/HAZ) of the AA6056 specimen [56].

Aluminum alloy plates made of AA7075–T7351 and AA6056 of 5 mm thickness were employed. Welds were produced using a Tricept TR805 robotic friction stir welder. The weld configuration had the AA7075 plate placed on the advancing side and the AA6056 plate on the retreating side of the friction stir welding tool (pin diameter, 5 mm; pin height, 4 mm; shoulder diameter, 15 mm). The welding parameters were: axial load, 11.5 kN; tool rotational speed, 900 rpm; tool traversing speed, 250mm/min. The welded specimens were subjected to slow strain rate tensile (SSRT) tests as per the ISO Standard 7359—Part 7 [57] at nominal strain rates of 10^{-6} and $10^{-7}\ s^{-1}$, in 3.5% sodium chloride solution, to assess their stress-corrosion cracking behavior [56].

The specimens tested both in air and in 3.5% NaCl solution exhibited a reduction in area value of 50% $\pm$ 2%, revealing that there was no stress-corrosion cracking in the weldment in 3.5% NaCl solution at a nominal strain rate of $10^{-6}\ s^{-1}$. Also, fracture was observed in the TMAZ/HAZ of the AA6056 specimen, with a behavior very similar to that observed in the tests in air. However, at lower nominal strain rate, $10^{-7}\ s^{-1}$, the TMAZ/HAZ region of AA7075 specimen was found to be susceptible to SCC, exhibiting intergranular fracture. It is thus concluded that although the weld nugget is resistant to SCC, the TMAZ/HAZ region of AA7075 in the weldment is prone to SCC in 3.5% chloride solutions at nominal strain rate levels on the order of $10^{-7}\ s^{-1}$ [56].

Macroscopic/microscopic examination of the failed specimens revealed the presence of pits in the AA7075 side of the welded specimen and also at the root of the weld nugget. However, the severity/extent of pitting/corrosion damage was found to be more pronounced in the AA7075 specimen than in the weld nugget. The optical micrograph of a representative region in the AA7075 specimen clearly indicates that the pits have propagated to a depth of around 200–300 μm in the base material during the SSRT-SCC test, which lasted about 15 h. Despite the development of such stress concentration risers in the gauge section of the weldment, the fracture occurred only in the TMAZ/HAZ of the AA6056 specimen, indicating that the mechanical stress, rather than the environment, governed the failure under these test conditions [56].

8.6.7. SCC of FSW of 7075-T651 and 7050-T451 Alloys

Aluminum 7075-T651 and 7050-T451 plates (300 × 100 mm × 10 mm) were friction stir welded at a travel speed of 140 mm/min and with a pin rotation speed of 240 rpm. The welding direction was parallel to the rolling direction of the alloy plate. The welds were investigated after at least 6 months of natural aging [58].

In constant extension rate testing (CERT) (at $5 \times 10^{-7}\ s^{-1}$), tensile samples prepared transverse to the weld suffered intergranular stress-corrosion cracking. In the as-welded condition, the reduction in strain to failure for 7075 welds was modest, from 7.2% (in air) to 5.8% (in aerated 3.5% NaCl). For as-welded 7050 samples, the reduction in strain to failure was significant, dropping from 10% to less than 2% [58].

In both 7075 and 7050 alloys, the weld nugget, the thermomechanically affected zone (TMAZ), and the heat-affected zone (HAZ) suffer intergranular corrosion during exposure to sodium chloride–hydrogen peroxide solution. Pitting associated with larger constituent particles occurs in all zones and does not appear to be strongly influenced by welding. FSW produced significant variations in microstructure and strength that affect corrosion resistance generally and that increase metal susceptibility to SCC. However, it has been found that little variations in the corrosion potential, pitting potential, or repassivation potential values were found on a zone-by-zone basis in deoxygenated chloride solutions [58].

The most vulnerable part of the weld showed environmentally assisted cracking (EAC) and this occurred regularly in the trailing side of the HAZ and TMAZ. The HAZs were found to be quite soft and these regions necked visibly during testing, suggesting a strong mechanical component to the cracking process since softening allows localization of plastic deformation. Coarse precipitation $Mg(ZnCu)_2$ and solute depletion along grain boundaries may have made boundaries more susceptible to cracking. Also, sweeping of the grain boundaries in the trailing side of the weld by plastic deformation during welding showed that a large number were oriented perpendicular to the load axis during CERT [58].

8.7. PREVENTION OF SCC

Although the cause of the SCC is the combination of several factors, the various approaches to prevent or reduce the risk of SCC can arbitrarily be divided as a function of the stresses, environmental considerations, metallurgical properties, surface modification, and hydrogen damage. Also, some attention has been given to the avoidence of hydrogen damage [13].

8.7.1. Design and Stresses

The sum of stresses in service and residual stresses, including the stress due to fabrication, should be below the threshold level, which, in the absence of reliable data, should be evaluated by testing as a percentage of the tensile yield strength. This can be achieved by a design that avoids concentration of stresses at the start, reducing operating stresses, relieving fabrication stresses by a heat treatment, and by choosing appropriate dimensions of the loaded alloy. Bolted or riveted joints can produce high local stresses that can cause SCC, so attention should be given to proper joint design and construction. Examples include using preformed parts, avoiding overtorquing of bolts, and providing adequate spacing and edge margins for rivets.

8.7.2. Environmental Considerations

Eliminate the critical ions that induce SCC—dissolved gases, heavy metals, and impurities—by chemical or physical methods such as degasification, demineralization, and distillation.

Avoid Intergranular Corrosion The heat treatments that provide high resistance to cracking are those that produce microstructures either free of precipitate along grain boundaries or with precipitate distributed as uniformly as possible within grains.

Other Forms of Corrosion and SCC During use, avoid stress concentrators through increasing susceptibility to different forms of degradation in service, especially localized corrosion, galvanic corrosion, and erosion corrosion. Accidental or nondesigned cyclic loading should also be avoided.

Temperature Avoid, if possible, the temperature level or range that causes SCC, depending on the microstructure and the ionic species.

8.7.3. Metallurgical Considerations

Select the appropriate resistant aluminum alloy with a convenient chemical composition and microstructure for the projected use environment.

Welding Stress relief at low appropriate temperatures is commonly used to lower susceptibility. Tensile residual stresses from welding can be particularly dangerous and a low-temperature thermal stress relief treatment is recommended for welded assemblies [59].

Tempers of 2xxx and 7xxx Alloys Direction and magnitude of stresses anticipated under conditions of assembly and service may govern alloy and temper selection. For products of thin sections, applied in ways that induce little or no tensile stress in the short-transverse (i.e., through-thickness) direction, the resistance of 2xxx alloys in T3 or T4 tempers or of 7xxx alloys in T6 tempers may suffice. Resistance in the short-transverse direction usually controls application of products that have a thick section or are machined or applied in ways that result in sustained tensile stresses in the short-transverse direction. More resistant tempers are preferred in these cases [22].

Stress Relief Through Stretching Residual stresses are induced in aluminum alloy products when they are solution heat treated and quenched. Figure 8.13a shows the typical distribution and magnitude or residual stresses in thick, high-strength material of constant cross section. Quenching places the surfaces in compression and the center in tension.

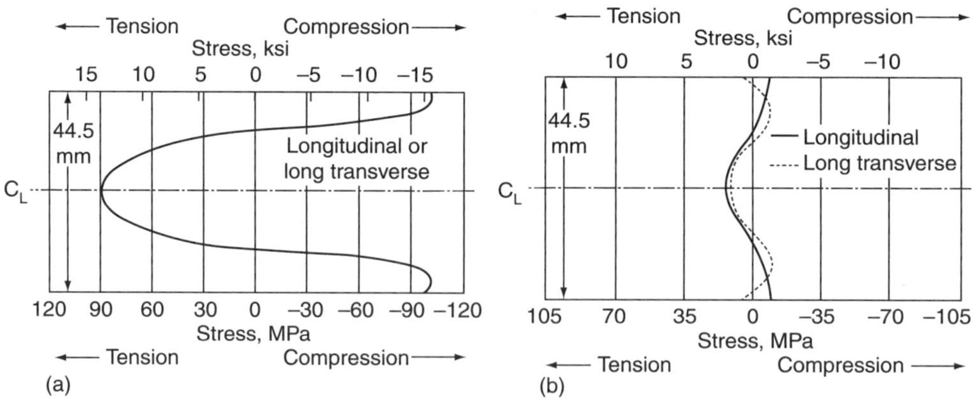

Figure 8.13 Comparison of residual stresses in a thick, constant cross section, 7075-T6 aluminum alloy plate before and after stress relief: (a) high residual stresses in the solution-treated and quenched alloy and (b) reduction in stresses after stretching 2% [22, 25, 27, 34].

If the compressive surface stresses are not disturbed by subsequent fabrication practices, the surface has an enhanced resistance to SCC because a sustained tensile stress is necessary to initiate and propagate this type of corrosion [22].

Aluminum products of constant cross section are stress relieved effectively and economically through mechanical stretching. The stretching operation must be done after quenching and, for most alloys, before artificial aging. Note the low magnitude of residual stresses after stretching in Figure 8.13b as compared to the as-quenched material in Figure 8.13a. Federal specifications for rolled and extruded products provide for stress relief by stretching on the order of 1–3%. The stress-relieved temper for heat-treated mill products minimizes SCC problems related to quenching stresses. The stress-relieved temper for most alloys is identified by the designation Tx5x or Tx5xx after the alloy number, for example, 2024-T351 or 7075-T6511 [22, 60].

8.7.4. Surface Modification

This includes surface treatments, conversion, coating, and cathodic protection that controls the surface in the domain of immunity.

- Shot peening and other mechanical processes that create compressive residual stresses at the surface are recommended.
- Surface conversion by oxidation, phosphating, and anodizing is helpful for certain alloys if it is followed by sealing with appropriate application of inhibitors or coatings.
- Coatings have been shown to extend life, but not to totally prevent SCC, with breaks in the coating reducing protection [31].
- Cladding a susceptible alloy with a nonsusceptible alloy is currently used for aluminum alloys.
- There should be nondestructive testing and inspection and maintenance programs to avoid SCC precursors, such as concentration of stresses by localized corrosion. In the case of protective coatings, routine maintenance is essential since scratches could create favorable sites for initiation of SCC.

Inhibitors Add efficient inhibitors for certain systems since it is believed to be associated with the formation of a stronger, more stable, or more readily repaired passive film [61]. Phosphates and other inorganic or organic inhibitors of corrosion, if they are used in fairly corrosive surroundings, decrease the effects of the SCC. Under certain conditions, inhibitors, especially oxidizing ones, should be added in appropriate quantities for best effect. A minimal critical concentration of some oxidizing inhibitors such as nitrites is absolutely necessary to avoid pitting. Actually, the application of more than one inhibitor for various reasons and good effect is recommended for corrosion control of some systems. Inhibitors can be chosen based on their mechanism of action (anodic, cathodic, or mixed). Capillary condensation can cause the formation of a liquid in the crevices at lower values of relative humidity. It would be prudent to fill crevices with corrosion-inhibited putty.

Cathodic Protection SCC can be greatly retarded, if not eliminated, by polarization to the level of cathodic protection potential. Cathodic polarization may reduce, or even prevent, SCC of some materials in certain aqueous solutions, but this should be recommended with great care for alloys that resist all types of critical hydrogen damage.

Corrosion Prevention of Welded and Nonwelded Structures Specific heat treatments for the chosen procedure of welding are fruitful depending on the alloy composition and microstructure, especially for the series AA2000 and AA7000. Sandblasting and shot peening are desirable for compressive stresses; however, they are not recommended for wrought alloys that have smooth skin-pass or finishing. Polishing does not show good efficiency. Surface preparation and surface conversion (phosphating or anodization, etc.), and organic and even metallic (e.g., Alclad) coatings, are viable options to consider, depending on the medium and the targeted required performance from the alloy.

8.7.5. Prevention of Hydrogen Damage

The methods of prevention could differ as a function of the type of hydrogen damage in using appropriate alloys that have good corrosion resistance for different forms of corrosion and especially to galvanic corrosion, which leads to hydrogen absorption. During welding, hydrogen embrittlement can be reduced by using dry conditions without humidity, since water and steam are the major sources of hydrogen. Low-hydrogen welding rods should be specified.

- Inhibitors and postprocessing bake-out treatments can also be used.
- Changing the environment can be very efficient. For example, blistering rarely occurs in pure acid corrosives without hydrogen-evolution poisons, such as sulfides, arsenic compounds, cyanides, and phosphorus-containing ions [13, 62].
- Similar to SCC prevention, one can add inhibitors to reduce the corrosion rate.
- Coatings are recommended just like for SCC, but they should be impervious to hydrogen penetration as well as resistant to the corrosive medium.
- Metallic, inorganic, and organic coatings, impermeable to hydrogen, are often used to prevent the formation of blisters of hydrogen [62].

REFERENCES

1. V. S. Sastri, E. Ghali, and M. Elboujdaini, *Corrosion Prevention and Protection—Practical Solutions*, Wiley, Chichester, West Sussex, UK, 2007, pp. 331–459.
2. T. Magnin and P. Combrade, in *Materials Science and Technology Series, Volume 1*, edited by R. W. Cahn, P. Haasen, and E. J. Kramer. Wiley-VCH, Weinheim, Germany, 2000, pp. 216–318, 537.
3. S. A. Shipilov, Fundamentals of physicochemical mechanics of fracture: Purposes and contents of the new educational course, in *Teaching and Education in Fracture and Fatigue*, edited by H. P. Rossmanith, E&FN Spon, London, 1996, pp. 293–299 (Chapter 28).
4. S. A. Shipilov, *Technology* **1**, 131–142 (1996).
5. M. G. S Fontana and N. D. S Greene, *Corrosion Engineering*. McGraw-Hill, New York, 1978, pp. 7–27.
6. R. H. Jones, in *ASM Handbook, Volume 13A, Corrosion: Fundamentals, Testing and Protection*, edited by S. D. Cramer and B. S. Covino, Jr. ASM International, Materials Park, OH, 2003, pp. 346–366.
7. Y.-Z. Wang, in *Uhlig's Corrosion Handbook*, 2nd edition, edited by R. W. Revie. Wiley, Hoboken, NJ, 2000, pp. 221–232.
8. H. D. Holroyd, N. J. Hardie, *Corrosion Science* **529** 533–535 (1983).
9. A. Turnbull, Embrittlement by the local crack environment, in *Progress in the Understanding of the Electrochemistry in Cracks,* edited by R. P. Gangloff. The Metallurgical Society, Warrendale, PA, 1984, p. 3.
10. M. J. Danielson, C. A. Oster, and R. H. Jones, *Corrosion Science* **32**, 1 (1991).
11. R. N. Parkins, in *Embrittlement by the Local Crack Environment*, edited by R. P. Gangloff. The Metallurgical Society, Warrendale, PA, 1984, p. 385.
12. R. H. Jones and R. E. Ricker, *Stress-Corrosion Cracking*. ASM International, Materials Park, OH, 1992, pp. 1–40.
13. *Surface Engineering for Corrosion and Wear Resistance*. ASM International, Materials Park, OH, 2001, pp. 1–81.

14. S. A. Shipilov, in *Catastrophic Failures Due to Environment-Assisted Cracking of Metals: Case Histories*, edited by M. Elboujdaini and E. Ghali. Proceedings of the International Symposium on Environmental Degradation of Materials and Corrosion Control in Metals, Montreal, 1999. METSOC, Montreal, 1999, pp. 225–242.
15. C. G. Interrante and L. Raymond, in *ASM Handbook, Volume 13A, Corrosion: Fundamentals, Testing and Protection*, edited by R. Baboian. ASM Internationnal, Materials Park, OH, pp. 367–380.
16. H. P. Van Leeuwen, *Engineering Fracture Mechanical*, 141 (1974).
17. B. Craig, S. D. Cramer, and B. S. Covino, in *ASM Handbook, Volume 13A Corrosion: Fundamentals, Testing, and Protection*, edited by S. D. Cramer and B. S. Covino, Jr. ASM International, Materials Park, OH, 2003, pp. 367–380.
18. B. Phull, in *ASM Handbook, Volume 13A, Corrosion: Fundamentals, Testing and Protection*, edited by S. D. Cramer and B. S. Covino, Jr. ASM International, Materials Park, OH, 2003, pp. 575–616.
19. H. H. Uhlig and R.W. Revie, *Corrosion and Corrosion Control: An Introduction to Corrosion Science and Engineering*, 3rd edition. Wiley, Hoboken, NJ, 1985, pp. 178–186.
20. D. G. Kolman, in *ASM Handbook, Volume 13A, Corrosion: Fundamentals, Testing, and Protection*, edited by S. D. Cramer and B. S. Covino, Jr. ASM International, Materials Park, OH, 2003, pp. 381–392.
21. M. H. Kamdar, in *ASM Handbook, Volume 13, Corrosion*, 9th edition, edited by L. J. Korb and D. L. Olson. ASM International, Materials Park, OH, 1987, pp. 171–187.
22. E. Ghali, in *Uhlig's Corrosion Handbook*, 2nd edition, edited by R. W. Revie. Wiley, Hoboken, NJ, 2000, pp. 677–715.
23. M. O. Spiedel, in *Hydrogen in Metals*, edited by ASM International Committee, American Society for Metals, Materials Park, OH, 1974, p. 249.
24. T. A. Marichev, *Werkstoffe und Korrosion* **34**, 300 (1983).
25. E. H. Hollingsworth and H. Y. Hunsicker, in *ASM Handbook, Volume 13, Corrosion*, 9th edition, edited by J. R. Davis (Senior Editor). ASM International, Materials Park, OH, 1987, pp. 583–609.
26. ASM Metals Handbook Committee, in *ASM Handbook, Volume 13, Corrosion*, edited by J. R. Davis (Senior Editor). ASM International, Materials Park, OH, 1987, pp. 93–196, 207–220, 231–233, 303–310, 596.
27. D. O. Sprowls and E. H. Spuhler, *Green Lettrer: Avoiding Stress-Corrosion Cracking in High Strength Aluminium Alloy Structures*, edited by Alcoa, Pittsburgh, PA, ASM International, Materials Park, OH, 1982.
28. B. Craig, *Hydrogen Damage*, Vol. **13**, 9th edition, ASM International, Materials Park, OH, 1987.
29. W.T. Tsai, A. MoccariI, Z. Szklarska-Smialovska, and D. D. Macdonald, *Corrosion* **40** (11), 573–583 (1984).
30. T. Magnin and P. Rieux, *Metallography*, 907 (1987).
31. L. L. Shreir, R. A. Jarman, and G. T. Burnstein, *Effect of Mechanical Factors on Corrosion*, Vol. **1**, 3rd edition, Butterworth-Heinemann, Woburn, MA, 1994, pp. 8: 3–242.
32. Z. Szklarska-Smialowska, *Corrosion Science* **19**, 753 (1979).
33. Z. Szklarska-Smialowska, *Hydrogen Embrittlement and stress Corrosion Cracking*. ASM International, Materials Park, OH, 1995.
34. J. G. Kaufman, in *ASM Handbook, Volume 13B, Corrosion: Materials*, edited by S. D. Cramer and B. S. Covino, Jr. ASM International, Materials Park, OH, 2005, pp. 95–124.
35. ASM International Handbook Committee, in *ASM Specialty Handbook, Aluminum and Aluminum Alloys*, edited by J. R. Davis. ASM International, Materials Park, OH, 1998, pp. 199–484, 579–731.
36. E. Ghali, *Aluminum Cast Alloys* [in English and French]. Centre Techniques des Industries de la Fonderie, Cedex, France, 2002, p. B 0430.
37. W. Wallace, D. W. Hoeppner, and P. V. Kandachar, in *Corrosion Handbook, Volume 1, Aircraft Corrosion: Causes and Case Histories*, AGARD Corrosion Handbook, AGARD, Neuilly-sur-Seine, France, 1985.
38. J. J. Bodu, M. Reboul, and D. Schuster, in *Corrosion Localisée*, edited by F. Dabosi, B. Beranger, and B. Baroux. Les Éditions de Physique, Les Ulis Cedex A, France, 1994, pp. 553–584.
39. ASM International Committee, in *Corrosion of Aluminum and Aluminum Alloys*, edited by J. R. Davis. ASM International, Materials Park, OH, 1999, pp. 63–74.
40. B. W. Lifka and D. O. Sprowls, *Signifiance of Intergranular in High-Strength Aluminum Alloy Products*. STP 516. ASTM, Philadelphia, PA, 1972.
41. E. N. Pugh, in *Encyclopedia of Materials Science and Engineering*, Vol. **2**. Pergamon Press, Elmsford, NY, 1986, pp. 889–890.
42. G. S. Frankel, *Introduction to Metallurgically Influenced Corrosion*, Vol **13A**. ASM International, Materials Park, OH, 2003, p. 257.
43. A. L. Greer, K. L. Rutherford, and I. M. Hutchings, *International Materials Reviews* **47** (2), 87–112 (2002).
44. D. J. Li, A. J. Akerman, K. J. Doherty, S. J. Poon, and G. J. Shiflet, *Scripta Materialia* **38** (4), 603–609 (1998).
45. A. Peker and W. L. Johnson, *Applied Physics Letters* **63** (17), 2342–2344 (1993).
46. ASM International Handbook Committee, in *Corrosion of Aluminum and Aluminum Alloy*, edited by J. R. Davis. ASM International, Materials Park, OH, 1999, pp. 63–74, 161–178.

47. K. F. Krysiak and ASM Committee on Corrosion of Weldments, in *Metals Handbook, Volume 13, Corrosion*, 9th edition, edited by J. R. Davis. ASM International, Materials Park, OH, 1987, pp. 344–368.
48. H. P. Godard, W. B. Jepson, M. R. Bothwell, and R. L. Kane, *The Corrosion of Light Metals*. Wiley, Hoboken, NJ, 1967, p. 269.
49. M. Pourbaix, *Atlas of Electrochemical Equilibria in Aqueous Solutions*. Pergamon Press, Elmsford, NY, 1966.
50. H. Kaesche, *Werkstoffe und Korrosion* **14**, 557 (1963).
51. R. Braun, *Materials Science and Engineering* **426** (1–2), 250–262 (2006).
52. R. Braun, *Materials Science Forum* **519–521**, 735–740 (2006).
53. G. Kotsikos, J. M. Sutcliffe, and N. J. H. Holroyd, *Corrosion Science* **42**, 17–33 (2000).
54. P. S. Pao, S. J. Gill, and C. R. Feng, *Scripta Materialia* **43**, 391 (2000).
55. D. Najjar, R. Magnin, and T. J. Warner, *Materials Science and Engineering A* **238**, 293 (1997).
56. P. Bala Srinivasan, W. Dietzel, R. Zettler, J. F. S Dos Satoz, and V. S Sivan, *Materials Science and Engineering A* **392**, 292–300 (2005).
57. International Standard ISO 7539, in *Part 7: Slow Strain Rate Stress Corrosion Tests*, Corrosion of Metals and Alloys—Stress Corrosion Testing. International Organization for Standardization, Geneva, 1989.
58. R. G. Buchheit and C. S. Paglia, in *Proceedings of the 2003 Meeting of the Electrochemical Society*, edited by R. G. Buchheit, R. G. Kelly, N. A. Missert and B. A. Shaw, Electrochemical Society, Pennington, NJ, 2004, pp. 94–103.
59. J. C. Scully, L. L. Shreir, R. A. Jarman, and G. T. Burstein, *SCC of Magnesium*, 3rd edition. Butterworth-Heinemann, Woburn, MA, 1994, p. 8, 127–129.
60. *Aluminum Standard.* Aluminum Association, Inc., Washington, DC, 1984, p. 12.
61. W. K. Miller, in *Stress-Corrosion Cracking*, edited by R. H. Jones. ASM International, Materials Park, OH, 1992, pp. 251–263.
62. M.G. Fontana and N.D. Greene, *Corrosion Engineering*. McGraw-Hill, New York, 1978.

Part Three

Performance and Corrosion Forms of Magnesium and Its Alloys

Chapter 9

Properties, Use, and Performance of Magnesium and Its Alloys

Overview

Magnesium crystallizes in the hexagonal close packed structure and is therefore not amenable to cold forming. Cast and wrought magnesium alloys, powder metallurg (P/M) prepared alloys, and metal matrix composites (MMCs) are described. Currently, the majority of large components of magnesium alloys are produced by high-pressure hot and cold chamber die casting, while an extensive number of ultra thin components are cast by Thixomolding, gravity sand, low-pressure sand and metal die molding, and semisolid and squeeze casting. An example of sheet and plate alloys is AZ31 (Mg–3Al–1Zn–0.3Mn), which is the most widely used. Magnesium composites containing boron, silicon carbide, and graphite are of increasing interest, particularly in the aerospace industry. The silicon-containing AS-based alloys show good creep properties at elevated temperatures and the Mg–Y–RE–Zr alloys show the highest creep performance. Corrosion due to poor design, flux inclusions, surface contamination, galvanic couples, and incorrectly applied or inadequate surface protection schemes is avoidable.

Magnesium alloys are used in the automotive, aerospace, electronics, and guided weapons industries because of their light weight and high strength-to-weight ratio. The use of magnesium alloys in structural applications is the most active area. Wrought msagnesium products, such as sheet and hydroformed extrusions, and a variety of other magnesium products, whether semi solid or forged, are being developed for automotive and other applications at an increasing rate. Composites based on aluminum and magnesium matrices are of great interest to the automotive and aerospace industries.

Corrosion resistance of alloys to natural and industrial atmospheres, and to aqueous solutions at different pH values and containing salts or organic compounds, is discussed. Performance in some dry gases or organic compound media at different temperatures is summarized. The effect of increasing the temperature is to increase the severity of attack. Mg is rapidly attacked by all mineral acids and alkalis except hydrofluoric and chromic acids. Severe corrosion may occur in neutral solutions of salts of heavy metals, such as copper, iron, and nickel. Moistened foreign materials on the surface can promote corrosion and pitting of some alloys unless the metal is protected by properly applied coatings. The presence of BF_3 or SF_6 in the ambient atmosphere is particularly effective in suppressing high-temperature oxidation up to and including the temperature at which the alloy normally ignites.

Corrosion Resistance of Aluminum and Magnesium Alloys: Understanding, Performance, and Testing By Edward Ghali

A. PROPERTIES OF MAGNESIUM ALLOYS

9.1. PHYSICAL AND GENERAL PROPERTIES OF MAGNESIUM

The perception of magnesium as a rapidly corroding material has been a major obstacle to its growth in structural applications despite its other obviously desirable physical properties. In fact, under normal environmental conditions, the corrosion resistance of magnesium alloys is comparable to or better than that of mild steel. It has been the uneducated use of magnesium in wet, salt-laden environments that has given rise to its poor corrosion reputation. Corrosion due to poor design, flux inclusions, surface contamination, galvanic couples, and incorrectly applied or inadequate surface protection schemes is avoidable not only for magnesium but for many other metals as well. Progressively, designers and engineers in the magnesium industry are establishing the correct use of magnesium in corrosive environments and are developing methods to improve the corrosion resistance of magnesium alloys by modifying alloy chemistry and improving surface protection technologies [1, 2].

Discovered in 1774, magnesium is the sixth most abundant element, constituting 2% of the total mass of the Earth's crust. The most important magnesium sources are magnesite $MgCO_3$ (27% Mg), dolomite $MgCO_3$–$CaCO_3$ (13% Mg), and carnallite KCl–$MgCl_2$–$6H_2O$ (8% Mg), as well as seawater, which contains 0.13% Mg or 1.1 kg Mgm^3 (third most abundant among the dissolved minerals in seawater)[3].

Magnesium and its alloys are different and have some unique and inherent corrosion behaviors, phenomena, and kinetics as compared to other metals. Under normal environmental conditions, the corrosion resistance of magnesium alloys is generally comparable to or better than that of mild steel. Magnesium is silvery white in appearance. It is a divalent metal. The atomic mass is 24.32 and the specific gravity of the pure metal 1.738 at 20 °C. The structure is hexagonal close packed. The lattice structure of magnesium has $c/a = 1.624$, and its atomic diameter (0.320 nm) is such that it enjoys favorable size factors with a diverse range of solute elements that have ±15% difference in atomic size [2, 4].

Magnesium crystallizes in the hexagonal close packed structure and is therefore not amenable to cold forming. Below 225 °C, only {0001} ⟨1120⟩ basal plane slipping is possible, along with pyramidal {1012} ⟨1011⟩ twinning. Pure magnesium and conventionally cast alloys show a tendency for brittleness due to intercrystalline failure and local transcrystalline fracture at twin zones or {0001} basal planes with big grains. Above 225 °C, new {1011} basal planes are formed and magnesium suddenly shows good strength [3]. Table 9.1 shows the physical data and the properties of magnesium. (See also aluminum properties as compared to magnesium, Chapter 4.)

The most widely used magnesium die casting alloy is AZ91 because of its superb castability even for the most complex and thin-walled parts. Currently, the majority of large components of magnesium alloys are produced by high- pressure hot and cold chamber die casting, while an extensive number of ultra thin components are cast by Thixomolding, gravity sand, low-pressure sand and metal die molding, and semisolid and squeeze casting. There are also processes that will be used more extensively in the future with Mg alloys, and processing parameters required to produce sound components. Magnesium has an important and growing future in the metalworking technologies of rolling, stamping, extrusion, and forging. The mechanical properties of cast and wrought magnesium components at ambient and elevated temperatures are greatly affected by their processing, and can differ considerably from test bar specimens. There are major efforts to improve the high-

Table 9.1 Physical Data and Properties of Magnesium

Property	Magnesium	References
Atomic mass	24.30506 g	Chapter 4 [18]
Tensile strength	No data	
Maximum oxidation number	2+.	48
Minimum oxidation number	0	Chapter 4 [1]
Crystal structure	hexagonal close packed	48
Spectroscopy	$(1s)^2$, $(2s)^2$, $(2p)^6$, $(3s)^2$	48
Ionization potential	I (7.64 eV), II(15.0 eV)	Chapter 4 [1]
Potential standard	−2.37 V	Chapter 4 [13]
Pauling electronegativity	1.2	Chapter 4 [1]
Density at 20 °C	1.741 g/cm	48
Melting point	649.5 °C	48
Boiling point	1107 °C	48
Specific heat at 20°C	1.030 kJ/kg·°C	4
Thermal conductivity at 20°C	157.5 W/m·°C	4
Van der Waals radius	0.16 nm	Chapter 4 [1]
Ionic radius	0.065 nm	Chapter 4 [1]
Number of common isotope	3 (isotopes 24, 25,26)	48

temperature mechanical properties of cast and worked magnesium alloys and extensive efforts to produce sheet, stampings, and extrusions [5].

A number of methods are available for the production of novel or nonequilibrium magnesium alloys with substantially improved corrosion resistance. Included in these methods are new rheocasting processes, rapid solidification processes, ion implantation, and vapor deposition. These processing methods typically enhance corrosion resistance by producing a more homogeneous microstructure, by increasing the solubility limits of alloying additions, or by some combination of both [6].

9.2. PROPERTIES OF CAST MAGNESIUM ALLOYS

Cast magnesium alloys have always predominated over wrought alloys, particularly in Europe, where, traditionally, cast alloys have comprised 85–90% of all magnesium products. The choice of a casting method for a particular part depends on factors such as the configuration of the proposed design, the application, the properties required, the total number of castings required, and the properties of the alloy [2].

9.2.1. Designation of Cast Magnesium Alloys

An international code for designating magnesium alloys does not exist, although there has been a tendency toward adopting the method used by the ASTM B275-94. In this system, the first two letters indicate the principal alloying elements according to the following code: A, aluminum; B, bismuth; C, copper; D, cadmium; E, rare earths; F, iron; H, thorium; K, zirconium; L, lithium; M, manganese; N, nickel; P, lead; Q, silver; R, chromium; S, silicon; T, tin; W, yttrium; Y, antimony; and Z, zinc. The letter corresponding to the element present in greater quantity in the alloy is used first; if they are equal in quantity, the letters are listed alphabetically. Letters are followed by numbers that represent the nominal compositions of these principal alloying elements in weight percent, rounded off to the nearest whole number; for example, AZ91 indicates the alloy Mg–9Al–lZn, the actual composition ranges being 8.3–9.7% Al and 0.4–1.0% Zn. Suffix letters A, B, C are chronologically assigned and usually refer to purity improvement. X is reserved for experimental alloys. For heat-treated or work-hardened conditions, the designations are specified by the same system as that used for aluminum alloys [7]. The two major systems of alloys, magnesium–aluminum and magnesium—zirconium are examined [2].

As an example, the designation AZ91B-F indicates the following:

AZ	The two principal alloying elements
B	This is the second alloy developed with the above aluminum and zinc compositions, being principally used in die casting. In this case, the "B" indicates that a higher residual copper level (0.35%) is permitted.
F	The alloy is used in its as-cast condition.

The wrought alloys may also be divided into two groups according to whether or not they contain zirconium. Specific alloys have been developed that are suitable for wrought products, most of which fall into the same categories as the casting alloys. The wrought alloys can be obtained in a number of tempers. The commonly used tempers are: T4, solution heat treatment only for 16 hours at 415 °C to homogenize the solution; T5, alloys artificially aged after casting; T6, alloys solu-tion treated, quenched, and artificially aged; and T7, alloys solution treated and stabilized [2, 8].

9.2.2. Alloying Elements

The earliest commercially used alloying elements were aluminum, zinc, and manganese and the Mg–Al–Zn system remains the most widely used for castings. Aluminum, zinc, cerium, yttrium, silver, thorium, and zirconium are examples of widely differing metals that may be present in commercial magnesium alloys. Apart from magnesium and cadmium, which form a continuous series of solid solutions, the magnesium-rich sections of binary-phase diagrams show peritectic or, more commonly, eutectic systems. Solubility data for binary magnesium alloys are given in Table 9.2; the first ten elements are those used in commercially available alloys [2].

Although early Mg–Al–Zn castings suffered severe corrosion in wet or moist conditions, the corrosion performance was significantly improved as a result of the discovery, in 1925, that small additions (0.2%) of manganese gave increased resistance. With this element, iron and certain other heavy metal impurities formed relatively harmless intermetallic compounds, some of which separate out during melting. In this regard, the classic work by Hanawalt et al. [9] showed that the corrosion rate increased abruptly once tolerance limits were exceeded; these tolerance limits are 5, 170, and 1300 ppm for nickel, iron and copper, respectively. The corrosion rate of pure magnesium as a function of iron content is shown in Figure 9.1, which clearly illustrates the tolerance limit for iron [2, 9].

Another problem with earlier magnesium alloy castings was that grain size tended to be large and variable, often resulting in poor mechanical properties, microporosity, and, in wrought products, excessive directionality of properties. Values of proof stress also tended to be low relative to tensile strength [2, 9].

Table 9.2 Solubility Data for Binary Magnesium Alloys

Element	Solid solubility		System
	Atomic %	Weight %	
Lithium	17.00	5.50	Eutectic
Aluminum	11.80	12.70	Eutectic
Silver	3.80	15.00	Eutectic
Yttrium	3.75	12.50	Eutectic
Zinc	2.40	6.20	Eutectic
Neodymium	~1.00	~3.00	Eutectic
Zirconium	1.00	3.80	Peritectic
Manganese	1.00	2.20	Peritectic
Thorium	0.52	4.75	Eutectic
Cerium	0.10	0.50	Eutectic
Cadmium	100.00	100.00	Complete solid solubility
Indium	19.40	53.20	Peritectic
Thallium	15.40	60.50	Eutectic
Scandium	~15.00	~24.50	Peritectic
Lead	7.75	41.90	Eutectic
Thulium	6.30	31.80	Eutectic
Terbium	4.60	24.00	Eutectic
Tin	3.35	14.50	Eutectic
Gallium	3.10	8.40	Eutectic
Ytterbium	1.20	8.00	Eutectic
Bismuth	1.10	8.90	Eutectic
Calcium	0.82	1.35	Eutectic
Samarium	~1.00	~6.40	Eutectic
Gold	0.10	0.80	Eutectic
Titanium	0.10	0.20	Peritectic

Sources: References 10 and 11.

9.2.3. Cast Magnesium Alloys Series

Pure magnesium ingots produced by the Pidgeon process or electrolytic process were utilized to produce ZM1 (Mg–4.5Zn–0.7Zr), ZM5 (Mg–8.2Al–0.5Zn–0.3Mn), and ZM6 (Mg–2.4Nd–0.7Zr–0.4Zn). The alloys ZM5-A and ZM6-A, which were prepared using the pure magnesium ingots produced by the Pidgeon process (i.e., distilled magnesium), showed better corrosion resistance than ZM5-B and ZM6-B, which were prepared using the pure magnesium ingots produced by the electrolytic process (i.e., electrolytic magnesium) [12]. For the same purity grade, the distilled magnesium may possess less impurity elements (Fe, Ni, Cu, and Cl) than electrolytic magnesium. The impurities, (e.g., Fe, Ni, Cu, chlorides) could accelerate the corrosion of these magnesium alloys, especially for ZM5, while alloying elements Zr and/or Nd could increase corrosion resistance of ZM1 and ZM6 [12].

The potential fields for magnesium alloys in the automotive industry are the segments of high-temperature applications. In the last 50 years, the main goal of alloy development has been to increase the high-temperature strength and creep resistance. Additions of silicon and rare earths as alloying elements improve the creep strength due to formation of

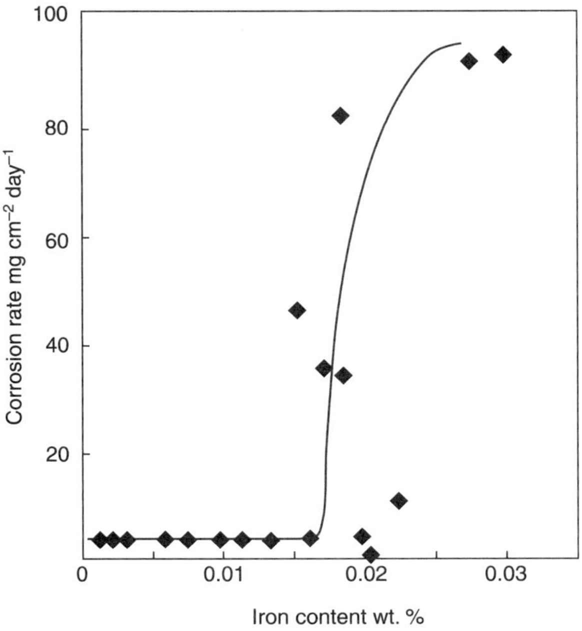

Figure 9.1 Effect of iron on corrosion of pure magnesium following on alternate immersion test in 3% NaCl [9].

high-temperature stable precipitates during solidification. Silicon containing AS-alloy, which had been used for gearbox housing in the Volkswagen Beetle® and AE-alloys, which show even better creep properties but are more costly, were the state of the art for high temperature creep-resistant magnesium alloys at the end of the last century. Advanced magnesium alloys such as QE- and WE-alloys containing silver (Q) and yttrium (W) in combination with rare earths (E) show even better creep properties at temperatures, that are suitable for applications in the power train or transmission housing of automobiles. Unfortunately, these alloys are not suitable for high-pressure die casting (HPDC). Magnesium producers and automobile companies developed a number of requirements for magnesium alloys to extend their applicability [13]:

- Room temperature properties as good as AZ91
- Elevated-temperature (>120°C) properties better than AZ91
- Castability comparable to AZ91
- Corrosion properties comparable to AZ91E
- Creep properties better than AE42
- Maximum increase in costs compared to AZ91

In order to reach these objectives, first tests with modifications of the commonly used magnesium alloys AZ91 and AM50, using additions of silicon, rare earths, tin, calcium, or strontium, have been done. Alloys were developed and patented but most of them were never brought to commercial application. Close to series production are some alloy developed by Volkswagen in cooperation with the Magnesium Research Institute. Mechanical properties at room and elevated temperatures of some common and newly developed alloys are listed in Table 9.3 [13].

Table 9.3 Mechanical Properties of Common and Newly Developed Alloys

Property	AZ91	AE42	ACM522	MRI153M	MRI230D	AJ62x
Ultimate tensile strength (UTS) at room temperature (RT) (MPa)	260	240	200	250	235	240
Yield Strength (YS) at RT (MPa)	160	135	158	170	180	143
Elongation at RT (%)	6	12	4	6	5	7
UTS at 150 °C (MPa)	160	160	175	190	205	166
YS at 150 °C (MPa)	105	100	138	135	150	116
Elongation at 150 °C (%)	18	22	—	17	16	27
Compressive YS at RT (MPa)	160	115	—	170	180	—
Compressive YS at 150 °C (MPa)	105	85	—	135	150	—
Impact strength (J)	8	12	—	8	6	—
Fatigue strength (MPa)	100	80	—	120	110	—
Corrosion rate (mg/cm^{2}day)	0.11	0.12	—	0.09	0.10	0.11

Source: Reference 13.

The silicon containing AS-alloys show good creep properties at elevated temperatures but limitations were in HPDC processes and also some corrosion problems. Norsk Hydro made progress in the development of AS-based alloys by using high-purity alloying elements and was able to show, that modern HPDC facilities are able to handle these alloys [13].

Noranda introduced strontium as another alloying element for magnesium alloys. They developed the AJ-alloys, which are used for the BMW hybrid crankcase, which is composed of an open deck insert made of Al–Si–17Cu–4Mg alloy. The insert is coated for better bonding with $AlSi_{12}$, which is applied by an arc spraying process, and the magnesium alloy AJ62-x is cast in a HPDC facility. Compared to an aluminum alloy crankcase, this hybrid one shows a weight reduction of approximately 25%, which means 10 kg saving on the front axle. The creep-resistant alloy AJ62-x shows good mechanical properties and recyclability. The part is used for BMW six-cylinder engines [13].

Mg–Al Based Alloys This system includes alloys containing 3–9 w % Al, combined with minor additions of zinc and manganese. Aluminum increases strength, castability, and corrosion resistance in saltwater. The maximum solubility of Al is 12.7 wt % at 437 °C. Commercial alloys usually don't exceed 10 wt % and the optimum alloys for strength and ductility is approximately 6 wt% Al [14]. Mg–Al is often alloyed with Zn to improve fluidity and room temperature strength and to overcome the corrosive effect of iron and nickel [4]. These alloys are widely available at moderate cost, and their mechanical properties are satisfactory from 95 to 120 °C (200–250 °F). At higher temperatures, the properties deteriorate. The most widely used of magnesium die-casting alloys is AZ91 because of its superb castability even for the most complex and thin-walled parts [2, 3].

Mg–Zr Based Alloys In 1937, it was discovered that zirconium had an intense grain-refining effect on magnesium. The lattice parameters of hexagonal zirconium are very close to those of magnesium. Paradoxically, zirconium could not be used in most existing alloys at that time because it was removed from solid solution owing to the formation of stable compounds with aluminum and manganese. This problem led to the evolution of a completely new series of cast and wrought zirconium-containing alloys with much improved mechanical properties at both room and elevated temperatures. Alloys containing zirconium as a grain-refining agent have the iron content reduced to about 0.004% because impurities separate during the alloying procedure. These alloys are now widely used in aerospace industries [2].

Since zirconium refines grains when added to magnesium, the result is greater casting integrity and improved mechanical properties. In addition, this alloy has more consistent properties through thin and thick sections. Zirconium can also be added to alloys containing zinc, rare earths, thorium, and silver, to refine the grains. The same phenomena happen with all common impurities found in magnesium alloy melts [4, 14]. These alloys generally possess much better elevated-temperature properties, but their more costly elemental additions, combined with the specialized manufacturing technology required, result in significantly higher costs. Many of the casting alloys are given simple heat treatments to improve their properties, while the wrought alloys can be obtained in a number of tempers [15].

Because of the particularly high solid solubility of yttrium in magnesium (12.5% maximum) and the amenability of Mg–Y alloys to age hardening, a series of Mg–Y–Nd–Zr alloys have been produced, which combine high strength at ambient temperatures with good creep resistance at temperatures up to 300 °C [16]. The heat-treated alloys have resistance to corrosion, which is superior to that of other high-temperature magnesium alloys and comparable to many aluminum-based casting alloys [17, 18]. Since pure yttrium is expensive and difficult to alloy with magnesium because of its high melting point (1500 °C) and its strong affinity for oxygen, a cheaper yttrium-containing (~75% Y) mischmetal together with heavy rare earth metals such as gadolinium and erbium could be substituted for pure yttrium (Table 9.4) [2, 19].

Since magnesium as well as cast aluminum alloys are in competition with other cast alloys such as that of cast iron, zinc, copper, and plastic, Table 4.5 gives a comparison of some important properties of these six mentioned cast materials.

9.3. PROPERTIES OF WROUGHT MAGNESIUM ALLOYS

Wrought materials are produced mainly by extrusion, rolling, and press forging at temperatures in the range of 300–500 °C. As with cast alloys, the wrought alloys may be divided into two groups according to whether or not they contain zirconium. Composition and mechanical properties of some wrought magnesium alloys are given in Table 9.5. Specific alloys have been developed that are suitable for wrought products, most of which fall into the same categories as the casting alloys [20]. Examples of sheet and plate alloys are AZ31 (Mg–3Al–1Zn–0.3Mn), which is the most widely used because it offers a good combination of strength, ductility, and corrosion resistance, and thorium-containing alloys such as HM21 (Mg–2Th–0.6Mn), which show good creep resistance at temperatures up to 350 °C. Magnesium alloys can be extruded at temperatures above 250 °C into either solid or hollow sections at speeds that depend on alloy content. Higher strength alloys such as AZ81 (Mg–8Al–1Zn–0.7 Mn), ZK61 (Mg–6Zn–0.7Zr), and the more recently developed ZCM711 (Mg–6.5Zn–1.25Cu–0.75Mn) all have strength/weight ratios comparable to those of the strongest wrought aluminum alloys. The alloy ZM21 (Mg–2Zn–1Mn) can be extruded at high speeds and is the lowest cost magnesium extrusion alloy available. Again, thorium-containing alloys, such as HM31 (Mg–3Th–1Mn), show optimal elevated-temperature properties. Magnesium forgings are less common and are almost always press-formed rather than hammer-forged [2].

Table 9.4 Nominal Composition, Typical Tensile Properties, and Characteristics of selected Cast Magnesium Alloys

Designation		Nominal composition, (wt %)												Tensile properties			
								Rare earth									
ASTM	British	Al	Zn	Mn	Si	Cu	Zr	MM[a]	Nd	Th	Y	Ag	Condition	0.2% Proof stress ($MN\ m^{-2}$)	Tensile strength ($MN\ m^{-2}$)	Elongation (%)	Characteristics
AZ63		6	3	0.3	—	—	—	—	—	—	—	—	As sand cast	75	180	4	Good room temperature strength and ductility
		—	—	—	—	—	—	—	—	—	—	—	T6	110	230	3	
AZ81	A8	8	0.5	0.3	—	—	—	—	—	—	—	—	As sand cast	80	140	3	Tough, leak tight casting with 0.0015% Be used for pressure die casting
		—	—	—	—	—	—	—	—	—	—	—	T4	80	220	5	
		9.5	0.5	0.3	—	—	—	—	—	—	—	—	As sand cast	95	135	2	General purpose, used for structural material
		—	—	—	—	—	—	—	—	—	—	—	T4	80	230	4	Alloy used for sand and die casting
AZ91	AZ91	—	—	—	—	—	—	—	—	—	—	—	T6	120	200	3	
		—	—	—	—	—	—	—	—	—	—	—	As chill cast	100	170	2	
		—	—	—	—	—	—	—	—	—	—	—	T4	80	215	5	
		—	—	—	—	—	—	—	—	—	—	—	T6	120	215	2	
AM60		6	—	0.2	—	—	—	—	—	—		—	As die cast	241	131	4	High-pressure die castings used for fans and wheels[b]
AM50		5	—	0.3	—	—	—	—	—	—	—		As die cast	125	200	7^c	High-pressure die castings
AM20		2	—	0.5	—	—	—	—	—	—	—	—	As die cast	105	135	10^c	Good ductility and impact strength
AS41		4	—	0.3	1	—	—	—	—	—	—	—	As die cast	135	225	4.5^c	Good creep properties up to 150 °C
AS21		2	—	0.4	1	—	—	—	—	—	—	—	As die cast	110	170	4^5	Good creep properties up to 150 °C
ZK51	Z5Z	—	4.5	—	—	—	0.7	—	—	—	—	—	T5	140	235	5	Sand castings, good room temperature strength and ductility
ZK61		—	6	—	—	—	0.7	—	—	—	—	—	T5	175	275	5	As for ZK51
ZE41	RZ5	—	4.2	—	—	—	0.7	1.3	—	—	—	—	T5	135	180	2	Sand castings, good room temperature. strength and castability

(continued)

Table 9.4 (*Continued*)

Designation		Nominal composition, (wt %)												Tensile properties			
								Rare earth						0.2% Proof stress	Tensile strength	Elongation	
ASTM	British	Al	Zn	Mn	Si	Cu	Zr	MM[a]	Nd	Th	Y	Ag	Condition	(MN m^{-2})	(MN m^{-2})	(%)	Characteristics
ZC63	ZC63	—	6	0.5	—	3	—	—	—	—	—	—	T6	145	240	5	Pressure tight castings, good elevated-temperature strength, weldable
EZ33	ZRE1	—	2.7	...	—	—	0.7	3.2	—	—	—	—	Sand cast T5	95	140	3	Good castability, pressure tight, weldable, creep resistant up to 250 °C
		—	—	—	—	—	—	—	—	—	—	—	Chill cast T5	100	155	3	
HK31	MTZ	—	—	—	—	—	0.7	—	—	3.2	—	—	Sand cast T6	90	185	4	Sand casting, good castability, weldable, creep resistant up to 350 °C
HZ32	ZT1	—	2.2	—	—	...	0.7	—	—	3.2	—	—	Sand or chill cast (T5)	90	185	4	As for HK31
QE22	MSR	—	—	—	—	—	0.7	—	2.5	—	—	2.5	Sand or chill cast (T6)	185	240	2	Pressure tight, weldable, high proof stress up to 250 °C
QH21	QH21	—	—	—	—	—	0.7	—	1	1	—	2.5	As sand cast T6	185	240	2	Pressure tight, weldable, good creep resistance and proof stress up to 300 °C
WE54	WE54	—	—	—	—	—	0.5	—	3.25[d]	—	5.1	—	T6	200	285	4[c]	High strength at room and elevated temperatures, good corrosion resistance, weldable
WE43	WE43	—	—	—	—	—	0.5	—	3.25[d]	—	4	—	T6	190	250	7[c]	

[a]MM, Mischmetal.

[b]www.magnesium.com/w3/data-bank/.

[c]Values quoted for tensile properties are for separately cast test bars and may not be realized in certain parts of castings.

[d]Contains some heavy metal rare earth elements.

Source: Reference 19.

Table 9.5 Nominal Composition, Typical Tensile Properties, and Characteristics of Selected Wrought Magnesium Alloys

Designation		Nominal composition, (wt %)								Tensile properties			
ASTM	British	Al	Zn	Mn	Zr	Th	Cu	Li	Condition	0.2% proof stress ($MN \cdot m^{-2}$)	Tensile strength ($MN \cdot m^{-2}$)	Elongation (%)	Characteristics
M1	AM503	—	—	1.5	—	—	—	—	Sheet, plate/F	70	200	4	Low-to medium-strength alloy, weldable, corrosion resistant
									Extrusions/F	130	230	4	
									Forgings/F	105	200	4	
AZ31	AZ31	3	1	0.3	—	—	—	—	Sheet, plate/O	120	240	11	Medium-strength alloy, weldable, good formability
				0.2[a]					Sheet, plate/H24	160	250	6	
									Extrusions/F	130	230	4	
									Forgings/F	105	200	4	
AZ61	AZM	6.5	1	0.3	—	—	—	—	Extrusions/F	180	260	7	High-strength alloy, weldable
				0.15[a]					Forgings/F	160	275	7	
AZ80	AZ80	8.5	0.5	0.2	—	—	—	—	Forgings/T6	200	290	6	High-strength alloy
				0.12[a]									
ZM21	ZM21	—	2	1	—	—	—	—	Sheet, plate/O	120	240	11	Medium-strength alloy, good formability, good damping capacity

(*continued*)

Table 9.5 (*Continued*)

Designation		Nominal composition, (wt %)								Tensile properties			
ASTM	British	Al	Zn	Mn	Zr	Th	Cu	Li	Condition	0.2% proof stress ($MN{\cdot}m^{-2}$)	Tensile strength ($MN{\cdot}m^{-2}$)	Elongation (%)	Characteristics
									Sheet, plate/H24	165	250	6	
									Extrusions/F	155	235	8	
									Forgings/F	125	200	9	
ZMC711		—	6.5	0.75	—	—	1.25	—	Extrusions/T6	300	325	3	High-strength alloy
LA 141		1.2	—	0.15[a]	—	—	—	14	Sheet, plate/T7	95	115	115	Ultralight weight (specific gravity 1.35)
ZK31	ZW3	—	3	—	0.6	—	—	—	Extrusions/T5	210	295	8	High-strength alloy, some weldability
									Forgings/T5	205	290	7	
ZK61		—	6	—	0.8	—	—	—	Extrusions/F	210	285	6	High-strength alloy
									Extrusions/T5	240	305	4	
									Forgings/T5	160	275	7	
HK31		—	—	—	0.7	3.2	—	—	Sheet, plate/H24	170	230	4	High-creep resistance up to 350 °C, weldable
									Extrusions/T5	180	255	4	
HM21		—	—	0.8	—	2	—	—	Sheet, plate/T8	135	215	6	High creep resistance up to 350 °C, short time exposure up to 425 °C, weldable
									Sheet, plate/T81	180	255	4	
									Forgings/T5	175	225	3	
HZ11	ZTY	—	0.6	—	0.6	0.8	—	—	Extrusions/F	120	215	7	Creep resistance up to 350 °C, weldable
									Forgings/F	130	230	6	

[a]Minimum.

Source: Reference 20.

9.4. MAGNESIUM POWDER

Magnesium powder has a wide range of applications in various chemical, pharmaceutical, metallurgical, and agricultural industries. Specific applications include steel desulfurization, pyrotechnics, manufacture of Grignard reagents used in pharmaceuticals and perfumes, effective chemical reductions in manufacture of beryllium and uranium, light source in flares and photoflash bombs, additives in electric welding electrode flux, and metal matrix composite fillers (www.magnesium.com). Magnesium stearate is a lubricant used widely in tabletting and capsule-filling processes. Magnesium stearate was found to exhibit polymorphism and to have various levels of lubricating ability [21].

Laser cladding and powder metallurgy are two nonconventional rapid solidification (RS) techniques. Laser cladding involves a high heat flux into a small area, resulting in very fast heating and cooling rates. Powder metallurgy alloys undergo RS during atomization of the source material to form metal powder. The powders are subsequently extruded, compacted, and sintered to form functional materials, often near the shape of the final product being manufactured [6].

9.5. MAGNESIUM COMPOSITES

There is considerable interest in the use of magnesium alloys in metal matrix composites (MMCs). This is a strenuous application for magnesium, considering the extreme galvanic nobility of many composite materials, such as graphite. In the case of AZ91 combined with alumina fibers, the corrosion rate of the MMC is 7 times higher than that of the bulk alloy, showing the paramount importance of galvanic effects. The microstructural as well as macroscopic effects of galvanic corrosion must be carefully regarded [6].

Low density, high elastic modulus, and increased thermal stability are some of the attractive attributes of magnesium MMCs. Consequently, magnesium composites containing boron, SiC, and graphite are of increasing interest, particularly in the aerospace industry. Vapor-deposited, corrosion-resistant magnesium–yttrium MMCs hold promise for the future [6].

It was found that, in general, less-ordered surface films show better performance (especially in terms of localized corrosion) due to better inherent breakdown resistance, higher ductility, and faster repassivation rates. In amorphous alloys, additionally, there are no grain boundaries to act as diffusion pathways to allow the ingress of oxygen or adverse solution species [6]. Rapid solidification offers a link between production technology and alloy development since the development of new alloys, cannot be accomplished by classical metallurgical procedures. Spray-forming is an attractive new technology for producing parts and near-complete products in almost final outline. Tubes, disks, rods, or sheets can technically be produced directly in one working step. The process is divided into the sputtering of a metal bath and the deposition of partly frozen drops on a substrate material [3].

MMC systems are currently under various stages of development. In the last two decades, much of the research has centered on cast aluminum-based MMCs, which have been the topic of discussion at many international forums. In contrast, research efforts on the processing and properties of magnesium-based MMCs have been rather limited. Magnesium and its alloys, with low density and high stiffness-to-weight ratios, are excellent candidates for matrix materials [22, 23].

By using appropriate procedures and alloy development, it could be possible to produce mechanical properties that are, in principle, comparable to those of aluminum and its alloys. However, this applies essentially only from room temperature up to approximately 150 °C.

Above these temperatures, only the relatively expensive Mg alloys of the QE series (Mg–Ag–RE) or WE series (Mg–Y–RE) show sufficient mechanical properties and creep resistance to be able to compete with aluminum alloys [24].

Metallic powders are manufactured mainly by the gas atomization of melts, where commercial hard materials are used as reinforcement. A clear improvement in the mechanical properties has been shown compared to the nonstrengthened basic alloys. Depending on the production process, an adjustment of whiskers and fibers could be detected, as well as the destruction of short fibers as soon as extrusion was used as a consolidation procedure. Another investigated consolidation is spray forming of Mg–MMCs, where SiC particles are brought into the atomization beam. However, these investigations are still in their infancy [24].

9.6. PARTICLES REINFORCING MAGNESIUM ALLOY MATRIX

9.6.1. SiC

Pure magnesium–30 vol % SiC particle composites are fabricated by a melt stir technique without the use of a flux or protective inert gas atmosphere. After hot extrusion with an extrusion ratio of 13, Mg–30 vol % SiC_P (p = particle) composites have been evaluated for their tensile properties at room and elevated temperatures (up to 400 °C). Composites in the as-cast conditions do not show any change in dendrite arm spacing: cell size compared to unreinforced pure magnesium. However, in the extruded conditions, average grain size of the composites is 20 mm compared to 50 mm in the pure magnesium. Microstructure shows no evidence of reaction product at the particle–matrix interface. At room temperature, stiffness and ultimate tensile strength (UTS) of the extruded composites are 40% and 30% higher, respectively, compared to unreinforced pure magnesium, signifying significant strengthening due to the presence of the SiC particles. Furthermore, up to temperatures of 400 °C, composites exhibit higher UTS compared to pure magnesium. Mg composites show a wear rate that is two orders of magnitude lower compared to pure Mg, when tested against a steel disk using a pin-on-disk machine [22].

Fracture behavior of pure magnesium reveals elongated dimples at room temperature and circular dimples at high temperatures.

9.6.2. Mg_2Si

Magnesium-based composites reinforced by Mg_2Si in situ formed via a mechanical milling process have been investigated. Characterization of the mechanical properties revealed that an increase in the amount of Si in Mg and mechanical milling duration led to an increase in yield and ultimate tensile strength. The increase in the mechanical properties is associated with the formation of Mg_2Si and refinement of microstructure. Microstructural analyses showed that the strengthening mechanisms were due to the dispersion strengthening of fine Mg_2Si particulates in the matrix as well as oxide dispersed particulates formed during mechanical milling [25].

Crystal sizes of milled powders decreased with increasing ball milling hours, while they slightly increased after the sintering and extrusion processes due to grain growth. It has been shown that Mg_2Si can effectively inhibit grain growth of magnesium matrix. An increase of yield strength with increasing ball milling duration and percentage of Mg_2Si was observed. Mg–Si alloys with addition of Al exhibit a higher tensile strength than Mg–Si alloys due to the solution strengthening effect of the aluminum in magnesium. Increase in elongation was due to homogenization of Mg_2Si after the first hour of mechanical milling. Further milling might lead to an oxidation of Mg powder [25].

9.6.3. Nanosized Alumina Particulates

Magnesium -based MMCs reinforced with only 1.11 vol. % of nanosized alumina particulates had exhibited mechanical properties comparable or even superior to similar composites containing much higher levels of micron-sized reinforcements. More surprisingly, the ductility of these composites exceeded even that of pure Mg. Studies had previously demonstrated that a smaller particulate size reduced wear in MMCs caused by delamination, a mechanism that has been shown to limit the advantages of the increased hardness and strength of the composites during sliding wear tests. The wear characteristics of Mg composites containing various amounts (up to 1.11 vol %) of nanosized alumina particulates in pin-on-disk dry sliding tests against hardened tool steel were examined, using a range of sliding speeds from 1 to 10 m/s, under a constant load of 10 N. The wear resistance of the composites improved with increasing amounts of reinforcement, which were particularly effective under the higher sliding speeds. Field-emission scanning electron microscopy (FESEM) identified the dominant wear mechanisms as abrasion, adhesion, and thermal softening. Reinforcement with only 1.11 vol % of nanosized alumina particulates was effective in increasing the wear resistance of pure magnesium by 1.8 times. Abrasion and adhesion were the dominant wear mechanisms, with a transition to thermal softening only under the highest sliding speed. Wear by delamination, which had been common in earlier work on Mg and Al MMCs with micron-sized reinforcements, was not evident [26].

B. USE OF MAGNESIUM AND MAGNESIUM ALLOYS

Magnesium alloys are used in the aircraft and guided weapons industries and in automotive construction because of their light weight and high strength-to-weight ratios. Another major application field is the electronic industry, due to the good electromagnetic interference shielding of the Mg [27]. New applications are emerging because of required properties, such as high stiffness-to-weight ratio, ease of machining, high damping capacity, and casting qualities. Magnesium is used as a canning material for uranium in gas-cooled reactors. Magnesium and its alloys can be used as sacrificial anodes for cathodic protection. Magnesium is itself used for alloying with other metals for different applications [2].

Wider availability of magnesium at a lower price than usual is leading to an increase in applications as well as research efforts. Magnesium has been used extensively as a sacrificial anode for cathodic protection. Currently, magnesium and its alloys are used in engineering structures such as in automotive components ($\sim$5 kg/vehicle), IT products (one-third of all laptops, many cameras, cell phones, and PDA bodies), and hand-held home and industrial equipment [5].

9.7. APPLICATIONS OF CAST MAGNESIUM ALLOYS

Die casting has seen the largerst use of magnesium alloys during the last few decades. In 1983, 36,100 tons were used for construction purposes, while in 2002, 138,638 tons were used. The use of magnesium alloys in structural applications is expected to see the most growth in the future. The main application is still high pressure die casting (HPDC) of magnesium alloys. The increase in usage from 27,000 tons in 1983 up to 152,000 tons in 2004 shows the potential growth of die casting within the last 20 years. Gravity castings as well as wrought materials are playing only a minor role, considering the total amount of material, but wrought alloys are expected to show a significant increase in applicability due to alloy development, process optimization, and identification of new potential applications [13].

9.7.1. Automotive and Aerospace Applications

The premier automotive use of magnesium is in the form of transmission casings; because these components are necessarily large structures, the weight savings realized with magnesium are substantial (20–25% over aluminum). Military uses for magnesium are extensive and include radar equipment, portable ground equipment, decoy flare ordnance, helicopter transmission and rotor housings, and in torpedoes [6].

Magnesium has recently emerged from obscurity as a reactive metal to become part of the suite of light material choices for modern industry. Over 10^8 kg of components are currently produced for the automotive sector; and there is a growing use of magnesium alloys in the aerospace, computer, camera, and consumer products industries. Automotive companies have increased their research support for magnesium. Several national and international car companies have started considering magnesium for critical applications such as cast engine blocks, transmission cases, and oil pans [5, 13]. Also, light metals are paving the way for the use of electric cars and other none or less polluting means of transport.

Magnesium alloys are expected to have a great impact in the automotive sector. Components include stampings in instrument panels, steering wheels, and steering column components; and the alloys are also used in window frames, seat structures, and carrier/support structures in automobiles [28]. Responsible for the increasing use of magnesium alloys in the last decade is the need to lower fuel consumption and emissions by reducing the weight of a vehicle; their use also gives designers the opportunity to balance and stabilize a car's center of gravity [13].

In the automotive industry, use of the AZ91 alloy instead of aluminum led to a total weight reduction of almost 25%, while the production equipment remained the same. Since the introduction of repetitive work in 1996, 600 parts are manufactured at Volkswagen in Kassel (Germany) per day; this trend can be increased and supported by several companies [3].

Military and civil helicopter gearbox casings are often made of magnesium, and BMW's new six-cylinder engine has a hybrid magnesium–aluminum block with a magnesium bedplate and cam cover. Daimler Chrysler's new seven-speed 7G-Tronic gearbox has a magnesium casing. General Dynamics latest military amphibious vehicle for the U.S. Marines, the Expeditionary Fighting Vehicle (EFV), includes magnesium in its complex transmission casings. The main magnesium alloy being evaluated for the EFV is Elektron 21, a new sand or investment casting alloy combining castability, corrosion performance, and the ability to operate at high temperature. The alloy chemistry consists of Nd–Gd (base)–Zn–Zr [29].

9.7.2. Application as Refractory Material

Magnesium compounds are used as refractory material in furnace linings for producing metals (iron and steel, nonferrous metals), glass, and cement. With a density of only two-thirds that of aluminum, it has countless applications in cases where weight reduction is important, (i.e., in aeroplane and missile construction). It also has many useful chemical and metallurgic properties, which make it appropriate for many other nonstructural applications. Magnesium components are widely used in industry and agriculture. Other uses include removal of sulfur from iron and steel, photoengraved plates in the printing industry, reducing agent for the production of pure uranium and other metals from their salts, flashlight photography, flares, and pyrotechnics [30].

9.7.3. Other Uses

Other Mg alloys used are listed in Table 9.6 while the weight of components composed of magnesium alloys are given in Table 9.7. ZE41A-T5 (Mg–4.2Zn–1.2Ce–0.7Zr), a sand cast Mg alloy, has been used in aircraft engine casings, auxiliary gearboxes, and gearbox casings [28].

9.8. APPLICATIONS OF WROUGHT MAGNESIUM ALLOYS

Today, not only cast magnesium parts, but many wrought magnesium products are being considered for wider application. Magnesium sheet, hydroformed extrusions, and a variety of other magnesium products, whether semisolid or forged, are being developed for automotive and other applications at an increasing rate [5].

Composites based on aluminum and magnesium matrices are of great interest to the automotive and aerospace industries [22]. Magnesium composites containing boron, SiC, and graphite are receiving increased attention, particularly from the aerospace industry. Vapor-deposited, corrosion-resistant magnesium–yttrium MMCs hold promise for the future. Because much of the bulk volume is taken up by the composite solute, only small

Table 9.6 Magnesium Alloy Uses

Use	Mg alloy	Use	Mg alloy
Airbag cover	AZ91	Front end	AM50/60
Airbag housing	AM60	Fuel tank cover	AM60
Armrest	AM60	Fuel tank support	AM60
Baffle plate	AZ91	Fuse box cover	AM50
Bulb fitting	AZ91	GPS frame	AZ91
Cabrio cover	AM50	Handbrake lever	AM50
Cabrio roof frame	AM50	Hinge on central console	AM50
Central console	AM60	Inner door frame	AM50
Compressor component	AZ91	Inner door handle	AZ91
Differential flange	AZ91	Intake manifold	AZ91
Door mechanism housing	AZ91	IP beam	AM50/60
Electronic box	AM50	Key-lock housing	AZ91
Engine blocks	AJ62	Mirror frame	AZ91
Engine brackets	AZ91	Oil pan	
Engine cradles	AE44	Oil pump housing	AZ91
Engine flange	AZ91	Pedal bracket	AM60
Fan clutch housing	AZ91	Radio frame	AZ91
Foot rest	AM50	Rear seat components	AM60

Source: Referencen 28.

Table 9.7 Components Containing Mg Alloys and Their Weights

Components containing Mg alloys	Weight (kg)
A-pillar	6.0–8.0
Airbag retainer	0.2–0.4
Armrest	0.6–0.8
B-pillar	6.0–8.0
C-pillar	7.0–9.0
Electronic box	0.4–0.8
Engine supports	0.4–0.8
Engine cradles	1.2–1.4
External mirror frame	0.5–0.6
Front end	3.0–5.0
Handbrake lever	0.6–1.0
Inner door frame	10.0–14.0
Intake manifold	1.5–3.0
IP beam	3.0–5.0
Key-lock housing	0.4–0.5
Oil pan	0.7–1.0
Pedal bracket	0.5–0.7
Rear deck lid	8.0–10.0

Source: Reference 28.

amounts of vapor-deposited material may be needed. Galvanic corrosion should be carefully addressed [6].

C. MAGNESIUM PERFORMANCE

Magnesium is shown to dissolve over a wide range of pH and potential as Mg^+ or Mg^{2+} in the absence of substances that can form soluble complexes, such as tartrate and metaphosphate, or insoluble salts, such as oxalate, carbonate, phosphate, and fluoride. Alloying different elements affects the nature of the protective film formed in presence of insoluble salts. At high pH, the corrosion product film of magnesium hydroxide (brucite) that forms on the surface is only semi protective. The pH values between 8.5 and 11.5 correspond to a relatively protective oxide or hydroxide film; however, above 11.5 a passive magnesium hydroxide layer dominates the electrochemical behavior of Mg [31, 32].

9.9. RESISTANCE OF MAGNESIUM ALLOYS TO ATMOSPHERIC CORROSION

Magnesium alloys are resistant to atmospheric corrosion because protective films form in a process similar to the formation of film in the active metal aluminum. When corrosion does occur, it is the result of the breakdown of this protective film. Corrosion of magnesium alloys increases with relative humidity (RH). At 9.5% RH, neither pure magnesium nor any of its alloys exhibit evidence of surface corrosion after 18 months. At 30% RH, only minor corrosion may occur. At 80% RH, the surface may exhibit considerable corrosion. In marine atmospheres heavily loaded with salt spray, magnesium alloys require protection for prolonged survival [6]. However, as humidity approaches 100%, more extensive tarnish films may form. While the thickness of such films is of only minor importance from an

Table 9.8 Corrosion Performance of Pure Magnesium in Some Artificial Media

Substance	Corrosion intensity
Ammonia	Resistant
Carbon dioxide and monoxide	Resistant
Dry chlorine	Little corrosion
Wet chlorine	Severe corrosion
Dry fluorine	Little corrosion
Wet fluorine	Negligible attack
Hydrogen peroxide	Little corrosion

Sources: Rerefernces 33 and 35.

appearance point of view, they may be very significant to the performance of a component, such as a computer disk drive made of die-cast magnesium alloy. Coatings or surface treatment can reduce the risk of such problems [33].

Composition of the corrosion products that form on magnesium alloys in an atmosphere varies from one location to another and from indoor to outdoor exposure. When humidity is high and a magnesium alloy has been coated or clad, local breakdown of the protective cladding/coating or film can promote pitting corrosion instead of general corrosion on the component [33].

Unprotected magnesium and magnesium alloy parts are resistant to rural atmospheres and moderately resistant to industrial and mild marine atmospheres, provided they do not contain joints or recesses that entrap water and thus promote the establishment of galvanic couples [6].

The oxide film on magnesium offers considerable surface protection in rural and some industrial environments, and the corrosion rate of magnesium lies between that of aluminum and that of low carbon steels. Tables 9.8 and 9.9 show the intensity of corrosion of pure magnesium in some dry and wet media [31, 33]. Table 9.10 shows the resistance of AZ31 alloy in the principal types of atmospheres [34].

Table 9.9 Corrosion Rate of Commercially Pure Magnesium in Various Media[a]

	Corrosion rate	
Medium	μm/yr	mils/yr
Humid air	1.0×10^{-5}	0.0004
Humid air with condensation	1.5×10^{-2}	0.6
Distilled water	1.5×10^{-2}	0.6
Distilled water exposed to acid gases	0.03–0.3	1.2–12
Hot deionized water (100 °C) (14 days stagnant immersion)	16	640
Hot deionized water inhibited with 0.25 NaF	5.5×10^{-2}	2.2
Seawater	0.25	10
3 M $MgCl_2$ solution	300	12,000
3 M NaCl (99.99% high-purity Mg with <10 ppm Fe)	0.3	12

[a]Grades 9980, 9990, 9991, 9995, 9998 except for NaCl solution.

Source: Reference 31.

Table 9.10 Results of 2.5 year Exposure Tests on Magnesium Alloy AZ31

Atmosphere	Corrosion rate (μm/yr)	Loss of tensile strength after 2–5 years (%)
Marine atmosphere	18.0	7.4
Industrial atmosphere	27.7	11.2
Rural atmosphere	13.0	5.9

Sources: Reference 34.

9.10. FACTORS AFFECTING ATMOSPHERIC CORROSION OF MAGNESIUM ALLOYS: EFFECT OF SULFITES AND SULFATES

SO_2, a frequent acidic gas present in polluted atmosphere, converts the insoluble hydroxide-carbonate films that form naturally on magnesium alloys into soluble bicarbonates, sulfites, and sulfates that can be washed away by rain [34]. Moreover, SO_4^{2-} may be formed by the reaction of acidic sulfur-bearing gases with $Mg(OH)_2$ or $MgCO_3$. Sulfates can also promote corrosion and pitting of some alloys unless the metal is protected by properly applied coatings [6].

If CO_2 is present on the surface of a magnesium–aluminum alloy, the protective film will be a mixture of hydrotalcite, $MgCO_3 \cdot 5Mg(OH)_2 \cdot 2Al(OH)_3 \cdot 4H_2O$, and hydromagnetite, $3MgCO_3 \cdot Mg(OH)_2 \cdot 3H_2O$. Due to hydrotalcite film, the magnesium–aluminum alloys are more tarnish resistant than other magnesium alloys when exposed to atmosphere [31]. Morphology of the corroded magnesium alloy with deposited NaCl was more homogeneous in the presence of CO_2 [36]. NaCl-induced corrosion was inhibited by CO_2, partly attributed to the formation of a slightly protective carbonate-containing film [37].

In the presence of CO_2, the surface is more passive due to the formation of a thick magnesium hydroxyl-carbonate film; grain boundaries and noble inclusion do not influence the distribution of corrosion products. The carbonated film inhibits both the anodic and the cathodic processes. The evidence from atomic force and electron microscopy and weight loss results from long-term exposures clearly show that ambient concentrations of CO_2 contribute to the corrosion resistance of magnesium in humid air [37].

It seems that the primary reaction in the corrosion of magnesium is the formation of magnesium hydroxide, $Mg(OH)_2$, followed by a secondary reaction with carbonic acid to convert the hydroxide to a hydrated carbonate. In an industrial atmosphere contaminated with sulfur compounds, hydrated and basic carbonates were found, together with magnesium sulfite ($MgSO_3 \cdot 6H_2O$) and magnesium sulfate ($MgSO_4 \cdot 7H_2O$). The sulfates may be formed by the reaction of acidic sulfur-bearing gases with $Mg(OH)_2$ or $MgCO_3$ [6].

9.11. WATER CORROSION

In stagnant distilled water at room temperature, magnesium alloys rapidly form a protective film that prevents further corrosion. Small amounts of dissolved salts in water, particularly chlorides or heavy metal salts, will break down the protective film locally, which usually results in pitting [6]. When agitation (erosion) destroys or depletes the surface film, corrosion can be increased significantly. The corrosion of magnesium alloys by pure water substantially increases as temperature increases [33].

Dissolved oxygen plays no major role in the corrosion of magnesium in either fresh water or saline solutions. However, agitation or any other means of destroying or preventing the formation of a protective film leads to corrosion. When magnesium is immersed in a small volume of stagnant water, its corrosion rate is negligible. The magnesium ions that initially dissolve increase the pH of the solution and cause the precipitation of insoluble $Mg(OH)_2$. When the water is constantly replenished so that the solubility limit of $Mg(OH)_2$ is never reached, the corrosion rate may increase [6].

The corrosion of magnesium alloys by pure water increases substantially with temperature. At 100 °C (212 °F), the aluminum–zinc (AZ) alloys typically corrode at 0.25–0.50 mm/yr (10–20 mils/yr). Pure magnesium and alloy ZK60A corrode excessively at 100 °C (212 °F), with rates up to 25 mm/yr (1000 mils/yr). At 150 °C (300 °F), all magnesium alloys corrode excessively [6]. Commercially pure magnesium in distilled water is reported to corrode at room temperature with a rate of 1.5×10^{-2} mm/yr (0.6 mil/yr) while at the boiling point the rate is 16 mm/yr (640 mils/yr) [34].

The effect of increasing the temperature is to increase the severity of attack. Magnesium alloys do not have adequate corrosion resistance for applications above ambient temperature [38]. Agitation or any other means of destroying or preventing the formation of a protective film leads to corrosion. When magnesium is immersed in a small volume of stagnant water, its corrosion rate is negligible. When the water is completely replenished, the solubility limit of $Mg(OH)_2$ is never reached and the corrosion rate may increase. In stagnant distilled water at room temperature, magnesium alloys rapidly form a protective film that prevents further corrosion [34].

Magnesium dissolution in aqueous environments generally proceeds by an electrochemical reaction with water to produce magnesium hydroxide and hydrogen gas, so that magnesium corrosion is relatively insensitive to the oxygen concentration [39, 40]. Dissolved oxygen plays no major role in the corrosion of magnesium in either fresh water or saline solutions generally [34]. Hanawalt et al. [9] stated that while the hydrogen overvoltage on pure magnesium is high it is greatly lowered by iron and there is no correlation of corrosion effect with hydrogen overvoltage. In acidic solution, and at relatively more negative potentials, it seems that hydrogen reduction is the main cathodic reaction [41].

Baril and Pebere [42] studied the corrosion behavior of pure magnesium in aerated and deaerated solutions (0.01 and 0.1 M) by steady-state current–voltage and electrochemical impedance measurements. It was shown that the anodic current densities were lower and the resistance values higher in deaerated media. They have stated that the presence of oxygen does not influence the cathodic reaction. The shift of the potential in the cathodic direction in aerated solutions and higher anodic corrosion current densities can be explained by the presence of bicarbonate ion in natural conditions (40 mg HCO_3^-/L). Around the corrosion potential, on the anodic side, the current should be partially controlled by diffusion [42].

Mg Alloys and Soils Except when used as galvanic anodes, magnesium alloys have good corrosion resistance in clay or nonsaline sandy soils but have poor resistance in saline sandy soils [6].

9.12. SALT SOLUTIONS

When exposed to magnesium salt solutions, the corrosiveness of chlorides was greater than that of bromides, which was greater than that of chlorates, as shown in Table 9.8. The lower corrosion rates in the 2 M solutions were attributed to concentration polarization [6]. Severe corrosion may occur in neutral solutions of salts of heavy metals, such as copper, iron, and nickel. Such corrosion occurs when the heavy metal, the heavy metal basic salts, or both plate out to form active cathodes on the anodic magnesium surface. Small amounts of dissolved salts of alkali or alkaline-earth metal (chlorides, bromides, iodides, and sulfates) in water will break the protective film locally, which usually leads to pitting [34]. In 4% NaCl, the corrosion rate of magnesium is 0.30 mg/cm^2· day at 35 °C, while in 48% caustic in the same solution, the corrosion rate drops to 0.01–0.02 mg/cm^2· day showing clearly the active–passive behavior and the quality of the passive layer of magnesium in alkaline medium [43].

Foreign materials that hold moisture on the surface can promote corrosion and pitting of some alloys unless the metal is protected by properly applied coatings. Fluorides form insoluble magnesium fluoride and consequently are not appreciably corrosive. Sodium fluoride has been reported to be the effective inhibitor in water, while sodium carbonate, sodium silicate, and sodium phosphate are not [31].

Oxidizing salts, especially those containing chlorine or sulfur atoms, are more corrosive than nonoxidizing salts [6]. Ion phosphates react to form phosphate layers suitable for paint application [43]. Chromates, fluorides, phosphates, silicates, vanadates, or nitrates cause little or no corrosion. The first four salts are frequently used in the chemical treatment and anodizing solutions for magnesium surfaces due to the formation of partial protective layers. Ammonium salts show higher attack than metal salts, very possibly because of their higher acidity.

Magnesium is rapidly attacked by all mineral acids, except hydrofluoric acid and chromic acid. However, pitting may occur in hydrofluoric acid solutions [33]. Wet organic matter or wet wood products produce attack since organic acids are produced during decay and corrode magnesium [34, 43].

Corrosion Mechanism Explaining Performance As the potential is lowered (more negative) due to the presence of anions, oxygen reduction becomes negligible relative to hydrogen evolution. The ability of an anion to reduce the magnesium potential appears to depend on the solubility of its magnesium salt. It has been suggested that anions are carried by electrochemical transport to anodic sites on the metal surface, where they form magnesium salts, which are acidic to the magnesium hydroxide film. The rapid uniform corrosion rate observed in 3 M $MgCl_2$ at a lower electrode potential supports this mechanism. Examples of these activating anions are Cl^-, Br^-, $SO_2{}^{-2}$, and $ClO_4{}^-$. In the presence of salt solutions of these anions, magnesium becomes several tenths of a volt active to the hydrogen electrode potential. Salts facilitate the cathodic reaction (hydrogen discharge) indirectly by potential shifting to less active levels; however, the hydrogen discharge still controls the corrosion rate. Alloying elements with low hydrogen overvoltage accelerates corrosion [44].

It has been shown also that oxygen plays a major role in the initiation of pitting of AZ91, HK31, and some Mg–Zn alloys in 5 wt% sodium chloride solution at room temperature at relatively high corrosion potentials. This concern of initiation of pitting by the oxygen reduction reaction can be extrapolated to other forms of localized corrosion [41].

A partially protective surface film plays an important role in the electrochemical dissolution processes for magnesium in NaCl, Na_2SO_4, and NaOH solutions. In the broken areas or on a film-free surface, the experimental data are consistent with the involvement of the intermediate species Mg^+ in the magnesium dissolution process. Magnesium is first oxidized to the intermediate species Mg^+; then the intermediate species chemically reacts with water to produce hydrogen and Mg^{2+}. The presence of Cl^- made the surface films more active or increased the broken area of the film, and also accelerated the electrochemical reaction rate from magnesium to magnesium univalent ions [45].

Outdoor Performance and Laboratory Testing Generally, the outdoor exposure corrosion rates were lower than any of the laboratory predictions. Especially in the automotive industry, complex accelerated laboratory tests have been developed to improve the quality of such predictions, with many car manufacturers having their in-house test procedures. None of these tests, however, can really simulate the complex outdoor conditions completely. This was also confirmed by a comprehensive outdoor study performed by Hydro Aluminum, in which magnesium AZ91D, AM50B, and AS21 test plates were exposed together with aluminum A380 test plates on trucks running in the Gothenburg area in Sweden for a total time of 3 years in comparison to accelerated laboratory tests. The field test revealed corrosion rates in the range of 0.012–0.032 mm/yr and pit depths from 0.2 to 0.65 mm after 3 years. The laboratory tests could predict the corrosion rates reasonably well but failed to estimate the pit depths. However, the most important finding was that magnesium components in applications such as transmission housings, engine blocks, and even engine cradles can be used without coatings for protection against general corrosion. Their performance was similar to that of the aluminum A380 alloy [13].

9.13. ACID AND ALKALINE SOLUTIONS

Magnesium is rapidly attacked by all mineral acids and alkalis except hydrofluoric acid (HF) and chromic acid (H_2CrO_4). Hydrofluoric acid does not attack magnesium to an appreciable extent, because it forms an insoluble, protective magnesium fluoride film on the magnesium; however, pitting develops at low acid concentrations. With increasing temperature, the rate of attack increases at the liquid line, but to a negligible extent elsewhere [6]. In acidic solution, hydrogen reduction is the main cathodic reaction [41].

Pure H_2CrO_4 attacks magnesium and its alloys at a very low rate. However, traces of chloride ion in the acid will markedly increase this rate. A boiling solution of 20% H_2CrO_4 in water is widely used to remove corrosion products from magnesium alloys without attacking the base metal. Magnesium resists dilute alkalis, and 10% caustic solution is commonly used for cleaning at temperatures up to the boiling point [6].

9.14. AQUEOUS ORGANIC COMPOUNDS

Aliphatic and aromatic hydrocarbons, ketones, ethers, glycols, and higher alcohols are not corrosive to magnesium and its alloys. Ethanol causes slight attack, but anhydrous methanol causes severe attack. The rate of attack in the latter is reduced by the presence of water. Gasoline–methanol fuel blends in which the water content equals or exceeds approximately 0.25 wt % of the methanol content do not attack magnesium. Pure halogenated organic compounds do not attack magnesium at ambient temperatures. At elevated temperatures or if water is present, such compounds may cause severe corrosion, particularly those compounds having acidic reaction products [6].

9.15. DRY ORGANIC COMPOUNDS

Dry fluorinated hydrocarbons, such as the Freon (E.I. Du Pont de Nemours, Inc.) refrigerants, usually do not attack magnesium alloys at room temperature, but when water is present, they may stimulate significant attack. While fluorinated hydrocarbons may react violently with magnesium alloys at elevated temperatures, newer processes involving fluorinated ketones are being developed to protect molten magnesium from burning [6].

In acidic foodstuffs, such as fruit juices and carbonated beverages, attack of magnesium is slow but measurable. Milk causes attack, particularly when souring [6, 33].

At room temperature, ethylene glycol solutions produce negligible corrosion of magnesium that is used alone or galvanically connected to steel; at elevated temperatures, such as 115 °C (240 °F), the rate increases, and corrosion occurs unless proper inhibitors are added. Measures should be taken to control the galvanic corrosion [6]. The corrosion of magnesium in ethylene glycol can be effectively inhibited by addition of fluorides that react with magnesium and form a protective film on the surface [46].

9.16. GASES AT AMBIENT TEMPERATURE UP TO ABOUT 100 °C

Dry chlorine, iodine, bromine, and fluorine cause little or no corrosion of magnesium at room or slightly elevated temperature. Even when it contains 0.02% H_2O, dry bromine causes no more attack at its boiling temperature (58 °C, or 136 °F) than at room temperature. The presence of a small amount of water causes pronounced attack by chlorine, some attack by iodine and bromine, and negligible attack by fluorine. Wet chlorine, iodine, or bromine below the dew point of any aqueous phase causes severe attack of magnesium. Water vapor in air or in oxygen sharply increases the rates of oxidation of magnesium and its alloys above 100 °C (212 °F), but boron trifluoride (BF_3), SO_2, and SF_6 are effective in reducing oxidation rates. The presence of BF_3 or SF_6 in the ambient atmosphere is particularly effective in suppressing high-temperature oxidation up to and including the temperature at which the alloy normally ignites [6].

The oxidation rate of magnesium in oxygen increases with temperature. At elevated temperature (approaching melting), the oxidation rate is a linear function of time. Cerium, lanthanum, calcium, and beryllium in the metal reduce the oxidation rate below that of pure magnesium. Beryllium additions have the most striking effects, protecting some alloys at temperatures up to the melting point over extended periods of time. Structural applications of magnesium alloys at elevated temperature are usually limited by creep strength rather than by oxidation [6].

Dry gases (Cl_2, I_2, Br_2, and F_2) cause little or no corrosion at room or slightly elevated temperature. Small amounts of water can cause pronounced attack by chlorine, iodine, or bromine and negligible attack by fluorine. Wet chlorine, bromine, or iodine below the dew point of any aqueous phase causes severe attack of magnesium. Dry sulfur dioxide is not harmful at ordinary temperatures, while wet sulfur dioxide is very corrosive due to the formation of sulfurous and sulfuric acids. Dry and wet ammonia causes no attack at ordinary temperatures [33, 34].

9.17. MAGNESIUM HIGH-TEMPERATURE CORROSION

Upon oxidation at 350 °C, magnesium develops dark, somewhat angular spots, some of them representing mounds 2 μm high, generally situated at grain boundaries or on scratches; reheating at 440 °C causes them to collapse. Elsewhere, uniform interference colors are observed, indicating a film of fairly uniform thickness. At 500 °C, rapid regrowth occurs at most of the points of collapse, with white powdery material resulting from the breakaway; at 520 °C, the white material spreads laterally over the whole surface. These observations suggest that breakaway is a phenomenon demanding nucleation energy [47].

At high temperatures, we observe a complete miscibility of Mg when its own halides are present on the metal surface [48]. In dry air up to 400 °C and to about 350 °C in moist air, magnesium corrosion resistance is acceptable. The oxidation rate of Mg in oxygen increases with temperature. At elevated temperature (approaching melting), the oxidation rate is a linear function of time. Cerium, lanthanum, calcium, and beryllium in the metal reduce the oxidation rate below that of pure magnesium. Beryllium additions of 5–19 ppm have the most striking effects, protecting some alloys at temperatures up to the melting point over extended periods of time [31]. Magnesium with small alloy additions of zirconium or beryllium has been used in gas-cooled reactors at temperatures above 350 °C. These alloys have adequate corrosion resistance in wet CO_2 and wet air at temperatures up to 500 °C [43].

At high temperatures, in the presence of water, chlorinated hydrocarbons may hydrolyze and form hydrochloric acid, causing corrosive attack of the magnesium [43]. In dry pure oxygen, it has been shown that at 400 °C and higher temperatures Mg evaporates and forms MgO nodules on the surface by reaction with oxygen [49]. Gol'dshleger and Amosov [50] have constructed a diagram of high-temperature oxidation of magnesium in a dry gas mixture containing oxygen and argon. They established a relation between the structural features of the oxide film and macrokinetics of high-temperature oxidation and mechanisms of transition between different oxidation regimes. The character of magnesium oxidation under isothermal conditions and the possibility of the existence of different combustion modes are primarily caused by a complicated dependence of properties and structural characteristics of the growing oxide film on temperature and oxygen concentration in the ambient medium. With increasing oxygen concentration (or temperature) in the gas phase, the characteristic size of magnesium oxide grains decreases, the film becomes denser, and its shedding is terminated. Gol'dshleger and Amosov found that the Mg corrosion rate at 21% oxygen concentration in dry argon was ~39.8 $g/cm^3 \cdot s$ at 555 °C and up to 65.3 $g/cm^3 \cdot s$ at 585 °C.

Water vapor in air sharply increases the rate of oxidation of magnesium at 100 °C, but boron trifluoride, boron hexafluoride, and sulfur dioxide are effective in reducing oxidation rates [34]. A negligible attack of pure magnesium is recorded with dry chlorine at ambient temperature while at a high temperature of 480 °C, the corrosion rate is 1.5 mm/yr and increased up to 10–20 times (15 and 30 mm/yr) at 540 and 565 °C, respectively. This behavior is very similar to that of aluminum; however, in the case of aluminum, the temperature of such attack is much lower (150–180 °C) [33].

Creep-Resistant Alloys The highest performance of Mg alloys commercially available today are the Mg–Y–RE–Zr alloys (e.g., WE54, WE43). While many elevated-temperature applications may be met by minor additions to AZ or AM type alloys, it is probable that some hotter engine or transmission applications may still require more "exotic" and costly

Table 9.11 Pure Mg High-Temperature Corrosion in Some Gas Media

Type of gas	Ambient or moderate temperature	High temperature
Dry air	Up to 400 °C, acceptable corrosion	
Moist air (30% humidity)	At about 350 °C, acceptable corrosion	
Wet CO_2 or wet air (80% humidity)	Up to 500 °C (+Zr, +Be), acceptable corrosion	
Halogenated organic compounds	No corrosion	Severe corrosion
Dry fluorinated hydrocarbons	No corrosion	Violent reaction Severe corrosion
Wet chlorinated hydrocarbons	No data available	Severe corrosion

Sources: References 31 and 35.

alloys if these can be justified. The ideal would be to develop a single high-temperature alloy to meet all requirement, since this would ensure significant volume production to minimize production and recycling costs and hence boost the commercial viability of the alloy [51]. Table 9.11 gives the corrosion resistance of pure magnesium as a function of humidity and halogen ion at different temperatures.

REFERENCES

1. D. S. Tawil, *Magnesium Technology*, Conference of the Institute of Metals, 1986, pp. 66–70.
2. E. Ghali, in *Uhlig's Corrosion Handbook*, 2nd edition, edited by R. W. Revie. Wiley, Hoboken, NJ, 2000, pp. 793–830.
3. K. U. Kainer, *Magnesium Alloys and Technology*. Wiley-VCH, Weinheim, Germany, 2003, pp. 1–22.
4. E. Aghion and D. Eliezer, in *Magnesium Alloys*, edited by C. E. Aghion and D. Eliezer. Consortium for Development of Magnesium Technologies, Beer Sheva, Israel, 2004.
5. M. O. Pekguleryuz and L. W. F. Mackenzie, in *International Symposium on Magnesium Technology in the Global Age*, Montreal, Canada, edited by M. O. Pekguleryuz and L. W. F. Mackenzie. Canadian Institute of Mining, Metallurgy and Petroleum, Montreal, 2006.
6. A. Shaw and C. Wolfe, in *ASM Handbook, Volume 13B, Corrosion,* edited by S. D. Cramer and B. S. Covino, Jr. ASM International, Materials Park, OH, 2005, pp. 205–227.
7. I. J. Polmear, *Light Alloys from Traditional Alloys to Nanocrystals*. Elsevier, Sydney, Australia, 2006.
8. ASTM B296 (reapproved 1990), in *Annual Book of ASTM Standards, Volume 02.02, Aluminum and Magnesium Alloys*. ASTM, Philadelphia, PA, 1994, pp. 288–289.
9. J. D. Hanawalt, C. E. Nelson, and J. A. Peloubet, *Transactions of American Society of Mining and Metallurgical Engineering* **147**, 273–299 (1942).
10. T. R. Massalski and H. Okamoto, in *Binary Phase Diagrams*, 2nd edition, Vols. 1–4, edited by T. R. Massalski and H. Okamoto. ASM International, Materials Park, OH, 1990.
11. A. A. Nayeb-Hashemi and J. B. Clark, in *Binary Alloy Phase Diagram*, edited by T. R. Massalski, H. Okamoto, P. R. Subramanian and L. Kacprzak. ASM International, Materials Park, OH, 1988.
12. P. Y. Li, H. J. Yu, S. C. Chen, and Y. M. Yu, in *Magnesium Technology 2003*, edited by H. I. Kaplan. TMS: The Minerals, Metals & Materials Society, Warrendale PA, USA, 2003, pp. 51–58.
13. K. U. Kainer, H. Dieringa, W. Dietzel, N. Hort, and C. Blawert, in *International Symposium on Magnesium Technology in the Global Age,* Montreal, Canada, edited by M. O. Pekguleryuz and L. W. F. Mackenzie. Canadian Institute of Mining, Metallurgy and Petroleum, Montreal, 2006, pp. 3–20.
14. I. J. Polmear, in *Magnesium and Magnesium Alloys*, edited by M. M. Avedesian and H. Baker. ASM International, Materials Park, OH, 1999, pp. 12–27.
15. K. G. Adamson, D. S. Tawil, Magnesium and Magnesium Alloys, in *Corrosion, Volume 1*, edited by L. L. Shreir, R. A. Jarman and G.T. Burstein, Butterworth-Heinemann Ltd, Oxford, UK, 1995, 1, 4.98–4.115.
16. J. P. King, *Advanced Materials Technology International*, Vol. 12 Sterling, London, 1990.
17. J. P. King, G. A. Iower, and P. Lyon, in *Light Weight Alloys for Aerospace Applications II*, edited by E. W. Lee and N. J. Kim. TMS, Warrendale, PA, 1991, p. 423.
18. W. Durako and L. Joesten, in *49th Annual World Magnesium Conference*, Chicago, International Magnesium Association, Wauconda, IL, 1992, pp. 87–92.
19. I. J. Polmear, *Magnesium Alloys and Applications, Materials Science and Technology*, The Institute of Materials, London, 1994, pp. 1–16. [http://www.ingentaconnect.com/content/maney/mst and www.maney.co.uk/journals/mst].
20. R. S. Busk, *Magnesium Products Desgin*. Marcel Dekker, New York, 1987.
21. Y. Wada and T. Matsubara, *Powder Technology* **78**, 109–114 (1994).
22. R. A. Saravanan and M. K. Surappa, *Materials Science and Engineering* **A276**, 108–116 (2000).
23. W. M. Chan, F. T. Cheng, L. K. Leung, R. J. Horylev, and T. M. Yue, *Corrosion Reviews* **16**, 43–52 (1998).
24. N. Hort and K. U. Kainer, in *Metal Matrix Composites, Custom–made Materials for Automotive and Aerospace Enginnering*, edited by K. U. Kainer. Wiley-VCH, Weinheim, Germany, 2006, pp. 243–276.
25. L. Lu, K. K. Thong, and M. Gupta, *Composites Science and Technology* **63**, 627–632 (2003).
26. C. Y. H. Lim, D. K. Leo, J. J. S. Ang, and M. Gupta, *Wear* **259**, 620–625 (2005).
27. E. Aghion and B. Bronfin, *Materials Science Forum* **350–351**, 19–28 (2000).
28. P. Maier, N. Hort, and K. U. Kainer, *Magnesium Verfahrenstechnik*, GKSS Workshop 2005. GKSS Forschungszentrum in der Helmholtz Gemeinschaft, Geesthacht, 2005.
29. N. Jeal, *Advanced Materials & Processes* **163**, 65 (2005).

30. Lenntech Periodic Table. Available at http://www.lenntech.com/Periodic-chart-elements/.

31. D. L. Hawke, J. Hillis, M. Pekguleryuz, and I. Nakatsugawa, in *ASM Specialty Handbook: Magnesium and Magnesium Alloys*, edited by M. M. Avedesian and H. Baker. ASM International, Materials Park, OH, 1999, pp. 194–210.

32. E. Ghali, Some aspects of corrosion resistance of magnesium alloys, in *International Symposium on Magnesium Technology in the Global Age*, Montreal, Canada, edited by M. O. Pekguleryuz and L. W. F. Mackenzie. Canadian Institute of Mining, Metallurgy and Petroleum, Montreal, 2006, pp. 271–293.

33. ASM International Handbook Committee, *Handbook of Corrosion Data*. 2nd edition. ASM International, Materials Park, OH, 1995.

34. A. F. Froats, T. Kr. Aune, D. Hawke, W. Unsworth, and J. Hillis, in *ASM Handbook, Volume 13, Corrosion*, edited by L. J. Korb, D. L. Olson, and J. R. Davis ASM International, Materials Park, OH, 1987, pp. 740–754.

35. J.-P. Gravel, Training Report in Chemical Engineering—MIC of Aluminum and Aluminum Alloys and Rational Biocorrosion of Magnesium and Magnesium Alloys. Chemical Engineering Department, Laval University, 2007.

36. L. Cui and L. Xiaogang, *Rare Metals* **25**, 190 (2006).

37. R. Lindström, L.-G. Johansson, G. E. Thompson, P. Skeldon, and J-E. Svensson, *Corrosion Science* **46**, 1141–1158 (2004).

38. M. J. Danielson, in *Environmental Effects on Engineered Materials*, edited by R. H. Jones. Marcel Dekker, New York, 2001, pp. 253–274.

39. G. L. Makar and J. Kruger, *Journal of the Electrochemical Society* **13**, 414–421 (1990).

40. B. R. Powell, V. Rezhets, M. P. Balogh, and R. A. Waldo, *Magnesium Technology 2001*. The Minerals, Metals and Materials Society/AIME, Warrendale, PA, 2001.

41. B. Y. Hur and K. W. Kim, *Corrosion Reviews* **16**, 85–94 (1998).

42. G. Baril and N. Pebere, *Corrosion Science* **43**, 471–484 (2001).

43. J. E. Hillis, in *ASTM Manual Series: MNL 20, Corrosion Testing and Standards: Application and Interpretation*, edited by R. Baboian. ASM International, Materials Park, OH, 1995, pp. 438–446.

44. W. A. Ferrando, *Journal of Materials Engineering* **11**, 299–313 (1989).

45. G. L. Song, A. Atrens, D. St. John, J. Nairn, and Y. Li, *Corrosion Science* **39**, 855–875 (1997).

46. G. Song and D. St. John, *Corrosion Science* **46**, 1381–1399 (2004).

47. U. R. Evans, *Corrosion and Oxidation of Metals*. Edward Arnold, London, 1968, pp. 3–11.

48. L. L. Shreir, R. A. Jarman, and G. T. Burstein, *Corrosion—Metal/Environment Reactions*, 3rd edition, Butterworth-Heinemann, Oxford, UK, 1995, pp. 1–18.

49. V. Fournier, P. Marcus, and I. Olefjord, *Surface and Interface Analysis* **34**, 494–497 (2002).

50. U. I. Gol'dshleger and S. D. Amosov, *Combustion, Explosion, and Shock Waves* **40**, 275–284 (2004).

51. J. F. King, in *Development of Pratical High Temperature Magnesium Casting Alloys*, edited by K. U. Kainer. Wiley-VCH, Weinheim, Germany, 2002.

Chapter 10

General, Galvanic, and Localized Corrosion of Magnesium and Its Alloys

Overview

General uniform corrosion is frequently observed and can lead to a passive surface under certain conditions. Nonuniform corrosion is a term not frequently used for magnesium alloys because, under some conditions, it is very close to pitting corrosion, at least in morphology. Exposure to humid air or to watr leads to the formation of a thick hydrated amorphous film that has an oxidation rate less than 0.01 μm/yr. Galvanic corrosion or bimetallic corrosion is important because most structural industrial metals and even the metallic phases in the microstructure create galvanic cells between them and/or the α-Mg anodic phase. Localized corrosion, such as pitting and filiform and crevice corrosion, of magnesium and its alloys has some differences in occurrence, kinetics, morphology, and mechanism as compared to other metals. Pitting and filiform corrosion can initiate simultaneously and filiform corrosion is observed frequently as the front runner. Filiform corrosion occurs on some uncoated extruded Mg alloys but not on bare pure Mg. Its occurrence on bare Mg–Al alloys indicates that highly resistant oxide films can be formed naturally.

Filiform corrosion propagation does not require the presence of dissolved oxygen in the environment; it is essentially fueled by hydrogen evolution occurring at the filament head and outside the filaments. If the potential of the passive metal or alloy is closer to that of the hydrogen evolution reaction, dissolved oxygen has an important effect on corrosion types such as pitting and filiform corrosion in pure water or upon exposure to atmospheric media. Anodic polarization and filiform corrosion studies in chloride solutions at neutral and alkaline pH values show that thixocast alloy has a lower corrosion rate than that of the die cast alloy.

A. GENERAL CORROSION

For engineering applications, magnesium is usually alloyed with one or more elements, including aluminum, manganese, rare earth metals, lithium, zinc, and zirconium. Under normal environmental conditions, the corrosion resistance of magnesium alloys to general corrosion is comparable to or better than that of mild steel.

Corrosion Resistance of Aluminum and Magnesium Alloys: Understanding, Performance, and Testing By Edward Ghali

10.1. CORROSION RESISTANCE OF PASSIVE MAGNESIUM

Although magnesium has a standard electrode potential at 25 °C of −2.37 V, its corrosion potential is more negative than −1.5 V in dilute chloride solution or a neutral solution with respect to the standard hydrogen electrode due to the polarization of the formed film of Mg $(OH)_2$. The oxide film on magnesium offers considerable surface protection in rural and some industrial environments, and the corrosion rate of magnesium lies between that of aluminum and that of low carbon steels (see Chapter 5).

The corrosion product film on magnesium starts at neutral pH and is stabilized with increasing pH. This causes the lustrous metal to assume a dull gray appearance when exposed to air. Exposure to humid air or to water leads to the formation of a thick hydrated amorphous film that has an oxidation rate less than 0.01 mm/yr. The corrosion rate of chemically pure Mg in salt water is in the range of 0.30 mm/yr. The corrosion resistance of commercial Mg alloys does not significantly exceed that of pure Mg. Within the Mg–Al alloy system, given that additional alloying elements are used in conjunction with Al and that tramp elements are present, manipulation of alloy chemistry and microstructure can significantly improve the corrosion behavior of these alloys. For example, AZ91E has one of the lowest corrosion rates of magnesium alloys, and AZ91D has good mechanical properties and low impurity level and behaves in an acceptable way for many applications in certain corrosive media, while WE43 has good mechanical properties and is widely used [1].

Cold working of magnesium alloys, such as stretching or bending, has no appreciable effect on corrosion rate. Shot- or grit-blasted surfaces often exhibit poor corrosion performance, not from induced cold work but from embedded contaminants. Acid pickling to a depth of 0.01–0.05 mm can be used to remove reactive contaminants, but unless the process is carefully controlled, reprecipitation of the contaminant is possible, particularly with steel shot residues. Therefore fluoride anodizing is often used when complete removal of the contaminant is essential [2].

Natural waters, depending on the source, may have variable concentrations of dissolved carbonates such as calcium, magnesium, and/or sodium, with the solubility controlled by the partial pressure of carbon dioxide and the solubility product of calcium carbonate that vary with alkalinity and temperature [3, 4].

Generally, differential aeration or oxygen cell and chloride ion concentration favor different types or forms of localized corrosion. This is not the case for magnesium alloys; however, oxygen influence becomes evident when the open circuit potential of the alloy is shifted to more noble or less negative values near that of the cathodic evolution reaction of hydrogen. Oxygen content has a decisive influence on passivation especially for aluminum-containing Mg alloys at $pH < 11$, while elevated temperatures above 40 °C and solutions with dissolved carbon compounds constrain passivation. A strong negative influence of the casting skin or rolling scale on passivation could be caused by incorporated impurities, as demonstrated by glow discharge optical emission spectroscopy (GDOES). A decreasing current density in the passive state and an increasing breakthrough potential with increasing aluminum content of the alloy were observed. Whenever passivation and breakthrough appeared, the specimens showed localized forms of corrosion, like pitting or filiform corrosion, and no repassivation under the open circuit condition occurs [5].

10.1.1. E_{corr} and Corrosion Rates in Natural and Aqueous Media

In natural atmospheres, the corrosion of magnesium can be localized depending on alloy composition, microgeometrical properties of the surface, and homogeneity of the

Table 10.1 Open Circuit Potential of Magnesium Electrodes Under Various Aqueous Solutions

Electrolyte	E_R (vs. NHE)
N NaCl	−1.72
N Na_2SO_4	−1.75
N Na_2CrO_4	−0.96
N HCl	−1.68
N HNO_3	−1.49
N NaOH	−1.47
N NH_3	−1.43
$Ca(OH)_2$ saturated	−0.95
$Ba(OH)_2$ saturated	−0.88

Source: Reference 2.

microstructure. This relies also on the conductivity, ionic species, and temperature of the electrolyte (see Chapter 9, Section C).

It should be mentioned that in stagnant distilled water, magnesium and magnesium alloys have a very low rate of corrosion. It is important then to characterize water chemistry such as pH, aeration, temperature, alkalinity, and specific dissolved ion concentrations. Contaminants may vary as a function of the source (well, river, estuary, storm drains) and of climatic conditions (wind direction, temperature, acid rain). Corrosion potentials of magnesium in normal solutions of five frequently found media are somewhat close while in saturated calcium or barium hydroxides, the potentials are nobler or less negative (Table 10.1). On the other hand, Table 10.2 shows some typical values of corrosion potential and general corrosion rates determined by the corrosion resistance polarization method for the most frequently used cast alloy, AZ91D.

Investigations of several Mg alloys with cyclic polarization curves and using limiting cathodic and anodic threshold currents showed interesting conclusions for the understanding of the passive properties of magnesium and its alloys [7–9] (see Chapter 18).

Table 10.2 Corrosion Potential E_c Versus Mercurous Sulfate Electrode (MSE)[a] of the Alloy AZ91D in Different Solutions at 25 °C Measuring the Higher Part of the Electrode

Solutions	pH	E_c (mV vs. MSE)	I_c (μA/cm^2)	Rp (Ω·cm^2)	R_c (μm/day)
100 ppm Cl^-	5.28	−1863	12.03	1805	0.75
100 ppm SO_4^{2-}	5.28	−1932	15.38	1412	0.96
100 ppm HCO_3^-	7.70	−1989	14.49	1499	0.90
100 ppm Cl^- + 100 ppm SO_4^{2-}	5.46	−1942	24.02	904	1.50
100 ppm Cl^- + 100 ppm HCO_3^-	7.88	−1987	22.49	966	1.40
100 ppm SO_4^{2-} + 100 ppm HCO_3^-	8.28	−2007	23.57	921	1.47
ASTM water: 100 ppm Cl^- + 100 ppm SO_4^{2-} + 100 ppm HCO_3^-	8.02	−1998	27.78	782	1.73
ASTM water saturated with $Mg(OH)_2$	8.93	−2019	22.56	963	1.40
300 ppm Cl^-	5.40	−1902	12.34	1760	0.77
500 ppm Cl^-	5.42	−1895	12.96	1676	0.81

[a] I_c, corrosion current; R_p, polarization resistance; and R_c, corrosion rate.
Source: Reference 6.

Saltwater corrosion studies are typically conducted in 3–5% sodium chloride solutions, following ASTM Standards G31–72 for immersion and ASTM B117-90 (ASTM, Philadephia, PA, USA) for salt-spray testing. In these test methods, a corrosive environment is simulated, as might be encountered in a marine or an automotive application (e.g., salty road splash). The chloride solutions, even in small amounts, usually break down the thin protective magnesium oxide film.

10.1.2. Corrosion Rate Methods of Mg–Al Alloys

10.1.2.1. Comparative Evaluation of Corrosion Rate Methods

The corrosion rates of Mg, AM50B, AZ91D, and eight experimental creep-resistant magnesium alloys were determined in 5% NaCl solution using salt-spray testing (ASTM B117). In parallel, other techniques are employed (weight loss, electrochemical dc polarization titration, and hydrogen evolution) [10]. A critical comparison of the employed techniques is given here. Specimens were machined from ingot or previouslycast plates. For immersion testing (e.g., titration, hydrogen evolution, and weight loss), rectangular specimens of approximate size 4 cm × 3.1cm × 0.3 cm were machined. The surface area of the specimens exposed to electrochemical testing was approximately 1 cm^2, which was a feature of the particular electrochemical cell used. Prior to measurements, the machined specimens were wet ground with SiC abrasive paper beginning with 400 grit size and successive grades to a final grit size of 1200, followed by rinsing with ethanol and air drying. Lacquer coating was applied to the specimen edges to avoid preferred corrosion in these regions [10].

The electrolyte for immersion tests was prepared by mixing 5 wt % sodium chloride, with the initial pH of the solution adjusted to pH 4 by adding either concentrated hydrochloric acid or sodium hydroxide solution. The solutions used for electrochemical polarization experiments were buffered solutions of either pH 4 or pH 6 with 5 wt % of sodium chloride added. All tests were conducted at 21 °C. Loss of noncorroded material during etching is always a concern for the weight loss method. Polarization experiments are relatively easy to conduct; however, a number of factors, such as scan rate, cell geometry, influence of the Nernst diffusion layer, agitation, or circulation of the electrolyte, may affect the corrosion process and therefore the repeatability and reliability of the corrosion rate determined by the polarization technique. In the titration method, the pH of the solution is held constant and it can be tailored to suit the requirements of the corrosion environment. In general, the titration test results are more reproducible. A lack of accuracy and experimental difficulty (e.g., proper gas sealing) are the main concerns with the hydrogen evolution method and hence the variability in the comparative corrosion rates between experiments is high [10].

10.1.2.2. Corrosion Rates of Mg–Al Creep-Resistant Alloys

There is an interest in the development of new high-temperature, creep-resistant alloys that can be considered more suitable for power train applications. Sivaraj et al. [10] examined the aqueous corrosion rate of eight creep-resistant magnesium alloys.

They consider three reference materials (pure magnesium, AM50B, and AZ91D cast alloys) and eight experimental Mg–Al alloys developed for high-temperature creep resistance. The casting processes for the experimental alloys were high-pressure die casting (HPDC) and sand casting. Creep resistance in these materials is generally attained through the addition of elements such as Ca, Sr, Nd, Y, or other rare earths and reductions in aluminum content, often with Si replacement.

The experimental alloys showed greater corrosion resistance than pure magnesium in most of the methods employed except dc polarization. The variation in the corrosion rates between the experimental alloys was not extraordinary, most rates falling within about an order of magnitude. The sand-cast alloys showed higher corrosion rates than the die-cast alloys, possibly due to processing and microstructural differences [10].

10.1.3. Critical Evaluation of the Passive Properties of Magnesium Alloys

The advantages of the mechanism of passivation of magnesium and its alloys are that the pH is alkaline, the passive film is stabilized, and serious types of corrosion such as pitting corrosion are self-limiting. Although the passive region is not perfect, especially at low buffered pH values [11], the passive film has the advantage of an effective stability for a broad range of pH values in the alkaline medium; this is not the case for aluminum alloys, which are vulnerable to cathodic corrosion and alkaline pitting. The solubility of the metal as related to the concentration of log Mg^{2+} in solution decreases linearly with pH starting at approximately pH 8.5.

However, there are some major defects in the film's quality of protection against general corrosion in the case of magnesium and its alloys.

1. The oxide film thickness on pure magnesium is relatively thin, ~2–3 nm, and continuing oxidation of the metal leads to the formation of a thicker hydrated film adjacent to magnesium. Even at high pH, the corrosion product film of magnesium hydroxide (brucite) that forms on the surface is only semiprotective. However, the protective film is of low density due to the plate-like structure of magnesium hydroxide, allowing for the ingress of electrolyte to the metal underneath. Its resistance is considered to be highly insensitive to the oxygen concentration but depends enormously on the sites of hydrogen discharge composed frequently of alloyed metals and different phases that can control the corrosion rate [2].
2. The Pilling–Bedworth ratio (PBR) values of the oxide and hydroxide are 0.81 and 1.77, respectively, and so the formation of $Mg(OH)_2$ inside the oxide film leads to the development of cracks. The stresses start accumulating during the oxide growth process (epitaxial stresses at the interface) and continue with increasing weight (see Chapter 2, Section 2.6). The film undergoes compressive rupture due to the higher molar volume of magnesium hydroxide compared to metallic magnesium and leads to spilling or flaking [12].

   The characteristics of the oxide film formed on a Mg-based WE43 alloy using ac/dc anodization techniques in an alkaline silicate solution at 30 mA/cm^2 for 5 min, followed by 25 min decreasing current, have been investigated. The anodic oxide film formed in alkaline silicate anodizing bath is composed of MgO, $Mg(OH)_2$, SiO_2, and MgF_2, with the molar ratio of MgO to $Mg(OH)_2$ being close to 2 : 1. As MgO readily reacts with water to form $Mg(OH)_2$, this could result in a partial blocking of the pores at the beginning, because the molar volume of $Mg(OH)_2$ is larger than that of MgO, hence increasing the film resistance. The further formation of $Mg(OH)_2$ inside the oxide film could change the mechanical stresses within the oxide film, causing some cracks to develop and the oxide film eventually fails (Figure 10.1). The evolution of the corrosion resistance using the ac impedance

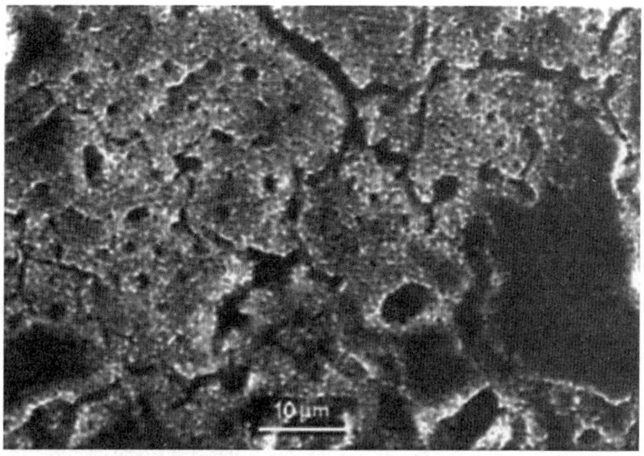

Figure 10.1 SEM image of anodized WE43 alloy after 50 hours of immersion in 0.86 M NaCl solution [13].

technique and open circuit potential measurements can be separated into three stages that could correspond to the initial hydration of MgO, the blocking of film pores as MgO begins hydrating, and then the formation of cracks and dissolution of $Mg(OH)_2$ into pores and the solution [2, 13].

3. There is the possibility of conversion of the protective surface film to soluble film such as bicarbonates, sulfites, and sulfates, that can be washed away by acid rain, agitation, or any other flowing liquids.

Possible Improvements of the Passive Behavior Improvements in corrosion resistance have been found to correlate with an increase in the concentration of the alloying element or its oxide in the passive film and to upgrade the passive behavior of Mg and its alloys. It is admitted that alloying affects the nature of this film [11]. Addition of substances that can form soluble complexes such as tartrate or metaphosphate or insoluble salts such as oxalate, carbonate, phosphate, and fluoride is efficient in reducing corrosion. Adding soluble chromates, neutral fluorides, or rare earth metal salts is effective in reducing magnesium-based metal corrosion [14]. Uniform corrosion on pure magnesium has been drastically reduced by exposure to chromate as well as dichromate, molybdate, and nitrate-containing solutions [11, 15].

10.2. THE NEGATIVE DIFFERENCE EFFECT (NDE)

The corrosion rate of chemically pure Mg in saltwater is in the range of 0.30 mm/yr. The corrosion resistance of commercial Mg alloys does not significantly exceed that of pure Mg. For most metal systems, there is usually good quantitative agreement between weight loss and electrochemical measurements and there is a decrease in hydrogen evolution during anodic polarization of most metals less active than magnesium. However, for magnesium metal, especially pure magnesium, this does not seem to be true. Over a restricted potential range of anodic polarization, the rate and amount of hydrogen evolution on magnesium actually increase as the potential of the metal is made more noble (hence the term negative difference effect, NDE). It is found experimentally that the hydrogen evolution reaction (HER) rate increases with increasing potential.

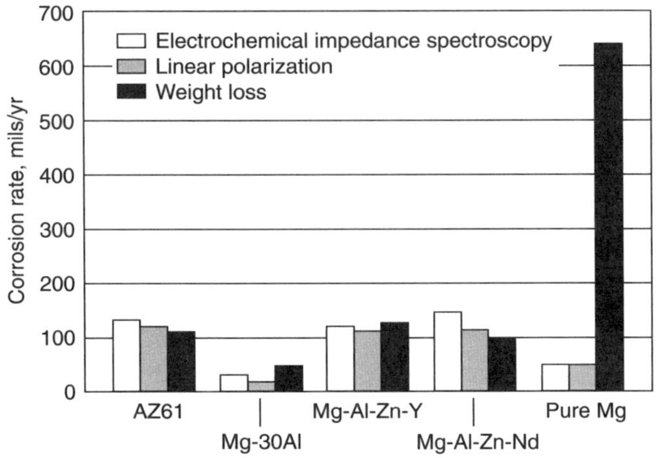

Figure 10.2 Comparison of corrosion rates of pure magnesium (99.98% pure) and magnesium alloys determined by electrochemical and gravimetric methods [11, 19].

Furthermore, more magnesium dissolution occurs than can be accounted for by integrating the charge passed and applying Faraday's law, assuming a valence of 2 for magnesium. In the absence of an added oxidant, the excess of magnesium that dissolves is approximately equal, in Faradaic terms, to the total amount of hydrogen evolved. For Mg, the negative difference effect is therefore not only of fundamental interest but also of practical importance since it contributes to the rapid corrosion of magnesium when it is galvanically coupled, in service, to less active metals [16].

Figure 10.2 gives a comparison of corrosion rates of alloys in pH 9.2 sodium borate solution determined from electrochemical techniques (polarization and electrochemical impedance spectroscopy) and the gravimetric method [11]. For pure magnesium, the weight loss values are roughly an order of magnitude higher than predicted by electrochemical techniques due to the negative difference effect [11, 17].

Figure 10.3 shows the effect of current density on the negative difference for anodic dissolution of magnesium in sulfuric acid obtained by the hydrogen evolution rate. When there is no negative difference effect, the voltage is proportional to $6.97I$, where I is the current input. The deviation from linearity becomes more pronounced with increasing current density and decreasing acid concentration [18, 19].

Mechanical and Chemical Attacks on the Protective Film Whitby [20] states that the protective film on magnesium exercises control of the corrosion rate. Evans [21] implies that destruction of this film accounts for the increased hydrogen evolution observed with anodic current flow. Robinson [22] points out that the increased corrosion rate of Mg at low pH values could be due both to depolarization of the local anode by breakdown of the $Mg(OH)_2$ protective film and to depolarization of the cathode by the increased availability of protons for discharge. He shows that the pH decreases appreciably with increase of magnesium ion concentration and suggests that the local corrosion rate of the magnesium anode might well be expected to rise with increasing applied current density since the magnesium ion concentration at the anode–solution interface is increased. McNulty and Hanawalt [23] propose that the dissolution of the metal resulting from current flow uncovers

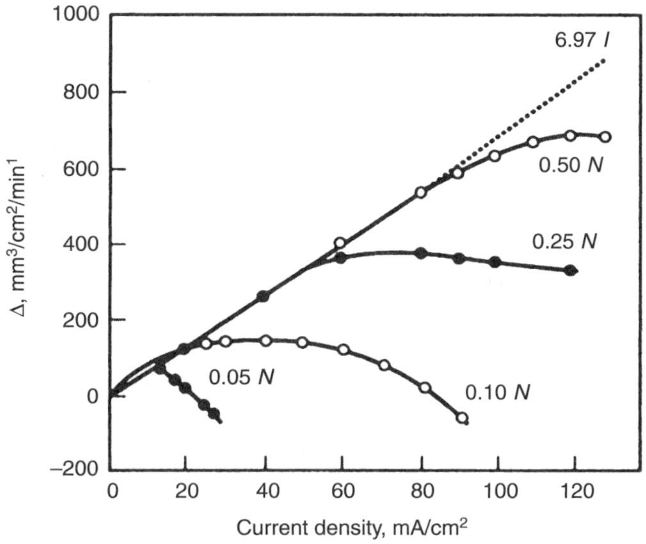

Figure 10.3 Effect of current density on the negative difference for anodic dissolution of pure magnesium for various concentrations of sulfuric acid [11]. Δ is the difference between hydrogenevolution rate V_1, from an electrode without current flowing, and hydrogenevolution rate V_2, from the same electrode with current flowing [18].

additional local cathodes, which increases the amount of local action current that can flow until a balance between the uncovering of new cathodes and the loss of old ones is reached.

Robinson and King [24] have proposed that the production of hydrogen occurs in the broken areas of the partially protective $Mg(OH)_2$ film formed on the magnesium surface or on a film-free surface and that magnesium reacts directly with water like sodium. Such damage occurs from applied current as a result of the buildup of magnesium ion concentration at the anode interface. Straumanis [25] also points out that "the very active" metals (Mg, Al, and Ti) react with the electrolyte (self-dissolution) in places where the protective scale is broken off from their surface. Glicksman [26] considers a model of proton transfer through the protective film along with film breakdown as an explanation of the phenomenon. Tomashov et al. [27] suggest that the excess corrosion represented by the negative difference effect may be due either to chemical reaction at the bare active surface where the film has broken down, or to the electrochemical reaction at microcathodes bared by the film destruction. They prefer the second possibility.

Loss of Metal by Disintegration (Chunk Effect) and Local Cell Actions Stampella et al. [16] consider that the excess hydrogen evolution and magnesium dissolution arise from a combination of two effects: first, spalling of magnesium metal followed by dissolution of the disintegrated particles by local action corrosion; and second, pitting of the magnesium followed by local cell action within the pits due to the highly acidic local environment there. However, no convincing mechanism has been proposed for the spalling of magnesium during anodic dissolution, although some form of environmentally induced cracking is an obvious possibility.

Mg^+ and Magnesium Hydride Formation MgH_2 is also considered as an intermediate of the anodic dissolution process. It was shown that autodissolution of metal is due to the participation of water in the anodic process, forming the hydroxonium ion on the active surface of the electrode. Subsequent reduction of the hydroxonium ion leads to hydride formation and autodissolution of the metal. The participation of anions of the electrolyte in the anodic process decreases the contribution of the reaction involving water and inhibits the autodissolution of Mg [28]. This does not exclude the direct dissolution of magnesium as the bivalent ion.

The anodic reaction is:

$$Mg + 2H_2O \rightarrow Mg(OH)_2 + 2H^+ + 2e, \quad E = -1.862 - 0.059\,\mathrm{pH}$$

The cathodic reactions can be [28]:

$$2H^+_{ads} + Mg + 2e \rightarrow MgH_2; \quad Mg + 2H_2O + 2e \rightarrow MgH_2 + 2OH^-, \quad E = 0.186 - 0.059\,\mathrm{pH}$$

Perrault [29] as well as Petty et al. [30] admitted that a metastable monovalent ion is produced as the intermediate species and reacts chemically with water to evolve hydrogen. Other investigators based their explanation on the anodic formation of monovalent magnesium ion, and its subsequent reaction with water to liberate hydrogen, followed by anodic oxidation of the magnesium hydride phase [24]:

$$MgH_2 + 2H_2O \rightarrow Mg(OH)_2 + H_2 + 2H^+ + 2e$$

or by hydrolysis,

$$MgH_2 + 2H_2O \rightarrow 2H_2 + Mg(OH)_2$$

This reaction occurs on the film-free magnesium surface and was not evident for magnesium with a surface film in 1 N NaOH solution, for example.

Song et al. [31, 32] propose that the area free of the surface film increases with increasing applied potential or current density. These film-free areas are crucial to the NDE behavior. Recently, Winzer et al. [33] explained the possibility of having the same overall reaction but through the electrochemical dissolution of magnesium as the monovalent ion (reaction 10.1) and electrochemical reduction (reaction 10.2) and this is completed through the oxidation–reduction chemical reaction (10.3) [32].

$$2Mg = 2Mg^+ + 2e^- \quad \text{anodic partial reaction} \tag{10.1}$$

$$2H^+ + 2e^- = H_2 \quad \text{cathodic partial reaction} \tag{10.2}$$

The intermediate species react with water chemically to produce hydrogen and Mg^{2+}:

$$2Mg^+ + 2H^+ \rightarrow 2Mg^{2+} + H_2$$

The product of the oxidation–reduction reaction can lead finally to hydroxide formation [34]:

$$2Mg^+ + 2H_2O = 2Mg^{2+} + 2OH^- + H_2 \quad \text{chemical reaction} \tag{10.3}$$

The overall reaction gives

$$2Mg + 2H^+ + 2H_2O = 2Mg^{2+} + 2OH^- + 2H_2$$

or

$$Mg + 2H_2O = Mg(OH)_2 + H_2$$

For (single–phase) Mg alloys, corrosion typically takes the form of localized corrosion since Mg is partially protected by the hydroxide film. Since the free corrosion potential is more positive than the pitting potential in 3% NaCl solution, for example, pitting spreads laterally in the case of magnesium alloys and deep pitting is not frequently observed. The chance development of areas of localized corrosion leads to undermining and release of Mg particles [33]. Winzer et al. [33] added that cathodic hydrogen production can still proceed on the surface film at such a negative potential, but the evolution rate decreases with increasing potential until the pitting potential is reached. At the pitting potential, the surface film begins to break down, and both H evolution and Mg dissolution become much easier on the film-free area. With increasing potential, the film-free area increases, so there is more H evolution. Also, the rates of cathodic and anodic reactions increase, causing more H to be produced at higher potentials. In the considered anodic dissolution reaction, there is only one electron involved—half the number of electrons expected. This means that for the same current density more Mg is dissolved than expected from the electrochemical direct divalent reaction. NDE is common in the corrosion of Mg and Mg alloys. Song et al. [33] found that α, β, and AZ91 die-cast alloys all showed NDE.

Schematic explanation of the NDE has been given by Song et al. [35] (Figure 10.4), showing that the anodic Mg dissolution current can increase faster than expected from the polarization curve due principally to the increase of the hydrogen evolution reaction (HER) at the applied potential. The normal anodic partial reaction and cathodic partial reaction are shown by the solid lines marked I_a and I_c, respectively, in a Tafel diagram (E versus log I). Both are assumed to obey Tafel kinetics. The rates of these two reactions are equal to I_0 at the corrosion potential E_{corr}.

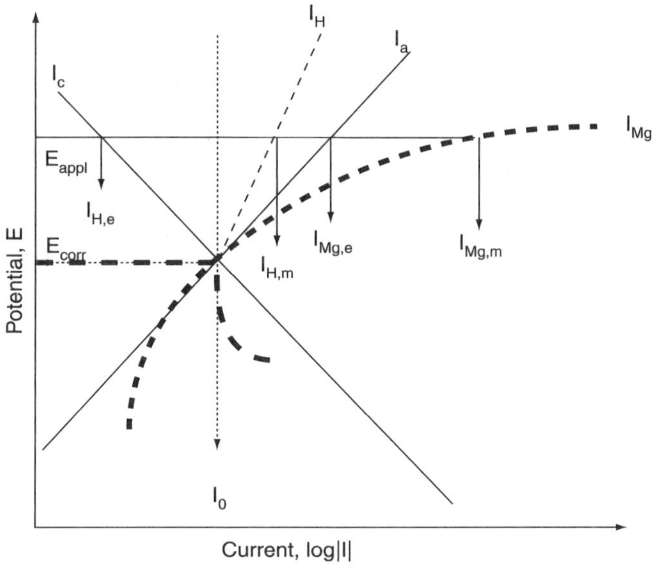

Figure 10.4 Schematic explanation of the negative difference effect [32].

When the potential is changed to a more positive value, E_{appl}, the rate of the anodic partial reaction increases along the curve marked I_a to the value $I_{Mg,e}$ and in the same time the cathodic reaction decreases along the curve I_c to the value $I_{H,e}$. Thus for an applied potential, E_{appl}, the actual rate corresponds to the value $I_{H,m}$ (which represents an HER current significantly greater than the expected current corresponding to $I_{H,e}$). Thus for an applied potential E_{appl}, the actual dissolution rate or the experimentally measured weight loss corresponds to the value $I_{Mg,m}$, which represents a corrosion current significantly greater than the expected current corresponding to $I_{Mg,e}$ [32].

In summary, it can be suggested that the damage of the protective film is essential for the two mechanisms of dissolution through monovalent magnesium ion with or without the help of hydride formation. The disintegration is a common factor that is observed in the presence of severe corrosion rate and gas evolution. Most of the time, these factors are related and probably two or more of them can predominate depending on the experimental conditions and the level of electrochemical polarization conditions. Complete understanding of these mechanisms can help in corrosion control and prevention for magnesium and its alloys [11, 17, 32].

10.3. KINETIC STUDIES OF GENERAL AND PITTING CORROSION OF MAGNESIUM ALLOYS

10.3.1. Electrochemical Noise Studies

10.3.1.1. The Potential

General corrosion and pitting corrosion of AZ91D-DC (die cast), AZ91D-ESTC (electromagnetically stirred thixocast), AZ91D-SFTC (stirring-free thixocast), and AJ62x-DC (die cast) specimens were studied in alkaline chloride medium (0.1 M NaOH + 0.05 M NaCl 2 mL H_2O_2) at 25 °C and pH 12.3 considering electrochemical noise (EN) measurements. The results confirmed to some extent the polarization results (passive zone, pitting current, and average corrosion rate) [36].

Noise and the current noise were recorded simultaneously for all the tested specimens. There are two distinct patterns showing the evolution of electrochemical potential and current fluctuations with time. Two distinct potential level values for the four specimens were observed: an initial potential noise around −1.43 V and a second, not reproducible, but generally ≥−1.20 V/Ag, AgCl/KCl saturated reference electrode. The first potential noise value changed a little with immersion time and with the type of specimen. This value became approximately −1.35 V after around 9 h and 3 h of immersion for AJ62x-DC and AZ91D-DC specimens, respectively.

The high current noise value corresponds to the first potential noise value and can be attributed to the active zone. The current noise is two times higher for AJ62x-DC (30 μA·cm^{-2}) than for the AZ91D-DC specimen. The low current noise value corresponds to the nobler potential and can be attributed to the passive zone. In the active zone, the AJ62x-DC specimen has the highest current noise followed by AZ91D-DC, AZ91D-SFTC, and AZ91D-ESTC specimens. Generally, for AZ91D specimens, after an immersion time of less than 10 min, there is a rise in the current of about 15 μA·cm^{-2} and a fall of about 200 mV for the potential. For the AJ62x specimen, after 1.5 h of immersion, there is a rise in the current of about 25 μA·cm^{-2} and fall of about 400 mV for the potential. The voltage and current noise values as a function of time are shown for AZ91D-DC in Figure 10.5.

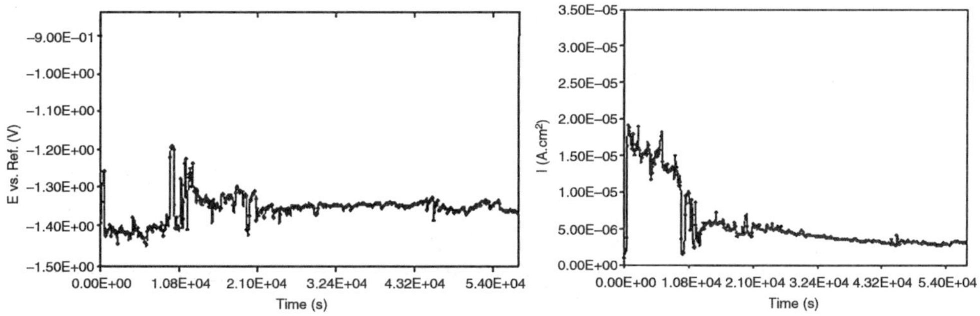

Figure 10.5 Potential and current noise evolution of Mg alloy AZ91D during 16 h immersion in 0.05 M NaCl solution saturated at pH 12, in atmospheric oxygen without agitation [36].

The time of transition from passive to active behavior is called the *induction time* and could correspond to the time observed before the appearance of pits. This value varies a little for the three AZ91D specimens (from 6 to 10 min) while it is on the order of 1.5 h for the AJ62x specimen. It has been found that the induction time and the number of transitions from active to active–passive behavior depend on the chemical composition and the microstructure of the magnesium alloy having the same surface preparation (mechanically polished down with 1200 grit silicon carbide abrasive paper). The induction time stays higher for the AJ62x-DC specimen than for the AZ91D specimens, but it is hard to make a distinction among the AZ91D specimens. AZ91D-SFTC and AZ91D-ESTC specimens present more transitions of active–passive behavior than do AZ91D-DC and AJ62x-DC specimens during a 16 h immersion time period. AJ62x-DC and AZ91D–ESTC specimens have a longer passive zone than AZ91D-DC and AZ91D-SFTC specimens [36].

In summary, an intense corrosion rate was observed at the beginning of the experiment and decreased with the immersion time. The best passive zone was observed for AJ62x-DC because of the intensive corrosion products formed at the surface. AZ91D-ESTC has shown the best corrosion resistance followed by AZ91D-SFTC and AZ91D-DC. Localized corrosion with dense pitted areas was observed during a 16 h immersion period for AZ91D-SFTC and AZ91D-ESTC specimens [36].

Zhang et al. [37] examined the behavior of die-cast AZ91D magnesium alloy with a mechanically polished surface in alkaline chloride solution (0.05 M NaCl at pH 12.0). Electrochemical noise (EN) data analysis in the time domain involves the calculation of the noise resistance R_n. Figure 10.6 shows that the evolution of $1/R_n$ is proportional to the corrosion rate, which was calculated instantaneously. With the increase of immersion time, the corrosion rate of AZ91D alloy exhibited a maximum at around 8000 s, with lower values at both shorter and longer immersion times, which is consistent with other investigations [36, 37].

10.3.1.2. *R_n and the Stages of Corrosion*

For the first stage during the first 90 min (5400 s), during the starting 1800 s, the corrosion rate of AZ91D alloy was low ($1/R_n$ was about 5.5 $\mu\Omega^{-1}\cdot cm^{-2}$) and increased gradually with the immersion time. After 30 min of immersion, a positive abrupt shift of $1/R_n$ value was recorded and the $1/R_n$ value stayed at a high level (10^{-12} $\mu\Omega^{-1}\cdot cm^{-2}$). For the second stage, for immersion time ranging from 90 to 180 min, the corrosion rate data points indicated that

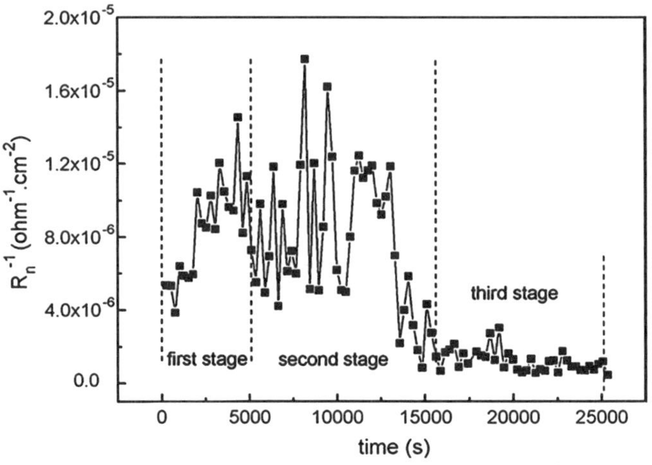

Figure 10.6 The active–passive behavior and pitting of AZ91D alloy in 0.05 M NaCl at pH 12.0 at open circuit potential [37].

the AZ91D alloy surface was experiencing a competition between pitting initiation and repassivation. After 180 min (10,800 s) of immersion, the corrosion rate of AZ91D alloy remained at a higher level (12 $\mu\Omega^{-1}\cdot cm^{-2}$) and was characterized by very small amplitude fluctuations, which suggested that frequent metastable pitting corrosion occurred on the alloy surface. When the immersion time increased from 13,500 to 18,400 s, a significant decline of corrosion rate overlapped large amplitude fluctuations and the corrosion rate decreased from 12 to 1.5 $\mu\Omega^{-1}\cdot cm^{-2}$ [37].

Analysis of EN data in the frequency domain involved the calculation of the noise power spectral density (PSD). The power spectral densities of current, PSD_i, were calculated as follows: $\log PSD_i = A_i + S_i \log f\,(i^2\ Hz^{-1})$ where A_i and S_i are, respectively, the noise intensity and the roll-off slope of current of PSD plots. The PSD results confirmed the values of R_n very well [37].

10.3.1.3. Wavelet Analysis Studies

For the first stage, the main feature was that the relative energy was mainly accumulated among some crystals and a maximum was found in one of them. Evolution of hydrogen bubbles on the AZ91D alloy surface might transiently increase the corrosion rate through removal of the corrosion product film and at the same time might shield the surface from the aggressive solution. During the second stage, at about 8960 s, a maximum was observed in one of the crystals, which demonstrated that the electrode surface was mainly undergoing metastable pitting corrosion. After about 11,800 s of immersion, eight successive crystals showed higher level values and the energy was mainly accumulated among some of them. Energy accumulation and the relatively higher weight of the crystals might reflect the fast fluctuations corresponding to the coexistence of metastable pitting corrosion and hydrogen bubble evolution. During the third stage, the inhibition of corrosion could be attributed to two aspects: the effect of hydrogen and the effect of aluminum enrichment [37].

The corrosion behavior of AZ91D magnesium alloy in alkaline chloride solution was investigated by electrochemical noise (EN). The noise resistance (R_n), power spectral density (PSD), and wavelet transform were considered in analyzing the EN data. There exist

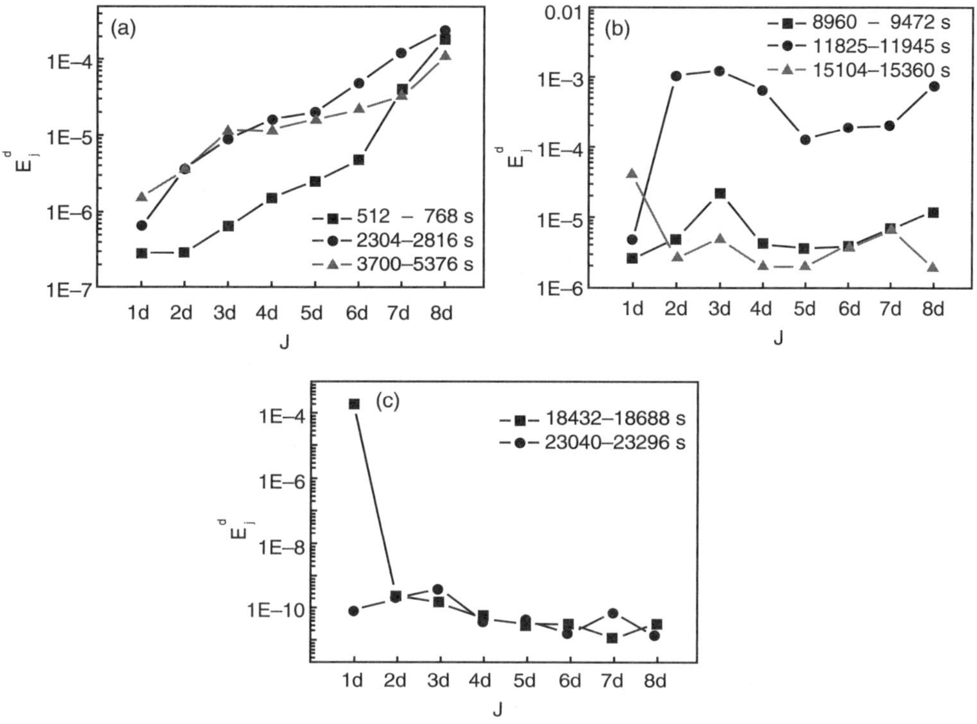

Figure 10.7 Variation of energy distribution plots (EDPs) corresponding to current noise of AZ91D alloy during the different immersion periods: (a) the first stage, (b) the second stage, and (c) the third stage [37].

three different stages of corrosion for AZ91D magnesium alloy in alkaline chloride solution. Three corrosion stages—the anodic dissolution process accompanying the growth, absorption, and desorption of hydrogen bubbles, the development of pitting corrosion, and the possible inhibition process by protective MgH_2 film—could clearly be distinguished by the analysis of EN time records, R_n, PSD, and the energy distribution plot (EDP) [37].

In summary, the results demonstrated that R_n determination accompanied by an EDP was a powerful tool to provide useful information about the dominant process for the different corrosion stages. The EDP was able to provide useful information about the dominant corrosion process of AZ91D alloy in chloride medium, according to the position of the maximum relative energy at different stages, and the evolution of different corrosion processes with time through the changes of relative energy of each crystal (Figure 10.7) [37].

10.4. CORROSION PREVENTION

Effective corrosion prevention for magnesium components and assemblies begins at the design stage. General corrosion attack in saltwater exposures can be minimized through the selection of high-purity magnesium alloys cast without introducing heavy metal contaminants and flux inclusions. Selective surface preparation, chemical treatment, and coatings are recommended. Oil application, wax coating, anodizing, electroplating, and painting are possible alternatives. Recently, it was found that a magnesium hydride layer, created on the magnesium surface by cathodic charging in aqueous solution, is a good base for painting.

Nakatsugawa [38] developed a method to coat a hydrogen-rich layer onto AZ91D by way of cathodic charging. MgH_2 is a reducing agent and decomposes gradually to form the hydroxide $Mg(OH)_2$ in an aqueous environment. The treated Mg or Mg–Al alloys show a pseudo-passive behavior in the anodic region in 5% NaCl solution and an increase of the Tafel slope in the cathodic region. The corrosion resistance of this coating is superior to Cr^{6+}-based conversion coating and has a fairly good adhesion to paint (see Chapter 15).

B. GALVANIC CORROSION

Galvanic coupling of magnesium alloys, high impurity content such as Ni, Fe, and Cu, as well as more noble secondary particles to the Mg matrix and surface contamination are detrimental for corrosion resistance of magnesium alloys. Bimetallic corrosion resistance has been and can be increased by fluxless melt protection, choice of compatible alloys, insulating materials, new high-purity alloys, and more homogeneous microstructures.

The chemical composition and the microstructural properties of the alloy as well as the conductivity and composition of the medium at the metal–medium interface are the controlling factors of the galvanic corrosion rate. For continuous outdoor use, where magnesium assemblies may be wet or subjected to salt splash or spray, attack resulting from galvanic corrosion is probably the most detrimental because magnesium is anodic (or sacrificial) to all other engineering metals. The degree to which the corrosion of magnesium is accelerated by bimetallic or galvanic corrosion can be predicted in considering the relative position of the two metals in the electrochemical series (Figure 10.8) and the open circuit potential (OCP) of every constituent in the considered solution and operating conditions. For example, in sodium chloride solutions (3–6%), the potential of magnesium alloys is ~ -1.67 V/SHE while the potentials of Al–12%Si and pure aluminum are -0.83 and -0.85 V/SHE, respectively. The OCPs of mild steel, nickel, and copper are much more positive (noble) in the same media and are ~ -0.78, -0.14, and -0.22, respectively [32, 39].

Galvanic corrosion of magnesium alloys can generally be attributed to some basic causes such as (1) poor alloy quality due to excessive levels of heavy metal or flux contamination, and (2) poor design and assembly practices, which can result in severe galvanic corrosion attack. Also, any heavy metal contamination of the aqueous solution at the interface can plate out on the metal surface and cause galvanic localized attack. The anodic reaction can be assumed to be the conventional divalent corrosion of magnesium

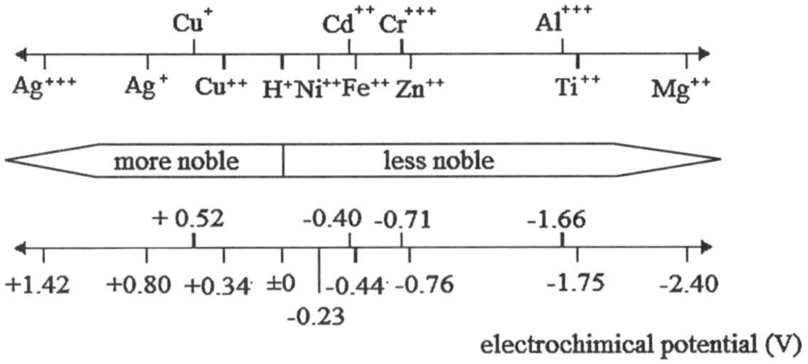

Figure 10.8 Standard reduction potentials of magnesium and some metals at standard conditions (atmospheric pressure and 25 °C) [39].

while the cathodic reaction can be the hydrogen ion reduction on the surface of more noble or less active phase of the magnesium alloy especially at high negative potentials:

$$Mg = Mg^{2+} + 2e^- \quad \text{anode reaction}$$

$$2H_2O + 2e^- = H_2 + 2OH^- \quad \text{cathode reaction}$$

The severity of galvanic activity is determined by the galvanic current, which flows in the completed circuit. This can be expressed as follows:

$$I = (E_k - E_a)/(R_m + R_e)$$

where E_k and E_a are the polarized measured potentials of the cathode and anode, respectively, and R_m and R_e are the resistance of the metal-to-metal contact and the electrolyte portions of the circuit, respectively. In many practical applications, R_m is negligibly small and is a function of electrical and mechanical factors and R_e (the electrolyte resistance) then becomes the controlling factor in the circuit resistance. The conductivity and composition of the medium in which the metal is immersed are controlling factors for the rate of galvanic corrosion.

The corrosion rate of pure magnesium after 60 days is ~1.5 mm/y (60 mils) in distilled water vented with carbon dioxide, almost 10 times that of the same solution vented to air through a caustic trap for carbon dioxide [2]. If the environment is rich in marine mists or deicing salts, the use of drain holes and sealants can help control corrosion by forcing the galvanic current in the electrolyte to flow through a thin and therefore highly resistive film (maximizing R_e). In practice, this limits the galvanic activity to a distance about 0.32–0.64 cm (1/8 to 1/4 inch) wide on either side of the magnesium–cathode interface. The magnesium alloy can still suffer severely, however, if the cathode does not polarize sufficiently to reduce or eliminate the effective potential difference ($E_k - E_a$) [40, 41].

10.5. HYDROGEN OVERPOTENTIALS

The degree to which the corrosion of magnesium is accelerated by the galvanic couple in a given environment (i.e., a given R_e) depends also in part on the polarization that reduces the electromotive force of the couple as the galvanic current develops. Because magnesium shows little, if any, anodic polarization in saltwater exposures, the reduction of the electromotive force of the galvanic cell typically results from polarization of the cathode, where water is reduced to hydrogen gas and hydroxyl ions. Some metals, such as iron, nickel, and copper, serve as efficient cathodes in what is thought to be the stepwise process of accepting and reducing the hydrogen ion to an atomic form (H), where it then combines to form the evolved hydrogen gas (H_2). It has been suggested that the intermediate electrochemical reaction controls the kinetics of the cathodic reaction and the whole corrosion rate: $2H^+ + 2e + H$ (adsorbed ion) $= H_2$. These metals have a low hydrogen overvoltage and can consequently cause severe galvanic corrosion of magnesium. Other metals, such as aluminum, zinc, cadmium, and tin, while equally cathodic to magnesium in some environments, serve as much less effective cathodes due to their tendency to inhibit the combination of atomic hydrogen on surfaces to form the hydrogen gas that evolves [42].

Equal areas of various cathodic materials and a magnesium alloy (Mg–6%Al–3%Zn–0.2%Mn alloy) were galvanically connected to steel, Cd-plated steel, Ni, Zn, Cu, brass, aluminum alloys 2024, 1100, and 5056, as well as Mg–1.5%Mn. These were tested by continuous immersion in 3% NaCl, in Midland tap water containing approximately 70 ppm chloride, and in distilled water. The duration of the test was 3 hours in 3% NaCl, 24 hours in Midland tap water, and 4 days in distilled water. All commonly used materials cause galvanic corrosion of magnesium in a strong chloride electrolyte [41, 43].

The relative areas of the magnesium anode and the dissimilar metal cathode have an important effect on the galvanic corrosion damage that occurs. A large cathode coupled to a small area of magnesium results in rapid penetration of the magnesium, because the galvanic current density at the small magnesium anode is very high, and anodic polarization in chloride solutions is very limited. Painted magnesium should not be coupled to an active cathodic metal if the couple will be exposed to saline or aggressive environments. A small break in the coating at the junction results in a high concentration of galvanic current at that point. Unfavorable area effects can also be seen in the behavior of some proprietary coatings using aluminum or zinc powder [41, 44].

In salt-spray tests using cast iron disks coupled to AZ91 die cast plates and separated by plastic spacers, it was found that a separation of about 4.45 mm (175 mils) was needed to ensure the absence of galvanic corrosion [41]. The salt-spray test is thought to produce a result biased against zinc due to the rapid cathodic attack on the zinc electroplate produced in this severe test exposure. This attack does not occur in many natural environments. The most compatible fastener coatings are based on zinc plating, with modifications to extend the life of the zinc. These modifications include chromating, silicate treatments, and alloying with tin [41, 45].

Figure 10.9 [11] indicates the relative severity of galvanic corrosion of die cast AZ91D caused by coupling with various dissimilar metals in salt spray (ASTM B117). These ratings provide appropriate guidelines for selection although the salt-spray test is severe and does not reflect real environmental conditions where one can observe the active–passive behavior of the alloy due to pH evolution to alkaline values in the absence of strong agitation [17, 46].

10.6. GALVANIC CORROSION OF PURE AND ALLOYED MAGNESIUM

In a highly conducting medium, such as 3% NaCl, most metals or intermetallic phases, cathodic to magnesium, will not polarize to the magnesium potential until a relatively high current density is reached, except the Al–5%Mg rivet alloy that normally polarizes at a very low current density. Aluminum alloys containing small percentages of copper (7000 and 2000 series and 380 die-cast alloy) may cause serious galvanic corrosion of magnesium in saline environments. Very pure aluminum is quite compatible, acting as a polarizable cathode; but when iron content exceeds 200 ppm, cathodic activity becomes significant (apparently because of the depolarizing effect of the intermetallic compound $FeAl_3$), and galvanic attack of magnesium increases rapidly with increasing iron content. The effect of iron is diminished by the presence of magnesium in the alloy. This agrees with the relatively compatible behavior of aluminum alloys 5052, 5056, and 6061 shown in the spray-salt fog test.

Data on galvanic corrosion of magnesium alloys were compiled in tests at Kure Beach, North Carolina, in which sheets of dissimilar metals were fastened to panels of AZ31B and AZ61A. Zinc, cadmium, or tin plating on steel all reduce galvanic attack of magnesium substantially when compared to that produced by uncoated steel. An inorganic chromate treatment on cadmium electroplate (and perhaps on other electroplates, and on metal

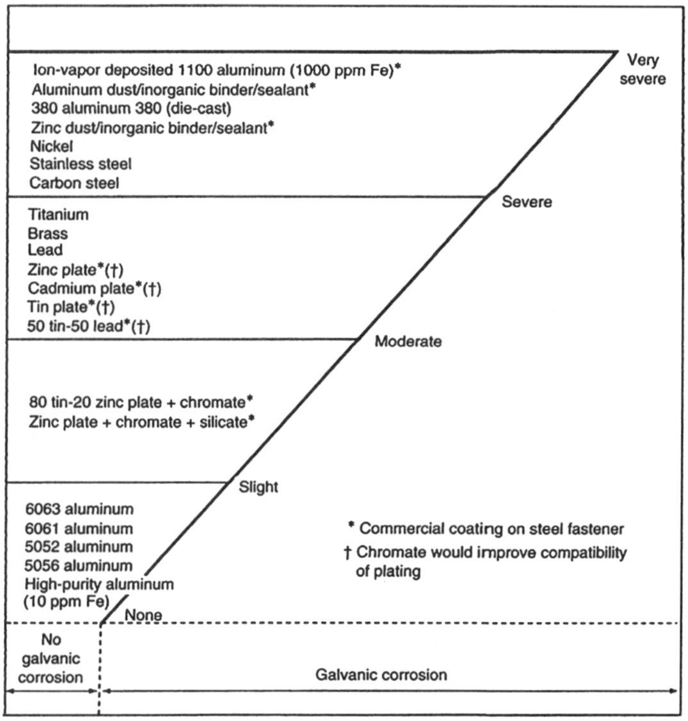

Figure 10.9 Relative galvanic corrosion produced by dissimilar fasteners attached to Mg alloy AZ91D (ASTM B117 salt-spray test) [11, 17, 46].

surfaces) was as effective in reducing the galvanic attack on magnesium as an epoxy coating. This observation is consistent with the known inhibitive effect of chromates on the cathodic reduction process [41, 43]. Certain zinc- and aluminum-filled polymer coatings on steel actually produced more damage to the magnesium than bare steel. This effect may be due to either an increase in the active cathode surface area resulting from the fine metal powders or flake employed in the coatings, or it may be due to the presence of a catalytic contaminant on the metal powder surface, such as iron.

Under conditions where the corrosion product is not continuously removed or under conditions of high cathodic current density, where the surroundings may become strongly alkaline, both the magnesium and an amphoteric contacting metal such as aluminum may suffer severe attack. Aluminum alloys containing appreciable magnesium, such as 5052, 6053, and 5056, are least severely attacked in chloride media when galvanically coupled. This fact was observed in galvanic couples of magnesium and aluminum alloys exposed to tide water and in the atmosphere at Hampton Roads, Virginia.

10.6.1. Cathodic Corrosion of Aluminum

Aluminum can be attacked by the strong alkali generated at the cathode when magnesium corrodes sacrificially in static NaCl solutions. Such attack destroys compatibility in alloys containing significant iron contamination, apparently by exposing fresh, cathodic active sites with low overvoltage. The aluminum alloys having substantial magnesium content

(5052 and 5056) are more resistant to this effect, but not completely so. A 5052 alloy would meet the essential requirement for a fully compatible aluminum alloy with a maximum of 200 ppm Fe or a 5056 alloy with a maximum of 1000 ppm Fe [41, 44]. Cathodic corrosion of aluminum is much less severe in seawater than in NaCl solution, because the buffering effect of magnesium ions reduces the equilibrium pH from 10.5 to about 8.8. The compatibility of aluminum with magnesium is, accordingly, better in seawater and is less sensitive to iron content [41].

Aluminum oxide is amphoteric (i.e., soluble in alkaline as well as acid solutions). The standard potentials of these two half-reactions are.

$$\text{Acid:} \qquad Al^{3+} + 3e^- = Al \quad (-1.66\ V/SHE) \tag{10.4}$$

$$\text{Alkaline:} \quad H_2AlO_3^- + H_2O + 3e^- = Al + 4OH^- \quad (-2.35\ V/SHE) \tag{10.5}$$

Reaction 10.5 has very close potential to the standard reduction potential of the Mg^{2+} half-reaction (−2.37 V). Commercial aluminum alloys contain several thousand parts per million of iron in the form of the intermetallic $FeAl_3$. The mutually destructive galvanic action between magnesium and commercial aluminum alloys in saltwater proceeds as follows:

1. Rise in the pH of the liquid in contact with the aluminum member. This is most likely the result of galvanic current flow between the magnesium and the initially passive aluminum.
2. Shift of the aluminum potential in the active direction in accordance with half-reaction (10.5).
3. Exposure of iron aluminum intermetallic particles (e.g., $FeAl_3$), which then engage in separate galvanic activity with the magnesium. This galvanic current flow accounts for the severe sacrificial corrosion of the magnesium, and the alkali generated at the cathode ensures continued corrosion of the aluminum in accordance with half reaction (10.5) [45].

10.6.2. Cathodic Damage to Coatings

Hydrogen evolution and strong alkalinity generated at the cathode can damage or destroy organic coatings applied to fasteners or other accessories coupled to magnesium. Alkali-resistant resins are necessary, but under severe conditions, such as salt spray or salt immersion, which do not simulate adequately a real application, the coatings may simply be blown off by hydrogen, starting at small voids or pores.

10.7. COMPOSITE COAT FOR MOLTEN MAGNESIUM

The corrosion behavior of silicon nitride bonding silicon carbide (Si_3N_4/SiC) composites in molten magnesium and AZ91 magnesium alloy were investigated through immersion tests. The microstructure and the component of the surface layer of the composites were characterized by scanning electron microscopy (SEM), energy dispersive spectroscopy (EDS), and X-ray diffraction (XRD). The corrosion attack of liquid magnesium or AZ91 magnesium alloy on Si_3N_4/SiC composite mainly took place at the initial stage of immersion. A reaction layer formed on the composite surface at the start, followed by a better corrosion resistance to liquid magnesium. After about 7 h, there was no further reaction with liquid magnesium because of the formed barrier layer. This composite can be considered as a coating material for handling the molten magnesium [47].

10.8. METAL MATRIX COMPOSITE GALVANIC CORROSION

Hihara and Kondepudi [48, 49] investigated the galvanic corrosion behavior of Mg MMCs of pure Mg and ZE41A–Mg alloy in contact with SiC monofilament (MF) or pure SiC particles. The results showed evidence of higher galvanic corrosion rates of the matrices in both cases; however, the ZE41A–Mg matrix showed a better corrosion resistance than pure Mg. Hall [50] observed the evidence of galvanic corrosion for carbon fiber/magnesium MMCs in a normal laboratory atmosphere of about 60% relative humidity at 20 °C. The determined rate of penetration was about 100 μm/yr. The magnesium matrix contained 1 wt % Al and Al carbides were formed at the fiber–matrix interface during squeeze casting but they had no significant effect on the atmospheric corrosion behavior.

Bakkar and Neubert [51] studied the corrosion behavior of carbon fiber reinforced magnesium metal matrix composite (MMC) with emphasis on the galvanic corrosion arising between the magnesium matrix alloy and the carbon fibers. Two magnesium alloys were used in this study—AS41 and AS41 (~0.5 Ca). The Mg alloys were reinforced with carbon short fibers (Sigrafil C-40) using the squeeze casting technique, in which the Mg alloy was forced into the carbon short performs of fibers. The carbon fibers (about 25 vol %) were distributed quasi-isotypically in the horizontal plane [51].

The corrosion behavior was examined in both neutral and alkaline aqueous solutions containing different concentrations of NaCl, using electrochemical techniques and the hydrogen evolution test. Two matrix alloys were used—AS41 and AS41 (0.5% Ca). Alkaline solution containing 100 ppm NaCl at pH 12 was considered by NaOH addition as well as neutral NaCl-containing solutions (pH ~7). The specimen was exposed to the NaCl solution for 5 minutes at the open circuit potential (OCP) prior to potentiodynamic polarization. The polarization was obtained by scanning 500 mV more negative than the OCP at a rate of 20 mV/min. Experiments were performed in an Avesta cell [51, 52].

The galvanic coupling with C fibers leads to severe corrosion of Mg matrix composites and invalidates the virtual effect of alloying elements on corrosion resistance. Applying electrochemical polarization on C/Mg MMCs creates crevice attack at perimeters of C fibers and hence leads to crevice corrosion at the C–Mg interface. As a result, the MMC shows high corrosion rates, regardless of the corrosion resistance of the Mg matrix alloy. In alkaline solutions, where Mg is passive, the corrosion potential (E_{corr}) of C/Mg MMCs is more noble than their monolithic matrix alloys, implying the eventual galvanic effect of C fibers by virtue of their high electrical conductivity. The two C/Mg MMCs studied monitor similar corrosion rates to be about 1.5 times that of the lower corrosion resistance Mg matrix alloy and 3 times that of the higher corrosion resistance one [51].

10.9. PREVENTION OF GALVANIC CORROSION

For indoor use, where condensation is not likely, no protection is necessary. Even in some sheltered outdoor environments, unprotected magnesium components can give good service life providing there are no water traps, and there is good ventilation, warm component temperature, or the presence of an oil film. Under corrosive conditions, the use of high-purity magnesium alloys will not reduce the effects of galvanic corrosion significantly. Magnesium is relatively insensitive to oxygen concentration [40, 41].

The new high-purity alloys, such as AZ91D, AZ91E, and AM60B, offer no defense against galvanic corrosion attack. Their improved performance in assemblies can only be realized if proper measures are taken to control the potential for galvanic attack through careful design, selection of compatible materials, and the selective use of coatings, sealants, and insulating materials. Alloying elements can form secondary particles that are noble to the Mg matrix, thereby facilitating corrosion, or enriching the corrosion product, thereby possibly inhibiting the corrosion rate (see Chapter 11, Section A).

Magnesium faying or mating surfaces should be assembled using "wet assembly" techniques. Inhibited primers or sealing compounds are placed between the surfaces at the time of assembly. Iron-based and to a lesser extent zinc-based, phosphate treatments are recommended as inhibitors to replace chromates. Sealing compounds, such as nonacidic silicone RTVs, polysulfide, epoxy resins, or plastic tapes, can be employed. Sealing/jointing compounds of the polymerizing or nonpolymerizing type are preferred as they will remain flexible and resist cracking. If possible, the compound or tape should extend beyond the joint interface by 3.2–9.5 mm. Polymerizing-type compounds are also used for caulking operations. For additional protection, mating surfaces can be primed prior to assembly and overpainted after assembly [41].

10.9.1. Joining Magnesium to Dissimilar Metal Assemblies

Good design can play a vital role in reducing galvanic corrosion. The elimination of a common electrolyte may be possible by the provision of a simple drain or shield to prevent liquid entrapment at the dissimilar metal junction. Alternatively, the location of screws or bolts on raised bosses may also help avoid common electrolyte contact, as would the use of nylon washers, spacers, or similar moisture-impermeable gaskets. The use of studs in place of bolts, provided the captive ends of the studs are located in blind holes, will reduce the area of dissimilar metal exposed by up to 50%.

The use of wet assembly techniques will eliminate galvanic corrosion crevices. Caulking the metal junctions will increase the electrical resistance (R_e) of the galvanic couple by lengthening the electrolytic path and so reduce the degree of attack should it occur. Vinyl tapes have also been used to separate magnesium from dissimilar metals or a common electrolyte and so prevent galvanic attack. Finally, overpainting the magnesium and more importantly the dissimilar metal after assembly will effectively insulate the two materials externally from any common electrolyte [40, 41].

Contacting components, fasteners, and inserts should be chosen for their compatibility; for example, a nonconductive, nonporous material; 5000 or 6000 series aluminum alloys; or tin- cadmium-, or zinc-plated ferrous alloys. The compatibility of plated fasteners can be further improved by the use of aluminum washers, organic coatings, or other inhibiting films.

Dissimilar metals that are compatible with magnesium are the aluminum–magnesium (5000 series) or aluminum–magnesium–silicon (6000 series) alloys, which should be used for washers, shims, fasteners (rivets and special bolts), and structural members where possible. Aluminum, zinc, cadmium, and tin are used to coat steel or brass components in order to reduce the galvanic couple with magnesium, under mild corrosive environments, but will have minimal effect in corrosive conditions, where additional precautions are required.

If painting is to be employed on only one of the contacting components, paint the cathodic material to avoid small anodic big cathodic relative areas. Painting both components is a better practice. Paints employed on cathodic components and complete magnesium assemblies should be chosen for resistance to alkalis in order to prevent stripping of the coating [40, 41].

10.9.2. Joining Magnesium to Nonmetallic Assemblies

Wood has a tendency to absorb water and then leach out natural acids that can attack magnesium. The wood, as well as magnesium–carbon fiber reinforced plastics, in the presence of a common electrolyte should first be sealed with paint or varnish and the faying surface of the magnesium should be treated as magnesium-to-magnesium assemblies [40].

C. LOCALIZED CORROSION

For immersed conditions, including corrosion under pools of condensate, attack may be, and usually is, irregular. Some areas become anodic to other and, as corrosion proceeds at the anodic areas, pitting develops. The unequal attack, which occurs in tap water, condensate, and other mild electrolytes, may lead to perforations of thin-gauge sheet and even to deep pitting of castings. In chloride solutions, such as seawater, attack on the metal usually results in pitting of some areas only; for reactive metallic surfaces, attack (e.g., sand blasting) may be so rapid that uniform dissolution is observed [32].

General corrosion can lead to localized corrosion, which is favored by a weak conductivity of the electrolyte and small anode/large cathode relative area ratios. Localized attack takes the form of pitting, filiform, and crevice corrosion. The corrosion of Mg–Al alloys in NaCl solutions is characterized by pit initiation and filiform corrosion, which develops into cellular corrosion.

10.10. PITTING CORROSION

Pitting, crevice corrosion, and filiform corrosion have been observed. Magnesium is relatively insensible to oxygen concentration. When corrosion occurs on a smooth machined magnesium alloy surface, this surface is roughened by the chemical action, and after the initial attack the degree of roughness does not change appreciably. In atmospheric attack, the roughening is really a microscopic form of pitting. In the usual industrial atmospheric conditions, the attack can correspond to uniform corrosion. There is a noticeable difference between the appearance of the aluminum-containing magnesium-rich alloys and the zinc/zirconium-containing magnesium alloys. In the former, the microscopic pits in the surface exposed to the weather tend to be narrow and relatively deep, whereas in the latter they are wider and tend to overlap, leading to a slightly wavy appearance [39, 41, 44, 53].

In immersed stagnant conditions, attack may be, and usually is, irregular. Galvanic cells develop and some areas become anodic to other areas, and as corrosion proceeds at the anodic areas, pitting develops. The unequal attack, which occurs in tap water, condensate, and other mild electrolytes, may lead to perforations of thin-gauge sheet and even to deep pitting of castings. In chloride solutions, such as seawater, attack on the metal usually results in pitting of some areas only; for reactive metallic surfaces attack (e.g., sand blasting) may be so rapid that uniform dissolution is observed [39, 44]

Stable corrosion pits initiate at flaws adjacent to a fraction of the intermetallic particles present as a result of the breakdown of passivity, and the resultant surface is very porous. Filiform corrosion of Mg–Al alloys in NaCl solutions, parallel to pit initiation, has frequently been observed. There is no evidence of initiation at particle-free areas, indicating that hydrogen evolution on the particle is the predominant controlling cathodic reaction (galvanic corrosion). Pitting and crevice corrosion are usually associated with the breakdown of passivity. Filiform corrosion was observed at critical concentrations before that of pitting. Pitting can also occur in nonpassivating alloys with protective coatings or in certain heterogeneous corrosive media [39, 41, 53, 54].

The AZ, AS, and AM type alloys maintain a bright and shiny appearance in the unattacked part of the corroded surface, whereas the AE alloys tend to become dull due to buildup of a relatively thick hydroxide film and formation of numerous small pits, only a few micrometers in depth. The corresponding backscattered electron image (BEI) and X-ray maps indicate a high chloride concentration in the pits and high aluminum concentration in the unpitted areas. The few studies of pitting of Mg and Mg alloys have been concerned with comparing the pitting behavior of cast to that of rapidly solidified Mg alloys. In these studies, two parameters indicative of pitting resistance were measured: (1) i_p, the passive current density, which is a measure of the protective quality of the passive film; and (2) E_b, the breakdown potential, which indicates the resistance to the breakdown of the passive film that results in pitting attack. The more positive the value of E_b, the more protective the film on the metal surface [41, 55].

10.10.1. The Pitting Potential Determination

Different possible solutions have been suggested for pitting investigations, most of them containing chloride ions at different concentrations and sometimes at alkaline pH for passive behavior or at buffered pH solutions (see Chapter 18, Section 18.1). It should be observed that pitting potentials can be nobler than the OCP. Magnesium alloy is apt to passivation as a function of the alkaline solution at pH $\geq$ 9.5 and the chloride ion concentration. During the active–passive behavior corrosion studies, two parameters indicative of pitting resistance are measured: i_p and E_b. The more positive the value of E_b, the more protective the film on the metal surface.

Cast and Rapidly Solidified AZ61 Alloy Makar and Kruger [19] showed that rapidly solidified AZ61 (Mg–6A1–1Zn) exhibited a breakdown potential that was around 200 mV higher than that of cast AZ61 in a buffered carbonate solution (pH 10) containing various levels of Cl^- (Figure 10.10). In a buffered borate solution (pH 9.2) containing various levels of Cl^-, there was no improvement in the E_b values observed for the rapidly solidified alloy. However, the pits formed at 1 V below E_b were hemispherical, apparently forming at defects in the black film that is observed when the cast AZ61 surface is at -1.5 V (SCE). No small hemispherical pits were found on the rapidly solidified AZ61 [41, 56]. It is interesting to note that the value of potential of the different microstructures of specimens in the borate buffered alkaline solution has an important influence on the morphology of the pits of this alloy.

Sr as Alloying Element Pitting corrosion potential of different magnesium alloys, AZ91D, AZ91E, ZA104, ZAC10403 (0.3Ca), and ZACS1040305 (0.3Ca and 0.5Sr), was studied in different aqueous corrosive media. The addition of strontium in

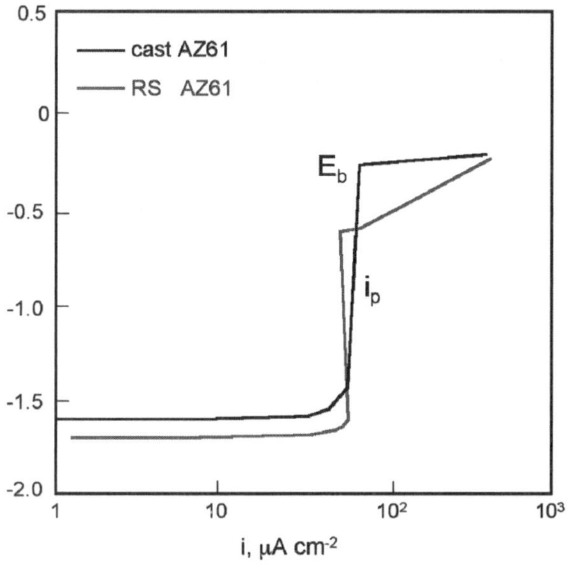

Figure 10.10 Anodic polarization scans for cast and rapidly solidified AZ61 (Mg–6Al–1Zn) in pH 10 sodium carbonate–sodium bicarbonate solution with 100 ppm NaCl [19, 41].

ZAC alloys shows a more beneficial effect on corrosion resistance against pitting. Generally, agitation of the solution was also favorable for prevention of pitting corrosion in magnesium alloys [57].

AZ91D Alloy and the α and β Phases Song et al. [32, 58] showed distinctive polarization curves and pitting potentials for α and β phases in 1 N NaCl solution. This shows the influence of microstructure on pitting that can be related to SCC [59]. Effectively, the pitting potential of the α phase was more negative (~15 mV) than the open circuit potential, while the pitting potential was more than 200 mV more noble (positive) than the corrosion potential for β phase (Figure 10.11). It is important to note that the corrosion rate at the open circuit potential was much higher for α phase than for β phase. It seems also that the general corrosion and pitting corrosion of magnesium α phase are vigorous at corrosion potential level and can be accelerated by superimposing the galvanic effect of the cathodic

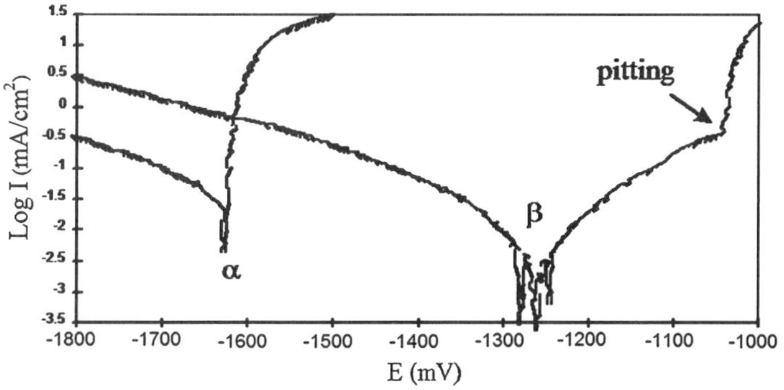

Figure 10.11 Polarization curves for the Mg α phase and the Al–Mg β phase in 1 N NaCl at pH 11 [32, 60].

β phase unless the nobler β phase can create an efficient protective barrier to isolate the surface from the aggressive medium [60].

Song et al. [61] examined the pitting potential that can correspond, under certain conditions, not only to pitting but also to filiform corrosion or as in strong chloride-containing solution (5%) at pH 11. The OCP or corrosion potential of the alloy with its $\alpha + \beta$ phases of the alloy was ~ -1.56 mV/SCE and is more positive (noble) than the "pitting" corrosion (10–15 mV). The OCP potential of the α phase alone was also 15 mV more positive than the pitting potential of this phase as can be deduced from Figure 10.11. It is important to state that this pitting potential, especially for the separate α phase, does not correspond to the conventional pitting potential that is determined due to the attack of a stable passive protective layer, as in the case of stainless steels or aluminum alloys, where the oxygen differential electrochemical cell and the autocatalytic mechanism of propagation are causing profound pits. The morphology of localized corrosion under these special conditions (negative to OCP and passive zone of the alloy) shows irregular shallow pits and in certain cases a specific type of filiform corrosion that starts with or before pitting. This filiform corrosion is not largely influenced by the oxygen differential cell [61].

10.10.2. Polarization Curves and Pitting Potential of AXJ Alloy

The AXJ530 alloy was developed to offer good creep resistance and contains mainly 4.4% Al, 2.6% Ca, 0.15% Sr, and 0.3% Mn. Specimens taken from the side wall (6.2 mm) from experimental box-like parts of AXJ530 alloy were prepared using three different routes. The first set of boxes was conventionally die cast (DC) from remelted billets. The second and third sets of boxes were thixocast (at 592 °C) either from semisolid billets solidified freely in a steel mold (SFTC), or from semisolid billets electromagnetically stirred during solidification in a copper mold (ESTC). The solution used for corrosion tests contains 0.05 M NaCl, 0.1 M NaOH, and 0.025 M H_2O_2 at a pH of 12.3. The hydrogen peroxide was added to promote good reproducibility of the corrosion potentials. All tests were performed at room temperature (25 °C) under atmospheric oxygen and without agitation. Potentiodynamic polarization tests were carried out at a scan rate of $10\ mV \cdot min^{-1}$. The starting potential was 250 mV more negative than the corrosion potential (E_{corr}). All potentials were referred to the Ag/AgCl electrode potential [62].

Figure 10.12 shows the polarization curves of AXJ530 specimens in alkaline solution. All specimens exhibit a self-passivation, which indicates that a protective passive film forms immediately after immersion. The polarization curves of all the AXJ530 specimens showed similar cathodic regions, which means that the same cathodic reactions occurred. The "current plateau" in the passive region covers a more important range of potentials for the two thixocast specimens than for die-cast ones. Also, the thixocast AXJ530 specimens displayed a relatively lower passivity current (i_p) than that of the die-cast type of specimens [62].

The pitting corrosion potential deduced from these curves does not correspond only to pitting, as for the conventional passive alloys of stainless steel or aluminum, but to pitting and/or filiform-like corrosion types of the localized corrosion form. The pitting potential of die-cast specimens was easier to determine since it was marked by an abrupt increase of the anodic current. When the pitting potential of die-cast specimens was reached, strong hydrogen evolution was clearly seen from the site of the localized attack. Thixocast specimens (AXJ530-SFTC and AXJ530-ESTC) do not show well-defined pitting potential, which usually marks the irreversible breakdown of the passive film. In this case, estimated

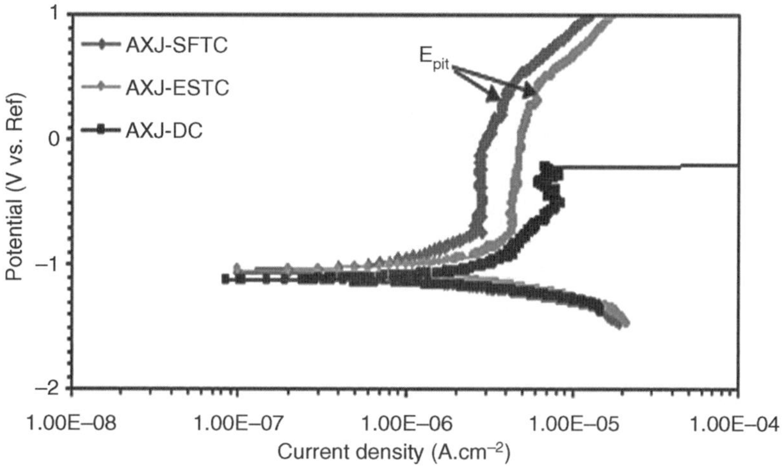

Figure 10.12 Polarization curves for AXJ530 specimens immersed in alkaline NaCl solution at pH 12.3 and 25 °C [62].

pitting potentials were determined as the potentials for which a slight but visible hydrogen evolution was observed; these points (E_{pit}) are indicated on each polarization curve in Figure 10.12 [62].

Most of AXJ530 specimens did not show visually clear signs of corrosion attack. The magnification of some regions of the surface of the AXJ530 alloys displayed numerous and short cracks. These cracks indicate metastable pits, that is, pits that passivated shortly after their initiation [63]. This behavior is confirmed by the weak current and potential noise fluctuations previously observed during immersion. Numerous and short cracks are found after 16 h of immersion and linked to the initiation and the repassivation of metastable pits. The density of these cracks, determined by quantitative metallography, is much higher for the die-cast specimens than for the two thixocast ones. It can be concluded that the resistance to pitting of the passive film of AXJ530-ESTC is the best among all the AXJ530 specimens for the examined conditions (Figure 10.13) [62].

Figure 10.13 Surface aspect of AXJ530-DC after 16 h of immersion in alkaline solution at pH 12.3 and 25 °C at two different magnifications [62].

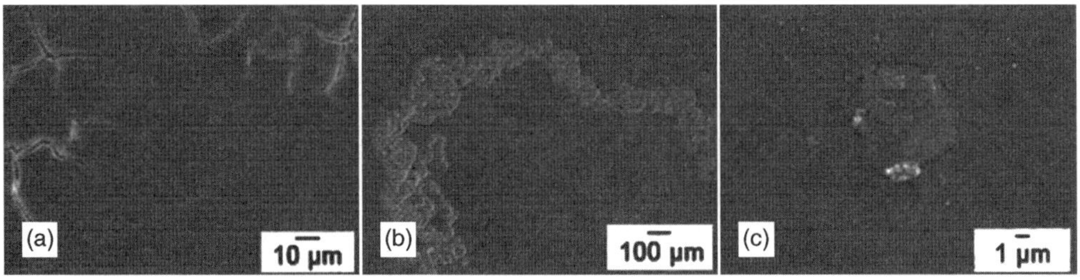

Figure 10.14 (a) Typical filiform-like corrosion of AXJ530 specimen; (b) aspect of a filament after cleaning with chromic acid; and (c) pitting corrosion initiating near an Al–Mn particle [62].

The observation of the three specimens of AXJ530 alloy in the alkaline solution (0.05 M NaCl, 0.1 M NaOH, and 0.025 M H_2O_2 at pH 12.3) showed also that some AXJ530 specimens sustain a type of corrosion similar or close to filiform corrosion, suggesting that highly resistant films can be naturally formed in this solution. The appearance of corrosion products was in the form of filaments, which were covered by a ruptured film (Figure 10.14a). After cleaning in a boiling chromic acid solution, the localized attack appears to be superficial and no pits were found inside the filaments (Figure 10.14b). Figure 10.14c shows the influence of the cathodic phase of an Al–Mn particle in pit initiation. The initiation of these filaments seems to occur mainly near the Mg–Al particles (i.e., simulated by a galvanic cell), while their propagation was found to be in random directions [62].

10.11. CREVICE CORROSION

Corrosion in crevices between Magnox A (Mg–0.18Al) and mild steel, and between Magnox A and PTFE, occurred in 200 g·m^{-3} NaOH (pH > 11.5) when the Cl^- concentration was about 1 g·m^{-3} or more [3]. Crevice corrosion is a well known type of corrosion that occurs at narrow gaps ("crevices"); however, it is somewhat different in mechanism for magnesium.

Generally, crevice corrosion is caused by the development of an anodic region within the crevice because of the exclusion of oxygen and a cathode region outside the crevice where the oxygen concentration is high. Oxygen differential cells could be established between cathode surfaces exposed, for example, to oxygenated seawater and anodic crevice areas, but several authors confirm that corrosion of magnesium is relatively insensitive to oxygen concentration differences. However, some authors consider the conventional mechanism of the oxygen differential cell for filiform corrosion and normally this can be extrapolated to crevice corrosion [39]. This approach can be considered where oxygen can accelerate the cathodic reaction at relatively less negative or more noble open circuit potentials; its influence can be admitted at least partially for certain alloys.

In crevices, where the differential aeration cell of oxygen does not play an essential role in corrosion, two other factors could initiate this type of crevice corrosion:

1. Hydrolysis reactions within crevices could produce changes in pH and chloride concentration in the crevice environment. It is very probable that crevice corrosion can be initiated because of the hydrolysis reaction. The formation of magnesium hydroxide should influence the properties of the magnesium–solution interface in the crevice.

2. The retention of moisture (which is unable to evaporate) in the crevice promotes the corrosion of the metal in the narrow recess over extended periods [39].

These two mechanisms for crevice corrosion can be extrapolated to filiform corrosion.

10.12. FILIFORM CORROSION

Filiform corrosion is typically associated with metal surfaces having an applied protective coating. Its occurrence on bare Mg–Al alloys indicates that highly resistant oxide films can be formed naturally. Filiform corrosion does not occur on bare pure Mg, indicating the strong influence of alloying elements on corrosion products and behavior. Extruded magnesium alloys with 3–8% Al and 0.5–0.8% Zn are susceptible to filiform corrosion and pitting corrosion in aqueous chloride solutions, depending on the chloride concentration [64, 65].

The overall variables of significance are temperature, material structure, and polarization of the microgalvanic cell. Figure 10.15 shows the mechanism and products of the filiform corrosion cell of magnesium [39, 66].

After the initiation period of corrosion pits, filiform corrosion dominates the morphology as narrow semicylindrical corrosion filaments project from the pit. Radial propagation is at a much slower rate than that of the filament tips projecting outward. Lunder et al. [66] observed that propagation of the filaments occurs with voluminous gas evolution at the head while the body immediately behind passivates. Electrochemical transport of chloride ions to the head of the filament appears to be an essential component, as is precipitation of insoluble $Mg(OH)_2$ by the anodic reaction with Mg^{2+} ions elsewhere along the filament. The corrosion products may vary because they depend on the environment [39].

Filiform corrosion initiates and then develops into cellular or pitting corrosion. Cellular corrosion occurs when a primary initiation site and secondary pits, formed along the filiform corrosion filaments, coalesce to form a corrosion cell with an epicenter at about the original pit initiation site. Growth proceeds at a steady radial rate independent of the material temper. Cellular corrosion continues until the cells impinge on one another, at which point they terminate, thereby forming clearly defined cell boundaries [39, 41, 66].

10.12.1. Initiation and Kinetics Parameters

Anodic polarization enhances filiform corrosion at the expense of the pitting process. Filiform corrosion is commonly observed and tends to occur at lower chloride concentration

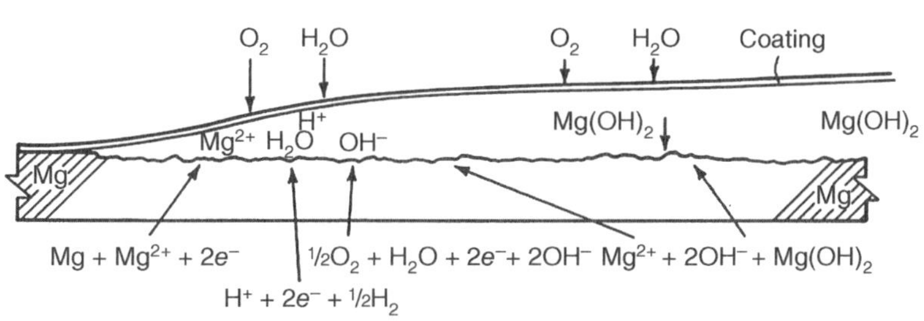

Figure 10.15 Diagram of the filiform corrosion cell in magnesium. Corrosion products and predominant reactions are identified. Filiform corrosion is a differential aeration cell driven by differences in oxygen concentration between the head and tail sections about 0.1–0.2 V [41, 66].

than pitting. The critical chloride concentration for initiation of localized corrosion (filiform and pits) was less than 0.05 M for several tested alloys. Weight loss increased with increased chloride concentration. Increases in temperature from 25 to 50 °C, did show a minor effect in promoting the initiation of localized corrosion. Magnesium without intentional alloying additions showed exfoliation in which individual grains were preferentially attacked along crystallographic planes [67].

10.12.2. Mechanism of Propagation

Filiform corrosion is thought to be a special form of tunneling, which appears to be the forerunner of regular pitting. Filiform propagation is characterized by unusually high rates under high anodic control at the surface. Morphology and directionality of filaments are determined by the material microstructure, such as compositional and crystallographic factors. Therefore the rate of filament propagation is independent of material temper, surface treatment, and presence of oxygen in the environment. It is controlled by mass transfer limitations resulting from the formation of a salt film at the filament tip [66].

Filiform corrosion tests were performed both in unstirred solutions and by exposure in a flow channel equipped with an optical cell for in situ microscopic observation of the corroding surface. These tests consisted of exposing the specimens to NaCl solutions of various concentrations (5% NaCl in most cases) at room temperature. Both deoxygenated solutions and solutions exposed to ambient conditions were employed [66, 68].

Filiform corrosion was observed in uncoated AZ31, while general corrosion mainly occurred in deposition coated AZ31, which seems to be suppressed as the immersion test proceeds. Morphologies and compositions of corrosion products formed on the uncoated and deposition coated AZ31 alloy are different from each other, which is believed to lead to the difference in corrosion behavior [68, 69].

In the as-cast condition, compositional variations orient the growth of filiform corrosion. In homogenized alloys, filiform corrosion propagates transgranularly along crystallographic directions. In Mg–Al alloys, precipitation heat treatment disperses the secondary $Mg_{17}Al_{12}$ precipitate, which blocks transgranular propagation of filiform corrosion, thereby reducing the corrosion rate [66].

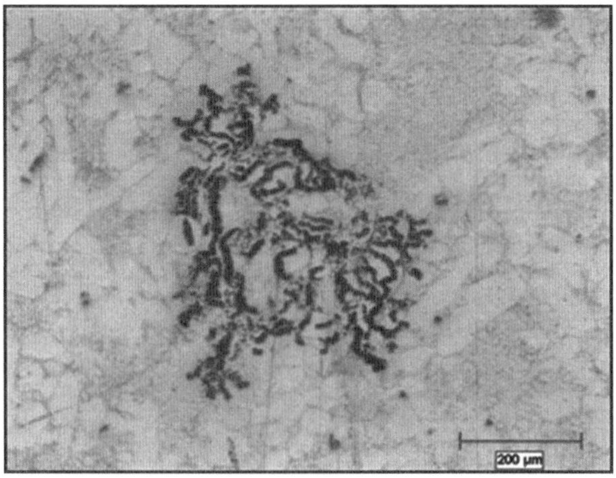

Figure 10.16 Filiform corrosion observed for AXJ530 thixocast in 0.05 M NaCl after anodic polarization [70].

Filiform corrosion observed on AZ91 is quite different in nature from the well-known mechanism of filiform corrosion and occurs on an uncoated surface with unusually high filament propagation velocities, very possibly due to the presence of a highly resistant air-formed oxide on the AZ91 alloy. Filiform propagation does not require the presence of dissolved oxygen in the environment; it is essentially fueled by hydrogen evolution occurring at the filament head and outside the filaments [66].

Anodic Polarization and Filiform Corrosion Experiments have shown that thixocast alloy shows a lower corrosion rate than that of die-cast alloy. The corrosion behavior of high-pressure die-cast and thixocast AXJ530 magnesium alloys has been investigated using polarization curves. Two solutions have been used, 0.05 M NaCl at pH 6.8 and 0.05 M NaCl + 0.1 M NaOH and 0.025 M H_2O_2 at pH 12.3. The alkaline medium seems to be an appropriate one to compare the active–passive behavior of magnesium alloys. Two types of localized corrosion have clearly been identified: intergranular pitting and filiform pitting. This latter results in dense pitted areas for the transition from active to passive behavior and is shown in Figure 10.16 [70].

REFERENCES

1. J. D. Hanawalt, C. E. Nelson, and J. A. Peloubet, *Transactions of the American Society of Mining and Metallurgical Engineering* **147**, 273–299 (1942).
2. A. Shaw, in *ASM Handbook, Volume 13A, Corrosion*, edited by S. D. Cramer and B. S. Covino Jr. ASM International, Materials Park, OH, 2003, pp. 692–696.
3. C. H. Baloun, in *ASM Metals Handbook*, Vol. 13, 9th edition, ASM International, Materials Park, OH, 1987, pp. 207–208.
4. E. Ghali, W. Dietzel, and K. U. Kainer, *Journal of Materials Engineering and Performance* **13**(1), 7–23 (2004).
5. T. Troβmann, K. Eppel, M. Gugau, and C. Berger, Investigation of the passivation behavior of magnesium alloys by means of cyclic current–potential curves, in *7th International Conference on Magnesium Alloys and Their Applications,* Weinheim, Germany, edited by K.-U. Kainer, Wiley-VCH, Weinheim, Germany, 2006, pp. 802–808.
6. C.-A. Loong, C. Q. Zheng, and J. D. Hwang, in *8th International Conference on Semi-solid Processing of Alloys and Composites (S2P)*. Limassol, Cyprus, Worcester Polytechnic Institute, Worcester, MA, 2004, Paper 19-03.
7. T. Troβmann, Verhalten von magnesium-Legierungen bei korrosiver und mechanisch korrosiver Beanspruchung, Dissertation, Darmstadt University of Technology "TU", Darmstadt, Germany, 2005.
8. S. Werdin, T. Troβmann, M. Gugau, and K. L. Kotte, *Materials Science and Engineering A Techniques* **36**, 659–668 (2005).
9. C. Berger, K. Eppel, J. Ellermeier, T. Troβmann, U. Dilthey, H. Masny, and K. Woeste, in *7th International Conference on Magnesium Alloys and Their Applications*, Weinheim, Germany, edited by K.-U Kainer. Wiley-VCH, Weinheim, Germany, 2006, pp. 734–742.
10. D. Sivaraj, R. McCune, and P. K. Mallick, Mohanty, in 2006 SAE World Congress, SAE Paper No. 2006-01-0257.
11. A. Shaw and C. Wolfe, in *ASM Handbook, Volume 13B Corrosion*, edited by S. D. Cramer and B. S. Covino, Jr. ASM International, Materials Park, OH, 2005, pp. 205–227.
12. F. H. Froes, Y. Kim, and S. Krishnamurthy, *Materials Science and Engineering* **117**, 19–32 (1989).
13. S. J. Xia, V. I. Birss, and R. G. Rateick, Jr., *Electrochemical Society Proceedings* **25**, 270–280 (2003).
14. P. Schmutz , V. Guillaumin, R. S. Lillard, J. A. Lillard, and G. S. Frankel, *Journal of the Electrochemical Society* **150**(4), B99–B110 (2003).
15. Q. Meng, T. Ramgopal, and G. S. Frankel, *Electrochemical and Solid State Letters* **5**(2), B1–B4 (2002).
16. R. S. Stampella, R. P. M. Procter, and V. Ashworth, *Corrosion Science* **24**(4), 325–341 (1984).
17. D. L. Hawke, J. Hillis, M. Pekguleryuz, and I. Nakatsugawa, in *ASM Specialty Handbook: Magnesium and Magnesium Alloys*, edited by M. M. Avedesian and H. Baker. ASM International, Materials Park, OH, 1999, pp. 194–210.
18. W. J. James, M. E. Straumanis, B. K. Bhota, and J. W. Johnson, *Electrochemical Society* **110**, 1117 (1963).
19. G. L. Makar and J. Kruger, *Journal of the Electrochemical Society* **13**(2), 414–421 (1990).
20. L. Whitby, *Transactions of the Faraday Society* **29**, 1318 (1933).
21. U. R. Evans, *Metallic Corrosion, Passivity, and Protection*, 2nd edition. Edward Arnold, London, 1946.

22. H. A. Robinson *Transactions of the Electrochemical Society* **96**, 499 (1946).
23. R. E. McNulty and J. D. Hanawalt, *Transactions of the Electrochemical Society* **81**, 423 (1942).
24. J. L. Robinson and P. F. King, *Journal of the Electrochemical Society* **108**, 36–41 (1961).
25. M. E. Straumanis, *Journal of the Electrochemical Society* **105**, 284 (1958).
26. R. Glicksman, *Journal of the Electrochemical Society* **106**, 85 (1959).
27. N. D. Tomashov, V. S. Komissarova, and M. A. Timonova, *Trudy Instituta Fizicheskoi Khimii, Academiya Nauk SSSR Korrozii Metal* **4**, 172 (1955).
28. A. P. Nazarov, T. A. Yurasova, V. V. Gubin, A. K. Buryak, and M. P. Glazunov, *Zashcita Metallow* **29**(3), 392–297 (1993).
29. G. G. Perrault, in *Encyclopedia of Electrochemistry of the Elements*, Vol. 8, edited by A. J. Bard. Marcel Dekker, New York, 1978, p. 263–319.
30. R. L. Petty, A. W. Davidson, and J. Kleinberg, *Journal of the American Chemical Society* **76**, 363 (1954).
31. G. Song and A. Atrens, *Advanced Engineering Materials*, **5**(12), 837–858 (2003).
32. G. L. Song and A. Atrens, *Advanced Engineering Materials* **1**(1), 11–33 (1999).
33. N. Winzer, A. Atrens, G. Song, E. Ghali, W. Dietzel, K.-U. Kainer, N. Hort, and C. Blawert, *Advanced Engineering Materials* **7**(8), 659–693 (2005).
34. G. L. Song, A. Atrens, D. St. John, J. Nairn, and Y. Li, *Corrosion Science* **39**(5), 855–875 (1997).
35. G. Song, A. Atrens, D. St-John, X. Wu, and J. Nairn, *Corrosion Science* **39**(10–11), 1981–2004 (1997).
36. A.-M. Lafront, W. Zhang, S. Jin, and R. Tremblay, *Electrochimica Acta* **51**(3), 489–501 (2005).
37. T. Zhang, Y. Shao, G. Meng, and F. Wang, *Electrochimica Acta* **53**, 561–568 (2007).
38. I. Nakatsugawa, Cathodic Protection Coating on Magnesium or Its Alloys: Methods of Production. Canada, Patent No. 218, 983; United States, Patent No. 6, 291,076 (2001).
39. E. Ghali, W. Dietzel, and K.-U. Kainer, *Journal of Materials Engineering and Performance* **13**(5), 517–529 (2004).
40. D. L. Hawke, J. E. Hillis, and W. Unsworth, *Technical Committee Report*. International Magnesium Association, Wauconda, IL, USA, 1988, p. 8.
41. E. Ghali, in *Uhlig's Corrosion Handbook*, 2nd edition, edited by R. W. Revie. Wiley, Hoboken, NJ, 2000, pp. 793–830.
42. E. Ghali, in *Uhlig's Corrosion Handbook*, 2nd edition, edited by R. W. Revie. Wiley, Hoboken, NJ, 2000, pp. 677–715.
43. H. P. Godard, W. B. Jepson, M. R. Bothwell, and R. L. Kane, *The Corrosion of Light Metals*. Wiley, Hoboken, NJ, 1967, p. 269.
44. A. F. Froats, T. Kr. Aune, D. Hawke, W. Unsworth, and J. Hillis, in *ASM Handbook, Volume 13, Corrosion*, edited by L. J. Korb, D. L. Olson and J. R. Davis. ASM International, Materials Park, OH, 1987, pp. 740–754.
45. D. Hawke and A. Olsen, *Corrosion Properties of New Magnesium Alloys*. Society of Automotive Engineers, Detroit, MI, USA, 1993, pp. 79–84.
46. D. L. Hawke, Galvanic Corrosion of Magnesium, SDCE 14th International Die Casting Congress and Exposition, Toronto, Canada, May 1987, Paper No. G-T87-004, Society of Die Casting Engineers, River Grove, IL, USA.
47. H. Chen, J. Liu, and W. Huang, *Materials Science and Engineering* **A415**, 291–296 (2006).
48. L. H. Hihara and P. K. Kondepudi, *Corrosion Science* **34**, 1761–1772 (1993).
49. L. H. Hihara and P. K. Kondepudi, *Corrosion Science* **36** 1585–1595 (1994).
50. I. W. Hall, *Scripta Metallurgica,* **21**, 1717–1721 (1987).
51. A. Bakkar and V. Neubert, *Electrochimica Acta* **54**, 1597–1606 (2009).
52. R. Qvarfort, *Corrosion Science* **28**, 135–140 (1988).
53. L. L. Shreir, R. A. Jarman, and G. T. Burstein, *Corrosion* **1**, 4.98–4.115 (1995).
54. D. O. Sprowls, in *ASM Handbook, Volume 13, Corrosion*, edited by J. R. Davis. ASM International, Materials Park, OH, 1987, pp. 231–233.
55. A. I. Asphahani and W. L. Silence, in *Metals Handbook, Corrosion*, 9th edition, edited by J. R. Davis (Senior Editor). ASM International, Materials Park, OH, 1987, pp. 113–122.
56. G. L. Makar, J. Kruger, and K. Sieradzki, *Corrosion Science* **34**, 1311 (1993).
57. G. Nussbaum, G. Regazzoni, and H. G. Jestland, in *Proceedings of the SAE International Congress and Exposition*, edited by Science and Engineering of Light Metals, RASELM 91, Tokyo, Japan, Oct. 1991.
58. G. Song and A. Atrens, in *Corrosion Behavior of Skin Layer and Interior of Die Cast AZ91D*, Werkstoff-Informationsgesellschaft, Wolfburg, Germany, 1998, pp. 415–419.
59. P. Adeva-Ramos, S. B. Dodd, P. Morgan, F. Hehmann, P. Steinmetz, and F. Sommer, *Advanced Engineering Materials* **3**(3), 147–152 (2001).
60. K. Nisancioglu, O. Lunder, and T. Aune, in *Proceedings of the 47th World Magnesium Conference*, Canne IMA, 1990, p. 43.
61. G. Song, L. Bowles, and D. H. St. John, *Materials Science and Engineering* **A366**, 74–86 (2004).
62. S. Amira, A.-M. Lafront, D. Dubé, R. Tremblay, and E. Ghali, *Advanced Engineering Materials* **9**(11), 973–980 (2007).
63. S. Mathieu, C. Rapin, J. Hazan, and P. Steinmetz, *Corrosion Science* **44**, 2737–2756 (2002).

64. K. Lubbert, J. Kopp, and E. Wendler-Kalsch, *Materials and Corrosion* **50**, 65–72 (1999).

65. R.-C. Zeng, Z. Jin, W. J. Huang, W. Dietzel, K.-U. Kainer, C. Blawert, and W. S Ke, *Transactions of Nonferrous Metals Society of China* **16**, 763–771 (2006).

66. O. Lunder et al., Filiform corrosion of a magnesium alloy, Paper presented at the 11th Annual Corrosion Congress, Florence, Italy, 1990, pp. 5.255–5.262.

67. V. Mitrovic-Scepanovic and R. J. Brigham, A fundamental corrosion study of magnesium, Progress Report No. 1, Metals Technology Laboratories, CANMET, Energy, Mines and Resources Canada, Ottawa, 1990, p. 10.

68. W. K. Miller and E. F. Ryntz, Jr., *Society of Automotive Engineers* (*SAE*) 830521 (1984).

69. A. Yamamoto, A. Wanatabe, K. Sugahara, S. Fukumoto, and H. Tsubakino, Applying a vapor deposition technique to improve corrosion resistance in magnesium alloys, in *Proceedings of the Second International Conference on Environment Sensitive Cracking and Corrosion Domage*, Hiroshima, Japan, edited by H. N. M. Matsumura, K. Nakasa, and Y. Isomoto. Nishiki Printing Ltd, Hiroshima, Japan, 2001, pp. 160–167.

70. S. Amira, M. Shehata, D. Dubé, R. Tremblay, and E. Ghali, Influence of the microstructure on the corrosion rate of AXJ530 magnesium alloy in 3.5% NaCl solution, in *24th Annual Conference of Egyptian Corrosion Society*, Hurghada. Egyptian Corrosion Society, Hurghada, Egypt, 2005.

Chapter 11

Metallurgically and Microbiologically Influenced Corrosion of Magnesium and Its Alloys

Overview

The influence of the composition, microstructure, and surface modification on the corrosion resistance mainly cast magnesium alloys is examined. The properties and influence of different phases of magnesium alloys on some types of corrosion, such as galvanic or bimetallic, intergranular, and exfoliation, are explained. The mechanical properties of magnesium and its alloys at high temperature are discussed with special reference to creep and "creep corrosion." Determination of the open circuit potential and polarization studies of every microstructure are explained. Post-heat treatment, rapid solidification, and the influence of joining or welding on the corrosion resistance of alloys are discussed. The definition, properties, and corrosion behavior of the exterior, the interior skin, and the bulk of cold chamber die-cast or thixocast alloys are detailed. Attention is given to the performance of hot chamber die-cast thin plates.

Magnesium undergoes two distinct corrosion phenomena when in the presence of a biological medium: microbiologically influenced corrosion (MIC) and rational biodegradation. The first phenomenon occurs everywhere in the biosphere, even in oxygen-free media where microorganisms are present. MIC of magnesium and its alloys has been carried out in nutrient broth solution to examine its performance as sacrificial anodes in cathodic protection. The second phenomenon occurs inside human or animal bodies, where the immune system prevents microorganism colonization. Physiological fluids containing water and large amounts of chloride are principally involved in this degradation process. Magnesium is used as a compatible biomaterial in the human body, where it undergoes rational biodegradation. However, for permanent implants, MIC of magnesium alloys should be controlled where bacteria are present, such as in the mouth (teeth) and digestive system.

Corrosion Resistance of Aluminum and Magnesium Alloys: Understanding, Performance, and Testing By Edward Ghali

A. METALLURGICALLY INFLUENCED CORROSION OF MAGNESIUM ALLOYS

Cast magnesium alloys have always predominated over wrought alloys. Two major cast magnesium alloys are available. The first group includes alloys containing 2–10% Al, combined with minor additions of zinc and manganese. Their mechanical properties are satisfactory from 95 to 120 °C. The second group consists of magnesium alloyed with various elements (rare earths, zinc, thorium, silver, etc.) except aluminum, all containing a small but important zirconium content that imparts a fine-grain structure that improves mechanical properties. These alloys generally possess much better elevated-temperature properties. Because of the particularly high solid solubility of yttrium in magnesium (12.5% maximum) and the amenability of Mg–Y alloys to age hardening, a series of Mg–Y–Nd–Zr alloys has been produced that combines high strength at ambient temperatures with better creep resistance at temperatures up to 300 °C. The heat-treated alloys have a resistance to corrosion, that is superior to that of other high-temperature magnesium alloys and comparable to many aluminum-based casting alloys. A yttrium-containing (~75% Y) mischmetal together with heavy rare earth metals such as gadolinium and erbium could be substituted for pure yttrium [1]. The wrought alloys are generally divided into two groups with or without zirconium, most of which fall into the same categories as the casting alloys, which can be obtained in a number of tempers (see Chapter 3).

Creep Resistance

Magnesium alloys, like aluminium alloys, do not exhibit a ductile–brittle transition at low temperatures. The elastic modulus and the notched and unnotched yield and tensile strengths remain constant or increase only slightly as the temperature decreases, and total elongation can increase or decrease slightly. The commonly used magnesium alloys have low performance at high temperatures. The elastic modulus of Mg can decrease significantly with increasing temperature; for example, the modulus of the AZ alloys can decrease by approximately 15% at 150 °C and 30–50% at 250 °C. For example, AS41 was developed to improve upon the creep resistance of AZ91, although it still does not perform as well as die-cast 380 aluminum alloy [2].

11.1. CASTING ALLOYS AND ALLOYING ELEMENTS

11.1.1. Casting Alloys

11.1.1.1. Magnesium–Aluminum Alloys

The AZ alloys, which contain zinc as a secondary alloying element, solidify with a sufficiently fine grain size to meet most property requirements. They are highly castable and have a minimum tendency toward hot cracking, this tendency increasing with increasing zinc content. These alloys, however, also have a tendency to develop microporosity. Alloy AZ91 is the most commonly used of all magnesium alloys due to its relatively low cost and generally adequate mechanical properties and processing characteristics. AM60, which contains manganese as a secondary alloying element, has a typical elongation of 6% in the as-cast condition and was developed to provide higher ductility and toughness required for die-cast automobile wheels. The second alloy, AS41, contains silicon and manganese as secondary alloying elements, and was developed to provide improved elevated-temperature properties for engine applications, such as the crankcase of air-cooled engines [2].

11.1.1.2. Magnesium–Zinc Alloys

The extremely effective precipitation hardening reactions of the Mg–Zn binary alloys and the grain-refining effect of zirconium combine to yield high strengths with good ductility. The Mg–Zn alloys with zirconium, thorium, or rare earths can provide very good combinations of room temperature yield strength and elongation. As the zinc content in these alloys is increased, microporosity and hot cracking become problems but these tend to be less severe in the alloys containing thorium or rare earths. All of these grades tend to be more costly than Mg–Al alloys [2].

11.1.2. Magnesium–Rare Earth, Magnesium–Thorium, and Magnesium–Silver Alloys

These alloys, being relatively expensive, are used selectively when service temperatures exceed about 150 °C (300 °F). Good elevated-temperature properties are obtained through the development of stable grain boundary precipitates, and they generally have good castability. The Mg–RE alloys are susceptible to oxidation while the Mg–Th alloys have more severe oxidation problems [2].

11.1.3. Alloying Elements and Tolerance Limit

Magnesium corrosion can be accelerated by galvanic coupling, high levels of certain impurities, especially nickel, copper, and iron, or contamination (especially of castings) by salts. The corrosion resistance of magnesium and its alloys is dependent on film formation in the medium to which they are exposed. The rate of formation, dissolution, or chemical change of the film varies with the medium, and also with the metallic alloying agents and impurities present in the magnesium. The principal rate-limiting factors in the atmospheric (aqueous) corrosion of magnesium alloys are associated with the breakdown of the magnesium hydroxide film and the rate of its reformation. Magnesium alloys are anodic to all other structural metals and will undergo galvanic attack if coupled to them. The effect of some important alloying elements on Mg alloy corrosion are summarized next.

Aluminum Increasing Al concentrations have a beneficial effect on the corrosion behavior of Mg–Al alloys, but the specific mechanism depends on the distribution of the Al within the magnesium matrix. Generally, alloying elements not only enhance the mechanical properties of Mg, but also impart a significant impact on the corrosion behavior of Mg–Al alloys. Alloying elements can form secondary particles, that are noble to the Mg matrix, thereby facilitating corrosion, or enrich the corrosion product, thereby possibly inhibiting the corrosion rate. Thus Mg–Al alloy corrosion behavior depends on the distribution of the alloying elements [3].

Increasing concentrations of 2–8 wt% Al in die-cast Mg–Al alloys decrease the corrosion rate, as shown in Figure 11.1. The corrosion rate of high-purity die-cast Mg

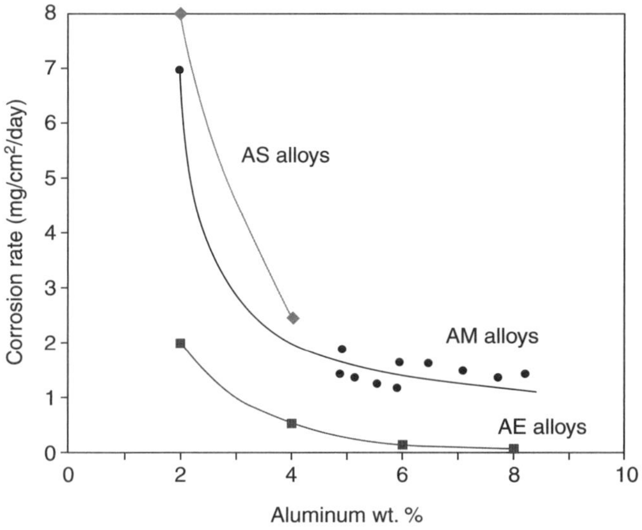

Figure 11.1 Corrosion rate of die-cast Mg alloy rods immersed in 5% NaCl solution as a function of Al content [6].

alloys in a chloride environment decreases rapidly with increasing aluminum content, up to about 4 wt%. Further Al additions, up to about 9%, give only a modest improvement in the corrosion resistance. Low Al additions, of approximately 2–4 wt%, result in α-Mg dendrites surrounded by the two-phase, $\alpha + \beta$, eutectic at grain boundaries, whereas higher AL additions, 6–9 wt%, tend to precipitate distinct β particles along grain boundaries, depending on solidification rates. Surrounding the Al-rich β phase are local concentrations of up to 10 wt% Al as a result of microsegregation during solidification [4]. The increasing presence of β particles, which begin to appear above 2 wt% Al, may cause, in part, the improved corrosion resistance of the higher Al-content alloys. The passivating effect of the Al-rich β phase, $Mg_{17}Al_{12}$, results in a low corrosion rate over a wide pH range. Auger depth profiling shows that, as the Al component dissolves, Mg-enriched film forms in alkaline media, and as the Mg component dissolves, an Al-enriched film forms in neutral and slightly acidic media. The synergistic effect of both components leads to the decreased corrosion rate of the β phase. During immersion testing, AE alloys exhibit a lower corrosion rate than AS, AM, and AZ alloys with similar Al content [5, 6].

Alloying with Al results in the precipitation of $Mg_{17}Al_{12}$. While $Mg_{17}Al_{12}$ precipitates, Al-rich coring zones act as a barrier against the extension of local corrosion, enhancing the corrosion resistance of Al-containing alloys. Alloying with at least 4 wt% Al is necessary to obtain an oxide with optimum corrosion properties. Only at this threshold does the Al_2O_3 component form a continuous passivating network, which could be a skeletal structure in the amorphous mixture of aluminum and magnesium hydroxides. It was also found that Al can have a detrimental effect on corrosion. Al was claimed to decrease the tolerance limit for Fe in an almost linear way and in small amounts (below 8%) to produce an anodically more active Mg solid solution [7].

The corrosion resistance of AZ91, AZ61, and AZ31 in 5% NaCl solution increased with increasing Al content. For AE42, ZAC8506, and AZ91D, the corrosion rate decreased in the sequence AE42 > ZAC8506 > AZ91D [8]. Additions of Al by rapid solidification processing results in decreased corrosion rates without precipitation of the β phase. Faster solidification rate of rapidly solidified (RS) alloys disperses fine $Mg_{17}Al_{12}$ particles, which increases the corrosion resistance, as does controlled precipitation of the β phase [9, 10].

In addition to the alloying ingredients added, certain other metals are usually present in small amounts. In the alloys containing aluminum, for example, iron usually amounts to about 0.02–0.05%. By using special techniques and care in melting, this level can be reduced to about one-tenth of this concentration. Such high-purity alloys have much better resistance to saltwater than do those of normal purity, but their corrosion behavior in industrial atmospheres is very similar. Furthermore, the practical value of the higher resistance to corrosion is largely offset when components are used in electrical contact with other more noble metals acting as cathodes. The effect of a steel bolt, for example, even when it has been zinc or cadmium plated, is much greater at the point of contact than that of the local cathodes in the impure alloys. Galvanic corrosion at joints with other metals is not markedly less in the case of the high-purity alloys. Nevertheless, such alloys have their place and, when they can be used without other metal attachments, provide better intrinsic resistance to corrosion by seawater than the alloys of normal purity [11].

Salt fog corrosion tests for Mn-containing Mg–Al alloys, such as AM50 and AM20, showed that corrosion pits initiated at low Al areas, and the matrix was attacked in the form of fissures. The fissures started from pitting locations and usually stopped in front of areas of high Al segregation. The continuous high Al segregation seemed to contribute much more in stopping the propagation of corrosion fissures than the discontinuous, more or less isolated, β-$Mg_{17}Al_{12}$ particles. However, the susceptibility to SCC increases as the Al content increases from 1% to 8% [8].

Zinc Zinc makes the Mg alloy electrochemically more noble, thereby minimizing the corrosion rate [12]. The low tolerance limits for the contaminants in AM60 alloy when compared to AZ91 alloy can be related to the absence of zinc. Zinc is thought to improve the tolerance of magnesium–aluminum alloys for some contaminants, but it is limited to 1–3% Zn because of its detrimental effects on corrosion above 3% [13].

Manganese Manganese can improve the corrosion resistance of Mg alloys, but this is not always the case. The corrosion rate of Mg alloys is related to iron content and the Fe/Mn ratio. Binary Al–Mn phase with lower Al/Mn ratio has a higher active potential. Therefore the corrosion rate increases when Mn is added into Mg–Al alloys to form A1–Mn and intermetallic phase Al–Mn–Fe [8].

Lithium Magnesium–lithium alloys have poor corrosion resistance [14].

Rare Earths The rare earths (REs) are typically added to Mg–Al alloys as cerium-based mischmetal (MM) containing lanthanum, neodymium, and praseodymium. The high corrosion resistance of the AE alloys appears to be related to the presence of passive Al-rich zones along the grain boundaries, acting as barriers against pit propagation. The Al_4MM phase particles precipitated in AE alloys exhibit a passive behavior and do not affect the corrosion process to a significant extent. A high resistance to localized corrosion is observed for the AE alloys with a high Al content [1]. All RE elements (including yttrium) form eutectic systems of limited solubility with magnesium. Therfore precipitation

Table 11.1 Proposed Tramp Element Tolerance Levels for Selected Mg–Al Die-Cast Alloys

Alloy	Fe/Mn	Fe(max)	Cu(max)	Ni(max)
AZ91B	0.032	0.0050	0.030	0.002
AM60B	0.021	0.0050	0.010	0.002
AS41B	0.010	0.0035	0.020	0.002
AE42X1	0.020	0.0050	0.050	0.005

Sources: References 5 and 22.

hardening is possible and appropriate. The precipitates are very stable and increase the creep resistance, corrosion resistance, and temperature strength. However, RE elements are affected by the medium and pH values [8, 15].

Rare earth elements are typically added to Mg–Al alloys as cerium-based mischmetal containing lanthanum, neodymium, and praseodymium. A typical composition of mischmetal is 50% Ce, 25% La, 20% Nd, and 3% Pr. These have very low solubilities in Mg (Ce, 0.09 at %; La, 0.14 at %; Nd, 0.10 at %; and Pr, 0.09 at %) [16] and react with Al to form Al_4RE intermetallics [17]. These intermetallics, with their high melting temperature, resist coarsening relative to Mg_2Si and provide enhanced creep resistance at higher temperatures. Compositions of solidified phases are given in Table 11.1 [18].

Mg–Th can undergo severe oxidation [14]. However, the normal saltwater corrosion resistance is only moderately reduced when compared to high-purity magnesium and Mg–Al alloys—0.5–0.76 mm/yr (20–30 mils/yr) as opposed to less than 0.25 mm/yr (10 mils/yr) in 5% salt spray [13]. Zirconium can stabilize the Mg matrix phase and reduce its corrosion rate. The beneficial effect of Zr cannot be extended to an alloy with too much Zr. The excess addition of Zr can lead to precipitation of Zr in the matrix, which is detrimental to the corrosion resistance [8].

Strontium Additions of Sr to Mg–Al alloys result in reduced grain size and a lower corrosion rate that is attributed not only to the reduced grain size but also to changes in the oxide layer structure and composition and in the electrochemical properties of the phases present [19]. A new family of creep-resistant Mg alloys is based on the Mg–Al–Sr system. The microstructure of the alloys is characterized by Al–Sr–(Mg) containing intermetallic second phases and the alloys exhibit better salt-spray corrosion resistance (0.09–0.15 mg/$cm^2 \cdot$ day) than other commercial Mg die-cast alloys such as AM60B, AS41, and AE42, and the Al die-cast alloy A380 [14].

Silver Silver, together with RE metals, strongly increases the high-temperature strength and creep resistance but also leads to low corrosion resistance [15].

Calcium The overall effect of an alloying element on corrosion rate depends on where and in what form and amount it is present in the alloy. Since both AC52 and AC53 were prepared by adding Ca to AM50, the addition of Ca first decreases the corrosion rate and then increases it. Calcium initially dissolves in Mg and improves its corrosion properties by lowering its activity. On further increasing the concentration of Ca in Mg, Ca forms an intermetallic compound $(Mg_2Al)_2Ca$ with Al and Mg, which probably creates more galvanic cells with Mg and increases the corrosion rate [20]. Addition of 0.3% Si in AC53 increases the corrosion rate of the alloy, but the rate decreases on adding Sr to AC53 + 0.3%Si. Both Ca and Sr are anodic or more active to Mg; therefore both in small amounts reduce the corrosion of magnesium. On the other hand, Si increases the corrosion because it is cathodic or less active to Mg.

Silicon Silicon is intentionally added to only the AS alloys, to combine with Mg, forming Mg_2Si, which precipitation strengthens the alloy and is relatively innocuous to the corrosion behavior. Alloying with Si does not have an important influence on the corrosion properties because phase formed Mg_2Si is a poor cathode. It has a corrosion potential of −1.65 V/SCE, close to the −1.66 V value for pure Mg in 5% NaCl solution saturated with $Mg(OH)_2$ at pH 10.5 [6].

Iron, Nickel, and Copper Noble tramp elements, like Fe, Ni, and Cu, are picked up during melting, handling, and pouring operations. Their influence can be seen in Figure 11.2 for die-cast AZ91 corrosion specimens in which the tramp elements were singularly increased. In general, the factor with the far strongest influence on the corrosion of Mg alloys is the amount of cathodic impurities, particularly of those with low hydrogen overvoltage. Noble impurities like Fe, Cu, and Ni promote microgalvanic corrosion and show tolerance limits above which corrosion rate rapidly increases (Figure 11.2). The solubility limit of Fe in Mg is 10 ppm and the eutectic phase of Mg–Fe corresponds to 60 ppm. The individual tolerance limits depend on the specific alloy and it has been observed that at ~90 ppm of Fe in the AZ91 alloy formation occurs of a fine dispersed $AlFe_3$ with one of the most noble constituents (−0.5 V) in 5% NaCl saturated with Mg $(OH)_2$, which could act as an effective cathodic site [7, 21].

The specific ASTM tramp element tolerances are shown in Table 11.1 (B94-92) [22] and they are typically the same or lower for ingots, as tramp elements are commonly picked up during the melting and pouring operations. In the recycling process, the Mg scrap is remelted and refined to a state in which it is free of internal impurities. Recently, a new

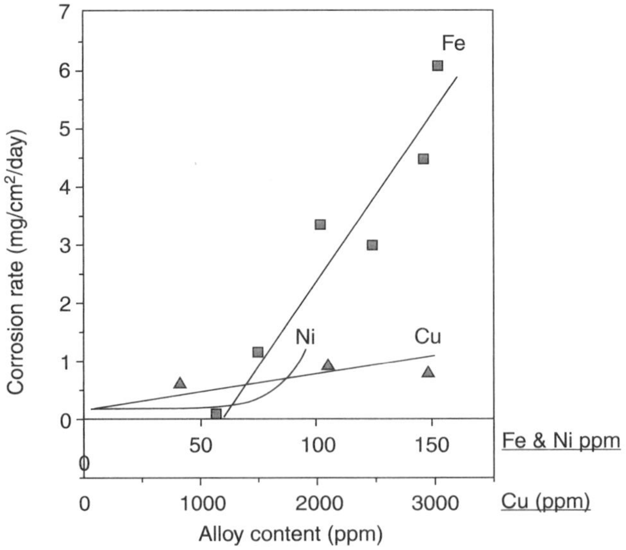

Figure 11.2 Die-cast AZ91 salt-spray performance versus tramp element content [5, 21].

process was developed that is based on SF_6 melt protection and a filter and/or argon gas sparging for nonmetallic impurities removal. This process differs from the one that has been used traditionally in both its method of melt production, where SF_6 is now used as opposed to flux, and its means of melt refining, where filter/sparging is now used instead of flux. One of the major problems of using recycled Mg alloys is their poor corrosion resistance. The presence of less active or more noble metal impurities such as Fe, Ni, and Cu represents the most detrimental factor in influencing the corrosion properties of Mg alloys.

Of these impurities, Fe is often the most problematic since it is introduced to the melt from steel pots and casting molds and is present as an impurity in the alloying elements. The detrimental effect of noble metals decreases as follows: Ni > Fe > Cu. Ni and Cu are usually not a problem because of their very low content in the primary production. Fe is very effective in catalyzing the reduction reactions, especially in hydrogen evolution, which is a cause of the corrosion process [7]. Intermetallic compounds containing more than a few percent iron are detrimental because they function as efficient cathodes. However, binary Al–Mn phases with a low Al/Mn ratio may also exhibit a relatively high cathodic current output, causing an increase in the overall corrosion rate.

Grain refinement increases the overall grain boundary area, thereby optimizing the distribution and minimizing the size of any possible detrimental intermetallics, such as Fe_3Al. The traditional grain refinement method in sand casting is to add an inoculent, which facilitates heterogeneous nucleation during solidification [19]. The iron tolerance for the Mg–Al alloys depends on the manganese present, a fact suggested many years ago but only recently proved. Effectively, additions of Mn make the Fe less efficient as a cathode. For AZ91 with a manganese content of 0.15%, this means that the iron tolerance would be 0.0048% ($0.032 \times 0.15\%$) [13]. For die-cast high-purity AZ91, the ASTM specification B94 recommends Fe < 50 ppm, Ni < 20 ppm, and Cu < 300 ppm [7]. The Mg–Fe phase diagram shows a very low solid solubility of Fe in Mg (9.9 ppm). In the absence of Mn, virtually all the Fe precipitates in Mg as Al_3Fe, which has a highly cathodic corrosion potential (Table 11.2).

Within an aggressive medium, Al_3Fe acts as an effective cathode, catalyzing the reduction reaction, especially hydrogen evolution, which controls the corrosion reaction. Due to the low solubility of Al_3Fe in Mg, increasing additions of Al result in smaller tolerance levels for Fe. Typically, up to 1 wt% of Mn is added to improve corrosion resistance by reducing the potential difference between iron-containing particles and the matrix. Its beneficial effect is attributed to either Mn combining with the Fe and precipitating to the bottom of the crucible and/or reacting with the Fe left in suspension during

Table 11.2 Corrosion Potentials of Synthetically Prepared Intermetallic Phases after Two Hours in Deaerated 5% NaCl Solution Saturated with $Mg(OH)_2$ (pH 10.5)

Compund	Corrosion potential (V/SHE)	Compound	Corrosion potential (V/SHE)
Al_3Fe	−0.50	$Mg_{17}Al_{12}(\beta)$	−0.96
$Al_3Fe(Mn)$	−0.71	Al_8Mn_5	−1.01
$Al_6(MnFe)$	−0.76	$Al_4(MnFe)$	−1.16
$Al_6(MnFe)$	−0.86	Al_4Mn	−1.21
Al_4MM	−0.91	Al_6Mn	−1.28
β-Mn	−0.93	Mg_2Si	−1.41
$Al_6Mn_5(Fe)$	−0.96	Mg 99.99%	−1.42

Source: Reference 6.

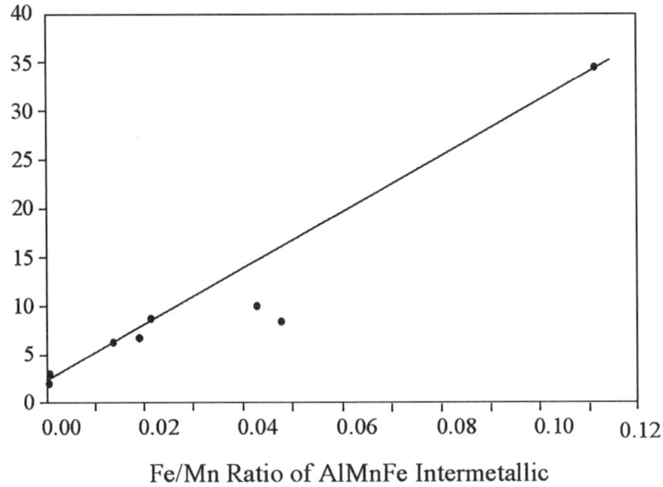

Figure 11.3 Relationship between the Fe/Mn ratio in the AlMnFe phase (up to 1 mm in size) and the corrosion rate [5, 23].

solidification [3]. The relationship between the Fe/Mn ratio in the AlMnFe phase and the corrosion rate is shown in Figure 11.3. Mn in excess of that needed to render the Fe content ineffective could be detrimental to corrosion resistance [5, 23].

The Ni tolerance depends strongly on the cast form, which influences grain size, with the low-pressure cast alloys showing just a 10 ppm tolerance for Ni in the as-cast (F) temper. Therefore alloys intended for low-pressure cast applications should have the lowest possible Ni level [13]. The tolerance limit for Ni is 5 ppm [4]. The influence of Cu content on microstructure and corrosion resistance of AZ91-based secondary Mg alloys was reported. Copper addition to the Mg alloy AZ91D results in grain refining and the formation of additional Mg–Al–Cu–Zn phases. With increasing Cu content in the intermetallics, the free corrosion potential will shift to more noble or less active values that create efficient local galvanic cells. If a uniform layer of Cu-rich intermetallics is formed as a barrier to prevent direct contact of the solution with the Mg matrix–Cu-rich intermetallics interface, corrosion resistance can be improved [8]. The tolerance limit for Cu is 1300 ppm [4].

11.2. CORROSION INFLUENCED BY METALLURGICAL PROPERTIES

11.2.1. Galvanic Corrosion and Secondary Phases

Metals with low hydrogen overvoltage, such as Ni, Fe, and Cu, constitute efficient cathodes for Mg and cause severe galvanic corrosion. The attack is especially severe if the other metal in the couple is passive or inert as, for example, stainless steels or copper-based alloys. Alloying metals that combine an active corrosion potential with a high hydrogen overvoltage (e.g., Al, Zn) are much less damaging [7]. Galvanic or bimetallic corrosion can be caused by impurities and secondary phases such as $Mg_{17}Al_{12}$, AlMn, Al_8Mn_5, $Mg_{12}Nd$, and Mg_2Pb, even when connected with Fe, Ni, and Cu [8]. Galvanic attack can be minimized by selecting high-purity alloys.

Corrosion behavior is optimized through alloy chemistry, by minimizing the cathodic sites, which evolve hydrogen gas, or by enriching the corrosion product film, which can

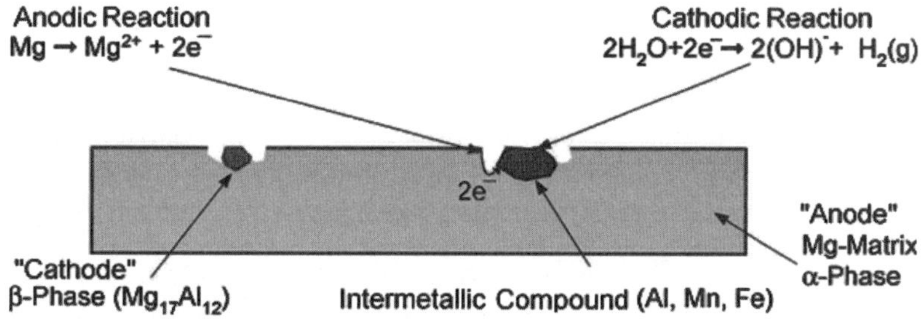

Figure 11.4 Schematic presentation of typical galvanic corrosion between some of the phases of Mg-Al alloys [14].

inhibit hydrogen gas evolution and decrease the corrosion rate. Microstructural enhancements, which refine the microstructure and homogenize the distribution of alloying elements, also disperse potentially deleterious elements, thereby enhancing corrosion resistance [3]. The potentials of intermetallic phases, prepared synthetically from the pure components by controlled solidification procedures, are given in Table 11.2 [6].

The β phase $Mg_{17}Al_{12}$ has an electrochemical polarization behavior different from the α matrix phase of the $\alpha - \beta$ binary phase alloys. The corrosion potential of the β phase is much more positive than the α phase and acts as a barrier to improve corrosion resistance. At the same time, it can also create a galvanic cell and accelerate corrosion, acting as a very effective cathode since its cathodic polarization curve shows high cathodic densities and low overpotentials if compared with that of the α phase. The amount and distribution of the β phase are the main factors that govern its beneficial or detrimental effect [24].

Figure 11.4 shows typical local galvanic cells that can lead to intergranular, stress corrosion cracking, or pitting corrosion depending on the properties of the solution, agitation, the microgeometry, and the microstructure of the surface. The different constituents of an AZ91 alloy (α, β, and MnAl phases) were synthesized and their corrosion resistance was studied by electrochemistry in ASTM D1384 water, pH 8.3. The pure phases were characterized through the corrosion potential, the polarization resistance, and polarization curves, then systematically coupled to assess the galvanic corrosion occurring in the AZ91 alloy. The aluminum content of the oxide film was obtained by X-ray photoelectron spectroscopy (XPS) measurements. The corrosion rate of the α solid solution alloys depends closely on their Al content. Aluminum enhances the corrosion resistance of the α phase through the formation of an A1-enriched superficial layer through a layer of a carbonate hydroxide of magnesium and aluminum. The β phase is 150 mV nobler than the α phase, but their corrosion rates are similar. The galvanic currents are low (below 20 $\mu A \cdot cm^{-2}$) whatever the implemented couples and close to the corrosion current previously measured for the AZ91 alloys [25].

Since the β phase is cathodic with respect to the matrix, it plays a dual role, depending on its volume fraction, $f = V_\beta / V_\alpha$, in the microstructure. It can be used as a corrosion barrier, and a cathode that causes galvanic corrosion. If f is lower, the β phase acts as a cathode that can accelerate the general corrosion of the α-Mg matrix; if f is higher, the β phase can be a barrier inhibiting general corrosion. Lunder et al. [26] studied the role of the β phase in the corrosion of AZ91; they suggest that the β phase has the better properties of the two metals (its corrosion resistance is similar to Mg in alkaline solution and Al in neutral solution); moreover, the corrosion resistance is better than that of Mg and A1 in alkaline solution. AlMn particles are commonly observed in the microstructure of Mg–Al alloys. Table 11.2

shows that the corrosion potential of AlMn phases is more active than that of $Mg_{17}Al_{12}$. They can create galvanic cells with different phases. A corrosion pit of as-extruded AM60 in 3.5% NaCl solution is caused by AlMn particles, rather than the β phase. Finally, the XPS analyses revealed that the corrosion layer of MnAl was mainly constituted of an aluminum hydroxide. Iron-rich phases, in particular, the FeAl phase (Table 11.2), is one of the most detrimental cathodic phases present in Mg–A1 alloys on the basis of its potential and its low hydrogen overvoltage. Mg_2Pb facilitates pitting and leads to a negative difference effect. $Mg_{12}Nd$ particles in the WE43 alloy also serve as cathodes with respect to the matrix. The formation of $Mg_{24}Y_5$ precipitates in hcp-Mg matrix during heat treatment caused the lowering of the corrosion resistance of the alloys. Mg_2Si seems to have no effect on the corrosion of Mg alloys [8].

Scanning Kelvin Probe (SKP) Jönsson and LeBozec [27] have studied the β-$Mg_{17}Al_{12}$ and η-Al_8Mn_5 phases of AZ91D by using different electrochemical techniques including the scanning Kelvin probe (SKP). The SKP is an electrochemical technique that measures the electrode potential at metal–polymer interfaces and thereby detects changes in the buried metal oxide structure and microstructure phases, and variations of the interfacial ionic conductivity with high spatial resolution of about 50 μm. Thus, one can distinguish between an ingression of ionic species into the polymer–metal interface, wet deadhesion, and a corrosive delamination [27].

It has been shown by Stratmann and Streckel [28] that the corrosion potential of a bare metal covered with a layer of electrolyte is linearly related to the Volta potential measured in air, according to the following equation:

$$E_{\text{corr}} = \frac{W_{\text{e}}^{\text{ref}}}{F} - \chi_{\text{gas}}^{\text{sol}} + E_{1/2} + \Delta\psi_{\text{sol}}^{\text{ref}}$$

where $W_{\text{e}}^{\text{ref}}$ is the electronic work function of the probe material, F is the Faraday constant, $\chi_{\text{sol}}^{\text{gas}}$ is the dipole potential of the solution–gas interface, $E_{1/2}$ is the half-cell potential of the reference electrode, and $\Delta\psi_{\text{ref}}^{\text{sol}}$ is the Volta potential difference [27]. The Volta potential is due to the charge on phases α and β. It is measurable or calculable by classical electrostatics from the charge distribution.

Both phases in AZ91D showed a more noble potential than the α-Mg phase under atmospheric weathering conditions. Moreover, they observed a clear relationship between the precipitation of Al-rich phases and Volta potential. The Volta potential values increased with the Al content in the phase and the Al-rich coring along the grain boundaries resulted in measurable changes. It can be added that a linear relationship was observed between the Volta potential measured by the powerful scanning Kelvin probe force microscopy (SKPFM) and that measured by SKP [27].

Thixomolding The SSP (thixomolding), which leads to Al-rich α phases, would be a way to reduce the AZ91 alloy corrosion. The galvanic corrosion currents are low whatever the implemented couples and close to the corrosion current measured on the SSP AZ91 alloy. They decrease with increasing Al content in the α phase when α is coupled either with β or with MnAl. Finally, the galvanic current of α/β is doubled if the β phase contains zinc, and this is the case in the AZ91 two-phase alloys. Galvanic corrosion occurs in die-cast and thixocast alloys between the two main phases (α and β). Its rate should be lower in the case of the semisolid cast alloy since both surface area ratio between cathodic and anodic sites, and differences between the A1 content of the α and β phases are smaller in this case. The better corrosion behavior of thixocast alloys is thus attributed mainly to the particular composition of the α phase resulting from the treatment of the semisolid alloy, prior to injection in the mold [25].

Rapid Solidification The rapid solidification process can refine the microstructure, which is beneficial to the corrosion properties. It can change the mechanism of corrosion, turning pitting corrosion of Mg–Al alloys into overall corrosion. The surface or skin layer of die-cast Mg–Al alloys with very fine grains, high β volume fraction, and continuous distribution of β phase along grain boundaries has a higher corrosion resistance than its core. It is also true that die-castings of Mg alloy AZ91D have better corrosion resistance than ingots [8].

11.2.2. Intergranular Corrosion

Intergranular corrosion (IGC) of magnesium alloys does not occur, because the grain-boundary constituent is invariably cathodic to the grain body. Corrosion of magnesium alloys is concentrated on the grains, and the grain-boundary constituent is not only more resistant to attack but is cathodically protected by the neighboring grain. Filiform corrosion initiates and then develops into cellular or pitting corrosion. However, in the early stages of immersion, a localized attack of magnesium and its alloys can be formed at the grain boundary at the interface of cathodic precipitates in mild corrosive media and can be considered as intergranular (intercrystalline) corrosion. Since IGC has much sharper tips than pitting corrosion, it is a more drastic stress riser and has a more damaging contribution to corrosion fatigue [6].

Intergranular corrosion is generally caused by localized attack of the α-Mg matrix, which corrodes preferentially, leaving the more noble intermetallics in relief along the grain boundaries, and is considered a metallurgically influenced corrosion form. The kinetics of the electrochemical cell is controlled by the hydrogen evolution reaction on predominant cathodic phases in the microstructure.

Figure 11.5 shows the corroded surface of alloy AE81 after the hydroxide film has been stripped off in chromic acid. The grain bodies with a low Al concentration corrode at a faster

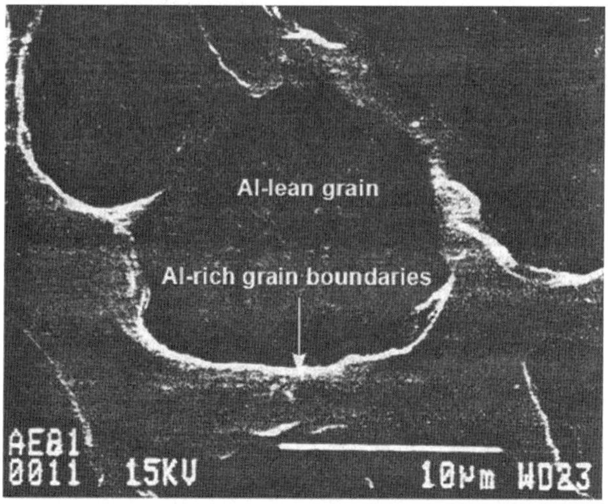

Figure 11.5 Morphology of corroded AE81 after removal of the hydroxide film. The grain boundaries with Al-rich areas (location A) are more resistant than the Al-lean grain (location B). [5, 6].

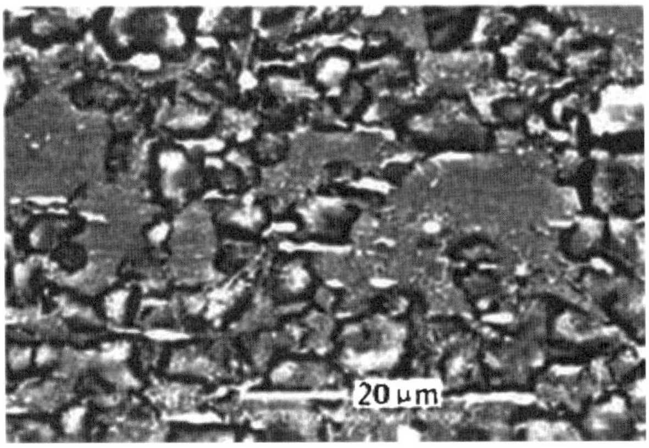

Figure 11.6 Intergranular corrosion morphology of AZ80-T5 in 3.5% NaCl aqueous solution after 1 h [8].

rate than Al-rich regions along the grain boundaries, as can also be seen in other Mg–Al alloys. However, on AE alloys, the pits do not easily penetrate the Al-rich zones. Good pitting resistance of the die-cast AE alloys is therefore attributed to the presence of these Al-rich zones, which appear to act as barriers against pit propagation. If these barriers are removed by homogenization heat treatment, the corrosion resistance is reduced. Homogenized AE81 exhibited corrosion rates more than 100 times higher than the as-cast material during a 3 day immersion test in 5% NaCl solution. It is not yet clear whether this unusual sensitivity of corrosion to heat treatment is related to the absence of Mn in this AE81 alloy. The corrosion rate of alloys AM80 and AZ91 were only moderately influenced by a similar heat treatment. In general, homogenized specimens exhibited deeper localized attack than the as-cast material [14].

Intergranular corrosion occurred after immersing aged AZ80 in 3.5% NaCl solution for 1 h (Figure 11.6). Corrosion appears along the grain boundaries and forms deep and narrow paths [8].

11.2.3. Exfoliation Corrosion

Exfoliation can be considered as a type of intergranular attack, and this is observed in unalloyed Mg above a critical chloride concentration. This morphology was not seen in Mg alloys, in which individual grains were preferentially attacked along certain crystallographic planes. The early stages of this form of attack caused swelling at points on the surface due to apparent delamination of the Mg crystals with interspersed corrosion products, but as attack proceeded, whole grains or parts of grains disintegrated and dropped out, leaving the equivalent of large irregularly shaped pits [14, 29]. Exfoliation morphology of corrosion should be distinguished from other localized corrosion. Since the naturally passive film on Mg metal is not so protective, it may suffer relatively more important localized corrosion. AZ91 in chloride-containing environments, for example, exhibits pitting at the outset (initiation sites are few and associated with intermetallic particles), filiform corrosion at early stages of propagation, and a cellular type of attack in the terminal stage.

11.2.4. High-Temperature Corrosion and Creep Deformation

The mechanical properties of the Mg alloys containing 2–10%Al, combined with minor additions of Zn and Mn, are maintained up to 95–120 °C. However, elevated temperature adversely affects their mechanical properties and the corrosion resistance deteriorates with increasing temperatures. Magnesium alloys containing various elements (rare earths, Zn, Th, and Ag) except Al, generally possess much better elevated-temperature properties, but the more costly elemental additions combined with the specialized manufacturing technology required result in significantly higher costs [13].

The highest performance Mg alloys commercially available today are the Mg–Y–RE–Zr alloys (e.g., WE54, WE43). While many elevated-temperature applications may be met by minor additions to AZ or AM type alloys, it is probable that some hotter engine or transmissions applications may still require more "exotic" and costly alloys if these can be justified [30]. The ideal would be to develop a single high-temperature alloy to meet all requirements, since this would ensure significant volume production to minimize production and recycling costs and hence the commercial viability of the alloy [30]. Studies showed that such alloying elements are rare earth elements (La, Ce, Pr, Nd, Th, Er, Gd), alkaline earth elements (Be, Ca, Sr), or 3d-transition elements (Y, Sc). This approach led to the development of many creep-resistant Mg alloys, such as AXJ530, AJ52, AJ62, AE42, MRI153M, and MRI230D, besides other experimental alloys, such as $MgSc_{10}$, $MgSc_{15}Mn$, and MgY_4ScMn (the numbers 10, 15 and 4 correspond to wt% of the corresponding element) [31, 32].

The insufficient creep strength of several commercial cast alloys was the original reason for developing experimental Mg alloys. Low range of high temperatures can cause poor bearing-housing contact, leading to an oil leak, increased noise and vibration, and/or more serious problems if used for manufacturing various housings. It is also worth while to mention that one of the principal requirements for new alloys is their price competitive ability with existing Mg and Al alloys. This requirement combined with die castability issues reduces possible options to alloy systems containing Al or Zn as major alloying elements, using Mn, Si, Ca, Sr, and Ce-based mischmetal as relatively small additions. Table 11.3 summarizes the mechanical properties and corrosion performance of some experimental alloys as compared to commercial ones [33].

11.2.5. Microstructure and Corrosion Creep of Magnesium Die-Cast Alloys

Structural applications of Mg alloys are limited by creep strength rather than by oxidation [34]. The highest sensitivity to creep in a corrosive environment is observed in the alloy with the highest Al content. Some Mg alloys have been developed in the past several years to meet the needs of structural applications.

The synergetic effect of corrosion and stress on the viscoelasticity of Mg alloys has been given the general name *corrosion creep*; however, its influence on stress corrosion cracking can be identified as *environment-enhanced creep* (see Chapter 13). Magnesium alloys show creep even at room temperature. Creep deformation of a brittle Mg–Al alloy in a corrosive environment leads to the surface film breakup and it has been stated that the borate anion acts as a corrosion inhibitor. Accelerated creep of Mg alloys in aggressive media can be due to removal of the protective barrier and metal dissolution [35].

Table 11.3 Ultimate Tensile Strength, Elongation, and Corrosion Rate of Some Experimental and Commercial Cast Magnesium Alloys

ASTM designation	Major alloying elements	UTS at 20 °C	MPa at 150 °C	Elongation at 20 °C	% at 150 °C	Corrosion rate [a] ($mg/cm^2 \cdot day$)	Reference
AZ91D	Commercial (Al, Zn)	260	160	6	18	0.11	33
AE42	Commercial (Al, RE)	240	160	12	22	0.12	33
AE21	Commercial (Al, Si)	230	120	16	27	0.34	33
MRI 153M	Experimental Al, Ca, Sr, RE (Be free)	250	190	6	17	0.09	33
MRI 230D	Experimental (Al, Ca, Sr, RE)	235	205	5	16	0.10	33
AJ62Lx	Experimental (5.6–6.4% Al, 1.7–2.21% Sr)	276		12		0.04	100
AJ62x	Experimental (5.6–6.4% Al, 1.7–2.21% Sr)	240		7		0.11	100
AE	Experimental (5–9% RE)	280		10–12		0.02–0.04 [b]	101
AM 60	Commercial (6% Al, 0.13% Mn)					0.055 [b]	101
AXJ	Experimental (0.87–2.6% Ca, up to 0.17% Sr) [c]		[d]		[d]	approx. 0.11	39,102

[a] The 200 h salt-spray test (ASTM Standard B117).

[b] The 10 day salt-spray test (Astme Standard B117).

[c] Optimum overall castability at 2% Ca.

[d] Twenty percent greater creep strength than AE42 at 150 °C.

Source: Reference 33.

11.2.6. The OCP, i_{corr}, and Corrosion Creep

An electrochemical testing setup for assessing the intrinsic corrosion resistance of creep-resistant Mg alloys in aqueous environments and the effects of passivating surface films anticipated to develop in the presence of engine coolants is under development [36]. This approach was found to provide a platform for the eventual assessment of the durability of certain passivating layers expected to develop during exposure of the Mg alloys to aqueous coolants. Five Mg alloys were examined starting with 99.98% Mg ingot as a reference material in the form of ingot stock. The specimens of the sand-cast alloys (MRI202S and SC-1) were 100 mm × 100 mm × 10 mm thick. The high-pressure die-cast alloys (MRI230D and AM50) had dimensions of 140 mm × 100 mm × 3 mm thick. The surface of a specimen was polished using 600 grit silicon-carbide papers with water lubricant [36].

Open circuit potential (OCP) and potentiodynamic polarization measurements permit the deduction of i_{corr} from polarization measurements. The relative corrosion resistance values of the base metals (AMC-SC1, MRI-202S, MRI-230D, AM50, and 99.98% Mg) in an appropriate chosen environment (Table 11.4) (pH 6 buffer solution of potassium phosphate, monobasic, disodium containing 1000 ppm NaCl) were reproducible and gave particularly stable values for the OCP or corrosion potential.

Estimations of corrosion rates of the treated base metal using direct current (dc) polarization (Table 11.4) were in general agreement with the ranking of alloys from cyclic testing. Figure 11.7 illustrates the time evolution of OCP for the various materials on replicate runs during 24 hours [36].

OCP and potentiodynamic scans are useful for distinguishing between corrosion behaviors of creep-resistant Mg alloys. The corrosion potential of HPDC MRI 230D showed a larger variation than the two sand-cast alloys, MRI 202S and AMC-SC1, in both tests, for the test duration applied. The creep-resistant alloys required longer times to reach stable OCPs compared to the AM50 and pure Mg. AMC-SC1 exhibited a fluctuating open circuit curve. Effectively, OCPs for alloys of different composition and microstructures do not correlate with the corrosion rate calculated from the potentiodynamic scans. AMC-SC1 exhibited the most noble OCP yet had the highest calculated corrosion rate of the alloys tested [36]. Fundamentally, there is no relationship between potential and corrosion current. However, in several situations for a group of specimens of the same alloy, and in the presence of certain variables (one or two most of the time), it can be found that more active potentials correspond to increasing current densities or corrosion rates.

Table 11.4 Corrosion Rate (CR) of Examined Alloys in 1000 ppm Cl^- Solution at pH 6

Alloy	CR (mmyr)	Comments
Mg 99.98%	7.5	Machined from ingot
AMC-SC1	2.65	Sand-cast alloy with rare earth content; predominantly in the form of mischmetal
Dead Sea MRI 202S	7.58	Sand-cast alloy with rare earth content; predominantly in the form of Nd
Dead Sea MRI 230D	1.4	High-pressure die-cast alloy containing Ca as the primary creep-resistant constituent; also contains approxoimatly 4% Al
AM50	1.33	High-pressure die-cast alloy; nominally 5% Al, <1% Mn

Source: Reference 36.

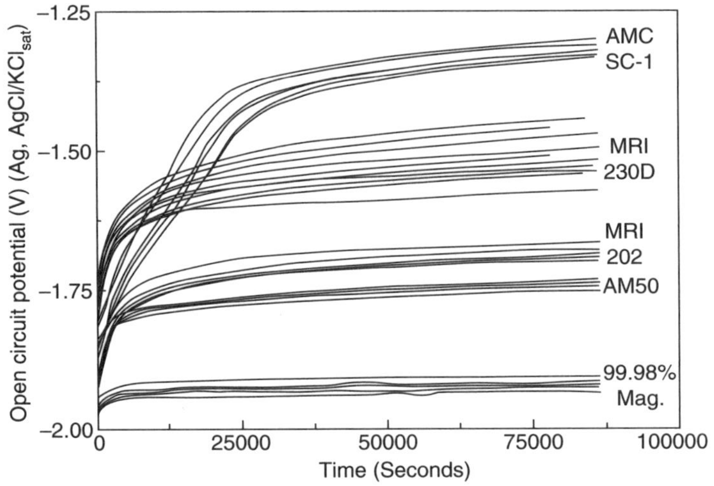

Figure 11.7 Time evolution of open-circuit potential (OCP) for the magnesium alloys with Ag, AgCl/KCl saturated 36.

This fragile relationship between corrosion rate and corrosion potential is hard to find in the case of Mg and its different alloys because of the formation of partially protective hydroxide in general and especially when this is joined with microstructures sensitive to creep.

11.2.7. Corrosion Creep and Aging

The surface of the Mg alloy AZ91D consists of an oxide-metallic film with a thickness of about 0.15 μm containing Mg and A1 oxides. If high-pressure die-cast AZ91D is exposed to 160 °C, aging will occur, which will result in reduction of the Al level in the α matrix and precipitation of more β phase. The variations of the level of Al solute and the amount of β phase could increase the corrosion resistance and decrease the mechanical strength of the alloy. Therefore aging time of the die-cast AZ91D should be closely controlled. The detrimental aging effect can influence both mechanical properties and corrosion performance. It has been noticed that if the aging time is properly controlled, aging can actually lead to improved strength and good corrosion resistance [24].

Prolonged aging in air at 200 °C promotes an increase in the $MgO/A1_2O_3$ ratio from 9 ± 2 to 14 ± 2. The Al concentration gradient in the surface layer is rather significant. The external layer of as-cast specimens contains up to 32 at % A1, whereas α-Mg grains in the bulk contain, on average, about 5% A1. Aging leads to a substantial increase in the surface concentration of Al at the expense of the acceleration of its diffusion and the intensification of supersaturated solid solution decomposition. During creep tests at elevated temperatures, the morphology of the β phase is significantly affected by strain and, to a lesser extent, by the casting temperature. Prolonged exposure of the alloy at 200 °C promotes a considerable acceleration of diffusion processes and a higher segregation of Al in the surface layer. The amount of β phase in the surface layer of as-cast specimens is substantially higher than in the center of the specimen and is equal to 40–50% and 15–25%, respectively. A long-term influence of elevated temperature and strain during creep tests leads to a substantial growth of the amount of β phase and to a change in its morphology [37].

Thus the microstructural instability of die-cast Mg–Al alloys, such as AZ91, at elevated temperatures is disastrous for their creep strength. There is a body of evidence showing that decreasing the Al content in Mg–Al alloys results in improved creep strength. It would seem that grain-boundary sliding plays a major role in the creep deformation of these fine-grained materials and that sliding is enhanced by the discontinuous precipitation reaction that occurs in regions of supersaturated α-Mg at elevated temperatures [38].

Reducing the volume fraction of β by lowering the Al content, such as in alloys AM50 and AM60, is then desired. The creep-resistant alloys AS21 and AE42 certainly have less supersaturated α-Mg in the die-cast condition than AZ91, and as a result there is less discontinuous precipitation during high-temperature creep. It has also been found that section thickness of the castings has a significant effect on the yield strength of the alloy [38].

11.2.8. Corrosion Creep of High-Strength AE42 and MEZ

The creep behavior of die-cast Mg alloys is examined for the high-temperature alloys AE42 and MEZ. Creep behavior in these fine-grained die castings is dependent on the stability of the near grain-boundary microstructure and is improved by rare earth element additions and reductions in Al content. These alloys, while having lower yield and tensile strengths than alloys with higher Al contents, exhibit considerable improvement in creep resistance, especially above 150 °C. The design of improved high-temperature Mg alloys must address both grain-boundary strength and stability as well as improved creep resistance by solid solution and precipitation strengthening [31].

Additional studies [39] have shown that in Al-containing alloys, $Al_{11}(RE)_3$ readily decomposes at high temperature to $Al_2(RE)$ and $Mg_{17}Al_{12}$. The higher tensile strength of AE42, which can be attributed in part to solid solution strengthening by Al, does not translate to higher long-time creep resistance when compared to MEZ. It has been shown that changes in near-grain boundary morphology can be observed after creep exposure for AE42, while no such changes are observed in MEZ. This indicates the potential importance of microstructural stability on creep resistance in these fine-grained die-cast alloys [31].

The general observations reported indicate that near-grain boundary microstructure is a controlling parameter in fine-grained die castings when little intragranular strengthening is derived from solid solution or from fine homogeneous precipitation in grain interiors. Because of this, the control of grain boundary microstructure is critical in the design of new high-temperature alloys. It is also important to consider additional matrix strengthening processes and further improvements in elevated-temperature grain boundary strength and stability if significant high-temperature uses for Mg alloys are to be realized [31].

11.3. INFLUENCE OF THE MICROSTRUCTURE, DIFFERENT PHASES, AND WELDING

11.3.1. Influence of Heat Treatments

Heat treatment can change the microstructure of Mg alloys. Aging makes Al atoms diffuse toward grain boundaries and form precipitation of the β phase, thus reducing the Al

Table 11.5 Heat-Treated AZ91 Corrosion Rates

Condition	Corrosion rate (mg/cm^2.day)
As-cast	10
T4	11
T6-aged at 120 °C	6
T6-aged at 205 °C	1

Sources: References 3 and 40.

concentration in the α-Mg matrix. For example, the Al content in the α-Mg matrix of Mg–9Al alloy decreases from 9% to 3% [8].

Heat treatment can drastically alter the size, amount, and distribution of the precipitated β phase, $Mg_{17}Al_{12}$, which in turn alters the corrosion behavior of IM Mg–Al alloys. A T4 heat treatment (solution heat treatment only for 16 h at 415 °C to homogenize the alloy) increases the corrosion rate slightly compared to the as-cast material, as shown in Table 11.5. Aune [9] attributed this increase to the resolution of β particles and release of elemental Fe. Aune [9] completed two T6 treatments with different aging times and temperatures—the T6 being a T4 followed by a T5 treatment. Both corrosion rates were well below those of as-cast samples and of samples given a T4 heat treatment (from 60% to 10% less for T6-aged at 120 and 205 °C, respectively). The lower temperature aging treatment formed a speckled precipitate of β particles, whereas the higher temperature treatment formed a discontinuous plate-like β precipitate with more surfaces. AZ91 and Mg–7.5Al were studied and it was found that the weight corrosion rate for the T6 treatment was only 1.1 ± 0.4 mm/y, and less than 3.2 and 4.7 mm/y, respectively, for T4 treatment immersion in 3.5% NaCl aqueous solution [8].

In the as-cast condition, compositional variations orient the growth of filiform corrosion. In homogenized alloys, filiform corrosion propagates transgranularly along crystallographic directions. In the T6 temper, filiform corrosion follows the same crystalline directionality and usually stops at or close to the grain boundary. Corrosion attacks the Al-depleted region between the β phase lamellae along the grain boundary, but passivates after a few minutes of propagation. The improved corrosion behavior results from the presence of β particles [40].

Eliezer et al. [41] and Lunder et al. [26]. investigated the corrosion of AZ91 by different heat treatments (F, T4, and T6 treatments) and found that the corrosion rate decreased in the sequence T4, F, and T6 treatments. Studies showed that corrosion resistance decreased in the sequence $Mg_{17}Al_{12}$ > after T6 treating > after T4 treatment.

Heating influences the salt-spray corrosion rate of die-cast commercial Mg–Al alloys. As shown in Figure. 11.8, alloys with higher residual-element (iron, nickel, and copper) concentrations were more negatively impacted by temperature. Using controlled-purity AZ91 alloy cast in both high-pressure and low-pressure forms, the contaminant tolerance limits have been defined as summarized in Table 11.6 for the as-cast (F), the solution treated (T4, held 16 h at 410 °C or 775 °F, and quenched), and the solution treated and aged (T6, held 16 h at 410 °C or 775 °F, quenched, and aged 4 h at 215 °C or 420 °F) [13].

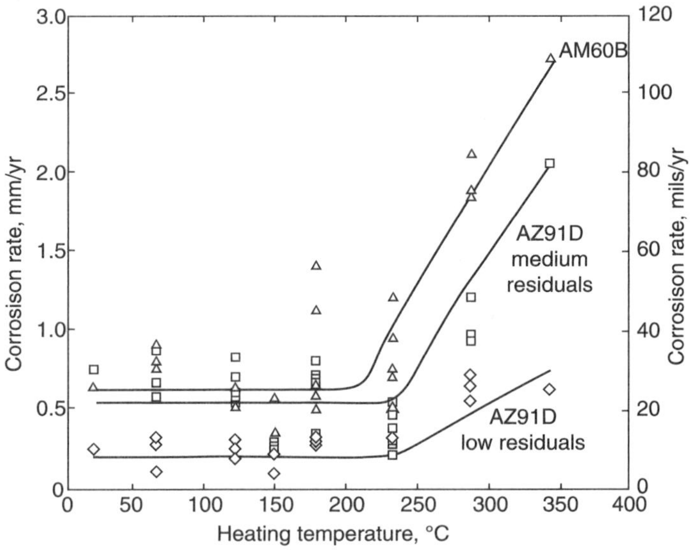

Figure 11.8 Contaminant tolerance limits versus temper and cast form for AZ91 alloy high-pressure die cast, 5–10 μm average grain size; low-pressure cast, 100–200 μm average grain size [13, 103].

11.3.2. Effect of Rapid Solidification

In rapid solidification (RS) technologies, including spray or droplet formation, continuous chill casting, and in situ melting, typical cooling rates are in the range of 10^5–10^7 °C/s [42]. Use of continuous chill casting typically produces a thin ribbon of metal, which is then broken into small particles. Then, as with the material formed by spray or droplet formation, the material is often consolidated and extruded. Improper processing can have a significant impact on corrosion behavior. The "chunk" effect [43] is caused by surface oxides on powder particles that lead to poor bonding within the final product [44]. Localized corrosion along these prior boundary oxides leads to particle-size pits and high corrosion rates [3]. Corrosion rates for atomized RS alloy are comparable to those of cast AZ91D, although those for melt spun RS alloys are significantly higher because of the "chunk" effect (Table 11.7).

Table 11.6 Contaminant Tolerance Limits

Contaminant (%)	Critical contaminant limit [a]			
	High pressure	Low pressure		
	F	F	T4	T6
Iron	0.032Mn	0.032Mn	0.035 Mn	0.046 Mn
Nickel	0.0050	0.0010	0.001	0.001
Copper	0.040	0.040	<0.010	0.040

[a]Tolerance limits expressed in wt% except for iron, which is expressed as the fraction of the manganese content (e.g., the iron tolerance of 0.2% Mn alloy = 0.0064% Fe in F temper).

Source: Reference 13.

Table 11.7 Corrosion Rates[a] of Selected Materials

Material (Wt%)	Atomized	Chill cast	Cast
Mg–7.7A1–2.9Zn–6.6Ce–0.35Mn	50		
Mg–10.1A1–2.7Zn–1.4Y–0.44Mn	60		
Mg–10.2A1–3.2Zn–5.8Ce–2.7Mn		350	
Mg–11.1A1–2.4Zn–3.2Y		430	
AZ91D		28	

[a]Rates in mpy (1 mpy = 25 μm/yr).
Sources: References 44 and 45.

Non equilibrium phases in RS alloys can influence the corrosion behavior; for example, Makar and Kruger [45] noted an increase in protection against pit initiation in RS AZ61 compared to IM cast AZ61. Corrosion resistance is improved because the more homogeneous microstructures tend to disperse elements and particles that normally act as cathodic centers, and because the extended solubility of various elements may shift the electrode potentials of light alloys to more noble values (Figure 11.9) [46–48].

Using rapid solidification processing, a number of Mg alloys have been produced in the form of melt spun ribbon, which is then usually mechanically ground to powder, sealed in cans, and extruded to produce bars. Alloy EA55RS (Mg–5Al–5Zn–5Nd) is now available commercially. Microstructures of the bulk products consist of fine grains 0.3–5 μm in size and dispersions of compounds, such as $Mg_{17}Al_{12}$, Al_2Ca, Mg_3Nd, and $Mg_{12}Ce$ [19]. Tensile strengths may exceed 500 MN · m^{-2}, which compares with maximum values of 250–300 MN · m^{-2} for conventionally cast magnesium alloys. Some alloys show improved creep resistance at moderately elevated temperatures, but others undergo accelerated creep deformation)[47].

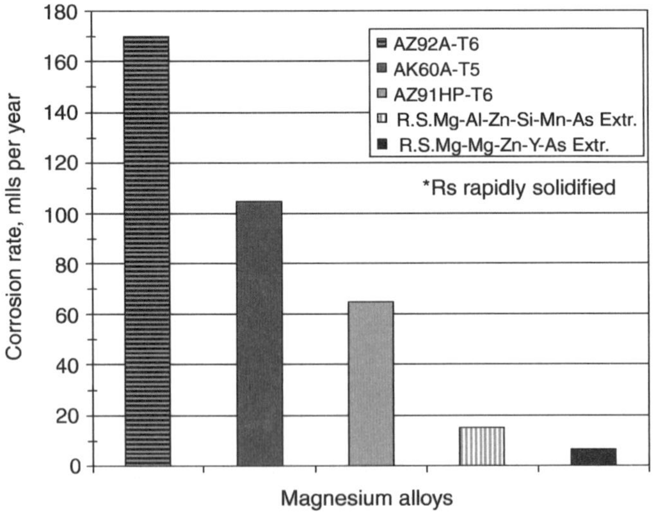

Figure 11.9 Corrosion rates (1 mpy ≈ 25 μm/yr) of rapidly solidified magnesium alloys tested in 3% NaCl at 21 °C compared with some commercial cast alloys (Extr = extruded) [47, 48]. [http://www.Ingentaconnect.com/content/maney/mst and www.maney.co.uk/journals/mst].

Table 11.8 Mg–Al Alloy Corrosion Rates[a]

Material	Corrosion Rate (mpy)
RS Mg–Al–Zn–Si–Mn	15
RS Mg–Zn–Al–Y	8
RS Mg–Zn–Al–Nd	11
AZ91HP-T6	82

[a]Rates in mpy (1 mpy = 25 μm/yr).
Source: Reference 3.

Effect of Alloying Elements During Rapid Solidification Hehmann et al. [49], experimentally measured the solid solubility extensions of 22 RS Mg alloys with extension factors ranging from 1.5 × to 1000 × . RS Mg–Al alloys with a maximum terminal solid solubility of 23.4 wt% have decreasing corrosion rates with increasing Al contents from 10 to 40 wt%. RS Mg–Al alloys, as do IM Mg–Al alloys, require further alloying elements to improve mechanical properties. Y, Mn, Nd, and Ce have been identified as beneficial to corrosion resistance, whereas Si, Zn, Ca, and Li have been identified as harmful to corrosion resistance [50]. The Mn acts in the same manner as for the IM alloys, by combining with the Al and Fe to form A1(Mn, Fe) intermetal [51]. This effect causes the low corrosion rate of RS Mg–Zn–Al–Si–Mn alloy (Table 11.8), despite the presence of detrimental Zn and Si.

Rare earth alloying elements (Y, Nd, Ce, and Pr) result in corrosion rates much lower than those of commercial AZ91HP-T6 alloy (Table 11.8). These elements form stable intermetallic particles in RS Mg–Al alloys, which, similar to the Mg_2Si particles, pin the grain boundaries and result in a refined microstructure. Because of the fast cooling rate, various forms of the intermetallics have been reported, such as $Mg_{17}Y_3$, Mg_3RE (RE = Ce, Nd, Pr), and Al_2Nd. The improved corrosion behavior of these alloys, compared to IM Mg–Al alloys, is attributed to the refined RS microstructure, formation of a protective film on the surface of the RS sample as a result of reaction of the saline solution with the rare earths, and the inertness of the second-phase particles [52].

Rapidly solidified $Mg_{97.16}Zn_{0.92}Y_{1.92}$ alloy exhibited high corrosion resistance, but the corrosion rate increased with increasing heat treatment temperature. Aging reduces corrosion resistance of Mg–RE alloys. For instance, the corrosion resistance of RS Mg–Zn–Y alloy decreased due to $Mg_{24}Y_5$ precipitation caused by heat treatment. Aging treatment of AZ80 has little influence on the fatigue life in air and in corrosive media at higher stress level, but remarkably improves the fatigue life in corrosive media at lower stress level [8].

11.3.3. Influence of the Microstructure of Some Mg Alloys

11.3.3.1. AZ and AM Alloys

Bender et al. [53] examined the corrosion rate of AZ, AM, and some Mg–Al–Mn–Li alloys in 0.01 M sodium chloride solution with special consideration of iron content. Polarization measurements were carried out with appropriate current densities at pH 9 (24 ± 3 °C) under argon atmosphere and RDE with 2000 rpm. The die-cast alloy AZ91-DC with a current density of 0.008 A/m^2 shows the best corrosion resistance followed by AZ91-PC with a current density of 0.009 A/m^2 and AZ31 with 0.012 A/m^2. The cast alloy AM20 shows a current density of approximately 0.01 A/m^2. The Mg–Al–Mn alloys alloyed with Li show higher values of the current density (approximately 0.018 A/m^2) than pure AM20 [53] (See Chapter 18, Section 18.1).

Regarding the different Fe concentrations of Mg–Al–Mn–Li alloys, no dependence of the polarization curves and polarization data on the Fe-concentration could be found. However, the anodic current density increases most for the AM20 alloy modified with Li, which has the lowest Al concentration (2%) and the highest Fe concentration ($< 0.034\%$). This shows clearly that by exceeding the maximal permissible value of Fe content, the corrosion behavior of the alloy is no longer dependent on the other alloying elements, but the Fe content determines the corrosion rate. The results obtained from noise electrochemical measurements were more convincing because of the high resolution of the method and reflects directly the real behavior of the metal without external polarization. The clear increase in the charge calculated from the noise measurements was considered to determine the influence of decreasing Al concentration and increasing Fe concentration on Mg alloys. A dependence of the charge on different Fe concentrations in the Mg–Al–Mn alloys modified with Li could be found. The charge increased with increasing Fe concentration. Furthermore, a dependence of the charge on Al concentration and fabrication process is found. The charge increased with decreasing Al concentration. The alloys AZ91-DC and AZ91-PC have nominally the same composition, but the charge for AZ91-PC is higher due to the macrograined structure in comparison to AZ91-DC, which has a fine-grained structure [53].

11.3.3.2. Influence of the Copper–Aluminum Phases

Copper is one of the critical elements that can enrich in Mg alloys during recycling of postconsumer scrap. Sources of copper are coatings on used Mg parts or Al alloy contaminations in Mg scrap. The tolerance limit for copper in AZ91 alloy to guarantee a sufficient corrosion resistance is ~0.07% for high-pressure die casting with a grain size of 5–10 μm and 0.04% for gravity die casting with a grain size of 100–200 μm. The influence of various alloying elements including copper on the corrosion resistance of magnesium (binary systems) has been studied [4]. More recent work suggests that some of the tolerance limits are too strict, especially for copper [54–56].

Copper contents have been varied between 0 and 2 wt% in gravity die-cast AZ91D alloy. The resulting microstructures were studied by LM, SEM + EDX, and XRD. Five different AZ91-based alloys were created by gravity die casting at a melt temperature between 745 and 782 °C. Starting with the AZ91D alloy (considered as reference), four alloys with increasing content of Cu were prepared—0.25, 0.5, 1.0, and 2%. Cross sections of 25 mm were taken from the ingots. The specimens were ground with silicon carbide paper to 2500 grit, and then polished with OPS™ before etching for microstructure observations [56].

All AZ91-based alloys, regardless of the Cu content, showed a typical dendritic cast microstructure. The addition of Cu showed a grain-refining effect, which was observed if the Cu content exceeded 0.5%. The reference alloy as well as the alloy with 0.25% Cu has an average grain (dendrite) size of about 500 μm. For the alloys with 0.5% and 1% Cu, the grain size was reduced to 250 μm. The smallest grain size of 200 μm was found for the alloy with 2% Cu. The microstructure of all AZ91 alloys consisted of three major phases. The primary Mg solid solution α (in dendritic morphology) was surrounded by an interdendritic secondary α and β'. The β' phase was mostly found as solid β (divorced eutectic $Mg_{17}Al_{12}$)

and only rarely as a coupled eutectic of α and β. The eutectic and secondary α have generally higher Al concentrations in solid solution than the primary α.

Already having low Cu content (0.25 wt% Cu), ternary Mg–Al–Cu intermetallic phases were observed.

Further increasing Cu concentrations resulted in more Cu-rich phases, and a significantly reduced corrosion resistance was found. The secondary phases formed due to the presence of an excess of Cu–Al and Zn surround the primary α dendrites. The dominant secondary phase is the β phase ($Mg_{17}Al_{12}$), which is on the one hand able to dissolve some of the zinc and can on the other hand embed some of the Cu-rich phases. With increasing Cu content in the alloy more Cu phases can be identified. At low contents (0–0.5% Cu), the dominating Cu phase was $Mg_6Al_7Cu_3$. In the medium Cu range (0.5–1%), more and more of the $(Mg,Al)_2Cu$ phase appeared, which was found in a Zn-free and a Zn-containing morphology. At the highest Cu concentration of 2%, another more Cu-rich precipitate of the type $(Mg,Al)Cu_2$ occurred. The diffraction pattern of the β phase reveals a decrease in intensity with increasing Cu content, indicating a reduced amount of β phase present in the alloys with higher Cu content [56].

EDX element concentration mappings across the various phases confirmed the higher Al concentrations of the interdendritic α in comparison to the primary α. There is a gradient of Al concentration from the grain boundaries toward the center of the grains, with the highest concentrations for the β phase, followed by the interdendritic secondary α and the primary α dendrites, which form the center of the grains. The amount of Al in solid solution varies from about 6% to 11% in the primary α and from 11% to 27% for the secondary interdendritic α. Randomly distributed in the alloys, but in a much smaller amount, further intermetallic phases were detected, such as Al_8Mn_5 and Mg_2Si. With the addition of Cu, additional phases formed, which were compounds of Mg, Al, Zn, and Cu in various combinations, depending on the amount of Cu added. Zn, however, seems to be solvated in the compounds without forming separate Zn-based phases [56].

Corrosion Forms and Resistance Every corrosion specimen was grinded with silicon carbide paper (1200 grit), cooled with deionized water, and cleaned in alcohol for corrosion resistance testing. Salt-spray testing for 48 h with 5% NaCl solution (pH 6.5) was performed according to DIN 50021. Potentiodynamic polarization measurements were conducted in 5% NaCl solution and the pH value of 11 was adjusted in certain manipulations using NaOH. Ag,AgCl/KCl saturated reference electrode was employed. After 30 min recording of the free corrosion potential, the polarization scan was started from −200 mV relative to the free corrosion potential with a scan rate of 0.2 mV/s. The test was terminated when a corrosion current of 10 mA was exceeded. From the cathodic branch of the polarization curve, the corrosion rate was determined using the Tafel slope. For determination of the polarization resistance from long-term corrosion tests, the specimens were polarized to ±10 mV relative to the free corrosion potential and the polarization curve was recorded. This was repeated every hour for 20 h. The polarization resistance was calculated from the slope of the polarization curve [56].

The major corrosion mechanism appeared to be a deep filiform type of corrosion, which spreads from certain starting points across the surface. With increasing Cu content in the AZ91 alloy, the surface area covered by filiform corrosion marks increased from about 5% to nearly 98% for the alloy with 2% Cu. However, for higher Cu contents, strong localized pitting corrosion was observed as well. The localized pitting corrosion was observed for Cu contents equal to or higher than 0.5% [56].

The effect of Cu additions on the local potential distribution on a specimen surface is estimated within the limits of electrochemical pen (EC-pen) equipment. The EC-pen was employed to measure the locally resolved free corrosion potential. A pen with 5% NaCl solution at pH 6.5 was used. A surface area of 6 mm $\times$ 6 mm was scanned using a step width of 0.5 mm and a measuring time of 3 s per step. The wetted surface area in contact with the electrolyte was about 1 mm^2. The reference electrode of the EC-pen had a potential of 0.037 V versus the Ag/AgCl electrode. The reference specimen AZ91D has a more or less uniform potential distribution (-1584.8 ± 10.9 mV), while a Cu-containing specimen revealed nobler potential with increasing variations (-1567.4 ± 33.2 mV). This means that the risk of localized galvanic corrosion was increasing [56].

The corrosion rates determined from the potentiodynamic polarization measurements at pH 6–7 showed that with increasing Cu content the corrosion rate increases. The lowest corrosion rate of 1.1 mm/y was measured for the reference alloy. With 0.25% Cu the corrosion rate was found to be 2.5 mm/y and increased further to 4.1 (0.5% Cu), 4.5 (1% Cu), and finally up to 12.1 mm/y for 2% Cu content. This trend is confirmed by the long-term polarization resistance measurements and the weight losses after 48 h of exposure to the salt-spray corrosion test. All corrosion tests performed at pH 11 allow the formation of more or less stable passive films. A moderate increase of the corrosion rate from 1 to 4 mm/y was observed for up to 1% Cu in the AZ91 alloy. This would suggest that although the $Mg_6Al_7Cu_3$ and $(Mg,Al)_2Cu$ phases are detrimental for the corrosion properties, they do not influence the passivation behavior as strongly. With the occurrence of the $(Mg,Al)Cu_2$ phase at 2% Cu content, the passivation behavior is strongly affected and the corrosion rate increases to 12 mm/y [56].

The real influence of Cu on the passivation behavior is better revealed in long-term corrosion tests. Measurement of the polarization resistance reveals that Cu-free AZ91 has a high and constant polarization resistance over long periods. Any Cu addition shifts the free corrosion potential to more noble potentials and reduces the polarization resistance and it can continuously decrease to rather low values if corrosion continues. It can be assumed that this is related to a gradual enrichment of Cu on the surface of the Mg alloy [56].

Looking at the above results, it seems reasonable to define the Cu tolerance limit at about 0.2% if no other modification of the alloy is done. This would give sufficient corrosion resistance in many applications. However, by increasing the amount of β phase, even higher tolerance limits up to 0.5% seem possible. If the latter is able to cover and embed the more critical Cu-rich phases, the corrosion rates are expected to remain low. To be more effective, this should include increasing the corrosion resistance and modification of the electrochemical potential of the β phase by alloying. Additionally, a more uniform distribution of Al in the α phase might contribute to more stable and uniform passive layers. Considering these aspects should allow developing a secondary AZ-based Mg alloy with reasonable high tolerance limits for impurities. However, Cu contents higher than 1% should be avoided as they are involved with the formation of intermetallics with Cu contents above 50 at %. Such intermetallics are too noble and are extremely detrimental to the corrosion resistance of the alloys [56].

11.3.3.3. Microstructure and Corrosion Resistance of AXJ530 Ingots and Billets

The AZ91D alloy is known to lose its creep resistance above 120 °C due mainly to the poor elevated-temperature properties of $Mg_{17}Al_{12}$ (β phase). Decreasing the Al content and adding alloying elements promotes the formation of intermetallic phases that are stable at elevated temperature. The AXJ530 magnesium alloy (4.4 wt% Al, 2.6 wt% Ca, and 0.15 wt% Sr) offers excellent castability, good creep resistance, and low cost, which meet material requirements for automotive power train applications. Indeed, the addition of Ca to the Mg–Al system leads to the formation of $(Mg,Al)_2Ca$ responsible for the improved creep resistance of the alloy. Specimens of AXJ530 alloy were obtained from billets cast in permanent molds without (AXJ530-SFTC) and with electromagnetic stirring (AXJ530-ESTC) and in large ingots, in order to produce relatively coarse phases and facilitate the observation of corrosion features. The liquidus and solidus of AXJ530 alloy used in this study were 607 °C and 524 °C, respectively, as determined by thermal analysis [57].

The corrosion behavior of AXJ530 magnesium alloy has been investigated at room temperature in a 3.5% NaCl solution at pH 6 and 25 °C. The corrosion behavior of the AXJ530 specimens was studied by potentiodynamic polarization, electrochemical noise analysis, and immersion tests. The phase composition, the morphology of the corroded surface, and its evolution with time were also analyzed. For potentiodynamic polarization experiments, the potential was scanned at a rate of 0.15 mV/s from −200 mV (with respect to the corrosion potential, E_{corr}) to an anodic potential corresponding to a current density of 1.0 mA/cm^2. Electrochemical noise measurements were carried out for specimens tested during 24 h immersion and the potential and current were simultaneously recorded at a frequency of 10 Hz [57].

The microstructure of specimens was studied with the scanning electron microscope (SEM). The spatial distribution of Al, Ca, and Mg in phases was characterized with an electron probe microanalyzer (EPMA), whereas the fraction of phases and porosity were determined using quantitative image analysis. The three examined phases are shown in Figure 11.10. In AXJ530 billets, fine lamellae in interdendritic spacing were observed. In addition, the higher volume fraction of $(Mg,Al)_2Ca$ phase leads to the formation of a more continuous, and then a more protective, network near the α-Mg grains. The fact that corrosion attack occurs first in eutectic α-Mg (Figure 11.10) can be attributed to geometric conditions favorable for a significant galvanic corrosion between the cathodic $(Mg,Al)_2Ca$ phase and the anodic α-Mg eutectic. Corrosion was found to spread within primary α-Mg grains after the eutectic α-Mg lamellae were completely attacked. Hence it is suggested that the preferential dissolution within the eutectic delays the onset of the corrosion attack of the primary α-grains [57].

In AXJ530-FSB (Frec strirring billet) and AXJ530-ESB specimens (electromagnetic stirring billet), the higher volume fraction of the cathodic $(Mg,Al)_2Ca$ phase and especially its lamellar form allow the formation of a galvanic cell with anodic α-Mg eutectic lamellae.

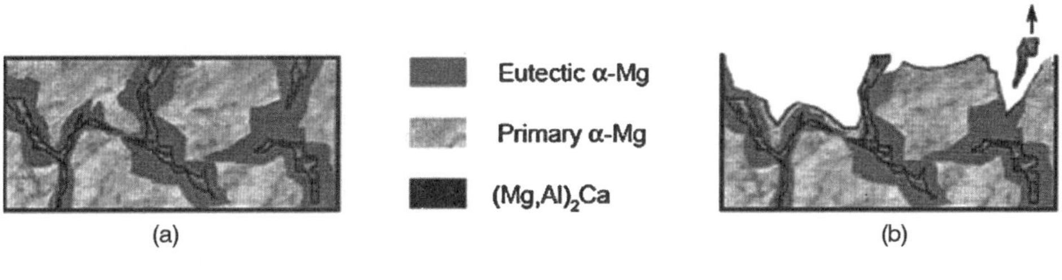

Figure 11.10 Schematic illustration of corrosion process at the surface ofAXJ530 ingot specimen: (a) before corrosion attack and (b) after corrosion attack [57].

Hence the corrosion attack initiates and propagates within the α-Mg eutectic, which delays the attack of primary α-Mg grains. Moreover, the cathodic $(Mg,Al)_2Ca$ phase forms a relatively continuous network, which prevents undermining by the preferential dissolution of the α-Mg eutectic[57].

Based on potentiodynamic polarization tests and electrochemical noise analysis, AXJ530 ingots showed lower corrosion resistance and higher activity, as compared to AXJ530-FSB and AXJ530-ESB specimens cut from billets. This behavior can be attributed to (1) the lower concentration of Al in primary α-Mg grains and (2) the low volume fraction of the cathodic $(Mg,Al)_2Ca$ phase, which inhibits the formation of a protective and continuous network around α-Mg grains. The presence of small and isolated $(Mg,Al)_2Ca$ particles and α-Mg eutectic also explains the low corrosion resistance of specimens cut from ingots. The sites left by undermined $(Mg,Al)_2Ca$ particles constitute possible occluded areas, which might favor localized corrosion. The mechanism explaining the corrosion of an AXJ530 ingot specimen, inspired from the work of Song et al. [58], is illustrated in Figure 11.10 [57].

11.3.3.4. Corrosion Resistance of AZ91D Die-Cast and Thixocast Alloys

Electrochemical studies of AZ91D-DC (die cast) (9% Al), AZ91D electromagnetically stirred billets thixocast (ESTC), AZ91D solidified freely thixocast (SFTC) billets, and AJ62x-DC (die cast) (4.4% Al and 2.6% Ca) specimens, were carried out for active–passive behavior and localized corrosion. Potentiodynamic and electrochemical noise measurements were made in alkaline chloride medium (0.1 M NaOH + 0.05 M NaCl + 2 mL H_2O_2) at 25 °C and pH 12.3 to investigate active–passive behavior and localized corrosion resistance [59].

Intense corrosion rate was observed generally at the beginning of the experiment and it decreased with immersion time. To identify clearly the active–passive behavior of Mg alloys, the current and potential noise values were varied from about 5 μA/cm^2 and −1230 mV/NHE for the active region to ∼ 0.5 μA/cm^2 and −1000 mV/NHE or more noble values for the passive region. The relative duration of the active and passive periods depends on the casting conditions. The best passive zone was observed for AJ62x-DC because of the corrosion products formed at the surface [59].

To compare the corrosion resistance of the Mg alloys, the spontaneous relative corrosion rate ($1/R_n$) is in accordance with the corrosion rate calculated from the polarization curves and decreases with immersion time. Analysis with the scanning reference electrode technique (SRET) shows the same tendency for the QEMF values recorded upon the specimen surface. The AZ91D-ESTC specimen has the best corrosion resistance followed by AZ91D-SFTC and AZ91D-DC specimens. SRET showed that the AJ62x specimen presented the biggest potential difference between the most active anode and the most active cathode and more numerous zones of intense localized corrosion [59].

To give additional information on the type of corrosion attack, an EN result analysis in a frequency domain shows two types of corrosion occurring on AJ62x-DC and AZ91D-DC specimens: localized dense pitted areas that turn to conventional pitting after 8 h of immersion, while over a 16 h immersion period, the two thixocast specimens (AZ91D-SFTC and AZ91D-ESTC) show only a localized corrosion with dense pitted areas, a kind of metastable pitting [59].

11.3.3.5. Corrosion Resistance of Die-Cast and Thixocast AXJ530 in Alkaline Medium

The corrosion behavior of AXJ530 magnesium alloy has been investigated in a solution containing 0.05 M NaCl, 0.1 M NaOH, and 0.025 M H_2O_2 at pH 12.3. Die-cast and thixocast specimens were tested using potentiodynamic polarization, electrochemical noise analysis (ENA), electrochemical impedance spectroscopy (EIS), and immersion tests. Specimens for corrosion tests were cut from the flat sidewall 6.2 mm thick and were mechanically polished to a depth of about 2 mm. These tests showed that all specimens were passive in that solution. However, the passive film formed on thixocast specimens was found to be more protective and resistant to localized corrosion than that of die-cast specimens [60].

Figure 11.11 shows the typical potential and current noise records of AXJ530-DC and AXJ530-ESTC after 16 h of immersion as an example. For the Die-cast specimen, the first period of 4 h corresponds to the beginning of a steady passive state. Indeed, the corrosion potentials of all specimens showed a pronounced drift over the first hours of exposure. This drift is typical when protective or partially protective corrosion films are produced. Thus after 4 h of immersion, sharp potential and current peaks appeared, showing variations of amplitude within 3 mV and 0.15 μA, respectively. After 16 h of immersion, similar fluctuations of potential (~4 mV) were observed while the current changed into a pattern with higher variation of amplitude (0.2 μA). These fluctuations of potential and current come from the instability of the passive film and are related to the film breakdown and repassivation processes associated with metastable pitting [60].

Potential and current noise records of AXJ530-ESTC after 4 h of immersion (Figure 11.11 b) also show variations similar to those of die-cast specimens, with low amplitudes of potential and current of 6 mV and 0.05 μA, respectively. After 16 h of immersion, AXJ530-ESTC showed variations of potential and current within 1 mV and 0.03 μA, respectively. By analogy with the behavior of active–passive alloys, the amplitudes of potential and current after 4 h and after 16 h of immersion for the two thixocast specimens, in particular, for AXJ530-ESTC, indicate a passive state. The performance of the AXJ530-SFTC specimen has the same pattern or shows the same trend and is closer to the best passive behavior of the ESTC specimen [60].

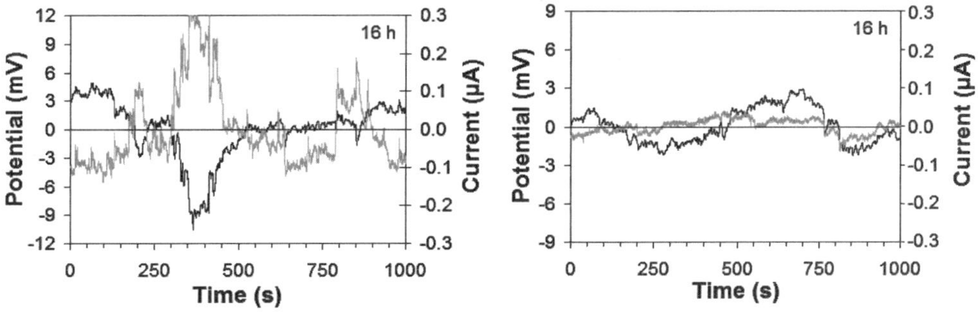

(a) (b)

Figure 11.11 Potential noise (black) and current noise (gray) for (a) AXJ530-DC and (b) AXJ530-ESTC for 1000 seconds after 16 h of immersion in alkaline solution [60].

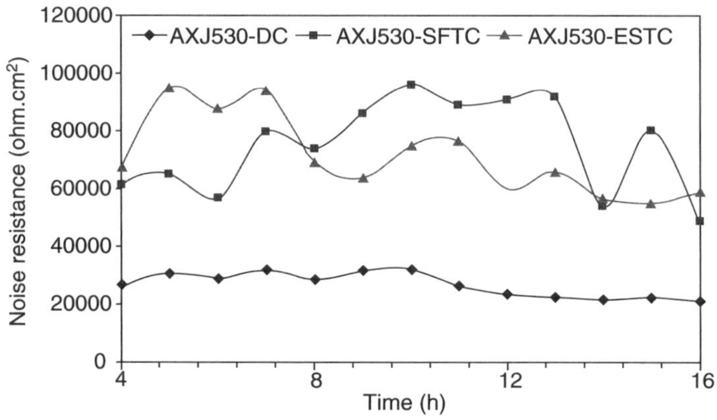

Figure 11.12 Noise resistance evolution for AXJ530 specimens for 4–16 hours of immersion in alkaline solution [60].

The evolution of the noise resistance R_n for the three AXJ530 specimens during the exposure to the alkaline solution was analyzed and is presented in Figure 11.12. The noise resistance R_n, which is defined as the ratio of the standard deviations of the potential noise (σ_E) and the current fluctuations (σ_I), has been often considered equivalent to the polarization resistance. Figure 11.12 shows clearly that the noise resistance R_n for the AXJ530-DC specimen is lower than that of the two thixocast specimens during immersion in the alkaline solution. The two thixocast specimens did not show any significant difference of behavior [60].

The electrochemical techniques used in this study show that die-cast and thixocast AXJ530 specimens are passive in the alkaline solution at pH 12.3 and 25 °C. Potentiodynamic polarization, ENA, and EIS indicate that the film formed on the two thixocast specimens is more protective and more resistant to localized corrosion than that formed on the die-cast specimens. According to the noise resistance, R_n, and the polarization resistance, the corrosion behavior of thixocast specimens is better than that of die-cast AXJ530 specimens. Numerous, short cracks are found after immersion tests and linked to the initiation and the repassivation of metastable pits. The density of these cracks is higher for the die-cast specimens than for the two thixocast ones. Few AXJ530 specimens sustain filiform corrosion, which suggests that highly resistant films can be formed naturally in he test solution [60].

11.3.4. Influence of Joining and Welding

In general, Mg alloys are difficult to weld due to the following reasons: oxidation, porosity and crack formation (especially when Mg alloys contain more than 6% Al and 1% Zn, or more than 3% Zn), and large fusion zone (FZ) and heat-affected zone (HAZ) as a consequence of excessive heat input. Therefore the conventional inert gas arc welding techniques are limited in the case of Mg alloys [61]. Welded Mg structures and localized corrosion and residual stresses, heated by welding, are found particularly dangerous to SCC resistance and so low-temperature thermal stress relief is the recommended practice. Shot peening and other mechanical processes that create favorable compressive surface residual stresses may also increase SCC resistance [62]. Welds on Mg–Al–Zn alloys should be aged or should be solution treated and aged to obtain good corrosion resistance in harsh environments and to reduce the risk of failure due to SCC [13].

Magnesium extruded alloys with approximately 3–8% Al and 0.5–0.8% Zn are susceptible to filiform corrosion and pitting corrosion in aqueous chloride solutions depending on the chloride concentration. The resistance of nonwelded alloys increases with Al content. On welding of the alloys, the corrosion resistance is determined by the Al/Mg proportion at the surface of the nonaffected material and the laser welding seam. The pits occur mainly in the HAZ of the welding seam [63].

Skar and Albright [64] investigated the corrosion properties of AM50A and AZ90D welds. The tests showed that the weld materials had poor corrosion resistance, and the corrosion rate of AM50A was higher than that of AZ91D. The AZ91D/AM50A welds showed corrosion rates comparable to AM50A alone. Embedded iron contaminants from wearing of the tool may be responsible for the poor corrosion properties of the welds. Partial melting of the β phase that easily dissolves iron may cause this wearing. As for dissimilar welding of Mg and Al alloys, Mg and weld seams neighboring the Mg side will corrode because of their lower potential [8].

Welding difficulties of Mg alloys can be reduced by applying low heat input with high power density and by applying shielding gas. The most suitable techniques that provide these characteristics are laser and electron beam welding processes. Since laser welding can be performed under ambient pressure, it is preferred over the electron beam technique. The effects of laser power and joint gap on the welding quality of 2 mm butt joints of ZE41A-T5 sand castings were investigated using a continuous wave 4 kW Nd: YAG laser system and 1.6 mm EZ33A-T5 filler wire at a welding speed of 6 m/min and surface defocusing. Smooth weld profiles with the minor welding defects were obtained at a laser power of 4 kW and joint gap of 0.3–0.4 mm. The hardness values in the fusion zone are similar to those of the base metal but there is a drop in the heat-affected zone after a natural aging of 18 months. A significant grain refinement was observed in the FZ due to high cooling rate and prolific Zr nuclei. No grain coarsening was observed in the HAZ [61].

Haferkamp et al. [65] made some investigations on the corrosion behavior of laser welded AZ91D in synthetic seawater. The corrosion morphology of the weld zone of AZ91D weld with gas tungsten arc (GTA) welding after a 48 h salt fog test was very evident. The corrosion morphology of the weld zone of AZ31 with laser beam welding after a 24 h salt fog test was much better than that with GTA welding. Figure 11.13 shows that the corrosion did not attack the weld zone. Also, the polarization curves of the AZ91D weld and its base material in 3.5% NaCl indicated that the corrosion rate of the weld zone is half that of the base material, very possibly because of the fine grain size, composition, and microstructural considerations. The laser beam welded material, AZ61HP, was found also to have excellent resistance. The high welding speed and fast cooling rate of the welds could improve the corrosion resistance of the weld zone connected with the same material because of its fine grain sizes and of solid solutions with higher Al content. The corrosion tendency of rapidly solidified welds, because of the high-power laser welding, is relatively low [8, 63].

The polarization curves of the AZ91D weld and its base materials (shown in Figure 11.14) indicate that the corrosion rate of the weld zone is half that of the base material due to its finer grain size. Considering the Tafel slope method for the cathodic branch of the polarization curves, the corrosion rate of the HAZ is the highest one, followed by the weld region, and finally, the best corrosion resistance is for the base material [8].

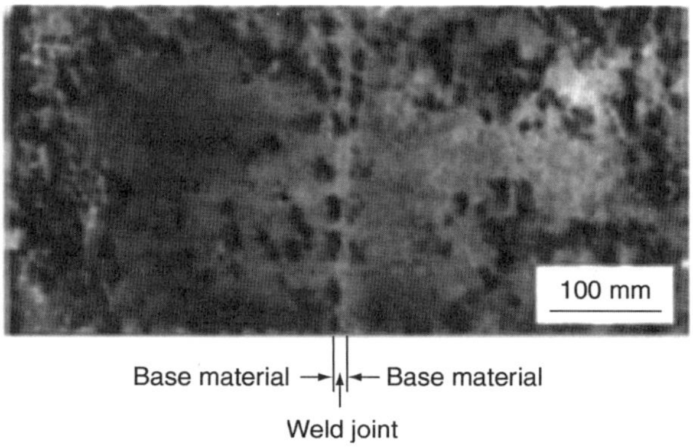

Figure 11.13 Corrosion morphology of AZ31 weld zone with laser beam welding. The polarization curves of the AZ91D weld and its base materials (in 3.5% NaCl solutions) indicated that the corrosion rate of the weld zone is half that of the base material [8].

Laser Butt Welding and SCC The use of laser butt welding (LBW) for cladding, alloying, and welding of Mg alloys is becoming an essential tool for successful component development. The LBW of Mg alloys shows joint efficiency (joint strength/base metal strtength), even up to 100%, whereas that of the convenient fusion welding processes such as tungsten inert gas is around 70–90%. Kannan et al. [66] examined the stress corrosion cracking (SCC) of laser beam welded AZ31 Mg alloy. This was carried out using slow strain rate tensile (SSRT) and constant load tests. The strain rate in air was 10^{-6}/s, and in the corrosive environment (148 mg/L Na_2SO_4, 165 mg/L NaCl, 138 mg/L $NaHCO_3$), the strain rate was 10^{-8}/s. The SCC tests showed that the laser beam welded AZ31 alloy failed in the fusion zone boundary when tested in the corrosive environment, in contrast to the failure in

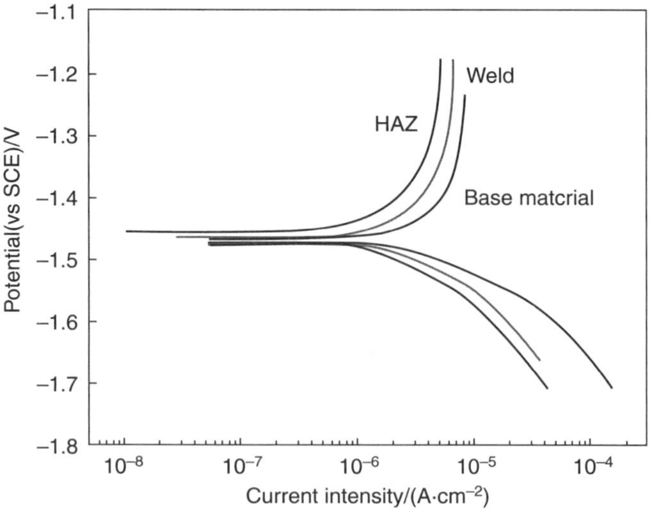

Figure 11.14 Polarization curves of the AZ91D weld and its base materials in 3.5% NaCl solutions [8].

the FZ base region observed in tests in air. During LBW, Mg vaporizes easily and this leads to a higher concentration of Al in the grain boundary of the finer part of the FZ. This can create local galvanic cell and initiate corrosion failure [66].

The Skin and the Bulk of Cast Alloys Hot chamber and cold chamber die-cast processes are differentiated by the metal-injection method. To date, most previous studies have investigated the corrosion properties of the cold chamber AZ91D die casting with a thick gauge section starting at 6 mm and going up to 40 mm in diameter. Blawert et al. [67] found that the skin of various cold chamber high-pressure die-cast (HPDC) Mg alloys had more inferior corrosion properties than did the bulk. In the absence of chemical conversion coating, removing the outermost skin by grinding clearly increases the corrosion resistance of the HPDC alloys.

11.3.5. Cold Chamber Processes

Considering arbitrarily the total skin between 0 and 100 μm and the outermost part of ~15–25 μm, the rest can be called interior skin and this can be followed by the bulk. It is important to define and describe the main parts in the thickness of the cast alloy (three) and relate that to the corrosion resistance of the alloy [68, 69].

Exterior Original Skin The following parameters and properties could be the cause of the low corrosion resistance of the outermost skin:

1. Contamination with Fe, Ni, and Cu can come from the die-casting mold and/or die lubricant. Certain fabrication processes, such as agitation during thixocasting, can accelerate the formation of some oxides at the skin–environment interface [57].
2. Organic surface contaminants should be removed with solvents from the surface of the specimens before polishing. Slight polishing exposes new surfaces with different microstructures and does not guarantee smooth and clean surfaces. However, it is difficult to remove superficial oxides and hydroxides in contact with the atmosphere without introducing another thin oxide layer (Beilby layer) during polishing [70].
3. Close to the surface of castings, the microstructure is fine and rapidly changing with depth. It is then important to mention that the skin depth is dependent on the casting conditions, the thickness of the casting wall, the mold temperature, and the alloy composition. Moreover, nonuniform cooling conditions over mold surfaces can lead to uneven microstructure at a given depth. The depth from the original surface is very important but difficult to evaluate with the routine practice of polishing [70].

Interior Skin There is often some microstructural discrepancy between the die-skin layer and the interior of the sample. Frequently, the microstructure of the interior skin is fine, homogeneous and less porous. Song et al. [58] inferred that a skin layer on the 6-mm-thick sample with a fine cast structure and a high fraction of $Al_{12}Mg_{17}$ β phase will exhibit better corrosion resistance than a sample with a coarser microstructure[71].

The Bulk Song et al. (1998) have studied the surface of specimens polished to remove 500 μm from the original casting surface and reach the bulk or its borders. Song et al found that when the cast skin was removed by 0.5 mm with abrasive paper, the obtained new surface had higher corrosion resistance than the deeper bulk (Song et al. 1998) [58]. It is suggested that the bulk that is close to the skin has fine microstructure because of a quicker solidification rate, then, it has a better corrosion property. Jin et al. (2007)[72] examined thixocast AZ91 alloy, the very deeper bulk specimen was not necessarily more susceptible to corrosion than that from the shallow depth bulk sections. Effectively, the bulks of many magnesium alloys are not homogeneous in composition and the microstructurec was more or less porous.

11.3.5.1. Skin and Bulk of HPDC AZ91D Alloy

The AZ91 alloy was processed in a cold chamber machine using a melt temperature of 690 °C and a pressure of 40 MPa. The alloy specimens were cast into step plates from which the 2 and 14 mm sections were used. Electrochemical experiments were performed in aqueous 5% NaCl solution, saturated with atmospheric oxygen and adjusted to pH 11 using NaOH in a stirred electrolyte at 22 °C. Every experiment consisted of three subsequent tests and the total test period was 23 hours [67]:

1. Thirty minute recordings of the free corrosion potential were taken.
2. Potentiodynamic polarization scan were done, starting from –200 mV relative to the free corrosion potential with a scans rate of 0.2 mV/s. The test was terminated when a corrosion current density of 0.1 mA/cm^2 was exceeded to minimize the damage on the specimen surface for further studies. The corrosion rate was calculated from the cathodic branch considering the current density at the intersection of the cathodic slope with the corrosion potential in the Tafel diagram.
3. Electrochemical impedance measurements at the free corrosion potential were carried out over the frequency range from 10 kHz to 0.01 Hz. The amplitude of the sinusoidal signals was 10 mV. The measurements were repeated every hour for a total of 22 hours. The charge transfer resistance was calculated from the intersections of the circle with the real axis (0° phase shifts) giving the solution resistance and the sum of the solution and the charge transfer resistance (assuming a simple Randles circuit model) [67].

The surface areas of specimens from the working electrode for scanning reference electrode technology (SRET) studies were 10 mm × 10 mm, cut from the 14 mm thick HPDC AZ91 alloy plates. The skins were ground, but only just to make the metal and the resin have the same level for the probe scan. The working surfaces of these electrodes were ground with SiC abrasive paper down to 1200 grit, washed with deionized water, and wiped immediately with tissue paper. The electrolyte was prepared using 500 ppm chloride ions added in the form of NaCl (about 0.014 M) [67]. Dilute nonagitated solutions are recommended for SRET pitting corrosion studies, although this shifts the pH to more alkaline values and may cause passivation.

Skin and Bulk Microstructural Regions Four different regions with different microstructures and corrosion performance can be distinguished: the outermost contaminated skin (0–15 μm), the fine globular subskin region (15–100 μm), the transition region with a mixture of globular grains and dendrites (100–500 μm), and finally the dendritic bulk (500 μm onwards). The corrosion resistance increases generally in the same order from the surface toward the center, except under certain conditions. The HPDC processing parameters and the cast component can influence the thickness of the skin layer that can be identified to a great extent by the presence of fine grains. The skin region in this case is ~100 μm thick.

In the center, especially in the thicker castings, larger variations of the corrosion resistance can be found. The boundary between the skin region and the bulk is not sharply defined. It is a smooth transition with increasing amount of primary dendritic α grains down to a depth of 500 μm. From 500 μm onwards, the microstructure is dominated by the larger primary α dendrites. Considering the microstructural properties of the 2 mm as well as the 14 mm thick plates, the wall thickness of the casting has no major influence on the extension of the skin region. This would suggest that the extension is controlled by the cooling on the die walls and not by the volume of the melt [67].

Corrosion Rate of the Exterior Skin Considering the skin as the outermost part of the casting and that nothing is removed from it, the corrosion resistance is low. The outermost skin is about 25 μm thick. The microstructure of the skin region with an average grain size of 5 μm is ten times finer in comparison to the center of the castings. Furthermore, the morphology is different. One can assume that the grain boundaries are the more active sites for a corrosion attack. Thus the higher fraction of them with smaller grain size causes higher corrosion rates. Present results suggest indirectly that only the presence of the β phase network is the determining aspect for good corrosion resistance. The enrichment of some impurities such as Fe in the near surface can increase corrosion rates [67].

Interior Skin If the outermost skin is removed, a better corrosion resistance can be obtained. This is often attribued to a finer microstructure, more uniform distribution of precipitates, and a dense protecting network of β phase along the grain boundaries in the case of Al-based alloys [67]. However, the corrosion measurements including potentiodynamic polarization and long-term electrochemical impedance spectroscopy (EIS) revealed a lower corrosion resistance of the original as-cast skin at about 100 μm deep. The thickness of this region is more or less independent of the thickness of the casting section. The skin region down to a depth of 100 μm is more or less globular, typical for an alloy solidified at the cold wall of a die. The surface region is free from any dendritic grain growth, which influences the amount and distribution of the β-phase containing the most important alloying element. This could explain the better corrosion behavior [67].

A higher number of β phase precipitates and smaller distances between them are believed to be the reason for a stronger galvanic attack in the skin region compared to the bulk. The strong uniform corrosion formed of numerous galvanic cells of this fine microstructure can result in a more uniform protective film. The interdendritic α phase, which forms a network along the grain boundaries, does not have the same capability to form a corrosion barrier as the β phase [67]. In the hot chamber for the die-cast process of thin plates, the corrosion performance of the skin region depends strongly on the amount and distribution of the α eutectic phase rich in Al. This latter has been found to be composed of fine β-$Al_{12}Mg_{17}$ particles and Al–Mn particles of more noble phases containing the matrix α (low Al) phase [71].

The Bulk Region The bulk is composed of much larger primary α dendrites and the interdendritic spacings are filled with a mixture of secondary interdendritic α

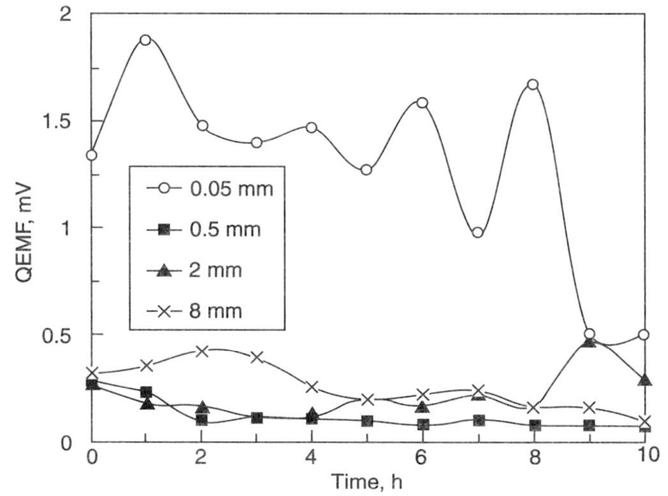

Figure 11.15 QEMF as a function of immersion time for different depths of HPDC AZ91alloy in 0.014 M chloride solution [67].

(also with a higher Al concentration) and β phase. The grain size of the interdendritic secondary α and the distribution of the β phase are similar to the skin region. However, the β phase precipitates are coarser than in the skin and the distances between them are larger [67].

To better interpret these results, the term *quasi-electromotive force* (QEMF), which describes the difference between the most positive potential and the most negative potential, was considered. The QEMF values as a function of immersion time for different depth specimens are plotted in Figure 11.15. It is seen that the 0.05 mm depth sample has the highest QEMF values and the 0.5 mm depth sample has the lowest QEMF values. However, the localized corrosion takes place on the anode zone [67].

A detailed study of the microstructure and the corrosion performance of HPDC AZ91 in the near-surface region down to a depth of 1 mm can be expressed in general terms by the scheme in Figure 11.16 [67].

11.3.5.2. Skin and Bulk Corrosion Resistance of Thixocast Specimens

The corrosion experiments were conducted using thixocast AZ91-TC alloy specimens that were cast into step plates at the GKSS Research Institute in Geesthacht, Germany. This was compared to AZ91-DC alloy specimens in a dilute chloride solution at room temperature. For exterior skin examination, the as-cast surface was ground just to make the metal and the resin surfaces have the same level for the probe scan; that is, the cast scale was not completely removed. All the other specimens for SRET in situ measurement (see Chapter 18) were 10 mm × 10 mm × 14 mm, ground with SiC abrasive paper down to 1200 grit. By observing the potential distribution on the metal surface, the corrosion form (general corrosion and/or localized corrosion) and corrosion rate can be estimated. Three-dimensional(3D) images of the potential distribution on the skins and at different bulk depths of 14 mm thick plates were obtained mainly in a neutral dilute chloride solution (0.014 M NaCl or 500 ppm Cl^-). The results showed that the as cast skins of AZ91-TC and AZ91-DC alloys had much higher QEMF values than the bulk samples [72].

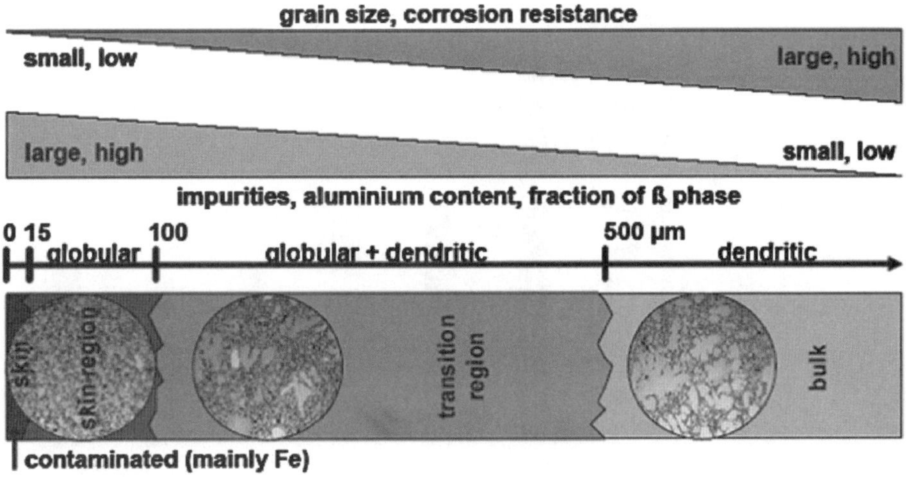

Figure 11.16 Correlation between microstructural features and corrosion properties of HPDC AZ91D alloy in chloride media [67].

Figure 11.17 shows the 10 hour average QEMF values of the skins and different bulk depth sections of the AZ91-TC sample across the whole thickness of the 14 mm thick sample. It is seen that the 6, 8, and 10 mm depth samples had lower QEMF values than that of the other depths [72].

The average value of the maximum local anode currents (MLACs) during the 10 hours was calculated for each depth of the bulk samples of AZ91-TC and AZ91-DC alloys; then the average values between the die and thixocast specimens were compared (Figure 11.18). It can be seen from Figure 11.18 that the average maximum local anode currents of AZ91-TC alloy were always lower than those of AZ91-DC alloy, meaning that the corrosion rate of AZ91-TC was lower than that of AZ91-DC for all depths. On average, the MLAC value (241 $\mu A \cdot cm^{-2}$) of AZ91-DC was twice as large as that of AZ91-TC (104.5 $\mu A \cdot cm^{-2}$). The

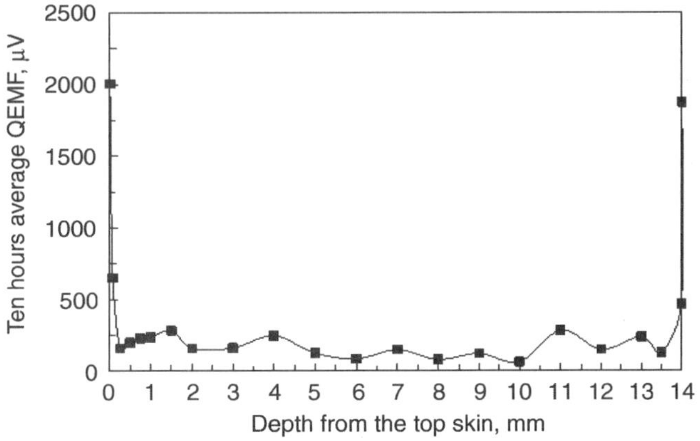

Figure 11.17 Ten hours average QEMF values as a function of depth for AZ91-TC alloy in 0.014 M NaCl solution at 23 °C [72].

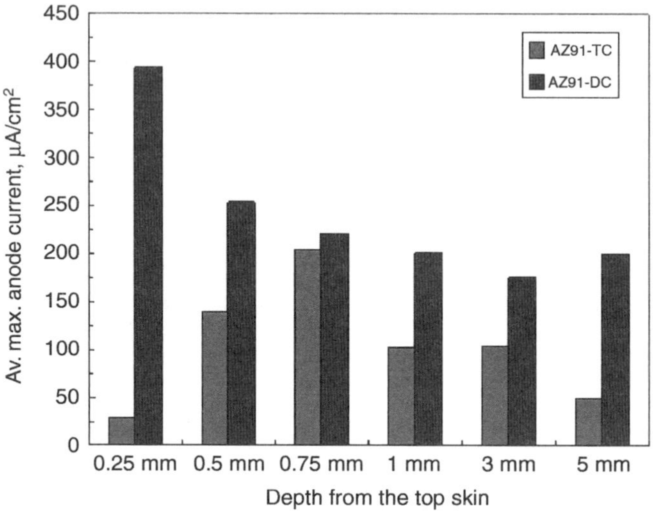

Figure 11.18 The 10 hour average MLAC values of AZ91-TC and AZ91-DC alloys in 0.014 M NaCl solution at room temperature (23 °C) [72].

MLACs of the skins were much higher than those of the bulk samples. Both the skin and bulk of the AZ91-TC alloy were more corrosion resistant than those of AZ91-DC alloy in this medium. Quantitative analysis, calculating the MLAC from the most negative anode potential, accurately determines the corrosion rate of AZ91-TC and AZ91-DC magnesium specimens [72].

The SRET results in 0.014 M NaCl were consistent with those obtained by electrochemical impedance spectroscopy (EIS) in 0.003 and 0.885 M NaCl solutions. Impedance tests were carried out for the skin of the as-cast surface without any grinding. It can be stated that the bulk of the thixocast AZ91 alloy had higher corrosion resistance than the skin. The 10 hour average MLAC was 40 times higher than that of the 8 mm depth bulk sample [72]. The thixocast AZ91 specimens had better corrosion resistance than the die-cast AZ91.

Skin and Bulk Thixo cast Performance in Alkaline Medium The corrosion resistance of die-cast and freely solidified or electromagnetically stirred thixocast AZ91D alloy has been studied using the electrochemical noise (EM) technique and EIS in dilute chloride solution to assess the influence of the microstructure on corrosion kinetics and morphology. In this work, EN and EIS were used to evaluate the corrosion depth profile of AZ91D alloy prepared using three different casting processes. Correlation between the corrosion behavior, the composition, and the microstructure is suggested. Cut specimens 1.0 cm^2 by 0.5 cm thick from the flat side walls of the boxes were examind after a mechanical polishing with a 1200 grit silicon carbide abrasive paper. The specimens were then cleaned with distilled water and ethanol. Six different depths from the surface of the specimens, were examined: 10 (skin), 50, 100, 200, 400, and 800 μm (±20 μm). All tests were performed in 0.05 M NaCl solution at 25 °C and pH 6.1, saturated with atmospheric oxygen and without stirring [70].

At depths between 10 and 50 μm (skin), all specimens showed general nonuniform corrosion with the lowest corrosion resistance. Between 100 and 200 μm (interior skin), the observed corrosion was accompanied by superficial undefined pits due to metastable pitting. The corrosion form in the interior skins gave the best corrosion resistance. Stable pitting corrosion was observed beyond 400 μm deep on the bulk specimens. The skin of all thixocast specimens prepared from both types of billets showed a more corroded surface than that of die-cast specimens. On the other hand, the interior skin as well as the bulk of thixocast specimens showed better corrosion performance than that of the die-cast specimens [70].

Corrosion Forms Independent of the casting process, the corrosion depth profile studies show three corrosion patterns: (1) general nonuniform corrosion in the skin (10–50 μm deep), (2) transition zone with shallow metastable pits (determined by EN) or 50% general nonuniform corrosion and 50% pitting corrosion (determined by EIS) in the interior skin (100–200 μm deep); and (3) pitting corrosion in the bulk (400 and 800 μm deep). Moreover, three specific frequency EN parameters (S_v, S_i, and S_r) have been identified for every type: (−2.5, −2.5, and 0), (−3, −2.5, and −0.5), and (−3.5, −3, and −0.5) (±0.5), respectively [70].

The surface regions of the studied castings generally have better corrosion resistance compared to the bulk of the material. In the skin, the preponderance of β phase particles and the smaller distance between them due to the fine microstructure promote a strong galvanic attack of primary α-Mg grains. However, the worst corrosion resistance observed in the skin of ESTC specimens could be ascribed mainly to oxygen contamination. The interior skin is a region that shows generally the best corrosion resistance, corresponding to low corrosion current and shallow unstable pits (metastable pitting) [70].

All material conditions in the test suffered from pitting corrosion, which occurs independently of the casting process and the region within the casting (surface or bulk). However, it appears that pitting occurs more readily in the bulk compared to the surface and more often on die-cast specimens compared to thixocast specimens. In the bulk, the heterogeneous microstructure containing large primary nonuniform α-Mg grains (die-cast) or round primary α-Mg grains (thixocast) and showing a large Al depletion gradient, is responsible for pitting attack, which occurs more often on die-cast specimens compared to thixocast specimens. Passive films on thixocast specimens appear to be more stable, but in the case of a defect, repassivation is more difficult compared to diecast specimens [70].

The change from general nonuniform corrosion to corrosion accompanied by shallow pits (metastable pitting) and finally stable pitting, with evident corrosion-protective films on top of cathodic areas, is observed for thixocast alloy specimens at a certain depth. Moreover, for all the alloys, the potential noise shifts, generally, to less noble values with the depth: from 10 to 800 μm, potential noise shifts from −1449 to −1487 mV versus Ag,AgCl/KCl_{sat} for die-cast specimens; from −1443 to −1510 mV for SFTC specimens; and from −1440 to −1515 mV for ESTC specimens. The potential of an alloy is determined by the potential of the constituent phase and the area fraction covered by this phase. The β phase is highly cathodic compared to the α phase in Mg alloys so a higher fraction of β phase shifts the potential toward more cathodic values, as was observed for the skin [70].

The skin (10 μm) of the three alloy specimens exhibits the lowest R_n and the skin of ESTC samples has been found to be the worst. Generally, R_n increases from 50 to 800 μm depth. However, there is a region between 100 and 200 μm where the resistance is the best, attributed to the metastable pitting mechanism of this microstructure. This is more evident for ESTC than for the other alloys. The bulk (800 μm depth) of ESTC specimens seems to have higher R_n than SFTC and DC specimens [70].

Between 10 and 50 μm depth, an interconnecting network at grain boundaries is more densely distributed. When the ratio of the surface area of the cathodic β phase to that of the anodic α phase increases, the potential shifts toward more noble or less negative values. This leads to a high galvanic current. There is a more homogeneous size of the α grains and a fine microstructure. In one study, for 75 minute immersion tests, the skin of ESTC specimens showed the highest corrosion rate, 68% and 51% higher than that of DC and SFTC specimens, respectively. Chemical analyses within the skin of the ESTC specimens indicate a high level of oxygen, likely as oxides. These oxides could explain the more intense corrosion observed for the skin in these specimens [70].

Between 100 and 200 μm depths, there is a transition zone—the *interior skin*. This transition microstructure, rarely identified, has the advantage of multiple fine electrochemical cells, leading very probably to a protective barrier or passive film. This is frequently controlled by the transient microstructure and the required critical current of passivation [70].

Starting at 400 μm and going deeper (bulk), the microstructure is more heterogeneous, containing large primary α grains (~50 μm) for the die-cast and preexisting rounded primary α-Mg grains for the thixocast specimens, both with low Al contents. One possible scenario for the start of pitting might be a loss of the local protective hydroxide layer. Once the protective layer has disappeared, the β phase promotes the galvanic corrosion of the less noble α-Mg grains. Thixocast specimens are more resistant to corrosion. The better corrosion behavior of thixocast specimens is attributed to their higher Al content in primary α-Mg phase. Moreover, the microstructure of thixocast specimens exhibits larger globules and less porosity than that of die-cast specimens. Eutectic α-Mg particles are embedded in a much denser interconnecting network of β phase. Generally, a higher corrosion rate has been found for the skin than for the bulk. Moreover, the interior skin shows better corrosion resistance than the bulk [70].

11.3.6. Hot Chamber Processes and Corrosion Resistance of Thin Plates

The hot chamber die-casting process has some advantages, such as rapid cycling, improved fluidity, and the use of a lower injection pressure, and offers a favorable method for the mass production of electronic appliance housings with thin section thicknesses. The sample thickness affects the cooling rate. Due to the high cooling rate of the die casting for a thin-gauge section, the nonequilibrium solidification structure in the die skin should differ from that of a thick die casting. The properties of the surface layer of an Mg alloy are important factors for the use of that alloy. Mechanical properties such as the yield stress of a nominally 1 mm thick Mg die casting falls from 186 to 160 MPa when 0.125 mm is removed from both surfaces. Since electrochemical corrosion is a surface phenomenon, the properties of the skin surface of a die-cast alloy control the corrosion resistance of the alloy upon exposure to a corrosive environment [71].

The corrosion properties of a hot chamber die-cast thin plate (e.g., 1.4 mm thick) have been examined. Additionally, during die casting, the mold, which is clamped together by hydraulic force, is rapidly filled (in 5–100 ms) by forcing the molten metal through a narrow gate. The metal solidifies with a high cooling rate (100–1000°C/s), yielding a fine-grained material. Grinding off the die skin enabled the observation of the as-cast microstructure in the interior of the die casting. The SEM photographs in Figure 11.19 illustrate the microstructures of the "without skin" (WOS) sample, viewed perpendicular to the surface.

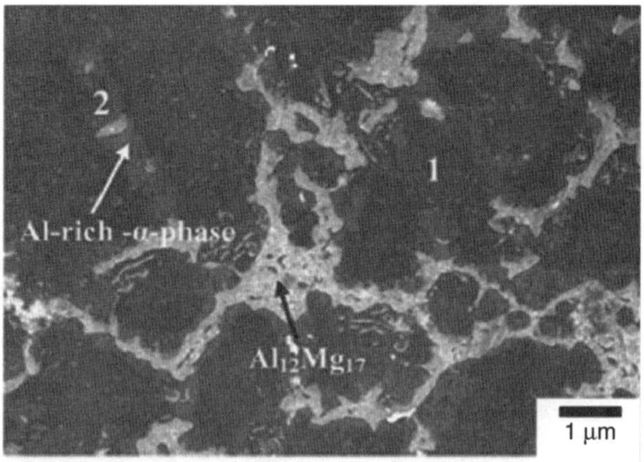

Wt pet / EDS positions	Mg	Al	Zn
1	95.7	4.3	0
2	83.9	13.1	3.0

Figure 11.19 Horizontal SEM view of the WOS sample (i.e., without die skin on surface), showing (a) the interdendritic networks of $Al_{12}Mg_{17}$ β phase (in bright contrast) and (b) the β network along the interdendritic boundary. The EDS compositions corresponding to positions 1 and 2 are indicated in the joined table [73].

Figure 11.19a shows that the as-cast structure contained interdendritic networks of $Al_{12}Mg_{17}$ β phase. The α-Mg grains were coarser than those in the die skin (Figure 11.19a). Figure 11.19b shows the β network distribution along the interdendritic boundary. Moreover, Figure 11.19b illustrates that the Al-rich β-Mg phase was observed near the β phase. The table accompanying Figure 11.19 shows the Al concentrations at the corresponding positions 1 and 2. The table shows that position 1 (the α-Mg) was approximately 4.3 wt% Al, while position 2 (the Al-rich α-Mg phase) was ~13.1 wt% Al. Moreover, the volume fraction of the Al-rich phase was 8.0 ± 1.4 vol % in the WOS sample, while the volume fraction of $Al_{12}Mg_{17}$ β phase reached up to 15.5 ± 1.7 vol %. Figure 11.19 shows that the main phase in the interdendritic network was the $Al_{12}Mg_{17}$ β phase.

The die-chill skin was composed of a thin layer of chill zone and a thick layer of interdendritic Al-rich α-Mg/$Al_{12}Mg_{17}$ β phase particle/α-Mg grain composite microstructures. The chill zone (4 ± 1 μm thick) had fine columnar and equiaxed grains and contained a distribution of submicron Mg–Al–Zn intermetallic particles. Beneath the chill zone, $Al_{12}Mg_{17}$ β particles were irregularly shaped but did not have interdendritic network morphology. Furthermore, an Al-rich α phase (also known as eutectic α) was in the interdendritic network, which occupied a higher volume fraction than the β phase in the die-cast skin layer [73].

The corrosion of the skin and bulk of a hot chamber die-cast AZ91D thin plate was studied to determimine the distinctive characteristics of this process. There is a structural inhomogeneity of the metal, caused by the formation of a chilled skin on the die-cast surface. Corrosion rates were determined by the electrochemical polarization test (0.166 mV/s) in 3.5 wt% NaCl solution and for constant immersion in the same solution for 120 hours at pH 6.1 and 3.5, adjusted by hydrochloric acid addition. The results showed severe corrosion of the sample with a die-cast skin while the sample without the die skin on the surface corroded more slowly [71].

With regard to the die skin microstructure of the hot chamber die-cast thin plate, rapid cooling of the die casting for a thin gauge section yields nonequilibrium solidification structures that comprise a large fraction of the Al-rich α phase and a relatively low fraction of $Al_{12}Mg_{17}$ β particles in the die skin of a thin die-cast plate. These findings differ from those for a thick die casting. Uan et al. [71] studied the role of interdendritic phases, primary β ($Al_{17}Mg_{12}$) and a surrounding α phase. They examined the effect on the corrosion rate of removing interdendritic phases (both the Al-rich α and the primary β) from the surface chill layer of the die-cast thin plate [71]. A hot chamber die-cast AZ91D thin plate with a die-chill skin on its surface was severely corroded in 5 wt% chloride solution ($I_{corr} \sim 1600\,\mu A/cm^2$), whereas a plate with a die-skin layer etched in an HF/H_2SO_4 aqueous solution to remove interdendritic phases had a substantially lower corrosion rate (3 to $\sim 16\,\mu A/cm^2$). The primary β phase in the die-chill layer was removed by Ar^+ etching, but the other main constituent phases remained on the surface of the sample. Although previous studies attributed intense galvanic corrosion to the cathodic phase (primary β), Van et al. [71] demonstrated that the removal of the primary β phase from the die-cast sample surface did not improve the corrosion performance of the sample [71].

The use of a transmission electron microscope (TEM) showed that the surrounding α-Mg phase (known as Al-rich α or eutectic α) actually did not contain a high concentration of Al solid solution. Instead, the Al-rich α was composed of fine $Al_{12}Mg_{17}$ β particles and Al–Mn-like particles (smaller than 0.5 μm) that were distributed in a low-Al-containing Mg matrix (~4 wt% Al). Such fine cathodic particles seem to participate strongly in the corrosion. Removing the primary β phase alone did not increase the corrosion resistance of the material, because many of the cathodic fine particles remained in the Al-rich α phase region, creating numerous galvanic cells. The primary β phase was then not the only (or dominant) cathodic phase that caused severe corrosion of the die-cast thin plate [71].

Samples sized $20 \times 20 \times 1.4\,mm^3$ were cut from the panels. A salt spray test and electrochemical polarization experiments were performed, to evaluate the corrosion resistance of the samples. The samples were kept in a chamber with salt spray at 35 °C for 168 hours (ASTM B117). A 5 wt% NaCl aqueous solution was used in the tests. At the end of the experiments, the samples were cleaned by dipping in a solution of 15 wt% CrO_3 + 1 wt% $AgCrO_4$ in 100 mL of boiling water. The corrosion rate was determined in millimeters per year (μm). Electrochemical polarization tests were conducted for $1\,cm^2$ in 5 wt% NaCl solution at room temperature at a scan rate of $0.5\,mV \cdot s^{-1}$. For electrochemical impedance spectra, the amplitude of the alternating current signal was 5 mV, and the range of measured frequencies was from 10 mHz to 100 kHz at 25 °C [71].

Figure 11.20a shows a surface FE-SEM image of the die skin. Figure 11.20b presents the concentration profile of the elements along the line displayed in Figure 11.20a. In Figure 11.20, the primary α-Mg grain, the Al-rich α phase, and the primary β phase corresponded to positions A, B (D), and C, respectively. The primary β phase (position C) and the Al-rich α phase (positions B and D) were composed of a high percentage of Al. In a primary α-Mg grain (e.g., position A), the Al concentration fell to ~4.5 wt% [71].

Figure 11.21 presents schematic cross sections of the microstructure at the die-cast surface and the corrosion caused by the galvanic effect. The primary α grains, primary β phase, and the Al-rich α eutectic phase presented in Figure 11.21 are greatly magnified. The figure emphasizes the effect of the Al-rich α microstructure on corrosion. This

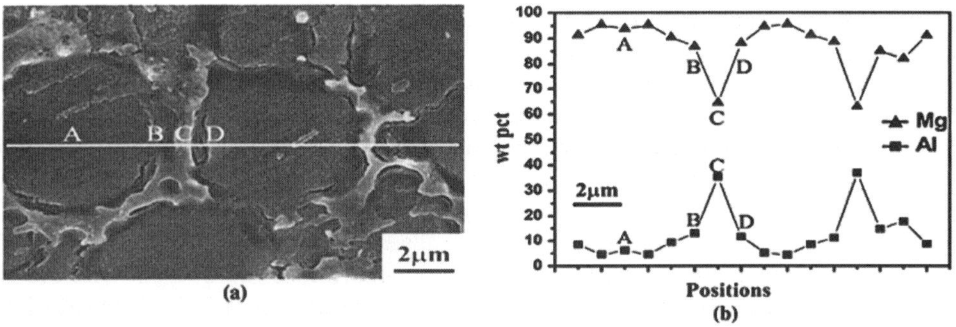

Figure 11.20 (a) Surface SEM observation of die skin on the as-die-cast sample, showing the SEM/EDS line scanning. (b) Concentration profiles for Mg and Al along the line in part (a), where position A is at primary a-Mg, positions B and D are located at Al-rich α, and position C is at primary β $Al_{12}Mg_{17}$ [71].

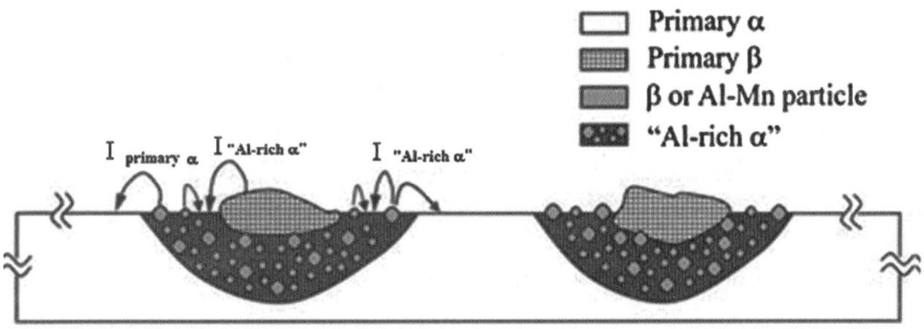

Figure 11.21 Schematic presentation of the possible galvanic corrosion cells between the different phases the primary β phase, the Al-rich α and its components (fine β, Al-Mn particles, and Al rich α phase), and the primary α-Mg phase [71].

schematic presentation shows the possible galvanic cells between the different phases of the alloy. It also shows that the primary β phase is not the major cathodic phase that caused the dissolution of the anodic phases due to the microgalvanic corrosion currents that flow between them. Rather, the intermetallic particles in Al-rich α participated markedly in the corrosion. For example, in this scheme, the primary β phase (the most noble microstructure) is not in close contact with the primary α phase (the most anodic phase) as are the other cathodic phases of the composite (mostly fine β particles, some Al–Mn/Al–Mn–Mg particles, and a low Al α-Mg matrix). These limits the EMF of the galvanic cell between the primary β and α phases due to the *RI* drop of the medium and this effect should be more evident in weaker electrolytes than the 5% NaCl solution examined [71].

B. MIC OF MAGNESIUM AND MAGNESIUM ALLOYS

MIC of magnesium alloys has been poorly studied and little information is actually available. Magnesium is one of the major cell components of bacteria, which implicates it in microbial attachment and growth at the start. Nandakumar et al. [74] studied the magnesium alloy AZ31B, which was covered with a biofilm-forming bacterium, *Pseudomonas* species. After 6 days, the increase in pH resulted in bacterial mortality and reduced the area covered by bacteria on the coupon surface. The formation of magnesium hydroxide and the dissolution of Mg shifted the pH of the coupon surface to higher alkaline values. Sawai [75] stated that MgO has an important antibacterial activity other than its influence in the shift of pH to high alkaline values. The antibacterial mechanism involving an oxygen radical was suggested [75]. Thus the formation of both MgO and $Mg(OH)_2$ adversely affects the growth and survival of bacteria.

The studies, done in relation to the rational degradation of magnesium sacrificial alloys for cathodic protection, were carried out in the absence of bacteria; however, for applications such as tooth implants that are in contact with the bacterial population of the mouth, the influence of bacteria on the corrosion resistance of the alloy should be investigated. Takaya et al. [76] reported that pure magnesium could have a good corrosion resistance in vitro. Kuwahara et al. [77] focused on the behavior of commercial pure magnesium and some magnesium alloys such as AZ91 in Hank's balanced salt solution (HBSS), and the loss of mass of each specimen was determined after immersion for various periods. HBSS is well known as an artificial nutritious liquid used for cultivating animal cells or bacteria. AZ91 alloy maintains a good corrosion resistance in HBSS. It was found that the corrosion resistance of pure Mg is improved by a surface modification at high temperature. MgO ($\sim$20 μm thick), formed on the specimen surface due to treatment at 530 °C for 9 hours, improved the corrosion resistance. A proposed crown or denture prepared by casting could integrate this heat treatment [77].

11.4. RATIONAL DEGRADATION

11.4.1. Behavior of Sacrificial Magnesium

Alloys of Al, Mg, and Zn (Table 11.9) are used as sacrificial anodes for metal protection in aqueous and soil environments. Al and Zn alloys are widely used for cathodic protection of steel in marine environments, while Mg-based alloys tend to be more appropriate in high-resistivity environments such as tap water. Mg and its alloys are less passivable than Zn and Al and the former ones are utilized in cathodic protection because of their theoretical active half-cell potential and adequate current capacity [78]. Magnesium is a widely used galvanic anode material for the protection of pipelines and other buried structures because its inherent negative potential and high current output per unit weight are desirable. However, magnesium anodes generally have a current efficiency below that of other galvanic anodes. In practice, current efficiency of magnesium anodes rarely exceeds 50%. This compares unfavorably with zinc and aluminum anodes, which have current efficiencies better than 90% [79].

The processes that contribute to the anode wastage are hydrogen evolution from microcathodic sites on the alloy surface, defined local cell action (LCA), and mechanical loss of metal, defined as the chunk effect (CE). Study of anode efficiency as a function of

Table 11.9 Chemical Composition (wt%) of Mg Anodes

Alloy	Mn	Ca	Al	Zn	Fe	Si	Ni	Cu
Mg–Mn–Ca	0.26	0.140	0.0091	0.0044	0.0025	0.018	<0.0005	0.0014

operating time and applied current densities suggested that a good functionality of the AZ63 alloy as sacrificial anodes in tap water (high-resistivity electrolyte) needed time and a relatively high impressed current (more than 0.44 mA/cm^2 for good cathodic protection of a steel structure). A heat treatment was used to dissolve the β phase in order to reduce the LCA effect. However, the performance of these anodes was relatively poor, indicating that β phase played an important role in the protection of as-cast Mg alloys. This result suggests that the intermetallic phase is more protective than pure Al in solution [78].

Addition of calcium to the Mg–Mn sacrificial anodes enhanced the anode efficiency by promoting uniformly distributed corrosion along the grain boundaries and increased the driving potential by intrinsic its intrinsic electronegative potential. The reason for the uniform attack is that Mg_2Ca precipitate (cathode) at grain boundaries is galvanically coupled to the α-Mg matrix (anode). Larger anodic area limits the localized nature of the intergranular attack [79].

11.4.2. Rational Biocorrosion of Mg and Its Alloys in the Human Body

Magnesium and some of its alloys could be considered as the most biocompatible structural metals for implants because of its minimal toxicity potential. Biocompatibility involves the exclusion of materials that exert toxic effect on cells [80]. Rational biocorrosion occurs inside the human body on biomaterial surfaces (implants and stents). The immune system prevents bacteria and fungi colonization in tissues like blood vessels and bones [81]. Thus rational biocorrosion involves oxidation reactions in electrolytic physiological media without the presence of bacteria. The rate of corrosion must be carefully controlled to avoid complete implant dissolution before the cure of the patient. Severe corrosion occurs in aqueous physiological environments where chloride ions are present at levels of 0.15 mol/L; for example, $Mg(OH)_2$ reacts with Cl^- to form highly soluble magnesium chloride and hydrogen gas (H_2) [80]. Pitting of magnesium is observed for Cl^- concentrations exceeding 0.03 mol/L [82].

Rational biocorrosion should lead to general uniform corrosion. Thus a corrosion rate can be monitored or calculated [81]. As an example, the reduction of implant volume can be determined by synchrotron-radiation based microtomography. Witte et al. [83], using this technique, have found that in vivo corrosion was about four orders of magnitude lower than in vitro corrosion of the tested alloys (two gravity-cast magnesium alloys—AZ91D and LAE442). It is desirable to reproduce in vitro electrolytic physiological solutions. A widely used solution is Hank's balanced salt solution (HBSS). A typical HBSS has a pH very close to that of blood, 7.4, and contains 8 g/L of NaCl, 0.4 g/L of KCl, 0.06 g/L of KH_2PO_4, 1 g/L of glucose, 0.01 g/L of phenol red, 0.048 g/L of Na_2HPO_4, 0.098 g/L of $MgSO_4$, 0.14 g/L of $CaCl_2$, and 0.35 g/L of $NaHCO_3$ [77].

The following reactions can be considered in an aqueous physiological environment:

$$Mg(s) + 2H_2O(l) \rightarrow Mg(OH)_2(s) + H_2(g) \quad \text{(hydrogen evolution)} \tag{11.1}$$

$$Mg(s) + 2\,Cl^-(aq) \rightarrow MgCl_2 + 2\,e^- \tag{11.2}$$

$$Mg(OH)_2(s) + 2\,Cl^-(aq) \rightarrow MgCl_2 + 2\,OH^- \tag{11.3}$$

11.5. STRESS CORROSION CRACKING AND IMPLANTS

Applications of magnesium alloys as biodegradable orthopedic implants are critically dependent on the mechanical integrity of the implant during service. Kannan and Raman [84] examined the stress-corrosion cracking (SCC) susceptibility of sand-cast Mg–Al–Zn alloy in modified-simulated body fluid (M-SBF), using the slow strain rate test (SSRT) method, with similar ion concentration to blood plasma and good Storage stability. The composition of m-SBF per liter is as follows: 5.403 g NaCl, 0.504 g NaHCO3, 0.426 g Na_2CO_3, 0.225 g KCl, 0.230 g $K_2HPO_4 \cdot 3H_2O$, 0.311 g $MgCl_2{\cdot}6H_2O$, 100 mL 0.2 M NaOH, 0.293 g $CaCl_2$, 0.072 g Na_2SO_4, and ~15 mL of 1 M NaOH. The solution was buffered with 2-(4-(2-hydroxyethyl)-1-piperazinyl)ethanesulfonic acid (HEPES) to maintain a physiological pH of 7.4 at 6.5 ± 0.5 °C.

The alloy exhibited 4.7% elongation to fracture (e_f) and 120 MPa ultimate tensile strength (UTS) in air. The mechanical properties of the alloy decreased by 17% (for UTS) and 21% (for e_f) when tested in m-SBF as compared to that in air. In this solution, the formation of hydroxyapatite film on magnesium alloy not only improves the biocompatibility but can also improve the corrosion resistance. However, the circumference of the tensile sample after the SSRT showed that pitting was not completely inhibited and that the primary and secondary cracks seem to have emanated from the localized attacked region. The crack propagated to the center of the sample, after which the sample failed due to overload. Considering the film-forming tendency of the alloy in m-SBF, it is suggested that the film-rupture model was a possible mechanism involved in the failure. This study suggests that the SCC susceptibility of sand-cast AZ91 magnesium alloy in m-SBF is not substantial and this aspect should not be a concern for its application in biodegradable orthopedic implants. However, if a high-strength magnesium alloy is projected for eventual use, it should be evaluated in the same medium for SCC prevention [84].

11.6. APPROACHES TO CONTROL BIODEGRADATION

To reduce the rapid corrosion rate of Mg in the human body, two basic approaches have been suggested: (1) alloying with biocompatible resisting elements and (2) surface treatment such as anodizing. Mg alloys that are actually used as biomaterials and that undergo rational biodegradation are described in Table 11.10.

11.6.1. Alloying

Heublein et al. [80] designed a new Mg alloy that was used as a sacrificial stent in blood vessels. This Mg alloy contained 2% Al and 1% rare earths (Ce, Pr, Nd). The complete degradation of this stent in a blood vessel is around 89 days (Figure 11.22). A new layer of endothelial cells (neointima) acting as a thin diffusion barrier grows over the stent, separates it from the bloodstream, and limits the formation of hydrogen. This layer of cells can be observed in Figure 11.23. Hydrogen gas bubbles are toxic for the vascular system. Thus a degradable Mg stent is unlikely to be used in the human vascular system due to the fatal hydrogen gas bubbles generated, even if the neointima restricts the hydrogen gas evolution at 10 days post-implantation of the stent [80, 87]. A similar natural protection of the Mg surface has been reported for a bone implant. A thick layer composed mainly of calcium ions grew on the Mg bone stent. Two to three weeks after surgery, no formation of hydrogen is

Table 11.10 Use of Some Mg Alloys for Medical Implants in the Human Body

Mg alloy	Use	Toxic potential	Reference
AE21	Cardiology: blood vessel implant	Moderate to highly toxic; contains 2 wt% Al, which can possibly induce Alzheimer's disease	80
AZ91D, AZ91, AZ61, AZ31	Bone surgery, dentistry	In low quantity, Zn is not toxic but Al (~2 wt%) is highly problematic.	83
LAE442	Bone surgery	Highly toxic since Li and Al are toxic; Li can induce some cancers	83
Lekton Magic Coronary Stent (Biotronik, Switzerland)	Cardiology: blood vessel implant	Low toxicity Zr (<5%), yttrium (<5%), and rare earths (<5%)	85
MgCa, pure Mg and 0.6–0.8 wt% Ca	Bone surgery	Minimal toxicity potential	86

Source: Reference 81.

observed [88, 89]. Hassel et al. [86] suggested that calcium alloying could be a base for promising alloys, since it has a minimal toxicity potential, good workability, and adequate mechanical properties and corrodes slower than some other alloys.

Case History Implantation of a biodegradable 3 mm magnesium stent was performed in a hybrid procedure in a baby weighing 1.7 kg. Reperfusion of the left lung was established and persisted throughout the 4 month follow-up period, during which the stent gradually degraded. The mechanical and degradation characteristics of the Mg stent proved to be adequate to secure reperfusion of the previously occluded left pulmonary artery. Bioabsorbable stents with different diameters may help develop new strategies in the therapy of vessel stenosis in pediatric patients [90].

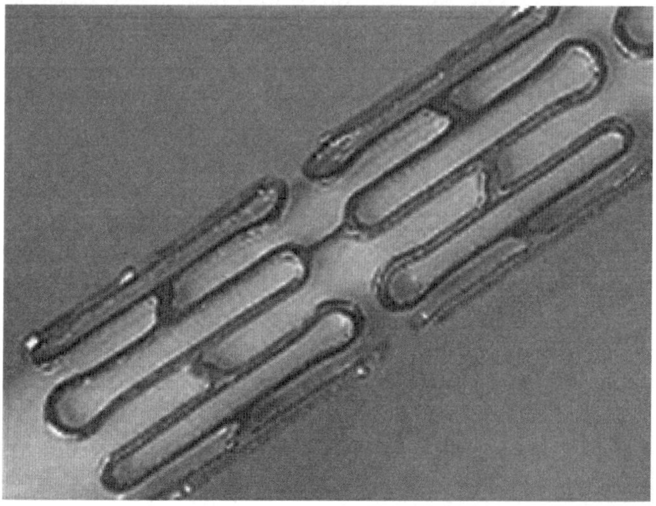

Figure 11.22 Sacrificial blood vessel stent.

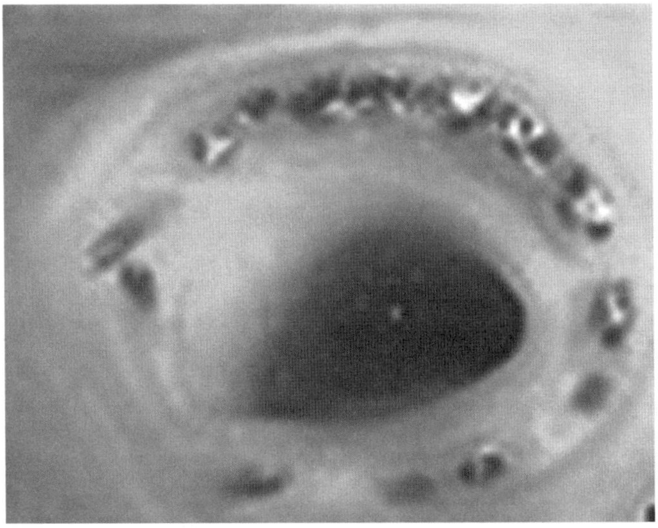

Figure 11.23 The proliferation of neointima over the stent was observed in an intermediate state 56 days after stent implantation [80].

11.6.2. Surface Treatment (Anodizing)

If the corrosion rate of a magnesium bone implant is properly controlled, hydrogen evolution will not be fast enough to lead to significant subcutaneous hydrogen gas bubbles, contrary to vascular stents. The alkalization effect resulting from the corrosion of magnesium could easily be balanced by metabolic mechanisms in the human body. Using anodized Mg coupons and a simulated body fluid as medium, Song and Song [87] found that the hydrogen evolution was less than 0.1 mL/cm^2/day with a shift in pH value (from 6 to 7) and nondetectable weight loss of the coupon after 100 hours of immersion. If the coating contains about 30% Si and is only about 4 μm thick, and if a 1 cm^2 surface area is dissolved, then only about 2.7 μg Si is dissolved into body fluid. Considering the high corrosion resistance of anodized Mg, a considerable length of time would be required to dissolve the 2.7 μg if there is 1 cm^2 of Mg surface left in contact with the blood. In fact, a trace amount of Si has been suggested to be essential in mammals. In this sense, the anodized coating should be nontoxic to the human body [87].

By utilizing magnesium's rapid corrosion reaction and controlling its corrosion process through Zn and Mn alloying, purification, and anodization, chemically active magnesium can be developed into a biodegradable, biocompatible implant material with a specific biodegradation process and tolerable hydrogen evolution rate, which may replace current problematic biodegradable polymers in applications[87].

11.6.3. Magnesium Implants and Bone Surgery

An optimal temporary bone implant material should have open pores to enable bone cells to migrate into the cartilage defect and to replace the degraded implant with new bone formation [91]. Magnesium is an osteoconductive metal that can create an environment where bone can grow. An osteoconductive metal doesn't induce or promote the synthesis of bone—it simply lets the growing bone cling to it. Yamasaki et al. [92] have investigated the action of magnesium and one of its corrosion products (Mg^{2+}) in promoting bone cell adhesion. The binding of cells to biomaterials plays an important role in the rapid restoration of a defective area. It should be noted that the content of magnesium in the body is 21–35 g (60% in the bone, 1% in the serum, and 39 % in the muscle) [88].

Biocompatible porous metals, including Ti and Mg, are successfully fabricated by an innovative powder metallurgy (P/M) process. The porous morphology, pore size, and porosity of the metals can be controlled exactly in this P/M process. An appropriate pore size distribution is in the range of 200–500 μm, which is specially managed for its appropriateness for porous bone substitutes. It is interesting to compare magnesium foam bone implant to that of titanium. The porosity is 50% for the magnesium implant and 78% for the titanium one. For titanium foam, the compressive strength is 35 MPa and Young's modulus is 5.3 Gpa, while, for magnesium foam the compressive strength is 2.33 MPa and Young's modulus is 0.35 Gpa; both the Ti foams and the Mg foams are presumably strong enough to resist handling during implantation and in vivo loading [93].

Osteoconductive bioactivity in Mg-based metals is suggested by observations of increased bone apposition around Mg-based implants compared to SR-PLA96 polymer implants [88]. Osteoconductive bioactivity in Mg-based metals can be explained by (1) the biochemical role for magnesium in the bone system, (2) the role of magnesium in the cell attachment process, and (3) the beneficial effects of magnesium-enriched materials on bone cell attachment and tissue growth [82].

Witte et al. [88] stated that magnesium alloy implants are completely degraded after 18 weeks of implantation. Small amounts of magnesium stimulate the cellular metabolism. New cortical bone formation can be seen during the biocorrosion of magnesium alloy implants. Orthopedic implants for bone, tendon, and ligament fixation have to maintain their stabilization properties over at least 12 weeks; magnesium alloys may represent an ideal alloy for future application in orthopedic implants, making a second surgery unnecessary (Figure 11.24). Witte et al. [88] also showed that in guinea pigs high magnesium ion

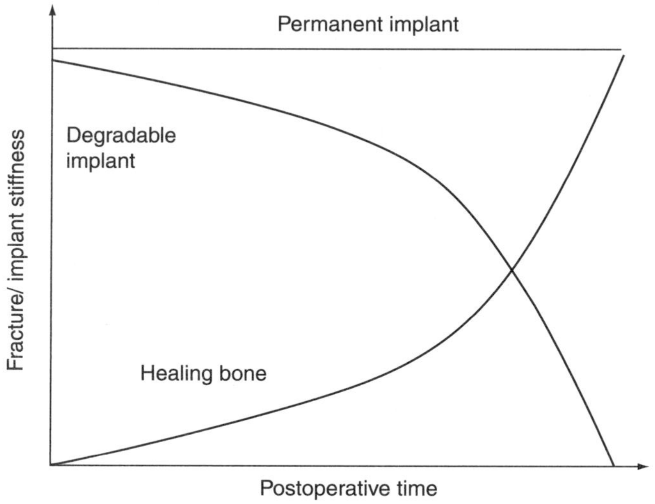

Figure 11.24 Basics of an ideal bone implant degradation compared with a permanent bone implant [89].

concentration could lead to bone cell activation and high mineral apposition rates. The bone mass increased around the magnesium rods and the surrounding soft tissue stayed intact [89].

As a lightweight metal with mechanical properties similar to natural bone, a natural ionic presence with significant functional roles in biological systems, and in vivo degradation via corrosion in the electrolytic environment of the body, magnesium-based implants have the potential to serve as biocompatible, osteoconductive, degradable implants for load-bearing applications. However, a great deal of research is still necessary. Modulation of the corrosion rate of magnesium-based materials in the physiological environment must be accomplished, possibly through the use of high-purity magnesium or experimenting with different compositions of alloys and surface treatments. The nontoxicity of magnesium-alloyed materials and more corrosion resistant variations must also be thoroughly evaluated. In vitro investigations of bone cell attachment, differentiation toward an osteoblast phenotype, and proliferation and formation of a mineralized matrix, and in vivo studies of bone apposition and tissue in-growth are necessary. The development, performance, and integration with bone tissue of porous magnesium-based implants are also topics requiring further investigation [82].

Two ways to slow down the degradation of magnesium alloy implants in the human body are by alloying with resistant biocompatible elements or by convenient surface treatment such as anodizing or surface conversion [94]. Xu et al. [95] investigated the corrosion of Mg–Mn–Zn rods in rat femora and calculated their corrosion rates according to the ratio of the cross-sectional area of the residual implant to the original one. The results showed that 54% of the samples had been degraded after 18 weeks. Li et al [96] employed Mg/1 wt% Ca alloy implant pins (10 mm in length, 2.5 mm in diameter) that degraded completely 3 months postoperative in a rabbit.

Among various surface treatment techniques, alkaline heat treatment was investigated extensively as one of the promising methods for metallic biomaterials. Gu et al. [94] considered a new Mg–Ca alloy (1.4 wt% Ca) treated for 24 h in Na_2HPO_4, Na_2CO_3, and $NaHCO_3$ alkaline solutions followed by a 12 h subsequent heat treatment at 773 K. Scanning electron microscopy (SEM) and energy-dispersive spectroscopy (EDS) results revealed that magnesium oxide layers with thicknesses of about 13 μm, 9 μm, and 26 μm were formed, respectively, on the surfaces of the Mg–Ca alloy. Atomic force microscopy (AFM) studies showed that the surfaces of the Mg–Ca alloy became rough after the three alkaline-heat treatments [97].

The electrochemical studies were carried out with in vitro corrosion tests in simulated body fluid (SBF) in an adjusted pH of 7.4 at 37 °C according to ASTM-G31-72. The open circuit potential (OCP), the pH value of the solution, and the hydrogen evolution rate were monitored during the immersion tests. Potentiodynamic polarization curves were obtained at a scanning rate of 1 mV/s. The results showed that corrosion rates of the Mg–Ca alloy were effectively decreased after alkaline heat treatments, with the following sequence: $NaHCO_3$ heated $<$ Na_2HPO_4 heated $<$ Na_2CO_3 heated. $NaHCO_3$ heated Mg–Ca alloy presented a uniform, dense, and thick surface, providing good protection for the substrate and the slowest corrosion rate. The results also indicated an accelerated calcium phosphate deposition rate on $NaHCO_3$ heated Mg–Ca alloy substrate. Therefore $NaHCO_3$ heat treatment might be a promising technique to improve the corrosion resistance and biocompatibility of biomedical Mg–Ca alloy. In summary, the rapid corrosion rate in the human body must be controlled to avoid high rates of toxic hydrogen gas bubbles and strong alkalization around the stent or implant, since the healing process of bone or blood vessel is actually slower than the complete biodegradation of the magnesium stent or implant [97].

Corrosion rates of the as-cast and heat-treated AZ91D samples were measured in SBF using a constant-immersion test for four different intervals of time (8 h, 1 day, 3 days, and 1 week). A potentiodynamic polarization test was carried out and a polarization scan was adjusted toward more noble values at a rate of 1 mV/s, after allowing a steady-state potential to develop. The results suggest that β-$Mg_{17}Al_{12}$ phase has significant influence on the corrosion behavior of the magnesium alloy. In the as-cast microstructure, the β-$Mg_{17}Al_{12}$ phase is highly cathodic to the α-Mg phase and can thus act as an effective cathode to cause microgalvanic corrosion [98].

Heat treatment was found to be effective in modifying the microstructures and thus electrochemical properties of magnesium alloy AZ91D. The homogeneous β-Mg microstructure led to uniform corrosion in the T4 treated samples. In the T6 microstructures, β-$Mg_{17}Al_{12}$ phase precipitated along the grain boundary and the inside grain provided the preferential site for intergranular corrosion and pitting. Solution treatment followed by 16 h aging (T6–16 h) gives the best corrosion resistance in SBF, although T4 treatment gives the lowest corrosion rate at the initial exposure to SBF [99].

REFERENCES

1. D. S. Tawil, Corrosion and surface protection developments, in *Magnesium Technology,* edited by C. Baker. The Institute of Metals, London, UK and The Institute of Metals, North American Publication Center, Brookfield, VT, 1986, pp. 66–70.
2. W. K. Miller and E.F. Ryntz Jr., *Magnesium for Automotive Applications: A State-of-the-Art Assessment*, Paper 830521. Society of Automotive Engineers, Warrendale, PA, 1984.
3. B. E. Carlson and J. W. Jones, in *Light Metals Processing and Applications*, METSOC, Quebec, 1993, pp. 833–847.
4. J. D. Hanawalt, C. E. Nelson, and J. A. Peloubet, *Transactions of the American Society of Mining and Metallurgical Engineering*, **147**, 273–299 (1942).
5. E. Ghali, in *Uhlig's Corrosion Handbook,* 2nd edition, edited by R. W. Revie. Wiley, Hoboken, NJ, 2000, pp. 793–830.
6. O. Lunder, K. Nisancioglu, and R. S. Hansen, in *Corrosion of Die Cast Magnesium-Aluminum Alloy, Paper No. 930755*, SAE International, Detroit, MI, 1993, pp. 117–126.
7. D. Eliezer, P. Uzan, and E. Aghion, Corrosion behavior, in *Magnesium Alloys, Science Technology and Applications*, S. Neaman Institute Technion, Haifa, Israel, 2004, pp. 231–257.
8. R.-C. Zeng, Z. Jin, W. J. Huang, W. Dietzel, K.-U. Kainer, C. Blawert, and W. Ke, *Transactions of Nonferrous Metals Society of China* **16**, 763–771 (2006).
9. T. K. Aune, Minimizing base metal corrosion on magnesium products. The effect of element distribution (structure) on corrosion behavior, Paper presented at 40th World Magnesium Conference. IMA, Dayton, OH, 1983.
10. T. K. Aune and H. Westengen, Property update on magnesium die casting alloys, *SAE Technical Paper 950242*. Society of Automotive Engineers, Warrendale, PA, 1995.
11. L. L. Shreir, R. A. Jarman, and G. T. Burstein, *Corrosion*, Vol. 1. Butterworth-Heineman, Woburn, MA, 1995, pp. 4.98–4.115.
12. J. E. Hillis and S. O. Shook, *SAE Technical Paper No. 890205*. Society of Automotive Engineers, Warrendale, PA, 1989.
13. C. Suman, Effect of direct aging on mechanical properties and corrosion resistance of die-cast magnesium alloys AZ91D and AM60B, Paper No. 900794, SAE International, Warrendale, PA, 1990.
14. E. Ghali, W. Dietzel, and K.-U. Kainer, *Journal of Materials Engineering and Performance* **13**(1), 7–23 (2004).
15. K. U. Kainer, *Magnesium Alloys and Technology*. Wiley-VCH, Weinheim, Germany, 2003, pp. 1–22.
16. A. A. Nayeb-Hashemi and J. B. Clark, in *Binary Alloy Phase Diagram*, 2nd Edition, edited by T. R. Massalski, H. Okamoto, P. R. Subramanian, and L. Kacprzak. ASM International, Materials Park, OH, 1988.
17. L. Y. Wei, Development of microstructure in cast magnesium alloys. Ph.D. thesis, Chalmers University of Technology, Goteborg, 1990.
18. O. Lunder, J. E. Lein, S. M. Hesjevik, T. Kr. Aune, and K. Nisancioglu, Filiform corrosion of a magnesium alloy, Paper presented at the 11th Annual Corrosion Congress, Florence, Associazione Italiana Di Metallurgia (AIM), Milano, Italy, 1990, pp. 5.255–5.262.
19. G. Nussbaum, G. Regazzoni, and H.G. Jestland, in *Proceedings of the SAE Internationnal Congress and*

Exposition, Society of Automotive Engineers, Warrendale, PA, 1990.

20. B. L. Tiwari and J. J. Bommarito, A novel technique to evaluate the corrosion behavior of magnesium alloys, in *Magnesium Technology 2002*, edited by H. I. Kaplan. The Minerals, Metals & Materials Society, Warrendale PA, 2002, pp. 269–275.
21. K. N. Reichek, K. J. Clark, and J. E. Hillis, *SAE Technical Paper No. 850417*. Society of Automotive Engineers, Warrendale, PA, 1985.
22. ASTM B94-92, in *Annual Book of ASTM Standards*. American Society for Testing and Materials, Philadelphia, PA, 1992.
23. O. Lunder, T. Kr. Aune, and K. Nisancioglu, *Corrosion*, **43**(5), 291 (1987).
24. G. L. Song, *Advanced Engineering Materials* **7**(7), 308–317, 563–586 (2005).
25. S. Mathieu, C. Rapin, J. Steinmetz, and P. Steinmetz, *Corrosion Science* **45**, 2741–2755 (2003).
26. O. Lunder, J. E. Lein, S. M. Hesjevik, T. Kr. Aune, and K. Nisancioglu, *Corrosion* **45**(9), 741–748 (1989).
27. M. T. D. Jönsson and N. LeBozec, *Corrosion Science* **48**, 1193–1208 (2006).
28. M. Stratmann and H. Streckel, *Corrosion Science* **30**(6–7), 697–714 (1990).
29. V. Mitrovic-Scepanovic and R. J. Brigham, *Corrosion* **48**(9), 780–784 (1992).
30. J. F. King, Develoment of pratical high temperature magnesium casting alloys, in *Proceedings of the National Congress, Magnesium Alloys and Their Applications*, Munich, Germany, edited by K. -U. Kainer. Wiley-VCH, Weinheim, Germany, 2000.
31. I. P. Moreno, T. K. Nandy, J. W. Jones, J. E. Allison, and T. M. Pollock, *Scripta Materialia* **48**(8), 1029–1034 (2003).
32. S. Amira, A.-M. Lafront, S. Jin, D. Dubé, R. Tremblay, and E. Ghali, Electrochemical behavior of AXJ530 magnesium alloy in neutral media, in *Magnesium Technology in the Global Age*, edited by M. O. Pekguleryuz and L. W. F. Mackenzie. Canadian Institute of Mining, Metallurgy and Petroleum, Montreal, Quebec, Canada. 2006, pp. 315–328.
33. E. Aghion, B. Bronfin, F. Von Buch, S. Schumann, and H. Friedrich, Dead sea magnesium alloys newly developed for high temperature applications, in *Magnesium Technology 2003*, edited by H. I. Kaplan. The Minerals, Metals & Materials Society, Warrendale, PA, 2003, pp. 177–182.
34. D. L. Hawke, J. Hillis, M. Pekguleryuz, and I. Nakatsugawa, in *ASM Specialty Handbook: Magnesium and Magnesium Alloys*, edited by M. M. Avedesian and H. Baker. ASM International, Materials Park, OH, 1999, pp. 194–210.
35. E. M. Gutman, A. Eliezer, Y. Unigovski, and E. Abramov, *Materials Science Engineering* **302**, 63–67 (2001).
36. B. J. Schneider, R. C. McCune, M. S. Ricketts, and J. A. Hines, Resistant magnesium alloys in aqueous environment, in *SAE International 2007 Proceedings*. Society of Automotive Engineers, Warrendale, PA, 2007, p. 7.
37. Y. B. Unigovski and E.M. Gutman, *Applied Surface Science* **153**, 47–52 (1999).
38. G. L. Dunlop, W. P. Sequeira, M. S. Dargusch, G. Song, A. Atrens, T. Kittel, D. St. Jhon, A. K. Dahle, and M. Murray, Microstructure of creep magnesium die alloys, in *Proceedings of the Annual World Magnesium Conference*, Hannover, Germany, International Magnesium Association Mclean, VA, 1998, pp. 68–73.
39. B. R. Powell, V. Rezhets, M. P. Balogh, and R. A. Waldo, *Magnesium Technology 2001*. The Minerals, Metals & Materials Society/AIME, Warrendale, PA, 2001.
40. T. K. Aune, O. Lunder, and K. Nisancioglu, *Microstructural Science* **17**, 231–241 (1989).
41. D. Eliezer, P. Uzan, and E. Aghion, *Materials Science Forum* **419–422**, 857–866 (2003).
42. F. Hehmann and H. Jones, *Magnesium Technology*. Institute of Metals, London, UK, 1987, p. 83.
43. A. Joshi, R. D. Adamson, and R. E. Lewis, in *Magnesium Alloys and Their Applications*, edited by B. L. Moredike and F. Hehmann. DGM Informationsgesellschaft, Oberursel, Germany, 1992, pp. 495–502.
44. A. Garboggini and H. B. McShane, in *Magnesium Alloys and Their Applications*, edited by B. L. Mordike and F. Hehmann. DGM Informationsgesellschaft, Oberursel, Germany, 1992, pp. 503–510.
45. G. L. Makar and J. Kruger, *Journal of the Electrochemical Society* **13**(2), 414–421 (1990).
46. S. K. Das and C. F. Chang, *Rapidly Solidified Crystalline Alloys*. Metallurgical Society of AIME, Warrendale, PA, 1985, pp. 137–156.
47. I. J. Polmear, *Materials Science and Technology* **10**, 1–16 (1994).
48. S. K. Das and C.F. Chang, Rapidly Solidified High Strength Corrosion Resistant Magnesium Base Alloys, U.S. Patent No. 4,853,035 (Aug. 1989).
49. F. Hehmann, F. Sommer, and B. Predel, *Materials Science and Engineering*, **A125**, 249 (1990).
50. A. Joshi and R.E. Lewis, in *Advances in Magnesium Alloys and Composites*, edited by H. Paris and W. H. Hunt. The Minerals, Metals & Materials Society, Warrendale, PA, 1988, pp. 89–103.
51. C. F. Chang, S. K. Das, D. Raybould, and A. Brown, *Metal Powder Report*, **41**(4), 302 (1986).
52. C. F. Chang, S. K. Das, D. Raybould, R. L. Bye, and E. V. Limoncelli, in *Advances in Powder Metallurgy*, edited by T. G. Gasbarre and W. F. Jandeska. MPIF/APMI, Princeton, NJ, 1989, pp. 331–346.
53. S. Bender, E. Boese, A. Heyn, and J. Goellner, Corrosion behaviour of magnesium alloys: material specific corrosion testing, in *Proceedings of the 7th International Conference on Magnesium Alloys and Their*

Applications, Dresden, Germany, edited by K.-U. Kainer. Wiley-VCH, Weinheim, Germany, 2007, pp. 721–726.

54. C. Scharf, C. Blawert, and A. Ditze Magnesium, in *6th International Conference on Magnesium Alloys and Their Applications*, Wolfsburg, Germany, edited by K.-U. Kainer. Wiley-VCH, Weinheim, Germany, 2004, pp. 980–987.
55. C. Blawert, E. Morales, W. Dietzel, N. Hort, K.-U. Kainer, C. Scharf, A. Ditze, and F. Endres, in *Magnesium Technology 2005*, edited by N. R. Neelameggham, H. I. Kaplan, and B. R. Powell. The Minerals, Metals & Materials Society, Warrendale, PA, 2005.
56. C. Scharf, A. Ditze, C. Blawert, E. Morales, W. Dietzel, and K.-U. Kainer, Influence of the copper content on microstructure and corrosion resistance of AZ91 based secondary magnesium alloys, in *SAE 2006 World Congress & Exhibition, Session: Magnesium Technologies*, Detroit, MI, SAE International, Detroit, MI, 2006.
57. S. Amira, D. Dubé, R. Tremblay, and E. Ghali, *Materials Characterisation* **59**, 1508–1517 (2008).
58. G. Song, A. Atrens, and M. Dargusch, *Corrosion Science* **41**, 249–273 (1998).
59. A.-M. Lafront, W. Zhang, S. Jin, and R. Tremblay, *Electrochimica Acta* **51**(3), 489–501 (2005).
60. S. Amira, A.-M. Lafront, D. Dubé, R. Tremblay, and E. Ghali, *Advanced Engineering Materials* **9**(11), 973–980 (2007).
61. H. Al-Kazzaz, M. Medraj, X. Cao, M. Xiao, and M. Jahazi, Effect of laser power and joint gap on weld quality of aerospace grade ZE41A-T5 magnesium alloy using ND:YAG laser, in *Proceedings of the Metallurgical Society of Canada*, Toronto, Canada, edited by M. O. Pekgulreyez and L.W. F. Mackenzie. METSOC, Toronto, Canada, 2006, pp. 503–518.
62. W. K. Miller, in *Stress-Corrosion Cracking*, edited by R. H. Jones. ASM International, Materials Park, OH, 1992, pp. 251–263.
63. K. Lubbert, J. Kopp, and E. Wendler-Kalsch, *Materials and Corrosion* **50**, 65–72 (1999).
64. J. I. Skar and D. Albright, in *Magnesium Technology 2002*, edited by H. I. Kaplan. The Minerals, Metals & Materials Society, Warrendale, PA, 2002, pp. 255–261.
65. H. Haferkamp, Fr. -W. Bach, P. Juchmann, M. Niemeyer and V. Kaese, in *Magnesium 2000*, edited by E. Aghion and D. Eliezer. Magnesium Research Institute, Beer-Sheva, Israel, 2000, pp. 449–455.
66. M. B. Kannan, W. Dietzel, C. Blawert, S. Riekehr, and M. Kocak, *Materials Science and Engineering A* **444**(1–2), 220–226 (2007).
67. C. Blawert, E. Morales, G. Wiese, V. Kree, W. Dietzel, S. Jin, and E. Ghali, Correlation between the microstructure of HPDC AZ91 and its corrosion properties, in *Proceedings of Magnesium Technology in the Global Age*, Montréal, Canada, edited by M. O. Pekguleryuz and L. W. F. Mackenzie. Canadian Institute of Mining, Metallurgy and Petroleum, Montréal, Canada 2006, pp. 295–314.
68. C. Blawert, E. Morales, V. Heitmann, W. Dietzel, S. Jin, E. Ghali, and S. Mandl, Corrosion properties of the skin of high pressure die cast AZ91 alloy, in *MET SOC Annual Conference*, Calgary, Alberta, Canada, edited by J.-P. Martin. CIM, Montreal, Canada, 2005, pp. 109–126.
69. S. Jin, N. Giguere, D. Dubé, R. Tremblay, and E. Ghali, Electrochemical impedance spectroscopy evaluation of the skin and bulk corrosion behavior of die-cast and thixo-cast AZ91D alloy in 5% (0.885 M) sodium chloride solution, in *Corrosion in Marine and Saltwater Environments II*, edited by D. A. Shifler. Electrochemical Society, Pennington, NJ, 2004, pp. 62–73.
70. A.-M. Lafront, D. Dubé, R. Tremblay, E. Ghali, C. Blawert, and W. Dietzel, *Canadian Metallurgical Quartely* **47**(4), 459–468 (2008).
71. C.-F. L. Jen-Yen Uan and B.-L. Yu, *Metallurgical and Materials Transactions A* **39A**, 703–715 (2008).
72. S. Jin, E. Ghali, C. Blawert, and W. Dietzel, in *Proceedings 210th ECS 2006 Meeting*, Cancun, Mexico, edited by N. Missert, A. Davenport, M. Ryan, and S. Virtanen. The Electrochemical Society, Pennington, NJ, 2007, pp. 295–311.
73. J.-Y. Uan and B.-L. Yu, *Matellurgical and Materials Transactions A* **36A**, 2245–2252 (2005).
74. K. Nandakumar, K. R. Sreekumari, and Y. Kikuchi, *Biofouling* **18**(2), 129–135 (2002).
75. J. Sawai, *Journal of Microbiological Methods* **54**(2), 177–182 (2003).
76. M. Takaya, K. Hashimoto, and Y. Toda, *Japan Institute of Light Metals* **48**, 352–353 (1998).
77. H. Kuwahara, Y. Al-Abdullat, M. Ohta, S. Tsutsumi, K. Lkeuchi, N. Mazaki, and T. Aizawa, *Materials Science Forum* **350–351**, 349–358 (2000).
78. M. Andrei, F. di Gabriele, P. L. Bonora, and D. Scantlebury, *Materials and Corrosion* **54**, 5–11 (2003).
79. G. Kim and Y.-W. Kim, *Materials and Corrosion* **52**, 137–139 (2001).
80. B. Heublein, R. Rohde, V. Kaese, M. Niemeyer, W. Hartung, and A. Haverich, *Heart* **89**, 651–656 (2003).
81. J.-P. Gravel, Training Report in Chemical Engineering—MIC of aluminum and aluminum alloys and rational biocorrosion of magnesium and magnesium alloys, Chemical Engineering Department, Laval University, Quebec City, 2007, pp. 1–67.
82. M. P. Staiger, A. M. Pietak, J. Huadmai, and G. Dias, *Biomaterials* **27**, 1728–1734 (2006).
83. F. Witte, J. Fischer, J. Nellesen, H.-A. Crostack, V. Kaese, A. Pisch, F. Beckmann, and H. Windhagen, *Biomaterials* **27**, 1013–1018
84. M. Bobby Kannan and R. K. Singh Raman, *Scripta Materialia* **59**(2), 175–178 (2008).

85. C. Di Mario, H. Griffiths, O. Goktekin, N. Peeters, J. Verbist, M. Bosiers, K. Deloose, B. Heublein, R. Rohde, V. Kasese, C. Ilsley, and R. Erbel, *Journal of Interventional Cardiology* **17**(6), 391–395 (2004).
86. T. Hassel, Fr.-W. Bach, A. Golovko, and Ch. Krause, Investigation of the mechanical properties and the corrosion behaviour of low alloyed magnesium–calcium alloys for use as absorbable biomaterial in the implant technique, in *Proceedings of 45th Annual Conference of Metallurgists of CIM*, Montreal, Quebec, Canada, edited by M. O. Pekguleryuz and L. W. F. Mackenzie. CIM, Montreal, Canada, 2006, pp. 359–370.
87. G. Song and S. Song, Magnesium as a possible degradable bio-compatible material, in *Magnesium Technology in the Gloal Age*, Montreal, Canada, edited by M. O. Pekguleryuz and L.W. F. Mackenzie. Metallurgical Society of CIM, Montreal, Canada, 2006.
88. F. Witte, V. Kaese, H. Haferkamp, E. Switzer, A. Meyer-Lindenberg, C. J. Wirth, and H. Windhgen, *Biomaterials* **26**, 3557–3563 (2005).
89. F. Witte, in *In-Vivo Korrosion von Magnesiumlegierungen und ihre biolokishen Effekte: Analysen von Tiermodellen*. GKSS, Workshop Magnesium in der Medizin, 2005.
90. P. Zartner, R. Cesnjevar, H. Singer, and M. Weyand, *Catheterization and Cardiovascular Interventions* **66**, 590–594 (2005).
91. J. Reifenrath, C. Palm, P. Müller, H. Hauser, H.-A. Crostack, J. Nellesen, F.-W. Bach, D. Besdo, M. Rudert, and F. Witte, Subchonrdal plate reconstruction by fast degrading magnesium scaffolds influence cartilage repair in osteochondrical defects, in *Proceedings of the 51st Meeting of the Orthopaedic Research Society*, Washington DC, Poster No. 1347, ORS, Rosemont, IL, 2005.
92. Y. Yamasaki, Y. Yoshida, M. Okazaki, A. Shimazu, T. Kubo, Y. Akagawa, and T. Uchida, *Biomaterials* **24**, 4913–4920 (2003).
93. C. E. Wen, M. Mabuchi, Y. Yamada, K. Shimojima, Y. Chino, and T. Asahina, *Scripta Materialia*, **45**, 1147–1153 (2001).
94. X. N. Gu, W. Zheng, Y. Cheng, and Y. F. Zheng, *Acta Biomaterialia* **5**(7), 2790–2799 (2009).
95. L. Xu, G. N. Yu, E. L. Zhang, F. Pan, K. Yang, *Journal of Biomedical Materials Research* **83A**(3), 703–711 (2007).
96. Z. Li, X. N. Gu, S. Q. Lou, and Y. F. Zheng, *Biomaterials* **29**(10), 1329–1344 (2008).
97. P. Gu, W. Zheng, M. Elboujdaini, and E. Ghali, Electrochemical study of magnesium hydride formation in aqueous solution, in *Degradation of Light Weight Alloys*, edited by D. Shifler. The Minerals, Metals & Materials Society, Warrendale, PA, 2007.
98. R. Ambat, N. N. Aung, and W. Zhou, *Corrosion Science* **42**, 1433–1455 (2000).
99. W. Zhou, N. N. Aung, and Y. S. Sun, *Corrosion Science* **51**(2), 403–408 (2009).
100. M. Pekguleryuz, P. Labelle, D. Argo, and E. Baril, Magnesium diecasting alloy AJ62X with superior creep resistance, ductility, and diecastability, in *Magnesium Technology 2003*, edited by H. I. Kaplan, The Minerals , Metals and Materials Society, Warrendale, PA, 2003, pp. 201–206.
101. P. Bakke, K. Petterson, and H. Westengen, Enhanced ductility and strength through RE addition to magnesium die casting alloys, *Magnesium Technology 2003*, edited by H. I. Kaplan, The Minerals, Metals and Materials Society, Warrendale, PA, 2003, pp. 171–176.
102. B. R. Powell, A. A. Luo, B. L. Tiwari, and V. Rezhets, The die cast ability of calcium-containing magnesium alloy, thin wall computer case, in *Magnesium Technology 2002*, edited by H. I. Kaplan, The Minerals, Metals and Materials Society, Warrendale, PA, 2002, pp. 123–129.
103. C. Suman, Effect of direct aging on mechanical properties and corrosion resistance of die-cast magnesium alloys AZ91D and AM60B, Paper No. 900794, SAE International, Warrendale, PA, 1990.

Chapter 12

Mechanically Assisted Corrosion of Magnesium and Its Alloys

Overview

Some studies concerning erosion corrosion and fretting fatigue corrosion are summarized. Preventive measures that can include inhibitors, surface treatments, and selected coatings are mentioned to improve wear resistance and wear corrosion of magnesium alloys. Testing the difference in performance of two types of coatings on light metals for erosion corrosion is discussed. It is important to examine the influence of erosion corrosion of road salts on car bodies and the efficiency of some preventive measures such as inhibitors, surface treatments, and selected coatings.

There are excellent studies on fatigue that are extremely appropriate but these do not address corrosion fatigue performance directly. Corrosion fatigue strengths of magnesium alloys can be as low as 10% of those in air and no fatigue limit could be determined because of accelerated crack kinetics of both initiation and propagation. Sea water has a greater corrosive effect than tap water because of chloride ions. Galvanic corrosion of magnesium decreases resistance to corrosion fatigue crack initiation. Mechanisms of corrosion fatigue are discussed. The influence of protective coatings on the condition of corrosion fatigue varied: anodic coatings showed small protective effect, but with certain organic materials added, they raised corrosion fatigue strength values to those obtained in air.

12.1. EROSION CORROSION AND FRETTING FATIGUE CORROSION

12.1.1. Erosion Corrosion

In waters free of passivators or dissolved oxygen, increased flow rates tend to erode any protective films that may have formed. Corrosion rates of most alloys have been examined over the acid pH range of water including even that of acid rain. Liquid impingement on the AZ91D alloy has been studied by Govender and Ferreira [1] using rectangular specimens, rotating disk electrodes, and two water jets at three impact velocities (9.53, 19, and 29 m/s approximately) in a 3.5% NaCl solution. The weight loss measurements were evaluated to be on the order of 25–80 times greater than those from immersion tests performed in the same solution for 24 h. The synergistic effect of the damage was a combination of corrosion of the α matrix and the mechanical removal of the corrosion product and erosion of the alloy.

Corrosion Resistance of Aluminum and Magnesium Alloys: Understanding, Performance, and Testing By Edward Ghali

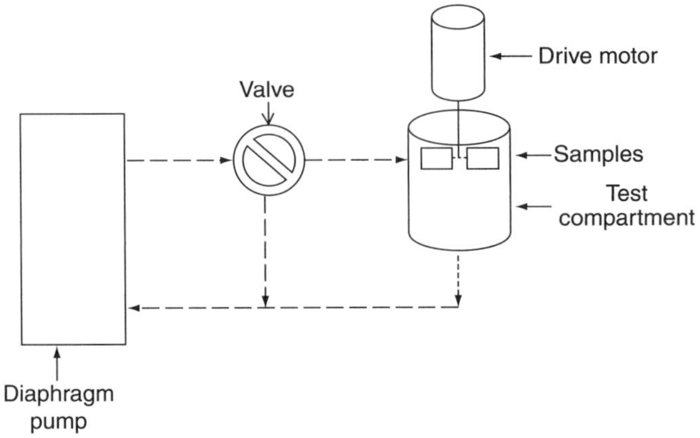

Figure 12.1 Schematic of erosion corrosion test equipment [2].

Ramsingh and DeRushie [2] determined the corrosion and erosion corrosion profiles of the following materials: Al 5052, Mg alloy AM60, 1052 commercial quality cold rolled (CQCR) steel, and several different types of protective coatings on the Mg alloy. They employed the equipment shown schematically in Figure 12.1, where the samples are mounted on the sample holder and rotated at a preset speed by the variable speed top-mounted electrical motor. The speed may be set up to 100 km/h by selecting the appropriate motor. Sand/salt slurry is poured into the slurry test compartment to fill the pump lines, and up to ~1 in. (2.5 cm) in the compartment. The temperature of the slurry may be controlled from 5 to 95 °C [2]. The 5 h test period, with the valve open for 10 s and closed for 90 s for each 100 s of the test, was established as the equivalent of driving 500 km (300 miles) at 35 km/h in the sand/salt slush present on urban roads after a snow or ice event. The results are given in Table 12.1.

If the weight loss results of the CQCR steel samples (4.31 mg/cm^2) are compared with the weight losses obtained on the same steel in under vehicle testing for 18 months [3], the equipment and procedure described above are more severe than the under vehicle test methods. The weight losses from the under vehicle test were in the range of 27.9–46.2 mg/cm^2 for the distance traveled range of 13,925–33,258 km. The weight loss in the erosion corrosion equipment was 4.31 mg/cm^2 for 500 km traveled, although in the case of the under vehicle tests, other factors such as crevice corrosion affected the results [2].

The equipment and procedures for assessing the erosion corrosion behavior of three alloys were appropriate and the weight loss results indicate good precision and reproducibility. The equipment and procedures are adequately suited to screen coatings for use on light metals in the automobile industry: for example, testing the difference in the performance of two types of coatings on light metals. The coated sample with a proprietary chemical conversion coating failed after 1 h of testing with 20% of the coating lost, while the epoxy powder and polyethylene coating showed no sign of wear after 10 h of testing [2].

Table 12.1 Average Weight Loss Results (mg/cm^2) After 5 h at 35 km/h

Material	3.5% NaCl	3.5% NaCl + 15% sand
CRCQ steel	2.06 ± 0.01	4.31 ± 0.02
Al 5052	0.06 ± 0.01	0.29 ± 0.01
Mg AM60	0.54 ± 0.03	4.14 ± 0.03

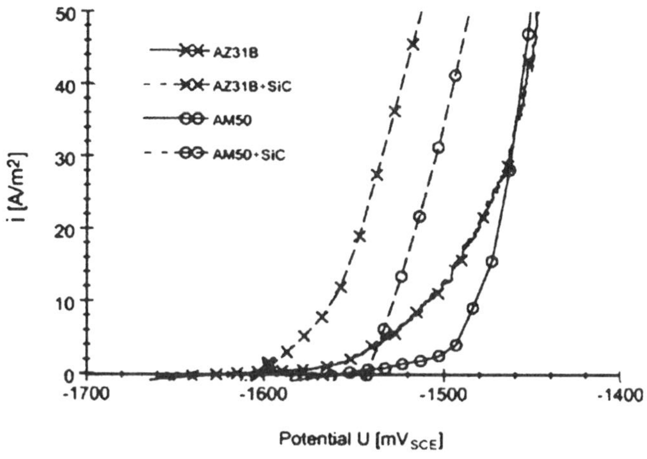

Figure 12.2 Polarization curves for nonreinforced and SiC-reinforced alloys of AZ31B and AM50 in 3% NaCl solution [4].

Further work in this area is badly needed to examine the influence of the environment, such as that of road salts, and the efficiency of some preventive measures that can include inhibitors, surface treatments, and selected coatings.

To improve wear resistance and wear corrosion of magnesium alloys, Gollner et al. [4] uniformly applied hard particles onto the magnesium surface using dispersion by plasma-arc powder weld surfacing with alternating current. The wear resistance of the two treated alloys, AZ31B and AM50, was excellent. After subjecting them to the Miller test (ASTM G75), the abrasive loss from samples reinforced with silicon carbide is very low, with values between 1% and 1.5%, while the abrasive loss was 25 times higher for magnesium alloy AM50 [4].

Figure 12.2 shows the polarization curves in a 3% sodium chloride solution for the two studied alloys with and without carbide-reinforced material, and it can be stated that the current increases at more positive potentials much more in the case of reinforced specimens compared to the two basic alloys. The composite surfaces showed pitting corrosion. In a mild solution like 0.01 N sodium hydroxide, the carbide-reinforced materials were also less resistant, showing dissolution of the material in the zone surrounding the carbide particles occurring from −1500 mV/SCE [4].

12.1.2. Fretting Fatigue Corrosion

Rotating bending fatigue strengths of Mg alloys were reduced by 50% in fretting conditions compared to those in air [5]. Reductions were less pronounced with Mg alloys characterized by high hardness and high fatigue strength values. Oxides and nitrides are often formed on the surfaces of Mg parts and are subjected to fretting fatigue conditions. Notch sensitivity of Mg alloys was insignificant when fretting corrosion predominated. Fretting can be reduced by surface rolling, sand blasting, or shot peening [6].

12.2. CORROSION FATIGUE OF MAGNESIUM ALLOYS

Magnesium alloys are seeing wider application in the transportation industry due to their high specific strength. Although they are mostly used in static parts (e.g., housing, bracket, panel), they have the potential to be used in load-bearing parts (e.g., wheel), which are subject to fatigue. The materials, however, usually have voids and other defects to some degree. Therefore it is of primary importance to examine the corrosion fatigue initiation and propagation behavior mechanisms [7]. For a given material, the fatigue strength (or fatigue life at a given maximum stress value) generally decreases in the presence of an aggressive environment. The effect varies widely, depending primarily on the particular metal–environment combination. The environment influences both fatigue crack initiation and fatigue crack growth rate. Expressing the crack growth rate, da/dN (where a is crack length and N is number of cycles), as a function of ΔK provides results that are independent of specimen geometry. The following factors must be considered for corrosion fatigue evaluation: alloy composition, microstructure, and yield strength; stress-intensity range, load frequency, and stress ratio; active–passive behavior and aggressive ions such as chloride ions; pH, electrode potential, and corrosion current in the aqueous environment [8, 9].

Fatigue tests generally follow ASTM E468-90 at ambient atmospheric conditions in spite of the possible influence of humidity and atmospheric oxygen accelerating galvanic corrosion cells. A rotating bending fatigue tester with a 125 mm cantilever length, operated at 1425 rpm at the level of 10^6 cycles, is frequently used [10].

12.2.1. Corrosion Fatigue of Cast Magnesium Alloys

Some available information concerns the fatigue behavior of sand-cast magnesium alloys; however, it has been claimed that similar comparative results are obtained with either sand-cast or die-cast materials. The stress versus life curves for machined laboratory fatigue specimens tend to become horizontal between 10^6 and 10^8 cycles and 30–50% of the ultimate tensile strength, suggesting an "endurance limit" similar to that of steel but differing from the continuously decreasing curve generally obtained with aluminum. Cast magnesium is very notch sensitive, with reported losses in fatigue strength being as great as 70% depending on the notch severity and stress level. Magnesium has been found to be more notch sensitive than cast aluminum, resulting in aluminum providing better notched fatigue resistance at high stress/short life and equal or better performance at low stress/long life. In general, magnesium is least competitive with aluminum in fatigue when there is a high applied stress, a severe stress concentration, or a high-temperature corrosive medium. Corrosion fatigue can also produce a significant reduction in fatigue strength, with the reduction being greatest at lower stress [11].

Morphology and Mechanism of the Fracture Under fatigue loading conditions, microcrack initiation in Mg alloys is related to slip in preferably oriented grains. Quasi-cleavage usually occurs in the initial stages of fatigue crack growth, which is common for hexagonal close packed (hcp) cells. Further crack growth micromechanisms can be brittle or ductile and trans- or intergranular, depending on the metallurgical structure and environmental influence. The corrosion environment is significantly detrimental relative to the air environment. Quasi-cleavage fatigue crack growth mechanisms have been identified in corrosion fatigue of cast magnesium alloy AZ91E-T6 in both air and 3.5% NaCl. Final fracture regions of samples in both environments were predominantly quasi-cleavage with some ductile dimples. Researchers observed crack initiation in MA12 in air and vacuum in the T2 condition; slip bands were short and thin and appeared after 15–20% of the total specimen life at the grain boundaries where the origin of microcracks occurred. In the T6 condition, in the same environments, slip bands were rare and occurred in individual grains after approximately 5% of the total life at high temperatures; intergranular slip, such as grain-boundary sliding, that controls cavity growth on grain boundaries for Magnox Al80 (0.8% Al) alloy at 430 °C, occurs [6].

There is no endurance limit for magnesium and its alloys in fatigue in corrosive media because of accelerated crack kinetics of both initiation and propagation. Galvanic corrosion of magnesium decreases resistance to corrosion fatigue crack initiation [6]. Corrosion pits are formed on the surface of both light weight alloys at a higher number of cycles: they become crack initiation sites [12]. The slope of the fatigue curve varies with the corrosive environment and the alloy composition.

Magnesium alloys are very susceptible to corrosion fatigue. Corrosion fatigue strengths can be as low as 10% of those in air, and no fatigue limit could be determined. Seawater has a greater corrosive effect than tap water because of chloride ions. The influence of a corrosion environment decreased with increasing frequency [6]. The Mg–1.5%Mn and Mg–2% Mn–0.5%Ce alloys are more resistant to corrosion fatigue than alloys containing aluminum and zinc; also, media such as 3% NaCl or seawater produce a much more rapid drop in the fatigue curve than does tap water [13]. Substantial reductions in fatigue strength of magnesium alloys have been studied in laboratory tests using NaCl spray or drops. Such tests are useful for comparing alloys and heat treatments.

In general, reduction of temperature increases the fatigue life of Mg alloys mainly by lengthening the crack initiation period [14]. Figure 12.3 shows data obtained on 1.6 mm (0.064 in.) Mg–6%Al–1%Zn–0.2%Mn sheet alloys tested with plate-type bending fatigue equipment in a chloride-containing spray of 0.01% NaCl. It also shows data obtained on protected and unprotected sheet to determine the effect of normal laboratory exposure. The two rates of spray shown in Figure 12.3 produced the same decrease in fatigue strength. Both alloys had approximately the same susceptibility to fatigue under corrosive conditions. Unprotected metal in the laboratory atmosphere had slightly lower fatigue strength than protected metal [15].

The corrosion environment was significantly detrimental relative to the air environment. Quasi-cleavage fatigue crack growth mechanisms have been identified in corrosion fatigue of cast magnesium alloy AZ91E-T6 in both air and 3.5% NaCl. Final fracture regions of samples in both environments were predominantly quasi-cleavage with some ductile dimples. Coatings that exclude the corrosive environment are considered to provide the primary defense against corrosion fatigue [16, 17].

Some coatings reduce the corrosion fatigue resistance of alloys for example, anodizing. However, the results of fatigue tests demonstrate that Keronite® coatings (plasma anodized surfaces) may cause no more than a 10% reduction in endurance limit of the Mg alloy studied, which is substantially lower than the effect from conventional anodizing. At the endurance limit, the transition to the nonfatigue region for the oxidized samples occurs substantially earlier than for the bare Mg alloy, due to the inhibition of the crack initiation by the oxide ceramic layer [10].

Corrosion Fatigue of Die-Cast Mg Alloy AZ91 It is generally admitted that the as-cast defects of Mg alloys such as voids or shrinkages often act as sites for initiation of fatigue cracks. The influence of corrosion on fatigue behavior and vice versa of the die-cast Mg alloy AZ91 was investigated by rotating beam testing. The samples were corroded by salt spray testing. Specimens taken from bulk material show poor fatigue properties. In die-cast bars (15 mm thick), large binding defects (pores) act as initiation sites for fatigue cracks limiting the 10^7 endurance limit to an amplitude of 50 MPa ($R = -1$). This effect of binding defects is dominant and determines the lifetime even after heavy corrosion (96 h salt spray

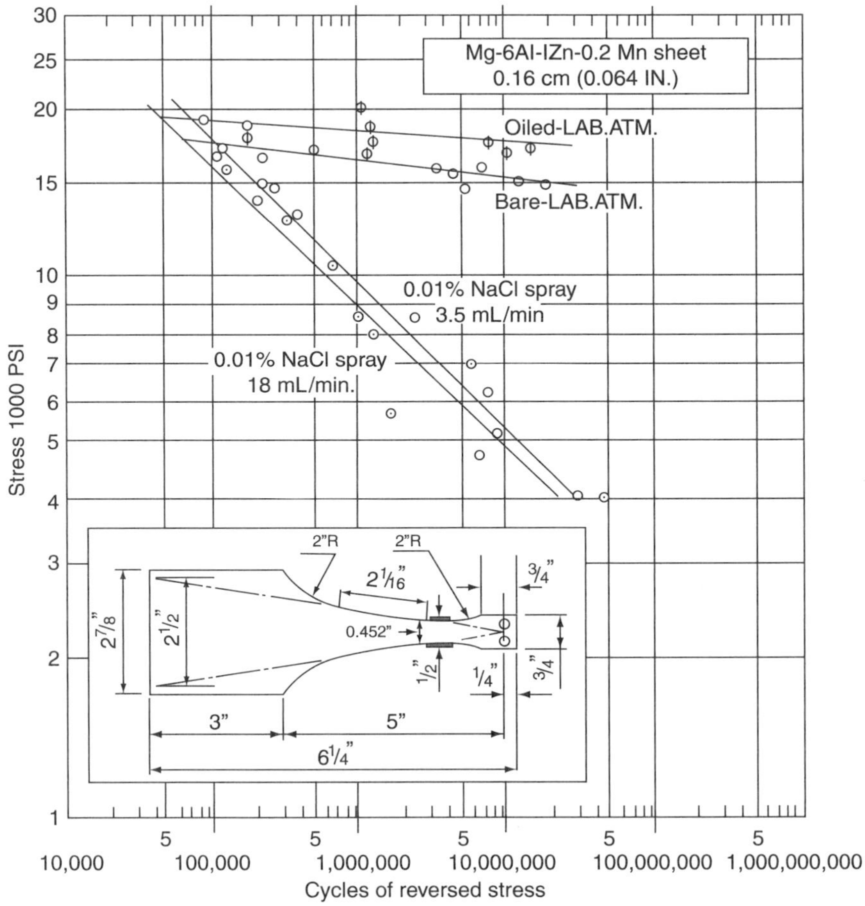

Figure 12.3 Effect of spray intensity of 0.01% sodium chloride on the resistance to fatigue of precipitated Mg–6%Al–1%Zn–0.2%Mn sheet. Specimen size—plate-type specimen, 1.6 mm (0.064 in.) thick; surface preparation—Aloxite ground; temperature—about 30 °C (90 °F) [17].

testing). Therefore precorrosion after salt spray that does not accumulate corrosion products or change the pH at the interface has no effect on the fatigue behavior under these conditions. Directly cast fatigue specimens (diameter 6 mm) show smaller pores and improved fatigue behavior with a 10^7 endurance limit of 100 MPa. Precorrosion in the case of cast specimens deteriorates the fatigue behavior and reduces the 10^7 endurance limit to 60 Mpa [18].

The corrosion behavior of AZ91 was measured by electrochemical polarization curves in 5% NaCl solution. In the unloaded state the die skin protects the alloy against corrosion. This protection effect is significantly reduced by applying a cyclic load. The rest potential is shifted to more active values by mechanical loading for specimens both with and without a die skin [18].

Magnesium alloys, which are used in the automotive industry, can suffer from dynamic loading in service, which can lead to fatigue and corrosion fatigue fractures. In order to determine the effect of a coating on the corrosion fatigue resistance of magnesium alloys, two die-cast alloys are considered—AM50 and AZ91D. The coatings were 5 and 25 μm thick anodized coats using the Anomag method (see Chapters 15 and 18). Three-point bending fatigue tests in natural seawater, tap water, and air as a reference were considered.

There seems to be no fatigue limit (endurance limit) for AM50 in tap water and for AZ91D in both tap water and seawater. Uncoated AM50 has a better corrosion fatigue resistance than AZ91D in chemical environments but not in air. It has been found that AZ91D has better corrosion fatigue resistance in tap water than in seawater, while AM50 has better corrosion resistance in seawater than tap water.

The comparison of fatigue life values at 110 MPa maximum stress level for coated and uncoated AM50 and AZ91D are given in Figure 12.4. It can be seen that AM50 always has better corrosion fatigue resistance than AZ91D and the coating generally improved the corrosion fatigue resistance in seawater. AM50 has excellent corrosion fatigue resistance in a seawater environment. The fatigue limit for 7×10^6 cycles in seawater is 105 MPa (maximum stress), which is higher than that in tap water and equal to that in air. Sealing did not show a marked favorable difference in spite of its known effect on increasing resistance to pitting corrosion. The resistance to corrosion fatigue was not reduced and even showed a slight improvement in the performance of AZ91D with the coating in this environment, since the coating acted as a protective barrier from corrosion under these conditions [19].

Corrosion Fatigue of Cast AM Alloy at Cold Temperatures To see the effect of corrosion on the fatigue life of Mg alloys, a simple facility was developed to immerse specimens of AM50 into 3.5 wt % NaCl water. This facility was installed in an environmental chamber of the AU-MTS set up. The fatigue test can be operated within an MTS or Instron frame under low and high temperatures without leakage. Fatigue testing results on rounded as-cast specimens of 6 mm radius showed serious corrosion-induced reduction of fatigue life. This reduction depends on the environmental temperature and the applied stress amplitude. For example, fatigue specimens under stress control with $R = 0.1$ and $\sigma_{max} = 120$ MPa show about 30% fatigue life reduction in saltwater compared to air at room temperature. It is important to underline that when the temperature was reduced from room temperature to −20 °C, this corrosion-induced reduction of fatigue life became more serious. It can be stated that low temperatures in this range accelerate fatigue corrosion. Extensive small pits were observed in the corroded surface and these pits were the locations for fatigue crack initiation and acted as the sources of crack coalescence.

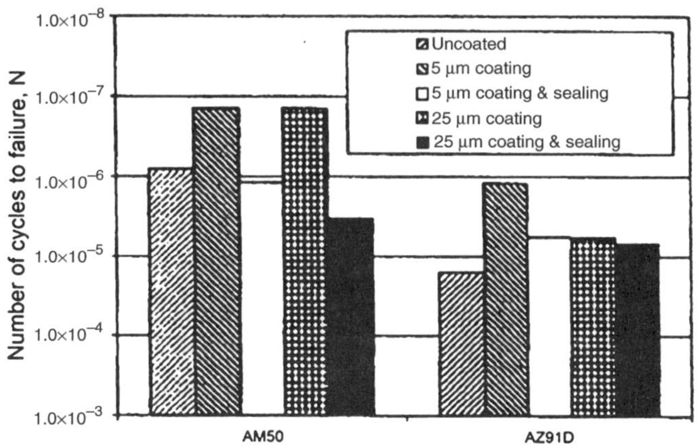

Figure 12.4 Fatigue life of uncoated and coated AM50 and AZ91D in seawater (30 Hz, $R = 0.25$) [19].

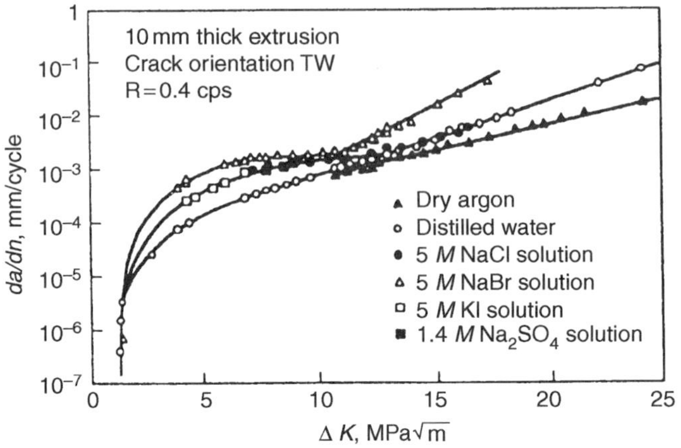

Figure 12.5 Corrosion fatigue crack growth curves of ZK60A-T5 in different environments [20].

12.2.2. Corrosion Fatigue of High-Strength Magnesium Alloys

Speidel et al. [20, 21] stated that all magnesium alloys behave similarly with respect to environmentally enhanced subcritical crack growth. Both stress-corrosion cracking (SCC) and corrosion fatigue (CF) cracks propagate in a mixed transgranular–intergranular mode. They measured the CF crack growth for all the aqueous environments in Figure 12.5 and compared the CF with SCC for this alloy. Microscopic studies showed that there is a distinct boundary between regions 11 and 111 for all the media. Sulfate as well as chloride ions accelerated the crack growth of both SCC and CF. The boundary between regions 11 and 111 in sodium bromide solution of the *da*/*dn* versus ΔK curve is higher than the stress-corrosion threshold (K_{ISCC}), which occurs at lower stress corrosion density [22].

12.2.3. Crack Propagation of Wrought Extruded Alloys

12.2.3.1. Surface Defects of Extruded AZ31 and AZ80 Alloys

It appears that the effects of composition and heat treatment are minor compared to the more significant and critical effect of surface condition [11]. Fatigue failure is usually initiated at a crack occurring at a surface defect or at some internal structural feature, such as a grain boundary [23]. Material defects may lead to a large scatter of the fatigue data for testing in air as well as in saltwater spray. The fatigue performance of two widely used wrought Mg alloys, AZ31 and AZ80, in extruded form was studied. Both shot peening and roller burnishing were found to improve fatigue life in air. When comparing air with aqueous NaCl solutions, AZ80 reacts more sensitively to an aggressive environment than AZ31. This influence becomes more pronounced if the surface is shot peened, but less pronounced after roller burnishing. This effect can be explained on the basis of differences in surface roughness induced by these mechanical surface treatments [24]. This can be due to overly severe peening that can reduce fatigue strength, since it has also been suggested that shot peening should be beneficial for fatigue and corrosion fatigue [25]. Shot peening the polished surface of Mg alloy AZ31 led to a decrease in fatigue strength. This was explained by the fact that compressive residual stresses on a polished surface are higher than a shot-peened one. Also, the plastic deformation induced by the shot peening treatment produced small cracks on the surface. It should be added that increased surface smoothness or cold working improved fatigue strength and fatigue time in air, but no improvement was observed in NaCl solution [26]. Concerning corrosion fatigue behavior, there should be an additional complex effect of the physicochemical, geometric properties of the surface combined with the electrochemical active–passive behavior at the interface of Mg alloys that control the beneficial effect of shot peening.

12.2.3.2. Crack Propagation of Extruded AZ80 and AZ61 Alloys

Relative Humidity (RH) The fatigue limit of as-extruded AZ61 is between 145 and 150 MPa at 20 °C and 50 °C. Corrosion fatigue occurred in 80% RH in air and corrosion pits were observed. The relative humidity influences not only the fatigue life but also the fatigue crack propagation rates in air. Atmospheric moisture accelerates fatigue crack propagation (FCP) rates of Mg alloys and decreases the ΔK_{th} to 55–75% of the respective values in vacuum [26].

Heat Treatment A solution heat treatment was found to increase the fatigue life. Mechanical and fatigue properties of sand-cast Mg–Al–Zn alloys showed an increase in the fatigue endurance limit using a T6 aging treatment. The FCP rate in a single-phase microstructure is slower than in a dual-phase one. The FCP rate of AZ91D alloy after the solution treatment was lower than that after an aging treatment, for example [26].

The influence of loading frequency, load ratio, aluminum content, heat treatment, temperature, and relative humidity on FCP of extruded Mg alloys AZ80 and AZ61 was examined. Using single-edge notched plate specimens, the effects of test parameters (e.g., loading frequency and load ratio), of materials (such as chemical composition, heat treatment state), and of environmental factors (temperature and relative humidity) were investigated [27].

It is demonstrated that the FCP rate increases to a great extent with a decrease in loading frequency from 10 Hz to 1 Hz. An increase in load ratio results in acceleration of the FCP rate. The effect of Al content on FCP rates depends on the combined effect of loading frequency, inhomogeneous microstructure and precipitates, and the thickness of the oxide film. The FCP rate of AZ80 is higher than that of AZ61 at 1 Hz, whereas it is lower at 10 Hz. Aging treatment facilitates the FCP rate. Increasing environmental temperature from room temperature to 60 °C and 120 °C, the FCP rate accelerates, particularly at 120 °C. A bend occurred in the curves of FCP rate versus stress-intensity factor at 120 °C. At first, the FCP rate increased sharply, and then went up slowly. Hydrogen embrittlement may be responsible for facilitation of FCP rate of Mg–Al alloys in higher relative humidity [27].

12.2.3.3. Corrosion Fatigue of Extruded AZ31 and AZ60 Magnesium Alloys

Wrought magnesium alloys show excellent mechanical properties and very good surface finish of a profile. For instance, die-cast and extruded AM50 alloys have ultimate tensile strength (UTS) values of 230 and 290 MPa, tensile yield strength (TYS0.2%) values of 125 and 180 Mpa, and elongation-to-fracture values of 12% and 18%, respectively (the specimen axis coincides with the extrusion direction). There is a practical reason for studying the corrosion fatigue of extruded Mg alloys, such as ZK60, AM50, and AZ31. The fatigue life of Mg alloys in corrosive solutions such as NaCl, for example, is always less than that in air. The degradation in fatigue strength for high-strength extruded AZ80 or die-cast

AZ91D alloys (8.59% A1, 1% Zn) due to NaCl is more pronounced than that for lower-strength AZ31 and AM20–AM40 (2–4% Al, 0.4% Mn) alloys, apparently due to a higher percentage of the second phase ($Mg_{17}Al_{12}$) in AZ80 or AZ91D alloys [28].

Corrosion fatigue tests were carried out on extruded AZ31 (3% Al, 1% Zn, 0.3% Mn), AM50 (5% Al, 0.4% Mn), and ZK60 (5% Zn, 0.5% Zr) Mg alloys in air, NaCl-based solution, and borate solution. Extruded hour glass-shaped specimens (minimum gauge diameter and length of 8.0 and 49 mm, respectively) (ASTM E466-82) were cut from extruded rods of AZ31, AM50, and ZK60 alloys by final turnery (roughness R_c below 3.2 mm). Rods 10 mm in diameter were produced from raw material 30 mm in diameter. Before the extrusion process, raw pieces were heated for 15 min at the working temperatures of 320 °C (ZK60), 350 °C (AZ31), and 375 °C (AM50). Extrusion rate was equal to 6 mm/s (ZK60, AM50) and 10 mm/s (AZ31). Fatigue tests were performed on a rotating beam type fatigue machine, employing complete reversible cycles ($R = -1$), equipped with a special electrolytic cell at 25 °C and at a frequency of about 30 Hz. The compositions of the electrolytes were as follows: 3.5% NaCl (pH ~5), 3.5% NaCl saturated with magnesium hydroxide $Mg(OH)_2$ (pH 10.5), 0.1 N $Na_2B_4O_7$ buffer solution (pH 9.3), and 0.1 N $Na_2B_4O_7$ saturated with magnesium hydroxide (pH 9.3) [28].

The relative fatigue life values (*N* solution/*N* air ratios) representing the number of cycles to failure in the solution and in air, respectively, were used for the analysis of the corrosion fatigue behavior of Mg alloys in various environments. These ratios vary in the same environment within a broad interval, depending on the processing and test conditions, for example, N solution/N air ratios vary for Mg alloys containing 8–9% Al from 0.01 to about 0.9–1 in 3.5–5% NaCl solutions. Immersion corrosion tests were carried out in a 2000 mL glass vessel during 72 h for both extruded alloys and die-cast AM50 and AZ91D alloys [28].

Figure 12.6 shows the fracture surface of an extruded AZ31 alloy specimen after a corrosion fatigue test in 3.5% NaCl in the crack origin area. The fatigue crack initiated from a corrosion pit since one can observe the fatigue crack origin connected with pitting. In the center of the micrograph, one can observe the presence of a cubic particle of Mn_3Mg_2 that can act as a cathode in a galvanic corrosion cell initiating pitting corrosion [28].

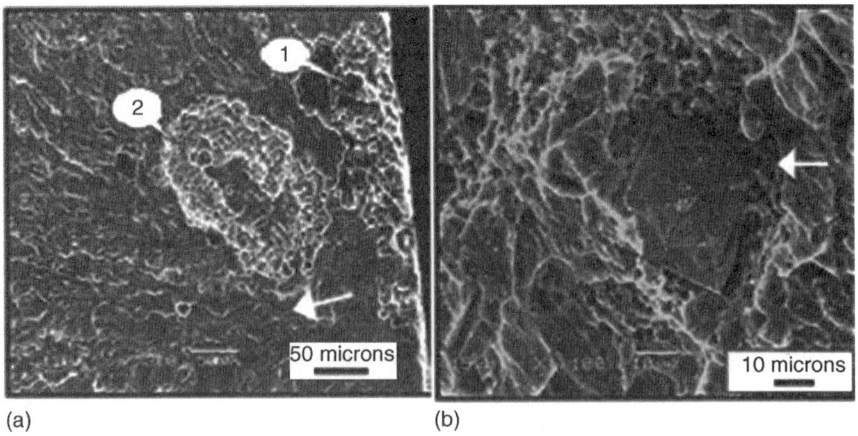

Figure 12.6 Fracture surface in crack origin area with pitting (a, b) in a sample of AZ31 alloy after a fatigue test in 3.5% NaCl. The direction of crack propagation is represented by the arrows: (a) ① and ② are pits on the surface of a specimen and on the fracture surface, respectively; (b) a cubic particle of Mn_3Mg_2 appears in the center of the micrograph [28].

Extruded alloys show a significantly longer fatigue life both in air and in NaCl-containing solutions in comparison with die-cast alloys. For instance, under the stress of 140 MPa, the lifetime of die-cast and extruded AM50 alloys was equal to 1.8×10^5 and 2.7×10^6 cycles in air, and 9.8×10^4 and 2.6×10^5 cycles in 3.5% NaCl, respectively. Extruded ZK60 alloy reveals very high fatigue and corrosion fatigue resistance properties in comparison with other alloys. For example, in air and in 3.5% NaCl saturated with $Mg(OH)_2$, fatigue fracture corresponding to $N = 10^5$ cycles was observed under the applied stresses of 215 and 190 MPa in ZK60; 180 and 150 MPa in AM50; and 180 and 165 MPa in AZ31 alloys. Under the same stress, the corrosion fatigue life of extruded alloys is significantly longer than that of die-cast alloys (*N* solution for extruded AM50 in NaCl is two to three times longer than that of die-cast AM50). Extruded alloys show a significantly higher sensitivity to the action of 3.5% NaCl solution in comparison with die-cast alloys. Extruded ZK60 has the lowest relative fatigue life (*N* solution/*N* air $\approx 10^{-3}$ to 10^{-2}) or the highest sensitivity to the action of NaCl-based solutions in comparison with that of AM50 and AZ31 alloys (*N* solution/*N* air $\approx 10^{-2}$ to 10^{-1}) [28].

Buffer borate solution represents a corrosive medium that decreases the fatigue life of extruded AZ31 alloy in contrast to die-cast alloys. Saturation of the borate buffer solution with magnesium hydroxide does not affect the corrosion fatigue behavior of extruded AZ31 alloy; while saturation of 3.5% NaCl solution with magnesium hydroxide leads to a marked increase in the corrosion fatigue life of this alloy. The dissolution rate of as-extruded ZK60, AZ31, and AM50 alloys in 3.5% NaCl increases with growing Al content. This fact agrees with the corrosion behavior of prestrained die-cast alloys AM50 and AZ91D [28].

12.2.3.4. Oil Environments as Inhibitors

Eliezer et al. [29] examined the fatigue and mechanochemical behavior of die-cast AZ91D alloy (Mg–95%Al–1%Zn) and AM (Mg–5%Al–0.4%Mn) alloys in air, transmission oil, and natural mineral oil material at 25 °C. Die-cast specimens were produced on a cold chamber machine with the locking force of 3450 kN. In addition, fatigue tests were carried out on hour glass-shaped specimens (67 mm long and 7.9 mm diamter). Die-cast specimens were examined with no mechanical treatment of the gauge. The Texaco Geartex EP-B SAE 85W90 commercial manual transmission oil containing extreme-pressure (EP) additives based on chloride or phosphorus components was considered and compared to light white mineral oil (0.84 g/mL) as a reference.

It was found that the mineral oil environment causes an increase in the lifetime of both alloys as compared to the gear oil. Mineral oil almost doubled the lifetime of AM50 alloy as compared to air, while gear oil increases the lifetime only by half; and AZ91 showed higher sensitivity to the environment if compared to AM50. It can be suggested that there was a reaction between the gear oil and the metallic surface—not just a physical type of adsorption but also a chemical one due to the presence of EP additives.

The initiation stage of the fatigue fracture in the two alloys is due to the Al content and the increase of the β phase in the AZ91D alloy. Higher Al content increased the dissolution rate of the Mg matrix, leading to the formation of pits; the area influenced by the corrosion reaction after the corrosion fatigue test was higher for AZ than for AM. Figure 12.7a shows the presence of pits and the initiation sites for corrosion fatigue failures in a borate solution saturated with magnesium hydroxide without stress. In the case of AM50, the corroded surface was more homogeneous and fewer corrosion pits were observed (Figure 12.6c). In the presence of stress or during the corrosion fatigue tests, the pits were initiated and grew

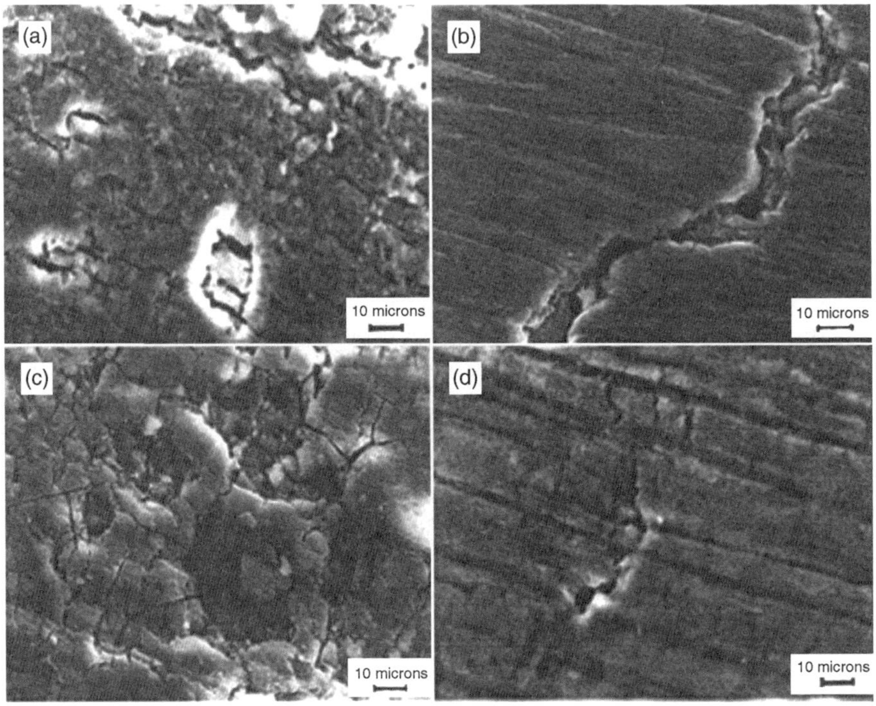

Figure 12.7 SEM micrographs of corroded surfaces in 0.1 N $Na_2B_4O_7$ solution saturated with $Mg(OH)_2$ without stress for (a) AZ91D and (c) AM50, and corroded surfaces under stress for (b) AZ91D and (d) AM50 [29].

into corrosion fatigue cracks. The quantity and importance of crack propagation are more evident for AZ91D (Figure 12.7b) than for AM50 (Figure 12.7d).

It can be stated that hard secondary phases promote strain hardening and thus increase the chemical potential of atoms; that is, they create the necessary conditions for mechanochemical dissolution. Consequently, the greatest resistance to develop a mechanochemical effect such as corrosion fatigue will be obtained for alloys with the lowest Al content. Indeed, deep pitting corrosion occurred and the lifetime of the alloy decreases according to the quantity of the β phase [29].

Corrosion Fatigue of the Skin and Bulk The corrosion fatigue of Mg alloy AZ91D under applied stress of 169 MPa in three different solutions—3.5% NaCl, gear oil, and gear oil with 3.5% NaCl—is presented in Figure 12.8. The die-cast skin (0.5 mm) was removed mechanically to examine the influence of the quantity and density of the β phase in the surface due to the die-casting process. It is evident that the specimen without the skin exhibits a higher fatigue lifetime in comparison to the die-cast surface due to the difference in Al content in the surface [29].

Surface Area and Thickness The fatigue limit decreases with increasing sample size. The FCP rate decreased significantly when the specimen thickness increased from 1.5 to 3 mm and was little changed when the thickness increased to 5 mm [6].

Grain Size The smaller the grain size at the surface, the higher the fatigue limit. The fatigue strength improved with the decrease in grain size for AZ91. However, this relation

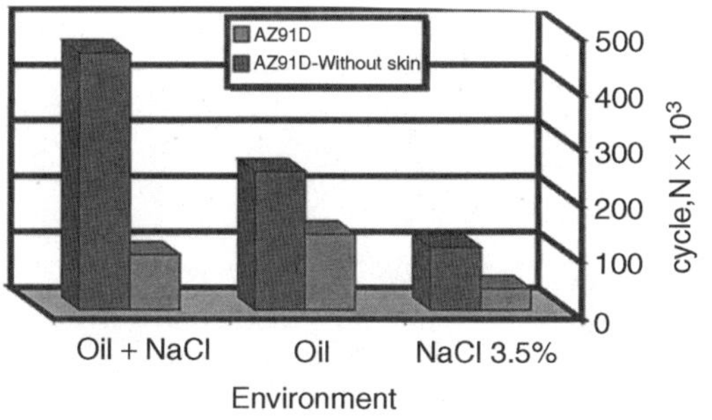

Figure 12.8 Corrosion fatigue resistance of the AZ91D alloy with and without the skin in three different solutions influenced very probably by the variation in Al content [29].

hardly relates to the center of the specimens. Also, the higher the applied stress load, the greater will be the grain size influence on fatigue life. Considering the kinetic parameters, the FCP rate increases with a decrease in grain size and the amplitude of the crack growth path deviation is greater for coarse grained material than a fine grained one [26]. Eliezer et al. [29] stated that the specimens of the cast alloys AZ91D and AM50 (6 mm in diameter) showed better stability in the *S–N* curves under the different environments (air and oil) than that of the 7.9 mm specimens because of the finer grain size; because of the less porous microstructure there was a smaller probability of finding a defect.

Crystallographic Textures The experimental results of Wagner et al. [30] and that of Zeng et al. [27] showed that the fatigue strength for extruded AZ31 and AZ80 in the longitudinal direction was significantly superior to that in the transverse direction.

Influence of the Frequency The lower the frequency, the faster the FCP rates of extruded Mg alloys AZ80 and AZ91 in air. Generally, in aqueous media and for frequencies less than 10 Hz, the lower the frequency the shorter the fatigue lives as observed for AM60 extrusions [26].

12.2.3.5. EAC of High-Strength Cast WE43 Rare Earth Alloy

Magnesium alloys are known to experience environmentally assisted cracking (EAC) in ambient atmospheric air. The greatest susceptibility has been found in Al containing alloys. EAC also occurs in pure Mg. Electrochemical measurements and observations of cracking in distilled water, moist air, and dry hydrogen gas suggest that EAC is promoted by hydrogen embrittlement. Transgranular stress corrosion crack propagation has been reported to occur by transgranular cleavage on the (0 0 01) basal plane. Intergranular cracking has been also reported. EAC of cast Mg alloy WE43 (4.2 wt %Yt, 2.3 wt %Nd, 0.7% Zr, and 0.8% HRE) in the T6 peak-aged condition was induced in ambient air in a notched specimen [31].

Cast plates of WE43 were solution treated at 525 °C. The specimen blanks were aged to the T6 condition at 250 °C in oil for 16 h before machining to their final shape. Constant-load, static fatigue tests were performed using cylindrical notched tensile specimens in ambient air at stresses between 250 and 3000 MPa. The test run-out was chosen at 100 h. Crack nucleation and propagation tests were performed using rectangular tested specimens in four-point bending. Before testing, the surface was polished and then etched with 30% nitric acid solution in methanol for about 30 s at − 30 °C and rinsed in methanol. The bend

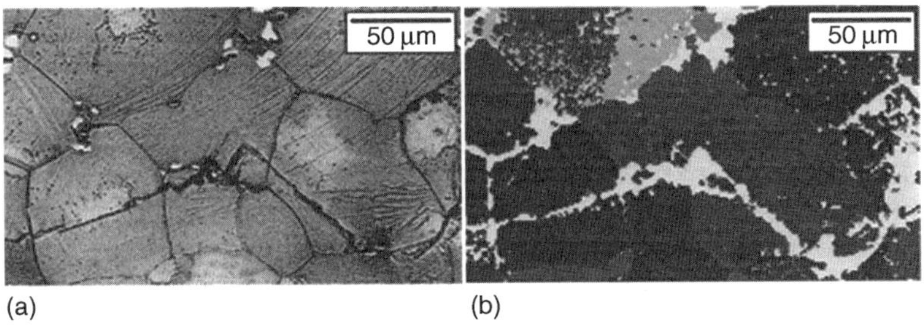

Figure 12.9 Transgranular and intergranular cracking: (a) backscatter electron image and (b) electron backscatter diffraction (EBSD) grain orientation map of the same region [31].

specimens were stressed at a rate of 1 MPa/s under displacement control, while the tensile surface was observed with an optical traveling microscope [31].

EAC or static fatigue occurs in WE43-T6 in atmospheric air, in a similar manner to other high-strength Mg alloys. Microstructure controls the nucleation of EAC. Cracks initiated at the intergranular brittle intermetallic and propagated by transgranular cleavage [31]. The transgranular cracks were stepped with crystallographic facets. Electron backscatter diffraction maps were used to determine the crystallographic orientation of the transgranularly cracked grains (Figure 12.9) [31].

A microstructural model for static fatigue limit in cast Mg alloys may be developed, which includes the effects of defects such as porosity [31].

12.2.4. Welding and Corrosion Fatigue of AZ31

Cross et al. [32] as well as Ellermeier et al. [33] have examined the influence of different welding processes (laser, electron beam, plasma, and gas-metal arc) on the corrosion resistance of Mg alloys exposed to saltwater immersion. Corrosion was found to initiate preferentially in the HAZ, and welds with a narrow HAZ showed higher corrosion resistance. It has been mentioned that the difference in Al concentration between filler and base materials as well as oxidation and evaporation processes taking place at the weld pool surface are important factors to control. Arc welding, including gas-metal arc and gas-tungsten arc processes, has been demonstrated to be a viable method for joining magnesium. Thate [34] reported that the relative corrosion resistance of base wrought alloys (AZ31 and AZ61) determines the corrosion behavior of the corresponding weldments and that the mentioned methods of welding have no observed influence on corrosion resistance.

Fatigue strength of Mg alloy weldments in air, for arc welds made on rolled plate, reaches ~50–60% that of the base alloy. Welded joints on extruded AZ31 have fatigue strength efficiencies around 80%. The notch effect of the weld toe determines the fatigue strength of the weldment, where use of filler metal improves the weld bead profile and fatigue properties [35].

Test specimens were 3 mm thick plate made from wrought AZ31 alloy (continuous cast and hot rolled) following ASTM E466-02. Welding was done using a variable polarity, gas-tungsten arc, cold wire feed process (VP-GTA-CWF) using AZ61 filler metal and a square, butt joint configuration. Welds were made full-penetration in a single pass, using appropriate parameters. Uniaxial fatigue tests were performed at 30 Hz in tension under load control (load ratio, 0.1) using a hydraulic MTS machine. Tests are performed at room temperature in both air and a 3.5% NaCl saturated solution without and with magnesium hydroxide acting as a buffer. The 3.5% NaCl solution used alone has a pH of 5 at the start and shifts gradually to 9.8. In the presence of NaOH, the alkaline salt solution has a constant pH of ~10.5. The cell was connected to a 10 L tank, and the solution was circulated at a speed of 0.3 L/min. It was observed that the weld metal picked up significant amounts of aluminum, zinc, and silicon from dilution with the filler wire, with a slight loss in manganese [32].

Cross et al. [32] have compared the fatigue and corrosion fatigue behavior of 3 mm thick rolled AZ31 plate and its corresponding weldments, tested in air and buffered salt solution. Use of the VP-GTA-CWF welding process has proved successful in generating welds of high quality using AZ61 filler wire [32].

Welds were observed to have a uniform bead along the weld length, with no undercutting and a few scattered pores under 0.5 mm in diameter. The grain growth in the HAZ increased in average diameter from $8 \pm 3\,\mu m$ to $38 \pm 16\,\mu m$. The coarse second phase in the base metal and HAZ has been identified as Al_8Mn_5. The weld metal consists of coarse equiaxed grains ($43 \pm 15\,\mu m$) with coarse β-$Al_{12}Mg_{17}$ eutectic phase constituent. This equilibrium β phase was almost continuous along the grain boundaries [32].

Figure 12.10 compares *S–N* curves for AZ31 wrought plate, in welded and unwelded conditions, for both air and buffered NaCl solutions. It is observed that the fatigue life of AZ31 base metal in salt solution is significantly reduced compared to that in air at high loads. This is likely due to the presence of corrosion pits serving to initiate fatigue cracks. At low loads, this difference becomes diminished, likely due to the interaction of corrosion at the crack tip. Weldments also have a reduced fatigue life compared to the base metal, with welds tested in salt solution displaying the lowest fatigue life. All welded specimens fractured through the weld metal, with the exception of two specimens (low load, high cycle), which failed in the base metal. Microhardness traverse measurements, across the weldment at 0.5 mm intervals, showed the weld metal to be slightly lower in hardness than the base metal: 40HV for the weld as compared to 50HV for the base metal [32].

Fatigue life of AZ31 weldments in air is significantly lower than the base metal for all loads examined. Fracture of weldments occurred predominantly in the weld metal, which is lower in strength and has a nearly continuous network of grain boundary β phase. Fatigue life of AZ31 base metal and weldments in buffered NaCl solution is lower than corresponding values in air, as expected. However, this difference becomes increasingly smaller in the low-load high-cycle regime, where corrosion is likely to play a role in crack propagation [32].

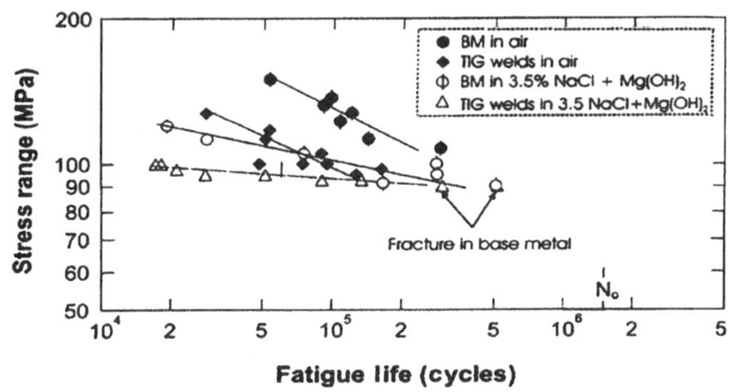

Figure 12.10 Comparison of *S–N* curves for AZ31 wrought plate in welded and unwelded conditions, and in air and buffered NaCl solution [32].

The weld metal with its higher Al content shows a corrosion potential that is more negative than either the HAZ or base metal on a freshly polished surface. However, after passivating, the weld metal shows better corrosion resistance and the base metal is found to have the most severe pitting [32].

12.2.5. Mechanisms of Corrosion Fatigue: Initiation and Propagation

Fatigue strength decreases with the increase in relative humidity (RH). Corrosion fatigue is observed for RH $\geq$ 80%. Fatigue cracks in the extruded Mg alloys are initiated at inclusions or oxide on the surface and subsurface of specimens in air, while in corrosive media, pits are the main cause of initiation.

Ando et al. [36] studied the crack propagation of single crystals and found that if the notch is perpendicular to the base plane, the fatigue cracks with basal slip grew along a zigzagged route, while in the case of the notch parallel to the base plane, fatigue cracks propagated in a straight line with [1012] twin. Horestemeyer et al. [37] stated that at low maximum crack-tip driving forces, the fatigue crack propagated preferentially through the α-Mg dendrite cells, while at high maximum crack-tip driving forces, the fatigue crack propagated preferentially through the β-particle laden interdendritic region [26].

Three mechanisms have been proposed for corrosion fatigue in air: (1) the absorbing mechanism, (2) oxide formation that formed films at the crack tip, which hinder the close of the crack surface, and (3) hydrogen embrittlement induced by water vapor. A possible mechanism for accelerated FCP rates in ambient air is the capillary condensation of water at the crack tip, which ensures exposure of the newly created surfaces. Generally, there are two mechanisms of corrosion fatigue in aqueous solutions—anodic dissolution and hydrogen embrittlement (HE). Experimental results by Eliezer et al. [38] showed that chlorine ions could penetrate or break through the oxide or hydroxide films on Mg surfaces and facilitate the dissolution of the bare surface [26].

It has been stated that anodic polarization accelerates crack growth of magnesium in sulfate and chloride solutions and plays an important role in corrosion cracking while cathodic polarization retards it by shifting the pH to more alkaline values, where a protective layer is formed. Fractographic and acoustic investigations showed that under certain circumstances HE plays an important role in crack propagation. Magnesium forms a stable hydride and Mg alloys are easily hydrogen-impregnated and their plasticity drops, that is, they are prone to internal HE [39].

Shipilov [40] suggested that in corrosion cracking and corrosion fatigue of Mg alloys, HE may be the primary mechanism of crack growth. Very high values of the parameter K (stress intensity) are necessary for HE in corrosion cracking and corrosion fatigue of VMD10 (Y, 7.6%; Zn, 1.7%; Cd, 1.5%; Zr, 0.3%). The addition of chloride ions to sodium hydroxide or CrO_3 solutions intensifies more strongly the local solution of the metal at the crack tip than HE. This causes a shift in the kinetic curves in the direction of lower values of K. The higher the chloride ion concentration, and the lower the passivator concentration, the greater the shift will be. This replaces the accelerating action of the cathodic polarization by a retarding one. The increase of strength and hardening of VDM10 alloy in aging (at 220 °C for 24 hours) eases its HE and leads to a reduction in the values of K_{HE} and K_{maxHE} corresponding to rapid growth rate. This reflects the general trend toward an increase in the HE effect with an increase in strength of constructional materials. However, the strength of an alloy is not the primary criterion of its tendency toward HE [40].

Among those mechanisms commonly used to identify corrosion fatigue crack growth (CFCG) are generally the technique based on the effect of applied potential on CFCG rates. This is accompanied by fractographic SEM studies currently used for this purpose. If the application of cathodic potential decreases the time to failure, whereas small anodic current increases the time to failure, this indicates that HE has no major effect on this mechanism.

Shipilov [41, 42] proposed a new approach for identifying the role of hydrogen-induced cracking (HIC) and stress-assisted dissolution in corrosion fatigue crack propagation. The approach intended to create short-term stimulation of electrochemical reactions within a growing crack. For this purpose, anodic or cathodic potential was applied to the specimen for a time period in which it is possible to register the change in the crack growth rate corresponding to the open circuit potential and to measure the crack growth rate under polarization. Due to the high resolution of the technique used to measure the crack extension, the time rarely exceeded 300 s. The approach made it possible to observe the non-single-mode effect of cathodic polarization on CFCG rates [41, 42].

Shipilov [42] stated that the effect of cathodic polarization on the CFCG behavior can be explained as follows, after evaluation of three different materials (low alloy steels, titanium alloys, and magnesium alloys). Cathodic polarization accelerated crack growth when the maximum stress intensity (K_{max}) exceeded a certain well-defined critical value characteristic for a metal–solution combination. When K_{max} was lower than the critical value, the same cathodic polarization, with all other conditions (specimen, solution, pH, loading frequency, stress ratio, temperature, etc.) being the same, retarded or had no influence on crack growth. The results and fractographic observations suggested that the acceleration in crack growth under cathodic polarization was due to HIC. Therefore the critical values of K_{max}, as well as the stress-intensity range (ΔK), were regarded as corresponding to the onset of CFCG according to the HIC mechanism and were designated as K_{HIC} and ΔK_{HIC}. HIC was the main cause of CFCG at $K_{max} > K_{HIC}$ ($\Delta K > \Delta K_{HIC}$) and $da/dN > da/dN_{cr}$ [42].

A mechanism of stress-assisted dissolution (SAD) at the crack tip played a dominant role in the corrosion fatigue crack propagation at $K_{max} < K_{HIC}$ and $(da/dN) < (da/dN)_{cr}$. In addition, SAD was the main mechanism of CFCG in titanium alloys and magnesium alloys in most solutions investigated in which cathodic polarization retarded and even stopped CFCG [42].

The relative contribution of HIC (in comparison with another possible mechanism such as SAD) to corrosion fatigue crack propagation increased with increasing R. Under static loading, HIC was more pronounced than under cyclic loading [42].

Shipilov [41] compared the HIC fatigue crack growth rates with critical distances ahead of the growing crack in high-strength low-alloy steels, titanium alloys, and magnesium alloys. He stated that six possible combinations of the relationship between HIC fatigue crack growth rates (increments of crack growth per cycle, Δa_p) and critical distances ahead of the crack were observed. Some of these combinations did not agree with well-known and conventional theoretical views on HIC and HIC fatigue crack growth mechanisms. The increment of HIC fatigue crack growth per cycle was lower than the cyclic plastic-zone size, for example, as observed for the magnesium alloy VMD10.

12.2.6. Prevention of Corrosion Fatigue

The influence of protective coatings on the corrosion fatigue (CF) condition varied: anodic coatings showed small protective effect, but with certain organic materials added, they raised CF strength values to those obtained in air [6].

It appears that the effects of composition and heat treatment are minor compared to the more significant and critical effect of surface condition. Machining, grinding, and shot peening can be very beneficial to fatigue resistance; however, shot peening must be properly controlled since overly severe peening can reduce fatigue strength [11].

Chromate conversion and anodic coatings can be used to improve environment resistance, although such coatings have been reported to be detrimental to fatigue resistance. More recent work, however, has shown this reduction results from prior acid cleaning, not the coating itself. Use of an alkaline cleaner or shot peening prior to acid cleaning can prevent this problem, with the coating then improving the fatigue and corrosion fatigue resistance [11].

There are also some conventional preventive methods that should be considered: avoid cyclic stresses, reduce maximum stress, use appropriate alloy selection, use inhibitors, achieve cathodic protection without hydrogen embrittlement, use surface treatments and organic coatings. Precautions should be taken when nobler metal coat is considered to avoid pores, scratches, and so on [9]. In the automotive industry, besides the appropriate conventional prevention methods, it has been noticed that the development of "road films" has a corrosion prevention measure that seems unpredictable, uncertain, and somewhat tenuous [11].

REFERENCES

1. G. Govender and J. H. Ferreira, Erosion corrosion of magnesium AZ91D alloy in the As-cast condition, in *Proceedings of the Second Israeli International Conference on Magnesium Science and Technology,* edited by E. Aghion and D. Eliezer. Magnesium Research Institute, Beer Sheva, Israel, 2000, pp. 371–376.
2. R. R. Ramsingh and C. DeRushie, *Materials Performance* **45**, 48–51 (2006).
3. R. MacDonald and R. R. Ramsingh, *Materials Performance* **24**, 48–53 (1985).
4. J. Göllner, S. Schultze, B. Bouaifi, and B. Ouaissa, *Materials and Corrosion* **53**, 13–22 (2002).
5. T. R. Beck, *Electrochimica Acta* **29**, 18–50 (1984).
6. V. V. Ogarevic and R. I. Stephens, *Annual Review of Material Science* **20**, 141–177 (1990).
7. Y.-Z. Wang, in *Uhlig's Corrosion Handbook*, edited by R. W. Revie Wiley-Interscience, Hoboken, NJ, 2000, p. 221–232.
8. W. Glaeser and I. G. Wright, in *ASM Handbook, Volume 13A, Corrosion*, edited by S. D. Cramer and B. S. Covino Jr. ASM International, Materials Park, OH, 2003, pp. 322–330.
9. V. S. Sastri, E. Ghali, and M. Elboujdaini, *Corrosion Prevention and Protection—Practical Solutions.* Wiley, Chichester, UK, 2007, pp. 109–176.
10. A. L. Yerokhin, A. Shatrov, V. Samsonov, P. Shashkov, A. Leyland, and A. Matthews, *Surface and Coatings Technology* **182**, 78–84 (2004).
11. W. K. Miller and E. F. Ryntz Jr., Paper 830521, Society of Automotive Engineers, 1984.
12. H. Mayer, M. Papakyriacou, S. Stanzl-Tschegg, E. Tschegg, B. Zettl, H. Lipowsky, R. Roesch, and A. Stich, *Materials and Corrosion* **50**, 81–89 (1999).
13. W. S. Loose, in *Corrosion Handbook*, edited by H. H. Uhlig. Wiley, Hoboken, NJ, 1976, pp. 218–252.
14. V. A. Serdyuk and N. M. Grinberg, *International Journal of Fatigue* **5**, 79–85 (1983).
15. A. F. Froats, T. Kr. Aune, D. Hawke, W. Unsworth, and J. Hillis, in *ASM Handbook, Volume 13, Corrosion*, edited by L. J. Korb, D. L. Olson, and J. R. Davis. ASM International, Materials Park, OH, 1987, pp. 740–754.
16. R. I. Stephens, C. D. Schrader, D. L. Goodenberger, K. B. Lease, V. V. Ogarevic, and S. N. Perov, SAE Technical Paper No. 930752, Society of Automotive Engineers, 1993.
17. A. Beck, *The Technology of Magnesium and Its Alloys.* F. A. Hughes and Co., London, 1940.
18. C. Mueller, R. Koch, and G. H. Deinzer, in *Magnesium Alloys and Their Applications*, edited by K.-U. Kainer. Wiley-VCH, Weinheim, Germany, 2000, p. 457.
19. W. G. Ferguson, W. Liu, P. Ross, and J. MacCulloch, Corrosion fatigue of high pressure die cast magnesium alloys, in *Proceedings of Magnesium Technology 2001*, edited by J. Hryn, The Minerals, Metals and Materials Society, Warrendale, PA, 2001, pp. 269–274.
20. M. O. Speidel, in *The Theory of Stress Corrosion Cracking in Alloys*, edited by J. C. Scully. North Atlantic Treaty Organization Scientific Affairs Division, Brussels, Belgium, 1971, pp. 289–355.

21. M. O. Speidel, M. J. Blackburn, T. R. Beck, and J. A. Feeney, in *Corrosion Fatigue: Chemistry, Mechanics and Microstructure*, edited by O. Devereux et al., National Association of Corrosion Engineers, Houston, Texas, 1986, p. 331.
22. A. Shaw and C. Wolfe, in *ASM Handbook, Volume 13B, Corrosion,* edited by S. D. Cramer and B. S. Covino, Jr. ASM International, Materials Park, OH, 2005, pp. 205–227.
23. P. N. Ross, J.-A. MacCulloch, and R. J. Esdaile, Anodizing Magnesium, Paper No. T99-063, 20th International Die Casting Congress and Exposition, Cleveland, 1999, pp. 213–218.
24. M. Hilpert and L. Wagner, Effect of mechanical surface treatment and environment on fatigue of wrought magnesium alloys, in *Proceedings of the International Congress on Magnesium*, Alloys and their applications, 2000, Munich, edited by K. U. Kainer, Wiley-VCH Weinheim, Germany, 2000, pp.304–311.
25. W. K. Miller, in *Stress-Corrosion Cracking*, edited by R. H. Jones. ASM International, Materials Park, OH, 1992, pp. 251–263.
26. Z. Rongchang, H. Enhou, and K. Wei, *Materials Science Forum* **488–489**, 721–724 (2005).
27. R.-C. Zeng, Z. Jin, W. J. Huang, W. Dietzel, K.-U. Kainer, C. Blawert, and W. Ke, *Transactions of Nonferrous Metals Society of China* **16**, 763–771 (2006).
28. Y. Unigovski, A. Eliezer, E. Abramov, Y. Snir, and E. M. Gutman, *Materials Science and Engineering A* **360**, 132–139 (2003).
29. A. Eliezer, O. Medlinsky, J. Haddad, and G. Ben-Hamu, *Materials Science and Engineering A* **477**, 129–136 (2007).
30. L. Wagner, M. Hilpert, and J. Wendt, *Materials Science Forum* **419–422**, 93–102 (2003).
31. T. J. Marrow, A. BinAhmad, I. N. Kan, S. M. A. Sim, and S. Torkamani, *Materials Science and Engineering A* **387–389**, 419–423 (2004).
32. C. E. Cross, G. Ben-Hamu, D. Eiezer, and P. Xu, *Corrosion-Fatigue of AZ31 Wrought Magnesium Weldments.* Wiley-VCH, Weinheim, Germany, 2006, pp. 727–733.
33. J. Ellermeier, K. Eppel, and C. Berger, *Magnesium-Bauteilfestigkeit: DVM-Tagunsband* 153–173 (2003).
34. W. J. Z. Thate, *Werkstofftechnik* **93**, (10) 699–704 (2003).
35. K. Krohn and S. Singh, SchweiBen von Magnesiumlegierung für dem Automobilbau, DVS Report 204, pp. 197–201.
36. A. Ando, N. Iwamoto, and T. Hori, *Journal of the Japan Institute of Metals*, **63**, 187 (2001).
37. M. F. Horestemeyer, N. Yang, and K. Gall, *Fatigue and Fracture of Engineering Materials and Structures* **25**, 1045 (2002).
38. A. Eliezer, E. M. Gutman, E. Abramov, et al., *Journal of Light Metals* **1**, 179–186 (2001).
39. T. A. Marichev, *Werkstoffe und Korrosion* **34**, 300 (1983).
40. S. A. Shipilov,in *Fiziko-Khimicheskaya Mekhanika Materialov*, Vol. **22**, Institute of Physical Chemistry, Academy of Sciences of the USSR, Moscow, 1986, pp. 21–25.
41. S. A. Shipilov, *Scripta Materialia* **47**, 301–305 (2002).
42. S. A. Shipilov, *Fatigue Fracture Engineering Material Structure* **25**, 243–259 (2002).

Chapter 13

Environmentally Induced Corrosion of Magnesium and Its Alloys

Overview

Magnesium alloys that contain neither aluminum nor zinc are the most stress-corrosion cracking (SCC) resistant. Aluminum content above a threshold level of 0.15–2.5% is reportedly required to induce SCC behavior in aluminum-containing magnesium alloys. Magnesium alloys are very susceptible to SCC through pitting. Slow cooling from the solution-treating temperature, along with certain thermomechanical treatments, can produce $Mg_{17}Al_{12}$ grain-boundary precipitates, which increase the susceptibility of an alloy to intergranular SCC.

SCC threshold stress is associated with the onset of plastic deformation. Both transgranular and intergranular crack propagation as well as mixed transgranular and intergranular crack propagation have been observed. Distilled water produces SCC in magnesium alloys during spraying or full, partial, or intermittent immersion, dissolved oxygen accelerates SCC, and rapid SCC has also been observed during seawater immersion. A film rupture or hydrogen-assisted cracking, or a combination of the two, would be consistent with the local cathodic condition that is produced by differential aeration or during bimetallic corrosion. The electrochemical dissolution and hydrogen embrittlement (HE) models are considered as propagation mechanisms of SCC. Some examples of laboratory failures (SCC–HE) of some magnesium alloys are given.

13.1. USE OF MAGNESIUM ALLOYS AND STRESS-CORROSION CRACKING FAILURES

The magnesium industry grew rapidly first in response to the need for lightweight military aircraft engines, airframes, wheels, and other parts. Then, after World War II, the nonmilitary market began using significant amounts of magnesium alloys in power tools, luggage, lawnmower decks, sporting equipment, camera cases, conveyors, and a variety of automotive components such as brackets, covers, housings, and wheels [1, 2]. Despite the SCC sensitivity shown by magnesium alloys in laboratory tests, it has often been reported that

Corrosion Resistance of Aluminum and Magnesium Alloys: Understanding, Performance, and Testing By Edward Ghali

service failures are rare and normally result from excessive residual stress produced during fabrication [3, 4]. In one laboratory study, an inorganic coating was found to accelerate SCC of a SCC-resistant alloy under certain conditions [5].

The apparent low incidence of SCC service failures of castings is attributable to low stresses actually applied or to stress relaxation by yielding or creep when a fixed deflection is imposed [6]. SCC will, however, be less likely to occur when the stress is produced by an imposed strain rather than a directly applied load. The room temperature creep in these alloys can produce stress reduction through the relief of both residual stresses and stresses induced by an imposed strain [1]. The only reports of SCC of pure magnesium have emanated from laboratory tests in which specimens were immersed in very severe SCC solutions [7–9].

Failures of wrought AZ80 aircraft components resulted from excessive assembly and residual stresses [3, 4]. More recently, however, failures of a number of forged AZ80 French aircraft components were reported, apparently resulting also from excessive assembly and residual stresses. Service failures of cast + forged South African magnesium aircraft wheels have also been described. Spiedel estimated that approximately 10–60 magnesium aerospace component SCC service failures occurred each year from 1960 to 1970, in a comprehensive review of more than 3000 unclassified failure reports from aerospace companies, government agencies, and research laboratories in the United States and five Western European countries. Of this total, more than 70% involved either cast alloy AZ91-T6 or wrought alloy AZ80-F [4].

13.2. KEY PARAMETERS

13.2.1. Alloy Composition and Magnesium Impurities

Pure magnesium is not susceptible to SCC when loaded up to its yield strength in atmospheric and most aqueous environments. Magnesium has a hexagonal close packed (hcp) structure, and in polycrystalline specimens, there is a threshold stress below which failure does not occur. In single crystals the yield strength point must be exceeded to produce plastic deformation, and this has been observed to be necessary for SCC to be observed [10]. The corrosion resistance of magnesium and its alloys is dependent on film formation in the medium to which they are exposed.

Hanawalt et al. [11] showed that the corrosion rate increased abruptly once tolerance limits were exceeded; these tolerance limits are 5, 170, and 1300 ppm for nickel, iron, and copper, respectively. Typically, up to 1 wt% of Mn is added to improve corrosion resistance by reducing the potential difference between iron-containing particles and the matrix (see Chapter 11). Mg–Al alloys have the greatest SCC susceptibility of all the magnesium alloys, and susceptibility increases with increasing aluminum content. Mg–Zn alloys have intermediate susceptibility, and the alloys that contain neither aluminum nor zinc are the most SCC resistant. So far no special heat treatments have been found that will reduce or eliminate SCC [12, 13]. An aluminum content above a threshold level of 0.15–2.5% is reportedly required to induce SCC behavior in aluminum-containing magnesium alloys, with the effect peaking at approximately 6% Al. Zinc also induces SCC susceptibility in magnesium alloys, so it is not surprising that the aluminum- and zinc-bearing AZ alloys, which are the most commonly used magnesium alloys, have the greatest susceptibility to SCC. Aluminum is the most susceptible element followed by zinc as the second susceptible metal for Mg alloys: for example, Mg–6% Al–1% Zn is susceptible in salt–chromate solution to intergranular crack propagation in the furnace cooled and transgranular crack propagation in the water quenched microstructure.

Increasing concentrations of 2–8 wt% Al in die-cast Mg–Al alloys decrease the corrosion rate. Low Al additions, of approximately 2–4 wt%, result in α-Mg dendrites surrounded by the two-phase, $\alpha + \beta$, eutectic at grain boundaries, whereas higher additions, 6–9 wt% Al, tend to precipitate distinct β particles along grain boundaries, depending on solidification rates. Surrounding the Al-rich α phase are local concentrations of up to 10 wt% Al as a result of microsegregation during solidification [14]. The increasing presence of β particles, which begin to appear above 2 wt% Al, may cause, in part, the improved corrosion resistance of the higher Al-content alloys [11]. However, higher Al-content alloys are frequently more pure and so this can contribute also to better performance [15].

Magnesium–manganese alloys, such as M1, are among the alloys with the highest resistance to SCC, and they are generally considered to be immune when loaded up to the yield strength in normal environments. Iron is known to reduce general corrosion resistance but its effect is to decrease SCC resistance or to have a minimal influence or no influence at all [1]. Mg–Mn alloys are immune in NaCl–dichromate solutions unless one adds on the order of 0.5 Ce [16]. Tests in humid air have resulted in SCC failures of Mg–Li–Al alloys, but SCC did not occur during testing of Mg–Li alloys strengthened with zinc, silicon, and/or silver instead of aluminum. Mg–14Li alloys, which have a body-centered cubic (bcc) structure, exhibit intergranular fracture in humid air although this can be prevented by a stabilizing treatment consisting of heating for 24 h at 149 °C after quenching [17].

Rare earth (RE) elements are typically added to Mg–Al alloys as cerium-based mischmetal (MM) containing lanthanum, neodymium, and praseodymium. The Al_4MM phase particles precipitated in AE alloys exhibit a passive behavior and do not affect the corrosion process to a significant extent. A high resistance to localized corrosion is observed for the AE alloys with a high Al content. Zn makes the Mg alloy electrochemically more noble, thereby minimizing the corrosion rate. Si is intentionally added to only the AS alloys, to combine with Mg, forming Mg_2Si, whose precipitation strengthens the alloy and is relatively innocuous to the corrosion behavior [14].

13.2.2. Microstructure and Crack Morphology

Although both transgranular and intergranular crack propagation have been observed (Figure 13.1), SCC in Mg alloys is usually transgranular, with significant secondary cracking (branching). Initiation of these cracks invariably occurs at corrosion pits. Very fine, parallel striations, spaced approximately 1 μm or less apart, have been observed during transmission electron microscopy (TEM) examination of transgranular SCC fracture surfaces. The striations are oriented perpendicular to the direction of cracking, occur within the coarser markings described above, and have been associated with successive positions of the crack during its discontinuous propagation. However, these striations do not correspond to crack-arrest fronts, but rather to the propagation of the crack along very small twin-boundary steps [1, 14].

Mixed transgranular and intergranular crack propagation (and occasionally totally intergranular cracking) have also been observed during magnesium SCC failures. Intergranular cracking is also reportedly promoted by a small grain size and/or a low residual iron content. Intergranular crack propagation may also be characteristic of other specific alloys or alloy–environment combinations. For example, SCC testing of a Mg–Zn–Zr alloy in water, a Mg–Li–Al alloy in humid air, and alloy AZ61 in a 75 °C 5% NaCl solution produced

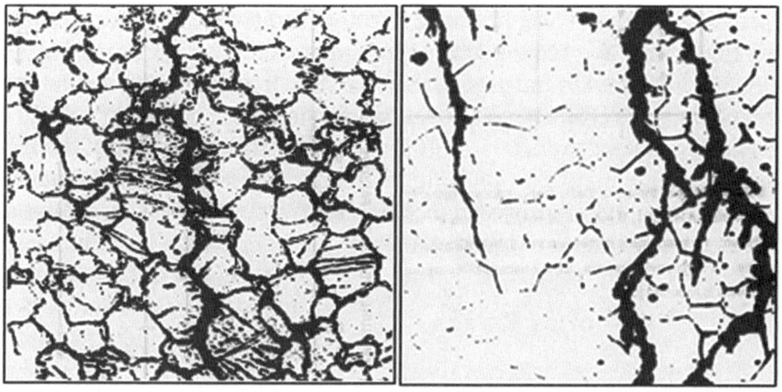

Figure 13.1 SCC in an extruded Mg–6Al–1Zn alloy tested in a salt–chromate solution, showing (a) intergranular crack in the furnace cooled alloy, and (b) transgranular propagation in the water quenched material [14].

predominantly intergranular fractures. It has been suggested that transgranular attack is not related to precipitates of any kind. Intergranular SCC has been related to localized galvanic attack on the matrix beside the cathodic $Mg_{17}Al_{12}$ grain-boundary precipitate [1].

Grain refinement increases the overall grain boundary area, thereby optimizing the distribution and minimizing the size of any possible detrimental intermetallics, such as Fe_3Al. Addition of strontium, as inoculants to Mg–Al alloys, results in reduced grain size and a lower corrosion rate that is attributed not only to the reduced grain size but also to changes in the oxide layer structure and composition and in the electrochemical properties of the phases present. Above a grain size of about 0.03 mm, cracking is transgranular irrespective of the heat treatment employed. In two-phase alloys, intergranular cracking occurs in specimens aged at ≤ 150 °C and transgranular cracking occurs in specimens aged at > 150 °C [16].

In rapid solidification technologies, including spray or droplet formation, continuous chill casting, and in situ melting, typical cooling rate are in the range of 10^5–10^7 °C/s. Improper processing can have a significant impact on corrosion behavior. The chunk effect is caused by surface oxides on powder particles that lead to poor bonding within the final product. Localized corrosion along these prior boundary oxides leads to particle-size pits and high corrosion rates. Corrosion rates for atomized RS alloy are comparable to those of cast AZ91D, although those for melt spun RS alloys are significantly higher because of the chunk effect [18].

Some studies show that heat treatment has an insignificant effect on SCC of Mg–Al–Zn alloys. Other contributions indicate that reduction in residual stress and the formation of a more resistant microstructure can reduce SCC susceptibility. Slow cooling from the solution-treating temperature, along with certain thermomechanical treatments, can produce $Mg_{17}Al_{12}$ grain-boundary precipitates, which increase the susceptibility of an alloy to intergranular SCC. The effect of heat treatment on SCC may also depend on the SCC environment. Artificial aging of cold-rolled AZ61 sheet improved SCC resistance when tested in salt–chromate solution or atmospheric exposure. However, the same treatment reduced SCC resistance in a 3% NaCl solution [1, 18].

Heat treatment can drastically alter the size, amount, and distribution of the precipitated β phase, $Mg_{17}Al_{12}$, which in turn alters the corrosion behavior of IM Mg–Al alloys including corrosion rate, filiform corrosion, and pitting. A T4 heat treatment (solution heat treatment only for 16 h at 415 °C to homogenize the alloy) increases the corrosion rate slightly compared to the as-cast material. The T6 treatment, being a T4 followed by a T5 treatment, gave corrosion rates well below those of as-cast samples and of samples given a T4 heat treatment (from 60% to 10% less for T6 aged at 120 and 205 °C, respectively). The lower temperature (120 °C) aging treatment formed a speckled precipitate of β particles, whereas the higher temperature treatment (205 °C) formed a discontinuous plate-like β precipitate with more surface [19].

13.2.3. Effect of Stress

Constant-load SCC tests have been shown to be more severe than constant-deflection tests. Under a constant load, stress increases as the cross section is reduced by cracking or corrosion. However, this condition produced decreasing stress when deflection is fixed. It has been suggested that SCC threshold stress is associated with the onset of plastic deformation, that is, the elastic limit of the alloy. The elastic limit is difficult to measure unambiguously, however, and the 0.2% yield strength (the stress at which 0.2% plastic deformation occurs) is generally used instead. It is interesting that the 30% yield strength limit recommended for die-cast alloy AZ91 correlates well with the elastic limit of this material, being equal to approximately one-third of the yield strength [20, 21]. The threshold stresses for monotonic or cyclic loading of initially plain specimens are essentially the same and are empirically related to the threshold stress-intensity factors from notched specimens subjected to constant load or constant strain. The severity of cracking shows a marked dependence on strain rate, with ductile failure at sufficiently fast or slow rates. The strain rate for most severe cracking is dependent on the composition of the solution and the applied potential [22].

Welding can introduce residual stresses, which are very dangerous for SCC resistance. More homogeneous microstructures tend to improve corrosion resistance. Under fatigue loading conditions, microcrack initiation in Mg alloys is related to slip in preferentially oriented grains. Some studies indicate that deformation, caused by cold work, reduces the SCC resistance of Mg alloys while other tests indicate the opposite. It seems that there has been no attempt to separate residual stress and microstructural effects as parameters influencing SCC resistance [1].

13.2.4. Effect of the Environment

Air Environments Exposure to normal atmospheric environments has been shown to produce SCC in susceptible Mg alloys. However, SCC severity is not necessarily associated with the corrosion severity of the environment. Rainfall, dew, and high humidity accelerate SCC of Mg alloys during atmospheric exposure. Laboratory tests have shown that SCC occurs in indoor air only when the relative humidity exceeds 85–98%, with additions of oxygen or carbon dioxide slightly reducing this threshold [1, 18].

Aqueous Media In general, the only solutions that do not induce SCC are either those that are nonactive to magnesium, such as dilute alkalis, concentrated hydrofluoric acid, and chromic acid, or those that are highly active, in which general corrosion predominates. Distilled water produces SCC in Mg alloys during spraying or full, partial, or intermittent, immersion. Dissolved oxygen accelerates SCC, and deoxygenating considerably retards or even prevents it. Rapid SCC has also been observed during seawater immersion [23]. Magnesium die-cast alloys are susceptible to SCC, requiring only partial immersion in distilled water. There is evidence that this SCC results from a cathodic process, with perhaps hydrogen-assisted cracking and film rupture both playing roles [23].

Magnesium SCC is also produced in many other dilute aqueous solutions, including (in approximate decreasing severity) NaBr, Na_2SO_4, NaCl, $NaNO_3$, Na_2CO_3, $NaC_2H_3O_2$, NaF, and Na_2HPO_4. Accelerated SCC has also been reported in dilute solutions of KF, KHF_2, HF, KCl, CsCl, NaI, KI, $MgCO_3$, NaOH, and H_2SO_4, HNO_3, and HCl acids. The desire for an accelerated test to study SCC of Mg alloys led to the development of an aqueous NaCl + K_2CrO_4 solution that produced very rapid cracking and helped the ranking of alloy susceptibility that appeared to correlate with atmospheric tests [24]. The aggressiveness of this solution is thought to result from the partial passivation of the magnesium by the chromate, which retards general corrosion but allows, or even promotes, local attack by the chloride. This electrolyte has remained very popular in laboratory studies of Mg SCC, even though it has since been found to correlate poorly with service experience. In one study, 40 g/L NaCl + 40 g/L Na_2CrO_4 has been applied to AZ80 and AZ61 under a stress of 10–14 ksi, while AZ91 needs more than 15 ksi for T4 or T6 in rural atmosphere. Another aggressive solution is 35 g/L NaCl + 20 g/L K_2CrO_4 for AZ31. In fact, no accelerated laboratory test has been developed that adequately predicts service life or the relative susceptibility of different alloys [1]. The addition of nitrate ions or carbonate ions to salt–chromate solutions inhibits SCC. This is believed to be associated with the formation of a stronger, more stable, or more readily repaired passive film [25].

KHF_2 This is very aggressive for SCC for almost all Mg alloys and is important in the context of anodizing and protective passive films containing fluorides. Pure Mg metal showed an intergranular cracking and Mg–Mn alloys were found to be susceptible to SCC in KHF_2 solution. Since F^- ion is an inhibitor for the corrosion of Mg, at least part of the electrochemical explanation may lie in inhibited film breakdown and repair kinetics [16].

Considering the lack of a specific standard or industrial practice for this purpose, a solution of 3.5% NaCl (pH 6.5) can be considered as testing medium for Mg alloys since this medium is recommended for all metals in ASTM G 44-88. Thus 3.5–5% NaCl saturated with magnesium hydroxide at pH 9 in the bulk solution has been used. A pH of 9 seems interesting because of partial passivation of Mg and possible pitting in a weak NaCl solution; however, the pH at the interface can rise to the range of 10.2–10.5. For SCC susceptibility, circulation of the electrolyte is recommended to avoid strong passivation [26].

pH, Cl^- and Distilled Water Although pH can affect the general corrosion of Mg alloys, a pH between 1.05 and 11.5 has been found to have no effect on SCC susceptibility in a salt solution. However, for pH values greater than 10–12, Mg alloys become very resistant to SCC [1]. The pH of the water in the subject tests increased during the first 4–5 days from an initial value of 5.5 to 8.0–8.5, where it remained during the rest of the testing period [23].

SCC is associated with environmental conditions leading to the local breakdown of a partially protective surface film. Film breakdown can be caused by chloride ion pitting, for example. However, it can be postulated that in the presence of relatively passive or barrier films, such as that expected for Mg in dilute sulfate solution or for some mg alloys in distilled water, film breakdown can be influenced by stresses [15]. It is also postulated that film rupture is unnecessary for SCC initiation in Mg alloys and can be substituted by a severe localized corrosion due to the cathodic phase of Mg_{17} Al_{12}, such as grain-boundary precipitates for AZ91 [27].

ASTM D1384-87 Water Immersion in ASTM D1384-87 water for 90 hours at pH 8.3, containing 165 mg/L NaCl, 138 mg/L $NaHCO_3$, and 148 mg/L Na_2SO_4 at the level of pitting potential has been studied [28] and could be an interesting medium as an SCC testing mechanism. Pitting potential can be determined by potentiodynamic polarization at an appropriate scan rate.

Effect of Temperature Susceptibility of Mg alloys to SCC has been found to increase with increasing temperature during tests in atmospheric environments, water, and a H_2SO_4 + NaCl solution. However, passivation kinetics is improved in a passivating solution; SCC can then be significantly reduced as temperature increases within a restricted range [1]. Temperature also affects the amount of creep deformation that occurs in a material. Therefore, under some conditions, SCC susceptibility may be reduced by the additional creep that occurs at higher temperatures. Creep can occur at low stresses at room temperature and can also produce stress reduction through relief of both residual stresses and stresses induced by an imposed strain [12, 23].

SCC and Liquid-Metal Embrittlement Early metallographic and fractographic observations of crack growth of pure Mg specimens were carried out in dry air (by surrounding specimens with a desiccant), a 3.3% NaCl + 2% K_2CrO_4 solution, and liquid alkali metals (Na, Rb, Cs) [7]. Specimens to be tested in dry air and the salt solution were first fatigue precracked in laboratory air and then were fractured by cantilever bending at deflection rates of 0.0005 up to 50°/s. High deflection rates produced high crack velocities that were measured from enlarged prints of cine films (taken at 50 frames/s) for the tests. Crack growth in dry ambient temperatures was macroscopically brittle and occurred parallel to {101X} planes or along grain boundaries, but fracture surfaces were microscopically fluted or dimpled. Fluted fracture surfaces parallel to {101X} planes and dimpled intercrystalline facets were also produced by SCC and liquid-metal embrittlement (LME) [9].

Cleavage-like {0001} fracture surfaces were observed after SCC and LME. Brittle fracture of Mg in inert environments occurs by a localized microvoid-coalescence process, which produces fluted transcrystalline facets and dimpled intercrystalline facets. SCC and LME occur by a more localized microvoid-coalescence process than that which occurs in inert environments and can be explained on the basis that adsorbed hydrogen/metal atoms weaken interatomic bonds at crack tips and thereby facilitate the nucleation of dislocations from crack tips [9].

The close similarities between SCC and adsorption-induced LME, and observations that embrittlement in aqueous environments could occur at crack velocities as high as 5 cm/s, suggested that adsorbed hydrogen atoms produced by dissociation of water molecules should be present at crack tips (rather than solute hydrogen, hydrides, or localized dissolution) and was responsible for SCC. Adsorbed hydrogen and metal atoms weaken interatomic bonds at crack tips and thereby facilitate the nucleation of dislocations from crack tips. Neither diffusion of hydrogen ahead of the crack nor localized dissolution at the crack tip would have time to occur. SCC at low velocities may also result from adsorbed hydrogen since the characteristics of SCC at low velocities are similar to those produced by rapid SCC and to those produced by adsorption-induced LME. Lynch and Trevena [9] proposed that transgranular SCC of pure Mg in NaCl + K_2CrO_4 solution occurred by adsorption-induced dislocation emission (AIDE) based on the similarity of the fracture surface morphology with that produced by LME, which involves adsorption of metal atoms at the crack tip.

13.3. INFLUENCE OF OTHER FORMS OR TYPES OF CORROSION ON SCC

Some types of corrosion are given here as example to show their relationship to SCC. Although in several situations, different types of corrosion have a synergetic accelerating effect on SCC, this is not always the case (general corrosion or corrosion creep).

13.3.1. Effect of General Corrosion

The corrosion rate of chemically pure Mg in saltwater is in the range of 0.30 mm/yr. The corrosion resistance of commercial Mg alloys does not significantly exceed that of pure Mg. In the long run, uniform and nonuniform corrosion cause thinning that can initiate and lead to SCC. However, indirectly, high uniform corrosion rates can retard pitting and SCC.

13.3.2. Bimetallic or Galvanic Corrosion

In a multiphase alloy, pitting is promoted by the galvanic cell between the cathodic $Mg_{17}Al_{12}$ phase and the matrix solid solution phase and that accelerates SCC [1]. A mechanically weak, tubular pitted surface is produced along active slip planes. It has been proposed that the role of corrosion is to produce pits or other stress concentrations that cause cracking by cleavage processes, and to remove obstacles that stop the crack. Fairman and Bray [25] proposed that the passage of dislocations on slip planes rupture the surface film, allowing a corrosion pit to develop, which then initiates cleavage [8, 9].

A film rupture or hydrogen-assisted cracking, or a combination of the two, would be consistent with the local cathodic condition that is produced by differential aeration or during bimetallic corrosion [23]. It is then necessary to consider the potential of the different phases of Mg. For example, the alloy AZ91 contains a high content of Al. β-$Mg_{17}Al_{12}$ in significant quantity along the grain boundary is cathodic with respect to the matrix of α-Mg and can initiate pitting in the alloy, assisted by the discontinuous barrier of this cathode created by the specific composition and microstructure of the alloy.

Mg has a negative free corrosion potential, E_{corr}, with a slightly more negative pitting potential, E_p, in solutions of practical importance like 3% NaCl. This is a very specific situation since pitting could start in certain aggressive solutions at open circuit potentials and that can pave the way to SCC without the necessary anodic polarization by oxidants to the level of pitting potential [29] (see Chapter 10). This can explain why Mg alloys are very susceptible to SCC through pitting while Al alloys are very susceptible to SCC through intergranular corrosion.

13.3.3. Pitting and Localized Corrosion

The oxide film on Mg offers considerable surface protection in rural and some industrial environments; the conductivity, ionic species, temperature of the electrolyte, alloy composition and homogeneity, differential aeration, and so on shift the corrosion morphology from general to localized corrosion, especially in the presence of a barrier layer. Pitting is a typical type of localized corrosion that can lead to SCC. Filiform corrosion initiates and then develops into cellular or pitting corrosion [14].

Interaction between a material and its environment can result in slow crack growth at stresses or stress-intensity factors markedly below those associated with fast fracture in the absence of environmental influences. The initiation of such slowly growing cracks has often been observed to occur at corrosion pits [30, 31]. The lower potential for cracking has been shown to be coincident with the pitting potential for some systems. The geometrical discontinuity introduced by a pit is often assumed to be important because of stress intensification in its vicinity, but the chemical conditions within a pit can be appreciably different from those without, and it may be that the latter is more important than the former in causing crack initiation. The association of stress-corrosion cracks with pits has been mentioned by a number of workers [32–34] studying Mg–7Al alloy in chloride–chromate solutions and commercial high–purity Mg exposed to sulfate solutions [8].

In atmospheric attack of a smooth machined Mg alloy surface, the roughening is really a microscopic form of pitting. There is a noticeable difference between the appearance of the Al-containing Mg-rich alloys and the Zn/Zr-containing Mg alloys. In the former, the microscopic pits in the surface exposed to the weather tend to be narrow and relatively deep, whereas in the latter they are wider and tend to overlap, leading to a slightly wavy appearance [14]. Song and Atrens [26, 35] showed distinctive polarization curves and pitting potentials for α and β phases in 1 N NaCl (see Chapter 10). This shows the influence of microstructure on pitting that can be related to SCC [2, 26, 35]. Stress-corrosion cracks invariably initiate at pits in numerous systems. The role of pitting is to disrupt films that otherwise prevent the ingress of hydrogen and, although there is some stress concentration associated with the pits, this is of lesser importance than the environmental implications of the presence of pits [22].

Susceptibility of age-hardened Mg alloys to environmentally induced cracking (EIC) appears to be related to alloying additions, particularly of aluminum, and cracking has not been reported for alloys containing less than about 1% Al. As in other alloy systems, the mechanism of EIC of Mg alloys has been variously ascribed either to continuous crack propagation as a result of preferential and enhanced anodic dissolution at the crack tip or to discontinuous crack propagation as a result of a series of hydrogen-induced brittle fractures at the crack tip. Although the experimental evidence in the literature strongly favors the latter, hydrogen-based mechanism, in aqueous environments cracking is generally preceded by pitting corrosion [8].

Pitting is a necessary precursor to hydrogen embrittlement and cracking because the pit walls provide the bare, active, film-free Mg surface that permits hydrogen evolved by the local cell cathodic reaction to enter the metal.

Prevention of pitting therefore prevents hydrogen entry and there is no embrittlement. The passive film on Mg is undoubtedly ruptured by slow strain rate tensile testing and bare metal would thereby be exposed to the environment. However, at -1500 and -2500 mV/SCE, even at a strain rate of $5.7 \times 10^{-6}\,s^{-1}$, repassivation of the Mg is sufficiently rapid to prevent sufficient hydrogen entering the metal to cause embrittlement. The crack path is exclusively transgranular in the finer-grained, commercial Mg. The coarser-grained, high-purity Mg exhibits a mixed inter- and transgranular crack morphology. Hydrogen-induced embrittlement and cracking provide a viable explanation for the spalling and disintegration of Mg anodes and for the negative difference effect that is observed when Mg is anodically polarized [8].

13.3.4. Welded Material and SCC

A low-temperature thermal stress relief is a recommended practice for welded Mg alloys. Residual stresses and pitting corrosion after welding are considered particularly detrimental to SCC resistance and so shot peening and other mechanical processes that create favorable compressive surface residual stresses may also increase SCC resistance. Extruded Mg alloys with approximately 3–8% Al and 0.5–0.8% Zn are susceptible to filiform corrosion and pitting corrosion in aqueous chloride solutions, depending on chloride concentration. The resistance of nonwelded alloys increases with Al content. Upon welding of the alloys, the corrosion resistance is determined by the Al/Mg proportion at the surface of the nonaffected material and the laser welding beam. The pits occur mainly in the heat affected zone (HAZ) of the welding beam. Laser beam-welded AZ61HP was found to have an excellent resistance [36].

13.3.5. Environment-Enhanced Creep and SCC of Mg Alloys

Environment-enhanced creep of Mg and its alloys can be considered as a part of the general term of *corrosion creep* that has been investigated for its strategic importance in service failures of Mg alloys in certain uses. The application of common die-cast Mg alloys in automotive drive train components, such as transmission housings and crankcases, is limited by their low creep strength at elevated temperatures in the range of 120–170 °C. The most common die-cast alloy is AZ91D with 9% Al content. Both AS21 and AE42 have been proposed as alternative die-cast alloys for such applications, and the creep strength of these two alloys is substantially better than AZ91—even the AM alloys are considered better [37–39]. This has been attributed to the presence of a large amount of low melting eutectic ($Mg_{17}Al_{12}$) [40].

It is believed that the decomposition of the microstructure is the prime cause of the weakness of die-cast AZ91 in creep and that the discontinuous precipitation reaction promotes grain-boundary migration and sliding, which are prominent creep deformation mechanisms in Mg alloys [41]. The intermetallic eutectic phase β-$Mg_{17}Al_{12}$ has an important role to play in imparting strength and stiffness to Mg–Al alloys but the inherent brittleness of this phase can limit the overall ductility unless it is controlled. Reducing the volume fraction of β by lowering the Al content, as in alloys AM50 and AM60, is then desired.

The combination of fluidity, strength, and ductility of an alloy with 9% Al is one of the major reasons that made the usage of AZ91D so widespread. However, while the material has good short-term strength at elevated temperatures, the creep resistance is poor. This has been attributed to the presence of a large amount of low melting eutectic ($Mg_{17}Al_{12}$). Another disadvantage of higher Al-content containing alloys is that ductility is usually sacrificed for grain strength [42]. However, when the volume fraction of the β-$Mg_{17}Al_{12}$ phase is increased in the surface layer (interior skin) because of highly nonequilibrium solidification, this provides for increased surface hardness and corrosion resistance. Corrosion rate of AZ91D has been reduced from 5.72 to 0.66 mm/y in 1 M NaCl [41].

The creep-resistant alloys AS21 and AE42 certainly have less supersaturated α-Mg in the die-cast condition and, as a result, there is less discontinuous precipitation during high-temperature creep. It has also been found that section thickness of the castings has a significant effect on the yield strength of the alloy [41].

Creep deformation of a brittle AZ91D alloy in a corrosive environment leads to the surface film breakup. This process results in an increase in the metal dissolution and corrosion rate on fresh surfaces. In general, it was reported that any environmental media that removed or altered the oxide film (e.g., acids or salt solution) increased the plasticity and accelerated the creep rate of some metals. But, in reality, such effects resulted from *both* the effects of metal dissolution and oxide removal. Indeed, there exists a theoretical and experimental background of chemomechanical effect, which consists of the environmentally assisted plasticization of metal in an active state without any passive film. This effect can promote an additional creep in the absence of surface barriers (e.g., oxide films or debris layers) [43].

The synergetic effect of corrosion and stress on the viscoelasticity of Mg alloys, named corrosion creep, has been studied in die-cast AZ91D (Mg–9%Al–1%Zn) and AM50 (Mg–5% Al–0.4%Mn) alloys in air and in a borate buffer solution. Gutman et al. [43] used a borate buffer solution (0.1 N $Na_2B_4O_7$) at pH 9.3 for the corrosion creep test. Stress values in the creep tests were 85–98% of the tensile yield strength. Corrosion rate measurements of Mg alloys in the 0.1 N $Na_2B_4O_7$ + $Mg(OH)_2$ saturated solution (pH = 10.5) versus tensile strain were made. This showed a higher corrosion rate of AZ91D in a deformed state than that of AM50, while in the underformed state (strain $\in = 0$) AZ91D has a lower corrosion rate than AM50.

The creep rate of AZ91D alloy in the buffer solution is significantly higher than in air. However, for AM50 alloy in the first stage of creep, the creep rate of the alloy in the buffer solution is lower than in air. The thickness of the oxide films formed on as-cast surfaces of AZ91D and AM50 alloys amounted to ~40 and ~100 nm in air, and ~1300 and ~1900 nm in the borate solution, respectively. The content of borate anions in the oxide film of specimens stressed in the buffer solution amounted to 17–25% for AZ91D and 20–45% for AM50 alloys. Apparently, a thicker film on the surface of a ductile AM50 alloy with a higher content of the borate anion $B_4O_7^{2-}$ decreases metal dissolution and corrosion rate. Here, the borate anion acts as a corrosion inhibitor at the first stage of creep [43].

Bonara et al. [40] performed electrochemical tests for AZ and AM alloys, using both dc potentiodynamic polarization and ac techniques (EIS) that were carried out in an aerated 0.05 M sodium tetraborate solution (pH = 9.7). Deformation increased the anodic current densities and shifted the potential to more active values. It was noted also for other alloys, AM20 and AM50, that the mechanical effect shows a maximum on the electrochemical activities when the strain hardening stage changes to the dynamic recovery stage. Also, AZ91D had a higher corrosion rate in the deformed state than AM50 under stress, while in the nonloading state, the corrosion rate was found to be higher for the AM50 alloy.

The plastic deformation effects on both potentiodynamic curves and impedance diagrams have shown that the anodic current density determined on potentiodynamic curves passes through a maximum, as predicted theoretically from the level of plastic deformation. Experimental confirmation of the correlation between the electrochemical characteristics due to metallic deformation and strain hardening stages (intensive strain hardening and dynamic recovery) was also related to the known change of dislocation during the plastic deformation. These features of mechanical behavior are the same in both active and pseudo-passive states, independent of surface film existence [40].

Pitting and environment-enhanced cracking or corrosion creep of pure Mg and die-cast Mg–Al alloys (AZ91D, AM50, and AS21) were examined in air, 0.1 N solution $Na_2B_4O_7$ (pH 9.3), and 3.5% NaCl solution (pH $\approx$ 5) at room temperature. In air, the creep behavior of pure Mg and its die-cast alloys demonstrates a decrease in the creep rate without rupture, typical of primary creep. In contrast to the data in air demonstrating only the first stage of the creep process, in corrosive solutions, secondary and tertiary creep due to plasticization was observed. The effect of environment on the creep behavior of Mg is connected, mainly, with plasticization of metal assisted by chemical reaction. Creep-stress-enhanced anodic dissolution, cracking, and finally creep rupture of pure Mg originate in a transgranular manner, while intercrystalline fracture was observed for alloys [44].

The lifetime of pure Mg increases from 12 h in sodium chloride solution to 170–320 h in the buffer solution under the same stress of 28 MPa. Elongation to fracture in tetraborate solution was two times higher than that in sodium chloride solution. In corrosive solutions, the creep life of Mg alloys at stress values of 120.5 MPa and the elongation-to-fracture values decrease with increasing Al content from 2.3% to 8.4% in AS21, AM50, and AZ91D alloys [44].

13.4. PROPAGATION MECHANISMS OF CORROSION

13.4.1. Electrochemical Dissolution Models

The electrochemical dissolution models concern the following possible mechanisms:

1. Intergranular SCC has been attributed to the galvanic corrosion of the matrix because of FeAl, $Mg_{17}Al_{12}$, and so on. Stress concentration can increase the local corrosion rate and/or rupturing of the protective surface film.
2. Film rupture leads to localized plastic deformation and creation of a galvanic cell. Using slow strain rate testing, Ebtehaj et al. [45] and Stampella et al. [8] indicated that a mechanism including strain-induced film rupture leading to corrosion and hydrogen production can advance crack propagation.
3. Fine corrosion tunnels in a SCC system are observed in specific dissolution processes based on the observations of Pickering and Swann [46].

It is important to examine the resistance to SCC as a function of polarization with respect to the open circuit potential. Slow strain rate tests were carried out on Mg–Al alloy (8.8 wt%) at various controlled potentials in a solution containing 5 g/L of both NaCl and K_2CrO_4 solutions. The reduction percentage in area "RA%," of a fracture, used as a convenient means of quantifying the cracking propensity, increased with decreasing potential (shifting to more negative values) until about −1.6 V/SCE, where it was barely distinguishable from the values obtained in a low strain rate test in dry air (Figure 13.2) [22].

The results show that at a chosen strain rate, the minimum susceptibility to SCC is attained as the potential shifts from the anodic region through the open circuit potential to more cathodic potentials until about −1.6 VSCE. Effectively, the reduction of SCC susceptibility attains the level of that in air in shifting the imposing potentials to very cathodic values. This means that anodic stimulation increased the severity of cracking, while cathodic current decreases susceptibility, compared to the cracking under open circuit or free corrosion potential. This is related to the fact that the anodic polarization of Mg, within a restricted range of potential, actually increases the amount of hydrogen discharged in comparison to that evolved when cathodic polarization is applied. The crack growth mechanism is almost certainly related to the ingress of hydrogen into the metal, supported by the generation of cracks by straining in gaseous hydrogen, having the same characteristics as those produced by immersion in aqueous solutions. The apparent contradiction is due to the negative difference effect observed with Mg-rich materials, that anodic stimulation facilitates hydrogen discharge to a greater extent than cathodic polarization [22].

Negative Difference Effect An approximately 1 μm thick brittle MgH_2 film has been observed on a magnesium SCC fracture [47]. It has been postulated that anodic polarization, within a restricted potential range, can discharge more hydrogen than

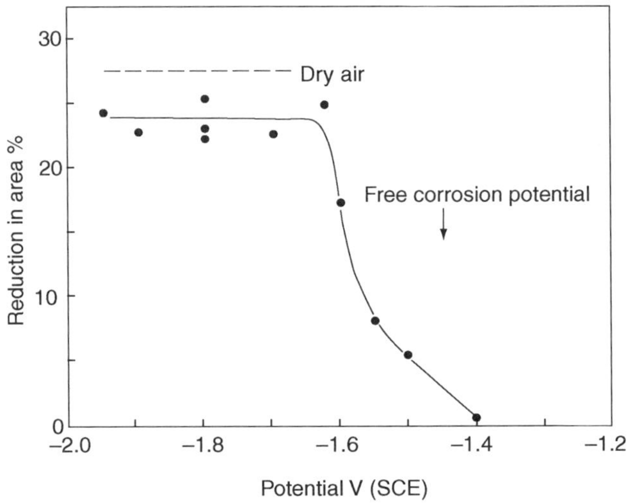

Figure 13.2 The effect of applied potential upon the reduction in area of fracture of initially plain specimens exposed to a solution containing 5 g/L sodium chloride and 5 g/L $KCrO_4$ and subjected to slow straining at $2 \times 10^{-6}\,s^{-1}$ [22].

cathodic polarization—a result of the *negative difference effect* (NDE). One explanation proposed for this phenomenon is that atomic hydrogen can enter the metal and cause embrittlement only when it is generated by local electrochemical reactions at active, film-free pit surfaces. Because cathodic polarization prevents pitting, SCC cannot occur. During anodic polarization, however, pitting is promoted, hydrogen is adsorbed/absorbed, and embrittlement proceeds [1]. It should be added that the hydrogen mechanism may not always be the primary mechanism for SCC growth (see NDE in Chapter 10).

Winzer et al. [15] stated that for most practical situations of Mg exposed to aqueous solutions, the hydrogen fugacity, f_H, is expected to be extremely large because the overpotential is large, over 1000 mV. Thus the hydrogen fugacity at the surface of Mg exposed to an aqueous solution is expected to be many orders of magnitude larger than that for steel exposed to an aqueous solution [15].

13.4.2. Hydrogen Embrittlement

Hydrogen is involved in the SCC propagation mechanism of Mg and Mg alloys; however, the concerned mechanisms are not evident and can be a function of the changing properties at the metal–solution interface of the initiated crack [15]. Atrens et al. [48] showed an estimate of the hydrogen diffusion coefficient at ambient temperature to be $10^{-5}\,cm^2/s$, which is sufficient to allow hydrogen transport ahead of a stress-corrosion crack in Mg at ambient temperature and confirms that the mechanism of SCC involves hydrogen, at least under certain circumstances. There are four admitted and related mechanisms and some of them act together: hydrogen-enhanced decohesion (HEDE), hydrogen-enhanced localized plasticity (HELP), adsorption-induced dislocation emission (AIDE), and formation of brittle stress-induced hydrides at the crack or delayed hydride cracking (DHC) [15, 49].

Hydrogen-Enhanced Decohesion HEDE is caused by weakening of the bonds between the adjacent metal atoms and it is proposed to be the most dominant mechanism for high-strength alloys that do not form hydrides. This involves a reduction in the electron charge density between metal atoms in the region ahead of the crack tip, where hydrogen accumulates by stress-assisted diffusion. The morphology of the fracture can be intergranular or transgranular. The transgranular fracture is expected to occur by conventional cleavage, resulting in river markings [50].

Hydrogen-Enhanced Localized Plasticity HELP is attributed to a decrease in the resistance to dislocation motion and an increase in the dislocation velocity due to the interaction between hydrogen atmospheres at mobile dislocations with stress fields and hydrogen atmospheres at microstructural features. HELP is then associated with the dislocation emission from the plastic zone ahead of the crack tip leading to ductile fracture surfaces [50].

Adsorption-Induced Dislocation Emission AIDE is attributed to dislocation emission, due to the weakening of metal–metal bonds by hydrogen atoms adsorbed at the crack tip and trapped within the first few atomic layers of metal. Since dislocations are emitted from the crack tip during AIDE, rather than from the plastic zone (per HELP), crack growth occurs by alternating slip on specific planes. The resulting fracture surfaces consist of low-index cleavage-like facets containing small shallow dimples [50].

Lynch and Trevena [9] suggested that HELP, HEDE, and AIDE can occur simultaneously, resulting in unique fracture surface morphologies. The coincidence of two HE mechanisms would occur only depending on the specific stress-corrosion crack range of velocities. The fractography for the HELP and AIDE mechanisms may be similar, since both mechanisms involve microvoids [50].

Delayed Hydride Cracking Makar et al. [51] underlined the brittle hydride model for crack propagation. It consists in the formation of brittle stress-induced hydrides. Winzer et al. [49, 50] admitted that hydride precipitation in Mg alloys is accompanied by plastic deformation, due to the volumetric misfit between the hydride and the substituted metal matrix. The DHC mechanism involves repeated stages of the following: (1) transient hydrogen diffusion toward the crack tip driven by stress and hydrogen concentration gradients; (2) hydride precipitation when the hydrogen solvus is exceeded; and (3) crack propagation through the extent of the hydride when it reaches a critical size of $\sim$0.8 μm. The stress-corrosion crack velocity, V_c, was calculated from the time for the hydride to reach the critical size. The model was implemented using a finite-element script developed in MATLAB [52].

Stress-corrosion crack velocities of $\sim 10^{-4}$ m/s, typical for Mg alloys in aqueous solutions, cannot be predicted by the DHC model based on the time to reach a critical hydride size for material with a low initial hydrogen concentration throughout. Such velocities might be predicted by a DHC model based on the time to reach the critical hydride size in steady state, when a significant hydrogen concentration would have built up at the crack tip. During steady-state stress-corrosion crack propagation for Mg in aqueous solutions, a high dynamic hydrogen concentration would be expected to build up just behind the crack tip. This is the most likely mechanism of DHC for transgranular stress-corrosion cracking (TGSCC) of Mg alloys and may be a feature of all cases of SCC where the crack propagation mechanism is hydrogen embrittlement assisted cracking (HEAC) [52].

Generally, *brittle fracture models* can account more for the high SCC cracking rates and the presence of flat, interlocking facets and shallow surface steps that are observed on TG SCC [1]. Pugh et al. [53] proposed that fracture occurs via an embrittled surface layer, either an oxide or a porous layer formed by selective dissolution of Mg or Al. The crack then becomes blunted and stops when it enters the ductile substrate. The process repeats after the film reforms. The crack growth mechanism is almost certainly related to the ingress of hydrogen into the metal, supported by the generation by straining in gaseous hydrogen of cracks having characteristics the same as those produced by immersion in aqueous solutions. It seems that anodic stimulation facilitates hydrogen discharge to a greater extent than cathodic polarization [22].

The hydrogen embrittlement mechanism is supported by certain experimental results. SCC crack initiation and propagation are accompanied by hydrogen evolution and crack propagation occurs at velocities at which only adsorbed hydrogen should be present at the crack tip. Immersion in a cracking solution before stress produces a fracture similar to a SCC fracture and the effect of preimmersion in a cracking solution is reversed by vacuum annealing. Testing in gaseous hydrogen results in the same crack characteristics produced in aqueous solution tests [22].

13.4.2.1. Cleavage Fracture Models: Electrochemical Attack and Mechanical Fracture

The majority of cleavage models involve alternating stages of electrochemical attack and mechanical fracture. Lynch and Travena [9] mentioned the presence of some plasticity, especially for high rate and faster crack velocities [15]. Fairman and West [27] proposed a model in which initiation of pitting is by film rupture due to basal slip. Stress concentration at the base of the pit initiates cleavage. Pugh and co-workers [53–57] showed evidence of a brittle cleavage-type mechanism involving hydrogen particularly with the stepped and faceted interlocking fracture surfaces. Pugh et al. [53] proposed the following sequence of alternating stages ahead of the crack tip: embrittlement of a thin film, film rupture, crack propagation, and crack arrest for transgranular cracking. This model was supported by crack propagation velocities of 0.35–2.5 mm/min [15].

13.4.2.2. Electrochemical Hydrogen Origin

It seems that the influence of atomic hydrogen on SCC is dependent on its origin, if it is created due to external polarization or local galvanic cells.

External Cathodic Polarization There was also acceleration of SCC by application of external cathodic polarization (0.2–10 mA). Liu [58] first suggested that cathodically generated hydrogen can be related to magnesium SCC. Experimental evidence supports this model, strengthened by the fact that SCC occurs under certain conditions at crack velocities at which only adsorbed/absorbed hydrogen should be present at the tip [9]. A weak, stress-induced magnesium hydride may form and has been observed on the surface of magnesium SCC fracture [12]. However, numerous studies reported that cathodic polarization inhibits or even prevents SCC [34].

Hydrogen from Active Galvanic Cells Cathodic effects can depend on the electrochemical characteristics and material properties [34]. Susceptibility of age-hardened

Mg alloys to environmentally induced cracking (EIC) appears to be related to alloying additions, particularly of aluminum, and cracking has not been reported for alloys containing less than about 1% Al. As in other alloy systems, the mechanism of EIC of Mg alloys has been variously ascribed either to continuous crack propagation as a result of preferential and enhanced anodic dissolution at the crack tip or to discontinuous crack propagation as a result of a series of hydrogen-induced brittle fractures at the crack tip. In aqueous environments hydrogen-based cracking is generally preceded by pitting corrosion [8].

Die-cast Mg alloys are shown to be susceptible to SCC, requiring only partial immersion in distilled water. Standard ASTM B557 die-cast tensile specimens in the as-cast F temper were used. The SCC tests were conducted in a dead-weight tension loading apparatus employing distilled water or a 3.5% NaCl solution covering the lower half of the specimen gauge length. There is evidence that this SCC results from a cathodic process, with perhaps hydrogen-assisted cracking and film rupture both playing roles. The tendency for SCC to occur at the air–water interface shows the effect of the galvanic cells created by different oxygen concentrations. A film rupture or hydrogen-assisted cracking, or a combination of the two, would be consistent with the local cathodic galvanic cell, which is produced by differential aeration [23].

13.5. SCC–HE OF SOME MAGNESIUM ALLOYS

The composition and microstructure of a Mg alloy determine its resistance to SCC in a specific corrosive solution and even distilled water. Magnesium alloys containing more than 1.5% Al are susceptible to SCC and wrought alloys are generally more susceptible than cast alloys. Tensile loads, less than 50% of yield stress, have been employed in SCC laboratory tests using slightly corrosive solutions [15].

Al–Zn Alloys A study of SCC and HE of Mg alloy AZ31 (3% Al, 1% Zn, and 0.3% Mn) has been carried by Song et al. [28]. The work was done to examine the influence of distilled water as compared to air and other aqueous solutions, such as sodium chloride (0.01 and 0.1 M) and ASTM D1387 (165 mg/L NaCl, 138 mg/L $NaHCO_3$, and 148 mg/L Na_2SO_4), on SCC. The time to fracture, reduction of area, elongation-to-fracture, and ultimate tensile strength were measured. The sample tested in air showed signs of ductility. In contrast, the samples tested in the various solutions had suffered corrosion and pitting, and there existed SCC. Both longitudinal and transverse orientations of AZ31 sheet Mg alloy are susceptible to SCC in various aqueous solutions and the SCC behaviors in the four tested solutions including distilled water are similar [28].

Figure 13.3 shows the stress–strain curves of longitudinal specimens of AZ31 in air, distilled water, and three tested solutions. Dramatic decreases of the studied parameters were observed in aqueous solutions compared to that in air. All the tested solutions induced SCC; the most susceptible was the 0.1 NaCl and the least was distilled water [28].

The fracture surfaces are given in Figure 13.4 for fractures in air, distilled water, ASTM solution, and 0.01 M NaCl solution. The surface fractured in air (Figure 13.4a) is interpreted as microvoid coalescence characteristic of a ductile rupture. Figures 13.4b–d show fracture surfaces in aqueous solutions with mixed intergranular and quasi-cleavage transgranular fractures [28]. The dominating cathodic reaction of the local galvanic cell can

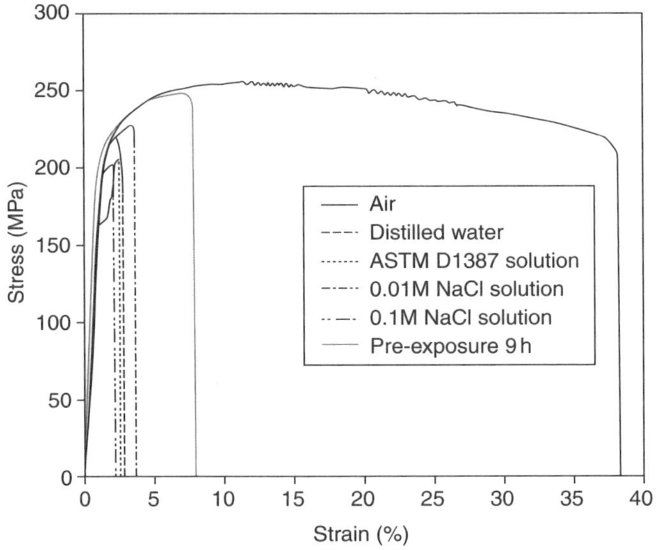

Figure 13.3 The stress–strain curves of longitudinal specimens of AZ31in air and the various tested solutions at a strain rate of $10^{-6}\,s^{-1}$ [28].

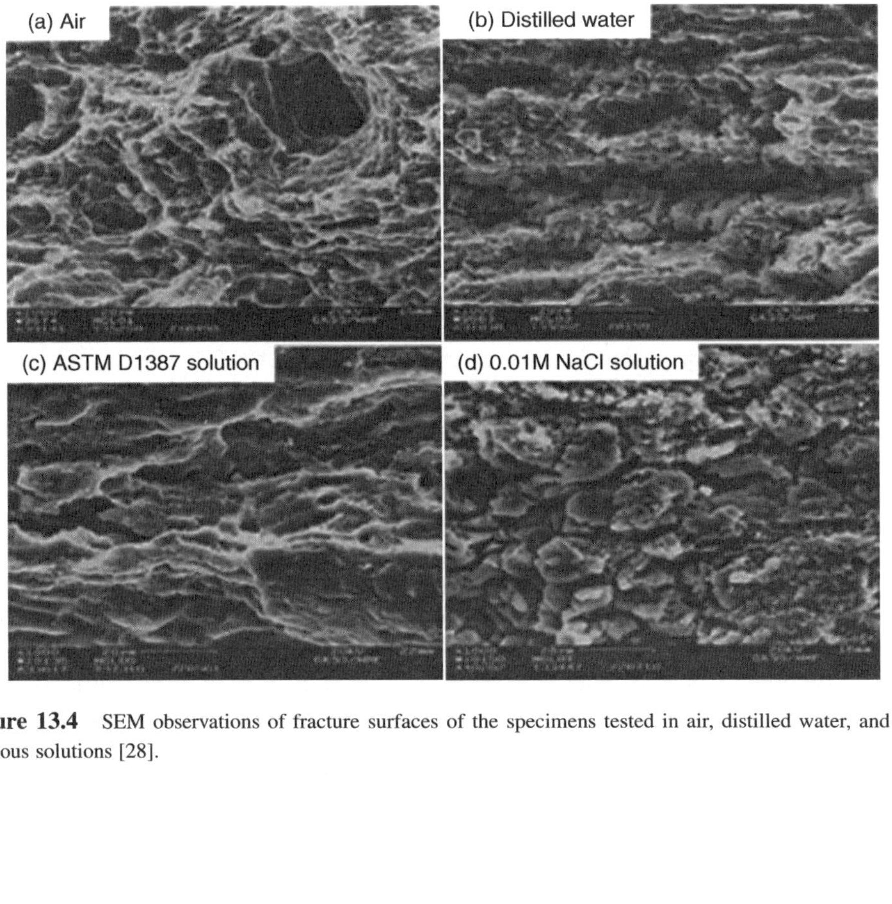

Figure 13.4 SEM observations of fracture surfaces of the specimens tested in air, distilled water, and two aqueous solutions [28].

be stated as

$$2H_2O + 2e^- = 2OH^- + 2(H) \text{ and } 2(H) = H_2$$

Exposure of Unstressed Specimens to HE Work was done to see if the exposure of unstressed specimens to the test solution can result in hydrogen absorption and subsequent embrittlement. Some specimens were exposed to 0.01 M NaCl solution for different times without being stressed and were then tested in air by loading using the constant strain rate method ($1 \times 10^{-6}\,s^{-1}$). There exists preexposure embrittlement in Mg alloy AZ31 and the extent of the exposure embrittlement increases with increasing preexposure time. Some of the hydrogen atoms would penetrate into the Mg alloy and then take part in HE. The role of the anodic dissolution is then to produce surface defects, which promote hydrogen production and its entry into the material. Both the preexposure embrittlement and SCC are mainly caused by HE [28].

Al–Mg Alloys AZ91, AZ31, and AM30 in Distilled H_2O Mechanisms for SCC of Mg–Al alloys have been investigated by scanning electron microscopy (SEM) of the fracture surfaces, for the two-phase alloy AZ91 and the single-phase alloys AZ31 and AM30 in distilled water. The AZ91 specimens were machined from as-cast ingots, whereas the AZ31 and AM30 specimens were machined from large extrusions such that their tensile axis was parallel with the extrusion direction [49]. The specimens were tested under linearly increasing stress test (LIST) and constant extension rate testing (CERT) conditions, in double-distilled H_2O or after precharging in gaseous H_2 at 30 bars and 300 °C for 15 hours, with control tests carried out in laboratory air. The SCC initiation was detected using the dc potential drop (DCPD) method [49].

The SCC mechanism of Mg alloys is inherently transgranular and somehow involves cathodically produced hydrogen entering the metal substrate. TGSCC of Mg is associated with conditions causing electrochemical breakdown or mechanical rupture of protective films at the crack surface, which allow hydrogen to enter the metal substrate [15]. Furthermore, it is likely that the mechanism for steady-state crack propagation in the TGSCC of Mg alloys involves stress-directed diffusion of hydrogen ahead of the crack tip. The most commonly proposed mechanism DHC, which involves repeated stages of stress-assisted diffusion of hydrogen ahead of the crack tip, leading to hydride formation and fracture [7, 59]. HELP, resulting from adsorption of hydrogen atoms at the crack tip, has also been suggested, particularly for higher crack velocities [9]. Both mechanisms should be considered and the predominant one is a function of the crack velocity [60].

The competition between DHC and HELP may be defined by the propensity for solute hydrogen to diffuse ahead of the crack tip, which is dependent on the metal substrate and stress distribution. Hydrogen ingress may occur as a result of mechanically or chemically induced film rupture. Thus mechanistic differences in TGSCC of Mg alloys under LIST and CERT conditions could be interpreted in terms of the difference in the crack tip stress fields over time and/or the breakdown/rupture of protective films [60].

Under LIST conditions, the load is relatively constant for consecutive stages of crack advance. Consequently, the stress intensity at the crack tip, and the driving force for hydrogen diffusion, would also be relatively constant (although the stress intensity at the crack tip may be reduced somewhat by blunting, a possible cause of crack arrest). In contrast, under CERT conditions, there is a reduction in load, with crack propagation. Thus there is a potential reduction in driving force for hydrogen diffusion and an increase in the time available for repassivation at the crack surface [60]. For the SCC of AZ91, AZ31, and AM30 in distilled water under CERT conditions, there was a decrease in SCC susceptibility with increasing strain rate [50]. This was characterized by an increasing difference between σ_{SCC} and the UTS measured in air, a decreasing elongation-to-failure, and a decreasing σ_{SCC} (for AZ91 and AZ31).

Winzer et al. [61] compared the fracture surfaces for AZ91 in distilled water under LIST and CERT conditions. The mechanical load on the specimen was steadily increased under LIST or CERT conditions. Under LIST conditions, complete fracture ensues shortly after σ_{SCC} is reached, resulting in a small SCC zone relative to that formed under CERT conditions. The morphologies of the SCC zones for AZ31 and AM30 under LIST conditions were consistent with those formed under CERT conditions: quasi-crystallographic markings containing ellipsoidal dimples for AZ31 and cleavage-like markings for AM30. For AZ91 [60], the SCC zones formed under LIST conditions were smaller than those formed under CERT conditions. In addition, under LIST conditions, there was no SCC initiation region for AZ31 and AM30 by localized (strain-assisted) dissolution as there was under CERT conditions [60, 62].

There was no evidence of irregular pitting in distilled water on the gauge surface for any alloy fractured in distilled water; however, AZ31 and AM30 specimens exhibited coarse markings along the length of the exposed surface and parallel to the axis of the specimen. These markings may be attributed to localized corrosion at the axially aligned Al–Mn particles [50].

SCC of AZ91 The fracture surface morphology for AZ91 in distilled water typically contained multiple thumbnail-shaped SCC zones, with the remaining fracture surface comprised of broken, jagged features, consistent with fracture in laboratory air [61]. The SCC zones were generally characterized by (1) parallel facets ~5 μm wide, (2) jogging at the edges of each facet, (3) extensive crack branching, and (4) cleavage through β particles (Figure 13.5). Examination at high magnification revealed that the parallel facets contained fine parallel markings (<0.5 μm apart) or ellipsoidal micro dimples. The orientation of the parallel facets and fine parallel markings was influenced by β particles (Figure 13.5). The fracture surfaces also contained pyramidal crevices (Figure 13.6) that have also been observed on fatigue fracture surfaces for AZ91 compact tensile specimens and are consequently attributed to an inert fracture process [49].

The fracture surfaces were generally consistent with those produced by fracture in air, but also contained small (<100 μm across) rounded regions adjacent to the gauge surface, which were attributed to SCC. Similar mechanisms for SCC initiation and propagation are proposed since the SCC initiation zones consisted of fine parallel markings or microdimples, similar to those observed within parallel facets for ongoing crack propagation under CERT conditions (Figure 13.5).

Figure 13.5 shows also that the fracture of β particles occurs by cleavage, which is characteristic of HEDE or DHC. There are no markings within the fractured β particles to indicate crack arrest or plastic deformation, which usually accompany DHC, although plastic deformation might be considered unlikely for the relatively brittle β phase. Moreover, there were no features within the fractured β particles that could be interpreted as magnesium hydride. Thus it is more likely that the mechanism for the fracture of β particles is HEDE [49].

The SCC propagation in AZ91 involves crack nucleation within β particles ahead of the primary crack tip. It was also shown that the stress-corrosion crack velocity for AM30

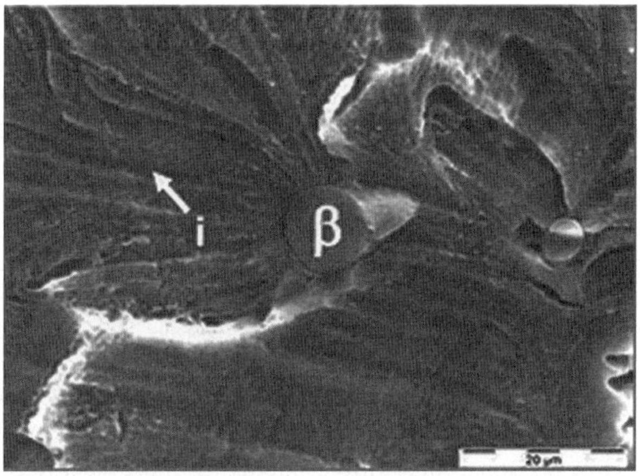

Figure 13.5 Fracture surface for AZ91 specimen tested in distilled water under CERT conditions showing the influence of β precipitates on the direction of parallel markings (i) [49].

is much slower than that for AZ91 and AZ31, indicating different crack propagation mechanisms [49]. The presence of β particles in AZ91 was associated with (1) a lower threshold stress, σ_{SCC}, for AZ91 (55–75 MPa) relative to AZ31 (105–170 MPa) and AM30 (130–140 MPa); and (2) a different SCC initiation mechanism for AZ91 relative to AZ31 and AM30 [49, 50]. The AZ31 and AM30 microstructures were relatively homogeneous and consisted of α matrix with small Al–Mn quadrilateral crystals (which were larger and more numerous for AZ31), small Mg–Si particles, and elongated Al–Mn plate-like crystals that were aligned collinearly in the extrusion direction.

A mechanism for crack initiation in AZ91 involving fracture of β particles close to the surface, with hydrogen ingress facilitated by mechanical rupture of the surface film, has been proposed [50]. The mechanism for SCC propagation in AZ91 involves (1) hydrogen trapping by β particles ahead of the primary crack tip; (2) fracture of β particles upon reaching some critical hydrogen concentration; (3) release of hydrogen trapped in β particles due to the reduction in internal hydrostatic stress; and (4) embrittlement of the surrounding matrix by the released hydrogen [50].

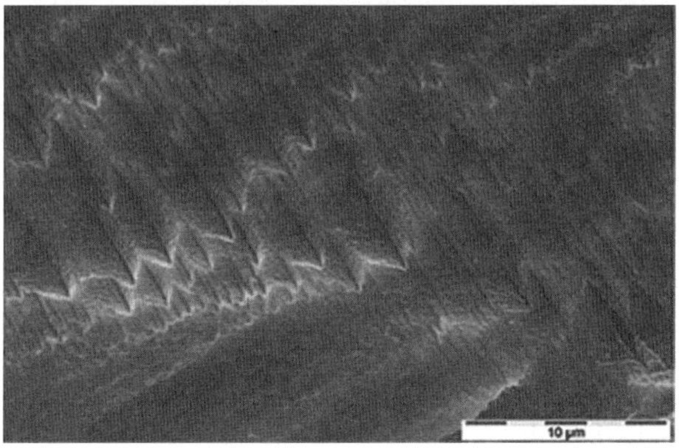

Figure 13.6 SEM micrograph of AZ91 specimen fractured in distilled water under CERT conditions showing pyramidal crevices [49].

The mechanism for crack propagation at moderate strain rates in AZ91 is similar to that in AZ31, with β particles acting as sources of hydrogen for mobile dislocations [49]. The mechanism for SCC propagation in AZ91 involves crack nucleation within β particles ahead of the primary crack tip. The fracture morphology for the β particles is indicative of HEDE. The fracture surface for AZ91 tested at the strain rate $3 \times 10^{-8}\,s^{-1}$ was similar to that for specimens precharged in gaseous H_2. The AZ91 specimens precharged in gaseous H_2 and the AZ91 specimens fractured in distilled water at $3 \times 10^{-8}\,s^{-1}$ have similar fracture surfaces, indicating that the same HE mechanism occurs under both conditions. This mechanism involves (1) the nucleation and growth of MgH_2 particles, (2) the sudden fracture of the MgH_2 particles at some critical stress, and (3) the decomposition of the MgH_2 particles after fracture [49].

The fracture surfaces for AZ31 and AM30 in distilled water under CERT conditions are indicative of stress-corrosion crack propagation on multiple parallel planes, with cliffs forming between consecutive planes by inert fracture. The stress-corrosion crack velocities for AM30 ($V_c = 3.6 \times 10^{-10}$ to 9.3×10^{-10} m/s) are much slower than for AZ91 ($V_c = 1.6 \times 10^{-9}$ to 1.2×10^{-8} m/s) and AZ31 ($V_c = 1.2 \times 10^{-9}$ to 6.7×10^{-9} m/s), indicating a different SCC mechanism. The results are consistent with a lower hydrogen diffusivity in the α phase in the absence of Zn, or the behavior of Al–Mn plate-like particles as sources of hydrogen for mobile dislocations (as per the β particles in AZ91) [50].

The mechanism for SCC initiation in AZ31 and AM30 involves highly localized dissolution. That the secondary cracks occur independently of microstructural features suggests that this mechanism is facilitated by mechanical rupture of the surface film (as per the mechanism for SCC initiation in AZ91) [27, 49]. The mechanism for SCC initiation in AZ91 appears to involve HE of β particles, whereas the mechanism for SCC initiation in AZ31 and AM30 involves localized dissolution of the α matrix [50].

The fractography for SCC propagation in AZ31 is characterized by elongated micro-dimples within quasi-crystallographic facets, indicative of a mechanism involving micro-void coalescence (HELP or AIDE). The occurrence of dimples within cleavage-like facets is a characteristic of AIDE; however, it is uncertain whether the facets observed on AZ31 fracture surfaces correspond to specific planes. The fractography for SCC propagation in AM30 is characterized by cleavage-like features, indicative of HEDE [50].

SCC Kinetics and Protective Films The kinetics of SCC of Mg alloy MA2-1 with various protective coatings was studied. Annealed alloy MA2-1 in the form of a 2 mm sheet (Al, 4.2%; Zn, 0.9%; Mn, 0.8%) was investigated. SCC tests were made under a constant load of $0.7\sigma_{0.2}$ at the start of the tests on a lever-type machine in 80 g/L $K_2Cr_2O_7$ + 5 g/L NaCl and 0.005 N NaOH + 0.01 N Na_2SO_4 media. The first solution caused rapid SCC and represented comparatively stringent test conditions, while the second permitted a more detailed study of the SCC kinetics since the time to failure was longer than in the first solution [5].

The SCC kinetics was investigated with simultaneous registration of the electrode potential and the elongation of the specimen (working part, 6 mm × 1.5 mm × 20 mm) to within + 1 μm. Protective coatings were obtained by chemical or electrochemical treatments. The protective films were 10 g/L NaOH at 323 K for 30 minutes; 40 g/L $K_2Cr_2O_7$ + 60 g/L $MnSO_4{\cdot}7H_2O$ for 30 minutes at 363 K; 50 g/L NaOH + 50 g/L Na_2CO_3 with an anodic current of 2 A/dm^2 at 303 K at a potential of 100 V for 30 minutes; and 50 g/L NaOH + 50 g/L Na_2CO_3, with $i_a = 3$ A/dm^2, temperature of 303 K, and tension ≤ 25 V for 30 minutes [5]. In SCC of the Mg alloy MA2-1 in chloride–chromate and alkali solutions, the elongation time curves have three characteristic parts. The first corresponds to inception of microcracks; the second to relatively slow development of one preferential crack; and the third to accelerated development of the latter. Protective films have a beneficial influence on the second part—development of the crack is slowed down by a factor of 2–5 [5].

13.6. SCC PREVENTION

1. Magnesium alloys that contain neither aluminum nor zinc are the most SCC resistant. Select a SCC resistant alloy for cladding a susceptible alloy. The stress should be between 30% and 50% of the tensile strength, including any additional stresses during service. It is recommended that external stress be 60–70% of imposed stresses.

2. Good design is needed for every structure to avoid increasing stress and pitting during service, especially for inserts. Bolted or riveted joints can also produce high local stresses that can cause SCC, so that attention should be given to proper joint design and construction. Examples include the use of preformed parts, avoiding overtorquing of bolts, and providing adequate spacing and edge margins for rivets. It has been recommended that inserts with a wall thickness greater than 1.25 mm be preheated before casting, because cast-in inserts may cause SCC due to local residual stresses created in the surrounding magnesium [63]. Susceptible alloys can be clad with nonsusceptible Mg alloys, but for exposed edges, wetting of both the alloy and the clad layer is important in order to achieve cathodic protection of the alloy [16].

3. Tensile residual stresses created by heating during welding were found to be particularly dangerous and, as a result, a low-temperature thermal stress relief treatment has become a recommended practice for welded assemblies [21]. Shot peening and other mechanical processes that create compressive surface residual stresses may be effective in increasing SCC resistance [18, 64].

4. Studies of the effectiveness of coatings have shown them to extend life but not to totally prevent SCC, with breaks in the coating producing an expected reduction in this protection [21, 63]. Inhibition, by nitrate or carbonate ions, of SCC in salt– chromate solutions is recommended for a more stable passive film [12]. Surface oxidation followed by anodizing is also reported to increase stress-corrosion life [16]. There should be appropriate inspection and maintenance programs generally and especially where protective coatings are used. Scratches could create favorable sites for initiation of SCC as well as pitting corrosion.

5. Cathodic polarization reduces or even prevents SCC as a general rule especially at high negative potentials.

REFERENCES

1. W. K. Miller, in *Stress-Corrosion Cracking*, edited by R. H. Jones. ASM International, Materials Park, OH, 1992, pp. 251–263.
2. P. Adeva-Ramos, S. B. Dodd, P. Morgan, F. Hehmann, P. Steinmetz, and F. Sommer, *Advanced Engineering Materials* **3**, 147–152 (2001).
3. M. Vialatte, *Study of the SCC Behavior of the Alloy Mg–8% Al.* NATO, Neuilly-sur-Seine, France, 1970 pp. 5.1–5.10.
4. M. O. Spiedel, *Metallurgical Transactions A* **6A**, 631–651 (1975).

5. Y. B. Yakovlev, L. P. Trutneva, N. I. Isaev, and G. Nemetch, *Protection of Metals* **20**, 300–306 (1984).
6. A. F. Froats, T. Kr. Aune, D. Hawke, W. Unsworth, and J. Hillis, in *ASM Handbook, Volume 13, Corrosion*, edited by L. J. Korb, D. L. Olson, and J. R. Davis. ASM International, Materials Park, OH, 1987, pp. 740–754.
7. E. J. Meletis and R. F. Hochman, *Corrosion* **40**, 39–45 (1984).
8. R. S. Stampella, R. P. M. Procter, and V. Ashworth, *Corrosion Science* **24**, 325–341 (1984).
9. S. P. Lynch and P. Trevena, *Corrosion* **44**, 113–124 (1988).
10. F. Meller and M. Metzger, U.S.N.A.C.A. Technical Note 4019 (1957).
11. J. D. Hanawalt, C. E. Nelson, and J. A. Peloubet, *Transactions of American Society of Mining and Metallurgical Engineering* **147**, 273–299 (1942).
12. W. K. Miller, *Metallurgical Transactions A* **22**, 873–877 (1991).
13. ASM Specialty Handbook Committee, in *ASM Specialty Handbook*, edited by M. M. Avedesian and H. Baker. ASM International, Materials Park, OH, 1999, pp. 138–162, 194–215.
14. E. Ghali, in *Uhlig's Corrosion Handbook*, 2nd edition, edited by R. W. Revie. Wiley, Hoboken, NJ, 2000, pp. 793–830.
15. N. Winzer, A. Atrens, G. Song, E. Ghali, W. Dietzel, K.-U. Kainer, N. Hort, and C. Blawert, *Advanced Engineering Materials* **7**, 659–693 (2005).
16. J. C. Scully, *SCC of Magnesium*, 3rd edition. Butterworth-Heinemann, Woburn, MA, 2000, pp. 127–129.
17. H. L. Logan, *Stress Corrosion Cracking of Metals*. Wiley, Hoboken, NJ, 1966.
18. W. S. Loose, in *The Corrosion Handbook*, edited by H. H. Uhlig. Wiley, Hoboken, NJ, 1948, pp. 232–250.
19. T. K. Aune, O. Lunder, and K. Nisancioglu, *Microstructural Science* **17**, 231–241 (1989).
20. G. Song, L. Bowles, and D. H. St.-John, *Materials Science and Engineering A* **366**, 74–86 (2004).
21. W. K. Miller and E. F. Ryntz, *SAE Transactions* **92**, 2.524–2.552 (1983).
22. K. Ebtehaj, D. Hardie, and R. N. Parkins, *Corrosion Science* **28**, 811–829 (1988).
23. W. K. Miller, Stress corrosion cracking of magnesium die casting alloys, *Materials Research Society Symposium Proceedings*, Vol. 125, edited by A. Barkatt, E. D. Verink, Jr, and L. R. Smith, Materials Research Society, Warrendale, PA, 1998, pp. 253–259.
24. G. F. Sager, R. H. Brown, and R. B. Mears, *Tests for Determining Susceptibility to Stress-Corrosion Cracking*. American Society for Testing and Materials, Philadelphia, 1945, pp. 255–272.
25. L. Fairman, H. J. Bray, and J. B. Corros, *British Corrosion Journal* **6**, 170–174 (1971).
26. G. L. Song and A. Atrens, *Advanced Engineering Materials* **1**, 11–33 (1999).
27. L. Fairman and J. M. West, *Corrosion Science* **5**, 711 (1965).
28. R. G. Song, C. Blawert, W. Dietzel, and A. Atrens, *Materials Science and Engineering A* **399**, 308–317 (2005).
29. K. Nisancioglu, O. Lunder, and T. Aune, in *Proceedings of the 47th World Magnesium Conference*, Canne, International Magnesium Association, IMA, Mclean, VA, 1990, p. 43.
30. R. N. Parkins, M. Elices, V. Sanchez-Galvez, and L. Caballero, *Corrosion Science* 379 (1982).
31. C. M. Rangel, and R. N. Parkins, in *Proceedings of the 6th European Symposium on Corrosion Inhibitors*, Vol.1. University of Ferrara, 1985, p. 655.
32. R. P. Frankenthal, *Corrosion Science* **7**, 61–62 (1967).
33. H. H. Uhlig, and J. R. Gilman, *Zeitschrift fuer Physikalische Chemie* **226**, 127 (1964).
34. W. R. Wearmouth, G. P. Dean, and R. N. Parkins, *Corrosion* **29**, 251–258 (1973).
35. G. Song and A. Atrens, Corrosion behavior of skin layer and interior of die cast AZ91D, in *Magnesium Alloys and Their Application*, edited by B. L. Mordike and K. U. Kainer, Mat Info, Frankfurt, Germany, 1998, pp. 415–419.
36. E. Ghali, W. Dietzel, and K.-U. Kainer, *Journal of Materials Engineering and Performance* **13**, 7–23 (2004).
37. T. K. Aune and H. Westengen, Property update on magnesium die casting alloys, SAE Technical Paper No. 950242, 1995.
38. W. E. Mercer II, Magnesium die cast alloys for elevated temperature applications, presented at the International Congress & Exposition, SAE Technical Paper, 1990.
39. M. S. Dargusch, G. L. Dunlop, and K. Pettersen, Elevated temperature creep and microstructure of die cast Mg-Al alloys, in *Proceedings of the Fourth International Conference on Magnesium Alloys and Their Applications*, Wolfsburg, edited by B.L. Mordike and K. U. Kainer, Mat Info, Frankfurt, Germany, 1998.
40. P. L. Bonora, M. Andrei, A. Eliezer, and E. M. Gutman, *Corrosion Science* **44**, 729–749 (2002).
41. G. L. Dunlop, W. P. Sequeira, M. S. Dargusch, G. Song, A. Atrens, T. Kittel, D. St. John, A. K. Dahle, and M. Murray, Microstructure of creep magnesium die alloys, in *Proceedings of the Annual World Magnesium Conference, Hannover Germany*, International Magnesium Association IMA, Mclean, VA, 1998, pp. 68–73.
42. A. A. Nayeb-Hashemi and J. B. Clark, in *Binary Alloy Phase Diagram*, 2nd edition, edited by T. R. Massalski, H. Okamoto, P. R. Subramanian, and L. Kacprzak. ASM International, Materials Park, OH, 1988.

43. E. M. Gutman, A. Eliezer, Y. Unigovski, and E. Abramov, *Materials Science and Engineering* **302**, 63–67 (2001).

44. Y. Unigovski, Z. Keren, A. Eliezer, and E. M. Gutman, *Materials Science and Engineering A* **398**, 188–197 (2005).

45. K. Ebtehaj, D. Hardie, and R. N. Parkins, *Corrosion Science* **28**, 811 (1993).

46. H. W. Pickering and P. R. Swann, *Corrosion* **19**, 373f–389f (1963).

47. L. Y. Wei, Development of microstructure in cast magnesium alloys, Thesis, Chalmers University of Technology, Goteborg, 1990.

48. A. Atrens, N. Winzer, G. Song, W. Dietzel, and C. Blawert, *Advanced Engineering Materials* **8**, 749–751 (2006).

49. N. Winzer, A. Atrens, W. Dietzel, R. G. Song, and K.-U. Kainer, *Minerals, Metals & Materials Transactions* **39A**, 1157–1173 (2008).

50. N. Winzer, A. Atrens, W. Dietzel, V. S. Raja, G. Song, and K.-U. Kainer, *Materials Science and Engineering*, **488**(1–2), 339–351 (2008).

51. G. L. Makar, J. Kruger, and K. Sieradzki, *Corrosion Science* **34**, 1311 (1993).

52. N. Winzer, G. Song, A. Atrens, W. Dietzel, C. Blawert, and K.-U. Kainer, Evaluation of Mg SCC using LIST and SSRT, in *Proceedings of the 7th International Conference on Magnesium Alloys and Their Applications*, Dresden, Germany, edited by K.-U. Kainer. Wiley-VCH, Weinheim, Germany, 2007, pp. 715–720.

53. E. H. Pugh, J. A. S. Green, and P. W. Slattery, On the propagation of stress-corrosion cracks in a magnesium–aluminum alloy, in *The Proceedings of the Second Internationnal Conference on Fracture, London*, edited by P. L. Pratt, Chapman and Hall Limited, London, 1969, pp. 387–395.

54. A. J. Bursle and E. N. Pugh, in *Mechanisms of Environment Sensitive Cracking of Materials*, edited by P. R. Swann, F. P. Ford and A. R. C. Westwood, Materials Society, London, 1977, pp. 471–481.

55. D. G. Chakrapani and E. N. Pugh, *Metallurgical Transactions A* **6**, 1155 (1975).

56. D. G. Chakrapani and E. N. Pugh, *Corrosion* **31**, 247 (1975).

57. D. G. Chakrapani and E. N. Pugh, *Metallurgical Transactions A* **7**, 173 (1976).

58. H. W. Liu, *Transactions of the ASME Journal of Basic Engineering* **92**, 633–638 (1970).

59. N. Winzer, A. Atrens, W. Dietzel, G. Song, and K.-U. Kainer, *Materials Science and Engineering A* **466**, 18–31 (2007).

60. N. Winzer, A. Atrens, W. Dietzel, G. Song, and K.-U. Kainer, *Materials Science and Engineering A* **472**, 97–106 (2008).

61. N. Winzer, A. Atrens, W. Dietzel, G. Song, and K.-U. Kainer, *Advanced Engineering Materials* **10**, 453–458 (2008).

62. T. L. Anderson, *Fracture Mechanics: Fundamentals and Applications*, 3rd edition. CRC, Boca Raton, FL, 2004.

63. E. Groshart, *Metal Finishing* **83**, 17–20 (1985).

64. M. A. Timonova, in *Corrosion Cracking of Magnesium Alloys and Methods of Protection Against It*, edited by I. A. Levin (translated from Russian). Consultant Bureau, New York, 1962, pp. 263–282.

Part Four

Coating and Testing

Chapter 14

Aluminum Coatings: Description and Testing

Overview

Pure aluminum, the 3xxx, 5xxx, and most 6xxx series alloys, are sufficiently resistant to be used in industrial atmospheres and waters without any protective coatings. Coatings are recommended for the higher strength 6xxx alloys, such as alloy 6013, and for all 2xxx and 7xxx alloys. The thickness of the natural oxide passive film can be increased by a factor of ~10 by prefilming in hot water and by a factor of 1000 or more by anodizing in sulfuric acid. Different options for mechanical and chemical surface preparation are available depending on choice of coating, appearance, and/or performance. Weak organic acids and their derivatives form insoluble salts and rely on the adsorption of the hydrophobic anions to provide a thin barrier layer. Chelating inhibitors create a thin tenacious passive layer (up to 20 nm). Aluminum alloys are protected by more active metals or by cathodic protection. Corrosion can be prevented or reduced by cladding. Some joint-sealing compounds that contain suitable soluble inhibitors are particularly recommended.

Aluminum can be protected by electroless or conventional plating. Aluminum and aluminum alloys in the active state act as a sacrificial anode in the form of plate or as a powder coating. Conversion layers can be created through physical vapor deposition, cathodic magnetron sputtering, high-energy ion beams, and laser ablation. Electrochemical anodization, plasma ablation, and chromate conversion coatings are frequently considered. There are thermoplastic coatings and converted coatings that are applied during or after processing and include principally three types of paints: epoxy, polyurethane, and moisture coatings. Corrosion monitoring is currently carried out by electrochemical impedance spectroscopy methods and electrochemical noise measurements.

General Considerations and Surface Preparation Pure aluminum and the 3xxx, 5xxx, and most 6xxx series alloys are sufficiently resistant to be used in industrial atmospheres and waters without any protective coatings. Examples of this are cookware, boats, and building products. Coatings are recommended for the higher strength 6xxx alloys, such as alloy 6013, and for all 2xxx and 7xxx alloys. One of the principal methods of protection is to enhance the thickness and quality of the natural oxide by prefilming in hot water, which can increase the thickness of the oxide passive film by about a factor of 10. The film can be thickened even more (to 1000 or more times the natural thickness) by

Corrosion Resistance of Aluminum and Magnesium Alloys: Understanding, Performance, and Testing By Edward Ghali

anodizing in sulfuric acid, for example. Chemical conversion treatments also provide protection, but to a much lesser degree, and are primarily used as a substrate for the subsequent application of organic films [1].

Conception, Alloy Selection, and Joint-Sealing Compounds During conception, the corrosion specialist should identify the different types of corrosion and prevention methods. Among the most common harmful effects are galvanic action, resulting from direct contact between aluminum and a dissimilar metal, such as copper, and indirect galvanic effects resulting from contact between aluminum and solutions containing reducible compounds of heavy metals. In some cases, design will prevent serious corrosion even though no other factors are altered. Similarly, since the various aluminum alloys differ widely in behavior, the selection of the most suitable alloy is important [2].

Aluminum-based alloys, such as 1100, 3300, 5052, 6053, Alclad 3300, Alclad 1017-T, and Alclad 2024-T, are highly resistant when freely exposed to most natural environments. They will all discolor or darken appreciably in most outdoor exposures, but will suffer no structurally appreciable changes in properties unless exposed in relatively thin sections below 0.076 mm (0.003 in.) thick [2].

Commercial aluminum alloys may contain other elements that provide special characteristics. Lead and bismuth are added to alloys 2011 and 6262 to improve chip breakage and other machining characteristics. Nickel is added to wrought alloys 2018, 2218, and 2618, which were developed for elevated-temperature service, and to certain 3xxx cast alloys used for pistons, cylinder blocks, and other engine parts subjected to high temperatures. Cast aluminum-bearing alloys may contain tin. In all cases, these alloying additions introduce microconstituent phases that are cathodic to the matrix and decrease resistance to corrosion in aqueous saline media. However, these alloys should be used in environments in which they are not subject to corrosion [2].

Joints, depressions, and other areas where moisture and dirt accumulate are more susceptible to corrosion than regions exposed to the atmosphere. Most plastic or semisolid joint-sealing compounds that conform and firmly adhere to adjacent metal surfaces are highly effective in preventing special attack in these regions. Some of these joint-sealing compounds that contain soluble inhibitors are particularly suitable [2].

In some cases, other mechanical factors, such as formability or hardness, may be of great importance in selecting an appropriate alloy for a specific application. Aluminum alloys such as 1100, 3300, or 5052, in the softer tempers, are readily formable and are also highly resistant to corrosion. If greater strength is required, alloy 6061 should be considered. This alloy combines good formability (in the W temper) with relatively high strength and good resistance to corrosion [2].

Surface Preparation by Plasma Ablation Surface preparation that includes cleaning first is well established for the currently used methods of aluminum finishing and coating for appearance and performance. Different mechanical and chemical options are available. Mechanical surface preparation includes abrasive blast cleaning, barrel finishing, polishing and buffing, and satin finishing [3]. Plasma ablation takes place by one of two methods—*sputtering* or *chemical etching*. Sputtering is used to remove organic contaminants and oxides from the metal surface. The surface is cleaned by bombarding it with inert gas plasma. Argon and hydrogen are used extensively for this procedure. Chemical etching involves chemical reaction between the impinging ion and the dislodged atoms. Oxygen is used with chemical etching to 'clean' the surface of a metal as well as deposit a synthetic oxide layer. The energy and quantity of oxygen control the properties of the oxide layer [4].

14.1. INHIBITORS

Most organic inhibitors come in the form of weak acids and their derivatives that form insoluble salts at the metal surface. There have been some reports on chelating inhibitors that formed an intimate bond between the organic complex and the metal surface or its oxide. The result was the creation of a thin (up to 20 nm) but tenacious passive layer. Organic salts rely on the adsorption of the hydrophobic anions to provide a thin barrier layer [4]. Solutions of sodium sebacate ($NaOOC(CH_2)_8COONa$), potassium hydrogen phthalate ($C_8H_5O_4K$), and sodium molybdate ($Na_2MoO_4 \cdot H_2O$) were added to a solution of $AlCl_3$ to test their effectiveness within a simulated aluminum pit. The sodium sebacate solution (0.1 M) immediately formed a thick white precipitate, while the other two inhibitors did not. Electrochemical evaluation of the sodium sebacate solution on fresh and already corroded aluminum found that it functions as an inhibitor to both pit nucleation and pit growth as long as the chloride concentration remained below 0.3 M.

Inhibitors and Control of the Environment Current knowledge makes possible the inhibition of aluminum in a wide range of both acidic and alkaline environments. Single materials and combinations have been identified that can be used with considerable confidence, frequently, however, within a narrow range of conditions. Inhibitors may be classified by surface reactivity as adsorptive or surface-reactive (where a precipitated film is formed to provide a barrier between the corrosive agent and the aluminum surface). Chromates, silicates, polyphosphates, soluble oils, and other inhibitors are commonly used to protect aluminum. Aluminum is concentration-sensitive to chromate solutions as well as to other anodic inhibitors. Combinations of polyphosphates, nitrites, nitrates, borates, silicates, and mercaptobenzothiazole are used in systems that include aluminum and other metals [2].

Composition differences among aluminum alloys often determine whether or not an alloy can be inhibited in a given environment, and so an understanding of the metallurgical variables is important. Investigations into the fundamental reactions at the aluminum–environment interface have added significant new understanding that now permits the selection of inhibitors to be made with greater precision and their application to proceed with fewer trial-and-error adjustments [2].

In a limited number of cases, removing some minor constituent from the contacting liquid or gas can prevent corrosion. For instance, copper compounds, which may make water corrosive to aluminum, can be removed by passing the water through a tower packed with aluminum chips. Finally, the use of periodic cleaning procedures may be highly beneficial in specific cases [2].

14.2. METALLIC COATINGS

Aluminum can be protected by electroless or by conventional plating. Aluminum and aluminum alloys act as a sacrificial anode in the form of plate (Alclad) or as an applied coat, or for long-duration sacrificial anodes on condition that they do not passivate in the specific medium. Aluminum alloys are protected by more active metals or by cathodic protection of Al alloys. Aluminum is used for metallic powder application as a protective coat. Also, magnesium can be used as a pigment in primers for a paint coating system (see Section 14.5).

14.2.1. Conventional Plating and Electroless Plating of Aluminum

Electroplating of aluminum and its alloys is occasionally used to provide resistance to corrosion since other anodic treatments of surface conversion are currently used and give efficient corrosion prevention. However, electrolytic metallic coatings are used to obtain a certain metallic appearance, increased electrical conductivity, improved solderability, or improved frictional properties. Aluminum-based alloys are more difficult to electroplate because of the highly passive aluminum oxide formed on the surface. Also, most metals are more noble than the alloy and any void, discontinuity, or scratch can lead to severe localized corrosion [5].

Electroless plating with nickel and copper seems to be the most established plating method and is currently used in aircraft/aerospace industry and electronics industry applications. Nickel is chemically plated on aluminum parts with complex shapes, such as that of integrated electronic circuits. A number of different reducing agents have been used in preparing electroless nickel baths such as sodium hypophosphite, amino boranes, sodium borohydride, and hydrazine. Sodium hypophosphite baths are currently used. Their principal advantages are lower cost, greater ease of control, and better corrosion resistance of the deposit [5].

Plating of aluminum becomes possible by the use of a zincate pretreatment to remove the aluminum oxide film. By the use of a double zincate treatment, the homogeneity of the zincate film can be increased. This consists of stripping the first deposited layer and performing the second treatment for the homogeneity of the zincate film. A commercial hypophosphite-based nickel electroless layer can be used for bumping. The pH value is 4.5 and the bath is operated at 90 °C. A plating rate of 25 μm/h is obtained. The deposits from this electrolyte contain 10% phosphorus [6].

Electroless Nickel Deposition on Al Integrated Circuits Electroless plating is an electrocatalytic process corresponding to the chemical reduction of metal ions on a base substrate in aqueous solution. Deposition progresses essentially linearly with time and quite heavy deposits can be produced analogous to those produced by conventional plating. However, electroless plating can produce very uniform thicknesses and layers since the nonuniform current distribution in the plating cell is avoided [6].

14.2.2. Surface Preparation for Thermal Spraying

Substrate surface preparation prior to thermal spraying is a key step to ensure good adhesion of the resulting coating. The most common approach consists of two successive stages:

1. Surface degreasing by solvent application to remove organic contaminants.
2. Surface roughening by grit blasting to ensure mechanical anchoring between the coating and the substrate.

This approach faces limitations and drawbacks for certain applications. One drawback is related to the use of solvents that are controlled by legislation protect peoples' health and for environmental considerations. Another issue is that grit blasting gives rise to residues that can be trapped in the material. These residues create some interface defects that either weaken the coating adhesion or create a severe indentation on the substrate [7].

Aluminum coatings can provide galvanic cathodic protection for several metals and alloys. In order to be a suitable protective solution on structural components, the mechanical integrity must be preserved. In particular, the fatigue properties are a challenge for thermal spray protective coatings on mechanical structures. To address the issue of the fatigue integrity of the aluminum alloy 7075 with an arc-sprayed protective coating, different surface preparations prior to arc spraying were considered. A feasibility study was performed using laser ablation as a surface preparation technique before or during arc spraying of coatings. Both fatigue and adhesive properties of aluminum coatings were evaluated in relation to substrate surface preparation techniques including laser ablation (PROTAL process), grit blasting, and shot peening [7].

Experimentation has shown that it is possible to thermally spray a metallic aluminum coating on to Al alloy 7075 without reducing the fatigue properties. Nitrogen as the atomizing gas for the arc spraying process provides significant improvement for the microstructural density of the aluminum coating. Moreover, laser ablation allows the deposition of aluminum coating on Al alloy 7075 using nitrogen as the atomizing gas with a clean interface and adequate adhesion. The combination of these two parameters on a shot-peened substrate allows for a significant improvement in fatigue properties to the level of an uncoated substrate and consequently maintains the material integrity and fatigue properties for structural applications [7].

It is feasible to use the PROTAL process with the arc spraying process in spite of the large plume and the significant overspray of the arc spray process. A similar bond strength to grit blasting can be achieved with the appropriate laser energy density for scouring and deoxidation modes. Finally, laser ablation allows a one-step process for both surface preparation and arc spraying on a robotic station [7].

14.2.3. Sacrificial Protection by Aluminum Alloys

Alclad Alloys These are duplex wrought products, supplied in the form of sheet, tubing, and wire, which have a core of one aluminum alloy and a coating on one or both sides of aluminum or another aluminum alloy. The cladding on each side is 2–5% of the total thickness. The coating is metallurgically bonded to the core over the entire area of contact. The coating is usually selected to be anodic to the core alloy in most natural environments and will galvanically protect the core where it is exposed at cut edges, rivet holes, or scratches. Such Alclad alloys are usually more resistant to penetration by neutral solutions than are any of the other aluminum-based alloys. Corrosion can be prevented or reduced by cladding with a more corrosion-resistant alloy, such as high-purity aluminum, a low magnesium–silicon alloy, or an alloy of 1% zinc. All of these cladding materials are frequently employed to give added corrosion protection to the 2000 and 7000 series alloys [8].

Sacrificial Aluminum Anodes Sacrificial aluminum anodes are used for cathodic protection of steel in seawater. Most aluminum sacrificial anodes are cast Al–Zn–Sn, Al–Zn–In, or Al–Zn–Hg alloys containing about 94–95% Al and 3.5–5% Zn. An addition of gallium (0.01–0.1%) or mercury ($\sim$0.05%) has been used in sacrificial anodes to avoid passivation. Al–Zn alloys with as little as 0.01% Sn in commercial-grade aluminum will cause surface darkening on annealing and increase susceptibility to corrosion, which appears to be due to migration of Sn to the surface. This effect may be reduced by small additions of copper (0.2%). These alloys are used in environments where aluminum remains active [8, 9].

Atmospheric corrosion and salt spray tests (ASTM Specification for Salt Spray Fog Testing B117-73) of aluminum–zinc alloy-coated steel showed that 55 w% aluminum–zinc is two to four times as corrosion resistant as a galvanized coating of similar thickness. Furthermore, for the galvanic protection of cut edges of sheet in some environments, this coating proved to be superior to aluminum coating [8].

Aluminum as Sacrificial Anode to Steel Structural aluminum displays potentials for the unalloyed metal (~99.5% pure) in the range of 0.6–0.8 and varies with the environment, while aluminum alloys exhibit a variety of potentials—some of them sufficiently active to be used as cladding materials for the more noble alloys. Copper has a dramatic effect on the anode alloy and it is suggested that a minimum of 0.025 of mercury is necessary for reasonable anode activity. With low iron and copper levels, consistently good performance can be achieved with 0.4% zinc and 0.04% mercury. A higher level of zinc is beneficial in anaerobic mud. Another developed group of alloys are all of a generic strain, which uses indium (~0.005–0.03 wt %) as the principal alloying metal. Usually, with high-purity aluminum and some zinc, the capacity of the anode is high and it operates close to 90% efficiency; it can exhibit higher driving force than the mercury anodes, usually at −1.10 V to the Ag,AgCl/KCl electrode or 0.3 V to the protected steel [10].

Sacrificial Aluminum Coat Spraying Aluminum spraying is a current practice to coat less-resistant alloys. For some composites, the corrosion behavior is governed by galvanic action between the aluminum matrix and the reinforcing material. Aluminum thermal spraying has been reported as a successful protection method for discontinuous silicon carbide–aluminum composites; for continuous graphite–aluminum or silicon carbide–aluminum composites, sulfuric acid (H_2SO_4) anodizing has provided protection, as have organic coatings or iron vapor deposited aluminum [2].

More recent work on corrosion prevention of aluminum alloys used for aerospace applications by sprayed sacrificial coatings has been reported. This can be applied in situ for maintenance. In one study, a coating 125 μm thick was sufficient since only 12.5 μm was consumed during 1000 hours of immersion of the coated alloy in agitated and aerated 3.5% sodium chloride solution at 25 °C. The Al–5Mg sprayed alloy offers similar protection as other sacrificial coatings based on Al, Zn, Al–Zn or Zn–15 Mg, although it gives the least sacrificial current. This coating can give better performance in alkaline environments since magnesium offers better resistance than aluminum for cathodic corrosion. This coating can prevent stress corrosion cracking; however, for pitting corrosion it is not so efficient since pitting potential is lower than the mixed potential of the coating and the substrate. Pitting, although retarded, can lead to transgranular ruptures and corrosion fatigue ruptures. It has also been shown that the corrosive medium can infiltrate through the pores of the coating to the alloy–coating interface similar to that observed with clad alloys [8, 11].

Stable Aluminum Anodes for Impressed Cathodic Protection Systems Most domestic water installations are protected using impressed cathodic current systems. A typical system uses aluminum anodes to protect the interior surface of the steel reservoir and the pipes. The principal reactions are

At the anode: $Al \rightarrow Al^{3+} + 3e^-; \quad Al^{3+} + 3H_2O \rightarrow Al(OH)_3 + 3H^+$

At the cathode: $O_2 + 2H_2O + 4e^- \rightarrow 4OH^-$

The anodic reaction produces aluminum hydroxide in a colloidal form while the cathodic one consumes oxygen. The formed aluminum hydroxide, on the order of 0.1–0.2 mm thick, acts as a protective layer for the interior cathodic surface of the reservoir that can be made of steel, galvanized steel, or copper. The pH of water should be controlled between 6 and 8.5, which is the zone of passivation of the aluminum anode.

Steel or aluminum can be used as consumable anodes in seawater for cathodic protection impressed current systems. Both operate at higher efficiencies than in fresh water, steel being about 90–100% efficient while aluminum is about 50–90%. However, permanent anodes made from lead alloys or platinum film electrodes are the two most suitable ones [10].

14.2.4. Aluminum Powder as a Coating

Cementation or diffusion of aluminum is used currently as a coating to protect metallic objects. A mixture of aluminum and appropriate flux are placed on the metallic surface at a high temperature in order to allow the diffusion of the coating material into the basic substrate. Zinc and chrome powders are used for the same purpose.

Electrophoretic Powder Deposits Electrophoretic powder deposits are obtained in galvanic baths, for example, alcohol, microadditives of $Mg(NO_3)_2$ or $Al(NO_3)_3$, and aluminum powder. A voltage of 100–120 V and a current density of 50–150 A/m^2 are recommended to deposit aluminum powder on steel substrate using the electrophoresis method. The electrostatic method permits one to deposit larger particles than the electrophoresis method [12]. The advantages of the hydroimpulse coating are its small grains, the possibility of applying the coating on parts of various shapes and configurations, and the fact that its adhesion to steel substrate is $\sim$100 MPa.

The coating is composed mainly of pure aluminum and some microalloying elements such as Mg, Na, and other trace elements. Hydroexplosive aluminum coating on steel substrate is comprised of three zones: an outwardly textured zone with grain sizes from 1 to 5 μm, the 10 μm thick inwardly (inner) ultrafine grain zone with grain diameters from 0.1 to 1 μm, and the 2–3 μm thick transition zone.

The coating does not have open porosity and the closed porosity is only 2–3%. However, there is a lack of structure in the brittle intermetalloids. This coating is not prone to SCC in wet sulfide media, but coated steel is prone to SCC in wet sulfide solutions containing NaCl additives. The coating is anodic in NaCl solutions while it is cathodic in hydrogen sulfide solutions. In NaCl and in wet sulfide solutions, corrosion rates are lower than that of pure aluminum. The coating is more resistant to pitting corrosion than pure Al [12, 13].

14.2.5. Cathodic Protection of Aluminum Alloys

Godard cited an early example of aluminum protection using sacrificial zinc anodes and Hatch mentioned the use of impressed current protection systems to protect painted aluminum ship hulls. Cathodic protection requires careful control to ensure that adequate protection is maintained without overprotection, which can lead to alkali attack (cathodic corrosion). Alclad alloys (layered aluminum products with one aluminum alloy integrally bonded to a more noble aluminum alloy core) may be viewed as having a self-contained cathodic protection system [2].

Buried aluminum pipelines are usually protected by sacrificial anodes: zinc for coated lines and magnesium for uncoated lines. Impressed current (rectifiers) systems are not used to protect aluminum pipes [8, 14]. In some applications, aluminum alloy parts, assemblies, structures, and pipelines are cathodically protected. The suggested practice is to shift the potential at least to −0.15 V but not beyond the value of −1.20 V versus a saturated $Cu/CuSO_4$ reference electrode. High cathodic currents can make the solution sufficiently alkaline to cause significant cathodic corrosion. In some soils, potentials as low as −1.4 V have been encountered without appreciable cathodic corrosion.

14.3. CONVERSION COATING

Conversion coatings are the most widely used prepaint treatment processes for metal substrates. Processes specifically designed for aluminum are of recent origin. Historically, however, phosphoric acid cleaners, wash primers, and iron and zinc phosphates and chromate-oxide coatings have all been utilized with satisfactory results. Accelerated chromate phosphates, chromate oxides, and anodizing and nonchromate formulations have recently been developed [15]. Conversion layers provide the ability to modify the aluminum surface to give better adhesion, a surface free of contaminants, or a coating layer that contains active corrosion inhibitors. Conversion coating can be selected based on corrosion protection functions and/or paint pretreatments. Corrosion prevention, at least partially, can be achieved through passivation of the metallic base of aluminum, by a barrier against moisture, by oxygen and other aggressive agents, by electrochemical insulation, and by protection against mechanical erosion. For paint pretreatments, an effective and continuous bonding, chemical stability, insolubility, imperviousness, and flexibility are essential properties, in addition to providing a wettable substrate for paint application and maintaining adhesive integrity between the metal and the paint film [15].

The U.S. aerospace industry places a high demand on coatings used to paint or repaint existing aircraft fleets. A typical coating system (pretreatment/primer/topcoat) is comprised of three individual coating layers. The first layer, a conversion coating, is the product of substrate pretreatment. The conversion coat is usually a very thin (10–60 nm) inorganic layer that provides corrosion protection and improved adhesion between the substrate and the primer, which is the second layer of the coating system. The primer provides similar functions to the conversion coat, but is comprised of a pigmented organic resin matrix. The application thickness of the primer can vary from 5 to 200 μm although a thickness of 25 μm (1 mil) is typically desired due to weight constraints on the aircraft. The primer is the principal provider of corrosion protection. Typical formulations consist of both chromated and nonchromated pigments enveloped in an epoxy resin. Finally, a top coat is applied that serves as the main barrier against environmental influences such as extreme climates and ultraviolet rays and it provides the aircraft with decoration and camouflage [4].

Conversion layers can be created through a variety of techniques, which include physical vapor deposition, cathodic magnetron sputtering, high-energy ion beams, and laser ablation. The most common conversion layers are from electrochemical anodization, plasma ablation, and chromate conversion coatings and modifications thereof [4].

14.3.1. Phosphates and/or Chromates

Corrosion coatings (chromates or phosphates) are recommended for the preparation of aluminum alloys. For milder environments, paint may be applied on the conversion coating, but a chromated primer should be applied for more aggressive media. Almost any type of paint (acrylic, alkyl, polyester, vinyl, etc.) is suitable. One or two coats of the finish paint should follow the primer. In general, a coating contains micropores, areas of low cross-link density or high pigment volume concentration (PVC), that provide a path for diffusion of corrosive agents such as water, oxygen, and chloride ions to the coating–metal interface. Therefore it is most often necessary to incorporate inorganic or organic inhibitors into a paint system for corrosion protection [4].

The chemical treatment of aluminum by immersion in acid chromate solutions is often employed to produce protective coatings on the metal surface. The coatings, known as chemical conversion coatings, increase the corrosion resistance of the metal and also serve as adhesive bases for organic finishes such as paints. The most effective corrosion inhibitor in use is hexavalent chromium (Cr^{6+}). A chromate conversion coating is a chemically grown oxide layer on the alloy substrate that provides an active barrier layer that reduces the rate of the cathodic oxygen reaction. The final result is a 10–60 nm thick film consisting of $Cr(OH)_3$ and $Al(OH)_3$ [4].

Brown et al. [16] examined the morphology and structure of the chemical conversion coating developed on annealed high-purity aluminum in an acid chromate–fluoride solution by transmission electron microscopy of stripped films and ultramicrotomed sections. A mechanism for conversion coating has been suggested. Deposition of hydrated chromium oxide at the cathodic sites takes place on the aluminum surface at preexisting metal ridges. Such material develops extensively over the ridged structure with time of immersion. The anodic sites exist between the preexisting metal ridges. At such regions, the fluoride-containing acid solution chemically thins the original air-formed film, resulting in a dynamic equilibrium between alumina film growth and its dissolution. The consequential loss of Al^{3+} ions to the solution results in local scalloping at the metal–conversion coating interface.

Growth of chemical conversion coatings on aluminum is strongly dependent on the purity of the metal. For specimens up to 99.99% purity, the aluminum surface is not homogeneous. Flaws are always present in the thin oxide layer covering the aluminum surface; such flaws are situated above the grain boundaries and cellular boundaries associated with impurity segregation in the aluminum substrate. These flaws are of the residual type and provide easy paths for electronic conduction in an otherwise insulating oxide layer. Consequently, coating growth (i.e., the reduction of the dichromate species to hydrated chromium oxide) occurs preferentially along the grain boundaries or cellular boundaries of the aluminum substrate [17].

However, for aluminum specimens of 99.9996% purity or higher, where the cellular structure is absent, the population density of the residual-type flaws in the oxide layer is considerably reduced. It is suggested then that coating growth proceeds by the tunneling of electrons through the thin, insulating, passive oxide layer to produce a coating of more uniform appearance than associated with substrates of reduced purity [17].

Growth of chemical conversion coatings on aluminum is also closely related to the heterogeneity of the aluminum surface. Even minor heterogeneities, such as grain boundaries or cellular boundaries, play an important role in the coating growth process. In practical situations, where various alloying additions are present within the metal, the coating growth process is expected to be considerably more complex than described here, due to the presence of precipitates and intermetallic particles of varying sizes and electrochemical behavior [17].

Zinc Phosphate Treatment Properly applied, this group of phosphates is good for corrosion protection and finds widespread use in mixed steel and aluminum product lines. The process of phosphating is used frequently before cold deformation of aluminum alloys for corrosion protection. The aluminum and its alloys can be deformed rapidly with appropriate lubricant and without the application of surface conversion treatment. In difficult situations of deformation of aluminum alloys, the use of crystalline zinc sulfate is indispensable after the application of a specific lubricant, as in the case of extrusion or dishing of a very thick plate of metal. An important preparation of the surface before phosphating is necessary to eliminate any leached metallic component of the alloy, such as copper or silicon [18].

Phosphate conversion coatings are popular because of low operational cost and low environmental toxicity [15]. Zinc phosphates were the initial replacement pigments for zinc chromates. They display no toxic effects and provide some corrosion protection to aluminum alloys by forming a $Zn_3(PO_4)_2 \cdot 4H_2O$ film. Bath composition typically consists of phosphoric acid, zinc dihydrogen phosphate, fluoride, and an oxidation accelerator such as NO_2 at pH 2–4 [4].

The zinc phosphate bath should have a balanced composition of salts and free acid. Its established total acid value should be maintained as well as the free-acid value for satisfactory performance. Accelerators such as sodium nitrite are used in some zinc phosphate solutions [3]. The metal phosphates are soluble in strong acids but crystallize out when the acidity is reduced. This mechanism occurs as the acid ions react with the aluminum surface, become neutralized, and produce an integral crystal growth on the metal surface. The aluminum surface is therefore converted to a finely crystalline phosphate film with acceptable texture for paint bonding. Crystalline phosphate films may be iridescent to gray and coating weights can be 0.108–0.538 g/m^2 (10–50 mg/ft^2) for iron phosphates, and 1.08–3.24 g/m^2 (100–300 mg/ft^2) for zinc phosphates [15].

Phosphating is an electrochemical and chemical reaction and gives rise to several passivating phenomena, namely, a precipitation to the amorphous state, crystallization and growth, and crystalline reorganization. The mechanism of iron and steel phosphating in zinc phosphate bath accelerated with nitric acid has been frequently examined. Effectively, the crystalline phosphate layers on steel are composed of metal oxide, iron phosphate, phosphophillite ($Zn_2Fe(PO_4)_2 \cdot 4H_2O$), and hopeite ($Zn_3PO_4 \cdot 4H_2O$). This can be extrapolated to another metal such as aluminum, except that the highly passive nature of aluminum oxide and the presence of fluoride in the bath as etcher and complexing agent of aluminum ions necessitate a fundamental explanation of the mechanism of aluminum phosphating that can lead to better understanding of the process [19].

The polarization curve of electrochemical reduction of hydrogen ions on the pure aluminum surface is situated at very low negative or active potentials. The phosphating solution contains phosphoric acid, zinc phosphates, nitrates, and fluorides or free hydrofluoric acid. As in chromate–phosphate treatments (see Section 14.3.2), the HF initiates a rapid homogeneous attack of the aluminum surface [18]:

$$3\,H_3PO_4 + Al \rightarrow Al(H_2PO_4)_3 + \tfrac{3}{2}H_2$$
$$3\,HF + Al \rightarrow AlF_3 + \tfrac{3}{2}H_2$$

The kinetics of the attack increases with fluoride ions and the oxygen saturation of the solution. The principal reaction of formation of insoluble tertiary phosphates of zinc (hopeite) starts at slightly higher pH due to the initial attack of the metal, since the acid bath contains the soluble zinc monophosphates:

$$3\,Me(H_2PO_4)_2 \rightarrow Me_3(PO_4)_2 + 4\,H_3PO_4$$

The formation of fluoride aluminum complexes, such as AlF_6^{3-}, occurs and these are critical for the support of the acid attack of the aluminum surface and can give rise to the precipitation of fluoride aluminum compounds [18]. This can lead gradually to the accumulation of fluorides, aluminates, and phosphates of aluminum in the bath. Sodium or potassium salts are then present to prevent the buildup of soluble aluminum in the bath, which inhibits coating formation. In the presence of sodium, aluminum, and fluoride ions, the following reactions are possible:

$$Al(H_2PO_4)_3 + 3Na^+ + 6F^- + 3H^+ \rightarrow Na_3AlF_6 + 3\,H_3PO_4$$
$$AlF_3 + 3\,F^- + 3\,Na^+ \rightarrow Na_3AlF_6$$

Layer Composition Simple analysis of the phosphate layers gives 34–37 wt% PO_4, 35–38 wt% Zn, 3–5 wt% Al, 4–6 wt% F, and water of crystallization [18]. It has been found that the phosphate layer is composed essentially of tertiary zinc phosphate (hopeite), with a thickness of 1–5 μm giving film weights of 2–6 g/m^2. In these phosphating solutions an underlayer of Al_2O_3 is formed. The addition of a ferrous salt can help the formation of the underlayer and this was found to consist of γ-Fe_2O_3 and γ-Al_2O_3. It has been suggested that the identical structure of the two oxides can lead to the formation of derivatives with molecular substitution [18].

Phosphate Application The application of phosphate coating for paint-based application normally comprises five successive operations: cleaning, rinsing, phosphating, rinsing, and chromic acid rinsing. Some of these operations may be omitted, combined, or integrated in one operation: for example, rinsing after cleaning with water at 71–82 °C and before phosphating since hot water is an additional cleaner. However, ambient temperature rinses are now often used. Hexavalent chromic acid rinse can be replaced by phosphoric acid or other recent alternatives [3].

Phosphating methods may be applied to an aluminum surface by either immersion or spray, or a combination of both. Application is by immersion at 125–140 °F for 1–4 minutes, or spraying at 125–160 °F for 30 seconds to 2 minutes. There is a trend to use zinc phosphating coating processes for automobile bodies using a combination of spray and immersion. Occasionally, a surface may be coated by brushing or wiping [3]. Product selections should be restricted to moderate service environments. Bath life is limited due to the low tolerance for accumulating aluminum [15]. Zinc phosphate coatings applied via spray are easily deposited on aluminum surfaces, provided fluoride ion is present in the bath [3].

Simultaneous Phosphating of Steel and Aluminum It is possible to phosphate steel and aluminum at the same time. In this case, rich solutions in zinc and nickel phosphate perform better than ordinary solutions. Also, the presence of HF and/or simple or complex fluorides is essential. The steel surface should be larger than that of aluminum in certain cases, and the aluminum surface should not exceed 10–15% of the total phosphate surface [18]. However, in processing a metal mix of aluminum, steel, and galvanized steel, separate fluoride additions to the bath are required if the metal mix consists of greater than 10% aluminum. Coating weights range from 0.27 to 2.2 g/m^2 (25 to 200 mg/ft^2) [3].

Phosphating of Al–Mg–Si Alloy Surface In vehicle body construction, extrusion alloys based on Al–Mg or Al–Mg–Si are most commonly used. Whereas the Al–Mg alloy does not suffer corrosion or paint adhesion problems using standard methods of application, subsurface corrosion can be troublesome with, for example, Al–0.4Mg–1.25Si alloys, especially when a machined surface is involved. This has been observed in the salt spray condensate water test and in atmospheric corrosion accelerated by NaCl. The underlayer filiform corrosion often starts in the direction of machining, after which it transforms into a broad frontally advanced action [20].

The performance of paint coated aluminum in the salt spray test is excellent. A major improvement in corrosion resistance to filiform corrosion is obtained when the Al–Mg–Si alloy surface, after machining, is etched in caustic soda by nitric–sulfuric acids before being phosphated to give a crystalline coating using a fluoride-containing, low-zinc method. The low-zinc phosphate coatings are, with respect to their corrosion resistance, markedly superior to normal zinc phosphating coatings. The low-zinc system produces coatings with a higher proportion of $Zn_2Fe(PO_4)_2$ (phosphophillite) [20].

14.3.2. Chromate–Phosphate Treatments

Chromate–phosphate coatings were the first pretreatment specifically developed for aluminum in 1945. Their products have performed excellently for the architectural metal and beverage can industries, since the coating does not contain Cr^{6+}. However, the demand for beverage cans is in decline and the high-performance top coats no longer require the strong performance of chromate-based pretreatments. Effectively, the excellent top coat paints, such as silicone fluoropolymers and powder coats, can minimize the performance requirements for the prepaint treatments. Advances in electocoat application technology are also proving beneficial [15].

The concentration limits of phosphate, fluoride, and hexavalent chromium ions, in the aluminum chromate–phosphating bath, can be expressed as: PO_4^{3-}, 20–100 g/L; F^-, 2–6 g/L; and CrO_3, 6–20 g/L. The F^-/CrO_3 (anhydrous) ratio should be controlled in the range 0.1–0.4. The pH is usually below 2.

Part of the aluminum can form F_3Al, which leads to formation of relatively soluble complexes such as $(AlF_6)^{3-}$ in combination with fluoride ions in the bath. Addition of HF is very critical to attack the aluminum surface, initiate the phosphating process, and precipitate the formed aluminum fluoride compounds as a mud for disposal. In every case, the aluminum compounds act as inhibitors of the process; they should be continuously removed, otherwise the zinc phosphate layer will not be as pure and homogeneous as desired [18].

Chromate–phosphate treatment can be applied by spray or immersion. Immersion times range from 30 seconds to 3 minutes at 110–130 °F, whereas spraying is done for 15–45 seconds at 95–130 °F. These baths produce crystalline or amorphous coatings of 15–1000 mg/ft^2. The film is iridescent to grayish green. Thickness can be significant, from 2.5 to 10 μm (0.1 to 0.4 mil). The chromium phosphate films are composed principally of hydrated chromium phosphate ($CrPO_4$), Cr_2O_3, and aluminum oxides. A typical air-dried coating is composed of 50–55% chromic phosphate, 17–23% aluminum phosphate, 22–23% water, and a trace amount of fluorides [15].

The corrosion performance of chromate–phosphate films is generally very close to chromic acid anodizing films and that of chromate oxide films. This treatment is recommended for severe and long-term service conditions. Because of its excellent performance, the American Architectural Manufacturers Association (AAMA) has designed it as a standard prepaint treatment and it is recommended also by U.S. military specifications [15]. In reality, this level of corrosion prevention is ideal for conversion coating.

Chromate Oxide Coatings with Superior Corrosion Performance These coatings were introduced in the early 1950s and are now widely used for domestic appliances, aircraft and electronic equipment, and continuous coil coating of architectural aluminum. Normally a chrome oxide bath consists of three principal constituents: acid chromates (H_2CrO_4), etchants (HF), and accelerators or complexing agents. The original accelerator, $Fe[(CN)_6]^{3-}$, was replaced by $(MoO_4)^{2-}$ because of environmental considerations. Paint-based coatings can be applied by spray, immersion, or brush at 25–60 °C (80–140 °F) for 15–45 seconds. Longer times are needed for bare corrosion protection coatings, which are applied by immersion. The formed iridescent yellow to brown color film is very adherent, amorphous, and mixed with metallic oxide corrosion products. The film thickness range is 0.13–1 μm (0.005–0.04 mil) and coating weights are from 15 to 100 mg/ft^2. The coating consists of chromium oxides covered with an absorbed monolayer of the accelerator [5, 15, 21].

Chromate oxide films are versatile and more widely used than the chromate–phosphate treatments, where parts are too long or assembled with dissimilar metals. They also have superior performance ratings compared to chromate–phosphate coatings. Without paint, the films have twice the salt spray resistance of a chromate–phosphate coating. They can stand very severe service conditions and comply with military specification MIL-C-554 and AAMA 605.2 [15].

Alkaline Chromates These treatments are as old as the described phosphate treatments. Alkaline chromates are primarily solutions composed of 2–3% sodium carbonate and 0.5% potassium dichromate. Immersion times range from 10 to 20 minutes at 95 °C (180–200 °F) at pH 10–11. This carbonate/dichromate ratio is critical to provide consistent coating action. Thin, gray, and porous films are 1–2.5 μm (0.04–0.1 mil) thick and coating weights are 1.077–5.38 g/m^2 (100–500 mg/ft^2), made up of aluminum oxide and dispersed chromate oxides. Maximum corrosion resistance is achieved by sealing in hot 5% potassium dichromate [15].

No-Rinse Chromate Processes These processes are finding application in the coil coating of aluminum. Adhesion and corrosion protection properties are equivalent to conventional processes. The applied solution is composed of Cr^{6+} and Cr^{3+} as well as other ingredients, such as F^- or PO_4^{3-}. Some formulations include organic compounds. For most paint applications, the coating weights range from 0.054 to 0.269 g/m^2 (5 to 25 mg/ft^2) [5].

14.3.3. Chromate Alternatives

Discharge of Cr(VI) into natural environments must be avoided; therefore molybdates should replace chromates in certain applications [2]. The advent of chromate replacements for aluminum alloys began in the late 1970s. As regulations became increasingly stricter, commercial, academic and, government facilities have relied on a collaborative effort to come up with novel innovations for the corrosion protection of aerospace aluminum alloys. More elaborate coating formulations have failed to produce the level of protection that chromates provide; therefore researchers have moved to more advanced technologies for improved protection. Some of the newer alternatives include low-temperature cationic plasma deposition, sol-gel and ceramic coatings, and various inorganic and organic inhibitors and conducting polymers, including some based on double helical structures. Although ultimate protection has not been achieved by one single technique, it is believed that a combination of the most promising alternatives will provide the desired protection [4].

The use of phosphochromatation instead of chromatation should be considered with great care since some Cr^{6+} in the phosphorchromatation layer has been proved to be present even with thorough rinsing of the treated material. A possible solution is a conversion coating using zirconium but the corrosion resistance is still unacceptable [22]. It has been postulated that a conversion coating based principally on potassium permanganate as the primary component is an acceptable alternative to chromate. Manganese oxides are by far the most closely related to chromic oxides in terms of similar chemical properties [23].

Nonchromate Systems A report has been issued by the U.S. Air Force comparing the corrosion resistance performance of selected fully nonchromate systems to the standard chromate-containing coating system. A total of 12 fully nonchromated systems, consisting of both experimental and commercially available products, were tested on 76 mm × 152 mm (3 in. × 6 in.) aluminum alloy AA2024-T3 panels and compared to the performance of the standard chromate system. The salt spray test (ASTM B117-94), filiform corrosion test (ASTM D2803), pull-off strength (PATTI) test (ASTM D4541), and electrochemical impedance spectroscopy (EIS) measurements were considered. The data identified two fully nonchromated systems—MSZ/WEP/APC and MSZ/EEP/APC—that demonstrated promising performance in all tested areas comparable to that of the standard chromated aircraft coating system [24].

The surfaces of the panels that were to receive an MSZ coating underwent the surface treatment method used at Warner-Robins ALC according to MIL-C-10578D. The CCC pretreatment was applied using the MIL-C5541E process specification by immersion for a period of 3–5 min to generate a coating weight between 0.430 and 0.645 g/m^2 (40–60 mg/ft^2). MSZ is a water-based, commercially available mixed silane–zirconate sol-gel, AC-131, with the Boegel EP-II trademark. The material was mixed according to the manufacturer's instructions (AC-131) and left for a 30 min dwell time. The application was spray-applied to upright coupons during several passes to ensure that the coupon was drenched for not < 30 s, but not > 2 min. The thickness of the films, measured by scanning electron microscopy (SEM), ranged from 0.6 to 0.9 μm [24].

The two evaluated nonchrome epoxy primers were: WEP, a commercial water-based nonchromated epoxy primer, PRC CF EW AE048A/B, meeting the specifications in MIL-PRF-85582, type I, Class N with the Eco Prime Trademark; and EEP, an experimental epoxy primer, 02-GN-083. The top coat was APC, product 99-GY-001 Color 3673, meeting MIL-PRF-85285D specifications. Analysis of the experimental data showed the ability of some of the fully tested nonchromate coating systems to provide corrosion protection comparable to that of the chromate control system (CCC/MIL-PRF-23377H/APC), but only when the criterion of a clean scribe is overlooked as stated in MIL-PRF-23377H for Class N primers. After 2000 h of exposure in the salt spray test, nonchromate systems with the MSZ surface treatment, and either WEP or EEP primer, demonstrated comparable corrosion protection [24].

Cerium Salts as Inhibitors Some of the most promising chromate replacement inhibitors are derived from cerium salts. They are believed to control the cathodic reaction by precipitating cerium hydroxide ($Ce(OH)_3$) at local regions of high pH. Aldykiewiczs et al. [25] studied the effects of cerium chloride on aluminum alloy 2024-T3 using in situ current density mapping. They concluded that there was a preferential Ce deposition over the copper-rich regions, which were associated with increases in pH due to O_2 reduction. Mansfeld et al. [26] modified a process that incorporated cerium, which was originally introduced by Hinton to inhibit corrosion of high-strength aluminum alloys. The process involves immersion of the alloy in hot $Ce(NO_3)_3$ followed by anodic polarization in Na_2MoO_4 and final immersion in hot $CeCl_3$ [4]. In 0.5 N NaCl, treated samples were reported to have pitting potential (E_{pit}) values 200 mV higher (more noble) than the untreated samples. The increase in E_{pit} is indicative of the system reaching a more noble state. The process yielded excellent corrosion resistance for the aluminum alloy 7000 series and marked improvements for the aluminum alloy 2000 series using this process. Cerium compounds are also being incorporated into sol-gel coating systems, which are new candidates for aluminum pretreatments and primers [4].

Sol-Gel Processes Sol-gel processing is a method in which thin oxide films can be deposited on a substrate at much lower temperatures than traditional ceramic processing methods. A variety of metal alkoxides, salts, or nitrates can be used as precursors for the oxide synthesis. Upon deposition, the coating undergoes hydrolysis and condensation to form a continuous three-dimensional oxide matrix. The oxide layer acts as an inert, hydrophobic barrier layer. The intrinsic stresses and processing conditions for one-coat systems limit the film thickness to less than 1 μm. Film thickness of >1 μm can be obtained by combining additives into the sol-gel or by creating multicoat systems. Due to the limited film thickness, sol-gel coatings are used as the substrate pretreatment [4].

Although one would expect a thicker film to provide better protection, the opposite is true for sol-gel films due to the tendency for cracking to occur in single-coat films greater than 1 μm. A more realistic approach for defect-free film formation is being taken by combining the various metal oxides with organic segments to form what are called *ceramers*. They are synthesized using organic functionalized metal alkoxides that promote reaction between an organic group and the inorganic alkoxide. Ceramer systems are potential chromate replacements because the increased flexibility of the coating thickness for ceramers is higher than that for traditional sol-gel coatings, which expands their use [4].

It has been stated by Hamdy et al. [27] that ceramic coatings based on salts like vanadia, ceria, silica, and molybdates prepared by sol-gel are promising future treatments for aluminum surfaces. The surface preparation (etching followed by oxide thickening in boiling water) prior to sol-gel treatment was found to have a marked influence on corrosion protection of AA6061-T6. Molybdate treatment showed the best corrosion resistance after 30 days in 3.5% sodium chloride solution due very probably to the formation of a compact film of molybdenum oxide with buffer action that rejects chloride ions [27].

Hamdy [28] studied the corrosion protection of aluminum composite AA6061-T6–10% Al_2O_3 by silicate/ceria treatments. One of the suggested treatments is oxide thickening in boiling water for 1 h followed by treatment in a solution containing a mixture of 10 mL sodium silicate (4 g/L) + 10 mL $CeCl_3$ (1000 ppm) for 2 h. It was found that the corrosion resistance was improved due to the formation of protective oxide films, which act as a barrier to oxygen diffusion to the metal surface. Cerate, and not silicates, plays the main part in the passivation process. EIS studies showed that etching before silicate/cerate treatment has a negative effect on the corrosion resistance that could be due to the increased distribution of silicon and that excludes the more efficient cerium oxide from the active centers [28].

Cerium Chloride Treatment Hamdy [29] stated that a surface preparation is essential for the corrosion protection of aluminum composite AA6061-T6–10% Al_2O_3. Preetching followed by oxidethickening was proposed before the treatment in pure solution of 1000 ppm of cerium chloride ($CeCl_3$ for 3 h or 1 day). Corrosion behavior was monitored using electrochemical impedance spectroscopy (EIS) and dc polarization techniques during immersion in 3.5% NaCl solution for up to 60 days. XPS and SEM techniques were used to examine the corrosion performance of different pretreatments.

Studies showed that preetching in 0.01 N KOH for 15 minutes plays an important role to inhibit the active sites, reject the chloride ions from the surface, form a silicon-rich surface, and increase the ability to distribute cerium uniformly on the surface. The oxide thickening step in distilled boiling water for 1 h following pickling is highly recommended. It is postulated that the mechanism of corrosion has been changed from simple adsorption on the surface treated directly with cerium chloride to absorption of cerium through the pores of the thick Al oxide layer [27].

Molybdate Solutions The effectiveness of the molybdate solutions was reported to be inhibited by the size and solubility of the oxide MoO_2 species. Sporadic pitting potentials of >100 mV led to the conclusion that the oxidizing power of MoO_4^{2-} was not as strong as the dichromate species ($Cr_2O_7^{2-}$). The effects of different inhibitors were tested, including molybdates, on the corrosion of aluminum using the power spectral density (PSD) of electrochemical noise measurement (ENM). PSD analysis proved to be an excellent method not only to determine the effectiveness of the inhibitor but also to provide some understanding of the mechanisms of the inhibitor. It was found that molybdates act as oxidizing inhibitors, and the main inhibiting effect was due to an adsorbed layer acting as a barrier to chloride ions [4].

Cerium, Manganese, Vanadium, and Molybdenum Pretreatments The influence of pretreatments of composite AA6061-T6–10% Al_2O_3 on the protection of aluminum by epoxy-treated and coated FLBZ 1074 fluoropolymer top coat was studied by Hamdy et al. [30]. The Ausimont coating system consists of solvent-based epoxy primer (80 μm) clear top coat of FLBZ 1074 (40 μm); FLBZ is the trademark of a fluoro-based top coat produced at Ausimont (Italy). The pretreatments were based on cerium, manganese, vanadium, and molybdenum chemical products [30].

The plastic materials, *Fluoropolymers*, are the most widely used when chemical resistance, stability at high and low temperatures, and good electrical properties are desired. Also, due to the very strong chemical bonding between carbon and fluorine, the fluoropolymers have some special properties, which make them very useful as coatings for equipment in the paint, varnish, and adhesives industries. Eight different types of pretreatmens were used as primer before applying the top coat of clear FLBZ 1074 [30].

The epoxy-treated fluoropolymer specimens showed a dramatic increase in corrosion rates under scratched conditions after less than 30 days of immersion in aerated 3.5% sodium chloride solution due to filiform corrosion. The new pretreatments showed outstanding durability using the salt spray test. No sign of corrosion was observed after 1140 h of exposure to the salt chamber while filiform corrosion took place for the epoxy-treated specimens after only 40 h of exposure. The mechanism of protection using the new treatments depends on formation of a highly protective oxide layer that is efficient to improve corrosion resistance and to maintain the adhesion performance within an acceptable range [30].

Silanes The organofunctional based silanes have been used recently as surface pretreatments for aluminum and other metals and have the capacity to be used in a variety of ways. Newer developments include the super-primer concept, which is formed of a composite consisting of silanes and organic resins, giving great flexibility in terms of composition and desired properties [31]. The silanes' action on corrosion has been enhanced by neutron scattering techniques. The main property is that they control the corrosion protection due to the hydrophobicity of the coating. If water reaches the surface, metallo-siloxane bonds are not resistant to water since silanes are not good in passivating treatment. They are thin, absorb water, and can be hydrolyzed at low temperatures. If an effective inhibitor package can be achieved that leaches out at a controllable rate, the superprimer can replace the chromate conversion coating and the chromate-containing primer [31].

The bis-silanes, containing OX as alkoxy groups and an amino or ureido as an organofunctional group, and especially silane mixtures have shown protection against both uniform and localized corrosion. Also, a modified silane system can contain an inhibitor with a defect-healing capability. Silanes reduce the corrosion rate of aluminum surface primarily because they replace hydrophilic hydroxyl groups with hydrophobic Al–O–Si groupings and with the action of the remainder of the cross-linked silane film. This is in addition to the improved adhesion to many paint systems because they are covalently bonded both to the metal oxide and to the paint polymer [32].

Huang et al. [33] proposed a novel, environment-protective, water-based metallic coating for aluminum alloys, which mainly contains metal flake, lithium silicate, and silane. The zinc flakes were 0.1–0.2 thick and 10 μm in diameter. The coating was sprayed on the finely polished aluminum surface, and then baked for 30 minutes at 200 °C. The film was actually formed by the lithium silicate and silane, which can form an interpenetrating polymer network (IPN) structure by forming Si–O–Si bonds in the larger molecules by means of cross-linking reaction of organosilicone and inorganic silicate. The lithium silicate water glass and silane have many advantages, such as good heat resistance and excellent water resistance. Adhesion and microhardness properties are excellent according to the standards.

In the salt spray test (ASTM B117-2003) (5% NaCl, pH 7 at 25 °C), the coating can endure for 250 h when the coating is 20 μm. The anticorrosion resistance increased with thickness and with the zinc flakes, which made an excellent filling material. Impedance studies found three kinds of electrochemical processes existing during the corrosion process. The zinc filler is attacked first, followed by the integrity of the film as related to the electric resistance and the capacitance of the coating. The last stage of attack is the meal itself, which should be controlled by the diffusion of the aggressive medium to the aluminum alloy–metal coating interface [33].

Cathodic Inhibitors Effective cathodic inhibitors of aluminum in neutral and alkaline solutions such as trivalent cerium acetate and an organic inhibitor such as tolyltriazole have shown promising performance on AA2024-T3 in 3.5% NaCl solution. To control water solubility of small particles of the chosen inhibitor, the inhibitor was encapsulated with a thin skin of an organic polymer by plasma polymerization. For instance, a plasma-polymerized skin made of a chosen organic compound, such as C_6F_{14}, was deposited on water-soluble salts and did not leach out in flowing water for at least 25 hours. A colorant can be integrated if selected from dyes that are water soluble but become insoluble upon curing [32].

14.4. ANODIZATION

Anodizing is an electrochemical method of converting aluminum into aluminum oxide on the surface of an aluminum piece. Anodized aluminum surfaces resist abrasion and anodizing improves the corrosion resistance of the alloy to weathering and other corrosive conditions. Anodizing is used in every area where aluminum items are produced. Actually, more aluminum is anodized than any other metal, such as titanium and magnesium. As an example, a film 5–7.6 μm thick is normally specified for bright automotive trim and 17–30 μm for architectural product finishes [34].

Anodizing is an electrolytic oxidation process in which the surface of the alloy becomes the anode and is converted into aluminum oxide—an amorphous, thick (~3–30 μm) layer, bound as tenaciously to the alloy as the natural oxide film (few nanometers thick). Anodic coatings, particularly those applied in a sulfuric acid electrolyte and suitably sealed, are highly effective in preventing discoloration or surface staining of the aluminum-based alloys mentioned previously. In addition, aluminum alloys that are used architecturally are more readily cleaned of atmospheric contaminants if they have been anodically coated. However, anodizing does not provide sufficient protection alone if the alloys themselves are unsuitable for the environment to which they are exposed. Anodic coatings are excellent paint bases [2].

Chromic acid anodization involves the electrochemical growth of an oxide layer where a thin, nonporous oxide layer is formed with a thicker porous layer on top of it. The thickness of the anodized layer is dependent on the applied voltage during film growth, but is usually 0.05–0.1 mil (1 mil = 25 μm) thick. Chromates are introduced in the final stage of anodization by sealing the porous layer with chromic acid (H_2CrO_4). Despite the superior corrosion protection offered by anodization, conversion coatings are preferentially used due to economic benefits [4].

The majority of anodizing processes are "soft". The current soft type of anodizing is done in chromic or sulfuric acid baths and is used in almost 90% of the production with oxide layers of 5–18 μm produced. Chromic acid anodizing produces a viable base for paint, but environmentally acceptable anodizing solutions are based on sulfuric and phosphoric acids. Hard coating is an excellent resistor and is used when wear resistance is requested and sometimes for corrosion resistance in many aircraft parts and food equipment. The hard type of anodizing (sulfuric acid bath alone or with some additives) produces a thickness on the order of 51 μm [17]. In hard coating, the part's surface is oxidized with oxygen as anode such that if the coat growth is on the order of 50 μm, 25 μm is below the original surface. It is important that penetration be on the same order as that of the exterior growth (Figure 14.1). Magoxid coat is a ceramic-like surface protection coating [34].

Anodic aluminum films are commonly composed of two layers [35]. The film possesses a special morphology corresponding to a hexagonal cell model structure. The porous-type anodic oxide film is shown in Figure 14.2 and can be obtained by anodizing in acid solution.

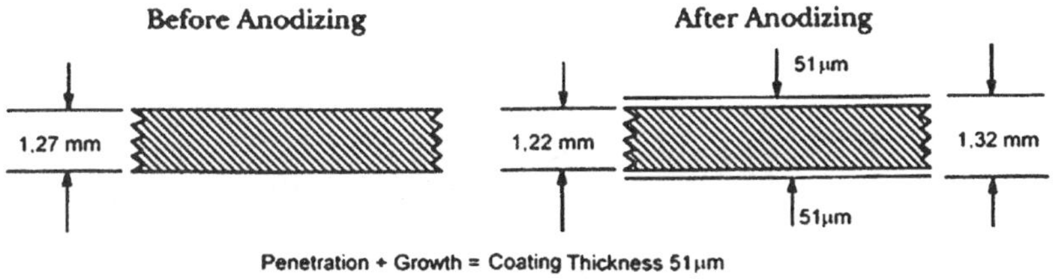

Figure 14.1 Schematic presentation of the effect of hard anodizing on thickness growth and penetration considered as a specific example [17].

The outer layer is characterized as a close packed array of columnar hexagonal cells that are perpendicular to the metal surface and each cell has a central pore that is normal to the substrate surface. At the base of the metal–metal oxide interface, there is the thin hemispherical barrier layer [35].

Morphology of the Film The morphology of the anodized oxide films can be observed by transmission electron spectroscopy and scanning electron microscopy. In TEM oxide films are removed from the metal substrate and thin slices of the vertical sections of oxide films are obtained by the ultrathin sectioning technique. Figure 14.3 shows the vertical section of a porous-type oxide film formed on aluminum in acidic medium [36].

The properties of the oxide layer obtained on aluminum in mixed electrolytes of oxalic acid–sulfuric acid are optimized using experimental design. For this purpose, a four-variable Doehlert design (bath temperature, anodic current density, sulfuric acid, and oxalic acid concentrations) was achieved. In order to maximize the growth rate and the microhardness of the anodic oxide layer and to minimize the effect of this speed on chemical and abrasion resistances of the anodized surface, a multicriteria optimization using a desirability function was conducted. Dissolution rate of the oxide in phosphochromic acid solution (ASTM B680-80) was used to express its chemical resistance. Under the determined optimal anodizing conditions ($C_{ox} = 12.6\,g{\cdot}L^{-1}$, 10 °C, 2.6 $A{\cdot}dm^{-2}$, C_{sul} $183.6\,g{\cdot}L^{-1}$), the estimated response values were $0.73\,\mu m \cdot min^{-1}$, $4.38\,g{\cdot}m^{-2} \cdot min^{-1}$, 481 HV, and $53.3\,g{\cdot}m^{-2}$ for growth rate, dissolution rate, microhardness, and weight loss after abrasion, respectively. The higher abrasion and chemical resistances of the optimum anodic layer can be correlated with its morphology revealed by SEM observations (Figure 14.4). The size of the pores (black spots on the images) is measured using a line

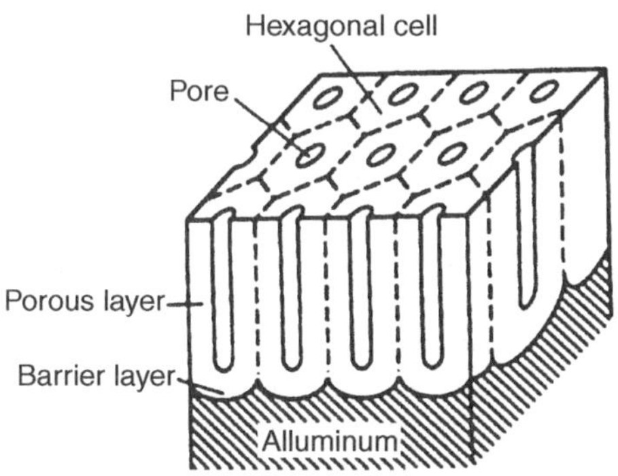

Figure 14.2 Structure of porous-type anodic oxide film formed on aluminum in acid solutions [35].

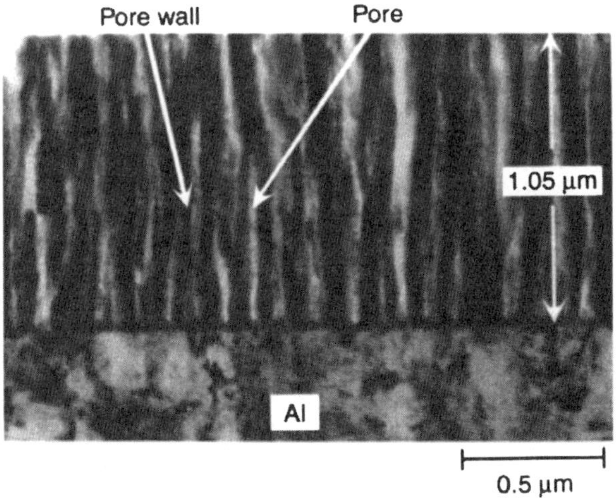

Figure 14.3 A vertical section of the anodized aluminum surface in 0.16 M oxalic acid by TEM showing the pore and its wall after 1 h of immersion at the open circuit potential in the same anodizing solution [35, 36].

profile. Their values range between 7 and 10 nm for the optimized layer. The less porous structure of the optimal anodic aluminum layer is confirmed by higher values of the coating ratios ($R = 1.71$) as compared to that for the nonoptimized samples (1.26 and 1.50) [37].

Coloring Colorings of anodic oxides can be achieved through three options (Figure 14.5): integral coloring, dyeing, and electrolytic coloring. Organic acids, such as oxalic, maleic, or sulfamic acids, can be used. The created anodic color (brown, gray, or black generally) is highly resistant to alteration. Many organic dyestuffs such as Alizaline Blue or Red-S give flashy colors such as gold, blue, or green. Inorganic dyeing through the process of precipitation of low-solubility salts such as PbS in the pores is much more resistant to heat and light but the color range is more limited than that of the organic ones. Electrolytic coloring consists of metal deposition at the bottom of the pore (Ni, Co, Sn, and Cu mainly to color from bronze to maroon to black). Low energy consumption, application for all alloys, and the light-fastness of the finishes are the reasons for using electrolytic coloring instead of integral coloring [35].

Electrolytic Coloring One of the currently used coloring methods of anodized aluminum is electrolytic coloring. During this process, aluminum is first anodized in sulfuric acid

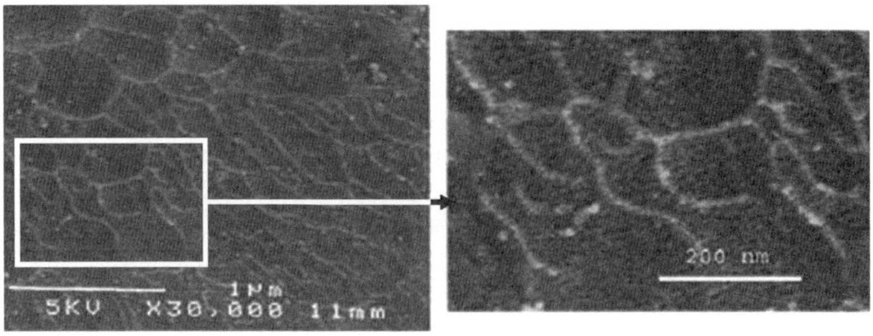

Figure 14.4 SEM top view images of anodic layers elaborated under optimal conditions [37].

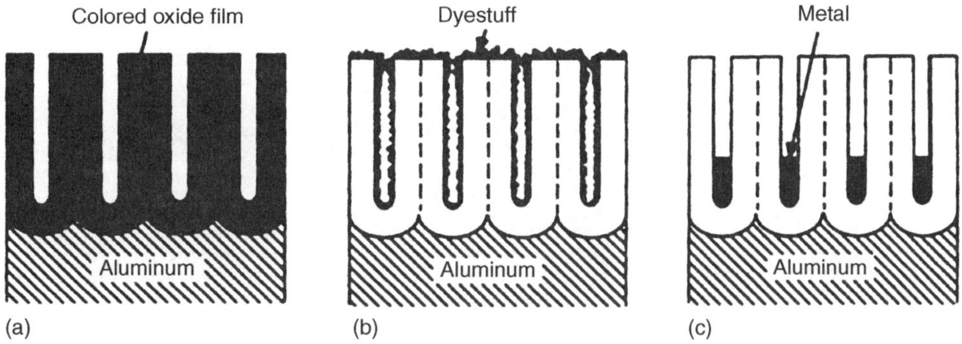

Figure 14.5 Coloring of porous-type anodic oxide films on aluminum by (a) integral coloring, (b) dyeing, and (c) electrolytic coloring [35].

solution, followed by an alternating current electrolytic deposition of a metal (tin, nickel, cobalt, etc.) at the base of the pores of the anodic coating. A very popular commercial bath is the $SnSO_4/H_2SO_4$ one. The ac electrolytic coloring process at 15 V_{rms} in acidic tin sulfate solutions for specimens of AA5083 and AA6111 unheated and heat treated was investigated and compared to pure aluminum. Specimens of AA5083 alloy and AA6111 unheated and heat-treated alloys were anodized in sulfuric acid baths, were electrolytically colored at 15 V_{rms} in acidic tin sulfate solutions, and were compared with those of pure aluminum [38].

Under standard electrolytic coloring conditions, the current efficiency for tin deposition was low for all examined materials. This bath provides good throwing power, has fewer tendencies to form complex compounds, is not sensitive to pH variations or bath contamination, and is relatively easy to operate. However, specimens are susceptible to atmospheric oxidation [38].

The current efficiency for tin deposition during electrolytic coloring at standard conditions is much higher for pure aluminum than for alloys. The anodizing voltage seems to influence the amount of tin deposited and the current passed and the effect is less for alloys than for pure aluminum, indicating that, for pure aluminum, the anodizing voltage affects to a greater extent the porosity of the film [38].

The alloy type affects the rate of tin deposition but certain qualitative characteristics and the stages of the electrolytic coloring process are similar for the alloys and pure aluminum. The temper of AA6111 did not affect the electrolytic coloring process, although it influenced the anodizing process. For AA5083, the increase of conductance of the oxide film resulted in an increase of hydrogen evolution with no improvement in tin deposition efficiency as compared with that of the pure aluminum [38].

Sealing The anodic oxidation of aluminum is followed immediately by sealing. The anodized coating can be sealed in hot water and a complete sealing corresponds to 15% weight gain. However, for better adhesion of primers, only a partial seal is recommended, on the order of 5% weight gain due to the adsorbed water. After anodizing, the specimen should be primed as soon as possible but before the freshly formed aluminum oxide adsorbs water and partially seals [39]. Manganese can be employed to alter the response in chemical finishing and anodizing [34].

Sealing of anodic alumina enhances the corrosion resistance of prepared coatings and increases the UV-light resistance of dyes in anodic coatings. Sealing is usually performed by dipping in boiling water. The anhydrous aluminum oxide is transferred to a hydrous one

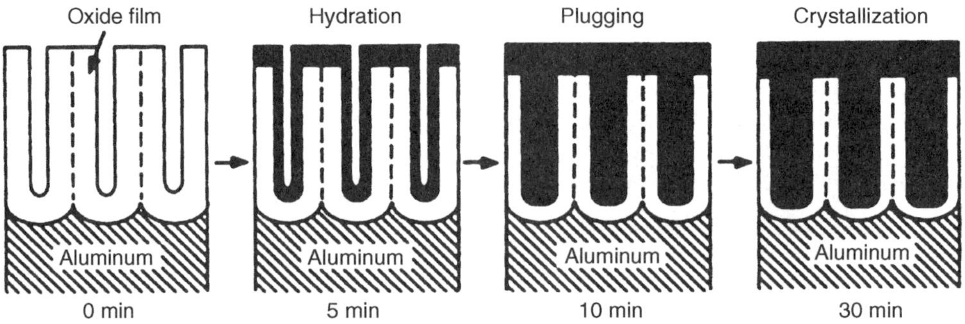

Figure 14.6 Process of pore sealing with hydroxide during dipping in boiling pure water [35].

gaining 1.5–2 molecules of H_2O for every molecule of Al_2O_3. The volume expansion by the formation of pseudoboehmite causes the pores to be sealed. Sealing for 10 minutes (Figure 14.6) is sufficient to close the pores completely and form a high crystallized form at the outermost part of the oxide film. Pressurized steam allows quick sealing, however, cold sealing in the presence of Ni^{2+} and F^- ions at ambient temperatures is currently used to save energy. Effectively, aluminum–nickel–fluoro complexes are deposited in the pores, blocking them [35].

Cerium Nitrate as Sealer Sealing is conducted in a solution composed of 3 g/L of Ce $(NO)_3$, 0.3 g/L H_2O_2, and 0.5 g/L H_3BO_3 having pH = 5, for 2 hours at 30 °C and then the specimen is rinsed with distilled water and dried with air. The thickness of the anodized coating increased approximately 3–5 μm after sealing as examined by an eddy-current thickness indicator; the color of the coating is yellow.

In NaCl solution, the Ce sealing of anodized aluminum alloy 2024 remained passive at the potential range at the open circuit potential in spite of a shift in the active direction of about 530 mV. The sealed anodized film on LY12 alloy was composed of outer and inner layers. By immersing the sample in 3.5 wt % NaCl solution for 6 days, the outer layer of the cerium conversion coat begins to lose its anticorrosive property. The inner sealed cerium layer of the anodized film is not corroded until 60 days of immersion. Thus the inner cerium sealed anodized layer plays a leading role in the corrosion protection of LY12 alloy [40].

Sol-Gel as Sealer Traditional sealing processes, such as hot water sealing, steam sealing, and cold nickel sealing, are well established while other sealing options are being investigated. Among the most promising new sealing methods are those focused on antismutting agents and electrochemical sealing based on the following requirements: corrosion resistance, abrasion resistance, and hardness without posing environmental problems. A promising approach for sealing anodic coatings is to form a protective layer with a sol-gel method. The glass-type layer is especially good for corrosion resistance of anodically oxidized aluminum [41].

The principle of the sol-gel process is to first prepare a sol (i.e., liquid colloidal dispersion) by hydrolysis of organometallic compounds and then have the coating gel (solidify) to form the hard coating. The sintering process finishes the production of the hard seal coat. The characteristics of the developed coating are a function of both the starting material and the conditions of the hydrolysis reaction. In this process, the coating is created using a colloidal method and a dip technique. The colloidal method of a sol-gel coating preparation means the hydrolysis of an organometallic compound (aluminum butoxide), which is obtained with an excess of stoichiometric water. A 20 minute anodic oxidation to reach a thickness of 10 μm is used for the following sol-gel type of sealing: the sealing is conducted at an operating temperature of 270 °C for 15 min with controlled heating and cooling of 10 °C/min. The samples are dipped into Al_2O_3 sol before the sealing process [41].

Corrosion resistance against atmospheric agents is comparable with hydrothermal sealing and even higher. The thickness of the sol-gel alumina coating, in combination with the other material component thicknesses, can influence the crack formation and the corrosion resistance of the material. A mixture of transition aluminas was identified for sol-gel type of coating, especially the transition δ-Al_2O_3. The disadvantage of this process is its high price and decreased abrasion resistance and hardness. In spite of this, for special purposes, this type of sealing is acceptable [41].

Anodization and Plasma Coatings Plasma coatings have two invaluable properties for corrosion protection of metals—a highly cross-linked matrix and excellent adhesion to the metal substrate. Keronite's plasma electrolytic oxidation process transforms the surface of aluminum alloys into a complex ceramic matrix by passing a pulsed, bipolar electrical current in a specific wave formation through a bath of low-concentration aqueous solution. A plasma discharge is formed on the surface of the substrate, transforming it into a hard, dense, ceramic oxide (mainly alumina), without subjecting the substrate itself to damaging thermal exposure. The process forms an ultrahard ceramic layer—from 800 to 2000 HV (Vickers Hardness Test) depending on the alloy and the coating's thickness [42].

The layer is attached to the substrate by a strong molecular bond, ensuring adhesion. The fused ceramic layers closest to the surface provide protection against corrosion and wear. The outer surfaces of the layer are porous and lend themselves well to the application of scratch-resistant, decorative top coats such as paints and lacquers, and can form composite coatings with PTFE (polytetrafluoroethylene, or Teflon), adhesives, or metals. The layer is typically between 10 and 150 μm thick and grows at a rate of around 1 μm minute—partly above the surface and partly below [42].

As an immersion process, it can be used to treat the inner surfaces of complex shapes. The ceramic layer can be adjusted for optimal performance in the chosen application. The process produces a completely uniform layer, even in the case of complex shapes or internal surfaces. The process is compatible with all known aluminum alloys, even those with a high copper content that cannot be treated using hard anodizing [4, 42]. The ceramic can withstand over 2000 hours in salt fog when sealed—a key test for corrosion resistance.

Keronite can be used in coating a range of exterior automotive parts, such as roof rails, door handles, door frames, and body panels; interior parts such as seat frames, instrument panel beams and supports, airbag retainers, and mirror brackets; and engine components such as piston crown and ring grooves and clutch rings. When used in engine components, the coating is designed to help improve powertrain performance and efficiency through reduced piston groove wear, better tolerance control, lower friction, and higher combustion chamber temperatures. The coating process reduces the temperature of aluminum pistons by approximately 85 °F [42].

Low-temperature cationic plasma deposition is currently used to create ultrathin hydrophobic barrier coatings on metals. Low-temperature cationic plasma deposition of inorganic or silicon monomers has displayed the best performance for corrosion protection with respect to steel substrates immersed in simulated seawater [4].

Modified Anodizing as Compared to Anodizing and Hard Anodizing Films Aluminum oxide coatings were deposited on Al–Si alloy substrates to produce hard and corrosion protective films using three different techniques: hard anodizing, anodizing, and modified anodizing. Rectangular coupons (25 mm × 25 mm × 5 mm) of a cast Al–Si alloy were used as the substrate for anodic coating deposition. The composition of this alloy in wt was 15.6% Si, 1.3% Fe, 0.12% other, and Al the balance. The substrates were ground and polished to a surface roughness of Ra = 0.1 μm before washing in water and then drying in air.

During the anodizing process, the substrates (exposed area 6.25 cm^2) were anodized at a constant current density of 0.016 A/cm^2 in 17% H_2SO_4 at 25 ± 1°C for 15 min and the voltage reached 20 V. The conditions remained the same except the temperature was controlled at 0–4 °C for the hard anodizing (HA) process with a final voltage of 25 V [43]. For the modified anodizing (MA), the substrates (exposed area 17.5 cm^2) were anodized at a constant current density of 0.012 A/cm^2 in 12% $NaHCO_3$ at 25 ± 1°C until the voltage increased to 340 V and then switched to a constant voltage (340 V) control mode. The whole process time was 30 min [43].

Oxide coatings were successfully deposited on a cast Al–Si alloy by hard anodizing, modified anodizing, and anodizing techniques. Small cracks and pores near second-phase particles caused by the internal stress are the result of the different film growth rates for the different alloy phases. The different coating features resulted in some differences between the coating hardness and surface hardness. Potentiodynamic polarization tests were conducted to assess the corrosion resistance of the coatings. A microhardness tester was used to measure the coatings' hardness. Scanning electron microscopy (SEM) and energy dispersive X-ray (EDX) analysis were used to investigate the coating microstructure and chemical composition both before and after corrosion [43].

It was found that the modified anodizing, an environmentally friendly coating method, could produce a hard oxide coating with good corrosion protection for the Al–Si alloy. All coatings provided effective protection for corrosion resistance, while the modified anodizing would be beneficial with respect to development of an environmental friendly process [43].

Pitting Corrosion of Anodized Coatings Moutarlier et al. [44] investigated the pitting initiated in different anodic films on AA2024 alloy. A polarization test in NaCl solution was used to initiate pitting corrosion in anodic layers produced in chromic acid, sulfuric acid, and sulfuric acid containing molybdate species. The mixed electrolyte containing sulfuric acid and molybdate species was studied as a substitute for the chromic acid electrolyte. Corrosion resistance of anodic films formed in sulfuric–molybdate was better when compared to films formed in sulfuric acid, but chromic anodic layers gave the best corrosion performance [44].

The Pitting Potential After anodizing and subsequent sealing, the pitting resistance of aluminum alloys is improved remarkably. The pitting behavior of anodized AA2024 in neutral NaCl solution was investigated using electrochemical methods and SEM [45]. Three stages were observed during potentiostatic polarization of anodized AA2024 in NaCl solution. In the first stage, current decreases with time, following the relation $\log i = -n \log t$, where parameter n indicates the passivation tendency of the alloy. The stage corresponds to the induction time for pitting and increased chloride concentration will shorten the stage. After the induction time, pitting occurs and the current begins to increase continuously. In the third stage, the current reaches a stable value, and the growth of pits is controlled by an ohmic drop [45]. The relationship between pitting potential of anodic film on AA2024 and chloride concentration in the solution follows the expression $E_P = A - B\log[a_{Cl^-}]$.

14.5. ORGANIC FINISHING

Thermoplastic coatings and *converted coatings*, applied during or after processing, include principally three types of paints: epoxy, polyurethane, and moisture coatings.

14.5.1. Thermoplastic Coatings or Liquors

The resin is in its final form and the coating dries solely by solvent evaporation. The film-forming process is merely the evaporation of the solvent. Examples are vinyls, acrylics, and chlorinated rubbers. If coatings are applied under high-humidity conditions, blushing of the coat occurs and the coat turns white. The blushed surface is porous with poorer resistance characteristics.

14.5.2. Converted Coating During or After Application

All such coatings undergo a chemical or physical change in the process of film formation before, during, and/or after application and they are different from the thermoplastic coatings in that they dry or react in a whole series of steps. There are some conversion coatings that require baking or heating, which are not practical when coatings are to be applied to large existing structures or equipment. The main types of converted coatings are the following [46].

Oil Paints These are very familiar paints that have a drying oil and a resinous varnish or resin as the binder. These usually dry more slowly than the thermoplastic ones, and the various drying stages are considerably more complex. These stages are solvent evaporation, oxidation, thickening, or polymerization. Gelation occurs when the polymers reach a size and concentration that form a continuous network. The paint appears as dry but effectively it contains a considerable quantity of liquid material and may be somewhat soft. The remaining film continues to cure or dry and becomes hard; this can be accelerated by a sunlight or heat mixture. When the films reach their ultimate hardness, they become more porous and loose resistance to moisture and chemicals [46].

Epoxy Coatings Epoxy coatings are created by a conversion process or cross-linking at ambient temperatures. The epoxy resin is mixed with an amine just prior to application. Its drying process consists of solvent evaporation followed by a chemical reaction of the amine and the epoxy resin resulting in cross-linkage. The amine becomes a part of the new polymer since it does not act as a true catalyst. Since this process is temperature-sensitive, and can occur in the absence of air where the cross-linkage takes place, the coating is called *thermoset* and it becomes neither soluble in its original solvents nor as sensitive to softening by heat as expected. There is another conversion reaction that occurs when an epoxy resin reacts with a second resin (e.g., a polyamide resin). In this case, the two resins (the epoxy and the polyamide) react and cross-link to form a solid resin film. The film is more resilient and elastic than the films formed using the amine epoxy reaction [46].

Polyurethane Coatings Polyurethanes (PUs) form a film through the chemical reaction of acrylic or polyester-modified urethane-based components with isocyanate reactive converter components. It is a conversion reaction and cross-links into a somewhat chemically resistant film. The main function of PU coatings is to improve the finished appearance. This process is not the same as epoxy; it is more humidity and temperature sensitive during the curing process. Excess humidity at this point can lead to loss of gloss, and the formation of a cheesy, nonuniform film, or wrinkling [46].

Moisture Coatings These are characterized by the fact that water from the atmosphere converts the film from a liquid to a solid. This is one of the processes by which moisture-cured PU coatings form. In this case, moisture from the air and/or substrate reacts with a PU resin during the initial evaporation stage, cross-linking it and increasing the molecular size until it becomes solid. The solvent-borne inorganic zinc (IOZ) coatings also require moisture from the air, whereas the waterborne IOZ coatings require carbon dioxide to change the silicate molecule (i.e., sodium, potassium, or ethyl silicate) into a continuous coating by reaction with the zinc pigment [46].

Aluminum is an excellent substrate for organic coatings if it is well cleaned and appropriately prepared. It has long been accepted that the durability of coatings on aluminum is determined first by the preparation of the metal base surface. A suitable preparation of the surface usually starts with degreasing, followed by eliminating existing oxides, forming a base layer, and applying a primer. Since the corrosion resistance of the aluminum base is very good, the resistance of the organic coatings is remarkable even after many years of exposure [47].

For indoor applications, the coating can be applied directly to a clean surface. However, a suitable primer coat, such as wash primer or a zinc chromate primer, usually improves the performance of the finish. For applications involving outdoor exposure, a surface treatment such as anodizing or chemical conversion coating is required prior to the application of a primer and a finish top coat, such as an epoxy urethane or polyurethane. Some new one-step, self-priming polyurethane top coats are also available, as are low volatile organic compound (VOC), high-performance primers such as epoxy polyamide [5]. Chromium-free conversion coatings are emerging to avoid the environmental problems caused by using Cr^{6+} compounds. Zinc phosphating provides a good base for organic compounds and has yielded very good results in the automobile industry with the cathodic electrodip coating [34].

The common antifouling paints for steels to prevent growth of algae, barnacles, and other sea organisms are not suited for aluminum. They may contain leachable heavy metals such as lead, arsenic, and copper that can plate on the aluminum surface and initiate galvanic corrosion, and so specific antifouling paints should be chosen for aluminum. For certain applications, the top coat can be replaced by adhesively bonded applied films. These flexible films provide a durable, weather-resistant finish when applied over standard, corrosion-resistant primers [5].

Clear protective coatings (lacquers) are used to provide protection while retaining glossy metallic appearance. All beverage and food containers are coated for prolonged shelf life and to prevent contamination of the food product. The absence of a hole or even a fine pore in the coating is highly recommended. Clear coatings are used also in the protection of anodized aluminum surfaces on commercial and residential buildings and ease cleaning procedures. Material in coil form can be coated very economically. The strip is first pretreated, rinsed, and dried, and then the paint is applied and baked in one continuous process. As a rule, a primer of about 5 μm is applied followed by a top coat of about 20 μm. Notable examples of organic coatings for aluminum are nonstick coatings on cooking utensils [5, 34], for example, polytetrafluorethylene (PTFE, or Teflon), and pressure-sensitive tapes and/or strippable plastic coatings for temporary protection of aluminum sheets or extrusions used in buildings [5].

Maximum protection depends on inspection and maintenance. Painted jet airliners are stripped of their coating and completely repainted as needed, usually for appearance purposes. Automobiles are repainted as needed usually for appearance or decoration. Dents and scratches in residential siding are rarely repaired, whereas rain-carrying systems such as gutters and down spouts are more frequently replaced than repaired [5].

14.5.3. Coatings Containing Metals More Active than Aluminum

Magnesium use as a pigment presents a possible alternative for the sacrificial protection of aluminum alloys. A Mg-based primer for the protection of aluminum structures, based on this concept, has been developed and tested [48]. An excellent performance of aluminum panels sprayed with this Mg-rich primer in Prohesion testing has been observed. This new class of metal-rich primer can be formulated for the protection of aluminum alloys in analogy to the Zn-rich primer devised for the protection of steel. The Mg-rich primer was made using a stabilized Mg particulate, 30–40 μm in average size, manufactured by Non Ferrum-Metallpulver GmbH, Salzburg, Austria. This particulate consists of Mg covered with a thin layer of MgO, intended to control the reactivity of magnesium and thus prevent further oxidation under dry conditions. Dispersion was made using a silane-modified multilayer/IPN polymer matrix [48].

To provide sacrificial protection, the Mg metal particles in the primer have to be in electrical contact with the substrate and also with each other. The Mg-rich primer was formulated at 50% PVC, approximately near the critical pigment volume concentration (CPVC) of the coating. The electrochemical behavior of Mg-rich primer on AA2024 and AA7075 has been studied via electrochemical impedance spectroscopy (EIS), open circuit potential (OCP), and potentiodynamic polarization. Electrochemical technique studies were complemented by scanning electron microscopy (SEM). Most of the electrochemical tests were made in 0.1 wt% NaCl distilled water. One of the experiments was conducted in Dilute Harrison's Solution, which emulates acid rain and consists of 0.35 wt % $(NH_4)_2SO_4$ and 0.05 wt % NaCl in distilled water [49].

Results showed that the Mg-rich primer provides sacrificial protection to the Al substrate by a two-stage mechanism. In a first stage, corrosion of aluminum is prevented by cathodic polarization, whereas at a later stage the precipitation of a porous barrier layer of magnesium oxide was observed. Magnesium had a very negative potential, compared to that of the aluminum alloy, and underwent fast corrosion, with visible bubbling on the surface. The high rate of the cathodic reaction was assessed by measuring the pH, which was approximately 11, on the Mg surface [49].

Corrosion of the Al alloys in 0.1% NaCl resulted in the formation of small pits. The solution pH was approximately 5. At this pH, Al becomes passive, as indicated in the Pourbaix diagram. The presence of inclusions and precipitates, however, can induce instability of the passive film and lead to localized corrosion, and so can the presence of aggressive ions, such as chlorides. The OCP measurements, however, have shown that the potential of the galvanic couple was lower than that of the bare alloys, and this has to be a result of cathodic protection.

The impedance of magnesium after 1 h of immersion was smaller than that of the bare aluminum substrate by approximately one order of magnitude—a conclusion that is in reasonably good agreement with the results from potentiodynamic polarization. When the Mg-rich coating was applied, the total impedance of the system increased significantly, again in good agreement with the potentiodynamic observations, showing that the sacrificial protection was manifested at a comparatively low Mg oxidation rate [49].

In order to assess the efficiency and protection mechanism at defective areas of the primer, a large scribe was induced with a knife on a coated sample, exposing an area of approximately 1 cm^2 prior to immersion in Dilute Harrison's Solution for 10 days of immersion. Inspection of the scribed area at the end of the experiment by SEM and EDAX revealed the aluminum surface was covered by a precipitate of magnesium oxides. This precipitate was porous and had several flaws; however, this did not correspond to sites of pitting corrosion of aluminum [49].

Unlike zinc, whose hydroxides precipitate at neutral pH, magnesium has a vast pH range over which it remains active. This range includes not only the regions of stability of $Al_3{}^+$ and Al_2O_3, but also overlaps the region of alkaline corrosion of aluminum, up to pH 11. Thus Mg becomes oxidized at a high rate, which could lead to exhaustion of the coating after a relatively short exposure period. Mixing magnesium with polymer significantly decreased the corrosion rate of the metal due to the barrier effect of the polymer and this can extend the lifetime of the coating [49].

In the case of zinc-rich primers for sacrificial protection, the zinc corrosion products precipitate inside the coating, around the zinc particles that originated them, blocking the pores of the coating and therefore increasing its barrier resistance. Magnesium acts in a somewhat different way. Because at the near-neutral pH of the solution the magnesium ions are soluble, they actually diffuse out of the primer layer. Furthermore, because of the high rate of electrochemical reactions, the pH can become quite alkaline at the cathodic sites, particularly if there is a relatively small defect in the primer. When these ions reach the cathodic areas, they will then precipitate as $Mg(OH)_2$ [49].

The Mg-rich coating used in this work has the capability of protecting alloys AA2024 and AA7075 against corrosion. The effect of magnesium starts with the polarization of aluminum, shifting its potential below the pitting corrosion potential. The consequence of this polarization can be either the prevention of pit nucleation at the exposed aluminum areas, or the inhibition of pit growth for the nucleated pits. During this stage, any defects on the surface will become cathodic, whereas the magnesium particles will be anodic. At the cathodic areas, reduction of hydrogen and possibly dissolved oxygen increases the pH above the threshold for the precipitation of magnesium oxide and the formation of magnesium hydroxide. This precipitation leads to the formation of a porous layer that further inhibits corrosion by a barrier mechanism. The typically high dissolution rate of magnesium is significantly decreased by its incorporation in the polymer [49].

14.5.4. Electrodeposited Coatings

An environmentally friendly system that has been proved effective by the automotive industry is electrodeposition of organic resins (e-coat). The unique advantage of electrodeposition is that a thick (up to several millimeters) coating can easily be deposited on a conducting substrate with good control of the thickness. The deposition process can be modified to undergo either cathodic or anodic deposition, depending on the resin functionality. Cathodic deposition is predominantly used today for corrosion applications because of some inherent disadvantages of the anodic process. Several types of resin have been successfully used for e-coat binders. These binders include acrylics, alkyds, epoxies, polyurethanes, polyamides, and polyesters. Electrochemical evaluation of e-coat systems has been used to demonstrate its excellent barrier properties. Twite and Bierwagen [4] presented data showing the superior barrier protection behavior of polyurethane/blocked isocyanate e-coats on aluminum alloys with and without a chromate conversion coat pretreatment.

14.6. CORROSION TESTING OF COATED METAL

14.6.1. Electrochemical Testing of Coatings

Corrosion monitoring of epoxy-coated aluminum 2024-T3 was carried out by electrochemical impedance spectroscopy methods and electrochemical noise measurements. The epoxy was electrodeposited on the surface of panels with one of following treatments: actone cleaned, alkaline cleaned, or plasma deposition of polytrimethylsilane. The epoxy coating was modified by one of the applied voltages: 100, 150, or 200 V (current-controlled coating). Six selected combinations of the surface treatment and the applied voltage were considered. The corrosive solution was 0.35 wt % ammonium sulfate and 0.05 wt % sodium chloride [50].

EIS and ENM (electrochemical noise measurement) data were collected over a 70 day immersion period on days 1, 7, 14, 60, and 70. For each sample pair, the impedance modulus at a frequency of 1 Hz was used. For ENM, time records of ZRA current and sample voltage were collected at a rate of 0.5 Hz for 128 s to provide the standard deviation of voltage and current. It was possible to conduct repeated measurements for EN over 5 h.

Linear regression analysis was used as the statistical treatment of the collected data to detect the individual contributions relating to the analysis technique. It has been concluded that EIS data can be used to monitor the protective quality of the coating as a function of surface treatment or applied voltage, while ENM data were too noisy. It was also found that a combination of 200 V with alkaline-cleaned aluminum would produce the highest impedance values [50]. Furthermore, there is considerable interest in potentiodynamic polarization techniques to rapidly assess the durability of coated aluminum surfaces that have been painted or given various polymeric or anodic surface treatments.

The use of electrochemical noise measurements to evaluate the corrosion protection as a probe for coated systems has already been studied; however, reproducibility of the results has been examined by Bierwagen et al. [51]. Monitoring is continuous and data can be gathered over a period of days, weeks, months, or even years. The backs and sides of the specimens were coated with a Colophony rosin/beeswax mixture, which is an inert high-resistance protective coating. Although this study considered grit-blasted painted steel with three-coat systems, the results can be extrapolated to coated aluminum or magnesium alloys with certain precautions. Each panel had 50 cm^2 of paint left exposed to better assess the influence of agitation and temperature. The solution was mainly NaCl (3%) or seawater. The Gaussian value determines the standard deviation based on the mean of all points of the sample, while the robust method bases the standard deviation on the median of the sample. The reproducibility of the ENMs was good and variations inherent in coating systems manifest themselves as variations in electrochemical properties in nominally identical samples. Different local variations in thickness leading to local structural variation are the most likely reason for the variability in the ENMs and corrosion results [51].

14.6.2. Conventional Testing

Evaluation of the protective ability of coatings is generally made in salt spray cabinets, such as those covered in ASTM B117 and G85 or in the 3.5% NaCl alternate immersion test (G44). Recent investigations, primarily on steel, have established the desirability of including ultraviolet (UV) light as part of the cyclic exposure, since UV light has a degrading effect on paints and other organic coatings. Various cabinet tests to provide UV light are under development, along with a corresponding Annex 5 to ASTM G85. Corrosion resistance of anodized aluminum is evaluated by conventional corrosion tests, such as CASS tests, salt spray tests, and other exposure tests. Galvanic corrosion tests are important because the corrosion rate of aluminum in the active state in contact with other less active or more noble metals can be more dangerous than that of aluminum alone [35] (see Chapter 15).

14.6.3. Corrosion Fatigue of Thermal Spraying of Aluminum as a Coating

We will examine the fatigue behavior of Al alloy 7075-T651 with ductile aluminum applied by a thermal spray process. The coating, deposited by four different commercial arc spray devices (guns), has been characterized. A thermal spray process—arc spraying—for ductile metallic materials was selected to apply 200 μm (nominal) thick aluminum coatings onto Al alloy 7075-T651 substrates. The aluminum coating was produced from four different commercial arc spray guns. Coated specimens, as well as polished and shot–peened specimens, were evaluated under fully reversed uniaxial loading ($R = -1$) at a constant amplitude of ±225 MPa in accordance with ASTM 466-82. A frequency of 20 Hz was selected to avoid potential frequency-induced heating with a sinusoidal loading wave form applied via a computer-controlled MTS servohydraulic load frame. The different coatings were evaluated also in terms of their microstructure [52].

While the shot peening pretreatment was observed to increase the fatigue resistance of polished specimens, application of the coatings subsequently reduced fatigue life to below that of the original polished coupons. Changes in the residual stress state of the shot-peened surface were identified as the most likely source of these reductions, even though no microstructural changes in the substrate were perceptible. Variations in fatigue life were also observed between the coatings resulting from the four spray guns. The roles of surface roughness and coating delamination in producing these decreases were investigated and stress concentrations resulting from coating delamination were identified as the primary detrimental factor affecting fatigue resistance. The effect of thermal spray coatings on the fatigue behavior of various substrate materials has attracted increased attention in recent years. With the absence of microgaps at the coating–substrate interface, even after the fatigue tests and with the lowest coating roughness, the equipment provided the best fatigue behavior for the Al-coated 7075-T651 Al alloy in this study [52].

14.6.4. Environmentally Assisted Cracking of Metallic Sprayed Coatings

Aluminum alloys play an important role in the manufacturing of air and ground vehicles. However, corrosion damage is often found on structural components subjected to fatigue loading. The alloying provides a large benefit, in terms of the high specific tensile strength (ultimate strength/density ratio), to the aluminum alloy but causes the aluminum alloy to be more sensitive to localized corrosion due to complex microscopic heterogeneities such as secondary phases. Arsenault and Ghali [54] studied thermal spray coating using arc spraying to evaluate the protection against environmentally assisted cracking (EAC) and localized corrosion on aircraft structural Al alloy 7075-T651. EAC and pitting corrosion at the coating–substrate interface are a challenge for thermal spray protective coatings on aluminum alloys under cyclic load and immersion. In this study, EAC was initiated on polished and shot-peened Al alloy 7075-T651 through a four-point bending test under cycling fluctuation load in 3.5 wt % NaCl solution kept at 25 °C in open air (ASTM G39-99) [53].

The applied load was kept under the yield strength (YS) (503 MPa) of 7075-T6 alloy and oscillated between 24% and 40% YS in tension ($R = 0.6$) at a frequency of 0.1 Hz. The selected stress level was sufficiently low to avoid premature coating damage and to allow a single failure mode. The failure mode validates the EAC mechanism such as SCC or corrosion fatigue. This approach has the benefit to initiate intergranular cracking in the aluminum alloy with a fast response, while maintaining the substrate material under elastic deformation. The samples used for EAC evaluation were rectangular in shape, having the dimension of 60.0 mm × 20.0 mm × 4.5 mm and machined in the short transverse direction to evaluate the coating protection performance in the most susceptible EAC direction of the 7075 aluminum alloy [54].

This study underlines the impact of coating material on the interface properties, in terms of interface quality (microgap) and adhesive strength. The Al coating shows, for five different surface properties, either lower microgap or higher bond strength than Al–5Mg. Moreover, the surface preparation on Al alloy substrate requires one to remove material mechanically, such as by grit blasting or by deoxidation, in order (1) to provide a low defect interface in order to avoid coating spalling under cyclic load test and (2) to avoid localized corrosion underneath the coating especially when under immersion applications. Both thermal spray anodic Al and Al–5Mg coatings did confer EAC protection to the Al alloy 7075-T651. However, the Al coating conferred the best protection against localized corrosion for the Al alloy 7075-T651 substrate [54].

REFERENCES

1. B. W. Lifka, in *Corrosion Testing and Standards: Application and Interpretation*, edited by R. Baboian. American Society for Testing and Materials, Philadelphia, PA, 1995, pp. 447–457.
2. E. Ghali, in *Uhlig's Corrosion Handbook*, 2nd edition, edited by R. W. Revie. Wiley, Hoboken, NJ, 2000, pp. 677–715.
3. ASM Metals Handbook Committee, in *ASM Handbook, Volume 3, Corrosion*, edited by L. J. Korb, D. L. Olson, and J. R. Davis. ASM International, Materials Park, OH, 1987, pp. 93–196, 207–220, 231–233, 303–310, 596.
4. R. L. Twite and G. P. Bierwagen, *Progress in Organic Coatings* **33**, 91–100 (1998).
5. ASM International Handbook Committee, in *Corrosion of Aluminum and Aluminum Alloys*, edited by J. R. Davis. ASM International, Materials Park, OH, 1999, pp. 63–74, 135–160.

6. J. Simon, E. Zakel, and H. Reichl, *Metal Finishing* **October**, 23–26 (1990).
7. B. Arsenault, P. Gougeon, M. Verdier, and D. L. Duquesnay, *Canadian Metallurgical Quarterly* **45**, 49–57 (2006).
8. B. W. Lifka, in *Corrosion Tests and Standards, Application and Interpretation*, 2nd edition, edited by R. Baboian. ASM International, Materials Park, OH, 2005, pp. 547–557.
9. ASM International Handbook Committee, in *Corrosion—Understanding the Basics*, edited by J. R. Davis. ASM International, Materials Park, OH, 2000, pp. 21–48.
10. J. Morgan, *Cathodic Protection*, 2nd ed. Cathodic Protection, Houston, TXs, 1993, pp. 135–140.
11. P. Marcoux, Prévention de la corrosion sous contrainte et de la corrosion localisée d'alliages d'aluminium aérospatiaux par des revêtements sacrificiels produits par projection thermique à l'arc électrique. Thesis, Université Laval, 2000.
12. V. Polyakov, *The Perspectives of Usage in Gas Industry Unporous Aluminium Coatings with High Resistance to Sulfide Cracking*. VNIIEGASPROM, Moscow, 1988, pp. 1–54.
13. V. Polyakov, in *The Recent Advances in Science and Engineering of Light Metals*, edited by K. Hirano and O. K. Ikeda. Tohoku University, Tokyo, Japan, 1991, pp. 371–376.
14. American Society of Metals, in *Aluminum and Aluminum Alloys*, edited by J. R. Davis. ASM International, Materials Park, OH, 1993, pp. 579–731.
15. A. O. Ita, *Organic Finish Guidebook and Directory*. Metal Finishing Magazine, NY, 2005, pp. 90–96.
16. G. M. Brown, K. Shimizu, K. Kobayashi, G. E. Thompson, and G. C. Wood, *Corrosion Science* **33**, 1371–1385 (1992).
17. G. M. Brown, K. Shimizu, K. Kobayashi, G. E. Thompson, and G. C. Wood, *Corrosion Science* **35**, 253–256 (1993).
18. G. Lorin, *La phosphatation des métaux*. Éditions Eyrolles, Saint-Germain, Paris, France, 1973.
19. E. Ghali and R. J. A. Potvin, *Corrosion Science* **12**, 583–594 (1972).
20. W. Rausch, *The Phosphating of Metals*, ASM International, Materials Park, OH, and Finishing Publications Ltd., Teddington, Middlesex, England, 2005, 416 pp.
21. B. W. Davis, P. J. Moran, and P. M. Natishan, in *Proceedings 98-17*, edited by R. G. Kelly, P. M. Natishan, G. S. Frankel, and R. C. Newman. Electrochemical Society, Pennington, NJ, 1999, pp. 215–222.
22. K. Lindsey, Gibbs High Speed Amphibian Technology, Nuneaton, England, 2006. http://www.gibbstech.co.uk/.
23. J. W. Bibber, *Metal Finishing* **91**, 46–47 (1993).
24. N. Voevodin, D. Buhrmaster, V. Balbyshev, A. Khramov, J. Johnson, and R. Mantz, *Materials Performance* 48–51 (2006).
25. A. J. Aldykiewiczs, H. Isaacs, and A. J. Davenport, *Journal of the Electrochemical Society* **142**, 3342 (1995).
26. F. Mansfeld, Y. Wang, and S. H. Lin, in *Electrochemical Society Extended Abstracts 95-2*. Electrochemical Society, Pennington, NJ 1995, p. 214.
27. A. S. Hamdy, D. P. Butt, and A. A. Ismail, *Electrochimica Acta* **52**, 3310–3316 (2007).
28. A. S. Hamdy, *Surface and Coatings Technology* **200**, 3786–3792 (2006).
29. A. S. Hamdy, *Materials Letters* **60**, 2633–2637 (2006).
30. A. S. Hamdy, A. M. Beccaria, and T. Temtchenko, *Surface and Coatings Technology* **155**, 184–189 (2002).
31. W. J. van Ooij, Potential of silane coupling agents to replace chromate metal pretreatments, in *Fifth International Symposium on Silane and Other Coupling Agents*, June 22–24, Toronto, Canada, 2005, Vol. 4, edited by K. L. Mittal, VSP, Leiden, The Netherlands, 2005.
32. W. J. van Ooij, V. Palanivel, H. Yang, and H. Mu, *A New Approach to Chromate-Free Coatings for Aerospace Aluminum Alloys*. NACE, San Diego, CA, 2003, pp. 53–55.
33. W. Huang, D. Li, T. Zheng, and M. Guo, *Materials Science Forum* **519–521**, 723–728 (2006).
34. ASM International Handbook Committee, *Corrosion—Understanding the Basics*. ASM International, Materials Park, OH, 2000, pp. 21–48, 100, 162, 214–215, 276, 286, 309, 513.
35. H. Takahashi, in *ASM Handbook, Volume 13A, Corrosion*, edited by S. D. Cramer and B. S. Covino, Jr. ASM International, Materials Park, OH, 2003, pp. 736–740.
36. S.-M. Moon, M. Sakairi, and H. Takahashi, *Behavior of Second-Phase Particles in Al5052 Alloy during Anodizing in a Sulfuric Acid Solution CSLM Observation*, Journal of the Electrochemical Society, Pennington, NJ, Vol. 151(7), 2004, pp. B399–B405.
37. W. Bensalah, K. Elleuch, M. Feki, M. Wery, and H. F. Ayedi, *Surface and Coatings Technology* **201**, 7855–7864 (2007).
38. I. Tsangaraki-Kaplanoglou, S. Theoharia, Th. Dimogerontakisa, N. Kallithrakas-Kontosb, Y.-M. Wang, H.-H. (Harry) Kuo, and S. Kia, *Surface and Coatings Technology* **200**, 3969–3979 (2006).
39. E. Groshart, *Metal Finishing* **98**, 80–81 (2000).
40. X. Yu and C. Cao, *Thin Solid Films* **423**, 252–256 (2003).
41. M. Zemanova and M. Chovancova, *Metal Finishing* **October**, 33–34 (2005).
42. S. Shrestha, A. Merstallinger, D. Sickert, and B. D. Dunn, *Some Preliminary Evaluations of Black Coating on Aluminium AA2219 Alloy Produced by Plasma Electrolytic Oxidation (PEO) Process for Space Applications*, Proceedings of the 9th International Symposium

on Materials in a Space Environment, Noordwijk, The Netherlands, 16–20 June 2003, pp. 57–65, Keronite International Ltd, 2008.

43. X. Li, X. Nie, L. Wang, and D. O. Northwood, *Surface and Coatings Technology* **200**, 1994–2000 (2005).
44. V. Moutarlier, M. P. Gigandet, and J. Pagetti, *Applied Surface Science* **206**, 237–249 (2003).
45. J. Ren and Y. Zuo, *Surface and Coating Technology* **182**, 237–241 (2004).
46. C. G. Munger, in *Corrosion Prevention by Protective Coatings*, edited by L. D. Vincent. NACE, Houston, TX, 1999, pp. 14–15.
47. C. Vargel, *Corrosion of Aluminium*. Elsevier, Boston, MA, 2004.
48. M. E. Nanna and G. P. Bierwagen, *Journal of Coatings Technology Research* 69–80 (2004).
49. D. Battocchi, A. M. Simones, D. E. Tallman, and G. P. Bierwagen, *Corrosion Science* **48**, 1292–1306 (2006).
50. R. L. De Rosa, D. A. Earl, and G. P. Bierwagen, *Corrosion Science* **44**, 1607–1620 (2002).
51. G. P. Pierwagen, D. J. Mills, D. E. Tallman, and B. S. Skerry, in *Electrochemical Behaviour Measurments for Corrosion Applications*, edited by J. R. Kearns, J. R. Schully, P. R. Roberge, D. L. Reicher, and J. D. Dawson. ASTM, Philadelphia, PA, 1996, pp. 427–445.
52. B. Arsenault, A. K. Lynn, and D. L. Duquesnay, *Canadian Metallurgical Quarterly Journal* **44**, 495–504 (2005).
53. Y. Z. Wang, R. W. Revie, M. T. Shehata, R. N. Parkins, and K. Krist, *Initiation of Environment Induced Cracking in Pipeline Steel: Microstructural Correlation*, The American Society of Mechanical Engineers, New York, 1998, pp. 529–542.
54. B. Arsenault and E. Ghali, in *Prevention of Environmentally Assisted Cracking of Structural Aluminum Alloys by Al and Al–5Mg Thermal Sprayed Coatings Using Different Surface Preparation Techniques*, edited by M. Jahazi, M. Elboujdaïni, and P. Patnaik, Canadian Institute of Mining, Metallurgy and Petroleum, Montreal, Quebec, Canada, 2006, pp. 61–74.

Chapter 15

Magnesium Coatings: Description and Testing

Overview

There are several surface treatments that can be used alone or in combination with oil, paint, or wax application depending on the concept and required duration, the projected use of the piece, and the aggressiveness of the medium. A new generation of chrome-free conversion treatments with good efficiency use phosphate permanganate ($KMnO_4 + MnHPO_4$), stannate ($NaOH + K_2SnO_3{\cdot}3H_2O + NaC_2H_3O_2{\cdot}\ 3H_2O + Na_4P_2O_7$), and cerium nitrate ($Ce(NO_3)_3$). Anodizing to form a nonconductive oxide layer (e.g., Dow17, HAE, Anomag, Keronite, Tagnite, Magoxid-Coat); galvanizing/electroplating (Zn, Cu, Ni, Cr) and electroless metal plating; and chemical vapor deposition, physical vapor deposition, flame or plasma spraying, and laser/electron beam surface treatments are possible options in spite of their high investment cost. Surface modification coatings like CrN and TiN can also enhance the wear resistance and offer improved corrosion resistance. The formation of a protective layer of "H-coat" is explained and the electrochemical conditions of Mg hydride formation–decomposition are discussed. Testing of anodized magnesium surfaces and coated metal by conventional and electrochemical procedures is described.

Accelerated laboratory testing aims to reproduce, in a much shorter time than in the field, natural corrosion and degradation processes of the paint system and the substrate without changing the corrosion/degradation mechanisms occurring in service. This is an excellent tool if it is well done and can predict the performance of the coated metal or alloy in service. Accelerated corrosion testing, which can have different testing modes in different corrosive aqueous media (immersion, alternating immersion, and immersion and spray) as well as in atmospheres with different relative humidities, are considered for corrosion of magnesium alloys with different coatings.

15.1. GENERAL APPROACH AND SURFACE PREPARATION

The very high strength-to-weight ratio, high stiffness, and mechanical castability of magnesium alloys besides their advantageous low densities are behind the recent intensive studies for innovative applications (the low density of pure Mg is 1.73 g/cm^3 and approximately 1.80 g/cm^3 for most Mg alloys, about two-thirds that of Al and its alloys). However, the major disadvantages of these Mg alloys are their poor corrosion and wear resistance and

Corrosion Resistance of Aluminum and Magnesium Alloys: Understanding, Performance, and Testing By Edward Ghali

high chemical reactivity. The development of high-purity Mg alloys, the most significant step toward improved corrosion resistance, has been achieved [1].

For certain applications, no protective treatment was required to obtain long and reliable service life because of the mitigating environmental factors outlined. Die-cast, investment-cast, and wrought magnesium components are used by computer and computer peripheral manufacturers for several applications where light weight, low inertia, rigidity, and heat sink requirements preclude the use of other metals or plastics. The noncorrosive operational environment within the computer eliminates the need for galvanic corrosion precautions [2]. An interesting finding was that magnesium components in applications such as transmission housings, engine blocks, and even engine cradles can be used without coatings for protection against general corrosion. Their performance was found similar to that of the aluminum alloy AA380 [1]. Modern automotive applications include air cleaner covers, retractable headlight assemblies, and clutch brake pedal supports. One of the more recent implementations of magnesium in the automotive industry is the use of alloy AM50 for the front-end support assembly for light-duty trucks. For some applications, where aesthetic appearance or corrosion protection is important, protective schemes for mildly corrosive environments should be considered [2].

The oxide film on magnesium offers considerable surface protection in rural and some industrial environments and the corrosion rate lies between that of aluminum and low carbon steels. Finishing of magnesium and magnesium alloys can include surface preparation, chemical treatment (surface conversion or modification), and coating practices for better performance in certain media and/or for decorative purposes. Capital investment, ease of manufacturing, coating performance, environmental issues, and projected use are important factors to consider. For military and helicopter applications for example, comprehensive protection schemes are required to achieve extended component life and to reduce maintenance costs. Schemes recommended for severely corrosive environments are required. Full wet assembly procedures and the coating of all exposed surfaces are essential. Good-quality aerospace paint systems should be used with wet assembly techniques. These schemes have given total protection for thousands of hours in various accelerated testing programs and will minimize corrosion spread. For civil and other less aggressive aerospace applications, protection schemes for moderately corrosive environments should be considered [2].

Coatings that exclude corrosive environments can provide the primary defense against different corrosion forms and even corrosion fatigue. A coating of magnesium components for protection against general corrosion and galvanic corrosion (e.g., surface chemical conversion, anodic oxidation) is frequently needed, especially for some parts with decorative functions [3]. Effective corrosion prevention for magnesium alloy components and assemblies should start at the design stage. Selective surface preparation, chemical treatment, and coatings are recommended. Oil application, wax coating, anodizing, electroplating, and painting are possible alternatives. Frequently, the designer should use the best combination of these methods to meet the functional need of the treated part. Recently, it has been found that a magnesium hydride layer, created on the magnesium surface by cathodic charging in aqueous solution, is a good base for painting. The degree of superficial corrosion that can be tolerated without affecting the performance of magnesium alloys and the severity of the service environment are determining factors in selecting an optimum finish [3, 4].

A thin oil or wax film is commonly used for the storage or shipping of sand-cast parts, which are treated before and during machining operations. A dry storage atmosphere is important. Chemical and electrochemical methods are used for the conversion of magnesium surfaces. A more corrosion-inhibiting and less alkaline to slightly acid film replaces the natural alkaline hydroxide-carbonate film on magnesium. The converted surface is generally more compatible with an organic coating.

Improved corrosion-preventive finishing systems can be achieved, for example, with a sealer, primer, and polyurethane top coat. Selecting a suitable surface treatment or finishing system necessitates consideration of surface conductivity, wear resistance, the service environment, alloy composition, and the possible presence of disimilar metals. Contacting components, fasteners, and inserts should be chosen for their compatibility, for example, a nonconductive, nonporous material; 5000 or 6000 series aluminum alloys; and tin-, cadmium-, or zinc-plated ferrous alloys [5]. There are several coating technologies available for protecting magnesium and its alloys. However, the widespread use of magnesium in the automotive industry, for example, is still deterred by the lack of appropriate protective coatings that can withstand harsh service conditions. A number of patents claim to have coating processes for magnesium and its alloys based on successful application for aluminum and its alloys; however, care should be taken since the active–passive behavior and film properties of magnesium alloys are different from that of aluminum alloys [6].

Cleaning and Surface Preparation Mechanical cleaning of magnesium alloy products is accomplished by grinding and rough polishing and buffiing, dry or wet abrasive blast cleaning, shot blasting, wire brushing or fiber brushing, and wet barrel or bowl abrading (vibratory finishing) [7]. Generally, mechanical cleaning is followed by chemical cleaning; however, under certain conditions one of them could be estimated to be sufficient. Chemical cleaning methods of magnesium alloys employ vapor degreasing, solvent cleaning, emulsion cleaning, alkaline cleaning, and acid pickling. Chemical cleaning starts normally by dissolving organic matter followed by the removal of corrosion oxides and even the passive layers if desired. Acid pickling is required for removal of impurities that are tightly bound to the surface or insoluble in solvents and alkalis. Ferric nitrate pickles deposits and the invisible chromium oxide passivating film. Acetic nitrate and phosphoric acid remove even invisible traces of other metals.

ASTM D1732 concerns the standard practices for preparation of magnesium alloy surfaces for painting where acid and alkaline cleaners, dip treatments, and anodizes are described.

15.2. METALLIC AND CONVERSION COATINGS

15.2.1. Metallic Coatings

Some special applications may require the metal plating of a magnesium component. For interior and mild exterior environments, especially in a marine atmosphere, a pore-free deposit is required for satisfactory corrosion resistance and this requires a nonporous base metal. In most situations, the plated surface is nobler than that of magnesium alloys, and so localized galvanic cells with important cathodic/anodic surface area ratios can be formed and lead to perforation.

15.2.1.1. Electroless or Galvanic Coatings

The most successful example of electroless plating technology for magnesium has a limited turnover rate of electrolyte. Further research is required to enhance the longevity of plating baths and to decrease waste generation. Another challenge associated with electroless plating on magnesium is the narrow window for operating conditions in order to obtain optimum coatings. However, this technology can produce uniform, corrosion-resistant, and wear-resistant coatings with good electrical conductivity and solderability, at a low cost [6].

The autocatalytic nature of the electroless Ni–P deposition process provides a uniform coating even on complex shapes; however, it has a short bath life due to sudden bath decomposition. This is further aggravated when plating Mg alloys due to the dissolution of active Mg nuclei. The inherent problem of the electroless deposition process is overcome by applying a layer of Ni–P alloys on the immersion-coated copper pretreatment film that is described later.

The application of electroless Ni–P on Cu immersion-coated Mg alloy AZ91 can be carried out in a bath containing 30 g/L $NiSO_4 \cdot 6H_2O$, 20 g/L CH_3COONa, 20 g/L $NaH_2PO_2 \cdot H_2O$, and 0.1–5.0 mg/L $(H_2N)_2CS$ (or 0–3.0 g/L $C_4H_2O_3$) with adjusted pH to 4.5 using HCl. The electrochemical corrosion performance of such a deposit was investigated in 5% NaCl solution. EDX analysis confirmed that no Mg or Cu dissolution was detected, indicating that the Ni–P coating successfully protected the inner layer from the long period of corrosion in 5% NaCl solution [6].

Copper Immersion Coating Before Electroless Treatments In order to achieve a quality coating on magnesium substrate, a process was designed that includes a chemical etching step to pretreat the magnesium surface, and an immersion coating process to form an underlayer for protecting the Mg surface for the subsequent electroless/electrodepositions in acidic and alkaline bath. The copper immersion coating (CIC) pretreatment on magnesium alloy AZ91D is a process involving several reactions taking place simultaneously, namely, Mg dissolution as an anodic reaction and hydrogen evolution and copper deposition as cathodic reactions. According to the *E*–pH diagrams of copper–water and magnesium–water, there is a broad pH region (between 2 and 9) within which the theoretical conditions for CIC are satisfied. The microstructure of the magnesium alloy AZ91D consists of α-phase matrix and intermetallic β-phase $Mg_{17}Al_{12}$ particles that are distributed at boundaries of small, cored grains of α matrix. Furthermore, the α-phase matrix, in turn, consists of primary α and eutectic α. Every phase has its characteristic potential and active or active–passive behavior. As an example, the β phase is much less active (acts as a cathode) if compared to the primary α or eutectic α, creating a galvanic corrosion cell of the α phase. This microstructural heterogeneity on the surface of the magnesium alloy substrate complicates the CIC process [6].

As an example, the pretreatment of substrates of magnesium alloy AZ91D coupons started with glass-beading for 10 seconds at a pressure of 450 kPa, followed by a sonication cleaning in isopropanol for 3–6 minutes. It was suggested that in an F^--containing aqueous solution, a surface film containing $Mg(OH)_nF_{2-n}$ may form and suppress anodic dissolution, thereby preventing the increase of copper coating coverage. Sonication is known for its capacity to erode/clean a substrate surface mainly through asymmetric cavitation. An alkaline degreasing process was subsequently performed in a solution containing 60 g/L NaOH + 10 g/L Na_3PO_4 at 75 °C for 3–6 minutes. After thoroughly rinsing in deionized water, the substrates were immediately transferred into a solution for CIC. The acidic bath contains 0.67 M $CuSO_4 \cdot 5H_2O$ + x M HF in various concentrations, while the alkaline bath consists of 100 g/L $K_4P_2O_7$, 30 g/L Na_2CO_3, 12.5 g/L $CuSO_4.5H_2O$ + x g/L NaF. All CIC processes can be performed at room temperature [6].

It has been demonstrated that, in an acidic bath, hydrofluoric acid can be employed to control the magnesium dissolution, and application of sonication could be used to modify the kinetic process. In the alkaline bath, both pH and the fluoride concentration play an important role in controlling the CIC process. In the acidic and the alkaline baths investigated, uniform and fine-grained CICs were achieved. The thermodynamically stable region of $Mg(OH)_2$ in the potential–pH diagram of magnesium–water gradually extends beyond 9. Accordingly, the dissolution of magnesium may be decreased, which may render CIC in an alkaline bath possible. With increasing pH, copper coverage first increases then decreases after a peak point at approximately pH = 10.3 [6].

15.2.1.2. Cathodic Treatments

Zinc and nickel can be plated directly onto magnesium from special electroless baths and are the only deposits used commercially as undercoatings upon which other commonly plated metals are deposited. The capital investment of electrochemical plating is relatively small. Standard electroplating magnesium involves surface conditioning, zinc immersion plating (zincate solution), and a cyanide copper strike ($\sim$8 μm), followed by a standard plating process of other metals. Porosity in the base metal promotes porosity in the deposit. Copper–nickel–chromium plating systems on magnesium satisfy decorative and protective requirements for interior and mild exterior environments for satisfactory corrosion resistance, especially in a marine atmosphere, where a pore-free deposit is required [8, 9].

Plated magnesium die castings are suitable, however, for such applications as interior automotive door handles and window cranks, where it is necessary to combine weight reduction with strength, durability, and good appearance. Metal plating of magnesium is not a corrosion-protective coating when magnesium and its alloys are exposed to severely corrosive environments. Recently, there have been some serious concerns over waste disposal [2].

15.2.1.3. Aluminum-Alloyed Powder Coating

An aluminum-alloyed coating was applied onto the surface of magnesium alloy AZ91D. The coating formed in aluminum powder at 420 °C for 1.5 h is rich in the β ($Mg_{17}Al_{12}$) phase. Polarization curve, ac impedance, salt immersion, and salt spray tests were carried out to investigate the corrosion behavior and assess the corrosion performance of the coated magnesium alloy. It was found that a coated AZ91D specimen was much more corrosion resistant and harder than the uncoated one. The improved corrosion resistance was mainly ascribed to the high volume fraction of β phase in the coating and this verifies the hypothesis that the corrosion performance of a magnesium alloy can be enhanced by a "skin" rich in β phase [10].

15.2.2. Chemical Conversion Surface Treatments

Chemical conversion is an important surface pretreatment for improving the corrosion resistance of an Mg alloy. Small cracks are distributed on the conversion coating layer during dehydration, improving the adhesion of subsequent paint layers or organic coatings to the surface of the Mg alloy substrate. There are a few applications in which Mg components may be subjected to prolonged immersion or contact with corrosive electrolyte. In some systems, it may be possible to add corrosion-inhibiting agents to the electrolyte. Maintaining electrolyte pH above 10.5 or adding soluble chromates or neutral fluorides is effective in reducing Mg-based metal corrosion [11].

Chemical and electrochemical finishing treatments can be used alone to provide short-term protection against corrosion and abrasion during shipment and storage, or as pretreatments for subsequent finishing methods. Conversion coatings do not provide adequate corrosion and wear protection from harsh service conditions when used alone. Although the anodized surface is more corrosion resistant and less porous than the phosphate one, for example, both are in need of a post-treatment to seal the porous surface. Conversion coating, which offers less wear resistance than anodizing, is still the state of the art as an effective measure to produce a primer for a subsequent organic coating because it is cheaper and more simple compared to anodizing processes, which need important investments and qualified operators.

The chemical finishing treatments, including the chrome pickle and chrome-free phosphate treatments, can be used also to provide a base for paint or short-term protection. The most widely used type of conversion coatings are chromate conversion coatings. However, use of the chromating process is strongly limited because of its serious health hazard and environmental risk due to the presence of leachable hexavalent chromium in the coatings.

Chrome Pickle The current steps of a chrome pickle are applications of an alkaline cleaner, cold rinse, chrome pickle (180 g/L of $Na_2Cr_2O_7{\cdot}2H_2O$ and 120–180 g/L HNO_3: specific gravity, 1.42), drying in air for 5 days, cold rinse, and hot rinse. A dichromate seal can be introduced between the cold rinse and hot rinse for better protection. Also, a dichromate treatment can replace the chrome pickle. A modified chrome pickle treatment introduces acid pickle and caustic dip or another acid pickle before the modified chrome pickle solution. The modified chrome pickle provides a uniform coating by optimizing the etching and passivating action of the chrome pickle bath and by thorough cleaning and washing. These treatments can be used generally to provide a base for paint or short-term protection [7, 9].

Dichromate as Inhibitor Treatment in a boiling dichromate solution (or the equivalent), followed by a slushy oil application, has been a satisfactory practice for a long time. Slight general uniform corrosion of Mg and its alloys can occur during this process. However, chromates have been shown to have very limited value over the less toxic and more economical commercial phosphates when applied over previously pickled surfaces [7].

Chrome-Free Phosphate Treatments The conversion coating on magnesium has traditionally been based on hexavalent chromium compounds. However, hexavalent chromium (Cr^{6+}) is a toxic substance that pollutes the environment and detrimentally affects people's health. Recently, investigations have succeeded in finding an alternative, chrome-free conversion coating process to protect magnesium against corrosion.

In the new conversion treatments, the Mg sample can be immersed in a bath of phosphate permanganate ($KMnO_4 + MnHPO_4$), stannate ($NaOH + K_2SnO_3{\cdot}3H_2O + NaC_2H_3O_2{\cdot}3H_2O + Na_4P_2O_7$), cerium nitrate ($Ce\ (NO_3)_3$), $CeCl_3/H_2O_2$, $La(NO3)_3$, $Pr(NO_3)_3$, or cobalt-III hexacoordinated complex ($Co(NO_3)_2 + NH_4NO_3 + NH_4OH + NH_3$). The Mg sample that is treated in the different baths is coated with various compounds. For example, the conversion coating films may be composed of $Mg_3(PO4)_2/MgMn_2O_8$, $MgSnO_3$, cerium oxide/hydroxide, lanthanum oxide/hydroxide, or praseodymium oxide/hydroxide [11].

A number of commercial phosphate treatments provide performance that is comparable to the best chromate-based surface treatments, particularly for new, high-purity die-cast alloys. Selected phosphate treatments can compete very effectively with chromates even in severe exposures, such as marine atmospheric environments. Promising alternatives offered by other conversion coatings are based on phosphate permanganate or fluorozirconate. A conversion coating based on an aqueous potassium permanganate electrolyte containing additions of alkali or ammonium salts with anions of the vanadate, molybdate, and wolframate group is also offered. Further alternatives based on organic acids, stannate, and salts of rare earth elements can also be considered. Some of these processes include a fluoride treatment (taking advantage of the relatively insoluble MgF_2).

It should be underlined that chemical treatment is recommended for paint formulations that are based on resins with low resistance to alkaline medium. Common chemical treatments alone do very little in agressive environments and may be unnecessary in mild environments. During the process of surface protection, in-process corrosion of Mg can occur [12–14].

Some alternatives are explained briefly here.

Phosphate-Based Baths The chemical conversion coating of magnesium alloy AZ91 from a bath containing 0.02 M $KMnO_4$ at 40 °C, with an adjustment of pH through the addition of 0.1 M nitric acid or HF or HCl and other additives, was examined by Umehara et al. [15]. The process starts with mechanical polishing of the Mg alloy surface, degreasing with acetone, activation with HF (200 mL/L) for 1 minute, a water rinse, and then the permanganate treatment for of 10–30 minutes, followed by a water rinse and drying. After electrochemical potentiodynamic studies and salt spray testing, the coated samples showed that the chemical-conversion examined baths give good corrosion resistance, almost equivalent to that of chromate treatment. The samples from the HF-added bath, in particular, showed a somewhat passive state in the anodic curve with greater anode polarization compared to the untreated surface of AZ91D. Also, the specimens treated in with nitric acid showed a greater anodic polarization. Most of the obtained layers showed the presence of magnesium oxides or hydroxides and manganese oxides.

Skar et al. [16] suggest that the phosphate permanganate treatment shows equivalent performance to a standard chrome pickle, both as stand-alone corrosion protection and as a base for subsequent coating. Phosphate permanganate should be accompanied by a deoxidizing process, either mechanical grinding or alkaline cleaning.

A new phosphate conversion coating that has excellent corrosion resistance compared to other coatings of this category is described by Han et al. [17, 18] on AZ91D. The AZ91D specimens were polished, pickled, and immersed in the phosphate bath at 40–90 °C for 5 minutes and then immersed in boiling solution to seal for 30 minutes. The black phosphate coating was in the amorphous state and quite likely composed mainly of $Mn_3(PO_4)_2$. This gave a macro smooth finish in spite of the presence of microcracks.

Corrosion resistance was carried out by salt immersion test in 5% NaCl solution for 48 h. Polarization curves were examined in 3.5% NaCl solution at pH 7 and 25 °C with a scan rate of 0.5 mV/s and starting potential of 100 mV cathodic to the open circuit potential. An obvious passive region exists in the anodic polarization curve and was at least two orders of magnitude less than for the other conventional methods. The corrosion rate measurements deduced from the polarization curves were much lower than for other organic acid, potassium permanganate, or chromate (Dow7) treatments [18].

Cerium-Containing Films The conversion coatings were formed by placing the magnesium electrodes in 50×10^{-3} mol·dm^{-3} solutions of $Ce(NO_3)_3$, $La(NO_3)_3$, or

$Pr(NO_3)_3$ at room temperature under gentle agitation for a 5 min period. This resulted in the formation of a visible coating on the electrode surface. These coatings were adherent, but could be removed by scratching the surface. A borate buffer solution (0.15 mol·dm^{-3} H_3BO_3/0.05 mol·dm^{-3} $Na_2B_4O_7$, pH 8.5) was used as the test electrolyte. The OCP increased from approximately −1500 to about −1410 mV/SCE as a cerium-containing film or coating formed on the magnesium surface. It was the best solution. For potentiodynamic polarization tests, the working electrode was immersed in the solution for 5 min and then polarized from the corrosion potential at a scan rate of 0.5 mV/s in the anodic or cathodic directions [19].

The coatings, which were formed by immersion in rare-earth salt-containing solutions, reduced significantly the dissolution of magnesium in a pH 8.5 buffer solution. With continued immersion of the treated electrodes in the aggressive pH 8.5 solution, the coatings first appeared to become more protective, but after periods exceeding 60 min they began to deteriorate. This is attributed to the formation of magnesium hydroxy corrosion products and mixed rare earth/magnesium oxide/hydroxide coatings, which on continued immersion became consumed by the formation of magnesium corrosion products. It has been found that the untreated Mg electrode dissolved, with the measured anodic current reaching values of 3 mA·cm^{-2} in the −1400 to −1200 mV/SCE range. The anodic current density was considerably reduced for the cerium-treated electrodes, with the lowest current, 25 μA·cm^{-2}, being recorded for the 5 minute treated electrode. Longer immersion times in the acidic cerium solution produced a less protective film, as was evident from the increase in the anodic current density recorded for the 80 min cerium-treated electrode. Sealing of these coatings should increase their protection quality and could result in more advantageous corrosion resistance [19].

The effects of a chemical cerium nitrate surface treatment for magnesium on the corrosion resistance and the mechanism of cerium compound layer formation have been studied by Ardelean et al. [20]. The chemical treatment was carried out in 0.05 M cerium nitrate or in 0.08 M cerium nitrate saturated with magnesium hydroxide at different temperatures—22, 40, and 80 °C. Some of the treated samples are annealed under oxygen at atmospheric pressure at 100, 150, and 200 °C. Effects of treatment parameters such as time, solution composition, and annealing temperature attributed to the formation of a homogeneous, uniform, and thick cerium compound on the magnesium surface. X-ray photoelectron spectroscopy (XPS) has shown that the deposited layer contains mainly CeO_2 with a low amount of $Ce(OH)_3$ and Ce_2O_3.

The corrosion resistance of the coated magnesium, with a cerium-containing film, has been investigated employing a rotating electrode (1500 rpm) in aerated 0.5 M Na_2SO_4 solution and using polarization curves, impedance spectroscopy, and galvanostatic reduction techniques. Cathodic and anodic polarization curves of the untreated, cerium-treated, and cerium-treated and annealed magnesium samples were recorded over the potential range of −2.5 to −1.8 V/MSE (saturated mercurous sulfate electrode) and showed a marked decrease of the anodic dissolution. The corrosion potentials (OCPs) were shifted toward more positive values and this shift was dependent on the solution composition and treatment time. The potential decay curves during galvanostatic reduction polarization in aerated 0.5 M Na_2SO_4 solution showed a remarkable low-end potential for the magnesium (cerium treated and annealed) electrode as compared to the untreated one. This indicates that the proton and water reduction reactions are markedly retarded. The electrochemical studies showed then (1) an inhibition of the cathodic reaction, (2) a marked decrease of the anodic dissolution, and (3) a shift of the corrosion potential and the dissolution reaction toward more positive values as a result of the cerium nitrate treatment [20].

Stannate Conversion Coatings Lin et al. [21] examined the effects of stannate ion content on the microstructure and corrosion resistance of stannate conversion coatings of AZ61 magnesium alloys. Stannate conversion coatings on magnesium alloys involve a nucleation and growth process. The dissolution of the substrate in the early stage of immersion is essential for the nucleation and growth of hemispherical particles of the coat. The conversion coating consisted of two layers: a relatively porous layer in contact with the substrate that was richer in aluminum compounds than that of the substrate, and a hemispherical particle layer as the major overlay covering the porous layer. During the coalescence of the hemispherical particles, some sites of discontinuity were inevitably left and influenced the corrosion resistance of the coat during the salt spray test. However, corrosion resistance was improved by increasing the stannate ion concentration and lowering the bath pH. This gave rise to finer particles, increased the population density of the hemispherical particles, and reduced the immersion time for optimal coating [21].

Potentiostatic polarization during the deposition of stannate chemical conversion coatings on magnesium alloy AZ91D accelerates dissolution of the magnesium alloy, promotes deposition of the coating film, and gives rise to a more uniform coating that improves corrosion protection compared to that deposited by the simple immersion method [22–24]. Before immersion in stannate solutions, the electrodes were pickled in a mixture of 0.25 wt % HF and 0.25 wt % HCl solutions. Composition of the stannate bath was 0.25 M $Na_2SnO_3{\cdot}3H_2O$, 0.073 M $CH_3COONa{\cdot}3H_2O$, 0.13M $Na_3PO_4{\cdot}12H_2O$, and 0.05 M NaOH. The conversion stannate coatings were formed under potentiostatic polarization at 40 °C between −0.8 and −1.4 V versus Ag,AgCl/KCl reference electrode. Potentiostatic polarization started immediately after immersion of the specimen in the bath. In order to ascertain the completion of the coating process under potentiostatic polarization, the current was traced during the coating process and the immersion was on the order of 50–60 minutes and was stopped when the current reached a minimum value of 1 μA.

For corrosion tests, the coated samples were first immersed in the borate buffer solution (0.15 M H_3BO_3 and 0.05 M $Na_2B_4O_7$, pH 8.5) for 10 minutes and then anodically polarized from the corrosion potential to a noble potential range of 500 mV with a scanning rate of 0.2 mV/s. The anodic polarization curves demonstrate a significant improvement in the corrosion performance of stannate coatings, showing a considerable reduction in anodic current. Among the different imposed potentials during film formation, the coated sample at −1.1 V was the best and gave also the lowest corrosion current value. The specimen coated by simple immersion at −1.4 V showed the worst corrosion behavior [22].

Phytic Acid Conversion Coating The formation of the phytic acid conversion coating, a new environmentally friendly chemical protective coating for magnesium alloys, was studied by Cui et al. [25, 26]. Phytic acid ($C_6H_{18}O_{24}P_6$) is an artificial and innocuous organic big molecule compound consisting of 12 hydroxyl groups and 6 phosphate radical groups. Commercial die-cast magnesium alloy AZ91D samples were used as the substrate. The phytic acid coating was deposited by the powerful chelating capability of the acid with magnesium and aluminum ions on the surface of the magnesium alloys.

The conversion aqueous solution was carried out with 5 g/L phytic acid at pH 8 and room temperature. Results show that the formation process can be divided into the following three steps: initial stage (0–15 min), metaphase stage (15 min to 1 h), and late stage (1–25 h). The gain in weight and the shift of potential versus more noble values were at their maximum values after 3–4 h of treatment. The composition of the coatings changed gradually with depth and the coatings contained hydroxyl, phosphate hydrogen radical, and phosphate radical, and this can improve the bond between the conversion coating and the substrate and facilitate later paint coating preparation [25, 26].

Electrochemical impedance spectroscopy (EIS) measurements were carried out in the frequency range of 10^{-2}–10^{5} Hz with a perturbation amplitude of 10 mV to examine the corrosion resistance properties of the coat. At the initial stage, as the immersion time increased, the thickness increased, which resulted in the increase of polarization resistance, R_p, and the decrease of the constant phase element (CPE). But with further growth of the coating, the coating degenerated, which resulted in the decrease of R_p and the increase of CPE. The microstructure and composition of the coating were affected by the phase of the substrate. The coating on β phase is always thinner but more compact than that on α phase [25, 26].

MAGPASS Coat The product produces a conversion layer consisting of oxides from both the passivation solution and base material itself and has been proposed as equal to the corrosion performance of current conventional chrome-containing conversion coatings on the market today [27]. Alodine 5200 is an organometallic titanium-based primer. For example, chromate-free conversion coatings, such as Alodine 5200 and MAGPASS, with an epoxy powder coat are available that can satisfy aggressive testing conditions established by the industry [28]. Another conversion surface treatment to consider is the Fluorozirconate treatment, which has shown a performance approaching that of the chrome pickle with respect to paint adhesion. This treatment gives adequate performance in mild corrosive environments and perhaps also in more severe exposures, providing the component is not exposed to stone chipping [16].

Magnesium Film by Vapor Deposition Recently, some studies have attempted to coat Mg-friendly metal on magnesium alloy to protect against corrosion. For instance, Yamamoto et al. [29] coated a thick, pure magnesium film on a Mg–Al–Zn alloy sample surface by vapor deposition. Their work concluded that a pure magnesium film may not only improve corrosion resistance but also increase the recyclability of alloys. Uan and Yu [11] coated a fine-grained Mg thin film on an AZ91D surface by vapor deposition. The film served as a sacrificial anode and cathodically protected the cathode (the AZ91D substrate).

Recycling Magnesium Scraps However, with respect to recycling magnesium scraps from automotive components or any postconsumed magnesium product, applications of chemical conversion coatings make it difficult to recycle the scraps into high-quality die-casting alloy ingots that meet ASTM specifications. Some of the main reasons are that the magnesium melt becomes contaminated by surface contamination and the formation of dross increases because of the presence of the conversion layer. According to Skar et al. [16], low recyclability and high toxicity are associated with the formation of conversion films on the surface. Thermal decoating of scraps of Mg alloy is an initial step toward recycling [11].

15.3. ANODIC TREATMENTS

15.3.1. Anodizing Description and Approaches

Anodizing of magnesium is an electrolytic process for producing a thick, stable oxide film on metals and alloys that can be used for paint adhesion, dyeing, passivation treatment, and better wear. Anodizing is less sensitive to the type of alloy being coated, has an acceptable cost of waste disposal, and can be considered as the most widely commercially used coating technology for magnesium and its alloys. However, the process is technologically more complex than electroless deposition, electroplating, or conversion coating—involving more capital investment and high operational cost [3].

Generally, the following steps are considered in practice: mechanical pretreatment, degreasing, cleaning and pickling, electrobrightening on polishing, anodizing using dc or ac current, dyeing or post-treatment, and sealing. Each anode (Mg or its alloy) is pretreated before anodization by abrasion of the surface with 600 grit paper, followed by a hot alkaline cleaner (a cleaner other than the current chromic acid), and a rinse in distilled water (stages a–c). The core part of this process (stage d) is the formation of an anodized layer that is normally about 5–30 μm thick, hard, dense, electrically insulating, and wear resistant. The films have a thin barrier layer at the metal–coating interface followed by a layer that has a cellular structure. Anodized coatings have varying degrees of porosity and must be scaled for use in agressive chloride media. The coatings contain pores whose dimensions and distribution are a function of the electrolyte properties, temperature, substrate, anodizing current density, and voltage. The coatings can be additionally colored, infused with various polymers to produce special properties including lubricating properties, and sealed (stages e and f). Depending on the aggressiveness of the environment, an anodized coating can also be painted for optimal protection [3].

Coloring Immediately after anodizing, organic dyes or inorganic pigments are frequently applied. Electrolytic coloring can be achieved also by electrolytic deposition of inorganic metal oxides and hydroxides into the pores of the film or by adding organic constituents to the anodizing electrolyte that decompose and form particles which become trapped as the anodic film grows [30]. The atmospheric corrosion performance of the newer color-anodized finishes is of interest. Based on the long-term weathering of dyed finishes, a limited range of special dyes for architectural applications is recommended [31]. Good performance is reported for the combined anodized and electrophoretically deposited clear laquer finishes, now used widely in Japan [32].

Sealing Sealing the pores of the oxides by hydrated base metal species is usually accomplished by boiling in hot water, steam treatment, dichromate sealing, and lacquer sealing. In a more severe environment, the open pore structures of the anodized layers on magnesium have to be sealed to give adequate corrosion resistance and consequently anodized magnesium alloys are generally designed to be sealed or covered by other protective layers. The anodizing standard (BS 1615, 1987) provides the necessary specification for an anodized product. Most of the testing data consider the sealed anodized layers [3]. Tests for quality of sealing of anodic coatings include dye spot tests with prior acid treatment of the surface (ISO 2143, 1981 and BS 6161—Part 5, 1982), measurement of admittance or impedance (ISO 2931, 1983 and BS 6161—Part 6, 1984), or measurement of weight loss after acid immersion (ISO 3210, 1983 and BS 6161—Part 3, 1984; and ISO 2932, 1981 and BS 6161—Part 4, 1981). The chromic-phosphoric acid immersion test (ISO 3210) has become the generally accepted reference test [3].

Influence of the Electrolyte Composition and the Voltage During the initial voltage ramp, before any sparking occurred on the anode, some electrolysis of water was always noted. The extent of this electrolysis varied widely among the different electrolyte mixtures. Bubble formation was often initially vigorous, but decreased with time as if the magnesium anode was somewhat passivated before sparking occurred. In a 3 M NaOH solution, extensive electrolysis occurred and continued during spark anodization. Addition of fluoride, aluminate, phospate, or tetraborate to the electrolyte solution reduced the electrolysis significantly from the level noted for 3 M NaOH. Aluminate ions are incorporated into the coating as a significant component, producing a coating of a micrometer in thickness. The coating had low surface roughness but was visually nonuniform. The form of the aluminum on the surface was likely to be magnesium aluminate, combined with magnesium oxide and hydroxide. In contrast, anodization in sodium fluoride resulted in improved surface texture, opacity, and color, but did not add significantly to the thickness [33].

All of the newer electrolytic plasma-based processes like Keronite (Keronite), Anomag (Magnesium Technology), Magoxid-Coat (AHC, Germany), and Tagnite use alkaline-based chromate-free solutions, which are more favorable from an environmental point of view in comparison with acidic chromate and/or fluoride-based older processes (e.g., Dow17, Cr22). The variations in the layer composition and the structure are a result of the different electrolytes and/or voltages used by the various processes.

15.3.2. Formation of Anodized Coatings

Anodizing can be accomplished by controlling either the voltage or the current. Under voltage control, the current drops with treatment time as the insulating oxide film is growing. Huber [34] showed the relationship between the applied voltage and the characteristics of a film formed on Mg in 1 M NaOH. At voltages up to 3 V, the current density remained low, and a light gray protective film of $Mg(OH)_2$ was formed. At intermediate voltages (i.e., 3–20 V), oxygen evolved, and a thick dark film of $Mg(OH)_2$ was found. Above 20 V, a thin protective coating was again produced. The formation of a compact anodic film was shown to be limited by the breakdown phenomenon accompanied by intensive sparking (above 50 V) [3].

Ono and Masuko [35] reported similar breakdown potentials for various fluoride-containing solutions. In the alkaline fluoride solution, the breakdown potential could vary with alloy and electrolyte from 50 to above 110 V. The barrier layer is composed mainly of MgF_2 and/or AlF_3, while the anodized layers at breakdown voltages are crystalline MgO. A peculiar phenomenon of high current density was observed at around 5 V that can correspond to the transpassive state; however, this state cannot be observed in acidic fluoride solutions such as Dow17 and ammonium fluoride for the AZ91 alloy [3].

The process parameters, the electrolyte composition, and the substrate can influence the corrosion properties. Mizutani et al. [36] reported a much better corrosion resistance for $Mg(OH)_2$ layers produced by anodizing at 3 V in NaOH rather than for MgO layers produced by anodizing at 10 and 80 V. The anodic polarization behavior of the anodic films in 0.1 M KCl solution was used for evaluating the corrosion resistance of the films. The anodic film formed in 3 M KOH solution with 1 M Na_2SiO_3 at 4 V for 60 minutes at 65 °C exhibited the highest corrosion resistance if compared to that formed in the presence of other concentrations of sodium silicate (0.5–5 M), correlating well with the quality of the film produced [3, 37]. When different concentrations of AlO_2^- ions were added in the electrolyte, the critical voltage remarkably increased and current density effectively decreased with increasing AlO_2^- content. The passivation effect of aluminate addition in the electrolyte was more effective than the addition of aluminum in the magnesium substrates [3, 35].

More uniform sparking has been found by adding $Al(NO_3)_3$ to an electrolyte of 3 M KOH + 0.21 M Na_3PO_4 + 0.6 M KF. However, the best corrosion resistance determined

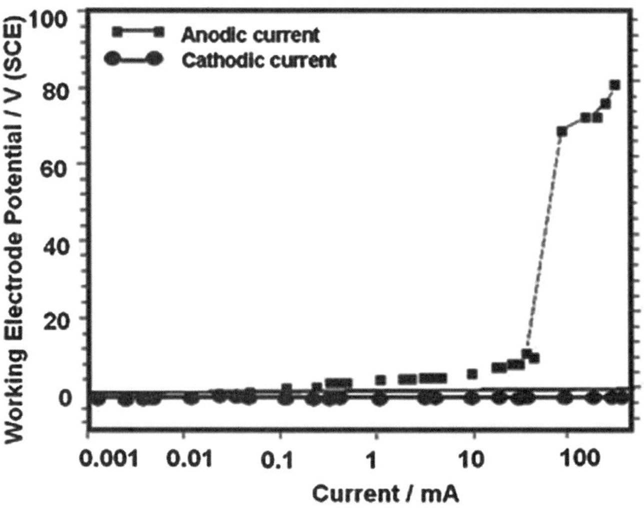

Figure 15.1 Electrochemical polarization curve for the magnesium alloy AZ91 in 3 mol/L NaOH at 24 °C [40].

by electrochemical impedance spectroscopy (EIS) was found for an addition of 0.15 M $Al(NO_3)_3$ [38, 39]. The electrochemical polarization curve shown in Figure 15.1 was obtained using a series of constant current steps and recording the stable current after 10 s in 3 M NaOH at 24 °C. A sudden jump to high potentials is observed when the anodic current exceeds ~20 mA/cm^2. The potential leaped to a value above 60 V and began to oscillate, and sparks appeared to move around [3, 40]. A collapse of the anodic film has been observed under a constant imposed current of 20 mA/cm^2 at 24 °C during a few seconds. The collapse of the barrier layer is due to the fact that the MgO initially formed has a molar volume of 11.3 cm^3/mol, which is smaller than the molar volume of the Mg base metal—14.0 cm^3/mol. This gives a Pilling–Bedworth ratio of 0.81 and 0.87 for pure magnesium and AZ9. The collapse is followed by repeated growth of the barrier layer with the same slope of dE/dT [40].

The relation between final current density after 10 min and formation voltage in 3 M KOH, 0.6 M KF, and 0.2 M Na_3PO_4 (pH 13) at 298 °C has been studied by Ono and Masuko [35]. They showed a difference between pure magnesium or AZ31 and AZ91 (Figure 15.2). The high current density at around 5 V can be explained by the transpassive state similar to that observed with sodium hydroxide solution. The peak current at 5 V decreased with increasing Al content of the substrate. The critical voltage of high current flow over more than 1000 A/m^2 accompanied by breakdown was relatively independent of substrate purity; namely, 60 V for 99.95% Mg, 99.6% Mg, and AZ31B, and 10 V for AZ91D [3, 35].

The formation of sparks during anodization on the magnesium surface has not been observed below 50 V. Initially, the sparks were very small and were extinguished very quickly. As the potential was increased, the sparks became larger and began to move over the surface of the anode. By stepping the potential during anodization, the anode would activate and passivate as the film grew, so the current fluctuated at any given potential. As the anode remained at a particular potential, the rate of formation of sparks would diminish. If the potential was then increased slightly, the sparks would start to form again and move about the surface [33].

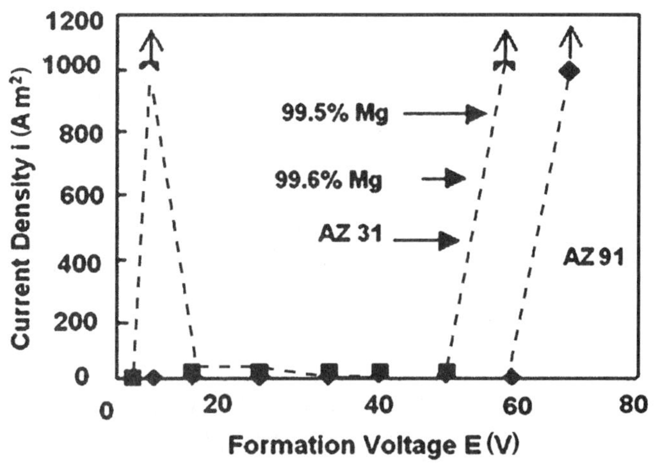

Figure 15.2 Influence of voltage–current characteristics of anodizing in KOH – KF – Na_3PO_4 solution on Mg substrates. Anodizing was performed for 10 min at 298 K [3, 35].

If a spark formed at a sharp edge or what were presumed to be point defects on the surface and did not move, there was inevitably a large pit burned into the anode. The behavior of the sparks did change with the electrolyte solution composition and concentration. Citrate in the electrolyte mixture produced a more controlled sparking process and prevented pit formation, which occurred under localized sparks. Iodide was a damaging electrolyte and led to uncontrolled pitting. Tetraborate contributed both to coating thickness and color, and lowered sparking voltage [33].

The thickness, chemical composition, and microstructure of anodized coatings formed on magnesium alloy AZ91D at various anodizing current densities were measured. It was found that anodizing current density influences the principal parameters of anodizing, and hence the coatings formed at different anodizing current densities had different corrosion resistances. This suggests that the corrosion performance of an anodized coating could be improved if a properly designed current waveform is used for anodizing. In order to improve the corrosion performance of an anodized coating, an optimized anodizing current waveform should be applied. A higher current density is generally required in the earlier stages for high production efficiency, and this should be followed in the later stages by low current density to minimize or seal the pores [41].

Anodic Film Composition at the Metal–Oxide Interface Metastable, solid-solution Mg–0.8 at% Cu and Mg–1.4 at% Zn alloys have been anodized up to 250 V at 10 mA/cm^2 in an alkaline phosphate electrolyte at 293 K in order to investigate the enriching of alloying elements beneath the anodic films. Rutherford backscattering spectroscopy (RBS) revealed enrichments to about 4.1×10^{15} Cu atoms/cm^2 and 5.2×10^{15} Zn atoms/cm^2, which correlate with the higher standard Gibbs free energies per equivalent for formation of copper and zinc oxides relative to that of MgO. The enriched layers were ~1.5–4.0 nm thick as measured by medium energy ion scattering (MEIS). The anodic films, composed mainly of magnesium hydroxide, contained copper and zinc species throughout their thicknesses; the Cu/Mg and Zn/Mg atomic ratios were about 18% and 25% of those of the alloys, respectively. Phosphorus species were present in most of the film regions, with a P/Mg atomic ratio of about 0.16. The magnesium ions in the film account for about 30% of the charge passed during anodizing [42].

15.3.3. Properties and Chemical Composition

The anodized coatings obtained by early anodizing processes in conventional electrolytes range from nearly transparent, colorless films to opaque gray. Properly abraded and cleaned surfaces generally give uniform colored coatings. More translucent coatings are not uniform in color because of poor surface pretreatment. Some coatings are very smooth; others were patchy, although all were porous on the microscopic scale. The center line average (Ra) for anodized coatings varies from ~ 0.5 to 1.1 μm in 3 M NaOH solution with or without the addition of 0.15 M hydroxide solution or 0.15 M fluoride or 0.05 M phosphate or 0.15 M tetraborate. The thickness was generally between 2 and 8 μm. It is a rougher surface for sodium tetraborate solution. Other electrolytes, such as phosphate and fluoride, did not produce a layer of sufficient thickness to be measured optically.

The films formed through the Dow17 and Anomag anodizing processes show a slight decrease in certain mechanical parameters such as the fatigue properties. However, the scattering of the *S–N* fatigue results was less for Anomag than for Dow17 treatments [3, 43].

The results of XPS analysis for some anodized samples by different processes have been studied by Barton and Johnson [33]. The outermost layer of the phosphate-anodized sample showed no phosphorus, nor did the tetraborate sample contain any boron. Visually, there were changes in the coating because of the presence of these electrolytes; however, any phosphorus or boron incorporated in the film might not be located at the surface, but may remain concentrated at the interface between the metal and the anodized layer. Fluoride was found in the fluoride-anodized surface, but in relatively small amounts. Aluminate showed up in abundant concentration, corresponding to the significant layer noted in thickness measurements. No aluminum was found in the fluoride, phosphate, or boron coatings.

15.3.4. Some Industrial and Developing Anodizing Processes

Galvanic anodizing is a low-voltage dc treatment that produces a thin black conversion coating, used mainly as a paint base (Chemical Treatment No. 9). A source of electric power is not required. Proper galvanic action requires the use of racks, made of stainless steel, Monel, or phosphor bronze. The galvanic anodizing is carried out in ammonium sulfate, sodium dichromate, and ammonium hydroxide solution at 49–60 °C for 10–30 minutes [7, 8]. A processing diagram of Chemical Treatment No. 9 is presented in Figure 15.3.

Proper galvanic action requires the use of racks, made of stainless steel, Monel, or phosphor bronze. When the work pieces are immersed in the anodizing solution, they are made the anodes, and the tank, if made of low-carbon steel, acts as the cathode. If the tank is equipped with a nonmetallic lining, separate steel cathodes must be used. This can be applied to all forms and alloys of magnesium to produce a protective black coating with good paint-base characteristics. Parts with attachments of other metals may also be treated. Because this process does not result in appreciable dimensional change, the parts are machined to close tolerances before treatment [8].

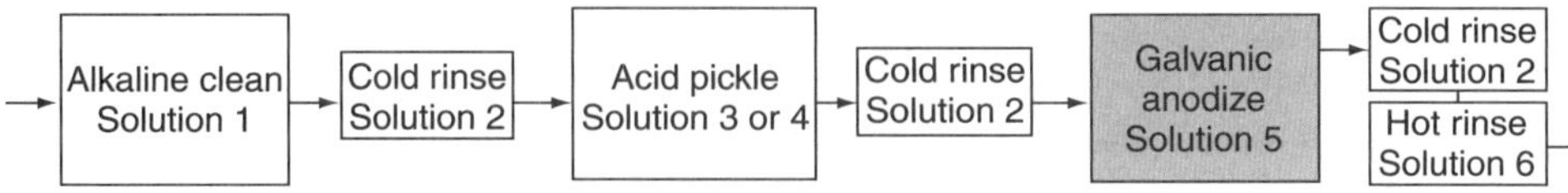

Figure 15.3 A galavanic anodizing diagram (Chemical Treatment No. 9 MIL-M-3171A) [8].

More substantial coatings (5–30 μm thick) require anodic polarization by external current. The Dow17 chemical treatment process and HAE treatment are currently used. For the two treatments, a two-layer coating is produced at lower and higher voltages, respectively. The first layer is about 5.0 μm thick, having a light green or greenish tan color. This is covered by a second, heavier coat about 30.0 μm thick and dark green in color. The second layer is vitreous, relatively brittle, and highly abrasive [9].

The anodizing bath of the Dow17 process is composed of an alkali metal hydroxide and a fluoride or iron salt or a mixture of the two at a high pH and can be applied for different alloys. The films formed on AZ91D alloy were uneven, which can be due to the presence of the intermetallic Mg–Al particles located at the grain boundaries. The electrolyte for HAE is composed of KOH, $Al(OH)_3$, K_2F_2, Na_3PO_4, and K_2MnO_4, and the current density is 1.5–2.5 A/dm^2. The terminating potential is 65–70 V after 7–10 minutes and 80–90 V after 60 minutes for the HAE treatment. The HAE process produces a dark brown coating that is hard, with good abrasion resistance, but it can adversely affect the fatigue strength of the underlying magnesium, particularly if it is thin. As an example, HAE is composed of six or seven steps: alkaline clean, cold rinse, anodize, cold rinse, dichromate bifluoride dip ($Na_2Cr_2O_7{\cdot}2H_2O + NH_4\ HF_2$), dry air, and possibly heated humidity again [30, 44].

Hard Anodizing The wear resistance and hardness of anodized films can be improved by operating at a decreased electrolyte temperature and increased current density (hard anodizing). The properties of the hard anodized film can be further improved by the incorporation of solid film lubricants, such as PTFE or molybdenum disulfide. Enhanced corrosion can occur if the coating contains defects and this can be due to an uneven film formation caused by metallurgical phase separations, mechanical pretreatments, or the geometry of the original metal. The fatigue strength of the base metal can be reduced during hard anodizing, especially in thicker films, and the produced ceramic material can be brittle and in adequate for certain applications [30].

Modified Acid Fluoride Anodizing The anodizing bath is composed of ammonium bifluoride, sodium dichromate, and phosphoric acid. The coating is formed by a chemical reaction, where magnesium is oxidized to Mg^{2+} and the Cr^{6+} is reduced to Cr^{3+}. An alternating current is necessary to ensure replenishment of the reactant concentrations at the interface. These coatings are highly stable if subjected to high humidity, high temperature, thermal cycling tests, and thermovacuum tests and are used for some space applications due to their high solar absorbance, high IR emittance, and good optical properties [30].

A heavier coating Cr22 treatment is a high-voltage process that is commercially available but not currently used. The terminating potential can be 320 or 350–380 V for heavier coatings. Green and black coatings can be produced on all alloys by varying the solution composition, temperature, and current density. The anodizing bath may contain chromate, vanadate, phosphate, and fluoride compounds [44]. These coatings provide excellent corrosion resistance in mild media or for unpainted parts of the structure when properly sealed. The sealing post-treatment consists of an immersion for 2 minutes in a solution of sodium silicate (10% by volume) at 85–100 °C [8, 9].

Plasma Anodizing Processes Other names such as microarc oxidation (MAO) or plasma electrolytic oxidation (PEO) processes can signify the same category of processes. In conventional anodizing of aluminum or magnesium, cell voltages can reach 50 V. In plasma anodizing, voltages are not usually less than 200 V and can be twice this. Besides these very high voltages, a clearly visible "glow" can be seen at and around components being plasma anodized. Although the electrolyte temperature might be typically 50 °C, the local temperature in the plasma zone would probably be in excess of 1000 °C, and it is this very high temperature that leads to formation of "glassy" or "ceramic" anodic coatings [45].

Magoxid-Coat Process In this process, the plasma is discharged by an external power source in a slightly alkaline electrolyte near the surface of the work piece (anode). The oxygen plasma generated causes partial short-term surface melting and ultimately the formation of an ceramic oxide layer. The anodizing bath for this process is free of chloride and may contain inorganic anions such as phosphate, borate, silicate, aluminate, or fluoride. The bath may contain organic acids such as citrate, oxalate, and acetate. A source of cations among alkali, alkaline earth, or aluminum ions is present. A stabilizer such as urea, hexamethylenediamine, hexamethylenetetramine, glycol, or glycerin is also added.

The coating is formed in a slightly alkaline bath, which gives $MgAlO_4$ and other beneficial compounds on the surface. The innermost or barrier layer is extremely thin, followed by a middle ceramic oxide, providing the majority of corrosion protection since it is almost nonporous. The outermost portion of the coating is a very porous ceramic layer [9, 46]. The coating consists of three layers, a 100 nm thin layer at the interface, followed by a low-porosity ceramic oxide layer, and finally a higher porosity ceramic layer (Figure 15.4). There are two options to make use of this porosity: one can impregnate the outer layer or one can grind it away in order to expose the harder and denser underlying layer [45].

Impregnation of the coating with particles of fluorine polymers has been shown to significantly improve the load-bearing properties of the coatings, while maintaining good adhesion and corrosion resistance. The produced coatings are uniform even on edges and cavities and provide wear and corrosion protection. Dyeing has been shown to result in a decrease in corrosion resistance [47, 48].

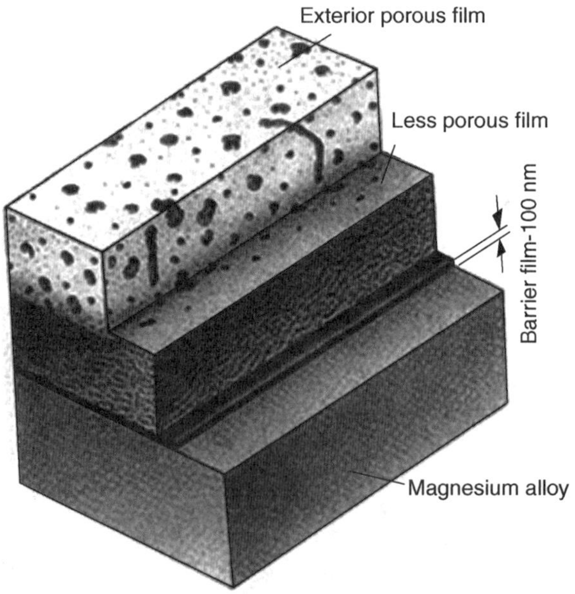

Figure 15.4 A Schematic cross section of anodized ceramic oxide Magoxid coat (~25 μm thick) [50].

Magoxid-Coat is effective for all wrought, die-cast, and pressure-cast Mg alloys. A number of applications are already in production such as supercharger rotors, housings, and handles for power tools, covers and brackets for high-performance automobiles, automobile and motorcycle rims, valve covers, transmission cases, control surfaces, pneumatic tools, slides, and carriages for data processing equipment [49, 50].

Anomag Process The anodizing bath consists of an aqueous solution of ammonia and sodium ammonium phosphate. The magnesium substrate is pickled in an aqueous hydrofluoric acid bath prior to anodizing in an aqueous bath composed of an alkali metal silicate and an alkali metal hydroxide. In a consecutive patent, the fluoride is integrated into the anodizing bath [51, 52]. The coating is formed when sparks are discharged at the surface. This melts the surface with a simultaneous deposition of a fluoride silicate coating [30].

The die-cast Mg samples treated by the Anomag process, followed by powder coating, had good paint adhesion properties and excellent corrosion protection. Tchervyakov et al. [53] examined the Anomag anodizing process of AZ91 and showed that the formed anodizing film is porous with a pore size of about 6 μm and porosity of 13%. Sealing and painting were shown to reduce the pore size and porosity to 3 μm and 4%, respectively, and increase the corrosion resistance [30].

The Keronite Process This is a plasma anodizing process that is occurs when sparks are discharged at the surface. This melts the surface with a simultaneous deposition of a ceramic layer. The Keronite process was conceived in Russia and uses alkaline chrome- and ammonia-free electrolyte. The layer is composed of $MgAl_2O_4$, together with SiO_2 and SiP. It can be stated that the thickness of the coatings produced by the original Keronite® technique leads to a moderate degradation of the fatigue performance [54]. An improved PEO method, utilizing a pulsed bipolar current, has recently been developed and made commercially available by Keronite Ltd., which allows coating growth rates of up to 10 μm/min. A substantial reduction in the thermal impact on the substrate can therefore be achieved [54].

The original Keronite oxidation treatment used amplitude-modulated ac mode at a frequency of 50 Hz. The waveform of the applied voltage (current) was fixed during the process, determining the relation between the duration and amplitude of positive and negative cycles. The improved Keronite® process (batch 3) utilizes a higher frequency ($\sim 10^3$ Hz), bipolar current mode with independent control of duration and amplitude for both positive and negative electrical pulses. Higher frequency current pulses enable the creation of sequences of shorter, yet more energetic, microdischarge events, ensuring a better balance between "oxidizing" and "fusing/recrystallizing" aspects of the coating formation process. Thus the high-frequency bipolar system is believed to allow a three to five times enhancement of the coating deposition rate combined with substantial improvement in the surface layer quality, with less porosity and roughness as well as different phase composition [54].

Keronite, as an example of a plasma anodized process, has an amorphous structure which does not crack on edges. Also, it maintains a largely uniform thickness all over a component and, in fact, is slightly thicker on the edges. The porous top surface of Keronite provides an excellent key for almost any type of top coat and has been used as a pretreatment to provide an A-class finish on auto body exterior panels. Although Keronite itself is an electrical insulator, it can still be electropainted or e-coated if certain precautions are taken and, indeed, a layer of Keronite followed by an epoxy-based e-coat can give very effective protection to magnesium at low cost [55].

While actual cost data are unavailable, a 10 μm Keronite coating, giving ample corrosion protection under most conditions, can typically be applied in 3 minutes, suggesting that in terms of "tank time costing" or power usage, this is not an exceptionally expensive process [45]. There are a variety of applications for Keronite, from the plasma anodized coatings within the electronics and electrical industries, to the pick-and-place nozzles for products that utilize the electrical insulating properties of Keronite. Magnesium bicycle frames can be cost effectively pretreated with Keronite. Keronite protects complex interior surfaces from corrosion, for example, grills of bicycles, wheel chairs and pushchairs, and sunglasses [55].

Tagnite Plasma Process This surface treatment is a chromate-free anodic electrode position surface treatment. The anodization bath consists of an aqueous solution containing hydroxide, fluoride, and silicate species. The abrasion resistance is improved compared to the Dow and HAE processes, even prior to organic finishing. Surface sealing of the coating further improved the corrosion resistance of the anodized alloys [30].

HAE and Plasma Anodizing Process HAE falls into the category of "plasma anodizing" processes, or plasma electrolytic oxidation (PEO). This transition corresponds to the moment when the surface of the alloy becomes totally covered by the PEO anodic film. The film obtained is composed essentially of a partially crystalline aluminum and magnesium oxide mixture, in which are incorporated all the elements of the electrolyte. The aluminum content of the film is a function of the aluminate concentration in the anodizing electrolyte [56, 57].

Microarc Oxidation or Microplasmic Ceramic Coatings The microplasmic process is an electrochemical microarc oxidation (MAO) process in a suitable electrolyte achieved by increasing the anodic voltage to a high stage, usually accompanied by intensive gas evolution and sparking phenomenon at the anode surface. Due to the high voltage and high current, intense plasma is created by microarcing at the specimen surface and this plasma in turn oxidizes the surface of the aluminum specimen. Thus the process is called a microplasmic process. The oxide film is produced by subsurface oxidation and considerably thicker coatings can be produced. A controlled high voltage ac power is applied to the metal part submerged in an electrolytic bath of proprietary composition. The process of the MAO treatment and the growth mechanism of the ceramic coating are similar; however, the preparation of MAO coatings on large aluminum and magnesium alloy workpieces, whose areas can reach 4 m^2, become possible using an average current density on the order of 0.7 A/dm^2 [58].

MAO is a promising surface treatment method on so-called valve metals, such as aluminum, magnesium, titanium, and their alloys. The die-cast magnesium alloy AZ91D was used as the substrate material. The cell voltage was varied in the range of 240–600 V; the current density was varied in the range of 0.5–5 A/dm^2; the MAO process was run for 5–60 min; the electrolyte temperature was controlled at 40 °C. Aqueous solutions of sodium aluminate and potassium fluoride were used as the constituents of the electrolyte. The process was divided into two stages. At the first stage, the cell voltage increased linearly at a very high rate of 80–300 V/min, the slope of the voltage–time response increased with the increase of the applied current densities and concentration of the electrolyte components. Approximately 3–20 min later, this process entered a second stage; a steady-state sparking was established on the anode surface and the cell voltage reached a relative stable value of 520–570 V. Variation of treatment time over the range of 10–40 min creates no obvious difference in the phase structure of the ceramic coatings [59].

The silvery white ceramic coatings fabricated at constant applied current densities on the surface of magnesium alloys by MAO are composed of spinel phase $MgAl_2O_4$ and intermetallic phase Al_2Mg. A few circular pores and microcracks are also observed to remain on the ceramic coating surface; the number of the pores decreases, while the diameter of the pores apparently increases with prolonged of treatment time. The corrosion resistance of ceramic coatings is improved more than 100 times compared to magnesium alloy substrate in chloride-containing solutions, as obtained by potentiodynamic and EIS investigations. The MAO process is fast, uses inexpensive equipment and raw materials, and produces no hazardous waste [59].

MAO in KOH Electrolyte Anodizing of magnesium alloy AM60 (6% Al + 0.27% Mn) was studied in a solution containing 1.5 M KOH + 0.5 M KF + 0.25 M $Na_2HPO_4 \cdot 12H_2O$ with addition of various $NaAlO_2$ concentrations. The experiments were carried out in the dc current galvanostatic mode. Observations of phenomena occurring at the sample surface plus voltage monitoring revealed three stages: traditional anodizing, followed by microarc anodizing, and finally arcing. The film was porous and cracked, with poor bonding to the substrate. It was composed of magnesium and aluminum oxide, and contained all the elements present in the electrolyte. The aluminum concentration in the film was dependent on the concentration of aluminate ions in the electrolyte. The transition from the microarc to arcing stage took place when the alloy surface was completely covered by the anodic film [56].

Magnesium Oxide Coating with Low-Temperature Electrolytic Plasma The synthesis of oxides in a low-temperature electrolytic plasma allows one to cover the surface of magnesium and its alloys with multifunctional ceramic oxide coatings in the same manner as previously shown for aluminum alloys. Remarkable increases in corrosion and wear resistance have been obtained. However, the commercial processes are limited to a coating thickness of generally 20–50 μm, mainly due to economical aspects (long treatment times) [60, 61]. For good wear protection, the layer should be sufficiently thick without becoming too brittle, as the load-bearing capacity of the magnesium substrate is low. For corrosion protection, it should be dense without through-going pores or defects.

The ceramic oxide coatings were produced on specimen plates (100 mm × 15 mm × 3 mm) of the magnesium alloy BMD10 (0.8% Zn, 7.1–7.9% Y, 0.63% Cd, 0.5% Zr, remainder Mg), which were immersed in an electrolyte containing potassium hydroxide and sodium silicate. The synthesis of the ceramic oxide layer was achieved using a cathodic-to-anodic current density ratio of one ($I_c/I_a = 1$). The treatment times were 20, 50, and 55 minutes, resulting in a layer thickness ranging from 40 to 120 μm. The dominating phase is MgO, followed by smaller amounts of Mg_2SiO_4 and Mg. The interface is fairly rough and it appears as if the roughness of the interface decreases with treatment time. From the SEM micrographs, the thickness can be estimated from nearly 50 μm for the 20 min treatment up to 125 μm for the 55 min treatment. The pore density is around 500 pores/mm^2 for all treatments, but only a very small fraction of all visible pores go through to the magnesium substrate, which is important for the corrosion resistance of the layer. Their density is 50 times lower than the total number of defects for the best performing coating after the 55 min treatment [62].

Electrolyte Agitation Power ultrasound enhances the growth rate of anodic coatings, possibly because of an increase in the rate of mass transfer of reactants and products through the film. It also plays an important role in the formation of ceranic coating structure and the distribution of coating composition. Power ultrasound at a constant frequency of 25 kHz was applied to magnesium alloy AZ31 and it was stated that an optimal acoustic power value should exist, under the ultrasound frequency of 25 kHz, to avoid layer fracture and obtain thick coatings [59].

The anodic coatings are composed of two phases, MgO and Al_2O_3, no matter whether ultrasound is applied or not. Results clearly show that ultrasound plays a very important role in the formation of the coating structure, the coating thickness, and the distribution of the coating composition. The coating consisted of two layers instead of one when the ultrasound field was applied and the acoustic power value increases to 400 W. The inner layer is compact and enriched in aluminum and fluorine. In contrast, the contents of aluminum and fluorine in the external layer are very low and its thickness is nonuniform [59].

Electrolysis Parameters and Bath Composition Gray and Luan [30] described some patents on the conditions of electrolysis and the bath composition. In order to obtain coatings that have little or no inherent color, that can easily be colored, and that provide a satisfactory adhesive base for lacquering or subsequent processing, a low-alkali aqueous electrolyte bath, containing borate or sulfate anions and phosphate and fluoride or chloride ions, adjusted to a pH of 5–11 and preferably 8–9, is employed. A direct current is applied and is either briefly turned off or its polarity is incompletely reversed to allow the formation of manganese phosphate and magnesium fluoride or magnesium chloride and optionally magnesium aluminate. Amines such as pyridine, β-picoline, piperidine, and piperazine or methanamine are especially appropriate for buffering the electrolytes and dissolve readily in water. It is recommended to anodize at 1–2 A/dm^2 with a voltage that increases preferably to 400 V [63]. Polybasic organic compound is used to get an almost nonporous insoluble metal-oxide–organic complex [64–66] or a multicomponent electrolyte is used for coating that gives superior decorative quality and corrosion and abrasion resistances [67]. Another anodizing process for forming a chemically stable and hard spinel compound of MgO–Al_2O_3 on magnesium surfaces has been invented [3, 68].

Environmentally Friendly Electrolyte A new anodizing process, based on an environmentally friendly electrolyte solution that contains no chromate, phosphate, or fluoride but can enhance the corrosion protection of magnesium alloy significantly, is currently being investigated. It uses potassium hydroxide and sodium carbonate, at a temperature of 5–85 °C, with a terminating voltage of 150 V; and sodium silicate and sodium borate, with a current density of 5–500 mA/cm^2, for a time of 10–80 min. Despite the presence of micropores that do not traverse the entire film and some flaws, the new film can form a relatively compact, intact, and uniform barrier layer that can protect the substrate against corrosion attack. The constant applied current can ensure that anodic films grow at an almost uniform rate. Higher current density and lower solution temperature benefit the film growth. Higher voltage achieves a thicker film when constant current is provided. By comparison with the films produced by the two classic processes (Dow17 and HAE), the new film can provide more effective corrosion protection to the substrate [69].

Low Potential Electrolysis The anodic behavior of Mg–Al–Zn alloy (AZ91D) under low potential electrolysis in 3 M KOH solutions was studied. Electrochemical measurements were carried out potentiostatically at 25 and 65 °C in 3 M KOH solutions, with and without addition of 0.5–5 M Na_2SiO_3, using a conventional cell with three electrodes. Anodic films incorporating silicon were formed during electrolysis, and the films formed under constant potential electrolysis at 4 V in 3 M KOH solution with Na_2SiO_3 were uniform and thicker than the films formed without Na_2SiO_3.The anodic film formed with 1 M Na_2SiO_3 for 60 min at 65 °C exhibited the highest corrosion resistance in the present study, correlating well with the quality of the film produced. A few atomic percent of silicon was present as Mg_2SiO_4 in the films, although the main compound was $Mg(OH)_2$. The anodic polarization behavior of the anodic films in 0.1 M KCl solution was used for evaluating the corrosion resistance of the films. The corrosion resistance of the films formed in solutions with Na_2SiO_3 increased in an anodic polarization test in 0.1 M KCl solution [37].

15.3.5. Forms of Surface Corrosion: Anodized or with Conversion Treatments

Generally, test results indicate that anodizing (e.g., Magoxide-Coat or Anomag) gives similar protection to powder coating. The mechanical strength of anodizing is better than standard pretreatment plus powder coating [70]. However, the physical misfit due to the smaller volume of the oxide as compared to the metal (11.3 versus 14.0 cm^3/mol) plays an important role during free corrosion. The corrosion properties of various anodized layers were studied in 5% NaCl solution by measuring electrochemical polarization curves (pH 6 and 10) and immersion tests at constant pH value (pH 6). The potential–current curves of anodized surfaces show that the corrosion current is dramatically reduced as compared to the untreated material. However, the form of corrosion is much more localized for anodized surfaces because of surface defects, and this can lead to deep pitting.

The defect density and especially open defects are the most important parameters for the performance of the plasma electrolytic hard ceramic coatings and is not influenced by the layer thickness even to more than 100 μm. The best performance for conventional anodized films was obtained for the thickest layer (125 μm) after 55 min of treatment, which also had the lowest defect density. With increasing layer thickness, the exchange of electrolyte in the defects becomes increasingly difficult so that the pH value can rise in the defects. The amount of $Mg(OH)_2$ in the pores and defects increases with time. As it expands, it may cause delamination of the interface in combination with evolved hydrogen and leads to failure. Therefore a sealing of the coating appears absolutely necessary to enable long-term exposure in aggressive environments [62].

15.3.5.1. General and Localized Corrosion of Anodized Surfaces

General corrosion of anodized surfaces is measured by polarization resistance and represents the corrosion of the metal–solution interface through the pores of the anodized film, a very close sort of localized corrosion of the anodic bare sites. Salt spray accelerated testing is used currently in spite of certain disadvantages (severity and stable pH; see Chapter 18). The corrosion forms obtained by the test can correspond to general and/or localized corrosion. Corrosion of anodized films on aluminum is evaluated after FACT (formerly ASTM B538, "Testing Anodized Aluminum Specimens"). The electrolyte is composed of 5 wt% NaCl similar to that of the sprayed solutions in the Salt Test (ASTM B117) and the copper accelerated Salt Spray Test (ASTM B368). The specimen is made the cathode to generate a high pH at the defects [71]. Although this test is especially critical for aluminum alloys because of their cathodic or alkaline pitting (see Chapter 14), the standard may also apply for anodized magnesium alloys.

Anodized magnesium coatings with good corrosion resistance can pass a 1 month immersion test in 3.5% NaCl or the 1000 hour salt spray test according to ASTM B117. The

Table 15.1 Corrosion Resistance of Some Anodized Coatings in Standard Salt Spray Test

Coating system	Corrosion resistance (h)
AZ91 HP untreated	0–10
AZ91 HP + Magoxid(MC) 25 μm	80–100
AZ91 HP + MC + sealing water glass	250–300
AZ91 HP + MC + sealing silane	430–600
AZ91 HP + MC + EP-powder paint 60–80 μm	750–1000
AZ91 HP + MC + silane + EP-powder paint 60–80 μm	>1000

Sources: References 3 and 72.

surface appearance should receive a rating of 9 according to ASTM D1654-92. An unsealed plasma Keronite surface treatment on AZ91 alloy (35 μm thick) showed similar resistance. Magoxid coatings with and without different sealing treatments were exposed to a standard salt spray test (DIN 50021). It was found that the corrosion resistance increased according to the treatments listed in Table 15.1 [3, 72].

Satoh et al. [73] have studied general and pitting corrosion of AZ91D after anodizing (Dow17). An alkaline degreasing pretreatment was done in sodium hydroxide and sodium phosphate solution, and an optimal sealing of the anodized surface was done in 50 g/L of Na_2SiO_3 solution at 25 °C for 900 s. The salt spray test was conducted according to JIS Z2371. The weight loss was independent of film thickness, but the estimated maximum penetration depth increased and the corrosion area ratio decreased with an increase in anodizing film thickness. The Dow17 and HAE coatings applied on test plates such as ZE41A alloy have little or no inherent salt spray resistance and corrosion sites were observed after 48 hours. However, after sealing even galvanic coupling cannot lead to corrosion after 165 h of salt spray testing [3].

The Anomag coating process (Magnesium Technology Ltd.) on the alloy AZ91 produces two different layers of 10–15 μm and 20–25 μm thickness. The porous film formed on AZ91 using the Anomag process consists mainly of magnesium phosphate such as $Mg_2(PO_4)_3$. The subsequent sealing and painting of the film affects the porosity, but not the composition of the film. The presence of an anodized film on the surface reduces general corrosion of the AZ91 alloy (ASTM B117). The most effective protection against general corrosion is obtained when the film is sealed and painted, with the corrosion rate reduced by 97% [53].

The corrosion behaviors of surface systems consisting of a chromating layer (MIL-M3173C, Type III), a chromate-free conversion treatment based on either fluoro complexes or phosphate permanganate, and a special hard anodizing layer (Magoxid-Coat), all with an identical coating system on top (silane sealing), were compared. The investigations were carried out on magnesium alloy AZ31B (Mg3AlZn), by performing the filiform corrosion test (DIN EN3665), the VDA-Wechseltest 621-415, and the salt spray test (DIN 50021SS). The best corrosion resistance results were obtained for the anodized layer, followed by the chromate-free passivation (MnO_2/MnO_3 principle) and the conventional MIL chromate treatment [3, 74].

The corrosion protection performance of selected anodized coatings on magnesium has been compared to that of traditional chromate conversion coating and vibratory finishing. Testing was carried out on die-cast AZ91D specimens for 80 days (General Motors GM9540P) and degradation was evaluated according to ASTM D1654. The performance of these anodizing coatings was better than that of the chromate treatment coatings. The best corrosion protection could be achieved by an Anomag/sealant system, followed by Magoxid/sealant and Cr-free passivation treatments [3, 75].

The Magoxid–Coat exhibited superior behavior in salt spray testing (DIN 50021) and also in the Taber abrasion test if compared to that of HAE and Dow17. Surface roughness after the treatment was lower for the Magoxid-Coat process ($R_Z = 7.66\,\mu m$) in comparison to HAE ($R_Z = 12.66\,\mu m$) and Dow17 ($R_Z = 18.26\,\mu m$). The original surface roughness of the AZ91 substrate was 2.25 μm for all treatments. It is important to state that even the relatively weak performing coatings, such as HAE, if sealed in a proper way, are able to provide years of service without problems, for example, in aircraft applications [3, 47, 75].

15.3.5.2. Galvanic Corrosion

Galvanic corrosion and pitting are the predominently examined forms of corrosion. Anodization of the samples was carried out using the Anomag process. Samples with two different film thicknesses were produced (10–15 μm and 20–25 μm thick). SEM examination showed that the film formed during the anodization process is porous. For a film thickness of 10–15 μm, the "anodization only" treatment produced a pore size of 6 μm and a porosity of 13%. Combined sealing and painting reduced the pore size to 3 μm and the porosity to 4%. The presence of an unsealed anodized film on the AZ91 alloy surface does not offer any protection against galvanic corrosion. However, when the film is sealed and painted, it offers superior protection against galvanic corrosion. During a 240 hour salt spray test, with a sealed film thickness of 20–25 μm, galvanic corrosion was completely eliminated even when the coating on the steel bolt was severely corroded and the steel bolt showed red rust over a large portion of the bolt head surface [53].

Electrochemical test results of AZ91 alloy showed that anodization, followed by sealing and painting, improves general corrosion resistance and significantly increases galvanic corrosion resistance better than base, anodized, and anodized/painted surfaces [53]. The internal galvanic corrosion caused by second phases or impurities could be avoided at least partially by careful surface treatments. Agitation or any other means of destroying or preventing the formation of a protective film leads to increasing corrosion kinetics. The better compatibility between magnesium and a second metal is determined by lower potential difference ($E_K - E_A$) and higher polarization resistance. The most compatible materials are the aluminum alloys of 5xxx and 6xxx series because of their relatively low potential difference, and the layers of 80Sn/20Zn due to their high polarization resistance. Steel, stainless steel, copper, nickel, and copper-containing aluminum alloys (e.g., A380) are incompatible [73].

Anodizing (Magoxide-Coat or Anomag) gave similar or better protection than standard pretreatment plus powder coating (60–80 μm thickness) against galvanic corrosion. To prevent galvanic corrosion, not only the anode (magnesium) but also the cathode and the electrical contact between anode and cathode should be isolated. It is well known that it is even better to coat the contact cathodic partner rather than the anodic one especially if the anode-to-cathode surface area ratio is low [3].

The Keronite coating shows improved galvanic corrosion resistance if sealed with JS500 top coat. There was no change in weight and appearance observed for the magnesium after a cyclic corrosion test according to General Motors test GM9540P. For the test, the Mg samples were fastened with zinc-plated and TriPass EVL 1000-M $10 \times 50\,mm^2$ steel bolt and tightened to 5 N.m, and subjected to 80 cycles. However, even unsealed Keronite coated (35 μm thick) AZ91 showed good corrosion resistance after a 1 month immersion test in 3.5% NaCl and after a 1000 hour salt spray test according to ASTM B117. The surface appearance achieved a rating of 9 according to ASTM D1654-92 [3].

Localized corrosion of the Tagnite-coated plates was measured using plates coated with thin (Type I) and with thick (Type II) coatings. The Tagnite-coated plates revealed substantially less corrosion and shallower pits than that on the HAE and Dow17 plates. For contact corrosion testing, sand-cast magnesium alloy AZ91E test plates coated with Tagnite-8200 Type I, Dow17 Type I, and HAE Type I were wet assembled using a cadmium-plated steel washer and bolt and placed in a salt spray chamber (ASTM B117 Test) for 1000 hours. A much greater galvanic attack was found on the Dow17 and HAE coated plates than on the Tagnite-8200 coated plate. The Tagnite coating can provide convenient corrosion resistance due to its uniform coating layer, better morphology, and smaller pores [3, 76].

15.3.5.3. Metallurgically Influenced Corrosion: Microstructure and Alloying Elements

The quality of the substrate showed important influence on corrosion initiation and kinetics. Two anodization waveforms (A and B) were employed on the WE43 magnesium alloy containing rare earth elements, developed by Magnesium Elektron Ltd. [77]: waveform A, the desired constant current density (i_1), was maintained for a period of time t_1, followed by a period of time (t_2) of decreasing current to produce an oxide film of a particular thickness. In waveform B, the current density (i_1) was kept constant throughout the anodization process for t_1 by gradually increasing the voltage from V_0 to V_2 at a ramp rate determined by the magnitude of i_1 (V_2 is reached more quickly at higher current densities). The oxide films thus formed during two types of waveforms employed in the ac/dc anodization are all composed of an underlying barrier layer and a thick, porous oxide film. The thickness of both the barrier and porous films depends linearly on the final voltage, but not on the current density and the type of waveform employed [4]. The anodic oxide film formed in the alkaline silicate anodizing bath is composed of MgO, $Mg(OH)_2$, SiO_2, and MgF_2, with the molar ratio of MgO to $Mg(OH)_2$ being close to 2 : 1 [7, 9, 78]. Due to the similarity in porous microstructure, composition, and thickness of the anodized coatings, the difference in corrosion resistance of the anodized commercial magnesium alloys is mainly due to that of the original substrate [79].

The porous characteristics of the anodized coatings should be considered and the presence of some "through holes" in the anodized coating will allow the aggressive medium to reach the substrate. In this case, the corrosion resistance of the substrate alloy at the bottom of the "through holes" is a determining factor [35, 80]. It has been found that the corrosion resistance of the same anodized coating can be different for different magnesium alloys and can therefore have different corrosion resistances for different magnesium alloys substrates such as AZ31B wrought sheet and AZ91B die-cast plate [81, 82].

Kotler et al. [83] and Blawert et al. [3] also showed that the same anodized coating had a different corrosion resistance if it was formed on different magnesium alloy substrates, such as AZ31B wrought sheet and AZ91B die-cast plate. Khaselev and Yahalom [84] have looked at the influence of the aluminum content and the amount of the β phase in the magnesium alloy substrate on the anodization process, and found that the aluminum content in the alloy is beneficial for passivity of Mg–Al alloys. The alloys, designated as Mg–1Al, Mg–5Al, Mg–10Al, Mg–22Al, and Mg–41Al, contained roughly 1 wt %, 5 wt %, 10 wt %, 22 wt %, and 41 wt % aluminum, respectively [3, 80].

The effect of the β phase in Mg–Al alloys on the corrosion performance of an anodized coating was studied. It was found that the corrosion resistance of the anodized coating was closely associated with the corrosion performance of the substrate alloy. In particular, Mg alloys with a dual phase microstructure of $\alpha + \beta$ with intermediate aluminum contents (i.e., 5%, 10%, and 22% Al) after anodization had the highest corrosion rate and the worst corrosion resistance provided by the anodized coating. The poor performance of an anodized coating was attributed partly to lower corrosion resistance of the substrate alloy and partly to the higher porosity of the anodized coating. The anodized coating on α phase had fewer pores and was smoother and more continuous; that on β phases had many tiny pores and large elongated curved defects. The anodized coatings on two-phase alloys were coarser in the areas close to the two-phase boundaries. The coarser areas were found to be around the boundaries of α and β phases. The low corrosion resistance of the anodized Mg–5Al and Mg–10Al alloys can be ascribed to the high density of through-holes in the coatings [80].

Influence of Alloying Elements For the same type of alloy, the impurity level (Fe, Ni, and Cu content) can determine the corrosion performance of an anodized coating applied on these. A primary HP AZ91 alloy coated with Magoxid-Coat showed much better corrosion resistance than a low-purity secondary AZ91 with the same coating [3].

A pretreatment anodization under controlled anodizing currents and potential waveform was used, and a post-treatment was applied systematically on three Mg–Zn alloys containing 0.5%, 1%, and 2% Zn, respectively. The anodized coating, Castanodise, contained pores measuring several micrometers deep. The exposure tests in 5% NaCl for 30 hours showed that the order of the corrosion resistance of the substrate alloys was Mg–2% Zn > Mg–1% Zn > Mg–0.5% Zn [85]. These alloys were mainly single phase, except for Mg–2Zn, which had few small zinc particles randomly present in the alloy.

Anodization of the samples was carried out on AZ91 using the Anomag process. Two different amounts of Mn, Al, Zn, and Li were used and the third containing Zn, Y, Cd, and Zr. The ceramic oxide coatings were produced at a cathodic/anodic current density ratio of 1. The corrosion properties were investigated in 0.3% HCl solution (pH 3.1) and 0.3% NaCl solution (pH 6.3). Generally, the corrosion potential becomes nobler with the formed ceramic oxide coatings and the current corrosion rates, deduced from potentiodynamic curves, decrease by one to two orders of magnitude. The corrosion resistance depended on the basic alloy, the pH, and the chloride concentration. Also, the coating thickness and the porosity influence the corrosion current values and this caused an anodic dissolution at the pore tips associated with cathodic sites at the side walls of the pores [61].

15.3.5.4. Corrosion Fatigue

Corrosion fatigue and fatigue properties are the most examined characteristics of anodized or coated alloys in the presence of external stresses, if compared to stress corrosion cracking of anodized surfaces. The influence of protective coatings on corrosion fatigue resistance varied depending on certain parameters. Anodic coatings showed small protective effect, but with certain organic materials added, they raised corrosion fatigue strength values to those obtained in air [3, 86]. Ogarevic and Stephens [86] mentioned that the anodic surface coating on some Mg alloys greatly increased the resistance to corrosion, but significantly reduced fatigue strength. However, it was found that anodizing the samples considerably reduced the range of values in air [78].

Not only anodization but also some conversion coatings can decrease corrosion fatigue resistance of a base Mg alloy. Chromate conversion and anodic coatings can be used to improve environment resistance, although such coatings have been reported to be detrimental to fatigue resistance. More recent work, however, has shown this reduction results from prior acid cleaning, not the coating itself. Use of an alkaline cleaner or shot peening prior to acid cleaning can prevent this problem, with the coating then improving the fatigue and corrosion fatigue resistance [87].

It has been suggested that localized surface heating effects during anodization may be responsible for introducing defects in the metal surface, which may then result in a degradation of mechanical properties. Although the residual surface stress could be reduced by adjustment of the film structure and chemical composition, and the substrate age softening can be reduced by shortening the treatment time, no adequate conventional anodizing techniques have so far been found to minimize the risk of premature fatigue failure for Mg alloys [3, 44].

Some anodizing processes such as Dow17 and Anomag show a slight decrease in certain mechanical parameters and do not reveal a clear difference between anodized and nonanodized surfaces, except that the scattering of the *S–N* fatigue results was less for Anomag than for Dow17 treatments [43]. AM60B samples were supplied for Anomag coating and these were coated to two thicknesses, 10 and 20 μm. The thicker coating always returned slightly lower mechanical properties than the thinner coating, but the thinner one exhibited an improved performance over the uncoated samples. This can be due in part to removal of surface crack initiation sites, as the coating itself has little ductility and would not be expected to contribute to the yield strength of the component [78]. The alloy AZ91 coated with 20 μm Magoxid-Coat showed no influence of the coating on the fatigue strength when compared to that of the pure alloy (DIN 50100). However, different results have also been reported, for example, a drop of 30–40% in fatigue strength at a higher loading of 125–150 N/mm^2 [3, 72].

The plasma electrolytic oxidation (PEO) technique could be an approach to reduce this risk [54, 57]. The improved Keronite process provides dense and uniform ceramic oxide layers with a fine-grained microstructure, which is more favorable for components experiencing fatigue loading. However, the fatigue strength of the base metal can be reduced during hard anodizing, especially in thicker films, and the produced ceramic layer can be brittle and in adequate for certain applications [3, 30].

A rotating bending fatigue tester at ambient atmospheric conditions showed that Keronite coatings for two thicknesses of ~7 and ~15 μm produced in different current regimes may cause no more than a 10% reduction in endurance limit of the studied Mg alloy (ASTM E468). At the endurance limit, the transition to the nonfatigue region for the oxidized samples occurs substantially earlier than for the bare Mg alloy, due to the inhibition of the crack initiation by the ceramic oxide layer. A probable cause of this reduction seems to be distortion of the metal subsurface layer rather than structural defects introduced by the oxide film. The results also suggest that the fatigue cracks in oxidized samples were formed and propagated at a higher stress intensity factor, which is probably caused by stress concentrations at the structural defects in the oxide layer and/or the heat treatment affected zone [3, 54].

Many anodizing processes for magnesium provide some increase in corrosion resistance, but often not a substantial one. This may be due in part to residual porosity in the anodic coating. In the case of spark anodizing, it may also be influenced by the characteristics of the process in which a high-energy disruptive event takes place close to the substrate surface. The Anomag process deliberately avoids the intense, localized heating that is believed to be associated with some other magnesium anodizing processes and that may adversely affect mechanical properties of the components to which it is applied [78].

15.3.5.5. Environmentally Influenced Corrosion

It may, however, be expected that improvements are possible as long as elastic or plastic deformation of the component does not result in cracking of the anodized surface. If the specimen is sealed with elastic material, the improved corrosion resistance of the layer system should prevent stress-corrosion cracking (SCC) or environmentally induced corrosion (EIC) [3].

15.4. SURFACE MODIFICATION

The different modification techniques are summarized in Table 15.2 and they are arranged by the approximate qualification of each technology [88].

15.4.1. Chemical and Physical Vapor Deposition

Some chemical vapor deposition (CVD) processes are not feasible for magnesium because of the high substrate temperature that cannot be allowed. There is, however, some low-temperature CVD processes such as the metal-organic (MO-CVD) method that has been adapted to magnesium, forming a TiO_2/Al_2O_3 film; a physical vapor deposition (PVD) technique was also used with SiC [89, 90]. In the case of plasma or thermally sprayed coatings, for example, adhesion can be a problem, unless the substrate surface is significantly heated, which is not always possible. In some cases, such coatings offer predominantly corrosion protection. Tribological behavior–exhibiting magnesium alloys have high wear rates in their untreated state and are subject to galling.

The use of gas-phase coating processes and laser surface melting/alloying/cladding to modify the surface or create coatings on magnesium are excellent alternatives with respect to environmental impact. These techniques produce very little and in some cases no hazardous waste. However, the capital cost associated with these techniques is much higher than solution phase coating technologies [6].

Aluminum Nitride Coatings by Arc-PVD Technique Generally, films produced by PVD methods have many applications, such as hard and thin layers in optics and microelectronics and as protective coatings against corrosion, in spite of the presence of defects in the ceramic coatings. PVD nitride coatings can show a better degree of corrosion protection of the substrates if care is taken to minimize these defects during and/or after the coating process [91]. Nitride coatings, such as TiN, CrN, AlN, and (Ti, Cr)N, have been the subject of intense studies [92, 93].

Aluminium nitride (AlN) films were coated on magnesium alloys (AZ31, AZ61, AZ63, and AZ91) using the PVD technique of dc magnetron sputtering, and the influence of the coating on the corrosion behavior of the magnesium alloys was examined. The substrate (magnesium alloys) to target (pure aluminum) distance was 90 mm [94]. The AlN films were deposited using the following process parameters: bias, −60 V, pressure, 0.4 Pa; and magnetron current, 5 A. The coating period was 90 min. An Al interlayer was deposited onto the substrates before the AlN coating to obtain good adhesion and corrosion resistance. The size of the defects was ~100–150 μm in diameter. The increase in the surface smoothness and the corrosion resistance of the substrate increased the protection ability of the AlN. Lower anodic current densities were observed for the alloys in 0.6 M NaCl solution as compared to that of the uncoated ones. A passive-like behavior state was observed for

Table 15.2 Properties of Several Surface Modification Techniques

Process	Modified/coated layer	Thickness, color	Surface roughness	Corrosion resistance	Installation cost	Toxic matter
Chemical conversion coating (Dow7)	Oxide/hydroxide of Mg and Cr	1–2 μm, light to dark brown	Negligible	Good	Low	$Na_2Cr_2O_7$
Hydride coating	MgH_2	1 μm, light to dark gray	Negligible	Good	Low	—
MO-CVD	SiO_2, Al_2O_3, CrO_3, TiO_2, etc.	1–5 μm, transparent	Negligible	Excellent	Low	—
Anodizing (HAE)	Oxide/hydroxide/ fluoride layer of Mg,Al,Cr	5–30 μm, light tan to dark brown	Small to medium	Excellent	Low	$Na_2Cr_2O_7$ for Post-treatment
Thermal spray coating	Mg–Mo, Mg–Co–WC, ZrO_2, etc.	100–2000 μm	Large	Good to Excellent	Average	Fused materials
PVD (ion plating)	TiN, etc	1–5 μm, transparent	Negligible	Excellent	High	—
Ion implantation	Mg containing N, C, Cr, Ta, etc.	<1 μm (effect zone is up to 100 μm), transparent	Negligible	Excellent	High	—
Laser surface melting	Fine-grained microstructure	<500 μm	Small to large	Good	High	Laser beam
Laser alloying	Mg–Si,Mg–Cu, etc.	100–2000 μrn	Large	Excellent	High	Laser beam

Source: Reference 88.

AlN-coated AZ91 alloy, and the highest improvement in the corrosion resistance was observed for AlN-coated AZ91 alloy due very probably to its smoother surface as compared to the other magnesium alloys [95].

After corrosion, it can be seen that the main corrosion source of the PVD-coated alloys are the defects in the coating (e.g., pores, pinholes). These defects are closely related to the surface quality of the substrates (e.g., surface morphology and defects) [96]. Also, inherent porosity arises from the columnar structure of the coatings. After corrosion experiments, some microcracks were formed in the coatings and these could increase the corrosion rate. The amount and the dimensions of defects depend on the kind of PVD process and on the deposition conditions, especially the substrate temperature and the bias voltage [92].

15.4.2. The H-Coat and Magnesium Hydrides

The presence of MgH_2 as an intermediate of corrosion reaction has been examined [97]. The application of magnesium alloys as a hydrogen storage material is also anticipated [98]. However, there seem to be few studies of either MgH_2 or other hydrides utilized as surface coatings. Magnesium-based hydrides are well known to have a high hydrogen-storage capacity [99]. Two different methods have been provided for hydrogen surface modification of high-purity magnesium (hp Mg) and magnesium alloy AZ91. One was electrochemical ion reduction (EIR) of hydrogen from an alkaline electrolyte on a Mg-based cathode. The other was plasma immersion ion implantation into Mg-based substrate. The Auger electron spectroscopy (AES) measurements provided a more accurate compositional depth profile. However, the presence of hydrogen in the bulk metal, which could not be identified by AES, was verified by secondary ion mass spectroscopy (SIMS) analysis.

15.4.2.1. The H-Coat

Nakatsugawa [100] developed a new surface treatment (hereafter called the H-coat) based on the formation of magnesium dihydride. The test specimen was etched in 5 wt % HF solution for 20 s. A cathodic intermittent current was applied to the specimen in the electrolyte containing 10 mol/m^3 NaOH and 200 mol/m^3 Na_2SO_4. The intermittent current has a frequency of 0.5 Hz with square-wave amplitude of 500 A/m^2, which was generated by a function generator and current source. During the electrolysis, hydrogen gas was evolved on the surface by the electrolytic decomposition of water, which is partially absorbed and diffuses into the interior of the specimen to form the metal hydride. The depth of hydrogen penetration into AZ91D specimen after 1.8 ks treatment is around 1–2 μm, confirmed by elastic recoil detection analysis. The treatment was applied up to 7.2 ks to study the effect of process time. In most cases, it was applied for 1.8 ks [101].

The Morphology of the Deposit Figure 15.5 shows the appearance of H-coated specimens with different treatment times. Prior to the treatment, the surface is etched by 5% HF solution, which darkens the grain boundary. Since the grain boundary contains more Al content and Mg, it is not attacked by the HF solution; the black compound may contain AlF_3 and Al_2O_3. With longer operation time, the darkened area spreads gradually into the grain, which indicates the formation of MgH_2. Therefore it is likely that the etched grain boundary offers a catalytic site for the hydrogen absorption. Fluorination has also been found to be effective for improving the initial activation of Mg-based hydrogen storage alloys. After etching, the state and composition of the grain boundary may have an important role as a catalytic site for hydrogen absorption [102].

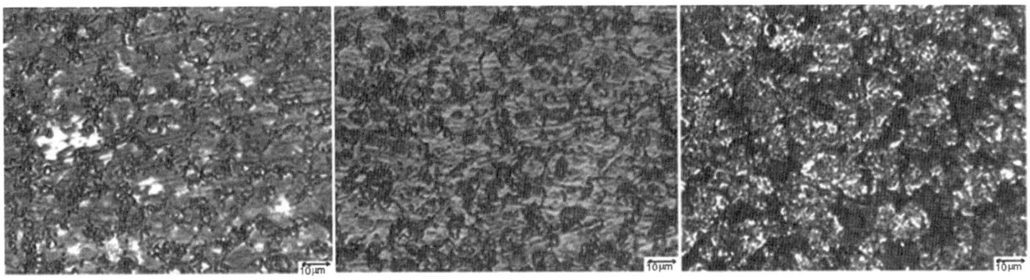

Figure 15.5 Appearance of die-cast AZ91D specimens after H-coat treatment [101].

Corrosion Resistance After electrochemical and conventional corrosion testing, the resistance and the stability of the coating proved excellent from the results of long-term immersion tests and baking treatments. The coating can be applied to pure Mg as well as Mg–Al alloys [101]. Corrosion testing was carried out by potentiodynamic polarization in chloride-containing aqueous solutions of pH 7 and 12. A significant improvement in the corrosion resistance of H-modified surfaces was verified. The improved corrosion resistance and the depth to which hydrogen penetrated the Mg-based substrates are strongly influenced by the alloying elements [103]. The hydride compound is a good base for painting. Because water is consumed during the process, maintenance of the bath requires the pH be kept in an appropriate range. As the process is easy to maintain with modest operating cost, its application as an industrial coating for Mg products is favored. Possible fragility, if any, of the Mg alloy due to hydrogen absorption during cathodic polarization should be addressed [100, 101].

The treated Mg alloys show a pseudo-passive behavior in the anodic region and an increase of the Tafel slope in the cathodic region. Longer treatment provides better passivity. H-coating shows better performance with AE42 alloy than the AZ and AM alloys. This is because the alloy contains 2%RE elements and Mg–RE compounds have a high affinity for hydrogen. RE elements also work as hydrogen storage material [98].

Baking Influence The paint adhesion of an H-coated surface is excellent. However, it is known that MgH_2 releases hydrogen at 563 K under 0.1 MPa. Thus the heat applied during the curing of painted layers may decompose MgH_2, which would induce blisters on the painted surface. To test for effects, an H-coated specimen was heated at 473 K for 1.2 ks in an oven and the effect was examined by polarization curve measurement. The effect of baking on the passive behavior of H-coated AZ91D in 5% NaCl solution was examined. It was found that heating under these conditions does not influence the electrochemical behavior appreciably except for the value of the OCP. The more negative value of OCP after baking was also observed for the untreated specimen, so we assume that the magnesium oxide/hydroxide film may be influenced by this heat condition. MgH_2 is thermally decomposed during the recycling stage, but the damage to melt quality would be negligible [101].

Effect of Alloying Element The anodic polarization curves of the treated pure Mg, die-cast AZ91D, AM60B, and AE42 alloys in the test solution show a pseudo-passive behavior for the tested alloys and the pure Mg and an increase in the Tafel slope in the cathodic region. The longer treatment provides better passivity. H-coating shows a better performance with the AE42 alloy since this alloy contains 2% RE elements and Mg–RE compounds have a high affinity for hydrogen and RE elements also work as hydrogen storage material. The

Mg–Al alloy with lower Al content shows larger $i_{passive}$ and less noble E_{break}. The kinetics of hydrogen absorption of Mg–Al compounds are low [101].

OCP and Polarization Studies of the Metal–Hydride and Hydroxide Interface The effect of process time, alloying element, and the stability of the coating are investigated by means of weight loss measurements and electrochemical analysis. The H-coat was applied according to the procedure explained in the patent. A hydrogen-rich or magnesium hydride layer can be created on the magnesium surface by cathodic electric charging of the magnesium alloy in aqueous solution [100, 101].

Samples were cut measuring 50 mm × 40 mm × 4 mm from pure magnesium and examined alloys and used as the test specimens. The test solution was 5% NaCl saturated with $Mg(OH)_2$ at pH ~10.5. The corrosion behavior of the H-coat was studied by immersing the coated specimen for 504 h (21 days) at 20 °C. The polarization was conducted from OCP to the anodic and cathodic directions with a scanning rate of 1 mV/s. The weight loss corrosion rate was calculated after removing the corrosion product in the cleaning solution containing 2 kmol/m^3 CrO_3 and 59 mol/m^3 $AgNO_3$. A correction was introduced to compensate for the active evolution of hydrogen gas during cleaning.

The potentiodynamic polarization curve was measured after the specimen was kept immersed in the test solution for 300 s at the OCP. Figure 15.6 shows the potentiodynamic anodic and cathodic polarization curves of untreated and H-coated AZ91D specimens in 5% NaCl solution [104].

The coated surface shows quasi-passive behavior and an excellent corrosion resistance in chloride solution. As the measurement was carried out at a different test area, the values of OCP of anodic and cathodic polarization curves are different. In the cathodic range, a Tafel region can be observed in both specimens. The slope of the polarization curve for the untreated specimen is –135 mV whereas the value for the H-coated specimen is –216 mV. The difference of the cathodic Tafel slope at the forward scan and at the reverse scan is reported for pure Mg in 1 mol/dm^3 NaOH solution [105]. As MgH_2 has an insulating nature, formation of the H-coat layer would create an increase in the *IR* drop. In addition, the accumulation of MgH_2 may also influence the reaction scheme [106].

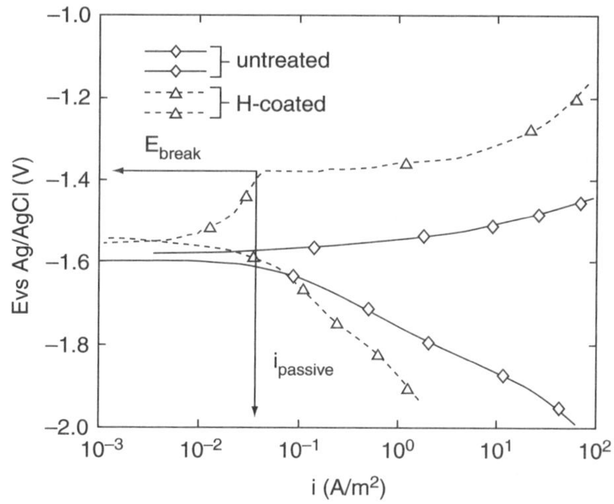

Figure 15.6 Potentiodynamic polarization curves of untreated and H-coated AZ91D in 5% NaCl solution [104].

In the anodic region, a pseudo-passive behavior, which is characteristic for H-coated specimen, is observed. Other chemical conversion coatings, such as Dow20 treatment, did not provide such a behavior and the current increased monotonously, which is similar to the untreated specimen in Figure 15.6. Similar pseudo-passive behaviors have been observed with Mg–Al alloy specimens prepared by rapid solidification processing [107] or the thixomolding technique. In these cases, the formation of a protective hydroxide film enriched with Al content is considered the principal reason. The relation with MgH_2 or the corrosion process is unknown. Hereafter, the pseudo-passive behavior will be characterized by the value of E_{break} from which the current increases rapidly [104].

The corrosion rate of untreated and H-coated AZ91D in 5% NaCl was calculated by the weight loss method after removing the corrosion product in the cleaning solution containing 2 kmol/m^3 CrO_3 and 59 mol/m^3 $AgNO_3$. It was found that the corrosion rate is around 0.02 g/m^2/h throughout the test period, which is generally one-third of that of the untreated specimen. If the accelerated corrosion occurring at the cut section were neglected, the corrosion rate would be much smaller.

Stability of H-Coat The phenomenon of a pseudo-passive behavior is unique to this surface treatment. However, MgH_2 is not thermodynamically stable in the presence of water, which would be gradually converted to $Mg(OH)_2$. So the stability of this film was studied by immersing the specimens in 5% NaCl solution for up to 21 days. With the lapse of time, the gray color of the H-coated surface gradually disappeared. In addition, a small amount of hydrogen bubbles were observed on the treated surface. The bubbles seemed to stay on the surface, which is clearly different from the evolution of hydrogen gas with Mg corrosion. Filiform corrosion, which is typical for Mg–Al alloys, was not observed at the coated specimen, except for the cut section. Because this section is more vulnerable to corrosion compared to the cast surface and difficult to treat, it might be preferentially attacked [104, 108].

Dependence of $i_{passive}$ and E_{break} on Treatment Time A pseudo-passive behavior was recognized with the 360 s treatment. The longer treatment leads to an E_{break} that is nobler and an $i_{passive}$ that is smaller, which means better passivity and an improved quality of the passive layer of the coated surface (Figure 15.7). It seems that $i_{passive}$ tends to saturate and attain a certain value. This is because the diffusion of H_2 through the solid Mg is limited to 10–20 μm. Further treatment may lead to a thinner coating layer, as MgH_2 (density, 1.45 mg/m^3) is more voluminous than Mg (1.74 mg/m^3) [104].

15.4.2.2. Stability of Mg Hydride

Understanding the role of magnesium hydrides in corrosion kinetics and corrosion prevention is of major importance in the area of corrosion kinetics and corrosion prevention. The formation of a magnesium hydride layer on the Mg–Al alloy surface under cathodic charging in aqueous solution has been investigated. The objective was to understand the hydride formation mechanism and to explore the role that magnesium hydride plays in the corrosion process. The investigations involved Mg–Al (~90% Mg, 9% Al) alloy using electrochemical techniques in aqueous solutions. The corrosion mechanism and hydride formation of the studied alloys were analyzed. The hydrogen evolution mechanisms at cathodic potentials were also examined [109].

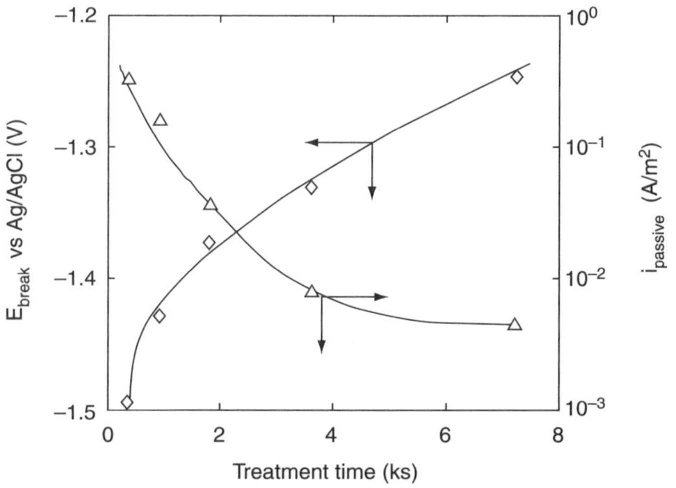

Figure 15.7 Effect of operation time on the pseudo-passive behavior of H-coated die-cast AZ91D [101, 104].

The theoretical Pourbaix [110] potential–pH diagram constructed by Perrault in 1974 indicates that a stable MgH_2 may exist if hydrogen overpotential exceeds 1 V. In the absence of O_2, MgH_2 could be produced cathodically as part of the corrosion reactions and it is important to examine and determine the presence and role of MgH_2 in the process of Mg corrosion. Gu et al. [109] conducted an investigation using cathodic polarization of the magnesium electrode. The approach was based on determination of Tafel polarization behavior coupled with examination and interpretation of ac impedance behavior of the cathodic polarization of Mg, where MgH_2 may be formed in parallel with, or as an intermediate, in cathodic H_2 evolution (the cathodic process in Mg corrosion in the absence of O_2), and finally formation of $Mg(OH)_2$ film during cathodic polarization. A mechanism has been then proposed to describe Mg hydride formation under a cathodic polarization condition [109].

Polarization measurements of steady-state currents provide useful information on the overall kinetics of the electrode process studied (Figure 15.8). The E versus log i relationship (the Tafel plot) and its slopes, (b), provide information about reaction mechanisms, i_o, exchange current density, which characterizes the kinetics of the electrode process at equilibrium. The cathodic potential $-E$ versus log i polarization curve for Mg in 0.1 N NaOH + saturated $Mg(OH)_2$ solution at 25 °C, for steady-state measurements recorded in the potential range of −1.0 to −2.4 V (vs. SCE), is given in Figure 15.6. The polarization measurement exhibits two distinguishable Tafel slopes as indicated by two straight lines. This kind of Tafel behavior implies a mechanism with parallel pathways. The solid line, with a slope of ~255 mV/decade, is attributed to reaction (a) of hydrogen evolution without hydride formation. It is the dominating reaction (indicated by solid line) until potential is more negative than −1.8 V (vs. SCE) [109]:

$$\text{(a)}\quad Mg + 2e^- + 2H_2O \rightarrow Mg + H_2 + 2OH^-$$

The dashed line, dominating in the potential range of −1.8 to −2.0 V (vs. SCE), is reaction (b) of hydride formation. The slope of this region is approximately ~130 mV/decade, which

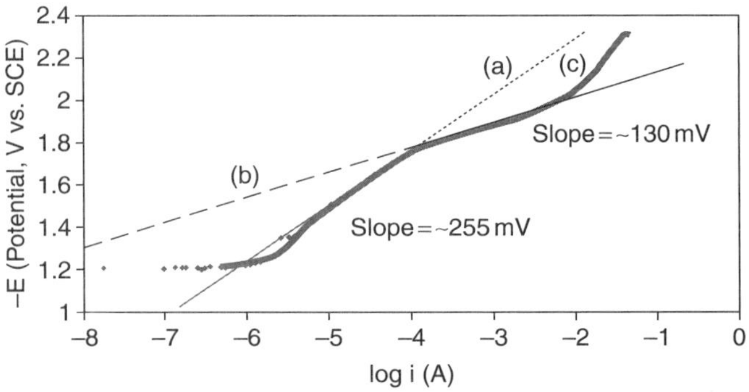

Figure 15.8 The potential ($-E$) versus log i polarization curve for Mg in 0.1 N NaOH + saturated $Mg(OH)_2$ solution, in the absence of oxygen, at 25 °C, recorded in the potential range −1.0 to −2.4 V. Scan speed was 0.167 mV/s and hydrogen gas was bubbled before and during the test [109].

is similar to that reported elsewhere and by Perreault (1978) [97].

$$\text{(b)} \quad Mg + 2e^- + 2H_2O \rightarrow MgH_2 + 2OH^-$$

The second linear region of the Tafel lines, beyond −1.6 V, can be attributed tentatively to Mg hydride formation. The ac impedance measurements are in excellent agreement with the steady-state polarization behavior. An interesting pseudo-inductive loop arises in the potential range beyond −1.6 V, and can be attributed to the hydride formation/decomposition process, with the former indicated at tested potentials. Cyclic voltammery studies confirmed these assumptions [109].

Beyond a potential of −2.0 V, reaction (c) eventually will be the rate-determining step, leading to a limiting current, since it is a potential-independent reaction. Rotating the electrode appears to have little effect and the polarization curve is *IR* drop corrected. This may indicate that such a limiting current is not under electrode kinetic control.

$$\text{(c)} \quad MgH_2 + 2H_2O \rightarrow Mg(OH)_2 + 2H_2\uparrow$$

Reactions (a) and (b) are competing reaction paths and reaction (c) is a path that consumes the hydride under the cathodic polarization condition and that forms $Mg(OH)_2$ by MgH_2 decomposition. Since reactions (a) and (b) are alternative paths, the path that proceeds with the greatest velocity characterizes the kinetics. It has been reported that reaction (b) takes place at a higher overpotential than that of reaction (a) [111]. Therefore it is logical to assume that reaction (a) dominates the Tafel behavior at the beginning of the cathodic polarization whereas reaction (b) becomes significant at high (more negative) polarization. It can be added that $Mg(OH)_2$ may be subjected to reduction as described by reaction (d) [109]:

$$\text{(d)} \quad Mg(OH)_2 + 2e^- \rightarrow Mg + 2OH^-$$

Formation of $Mg(OH)_2$ in Alkaline Solution The behavior of magnesium in aqueous solutions over the pH region 12–13 at 25 °C has been examined. A magnesium alloy polished specimen containing 9% Al and other minor elements, in the shape of a cylinder with geometric surface area of 2.0 cm^2, was used as the working or experimental test electrode. The sample was mounted in a rotating shed, allowing various rotating speeds

Table 15.3 List of Holding Time, Open Circuit Potential, and Polarization Resistance

Holding time (min)	E_{OCP} (V vs. SCE)	R_p (kΩ)
10	−1.196	22.1
20	−1.101	40.7
30	−1.053	34.4

Source: Reference 109.

during the experiment. A Pt electrode, having a large (2.54 cm × 2.54 cm) surface area, was used as the counterelectrode in all experiments. The large surface area of the Pt electrode minimizes the Faradaic current density of O_2 evolution at the counterelectrode during the polarization experiment. One compartment cell was used for the experiments with 0.1 N NaOH + saturated $Mg(OH)_2$ solution. Hydrogen gas was bubbled before and during the experiments [109].

To examine the $Mg(OH)_2$ film formation associated with MgH_2 decomposition, the working electrode was held at the potential of −1.0 V (versus the open circuit potential) for various times from 10 to 30 minutes. After that, the potential control is switched off to allow the electrode to settle to its natural corrosion potential. Polarization (Tafel) measurement was then conducted to determine the polarization resistance. Table 15.3 lists the values of potential and the polarization resistance. It is clear that the rest potential (or open circuit potential) shifted toward the positive direction.

During cathodic polarization, the hydride formation is formed according to the reaction

$$Mg + 2e^- + 2H_2O \rightarrow MgH_2 + 2OH^-$$

The potential shift toward the positive direction is an indication of $Mg(OH)_2$ formation as a result of the following reaction:

$$MgH_2 + 2H_2O \rightarrow Mg(OH)_2 + 2H_2\uparrow$$

The estimated polarization resistance is also increased as the holding time increases but the result is not conclusive. The steady-state test condition here allows the generation of Mg $(OH)_2$. As indicated in the above reaction, $Mg(OH)_2$ film formation would be the result of hydride deformation. However, under the cathodic condition, the $Mg(OH)_2$ film can also be reduced as indicated. The observation of no further increase of polarization resistance for longer holding times may support the proposed mechanism: hydride formation and decomposition resulting in oxide formation, while the following reaction reduces it:

$$Mg(OH)_2 + 2e^- \rightarrow Mg + 2OH^-$$

Cyclic Voltammetry of Hydroxide Film Growth This method is used in the study of the oxide film growth. A linearly swept potential signal, $V(t)$, is applied to the electrochemical system and the response signal, for example, the current $i(t)$, is recorded. In this study, the method used is to investigate the hysteresis behavior of the magnesium electrode involving hydride formation and deformation. The scan speed of 5 mV/s was applied. To further examine the polarization behavior of magnesium hydride formation, a cyclic

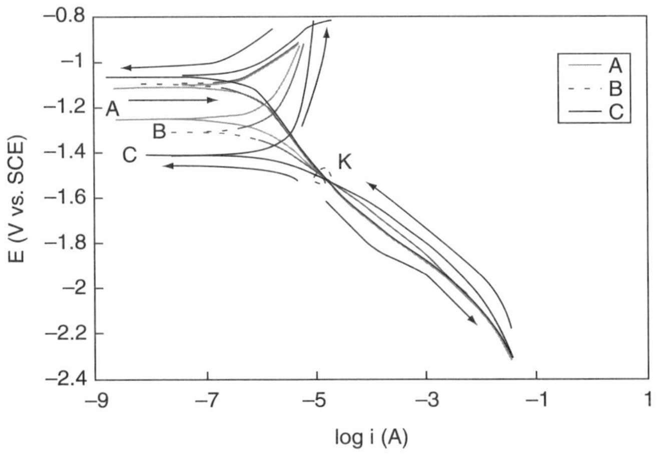

Figure 15.9 Cyclic voltammetry curves (scan direction is indicated by arrows). Potential reverses at (A) −1.7 V, (B) −2.0 V, and (C) −2.3 V versus. SCE. Curves are *IR* corrected. Scan rate was 5 mV/s and hydrogen gas was bubbled before and during the test [109].

voltammetry technique was applied. The potential was cycled from 0.3 V (vs. open circuit potential) to various potentials in the cathodic area [109].

Figure 15.9 shows both ascending and descending curves (indicated by arrows) of potential changes. Scans start at an anodic potential, go to various cathodic potentials ranging from −1.4 to −2.3 V versus SCE (only three curves are shown in Figure 15.9), and then return back to the starting potential [109]. The following observations can be deduced from this figure:

1. There is an equal current point that has been identified. At the constant potential scan rate, regardless of the various degrees of cathodic polarization, there is a point (indicated as K in Figure 15.9) where the forward scan current is equal to the returning scan current. The corresponding potential is approximately −1.55 V (vs. SCE). Such a potential could be the MgH_2 formation potential, below which there is no MgH_2 formation [109].
2. A hysteresis phenomenon does not appear if the potential scan is less negative than −1.6 V (curve A in Figure 15.9). However, significant hysteresis in the polarization behavior is observed in curves B and C, where potential goes beyond −1.6 V. The hysteresis is related to magnesium hydride formation and decomposition processes. A slope change in forwarding scans (curves B and C) is clearly seen, which denotes the formation of magnesium hydride. Interestingly, the current of the reverse scan is larger than that of the forward scan. This indicates very probably oxidation (or decomposition) of magnesium hydride (the reverse of reaction (c)). It appears that the reaction of magnesium hydroxide formation is slow and the magnesium hydride formed in the forward scan is still intact; otherwise, the current of the returning scan should be smaller.
3. The reverse scan rest potential, after a more negative polarization, is more negative than those that have less negative polarization; for example, the reverse scan rest potential in more negative sequence is C, B, and A. The decomposition of metal hydride to form the surface oxide film and significant $Mg(OH)_2$ formation can result in a less negative rest potential [109].

15.5. ELECTROCHEMICAL CHARACTERIZATION OF THE METAL–FILM INTERFACE

15.5.1. OCP and Polarization Studies of the Metal–Oxide Interface

Although it has been found by some researchers that dissolved oxygen has no influence on magnesium corrosion or the polarization curve under certain circumstamces, it is always necessary to examine the state of saturation of the solution with bubbled oxygen or with atmospheric oxygen or without oxygen at all by bubbling argon or nitrogen. It should also be determined if the electrode is steady or rotating and if the solution is stagnant or not; and if agitated or circulated, quantitative description is necessary.

OCP and Corrosion Resistance The corrosion resistance of anodized films on the WE43 alloy containing 1.1% RE elements (except Nd) in an alkaline silicate solution have been investigated [77]. Anodization of the WE43 alloy significantly improves its corrosion resistance and greatly increases the time to pitting in 0.86 M NaC1 solution. No pitting of the anodized WE43 alloy surface in 0.86 M NaC1 solution has been observed visually, even after more than 140 hours of immersion. Excellent corrosion protection of the WE43 alloy can be achieved by ac/dc anodization in an alkaline silicate solution at high voltages (>350 V). The characteristics of the oxide film formed at 30 mA/cm^2 for 5 min, followed by 25 min decreasing current, have been investigated. The change of the corrosion resistance and open circuit potential as well as the ac impedance studies suggested a sequence of three stages (Figure 15.10) [4].

The first stage in Figure 15.10 corresponds to the decrease in resistance and OCP and can reflect the hydration of the oxide film and penetration of water and sodium chloride. In the first period, impedance measurements show a decrease in R_1 and CPE1, followed by a decrease in R_2 and CPE$_2$, which could signify the trend of a thinner barrier with tiny pores and cracks leading to an increase in the conductivity of the film through the pores and the infiltration of sodium chloride. During the second stage, it is suggested that MgO is being

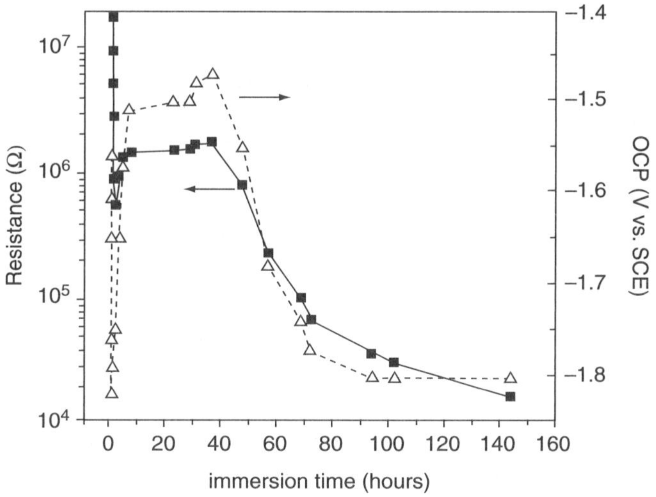

Figure 15.10 Corrosion resistance (solid line) and OCP (dotted line) as a function of immersion time in 0.86 M NaC1 for WE43 alloy, ac/dc anodized in alkaline silicate solution at 30 mA/cm^2 for 5 and 25 min [4].

converted to the lower density $Mg(OH)_2$ film at the surface inside the pores of the oxide film, since thermodynamically $Mg(OH)_2$ is more stable than MgO in aqueous solutions. This could result in a partial blocking of the pores, because the molar volume of $Mg(OH)_2$ is larger than that of MgO, hence increasing the film resistance. This indicates that the effective surface area of the film increases, and the barrier layer will become increasingly exposed to solution and hydrated with the formation of cracks. The further formation of $Mg(OH)_2$ inside the oxide film could change the mechanical stresses within the oxide film, causing some cracks to develop, as has been stated by SEM observations after 5 and 50 hours of immersion in NaC1 [4].

Corrosion Rate Measurements Specimens from two different magnesium alloys were coated with a ceramic oxide layer produced at a cathodic/anodic current density ratio of 1. The corrosion properties were investigated in 0.3% HCl solution (pH 3.1) and 0.3% NaCl solution (pH 6.3). The results revealed that the corrosion potential became nobler with the ceramic oxide coatings formed and the corrosion rate, deduced from potentiodynamic curves, was decreased by one to two orders of magnitude. The corrosion resistance depended on the basic alloy, the pH, and the chloride concentration. Also, the coating thickness and the porosity influenced the corrosion rate [3, 61].

Corrosion tests on rather thick anodized specimens revealed a clear correlation between pore density and corrosion resistance. Potentiodynamic polarization curves obtained in 5% NaCl aqueous solution (pH 10 adjusted by NaOH, scanning rate 0.2 mV/s) gave a corrosion rate of 3.2 mm/yr for the substrate and 0.001 mm/yr for the 55 min treated sample. It is interesting to note that the thicker layer obtained after 50 min treatment offered less protection than the layer obtained through 20 min anodizing. This indicates that the defect density is the dominating influence rather than the layer thickness. Increasing the layer thickness to more than 100 μm did not reduce the amount of open defects that provided detrimental contact between the magnesium substrate and the electrolyte. Sealing the coating appeared absolutely necessary to enable long-term exposure in aggressive environments [3, 61].

Potentiodynamic Polarization (Breakdown Potential) Accelerated electrochemical test (anodic polarization) results of AZ91D uncoated and coated with 35 μm Keronite are shown in Figure 15.11. The coated specimen displayed a more noble rest potential, E_{corr}, compared to that of the uncoated specimen. A more noble rest potential in deaerated electrolyte generally signifies less susceptibility to corrosion attack. Moreover, the coated specimen displayed greater resistance to initiation of corrosion attack shown by the presence of a breakdown potential, E_b, and a passive potential range, $E_b - E_{corr}$ [3, 112].

15.5.2. Impedance Measurements

Electrochemical impedance measurements (EIMs) are carried out in a tetraborate buffer solution (0.05 M at pH 9.7) for two alloys, AM50 and AZ91D. The scanned frequency ranged from 6 mHz to 100 Hz. The EIS spectra obtained under anodic polarization inside the potential range of the MgO formation exhibit one capacitive loop followed by a linear part for both magnesium alloys. The Nyquist plots of both magnesium alloys at OCP exhibit two capacitive loops, one for high and intermediate frequencies and the other, the small one, for low frequencies. The first capacitive loop is attributed to the charge transfer process. Thus for frequencies higher than 1 Hz, a resistor R_p and a capacitor C_{dl} in parallel can model the electrode–electrolyte interface [87].

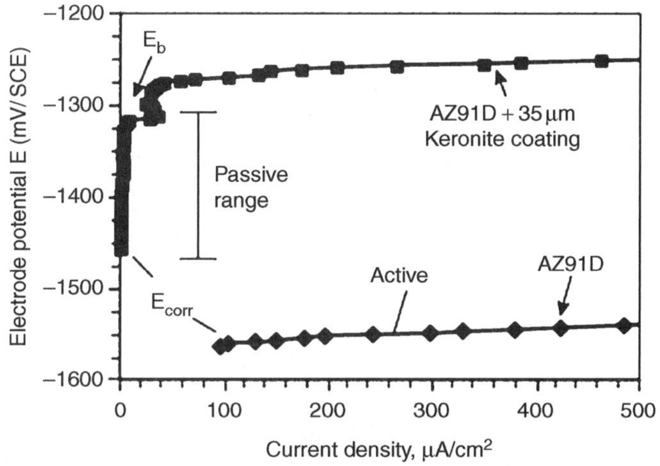

Figure 15.11 Accelerated electrochemical test (anodic polarization) results for AZ91D uncoated and coated with 35 μm Keronite [3, 112].

Partial data fitting, done with Boukamp circuit equivalent software [113] for the charge transfer process, produced the R_p (polarization resistance) and C_{dl} (double-layer capacitance) values. The R_p of the charge transfer process is 207.7 and 374 $\Omega \cdot cm^2$ for AM50 and AZ91D alloys, respectively. The obtained capacitance values are 22.6 and 68 $\mu F\,cm^2$, for AM50 and AZ91D, respectively. The slightly lower value of C_{dl} for the AM50 alloy implies the formation of a thick, protective film on the electrode surface—much lower C_{dl} values being already reported for other Mg-based alloys [114]. The second small capacitive loop is generally attributed to mass transfer in the solid phase [115], which consists of the oxide/hydroxide layers (see Chapters 16 and 18).

An increase of R_p, which is significant in the case of AZ91D alloy, suggests that the layer is growing on the electrode surface. The equivalent circuit consists of a resistor (R_p) in series with a constant phase element (CPE), the two being connected with a capacitor (C_{dl}) in parallel. The CPE can be assumed to be Warburg diffusion according to the n values that are close to 0.5. Thus under anodic polarization, the corrosion process is controlled by mass transfer of the corrosion products through the oxide layers [116]. The Nyquist plots for both Mg alloys obtained under cathodic polarization show one capacitive loop, which is attributed to water reduction [115].

For anodized surfaces produced in sodium fluoride and phosphate electrolytes, the corrosion resistance was markedly improved, compared to the nonanodized material, as measured by the polarization resistance. A value for the polarization resistance of 90 $k\Omega \cdot cm^2$ for the sodium fluoride-anodized film corresponds to only a few microamperes of corrosion current, while the corrosion resistance of the nonanodized surface was 6.6 $k\Omega \cdot cm^2$. Phosphate behaved similar to fluoride by incorporating at the metal–coating interface but did not provide the protective nature of the fluoride layer as determined by impedance spectroscopy [33]. The corrosion experiments were carried out in 0.1 M sodium perchlorate at the open circuit potential of the electrodes, generally near −1.5 V (vs. SCE) in a frequency range of 10^5 Hz to 0.5 mHz. The impedance spectroscopy suggested that the electrical properties of the anodized films were affected by both the extent of the coverage and the barrier characteristics of the film at the bottom of the pores. The capacitance value measured for the nonanodized magnesium electrode was 42 $\mu F/cm^2$.

This falls in the range of normal metal–solution interfaces from 10 to 100 μF/cm^2. The capacitances measured for the anodized surfaces were significantly below this value, in a range of 0.65–9.5 μF/cm^2 [3].

EIS of Magnesium Surface Film and the Inductive Loop Nakatsugawa and Ghali [117] carried out EIS measurements on the surface film formed on AZ91D in 5% NaCl solution saturated with magnesium hydroxide (pH 10.5). Corrosion has been monitored using the square-wave current method. Effectively, a function generator was connected to a galvanostat that poroduced a repetitive symmetrical square-wave current with respect to the zero current that was applied. Figure 15.12 shows an exemple of a Nyquist diagram of EIS obtained for the corroding surface in a chloride alkaline solution. In addition to a *RC* semicircle, an inductive loop is observed in the low-frequency region. The solution resistance can be neglected and two interesting parameters can be deduced, R_{peak} and R_{steady} [117].

The input of square rate current with lower frequency gave a response with an overshooting shape (Figure 15.13) and Figure 15.14 shows the time variation of R_{peak} and R_{steady}. The current inputs with frequency 1 and 0.1 Hz were chosen to measure R_{peak} and R_{steady}. These two resistances increased during the initial period and overshooting was observed only after 30 minutes of immersion and this was accompanied by potential fluctuations. This gave rise to the overshooting and to the inductive loop. R_{steady} decreased with time and several pits accompanied with hydrogen gas evolution were observed on the electrode surface. It is then suggested that R_{steady} is related to the resistance at the corrosion sites while R_{peak} may represent the film resistance [117].

Mansfeld et al. [118] have observed an inductive loop during the corrosion of Mg–Al alloy in salt solution. Shi et al [119] further confirmed the corrosion performance of anodized magnesium alloys as revealed by anodic polarization curves using ac

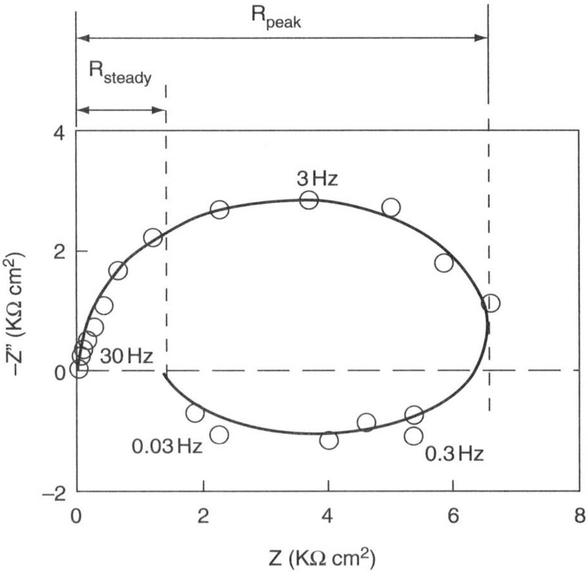

Figure 15.12 Impedance diagram of corroding AZ91D in 5 wt % NaCl solution (pH = 10.5) after 5 hours [117].

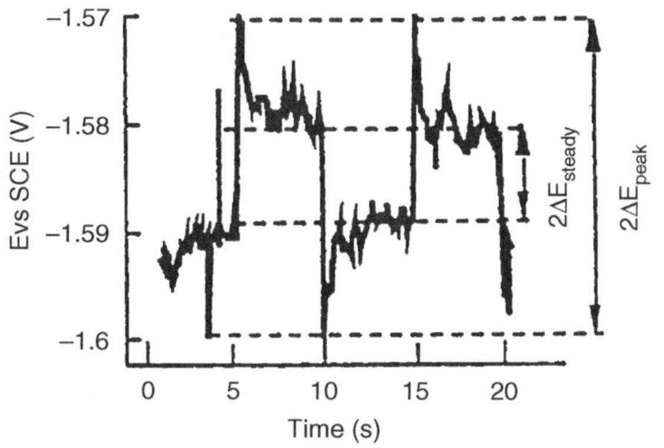

Figure 15.13 Potential response of corroding AZ91D polarized by square-wave current input [117].

electrochemical impedance spectroscopy (EIS) and found that the breakdown of an anodized coating on magnesium alloys corresponded to an inductive loop on the Nyquist plots. The occurrence of an inductive loop during EIS measurements was then observed and attributed to the formation of a surface film on Mg–Al alloy in alkaline chloride solution [117].

EIS, performed on AZ91 magnesium alloys, suggested that the electrical properties of the anodized films were affected by both the extent of the coverage and the barrier characteristics of the film at the bottom of the pores. Experiments were carried out in 0.1 M sodium perchlorate at the open circuit potential of the electrodes, generally near −1.5 V (vs. SCE) after equilibration for 1 hour and a frequency range of 10^5 Hz to 0.5 mHz. The capacitance value measured for the nonanodized magnesium electrode was 42 $\mu F/cm^2$, which is in the normal range for metal–solution interfaces, from 10 to 100 $\mu F/cm^2$. The capacitances measured for the anodized surfaces were significantly below this value, in a range of 0.65–9.5 $\mu F/cm^2$. Polarization resistances of fluoride or phosphate anodized alloys showed a tenfold increase in protection compared to the nonanodized material [33].

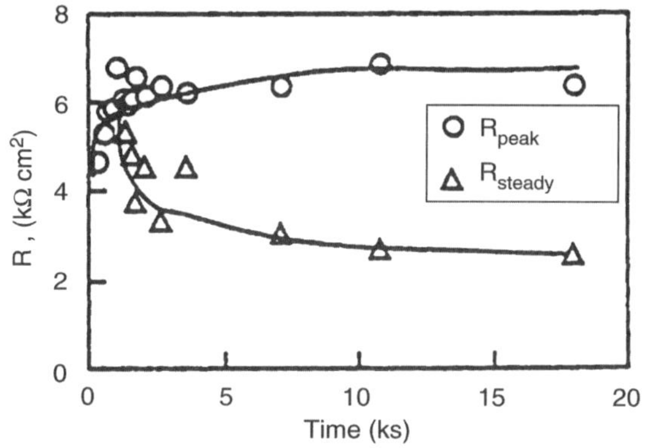

Figure 15.14 Time variation of R_{peak} and R_{steady} of corroding AZ91D [117].

15.6. ORGANIC FINISHING AND CORROSION TESTING OF COATED MATERIAL

15.6.1. Organic Coatings

Organic coatings are extremely versatile and can be applied to many metals, provided an appropriate pretreatment can be developed for the substrate. The adhesion and corrosion resistance of these coatings are inadequate without pretreatment. Organic coatings are typically the last step(s) in a coating system. They may be applied for a purely decorative effect or to enhance the corrosion resistance of the overall coating system. There are some environmental concerns with the use of solvent-borne organic coatings but the development of water-borne and powder coating technologies has led to a decrease in the use of these chemicals [6].

The mechanism of corrosion in aqueous media consists generally of an anodic dissolution of the metal and a cathodic reduction of hydrogen or oxygen if the coating is permeable to oxygen. This alternative cathodic reaction is capable of producing high local pH values and chemical degradation of the polymer or adhesive bonds. Some tests exploit the electrochemical mechanism to determine specific corrosion succeptibilities of coatings system such as cathodic disbondment and pinhole detection [71].

Generally, the coating should be thick, resistant to wear and scratching, and flexible enough to follow the deformation of the substrate without cracking or delamination. For die-cast magnesium components, the coating processes should be able to cope with some surface imperfections such as pores, impurities, and composition and microstructure variations. However, a defect of a surface treatment, especially in the absence of a further covering film, can lead to more severe localized corrosion [3]. For optimum corrosion resistance, a chemical conversion coating or anodizing treatment is required prior to the application of the organic finishing system. It has been found that anodized components provide greatly improved corrosion durability, if the anodized coat is sealed with a penetrating resin sealer prior to the application of primer and top coat [7, 13]. Finishing with a single layer $\sim$13 μm thick depends on the corrosive medium and the tolerance in corrosion rate.

Primers (13 μm thick) for magnesium should be based on alkali-resistant vehicles, such as polyvinyl butyral, acrylic, polyurethane, vinyl epoxy, and baked phenolic. Titanium dioxide or zinc chromate pigments are used as inhibitors in these vehicles. Finish coats should be compatible with the primer. Vinyl alkyds are resistant to alkalis; acrylics are resistant to chloride environments; alkyd enamels are used for exterior durability; and epoxies have good abrasion resistance [7, 9, 13]. Oils, alkyds, and nitrocellulose are best avoided, except for use in mild exposures. Baked paints are harder and more resistant to attack by solvents and are preferable to air-dried under these conditions. For anodized films, the most effective protection against general corrosion is obtained when the film is sealed and painted [53].

High-Temperature Resistance These finishes have an increasing temperature resistance in the following order: vinyls, epoxies, modified epoxies, epoxy-silicones, and silicone vinyls at 150 °C; epoxies and modified epoxies at higher temperatures are recommended. Above 200 °C applications, silicone, epoxy-silicone, and polyimide based paints will provide effective corrosion-resistant coatings on magnesium. In environments where lubricants may be present, polyimide coatings are preferred, particularly when applied onto No. 17 anodic pretreatment [2]. A current example of a paint system can consist of an epoxy-based primer or sealer and an acrylic top coat. With epoxy and epoxy-polyester, electrostatic powder deposition is used successfully on magnesium alloys. Cathodic electrodeposition of a resin from an aqueous emulsion can give a more uniform coverage than spray or dip systems of painting [9, 120, 121].

Polymer Development Polyoxadiazoles (PODs) or oxadiazole-based polymers were designed by attaching functional groups, that is, diphenyl ether (DPE) and diphenylhexafluoropropane (6FP), to the main polymer chain for the purpose of low water permeability and eventually for high corrosion protection of magnesium alloys [122]. Electrochemical experiments showed that the POD–6FP coated AM50 magnesium alloy exhibited three to four orders of magnitude higher corrosion/coating resistance compared to the POD–DPE coated alloy, which is attributed to the low water affinity of the former.

Some Schemes of Protection Shaw and Wolfe [2] gave some interesting examples of uses and corresponding schemes of required successive treatments and coatings.

Housing For exterior housings, decorative surface treatments consisting of chromate pretreatment followed by textured epoxy powder coatings have proved satisfactory. Within disk drive units, where minute dust or other particles could cause disk or head failure, very thin (0.003 mm, or 0.0001 in.) specialized conformal coatings are applied to protect against atmospheric oxidation. Other similar commercial applications include housing or support frames in portable video equipment and a range of optical and medical electronic equipment [2].

Wear and Corrosion Protection Some magnesium applications may require surface protection against wear and corrosion. The use of resin-sealed hard-anodizing treatments, particularly thick HAE treatments, will provide excellent abrasion and corrosion resistance, but some applications (e.g., pulley wheels) will themselves cause excessive wear on other components. Under these conditions, dry-lubricant coatings are required. Nylon coatings applied by electrostatic powder or fluidized bed techniques onto chromated and suitably primed magnesium surfaces have proved effective for use in aircraft pulley control systems. Thick nylon coatings also offer good damage and erosion resistance. Other systems, based on fluorocarbon (Teflon, E.I. Du Pont de Nemours & Company) impregnation of anodic pretreatments or resin-bonded solvent-based fluoropolymer coatings, are also effective in providing combined corrosion resistance and lubricity [2].

Underbody and Wheel Applications For structural underbody and wheel applications, in which frequent exposure to water splash, stone impact damage, and the absence of mitigating environmental factors are problems, more comprehensive protection schemes may be required, together with protection against galvanic corrosion. Where applicable, use of underbody wax-type coatings can provide additional protection. It is expected that use of high-purity alloys will enable designers to specify many more magnesium automotive applications in the future. Cathodic electrophoretic epoxy primers applied to chromic acid pickled or dichromate pretreatments have proved effective, particularly in conjunction with high-purity AZ91D die castings. Electrostatic powder spray is used for top coating. Specially designed fasteners incorporating nylon or plastic washers, sleeves, shims, and so on help reduce the risk of galvanic corrosion on exposed parts [2].

Diving Sea Suits For atmospheric pressure deep-sea diving suits, the body and helmet are preferably made of a magnesium alloy. Surface protection consists of a thick HAE anodic film that is surface sealed with high-temperature stoving epoxy resin, full wet assembly procedures, and final painting with primer and top coat. Coupled with good maintenance, this protection scheme has given satisfactory service between major overhauls for intervals of 4 years [2].

15.6.2. Conventional Corrosion Testing of Coated Metal

15.6.2.1. Physical and Mechanical Testing Methods

Film hardness is best measured by making microhardness indents on a cross section of a film, but a minimum film thickness of 25 μm is required (BS 1981). One of the side benefits of magnesium's hardness is that it has a high scratch resistance. For example, 10 μm of Keronite on its own requires three times more force than an equivalent thickness of conventional anodizing to break through to the magnesium substrate. This can then be increased tenfold when the Keronite layer is impregnated with an organic top coat such as e-coat, PTFE, or powder coat [55].

A typical test of adhesion is to determine the possible nondamaging pull of a tape over the painted material after x hours at 100% humidity exposure at a certain temperature. For example, the tape pull test registered a pass after 100 hours of 100% RH at 50 °C for cast magnesium parts treated with Keronite and then sprayed with three different powder top coats [55]. The adhesion of the paint films can be measured by a crosscut tape test in accordance with the Japanese standards for evaluation of paint film [123]. Standard practices for preparation of magnesium alloy surfaces for adhesive bonding can be found in ASTM D265.

For abrasion resistance measurements, a test based on a loaded abrasive wheel, which moves backward and forward over the film surface, has improved the sensitivity of such measurements [124]. A magnesium alloy coated with a protective coating, containing magnesium phosphate and magnesium fluoride, that is 15–30 μm thick resists wear with a loss of mass measuring less than about 20 mg following 10,000 revolutions in a Taber abraser (CS 10, 10 N).

15.6.2.2. Atmospheric Corrosion Performance

One of the currently accepted tests for judging the ability of a paint base to sustain its protectiveness after damage is to scribe the test panels with a sharp instrument, which penetrates the paint and coating layer, leaving only the bare substrate exposed. Once scribed, the panels are placed in salt spray and evaluations are taken on regular intervals to determine how far corrosion has migrated from the scribe line to undamaged areas (ASTM D1654) [3].

The performance of the magnesium alloy AZ91D with different surface treatments, some paint coated and some without paint, was studied by Umehara et al. [74] and compared to that of a die-cast aluminum alloy (ADC12). The conversion and anodizing surface treatments, described in Table 15.4, were applied to the two alloys. In addition, a 100 mg/m^2 coating of Alodine #1000 and a 20 mg/m^2 of Alodine #1200 were applied to the die-cast aluminum panel as the control (Alodine is a brand name of the aluminum surface treatment

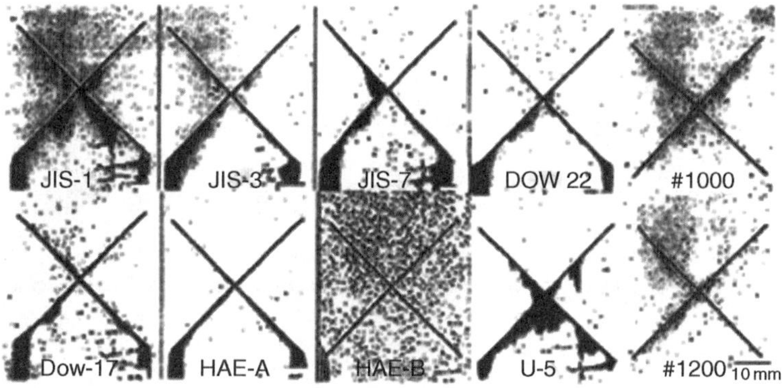

Figure 15.15 Corrosion performance of paint coated AZ91D magnesium and ADC12 alumium die-cast panels exposed to the atmosphere (Choshi, Japan) for 3 years [74].

agent of the Nihon Parkerizing Co., Ltd. and is known as Alocrom in Europe and America). The paint film was applied on both magnesium and aluminum alloys in two layers: a primer containing epoxy resin (25–30 μm thick), setting at 170 °C for 20 minutes, and acrylic paint resin (25–30 μm thick), setting at 150 °C for 20 minutes as the finish coat.

Figure 15.15 shows the surface appearance of the magnesium and aluminum alloys after a 3 year atmospheric exposure test in Choshi, Japan. Little corrosion was observed in the areas other than the crosscut section, which resembled general corrosion, and no large difference was observed between the two materials. The pitting corrosion depth in the crosscut section, determined by using an optical microscope, was measured when the paint had been peeled off and the corrosion products removed [74]. Table 15.4 shows the width of corrosion and peeling in the areas around the crosscut section for some surface treatments. Judging from these results, the surface treatments generated from the JIS-7 type of bath had a corrosion resistance inferior to that of conversion film chromate JIS-3 or the anodizing film HAE-B.

Table 15.4 also gives the performance of the painted surface treated alloys after 36 months of exposure in Miyakojima, Japan, based on the appearance and adhesion of the paint film. The conversion film specimens containing chromate JIS-1 and JIS-3, and the anodized Dow17, HAE-B, and U-5 type specimens, demonstrated a corrosion resistance equal or superior to that on the ADC12 die-cast panel whether or not the substrate is machined. Moreover, Dow17, HAE, and U-5 coating without a paint top coat also showed good corrosion resistance in the atmospheric exposure test [3]. In the crosscut section, the depth of pitting corrosion on AZ91D was greater than that on ADC12 material, indicating that the bare or noncoated magnesium alloy has poorer corrosion resistance than the aluminum one [74].

It should be added that a 3 year test is considered a short-term exposure test and it is possible that some other important observations could occure in the long-term atmospheric corrosion studies.

15.6.2.3. Accelerated Corrosion Testing Methods

Accelerated testing for coated metal is an excellent tool if it is well done and can predict the performance of the material in practice. Accelerated corrosion testing that can have

Table 15.4 Description of Surface Treatments and Corrosion Performance of Painted and Coated Materials Based on Japanese Industrial Standards of the Japanese Standards Association

Treatment	Bath	Atmospheric corrosion rate (mm/36 months) (Choshi, Japan)	Atmospheric corrosion area ratio (Miyakojima, Japan)
JIS-1 conversion	$Na_2Cr_2O_7$, HNO_3	No data	<0.01% (cast) <0.01% (machined)
JIS-3 conversion	$Na_2Cr_2O_7$, CaF_2	0 (cast) 0 (machined)	<0.01% (cast) <0.01% (machined)
JIS-7 conversion	$Na_2Cr_2O_7$, $Mn(H_2PO_4)$, NaF	0.05 (cast) 0,08 (machined)	0.01 to ~0.02% (cast) <0.01% (machined)
Dow22 conversion	$Na_2Cr_2O_7$, $KMnO_4$, H_2SO_4	No data	<0.01% (cast) <0.01% (machined)
Dow 17 anodizing	H_3PO_4, $Na_2Cr_2O_7$, NH_4HF_2,	No data	<0.01% (cast) <0.01% (machined)
HAE-A anodizing	KOH, KF, $Al(OH)_3$, $KMnO_4$, Na_3PO_4	No data	<0.01% (cast) <0.01% (machined)
HAE-B anodizing	KOH, KF, $Al(OH)_3$, $KMnO_4$, Na_3PO_4	0.03 (cast) 0 (machined)	<0.01% (cast) 0% (machined)
U-5 anodizing	Na_2SiO_3, carboxylate, fluoride	No data	<0.01% (cast) <0.01% (machined)
Alodine 1000	—	0 (cast)	<0.01% (cast) <0.01% (machined)
Alodine 1200	—	0 (cast)	<0.01% (cast) <0.01% (machined)

Source: References 74.

different testing modes in different corrosive aqueous media (immersion, alternating immersion, and immersion and spray), as well as in atmospheres with different relative humidity percentages, is considered for corrosion of magnesium alloys with and without coatings. Accelerated laboratory testing aims to reproduce, in a much shorter time than in the field, natural corrosion and degradation processes of the paint system and the substrate without changing the corrosion/degradation mechanisms occurring in practice.

Salt Spray Testing The alternate intermittent immersion in salt water or salt spray is often used to compare the corrosion resistance of magnesium alloys to each other and to other materials [125, 126]. Evaluation of painted or coated specimens subjected to corrosive environments is done according to ASTM D1654, which employs the ASTM B117 exposure for acceleration of corrosive disbanding of the organic coating from the metal substrate. A test period of 28 days has been found to be more severe than 5 years of atmospheric exposure [44].

Magnesium is subjected to various saltwater-based accelerated corrosion tests developed for special purposes. An important example is the proving ground cycle test used by automobile manufacturers. Correlation has been made between the test and the long-term service of the vehicle. Corrosion rates of die-cast AZ91 alloy in 5% NaCl salt spray for 10 days (ASTM B117) showed corrosion rates 200 times larger than samples experiencing a marine atmosphere exposure for 2 years on the Texas Gulf Coast, yet some parallel conclusions can be drawn. For some unprotected magnesium alloys (sheet or sand cast) in salt spray (20% NaCl), tidal immersion, and marine atmospheres, no fundamental differences have been found between spray and immersion tests for the alloys (AZ31-H24, AZ63-F, AZ91C-F) [125].

The pH of the salt solution was such that when atomized at 35 °C, the collected solution was in the pH range of 6.5–7.2 (ASTM B117). The accelerated test used most often is the salt spray test, which involves continuous spraying of 5% sodium chloride (NaCl) in distilled water at 35 °C. It can be performed in accordance with ASTM B117, ISO 7253, ISO 9227, DIN 53167, or BS 3900. There are also the acidified salt spray or AASS test (standard salt spray acidified with acetic acid, ASTM G85) and the copper-accelerated or CAS test (standard salt spray acidified with acetic acid plus the addition of copper chloride, ASTM B368).

In 1939, the neutral salt spray test was standardized in ASTM B117 [127]. In the early 1980s, it was concluded in the literature that salt spray tests based on continuous NaCl spray alone are particularly unreliable for simulating accelerated corrosion in industrial atmospheres [128]. The generally recognized four largest failings of the standard salt spray test are the lack of a wet–dry cycle, the high concentration of electrolyte, the type of electrolyte, and the lack of ultraviolet (UV) or infrared (IR) radiation. However, in spite of salt spray test limitations, for many paint systems and coated steel products, performance requirements are described in terms of salt spray resistance for a specified number of hours.

For many years, research and test equipment laboratories have looked at varying the salt concentration, combining sodium chloride and other salts, and adding sulfur dioxide and other gases. Other investigations have examined the effect of different intensities and energy levels of lighting (or radiation), the effect of condensing and noncondensing humidity load, plus variations in temperature, mechanical load, and other factors. In the last decade, more attention has been given to the effect of cyclic load—the combination of various factors in a specific cyclic order [128].

Humidity or Condensation Test This is a simple variation of the salt spray test. Test panels are exposed to a climate with very high relative humidity (RH) and no electrolytes, usually at 40–50 °C. The basis for the test is the same as the salt spray test: coated panels are exposed so that moisture condenses on the test face. This happens at very high RH and where there is a small temperature difference across the panel. Available standards include ISO 6270, ISO 11503, BS 3900, ASTM D2247, ASTM D4585, and DIN 50017. The basic procedures for outdoor exposure (in Florida, the "benchmark" location) are the exposure angle, the methods of mounting the panels, the effect of mounting on moisture, and the effect of seasonal variability [129].

Cyclic Corrosion Testing The cyclic "Prohesion test" was developed in the United Kingdom in the 1970s for industrial maintenance coatings applications and is described in ASTM G85, together with other modifications to the basic salt spray test. In 1996, ASTM adopted a new test procedure (ASTM D5894) for the accelerated evaluation of corrosion-resistant coatings. In 1999, Jarrald [130] gave an updated overview of the available climates that can be used in cyclic corrosion testing, considering time, temperature, relative humidity, composition of electrolyte solution, UV radiation, and so on.

Since cyclic tests are performed generally with a milder salt solution than 5% NaCl, tests are less aggressive than that of the traditional salt spray, so they must be extended for more than 2000 h. The length of each time period within a cycle in accelerated testing must be optimized to obtain the best correlation between accelerated tests and practice. In addition, the dry periods must be long enough so the effects of drying out are realized. The best predictions from accelerated tests are obtained if paint systems of the same type of binder are compared [128].

Cyclic corrosion testing (CCT) can produce failures representative of the type found in outdoor corrosive environments. It is important to pattern accelerated laboratory tests after these neutral cyclic conditions, because actual atmospheric exposures usually include both wet and dry conditions. Research indicates that, with CCT, the relative corrosion rates, structure, and morphology are more similar to those seen outdoors. Simple exposures like Prohesion may consist of cycles between salt fog and dry conditions. More sophisticated automotive methods call for multistep cycles that may incorporate immersion, humidity, and condensation along with salt fog and dry-off events [130].

Prohesion tests are effective for evaluation of a variety of corrosion mechanisms, including general, galvanic, and crevice corrosion. Prohesion has a reputation as a good test for filiform corrosion (equivalent to ASTM G85/A4). The Prohesion electrolyte solution is much more dilute than traditional salt fog. In addition, the spray atomizing air is not saturated with water [130]. The electrolyte solutions in this case consist of 0.05% sodium chloride and 0.35% ammonium sulfate having a pH between 5.0 and 5.4. The Prohesion exposure cycle is 1 hour salt fog application at 25 °C (for ambient) and then 1 hour dry-off at 35 °C (the dry-off is achieved by purging the chamber with fresh air, such that within 45 min all visible droplets are dried off the specimens).

In a Prohesion test (with conditions like those described by Brennan [127]), the corrosion characteristics of the coated steels showed distinct differences related to the microstructural characteristics and/or to minor compositional differences of substrates. A clear alkyd lacquer can be chosen to facilitate the monitoring of the corrosion features [131]. The results from an acid rain test (2 hours of spraying with acid rain of pH 3.5 plus 1 hour drying) did not distinguish the various coated specimens in terms of corrosion behavior. The Prohesion test was more corrosive than the acid rain test [128].

REFERENCES

1. K.-U. Kainer, H. Dieringa, W. Dietzel, N. Hort, and C. Blawert, in *International Symposium on Magnesium Technology in the Global Age: Magnesium in the Global Age*, Montreal, Canada, edited by M. O. Pekguleryuz and L. W. F. Mackenzie. Canadian Institute of Mining, Metallurgy and Petroleum, Montreal, 2006, pp. 3–20.
2. A. Shaw and C. Wolfe, in *ASM Handbook, Volume 13B, Corrosion*, edited by S. D. Cramer and B. S. Covino, Jr. ASM International, Materials Park, OH, 2005, pp. 205–227.
3. C. Blawert, W. Dietzel, E. Ghali, and G. Song, *Advanced Engineering Materials* **8**(6), 511–533 (2006).
4. S. J. Xia, V. I. Birss, and R. G. Rateick, Jr., *Electrochemical Society Proceedings* **25**, 270–280 (2003).
5. D. L. Hawke, J. E. Hillis, and W. Unsworth, *Technical Committee Report*. International Magnesium Association, Washington, DC, 1988, p. 8.
6. B. L. Luan, J. Gray, L. Yang, W.-J. Cheong, and D. Shoesmith, *Materials Science Forum* **546–549**, 513–518 (2007).
7. J. E. Hillis, in *Metals Handbook, Surface Cleaning, Finishing, and Coating*, *volume 5*, 9th edition, edited by S. R. Lampan and F. Reidenbach. American Society for Metals, Materials Park, OH, 1982, pp. 819–834.
8. J. E. Hillis, in *ASTM Handbook, Volume 5, Surface Engineering*, edited by C. M. Cotell, J. A. Sprague, and F. A. Smidt Jr. ASM International, Materials Park, OH, 1990, pp. 819–832.
9. E. Ghali, in *Uhlig's Corrosion Handbook*, 2nd edition, edited by R. W. Revie. Wiley, Hoboken, NJ, 2000, pp. 793–830.
10. L. Zhu and G. Song, *Surface and Coating Technology* **200**, 2834–2840 (2006).
11. C.-F. L. Jen-Yen Uan and B.-L. Yu, *Metallurgical and Materials Transactions A* **39A**, 703–715 (2008).
12. J. E. Hillis and R. W. Murray, Finishing alternatives for high purity magnesium alloys, in *SDCE 14th International Symposium*. Society of Diecasting Engineers, Paper No. G-T87-003, Rosemont, IL, 1987, pp. 1–7.
13. J. H. Hawkins, *Assessment of Protective Finishing Systems for Magnesium*. International Magnesium Association, Washington, DC, 1993.
14. *Corrosion and Protection of Magnesium*. Amax Magnesium, Salt Lake City, UT, 1984, pp. 1–30.
15. H. Umehara, M. Takaya, and S. Terauchi, *Material Science Forum* **419–422**, 883–888 (2003).
16. J. I. Skar, M. Wlter, and D. Albright, *Journal of the Society of Automotive Engineers*, **970324**, 7–11 (1997).
17. E.-H. Han, W. Zhou, D. Shan, and W. Ke, *Materials Science Forum* **419–422**, 879–882 (2003).
18. E.-H. Han, W. Zhou, D. Shan, and W. Ke, in *Chemical Conversion Coating on AZ91D and Its Corrosion Resistance*, edited by A. A. Luo. Symposium sponsored by the Magnesium Committee of the Light Metals Division of the Minerals, Metals & Materials Society with the International Magnesium Association, Charlotte, NC 2004.
19. A. L. Rudd, B. Breslin, and F. Mansfeld, *Corrosion Science* **42**, 275–288 (2000).
20. H. Ardelean, M. B. Yves, C. Fiaud, and P. Marcus, Electrochemical study of the effects of a chemical cerium nitrate treatment on the corrosion behaviour of magnesium, in *Environmental Degradation of Materials and Corrosion Control in Metals*, edited by M. Elboujdaini and E. Ghali. METSOC, Montreal, Quebec, 1999, pp. 85–102.
21. C. S. Lin, H. C. Lin, K. M. Lin, and W. C. Lai, *Corrosion Science* **46**(1), 93–109 (2006).
22. H. H. Elsentriecy, K. Azumi, and H. Konno, *Electrochimica Acta* **53**(2), 1006–1012 (2007).
23. M. A. Gonzalez-Nunez, C. A. Nunez-Lopez, P. Skeldon, G. E. Thompson, H. Karimzadeh, P. Lyon, and T. E. Wilks, *Corrosion Science* **37**(11), 1763–1772 (1995).
24. M. A. Gonzalez-Nunez, P. Skeldon, G. E. Thompson, and H. Karimzadeh, *Composites* **55**(12), 1136–1143 (1999).
25. X.-F. Cui and Q.-F. Li, *Corrosion Science and Protection Technology* **19**(3), 203–206 (2007).
26. X.-F. Cui and Q.-F. Li, *Corrosion Science and Protection Technology* **19**(4), 275–277 (2007).
27. R. Lindemann, *MAGPASS-COAT. Chrome-Free Passivation for Magnesium-Based Materials.* AHC Oberflächentechnik GmbH, Göppingen-Voralb, or Eschenbach, Germany, 2003.
28. W. Zheng, C. DeruShi, R. Zhang, J. Lo, and S. Dionne, in *Corrosion Mitigation of Mg Alloys for Structural Automotive Applications*, NACE International, Northern Area Meeting, Ottawa, 2003.
29. A. Yamamoto, A. Wanatabe, K. Sugahara, S. Fukumoto, and H. Tsubakino, Applying a vapor deposition technique to improve corrosion resistance in magnesium alloys, in *Proceedings of the Second International Conference on Environment Sensitive Cracking and Corrosion Domage*, Hiroshima, Japan, edited by H. N. M. Matsumura, K. Nakasa, and Y. Isomoto. Nishiki Hiroshima, Japan, 2001, pp. 160–167.
30. J. E. Gray and B. Luan, *Journal of Alloys and Compounds* **336**, 88–113 (2002).
31. C. T. Speiser and H. Schenkel, *Aluminium* **50**, 159 (1974).
32. K. Shibata, *Light Metal Age* **41**, 22 (1983).
33. T. F. Barton, and C. B. Johnson, *Plating and Surface Finishing* 138–141 (1995).

34. K. Huber, *Journal of the Electrochemical Society* **100** (8), 376–382 (1953).
35. S. Ono and N. Masuko, *Materials Science Forum* **419–422**, 897–902 (2003).
36. Y. Mizutani, S. J. Kim, R. Ichino, and M. Okido, *Surface and Coatings Technology* **169–170**, 143–146 (2003).
37. H. Fukuda and Y. Matsumoto, *Corrosion Science* **46**, 2135–2142 (2004).
38. Y. Zhang, C. Yan, F. Wang, H. Lou, and H. Cao, *Surface and Coatings Technology* **161**, 36–43 (2002).
39. H.-Y. Hsiao and W.-T. Tsai, *Surface and Coatings Technology* **190**, 299–308 (2005).
40. E. A. Carter, T. F. Barton, and G. A. Wright, *Anodisation of Magnesium at High Voltages*, Surface Treatment 99: 4th International Conference on Computer Methods and Experimental Measurements for Surface Treatment Effects, Assisi, Italy, 20–22 Sept. 1999, pp. 169–177.
41. Z. Shi, G. Song, and A. Atrens, *Corrosion Science* **48** (8), 1939–1959 (2006).
42. M. Abulsain, A. Berkani, F. A. Bonilla, Y. Liu, M. A. Arenas, P. Skeldon, G. E. Thompson, P. Bailey, T. C. Q. Noakes, K. Shimizu, and H. Habazaki, *Electrochimica Acta* **49**, 899–904 (2004).
43. T. Kanadi, A. Saigo, M. Hino, and K. Sakaj, *Journal of Japan Institute of Light Metals* **52**(10), 477–478 (2002).
44. J. E. Hillis, in *ASTM Manual Series: MNL 20, Corrosion Testing and Standards: Application and Interpretation*, edited by R. Baboian. ASM International, Materials Park, OH, 1995, pp. 438–446.
45. A. Kuhn, *Metal Finishing* **44-**49 (2003).
46. C. Jurey, Magoxid-Coat, A hard anodic coating for magnesium, in *Proceedings of the International Magnesium Association*, Washington, DC, 1993, pp. 80–88.
47. P. Kurze, D. Banerye, and H.-J. Kletze, Method of producing oxide ceramic layers on barrier layer-forming metals and articles produced by the method. U.S. Patent No. 5385662, 1995.
48. P. Kurze and H.-J. Kletze, Method of producing articles of aluminium, magnesium or titanium with an oxide ceramic layer filled with fluorine polymers. U.S. Patent No. 5487825, 1996.
49. E. Brown, *Light Metal Age* 14–17 (1993).
50. P. Kurze, in *Korrosion und Korrosionsschutz von Magnesiumwerkstoffen*, Fortbildungsseminars der Deutschen Gesellschaft fuer Materialkunde e.V. in Zusammenheit mit GKSS Geesthacht, Germany, edited by K.-U. Kainer. Wiley-VCH, Weinheim Germany, 2000, pp. 236–243.
51. O. Kozak, Method of coating articles of magnesium and an electrolytic bath. U.S. Patent No. 4620904, 1986.
52. O. Kozak, Anticorrosive coating on magnesium and its alloys. U.S. Patent No. 4184926, 1980.
53. V. Tchervyakov, G. Gao, J. Bomback, A. P. Pchelnikov, and G. Cole, in *Laboratory Evaluation of Corrosion Resistance of Anodized Film on Magnesium*, edited by H. I. Kaplan, J. Hryn, and B. Clow. The Minerals, Metals and Material Society, Warrendale, PA, 2000, pp. 143–152.
54. A. L. Yerokhin, A. Shatrov, V. Samsonov, P. Shashkov, A. Leyland, and A. Matthews, *Surface and Coatings Technology* **182**, 78–84 (2004).
55. Keronite International Ltd. http://www.keronite.com/.
56. S. Verdier, M. Boinet, S. Maximovitch, and F. Dalard, *Corrosion Science* **47**, 1429–1444 (2005).
57. A. L. Yerokhin, X. Nie, A. Leyland, A. Matthews, and S. J. Dowey, *Surface Coating Technology* **122**, 73 (1999).
58. J. L. Patel and N. Saka, *Interceram* **50**(5), 398–401 (2001).
59. H. F. Guo and M. Z. An, *Applied Surface Science* **246**, 229–238 (2005).
60. M. D. Klapkiv, H. M. Nykyforchyn, and V. M. Posuvailo, *Surface and Coating Technology* **100–101**, 219 (1998).
61. H. M. Nykyforchyn, W. Dietzel, M. D. Klapkiv, and C. Blawert, Corrosion properties of conversion plasma coated magnesium alloys, in *6th International Conference on Magnesium Alloys and Their Applications*, Wolfsburg, Germany, edited by K.-U. Kainer. Wiley-VCH, Weinheim, Germany, 2003, p. 176.
62. C. Blawert, V. Heitmann, W. Dietzel, H. M. Nykyforchyn, and M. D. Klapkiv, *Surface and Coating Technology* **200**, 68–72 (2005).
63. E. L. Schmeling, B. Roschenbleck, and M. H. Weidemen, Method of producing protective coatings that are resistant to corrosion and wear on magnesium and magnesium alloys. U.S. Patent No. 4978432, 1990.
64. S. J. W. Plazter, Electrolytes for electrochemically treating metal plates. U.S. Patent No. 4578156, 1986.
65. T. M. Gillich, J. E. Walls, S. F. Wanat, and W. J. Rozell, Anodic aluminum (alloy) oxidation by pulse plating, Patent number DE3305354A (03305354/DE-A), West Germany, 17 February, 1983.
66. T. M. Gillich, J. E. Walls, S. F. Wanat, and W. J. Rozell, Electrochemically treated metal plates. U.S. Patent No. 4448647, 1984.
67. W. Kobayashi, K. Uehori, and M. Furuta. Anodising solution for anodic oxidation of magnesium and its alloys, U.S. Patent No. 4744872, 1988.
68. W. Kobayashi and S. Takahata, Aqueous anodising solution and process for colouring article of magnesium or magnesium-base alloy. U.S. Patent No. 4551211, 1985.
69. Z. Zhang, Q. Zhong, J. Zhang, and C. Cao, Electrochemical noise analysis of LY 12 in sodium chloride solution with wavelet transform technique, in *Corrosion 2002*, NACE International, Houston, TX, 2002.
70. J. I. Skar and D. Albright, in *Magnesium Technology 2002*, edited by H. I. Kaplan. The Minerals, Metals & Materials Society, Warrendale, PA, 2002, pp. 255–261.

71. R. D. Granata, in *Corrosion Tests and Standards: Application and Interpretation*, edited by R. Baboian. ASTM, Philadelphia, PA, 1995, pp. 525–530.
72. F. Hayduk and P. Kurze, in *Magnesium 2000*, Second Israeli International Conference on Magnesium Science & Technology, Dead Sea, Israel, edited by E. Aghion and D. Eliezer, the Magnesium Research Institute (MRI) Beer-Sheva, Israel, 2000, 431–440.
73. F. Satoh, Y. Asakawa, T. Nakayama, and H. Satoh, *Surface Technologies* **43**(3), 67–70 (1993).
74. H. Umehara, M. Takaya, and T. I. Tsukuba, *Aluminium International Journal* **75**(7/8), 634–641 (1999).
75. G. G. Wang, K. Stewart, X. Berkmortel, and J. I. Skar, Corrosion prevention for external magnesium automotive components, SAE Technical Paper 2001-01-0421, SAE International, Warrendale, PA, USA.
76. Technology Application Group. http://www.tagnite.com/.
77. Magnesium Elektron Ltd., MEL Data Sheet 467A, Electron WE43, 1997.
78. P. N. Ross, J.-A. MacCulloch, and R. J. Esdaile, Paper No. T99-063, presented at 20th International Die Casting Congress and Exposition, Cleveland, 1999, pp. 213–218.
79. Z. Shi, G. Song, and A. Atrens, *Corrosion Science* **48** (11), 3531–3546 (2006).
80. Z. Shi, G. Song, and A. Atrens, *Corrosion Science* **47** (11), 2760–2777 (2005).
81. O. Khaselev, D. Weiss, and J. Yahalom, *Corrosion Science* **43**, 1295–1307 (2001).
82. A. J. Zozulin and D. E. Bartok, *Metal Finishing* **92**, 39–44 (1994).
83. G. R. Kotler, E. N. Aqua, and D. L. Hawke, *Light Metal Age Journal*, **34**(7–8), 20–21 (1976).
84. O. Khaselev and J. Yahalom, *Corrosion Science* **40**(7), 1149–1160 (1998).
85. Z. Shi, G. Song, and A. Atrens, Effects of zinc on the corrosion resistance of anodised coating on Mg-Zn alloys, in *1st International Light Metals Technology Conference*, edited by A. Dahle. CRC for Cast Metals Manufacturing, Brisbane, Australia, 2003, pp. 393–396.
86. V. V. Ogarevic and R. I. Stephens, *Annual Review of Material Science* **20**, 141–177 (1990).
87. W. K. Miller and E. F. Ryntz, Jr., *Society of Automotive Engineers (SAE)* 830521 (1984).
88. I. Nakatsugawa, Surface modification technology for magnesium products, in *Magnesium—A Material Advancing to the 21st Century*, International Magnesium Association, Ube City, Japan, 1996 pp. 24–29.
89. M. Benmalek, P. Gimenez, and G. Regazzoni, in *Proceedings of the 47th Annual World Magnesium Conference* Cannes, IMA, Cannes, 1990, p. 117.
90. F. Dabosi, R. Morancho, N. Pebere, and D. Pouteau, CVD, in *Proceedings of the 47th Annual World Magnesium Conference*, Cannes, IMA, Cannes, 1990, pp. 51–55.
91. B. S. Yilbas, A. Z. Sahin, Z. Ahmad, and B. J. Abdulaleem, *Corrosion Science* **37**(10), 1627 (1995).
92. F. Hollstein, R. Wiedemann, and J. Scholz, *Surface and Coatings Technology* **162**, 261 (2003).
93. F. Vacandio, Y. Massiani, P. Gravier, S. Rossi, P. L. Bonora, and L. Fedrizzi, *Electrochimica Acta* **46**, 3827 (2001).
94. J. Y. Chen, G. P. Yu, and J. H. Huang, *Materials Chemistry and Physics* **65**, 310 (2000).
95. H. Altun and S. Sen, *Surface and Coatings Technology* **197**, 193–200 (2005).
96. M. Urgen and A. F. Cakir, *Surface Coatings Technology* **96**, 236 (1997).
97. G. G. Perrault, in *Encyclopedia of Electrochemistry of the Elements*, Vol. 8, edited by A. J. Bard. Marcel Dekker, New York, 1978, pp. 263–319.
98. Y. Osumi, in *Hydrogen Absorbing Alloys–Its Properties and Applications*, Agune, Tokyo, Japan, 1993.
99. A. P. Nazarov, T. A. Yurasova, V. V. Gubin, A. K. Buryak, and M. P. Glazunov, *Zashcita Metallow* **29**(3), 392–397 (1993).
100. I. Nakatsugawa, Cathodic protection coating on magnesium or its alloys methods of production, Canadian Patent No. 218,983; U.S. Patent No. 6,291,076, 2001.
101. I. Nakatsugawa, J. Renaud, and E. Ghali, Cathodic protective coating for Mg alloys, in *Environmental Degradation of Materials and Corrosion Control of Metals*, 38th Annual Meeting of Metallurgists of CIM, edited by M. Elboujdaini and E. Ghali. CIM, Montreal, 1999, pp. 103–112.
102. X. L. Wang, N. Haraikawa, and S. Suda, *Journal of Alloys and Compounds* **231**, 397–402 (1995).
103. A. Bakkar and V. Neubert, *Corrosion Science* **47**, 1211–1225 (2005).
104. I. Nakatsugawa, F. Yamada, H. Takaysu, T. Tsukeda, and K. Saito, Corrosion behavior of thixomolded Mg–Al alloys, 38th Annual Conference of METSOC, in *Environnemental Degradation of Materials and Corrosion Control in Metals*, Quebec City, Canada, edited by M. Elboujdaini and E. Ghali. METSOC, Quebec City, Canada, 1999, pp. 113–123.
105. B. E. Conway, Development of electrochemical methodologies for studying complex corrosion of magnesium: Final report. Canadian Government Research Announcement, 1993; p. Index, 23.
106. I. Nakatsugawa, J. Renaud, E. Ghali, and E. J. Knystautas, Electrochemical formation of magnesium hydride and its application to surface coating, in *First Israeli International Conference on Magnesium Science and Technolgy*, 1997, "Magnesium 97", edited by E. Aghion and D. Eliezer, published by the Magnesium Research Institute (MRI) Beer-Sheva, Israel, 1998, pp. 220–225.

107. D. Daloz, P. Steinmetz, and G. Michot, *Corrosion* **53** (12), 944–954 (1997).

108. G. L. Song and A. Atrens, *Advanced Engineering Materials* **1**(1), 11–33 (1999).

109. P. Gu, W. Zheng, M. Elboujdaini, and E. Ghali, Electrochemical study of magnesium hydride formation in aqueous solution, in *Degradation of Light Weight Alloys*, edited by D. Shifler. The Minerals, Metals & Materials Society, Warrendale, PA, 2007.

110. M. Pourbaix, *Atlas of Electrochemical Equilibria in Aqueous Solutions*. Pergamon Press, New York, 1966.

111. R. P. Simpraga, B. E. Conway, and V. M. Scepanovic, *Electrochemical Society Proceedings* **94,** 21(1994).

112. S. Shrestha, A. Sturgeon, P. Shashkov, and A. Shatrov, *Magnesium Technology* 283–287 (2002).

113. B. A. Boukamp, *Equivalent Circuit (EQUIVCRT.PAS) Users Manual*, The Netherlands University of Twente, Twente, 1988/1989, pp. 6–26.

114. R. Udhayan and D. P. Prokash Bhatt, *Journal of Power Sources* **63**, 103 (1996).

115. S. Turgoose and R. A. Cottis, in *Electrochemical Impedance: Analysis and Interpretation*, edited by J. R. Scully, D. C. Silverman, and M. W. Kendig. ASTM, Philadelphia, PA, 1993, pp. 173–191.

116. C. H. R. Tsai, in *Electrochemical Impedance: Analysis and Interpretation*, edited by J. R. Scully, D. C. Silverman, and M. W. Kendig, ASTM, Philadelphia, PA, 1993, pp. 37–52.

117. I. Nakatsugawa and E. Ghali, *Extended Abstracts of The Electrochemical Society* **97–1**, 140 (1997).

118. F. Mansfeld, Y. Wang, X. Xiao, and H. Shih, Monitoring of localized corrosion of aluminum alloys with electrochemical impedance spectroscopy (EIS), in *Symposium on Critical Factors in Localized Corrosion* The Electrochemical Society, Pennington, NJ, 1992, p. 469.

119. Z. Shi, G. Song, and D. St John, A method for evaluating the corrosion resistance of anodised magnesium alloys, *Corrosion and Prevention 2001*, Australian Corrosion Association, Newcastle, Paper #058, 2001.

120. R. W. Murray and J. E. Hillis, *Magnesium Finishing: Chemical Treatment and Coating Practices*. Society of Automotive Engineers, Warrendale, PA, 1990.

121. P. L. Hagans, *Surface Modifications of Magnesium for Corrosion Protection*. International Magnesium Association, London, UK, 1984.

122. M. Bobby Kannan and R. K. Singh Raman, *Scripta Materialia* **59**(2), 175–178 (2008).

123. *Standards for Evaluation of Paint Systems*. Paint Inspection Association, xxx, Japan, 1970.

124. L. L. Shreir, R. A. Jarman, and G. T. Burstein, *Corrosion, Corrosion Control*. Butterworth-Heinemann, Oxford, UK, 1995, pp. 19:1–19:48.

125. A. F. Froats, T. Kr. Aune, D. Hawke, W. Unsworth, and J. Hillis, in *ASM Handbook, Volume 13, Corrosion*, edited by L. J. Korb, D. L. Olson, and J. R. Davis. ASM International, Materials Park, OH, 1987, pp. 740–754.

126. J. D. Hanawalt, C. E. Nelson, and J. A. Peloubet, *Transactions of American Society of Mining and Metallurgical Engineering* **147**, 273–299 (1942).

127. P. J. Brennan and S. J. Grossman, *Combined Corrosion/Weathering Exposures: A Status Report*, Paint and Ink International Journal, November/December, 1996, pp. 30–32.

128. B. P. Alblas and J. J. Kettenis, *Protective Coatings Europe* **February**, 49–59 (2000).

129. M. J. Crewdson and P. Brennan, *JPCL* **12**(9), 17–25 (1995).

130. B. J. Jarrald, *Polymers Paint Colour Journal* **189**(1), 10–12 (1999).

131. I. Costa, S. E. Faidi, and J. D. Scantlebury, *Corrosion Science* **35**(5–8), 1367–1374 (1993).

Part Five

Evaluation and Testing

Chapter 16

Conventional and Electrochemical Methods of Investigation

Overview

Different categories of testing are summarized. Important and promising electrochemical testing methods of the corrosion resistance of aluminum and magnesium alloys and their coatings are given in more detail. Field, pilot-plant, and laboratory tests are the first-choice methods; then the test mode can be adapted for total immersion, partial immersion, alternate or intermittent immersion, or a cyclic exposure to wet/dry conditions to simulate service conditions. Visual observation usually can define areas of severe corrosion and metallographic examination generally is limited to such regions. The removal of corrosion products from metal specimens can be done by mechanical, chemical, or electrolytic cathodic cleaning.

Investigations usually progress from light microscopy, to scanning electron microscopy (SEM), and then to transmission electron microscopy (TEM). Nondestructive evaluation techniques (electrical resistance measurement, eddy-current testing, ultrasonic flaw detection, and radiography examination) are discussed. Macro- and microhardness testing and tensile, fatigue, or impact tests are carried out for certain applications. Chemical and physical methods of surface testing are described. Published and previous data on the performance of metals or alloys under similar conditions of service are of major importance. Electrochemical polarization test methods (cyclic voltammetric, potentiodynamic, potentiostatic, and galvanostatic techniques) can establish criteria for susceptibility to several forms of corrosion and anodic or cathodic protection.

The basics of electrochemical impedance spectroscopy (EIS) are explained with an introduction to equivalent circuits, Nyquist and Bode plots, and their relevance to corrosion studies. Analysis of electrochemical noise measurements (ENMs) have been explained in the time and frequency domains. Spectral analysis of the noise in the frequency domain allows calculation of the power spectral density (PSD) using the fast Fourier transform (FFT). The spectral noise resistance PSD, R_n, can be equal to the impedance modulus, Z. Chaos analyses (rescaled range and stochastic process) serve to highlight the features present in the frequency domain of the transformed data record. Wavelet analysis distinguishes periodic and nonperiodic variation in the signal power in the time and frequency domains.

Corrosion Resistance of Aluminum and Magnesium Alloys: Understanding, Performance, and Testing By Edward Ghali

The scanning reference electrode technique (SRET) is designed to study localized corrosion such as pitting or intergranular corrosion. This is also associated with the scanning vibrating electrode technique (SVET), used frequently for coated materials. Microelectrodes and scanning electrochemical potential microsystems are identified as important emerging methods of investigation for active–passive behaviors of metals accompanied occasionally by atomic force microscopy (AFM). The wire beam electrode (WBE) is a multipiece electrode constructed with a variable number of metal wires embedded in insulating material and is used in conjunction with electrochemical noise measurement (ENM) and SRET.

16.1. CORROSION TESTING APPROACHES AND METHODS OF INVESTIGATION

Newly developed alloys and composite materials need extensive testing before use, as well as traditional alloys that are projected for use in new media or for new applications. Moreover, it is very hard, for example, to predict theoretically the performance of an aluminum alloy in a mixture of mineral acids or certain organic chemical products.

16.1.1. Testing Approach

The degradation of materials generally occurs via three well-known avenues frequently interconnected: corrosion, fracture, and wear. This complete approach considers the following steps: environmental definition; material definition; mode definition (morphology of corrosion); superposition and comparison; failure definition; development of a statistical framework; accelerated testing; prediction; monitoring, inspection, and feedback; and finally modification of the system based on data taken from operation [1].

The main purposes of corrosion testing, inspection, or monitoring are generally evaluation and selection of materials for a given application; evaluation and design of new alloys; determination of the aggressiveness of the interfacial medium for a certain application; control of corrosion resistance of the material; inspection and diagnosis of causes of corrosion failures; obtaining reference or database information; and study of the corrosion kinetics and mechanisms. Inspection is the safe way to secure the duration of service of a structure and sometimes avoid catastrophic failures. Monitoring allows workers to follow the effectiveness of a corrosion-control system and provides early warning when damaging conditions arise [1]. Tests of the bare, unprotected metal are useful to establish a minimum basis of performance and comparative ranking. However, the applicability to the end use and prediction of service life require evaluation of the protective coatings and maintenance procedures during service [2].

16.1.2. Categories of Corrosion Testing

There are three possible categories of corrosion testing: field, pilot-plant, and laboratory tests.

1. *Field Tests.* These tests are currently referred to as simulated-service testing and could be considered to be the most reliable predictor of corrosion behavior in the absence of actual service experience. Examples of field tests are atmospheric exposure of a large number of specimens in racks at one or more geographical locations and similar tests in soils or seawater. Simulated-service tests can be more helpful than accelerated tests at the design stage and in failure investigations based on environmental and mechanical conditions that are as close as possible to the service conditions [1].

2. ***Pilot-Plant Tests.*** These tests are usually more desirable than laboratory tests, but they are also more expensive. Here the tests are made in a small-scale plant that essentially duplicates intended large-scale operation [1].

3. ***Laboratory Tests.*** These tests are used to predict corrosion behavior when service history is lacking and time or budget constraints prohibit field testing. Laboratory tests cover immersion tests, cabinet-controlled or autoclave-controlled environments, and different electrochemical procedures [1].

16.1.3. Testing Duration

The typical time allowed for testing of candidate materials to support a materials selection process is usually measured in weeks and months—a relatively short period of time compared to the desired useful life of different equipment, which can vary from approximately 7–10 years for automobiles, up to 1000 years (without appreciable leakage) in high-level nuclear waste containers. Typically, accelerated laboratory corrosion tests increase the severity of the environment by exposing the materials to more concentrated solutions, higher-temperature solutions, or increased periods of wetness and possible certain other parameters depending on the tested type of corrosion.

Accelerated tests should produce the types of corrosion, pitting, or intergranular corrosion (IGC) that occur in service and cover at least the major contaminants and environmental variables expected. This can produce the essential mechanism of corrosion and keep the major ties to reality and service performance. Engineering requirements generally dictate the type of corrosion of most concern and the most significant method of evaluation. For example, for liquid containers, the primary concern may simply be to avoid perforation and leaking. Depth of corrosion is another major concern. However, if the liquid is under pressure, the type of corrosion can be of importance, as this can affect burst strength [1, 2].

The observed performance of alloys in service can vary from that based on some accelerated tests because of the following conditions:

- Different microstructures of the metal or alloy due to impurities, heat treatment, surface conditioning, or welding can exist.
- Aeration and oxidants such as ferric and cupric ions can affect tests; however, the most frequent cell is the oxygen differential aeration cell.
- Stagnation of the solution can change the interfacial properties such as pH and concentration of different ions and species, thus leading to a change in the electrochemical potential. Consequently, agitation or circulation of the electrolyte on the laboratory scale should match the projected conditions in service.
- Galvanic corrosion could be created in many ways especially for metals such as aluminum or magnesium in the active state. For example, impurities in the environment (natural or industrial), such as small amounts of more noble metals like copper in solution, can deposit as metallic copper on aluminum or magnesium alloy surfaces and accelerate corrosion.
- The action of microorganisms is frequently an important factor that has been overlooked in accelerated tests and by investigators and designers.

- Suspended solids can give rise to mechanically influenced corrosion, a factor frequently neglected in accelerated testing [3].

16.1.4. Testing Modes

The simulation of corrosion media and immersion techniques is now described [1]. The current test modes can be grouped into the following:

- ***Wetting.*** Wetting the metallic surface in a medium saturated with water vapor can simulate acid rain or industrial polluted atmospheres.
- ***Spraying of Aggressive Media.*** These tests can simulate ocean climate corrosion by spraying NaCl solution and NaCl + acetic acid to simulate salted-road attack.
- ***Erosion Corrosion.*** Impingement, cavitation corrosion, and fretting corrosion due to the impact of liquid and/or solid particles are critical factors to consider.
- ***Wear Corrosion.*** Wear corrosion in an aggressive medium containing, for example, HCl or HNO_3, must also be considered.

Considering the environmental conditions that must be simulated, the test mode should be adapted for total permanent immersion, partial immersion, alternate or intermittent immersion, or a cyclic exposure to wet/dry conditions.

- ***Total Immersion Tests.*** The ASTM G31 standard and NACE TM0169 standard give general laboratory guidance on how to carryout these studies.
- ***Partial Immersion Tests.*** Partial immersion conditions provide a very suitable accelerated test for metals such as aluminum alloys and others that develop concentrated attack at the liquid line (or splash zone, in certain equipment) [1].
- ***Alternate Immersion Tests.*** In these tests a thin film of the solution, frequently renewed and almost saturated with oxygen, can be maintained on the test specimen during most of the exposure period, even when the shape of the specimen is complex [1] and this can accelerate different forms of corrosion. This simulates, for example, the effects of the rise and fall of tidal waters, the changes from humid to dry weather, and the movements of corrosive liquids in chemical plants. In addition, these conditions can provide a relatively rapid test for the effect of aqueous solutions.

16.1.5. Removal of Corrosion Products

The removal of corrosion products from metal specimens is described in ASTM G1, which covers the suggested procedures for preparing bare, solid metal specimens for tests, for removing corrosion products after the test has been completed, and for evaluating the corrosion damage that has occurred [4]. Emphasis is placed on procedures related to the evaluation of corrosion by mass loss and pitting measurements (e.g., ASTM standards [5]). This is also covered in ISO/DIS 8407.2.

The various methods may be classified as follows:

1. ***Mechanical Treatments.*** Scrubbing with a bristle brush, scraping, and wire brushing are currently used methods. Grit and shot sand blasting are good means of abrasion, but the surface could be contaminated with corrosion products (e.g., sand) and the metallic surface could become very rough [4], having some residual compressive stresses.

2. *Chemical Treatments.* Organic solvents are currently used to remove organic deposits from the metallic surface during chemical cleaning [4].

3. *Electrolytic Cathodic Cleaning.* This can be applied to aluminum and magnesium as well as their respective alloys. The metal could be polarized as a cathode in sulfuric acid usually containing an appropriate inhibitor such as citric acid, potassium cyanide, or caustic soda. After scrubbing to remove loosely attached corrosion products, the metal is cathodically polarized in hot dilute sulfuric acid under the following conditions: electrolyte–sulfuric acid (5 wt %) plus an appropriate inhibitor (0.5 $kg{\cdot}m^{-3}$). The temperature should be 75 °C, the cathode current density 2000 $A{\cdot}m^{-2}$, and the time of cathodic polarization 3 min. The anode should be carbon or an inert electrode. After the electrolytic treatment, the specimen is scrubbed with a brush, rinsed thoroughly, especially for aluminum alloys to avoid aggressive alkaline medium, and then dried. Care should be taken in the case of magnesium alloys to avoid high cathodic currents that could form magnesium hydrides. Electric treatment may result in the redeposition of a metal, such as copper, from reducible corrosion products, and thus decrease the apparent weight loss [4].

4. *Chemical Cleaning of Al, Mg and Their Alloys.* *Aluminum alloys* are dipped for 5–10 min in an aqueous solution containing 2 wt% chromic acid (CrO_3) plus 5 vol% orthophosphoric acid (H_3PO_4, 85%) maintained at 80 °C. Ultrasonic agitation will facilitate this procedure. The specimen is rinsed in water to remove the acid, brushed very lightly with a soft bristle brush to remove any loose film, and rinsed again. If film remains, the specimen is immersed for 1 min in concentrated nitric acid and the previous steps are repeated. Nitric acid may be used alone if there are no deposits [4]. Cleaning in concentrated nitric acid, as given in ASTM G1, is the procedure normally used for aluminum because it removes most corrosion products effectively [2].

Magnesium alloys are dipped for approximately 1 min in boiling 15% chromic acid to which has been added with agitation 1% silver chromate solution [4].

16.2. PHYSICAL AND MECHANICAL TESTING OF CORRODED MATERIALS

16.2.1. Visual and Microscopic Techniques of Testing

16.2.1.1. Visual Techniques of Testing

Samples are first examined visually, preferably with the aid of a hand magnifier or other suitable viewing aid. At this stage, such features as the extent of damage, general appearance of the damage zone, color, texture, and quantity of surface residues are of primary interest. If substantial amounts of foreign matter are visible, cleaning is necessary before further examination. The residues can be removed in some areas, leaving portions of the failure region in the as-received condition to preserve evidence. When only small amounts of foreign matter are present, it is sometimes preferable to defer cleaning so that the surface can be examined microscopically first or to defer cleaning until necessary for surface examination at higher magnifications [1].

Washing with water or solvent, with or without the aid of an ultrasonic bath, is usually adequate to remove soft residues that obscure the view. Inhibited pickling solution or electrolytic cathodic cleaning could remove adherent oxides. Alternatively, plastic replicas can be used in cleaning; the replicas also retain and preserve surface contaminants, thus making them available for analysis. Dilute acid cleaning or pickling is a possible avenue if the metallic piece is at least partially protected by cathodic polarization [1].

Regardless of the type of corrosion, the initial, basic evaluation should be a visual inspection in the "as-corroded" condition, assisted by low-power (3–10 ×) magnification. If the sample is disassembled or cleaned, the visual/photographic inspection should be repeated [2]. Macroscopic forms of corrosion affect greater areas of corroded metal and are generally observable with the naked eye or can be viewed with the aid of a low-power magnifying device. Macroscopic examination can identify the following forms: galvanic corrosion, erosion corrosion, crevice or pitting corrosion, exfoliation corrosion, and dealloying. Microscopic examination can identify intergranular corrosion, stress corrosion cracking, corrosion fatigue, and subsurface corrosion (observed frequently at high temperature) [1].

16.2.1.2. Microscopic Examination

Visual examination usually can define areas of most severe corrosion and metallographic examination generally is limited to such regions. Investigations progress from light microscopy, to scanning electron microscopy (SEM), and then to transmission electron microscopy (TEM) depending on the need. It is important to observe minute features on corroded surfaces and to evaluate the influence of the microstructure on the initiation and propagation of corrosion by the corrosive environment such as grain-boundary attack or selective leaching. This is achieved by polishing or polishing followed by etching. The corrosion products should be retained if they possess sufficient coherence and hardness to be polished. One method of keeping the surface material in place is to impregnate the sample with a castable resin, which is allowed to harden before samples are cut. Polishing on napless cloths with diamond abrasives is recommended for maximum edge retention [1].

Optical Microscopy Metallographic examination should begin with optical microscopy at appropriate magnifications of 50–500 to determine the type, depth, and extent of corrosion. Examination of a small region or area of either unprepared surface or polished and etched surface by optical (light) microscopy of 25–100 × is currently carried out. Normally examination is of a sample cut from the bulk, but on-site examination and replication techniques are possible. High-temperature corrosion needs special attention: care should be given to the examination of subsurface corrosion, since observation by the naked eye of the surface could lead to a faulty conclusion [1].

Both pitting and intergranular corrosion of aluminum tend to be localized and nonuniform over the surface. Both forms of corrosion tend to develop in hemispherical shaped sites, and several (typically three) random metallographic sections of about 20 mm in length are examined to ensure a better chance of the plane of polish being at the deepest portion of the hemisphere. Repetitive polishing and measurement of the metallographic sections provide another means of assuring the maximum depth of penetration has been measured and calculating the average depth for statistical evaluations [1].

Scanning Electron Microscopy (SEM) It is frequently important to examine unprepared surface (e.g., fracture face) or prepared surface (e.g., polished or etched) by SEM since resolution could go up to 20 nm to identify the corrosion form or type (pitting, intergranular, selective leaching) as well as intergranular or transgranular ruptures (corrosion failures). A scanning electron microscope is a valuable tool to differentiate intergranular stress corrosion cracking from hot-short cracking, and transgranular stress corrosion cracking from corrosion fatigue [1]. It has been shown that for more than half of the examined samples, the depth of corrosion measured by SEM examination of precorroded fatigue specimens was greater than that determined by optical microscopy. Investigators often employ a mechanical test (tension or fatigue test) of a corroded specimen, which inherently fractures at the deepest corrosion site. The fractured surface is then examined to measure the depth of corrosion at the fracture initiation site [2].

Transmission Electron Microscopy (TEM) TEM requires a very thin foil or surface replica through which electrons are transmitted. Magnifications can be 2000–4000×. Resolution can be about 2 nm, which provides the highest microstructural details; however, the maximum area that can be examined is limited at between 30 μm and 3 mm in diameter [1].

16.2.1.3. Corroded Surfaces

Examination of the cleaned surface should start at relatively low magnifications by a stereomicroscope, for example. Gross topographic features such as pitting, erosion corrosion, and wear should be sought out [1]. The form of corrosion could dictate special testing techniques in conjunction with microscopy. For example, a biochemical analysis is required to identify the organism in microbiologically influenced corrosion studies. In erosion corrosion studies, determination of the effects of particulate matter in a stream may require the flow pattern of water combined with microscopic examination of the metallic surface [3, 6, 7].

16.2.2. Nondestructive Evaluation Techniques

For a part in which the internal damage may have resulted from corrosion or the combined effects of corrosion, stress, and imperfections in the metal, the application of a nodestructive detection method before cutting may be desirable. Some of the following techniques may introduce contaminants into the test specimen (e.g., dye penetration inspection) and this possibility must be decided carefully for every case [1]:

1. ***Dye Penetration Inspection.*** This is currently used for corrosion detection or failures. Some substances found in liquid penetrants can be chemically similar to corrosion products or to those that may cause stress-corrosion cracking (SCC) [1].

2. ***Eddy-Current Inspection.*** This allows detection of cracks or defects and determination of the thickness of coatings, which cause variation in eddy currents (electromagmetic) induced by an applied alternating magnetic field.

3. ***Acoustic Emission.*** This allows detection of corrosion cracks online in particularly large vessels or systems that are generally too complex to inspect using other nondestructive evaluation techniques. However, operational or extraneous noise is often difficult to distinguish from crack growth [1]. Acoustic emission is recommended, for example, for aircraft applications to detect and locate active fatigue cracks inside fastener holes without the removal of the fasteners. Visible crack detection takes place generally at 75% of the fatigue life of the rivet row of the longitudinal fuselage skin in a multisite environment, while acoustic emission could detect it at ~20–40%.

4. ***Electrical Resistance.*** This is good for on-site or in-situ testing and monitoring, applied for integrity of coatings, corrosion rate, surface-breaking crack, or defect depth determinations.

5. *Ultrasonic Inspection.* This can be used for online monitoring; however, it is often impractical for thoroughly inspecting large vessels and extensive piping systems. There are three kinds of waves to consider. Longitudinal waves are applied normal to the surface and one detects the reflection of pulsed (typically $1000\,s^{-1}$), high-frequency waves (typically 5–10 MHz). These waves can detect a corrosion loss or wastage of material up to 250 mm thick. Shear waves are angled to the surface and are used to examine the location and orientation of any cracks or defects in the material up to 250 mm thick; this is particularly applied for welds. Surface waves give more information on the metal–solution interface and this technique is considered as fair practice for aluminum alloys (Table 16.1).

6. *Radiography Techniques.* These techniques use X-rays or gamma rays to penetrate the sample/structure (with subsequent photographic recording of findings). Extent of penetration depends on the thickness and on the material and its contained cracks and defects.

7. *Temperature Measurements.* These could be carried out using simple indicators applied to the surface, such as crayon/paints/lacquers for surface temperature, which undergo a color change or soften over a specific temperature range. The upper temperature limit is less than 1000 °C and has a poor resolution of temperature (typically ±50 °C). Radiation pyrometry implies matching by color a heated electrical filament and the target using emitted visible radiation (optical pyrometry). Both infrared radiation and visible radiation emitted by the target (total radiation pyrometry) are detected. Scanning detection techniques using infrared are subject to error in the presence of water vapor and CO_2, which absorb radiation.

8. *Pressure Measurements.* These could be made using a pressure transducer "plumbed" in or temporarily attached to a pressure line or vessel, which gives good resolution in the range of 0–45 MPa.

Table 16.1 shows the application of these techniques to aluminum and its alloys [1].

Table 16.1 Nondestructive Evaluation Techniques for Aluminum and Its Alloys for Evaluating Suspected Damage Due to Stress-Corrosion Cracking

Technique	Detectability
Visual	Poor
Penetrant: visible	Good
Penetrant: fluorescent	Excellent
In situ metallography	Fair
Radiographic: gamma ray	Poor
Radiographic: X-ray	Good
Ultrasonic: shear wave	Poor
Ultrasonic: longitudinal wave	Poor
Ultrasonic: surface wave	Fair
Eddy current: standard	Fair
Eddy current: remote field	Fair
Acoustic emission	Fair

Source: Reference 1.

16.2.3. Mechanical Testing

Hardness testing can be used to assist in evaluating heat treatment in comparing the hardness of the failed component with that prescribed by specification. Tensile, fatigue, or impact tests should be carried out as justified for certain applications. The determination of plane-strain fracture-toughness values may also be considered. Some tests on specimens at different temperatures are currently considered [1].

1. ***Macro- and Microhardness.*** The bulk (macro) indentation of a specimen surface by standard indenter (pyramid, ball, or cone) under known load, normally in the range of 1–3000 kgf, is performed.

Microhardness is determined in a similar fashion to bulk indentation but with pyramid diamond indenters used only in the range of 0.001–3.5 kgf (kilogram force; 1 kgf = 9.80665 Newton).

2. ***Tensile Test.*** The specimen is loaded under tension at a known loading or deflection rate. Normally, the test is carried out at a low strain rate and ambient pressure and temperature. The elastic limit and modulus are determined [1, 8].

3. ***Impact Test.*** The specimen (usually prenotched) is loaded at a high strain rate. Tests may be carried out for a range of temperatures (low). The energy absorbed in impact failure is recorded (upper and lower-shelf impact energies and impact energy transition from ductile to brittle).

4. ***Creep and Stresses.*** A load is maintained on a specimen subjected to high temperature (relative) for periods of up to 100,000 h. Creep strength (stress rupture) is examined at a given temperature and time. Creep strain, at a given load, temperature, and time, or apparent creep modulus (applied stress divided by creep strain) are also considered.

5. ***Stress-Corrosion Cracking.*** Performance is very sensitive to the orientation of the specimen with respect to wrought direction, for example. Data from tests on smooth specimens have limited use, often only for relative ranking or quality assurance purposes. A typical test for determining K_{ISCC} is to maintain the load on the specimen, often notched or precracked and subjected to a specific environment. The applied load is normally up to 0.9 of the yield strength and exposure is often limited to 1000 h. Susceptibility (e.g., threshold stress) to stress-corrosion cracking (SCC), critical stress-intensity factor for SCC, K_{ISCC}, and crack growth rate versus stress-intensity factor, *da/dn* versus $K_{ISCC,}$ are determined.

6. ***Cyclic Load.*** A cyclic load is applied to smooth or prenotched or precracked specimens, sometimes subjected to specific environments. Normally a range of loads is applied. Fatigue stress, S, versus number of cycles to failure, N (from replicate tests at a range of loads), is considered. Fatigue crack growth rate (normally versus stress-intensity range, ΔK) and threshold stress intensity for fatigue crack growth are measured. S–N data are susceptible to considerable scatter and a statistical approach is recommended.

7. ***Fracture Toughness.*** The tensile load is applied to a prenotched or a cracked specimen of known precise dimensions. Maximum load and/or crack opening displacement (COD) are recorded. Linear elastic fracture mechanics properties or elastic–plastic fracture properties (i.e., K_c, $K_{Ic,}$ etc.) are determined.

16.2.4. Chemical Analysis

Chemical analysis of any deposit, scale, or corrosion product, or of the medium with which the affected material has been in contact, is required. Certain gaseous elements or interstitials need special care and are normally not reported in a chemical analysis but they could have profound effects on the mechanical properties of some metals. Oxygen and nitrogen may give rise to strain aging and quench aging. Hydrogen may induce brittleness, particularly when absorbed during welding, cathodic cleaning, electroplating, or pickling. Within limits, the distribution of the microstructural constituents in a material is of more importance than their exact proportions. In a failure investigation, chemical analysis is recommended to ensure that the metal matches that specified for the application [1].

16.2.4.1. Spot Test

This concerns the application of selected reagent(s) to the surface and detection, by eye or with the aid of a microscope, of subsequent reaction. It indicates the presence or absence of specific elements, with detection limits depending on the nature of the metal. This is suitable for on-site use and semiquantitative analysis is possible in certain cases [1, 8].

16.2.4.2. Conventional Wet Analytical Chemistry

Quantitative chemical composition is time consuming and requires dissolution of solid samples for most elemental analysis. Gravimetric/volumetric/colorimetric/electrochemical and atomic absorption techniques could be used depending mostly on the nature of the element and required analytical precision. Complete analysis of major elements in one systematic consecutive separation of the desired elements is possible. Detection limits vary from 0.001% to 100% with an accuracy of $\pm 1\%$ of the detected level [1, 8].

16.2.4.3. Alternative Approaches of Analysis

Other techniques are very appropriate to achieve either a complete quantitative analysis of major elements and trace elements or a semiquantitative analysis.

Mass Spectrography This involves recording on a photographic plate the spectrum produced after ionizing a solid sample and accelerating the ions through a magnetic field. Spectral line densities are compared, subsequently, with a standard. Determination of quantitative bulk or local chemical composition including minor and trace constituents is possible. Accuracy is $\pm 5\%$ of detected level. Detection limits are from 10 ppb to 1% and one can analyze all elements in the metal from lithium (atomic weight 7) to uranium (atomic weight 238) and, with specialized equipment, H, N, O, and so on [1, 8].

Emission Spectrography This involves recording on a photographic plate the visual and ultraviolet spectrum produced by sparking a solid sample or by introduction of a solution into a plasma. Spectral line densities are compared, subsequently, with a standard. Determination of qualitative/quantitative bulk or local chemical composition is possible. Detection limits are from 1 ppm to 30%. Portable versions exist for semiquantitative on-site use, with detection limit not less than 0.1%. Accuracy is $\pm 5\%$ of detected level. Light elements, H, C, N, O, S, P, and Cl, are not detected [1, 8].

Mass/Emission Spectrometry and Mass Spectrography The output is converted using photomultipliers to direct reading of elemental quantitative concentrations, up to 20 elements during each analysis. It should be underlined that C, S, and P in metals may be detected by vacuum emission equipment. A full range of photomultipliers may not be available [1, 8].

Electron Diffraction Analysis Electrons from an electron microscope are scattered and transmitted through a thin film or reflected from a solid surface (from depths up to 50 Å) by the crystal lattice in a 1 μm diameter sampled area. It is limited to the identification of crystal structure or crystalline phases of solids. Diffraction techniques identify crystal structure only [1, 8].

X-Ray Diffraction Analysis X-rays are transmitted through or reflected from a solid sample. Samples in powder form are ideal for analysis of corrosion products. Quantitative crystalline phase analysis is possible but is limited to crystalline microstructures. Detection limits are 1–5% and accuracy is between ±1% and ±10% of detected level [1, 8].

Electron Probe Microanalysis Analysis is done by crystal spectrometry or energy dispersion of X-rays emitted as a result of applying a focused (1 μm diameter) electron beam to a surface. Qualitative and quantitative elemental composition can be determined for an excited volume of a few cubic micrometers. Specialized equipment in conjunction with an electron microscope is necessary and electron penetration depth is about 10 μm, and so it is limited to the analysis of surface films. This is used in conjunction with scanning electron microscopy. X-ray mapping of a 0.25 mm^2 area or line scanning is possible analysis [1, 8].

Electron Spectroscopy This involves analysis of either photoelectrons or Auger electrons emitted from a surface excited by X-ray photoelectron spectroscopy (XPS) or Auger electron spectroscopy (AES). Quantitative chemical analysis of the outermost atomic layers of surfaces is attained. This requires relatively large surface area (cm^2 for XPS). It is suitable to a depth of about 2 nm and analyzes all chemical compounds and detects all elements except hydrogen and helium [1, 8].

16.2.5. Surface Chemical Analysis

Energy-dispersive and wavelength-dispersive X-ray spectrometers are employed as accessories for scanning electron microscopes and permit simultaneous viewing and chemical analysis of a surface. To detect the elements in extremely thin surface layers, AES, XPS, Mössbauer spectroscopy, secondary ion mass spectroscopy (SIMS), low-energy ion scattering spectroscopy (LEISS), and several other techniques also are useful. Figure 16.1 shows the depths to which several of these techniques are capable of analysis. AES can provide qualitative and semiquantitative determinations of elements with atomic numbers of 3 or higher. The size of the area examined varies greatly with test conditions and may be from 1 to 50 μm in diameter [1, 8].

16.2.6. Published Data of Performance and Corrosion Resistance

Besides the chosen testing procedures, thermodynamic data (*E*–pH diagrams), recent published contributions, industrial reports, and manufacturer's data are essential to consult.

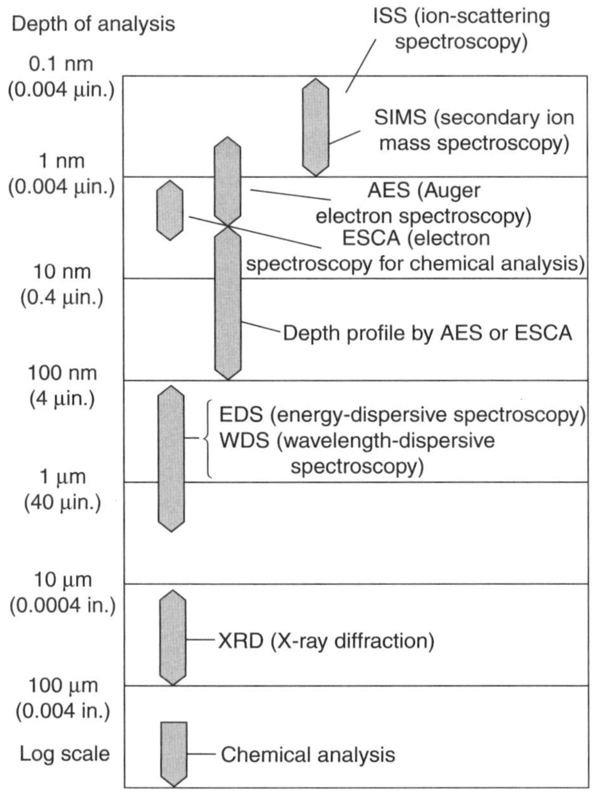

Figure 16.1 Relative depth of penetration of various surface analysis techniques [1].

The most desirable data are those obtained for the material of interest for the intended conditions of use. Such data are generally not easily available; however, examples of corrosion resistance of many metals and alloys in a variety of media and conditions are frequently and periodically published (see Part Six, Appendixes). Also, detailed information can be found in standards issued by the ASTM, NACE International, International Organization for Standardization (ISO), and the Materials Technology Institute of the Chemical Process Industries (MTI) [1].

Aluminum and aluminum alloys are more abundant and have been tested and used in different media and environments for long periods. Much less data are available concerning magnesium alloys. It is almost obligatory to look at the corrosion data before testing, since it gives a thorough background on the important parameters for testing for a certain alloy in a specific medium such as pH, temperature, oxygen content, and agitation. The following books have been chosen arbitrarily, essentially because of their basic and complimentary data (see Part Six, Appendixes).

The *Corrosion Data Survey: Metals Section* published by NACE International [9] is very useful for identifying unsuitable materials, locating which may have satisfactory performance, and giving corrosion rates and limited information on other types of corrosion. Aluminum alloys are considered, limited to some alloys of the series 1xxx, 3xxx, 5xxx, 6xxx, and cast 43, B214, 356, and 406. No aluminum alloy containing more than 1% Cu should be considered to have an equivalent corrosion resistance. The ideal rating has been assigned when corrosion is less than 50 μm (2 mils) per year and a secondary rating represents less than 508 μm (20 mils) per year. A third classification is provided to consider a corrosion rate between 508 and 1270 μm (20 and 50 mils) per year. The final rating is given when the corrosion rate is probably too high (>1270 μm/y) to consider for structural purposes.

The *Handbook of Corrosion Data* [10] is organized on the basis of environment (chemical compound) and is complementary to the book published by NACE. It gives comparative corrosion characteristics of aluminum alloys for general and stress-corrosion cracking and describes the general and past performance of Al alloys in chemical and natural environments (seawater, water, organic and inorganic chemicals, etc.).

The *Corrosion Resistant Materials Handbook* [11] considers the performance of metals, plastics, nonmetallics, and rubbers in enormous varieties of corrosive media for industrial, food processes, and so on. Pure aluminum is considered the standard for the behavior of all alloys. The tables consider saturated aqueous solutions and the influence of temperature on corrosion upto 460–560 °F (238–293 °C). The classification of performance as E (excellent), G (good), S (satisfactory), and U (unacceptable) is almost along the same lines as that by NACE International.

The *Corrosion Resistance Tables: Metals, Plastics, Nonmetallics and Rubbers* [12] gives the performance of aluminum alloy as unity and describes the performance of aluminum alloys in different corrosive solutions or organic colorants as a function of the concentration and temperature. The ratings are similar to that of NACE International. This book considers some information for selection of materials with better resistance to corrosion, especially for targeted applications. The comparative resistance of construction materials like tanks and gates is given in a series of tables.

Although it is strongly recommended to consult these references, equivalent or more up to date ones, care should be taken in the extrapolation of the information to specific situations. Available and published data for atmospheric exposure for long periods are valuable. However, the obtained results should be revised since atmospheric conditions are changing with time, such as for acid rain, and that can influence the performance of the material. The chemical composition, microgeometry of the surface, impurities, and the microstructure of the alloy as well as properties of the environment at the interface (agitation, circulation of the electrolyte, oxidant chemicals or local oxygen, erosion, suspended materials, microorganisms, temperature, etc.) are of concern in the selection of materials and evaluation of corrosion forms [7].

16.3. ELECTROCHEMICAL POLARIZATION STUDIES

Electrochemical polarization test methods are extremely pertinent for understanding and evaluating the corrosion resistance of materials and the effect of changes in the corrosive environment. The dc electrochemical techniques (cyclic voltammetric, potentiodynamic, potentiostatic, and galvanostatic techniques) are basic rapid tools of investigation [13]. They can establish criteria for anodic or cathodic protection and susceptibility to several forms of corrosion. These are particularly powerful for characterizing aluminum alloys (different phases, active–passive behavior, pitting, crevices, etc.) when used in conjunction with some of the previously mentioned methods. As an example, determination of critical pitting potential in a certain chloride medium, study of intergranular attack, and relating some of these measures to stress-corrosion cracking of aluminum alloys are advanced subjects in controlling these types of corrosion [14]. The dc techniques have important and appropriate possibilities for exploring the influence of alloying elements on the quality of the passive aluminum oxide and on the performance of these alloys in different electrolytes [15].

16.3.1. Measurements of the Corrosion Potential

The determination of corrosion potential of the corroded metal and its evolution is an important and simple type of measurement. This is frequently called open circuit potential (OCP), dissolution potential, or stationary potential depending on its equilibrium state. Potential–time relationships have widely been used for studying film formation and film breakdown, as indicated by an increase or decrease in the corrosion potential, respectively. This could show the active behavior of the metal as well as the active–passive behavior of Al, Mg, and their alloys. This depends on the solution properties at the metal–solution interface. Application of the Nernst equation based on thermodynamic data can give the metal potential as a function of the ion activity of the metal in the medium. An activity of 10^{-6} mole/L (or M) is generally considered as an approximation when the activity of the corroded ions in solution is not well known. The corrosion potential deduced from the intersection of anodic and cathodic Tafel slopes after a certain scan rate dE/dt can be different from that determined at the OCP depending on immersion time and steady or stationary states of the metal–solution interface (see Chapter 3).

16.3.2. Potentiodynamic Methods

A potentiostat that will maintain an electrode potential within 1 mV of a preset value over a wide range of applied currents should be used. For the type and size of standard specimen supplied, the potentiostat should have a potential range from −0.6 to 1.6 V with respect to corrosion potential and a current output range from 1.0 to 10^5 μA. In the potentiodynamic method, a potentiostatic potential sweep rate of 0.3 mV/s is generally used, depending on the metal/solution properties and steady states, and a continuous recording of the current is made with change in potential from the cathodic zone to corrosion potential and then to the anodic zone [16]. Scan rate is very important to potentiodynamic polarization. In general, there is less distortion of the true polarization curves at the lowest scan rates. Scan rates from 0.05 to 0.2 mV/s may give the standard active–passive behavior, otherwise 0.01–0.04 mV/s rates can be tried. In ASTM G5, a scan rate of 600 mV/h is recommended. Argon or nitrogen should be bubbled to remove oxygen from the solution. The data collected can be used to construct a polarization curve, which usually is plotted as E_{appl} versus the logarithm of the measured current density [16].

ASTM G5 outlines the standard methods more precisely for making potentiostatic and potentiodynamic anodic polarization measurements. Potentiodynamic polarization curves can be obtained as a single sweep, and sometimes the scan is reversed when E_{final} or a preset maximum current density is reached, and stopped when previous E_{corr} or a new value of E_{corr} is reached. Similarly, infrequently changing current can be imposed on the metal and the corresponding potential is recorded (ASTM G5 2006).

Plots of E_{appl} versus $\log I$ are called *Tafel plots*. The corrosion current density, i_{corr}, can be determined using two equations by extrapolation of the Tafel lines [7]: a complete polarization curve consists of a cathodic part and an anodic part. The cathodic portion of the polarization curve contains information concerning the kinetics of the reduction reaction(s) occurring for a particular system. Depending on the solution composition, a mass-transport-controlled region can be reached at more negative potentials, at which the reaction rate (the measured current density) depends only on the solution composition and the hydrodynamic conditions. The particular features of the anodic part of the polarization curve depend strongly on the metal–electrolyte system. Usually, a charge-transfer-controlled region occurs at potentials close to E_{corr}. So-called passive metals show an active–passive transition followed by a passive region and a region of oxygen evolution at the highest applied potentials. For those metals, which are susceptible to localized corrosion, a large increase of the current occurs in the passive region when the pitting potential, $E_{pit,}$ has been exceeded [16].

The relationship between the applied potential, $E_{appl,}$ and the measured current density, i, in the Tafel region is given by [17, 18]

$$E_{appl} = E_{corr} + \beta_a \log(i/i_{corr}) \quad \text{for the anodic polarization curve}$$

$$E_{appl} = E_{corr} - \beta_c \log(i/i_{corr}) \quad \text{for the cathodic polarization curve}$$

where β_a and β_c are positive constants (b_a and b_c are equally used). From i_{corr}, the corrosion rate can be calculated using Faraday's law [7, 16]. This method of obtaining corrosion current density is called the *Tafel extrapolation method.* It is also possible to determine i_{corr} from either the anodic or the cathodic Tafel plot or from the intersection of both plots. It is advantageous to use computer software to record polarization curves and analyze the experimental data in terms of parameters such as β_a, $\beta_{c,}$ and i_{corr}. From the numerical values of β_a and β_c, conclusions concerning the rate-determining step in the reaction mechanism can be made.

The determination of the Tafel constants from $\eta - i$ curves for each system studied is time consuming and may not be particularly accurate owing to resistance and mass transfer effects. Hoar has criticized the method on these grounds and has pointed out that the complete Tafel equations for the anodic and cathodic reactions, which have to be determined to evaluate the Tafel slopes, can be used to calculate i_{corr} without resorting to the polarization resistance technique. Mansfeld suggests that polarization curves obtained in the R_p region can be fitted to various theoretical curves, preferably by computer analysis, to give separate values of both β_a and β_c, which may change substantially during the corrosion test. Makrides uses a single mass-loss determination at the end of the test to obtain constant. Once the constant has been determined it can be used throughout the tests, providing that there is no significant change in the nature of the solution that would lead to markedly different values of the Tafel constants. The polarization resistance has the advantage that the small changes in potential required in the determination do not change the interface and the system significantly [4].

Polarization Resistance (i_{corr}) In measurements of the polarization resistance, only a small electrochemical signal is applied in order to ensure that the system under investigation remains linear. This requirement also ensures that these measurements are nondestructive and therefore can be repeated many times on the same system [19, 20]. Stern and co-workers used the term *linear polarization* to describe the linearity of the η–i curve in the region of E_{corr}, the corrosion potential. The slope of this linear curve, ΔE–Δi, is termed the *polarization resistance*, R_p, since it has dimensions of ohms, and this term is synonymous with linear polarization in describing the *Stern–Geary technique* for evaluating corrosion rates [4].

$$\frac{1}{R_p} = \left(\frac{\Delta i}{\Delta E}\right)_{E_{corr}} = 2.3\left(\frac{\beta_a + |\beta_c|}{\beta_a|\beta_c|}\right) i_{corr}$$

where R_p is the polarization resistance determined at potentials close to E_{corr}, and β_a and β_c are the Tafel constants; note that in the case of β_c the negative sign is disregarded. This equation shows that the corrosion rate is inversely proportional to R_p (or directly proportional to the reciprocal slope of the ΔE–Δi curve) at potentials close to E_{corr} (>10 mV), and that i_{corr} can be evaluated providing the Tafel constants are known. For a process that is controlled by diffusion of the cathode reactant (transport control) and in which the anodic process is under activation control, a similar linear relationship applies [4]:

$$\frac{1}{R_p} = \left(\frac{\Delta i}{\Delta E}\right)_{E_{corr}} = \frac{2.3\, i_L}{\beta_a} = \frac{2.3\, i_{corr}}{\beta_a}$$

where i_L is the limiting current density of the cathodic reaction and it is assumed that $i_L = i_{corr}$. Taking arbitrary values of the Tafel constants showed that corrosion rates determined by the polarization resistance techniques are in good agreement with corrosion rates obtained by mass loss methods [4]. Especially when only a small polarization range ($E - E_{corr}$) is applied (usually ±30 mV or less), this approach to obtain corrosion rates and Tafel slopes in the non-Tafel region is nondestructive and can be applied repeatedly without danger of changing the surface under investigation [16].

The polarization resistance (R_p) is defined as $R_p = (dE/di)_{E_{corr}}$ and is determined as the slope of the polarization curve when $I = 0$ or $i_a = i_c$ in absolute values (ASTM G59) [7]. R_p is a constituent of the total resistance in the circuit (R_p and R_s). Elimination of R_s is highly recommended (see Chapter 3). Figure 16.2 gives a standard polarization curve for type 430

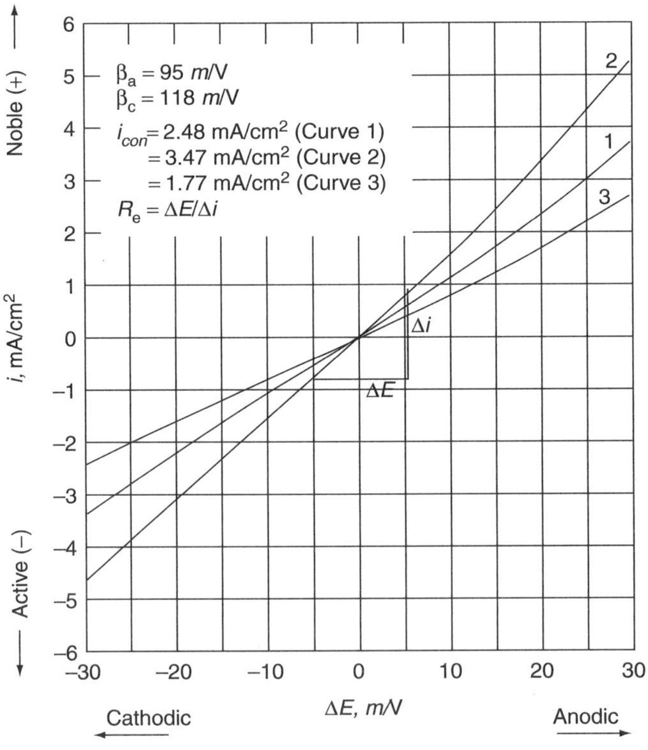

Figure 16.2 Standard potentiodynamic polarization curves in the vicinity of E_{corr} for type 430 stainless steel, in 1 N H_2SO_4 at 30 °C [7, 16].

stainless steel in 1.0 N H_2SO_4, purged with hydrogen, nitrogen, or argon at 30 °C. Experimental curves should lie between curves 1 and 3; otherwise experimental errors occur (ASTM G59) [7, 16].

The ASTM G102 "Electrochemical Measurements" guidance covers converting the results of electrochemical measurements into rates of uniform corrosion. Calculation methods for converting corrosion current density values into either mass loss rates or average penetration rates are given for most engineering alloys. In addition, some guidelines for converting polarization resistance values into corrosion rates are provided.

Most authors emphasize the importance of making polarization resistance measurements at both positive and negative overpotentials. Oldham and Mansfeld conclude that although linearity is frequently achieved this is due to three possible causes: ohmic control due to the *IR* drop rather than control according to linear polarization; the similarity of the β_a and β_c values; and the predisposition by the experimenter to assume that the ΔE–Δi curves near E_{corr} must be linear. Oldham and Mansfeld showed that linearity of the ΔE–Δi curve is not essential and that i_{corr} can be evaluated from the slopes of the tangents of the nonlinear curve determined at potentials of about 20–30 mV more positive and negative than E_{corr} [16].

16.3.3. Cyclovoltammetry Techniques and Pitting

Pitting scans are carried out frequently to determine E_{pit} and sometimes also the protection potential, E_{prot}. ASTM G61 gives a procedure for conducting cyclic potentiodynamic measurements to evaluate relative susceptibility to localized corrosion and to determine the E_{pit} at which the anodic current increases rapidly from the low values in the passive region. The more noble E_{pit} is with respect to $E_{corr,}$ the less susceptible the alloy is to initiation of localized corrosion. The intersection of the returning scan with the starting one gives the E_{prot} value below which pits do not initiate in the considered medium and experimental conditions. The critical pitting potential, E_{cpr} lies between the breakdown potential and the protection potential and can be determined by the scratch repassivation method (see Chapter 17) [7,16, 21].

16.3.4. Potentiostatic, Galvanostatic, and Galvanodynamic Methods

Potentiostatic Techniques Recording of the current as a function of time at a constant applied potential provides information about the time dependence of the rate of an electrochemical reaction at a given potential. This can be applied to determine pitting potentials. The breakdown of the films can be determined by holding the sample in the electrolyte at a higher potential than E_b until localized corrosion occurs (see Chapter 17) [3, 7].

Galvanostatic and Galvanodynamic Techniques These measurements are frequently closer to some industrial processes than the potentiostatic and potentiodynamic techniques. In galvanostatic techniques, a constant current is applied, and the potential is monitored as a function of time until the time rate of change in potential approaches zero. It is also possible to record an entire polarization curve in galvanostatic mode [16].

16.4. THE ac ELECTROCHEMICAL IMPEDANCE SPECTROSCOPY TECHNIQUE

16.4.1. Introduction

Impedance is the term used to describe the alternating current (ac) equivalent of direct current (dc) resistance. It is a totally complex resistance encountered when a current flows through a circuit made of resistors, capacitors, or inductors, or any combination of these. The ac electrochemical impedance spectroscopy (EIS) is an electrochemical method in which an ac signal (hence the alternative name of ac impedance) is used. The impedance is measured as a function of the frequency of the ac source. The technique where the cell electrode impedance is plotted versus frequency is called electrochemical impedance spectroscopy. Modeling of impedance measurement aims to deduce a relationship between electrical components like resistance (R) and capacitance (C) to the phenomena occurring at the metal–solid interface such as charge transfer, diffusion, and adsorption. In the case of alternating applied voltages, there is another component of the resulting current that must be allowed for and this is due to the capacitance, C_{dl}, of the surface. Because the current can either flow through the interface, causing oxidation or reduction of the species, or charge or discharge the capacitance, these two components of the surface are in parallel [22].

The ac impedance is actually a conventional technique for the study of corrosion phenomena, performance of semiconductors, electrolytic phenomena, electro-organic synthesis, and coatings evaluation. The impedance measurements are most useful for studying complex systems such as the anodic behaviors of metals and other composite materials, corrosion, and states of electrodes during charging/discharging cycles of batteries, surface characterization of polymer-modified electrodes, and many other complex electrochemical phenomena [23]. It has become a routine tool for practical corrosion prediction and can give a rapid assessment of the action of an inhibitor or the protection performance of paint and a rapid estimation of corrosion rates even in low conductivity media [24]. Since this technique permits the rejection of noise, EIS measurements could be considered for electroanalytical measurements [16].

16.4.1.1. Some Advantages of EIS over dc Techniques

1. The EIS technique applies very small signals that are nondestructive. Frequently, a small voltage signal is applied to a corroding metal and the resulting current is measured. Since the applied signal is generally in the range of 5–10 mV peak-to-peak (very low amplitude), the tests are nondestructive for the properties of the metal–solution interface. The ac impedance excitation amplitudes of this magnitude cause only minimal perturbation of the electrochemical test system, thus reducing errors caused by the measurement technique. The possibility exists for studying corrosion reactions and measuring corrosion rates in low-conductivity media for surfaces coated with paint, where traditional dc methods fail. In addition, polarization resistance as well as double-layer capacitance data can be obtained in the same measurements [24]. However, impedance methods for the evaluation of corrosion performance of painted metals are not without their limitations. These limitations are caused by the measuring technique itself—the accuracy, frequency range, and the time taken to obtain the impedance data being dependent on the equipment configuration.

2. The technique offers valuable mechanistic information because of the possibility of obtaining data on the capacitance of the electrode, on the charge transfer kinetics, on the many parameters of the electrochemical cell, such as R_{ct}, C_{dl}, Z, phasor angle ϕ, and Warburg impedance $Z\omega$, and even on the equivalent circuit of the examined electrochemical cell [16].

3. The advantage of impedance measurements for corrosion studies at or close to open circuit potentials is that it provides a small potential perturbation on the corroding system. It has been shown that some types of electrical interference with the polarization resistance technique of dc methods have much less influence when ac techniques are used [42]

4. It was demonstrated that uniform corrosion occurs when anodic and cathodic reactions take place with a constant change of location and time of the individual processes. The analysis of impedance data can lead to the value of the polarization resistance. However, the values of the Tafel slopes, which have to be known as a function of time, electrode, and electrolyte parameters in order to calculate the corrosion current density (CCD), i_{corr}, or the corrosion rate, have to be obtained by conventional electrochemical dc polarization techniques [24].

5. EIS is still being used for localized corrosion studies involving pitting. The statistical variation of pit nucleation and the absence of steady states prevent long-time measurements in the low-frequency region. In addition, in the pitting region, a complicated Nyquist plot is obtained and is difficult to interpret. However, characteristic changes have been discovered in the low-frequency region. It should be noted that the impedance spectra for pits in stainless steels and magnesium are different from those in aluminum [16, 25–27].

6. A corrosion system can be examined by EIS at fixed potential or current. By imposing potential sweeps, potential steps, or current steps, the electrode is typically driven to a condition far from equilibrium, and the response is usually a transient signal. At an examined potential, for example, the properties of the system can be examined through analysis of the frequency dependence of the impedance [20].

16.4.1.2. The EIS Standards

ASTM G106, for example, provides a standard material, electrolyte, and procedure for collecting electrochemical impedance data at the open circuit or corrosion potential that should reproduce data determined by others at different times and in different laboratories. Also, it provides an experimental procedure to check one's instrumentation and a technique for collecting and presenting electrochemical impedance data.

16.4.2. EIS Terms and Equivalent Circuits

The terms resistance and impedance both denote an opposition in the flow of electrons or current. In direct current, only resistors produce this effect while for ac circuits, two other circuit elements, capacitors and inductors, impede the flow of electrons. The total impedance of the circuit is the combined opposition of all its resistors, capacitors, and inductors to the flow of the electrons. The opposition of capacitors and inductors is given the name *reactance*, symbolized by X and measured in ohms (X_C for capacitive reactance and X_L for inductive reactance) [23].

Consider Ohm's law, $V = IR$, where V is the voltage across a resistor in volts, R is the resistance in ohms, and I is the current in amperes, and conductivity $k = 1/R$. For ac signals, $E = IZ$, where E and I are waveform amplitudes for potential and current, respectively, and Z is the *impedance* while *admittance* $Y = 1/Z$. The impedance of a system at a given frequency is defined by two terms that relate the current output to the input voltage: these are the amplitude of the ac current divided by the amplitude of ac voltage and the phase angle (proportional to the shift in time between peak current and peak voltage). The collection of these parameters for different frequencies is the impedance spectrum [23].

Expression of $|Z|$ Consider the case where a resistance R and a capacitance C are in series and a voltage E is applied across them. This voltage must be equal to the sum of the individual voltage drops across the resistor and capacitor [28]: $E = E_R + E_C$ and $E = IZ$ and the voltage is linked to the current in this case through a vector, $Z = R - jX_C$, called the impedance.

If a voltage E is applied to a system having a variable frequency under the form

$$E = E_0 \sin(\omega t)$$

where E_0 is the amplitude of the signal and ω is the angular frequency ($2\pi f$) and t is the time in seconds, then the current response will be in the same form:

$$I = I_0 \sin(\omega t + \phi)$$

where I_0 is the amplitude of the current and ϕ is the phase angle. The ratio of voltage amplitude to that of the current defines the impedance term:

$$|Z| = E_0 / I_0$$

Impedance can be expressed as a complex number, where the resistance is the real component and the combined capacitance and inductance is the imaginary component. Vector analysis provides a convenient description of the wave in terms of its amplitude and its phase characteristics. The real and imaginary components of an ac waveform are defined with respect to some reference waveform. The real component is in phase with the reference waveform, and the imaginary component, also referred to as the *quadrature component* is exactly 90° out of phase. The reference waveform allows expression of the current and voltage waveform as vectors with respect to the same coordinate axes, which facilitates mathematical manipulation of the vector quantities [23].

In general, the impedance can be represented as [28].

$$Z(\omega) = Z_{Re} - jZ_{Im}$$

where Z_{Re} and Z_{Im} are the real and imaginary parts of the impedance, for example, $Z_{Re} = R$ and $Z_{Im} = X_C = 1/\omega C$. The magnitude of Z, written as $|Z|$, is given by

$$|Z|^2 = R^2 + X_C^2 = (Z_{Re})^2 + (Z_{Im})^2$$

The phase angle is expressed as ϕ and is given by

$$\tan\phi = Z_{Im}/Z_{Re} = X_C/R = 1/\omega RC$$

A vector can be expressed in terms of real (I') and imaginary (I'') coordinates. Components along the abscissa are real while components along the ordinate are assigned as imaginary and are multiplied by $j = \sqrt{-1}$. Any ac current vector can be defined as the sum of its real and imaginary components [23]:

$$I_{\mathrm{TOTAL}} = I' + I''\mathrm{j} \quad \text{where} \quad j = \sqrt{-1} \quad \text{and} \quad E_{\mathrm{TOTAL}} = E' + E''j$$

Then the impedance vector could be expressed as

$$Z_{\mathrm{TOTAL}} = \frac{E' + E''j}{I' + I''j} = Z' + Z''j$$

The absolute magnitude of the impedance vector and the phase angle can then be expressed as

$$|Z| = \sqrt{(Z')^2 + (Z'')^2}, \quad \tan\theta = \frac{Z''}{Z'}, \quad Z' = |Z|\cos\theta \text{ and } Z'' = |Z|\sin\theta$$

Equivalent Electrical Circuits Table 16.2 shows that the impedance of a resistor has no imaginary component at all. The phase shift is zero degree; that is, the current is in phase with the voltage. The ac impedance of a capacitor, on the other hand, has no real component and an imaginary component that is a function of both capacitance and frequency. Similar expressions can be derived for other circuit elements, and the impedance for any combination of circuit elements can be defined by a linear combination of the elemental expressions. For example, the parallel resistor–capacitor network impedance has an expression that contains both real and imaginary components as shown in Table 16.2 [23].

Electrochemical systems can be examined with respect to their equivalent electrical circuits. R_{Ω} is the uncompensated resistance between the working electrode and the reference electrode. R_p or R_{ct} is the polarization resistance at the electrode–solution interface. C_{DL} could represent the double-layer capacitance at this interface. Knowledge of R_p permits the calculation of electrochemical reaction rates. Capacitance measurements can provide information on adsorption and desorption phenomena, film formation processes at the electrode, and the integrity of the organic coating.

Table 16.2 Applications Amenable to Impedance Studies

Circuit element	Impedance equation
	$Z = R + 0j \qquad j = \sqrt{-1}$
	$Z = 0 - j/\omega C \qquad \omega = 2\pi f$
	$Z = 0 + j\omega L \qquad \omega = 2\pi f$
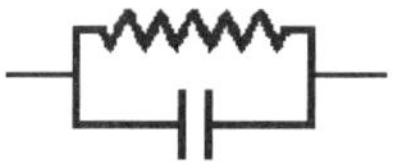	$Z = \frac{R}{1 + \omega^2 C^2 R^2} - \frac{j\omega C R^2}{1 + \omega^2 C^2 R^2}$

Source: Reference 23.

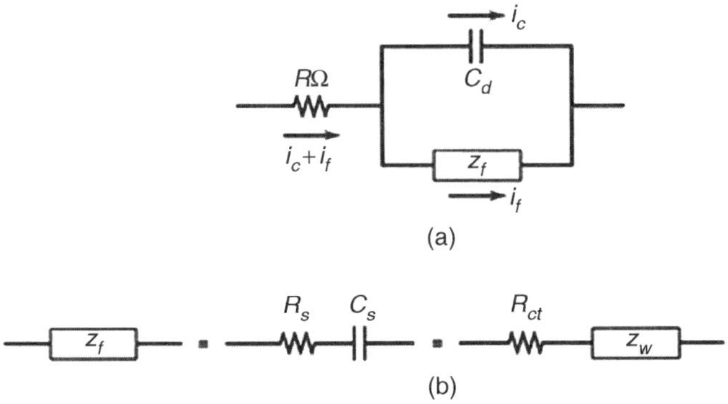

Figure 16.3 (a) Equivalent circuits of an electrochemical cell. (b) Subdivision of Z_f into R_s and C_s, or into R_{ct} and Z_w [28].

A frequently used circuit, called the *Randles equivalent circuit*, is shown in Figure 16.3a. The parallel elements are introduced because the total current through the working interface is the sum of distinct contributions from the Faradaic process, i_f, and double-layer charging, i_c. The double-layer capacitance is represented by a simple linear circuit by the elements C_{DL}. The Faradaic process cannot be represented by an equivalent circuit element like R and C, whose values are independent of frequency. It must be considered as general impedance, Z_f. Since the current must pass through the solution resistance, R_Ω is inserted as a series element to represent this effect in the equivalent circuit [28, 29].

The Faradaic impedance has been considered in the literature in various ways. Figure 16.3b shows two equivalents that have been made. The simplest representation is to take Faradaic impedance as a series combination comprising the series resistance, R_s, and the pseudocapacity, C_s [30]. An alternative is to separate a pure resistance, R_{ct}, the charge-transfer resistance, from another general impedance, Z_W, the Warburg impedance, which represents a kind of resistance to mass transfer. In contrast to R_Ω and C_d, which are nearly ideal circuit elements, the components of the Faradaic impedance are not ideal, because they change with frequency ω. The chief objective of a Faradaic impedance experiment is to discover the frequency dependency of R_s and C_s or R_{ct} and Z_W. Theory is then applied to transform these functions into chemical information [28]. Actually, the frequently used symbols for the equivalent circuit are R_Ω for solution resistance and R_{ct} for charge-transfer resistance and the latter is frequently compared to R_p. Models that account for diffusion-controlled electrochemical processes usually employ an element for charge-transfer resistance R_t or R_{ct}, rather than a polarization resistance, R_p.

The circuit in Figure 16.4 is analogous to an electrochemical reaction coupled to a chemical reaction. Here R_{CR} and C_{CR} represent, respectively, the resistive and capacitive effects of the chemical reaction. The box enclosing R_W and C_W is actually a simplification of the circuit representing the Warburg impedance, which accounts for mass transfer limitations due to diffusion processes adjacent to the electrode. Effectively, if, for example, the chemical reaction produces the electrochemical active species and its rate is slower than that of the electrochemical reaction, the slower rate of the chemical reaction will control the kinetics of the whole reaction in a similar way to that of the diffusion process [23].

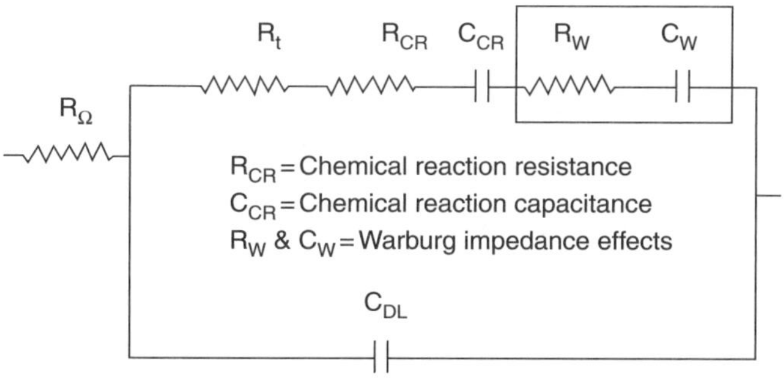

Figure 16.4 Equivalent circuit for an electrochemical reaction coupled to a chemical reaction [23].

The rate of an electrochemical reaction can be strongly influenced by the diffusion of one or more reactants to the electrode surface. This is often the case when a solution species, prior to reaction, must diffuse through a surface film on the electrode. This situation can exist when the electrode surface is covered with reaction products, adsorbed solution components, or a prepared coating. In the case of a painted metal–solution interface, one modification takes into account diffusion processes within pores in the paint film, which are modeled by the inclusion of a Warburg or pseudo-impedance [23], where Z_W can be expressed as

$$Z_W = \sigma\omega^{-1/2}(1 - j)$$

where σ is the Warburg impedance coefficient (ohm·s$^{1/2}$) and $\omega = 2\pi f$ (rad·s^{-1}).

Whenever diffusion effects completely dominate the electrochemical reaction mechanism, the impedance is called *Warburg impedance*. For a diffusion-controlled electrochemical reaction the current is 45° out of phase with the potential excitation. This means that the real and imaginary components of the impedance vector are equal at all frequencies. The value of $|Z|$ for a diffusion-controlled system varies with frequency as $\omega^{-1/2}$.

The electrochemical reaction states are better understood using EIS and the electronic components are determined by impedance measurements. Figure 16.5a shows the model of the interface proposed by Hemholtz [31]. In a simple case, the interface can be modeled by an equivalent circuit as shown in Figure 16.5b. This is also called a Randles circuit, which is made of a double-layer capacitor in parallel with a polarization resistor (also known as a charge-transfer resistor, R_{ct}, with certain constraints) and Warburg impedance, Z_W, connected in series with a resistor that measures the resistance of the electrolyte, R_s [32].

16.4.3. Impedance Plots

EIS data are recorded as a function of frequency of an applied ac signal at a fixed working point (E, i) of the polarization curve. In corrosion studies, this working point is often E_{corr} ($E = E_{corr}$, $I = 0$). Usually, a very large frequency range has to be investigated to obtain the complete impedance spectrum using the frequency response analyzers (FRAs). In most corrosion studies, this frequency range extends from 100 kHz to 1 MHz. Measurements at lower frequencies are very time consuming [16].

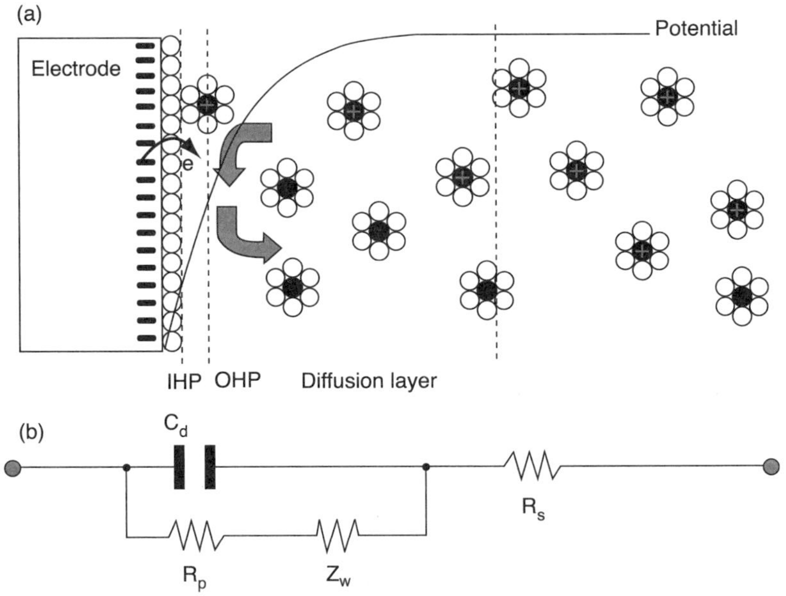

Figure 16.5 A simple metal–aqueous solution interface in which the vertical dotted lines in (a) can match the electronic components determined after EIS studies [32].

Impedance data are usually determined with a three-electrode system, although it is also possible to use a two-electrode system in which both electrodes are of the same material. In this case, the EIS data are collected at $E = 0$ V. A potentiostat is used to apply the potential at which the data are to be collected. The FRA is programmed to apply a series of sine waves of constant amplitude, small enough to remain in the linear potential region, and varying the frequency. Impedance data are determined by FRA at each frequency and stored in its memory. Because a very large number of data points have to be collected, displayed, and analyzed, it is essential to use adequate software for these purposes. Such software is available from all major manufactures of FRAs [16].

Kramers–Kronig (KK) transforms or algorithms have been used in electrochemistry for two purposes: first, for obtaining the polarization resistance from the frequency-dependent imaginary component, second, to assess the quality of the measured impedance data and address their validation. The validation of the experimental data of impedance measurements should follow basic conditions [33]:

1. The response of the system must be the result only of the applied perturbation.
2. The relationship between the perturbation and response is independent of the magnitude of the perturbation.
3. The system returns to its starting state after the perturbation is removed.
4. The transfer function (impedance) must be finite as the frequency approaches both 0 and ∞ and is continuous and finite valued at all intermediate frequencies.

The data are usually provided by EIS equipment. Nyquist and Bode plots are commonly used for the graphical presentation of impedance data obtained over a wide frequency range.

16.4.3.1. The Nyquist Plot

The *Nyquist plot*, known as a Cole–Cole plot or a complex impedance plane plot, is a common technique for evaluating ac impedance data (real Z' or Z_{Re} on the x axis and imaginary Z'' or Z_{Im} on the y axis). In this technique, the imaginary part of the impedance Z'' is plotted against the real part Z'. Because the majority of the responses of corroding metals have negative Z'', it is conventional, for corrosion studies, to plot $-Z''$ against Z'. So it should be noted that in this plot the y-axis is negative and each point on the Nyquist plot is the impedance at one frequency [23].

On the Nyquist plot, the impedance can be represented as a vector of length $|Z|$. The Nyquist plot is popular since it shows easily the ohmic resistance. The shape of the curve (often a semicircle) does not change when the ohmic resistance changes. The plot emphasizes circuit components that are in series, such as R_W. The ohmic resistance and polarization resistance can easily be read directly from the Nyquist plot; however, frequency does not appear explicitly for every point and the electrode capacitance can be calculated only after the frequency information is known [23].

Figure 16.6 shows the high- and low-frequency limits as well as the intermediate zone of frequencies. An actual plot of impedance in the complex plane will combine the features of the two precedent limiting cases as well as that corresponding to the intermediate frequencies. The low limit frequency is the zone characterized by the mass transfer control (on the extreme right), while on the extreme left, the reaction is kinetically controlled by the rate of electron transfer [28].

When we consider the high-frequency limit, the Warburg impedance becomes unimportant compared to R_{ct}. For low frequency, as $\omega \to 0$, this leads to

$$Z_{Re} = R_\Omega + R_{ct} + \sigma\omega^{-1/2} \quad \text{and} \quad Z_{Im} = \sigma\omega^{-1/2} + 2\sigma^2 C_d$$

Elimination of ω between these two equations gives

$$Z_{Im} = Z_{Re} - R_\Omega - R_{ct} + 2\sigma^2 C_d$$

Typical multiple time-constant Nyquist plots are obtained frequently and close inspection is necessary when one of the semicircles is much smaller than the other(s) (www.princeton

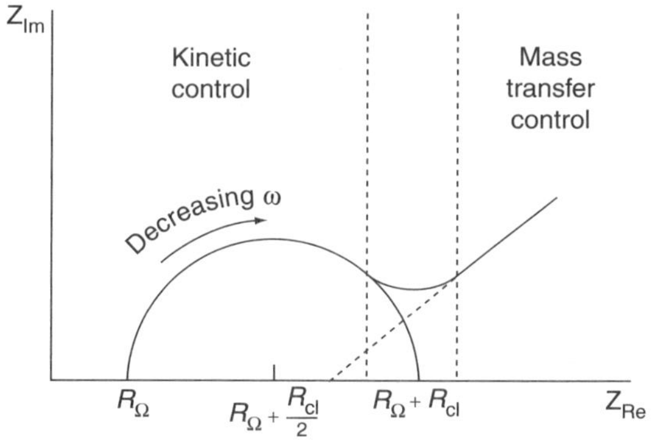

Figure 16.6 Impedance plot for an electrochemical system showing that the regions of mass transfer and kinetic control are found at low and high frequencies, respectively [28].

appliedresearch.com). EIS is also a good technique to investigate the anodic behaviors of metals and composites. In most complex systems, the impedance response involves more than one semicircle. The second or third semicircles in the lower frequency domains in Nyquist impedance plots, when present, represent the electrochemical reactions of slower rise times. When the interface is complex because of adsorption or chemical reactions, more than one semicircle can be observed. The i–η curve follows the Butler–Volmer equation until the current is limited by mass transfer when the overpotential η becomes large. The I–η curve constructed from the impedance data can provide information not available from impedance results alone [32].

Frequency Dispersion In most corrosion systems, the capacitive semicircle exhibits significant deviation from the ideal semicircle. This has often been referred to as *frequency dispersion* attributed to circuit elements. Detailed analysis of the experimental data shows that this deviation can be described by a rotation of the semicircle below the real axis by a certain angle. A good approximation, which is often used in corrosion studies to describe this effect, is a well-known dispersion formula [34–37]. In general, the physical meaning of the frequency dispersion is not yet fully understood [38]. The occurrence of the inductive loop, which becomes more pronounced in the presence of inhibitors, has created an additional problem concerning the significance of the inductive loop and its relationship to polarization resistance [24].

Theoretical models, which describe the roughness of the electrode surface in terms of fractal geometry, predict a correlation between the exponent σ and the fractal dimension D_F [39]. Experiments on model electrodes of fractal geometry were found to be in agreement with the theoretical prediction [40]. The effect of the surface roughness and porosity is of great importance, and several impedance models were derived to explain the effect of three-dimensional electrode structures, as reviewed by De Levie [41].

16.4.3.2. The Bode Plot

The Bode plot has some distinct advantages over the Nyquist plot, as shown in Figure 16.7. A typical plot considers log of frequency on the x axis and both the absolute value of the impedance ($|Z| = Z_0$) and phase shift on the y axis. The plot uses the logarithm of frequency to allow a very wide frequency range to be plotted on one graph, but with each decade given equal weight. The Bode plot provides a clearer description of the electrochemical system's frequency-dependent behavior than does the Nyquist plot. The Bode plot also shows the magnitude ($|Z|$) on a log axis so that you can easily plot wide impedance ranges on the same set of axes.

The Bode plots easily shows the frequency break points associated with each limiting step. In some electrochemical processes, there is more than one rate-determining step. Each step represents a system-impedance component and contributes to the overall reaction-rate constant. The ac impedance experiment can often distinguish among these steps and provide information on their respective rates or *relaxation times.* The Bode plot also has some disadvantages. The greatest one is that the shape of the curves can change if the circuit values change [23].

The log $|Z|$ versus log ω curve can yield values of R_Ω and R_p or R_{ct}. At the highest frequencies shown in Figure 16.7, the ohmic resistance dominates the impedance and log (R_Ω) can be read from the high-frequency horizontal plateau. At the lowest frequencies, polarization resistance also contributes and $\log(R_\Omega + R_p)$ can be read from the low-frequency horizontal plateau. At intermediate frequencies, the curve should be a straight

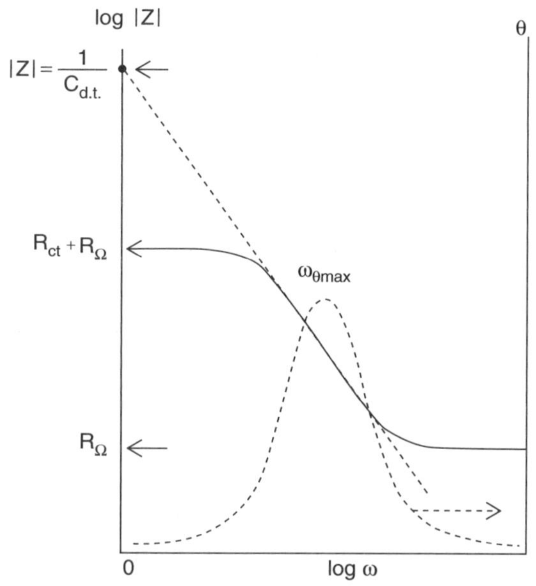

Figure 16.7 Bode plot for a simple electrochemical system [23].

line with a slope of -1. Extrapolating this line to the log $|Z|$ axis at $\omega=1$ (log $\omega=0$, $f=0.16$ Hz) yields the value of C_{DL} from the relationship

$$|Z| = 1/C_{DL}, \quad \text{where } \omega = 2\pi f$$

The Bode plot format also shows the phase angle $\ominus$. At the high- and low-frequency limits, where the behavior of the Randles cell is resistor-like, the phase angle is nearly zero. At intermediate frequencies, $\ominus$ increases as the imaginary component of the impedance increases. The $\ominus$ versus log ω plot yields a peak at $\omega_{(\ominus=\max)}$, the frequency, in radians, at which the phase shift of the response is maximum. The double-layer capacitance, C_{DL}, can be calculated from the following equation [23]:

$$\omega_{\ominus\max} = \frac{1}{CR_p}\sqrt{1+\frac{R_p}{R_\Omega}}, \quad \text{where } \omega_{\ominus\max} = 2\pi f_{\ominus\max}$$

16.4.3.3. Idealized Randles Plot

The Randles plot is useful in determining whether the Warburg impedance is a pronounced component of the equivalent circuit model. Identifying this condition experimentally is an aid in describing a reaction mechanism. Diffusion control is indicated by a slope of $-1/2$ or $-1/4$ in the linear portion of the Bode plot. This slope is -1 for a simple, kinetically controlled reaction. Figure 16.8 shows an idealized Randles plot of Z' versus $\omega^{-1/2}$ for a diffusion-controlled system. In this case, Z' and Z'' are equal and are linear functions in $\omega^{-1/2}$. The ordinate intercept of this line yields the Warburg impedance, $|Z_w|$. Thus the linearity and slope of the plot can be used as a test of diffusion control, and in certain cases the Warburg diffusion coefficient can also be calculated [42] (www.princeton appliedresearch.com [23]).

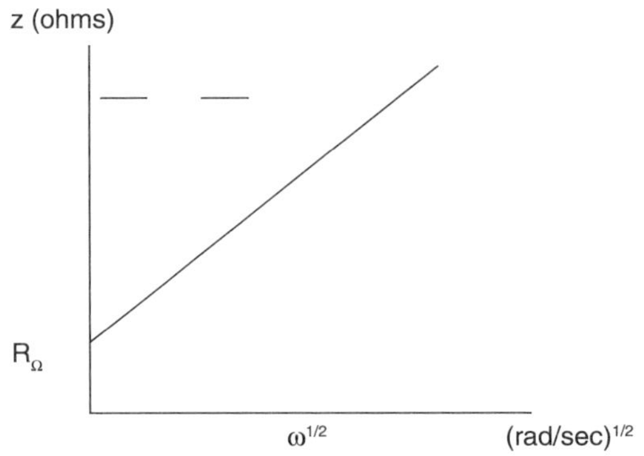

Figure 16.8 Idealized Randles plot of Z' versus $\sqrt{\omega}$ [23].

16.5. ELECTROCHEMICAL NOISE MEASUREMENTS

16.5.1. Historical and Electrochemical Noise Definition

Electrochemical noise (EN) is relatively an emerging tool to monitor almost all forms of corrosion. When corrosion is initiated on the surface of a metal, a spectrum of EN is concurrently created, and the characterization of this spectrum is an indication of the severity of general corrosion and the form of corrosion. It is especially sensitive to localized corrosion such as pitting and crevice corrosion and even metastable pitting [25]. The EN technology has recently been practically applied to corrosion-related problems in major industrial sectors, such as oil and gas, nuclear power, and aerospace industries. Specific studies show it to be a promising technology for continuous monitoring of the performance of inhibitors in a plant, for example [43].

Blanc et al. [47] undertook some EN studies as an interesting tool to monitor almost all forms of corrosion. When corrosion is initiated on the surface of a metal, a spectrum of EN is concurrently created, and the characterization of this spectrum is an indication of the severity of general corrosion and/or other forms of corrosion. These are related to instantaneous changes due to a gain or loss of electrons at the metal–electrolyte interface, giving rise to minute transients in the electrical charges on the electrode during a corrosion process in the form of fluctuations of potential or current, typically of low frequency (< 10 Hz) and low amplitude.

EN originates, in part, from natural variations in electrochemical rate kinetics during a corrosion process. EN is often regarded as a random (stochastic) phenomena coupled to deterministic kinetics. Electrochemical noise measurement (ENM) provides some of the most fundamental information on the basic (stochastic) processes involved in the electrochemical corrosion reactions of anodic dissolution and electrodeposition processes. In parallel, the phenomenon of coupling current and associated current noise signals arising from the galvanic coupling of nominally identical materials were studied and examined successfully [44].

Hladky [45] made an original EN patent application concerning the monitoring of corrosion using potential measurements between the examined metallic and reference electrodes. Later, Eden et al. [43] authored a patent and published a paper that described a method and apparatus for the detection of localized corrosion using current noise measurements. Actually, the EN technology is applied to solve corrosion-related problems in major industrial sectors, such as oil and gas, fossil fuel power generation, petrochemicals, nuclear power, and aerospace industries and help corrosion prevention with continuous monitoring of the action of inhibitors in a plant [43].

There are two types of ENMs: the electrode potential and cell current noise. The potential can be measured by recording the potential difference between two identical electrodes or by measuring the potential difference between a reference electrode and the examined metal. The former has the advantage of eliminating the effect of other factors such as the temperature and the variation of the electrochemical properties at the metal–solution interface since the potentials of the two identical electrodes change at the same time. The current noise is measured by using a zero resistance ammeter, and frequently the potential noise between the two identical electrodes can also be monitored at the same time using an extra reference electrode [46].

ENM gives instantaneous data that can show the relative corrosion resistance of different alloys in the same medium over long periods. However, agitation or convection can mask the noise signals, making it difficult to simulate operational conditions in certain situations. Zero-resistance ammeters (ZRAs) include an electronic instrument designed to measure the current flowing in a circuit without introducing the additional voltage drop associated with a standard ammeter. As with the potentiostat, the main functional component is an operational electronic amplifier that supplies current necessary at its output to maintain a zero potential difference between the two input potentials and that no current flows into or out of its input terminals [13].

Several parameters can be derived and calculated from the EN; however, the theoretical basics of some of these parameters are still being developed. The EN resistance, R_n, which is the ratio of the standard deviations of potential noise and current noise, is generally accepted as equivalent to the polarization resistance. The *pitting index* is defined as the standard deviation of current noise divided by the mean current. The power spectral density of a noise can be calculated using the maximum entropy method (MEM) and the fast Fourier transform (FFT) [46]. The concept of noise impedance was further developed, leading to the production of electrochemical impedance spectra using the spontaneous electrochemical potential and current noise signals [7, 43].

This technology can be used without disturbing the system under investigation. The corrosion phenomena that could be examined by EN technology include adsorption, inhibitor activity, passive film stability, chemical cleaning, coating failure, and gas generation. EN is good for characterizing general corrosion continuously and is especially sensitive to several different types of localized corrosion [25]. At a practical level, ENM studies have been related to chemical surface cleaning, general corrosion, pitting corrosion, crevice corrosion, preferential weld corrosion, microbiologically induced corrosion (MIC), corrosion under fouling, cavitation and erosion corrosion, intergranular stress-corrosion cracking (IGSCC), transgranular stress-corrosion cracking (TGSCC), and coatings degradation. It is considered on-site for online corrosion monitoring, process control, surveillance, and troubleshooting. On the other hand, agitation or convection can mask the noise signals, making it difficult to simulate operational conditions in certain situations [46]. However, Kearns et al. [47] stated that there are no established test methods and no consensus on a theoretical framework for interpreting electrochemical noise measurement data at this time.

System Sources of Noise Mass transport fluctuations and bubble nucleation are important to consider since these can change The pattern of corrosion types such as general and pitting corrosion (examined later) should be observed with care during data analyses and compared to that from mass transport and gas bubble formation [22].

Mass Transport Fluctuations In nominally laminar conditions, it is expected that some fluctuations will occur. This could be observed more clearly in strongly agitated and turbulent conditions. The variation in the diffusion boundary layer thickness at the interface will give rise to fluctuations in the current. It is not clear if this source of noise can be analyzed as a Poisson process; the relationships between the measured noise and the diffusion processes are not well defined.

Bubble Nucleation, Growth, and Detachment This source of noise is unlikely to behave according to the treatment discussed earlier. When hydrogen evolution is the predominant cathodic reaction, it is expected that the growth and detachment of hydrogen bubbles will tend to give rise to certain fluctuations of the current [48].

16.5.2. EN Generation and Data Acquisition Systems

The starting point for the development of a theory of EN is a theoretical analysis of the noise associated with a randomly occurring, brief pulse of charge, with the occurrence of each event being independent of any other event. This is known as a Poisson process, and the simplest example is the flow of electronic current, in which case each event is the passage of an individual electron through the measuring circuit. If we define the noise current, I_n, as the instantaneous current minus the mean current, it can be shown that the noise current is given by

$$\bar{I_n^2} = 2eIb$$

where $\bar{I_n^2}$ is the mean squared noise current, e is the charge on the electron, I is the average current flowing, and b is the bandwidth of measurement. The result of this process is known as *shot noise* and is an unavoidable minimum noise current associated with the flow of current. When we consider an electrochemical reaction, providing we can treat the dissolution process as a series of brief events, we can use a similar analysis to predict the noise current:

$$\bar{I_n^2} = 2qIb$$

where q is the charge in each dissolution event. If the dissolution event has a significant duration, the noise at high frequencies (where the period becomes less than the duration of the event) will fall to that due to the individual reactions (i.e., q will become the charge on the electron times the number of electrons involved in the reaction). The slope of the power spectrum between the low- and high-frequency limits will be a function of the shape of the current transients associated with the individual events, although the slope will only be clearly distinguishable if q corresponds to a large number of electrons [48].

Instrumentation and Correlating Trends There are many commercially available data acquisition systems (DASs), a few of which are suited to ENMs. In terms of the data acquisition rate, there does not appear to be any advantage in taking measurements at high rates, and a minimum logging period of 1 second has been found to be satisfactory for most circumstances. At higher rates of data acquisition, the amplitude of instrumentation noise approaches the EN levels. In addition, electromagnetic interference at power supply frequencies becomes problematical. By measuring at low frequencies, it is possible to reject most spurious noise sources [49].

The importance of DAS equipment has been shown by laboratory round-robin testing of EN standards. In these tests, reproducible information was obtained only when there was sufficient standardization in techniques. The round-robin testing resulted in the following recommendations:

1. The equipment should allow experimenter to choose the appropriate frequency range (0.01–10 Hz).
2. The complete measuring apparatus must exhibit small amounts of background noise less than $\pm 100\,\mu V$ or $\pm 10\,nA$.
3. Continuous measurements of at least 10 minutes duration should be possible.
4. Greater attention should be paid to the interactions between the ways that ENMs are made, such as sampling rate, filtering, data processing, and background noise.
5. There should be better standardization in the methods used for data evaluation.
6. Experimenters should confirm trends by different parameters and analyses of data [49, 50].

Most high-resolution measuring instruments involve electronic filtering to avoid problems such as aliasing. Beyond the use of filtering to enhance resolution, it can be used as an online data-processing technique. For instance, the signal can be filtered through a bandpass system centered at 50 ± 10 mHz, and, subsequently, the root mean square (rms) of the filtered signal can be calculated. The type (e.g., lower cutoff frequency, upper cutoff frequency), format (e.g., Bessel, Butterworth, constant phase), and order of the bandpass filter have to be determined for the system of interest. The selection of the correct filter for a given application is a complex topic. Alternatively, the rms values of the electrochemical potential noise measurement (EPNM) and electrochemical current noise measurement (ECNM) over time can be used directly to measure corrosion processes [51]. The most basic type of data analysis from a three-electrode system for which current and potential are measured simultaneously is recognizing a pattern or correlating trends in the current and potential signals. In the case of a general mode of corrosion, the noise signal is fairly random. There may be good correlation (>90%) between the coupling current noise and the corrosion potential noise, although there is no distinctive shape or any real evidence of individual transients [51].

In the case of a stable passive alloy, the EN, both current and potential, observed in the presence of passive films is typically of small amplitude, and a slow drift in the corrosion potential is typically observed. Although the current and potential correlate, the passive current characteristic of the system has a low-level noise signal. The fine structure observed in many time records suggests that the current flow is random because of electronic or ionic diffusion within the film [50, 51].

Apparent plateaus at low frequencies can arise from several sources, such as the trend removal or windowing or the extrapolation of the maximum entropy method (MEM) below the bandwidth of the analysis. In certain circumstances, it is suggested that the lower valid frequency should be taken as something like 3/time record period. If in doubt, a longer time record is considered with less frequent sampling in order to test that a plateau remains the same. On the other hand, an artificial high-frequency plateau can be obtained, for example, due to aliasing of higher frequencies or instrumentation noise. The validity of a high-frequency plateau can be checked by taking samples more frequently and reducing the length of time record if necessary. Noise due to instrumentation is more likely at higher frequencies (and correspondingly lower amplitudes), and it is recommended to check for instrumental noise by replacing the electrochemical cell by a dummy cell of comparable impedance to that of the electrolytic cell [52]. For improving the performance of the system when used for impedance analysis, some ENM models could use a computer-generated white noise signal [53].

Acquisition of Noise Signals There are two types of ENM: the electrode potential and cell current noise. The potential can be measured by recording the potential difference between the two identical electrodes or by measuring the potential difference between reference electrode and the examined metal. The former has the advantage of eliminating the effect of other factors such as the temperature and the variation of the electrochemical properties at the metal–solution interface since the potentials of the two identical electrodes change at the same time. The current noise is measured by using a zero-resistance ammeter, and frequently the potential noise between the two identical electrodes can also be monitored at the same time using an extra reference electrode [46].

Noise signals may be acquired in several ways. There are basically three different interfaces for ENM that can be used to follow the evolution of the electrochemical processes: the freely corroding mode, the potentiostatic mode, and the galvanostatic mode.

The Freely Corroding Mode For measurements at the free corrosion potential, it is possible to measure potential and current fluctuations together, or independently, using an interface arrangement. This is commonly called zero-resistance ammeter/electrometer (ZRA) and is carried out without any polarization control. By measuring EN with an open circuit, the corrosion system is not disturbed by an external voltage or current source and hence no additional corrosion effects are induced. The current between the coupled working electrode pair is converted to a voltage by the current amplifier. Performing the experiment with two identical electrodes under open circuit conditions with a ZRA allows a measurement to be made with no external perturbation, closely simulating ambient real-use conditions. Both potential and current can be measured simultaneously (Figure 16.9). The electrochemical current noise is then measured as the galvanic coupling current between two nominally identical working electrodes (WEs) kept at the same potential, and with modern devices the potential difference can be maintained within a microvolt. This is widely used in plant monitoring/surveillance [43, 48].

The ZRA is an electronic instrument designed to measure the current flowing in a circuit without introducing the additional voltage drop associated with a standard ammeter. As with the potentiostat, the main functional component is an operational electronic amplifier that supplies the current necessary at its output to maintain a zero potential difference between the two input potentials; no current flows into or out of its input terminals [13]. The ZRA acts to maintain the potentials of the two WEs at the same level. The current required to maintain the two WEs at the same potential flows through the feedback resistor R, and the voltage output of the ZRA is related to current flowing through the system by $V = IR$. The potential buffer is often used to ensure that there is no current drain on the reference electrode. Ideally, with no cells connected, the ZRA will give a signal of 0 V; however, intrinsic noise sources and offsets will inevitably be present. It is important to understand the limitations of such interfaces, and the baseline offset and intrinsic noise parameters should be measured as a matter of course [43].

The coupled potential is measured via a reference electrode such as a saturated silver chloride electrode Ag,AgCl/KCl_{sat} or calomel electrode. A potentiostat and a

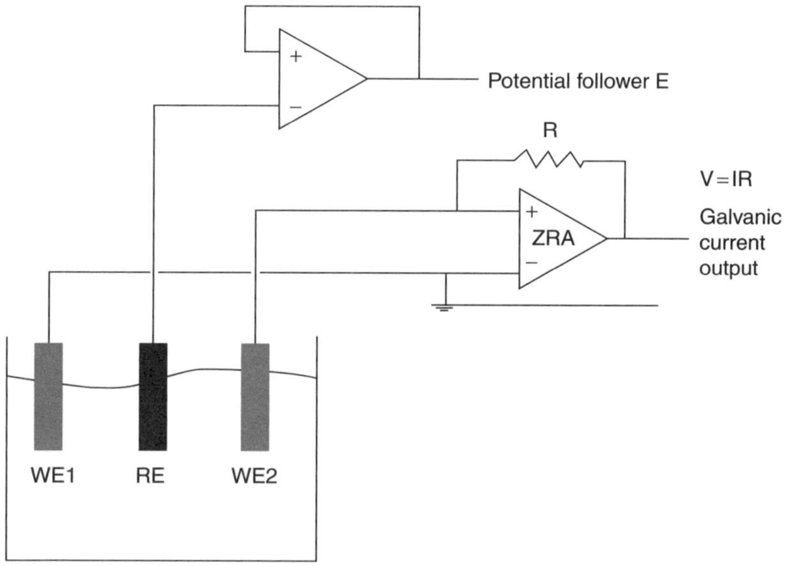

Figure 16.9 Simultaneous monitoring of potential and current noise at the free corrosion potential using an interface setup in zero-resistance ammeter (ZRA) mode and a potential/buffer follower [43].

computer is generally used to log potential and current values sampled at a scan rate of 10 Hz for 204.8 s, giving a total of 2048 readings. The noise data for corrosion studies could be collected for 2, 12, 24 h, or longer immersion periods depending on the stability of the interface.

The systems used for measuring EN signals have several key components that depend on the type of noise measurements being undertaken. It is common practice to use a three-electrode cell (sensor). The noise signals being measured may require some form of signal conditioning, and this is included in the interface. For the purpose of logging data from a series of sensors, a multiplexer is often incorporated. This may configure in two particular ways, either before or after the signal conditioning units. The data acquisition unit and the computer controller may be either an integrated subsystem or separate devices.

These transients manifest in the form of potential and current noise, which can be exploited to describe a corrosion event. Various physical and chemical processes can give rise to seemingly random low-frequency signals. The potential and/or current spontaneous fluctuations of a metal in a solution as a function of time from these stochastic processes, taken as a group, are referred to as EN. The foundations of EN technology lie in original work undertaken by Iverson [54], who studied transient voltage changes produced in corroding metals using elementary instrumentation [54].

Under certain circumstances, it is possible to engineer a differential WE arrangement such that effects of, for example, stress, differential aeration, and sensitization can be simulated. One such case would be the study of susceptibility of materials to stress-corrosion cracking; in this instance, the sensor arrangement would be a sample of unstressed material as one WE, and a sample of stressed material as the second WE [43, 50].

Potentiodynamic and Galvanostatic Modes Three-electrode cell configurations composed of working, auxiliary, and reference electrodes could be used for potentiodynamic or galvanostatic control monitoring current or potential fluctuations, respectively. In the potentiostatic mode, the potential of the working electrode with respect to the reference electrode is controlled to the set potential. The current required to maintain the controlled potential is simply measured as the voltage drop across the sensing resistor R in the circuit. Some potentiostats may incorporate current followers. Potentiostatic arrangement is often used for laboratory studies, particularly for accelerated testing of material susceptibility to a variety of failure mechanisms such as that for pitting or stress-corrosion cracking within defined potential levels [43]. In the galvanostatic mode, the current between the working electrode and auxiliary electrode is kept constant and the variations in potential of the working electrode are measured directly with respect to a reference electrode. This mode is a good simulation of some refining and electrowinning processes that are carried out at constant current [22].

16.5.3. Analysis of ENM Data

Analysis of ENM data involves the characterization of low-amplitude, low-frequency, random fluctuations of current and potential signals from electrochemical systems. Observation of the short-term transients in the potential–time and current–time records is useful to identify the corrosion resistance at the metal–solution interface, especially for previously studied systems. The absolute magnitude of the current signal can be used qualitatively to assess corrosion rate. These transients are usually indicative of spontaneous changes in the corrosion mechanism, such as passive film breakdown due to pit initiation processes, cavitation attack, and certain types of SCC. Specifically, the amplitude of the fluctuations observed in EN records can correspond to the intensity of the corrosion process while the fluctuation shape observed in these records can be correlated with a certain form or type of corrosion [55].

EN usage has grown alongside more conventional electrochemical techniques, such as linear polarization resistance, electrochemical impedance, and harmonic analysis for 15 years [43]. The promise of EN data is that it can yield information regarding both the uniform rate and localized corrosion processes for the system such as for iron, aluminum, magnesium, and their alloys [56–58]. EN originates in part from natural variations in electrochemical rate kinetics during corrosion processes and is often regarded as a random, low-frequency (stochastic) phenomenon coupled to deterministic kinetics. All general and localized corrosion processes result in fluctuations in the free corrosion potential and current. The fluctuations are thought to contain significant information about both the rate and the mechanism of corrosion [59]. Specifically, the amplitude of the fluctuations observed in EN records can be correlated with the intensity of the corrosion process while the fluctuation shape observed in these records can be correlated with the type of corrosion process [60].

Statistical analyses that treat the data as a set of values, without regard for the order in which the data were collected, are termed *sequence-independent*. Data analyses in the time domain are statistical analyses that take into account the order in which the data were collected and are termed *sequence-dependent*. These methods retain more of the information in the data than the sequence-independent methods, but are much more difficult analyses to perform in terms of computational requirements and in error propagation. These methods include the autocorrelation function, power spectra, higher-order spectra, and wavelet methods [50].

The spectral analysis of the noise in the frequency domain allows calculation of the power spectral density (PSD) using the fast Fourier transform (FFT). It can be used to characterize the type of corrosion attack. Low-frequency amplitude, slope of the PSD, and the change in the slope (roll-off frequency) contain information about the type of corrosion process occurring [57]. There is no universal table of slopes for a specific corrosion type. The relevant slopes for each corrosion system should be established by experiment. The PSD noise resistance, R_n, is an alternative parameter that denotes the square root of the ratio of the potential PSD to the current PSD, which can be equal to the impedance modulus Z [61]. The chaos analyses (rescaled range and stochastic process) serve to highlight the features present in the frequency domain of the transformed data record [62]. It is known to reveal the persistent or nonpersistent nature of the process [63]. The wavelet analysis presents its advantages with regards to the Fourier analysis in distinguishing periodical and nonperiodical variation in the signal power in time and frequency as opposed to the Fourier analysis that only considers frequency [60].

The different approaches have been proposed for processing the experimental records, the statistical, the spectral, and the chaos –theory (non random) and the wavelet transform-based method (for non stationary phenomena) is considered in some detail as follows:

16.5.3.1. Data Analysis in the Time Domain

During a corrosion process, the cathodic and anodic reactions can create minute transients in the electrical charges at the metal–liquid interface and this causes the fluctuations of the potential and current of the electrode. These transients manifest in the form of potential and current noise, which can be exploited to map a corrosion event. By measuring EN under open circuit condition, the corrosion system is not influenced or disturbed by an external field and so ENM reflects the real performance of the electrode [48]. It allows calculating the noise resistance, R_n, which has been found equivalent to the polarization resistance, R_p, in potentiodynamic studies and is associated with corrosion rate.

The simpler statistical methods treat the time record as a collection of individual potentials or currents (or, in statistical terms, as a sample from the population of all values that will occur over infinite time) and ignore the relationship between one value and the next [61]. Statistical analysis is typically undertaken on batch data, although modifications can be made to make this semicontinuous by data blocking or by discounting of the past. The following eight sample statistics parameters are frequently considered [43, 50].

1. The *mean* can be expressed by

$$\bar{x}_n = \frac{1}{n}\sum_{i=1}^{n} x_i$$

The average current is usually zero between two identical electrodes but this could be very valuable in the case of dissimilar electrodes. The mean potential corresponds to the average open circuit potential and can be interpreted in conventional ways and the fluctuations of E could be related to different corrosion processes. The mean may carry more useful information for dissimilar electrodes [22, 48].

The variance is expressed as

$$m_{2,n} = \frac{1}{n}\sum_{i=1}^{n} (x_i - \bar{x})^2$$

The variance of the signal depends on the range of frequencies included in the signal and is often related to the power in the data. Generally, the variance of the current increases as the rate of the corrosion process increases and as the corrosion becomes more localized. In contrast, the variance of the potential is expected to decrease as the rate of the corrosion process increases, but to increase as the corrosion becomes more localized. It is more usual to use the standard deviation of the signal.

2. *The third central moment* is expressed as

$$m_{3,n} = \frac{1}{n}\sum_{i=1}^{n}(x_i - \bar{x})^3$$

This is a measure of the asymmetry of the data around the mean value and permits one to calculate skewness.

3. *The skewness* (*or skew*) is given by

$$g_1 = \frac{m_{3,n}}{(m_{2,n})^{2/3}}$$

A normal distribution will have zero skewness, meaning that the distribution is symmetrical about the mean; while a distribution with a tail in the positive direction will have a positive g_1 and a distribution with a tail in the negative direction will a negative g_1. EN analyses of the potential signal have been done in the time domain:

$$\text{Skewness} = \frac{1}{N}\sum_{k=1}^{N}\left(\frac{E_n[k] - \bar{E}}{\sqrt{E_n[k]^2}}\right)^3$$

4. *The fourth moment* is given by

$$m_{4,n} = \frac{1}{n}\sum_{I=1}^{n}(x_i - \bar{x})^4$$

This is used to derive a value for kurtosis.

5. *Kurtosis* is given by

$$g_2 = \frac{m_{4,n}}{(m_{2,n})^2}$$

The kurtosis is a measure of the flatness or peakness. A normal distribution gives a g_2 of zero. A positive g_2 reflects a more peaked distribution and a negative g_2 a less peaked distribution. A g_2 above 2 is typical of localized corrosion.

$$\text{Kurtosis} = \frac{1}{N}\sum_{k=1}^{N}\left(\frac{E_n[k] - \bar{E}}{\sqrt{E_n[k]^2}}\right)^4 - 3$$

Three is subtracted to bring the kurtosis for a normal distribution to zero [56].

6. The *standard deviation* is expressed as

$$\sigma = (m_{2,n})^{1/2}$$

The standard deviation of the signal (σ) is a measure of the spread of the data around the mean value and is related to the broadband ac component of the signal.

7. The *coefficient of variance* (CoV) is given by

$$\mathrm{CoV} = \frac{\sigma}{\bar{x}}$$

The coefficient of variance corresponds to the standard deviation divided by the mean value.

8. The *root mean square* (rms) value is expressed as

$$\mathrm{rms} = \frac{1}{n}\sum_{i=1}^{n} x_i^2$$

In performing rms calculations, the number of data points, acquisition rate, length of time, the standard deviation formula, and profiteering are to be determined by the user. The rms values, standard deviation values, or bandpass-filtered noise from digital instrumentation for both V_{n} and I_{n} can be related using Ohm's law to produce a resistance noise value (termed R_{n0}) as the frequency approaches 0 Hz.

16.5.3.2. Nondimensional Skewness and Kurtosis Values

These values can identify changes in the distribution of the noise signals. Skew (or skewness) is a measure of the symmetry of the distribution. A value of zero implies that the distribution is symmetrical about the mean; a positive skew implies that there is a tail in the positive direction, and a negative skew implies that there is a tail in the negative direction. The value of skew may be used to identify whether the data have a uniform distribution, such as may occur, for example, during pitting attack, cavitation, and SCC. If the current noise is measured between identical electrodes, the time record consists of unidirectional transients if one electrode happens to be active and will be typically heavily skewed. This may be a useful measure for detecting such transients that are considered to describe metastable pitting; however, the skew will be difficult to interpret if both electrodes are active (bidimensional) [48].

The value of kurtosis reflects the distribution of the signals. A value of zero indicates a normal distribution. A positive kurtosis implies a more spiky distribution, whereas a negative kurtosis implies a flatter distribution [48]. The value of kurtosis for the potential and current signals is a sensitive indicator of changes in corrosion rate and mechanism. For data that exhibit spontaneous sudden changes in corrosion rate that may occur due to changes in amplitude distribution, the kurtosis value will typically be >5. Kurtosis will also reflect sudden changes in corrosion rate that may occur due to changes in flow rate, pH, and so on. This is probably a better approach to describe localized attack (pit initiation/propagation, intergranular corrosion) than the use of the localization index and should be sensitive to stress-corrosion cracking phenomenon [46].

16.5.3.3. Electrochemical Noise Resistance (R_n) and Uniform Corrosion

Uniform corrosion might be expected to be free of noise, with atoms leaving the metal surface at a uniform rate. However, even a perfectly homogeneous process will give some fluctuations in rate, akin to Brownian motion. Furthermore, there are a number of mechanisms by which it might be expected that even a uniform dissolution process will occur as a series of bursts. For uniform dissolution processes, it is expected that the value of q will correspond to the charge liberated by 10^2–10^6 atoms. R_n is the currently used parameter and presents the ratio of the standard deviations of the potential noise (σ_E) and current noise (σ_I): $R_n = \sigma_E/\sigma_I$. This could be equivalent to the ac charge transfer resistance, R_{ct}, or the polarization resistance (R_p), and the ratio $1/R_n$ is proportional to the corrosion rate [43]. The R_n value is determined by a polarization method in which the perturbation is generated by the corroding system rather than an external instrument. Naturally occurring transients are used to compute an electrochemical interfacial resistance instead of the imposed galvanostatic or potentiostatic signals used for the determination of the linear polarization resistance (the Stern–Geary approach) for corrosion rate measurements, R_p.

Chen and Bogaerts [64] theoretically proved that the noise resistance is indeed equivalent to the polarization resistance in the case of uniform corrosion. However, in their derivation, they assumed that steady states are attained and that the system is totally under activation control following the Butler–Volmer equation; that is, diffusion or intermediate chemical reactions (concentration control) as well as resistance control are not taking place. Also, the potential of the working electrode should be far away from the equilibrium potentials of the individual anodic and cathodic reactions. If instantaneous open circuit or corrosion potential does change with time, especially at the very beginning of immersion, it is advisable to select a shorter time duration or remove the drift of the corrosion potential by using curve-fitting techniques, such as the linear curve-fitting method or polynomial curve-fitting method [64].

However, it has been shown for passive systems, such as stainless steel or titanium alloys exposed to natural media, that R_n is several orders of magnitude lower than R_p. Moreover, R_n was found to depend on bandwidth Δf of the noise measurement. Generally, R_n is close to R_p only in those cases where the impedance is independent of frequency within Δf [65].

The following application is an example of ENM used for uniform and pitting corrosion monitoring of stainless steels tanks [66]. Three corrosion monitoring systems utilizing electrochemical noise and linear polarization resistance (LPR) technologies were employed for a radioactive waste storage site. The data showed mostly that the corrosion was uniform at a very low rate. The only off-normal corrosion behavior was the appearance of a few weeks of data indicative of pit initiation. The pitting lasted for a short period and returned to the uniform corrosion form. The EN-based corrosion monitoring system could become a valuable tank integrity management tool for nuclear waste tanks, particularly as these tanks meet or exceed their design life.

16.5.3.4. Localization Index and Pitting Corrosion

The spatial separation of the anodic processes in the pit and the cathodic processes on the surrounding surfaces necessitates the passage of current, which gives rise to the measured noise signals. Since localized corrosion sites are typically very small, on the order of 100 μm in diameter or less, the current densities inside these cavities can be on the order of 1 A/cm^2. EN studies can be performed under open circuit potential conditions, and close to the natural conditions of pitting. It has been suggested that current transients below the pitting potential may be attributed to the formation of pit embryos.

Pistorius and Burstein [67] discussed several factors that can influence the proper interpretation of ENMs in the study of metastable pits. These factors can be probe size, sampling rate, and system noise. Current measurements seem to give more clear information on the corroding system than that of the potential [7, 25, 67, 68]. The pitting initiation process is often found to result in metastable pit nucleation and propagation, giving rise to current transients lasting for a time of about 1 second, and involving a charge of about 10^{-6} C (corresponding to around 10^{12} atoms). Thus the noise associated with pitting corrosion is much larger than that observed for general corrosion.

Much of the initial research on pitting using ENM involved the potentiostatic technique. Current transients were obtained by sweeping anodically into the region of pitting potential. The pitting and metastable pitting processes were then characterized by extensive statistical analysis to identify the influence of alloying additions such as molybdenum on the corrosion resistance of stainless steel [50].

The distribution of the current around the rms has been considered to show localized attack and is frequently used to describe pitting corrosion. The localization index (LI) deduced from ENMs is defined as the ratio of the standard deviation to the rms value of the current: $LI = \sigma_I / I_{rms}$. As a general rule, if the LI has a value approaching 1 (e.g., pitting index), the corrosion process is unstable and therefore more likely to be stochastic, indicating a localized form of corrosion. More uniform corrosion processes, on the other hand, have LI values that are typically on order of 10^{-3}. It should be noted that a slow drift in the current through zero may give an artificially high value of LI, and therefore indicate localized corrosion even though the corrosion process has a Gaussian distribution that is symptomatic of general corrosion [43].

The suggestion that LI can be used to determine the prevailing corrosion mechanism should be considered with great caution. Statistical analysis of the noise signals indicates that the noise data generated during general corrosion has a relatively normal Gaussian distribution and will exhibit few rapid transients. In contrast, localized corrosion processes such as pitting and SCC have transients in the time record trace, and characteristics that help distinguish between them. Localized corrosion leads to deviation from a normal distribution (Poisson distribution) and is frequently characterized by a high number of overlapping transients, skewness, and kurtosis values and other types of signal analyses [16, 50].

16.5.3.5. Spectral Analysis

Spectral analysis–power spectral density (PSD) plots are frequently used by electrical engineers to analyze the noise of a signal. In performing spectral density calculations, parameters such as the number of data points, acquisition rate, length of acquisition time, the standard deviation formula, and prefiltering are to be determined by the user, who in this way influences the calculated result. Mansfeld et al. [61] and Bertocci et al. [69, 70] introduced the spectral noise resistance. Transformation of the experimental EN data from the time domain to the frequency domain leads to the estimation of its power spectral density (PSD) or spectrum. Spectral estimation is the process of estimating the power present at the various frequencies and PSD is the distribution of the power of the signal in the frequency domain. A certain type of PSD plot that shows a continuous increase in PSD with decreasing frequency is referred to as a $1/f$ plot and is generally taken to represent the random noise associated with a signal.

In corrosion studies, $1/f$ type behavior has been interpreted as evidence of a process occurring randomly over a specific period of time and a given space (such as a working electrode surface area) coupled to a process described by chemical kinetics. This leads to a caution about the use and interpretation of PSD plots: since the PSD is based on a signal-

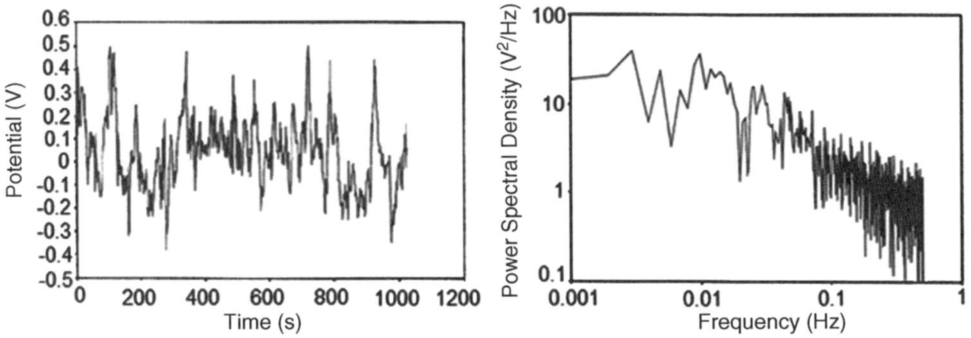

Figure 16.10 Typical power spectra of (a) potential–time record and (b) PSD–frequency record [22].

averaging method (such as FFT), it represents an average of system behavior during the selected data acquisition period rather than discrete or localized events. For this reason, a PSD is not particularly useful for corrosion monitoring in comparison to a ENM–time plot. However, a PSD plot is very useful for determining and/or visualizing how the amplitude of an ENM varies as a function of frequency. From this, one might deduce the relative contribution to the overall ENM of a slowly occurring corrosion process versus that of a more frequently occurring event. In the studies of passivity and localized corrosion, workers have used potentiostatic measurements to obtain PSDs of ECN and galvanostatic measurements to obtain PSDs of EPN [50].

Two methods are commonly used to estimate power spectra in EN studies: standard fast Fourier transform (FFT) or maximum entropy method (MEM) [46]. FFT performs a spectral analysis of the random transient of the EN signal in a frequency range for a chosen sampling time and data recording time [71]. The noise data FFT are presented generally in the frequency domain from 5×10^{-3} to 5 Hz as PSD. Figure 16.10 shows typical plots. On the y axis of the power spectrum plot, the PSD is obtained by dividing the power by the width of the frequency band [22].

The PSD can be obtained through the periodogram method based on the Cooley–Tukey FFT algorithm:

$$\phi(f) = \frac{2}{T}\frac{1}{N}\sum_{j=1}^{N} |I_j(f)|^2$$

where f is the frequency, T is the acquisition time, $I_j(f)$ is the Fourier transform of the elementary time recording, and N is the number of time recordings [72, 73].

The acquisition or sampling time determines the low-frequency limit and the sampling rate determines the high-frequency limit of the resulting information. Thus the bandwidth of the noise spectra (Δf) was defined by ($\Delta f = f_{max} - f_{min}$, where $f_{max} = \frac{1}{2} f_s$ (f_s is the sampling frequency) and $f_{min} = 1/T$ (T is the measurement time). Usually, f_s is 2 Hz and T is 1024 seconds [74]. The resistance EN results were reported to be very sensitive to the variation of the sampling frequency. There are many restrictions that limit the bandwidth of the noise spectra (Δf), such as equipment stabilities that affect the spectrum shape at low frequencies and high-frequency noises from the power source [65, 69–71].

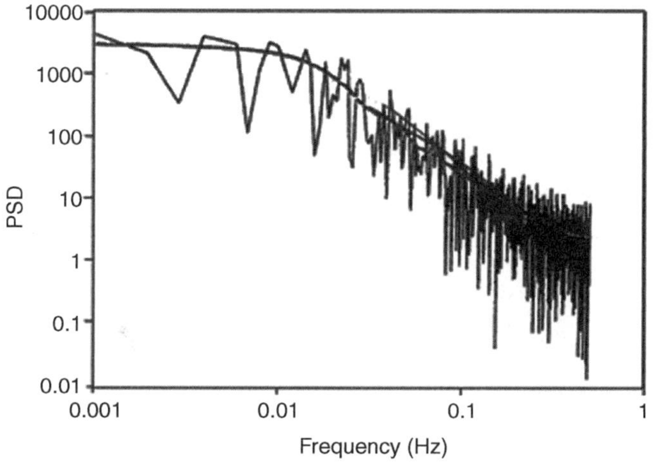

Figure 16.11 Power spectrum density for total signal or overall time record (noisy curve) as compared to the sum of the spectra for the individual events or transients (smooth curve) [22].

For the MEM, the PSD can also be estimated by

$$\phi(f) = \frac{2\sigma^2 \Delta t}{\left| 1 + \sum_{k=1}^{M} a_k e^{-2i\pi f k \Delta t} \right|^2}$$

where $a_1, \ldots, a_M$ are $M + 1$ coefficients dependent on M, which are calculated from the $(M + 1) \times (M + 1)$ matrix equation, and M is the order of the model. Therefore the MEM will be called MEM(M) [72, 73].

However, the FFT produces noisy spectra, whereas the MEM produces smoother spectra. The order of M can have a marked influence on the ability of the MEM to model a spectrum, and the order should always be reported. There is a tendency to use a small order to give a smoother spectrum, but it must be appreciated that this also leads to a loss of detail at low frequencies. It is recommended to start with an order of ~1–2% of the number of points in the time record and then compare the MEM with the FFT to ensure that they give a good match, particularly in the low-frequency region. An unusually smooth FFT power spectrum or one showing periodic behavior typically indicates that the spectrum is dominated by one or two features in the time record, such as single large-amplitude transient (Figure 16.11) [22, 71, 75].

For the noise PSD type of spectral analysis, a noise impedance ($|Z_n|$) can be described as a function of frequency by correlating I_n and V_n using frequency response analysis. The Z_n value has been related to both the charge transfer resistance and polarization resistance. Transient analysis mechanistic information can also be obtained by analysis of the individual transients making up the noise signal, because these are influenced by the electrochemical parameters controlling the metal dissolution and cathodic reduction reactions. The shape of a transient can be shown to be the result of specific events associated with localized corrosion [50].

Analysis of EN data in the frequency domain involves the calculation of the PSD. Potential and power PSD plots have the following general frequency dependence:

$$\log V_{\text{PSD}} = A_V + S_V \log f$$

$$\log I_{\text{PSD}} = A_I + S_I \log f$$

where S_v and S_I are the slopes of potential and current and A_V and A_I are the noise intensities of potential and current of PSD plots, respectively. It follows that the slope of the spectral noise plot is $S_{\text{rsn}} = 0.5\ (S_V - S_I)$ [74].

The expected slopes of the PSD in the frequency domain (of the current signal) are 0 for stochastic (Poisson) processes, −0.5 for diffusion-controlled processes, and −1 for processes having a uniform corrosion (Gaussian distribution). If the process is Gaussian, the corrosion rate may be estimated from the resistance noise value or, alternatively, from the current noise value. The potential noise signal (B) and the current noise due to the corrosion current could be related using the following relationships [43]:

$$V_{\text{n}} = K\sqrt{\frac{1}{f^{\alpha}}} \times I_{\text{corr}} \times R_{\text{ct}}$$

where the Stern–Geary constant $B = I_{\text{corr}} \times R_{\text{ct}}$ and the corrosion noise is given by

$$i_{\text{n}} = K\sqrt{\frac{1}{f^{\alpha}}} \times I_{\text{corr}}$$

For the current signal we would expect three possible values for α: namely, 0, 1, and 2, respectively. In the frequency domain PSD estimate, this would equate to slopes of 0, −1, and −2, respectively. A slope of 0 equates to stochastic, Poisson processes; a slope of −1 to diffusion-controlled processes; and a slope of −2 to Gaussian processes. The factor K could be estimated from knowledge of a general corrosion process, where $\alpha = 2$, and a knowledge of the measured potential noise and the B value [43]. Moreover, the power level of the signal as calculated from the PSD in the frequency-independent region seems to be related to the corrosion rate. The concept of noise impedance was further developed leading to the production of electrochemical impedance spectra using the spontaneous electrochemical potential and current noise signals [7, 43].

In the ZRA mode, when potential and current data are determined simultaneously, spectral noise plots can be constructed in which R_{sn} is plotted versus the frequency f [74]. Figure 16.12 shows experimental potential V (part a) and current I (part b) fluctuations for mild steel exposed to aerated 0.5 N NaCl as well as the PSD plots for these data (parts c and d). The drift in the raw materials data that causes erroneous PSD was removed using a linear method. From the PSD plots for potential and current fluctuations, the spectral noise or noise impedance plot (R_{sn}) (Figure 16.12e) can be determined as the ratio of the FFT of potential and current fluctuations, or the square root of the ratio of the potential and current PSD plots. Good agreement was observed between the impedance spectra and the spectral noise plots [16, 20].

16.5.3.6. Wavelet Transform-Based Analysis

Wavelet analysis is a relatively new mathematical tool used to supplement conventional Fourier analysis. The statistical analysis in the time domain and the spectral method in

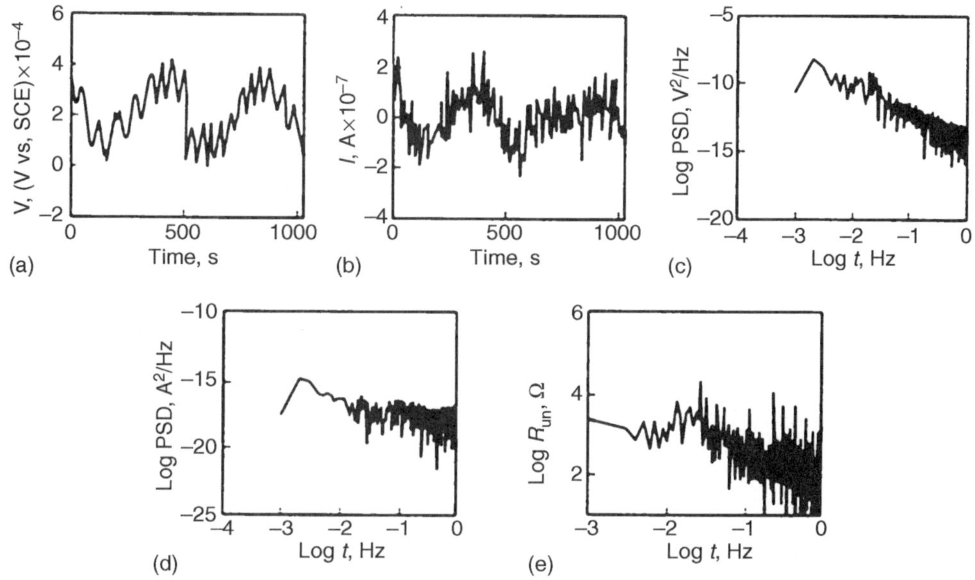

Figure 16.12 EN data after trend removal for mild steel exposed to 0.5 N NaCl: (a) voltage (versus saturated calomel electrode) versus time history; (b) current versus time history; (c) power spectral density (PSD) plot from voltage data; (d) PSD plot from current data; and (e) spectral noise impedance, $R_{sn,}$ derived from PSD [16, 20, 75].

the frequency domain are the main two mathematical treatments of EN. They are suitable for analyzing stationary phenomena, but they are limited in treating nonstationary signals. The FFT has some inherent disadvantages; for example, the transform results do not provide time resolution, and detrending and windowing have an influence on the transform results. Also, spectral results require a stationary signal having statistical properties that are stable with time. Typical potential noise drifts with time, and so removing this drift by subtracting a linear regression line from the data could introduce errors in the power spectrum while windowing preanalysis could induce errors with the ends of the data sequence [76].

Generally, the amplitude of the fluctuations in ENM can be correlated with the intensity of the corrosion process, while the fluctuation shape can be correlated with the type of corrosion (morphology). Since most ENMs are characterized by a high number of overlapped transients, wavelet analysis is proposed as an alternative to the FFT when a time-scale analysis is required or when we study transients into a signal. Aballe et al. [60] stated that wavelet transform allows analysis of the signal at different scales and over time simultaneously and suggested the orthogonal wavelet transform. In practice, the orthogonal wavelet transform is computed by the fast wavelet transform algorithm and includes lowpass filter, highpass filter, and down-sampling. The interesting feature of this method is the decomposition of EN records into different sets of wavelet coefficients that contain information about corrosion events at a determined time scale. As an example, wavelet coefficients resulting from the orthogonal wavelet transform of a current noise record corresponding to a 2024 aluminum alloy specimen after 5 h of immersion in a 0.6 M NaCl + 1.3 mM $CeCl_3$ solution have been given.

Aballe et al. [55] explained the theoretical basis of wavelet analysis and considered a wide range of corroding systems involving Al alloys AA5083 and AA2024 and 304 stainless steel in different aqueous solutions. ENM records, containing 2048 data points collected at a rate of 2.15 points/s, could show the necessary conditions that could define a frequency window where most usual corrosion processes can be detected. ENM analyses were carried on using both Fourier and wavelet transforms for comparative purposes. The wavelet analysis technique allows satisfactory analysis of ENMs in which no strong periodicity is observed, since it deals with signals formed by discrete events of a particular scale randomly distributed in time. Although the considered examples were current–time records, potential–time records can be analyzed in the same way.

Aballe et al. [77] proposed an algorithm to detect and classify transients in ENMs. They proposed a nondecimate discrete wavelet transform followed by a subsequent intercalary analysis of the signals that could isolate transients while neglecting other components. This can reveal the existence of transients and provide a comparative characterization of their size and scale. The proposed method can be adapted to several kinds of signals since one can choose different wavelet bases or change the particular crystals to be correlated.

Zhang et al. [78] summarized their practical approach for magnesium corrosion studies in the following six equations. Consider a time record x_n $(n = 1,\ldots, N) \in L^2(R)$ expressed on a Cartesian basis. The essential approach of wavelet analysis consists in representing the time record x_n using a linear combination of basic functions $\phi_{j,k}$ and $\varphi_{j,k}$.

$$x(t) = \sum_k s_{j,k}\phi_{j,k}(t) + \sum_k d_{j,k}\varphi_{j,k}(t) + \sum_k d_{j-1,k}\varphi_{j-1,k}(t) + \cdots + \sum_k d_{1,k}\varphi_{1,k}(t) \tag{16.1}$$

where $s_{J,k}$ and $d_{j,k},\ldots, d_{1,k}$ are the so-called wavelet coefficients; J is a small natural number that depends mainly on N and the basis function; and k ranges from 1 to the number of coefficients in the specified component. The basis for this decomposition is formed from mother-wavelet $\psi(t)$ and father-wavelet $\psi(t)$, by translating in time and dilating in scale [78, 79].

$$\psi_{j,k}(t) = 2^{-j/2}\psi(2^{-j}t - k) \tag{16.2}$$

$$\varphi_{j,k}(t) = 2^{-j/2}\varphi(2^{-j} - k), \quad j,k \in Z \tag{16.3}$$

where $k = 1, 2,\ldots, N/2$, N is the number of data records, $j = 1, 2,\ldots, J$, J is often a small natural number, and Z is the set of integers. Meanwhile, the wavelet scale j may be interpreted as "$2 - j$ frequency"; consequently, the "frequency" decreases with increasing j.

According to formula (16.1), in the first step of this algorithm, the discrete signal, $x(x_1,x_2,\ldots, x_N)$, is decomposed into two sets of coefficients: one low-frequency set that contains information about the general trend of the signal and a high-frequency set that contains information about the local fluctuations in the signal. At the end of the last step of the algorithm, the detail coefficients, d_1, d_2, d_J and the general coefficients, s_J are saved, and they will encode information regarding the different features present in the signal. The necessary calculations for decomposition of the analyzed signal were performed using function wavedec that was included in the MATLAB wavelet toolbox [23]. The orthogonal function $\varphi(t)$, which was called sym of the eighth order was applied. The main property of the chosen orthogonal $\varphi(t)$ is that energy of the analyzed signal $x(n)$ is equal to the sum of energies of all components obtained by the wavelet transform [78].

To characterize the fast wavelet transform results in more detail, the energy, E_j, named here as a relative energy of a crystal, was calculated and plotted versus the crystal name. This plot is currently referred to as the EDP. The overall energy of the signal is

$$E = \sum_{n=1}^{N} x_n^2, \quad n = 1, 2, \ldots, N \tag{16.4}$$

Then the relative energy of a crystal, which estimates the contribution of every crystal to the overall signal, can be calculated as follows:

$$E_j^d = \frac{1}{E} \sum_{k=1}^{n/2^J} d_{j,k}^2, \quad j = 1, 2, \ldots, J \tag{16.5}$$

$$E_J^S = \frac{1}{E} \sum_{k=1}^{n/2^J} S_{j,k}^2 \tag{16.6}$$

16.5.3.7. Chaos-Based Methods and Wavelet Fractal Analysis

Certain researchers have applied techniques developed from chaos theory and fractal analysis to ENM data. Wavelet transforms from neural network analysis would appear to be useful for identifying the significant segments within long, complex data sets. The use of any of these techniques for practical corrosion applications is an ongoing effort directed, in part, toward reducing the vast amount of data obtained during continuous monitoring to a few sensitive parameters [50].

The chaos approach implies that noise generation processes may cause fluctuations by virtue of the instability of the governing equations rather than purely random processes. Several investigators proved that chaos analysis of EN data showed a chaotic behavior. It has been shown that localized corrosion is generated by a deterministic chaotic process, while uniform corrosion is a random process. In this way, different chaos-based mathematical treatments could be applied to ENM data. Rescaled range analysis has been used to broaden the statistical treatment of the electrochemical potential noise [22, 55].

For a time series of EN of a corrosion process, the noise signal is usually nonstationary. Its features change frequently as a function of the time variation of the interfacial environment and the electrode material and this changes its fractal characteristics [76]. The fractal dimension is an important characteristic because it contains dimensions of geometrical properties. In practice, EN could be divided into two kinds of time series: the fractional Gaussian noise and fractional Brownian motion. The analysis of their fractal characteristics is often related to their persistence and stationary characteristics. Persistence measures the correlation between adjacent values within a certain time and nonexisting correlation indicates white noise. Although a general fractal parameter could give useful information, a local fractal parameter is more interesting since it gives the characteristics at a certain interval [76].

Large-amplitude localized transients are considered to be nonstationary in character. For an EN signal originating from a corrosion process, the wavelet transform could decompose it into a series of components at different scales and locations. Each component is defined by a set of wavelet coefficients, which is the similarity between the noise signal and wavelet function, and thus contains information about the time scale characteristic of the associated corrosion event. Wavelet transform is considered suitable for analyzing the self-similarity of a time series with fractal characteristics, since some fractal characteristics were analyzed by chaos theory, but its computation is complex and difficult to understand [76].

For simplification, the Hurster index (Hu) and the Hausdorff exponent (Ha), with a close relationship to the fractal characteristic, are generally considered. Compared with other fractal components, the wavelet standard deviation (STD) exponent Hws has smaller deviation and could describe the fractal characteristics of ENMs over a wide range. Hws calculation avoids the inherent defects of detrending and windowing if compared to the spectral power β from FFT analysis [76].

The Hws was chosen to examine general and local irregularity of potential signals generated from aluminum alloy 7075-T76 in 3.5% NaCl solution without and with added corrosion inhibitors. After the specimen was immersed in the test solution for 24 h, its potential noise signal at open circuit was sampled with an instrument for measuring and analyzing EN. The sampling interval was 2 s and the length of the measurement was 1024 points. It has been found that the smaller general Hws value corresponds more frequently to the potential fluctuation on the surface and less to the effect of the inhibitor in such a way that the difference between the general Hws and 1 reflected the condition of surface passivity. Local Hws was more suitable to describe the varying features of a frequently fluctuating signal and could quantitatively describe the fractal characteristic with time [76].

16.5.4. Potentiodynamic, Potentiostatic, and Galvanostatic EN Studies

Potentiodynamic and Potentiostatic Studies The potentiodynamic polarization method has been used by Keddam et al. [80]. Pit nucleation is described by the Poisson process. The amplitude and the frequency of the current noise decreased with aging of the passive film. The capacitance measurements correlated with the passive current and suggested increasing charge storage with completion of the full passive state. The potentiostatic technique is particularly useful for investigation of passivity [51].

Electrocrystallization Noise It has been shown that ENMs during electrodeposition of metals show fluctuations of the current passing through the interface that reflect the morphology and structure of the deposit [81]. Electrodeposition of nickel and zinc was considered and it has been shown that, during electrodeposition, the relationship between the noise power and current density depends on preferred orientation of the nickel deposit, while in the case of zinc, the noise power strongly depends on the deposit morphology and seems an increasing function of the surface roughness [44].

Yang et al. [82] stated that when the two-dimansional (2D) growth of nickel deposit was compact, this gave EN features of a slowly positive potential drift, and the maximum relative energy accumulated in the region of the replotted relative energy distribution plot (RP-EDP) showed smaller scales. While under diffusion control, the three-dimensional (3D) growth of nickel showed fast positive potential drift and subsequent remarkable negative shift and this gave a dendritic deposit having the maximum relative energy accumulated in the RP-EDP with larger scales [73].

It has also been found that the EN generated during the electrodeposition of dentritic or a large conglomerate of zinc shows large potential amplitude and positive potential drift, while the compact deposit of zinc possesses small noise amplitude and little potential drift. The maximum relative energy of the energy distribution plot obtained from wavelet analysis in the time domain records shifted from larger scales to smaller ones, corresponding with the change of a dendritic microstructure to a compact one [83].

Budevski et al. [84] studied the silver deposition by transient techniques on quasi-perfect silver single-crystal faces. Simulation of simplified 2D nucleation and the mono-nuclear growth mechanism showed that nucleation and propagation rates were reflected in the power spectrum and the autocorrelation function of EN analysis. If the rates for both phenomena are Poissonian distributed, the analysis of data becomes difficult.

Oxygen Evolution Reaction on Lead Anodes Huet et al. [85] performed ENM on the oxygen evolution reaction on lead dioxide and lead dioxide-composite materials. They showed that the PSD of the electrolyte resistance fluctuations is very similar for the two materials, suggesting a similar gas profile evolution. In fact, the potential PSD fluctuations were different, showing an ohmic effect for composites containing catalytic particles, while nonohmic effects are clearly dominant for the pure lead dioxide, probably because of the much higher activation potential. It is interesting to note that a plateau was observed at intermediate frequencies and this could correspond to bubble coalescence phenomena.

16.6. SCANNING REFERENCE ELECTRODE TECHNIQUE

The scanning reference electrode technique (SRET) is a powerful tool to monitor the initiation and propagation of localized corrosion of metals such as Al, Mg, and their alloys in dilute aqueous chloride media without agitation, saturated with atmospheric oxygen, to follow up the breakthrough of the passive layers. It has been used mainly for investigation of local corrosion processes, such as pitting and crevice corrosion. More recently, SRET has also been used for the evaluation of delamination of organic coating. The SRET is based on a microprobe that is capable of detecting local variation of the electrochemical properties of a metal surface with high lateral resolution. An optimized SRET has been used to evaluate the performance of several clear coatings, with good correlation between SRET signals and permeability as well as wet adhesion of these coatings.

The SRET is extensively used to study localized corrosion such as pitting or intergranular corrosion. It enables the measurement of localized current densities in the vicinity of pits in aqueous solutions. Applied potentiodynamic pitting scans can be obtained for localized areas immediately adjacent to accurately defined regions of the electrode surface [86, 87]. SRET is also associated with the scanning vibrating electrode technique (SVET), in which the probe is mounted on a biomorph piezoelectric reed, which vibrates the tip normal to the electrode at a characteristic frequency [87]. Another variation of the technique has been used in localized measurement of electrochemical impedance spectra [88].

The commercially fabricated SRET consists of a pair of platinum electrodes made of a wire 0.2 mm in diameter. The probes have electrochemically sharpened tips with an approximate radius of 1 μm and, apart from the tips, have as little platinum exposed to the electrolyte as possible. One tip is close to the metal surface and the electric field created by the ion flux (10–20 μm from the metal). The other probe, a few millimeters behind the first, samples the noise in the bulk electrolyte. The output from the platinum electrodes is directed to an ac coupled differential amplifier before being digitized into a form that the computer can use and display. Quantifying the degree of localized corrosion in terms of current density, as opposed to simply measuring the amplitude of the electric field adjacent to the source of activity, is a challenging goal for SRET. The "point in space" (PIS) is a reference calibration signal for a known current derived from an electrode of known surface area. Every specific metal/interface with operational conditions should have certain characteristics for more sensitive signals [86, 87].

The SVET is similar to the SRET and has been used for testing the integrity of coated metallic surfaces. SVET uses a microelectrode and scanning and measurement systems in the following typical conditions: distance from the surface of specimen to the microelectrode, 8 μm; oscillation frequency, 625 Hz; amplitude of vibration of the microelectrode, 0.5 μm; and scanning area, 1.5 mm^2. If the pores are very small, information on locations of the distributed potential cannot be obtained. Using a Vickers hardness tester, indentations can be made on the coating with ~50 kg weight and for a diagonal line length of ~405 μm to induce an artificial defect. Such studies can be extrapolated to examine active–passive behaviors of metals such as that of aluminum and magnesium with changes in chemical and coating characteristics [86, 89].

The SRET measures microgalvanic potentials existing locally on the surface of material under examination using a uniquely designed vibrating probe scanning the surface under microcomputer control. It works with an *x-y-z* scanning stage, which has maximum movement of 100 mm in the *x* direction and 75 mm in the *y* direction. However, it is also possible to scan larger samples by simply positioning the scanning head above different areas of the sample. A vibrating probe is used with a lock-in amplifier to eliminate noise occurring at any frequency other than the frequency of probe vibration. Probe vibration is controlled by a piezoceramic displacement device. Vibration amplitudes from 1 to 60 μm in a direction perpendicular to the sample surface are possible. Spatial resolution of probe is 5–10 μm and probe tip dimensions vary from 2 to 5 μm in diameter. The distance between probe and sample surface should be below 100 μm. The SVET measures localized electrochemical activity of less than 5 $\mu A/cm^2$. The vibrating probe is under full control of the computer and the resultant data can be displayed in the form of line scans, 2D or 3D area maps, and collared models [87].

The chosen vibrating probe amplitude in a typical example is adjusted at 45 μm, and the data collection is scanned at 32 points per scan line on the *x*-axis and for a total of 24 lines on the *y* axis; the signals over the metal specimen of 10 mm by 10 mm are considered. Each vibrating probe scan lasted approximately 19 min and 1 min of rest was allocated between two scans. All the experiments were conducted over a period of 10 hours. After ensuring that the surface of the specimen is parallel to the surface on the Perspex tripod and level, the probe is lowered to a height ~100 μm above the specimen surface, and the solution is added into the cell [90].

The electrolytic cell for the SRET measurements is a 5 liter cell containing about 4 liters of NaCl solution. The electrolyte temperature is generally kept at ambient temperature. The electrolyte is naturally saturated with atmospheric oxygen since no mechanical or magnetic stirring is recommended. It should be added that the vibrating scanning probe gives a slight agitation to the solution [91]. After calibration, the chosen rectangular area overlapping the acrylic resin is scanned. The overall SRET data were used to reproduce a 3D image mapping of the surface potential over the specimen. The tested specimen is mounted in the specimen

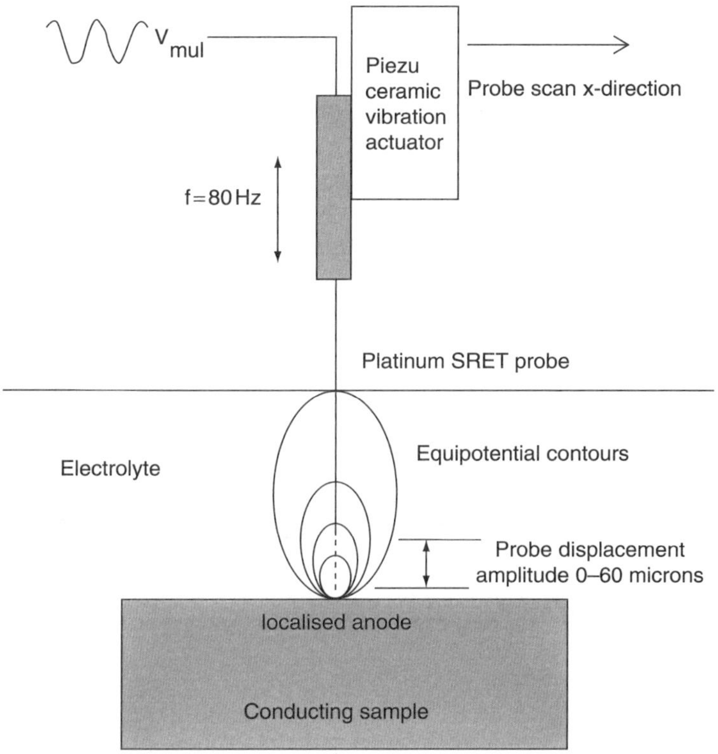

Figure 16.13 Scheme of SRET operating system; sensing probe is ~100 μm from the specimen surface [92].

holder of a SRET apparatus model SVP100 for free corrosion measurement in test solution (Figure 16.13) [92, 93].

The 3D SRET map of potential differences measured over the exterior skin of the AZ91D-DC specimen for an immersion time of 3.5 h is shown in Figure 16.14. The potential differences were distributed into different zones corresponding to neutral, low, or intense cathodic and

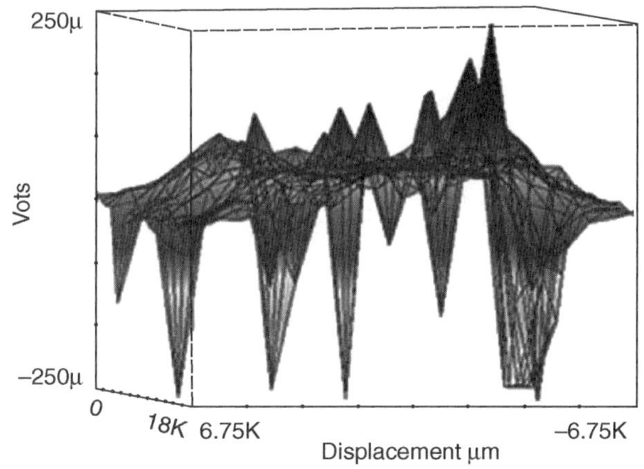

Figure 16.14 Three-dimensional potential images at different immersion times in 0.05 M NaCl solution at pH 6.1 and 23 °C at $t = 3.5$ h from the exterior skin of an AZ91D-DC specimen; cathodic ↑ and anodic ↓ [93].

anodic activities. The down ward-pointing cones (↓) correspond to anodic potentials (inside pits) and the up ward-pointing cones (↑) represent the cathodic potentials [93].

The most anodic (active) potential is the lowest point in the down ward cones and vice versa. From the 3D SRET maps, the AZ91D specimen presented many pits and also showed intense cathodic sites. The potential difference between the most active cathode and the most active anode on the surface of AZ91D-DC may be considered as the quasi-electromotive force (QEMF) of the most active corrosion cell. The QEMF of the corrosion cell for the exterior skin of AZ91D-DC can then be plotted as a function of immersion time [93]. The term QEMF describes well the corrosion tendency and intensity of thixocast and die-cast AZ91 magnesium alloys since the current flows between local galvanic cells and the *RI* drop at the interface can be considered the same for the same electrolyte for the sake of comparison (see Chapter 18) [91].

16.7. MICROSYSTEMS AND WIRE BEAM ELECTRODE

16.7.1. Microsystems and Atomic Force Microscopy

The effect of heterogeneities in alloys on corrosion has been investigated by measuring the corrosion potentials of selected small and/or minute areas of the surface of the metal. If the surface of the metal is coated, it is possible to produce punctures in the film down to 35 μm diameter. Microelectrodes of Ag,AgCl/Cl^- can be used with capillary tips down to 15 μm for potential measurements and microglass electrodes of approximately 30 μm diameter for the determination of pH. Recently, microelectrodes with an internal diameter at the tip of about 0.2 μm to measure localized corrosion potentials in the grain boundary regions of aged Al–Mg and Al–Cu alloys, and microelectrodes with a tip diameter of 1 μm to measure pH during stress corrosion of aluminum alloys were realized for corrosion studies [4].

A microelectrochemical technique applying microcapillaries as electrochemical cells has been developed, allowing the study of local processes on passive metal surfaces. Only small surface areas a few micrometers or even nanometers in diameter are exposed to the electrolyte. Due to enhanced current resolution down to picoamperes and femtoamperes processes in the micro- and nanometer range can be studied. Microelectrochemical techniques, combined with statistical evaluation of the experimental results, allow more insight into the mechanism of these processes [26].

A precision scanning microelectrode system of the scanning electrochemical potential microscope (SECPM) is available and it can be used to monitor or impose current flowing between a microelectrode and a specimen surface in a solution with extremely high spatial resolution (Uniscan model SECM 270). The closed loop microelectrode can position a specimen of 70 × 70 × 70 mm and directly display the microelectrode position in all axes. Use of the scanning electrochemical microscope is recommended for the study of metals showing active–passive behaviors that can be vulnerable to localized corrosion such as pitting of aluminum or magnesium alloys in aggressive chloride solutions. The initiation and propagation of corrosion pits are better examined by high-resolution electrochemical maps.

Surface analysis at the atomic level is of critical importance especially in the nanocomposite field. While scanning or transmission electron microscopy offers excellent 2D representations of samples, these techniques do not allow quantifying surface profiles. In contrast, atomic force microscopy (AFM) is able to establish the surface topography precisely, determine the roughness, and give a better description of surface passive films.

Furthermore, lateral force measurements (LFMs) would yield additional properties of materials by investigating the force–distance relationship between sample and tip. Since the force gradient depends on the material's nature and the geometry of both the surface and tip, the identification of various materials in thin-film form is possible. In LFM mode, it is also possible to study the adhesion forces related to metallic, covalent, or ionic bonds. AFM can be adapted by the addition of specific accessories for electrochemical measurements such as surface conductivity or electrochemical potential simultaneously to the topographical mapping of the surface. This enables more profound studies of corrosion and passivation, surface modification, electrodeposition, and the influence of biological systems on corrosion [94].

16.7.2. Wire Beam Electrode

When a metal substrate surface has a nonuniform microstructure, heterogeneous electrochemical processes occur. The wire beam electrode (WBE) is a multipiece electrode constructed with a variable number of metal wires embedded in insulating material. Each wire surface is much smaller than the total electrode surface and for most surfaces its corrosion and other electrochemical parameters can be assumed to be uniform even if the process on the whole electrode surface is not. An aluminum WBE has been constructed and tested on the bare metal. The electrolyte used was an aqueous solution with 0.35 wt % $(NH_4)_2SO_4$ and 0.05 wt % NaCl (dilute Harrison's solution). The data acquired indicate that the behavior of a continuous surface under corrosion can be emulated using a multipiece electrode. Differences between the behavior of a plate and wire electrode still remain and the SVET has the capability to address them [95]. There is interest in applying the WBE in a novel experimental setup to simultaneously measure electrode potential noise and create WBE current distribution maps for direct comparison of the two methods and correlation of electrode noise and corrosion behavior. Experiments have been carried out using a stainless steel WBE exposed to a solution containing $FeCl_3$. The obtained results suggest that the WBE method could be used in combination with the noise signatures to achieve early recognition, detection, and prediction of localized corrosion [96].

The WBE and the SRET have been used in combination to study the anodic dissolution of aluminum (AA1100) in 0.5 M NaCl solution, and the inhibiting effects of potassium dichromate, and to demonstrate the applicability of the WBE for investigating corrosion processes under anodic polarization. Both techniques were used successfully in determining corrosion behavior and profile below and above the pitting potential as well as the effects of the inhibitor. The pitting potential determined using the WBE method was found to correlate well with that determined using the conventional pitting scan method; and the anodic dissolution profile determined using the WBE method was found to correlate with maps using the SRET. Moreover, the WBE appeared to be more sensitive and could provide more detailed information of corrosion processes than the SRET. Two mechanisms of localized corrosion initiation of aluminum have been suggested based on the results of this study: the initiation of localized corrosion due to the disappearance of minor anodes, and the initiation of localized corrosion due to the initiation of new anodic sites [97].

Finally, it was demonstrated that the WBE is appropriate for understanding the electrodeposition of polyaniline coatings on AA1100. Effectively, if the aluminum surface is not treated, the anodic polarization current maps showed localized anodic current distribution resulting in nonuniform electrodeposition. This nonuniform coating has been found to accelerate general corrosion of AA1100. In contrast, if the AA1100 electrode was pretreated by a cathodic polarization process, anodic polarization current maps would show a random anodic current distribution, leading to uniform polyaniline coating on the whole surface. This applied film prevents localized corrosion of AA1100 primarily by enhancing the passive film rather than by acting as a barrier film [98].

REFERENCES

1. ASM International Handbook Committee, in *Corrosion—Understanding the Basics*, edited by J. R. Davis. ASM International, Materials Park,OH, 2000, pp. 427–491.
2. B. W. Lifka, in *Corrosion Tests and Standards, Application and Interpretation*, 2nd edition, edited by R. Baboian. ASM International, Materials Park, OH, 2005, pp. 547–557.
3. V. S. Sastri, E. Ghali, and M. Elboujdaini, *Corrosion Prevention and Protection— Practical Solutions*. Wiley, Chichester, UK, 2007, pp. 109–176.
4. L. L. Shreir, R. A. Jarman, and G. T. Burstein, *Corrosion, Corrosion Control*. Butterworth-Heinemann, Oxford, UK, 1995, pp. 19:1–19:48.
5. *Annual Book of ASTM Standards*. American Society for Testing Materials, Philadelphia, PA, 2006; pp. D1384-01;G1-90; G5-94; G16-95; G46-94; G78-01; G61-86; 69-01; G102-89, and G110-92.
6. C. P. Dillon, *Forms of Corrosion Recognition and Prevention*. International Association of Corrosion Engineers, Houston, TX, 1982.
7. V. S. Sastri, E. Ghali, and M. Elboujdaini, *Corrosion Prevention and Protection— Practical Solutions*. Wiley, Chichester, UK, 2007, pp. 331–459.
8. L. M. Wyatt, D. S. Bagley, and M. A. Moore, *An Atlas of Corrosion and Related Failures*, MTI Publication 18, Materials Technology Institute of the Chemical Process Industries Inc., New York, 1987.
9. *Corrosion Data Survey: Metals Section*, 6th edition. National Association of Corrosion Engineers, Houston, TX, 1985.
10. ASM International Handbook Committee, *Handbook of Corrosion Data*, 2nd edition, edited by B. C. Craig and D. S. Anderson. ASM International, Materials Park, OH, 1995.
11. D. J. De Renzo, *Corrosion Resistant Materials Handbook*, 4th edition. Noyes Data Corporation, Park Ridge, NJ, 1985.
12. P. A. Schweitzer, *Corrosion Resistance Tables: Metals, Plastics, Nonmetallics and Rubbers*, 2nd edition. Marcel Dekker, New York, 1986.
13. N. G. Thompson and J. H. Payer, in *Corrosion Testing Made Easy*, edited by B. C. Syrett. NACE International, Houston, TX, 1998, pp. 72–77.
14. M. Elboujdaini, M. T. Shehata, and E. Ghali, An electrochemical investigation and corrosion performance of some aluminum alloys in chloride media, in *Proceedings of the 31st Annual International Metallographic Society Meeting, Volume 26, Analysis of In-service Failures and Advances in Microstructural Characterization*, Ottawa, Canada, edited by D. O. Northwood, M. T. Shehata, and J. Wylie. Metallographic Society, Ottawa, Canada, 1998, pp. 91–97.
15. M. Elboujdaini, E. Ghali, R. G. Barradas, and M. Girgis, *Corrosion Science* **30**(8–9), 855–867 (1990).
16. *ASM Handbook*, Vol. 13A. ASM International, Materials Park, OH, 2003, 446–462.
17. D. A. Jones, *Principles and Prevention of Corrosion*. MacMillan Publishing, New York, 1992.
18. H. Kaesche, *The Corrosion of Metals*. NACE, Houston, TX, 1985.
19. F. Mansfeld, H. Xiao, L. T. Han, and C. C. Lee, *Progress in Organic Coatings* **30**, 89 (1997).
20. S. Mansfeld, Z. Sun, C. H. Hsu, and A. Nagiub, *Corrosion Science* **43**, 341 (2001).
21. N. Sato and G. Okamoto, in *Comprehensive Treatise of Electrochemistry*, edited by J. O. M. Bockris, B. E. Conway, E. Yeager, and R. E. White. Plenum Press, New York, 1981, pp. 193–245.
22. R. Cottis and S. Turgoose, *Testing Made Easy: Electrochemical Impedance and Noise*. NACE International, Houston, TX, 1999, pp. 9–49, 71–93.
23. *Basics of Electrochemical Impedance Spectroscopy*. Princeton Applied Research Co., Princeton, NJ, 2007, p. 13.
24. F. Mansfeld, *Corrosion-NACE* **36**(5), 301–307 (1981).
25. Z. Szklarska-Smialowska, *Pitting and Crevice Corrosion*. NACE International, Houston, TX, 2005, pp. 327–329.
26. H. Böhni, in *Uhlig's Corrosion Handbook*, edited by R. W. Revie. Wiley, Hoboken, NJ, 2000, pp. 173–190.
27. F. Mansfeld, Y. Wang, X. Xiao, and H. Shih, Monitoring of localized corrosion of aluminum alloys with electrochemical impedance spectroscopy (EIS), in *Symposium on the Critical Factors in Localized Corrosion*, Electrochemical Society, Pennington, NJ, 1992, p. 469.
28. R. Bard and L. R. Faulkner, *Electrochemical Methods*. Wiley, Hoboken, NJ, 2001, pp. 368–416.
29. D. E. Smith, AC Polarography and related techniques: Theory and practice, in *Elecroanalytical Chemistry*, Vol. 1, edited by A. J. Bard, Marcel Dekker, Inc., NY, 1966, pp. 102–132.

30. M. Sluyters-Rehbach and J. H. Sluyters, *Electroanalytical Chemistry* **4**(1) (1970).
31. R. Parsons, *Journal of Chemistry Reviews* **90**, 813 (1990).
32. J.-S.Y. Su-Moon Park, *Analytical Chemistry*. Amercan Chemical Society, Washington DC, 2003, pp. 451A–455A.
33. D. C. Silverman, in *Uhlig Corrosion Handbook*, 2nd edition, edited by R. W. Revie. Wiley, Hoboken, NJ, 2000, pp. 1179–1225.
34. K. S. Cole and R. H. Cole, *Chemical Physics* **9**, 34 (1941).
35. K. Juttner, *Electrochimica Acta* **35**, 1501 (1990).
36. G. Grundmeier, K. M. Juttner, and M. Stratmann, in *Corrosion and Environmental Degradation*, Vol. 19/I, edited by M. Shutze. Wiley-VCH, Weinheim, Germany, 2000, pp. 285–308.
37. M. Stratmann, W. Furbeth, G. Grundmeier, R. Losch, and C. R. Reinartz, in *Corrosion Mechanisms in Theory and Practice*, edited by P. Marcus and J. Oudar. Marcel Dekker, New York, 1995, pp. 373–419.
38. J. R. MacDonald, *Impedance Spectroscopy*. Wiley, Hoboken, NJ, 1987.
39. L. Nyikos and T. Pajkossy, *Journal of the Electrochemical Society* **30**, 1533 (1985).
40. T. Pajkossy and L. Nyikos, *Electrochimica Acta* **34**, 181 (1989).
41. R. De Levie, *Annual Biomedical Engineering* **20**, 337 (1992).
42. J. E. B. Randles, *Discussions of the Faraday Society* **1**, 11 (1947).
43. D. A. Eden, in *Uhlig's Corrosion Handbook*, edited by R. W. Revie. Wiley, Hoboken, NJ, 2000, pp. 1227–1238.
44. D. A. Blanc, C. Gabrielli, M. Ksouri, and R. Wiart, *Electrochimica Acta* **23**(4), 337–340 (1976).
45. K. Hladky,U.S. Patent No. 4,576,678, 1981.
46. B. Moniz and H. Zhang, in *Chemical Processing*, edited by R. Baboian. ASTM International, West Conshohocken, PA, 2005, pp. 779–794.
47. J. L. Kearns, D. A. Eden, M. R. Yaffe, J. V. Fahey, D. L. Reichert, and D. C. Silverman, in *Electrochemical Noise Measurement for Corrosion Applications*, edited by J. R. Kearns, J. R. Scully, P. R. Roberge, D. L. Reichert, and J. L. Dawson. ASTM International, West Conshohocken, PA, 1996, Vol. ASTMSTP 1277, pp. 446–470.
48. R. A. Cottis and A. Llewellyn, Electrochemical noise, electrochemistry for corrosion. Available at http://www.cp.umis.ac.uk/JCSE/vol1.
49. J. Hickling, J. Goellner, A. Burkert, and A. Heyn *Corrosion* **385**, 22 (1989).
50. B. Arsenault, *Overview of the Electrochemical Noise Technique*. Laval University, Quebec, 2007, p. 16.
51. J. L. Dawson, *Electrochemical Noise Measurement: The Definitive In-Situ Technique for Corrosion Applications*. ASTM, West Conshohocken, PA, 1996, pp. 3–35.
52. R. A. Cottis and A. Llewellyn, Electrochemistry for corrosion, electrochemical noise. Available at http://corrosiontest-its.Manchester.qc.uk/lecture notes/echem/noise file.html.
53. Gamry Electrochemical Noise Software, Electrochemical Signal Analyzer Software, Gamry Instruments, Edmonton, Canada. http:// www. gamry.com/Products/ESA400/ESA400.htm.
54. W. P. Iverson, *Journal of Electrochemical Science* **6**, 617 (1968).
55. A. Aballe, M. Bethencourt, F. J. Botana, and M. Marcos, *Electrochimica Acta* **44**, 4805–4816 (1999).
56. R. D. Klassen, P. R. Roberge, A.-M. Lafront, M. O. Oteyaka, and E. Ghali, *Canadian Metallurgical Quarterly* **44**(1), 47–52 (2005).
57. E. Slavcheva and G. Schmitt, *Materials and Corrosion* **53**(9), 647–655 (2002).
58. J. A. Wharton, R. J. K. Wood, and B. G. Mellor, *Corrosion Science* **45**(1), 97–122 (2003).
59. R. J. K. Wood, J. A. Wharton, A. J. Speyer, and K. S. Tan, *Tribology International* **35**(10), 631–641 (2002).
60. A. Aballe, M. Bethencourt, F. J. Botana, and M. Marcos, *Electrochemistry Communications* **1**, 266–270 (1999).
61. F. Mansfeld, C. Chen, C. C. Lee, and H. J. Xiao, *Corrosion Science* **38**, 497 (1996).
62. P. Roberge, *ASTM Publication* **STP 1277**, 142–155 (1996).
63. A. Horvath and R. Schiller, *Corrosion Science* **45**(3), 597–609 (2003).
64. J. F. Chen and W. F. Bogaerts, *Corrosion Science* **37**(11), 1839–1995 (1995).
65. F. Mansfeld, L. T. Han, C. C. Lee, and G. Zhang, *Electrochimica Acta* **43**, 2933 (1998).
66. G. L. Edgemon, *Materials Performance* 52–55 (2005).
67. P. C. Pistorius and G. T. P. Burstein, *Transactions of the Royal Society of London Series A* **341**, 531 (1992).
68. T. Suter and H. Bohni, *Electrochimica Acta* **43**, 2843 (1998).
69. U. Bertocci, C. Gabrielli, F. Huet, and M. Keddam, *Journal of the Electrochemical Society* **144**(1), 31 (1997).
70. U. Bertocci, C. Gabrielli, F. Huet, M. Keddam, and P. Rousseau, *Journal of the Electrochemical Society* **144** (1), 37 (1997).
71. G. Gusmano, G. Montesperelli, S. Pacetti, A. Petitti, and A. D'amico, *Corrosion* **53**(11), 860–868 (1997).
72. L. Beaunier, J. Frydman, C. Gabrielli, F. Huet, and M. Keddam, in *Electrochemical Noise Measurement for Corrosion Applications*, edited by R. Kearns, J. R. Scully, P. R. Roberge, D. L. Reichert, and J. L. Dawson.

ASTM, West Conshohocken, PA, 1996, Vol. ASTM STP 1277, pp. 114–128.

73. F. Safizadeh, *Study of Pitting Corrosion of Aluminum by Electrochemical Noise Measurements*. Laval University, Quebec, 2007, p. 25.

74. C. C. Lee and F. Mansfeld, *Corrosion Science* **40**, 959 (1998).

75. K. Hladky and J. L. Dawson, *Corrosion Science* **22** (3), 231 (1982).

76. X. F. Liu, H. G. Wang, and H. C. Gu, *Corrosion Science* **48**, 1337–1367 (2006).

77. A. Aballe, M. Bethencourt, F. J. Botana, and M. Marcos, *Electrochimica Acta* **46**, 2353–2361 (2001).

78. T. Zhang, Y. Shao, G. Meng, and F. Wang, *Electrochimica Acta* **53**, 561–568 (2007).

79. M. W. Kending, S. Jeanjaquet, and J. Lumsden, Electrochemical impedance of coated metal undergoing loss of adhesion, Conference article, November 1991, San Diego, CA, *ASTM Special Technical Publication*, Philadelphia, PA, Issue No. 1188, 1993, pp. 407–427.

80. M. Keddam, M. Krarti, and C. Pallotta, *Corrosion* **43**, 454 (1987).

81. C. Gabrielli, M. Ksouri, and R. Wiart, *Journal of Electroanalytical Chemistry* **86**, 233–239 (1978).

82. Z.-N. Yang, Z. Zhang, W. H. Leng, K. Ling, and J.-Q. Zhang, *Transactions of the Nonferrous Metallurgical Society of China* **16**, 209–216 (2006).

83. Z. Zhang, W. H. Leng, Q. Y. Cai, F. H. Cao, and J. Q. Zhang, *Journal of Electroanalytical Chemistry* **578**, 357–367 (2001).

84. E. Budevski, W. Obretenov, W. Bostanov, and G. Staikov, *Electrochimica Acta* **34**, 1023–1029 (1989).

85. F. Huet, M. Musiani, and R. P. Nogueira, *Electrochimica Acta* **48**, 3981–3989 (2003).

86. E. Ghali, W. Dietzel, and K.-U. Kainer, *Journal of Materials Engineering and Performance* **13**(5), 517–529 (2004).

87. K. R. Trethewey, D. A. Sargent, D. J. Marsh, and S. Haines, in *Modelling Aqueous Corrosion from Individual Pits to System Management, Volume 266, Applied Sciences*, edited by K. R. Trethewey and P. R. Roberge. NATO ASI Series E, Academic Publishers, Boston, No. 266, 1994, pp. 417–442.

88. H. S. Isaacs and M. W. Kendig, *Corrosion* **36**, 269 (1980).

89. N. Morihiro, M. Takeda, G. Katayama, M. Kido, and Y. Harada, Corrosion behaviour of WC thermal spray materials, in *Second International Conference on Environment Sensitive Cracking and Corrosion Damage*, Hiroshima, Japan, edited by ESCCD, Hiroshima, Japan, 2001.

90. W. Zhang, S. Jin, E. Ghali, R. Tremblay, M. Shehata, and E. Es-Sadiqi, *Advanced Engineering Materials* **8**(10), 973–980 (2006).

91. S. Jin, E. Ghali, C. Blawert, and W. Dietzel, in *Proceedings of 210th ECS Meeting*, Cancun, Mexico, edited by N. Missert, A. Davenport, M. Ryan, and S. Virtanen. ECS Transactions, Pennington, NJ, **3**(31), 2007, pp. 295–311.

92. J. N. Murray, D. C. Hansen, and R. A. Brizzolara, in *Corrosion 2000*, NACE International, Houston, TX, 2000.

93. W. Zhang, S. Jin, E. Ghali, and R. Tremblay, *Canadian Metallurigical Quartely* **45**(2), 181–188 (2006).

94. T. Zhao, D. Zagidulin, G. Szymanski, and J. Lipkowski, *Electrochimica Acta* **51**, 2255–2260 (2006).

95. D. Battocchi, J. He, G. P. Bierwagen, and D. E. Tallman, *Corrosion Science* **47**, 1165–1176 (2005).

96. Y.-J. Tan, N. N. Aung, and T. Liu, *Corrosion Science* **48**, 23–38 (2006).

97. T. Liu, Y.-J. Tan, B. Z. M. Lin, and N. N. Aung, *Corrosion Science* **48**, 67–78 (2006).

98. T. Wang and Y.-J. Tan, *Corrosion Science* **48**, 2274–2290 (2006).

Chapter 17

Evaluation of Corrosion Forms of Aluminum and Its Alloys

Overview

Generally, the essential requirements of accelerated laboratory testing are that the acceleration should produce the same form and type of corrosion and/or similar mode of failure or cracking and reflect at least a known order of resistance of some alloys in service media. Nonheat or nonheat-treatable alloys have better corrosion resistance than heat-treatable ones. Brief descriptions of the corrosion resistance of aluminum alloys to seven forms of corrosion are given: general uniform corrosion, nonuniform corrosion, galvanic corrosion, metallurgically influenced corrosion (METIC), microbiologically influenced corrosion (MIC), mechanically assisted corrosion (MECIC) and environmentally induced cracking (EIC) or stress-corrosion cracking (SCC).

Silicon carbide, graphite, and boron are cathodic to aluminum and do not polarize easily. Al/SiC metal matrix composites (MMCs) are susceptible to localized galvanic corrosion in marine environments. The value of the pitting potential of Al 1199 (99.99% Al) in sodium chloride solutions can vary from approximately −0.35 to 0.54 V/SHE depending on the activity of Cl^- at 25 °C. The critical pitting potential, $E_{SCR,}$ which corresponds to a more stationary state, lies between the breakdown potential and the protection potential and should also be considered. Dynamic polarization curves show, for example, superior pitting corrosion behavior of an electron beam weld of AA2219 alloy as compared to a gas tungsten arc weld at 0.166 mV/s.

Testing for intergranular corrosion susceptibility varies with the alloy family and the fabrication process that can occur in most environments. Periodic removal of sample plugs could monitor general biofouling tendencies, but the value of this method is limited because of the patchy nature of biofilms.

The strongest aluminum alloys have the greatest corrosion resistance to erosion corrosion. SCC is observed only for alloys with soluble alloying elements, primarily Cu, Mg, Si, and Zn, in appreciable quantities. SCC in aluminum alloys is typically intergranular, although transgranular SCC has been observed for a few alloys under highly specific environmental conditions and may be part of the propagation mechanism in some 7xxx alloys. Stress-corrosion branched cracks can be seen under specific conditions for some aluminum alloys. Microbranching can occur under less restrictive conditions, for example.

Corrosion Resistance of Aluminum and Magnesium Alloys: Understanding, Performance, and Testing By Edward Ghali

During the design of a certain structure, selection of the material with appropriate mechanical properties should determine whether the material requires surface protection for use in the intended environment and if a coating is appropriate for this use. Generally, alloys of the high-strength 2xxx and 7xxx series alloys necessitate a certain type of corrosion prevention or coating depending on the requested duration or useful life of the structure [1]. A brief description of corrosion forms that are closely related to the different series of cast and wrought aluminum alloys is imperative in evaluation of the corrosion resistance of the chosen alloy since testing differs generally with the corrosion type. The corrosion resistance of aluminum alloys to different corrosion forms is then summarized here.

Cast Alloys Nonheat or heat-treatable alloys have better corrosion resistance than heat-treatable ones. Cu is the element most prone to general corrosion. Heat-treatable cast alloys (F) without Cu such as 356 or 357 have high corrosion resistance like nonheat-treatable wrought alloys. Heat-treatable cast alloys (T) with Cu have lower corrosion resistance with increasing Cu percentage and more so than with wrought alloys. This lower resistance is compensated by the thicker sections of cast alloys. The Corrosion resistance of heat-treatable alloys depends not only on Cu but also on Zn content. It has been shown that addition of inhibitors is efficient in controlling the influence of Zn on corrosion acceleration. As predicted from *E*–pH diagrams, Al alloys show good passive behavior between pH 4 and 8 [1].

Wrought Alloys Nonheat-treatable alloys of the 1xxx, 3xxx, 4xxx, and 5xxx series have a high resistance to general corrosion:

1xxx $\geq$ Al 99%: excellent resistance to corrosion.

3xxx Al–Mn: good corrosion resistance improved by inhibitors, most widely used.

4xxx Al–Si: good corrosion resistance improved by inhibitors, mainly for welding.

5xxx Al–Mg: good corrosion resistance that can be improved by inhibitors, possibility of stress-corrosion cracking; widely used in marine atmospheres.

Heat-treatable alloys of the 2xxx, 6xxx, and 7xxx series have different corrosion behaviors:

2xxx Al–Cu, Al–Cu–Mg, Al–Cu–Si–Mg: low resistance to general corrosion, possibility of intergranular attack, difficult to inhibit; currently used in the aircraft industry.

6xxx Al–Mg_2Si: high resistance to general corrosion.

7xxx no copper: high resistance to general corrosion.

7xxx Al–Zn–Mg–Cu: inhibitor may be used; can have very high strengths.

General Corrosion Al is active in strong acidic or alkaline media and this can lead to general corrosion. In the presence of strong agitation and numerous local galvanic cells, the morphology of the attack frequently leads to general corrosion. In this case, polarization resistance and conventional weight loss methods are good to predict general corrosion rates (see Chapter 16).

Galvanic Corrosion Dissimilar metal corrosion (galvanic corrosion) is evident when the areas of anodic sites are much smaller than other common nobler structural materials.

Avoid copper especially in this situation. Al cements (reduces) or deposits nobler metal ions (deposition corrosion), which initiates dissimilar metal corrosion. Hg (parts per billion) should be avoided. When an electric current leaves the Al surface to enter an environment such as water, soil, or concrete, the Al is corroded by stray current corrosion as in concrete (roadways) and soils (pipelines and drainage systems). Cathodic protection or metal overprotection of Al alloys could lead to alkali attack (aluminate ion formation) since Al is amphoteric.

Localized Corrosion Imperfections can be 2–4 nm deep at room temperature; pitting is observed principally because of the oxygen concentration differential cell. Crevice corrosion is observed for stagnant depleted zones of oxygen. The volume of corrosion product can reach five times that of the subjacent aluminum, and this ratio (oxide volume /metal volume) is double that of rust on steel. This can damage, for example, packages of aluminum sheets or aluminum wraps of coil by water staining (crevice corrosion). Filiform corrosion (e.g., under paint) is due to the active anodic head and an oxygenated tail (cathode). For example, filiform corrosion of 2024 and 7000 series alloys coated with polyurethane is frequent and blisters caused by H_2 are observed in the head region. Also, filiform corrosion of Al alloys of the 3xxx series has been observed.

Pitting of Al matrix composites 1050 and 2124, each reinforced with silicon carbide particles (SiC_p) in the size range 3–40 μm, has been studied in 1 N NaCl solution. Pores and crevices at SiC_p–matrix interfaces strongly influence pit initiation, which is further aided by the cracking of large SiC_p during processing. The presence of $CuAl_2$ and $CuMgAl_2$ precipitates in 2124-SiC_p composite also promotes pitting attack at SiC_p–matrix and intermetallic–matrix interfaces [1].

Metallurgically Influenced Corrosion (METIC) Intergranular corrosion is caused by a direct corrosion of a precipitate that is less corrosion resistant or by corrosion of a denuded zone adjacent to a noble phase. Exfoliation corrosion is a lamellar selective subsurface attack of the plate and appears as striations in parallel planes in the direction of working. This occurs in slightly acid medium and a nobler metal accelerates exfoliation. It needs some residual stresses such as that left after lamination.

Weldments can crack because of mercury–zinc amalgam and residual stresses. Soldered joints are good in milder environments, but not in more aggressive ones. Brazed joints in Al alloys have good resistance to corrosion but excessive flux should be removed. Weldments of nonheat-treatable alloys have good resistance to corrosion, while for heat-treatable alloys, corrosion is selective in the weld or in the heat-affected zone (HAZ). Stress-corrosion cracking in weldments is due to residual introduced stresses during welding but is rare.

Microbiologically Influenced Corrosion (MIC) Biological attack is generally localized; for example, *Cladosporium resinae* (principal organism) produces organic acids with pH values $\leq$3–4. Pseudomonads produce corrosion product (slime-forming) that leads to the formation of a tubercle, with the inside region being highly anodic and the outside being cathodic due to dissolved oxygen.

Mechanically Assisted Corrosion The strongest Al alloys have the greatest resistance to erosion corrosion. Fretting corrosion causes localized corrosion and fatigue corrosion. This can be prevented by lubrication, restricting movement, and selecting corrosion-resistant alloys. Corrosion fatigue is generally transgranular, simulated by localized attack (pitting or intergranular). Prevention could be achieved by appropriate peening, inhibitors, painting, and cathodic protection. For welds, peening and coating are a good practice to increase resistance to corrosion fatigue.

Environmentally Induced Corrosion Stress-corrosion cracking is observed only for alloys with soluble alloying elements, primarily Cu, Mg, Si, and Zn, in appreciable quantities. As for hydrogen damage, humid hydrogen gas causes minimum ductility below 0 °C, as in other face-centered cubic (fcc) structures. Intergranular and/or transgranular embrittlements are observed. Blistering is observed near surface voids that coalesce to produce a large blister.

17.1. GENERAL CORROSION OF ALUMINUM AND ITS ALLOYS

Uniform corrosion of aluminum in practice is rare, except in special, highly acidic, or alkaline aggressive media, as can be deduced from *E*–pH Pourbaix diagrams. Etching an Al alloy specimen for 1 min in 5 wt% sodium hydroxide at 80 °C is an example of a general uniform corrosion that is used for atmospheric exposure studies (ASTM G66). Under controlled anodic polarization conditions and/or nonpassivating media (e.g., one containing chloride ions), uniform general corrosion is observed and depends mainly on the chemical concentration and temperature of the electrolyte. Several industrial processes use this principle to electrochemically machine or size parts.

Gravimetric methods explained in ASTM G1 are the most conventional way to estimate general corrosion. For steady dissolution rate applications, thickness-measuring devices or gravimetric procedures can be used [1, 2]. The standard practice for laboratory immersion corrosion testing of metals, ASTM G31 "Immersion Corrosion Testing of Metals", serves to quantify gravimetrically the attack of a certain agent. This practice describes accepted procedures and factors that influence laboratory immersion corrosion tests, particularly mass loss tests. These factors include specimen preparation, apparatus, testing conditions, methods of cleaning specimens, evaluation of results, and calculation and reporting of corrosion rates.

Atmospheric Corrosion The standard practice for conducting atmospheric corrosion tests on metals (ASTM G50) deals with the description of general and localized corrosion. Generic types of atmospheres used are seacoast, industrial, urban, and rural. Sometimes specific geographical locations or local chemical conditions are important because they can produce unique results [1]. Corrosion rates can be expressed in either penetration per unit time or loss in thickness over the exposure period. If the corrosive attack is nonuniform, the corrosion data can be misleading. In cases where corrosion is in the form of pitting, a pitting factor should be reported. The depth of the deepest pit or the average depth of ten deepest pits should be measured [3, 4].

The extent of attack that occurs on Al alloys buried underground varies greatly, depending on the soil composition and climatic conditions. In dry, sandy soil, corrosion is negligible. In wet, acid or alkaline soils, attack may be severe [3]. Chemical dip and sulfuric acid anodic coatings are generally protective for 6053-T and presumably for other Al alloys. The depth of attack is determined by microscopic examination of cross sections and by the change in tensile strength of a specimen.

Influence of High Temperature or High Pressure or Both ASTM G111 is commonly used to evaluate the corrosion performance of metallic materials under conditions that attempt to simulate service conditions that involve high temperature (HT), high pressure (HP), or HTHP combination. Clinical processing, food processing, pressurized cooling water, electric power systems, and petroleum production and refining are examples where this type of testing is important and useful. Both self-stressed and externally stressed specimens are acceptable. For electrochemical measurements or control, cylindrical electrode specimens, where only the lower portion of the electrode is exposed to the liquid phase of the test environment and where the electrical connections are made externally to the test cell, are convenient.

A common practice is to conduct tests in environments that have been sampled and retrieved from field and plant locations. Depletion of reactive constituents or concentration of constituents may occur and their monitoring in these cases is necessary. The ratio of solution volume to specimen surface area is important and a minimum ratio of 30 mL/cm^2 should be maintained.

Corrosion in Engine Coolants ASTM D4340 is the standard test method for corrosion of cast Al alloys in engine coolants under heat-treating conditions. The test covers a laboratory screening procedure for evaluating the effectiveness of engine coolants in combating corrosion of cast Al alloys used for engine cylinder heads while exposed to an engine coolant under a pressure of 28 kPa. The temperature of the specimen is maintained at 135 °C and the test is continued for 1 week. The effectiveness of the coolant for preventing corrosion of the Al under heat transfer conditions is evaluated by the loss of weight of the specimen in mg/cm^2/week.

Rational Corrosion of Al Coatings Al–Zn based coatings are used more and more as sacrificial anodes because of their low rate of uniform general corrosion. Atmospheric corrosion and ASTM salt spray tests of Al–Zn alloy-coated steel showed that 55 wt % Al–Zn is two to four times as corrosion resistant compared to a conventional galvanized coating of similar thickness. Furthermore, for the galvanic protection of cut edges of sheet in some environments, these coatings proved to be superior to Al coating [5].

17.2. GALVANIC CORROSION

17.2.1. General Considerations

Dissimilar metallic corrosion (frequently called galvanic corrosion) can cause general or localized corrosion depending on the relative surface areas of anode to cathode, the geometry of the corrosion cell, agitation, the properties of the material, the conductivity, and the composition of the corrosive media. Evaluation of galvanic corrosion throws light on compatibility of materials and predicts the extent and spatial distribution of corrosion damage. Galvanic corrosion can be uniform and/or localized. For example, pitting is observed in 5% sodium chloride solutions at an acidic pH of 1.25. However, well-defined pitting corrosion is common at neutral and alkaline solutions. The geometry of a sample plays an important role in galvanic corrosion, but it is not possible to define a particular geometric configuration as a standard for testing. Thus the experimental design for galvanic corrosion testing is unique and variable from one situation to another [6]. The extent of galvanic effects is influenced by the potential relationships of the metals involved, their polarization characteristics, the relative areas of anode and cathode, and the internal and external resistances in the galvanic circuit.

The simplest procedure in studying galvanic corrosion is a measurement of the open circuit potential difference between the metals in a couple in the environment under consideration. A better procedure is to make similar open circuit potential measurements between every individual metal and an appropriate reference electrode, which will yield the same information and will also permit observations of any changes in potential of the individual metals with time that will affect the overall potential difference in the couple. For most practical laboratory testing, the saturated calomel half-cell is the most convenient. The precision of the determinations is adequate and it is easy to maintain a constant concentration of potassium chloride.

The preferred potential-measuring instruments are potentiometers or electrometers, either of which permits measurements to be made without flow of sufficient current to polarize the electrodes during the determinations. It is also possible to use millivoltmeters if the internal resistance of the instrument is high enough to avoid any appreciable flow of current. Instruments (zero-resistance ammeters) that permit current measurements to be made with zero resistance in the measuring circuit are available. This may also be achieved by connecting the two metals to the input of a potentiostat and setting the control potential to 0 volts; the output current of the potentiostat will then be equivalent to the galvanic current flowing between the two metals. A convenient method of carrying out a galvanic test is to use two assemblies. Each assembly includes two pairs of dissimilar metals—one pair coupled galvanically while the other pair is left uncoupled in order to determine the normal corrosion rates under the same environmental conditions [7, 8].

Galvanic corrosion tests between different metals or microstructures can be accomplished in acid or neutral chloride solutions containing hundreds of ppm Cl^- and up to 5% Cl^- at ambient laboratory temperature or at the desired temperature to examine the galvanic effect of different materials. The ratio of the anode to the cathode areas of the specimen is generally 1:1 or adjusted to the projected use. Agitation or circulation of the electrolyte should match the conditions in practice [6].

Electrochemical and nonelectrochemical methods can be used to measure the parameters of interest. Electrochemical methods include measurements of potential, under coupled or noncoupled conditions, to provide information on the polarity of a bimetallic couple and extent of anodic and cathodic polarization; or measurement of galvanic current to indicate quantitatively the intensity of galvanic corrosion; or measurement of the current–potential relationship of each metal with a potentiostat to understand the kinetics. The nonelectrochemical methods include weight-loss or thickness-loss determination and visual or instrumental examination of the corroded surface [9].

There are two ASTM guidelines for galvanic corrosion testing: ASTM G71 and ASTM G82, concerning the development and use of a galvanic series for predicting galvanic corrosion performance. Under atmospheric conditions, two different types of testing methods have been standardized for determination of the weight loss due to galvanic corrosion: ISO 7441 and ASTM G104 for plate testing, and ASTM G116 for wire-on-bolt testing. In the plate type of assembly, a strip of one metal is attached by bolts to a panel of another metal to ensure electrical contact. The assembly is exposed to the atmosphere and compared to that under noncorrosive conditions. The galvanic corrosion is evaluated by visual examination or by weight-loss measurements of the strip or panel. In the wire-on-bolt type of assembly, a wire of the metal to be tested is tightly wound around the threads of a bolt of the other metal in the couple, exposing the assembly to the atmosphere. The galvanic corrosion can be estimated quantitatively by comparing the weight loss of the coupled wire to that wound on the threads of a plastic bolt [9].

17.2.2. Influence of the Composition and Microstructure

Aluminum-based alloys are anodic to many other common structural alloys. Galvanic action is much more pronounced in marine or seacoast atmospheres than in rural or industrial locations. The galvanic series of Al alloys and other structural metals in seawater shows that only Mg and Zn are more anodic and corrode to protect Al. This type of corrosion can be found in strong acidic or strong basic solutions. The rate of corrosion can vary from several micrometers per year to several micrometers per hour.

Contact with Cu or Cu-based alloys causes more pronounced galvanic attack than does contact with most other metals like stainless steels or steel. Zc is anodic to Al in most neutral or acidic solutions; while in alkaline solutions, the potentials reverse so that, in these media, contact with Zn can cause accelerated attack of Al [1, 10].

Geometry is important when the intent is to consider the influence of microstructural parameters, alloying, or mechanical working on reaction kinetics. The geometry of the design should result in a uniform potential distribution on the surface of the anode and cathode; to meet this requirement, the distance between the two metals should be larger than the dimensions of the samples [9].

Generally, the corrosion behavior of composites is governed by galvanic action between the Al matrix and the reinforcing material in an aggressive medium. Silicon carbide, graphite, and boron are cathodic to Al and do not polarize easily. Al/SiC MMCs are susceptible to localized galvanic corrosion in marine environments. Pit initiation occurs in a similar way to that of the unreinforced alloys, but the rate of pit propagation is higher for composites due to the galvanic effect [11]. These composites are marginally inferior to their matrix alloys in corrosion fatigue resistance [12].

Aluminum and its alloys are sensitive to some inorganic and organic products at certain operational conditions. The standard practice for laboratory immersion corrosion testing of metals, ASTM G31 serves to quantify gravimetrically the attack of a certain agent [2]. Coupon corrosion testing is predominantly designed to investigate general corrosion. Dissimilar metallic corrosion of Al alloys can be estimated by coupling one coupon to another electrically and comparing with that of insulated samples. Similarly, crevice corrosion can be quantified if the metal surface is blocked by a spacer or supporting hook, for example [1].

17.2.3. Electrochemical Testing

Electrochemical methods, such as measurement of the corrosion potential (open circuit) and anodic polarization curves, are primarily laboratory tests. Testing follwing ASTM G3 is recommended.

Corrosion Potential Measurements Testing for galvanic corrosion can follow ASTM standards in the form of potential measurements. It is important to follow a revised procedure to determine the free or open circuit potential such as the standard practice for measurement of corrosion potentials of Al alloys (ASTM G69). The test solution is IM sodium chloride and 9 mL of 30% hydrogen peroxide reagent per liter of aqueous solution at 25 °C to facilitate the cathodic reaction and favor passivation. The period of immersion is on the order of 1 h and the average value for the last-half hour should be within ± 5 and ± 10 mV for duplicate specimens.

The corrosion potential of an Al alloy depends on the amounts of certain alloying elements that the alloy contains in solid solution, and so it is used to determine the amount of material in the solid solution, the heat-affected areas from welding, and the validity of thermal treatments. Commercially unalloyed Al (1100 alloy) has a potential of −750 mV; 2024-T3 alloy, with its almost nominal 4.3% Cu in solid solution, has a potential of about −610 mV; while 7072 alloy with its nearly nominal 1.0%, Zn in solid solution, has a potential of −885 mV/SCE (ASTM G69) [1]. Because of the active–passive behavior of Al in numerous media and in different applications, instantaneous and stationary potential values are influenced by the composition and renewal of the species at the metal–solution interface. These values and potential–time curves during the first hour of immersion generally could indicate the tendency of the corrosion resistance of the alloy.

Polarization Measurements Electrochemical testing and determination of polarization characteristics of every component are recommended. ASTM G5, the standard reference test method for making potentiostatic and potentiodynamic anodic polarization measurements, is recommended. If one of the metals has an active–passive behavior, the state of the contact material should be considered for the expected active and passive states.

17.3. LOCALIZED CORROSION OF ALUMINUM AND ITS ALLOYS

The susceptibility of alloys to localized corrosion can be evaluated, such as that of an alloy having an active–passive behavior in certain media, by measurement of the corrosion potential (open circuit or rest potential), cyclic voltammetryc, potentiodynamic methods, anodic polarization, galvanostatic studies, scratch potentiostatic methods, triboellipsometric methods, pit-propagation rate curves, electrochemical impedance spectroscopy (EIS) studies, and electrochemical noise measurements (ENMs). Time-to-perforation data can be obtained by designing a specimen that is pressurized with air. This pressure is monitored over a period of time until failure is indicated by a decrease in pressure [13].

17.3.1. Pitting Corrosion

Pitting corrosion is normally encountered on all Al alloys and in most environments, particularly those containing the chloride ion (i.e., laboratory tests, field tests, and service applications). It occurs randomly, but not necessarily uniformly, over the entire surface. The depth and frequency of pitting depend on the alloy and environment. Pitting can be locally accelerated by crevices and contact with dissimilar metals. Pitting in Al tends to be self-stopping, following a polynomial or power function and depending on the environment [1].

Depending on the use and the urgency of the situation, the following avenues of pitting evaluation should be to considered. Measurement of mass loss, along with visual comparison of pitted surfaces, may be sufficient to rank the relative resistance of alloys in laboratory tests. Visual examination of pits should determine the size, shape, and density of pits as far as possible. Pitting assessment needs micrometer gauges. Metallographic examination is very important to show the correlation with the microstructure and to distinguish between pitting, intergranular corrosion, and dealloying. Pitting is evaluated on flat panels that are used when this is the principal purpose of the test. Determination of pitting effect on mechanical properties is frequently done.

The testing procedure ASTM G1 describes surface preparation of specimens for pitting corrosion evaluation, while ASTM G46 covers specimen evaluation after testing. ASTM G46 gives the standard rating chart for pits and the different methods of examination of pits and also describes nondestructive inspection of pitting that can include dye penetration inspection and electromagnetic, ultrasonic ray, and radiographic examinations. When pitting is involved, it is mandatory to use large samples and obtain statistically valid data [14]. It is recommended to rate the pits in terms of density (x pits/m^2), size (average pit opening in mm^2), and average pit depth (mm). The deepest pits should be measured along with the maximum pit depth or the average of the ten deepest pits, preferably both (ASTM G46). The metal penetration can be expressed as a pitting factor, which is the ratio of the deepest metal penetration to the average metal penetration. However, the *pitting factor* is not useful for Al since pitting tends to be nonuniform [1]. Neutral 5% sodium chloride salt spray testing (ASTM B117) and 3.5% sodium chloride by alternate immersion testing (ASTM G 44) are generally recommended.

17.3.1.1. Cyclic Voltammetry

Cyclic Voltammetric Studies These are quick, generally reproducible tests to show the comparative performance of Al and different alloys under chosen conditions. For some situations, a combination with other techniques, such as potential step measurements, microscopic exams, and X-ray photoelectron spectra (XPS), can help one to understand the mechanism of corrosion and to determine the composition of the passive film of an alloy. The anodic polarization behavior was determined for two Al alloys (Al–Mg and Al–Si–Mg) and pure Al. Pits initiate at precipitates of intermetallic compounds due to anodic polarization.

A typical cyclic voltammogram in 3% NaCl employing a slow linear potential scan (5 mV/s) is shown in Figure 17.1 for the Al–Mg alloy AA5083. One can observe an evident peak (P) prior to breakdown potential in the chloride solution that is attributed to the

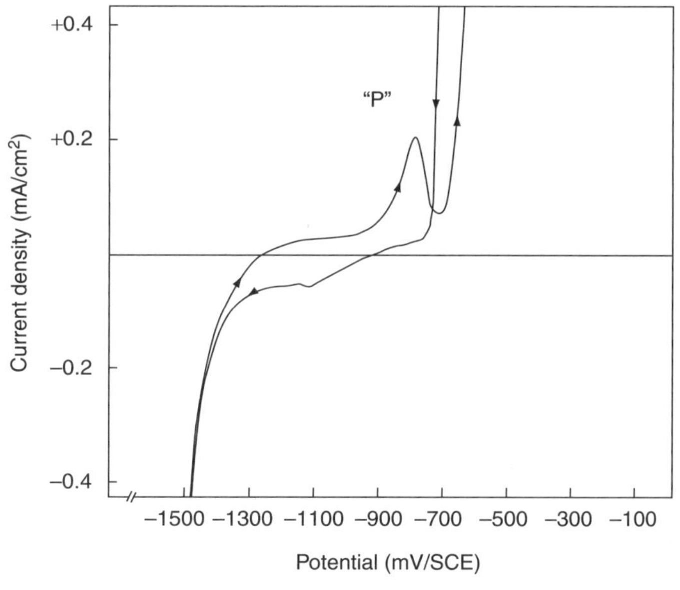

Figure 17.1 Cyclic voltammogram for AA5083 electrodes in 3% NaCl at room temperature and cycling at V = 5 mV/s (under stirred condition) [15].

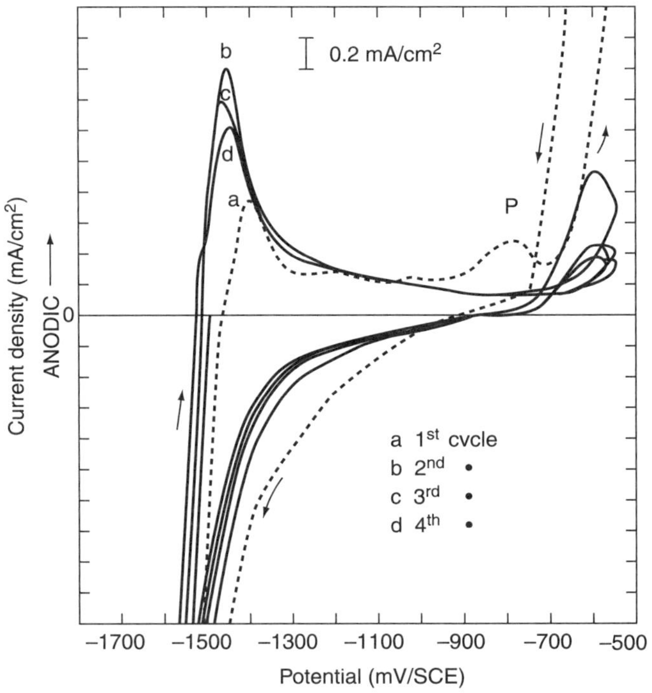

Figure 17.2 The effect of continuous cycling at 100 mV/s upon the *I–E* profiles of AA 5083 in 3% NaCl and under unstirred condition [15].

influence of Mg and/or to the presence of Fe-rich phases. This peak was not found in the voltammogram of pure Al for the same conditions [15].

At high sweep rates (e.g., 100 mV/s), the voltammogram exhibited a well-defined peak (Figure 17.2). During continuous cycling over the chosen potential limits, a marked increase in the peak current is observed in the second cycle due to metal attack and hydrogen evolution, followed by a gradual decrease in current in the following cycles that correspond to the formation of the protective film [15].

The presence of a larger quantity of Mg in the matrix of AA5083 has an effect on the electrochemical behavior and consequently on the Al_2O_3 passive oxide film as confirmed by XPS analysis of the film. A porous film of MgO is formed on the electrode surface in addition to Al_2O_3 [15].

Potentiostatic and Galvanostatic Studies Cyclic voltammetry results can be better explained by potentiostatic and galvanostatic studies. To elucidate the process of nucleation and growth of the film in the region of the peak (P), for example, the working electrode has been investigated using the potential step method. The potential was held at an initial potential of −1500 mV/SCE and then, by stepping the system to −800 mV, potentiostatic transients (current as a function of time) of less than 1 minute to minimize the influence of impurities in the two alloys were obtained for the pure Al electrode and the AA5083. The amount of charge was calculated after pulsing the electrodes to different anodic potentials. This showed a similarity in behavior between the two alloys [15]. In the same manner, galvanostatic methods for localized corrosion can be carried out at constant chosen currents, and the potential evolution as a function of time is recorded until the time rate of change in potential approaches zero [13].

17.3.1.2. Determination of the Pitting Potential (E_p)

Electrochemical studies of pitting corrosion usually indicate that stable pitting occurs only above a critical potential or within a specific potential range. Since the mechanism of autocatalytic propagation of localized corrosion in Al alloys is valid for pitting as well as crevice and filiform corrosion, the determination of the pitting potential expresses the corrosion resistance of the Al alloy for several types of corrosion. Most commonly, potential–current curves are measured either by applying stepwise potentials (potentiostatically) or by applying a constant potential sweep rate (potentiodynamically) and recording the resulting current. The cyclic potentiodynamic polarization method is the conventional way to determine pitting potential in scanning the potential with an appropriate rate (~0.05–0.2 mV/s) from cathodic to more anodic ones and protection potentials during the scan return. It is the currently used method to determine E_p, in which the potential is made more positive and the resulting current density is measured for a specimen immersed in a deoxygenated electrolyte by bubbling argon or nitrogen (Figure 17.3). The critical current density, i_{crit} characterizes the active–passive transition. The pitting potential E_p (E_b, E_{bd}, E_{pit}) corresponds to the formation of possible stable pits that start to grow and is a function of a certain medium with a certain chloride concentration.

The more noble the breakdown potential, the more resistant will the alloy be to pitting and crevice corrosion. The protection potential E_{prot} (or repassivation, E_R, E_{rep}, E_{pp}) after reversal of the potential sweep direction is the value below which the already growing pits are repassivated and the growth is blocked [16, 17]. The potential at which the hysteresis loop is completed upon reverse polarization scan determines the potential below which, there is no localized attack for this scan rate (ASTM G5, ASTM G61). Also, allowing too much pitting propagation to occur along with the accompanying chemistry changes can influence the reversal in the scan rate [13].

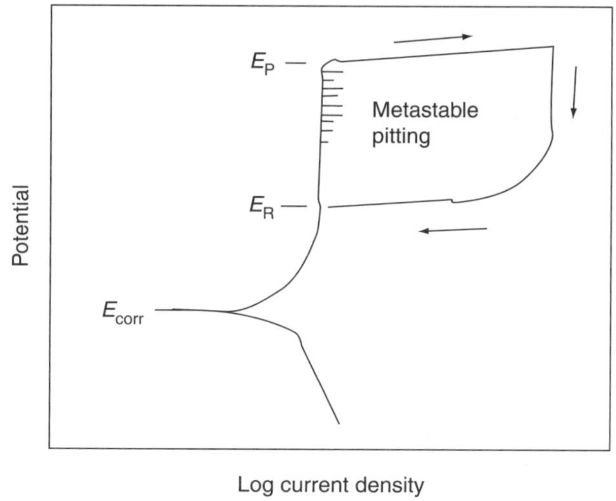

Figure 17.3 Schematic polarization curve for a metal showing active–passive transition as well as pitting in the passive potential range: for $E > E_p$ pitting will occur; for $E < E_R$ growing pits will repassivate [18, 19].

It is also recommended to polarize the individual alloy surface at potentials above and below the determined pitting potential value to validate the chosen scan rate and give some kinetic data on the initiation and propagation of pits at different levels. Another possibility is to initiate pits above the pitting or breakdown potential and then shift to lower values above or below the protection potential. It is assumed that at imposed values below the protection potential, one should observe current decrease until complete repassivation.

A relatively interesting new term called the *critical pitting potential,* E_{SCR}, that corresponds to a more stationary state and lies between the breakdown potential and the protection potential, should also be considered. The E_{SCR} is determined by scratching the alloy surface and exposing the exposed surface to a constant potential after the scratch repassivation method for localized corrosion. The current change is monitored as a function of time and this will show the influence of potential on the induction time and the repassivation time [13, 20]. Recent experimental work suggests that E_{pit} and E_{prot} would converge to a unique pitting potential [21].

Thompson and Payer [21] introduced a unique pitting potential, which corresponds to both the most active value of E_p determined after a long incubation time and the most noble value of E_r measured following only minimal pit growth, and which can correspond to the stationary pitting potential between the pitting and protection potential values [17]. The region between the protection and pitting potentials can correspond to *metastable pits* that are typically on the order of micrometers in size, at most, with a lifetime on the order of seconds or less; under certain conditions, they can continue to grow to form large pits. Thus the a term metastable is used to express the region where pits initiate or grow for a limited time and close because of repassivation. This is different from the fact that large pits can stop growing for different reasons, such as accumulation of corrosion products or pH change at the interface etc [18, 21].

ASTM G61 is a standard immersion test using cyclic potentiodynamic polarization (hysteresis) for pitting corrosion testing [4]. It is interesting to compare the resistance to pitting as a function of the critical concentration of chloride, which causes pit initiation on different alloys, and as a function of temperature. Increasing temperature usually increases the pitting tendency of metals and alloys, showing high pitting potentials at low temperature and decreasing potentials in the temperature range between 0 and 70 °C [6].

Pitting Potential of Aluminum Alloys The value of the pitting potential of AA1199 (99.99% Al) in sodium chloride solutions can vary from −0.35 to 0.54 V with respect to SHE depending on the activity of the Cl^- ion [17, 22].

Potentiodynamic and potentiostatic studies of the alloys 2024 and 7075 were conducted to determine the relative susceptibility of these two alloys to pitting, which can influence stress-corrosion cracking and corrosion fatigue [23, 24] (ASTM 101). The pitting potential in a nonaerated solution of 3% NaCl at 20 °C was −625 mV/SCE for 2024-T3, −585 mV/SCE for 2024-T4, and −657 mV/SCEF for alloy 7075-T6. Potentiostatic studies showed higher passive currents for 7075 than for alloy 2024. Analysis of the products in the pit showed high concentrations of chloride ions [25, 26].

Koteswara Rao et al. [27] have developed tests for comparing the corrosion performance of two different types of welding of Al alloys. These tests were conducted according to the ASTM G3. All the experiments were conducted in 3.5% NaCl solutions with pH adjusted to 10. A saturated calomel electrode (SCE) and a carbon electrode were used as reference and auxiliary electrodes, respectively. The potential scan was carried out at 0.166 mV/s. Figure 17.4 shows a higher corrosion current density during passivation for the

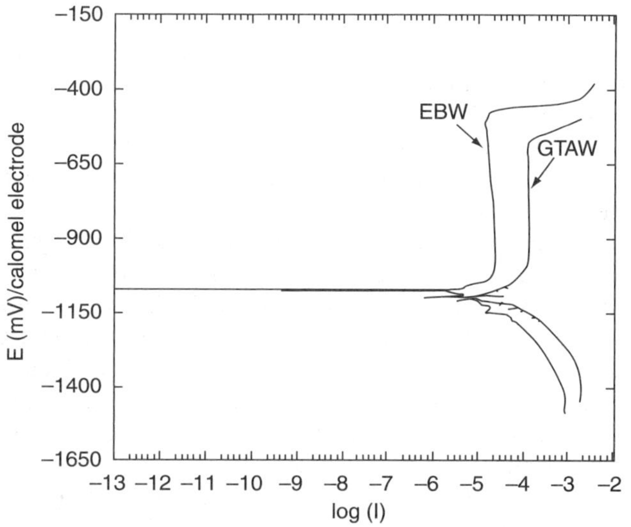

Figure 17.4 Dynamic polarization curves showing superior pitting corrosion behavior of an electron beam weld of AA2219 Specimen as compared to a gas tungsten arc weld at 0.166 mV/s [27].

gas tungsten arc welded (GTAW) alloy. Thus the electron beam weld is less susceptible to pitting corrosion than that of the gas tingsten are weld [27].

A microelectrochemical technique, applying microcapillaries for electrochemical cell measurements, has been developed in which small surface areas of a few micrometers or even nanometers in diameter are exposed to the electrolyte. This leads to strongly enhanced current resolution, down to picoamperes. Microelectrochemical techniques and statistical evaluation of the experimental results allow one to gain more insight into the mechanism of these processes [17].

17.3.1.3. EIS and Localized Corrosion

EIS is still being refined for localized corrosion studies generally and for pitting in particular. The statistical variation of pit nucleation and the absence of steady states prevent long measurements in the low-frequency region. In addition, in the pitting region, a complicated Nyquist plot is obtained and difficult to interpret. However, Lee and Mansfeld [28] demonstrated that characteristic changes exist in the low-frequency region. It should be underlined that the impedance spectra for pits in stainless steels and magnesium are different from those of aluminum [29].

In addition to the local activation or pit initiation process, the stability of the passive film is decisive for the corrosion resistance of passive metals and alloys. Fast and effective repassivation, necessary for highly corrosion-resistant alloys, may only occur if highly stable films are formed during repassivation. The stability of passive films is often reflected by the semiconductive properties of these films. Therefore EIS, photoelectrochemical methods, and in situ analytical techniques are very valuable tools to study the chemical and electrochemical behavior of these passivating oxide films [17].

The conventional EIS method involves some difficulties, one of which is that the equivalent circuit model can be designed in more than one way, so that the impedance spectra are difficult to analyze. Another difficulty is that conventional EIS requires a very long time to keep track of the test system, such as Al alloy–NaCl. EIS using the time domain method is a transient technique and can be used to measure unstable systems precisely.

EIS and Localized Corrosion of AA2024-T3 The evaluation of localized corrosion of aluminum alloy 2024-T3 following a corrosion test was performed by EIS in the time domain (quantitative and nondestructive technique). Since localized corrosion is a stochastic phenomenon, a statistical approach was considered and 50 parallel specimens were submitted to an alternating immersion test in 5% NaCl + 0.5% $(NH_4)_2SO_4$ solution for 28 days. Analysis of data by the time domain consists of a transient that can be used to measure an unstable system precisely. A special parameter at low-frequency impedance R_d ($f \leq 0.1$ Hz) was extracted from the impedance spectra to reflect the degree of corrosion [29]. Meanwhile, maximum corrosion depth (d_m) was also determined microscopically for the sake of comparison [30].

It has been found that both $1/R_d$ and d_m have similar statistical distribution and time-dependent growth. The dynamic variation curves of $1/R_d$ and d_m versus the test period were both somewhat S shaped. For simulation, the dynamic variation curve of AA2024-T3 should be divided into two parts to treat the data: pitting and intergranular corrosion corresponding to a sigmoidal curve and exfoliation corrosion giving a line. A quantitative relationship between $1/R_d$ and the maximum depth has been established [30].

17.3.1.4. Electrochemical Noise Measurements

Metastable and Stable Pitting Electrochemical noise measurements (ENMs) are required to investigate the rather fast kinetics of localized corrosion. Electrochemical noise (EN) is an interesting tool to monitor the metastable pitting by recording the galvanic current between nominally identical electrodes on the corrosion potential of a single electrode. Since localized corrosion sites are typically very small, on the order of 100 μm in diameter or less, the current densities inside these cavities can be on the order of 1 A/cm^2, and that can be detected [29, 31]. Sato [32] observed current transients below the pitting potential and attributed them to the formation of pit embryos. The measured current noise seems to give clearer information on the corroding system than the potential noise [21].

Noise analysis obtained from microelectrochemical investigations under potentiostatic conditions are recommended, as well as microscopic studies to distinguish between repassivating superficial pits and penetrating ones [17, 33]. The scanning reference electrode technique (SRET) is an appropriate complementary tool [29].

Pride et al. [34] focused on the analysis of EN signals associated with metastable pitting and the transition from metastable to stable pitting of Al-based materials. To achieve these goals, the susceptibility to pitting corrosion was varied through control of environmental variables (halide concentration, inhibitor addition, aeration), and metallurgical factors were examined using progressively more susceptible materials: high-purity Al, aged Al–2%Cu, and AA2024-T3. Two complementary approaches were carried out for EN studies: potentiostatic polarization and open circuit setups.

1. High-purity Al, roughly simulating Cu-depleted grain boundary zones in aged Al–Cu alloys, was potentiostatically polarized so that current spikes associated with individual pitting events could be analyzed.
2. ENMs of current and potential were carried out between nominally identical galvanically coupled Al, aged A1–2%Cu, and AA2024-T3 electrodes.

The capability of various single-event, statistical, and spectral analysis methods to forecast the transition from metastable to stable pit growth is investigated. In the case of the statistical and spectral analyses methods, a large population of metastable pitting events detected over a given time period were analyzed to gain insight into the pit stabilization process. The following colclusions were reached:

- In the case of the single-event approach, the growth rate, anodic charge, and pit current density for individual pitting events were determined for a large population of pitting current spikes obtained potentiostatically. Resulting pit properties are compared to electrochemical criteria for pit stabilization without taking into consideration the existence of any relationship between one pit and the next.
- Pit stabilization occurred when individual pits exceeded a threshold of $I_{pit}/r_{pit} > 10^{-2}$ A/cm at all times during pit growth as established from potentiostatic measurements. The magnitude of this ratio is linked directly to the concentration of the aggressive solution within pits. Two related statistical pit stabilization factors ($I_{rms}/r_{pit\ total}$ from ECN data and the mean of $(I_{peak} - I_{ox})/r_{pit}$ values from each pit current spike) were obtained from galvanic ECN data containing a large number of pit current spikes. These parameters provided a better indication of the transition to stable pitting than the pitting index or noise resistance but also had shortcomings.
- Spectral analysis using current and potential power spectral density (PSD) data provided qualitative information on pit susceptibility. However, the transition to stable pitting could not be accurately defined because of a lack of information on pit sizes in spectral data [34, 35].

Aballe et al. [36] showed that the exposure of AA5083 in oxygen-saturated 3.5% NaCl solution led to a localized alkaline corrosion (LAC), which was the principal corrosion process using weight loss, linear polarization, and EN methods, and surface observations were carried out after 15 days of immersion by optical microscope and SEM. They stated that the cathodic precipitates of Al(Mn, Fe, Cr) surrounding the pit are more effective depending on the degree of polishing from SiC 80 grit (Ra 2.02 μm) to 1200 grit (Ra 0.13 μm). The greater susceptibility of the more finely polished sample was interpreted as a function of the density of the cathodic intermetallic particles exposed on the smoother surface.

Electrochemical Noise and Impedance Studies Gouveia-Caridade et al. [37] examined the influence of chloride ion induced corrosion on the performance of pure Al at OCP conditions by ENM and EIS techniques as a function of pH. Solutions were prepared with and without acetic acid/sodium acetate buffer. The considered solutions were 0.1 M KCl at pH 5.4 and two similar solutions buffered at pH 5.4 and pH 4.3, respectively. The potential and current noise measurements were recorded simultaneously and the data were analyzed in the time and frequency domains. After 1 h immersion of the electrodes, the current and potential noise was sampled at a frequency of $f_s = 100$ Hz for 20.48 s, giving a total of 2048 points. This procedure was repeated for longer times up to 4 h. Electrochemical impedance spectra were recorded for different immersion times where a sinusoidal perturbation of 10 mV rms was applied at the cell rest potential over the frequency range from 65.5 kHz to 0.1 Hz in 10 steps per decade.

The calculated values for R_n from ENMs were larger for the solution with pH 4.3 than the other two solutions, which should give a lower general uniform corrosion rate for the lower pH. This can be due to the formation/dissolution of aluminum oxide and the presence of metastable pits on the surface because of the less stable passive film. The PSD plots and the spectral noise plots were fitted to the considered data in the literature [38] (see Chapter 16), thus giving the following slope relation: $S_{rsn} = 0.5(S_V - S_I)$. It has been suggested that the absolute magnitudes of the slopes of PSD are characteristic of the type of corrosion.

The analysis in the frequency domain showed that the much larger current slope of PSD was that of pH 4.3, and this can be explained by the more corroded surface of Al and the presence of metastable pitting. The *characteristic frequency* is inversely proportional to the PSD and independent of the current noise and has generally been applied as a parameter to evaluate the degree of localized corrosion. The lower the characteristic frequency, the more localized corrosion was found for the solution having the lower pH (although the difference between it and that of the buffered solution at pH 5.4 was not large). However, this was confirmed by visual observation and calculated corrosion rates determined by Tafel plots [37].

Impedance spectra of Al obtained for the different immersion times of the three solutions have been fitted by an equivalent circuit, which comprises the cell resistance in series with a constant phase element modeled as a capacitor. Comparison of the results of the impedance modulus and spectral noise plots for Al exposed to the three solutions gave the same trend after 1 h as that after longer periods up to 4 h. A model has been developed based on localized corrosion, which equates the noise impedance and the electrochemical impedance.

The R_{sn} values from ENMs were found to be higher than those of the impedance modulus Z with a vertical line for every side. A possible explanation for this difference is that in the experimental impedance setup, a fixed potential is imposed that is equal to the OCP at the beginning of the experiment, and this may contribute artificially to a more stable passive film. Under ENM conditions, the OCP varies as under natural exposure conditions of the materials; the local breakdown could give rise to the formation of fresh aluminum oxides, and so increase the resistance of the surface with a larger oxide film [37].

Nagiub [39] examined the localized corrosion of AA2024 in artificial seawater (VNSS) at pH 7.5 and 0.5 M NaCl with EIS and ENM techniques. A Pit growth law was obtained for Al alloy exposed to both solutions. It has been stated that sodium chloride solution is more aggressive than the artificial seawater (VNSS), something that has been observed for other alloys with other techniques under somewhat different experimental conditions. Also, the values of the noise resistance (R_n) from the ENMs are in agreement with the polarization resistance R_p corresponding to the pitted area (R_{pit}) obtained from EIS studies [39, 40].

The noise resistance R_n was assumed to be close to the polarization resistance R_p and so corrosion rate can be calculated by inserting R_n in the Stern–Geary equation as $I_{corr} = B/R_n$ [41]. Lee and Mansfeld [28] showed that R_n will be equal to R_p for a system with high corrosion rates for which the impedance reached the dc limit within the frequency range. Also, some factors should be avoided that can influence the value of R_n, such as the dc drift, sampling rate, and instrument noise [42, 43].

Inhibitors and Weibull Probability Plot Na and Pyun [44] studied sulfate and molybdate ions as inhibitors for pitting corrosion of pure Al in 0.1 M NaCl solution by using potentiodynamic polarization measurement and ENM. Various amounts of sulfate or molybdate ions (0.00, 0.01, 0.05, and 0.1 M) were considered. ENMs were made using the OPC setup and two galvanically coupled Al specimens. V and I were simultaneously recorded for 48 h and the results were supported by conventional potentiodynamic polarization studies from –1.2 to 0.4 V/SCE at a scan rate of 0.5 mV/s.

The sampling interval was 0.5 s and each time record consisted of 2048 data points. For every solution, 174 time records were analyzed. A Weibull probability plot was constructed on the basis of a stochastic theory to resolve the noise signal. The change in the linear slope in one Weibull probability plot with the frequency of events has been discussed in terms of the coupled processes of uniform and pitting corrosion. It has been found from the two linear regions in one Weibull probability plot that pitting corrosion was clearly distinguished from uniform corrosion in the stochastic analysis [44]. From the statistical variation in the frequency of events, it was found that the width of the distribution of the frequency of events was broader in the presence of sulfate ions. On the other hand, the distribution of the frequency of events shifted to a lower frequency in the presence of molybdate. It was then concluded that the two examined inhibitors influence the pit initiation differently. Sulfate ions can retard pit initiation by competitive adsorption with chloride ions, but accelerate the growth of preexisting pits while molybdate ions can create thick insoluble films of molybdenum species over the aluminum oxide passive film and/or retard the diffusion or the ingress of Cl^- through the oxide protective film to the Al–electrolyte interface [44].

The analysis of ENMs obtained from pure aluminum during the breakdown of oxide film in aqueous neutral chloride solution was carried out and hydrogen evolution in alkaline solution was considered [45]. Pyun and Lee [46] have used the conditional probability density to predict a pitting failure based on the stochastic theory. In this method, the Weibull probability was introduced based on the stochastic theory as a method for analyzing EN data rather than the PSD usually used in the conventional method; however, additional analysis is still needed [45].

The plot of the cumulative probability numerically calculated was first transformed from the domain of the frequency of events to the mean free time domain, and then the Weibull probability plot was constructed by fitting the Weibull distribution function to the calculated cumulative probability. Finally, the conditional event generation rate was numerically determined as a function of time [45]. Among noise-generating processes, shot noise is caused by the fact that the current is carried by discrete charge carriers. Shot noise is an individual event independent of other events, like the stochastic processes. The shot noise produced during breakdown of the oxide film, pit initiation, and hydrogen evolution can be related to the average mean corrosion current:

$$I_{\mathrm{corr}} = q f_{\mathrm{n}}$$

where q is the average charge in each event and f_{n} is the frequency of events. Also,

$$\bar{I}_{\mathrm{corr}} = B/R_{\mathrm{n}} \quad \text{and} \quad \bar{q} = \sigma_I \sigma_E / Bb$$

where B is the Stern–Geary coefficient, R_{n} is the noise resistance parameter, σ_I is the standard deviation of current, σ_E is the standard deviation of potential, and b is the bandwidth of measurement. The value of f_{n} can then be deduced:

$$f_{\mathrm{n}} = \bar{I}_{\mathrm{corr}}/\bar{q} = B^2 b/{\sigma_E}^2$$

The cumulative probability $F(f_{\mathrm{n}})$ at each f_{n} can then be numerically determined from the set of f_{n} data. The Weibull distribution function is a widely used cumulative probability function for predicting lifetime in reliability testing and has the advantage of being able to analyze data even where two or more failure modes are present. The cumulative probability $F(t)$ of a failure system can be introduced as the Weibull distribution function based on a "weakest link" [46] and the following equation can be deduced:

$$\ln\{\ln[1/(1-F(t))]\} = m \ln t - \ln n$$

where m and n are the shape and scale parameters, respectively, m is a dimensionless parameter, and n is expressed in s^m. These two parameters can be determined from the slope of the linear $\ln\{\ln[1/(1-F(t))]\}$versus ln t plots (Weibull probability plot) and from the intercept on the $\ln\{\ln[1/(1-F(t))]\}$ axis.

The conditional event generation rate $r(t)$ is employed as a kind of failure rate in reliability engineering corresponding to $r(t) = m/n\, t^{m-1}$ [45]. The value of $r(t)\, dt$ represents the generation probability of events in the next unit time dt for the specimens in which events have not yet been generated when t has elapsed. This just corresponds to the pit embryo formation rate in pitting corrosion [46].

From two linear regions in one Weibull probability plot, it was concluded that breakdown of the oxide film and hydrogen evolution, respectively, were clearly distinguished from uniform corrosion in the Weibull probability plot. As an example, the role of anion additives as inhibitors in the breakdown of the oxide film was discussed in terms of distribution of the mean free time and the conditional event generation rate [45].

The Weibull probability plots for pure aluminum in aqueous 0.1 M NaCl solution containing various concentrations of sulfate or molybdate ions showed two satisfactory linear regions for every plot. These are likely to indicate the existence of two modes of corrosion failure. It is suggested that the slopes in the relatively shorter t_m are associated with uniform corrosion, while the slopes in the longer t_m range correspond to localized corrosion such as metastable pitting or pitting initiation. It is stressed that the two slopes do not rigorously represent pure uniform corrosion or pitting corrosion, respectively. Also, the conditional event generation rate was calculated as a function of time t and it was observed that the value of $r(t)$ in the presence of sulfate or molybdate ions was remarkably lower compared to the reference solution without additive at a given time t. This indicates that metastable pitting or pit initiation is suppressed in the presence of these inhibitors [45].

Na and Pyun [47] used ENMs to examine the susceptibility to pitting corrosion of three aluminum alloys, AA2024-T4, AA7075-T651, and AA7075-T761, in 0.05, 0.1, 0.5, and 1 M NaCl solutions at room temperature. The EN data were analyzed based on the combined stochastic and shot noise theories using the Weibull distribution function [44]. It was suggested that two stochastic processes of uniform corrosion and pitting corrosion exist. Pitting corrosion was distinguished from general uniform corrosion in terms of the frequency of events in the stochastic analysis. Based on the conditional probability concept, the susceptibility to pitting corrosion was appropriately evaluated by determining pit embryo formation rate in the stochastic analysis. Parallel and as a support to ENMs and analysis, potentiodynamic polarization and electrochemical impedance studies were carried out. It has been concluded that the susceptibility to pitting corrosion was decreased in the following order: AA2024-T4 (the naturally aged condition), AA7075-T761 (the overaged condition), and finally AA7075-T651 (the near-peak-aged condition) [47].

Wavelet Transform and Transient Shapes of Pitting Types The development of a passive film and two types of pitting (crystallographic pitting and alkaline pitting) are examined. Crystallographic pitting has already been mentioned to occur in the 100 crystallographic planes at −0.720 V versus Ag,AgCl/KCl reference electrode, while alkaline pitting is considered to be a consequence of the galvanic cell between the Al(Mn, Fe, Cr) and the matrix. The intermetallic particles facilitate or depolarize the cathodic reduction of oxygen. Aballe et al. [48] studied AA5083 at the OCP in 3.5% NaCl solution using ENM statistical parameters, wavelet transform, and transient shapes to examine the influence of 500 ppm Ce^{3+} ion addition in the form of cerium chloride. The samples were exposed for 4 days in order to follow the time evolution of the system. Six consecutive voltage-time and current–time records were taken every 10 hours and an average value of each parameter was considered for each 10 h record. Both potential and current for an open circuit were acquired at a sampling of 2.15 points per second for about 15 minutes, giving 2048 data points for every register [48].

A study of the noise fluctuation magnitude indicates that the activity was much lower in the presence of cerium ions. This has been explained by the loss of intermetallic particles from the alloy surface in the corrosive medium, while these particles were blocked in place in the presence of Ce^{3+}. The shape of transients suggests that the inhibiting effect of cerium chloride was hindering alkaline pitting before and during the attack. It was observed that the cerium cations precipitate onto the Al(Mn, Fe, Cr) particles, probably hindering the cathodic reaction from occurring [48]. The voltage median evolution of AA5083 samples exposed to a 3.5% NaCl solution suggests that the main corrosion process is crystallographic pitting for about the first day, followed by alkaline pitting. The evolution of OCP, current, and potential noise parameters showed that the corrosion activity decreases with time with or without cerium ions [48].

The Energy Distribution Plot and Corrosion Types Rod specimens from commercial alloy 2024-T3 were studied in 2% sodium chloride solution by EMNs [9, 49]. The potential noise fluctuations during at the OCP of the alloy were recorded and analyzed using the wavelet transform technique. Between EN experiments, EIS measurements were performed in the direction of decreasing frequency for the range of 10^5 to 0.015 Hz.

It has been found that the noise signal is composed of distinct types of events, which can be classified according to their scales (their time constants). The energy distribution plot can be used as the fingerprints of the EN signal to differentiate between the corrosion types. During general corrosion, all the transients of the Al alloy have some contributions to the original signal while the event with the largest time constant makes the largest contribution. Concerning EIS studies, two depressed semicircles can be observed in the Nyquist plot within the passive region. The high-frequency one is related to the dielectric performance of the surface film, while the other can be associated with the oxide film on the AA12024-T3 surface. A third depressed semicircle appeared with prolonged immersion time, very probably due to the formation of corrosion products [9, 49–51].

Zhang et al. [9] studied the magnesium alloy LY12 in 1 N sodium chloride solution with EN and impedance techniques. They found that the fractal dimension (D_n) obtained from power spectral density (PSD) was directly proportional to the intensity of pitting corrosion or the value of the pitting parameter (S_E) derived from dimensional analysis, while the fractal dimension (D_E) obtained from EIS is mainly related to the uniform corrosion [9, 52].

Shi et al. [53] studied corrosion of AA2024-T3 in simulated acid rain under cyclic wet–dry conditions by ENM, potentiodynamic polarization, and SEM techniques. Average and standard deviation as well as wavelet transformation have been employed for EN analysis. The results showed that the main cathodic reaction changes from the reduction of protons to that of oxygen molecules with the increase of pH from 3.5 to 4.5 or to 6. The results showed also that the corrosion was much more vigorous at pH 3.5 than that at pH 4.5.

The potential standard deviation showed that, with the increase of pH, the number and the frequency of occurrence of peaks above the baseline dramatically decrease. This was interpreted because of high diffusion and migration rate of hydrogen ions in solution and through the less stable passive film at pH 3.5.

The time evolution of potential three-dimensional plots (EDP) at the three considered pH values gives interesting assumptions on the kinetics and mechanism of pitting. At pH 3.5, three consecutive stages were observed that can indicate a balance region between the pitting process and the diffusion of cathodic reagents, followed by a region controlled by diffusion of reagent through the formed corrosion products; the final stage mainly concerns the initiation of new pits in the already formed ones. For higher pH systems (4.5 and 6), the energy is mainly accumulated in the finer crystals all through the tests except at the beginning. This is because pitting is the main corrosion type, although quite limited, and the diffusion is not a controlling process due to the presence of rare corrosion products [53].

17.3.2. Crevice Corrosion

Crevice corrosion due to corrosion products can also be localized easily, and physical and chemical methods of investigation could be better for this type of corrosion. Staining of Al sheets of several wrought alloys, because of crevice corrosion during transport or storage (e.g., 3xxx; see Chapter 5), is observed frequently for Al sheets and can be identified by simple or magnified visual examination and photographs. However, measurement of reflectivity or image clarity is a more quantitative evaluation [1].

ASTM G78 describes crevice corrosion testing using a number of geometries. It tests iron- and nickel-based alloys in seawater, which could also be considered for aluminum or magnesium and their alloys since specimen preparation and the crevice assembly can be used in any environment [54]. Similar to that mentioned for pitting corrosion, two identical potentials are considered for crevice corrosion resistance: crevice/pit initiation potential (E_{crev} or E_p) and crevice/pit repassivation potential (E_{rcrev} or E_{rp}). Also, the pH below which an alloy does not exhibit any passivity is called the *depassivation* pH_d, and this can be measured by generating polarization curves in various simulated crevice solutions with decreasing pH using ASTM G5 or G61. The pH_d can also be determined under galvanostatic polarization while simultaneously decreasing the pH by HCl addition [55]. In the latter method, there is also the influence of the chloride or halide ion addition; however, there has been no standard for the determination of pH_d until now.

ASTM G61 could be used to determine the repassivation potential using the cyclic potentiodynamic method. However, when hysteresis between the forward and back scans is observed, care should be taken to prove that this corresponds to crevice corrosion, since, for example, hysteresis could be observed in highly corrosion-resistant alloys due to transpassive dissolution during continuous scanning at potentials higher than 1 V/SCE [55].

The Materials Technology Institute (MTI) of the Chemical Process Industry has identified five corrosion tests for iron- and nickel-based alloys, from which MTI-4 could be the most appropriate for Al or Mg alloys. The MTI-4 method uses an increase in neutral bulk Cl^- concentration at eight levels, ranging from 0.190 to 3% NaCl, to establish the minimum critical Cl^- concentration that produces crevice corrosion at room temperature (20–24 °C) [56].

17.3.3. Filiform Corrosion Testing of Al Alloys

Generally, this depends more on the type and thickness of the coating, surface preparation, and adhesion than on the electrochemical properties of the metallic substrate. However, it is found that higher Cu content Al alloys are more susceptible, showing very possibly the influence of galvanic cells [1]. Humidity, chloride concentration, and acidity are environmental accelerating parameters for initiation and propagation. The preferred initiator of corrosion is the salt fog atmosphere described in ASTM B117 (for 4–24 h) or dipping in a salt solution without rinsing. If susceptible, filiform filaments will gradually grow out perpendicularly from the scratch. Many of these filaments will later orient themselves in the rolling direction of the panel. The susceptibility of Al to filiform corrosion can be determined by placing several coated and scratched panels in a salt fog chamber as described in ASTM D2803 [57].

This form of corrosion can occur during exposure to seacoast atmosphere, and it has been developed at inland atmospheric exposure sites by spraying the specimens periodically (about three times per week) with a 3–5% solution of sodium chloride. There is a growing interest in Al for auto body sheet, together with the need to maintain an aesthetically pleasing painted surface [1]. Laboratory tests that has been used are the 3.5% NaCl alternating immersion test, ASTM G44, or exposure to hydrochloric acid vapors for 24 h followed by prolonged exposure to high humidity at slightly elevated temperatures of about 50–65 °C [58].

ASTM D2803 is the standard guide for testing filiform corrosion resistance of organic coatings on metals. Coated metal specimens are scribed and placed in a corrosive atmosphere to initiate corrosion. The specimens are then exposed to controlled temperature and humidity conditions (70–95% range) for 6 weeks.

17.4. METALLURGICALLY INFLUENCED CORROSION

17.4.1. Intergranular Corrosion Testing

Intergranular corrosion testing susceptibility depends primarily on the type of alloy and fabrication process and can occur in most environments. Testing for intergranular corrosion susceptibility varies with the alloy family and the fabrication process and can occur in most environments [1]. ASTM 669 for measurement of corrosion potentials of Al alloys is useful; however, there is limited need for electrochemical methods for predicting intergranular corrosion susceptibility, considering that the process is fundamentally related to the presence of galvanic cells [59].

An immersion period of $\sim$ 6 h in a solution containing 57g of sodium chloride and 10 mL of hydrogen peroxide (30%) per liter following ASTM G110 ("Practice for Evaluating Intergranular Corrosion Resistance of Heat Treatable Aluminum Alloys by Immersion in Sodium Chloride + Hydrogen Peroxide Solution") is most often used to determine inherent susceptibility to this form of corrosion. It is primarily for testing wrought heat-treatable alloys (2xxx, 7xxx, and 6xxx) but may be used for other Al alloys including castings but the degree of susceptibility in this accelerated test may not be representative of performance in outdoor atmospheres [10, 59]. A metallographic cross section approximately 20 mm in length, preferably through a corroded area, should be examined. The type, extent, and depth of intergranular corrosion can then be determined. Determination of the effect of intergraular corrosion on mechanical properties could be carried out (ASTM G110) [1].

A rather unique mass-loss test method, ASTM G67 ("Test Method for Determining the Susceptibility of Intergranular Corrosion of 5xxx Series Aluminum Alloys by Mass Loss After Exposure to Nitric Acid, NAMLT Test"), has been developed to provide a quantitative measure of the intergranular susceptibility of 5xxx series Al–Mg and Al–Mg–Mn alloys. The technique is useful to determine whether material has become sensitized by exposure to elevated temperature. Reproducibility of the mass-loss results decreases with increased susceptibility to intergranular attack [10].

17.4.2. Exfoliation Testing

Exfoliation corrosion occurs only for certain tempers and alloys, and protective coatings do little to prevent this type of corrosion. Resistance to this form of corrosion should be considered because the resultant damage can be quite severe [10]. Exfoliation corrosion occurs in products having highly directional grain structure. The corrosion begins as lateral intergranular corrosion on subsurface grain boundaries parallel to the metal surface, but entrapped corrosion products produce internal stresses that tend to lift off the overlying metal. This spalling off of the metal creates fresh metal surfaces for continued corrosion; consequently exfoliation corrosion does not become self-limiting.

The standard guide for conducting exfoliation corrosion tests on aluminum alloys, ASTM G112, provides guidelines regarding specimen preparation, significant exposure periods, and inspection and evaluation techniques in both laboratory accelerated environments and in natural, outdoor atmospheres. The degree of susceptibility to exfoliation of some materials will vary in the different accelerated tests. Performance in seacoast atmosphere is generally regarded as the baseline against which the appropriateness of any accelerated test should be judged. Two years of exposure at the seacoast is considered the minimum exposure period for this purpose, and longer seacoast experience of 4 years or more is desirable [1, 10, 60, 61].

ASTM standards have been prepared for several short-time-laboratory tests that range from 1 day to 2 weeks of exposure; namely, ASTM G34 for 2xxx and 7xxx alloys, ASTM G66 for 5xxx alloys, and three methods of salt spray (fog) testing, ASTM G85. The standard test method for exfoliation corrosion susceptibility in 2xxx and 7xxx series Al alloys (EXCO Test) is ASTM G34. Exfoliation corrosion testing is intended for use with high-strength 2xxx and 7xxx ingot metallurgy alloys. It describes a procedure for constant immersion exfoliation corrosion testing and applies to all wrought products such as sheet, plate, extrusions, and forgings produced from conventional ingot metallurgy process. The test solution is 4.0 M NaCl, 0.5 M potassium nitrate (KNO_3), and 0.1 M nitric acid (HNO_3) and the test period is 96 h for 2xxx and 48 h for 7xxx alloys [62].

The EXCO test provides a useful prediction of the exfoliation corrosion behavior in marine and industrial environments, for example. The visual rating of corroded specimens after 48–96 h could show no appreciable attack, pitting, or exfoliation. EXCO test kinetics were studied by interrupting the test and by nothing the corrosion appearance and characterizing the corrosion morphology on metallographical and FEG/SEM cross sections determined at mid-thickness (ASTM G34).

The standard test method for visual assessment of exfoliation corrosion susceptibility of 5xxx series Al alloys (ASSET Test) is ASTM G66. The test covers a procedure for continuous immersion exfoliation corrosion testing of 5xxx series Al–Mg alloys containing 2.0 % or more Mg and applies only to wrought alloys. The specimens are immersed for 24 h at 65 ± 1 °C in a solution containing 1.0 M ammonium chloride (NH_4Cl),

0.25 M ammonium nitrate (NH_4NO_3), 0.01 M ammonium tartrate ($(NH4)_2C_4H_2O$), and 0.09 M hydrogen peroxide (H_2O_2).

The visual appearance of corroded specimens can show the absence or presence of appreciable attack, pitting, and exfoliation. The susceptibility to exfoliation is determined by visual examination using performance ratings established by reference to standard photographs. Highly cold-worked tempers of certain 3xxx and 5xxx alloys can incur a less aggressive form of exfoliation that proceeds in a transgranular mode, following selective precipitation along slip bands. Standard practice for modified salt spray (fog) testing, ASTM G85, is applicable to Al alloys. The variations described in this test are useful when a different or more corrosive environment than the salt fog described in ASTM B117 is desired [62].

SCC laboratory tests that have been used are the 3.5% NaCl alternating immersion test, ASTM G44, or exposure to hydrochloric acid vapors for 24 h followed by prolonged exposure to high humidity at slightly elevated temperatures of about 50–65 °C [1].

17.4.3. Joining and Testing

The corrosion behavior of a friction stir welded AA7108-T79 specimen has been investigated using a modified ASTM G34 EXCO accelerating test accompanied by electrochemical measurements for evaluating intergranular corrosion. The modified EXCO test employs 15 vol% dilution of a solution of 4.0 M NaCl, 0.5 M KNO_3, and 0.1 M HNO_3. Corrosion testing revealed that the edge regions of the thermomechanically affected zone were most susceptible to exfoliation (intergranular) corrosion and extended into the heat affected zone of the weld [63] (see Chapter 6).

17.5. MIC AND BIODEGRADATION EVALUATION

Evaluating microbiologically influenced corrosion (MIC) is expected to determine whether a particular metal/microbe is dangerous enough that a corrosion control program has to be employed. MIC results in an unacceptable level of corrosion and could lead to specific failures. Sampling is critical for appropriate MIC studies. Since there are no definitive tests for MIC, several types of circumstantial evidence–metallurgical, microbiological, chemical, and electrochemical—are generally used [64].

It should be underlined that sampling of MIC requires clean, sterile, sealable, carefully labeled containers for corrosion products and electrolyte close to corrosion sites. Labels on the sample containers should indicate, for example, sample origin, date, sampling time, and tests to be performed on the sample. Most sampling containers are glass or plastic. Sterile plastic bags provide lightweight sample containers. Commercially available devices such as autoclaves can be controlled manually or automatically to produce temperatures and pressures required for sterilization of liquids and solids. Water or process fluid samples should be collected without introducing microorganisms using a sterile needle and syringe [65]. High-temperature values of aqueous solutions should be avoided during sampling since certain microorganisms do not survive at temperatures above ~40 °C.

Deposits from the corroded areas are primordial for testing purposes because causative organisms are expected to be found in relatively high numbers at the corrosion site. In liquid phase, planktonic organisms are microscopic organisms that float or drift in water while sessile microorganisms grow on the surface. A given bacteria species can be a part of the plankton and become sessile after binding to a given surface. Liquid samples have one major drawback: numbers and types of organisms present in the fluid (planktonic) may have little correlation to those on surfaces (sessile). Since biocides are more effective in killing planktonic organisms than sessile organisms, samples of systems treated with biocides may indicate low numbers in water samples while there are high numbers growing on metal–liquid interfaces. Periodic removal of sample plugs could monitor general biofouling tendencies, but its worth is limited because of the patchy nature of biofilms. Generally, sample plugs should either be preserved for SEM examination for solid deposits or swabbed for microbiological characterization [65].

Metallographic Evidence Metallurgical testing has been used extensively to identify potential MIC. Some types of MIC are recognizable, in part, by the pattern of corrosion products on the surface, which can easily be examined with a low-power microscope (5× to 60×) [64, 65].

Microbiological Evidence Data must be gathered while the corrosion site is still wet. It is important to photograph the initial appearance of the corrosion site while the organisms are still alive [64]. Biofilm community structures can be analyzed using cluster analysis of the phospholipid fatty acid (PLFA) profiles. Direct molecular characterization of natural microbial populations can be accomplished with sequence analysis of 5SrRNA (to identify with high precision the types of microorganisms involved in the corrosion process). More recently, fluorescent dye-labeled oligonucleotide probes have been used for microscopic identification of single cells and characterization of mixed populations. Polymerase chain reaction amplification, comparative sequencing, and whole cell hybridization have been combined to selectively identify and visualize sulfate-reducing bacteria (SRB) in both established and developing multispecies biofilms [65].

Techniques for analyzing microbial metabolic activity at localized sites are being developed. Little et al. [65] incubated microbial biofilms with C-metabolic precursors and autoradiographed the biofilms to localize biosynthetic activity on corroding metal surfaces. The localized uptake of labeled compounds was related to localized electrochemical activities associated with corrosion reaction [65]. In complement, some laboratory methods in microbiology can be used to demonstrate the presence of MIC or to study the MIC process: culture methods, adenosine-triphosphate (ATP photometry) examination, fluorescently labeled rRNA-targeted nucleic probes, enzyme methods, immunological techniques, adenylyl sulfate (APS) reductase antibodies to detect SRB, monoclonal antibody probes, and hydrogenase assessment [66, 67].

Chemical and Spectral Evidence Detailed chemical analysis should be done for the corrosion products and any biological mounds present at or near the corrosion site. Evaluation of the chemistry of the liquid phase and its variability, both spatially and with time in relation to the observed corrosive attack, is necessary. Details should include the color, texture, odor, and distribution of the materials as well as their organic and inorganic chemistries [64]. For example, the color change of corrosion deposits from black to brown is, in itself, a good indication of sulfide corrosion product. Tests for sulfate, total organic carbon (TOC), pH, sulfide, and oxygen concentration are also useful indicators of the potential for SRB growth [64, 68].

Infrared spectroscopy has been used for many years as an analytical technique in microbiology and can be used to detect bacteria and biomolecules at a metal surface. Time-of-flight secondary ion mass spectrometry and FFT impedance spectroscopy can also be used to study the biofilm and biofouling [66, 69].

Electrochemical Evidence Since microorganisms change rates of electrochemical reactions leading to corrosion, electrochemical techniques for monitoring in field applications and kinetic laboratory studies can be used to evaluate MIC influence. Common methods of accelerating corrosion reactions, such as increasing temperature or concentration or aggressive species, cannot be used. Conventional anodic or cathodic polarization techniques can produce misleading information because the very high fields produced at the metal surface during polarization are incompatible with viable microorganisms. The methods applying very small imposed changes in corrosion potentials and currents (polarization resistance), or at free corrosion potentials (electrochemical impedance and electrochemical noise measurements) are appropriate [65].

Microscopic Evidence Microscopic techniques such as fluorescence microscopy, scanning electron microscopy (SEM), atomic force microscopy (AFM), environmental scanning electron microscopy, and confocal laser scanning microscopy can be used to monitor MIC [66]. For example, the relationship of microorganisms to corrosion products can be examined with a scanning electron microscope SEM [65]. AFM images could afford more information on slight changes of the surface due to corrosion. This technique has good resolution efficiency. Liu et al. [70] have used AFM for the investigation of MIC caused by SRB on LY12 aluminum alloys.

If biosensor and electrochemical techniques are combined together, they can play a significant role in monitoring microbial corrosion. A monitoring scheme for controlling both biofouling and biocorrosion should include the generation of as many of the following types of data as possible [64, 67, 71, 72].

Sessile bacterial counts of the organisms in the biofilm on the metal surface can be made by either conventional biological techniques or optical microscopy accompanied by direct observation of the community structure of the biofilm. This can be done on metal coupons made from the same alloy used for the system. Sampling devices may be either directly implanted or side-stream implanted. Metal coupons, generally made with the same structural material of the system, have a known surface area, which enables an accurate count of sessile bacteria per square centimeter after biofilm detachment. Coupons are mounted in holding assemblies that are inserted in the pipework of the laboratory or industrial system [62].

Identification of the microorganisms should be carried out in both the process water and on the metal surface. Monitoring methods must provide information on well-established biofilms such as those that develop in water systems. Identification and analysis of corrosion products and biofilms should be followed by the evaluation of the morphology, form, and type of corrosion after removal of biological and corrosion product deposits [68, 73].

An effective monitoring program, either for the laboratory or for the field, must supply information on water quality and corrosive attack. Researchers at Shell Petroleum Company (Calgary, Canada) have developed an approach to risk assessment of carbon steel pipelines, based on oxidation–reduction potential measurements, water analysis, and operation parameters [68, 74].

Tests for nitrite-utilizing bacteria or sulfur oxidizers are complex. A microbial biosensor was developed for monitoring MIC of metallic materials in industrial systems. The *Pseudomonas* species isolated from corroded metal surfaces was immobilized on acetylcellulose membrane. The microbial biosensor response was observed by measuring the concentration of oxygen consumed by the microorganism for respiration. The biosensor showed short response time, high sensitivity, and, high specificity. It is easy to handle and is cheaper than enzyme-based biosensors.

The microbial biosensor was used also for the measurement of sulfuric acid in a batch culture medium contaminated by microorganisms. A linear relationship between the microbial sensor response and the concentration of sulfuric acid was observed. The response time of the biosensor was 5 min and was dependent on the immobilized cell loading of *Pseudomonas* species, pH, temperature, and corrosive environments. The microbial biosensor response was stable, reproducible, and specific for sensing of sulfur-oxidizing bacterial activity. Because of specificity, it showed poor response in the presence of other groups of microorganisms [75].

Electrochemical corrosion measurements were made using electrical resistance or polarization resistance type probes. For corrosion assessment, the electrical resistance method, widely used in the industry, is only appropriate for indicating a change in the general corrosion rate, but the results are difficult to interpret in the presence of localized corrosion such as pitting, the most frequent form of attack found in MIC. If biofilm or localized corrosion occurs, the polarization resistance reveals that something is happening but may not give an accurate determination of the corrosion rate. Monitoring programs for MIC have focused mainly on the assessment of planktonic populations in water samples and generalized corrosion by using corrosion coupons or some kinds of resistance or polarization resistance probes. However, as mentioned in sampling, planktonic populations do not properly reflect the types and numbers of organisms living in biofilms since a sessile community is built through the formation of a secreted gel containing 95% or even more water and a matrix of exopolysaccharidic substances EPSs that are partially protective [67].

It is appropriate to combine biosensoring with electrochemical methods. MIC is difficult to monitorize since it is initiated by the inhomogeneity of the environment created by the microorganisms immediately adjacent to the metal at the interface. In a bacterial *Thiobacillus* culture, both oxygen and pH change during growth. Oxygen is consumed by the bacteria and sulfuric acid is generated. Dubey and Upadhyay [75] showed that mild steel potential shifts to more active values (negative) due to the presence of *Thiobacillus*. Open circuit potential evolution and biosensoring are frequently very useful when employed together [75].

17.6. MECHANICALLY INFLUENCED CORROSION OF ALUMINUM AND ITS ALLOYS

17.6.1. Erosion Corrosion Testing

Exposure to various outdoor atmospheres is contained in ASTM G50. Generic types of atmospheres used are seacoast, industrial, urban, and rural. Sometimes specific geographical locations or local chemical conditions are important because they can produce unique results [76]. Field tests in outdoor atmospheres, especially seacoast and industrial atmospheres, are essential in certain situations and form the primary baseline for determination of long-term service performance [1].

ASTM G73 considers that liquid impingement erosion testing is essential to consider for aluminum alloys. The standard test method for cavitation corrosion and erosion corrosion characteristics covers aluminum pumps with engine coolants. An aluminum automotive water pump, driven at 4600 V/min by an electric motor, is used to pump the solution at 113 °C for 100 h and to serve as the object specimen in evaluating the cavitation erosion corrosion effect of the coolant under test. The test coolant is prepared by adding one part engine coolant concentrate for five parts corrosive water by volume. The water should contain 100 ppm each of sulfate, chloride, and bicarbonate ions added as sodium salts. This is to evaluate depressions, random pits, grooves, clusters of pits or scalloping, or both, and localized areas of metal removal in high-impingement regions. Pump case failure or pump case leaking are then evaluated (ASTM D2809).

The measurement of corrosion and wear, as well as erosion corrosion and corrosion–wear interactions, is a multistep process. Each component of the interaction must be measured separately. The results may then be combined to identify the synergistic effects and create a complete picture of the damage process. The standard ASTM G119 applies to systems in liquid solutions or slurries and some aspects of it can be adapted to dry corrosion and wear interactions as well [6].

Jet and whirling-arm tests are currently used in the field of erosion testing (ASTM G76). In the whirling-arm test, the impact velocity is well known, and an entire face of the specimen can be eroded, producing a more uniform surface. Machining is most commonly used method in the high-temperature abrasive test, because the process of machining produces an elevated temperature. One of the most commonly used slow abrasive elevated-temperature tests is the high-temperature ring-on-disk test [6, 77].

Equipment and procedure for assessing the erosion-corrosion behavior for aluminum and magnesium alloys as well as for steel have been given studied by Ramsingh and DeRushie [78]. These are adequately suited to screen coatings for use on light metals in the automobile industry (see erosion corrosion of magnesium alloys, Chapter 12).

Galling Stress Wear galling appears as a groove, or score mark, terminating in a mound of metal and this is a good measure of wear resistance of a given material pair [79].

17.6.2. Corrosion Fatigue Testing

Despite the estimated billions of dollars spent combating fatigue, a complete understanding of the scientific basis for reliable estimates of fatigue life for all conceivable load and environmental combinations is hard to come by. The following factors must be considered in corrosion fatigue testing:

- Stress-intensity range, load frequency, and stress ratio.
- Electrode potential in aqueous environment and intended environment composition.
- Chemical alloy composition, mechanical properties (e.g., yield strength), and microstructure [80].

On the basis of the limited evidence available, it seems that resistance-to-failure processes (fatigue, stress corrosion, and hydrogen embrittlement) decrease as specimen size increases and consequently tests on small specimens may not reflect the results that would be obtained with a much larger structure or component in service. For example, it is possible for brittle fracture of a large structure to occur at or near ordinary temperature, while subsequent laboratory tests of Charpy or Izod specimens show a transition temperature well below −18 °C. Because of the complexity of service environments, a good correlation of normal laboratory corrosion fatigue data with actual service performance is difficult. Low-stress, long-duration laboratory tests lower the fatigue strength of aluminum alloys [81].

Laboratory fatigue tests are carried out in ambient air and the cyclic stress may be axial, torsional, or reverse bending. Tests of fatigue consist of submitting the material to a certain frequency of alternate cyclic compression–tension stresses of different values and plotting, the *C–N* curve. When testing polished cylindrical specimens rotating under flex in a Moore machine, the stress just below that which causes failure at 500×10^6 cycles is taken as the fatigue strength. For sheet specimens subjected to reverse bending in a Krouse machine, the value is usually taken for survival at 50×10^6 cycles [6, 81, 82].

Corrosion fatigue tests are carried out with the same apparatus, with provisions made to subject the specimen to a corrosive environment by spraying or dripping a solution onto the specimen, creating a mist in an enclosure about the specimen, or immersing the specimen in a solution. The corrosive environments used most frequently are distilled or demineralized water, tap water, and brines (including natural or synthetic seawater).

Types of Tests The complete fracture approach considers cycles-to-failure tests and the crack growth test for the kinetics of crack propagation. In cycles-to-failure testing, specimens or parts are subjected to a sufficient number of stress cycles to initiate and propagate cracks until complete fracture occurs. Such data are usually obtained by testing smooth or notched specimens. In crack propagation testing, preexisting cracks or sharp defects in a material reduce or eliminate the crack initiation portion of the fatigue life of the component. The voluntary recommended practice ASTM F1801 and the standard ASTM F1160 are recommended for corrosion fatigue testing of metallic implant materials [6, 80].

Specimen Configuration A typical fatigue test specimen has three areas: the test section and two grip ends. The test section in the specimen is reduced in cross section to prevent failure in the grips ends. Round specimens for axial fatigue machines may be threaded, button-head, or constant-diameter types for clamping in V-wedge pressure grips. For rotating-beam machines, short, tapered grip ends with internal threads are used, and the specimen is pulled into the grip by a draw bar. Torsional fatigue specimens are generally cylindrical. Flat specimens for either axial or bending fatigue tests are generally reduced in width in the rest of the section but may also have thickness reductions [6, 80].

Fracture Mechanics Approach Fracture mechanics provides the basis for many modern fatigue crack-growth studies. ΔK is the stress-intensity range $(K_{\max} - K_{\min})$, where K is the magnitude of the mathematically ideal crack-tip stress field in a homogeneous linear-elastic body and is a function of applied load and crack geometry:

$$\Delta K = \alpha\sqrt{\pi\alpha(\sigma_{\max} - \sigma_{\min})}$$

where α is a function of the geometry of the rupture and the test sample, and σ is the stress.

Most corrosion fatigue crack growth (CFCG) rate investigations attempt to follow the general provisions of standard test method ASTM E647. In this constant load-amplitude method, crack length is measured visually or by an equivalent method as a function of elapsed cycles, and these data are subjected to numerical analysis to establish the rate of crack growth. Crack growth rates are then expressed as a function of crack-tip intensity range, ΔK, which is calculated from expressions based on linear-elastic stress analysis [6, 80].

Expressing the crack growth rate, da/dN (where a is crack length and N is number of cycles), as a function of ΔK provides results that are independent of specimen geometry, and this enables the comparison of data obtained from a variety of specimen configurations and loading conditions [80].

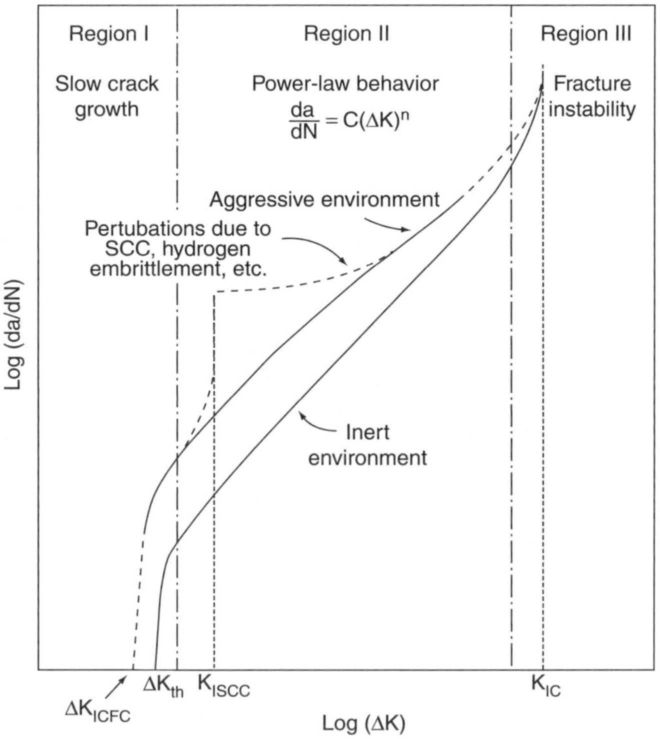

Figure 17.5 Corrosion fatigue crack grawth rate (*da*/*dN*) as a function of the cyclic crack- tip stress-intensity range (ΔK) [6, 80, 84].

Results of fatigue crack growth rate tests for many metallic structural materials have shown that complete *da*/*dN* versus ΔK curves have three distinct regions of behavior, as shown in Figure 17.5. In an inert (or benign) environment, the rate of crack growth depends strongly on K at K levels approaching K_{Ic} (plane-strain fracture toughness) at the high end (region III) and at levels approaching an apparent threshold, ΔK_{th}, at the lower end (region I), with an intermediate region II that depends on some power of K or ΔK on the order of 2–10 [83]. This is described by the power-law relationship [6, 80].

$$\frac{da}{dN} = C(\Delta K)^n$$

where C and n are constants for a given material and stress ratio. In an aggressive environment, the CFCG curve can be quite different from the pure fatigue curve. The growth or extension of a fatigue crack under cyclic loading is principally controlled by maximum load and stress ratio (minimum/maximum stress). However, as in crack initiation, there are a number of additional factors that may exert a strong influence, especially with the presence of an aggressive environment [6]. The environmental effects are quite strong above some threshold for SCC (K_{ISCC}) and may be negligible below this level (K_{ISCC} is the stress-intensity threshold for plane-strain environment-assisted cracking). In addition, certain loading factors, such as frequency, stress ratio, and stress waveform, can have marked effects on the crack growth curves in aggressive environments [6, 80, 84].

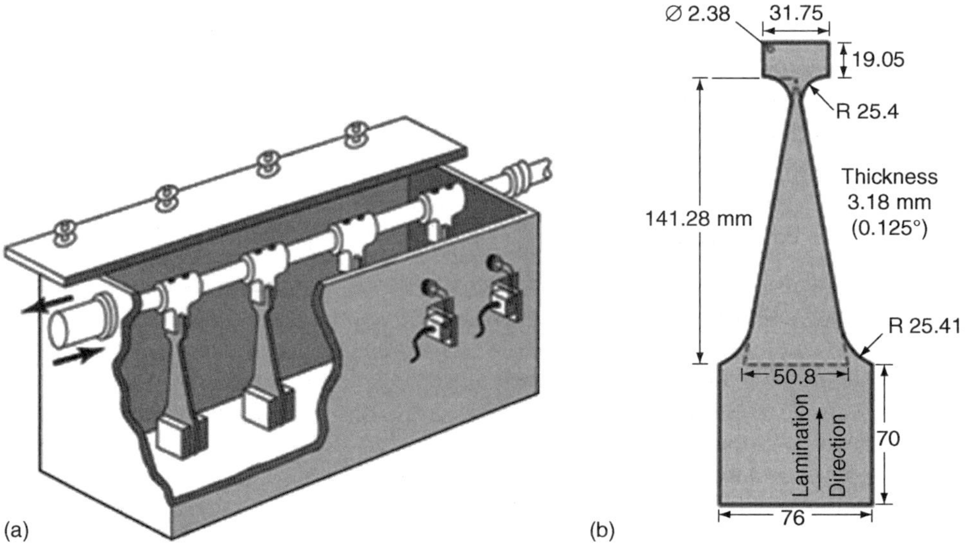

Figure 17.6 Corrosion fatigue test sample and test apparatus [6, 81].

Electrochemical Measurements During Corrosion Fatigue A bending fatigue test where the samples are immersed in the corrosive medium is generally more convenient than the tensile fatigue test. The advantage of this apparatus for corrosion fatigue Studies of cast or wrought alloys is that four samples can be tested at the same time, allowing a statistical approach for electrochemical measurements for open circuit potential or imposed potential or current during fatigue experiments. Corrosion fatigue (CF) test apparatus is schematized in Figure 17.6 [15, 85].

The preliminary step in the experiment is to determine the displacement caused by the desired applied load and the exact stresses are determined by strain gauges. The frequency of the applied cyclic bending is generally 60 cycles/min and the stress amplitude is $R = -1$. The number of cycles per minute can be lowered to just several cycles per minute for observing a dominant corrosion influence of the CF testing. The electrolyte is generally an aqueous solution of 3.5% sodium chloride utilized at ambient temperature and deoxygenated by the bubbling of argon for 1 hour before and during the test. Samples can be polished, sandblasted, shot peened, and so on.

The behavior of the alloy in CF testing can be studied at the free corrosion potential under different percentages of stress amplitude of the elastic limit. From potentiokinetic curves, $I = f(E)$, the protection or the pitting potential can be deduced and maintained for the entire duration of the stress test. Other levels of potentials between the pitting and protection potentials can also be considered [86]. The average difference between triplicates or more did not exceed 15% [81].

17.7. ENVIRONMENTALLY INFLUENCED CORROSION

The objectives of SCC testing programs include the following [87]:

- Determination of the risk of SCC for a given application and comparison of alloys and mill products.

- Examination of the influence of chemical composition, metallurgical processing, and fabrication practices for structural components and prediction of service life.
- Evaluation of protective systems and their effect on service life of the structure.
- Evaluation of claims for SCC performance of improved mill products.
- Studies of SCC mechanisms and development of new alloys that can offer longer, more safe, and more efficient service for certain environments and at the same time considering cost-effective evaluation of the material.

The most significant type of SCC data is the determination of the safe stress level below which failure will not occur in the concerned environment. However, stress threshold values for aluminum alloys are difficult to determine. The corrosion performance predictions should be obtained from previous data and by testing. The initiation period increases as the applied stress is decreased and can vary from a few hours to many years. Once cracking has begun, the crack grows linearly at a constant velocity independent of the applied stress. This continues until unstable crack growth occurs, leading to rapid failure. SCC failure times are not normally distributed, but the logarithm of failure times tends to have a normal distribution. This permits probability plots of cumulative percent failure (or survival) versus exposure time [1].

SCC testing can be divided into those conducted in natural environments, such as atmospheric exposure tests and seawater immersion tests, and those conducted under laboratory conditions or other fabricating operations [88].

17.7.1. SCC Testing Procedures of Aluminum Alloys

The essential requirements of accelerated laboratory testing are that the acceleration should produce the same mode of failure and reflect at least a known order of resistance of some alloys in service media [58]. The most common approaches employed to achieve SCC testing objectives are the use of high stresses, slow continuous straining, precracked specimens, higher concentration of species in the test environment than in the service environment, increased temperature, and electrochemical stimulation [87]. For electrochemical corrosion, the properties of the medium at the interface should be considered in accelerated tests.

Several aluminum alloy product specifications require defined levels of performance with respect to resistance to SCC. Standard tests used to measure such performance are described in standard methods and are referenced in materials specifications. For spray testing, ASTM B117 states the relevant laboratory testing conditions for the evaluation of SCC in neutral 5% NaCl. For alternate immersion, ASTM G44 recommends a widely used solution (3.5% sodium chloride at $\sim$ pH 6.5) for testing smooth specimens of aluminum alloys with improved resistance to SCC. This is considered in ASTM G47 for determination of the susceptibility to SCC which covers method of preparation and exposure of high-strength 2xxx series alloys (1.8 to 7% Cu) and 7xxx series alloys (0.4–2.8% Cu). Alternate immersion in 3.5% NaCl solution is given in ASTM G64 for aluminum heat-treatable alloys [89].

The test ASTM G64-91 describes the standard classification of resistance to stress-corrosion cracking of heat-treatable alloys for the 2xxx, 6xxx, and 7xxx series. The test concerns the intergranular path leading to ultimate fracture. The stress-corrosion ratings are based on laboratory tests of standard smooth specimens for susceptibility at specified stress levels. These stress levels are not to be interpreted as threshold stresses and are not recommended for design. The 3.5% NaCl alternate immersion test (ASTM G44) is usually chosen. Pitting or pitting plus transgranular cracking is not considered SCC failure. The ratings do not apply to metal in which the metallurgical structure has been altered by welding, forming, or other fabrication processes [6, 10].

Several criteria for products of 7xxx copper-containing alloys in T76, T73, and T736 tempers are based on combined requirements for tensile strength and electrical conductivity [90]. It has been shown that the relative performance of three different tempers of 7075 alloy plate can be influenced by the choice of a test specimen, differences in the outdoor atmosphere, and interlaboratory variations in performing the standard 3.5% sodium chloride alternate immersion test. A system of rating SCC resistance of high-strength aluminum alloy products has been incorporated into ASTM G64 [1].

For 7xxx series alloys containing less than 0.25% copper, the ASTM G103 method for performing a stress-corrosion test for low-copper-containing Al–Zn–Mg alloys in boiling 6% sodium chloride solution is used. Evidence shows that boiling 6% sodium chloride medium correlates better with outdoor atmospheric exposure than alternate immersion in 3.5% sodium chloride solution (ASTM G44) for the 7xxx series alloys containing little or no copper. This procedure is intended for a statically loaded, smooth, nonwelded or welded specimen of the 7xxx series containing less than 0.26% copper. The specimens are totally immersed in boiling 6% NaCl for up to 168 hours [1].

The test ASTM G139-96 determines SCC resistance of heat-treatable aluminum alloys using a breaking load method. The concept uses residual strength as the measure of damage evolution. Continuous immersion in boiling 6% NaCl solution for 4 days is widely used by U.S aluminum producers for testing smooth specimens of 7xxx series alloys containing no more than 0.26% Cu. It is intended for nonwelded or welded samples of Al–Zn–Mg–Cu alloys containing less than 0.26% Cu [89].

The test ASTM G58 is specific for welding and concerns the standard practice for preparation of stress-corrosion test specimens for weldments. The test considers the weld heat affected zone and the parent metal as produced by a specific welding process and the resistance of a deposited weld metal. The test determines the stress level for failure and identifies the zone of the SSC failure. It evaluates also the effect of notches and stress raisers. The test media and exposure times may vary from long-term tests in plant equipment or in outdoor environments to various laboratory tests. In the development of high-strength, welded aluminum alloys, it is necessary to determine the resistance to stress corrosion of the experimental combination of parent metal and filler alloys. An exhaustive evaluation was made on beam specimens, which were loaded both by constant deformation and by constant load [1, 91].

The standard test method for determination of susceptibility of metals to embrittlement in hydrogen-containing environments at high pressure, high temperature, or both is ASTM G142-96, which applies to all materials including wrought and cast alloys. The specimens of selected material are exposed to hydrogen-containing environment at high pressure or high temperature, or both, while being pulled to failure in uniaxial tension. The susceptibility to hydrogen embrittlement is evaluated through the determination of standard mechanical properties in tension such as yield strength, ultimate tensile strength, notched tensile strength, reduction in area, and elongation [1].

Testing in Liquid Metals or Melting Salts Tests in liquid or melting salt media such as alkali metals should be conducted in a glove box filled with high-purity argon. Immersion, alternating immersion, and spray testing can also be conducted. Frequently, after partial cracking, the specimens should immediately be washed in ethanol, dried, and then cracked in dry air so that the appearance of fracture surfaces produced in air and liquid or melted salt environments could be compared for the same grain or same grain boundary [92].

17.7.2. Test Specimens

ASTM B557 is currently used for the preparation of die-cast tensile specimens [93]. All tests are conducted with the material in the as-cast, F temper since die castings are not normally heat treated. The use of unmachined, as-cast specimens has the disadvantage of providing little control of property and specimen uniformity. However, it is deemed important to fully incorporate the gradients of structure, chemistry, and porosity, which are inherent in die-cast material. The SCC tests are conducted in a dead-weight tension loading apparatus with the test specimen mounted through a sealed hole in the bottom of a Plexiglas container [94].

For testing cylindrical tensile specimens, the solutions are contained in cylindrical glass tubes closed at each end with rubber stoppers and with facilities for measuring or controlling the potential with respect to an external reference electrode, generally saturated calomel. For tests on bar specimens, the test solution is contained in the region of the notch by pieces of adhesive tape attached to opposite sides of the specimen, the solution level in the notch being below the position of attachment of the gauge arms [94].

Smooth Specimens There are various types of smooth test specimens and different methods of stressing are available; however, standard ones should be preferred. Performance is based on visual cracking (ASTM G34). The commonly used types of specimens for tests under elastic-range stress are bend-beam specimens (ASTM G30 and G39), C-ring specimens (ASTM G38), O-ring specimens, tension specimens, and tuning fork specimens. Plastic strain specimens and residual stress specimens are also used for certain conditions. Static loading of precracked specimens as well as slow-strain-rate testing should be considered. Stressed O-rings have also been used to evaluate protective treatments for SCC prevention [95].

ASTM G49 ("Standard Practice for Preparation and Use of Direct Tension Stress Corrosion Test Specimens") is often used for study of the SSC mechanisms and covers the use of axially loaded, quantitatively stressed ASTM standard tension test specimens. The tensile bar is tested according to ASTM G49 and G44, and before it fails, the bar is tension tested to determine the amount of corrosion damage. ASTM G58 shows the instructions relative to welded specimens.

The specimens can be subjected to various loading conditions involving constant load, constant strain, or monotonically increasing strain to total failure in some of the slow strain rate tests. Other tests include cyclic loading as well as slow straining over a limited stress range. Care must be taken to electrically insulate the specimen from any metallic loading fixtures [6, 94, 95].

Precracked Specimens The use of precracked SCC specimens is necessary for some metals in which an SCC flaw will not initiate, but in which other types of flaws such as weld cracks can propagate by SCC. This is not necessary for aluminum alloys, which will initiate a SCC flaw on unflawed smooth specimen; however, the period of initiation can be long and variable. Most SCC testing of aluminum has been done using smooth specimens because these specimens and test methods are faster and less expensive.

Specimens are loaded to a percentage of the fracture toughness and the loading method can result in a constant, an increasing, or a decreasing level of stress intensity as the crack progresses. The last method is the usual one, using double-cantilever notched, precracked specimens. The data measured are crack lengths, from which crack velocity and the limiting stress intensity for crack growth can be calculated, and an international standard exists (ISO 7539-6). However, both methods tend to rank aluminum alloys in the same relative order of increasing resistance to SCC [1, 89].

Bolt-loaded double-beam type of precracked specimen testing has been performed extensively, and the ranking of materials by this method is generally in good agreement with that obtained with smooth samples. The bolt-loaded *K*-decreasing type of test is simple to perform and the results appear to be helpful to the SCC control in engineering structures [89]

17.7.3. Stressors

Corrosion acceleration for testing alloys is achieved through the use of various "stressors," such as cold work of the material, higher concentration of the aggressive ion, lower pH, higher temperature, or higher stress. Externally applied loads are easier to evaluate and to control, but residual stresses are those that normally are responsible for stress-corrosion failures under service conditions. It is good practice to employ both methods of stressing. Although laboratory tests are useful in encouraging conservative design of magnesium alloy structures, results of long-term atmospheric tests of tensile-loaded specimens are considered to be very important [88, 95, 96].

The specimens can be subjected to various loading conditions involving constant load, constant strain, or monotonically increasing strain to total failure in some of the slow strain rate tests. Other tests include cyclic loading as well as slow straining over a limited stress range [94]. For example, Arsenault et al. [97] examined the SCC mitigation of aluminum alloy 7075-T6 by thermal spray coatings. The accelerated SCC mechanical tests were performed using a fluctuating load set up at a low frequency of 0.1 Hz through a four-point bent beam apparatus.

Constant Load SCC tests have been shown to be more severe than constant-deflection tests. Under a constant load, stress increases as the cross section is reduced by cracking or corrosion. However, this condition produces decreasing stress when deflection is fixed. It has been suggested that SCC threshold stress is associated with the onset of plastic deformation, that is, the elastic limit of the alloy. The elastic limit is difficult to measure unambiguously; however, the stress at which 0.2% plastic deformation occurs is generally used[6].

Constant Deflection A recent technique for use with constant-deflection specimens is the determination of the residual "breaking strength" (load-carrying ability) after short exposures to an environment such as that given in ASTM G44, rather than use of pass/fail data. Both unstressed and stressed specimens are exposed, providing a measure of the effect of applied stress on corrosion and on resistance to SCC. An adequate number of replicate specimens (typically five), two or more stress levels (one being no stress), and three of longer exposure periods (one being no exposure) are required. Such a method is in use for high-strength aluminum alloys and the advantages of the method are considerably shorter exposure periods and statistical analysis of results [1]. Constant-load SCC tests have been shown to be more severe than constant-deflection tests since under constant load, the stress increases as the cross section is reduced by crack propagation. However, this condition produced decreasing stress when deflection is fixed [98].

Slow Strain Rate Test In this test, the sample is pulled to failure at a preselected strain rate in a tensile machine surrounded by the aggressive medium. If SCC failures occur, there are many secondary cracks associated with the primary fracture and there will be a significant reduction in ductility that can be expressed by the elongation percentage and the percent reduction in area; otherwise the sample will fail by ductile tensile overload and the classical dimpled fracture surface can be identified by scanning electron microscopy. The interest in these tests, for aluminum or metals that possess active–passive behaviors, is that by straining the sample at a selected slow strain rate, the passive film is disrupted. This disruption in the passive film gives the occasion to initiate cracks quickly in the susceptible alloy by the medium that contains the specific aggressive ion. Results may be obtained in 1–5 days depending on the strain rate, while the U-bend test takes much longer to run since the breakdown of the passive film with possible crack initiation is a very slow process and could take more than a month [99].

Standard tension specimens (ASTM E8) are generally recommended for use with the specified conditions of gauge lengths, radii, and so on, unless specialized studies are being conducted. The strain rate chosen frequently for the tests, based on several studies, indicates important susceptibility to cracking at about $2 \times 10^{-6}\,s^{-1}$ for steels, aluminum, and magnesium alloys. However, the tests refer to open circuit conditions and the strain rate sensitivity of cracking is dependent on potential as well as solution composition. Where necessary, the potential of the specimens can be controlled using a potentiostat during slow strain rate tensile testing [100]. The reduction of area is a simple and appropriate way to quantify the susceptibility to SCC.

The specimens are polished generally to a 1200 grit finish, ultrasonically cleaned in acetone and alcohol, and then dried in hot air before testing. Figure 17.7 shows a SCC specimen mounted in the slow strain rate machine test cell. Using the potentiostat and auxiliary platinum electrode, anodic or cathodic polarization for galvanostatic or potentiostatic control can be carried out during the test [101]. A candidate solution for the Slow strain rate tensile (SSRT) test is 3% NaCl + 0.3% H_2O_2.

The reduction of area is a simple and appropriate way to quantify the susceptibility to SCC. Both the ac and dc potential drop methods are well-established techniques for monitoring subcritical crack growth as well as a combined ac/dc potential drop measuring technique [6, 102].

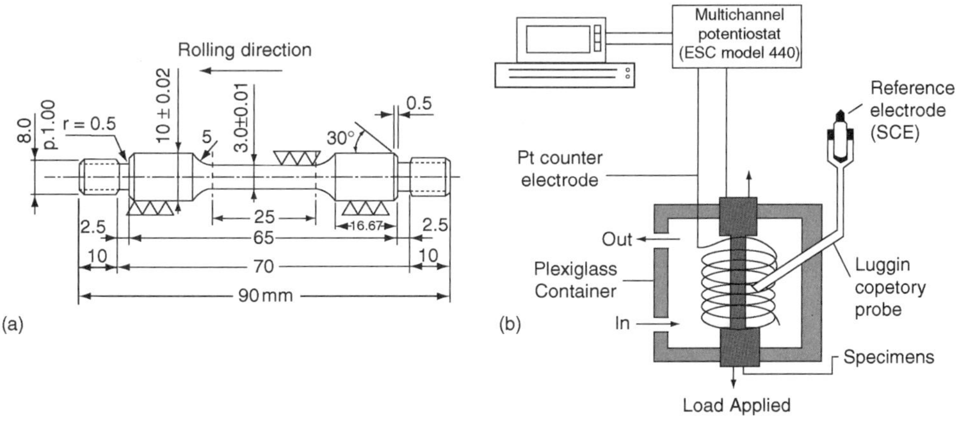

Figure 17.7 SCC experiment cell apparatus for slow strain rate experiments that can be operated with or without polarization: (a) the specimen and (b) the cell setup [101].

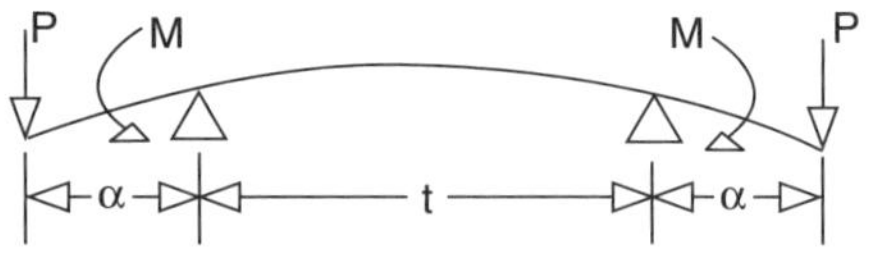

Figure 17.8 Schematics of SCC specimen in pure bending [23, 103].

The slow strain rate (SSR) test in ASTM G129 can apply to wrought or cast aluminum alloys to evaluate the susceptibility of metallic materials to environmentally assisted cracking. This test concerns the use of axially loaded specimens submitted to a SSR in an aggressive environment. In the absence of a designed environment, 3.5% NaCl is often considered. In many cases, the initiation of environmentally assisted cracking (EAC) is accelerated through the application of a dynamic strain in the gauge section or at a notch tip or crack tip, or both, of a specimen. Due to the accelerated nature of this test, the results are not intended to necessarily represent service performance, but rather to provide a basis for screening, for detection of an environmental interaction with a material, and for comparative evaluation of the effects of metallurgical and environmental variables on sensitivity to known environmental cracking problems [1, 62] and ISO 7539.

The three-point and four-point samples are frequently used for SCC and SSRT tests. Figure 17.8 shows a four-point stress–strain sample, where the maximum tension and compression are produced in points furthest from the neutral axis. Since the sample is symmetrical, the neutral axis lies along the mid way point of the sample thickness and the compressive and tension stresses are equal at the extremities [103]. Figure 17.9 shows a

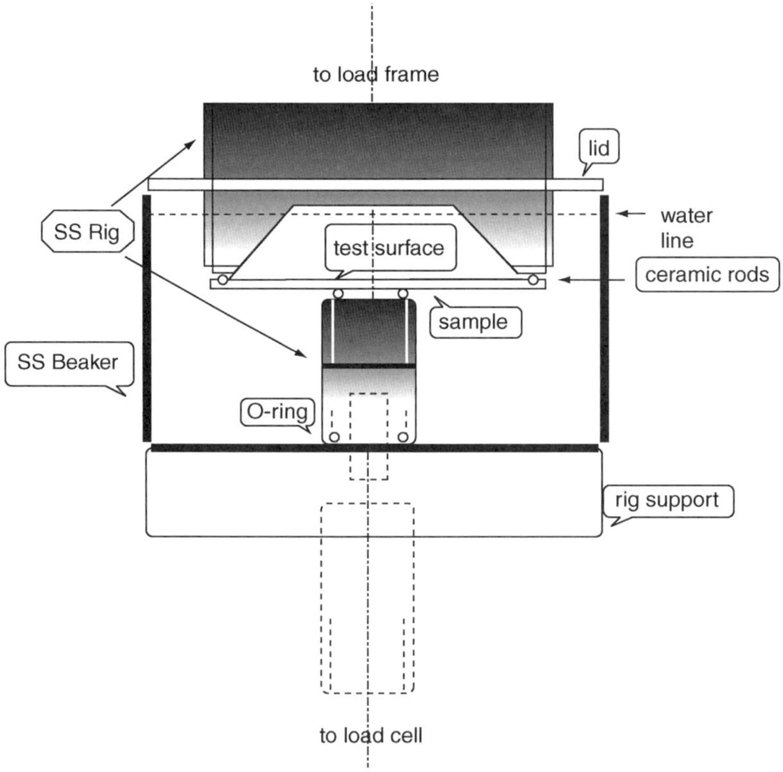

Figure 17.9 Test method of four-point specimen connected to a SSRT controlling machine together with a potentiostat [101, 103].

four-point SSC sample cell, loaded in a pneumatically operated computer-controlled machine using four-point bend tests together with a potentiostat. Cyclic loading could also be applied [97, 103](see Chapter 14 for more details).

17.7.4. Fracture Morphology and SCC of Aluminum Alloys

SCC of aluminum and its alloys requires the interaction of a metallurgically susceptible material, a corrosive environment (water vapor may be sufficient), and sustained (but not necessarily continuous) tensile stress [1]. Generally, propagation follows three distinct regions in SCC (Figure 17.10). It should be noted that true threshold stress intensity may not even exist for initiation of cracks in aluminum alloys, and K_{ISCC} could be defined by convention as the stress intensity corresponding to a crack growth rate of 10^{-9} or 10^{-10} m/s. It is also assumed that intergranular corrosion could be behind the increase of localized stresses that initiate the crack [104]. For macrobranching, the crack velocity must be independent of the crack-tip stress intensity (region II) and the crack-tip stress intensity K_B necessary for macrobranching is equal to or larger than 1.4 times the stress intensity K_P at the beginning of the plateau. Microbranching can occur in a broad zone of stresses starting at the level of K_M in the middle of region I.

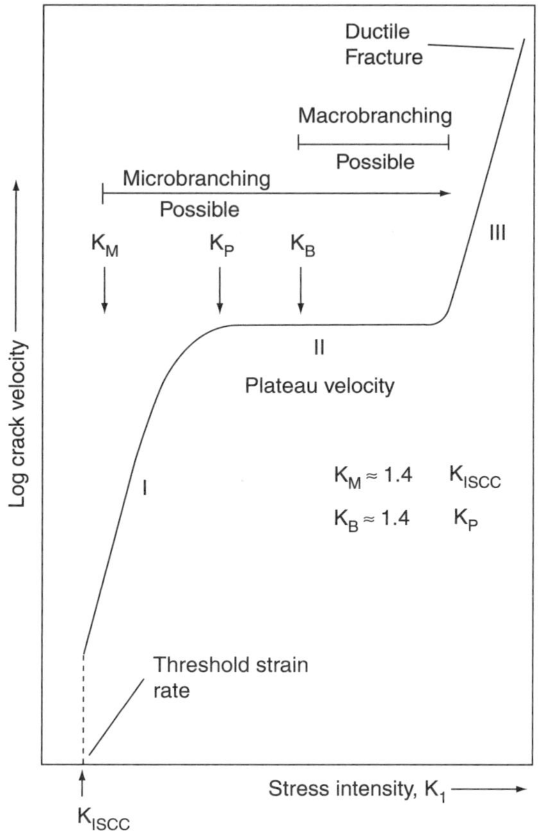

Figure 17.10 Schematic representation of the dependence of stress-corrosion crack growth rate, microbranching, and macrobranching on the crack-tip stress intensity [104].

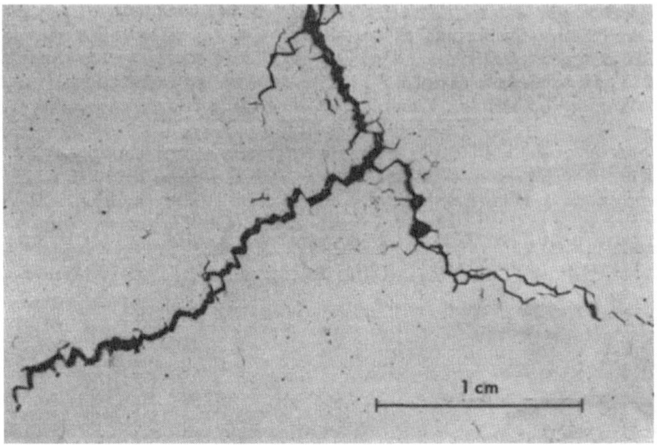

Figure 17.11 Macroscopic branching of a stress-corrosion crack in a ternary Al–Mg–Zn alloy exposed to aqueous halide solutions at high stress intensity [104].

Generally, failed specimens appear macroscopically brittle and exhibit highly branched cracks that propagate intergranularly and/or transgranularly, depending on the metal properties and chemical properties of the environment. Intergranular cracks are believed to propagate continuously or discontinuously, depending on the system while transgranular SCC propagation is often discontinuous, corresponding to periodic jumps on the order of a micrometer. In stress corrosion, ruptures are fragile and are sometimes characterized by the presence of cleavages, notably in the case of the hydrogen embrittlement [6, 82, 105].

Cleavage is a brittle fracture that occurs along specific crystallographic planes. Cleavage has a well-defined crystallographic orientation and it is easy to recognize its occurrence by optical microscopy as it exhibits brilliant and flat fracture facets that are related to the dimension of the grain size of the material under study. Under a scanning electron microscope, flat fracture facets exhibit cleavage steps and a river pattern that are caused by the crack moving through the crystal along a number of parallel planes, which form a series plateaus and connecting ledges [6, 82].

SCC in aluminum alloys is typically intergranular (Figure 17.11), although transgranular SCC has been observed for a few alloys under highly specific environmental conditions and may be part of the propagation mechanism in some 7xxx alloys [104]. Stress-corrosion branched cracks can be seen under certain conditions for some aluminum alloys. In macrobranching, a crack separates into two or more cracks that tend to diverge, while in microbranching, the crack front splits into several local cracks with separation distances in which the crack is close to a grain diameter. Microbranching can occur under less restrictive conditions, for example, at a wide zone of stress intensities at the crack tip, while macrobranching occurs at a more narrow zone of high values of stresses [104].

Macrobranching of stress-corrosion cracks usually does not occur in commercial high-strength aluminum alloys exposed to moist gases or aqueous solutions because the fracture path follows the grain boundaries that have strong preferred orientation. However, during the heat treatment, recrystallization occurs and locally an equiaxial grain structure is developed. For example, macrobranching was observed for forged specimens of aluminum alloy 7079-T6 and high-purity ternary Al–Mg–Zn alloy. The latter showed two macrobranches and every macrobranch had a considerable amount of microbranching (Figure 17.12).

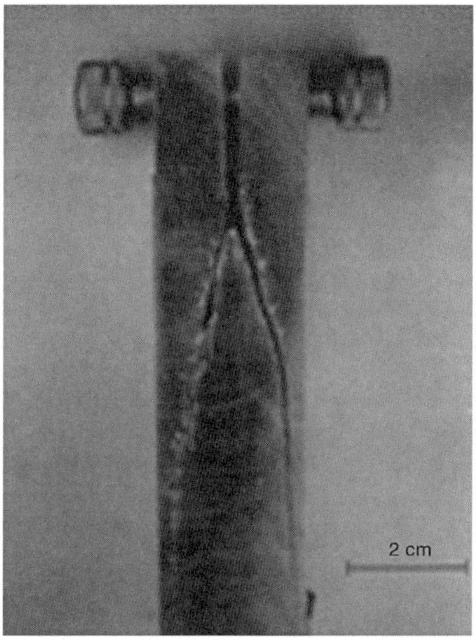

Figure 17.12 Macroscopic branching of a subcritical crack filled with mercury in high-strength aluminum alloy 7178-T651 [104].

In the case of commercial high-strength aluminum alloys with strongly preferred grain boundary orientation, specimens exposed to liquid metals such as mercury showed macroscopic branching. As the branches grew longer and the stress intensity dropped, the transgranular component of the crack growth was reduced and the last stages of the two macrobranches became mostly intergraular (Figure 17.12). It is common to observe that the path of SCC can change from transgranular at high stress intensities to predominantly intergranular at low stress intensities [104].

REFERENCES

1. B. W. Lifka, in *Corrosion Tests and Standards, Application and Interpretation*, 2nd edition, edited by R. Baboian. ASM International, Materials Park, OH, 2005, pp. 547–557.
2. C. Vargel, in *Corrosion de l'aluminium*, edited by C. Vargel. Dunod, Paris, France, 1999, pp. 199–217.
3. R. B. Mears, in *Corrosion Handbook*, edited by H. H. Uhlig. Wiley, Hoboken, NJ, 1976, pp. 39–55.
4. ASM International Handbook Committee, in *Corrosion—Understanding the Basics*, edited by J. R. Davis. ASM International, Materials Park, OH, 2000, pp. 427–491.
5. J. C. Zoccola, H. E. Townsend, A. R. Borxillo, and J. B. Horton, *Atmospheric Corrosion Behavior of Aluminun Zinc Alloy-Coated Steel*. American Society for Testing and Materials, Philadepphia, PA, 1978, pp. 165–184.
6. V. S. Sastri, E. Ghali, and M. Elboujdaini, *Corrosion Prevention and Protection—Practical Solutions*. Wiley, Chichester, UK, 2007, pp. 331–459.
7. L. L. Shreir, R. A. Jarman and G. T. Burstein, *Corrosion, Corrosion Control*. Butterworth-Heinemann, Oxford, UK, 1995, pp. 19:1–19:48.
8. L. L. Shreir, R. A. Jarman, and G. T. Burstein, *Corrosion—Metal/Environment Reactions*, 3rd edition. Butterworth-Heinemann, Oxford, UK, 1995, pp. 1–18.
9. J. Q. Zhang, Z. Zhang, J. M. Wang, H. B. Shao, and C. N. Cao, *Acta Metallurgica Sinica* **14**(2), 91–96 (2001).
10. E. Ghali, *Uhlig's Corrosion Handbook*, 2nd Edition, edited by R. W. Revie. Wiley, Hoboken, NJ, 2000, pp. 677–715.

11. ASM International Handbook Commitee, in *Aluminum and Aluminum Alloys*, edited by J. R. Davis. ASM International, Materials Park, OH, 1993, pp. 3–55, 88–120, 579–731.
12. K. A. Lucas and H. Clarke, *Corrosion of Aluminium-Based Metal Matrix Composites*. Research Studies Press Ltd, Taunton, Somerset, UK, 1993.
13. J. R. Scully, in *Corrosion*, Vol. 13, edited by J. R. Davis. ASM International, Materials Park, OH, 1987, pp. 212–220.
14. V. S. Sastri, E. Ghali, and M. Elboujdaini, *Corrosion Prevention and Protection—Practical Solutions*. Wiley, Chichester, UK, 2007, pp. 109–176.
15. M. Elboujdaini, M. T. Shehata, and E. Ghali, in An electrochemical investigation and corrosion performance of some aluminum alloys in chloride media, in *Proceedings of the 31st Annual International Metallographic Society, Volume 26, Analysis of In-Service Failures and Advances in Microstructural Characterization*, Ottawa, Canada, edited by E. A. D. O. Northwood, M. T. Shehata, and J. Wylie. Metallographic Society, Ottawa, Canada 1998, pp, 91–97.
16. Z. Szklarska-Smialowska, *Pitting Corrosion of Metals*. NACE, Houston, TX, 1986, pp. 39–69.
17. H. Böhni, in *Uhlig's Corrosion Handbook*, edited by R. W. Revie. Wiley, Hoboken, NJ, 2000, pp. 173–190.
18. G. S. Frankel, in *Corrosion: Fundamentals, Testing and Protection*, Vol. 13A, edited by D. Cramer and B. S. Covino, Jr. ASM International, Materials Park, OH, 2003, pp. 236–241.
19. H. P. Leckie and H. H Uhlig, *Journal of the Electrochemical Society* **113**, 1262 (1966).
20. N. Sato and G. Okamoto, in *Comprehensive Treatise of Electrochemistry*, edited by J. O. M. Bockris, B. E. Conway, E. Yeager, and R. E. White. Plenum Press, New York, 1981, pp. 193–245.
21. N. G. Thompson and J. H. Payer, in *Corrosion Testing Made Easy*, edited by B. C. Syrett. NACE International, Houston, TX, 1998, pp. 72–77.
22. R. B. Inturi and Z. Szklarska-Smialowska, *Corrosion Science* **34**(12), 1973–1989 (1993).
23. T. J. Summerson and D. O. Sprowls, Corrosion behavior of aluminum alloys, in *Proceedings of the International Conference on the Hall–Heroult Process*, Engineering Materials Advisory Services, University of Virginia, 1986, pp. 1576–1662.
24. P. Marcoux, Prévention de la corrosion sous contrainte et de la corrosion localisée d'alliages d'aluminium aérospatiaux par revêtements sacrificiels produits par projection thermique à l'arc électrique. Master's thesis, Université Laval, Québec, Canada, 2000.
25. S. Berrada, M. Elboujdaini, and E. Ghali, *Journal of Applied Electrochemistry*, **22**, 1065–1071 (1992).
26. F. Mansfeld and M. W. Kendig, *Journal of the Electrochemical Society* **135** (4), 828–833 (1988).
27. S. R. Koteswara Rao, G. Madhusudhan Reddy, K. Srinivasa Rao, M. Kamaraj, and K. Prasad Rao, *Materials Characterization* **55**(4–5), 345–354 (2005).
28. C. C. Lee and F. Mansfeld, in *Corrosion/99*, Paper No.195. NACE, Houston, TX, 1999.
29. Z. Szklarska-Smialowska, *Pitting and Crevice Corrosion*. NACE International, Houston, TX, 2005, pp. 327–329.
30. D. Li, Y. Hu, and B. Guo, *Materials Science and Engineering* 173–176 (2000).
31. D. A. Eden, in *Uhlig's Corrosion Handbook*, edited by R. W. Revie. Wiley, Hoboken, NJ, 2000, pp. 1227–1238.
32. N. Sato, *Journal of the Electrochemical Society* **129**, 260 (1982).
33. T. Suter and H. Bohni, *Electrochimica Acta* **43**, 2843 (1998).
34. S. T. Pride, J. R., Scully, and J. L. Hudson, *Analysis of EN from Metastable Pitting in Aluminum, Aged Al–2%Cu and AA 2024-T3*. in Electrochemical Noise Measurement for Corrosion Applications, ASTM STP 1277, Philadelphia, PA, USA, 1996, pp. 307–331.
35. B. Arsenault, *Overview of the Electrochemical Noise Technique*. Laval University, Quebec, Canada, 2007, p. 16.
36. A. Aballe, M. Bethencort, F. J. Botana, M. Marcos, and J. M. Sanchez-Amaya, *Corrosion Science* **46**, 1909–1920 (2004).
37. C. Gouveia-Caridade, M. I. S. Pereira, and C. M. A. Brett, *Electrochimica Acta* **49**, 785–793 (2004).
38. F. Mansfeld, L. T. Han, C. C. Lee, and G. Zhang, *Electrochimica Acta* **43**, 2933 (1998).
39. A. M. Nagiub, Evaluation of localized corrosion for Al-2024 with electrochemical impedance spectroscopy (EIS) and electrochemical noise analysis (ENA), in *Proceedings of the Egyptian Corrosion Society Conference*, Ain El-sukna, Red Sea, Egypt. Egyptian Corrosion Society, Ain El-sukna, Red Sea, Egypt, 2003.
40. Y. Mansfeld, S. H. Wang, S. H. Lin, H. Xiao, and H. Shih, *ASTM Publication* **STP 1181**, 297 (1993).
41. J. L. Kearns, D. A. Eden, M. R. Yaffe, J. V. Fahey, D. L. Reichert, and D. C. Silverman, in *Electrochemical Noise Measurement for Corrosion Application*, edited by J. R. Kearns, J. R. Scully, P. R. Roberge, D. L. Reichert, and J. L. Dawson. ASTM, West Conshohocken, PA, 1996, Vol. ASTM STP 1277, pp. 446–470.
42. S. Mansfeld, Z. Sun, C. H. Hsu, and A. Nagiub, *Corrosion Science* **43**, 341 (2001).
43. A. Nagiub and F. Mansfeld, *Corrosion Science* **43**(11), 2001–2009 (2001).
44. K. H. Na and Su-Il Pyun, *Journal of Electroanalytical Chemistry* **596**, 7–12 (2006).
45. K. H. Na, Su-Il Pyun, and H.-P. Kim, *Corrosion Science* **49**(1), 220–230 (2007).
46. S.-I. Pyun and E. J. Lee, *Surface Coating Technology* **62**, 480–485 (1993).

47. K. H. Na and Su-Il Pyun, *Corrosion Science* **50**(1), 248–258 (2008).

48. A. Aballe, M. Bethencourt, F. J. Botana, M. Marcos, and R. M. Osuna, *Electrochimica Acta* **47**, 1415–1422 (2002).

49. Z. Zhang, Q. Zhong, J. Zhang, and C. Cao, Electrochemical noise analysis of LY 12 in sodium chloride solution with wavelet transform technique, in *Corrosion 2002*, NACE International, Houston, TX, 2002, 11 pp.

50. F. H. Cao, Z. Zhang, J. X. Su, Y. Y. Shi, and J. Q. Zhang, *Electrochimica Acta* **51**, 1359–1364 (2006).

51. J. D. Damborenea and B. Fernandez, *ASTM Publication* **STP 1277**, 398–410 (1996).

52. E. Ghali, W. Dietzel, and K.-U. Kainer, *Journal of Materials Engineering and Performance* **13**(5), 517–529 (2004).

53. Y. Shi, Z. Zhang, J. Su, F. Cao, and J. Zhang, *Electrochimica Acta* **51**, 4977–4986 (2006).

54. ASM Metals Handbook Committee, in *Corrosion*, Vol. 13, edited by L. J. Korb, D. L. Olson, and J. R. Davis. ASM International, Materials Park, OH, 1987, pp. 93–196, 207–220, 231–233, 303–310, 596.

55. N. Sridhar, D. S. Dunn, C. S. Brossia, G. A. Cragnolino, and J. R. Kearns, in *Corrosion Tests and Standards: Application and Interpretation*, 2nd edition, edited by R. Baboian. ASM International, Materials Park, OH, 2005, pp. 221–232.

56. R. M. Kain, *Corrosion* **40**(6), 313–321 (1984).

57. ASM International Handbook Committee, in *Corrosion of Aluminum Alloys*, edited by J. R. Davis. ASM International, Materials Park, OH, 1999, pp. 45–62.

58. B. W. Lifka, in *Corrosion Testing and Standards: Application and Interpretation*, edited by R. Baboian. American Society for Testing and Materials, Philadelphia, PA, 1995, pp. 447–457.

59. J. E. Hatch, *Aluminum: Properties and Physical Metallurgy*. ASM International, Materials Park, OH, 1984.

60. S. Lee and B. W. Lifka, in *New Methods for Corrosion Testing of Aluminum Alloys, ASTM STP 1134*, edited by V. S. Agarwala and G. M. Ugiansky. American Society for Testing and Materials, Philadelphia, PA, 1992.

61. B. W. Lifka and D. O. Sprowls, in *Corrosion in Natural Environments, ASTM STP 558*, edited by ASTM Committee. G-1. American Society for Testing and Materials, Philadelphia, PA, 1974.

62. Annual Book of ASTM Standards Committee, *Annual Book of ASTM Standards*. American Society for Testing Materials, Philadelphia, PA, 2006, pp. D1384–01, G1-90, G5-94, G16-95, G46-94, G78-01, G61-86, G69-01, G102-89, and G110-92.

63. D. A. Wadeson, X. Zhou, G. E. Thompson, P. Skeldon, L. Djapic Oosterkamp, and G. Scamans, *Corrosion Science* **48**, 887–897 (2006).

64. S. C. Dexter, in *ASM Metals Handbook, Volume 13A, Corrosion*, edited by S. D. Cramer and B. S. Covino, Jr., ASM Internationals, Materials Park, OH, 2003, pp. 398–416.

65. B. J. Little, P. A. Wagner, and F. Mansfeld, *Microbiologically Influenced Corrosion, Corrosion Testing Made Easy*, NACE International, Houston, TX, 1997, Vol. 5.

66. W. Wang, J. Wang, H. Xu, and X. Li, *Materials and Corrosion* **57**(7), 531–537 (2006).

67. H. A. Videla and L. K. Herrera, *International Microbiology* **8**, 169–180 (2005).

68. J. F. D. Stott, in *ASM Handbook, Volume 13A, Corrosion: Fundamentals, Testing, and Protection*, edited by S. D. Cramer and B. S. Covino, Jr. ASM International, Materials Park, OH, 2003, pp. 644–649.

69. P. Norouzi, J. Hamedi, M. R. Ganjali, and A. Ahmadi, *Materials and Corrosion* **57**(7), 538–542 (2006).

70. J. Liu, H. Wu, and S. Li, in *Proceedings of the Second International Conference on Environment Sensitive Cracking and Corrosion Damage*, edited by H. N. M. Matsumura, K. Nakasa, and Y. Isomoto. ESCCD; Hiroshima, Japan, 2001, pp. 109–112.

71. H. A. Videla, in *Manual of Biocorrosion*, CRC Press, Boca Raton, FL, 1996, pp. 13–45, 121–133, 160–173, 179–185, 228–237.

72. T. P. Zintel, G. J. Licina, and T. R. Jack, in *A Practical Manual on Microbiologically Influenced Corrosion*, Vol. 2, edited by J. G. Stoecker. NACE International, Houston, TX, 2001, p. 10.1.

73. N. S. T. Method, *Field Monitoring of Bacterial Growth in Oilfield Systems*. National Association of Corrosion Engineering International, Houston, TX, 1994.

74. R. C. John, I. J. Rippon, J. J. Maarten, S. Thomas, S. D. Kapusta, M. M. Girgis, and T. Whitham, B"F"M" Post" NACE International, Houston, TX, 2002, Paper No. 02235.

75. R. S. Dubey and S. N. Upadhyay, *Biosensors & Bioelectronics* **16**, 995–1000 (2001).

76. B. W. Lifka, *Aluminum*, **53**(12), 750–752 (1977).

77. H. Tylczak and T. A. Alder, *Gaseous Corrosion–Wear Interactions*, Vol. 13A. ASM International, Materials Park, OH, 2003.

78. R. R. Ramsingh and C. DeRushie, *Materials Performance* **45**(4), 48–51 (2006).

79. B. Bhushan, *Introduction to Tribology*. Wiley, Hoboken, NJ, 2002, pp. 331–420.

80. B. Phull, in *ASM Handbook, Volume 13A, Corrosion*, edited by S. D. Cramer and B. S. Covino, Jr., ASM International, Materials Park, OH, 2003, pp. 625–638.

81. M. Elboujdaini and E. Ghali, *Materials Science Forum* **44–45**, 153–168 (1989).

82. M. G. Fontana and N. D. Greene, *Corrosion Engineering*. McGraw-Hill, New York, 1978, pp. 7–27.

83. A. J. Mc Evily and R.P., Wei, *Fracture Mechanics and Corrosion Fatigue*. NACE, Houston, TX, 1973, pp. 381–395.

84. P. M. Scott, in *Corrosion Fatigue: Mechanics, Metallurgy, Electrochemistry and Engineering*, Vol. STP 801, edited by T. W. Crooker and B. N. Leis. American Society for Testing and Materials, Philadelphia, PA, 1983, pp. 319–350.
85. P. Mehdizadeh, R. L. McGlasson, and J. E. Landers, *Corrosion* **22**(12), 325 (1965).
86. M. Elboujdaini, M. T. Shehata, and E. Ghali, *Microstructure Science* **25**, 41–49 (1997).
87. J. E. Hillis, in *ASTM Manual Series: MNL 20, Corrosion Testing and Standards: Application and Interpretation*, edited by R. Baboian. ASM International, Materials Park, OH, 1995, pp. 438–446.
88. A. F. Froats, T. Kr. Aune, D. Hawke, W. Unsworth, and J. Hillis, in *ASM Handbook, Volume 13, Corrosion*, edited by L. J. Korb, D. L. Olson, and J. R. Davis. ASM International, Materials Park, OH, 1987, pp. 740–754.
89. B. Phull, in *ASM Handbook, Volume 13A Corrosion* edited by S. D. Cramer and B. S. Covino, Jr. ASM International, Materials Park, OH, 2003, pp. 575–616.
90. E. H. Hollingsworth and H. Y. Hunsicker, in *ASM Metals Handbook, Volume 13, Corrosion*, 9th ed., edited by L. J. Korband and D. L. Olson. ASM International, Materials Park, OH, 1987, pp. 583–609.
91. P. R. Roberge, *Handbook of Corrosion Engineering*, McGraw-Hill, New York, 1999, pp. 598–600.
92. S. P. Lynch and P. Trevena, *Corrosion* **44**(2), 113–124 (1988).
93. W. K. Miller, Stress corrosion cracking of magnesium die casting alloys, in *Materials Stability and Environmental Degradation*, Vol. 125, edited by A. Barkatt, E. D. Verink, Jr, and L. R. Smith, Materials Research Society, Warrendale, PA, 1988, pp. 253–259.
94. R. N. Parkins, and Y. Suzuki, *Corrosion Science* **23**, 577 (1983).
95. D. Sprowls, in *Stress Corrosion Cracking*, edited by R. H. Jones. ASM International, Materials Park, OH, 1992, pp. 336–415.
96. G. F. Sager, R. H. Brown, and R. B. Mears, *Tests for Determining Susceptibility to Stress-Corrosion Cracking*. American Society for Testing and Materials, Philadelphia, PA, 1945, pp. 255–272.
97. B. Arsenault, P. Marcoux, S. Simard, and E. Ghali, *Stress Corrosion Cracking Mitigation of 7075-T6 Aluminum Alloy By Thermal Spray Coatings,* Paper 06496-NACE, in Corrosion 2006 Conference, San Diego, CA, NACE International, Houston, Texas, 2006.
98. W. E. Mercer and J.E. Hillis, The Critical Contaminant Limits and Salt Water Corrosion Performance of Magnesium AE42 Alloy (Paper 920073), Society of Automotive Engineers, Detroit, MI, 1992.
99. B. Moniz and H. Zhang, in *Chemical Processing*, edited by R. Baboian. ASTM International, West Conshohocken, PA, 2005, pp. 779–794.
100. K. Ebtehaj, D. Hardie, and R. N. Parkins, *Corrosion Science* **28**, 811–829 (1988).
101. M. Elboujdaini, E. Ghali, and M. T. Shehata, Effect of applied stress and potential on SCC and corrosion fatigue behaviour of aluminum in chloride media, in *Proceeding of 23rd Annual Conference of Egyptian Corrosion Society*, Nozha Beach Resort, Red Sea, Egypt, Egyptian Corrosion Society, Cairo, 2004, p. 14.
102. W. Dietzel and K.-H. Schwalbe, *Materialprüfung* **28** (11), 368–372 (1986).
103. Y. Z. Wang, R. W. Revie, M. T. Shehata, R. N. Parkins, and K. Krist, in, *Initiation of Environment Induced Cracking in Pipeline Steel: Microstructural Correlation, Proceedings of the International Pipeline Conference* 1998, edited by International Pipeline Conference. The American Society of Mechanical Engineers, New York, 1998, pp. 529–542.
104. M. O. Speidel, in *The Theory of Stress Corrosion Cracking in Alloys*, edited by J.C. Scully. North Atlantic Treaty Organization Scientific Affairs Division, Brussels, Belgium, 1971, pp. 289–355.
105. T. Magnin and P. Combrade, in *Materials Science and Technology Series*, Vol. 1, edited by M. Shutze. Wiley-VCH, Weinheim, Germany 2000, pp. 216–318, 537.

Chapter 18

Evaluation of Corrosion Forms of Magnesium and Its Alloys

Overview

Most magnesium testing solutions include chloride or sulfate ions in neutral or alkaline media and are used to evaluate general and localized corrosion. A sodium chloride solution (3%) is used currently. Polarization curves obtained in 1 N NaCl solution adjusted to pH 11 by NaOH showed different pitting potentials as a function of the microstructure. A 1 N sodium hydroxide solution saturated with Mg_2SO_4 shows the influence of the aggressive sulfate ion with reference to industrial media. With pH increasing above 10.2—the point at which magnesium hydroxide is formed—the effect of impurities both in the metal and in the solution media is apparently overshadowed by the high tendency of film formation. Buffer solutions and saturation of the solution with magnesium hydroxide after thorough surface cleaning have been adopted frequently.

Besides weight loss, poteniodynamic polarization methods, estimates of the polarization resistance R_p, and the volume measurement of evolved hydrogen can give the instantaneous corrosion rate, kinetics, and the evolution of corrosion behavior of magnesium and its alloys. Alternate intermittent immersion in salt water or salt spray is often used to compare the corrosion resistance of different magnesium alloys.

During cyclovoltammetric sweeps, changes in the passive films and the solution composition at the metal–electrolyte interface (e.g., hydroxide formation and pH increase), as well as development of localized forms of corrosion such as pitting, take place. Impedance measurements are usually performed at open circuit potentials and under potentiostatic conditions. In many magnesium alloys, the open circuit potential is higher than the pitting potential, and thus the pitting potential cannot be determined with traditional methods.

Three approaches are possible for obtaining instantaneous values of R_p from the EN current and potential measurements: namely, noise resistance, spectral noise resistance, and self-linear polarization resistance (SLPR). The term-quasi electromotive force (QEMF), corresponding to the difference between the most active cathode potential and the most active anode potential, is used to interpret the scanning reference electrode technique (SRET) results. The strain rate chosen for the tests indicating maximum susceptibility to cracking was about $2 \times 10^{-6} s^{-1}$ for a solution containing $5 g{\cdot}L^{-1}$ sodium chloride and $5 g{\cdot}L^{-1}$ $KCrO_4$. Two frequently used laboratory methods for characterizing stress-corrosion

Corrosion Resistance of Aluminum and Magnesium Alloys: Understanding, Performance, and Testing By Edward Ghali

cracking (SCC) and mechanism considerations are the constant extension rate test (CERT) and the linearly increasing stress test (LIST).

Previous service failure data for magnesium alloys are not so numerous and in certain cases, access to this data is somewhat difficult. Corrosion acceleration for testing magnesium alloys is achieved through the use of various "stressors" such as cold working of the material, higher chloride concentration, lower pH, higher temperature, or higher stress. Accelerated tests of unprotected metal surfaces are appropriate to establish the minimum required basis of performance, while prediction of service life requires evaluation of the coatings and selection of maintenance procedures. Corrosion rates tend generally to decrease as the electrolyte becomes spent and saturated with magnesium ions, depending on the quality of the passive layer and pH shift. Since electrochemical corrosion is a function of the metal–solution interface, flow rates and ratio of area of metal surface to volume of solution should be considered. Accelerated testing should produce the same mode of failure and reflect at least a known order of resistance of some alloys in service media [1].

18.1. TESTING SOLUTIONS

Most magnesium testing solutions include chloride or sulfate ions in neutral or alkaline media and are used to evaluate general and localized corrosion. Buffer solutions and saturation of the solution with magnesium hydroxide as well as surface cleaning have been adopted by some researchers or professionals. All organic materials should be degreased thoroughly and if desired a 10% caustic solution is frequently used for cleaning magnesium at temperatures up to the boiling point before testing and in practice. Soak cleaners are used as alkaline cleaning in concentrations of 30–75 g/L (4–10 oz/gal) and at 71–100 °C [2].

A deliberate increase of pH to about 10.5 by the addition of CaO or NaOH suppresses corrosion in a large volume of water. The beneficial effects of a calcium oxide treatment are the possible precipitation of undesirable metals that can initiate galvanic corrosion, such as copper, in addition to the formation of a protective hydroxide. It has been observed that once attack is inhibited, and if it is uniformly distributed, an increase of the critical pitting potential and a delay in the onset of pitting usually occurs [3].

18.1.1. Hydroxide Solutions

With pH increasing above 10.2—the point at which magnesium hydroxide is formed—the effect of impurities both in the metal and in the solution media is apparently overshadowed by the high tendency of film formation. When the pH exceeds 10.5, a value that corresponds to the pH of saturation with magnesium hydroxide, a magnesium hydroxide film is formed on the surface, and magnesium becomes very resistant to corrosion in alkaline solutions [4]. In deaerated 0.1 N sodium hydroxide solution (pH 13), even without sodium chloride additions, the corrosion potential of magnesium FSI alloy (3.1% Al and 1.3% Zn) oscillated from passive to active values periodically. The addition of a strong oxidizing agent, such as hydrogen peroxide (10 mL), maintained the potential in the passive region [1]. In NaOH solutions adjusted to pH 12, the protective properties of the passive layer were shown to be much better than that of the basic metal. However, the properties of the formed passive layer are generally a function of the geometrical properties (roughness and porosity) of the base metal, alloying elements, and microstructure. In 1 M NaOH solution (pH 14), magnesium alloys should show generally a good passive behavior [2].

18.1.2. Chloride, Sulfate, and Hydroxide Solutions

A sodium chloride solution (3%) has been used by Hanawalt et al. [5] to generate important data on the tolerance limits and individual or combined effects of impurities in magnesium alloys. A 1 N NaCl solution with a normal pH around 7 is recommended and used by some authors and compared to that of 1 N NaCl adjusted to pH 11 by NaOH to show the influence of pH, the passivation aptitude, and the passivation quality of the corrosion products formed. A solution of 5% NaCl saturated with $Mg(OH)_2$ at pH 9 can distinguish between the behavior of different Mg samples and is frequently used but not standardized. This can be accompanied by experiments in 5% NaCl solution to show the influence of saturation with magnesium hydroxide [5].

Polarization curves obtained in 1 N NaCl solution adjusted to pH 11 by NaOH showed different pitting potentials as a function of the microstructure. A 2 N NaCl solution alone has also been used, but it seems to be aggressive, especially if magnesium composites are tested. Solutions of 0.1 M NaOH with additions of 0.005, 0.01, 0.02, and 0.03 M NaCl are used for determination of pitting or filiform corrosion during a 24 h immersion period. Passivation and reproducible results are obtained by using an addition of 10 mL hydrogen peroxide (30%) in corrosion pitting studies, preferably with pH adjustment to 11 [11]. A 1 N sodium hydroxide solution saturated with Mg_2SO_4 can show the influence of the aggressive sulfate ion with reference to industrial media [6–8].

18.1.3. ASTM D1384-96 Corrosive Water

Corrosive water according to ASTM D1384-96 has been used, either alone or saturated with magnesium hydroxide, to simulate the Mg alloy–solution interface that may contain important concentrations of magnesium hydroxide in stagnant solutions. The ASTM corrosive water has been designed to distinguish between coolants that are definitely deleterious from the corrosion point of view. The corrosive water contains 100 ppm each of sulfate, chloride, and bicarbonate ions introduced as sodium salts, and is prepared using the anhydrous form of 148 mg of sodium sulfate, 165 mg of sodium chloride, and 138 mg of sodium bicarbonate. It has been shown that the ASTM corrosive water alone is moderately aggressive, while that saturated with magnesium hydroxide is less aggressive and simulates the Mg metal–solution interface, giving lower corrosion rates.

The samples (rods of 40 mm diameter and 10 mm thickness) were exposed for 7 days to ASTM D1384-96 corrosive water (pH 8.2) without and with saturated magnesium hydroxide solution at pH 10.6, at room temperature, without stirring for a solution volume of 5 times that of every exposed square centimeter [9]. The solution was exposed to atmospheric oxygen and not agitated. This was appropriate to distinguish between the electrochemical corrosion resistance of high-pressure die-cast and semisolid cast AZ91D magnesium alloys [9]. Also, Adeva-Ramos et al. [10] examined pitting corrosion by immersing samples in the ASTM D1384-01 corrosive medium at pH 8.2. This was followed by potentiodynamic polarization for the determination of pitting potential in the same solution but saturated with $Mg(OH)_2$ at pH 10.6 [2, 10].

18.1.4. Buffered Solutions

Buffered solutions generally do not correspond to practical aspects of magnesium corrosion, except for those buffered with high magnesium hydroxide concentrations since this can simulate the interfacial corrosion of some magnesium alloys in a nonstirred solution. Generally, in solutions with low concentrations and initial pH values of 7, 9, and 11, the pH value had changed to a value of approximately 10.5 [11]. Borated boric acid buffer (pH 8.4) containing 0.001 N NaCl is advantageous for materials that are susceptible to pitting. The low percentage of sodium chloride and sodium hydroxide can give a controlled attack and reproducible results. Higher quantities of NaCl can also give reproducible results for situations less sensitive to agitation [2, 12].

A 0.1 N sodium borate solution buffer (pH 9.3) saturated with $Mg(OH)_2$ at pH 10.5 was used to measure corrosion rates at different strain percentages by the linear polarization method. It has been found that the borate anion acts as a corrosion inhibitor. For strained specimens, an aerated sodium tetraborate (0.05 M) with a pH of 9.7 has been chosen since the magnesium surface can be covered by a protective layer and the pH is stable. It has been shown by Inoue et al. [13] that corrosion rates of cast pure or highly pure magnesium AZ31 and AZ91 in deaerated solutions containing 10 g/L NaCl at pH 6.5 and borate buffer at pH 9 depended solely on the pH of the solution. The corrosion rates were determined gravimetrically. This electrochemical behavior corresponds to the resistivity of the passive layer to anodic reactions. AZ91E was an exception. Higher buffer capacity masked the detrimental effect of the cathodic impurities by reducing the difference between local pH at cathodic and anodic sites [13]. However, Gutman et al. [14] showed that the highest sensitivity to creep in the corrosive environment is observed in the alloy with the highest Al content—AZ91D and AM50 magnesium alloys, tested in 0.1 N sodium borate aqueous buffer solution at pH 9.3 or in the same solution saturated with magnesium hydroxide at pH 10.5 [2, 14].

18.2. GENERAL CORROSION FORM

18.2.1. Immersion Testing and Corrosion Rate

Saltwater corrosion studies are typically conducted in 3–5% sodium chloride solutions, following ASTM G31 [15] for immersion [16, 17]. ASTM G31 ("Standard Practice for Laboratory Immersion Corrosion Testing of Metals") suggests a reasonable correlation with salt spray performance; however, the measured weight loss corrosion rates typically run two or three times those measured in the ASTM B117 salt spray test for the high-purity alloys [18]. Corrosion products may have an effect on the pH at the interface and the form of corrosion, and so it is important to define the parameters concerning agitation or circulation of the solution.

18.2.1.1. Weight Loss

Hanawalt et al. [5] have successfully compared some selected magnesium alloys to pure magnesium using weight loss measurements and samples measuring 25.4 mm × 38.1 mm with a thickness of 6.35 mm. Complete removal of the corrosion products was performed by a 1 minute immersion in boiling 20% CrO_3 solution containing 1% silver nitrate [5]. The usual boiling 20% $H_2Cr_2O_4$ solution in water is used to remove corrosion products from magnesium and magnesium alloys without attacking the base metal, while the silver nitrate is expected to form a fine precipitate of Ag_2CrO_4 that can react in turn to precipitate chloride ions that can be carried away from the corrosion products [19]. Molded samples with limited surface are preferred to control the surface conditioning of the surface and the potential recording for the first hour and at the end. Triplicates are recommended.

To reduce the error caused by the dissolution of the uncorroded areas and substrate under the corrosion products during removal of the corrosion products, an uncorroded specimen is usually employed as a reference in the chromic cleaning process [20]. Weight loss increased with increased chloride concentration [11]. Although gravimetric studies have been used by several investigators, the weight loss method is not capable of measuring corrosion rates of magnesium over short periods of time since, in saltwater, as the dissolution reaction proceeds, the pH of the solution increases and $Mg(OH)_2$ precipitates on the surface of the sample often causing a weight gain [21].

Other more sensitive methods like atomic absorption, spectrophotometric, and calorimetric methods to determine magnesium and alloying elements in solution are available. The use of highly pure analytical quality reagents is recommended, and the magnesium hydroxide reagent can be avoided as an electrolyte [2]. The weight loss of a specimen is currently measured by dissolving the corrosion products in a hot chromic solution. The specimen is washed with distilled water and dried quickly in hot flowing air. Then it is put into a 100 g/L CrO_3 solution at about 90–95 °C for about 5–10 min. Following that, the specimen is washed with distilled water, dried, and weighed [20].

Tchervyakov et al. [22] considered the corrosion rates of AZ91 alloy by immersion tests. The samples (surface area, 5 cm^2) were immersed in 3% NaCl (100 mL) for 5 hours without polarization. Every half hour, a solution sample of 5 mL was taken out for atomic absorption analysis that was compensated from the fresh solution. The corrosion rate of the alloy was determined as the sum of corrosion rates of Al and Mg.

18.2.1.2. Hydrogen Measurements

Volume Measurement of Evolved Hydrogen This method gives the instantaneous corrosion rate, kinetics, and the evolution of corrosion behavior of magnesium and its alloys. The evolution of 1 mole of hydrogen gas corresponds to the dissolution of 1 mole of magnesium according to the following corrosion reaction:

$$Mg + 2H_2O = Mg^{2+} + 2OH + H_2$$

This is based on the belief that, particularly in aggressive solutions such as NaCl, the cathodic reaction is mainly hydrogen evolution and the contribution of oxygen reduction is practically negligible. The experimental setup can be designed in such a way that hydrogen evolution from the undermined particles can also be collected together with the hydrogen from the magnesium specimen. The dissolution of some alloying elements can cause error since the corrosion rate is based on the magnesium dissolution reaction. Some alloying elements, however, can also produce hydrogen, which to some extent reduces the error [20].

The accuracy of this method could be affected by some corrosion products stuck on the specimen surface, so it is best used for short-term immersion experiments. It is difficult to determine whether the hydrogen evolution rate is more accurate than the weight loss measurement. However, compared with the estimation of corrosion rates based on polarization curves, hydrogen evolution collection is undoubtedly very reliable [20].

Evolution and Control of pH Measurements The principle of the method is based on measuring specific ions (such as hydrogen or magnesium cation) by an ion selective electrode in a corrosive environment such as 5 wt % NaCl solution in distilled water. The variation of ion concentration with time is recorded, and the corrosion rate is calculated.

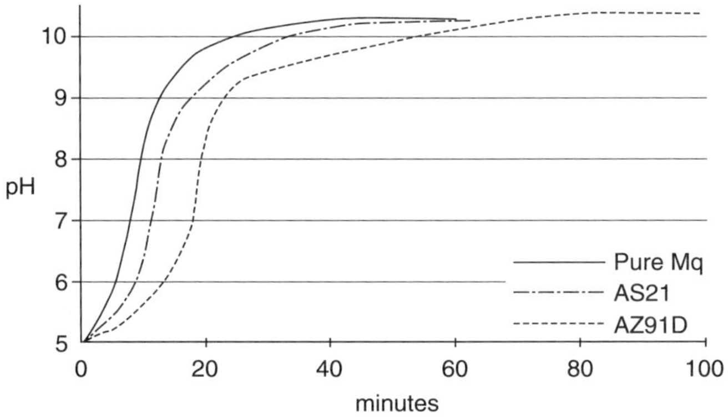

Figure 18.1 A schematic presentation of pH evolution as a function of time for Mg and two commercial alloys in 5% NaCl solution [2, 23, 24].

Weiss et al. [23] gave the following experimental details, The sample size can be about 40 mm × 30 mm × 15 mm, machined, thoroughly washed with acetone, and dried in air. A solution of 5 wt % NaCl in distilled water is generally considered. The solution volume is recommended to be 1 L and should be boiled for CO_2 degassing and kept under argon bubbling for the entire test [2, 23, 24].

Previous measurements based on the corrosion rates of each sample were calculated from several measurements (pH between 10 and 15) which were taken at an interval of 1 minute and showed good agreement with that obtained by weight loss methods. Figure 18.1 gives a schematic presentation of the evolution of pH as a function of time for magnesium and some commercial alloys. As in the volumetric method of hydrogen determination, the oxygen catalysis of the cathodic reaction, the dissolution of oxides, or alloying elements can cause some error in the calculation of the corrosion rate [2, 23].

Controlled pH Measurements In the last two methods, the hydrogen concentration of the salt solution decreases significantly with time, causing the precipitation of $Mg(OH)_2$ on the surface of the samples. This affects the intrinsic corrosion rate. Tiwari and Bommarito [21] studied the corrosion behavior of newly developed creep resistant Mg–Al–Ca (AC) alloys with several known magnesium alloys. They succeeded in determining the intrinsic corrosion rate of these alloys and to thus evaluate the performance of new or experimental alloys.

The method measures the dissolution rate of magnesium alloy samples in 5% NaCl solution under a controlled pH. It is fast (<5 hours) and capable of discerning the effect of small compositional changes or different tempers on the corrosion behavior of selected magnesium alloys. As a magnesium alloy sample dissolves in a 5% NaCl solution, the dissolution rate is determined by measuring the amount of HCl added to the NaCl solution to control the pH between 5 and 7. The corrosion curve, showing the amount of HCl added (in milliliters) as a function of time, generally has three portions (Figure 18.2).

The first portion corresponds to the reaction of magnesium oxide with hydrogen ions, producing water and magnesium ions. The corrosion rate can be constant from the beginning if an excessive oxide layer on the surface is not present and this portion is not observed [2, 21]. In the middle or second portion of the corrosion curve, the rate is controlled exclusively by the reaction $Mg + 2H^+ = Mg^{2+} + H_2$. The corrosion rate is determined from the slope of the curve where the dissolution rate reaches a steady state. This represents

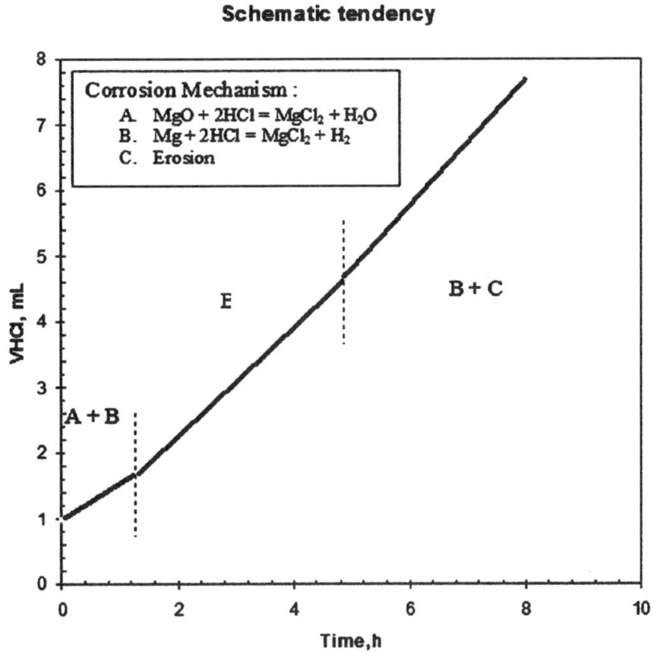

Figure 18.2 A schematic presentation of the three regions of attack of a commercial alloy AC53 in a controlled pH between 5 and 7 using 1 M HCl [2, 21].

the intrinsic corrosion rate of the alloy. In the third portion of the corrosion curve, disintegration of Mg caused by the H_2 bubbles in the form of fine particles was observed, increasing the effective surface area and thus the corrosion rate. The rate of corrosion seems to be closer in the second and third portions of the curve; however, this can depend on the microstructure and the composition of the alloy [2, 21].

This method was capable of discerning the effect of small changes in alloy composition on corrosion rates. An addition of small amounts of Si (0.3%) to AC alloys increases the corrosion rate, while Sr (0.07%) reduces it. It was also shown that the corrosion rate of AC52 (Mg–5%Al–2%Ca) is comparable to AZ91, meeting one of the several design criteria for new, creep-resistant alloys [2, 21].

18.2.2. Salt Spray Corrosion Test

The alternate intermittent immersion in saltwater or salt spray is often used to compare the corrosion resistance of magnesium alloys to each other and to other materials [5, 19]. In these test methods, a corrosive environment is simulated, as might be encountered in a marine or an automotive application (e.g., from salty road splash). The chloride solutions, even in small amounts, usually break down the thin protective magnesium oxide film. The pH of the salt solution should be such that when atomized at 35 °C, the collected solution is in the pH range of 6.5–7.2. The accelerated test used most often is the salt spray test, which involves continuous spraying of 5% sodium chloride (NaCl) in distilled water at 35 °C. It can be performed in accordance with ASTM B117, ISO 7253, ISO 9227, DIN 53167, or BS 3900.

For ASTM B117, salt spray testing exposure periods of 7–10 days are commonly used to evaluate the effects of trace contaminants within the alloy microstructure, for untreated or uncoated surfaces. Chromic acid solution and 1% silver dichromate are used to remove corrosion products [18]. The corrosion rate is based on measuring the weight loss of the specimen [6].

The marine atmosphere corrosion rate of Mg–Al alloys is much smaller than the salt spray rate, but both rates are affected by the impurity content. The effects of nickel, iron, and copper content on the corrosion of controlled-purity AZ91 die-cast alloy in a 10 day salt spray (5% NaC1, ASTM B117) test are equivalent to that in a 2 year atmospheric exposure on the Texas Gulf Coast. Although the salt spray corrosion scale is 200 times larger than the marine atmospheric corrosion scale, yet, some parallel conclusions can be drawn [19].

- Breakpoints relative to nickel and copper contamination are the same in both exposures (50 and 550 ppm, respectively).
- AZ91 corrosion remains low in both exposures over the iron contamination range studied (0.31% Mn alloy), confirming that the critical iron/manganese ratio (0.032 for this alloy) was not significantly exceeded.
- High-purity AZ91 alloy is shown in both exposures to have lower overall corrosion rates than carbon steel or die-cast aluminum 380 [25].

The salt spray method is slow (3–10 days) and labor-intensive, and it is error-prone due to the cleaning procedure used. The corrosion samples also lose weight due to crumbling during the tests. The method could result in misleading data depending not only on corrosion rate but also on the aptitude of the alloys to disintegrate [21]. For some unprotected magnesium alloys (sheet or sand cast) in salt spray (20% NaCl), tidal-immersion, and marine atmospheres, no fundamental differences have been found between spray and immersion tests for the alloys (AZ31-H24, AZ63-F, AZ91C-F) [19]. However, when the ASTM B117 salt spray test was carried out for 100 hours in a milder solution of 0.86 M NaCl, a corrosion rate of 0.018 MCD ($mg/cm^2 \cdot$ day) was obtained for an AZ91D alloy prepared by thixomolding with 0.8 mm thickness. This rate is almost 40% that of the conventional cast alloy [26].

Salt spray with a highly conductive electrolyte and continuous wetting of the sample surface is very aggressive for galvanic corrosion and consequently also for corrosion testing of magnesium. The corrosion mechanism of magnesium alloys is attributed to micro-galvanic corrosion between the matrix and the noble intermetallic particles and secondary phases. This corrosion is strongly promoted in salt spray due to the very conductive electrolyte and continuous wetting of the surface [27].

It is admitted that salt spray tests based on continuous NaCl spray alone are particularly unreliable for simulating accelerated corrosion in industrial atmospheres [28]. The generally recognized four largest failings of the standard salt spray test are the lack of a wet/dry cycle, the high concentration of electrolyte, the type of electrolyte, and the lack of radiation (ultraviolet (UV) or infrared (IR)). Two alternatives to ASTM B117 exist: alternate immersion testing (30 s immersion in 3% NaCl followed by 2 min in air) or continuous immersion in 3% or 5% NaCl for a period of 3–7 days or more. Existing variations to the standardized (neutral) salt spray test are the acidified salt spray or AASS test (ASTM G85) and the copper-accelerated or CAS test (ASTM B368) [18].

Humidity Tests Indoor corrosion resistance and influence of potential flux contamination are better evaluated by the mild humidity test than saltwater immersion or salt spray testing. The exposure of a freshly machined surface to an atmosphere of 60–80% relative humidity at 20–30 °C will reveal the presence of flux inclusions within 24–48 hours [18]. A simple test chamber employing a saturated sodium thiosulfate solution as a controlled source of 70% relative humidity has been proposed by Emley [29].

The published data has focused on automotive applications, where the standard testing has been ASTM B117 salt spray and automotive cycle tests. The high-purity magnesium alloy components typically perform very well in such tests, since the formed differential oxygen concentration cell does not appear to play a role in magnesium corrosion or coating adhesion. The cycle tests are specific to each of the automotive companies but involve cyclic exposure to saltwater, drying, and high humidity and this is more realistic than the standard salt spray exposure alone. An important example is the proving ground cycle test used by automobile manufacturers. Correlation has been made between the test and the long-term service of the vehicle [16].

Cyclic Corrosion Tests For alternating immersion testing, Hanawalt et al. [5] carried out experiments in 3% NaCl solution for 16 weeks. The cycles consisted of 1/2 minute in the solution followed by 2 minutes in air (Figure 18.3). During the latter time, the specimens do not dry completely. The cyclic corrosion test currently used is that of GM9540P. The duration of several days or more depends on the corrosion resistance of the examined bare alloy or the tested chosen coat [5, 27].

However, cyclic corrosion testing gives far better correlation to service than salt spray testing. The cyclic corrosion test uses a less conductive electrolyte and combines drying/wetting of the surface; it is thus less aggressive. For some applications, this test may be too mild due to the short time of the salt mist application and the lack of deposits of mud and dirt with hygroscopic salts [2, 27].

18.2.3. Some Electrochemical Methods of Investigation

18.2.3.1. Open Circuit Potential Determinations

During the immersion of magnesium and its alloys in aqueous solutions containing aggressive ions, the active and passive behaviors necessitate a time interval between 20 and 60 minutes to obtain rather stable corrosion potential values [9, 30, 31]. For some alloys in certain media, a relatively varying nonstationary potential can be obtained even after 1 hour. The potential of the α phase Mg_4Al is $-1.82\ V_{ECS}$ ($-1.58\ V_{SHE}$) and that of the β

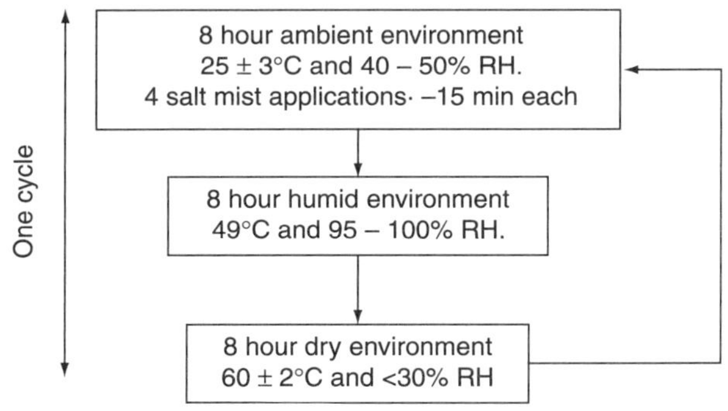

Figure 18.3 General Motors adopted cyclic testing GM9540P: one cycle represents 1 day of exposure [27].

phase $Mg_{17}Al_{11}Zn_1$ is $-1.23\ V_{ECS}$ ($-0.99\ V_{SHE}$) in ASTM D1384 water at pH 8.2 and room temperature. The aluminum concentration in the primary α phase is 3 wt%, whereas it is only 1.8 wt % in a die-cast alloy. Lunder et al. [32] and Beldjoudi et al. [33] showed a decrease of both the corrosion current density i_{corr} and the corrosion potential with the increase in Al content in a solution of 5% NaCl saturated with $Mg(OH)_2$.

In a deaerated solution, Baril and Pebere [34] found that as the concentration of sodium sulfate decreases from 0.1 to 0.01 M Na_2SO_4, the corrosion potential shifts toward more noble (positive) potentials of about 100 mV. The current densities were equal to, or greater than, 10 mA/cm^2 for 0.1 M Na_2SO_4 and were halved for 0.01 M aerated solutions. In deaerated solutions, the corrosion potential of pure magnesium is more noble (positive) and the anodic current densities are lower when compared to that of the aerated solutions. This was attributed to the absence of bicarbonate ions and carbon dioxide gas, which is present in the natural environment. An oxidant such as chromate shifts the potential to less negative potentials (i.e., more noble values) and reduces the corrosion current density [2].

18.2.3.2. Polarization Measurements

Potentiodynamic Polarization Studies The working electrode generally consists of a cylindrical rod on the order of 1 cm^2 that gives a good representation for magnesium alloys. It is more advisable to have a round surface to avoid the phenomenon of corrosion around the edges (concentration of corrosion current and products). The solution volume can be on the order of 100 mL per square centimeter of the metal surface. It is recommended to have a rather stable open circuit potential before scanning. The potentiodynamic scan can start at -250 mV versus corrosion potential (E_{corr}) and proceed to an anodic potential where the current density does not exceed 1 mA/cm^2. The potential scan rate can be on the order of a 0.1–0.2 mV/s; however, depending on the composition, microstructure of the examined alloy, and solution characteristics, slower scan rates could be advantageous.

Cathodic cleaning of the surface at moderate or high cathodic potentials is avoided as well as very slow stationary potential scanning due to the possible formation of magnesium hydride compounds that can change the properties of the natural corroded surfaces. It has been found that the level of cathodic polarization, adopted normally to clean the metallic surface of oxide films, can influence the form and characteristics of the polarization curves [11]. However, higher cathodic potentials have been used, even a polarization of -4 V. Recent data can show at what potentials magnesium hydrides may form, and thus change the electrochemical properties of the magnesium surface [2].

The polarization resistance method has been used successfully for comparison purposes to determine i_{corr} as a function of the elongation for AM50 and AZ91D commercial alloys. In an aerated borate solution at pH 9.7 and room temperature, the corrosion rate passes over a maximum in relation to the increase in plastic deformation. Also, the most negative value of the open circuit potential corresponds to the maximum value of the corrosion rate [9, 30, 35]. The electrochemical impedance spectroscopy at the open circuit potential or at imposed potentials gives R_p values when one is examining the solution resistance. The $1/R_p$ corrosion current calculations for AM50 and AZ91D were found to give a parallel classification to that obtained from potentiodynamic measurements, using the linear polarization method (0.2 mV/s) in aerated buffer solution at pH 9.7 [30].

Rotating Disk Electrode It has been reported that magnesium has abnormal polarization behavior. Under anodic polarization conditions, the broken areas in the surface film and other parameters could change with applied polarization potential or current density, giving nonreproducible results. During polarization of magnesium alloy, the influence of unstable hydroxide layers typically formed on magnesium and magnesium alloys may be suppressed with the help of a rotating disk electrode (RDE). Considering a less aggressive electrolyte such as 0.01 M NaCl under argon atmosphere and a RDE with 2000 rpm at pH 9 ($24 \pm 3\,°C$), polarization measurements are suited to characterize different magnesium alloys with appropriate current densities. Experiments can be carried out also in a 0.05 M Na_2SO_4 solution to avoid chloride ions at different pH values [36].

Polarization curves were carried out in a three-electrode cell, the working electrode was a RDE; a saturated Ag/AgCl (197 mV/E_{SHE}) electrode was used as the reference electrode and a platinum electrode as counterelectrode. The scan rate of 60 mV/min started from cathodic potentials from −2000 mV to about 0 mV or until the current density reaches the threshold of 5 mA/cm^2 [36]. A rotation speed of 2000 rpm was chosen because at this speed a stable flow field was developed, which removed the reaction products and therefore gave reproducible corrosion data. Aqueous electrolytes with pH 9 were found suitable to slow the dissolution rate (slower than at pH 7) and prevent formation of dense layers on the surface of the magnesium alloys, which occurs at higher pH values. During the polarization measurements, argon was bubbled into the electrolyte to avoid the adsorption of CO_2 from ambient air that can cause a decrease of pH value.

The polarization measurements were found suitable for characterizing the corrosion resistance of different magnesium alloys and were confirmed by electrochemical noise measurements (ENMs) without external polarization. The salt spray test also has the relative advantage of sweeping the corrosion products from the surface; however, it was originally developed for materials with coatings and does not adequately take into account the specific corrosion behavior of magnesium and magnesium alloys. The advantage of polarization measurements with the RDE as compared to the typically used salt spray test is the ease of obtaining an instantaneous corrosion rate and the removal of water-soluble and not strongly adherent corrosion products from the exposed surface. Some ENMs at open circuit potential confirmed the trend of results obtained using the RDE polarization method [36].

18.2.3.3. Cyclic Current Potential Measurements

During cyclovoltammetric sweeps, changes in the passive films and the solution composition at the metal–electrolyte interface (e.g., hydroxide formation and pH increase), as well as development of localized forms of corrosion such as pitting, take place. These can be identified on the cyclic current–potential curves and can be amplified by the choice of test conditions. A test method was designed based on current-limited cyclic current–potential curves. The solution was saturated with gases (air, nitrogen, oxygen, or carbon dioxide) for 1 hour before starting the test (meanwhile specimens were kept in the gas phase of the cell) and aerated with the respective gases during the whole test [37].

Following anodic polarization (scan rate 1000 mV/h) beyond the range of the pitting potential, after passing a threshold current of +0.05 mA/cm^2, a reversal of the polarization direction to more negative potentials is initiated to enable the repassivation of the material surface. The chosen current threshold prevents massive damage to the surface, on the one hand side, but also ensures a breakthrough at the pitting potential. After another reversal of the polarization direction at a negative current density of −0.05 mA/cm^2, further complete polarization sweeps are conducted. Observation of the tested surface during the experiment, as well as after the test, normally reveals that the total number of pitting sites does not increase significantly with rising number of polarization sweeps. That means pits initiated at the beginning tend to be deactivated during the cathodic polarization (until $-0.05\,mA/cm^2$ is reached) and are reactivated after the reversal of the polarization direction. A cathodic threshold of only $-0.01\,mA/cm^2$ is not sufficient to achieve this effect of repassivation [37].

These conditions enable one to examine and differentiate between the active and passive behaviors of the AZ91 alloy with its casting skin and ground surface. The ground specimen showed active–passive behavior; the open circuit potential of the ground specimen was shifted toward more noble values with increasing number of completed sweeps. It was also possible to recognize a limited passivation of the magnesium alloy. After the breakdown potential is exceeded, the current density rises steeply at the cyclic pitting potential. After two sweeps, a passive area up to $-1100\,mV$ (NHE) sets in. After further sweeps, the pitting potential will even shift toward the range of $-750\,mV$ (NHE) (see the arrows on Figure 18.4) with a decrease of the current density in passive state, whereas the maximum pitting potential may not appear in the last sweeps. In contrast to the ground specimen, no passivation is identified on a surface with a casting skin, indicating rapid breakdown of passivity at or near the open circuit potential (E_{oc}). In this case, merely a shift of the open circuit potential is taking place, presumably due to the alkalization of the boundary layer (Figure 18.4) [37–40].

18.2.3.4. Impedance Measurements

Impedance measurements are usually performed at open circuit potentials (E_{oc}) and under potentiostatic conditions.

EIS of Uniform and Pitting Corrosion Electrochemical impedance experiments were performed in aqueous 0.003 M NaCl solution, saturated with atmospheric oxygen. A corrosion cell (333 mL) consisting of an Ag/AgCl reference electrode, a Pt counterelec-

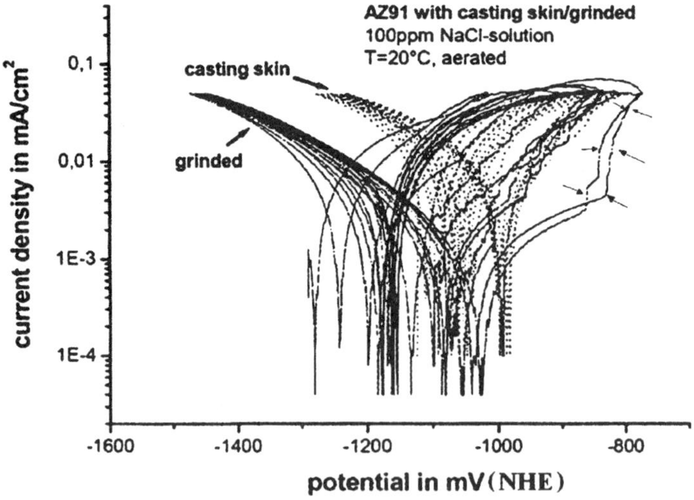

Figure 18.4 Cyclic polarization curves of AZ91 for cast (dotted line) and ground (solid line) specimens in chloride solution at ambient temperature using threshold limiting current densities of $0.05\,mA/cm^2$. The arrows show the passive behavior of the ground sample using chosen current thresholds [37].

trode, and the specimen as the working electrode (1.02 cm^2) was used. The electrolyte was not stirred and the temperature was controlled at (22 ± 0.5 °C) during the experiments. After 1 hour recording of the free corrosion potential, electrochemical impedance measurements at the free corrosion potential were carried out over the frequency range from 10 kHz to 0.1 Hz. The amplitude of the sinusoidal signals was 10 mV. The measurements were repeated every hour for a total of 22 hours and the total test period was 23 hours [41].

Two general types of impedance spectra were observed (Figure 18.5). Type I with the diffusive part at low frequencies was obtained when no sign of pitting or active corrosion was visible on the surface of the specimens. In that case the charge transfer resistance was determined using a Randles circuit. Type II was only obtained when pitting marks were visible at the surface (active corrosion). Here, the charge transfer resistance was calculated from the two intersections of the experimental curve with the real axis (0° phase shifts) giving the ohmic resistances of the solution (R_s) and the sum of the solution and charge transfer resistance (R_{ct}) [41].

Impedance measurements were carried out in a tetraborate buffer solution (0.05 M at pH 9.7). The scanned frequencies ranged from 6 mHz to 100 Hz. The Nyquist plots of magnesium alloys at open circuit potential exhibit two capacitive loops, one for high and intermediate frequencies and the other, the small one, for low frequencies. The first capacitive loop is attributed to the charge transfer process. Thus for frequencies higher than 1 Hz, a resistor R_p and a capacitor C_{dl} in parallel can model the electrode–electrolyte interface. A partial data fitting made with the Boukamp circuit equivalent software for the charge transfer process produced R_p (polarization resistance) and C_{dl} (double-layer capacitance) values [2, 42]. The R_p of the charge transfer process was 207.7 and 374.0 Ω.cm^2 for AM50 and AZ91D alloys, respectively. The obtained capacitance values were 22.6 and 68.0 μF/cm^2 for AM50 and AZ91D, respectively. The slightly lower value of C_{dl} for the AM50 alloy implies the formation of a thick protective film on the electrode surface; much lower C_{dl} values have been reported for other Mg-based alloys [43]. The second small capacitive loop is generally attributed to the mass transfer in the solid phase, which consists of the oxide/hydroxide layers [2, 44].

Over a period of 72 hours immersion of the die cast AZ91D in ASTM corrosive water saturated with magnesium hydroxide (pH 10.6), the R_t (transfer resistance) value increased gradually due to the formation of a protective corrosion film. A discontinuity during the first 10 hours is linked to a partial degradation of the coating, and this process is irreversible. This behavior is closely related to the microstructure and composition of this alloy. Maximum R_t

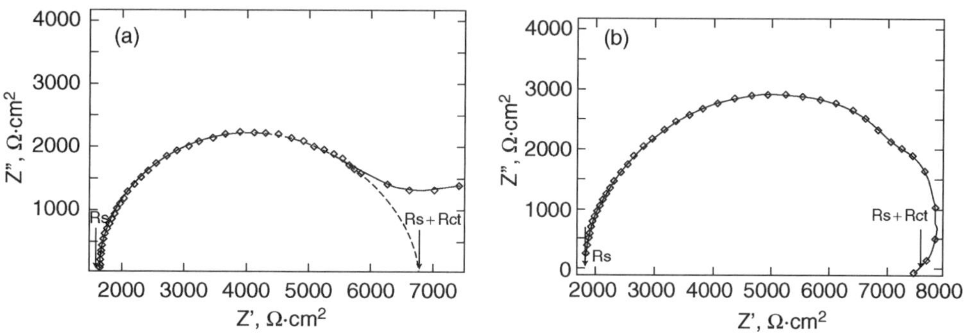

Figure 18.5 Typical impedance spectra obtained for AZ91D in 0.003 M NaCl: (a) type I, uniform corrosion with diffusive part at low frequencies; and (b) type II, pitting or active corrosion with inductive part at low frequencies [41].

values are on the order of 17 kΩ.cm^2 but dropped to about 5 kΩ.cm^2 during the 72 hour immersion. Mathieu et al. [9] reported that the potential amplitude is set normally to 5 mV, and the frequency range to 10 kHz to 5 MHz.

Generally, high- and low-frequency capacitive loops are seen. The high-frequency capacitive loop can be attributed to charge transfer reactions and the diameter of the loop to the transfer resistance. The capacitance values for this loop were always below 50 μF/cm^2 and can be attributed to the double-layer capacitance (C_{dl}) of the partially covered surface. C_{dl} values were on the order of 5–10 μF/cm^2 and increased progressively to about 20 μF/cm^2. The second capacitive loop is generally attributed to the diffusion of ions through the hydroxide or oxide coating. Generally, C_{dl} should decrease with the increase of passivity, but C_{dl} is not in agreement with the value obtained when die-cast AZ91D underwent the 72 hour immersion in ASTM D1384 corrosive water saturated with magnesium hydroxide. It is then stated that the high frequency loop can be related to the charge transfer and the surface film formation, as in the case of pure magnesium [2, 15, 34, 45].

EIS Spectra Under Anodic Polarization Impedance measurements are appropriate to show the influence of mechanical deformation on surface electrochemical reactions [30]. EIS spectra obtained under anodic polarization inside the potential range of the MgO formation exhibit one capacitive loop followed by a linear part for both magnesium alloys. An increase of R_p, which is significant in the case of AZ91D alloy, suggests that the layer is growing on the electrode surface. The equivalent circuit consists of a resistor (R_p) in series with a constant phase element (CPE), the two being connected with a capacitor (C_{d1}) in parallel. The CPE can be assumed to be Warburg diffusion according to the n values close to 0.5. Thus, under anodic polarization, the corrosion process is controlled by the mass transfer of the corrosion products through the oxide layers [46]. The Nyquist plots for both Mg alloys obtained under cathodic polarization show one capacitive loop, which is attributed to water reduction [2, 30, 46].

Cathodic EIS Measurements The following results show the conventional diagram for the 90%Mg–9%Al alloy in alkaline solution saturated with magnesium hydroxide and also describe the pseudo-inductance loop that can be formed by $Mg(OH)_{ads}$, MgH_2, or both species on the surface. Complex-plane plots exhibit a single semicircle in the case of a simple, one-step Faradaic process having a potential-dependent Faradaic resistance, R_F, in parallel with the double-layer capacitance. The diameter of such plots is a measure of R_F, which is inversely related to the reaction current or rate constant, at a given potential, and is hence usually exponentially related to potential (as for Tafel relation) of the process at the given dc current passing. In the case of two-step or more complex processes (as in corrosion), the complex-plane diagram can exhibit two or more semicircles but the analysis for this situation is quite complex [47].

The setup consists of a programmable sinusoidal generator and analyzer. The generator provides an output signal of known amplitude (10 mV in the present case) at various frequencies. This signal is applied to the electrochemical system to be examined, through the potentiostat [47]. The response of the electrochemical system to a sinusoidal input signal has a fundamental ac component of the system's responses. The function of the analyzer is to accept the electrochemical system's response to the stimulus signal of the generator using a correlation process, resolving it in terms of real and imaginary components of the overall impedance for plotting in the complex plane. Relations of phase angle and log |Z|, the impedance vector, to log frequency can also be analyzed and displayed simultaneously. The ac impedance measurements can be made using a Solartron 1260 Frequency Response

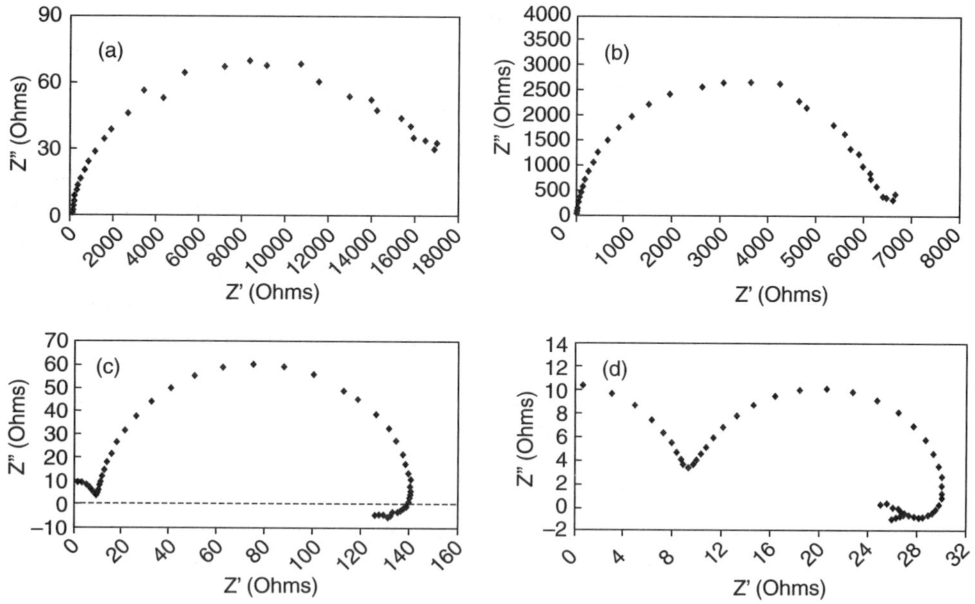

Figure 18.6 Complex plots were taken at potential values of (a) −1.2, (b) −1.4, (c) −1.7, and (d) −1.9 V/SCE [47].

Analyzer (FRA), equipped with a Solartron Potentiostat and a 1287 Electrochemical Interface (ECI), which controlled the electrode system. The data are generally stored on a $3\frac{1}{4}$ inch disk and then transferred to a personal computer for further data fitting treatment analysis [47].

The test was conducted in the potential range of −1.2 to −1.9 V (vs. SCE) (Figure 18.6). In Figure 18.6a,b, a single semicircle was observed in the complex-plane plots. The potential-dependent polarization resistance, R_R, decreases with increase of the cathodic potential, which indicates that a simple Faradaic process takes place over this potential range [47].

For plots in Figure 18.6c,d, the ac impedance behavior differs from that shown in Figure 18.6a, b. The polarization resistance for this case decreases with increase of cathodic potential and the loop that is characteristic of pseudo-inductance [48] appears in the complex-plane plot. The observation of the pseudo-inductance loop at the potential range of MgH_2 formation is in good agreement with results reported in Simpraga et al. [49], and it is attributed to possible H_2 evolution at the hydride phase for the tested potentials. Such a pseudo-inductance loop was also observed in Mg corrosion as reported elsewhere, where the Mg corrosion was tested in aerated sodium sulfate solutions. Baril and Pebere [50] attributed the effect to adsorbed intermediate, probably $Mg(OH)_{ads}$. It can then be suggested from EIS studies that the pseudo-inductance loop is due to formation or decomposition of MgH_2 or a combination of both processes [47].

18.3. GALVANIC OR BIMETALLIC CORROSION OF MAGNESIUM AND ITS ALLOYS

In general, the corrosion potential difference between anode and cathode becomes the driving force for galvanic corrosion. A galvanic corrosion test between low-carbon steel and magnesium alloys (NA42/AZ91D) was accomplished in acid–chloride solution (pH 5.4, 200 ppm Cl^-) at ambient laboratory temperature ($22 \pm 2\,°C$) to compare the galvanic effect. The galvanic currents were measured using a zero-resistance ammeter (ZRA) for 7 hours. The ratio of the anode to the cathode area of the specimen was 1 : 1. Although the Mg–4Ni–xAl alloys had lower corrosion resistance than AZ91D, the results of galvanic corrosion tests between the low–carbon steel and the Mg alloys indicated that the galvanic corrosion resistance was higher in Mg–4Ni–2Al (NA42) than in AZ91D [51]. It seems that the effect of nickel as the alloying element is shifting the magnesium alloy potential to less negative or more positive potentials than that of aluminum [2].

18.4. LOCALIZED CORROSION OF MAGNESIUM AND ITS ALLOYS

In natural atmospheres, the corrosion of magnesium can be localized. The conductivity, ionic species, differential oxygen concentration cell, temperature of the electrolyte, alloy composition, and homogeneity of the microstructure can influence the corrosion morphology. The susceptibility of magnesium alloys to localized corrosion can be evaluated as that of an alloy having an active–passive behavior in certain media by cyclic voltammetric, potentiodynamic, galvanostatic, scratch potentiostatic, and triboellipsometric methods, pit propagation rate curves, impedance spectroscopic studies, and electrochemical noise measurements. The appearance, morphology, distribution, and depth of pits (average penetration of the 10 deepest pits and the deepest one) should be determined in parallel with pitting potential determinations. Microscopic studies are especially valuable in addition to electrochemical studies. Time to perforation data can be obtained by designing a specimen that is pressurized with air. This pressure is monitored over a period of time until failure is indicated by a decrease in pressure [2, 4, 52].

18.4.1. Open Circuit Potential and Pitting Corrosion Studies

In a sodium chloride, air-saturated solution without peroxide additions, the free corrosion potential of Mg and some of its alloys were measured in the range of -1500 to -1600 mV/SCE for chloride concentrations of 5×10^{-3} to 5×10^{-4} M. In these solutions, no visible localized attack occurred, but a weight loss of approximately 5 mdd (milligram per square decimeter per day) (equivalent to a current density of 5 $\mu A/cm^2$) was measured. On the other hand, with the addition of peroxide, the potential becomes much nobler, and the effect of the critical chloride concentration becomes clear. With 0.005 M NaCl, localized corrosion occurs, and the potential remains below the hydrogen line (≈ -1.009 mV/SCE at pH 13), indicative of a corrosion mechanism in which hydrogen reduction is the cathodic reaction. With 0.0005 M NaCl, no localized attack occurs, and the potential shifts toward very noble values. Localized attack is observed only when the free corrosion potential adopts a value active to the hydrogen line (Figure 18.7) [2, 52].

An increase in the chloride level to the critical value for each alloy leading to the onset of localized corrosion is accompanied by a shift in free corrosion potential to a value active to the hydrogen line, thereby making the production of hydrogen thermodynamically possible. This means that hydrogen is a prerequisite for localized corrosion [2, 4, 52].

It is possible to use the pH change with time as a measure of pitting corrosion resistance of magnesium alloys. In acidic solutions, pitting is initiated mainly by the reduction of hydrogen ions, but in neutral solutions, the reduction of dissolved oxygen plays a major role during pitting. At a relatively high corrosion potential, at the initiation step, oxygen

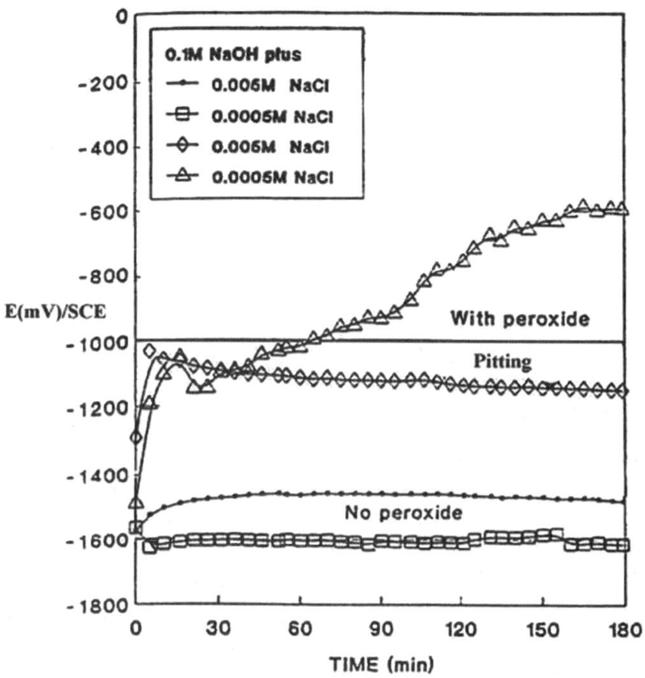

Figure 18.7 Corrosion potential of magnesium in 0.1 N NaOH solution. The hydrogen evolution reaction is around −1.009 V/SCE at pH 13 [2, 52].

consumes hydrogen ions. In neutral or weak acidic solutions, two steps of pitting were clear—initiation and propagation. Between these two steps, there is a retention period that can vary in length according to the corrosion resistance of the alloy and disappears completely at pH 1.25 in 5% aerated sodium chloride solution [2].

In many magnesium alloys, the open circuit potential is higher than the pitting potential, and thus the pitting potential cannot be determined with traditional methods. However, when the solution is mild and the alloy shows a good active–passive behavior, cyclic voltammetry can be used. By using the polarization curve showing the active–passive behavior for a magnesium alloy (with or without a coating), and as a function of increasing chloride concentration, E_{bd} and $E_{protection}$ can be determined. The breakdown potential corresponding to the considerable increase of the anodic current at a certain scan rate gives the susceptible condition for the initiation of localized attack. The nobler the breakdown potential, the more resistant will be the alloy to pitting and crevice corrosion. The potential at which the loop is completed upon reverse polarization determines the potential below which there is no localized attack for this scan rate. The obtained values are dependent on the scan rate. Also, allowing too much pitting propagation to occur along with the accompanying chemistry changes can influence the reversal in the scan rate [53–55].

The quality of passivation depends on the solution parameters, pH, chloride concentration, agitation (if any), and temperature. For example, in ASTM D1384 corrosive water saturated with $Mg(OH)_2$ at pH 10.6, and 25 °C, the critical passivation current and the passivation current are higher than 100 $\mu A/cm^2$ for pure Mg and lower than 10 $\mu A/cm^2$ for AZ91D (die cast). It is not only the absolute value of the passivation potential that counts for easy passivation, but also the difference between the passivation potential and the corrosion potential; this indicates the potential polarization required to attain passivation [9]. The good corrosion resistance of thixomolded AZ91D was electrochemically verified in 0.17 mol/L NaCl solution saturated with $Mg(OH)_2$ by the presence of pseudo-passivation with a small $i_{passive}$ and a noble E_{bd} [2, 53, 56].

For sodium chloride solution deaerated at 25 °C, it is believed that concentrations up to 0.05–0.1 M should be sufficient to identify relative resistance of magnesium alloys in an alkaline solution of magnesium or sodium hydroxides at about pH 11–13 [11]. The scan rate should be around 0.1–0.2 mV/s, or much less than 6 mV/s for iron-, nickel-, or cobalt-based alloys [2, 15]. Mitrovic-Scepanovic and Brigham [52] found that the critical concentration of chloride that causes pit initiation on a number of magnesium alloys falls in the range of 2×10^{-3} to 2×10^{-2} M NaCl.

Another solution is 0.1 M NaOH + 0.005 M NaCl or 0.01 M NaCl, or 0.02 M NaCl [4, 15]. Immersion takes place in ASTM D1384-87 water for 90 hours at pH 8.3 containing 165 mg/L NaCl, 138 mg/L $NaHCO_3$, and 148 mg/L Na_2SO_4. This is followed by potentiodynamic polarization for the determination of pitting potential in the same solution but saturated with Mg $(OH)_2$ [10].

In the *scratch-repassivation method* for localized corrosion, the alloy surface is scratched and exposed to a constant potential. The current change is monitored as a function of time and this will show the influence of potential on the induction time and the repassivation time. A careful choice of the level of potential between the breakdown potential and the critical pitting potential can give the critical pitting potential for a certain material under specific environmental conditions [2, 4].

ASTM F746-87 is designed to determine comparative laboratory indices of performance and can be used to rank materials in the order of increasing resistance to pitting and crevice corrosion. The resistance of surgical implants to localized corrosion is carried out in dilute sodium chloride solution (9 g/L) under the specific conditions of the potentiodynamic test method. Typical transient decay curves under potentiostatic polarization should monitor susceptibility to localized corrosion. Alloys are ranked in terms of the critical pitting potential: the higher (more noble) this potential, the more resistant the alloy is to passive film breakdown and to localized corrosion [57].

The corrosion potential of the working electrode (specimen) is recorded for 1 hour in the NaCl solution (E_1). The current is recorded at +0.8 V/SCE for a period of time that depends on the reaction. If localized corrosion is not stimulated in the initial 20 seconds, the polarizing currents will remain very small or decrease rapidly with time. Stimulation of localized corrosion is marked by increasing polarization current with time or by current densities that exceed 500 μA/cm^2. The test consists of alternating between stimulation at 0.8 V/SCE and returning to a preselected potential (E_1 + a jump of + 50 mV) until continuous increases or large fluctuations in current occur during the 15 minute observation period. Evidence of pitting and crevice corrosion should be noted. This procedure is strongly recommended for magnesium alloys in development for human body implants [2].

18.4.2. Electrochemical Noise Measurements

Electrochemical noise measurements (ENMs) give the electrochemical current noise that corresponds to the galvanic coupling current between two identical working electrodes (WEs), frequently with an exposed area of 1.0 cm^2 for each electrode kept at the same potential. Potential and current can be obtained during a time period of 1024 seconds, which fixes the frequency range (Δf) in the region between 1 Hz and about 1 mHz.

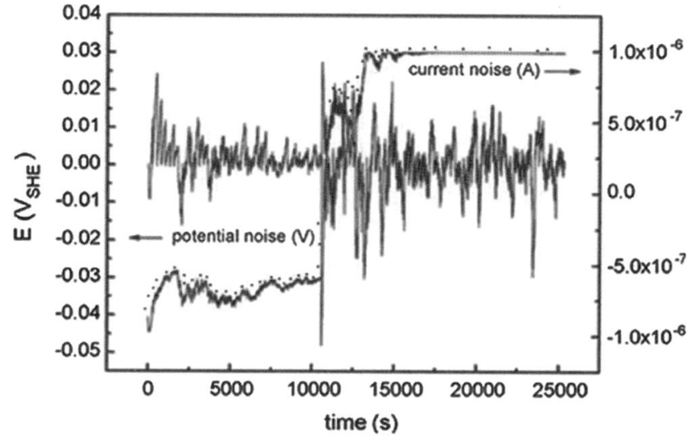

Figure 18.8 Electrochemical noise of AZ91D alloy during a 256,000 s immersion period in 0.05 M NaCl aqueous solution with pH 12.0 [58].

Figure 18.8 shows the collected electrochemical noise data of the AZ91D alloy for a 25,600 s immersion period. Every set of EN records, containing 1024 data points, was registered with a data-sampling interval of 0.25 s. The data were recorded immediately after the immersion of the working electrodes in the solution [58].

The analysis of EN data can be performed in both time and frequency domains. Generally, the low current noise corresponds to a more noble potential and can be attributed to the "passive zone." The high current noise corresponds to the first potential noise and can be attributed to the "active zone." In the time domain, the most interesting parameter of the statistical analysis is the noise resistance (R_n), defined as the ratio of a standard deviation of the potential noise to that of current noise, which can be associated with the polarization resistance (R_p). The ratio $1/R_n$ is proportional to the corrosion rate [59].

The corrosion rate can be obtained from an estimate of the polarization resistance, R_p, which is inversely related to the linear corrosion rate by the Stern–Geary approximation. Three approaches are possible for obtaining instantaneous values of R_p from the EN current and potential measurements: namely, noise resistance, spectral noise resistance, and self-linear polarization resistance (SLPR). Considering the SLPR method, noise measurements were carried out for two magnesium alloys, AZ91D and ZA104, in a 5% sodium chloride solution saturated with magnesium hydroxide. The samples were not subjected to an external applied current.

A measure of the localized corrosion behavior, according to SLPR theory, is the probability density of anodic transients. This can be obtained by plotting the frequency of anodic transients versus the intensity of anodic transients as shown in Figure 18.9. Most of the current transients for the ZA104 sample were much larger than those from the AZ91D sample. The frequency of transients from the ZA104 sample was also much higher than that for the AZ91D sample. The corrosion rates, as measured from SLPR theory, showed corrosion ratings similar to that obtained from the conventional polarization resistance method. Also, the scanning reference electrode technique (SRET) illustrated that localized corrosion occurred more frequently on ZA104 rather than on AZ 91D [60]. In a consecutive study, it was shown that increasing the level of zinc and lowering the level of aluminum relative to the AZ91D sample does not improve resistance to pitting corrosion [61].

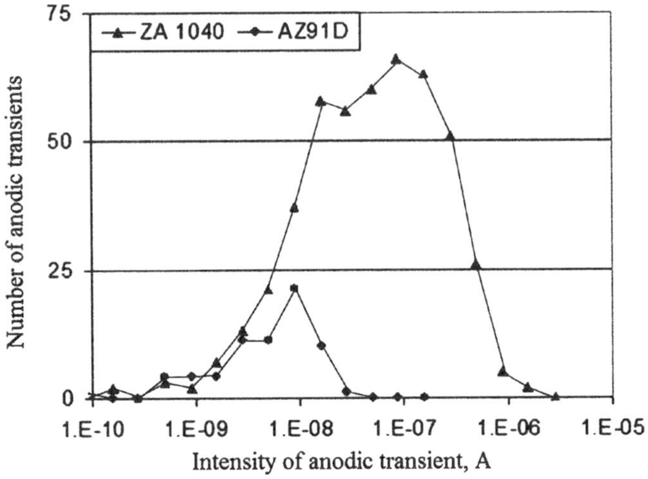

Figure 18.9 Noise electrochemistry using SLPR theory: number of anodic transients versus the intensity of transients for AZ91D and ZA104 samples during an 8 h immersion in 5% NaCl solution saturated with $Mg(OH)_2$ [60].

Power Spectral Density Parameter Noise data can be transformed into the frequency domain using a fast Fourier transform (FFT) algorithm and a fast wavelet transform (FWT) algorithm. For FFT, the power spectral density (PSD) can be expressed in a frequency domain from 5×10^{-3} to 4 Hz. The power spectral densities of potential (PSD_v), of current (PSD_i), and of the ratio ($PSDR_{sn}$) are calculated as follows:

$$\log \mathrm{PSD}_v = A_v + S_v \log f \quad (\mathrm{V}^2/\mathrm{Hz})$$
$$\log \mathrm{PSD}_i = A_i + S_i \log f \quad (\mathrm{A}^2/\mathrm{Hz})$$
$$\log \mathrm{PSDR}_{\mathrm{sn}} = A_\mathrm{r} + S_\mathrm{r} \log f \quad (\Omega^2)$$

where S_v and A_v, S_i and A_i, and S_r and A_r are, respectively, the slope and the noise intensity of potential, current, and ratio of PSD plots [62].

Corresponding PSD curves for magnesium alloys in 0.05 M NaCl solution at pH 6.1 [59] are ahown in Figure 18.10. The slope of potential is on the order of -3 for the PSD of potential in 0.05 M NaCl solution at pH 6.1; the slope of current is on the order -2.5 for the PSD of current in 0.05 M NaCl solution at pH 6.1. The slope can be on the order of -0.25 [$S_{\mathrm{Rsn}} = 0.5(S_v - S_i$]. The influence of pitting corrosion on this parameter can then be studied as a function of immersion period, chloride concentration, or microstructural variations [59].

It was also found that the pit lifetime analysis of EN data in the frequency domain involves the calculation of the noise PSD. Pitting corrosion of AZ91D-DC (die cast), AZ91D-ESTC (electromagnetically stirred billets; thixocast), AZ91D-SFTC (billets solidified freely; thixocast), and AJ62x-DC (die cast) specimens was studied in alkaline chloride medium (0.1 M NaOH + 0.05 M NaCl + 2 mL 30% H_2O_2) at 25 °C and pH 12.3. The passive behavior of four specimens of AZ91D-DC (die cast) magnesium alloys correspond to a lower value of A_v and A_i as compared to that obtained during active behavior ($A_v < -9.19$ and $A_i < -17.44$ against $A_v > -6.86$ and $A_i > -14.76$, respectively). Effectively,

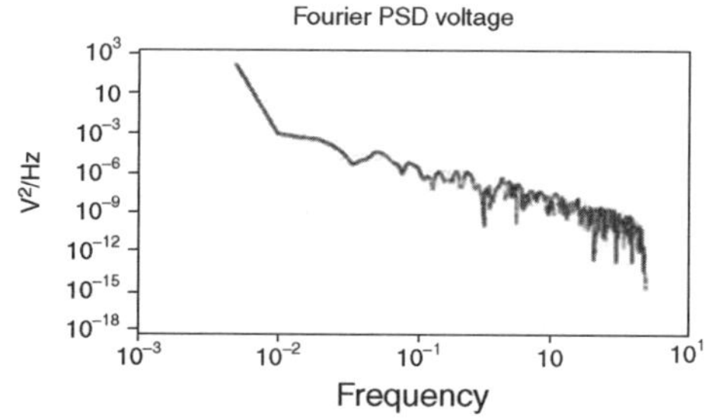

Figure 18.10 Power spectral density (PSD) plots for magnesium alloys in 0.05 M NaCl solution at pH 6.1 [59].

passive behavior involves a high value of A_r. For both die-cast specimens, A_r increases with time, meaning that the intensity of the corrosion phenomenon decreases. Also, EN analyses in the frequency domain were able to show even a change in the morphology or the submode of pitting, becoming a classical pitting type, for AJ62x and AZ91D die-cast specimens [62].

Typical PSD current plots of passive states for three different AZ91D specimens and AJ62 are shown in Figure 18.11 [62].

In the active state, the change in the PSD of the current plots after 2, 5, 11, and 16 h of immersion in the alkaline chloride medium were examined. AZ91D-SFTC and AZ91D-ESTC specimens present similar noise intensity for all periods of immersion. This type of corrosion has specific frequency parameters: A_v, A_i, S_v, and S_i are approximately −6.5, −15.0, −3.0, and −1.8, respectively. For both specimens, A_r increases with time, meaning that the intensity of the corrosion phenomenon decreases. For AJ62x-DC and AZ91D-DC specimens, both the noise intensity and the PSD slopes decrease with the period of immersion. A_v changes from −6.8 to −8.5, A_i from −14.0 to −16.0, S_v from −2.6 to −3.6, and S_i from −2.0 to −2.5. The shift of the noise toward lower levels and the increase of the negative PSD_i slope with time reflect a change in corrosion type. This change is more significant for AJ62x-DC than for AZ91D-DC. AJ62x-DC presents the lowest A_r value (9.4) and AZ91D-ESTC the highest (10.9).

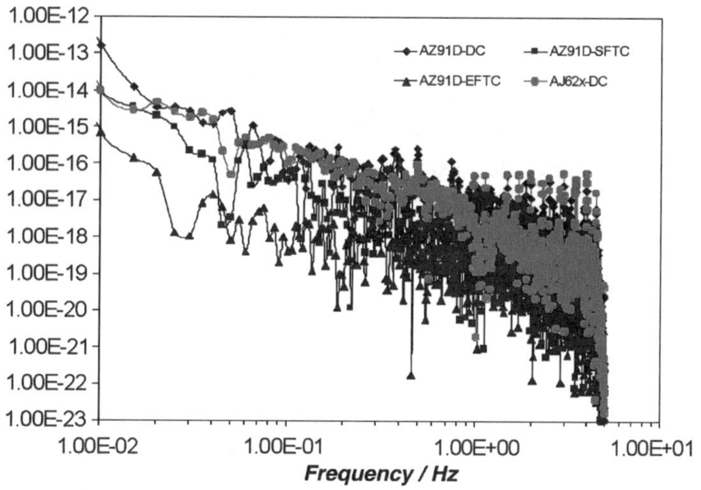

Figure 18.11 PSD current plots versus frequency showing a passive zone for different magnesium alloy specimens in alkaline solution [62].

EN time records, R_n, PSD, and the energy distribution plot (EDP) were able to provide useful information about the three corrosion stages dominant in the corrosion processes of AZ91D alloy in chloride medium. EDP can be deduced from wavelet analysis using WT sym8 for the EN data [58]. See Chapter 10 for more details.

EN Galvanostatic Mode The corrosion behavior of the magnesium alloy AM60 has been investigated by ENMs in chloride and sulfate solutions. Galvanostatic tests in combination with the measurements of the current noise were useful in determining the breakthrough potential in 0.01 M aerated sodium sulfate solution. In this study, the potential was stabilized for 30 minutes in the open circuit and then subjected step by step to an increasing gradual imposed current (galvanostatic mode). The potential–time noise diagram shows oscillations characteristic of the pitting phenomenon at a certain level. The potentials determined from this method were found to be very similar to those obtained by cyclic voltammetry for the same conditions using a scan rate of 0.1 mV/s for a current density of 1 mA/cm^2. This technology helped to examine the influence of different corrosion inhibitors [2, 31].

18.4.3. Magnesium SRET Studies

SRET measurements were carried out for two magnesium alloys, AZ91D and ZA1040, in a solution of 5% sodium chloride saturated with magnesium hydroxide. In this approach, the micropotential gradients in solution just above the surface of the material under study are recorded. The micropotential gradient in the solution that is perpendicular to the surface is proportional to the ionic current either leaving from or going toward the surface. The potential difference is equal to the ionic current times the solution resistance. The SRET maps of potential difference, or equivalently ionic current, over both samples are examined at two different times: that is, 20 minutes and 8 hours of immersion. An example of a SRET map of potential difference, or equivalently ionic current, over the AZ91D sample after 40 minutes is shown in Figure 18.12. Most of the AZ91D surface is cathodic. The potential differences were discretized into zones corresponding to neutral, cathodic, first-stage anodic, second-stage anodic, and third-stage anodic. There are fisrt-stage anodic zones with neutral regions separating the cathodic regions. In contrast, most of the ZA1040 surface of the other alloy was at first-stage anodic with only small regions of cathodic activity. The ZA1040 sample also had small zones of intense localized corrosion up to third-stage anodic [63].

Figure 18.13 gives a two-dimensional SRET map of a ZACS1040305 and shows the distribution of pits on the surface after only 48 minutes in a solution of 5% NaCl saturated with magnesium hydroxide. As a function of time, some of the pits were blocked, giving rise to growing pits in other sites. This behavior was closely related to the microstructure and the alloying elements since the commercial AZ91D showed a different pattern of pitting corrosion [2, 63].

The SRET studies illustrated that localized corrosion occurred more frequently on the ZA1040 sample than on the AZ91D sample. The location of localized corrosion was also much more likely to change on the ZA1040 sample than on the AZ91D sample [2, 63]. This is almost a semiquantitative analysis of the distribution and the location of the anodic and cathodic sites and a clear association of the corrosion current to pitting corrosion. Combination of the surface analysis techniques of electron microscopy, SRET, and EN provides interesting and useful comparisons. It was found that increasing the level of Zn and

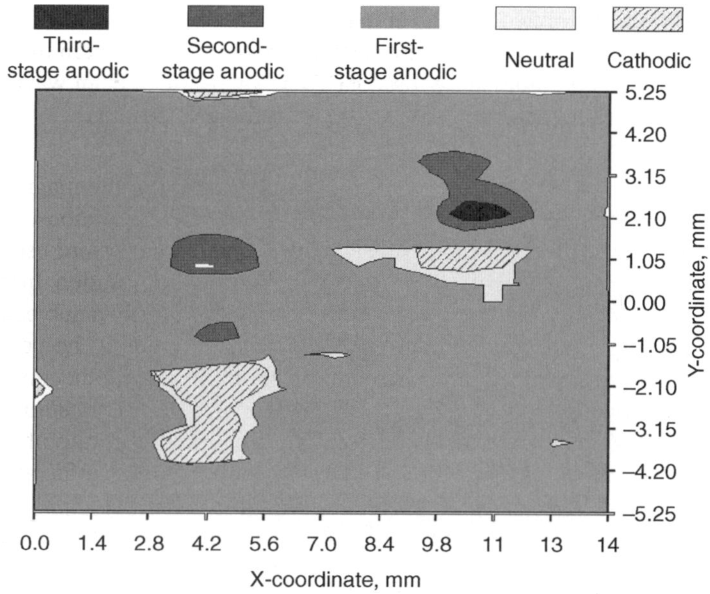

Figure 18.12 Scanning reference electrode technology (SRET) map of ionic current (potential difference in solution above sample) for the AZ91D sample after 40 minutes immersion in 5% NaCl solution saturated with $Mg(OH)_2$ [63].

lowering that of Al relative to their levels in the magnesium alloy AZ91D does not seem to improve corrosion resistance.

Three different casting methods were used to produce box-like parts from which corrosion specimens were prepared. AZ91D alloy was die cast (AZ91D-DC), thixocast

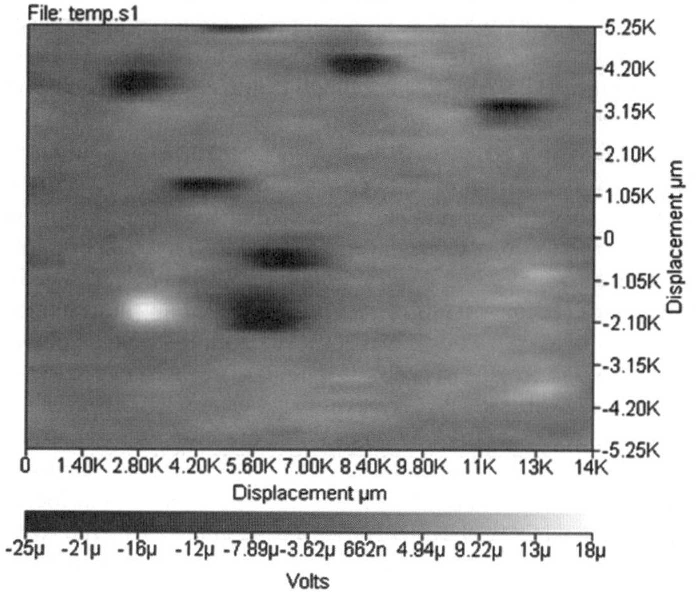

Figure 18.13 Two-dimensional SRET map of a ZACS after 48 minutes in a solution of 5% NaCl saturated with $Mg(OH)_2$ [63].

from billets solidified freely in permanent mold (AZ91D-SFTC), or thixocast from electromagnetically stirred billets solidified in a water-cooled and coated copper mold (AZ91D-ESTC); AJ62x alloy was only die cast (AJ62x-DC). Corrosion test specimens of 10 mm × 10 mm × 5.4 mm in size were embedded in acrylic resin, polished with SiC abrasive paper down to 1200 grit, cleaned with alcohol, and then dried before exposure to the electrolyte [64]. The testing solution was 0.05 M NaCl solution at pH 6.1.

In SRET measurements, mechanical or magnetic stirring is generally not used; however, the scanning vibrating probe (SVP) could give the solution over the specimen a weak agitation during the potential measurements. SRET analyses and polarization tests were performed at 23 and 25 °C, respectively, under atmospheric oxygen. The test specimen was mounted horizontally in the cell of the SRET apparatus (EG&G Instruments—Model SVP100) for free corrosion potential measurement. After ensuring that the surface of the specimen was parallel to the surface on the Perspex tripod and leveled, the probe was lowered to a distance of 100 μm above the sample surface, and the chloride solution was added. The basic configuration consists of connecting one positive input to the vibrating probe and the other negative input to a separate reference probe (carbon electrode in this case) placed in the solution close to the surface to be measured.

After conditioning of the signal, a rectangular area (18.0 mm × 13.5 mm) was scanned, overlapping slightly the acrylic resin. The vibrating probe amplitude was adjusted to 45 μm, and the data collection was scanned at 32 points per scan line on the X-axis and for a total of 24 lines on the Y-axis. For each data point, 500 potential measurements were taken at a frequency of 81 Hz. Each SVP scan lasted approximately 19 minutes and 1 min of rest was allocated between each scan. All the experiments were conducted over a period of 16 h. The overall SRET data were used to reproduce a three-dimensional image mapping of the surface potential over the specimen [64].

The term quasi-electromotive force (QEMF), corresponding to the difference between the most active cathode potential and the most active anode potential, is used to interpret the SRET results. It is believed that the corrosion rates and tendency of metals and alloys can be related at least indirectly to the evolution of the electromotive force of the corrosion cell. This difference could normally be determined for every complete 12 min scan of the whole surface and should have the same trend as the electromotive force (EMF) of the most active corrosion cell [41].

The QEMF values of the corrosion cell as a function of immersion time for four specimens of magnesium alloys in 0.05 M NaCl solution are plotted in Figure 18.14. The average QEMF values of corrosion cell during a 16 h immersion period were 402 μV for AJ62x-DC, 398 μV for AZ91D-DC, 188 μV for AZ91D-SFTC, and 158 μV for AZ91D-ESTC. It is clear that AZ91D-ESTC had the lowest QEMF among the four specimens and had the best corrosion resistance, whereas AJ62x-DC had the highest QEMF among the four specimens and had the worst corrosion resistance. Considering the QEMF of the different samples and measuring the depth of the pits of the four samples, it can be stated that the AJ62x-DC specimen has the deepest pits. It is then assumed that there is a direct relation between the magnitude of QEMF and the corrosion rate of every specimen [64].

SRET studies for magnesium alloy AZ91D were carried out for the skin and the 0.25 mm depth bulk samples as a function of immersion time. The QEMF values of the skin were always much higher than that of 0.25 mm depth bulk. Mostly, the difference reached a factor of 10. This means that the skin sample would undergo much more severe corrosion than the bulk sample. Actually, many deep pits formed on the skin sample, while there was no visible pitting on the bulk sample surface after 10 hours of experiment. It was reported that the high corrosion rate of the skin samples was due possibly to some contaminants with

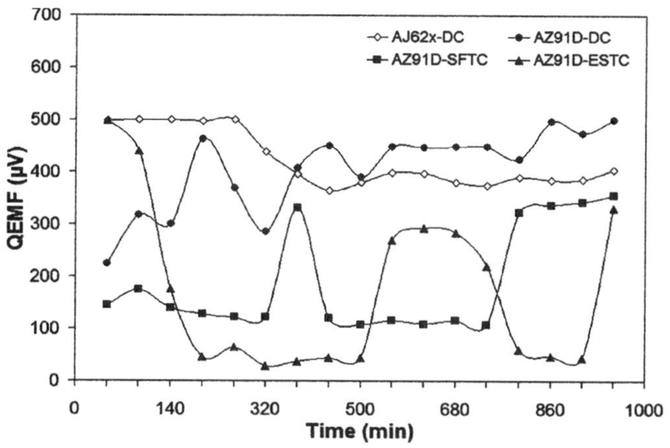

Figure 18.14 QEMF values of corrosion cell versus immersion time in 0.05 M NaCl solution at pH 6.1 and 23 °C for each specimen [64].

low hydrogen overpotential such as iron and nickel from the mold during the casting process [41].

The maximum local anode currents (MLACs) can be obtained if the measured anode potential values are multiplied by the calibration factor (2.5 μA μV). Figure 18.15 shows the MLAC values as a function of immersion time for skin sample and 0.02, 6, 8, and 10 mm depth bulk samples of AZ91-TC alloy in 0.014 M NaCl solution. It is seen that the MLAC values of the bulk samples are much lower than that of the skin sample. However, the 0.02 mm depth sample had rather high MLAC values, though lower than that of skin. It can be assumed that the 0.02 mm depth bulk sample was not completely polished and that it still kept a small fraction of the as-cast skin. It is clear that the 8 mm depth bulk sample had very low anode current; at the end of the 10 h immersion, the MLAC was only 17.5 μA/cm^2. The average MLAC value during the 10 h was 66.6 μA/cm^2. The maximum corrosion rate was less than 0.2 μm/h. So, after the immersion experiment, the surface was

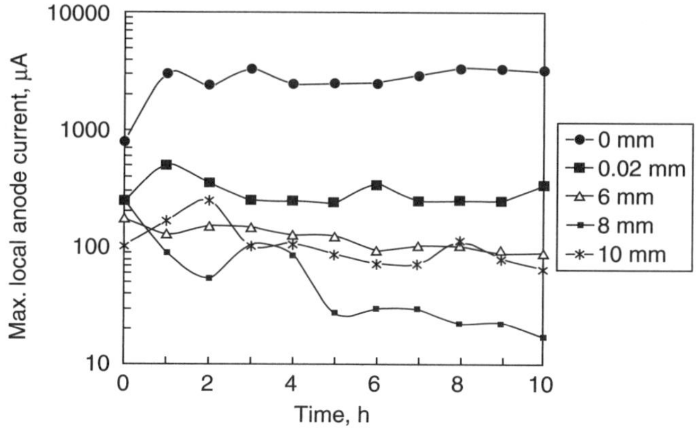

Figure 18.15 Maximum local anode current (MLAC) as a function of immersion time for skin sample (0 mm) and 0.02, 6, 8, and 10 mm depth bulk samples of AZ91-TC alloy in 0.014 M NaCl solution at 23 °C [41].

intact without any pitting. As compared to the skin, the 10 h average MLAC value for the skin sample was 2696 μA/cm^2, that is, 40 times that of the 8 mm depth bulk sample [41].

18.5. METALLURGICALLY INFLUENCED CORROSION OF MAGNESIUM AND ITS ALLOYS

There are rare instances of reported intergranular corrosion (IGC) of magnesium alloys, as in the case of chromic acid solution contaminated with chlorides or sulfates. Testing of intergranular attack necessitates metallographic examination. Certain media have been commonly used for evaluating the susceptibility to IGC of magnesium, copper, lead, and zinc alloys [65]. An innovative procedure for IGC of magnesium should be considered since there is no special targeted ASTM procedure. Susceptibility to this type of attack depends primarily on the type of alloy, fabrication process, and the presence of efficient cathodic sites at the interface. The standard practice ASTM G110-92 that evaluates IGC resistance of heat-treatable aluminum alloys by immersion in sodium chloride and hydrogen peroxide solution could be the most suitable standard for testing intergranular attack in magnesium alloys [15]. According to this standard, an immersion period of ~ 6 h in a solution containing 57 g of sodium chloride and 10 mL of hydrogen peroxide (30%) per liter is recommended. A metallographic cross section approximately 20 mm in length, preferably through a corroded area, should be examined. The type, extent, and depth of IGC can be determined [1, 15]. It is very possible that a less concentrated sodium chloride test solution could be appropriate for magnesium alloys. For simple etching, for surface cleaning, and for microscopic observation or testing purposes, an immersion for 1 minute into a boiling 20% CrO_3 solution with 1% silver nitrate is usually sufficient [2].

18.6. MIC AND BIODEGRADATION OF MAGNESIUM AND ITS ALLOYS

Evaluation of MIC of magnesium should be similar to that of other metals and testing standards of Al in this respect could be extended to Mg alloys (see Chapter 17). The behavior of magnesium alloys in 0.9% NaCl solution and Hank's balanced salt solution (HBSS) was studied using EIS and cyclic polarization (CP). The purpose of the research was to find the correlation between biological environments, corrosion exposure, and magnesium alloys. During the experimental procedure, different physiological solutions, including fetal calf serum (FCS), at different temperatures as well as an external pH control were used. A system formed of bioreactor, in which key parameters of a solution were adjusted, and electrochemical cell, in which measurements were made, was realized. The following parameters inspired from living organisms are considered: temperature of solution, 36.5–40 °C; pH, 7.35–7.45; volumetric speed of movement of a solution between a reactor (500 mL) and an electrochemical cell (500 mL), 100 mL/min; speed of circulation of a solution inside the electrochemical cell, 300 mL/min. Measurements were taken using a potentiostat linked to a PC (Table 18.1) [66].

It can be shown that the impedance differs from one solution to another and that HBSS is less aggressive than the NaCl solution only (Figure 18.16). From these data, the best anticorrosive properties of alloy AZ91 can be seen more evidently as compared to AM50. The HBSS solution acts as a stable buffer solution and is less reactive compared to saline 0.9% NaCl. It was stated that the presence of FCS combined with HBSS in the tested

Table 18.1 Results of Electrochemical Measurements of Magnesium Alloys AM50 and AZ91 in Solution at $T = 37$ °C and pH = 7.4

Alloy	Corrosion rate before CP (μm/yr)	Impedance before CP (ohms)	Solution
AM50	57.15	900	0.9% NaCl
AZ91	15.24	1800	0.9% NaCl
AM50	22.86	2600	HBSS
AZ91	12.7	3500	HBSS

Source: Reference 66.

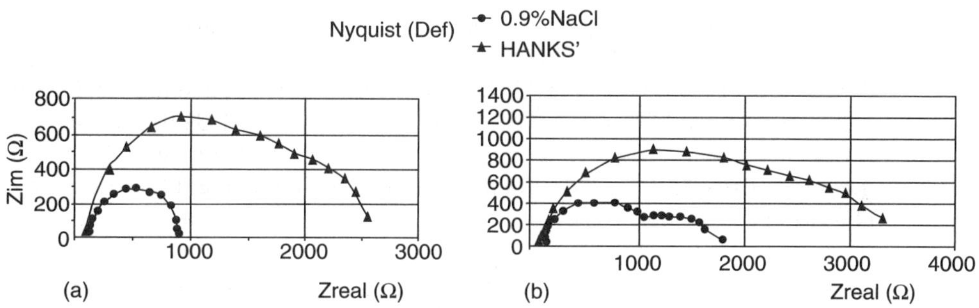

Figure 18.16 EIS for (a) AM50 and (b) AZ91 before cyclic polarization, $T = 37$ °C and pH = 7.4 [66].

solutions influences or could decrease the corrosion rate for both AZ91 and AM50, which should promote future magnesium implants [66].

18.7. CORROSION FATIGUE

The general characteristics of magnesium alloy fatigue are that the curves generally feature a fatigue limit, and this is similar to that of most mild steel alloys, and in contrast with the behavior of most nonferrous alloys, particularly aluminum alloys. Under conditions where the dead-load component is relatively high, $R > 0.25$, it is common for the tests to be conducted within the plastic region of the stress–strain curve, that is, at maximum loads, exceeding the 0.2% proof stress of the alloy [67]. The threshold stresses for monotonic or cyclic loading of initially plain specimens are essentially the same and are empirically related to the threshold stress-intensity factors from notched specimens subjected to constant load or constant strain.

Air, deionized water (pH 7.0), tap water generally with pH 5.0, owing to the absorption of carbon dioxide, 3.5% sodium chloride solution, and natural seawater have frequently been employed. At 7,000,000 cycles, the tests were terminated in cases where the samples had not failed prior to this. It appears that a good quality die casting may be expected to contain about 1% porosity, present as fine micropores and often distributed in bands within the casting. However, the wide disparity between separate die castings introduces an element of randomness for fatigue and corrosion fatigue testing. A fatigue crack may enter a pore, or perhaps several pores in its propagation phase, and while under some circumstances this may accelerate the crack progression, it is reasonably likely to retard the crack if the pore is relatively small, by virtue of crack blunting effects [67].

18.8. SCC TESTING AND EVALUATION OF MAGNESIUM ALLOYS

The methods of stressing employed in the stress-corrosion testing of magnesium alloys are, in general, the same as those employed in the testing of aluminum and other metals. In rough preliminary surveys, however, it has frequently been found expedient to stress sheet specimens by bending them in an arc sufficient to produce a stress somewhat in excess of the yield strength and retain them in that position by means of a waxed cord. In a modified form of this method, a holding member machined from the same sheet as the specimen is also used in place of the string. The only criticism of this suggested test (developed in the year 1945) is that it is now known that 30% of the elastic limit is the recommended practice for SCC prevention [17, 68].

SCC in magnesium alloys is usually transgranular with significant secondary cracking (branching). Initiation of these cracks has been found to occur invariably at corrosion pits. Mixed transgranular and intergranular crack propagation, and occasionally totally intergranular cracking, have also been observed during SCC in magnesium.

18.8.1. Static Loading of Smooth Specimens and General Considerations

Tests for predicting the stress-corrosion performance of an alloy in a particular service application should be conducted with a stress system similar to that anticipated in service. Table 18.2 lists the numerous sources of sustained tension that are known to have initiated SCC in service and the applicable methods of stressing [69].

Results are strongly influenced by mechanical aspects of the test, such as method of loading and specimen size. These mechanical aspects can control initiation and propagation lifetimes and can influence estimates of a threshold stress. Therefore an apparent threshold stress for SCC is not a material property, and threshold estimates must be qualified with regard to the test conditions and the level of statistical significance [69].

Table 18.2 Stressing Methods Applicable to Various Sources of Sustained Tension in Service

Source of sustained tension in service	Constant strain	Constant load
Residual stress		
Quenching after heat treatment	X	
Forming	X	
Welding	X	
Misalignment (fit-up stresses)	X	
Interference fasteners	X	
Interference bushings		
Rigid	X	
Flexible		X
Flareless fittings	X	
Clamps	X	
Hydraulic pressure	X	X
Dead weight		X
Faying surface corrosion	X	X

Note: The greatest hazard arises when residual, assembly, and operating stresses are additive.

Source: Reference 69.

18.8.2. Stresses

Effect of Strain Rate for Mg Magnesium castings have been shown to fail in laboratory tests under tensile loads as low as 50% of yield strength in environments causing negligible general corrosion. Under certain conditions of testing, stressed specimens may (1) be as resistant to attack as unstressed specimens, (2) show acceleration of corrosion when stressed, or (3) develop stress-corrosion cracks without showing appreciable acceleration of general corrosion [68].

SCC Testing of Magnesium Alloys Atmospheric exposures are used extensively in the stress-corrosion testing of magnesium alloys and give very reliable results. The principal disadvantage of such tests is the comparatively long time required for their completion. In the investigation of experimental alloys, thermal treatments, and other means of preventing SCC in magnesium alloys, accelerated tests can give results in comparatively short times and are practically essential [68].

Dry Air and Liquid Alkali Metals Specimens can be tested in dry air (by surrounding specimens with a dessicant), a 3.3% NaCl + 2% K_2CrO_4 solution, and liquid alkali metals (Na, Rb, Cs) [70]. Specimens to be tested in dry air and the salt solution can be first fatigue precracked in laboratory air and then fractured by cantilever bending at deflection rates of 0.0005 up to 50 degrees/s. High deflection rates produced high crack velocities that can be measured from enlarged prints of cine films (taken at 50 frames/s) of the tests. It is recommended to keep a few specimens for partial cracking in the salt solution, immediately washing in ethanol, drying, and then cracking in dry air so that the appearance of fracture surfaces produced in the two environments could be compared for the same grain or same grain boundary. Tests in liquid alkali metals should be conducted in a glove box filled with high-purity argon [71].

18.8.3. Solutions and Operational Conditions

Exposure of SCC specimens to different types of atmospheres and solutions is currently considered because the chloride-containing solutions tend to show a general corrosion form rather than material cracking. Cracking of highly stressed susceptible alloy specimens can occur within a few hours, while exposure of the specimens under currently used stresses in an industrial atmosphere requires 1000 h of testing without incurring excessive pitting for crack initiation and propagation. There is no standard solution accelerated test for magnesium alloys; however, the following solutions are suggested accompanied with some appropriate comments:

1. Interrupted slow straining, constant strain, and constant load testing of a Mg–Al alloy in solutions containing various concentrations of sodium chloride and potassium chromate has revealed a susceptibility to stress-corrosion failure that is maximal at a chloride/chromate ratio around 1. This is reflected in both the threshold stress for failure and in the crack velocity [72]. Short-term accelerated tests, such as sodium chloride/potassium chromate ($NaCl/K_2CrO_4$) tests are recommended, although they do not predict SCC behavior reliably in practice [19]. However, it has recently been reported that good correlation has been found between the SCC behavior of Mg–Al–Zn in industrial atmosphere and SCC susceptibility observed in 35 g/L sodium chloride and 20 g/L potassium chromate at pH 8 [65]. The strain rate chosen for the tests indicating maximum susceptibility to cracking was about $2 \times 10^{-6}\ s^{-1}$ for a solution containing 5 g/L sodium chloride and 5 g/L

$KCrO_4$. However, those tests refer to open circuit conditions and the strain rate sensitivity of cracking is dependent on potential as well as solution composition [72]. The greater tendency for fracture as the chloride concentration is increased is readily explained by the greater amount of hydrogen produced in the pits at which crack initiation invariably occurs. However, the more extensive pitting that is associated with the more aggressive solutions means that initiation of stress-corrosion cracks becomes more difficult, with a consequent increase in threshold stress and a sharp decrease in crack velocity [72].

The effects of varying the concentration of sodium chloride content with fixed potassium chromate content or of varying the concentration of potassium chromate with fixed sodium chloride content on the time to SCC for magnesium-based alloy (AM–D_{57}S–H) specimens have been examined. It has been shown that addition of 35–80 g/LNacl to 5 g/L of K_2CrO_4 has reduced the time to failure approximately to 15 and 5 hours, respectively. Addition of 10 g of potassium chromate to 35 g/L sodium chloride has reduced the time to failure to below 5 minutes of $\sim$2.5 mm thick sample stressed to 100% of the yield strength by dead load in a simple beam test. This shows the importance of the relative quantities of chromates and chloride solutions in a severe test of 100% of the elastic limit [68].

Mg–7Al alloy exposed to chloride/chromate solutions displays transgranular stress-corrosion cracking with minimal threshold stresses and maximal crack velocities at intermediate chloride chromate concentration ratios, and with no cracking in chloride-free or chromate-free solutions. The severity of cracking shows a marked dependence on strain rate, with ductile failure at sufficiently fast or slow rates [72].

Increasing the chloride/chromate concentration ratio of solutions facilitates discharge of hydrogen and also increases the frequency of pitting. Up to chloride/chromate ratios of 1–2, these trends result in increased susceptibility to cracking. At ratios higher than 2, cracking susceptibility diminishes again, despite the enhanced discharge of hydrogen. However, the greater dissolution at such ratios results in blunting of the cracks or their disappearance, so that cracking susceptibility decreases at the higher chloride/chromate ratios [72].

It was also observed that the more extensive pitting in the more aggressive solutions is associated with more difficult initiation of stress-corrosion cracks, with a consequent increase in threshold stress and a sharp decrease in crack velocity. Stress-corrosion cracks invariably initiate at pits in this system. The role of pitting is to disrupt films that otherwise prevent the ingress of hydrogen and, although there is some stress concentration associated with the pits, this is of lesser importance than the environmental implications of the presence of pits [72].

2. A solution of 3.5% NaCl pH 6.5 is recommended after ASTM G44-88 for all metals (ASTM G69-81). However, it seems that this solution can favor other forms of corrosion, such as general and localized corrosion, more than SCC for certain magnesium alloys. Cracking resistance of metals and alloys by alternate immersion in 3.5% sodium chloride solution is also used frequently. For a spray test, ASTM B117-95 states the relevant conditions for conducting the test. However, it seems that some of these generally accepted tests are somewhat severe for some uncoated magnesium alloys.

3. Laboratory tests have been conducted using potassium hydrogen fluoride (medium frequently used during magnesium anodizing) and a dilute solution of sodium chloride plus sodium bicarbonate [65].

4. The test environment can be a deaerated, aqueous 10^{-3} M Na_2SO_4 solution with the pH adjusted to 10.0 by additions of sodium hydroxide for slow strain rate tensile testing. However, lower tensile strength and lower elongation rate are found, but no reduction in area is observed, at least for certain magnesium alloys [73].

5. Test method ASTM D1384-01, which has been used for some electrochemical measurements for pitting corrosion, covers a simple beaker-type procedure for evaluating the effects of engine coolants on metal specimens under controlled laboratory conditions. This can constitute a mild interesting corrosion medium for SCC testing.

Partial Immersion and Differential Oxygen Cell Distilled water or a 3.5% NaCl–distilled water solution covers the lower half of the specimen gauge length; this partial immersion is used to expose the magnesium to moisture and oxygen gradients. As a control, an unstressed, partially immersed companion test bar can be included in each container. The pH of the water in the subject tests increases during the first 4–5 days from an initial value of 5.5 to 8.0–8.5, where it remained during the rest of the testing period. The solution was removed following fracture of the stressed specimen, which was then tension tested to determine the effects that immersion alone had on the tensile properties. SCC is inhibited in high (greater than 10–12) pH aqueous solutions [74].

Effect of the Applied Potential The strain rate for most severe cracking is dependent on the composition of the solution and the applied potential. Where necessary, the potential of the specimens can be controlled potentiostatically during slow strain rate tensile testing, using a potentiostat, a platinum counterelectrode, and saturated calomel reference electrode. The conventional slow strain rate tensile tests were carried out at a cross-head speed of 1.14×10^{-4} mm/s, which is equivalent to a strain rate of $5.7 \times 10^{-6}\ s^{-1}$, under the different environmental conditions reported. Load-elongation curves were recorded for each test [73].

Raising the potential above the open circuit value increases cracking susceptibility and lowering the potential decreases susceptibility as compared to the cracking observed under open circuit conditions [72] (see Chapters 8 and 13). After testing, all the specimen fracture surfaces can be examined using scanning electron microscopy. In addition, longitudinal sections through the fractures can be mounted, polished, and etched and examined using optical microscopy [73].

18.8.4. Constant Extension Rate and Linearly Increasing Stress Tests

The two frequently used laboratory methods for characterizing SCC and mechanisms are the constant extension rate test (CERT) and the linearly increasing stress test (LIST). In the CERT and LIST, the nominal load and extension, respectively, are increased until specimen fracture. They allow more rapid characterization of SCC susceptibility compared to constant load tests and permit direct measurement of critical parameters such as the threshold stress (σ_{SCC}) and crack velocity (CV) when coupled with equipment to measure crack extension, although both methods have their inherent disadvantages in reflecting the close operating cracking processes [75].

18.8.4.1. Constant Extension Rate Test

The mechanical load on the specimen is steadily increased under constant extension rate test (CERT) conditions. The slow strain rate test (SSRT) or CERT setup (Figure 18.17) incorporates a constant strain rate by means of an open-loop control system; the average strain is measured by two high-resolution linear variable displacement transducers (LVDTs)

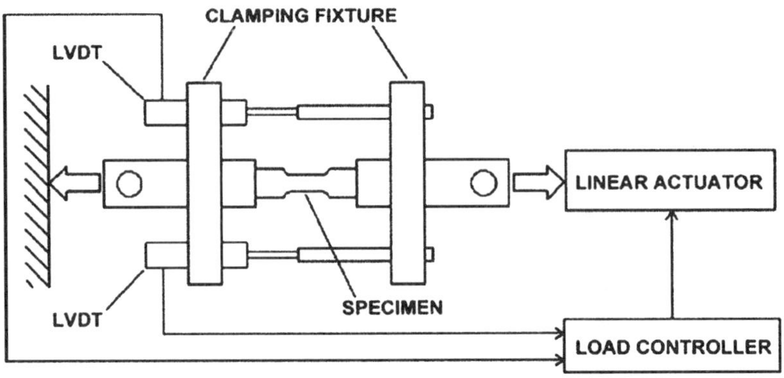

Figure 18.17 Schematic illustration of the SSRT apparatus [75, 79].

in parallel with the specimen while a geared synchronous motor increases the strain and the load accordingly [75]. In the CERT, a pair of LVDTs connected to the load train and in parallel with the specimen LVDTs are used to control the cross-head speed of the test machine via a feedback circuit [76].

By varying the initial and final stresses from test to test, this approach allows a threshold stress for cracking to be defined as a function of strain rate [77]. A combined ac/dc potential drop measuring technique in some cases can help in obtaining more information from a single test, in particular, for crack propagation or the onset of stable crack growth [78]. The direct current potential drop (DCPD) method, for example, involves applying a constant current (~5 A) to the specimen and measuring the increase in specimen resistance, ΔR, as the cross section is reduced, primarily by crack propagation, according to the relationship $\Delta R = rl/\Delta A$, where ΔA is the change in cross-sectional area, r is the resistivity, and l is the length of the specimen [76].

Actually, both the ac and dc potential drop methods are well established techniques for monitoring subcritical crack growth. SCC testing in distilled water and 5 g/L NaCl under SSRT conditions showed that maximum susceptibility occurred in distilled water at the slowest strain rate [75]. SCC initiation can be detected using the DCPD method [79, 80].

Early CERTs or SSRTs of magnesium alloys were done by Parkins and Suzuki [77] based on maximum susceptibility in the solution containing the relative concentrations of 5 g/L sodium chloride and 5 g/L $KCrO_4$ [72]. The strain rate chosen for the tests was based on maximum susceptibility to cracking at about $2 \times 10^{-6}\,s^{-1}$. In tests employing slow straining over a limited stress range, the specimens were rapidly strained in air at a selected applied stress and then left to relax under a constant strain. Dynamic straining was then applied at various strain rates, with the corrosive solution present, until the applied stress reached an upper limit at which the test was stopped. The specimen was then broken rapidly in air, so that the extent of slow crack growth could be measured on the fracture surface by scanning electron microscopy.

18.8.4.2. Linearly Increasing Stress Tests

The linearly increasing stress test (LIST) apparatus steadily increases the nominal load.The LIST apparatus is illustrated in Figure 18.18, where the specimen is attached to one end of a lever arm while the opposite end of the arm is attached such that the tensile load applied to the specimen increases linearly as the distance between the fulcrum and the mass is increased by

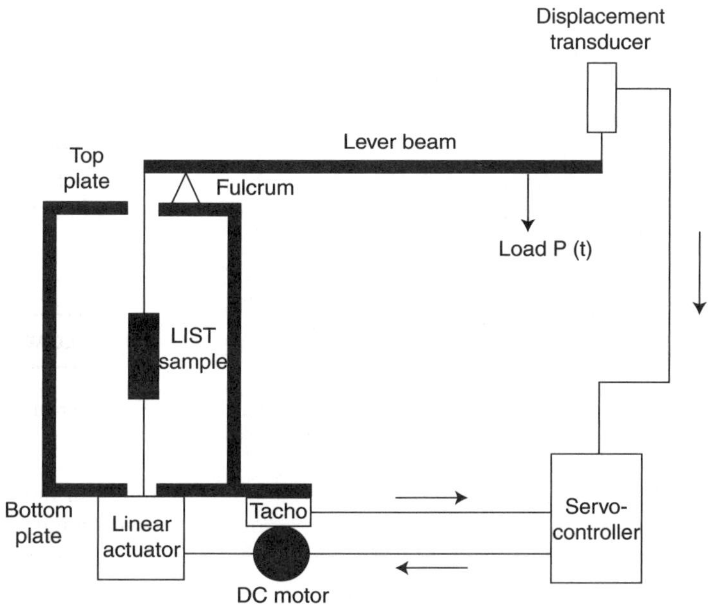

Figure 18.18 Schematic illustration of the LIST apparatus [75].

means of a screw thread and synchronous motor. It has been shown by Winzer et al. [75] that the LIST method provides more rapid determination of the threshold stress, whereas the SSRT method provides a larger fracture surface area for analysis and increased crack propagation time for determining the overall steady-state crack velocity. This latter trait may be particularly important in determining the underlying mechanism for SCC in Mg alloys, since the predominant mechanism is somehow related to the crack velocity [75].

18.8.5. SCC CERT Versus LIST Techniques

During a LIST, fast fracture occurs a relatively short time after the threshold stress is attained; whereas during CERT, crack growth over a much longer time period is facilitated by a reduction in stress. The loading LIST and CERT conditions control the stress or strain and their distribution near the crack tip and influence the propensity for mechanically or chemically induced film rupture and repair at a crack surface. Thus the choice of test method may affect the tendency for H to enter the substrate and cause crack initiation [75, 81]. Transgranular stress-corrosion cracking (TGSCC) of the Mg alloy AZ91 in distilled water and 5 g/L NaCl solution has been evaluated using the LIST and CERT. The LIST is typically 30–50% shorter in duration, whereas the CERT produces a larger SCC fracture surface [75].

Fractography as Related to Mechanisms of CERT and LIST Cracks LISTs in SCC-causing environments resulting in small, localized regions of parallel markings suggest that when the applied stress reaches σ_{SCC}, H embrittlement (HE) causes the crack to propagate until some critical crack length is reached a short distance from the surface. In this case, the crack propagation mechanism is overwhelmed by some non-H-assisted process such as cleavage or microvoid coalescence. In contrast, CERTs result in considerably larger regions of parallel markings. This may be attributed to a tendency for stress relaxation coupled with crack tip opening displacement; as the crack propagates some distance, stress may be reduced below the threshold or the local stress at the crack tip may be reduced such that the HE processes are retarded.

The SCC fracture surfaces resulting from the interrupted CERTs are macroscopically different from those corresponding to later-stage crack propagation under CERT conditions. This is indicative of a load dependence of the predominant mechanism [81]. The increased crack propagation time under CERT conditions may be particularly important in determining the underlying mechanism for SCC in Mg alloys, since the predominant mechanism is somehow related to the crack velocity. The LIST method may be more suitable for investigating SCC in aggressive environments, where pitting may be a consideration [81].

REFERENCES

1. B. W. Lifka, in *Corrosion Testing and Standards: Application and Interpretation*, edited by R. Baboian. American Society for Testing and Materials, Philadelphia, PA, 1995, pp. 447–457.
2. E. Ghali, W. Dietzel, and K.-U. Kainer, *Journal of Materials Engineering and Performance* **13**, 517–529 (2004).
3. ASM Specialty Handbook Committee, in *ASM Specialty Handbook*, edited by M. M. Avedesian and H. Baker. ASM International, Materials Park, OH, 1999, pp. 138–162, 194–215.
4. ASM Metals Handbook Committee, in *ASM Handbook, Volume 13, Corrosion*, edited by L. J. Korb, D. L. Olson, and J. R. Davis. ASM International, Materials Park, OH, 1987, pp. 93–196, 207–220, 231–233, 303–310, 596.
5. J. D. Hanawalt, C. E. Nelson, and J. A. Peloubet, *Transactions of the American Society of Mining and Metallurgical Engineering* **147**, 273–299 (1942).
6. G. L. Song and A. Atrens, *Advanced Engineering Materials* **1**, 11–33 (1999).
7. G. L. Song, A. Atrens, D. St. John, J. Nairn, and Y. Li, *Corrosion Science* **39**, 855–875 (1997).
8. G. Song, A. Atrens, D. St.-John, X. Wu, and J. Nairn, *Corrosion Science* **39**, 1981–2004 (1997).
9. S. Mathieu, C. Rapin, J. Hazan, and P. Steinmetz, *Corrosion Science* **44**, 2737–2756 (2002).
10. P. Adeva-Ramos, S. B. Dodd, P. Morgan, F. Hehmann, P. Steinmetz, and F. Sommer, *Advanced Engineering Materials* **3**, 147–152 (2001).
11. V. Mitrovic-Scepanovic and R. J. Brigham, *Corrosion*, **48**(9), 780–784 (1992).
12. P. P. Trzaskoma, *Corrosion* **42**, 609–613 (1986).
13. H. Inoue, K. Sugahara, A. Yamamoto, and H. Tsubakino, *Corrosion Science* **44**, 603–610 (2002).
14. E. M. Gutman, A. Eliezer, Y. Unigovski, and E. Abramov, *Materials Science Engineering* **302**, 63–67 (2001).
15. Annual Book of ASTM Standards Committee, *Annual Book of ASTM Standards*. American Society for Testing Materials, Philadelphia, PA, 2006, pp. D1384-01, G1-90, G5-94, G16-95, G46-94, G78-01, G61-86, G69-01, G102-89, and G110-92.
16. J. E. Hillis, Private communication. The Dow Chemical Company, 1998.
17. E. Ghali, in *Uhlig's Corrosion Handbook*, 2nd edition, edited by R. W. Revie. Wiley, Hoboken, NJ, 2000, pp. 793–830.
18. J. E. Hillis, in *ASTM Manual Series: Volume MNL 20, Corrosion Testing and Standards: Application and Interpretation*, edited by R. Baboian. ASM International, Materials Park, OH, 1995, pp. 438–446.
19. A. F. Froats, T. Kr. Aune, D. Hawke, W. Unsworth, and J. Hillis, in *ASM Handbook, Volume 13, Corrosion*, edited by L. J. Korb, D. L. Olson and J. R. Davis. ASM International, Materials Park, OH, 1987, pp. 740–754.
20. G. Song, A. Atrens, and D. St. John, in *Magnesium Technology*, edited by J. Hryn. The Minerals, Metals & Materials Society, Warrendale, PA, 2001, pp. 255–262.
21. B. L. Tiwari and J. J. Bommarito, A novel technique to evaluate the corrosion behavior of magnesium alloys, in *Magnesium Technology 2002*, edited by H. I. Kaplan. The Minerals, Metals & Materials Society, Warrendale, PA, 2002, pp. 269–275.
22. G. G. Tchervyakov, J. Bomback, A. P. Pchelmikov, and G. Cole, in *Laboratory Evaluation of Corrosion Resistance of Anodized Film on Magnesium*, edited by H. I. Kaplan, J. Hryn, and B. Clow. The Minerals, Metals & Material Society, Warrendale, PA, 2000, pp. 143–152.
23. D. Weiss, B. Bronfin, G. Golub, and E. Aghion, Corrosion resistance evaluation of magnesium and magnesium alloys by an ion selective electrode, in *Magnesium 97, First Israeli International Conference on Magnesium Science & Technology*, Dead Sea, Israel, edited by E.

Aghion and D. Eliezer. Magnesium Research Institute (MRI), Beer-Sheva, Israel, 1998, pp. 208–213.

24. E. Ghali, W. Dietzel, and K.-U. Kainer, *Journal of Materials Engineering and Performance* **13**, 7–23 (2004).
25. A. Shaw and C. Wolfe, in *ASM Handbook, Volume 13B Corrosion*, edited by S. D. Cramer and B. S. Covino, Jr. ASM International, Materials Park, OH, 2005, pp. 205–227.
26. A. Luo, M. Balogh, and B. R. Powell, *Tensile Creep and Microstructure of Magnesium–Aluminium–Calcium Based Alloys for Powertrain Applications: Part 2*. Society of Automotive Engineers, Warrendale, PA, 2001.
27. J. I. Skar and D. Albright, Corrosion behavior of die-cast magnesium in ASTM B117 salt spray and GM9540P cyclic corrosion test, in *Magnesium Technology 2003*, edited by H. I. Kaplan. The Minerals, Metals & Materials Society, Warrendale, PA, 2003, pp. 59–63.
28. B. P. Alblas and J. J. Kettenis, *Protective Coatings Europe* **February,** 49–59 (2000).
29. E. F. Emley, *Principles of Magnesium Technology*, Pergamon Press, New York, 1966, pp. 183–185.
30. P. L. Bonora, M. Andrei, A. Eliezer, and E. M. Gutman, *Corrosion Science* **44**, 729–749 (2002).
31. R. Feser and J. W. Erning, *Materials and Corrosion* **52**, 456–461 (2001).
32. O. Lunder, T. Kr. Aune, and K. Nisancioglu, *Corrosion* **43**, 291 (1987).
33. T. Beldjoudi, C. Fiaud, and L. Robbiola, *Corrosion* **49**, 738–745 (1993).
34. G. Baril and N. Pebere, *Corrosion Science* **43**, 471–484 (2001).
35. C. F. Baker, in *Magnesium Alloys and Their Applications*, edited by B. L. Mordike and F. Hehman. Informationsgesellschaft, Garmisch-Partenkirchen, Germany, 1992, pp. 77–87.
36. S. Bender, E. Boese, A. Heyn, and J. Goellner, Corrosion behaviour of magnesium alloys: material specific corrosion testing, in *Proceedings of the 7th International Conference on Magnesium Alloys and Their Applications*, Weinheim, Germany, 2006 edited by K.-U Kainer. Wiley-VCH, Weinheim, Germany, 2007, pp. 721–726.
37. T. Troβmann, K. Eppel, M. Gugau, and C. Berger, Investigation of the passivation behavior of magnesium alloys by means of cyclic current–potential-curves, in *Proceedings of the the 7th International Conference on Magnesium Alloys and their Applications*, Weinheim, Germany, 2006 edited by K.-U. Kainer. Wiley-VCH, Weinheim, Germany, 2007, pp. 802–808.
38. T. Troβmann, Verhalten von magnesium-Legierungen bei korrosiver und mechanisch korrosiver Beanspruchung, Dissertation D17, Darmstadt University of Technology "TU", Darmstadt, Germany, 2005.
39. S. Werdin, T. Troβmann, M. Gugau, and K. L. Kotte, *Materialwissenschaft und Werkstofftechnik Journal*, **36**, 659–668 (2005).
40. K. Eppel, J. Ellermeier, T. Troβmann, C. Berger, H. Masny, K. Woeste, and U. Dilthey, in *Proceedings of the 7th International Conference on Magnesium Alloys and Their Applications*, Weinheim, Germany, 2006 edited by K.-U. Kainer. Wiley-VCH, Weinheim, Germany, 2007.
41. S. Jin, E. Ghali, C. Blawert, and W. Dietzel, SRET evaluation of the corrosion behavior of thixocast AZ91 magnesium alloy in dilute NaCl solution at room temperature, in *Proceedings of the 210th ECS 2006 Meeting*, Cancun, Mexico, edited by N. Missert, A. Davenport, M. Ryan, and S. Virtanen. The Electrochemical Society, Inc., Pennington, NJ, Vol. 3, No. 31, 2007, pp. 295–311.
42. B. A. Boukamp, in *Equivalent Circuit (EQUIVCRT. PAS) Users Manual*. University of Twente, The Netherlands, 1988/1989, pp. 6–26.
43. R. Udhayan and D. P. Prokash Bhatt, *Journal of Power Sources* **63**, 103 (1996).
44. S. Turgoose and R. A. Cottis, in *Electrochemical Impedance: Analysis and Interpretation*, edited by J. R. Scully, D. C. Silverman, and M. W. Kendig. ASTM, Philadelphia, PA, 1993, pp. 173–191.
45. N. Pebere, C. Riera, and F. Dabosi, *Electrochimica Acta* **35**, 555–561 (1990).
46. C. H. R. Tsai, in *Electrochemical Impedance: Analysis and Interpretation*, edited by J. R. Scully, D. C. Silverman, and M. W. Kendig. ASTM, Philadelphia, PA, 1993, pp. 37–52.
47. P. Gu, W. Zheng, M. Elboujdaini, and E. Ghali, Electrochemical study of magnesium hydride formation in aqueous solution, in *Degradation of Light Weight Alloys*, edited by D. Shifler. The Minerals, Metals & Materials Society, Warrendale, PA, 2007.
48. S. Chechirlian, P. Eichner, M. Keddam, H. Takenouti, and H. Mazille, *Electrochimica Acta* **35**, 1125 (1990).
49. R. Simpraga, B. E. Conway, Role of processes involving hydrogen and H_2 in the corrosion of magnesium in aqueous solutions, in *Proceedings of the Symposium on Electrochemistry and Materials Science of Cathodic Hydrogen Absorption and Adsorption,* San Francisco, CA, edited by B. E. Conway and G. Jerkiewicz, 1994, Electrochemical Society, Pennington, NJ, 1995, vol. 94–2, pp. 427–439.
50. G. Baril and N. Pebere, in *Environmental Degradation of Materials and Corrosion Control in Metals*, edited by M. Elboujdaini and E. Ghali. The Canadian Institute of Mining, Metallurgy and Petroleum, Montreal, Quebec, Canada 1999, pp. 55–68.
51. Y.-S. Choi, J-G Kim, S. K. Kim, and Y.-J. Kim, Aqueous corrosion characteristics of Mg–4Ni–xAl alloys in acid–chloride solution, in *Magnesium Technology 2002*, edited by H. I. Kaplan. The Minerals, Metals & Materials Society, Warrendale, PA, 2002 pp. 277–282.
52. V. Mitrovic-Scepanovic and R. J. Brigham, *Corrosion* **48**, 780–784 (1992).
53. B. Y. Hur and K. W. Kim, *Corrosion Reviews* **16**, 85–94 (1998).

54. I. Nakatsugawa, J. Renaud, and E. Ghali, Cathodic protective coating for Mg alloys, in *Environmental Degradation of Materials and Corrosion Control of Metals*, edited by M. Elboujdaini and E. Ghali. METSOC, Montreal, Canada, 1999.
55. I. Nakatsugawa, F. Yamada, H. Takaysu, T. Tsukeda, and K. Saito, Corrosion behavior of thixomolded Mg–Al alloys, in *Environmental Degradation of Materials and Corrosion control in Metals*, edited by M. Elboujdaini and E. Ghali. METSOC, Quebec, Canada, 1999, pp. 113–123.
56. I. Nakatsugawa, H. Takayasu, and K. Saito, Corrosion behavior of thin wall magnesium products molded by thixomolding, in *Magnesium Alloys and Their Applications*, edited by K.-U. Kainer. Wiley-VCH, Weinheim, Germany, 2000, pp. 445–450.
57. D. O. Sprowls, in *ASM Handbook, Volume 13, Corrosion*, edited by J. R. Davis. ASM International, Materials Park, OH, 1987, pp. 231–233.
58. T. Zhang, Y. Shao, G. Meng, and F. Wang, *Electrochimica Acta* **53**, 561–568 (2007).
59. A.-M. Lafront, M. Ö. Öteyaka, R. D. Klassen, P. R. Roberge, and E. Ghali, Study of the corrosion of zinc and aluminium magnesium alloys by electrochemical noise (EN) and scanning reference electrode technique (SRET), presented at the *NACE International Seminar, Session 2, Montreal, Canada*, 2002.
60. R. D. Klassen, P. R. Roberge, A.-M. Lafront, M. Ö. Öteyaka, and E. Ghali, Comparison of the corrosion behaviour between two magnesium alloys by electrochemical noise and scanning reference electrode technique, in *Magnesium, Proceedings of the 6th International Conference, Magnesium Alloys and their Applications,* edited by K.-U. Kainer. Wiley-VCH, Weinheim, Germany, 2003, pp. 655–660.
61. R. D. Klassen, P. R. Roberge, A.-M. Lafront, M. Ö. Öteyaka, and E. Ghali, *Canadian Metallurgical Quarterly* **44**, 47–52 (2005).
62. A.-M. Lafront, W. Zhang, S. Jin, and R. Tremblay, *Electrochimica Acta* **51**, 489–501 (2005).
63. M. Ö. Öteyaka, A.-M. Lafront, E. Ghali, and R. Tremblay, Pitting corrosion of some magnesium alloys by scanning reference electrode technique (SRET), in *Proceedings of the Conference of NACE International*, Québec, Canada, NACE International, Houston, Texas, USA 2002, pp. 26–27.
64. W. Zhang, S. Jin, E. Ghali, R. Tremblay, M. Shehata, and E. Es-Sadiqi, *Advanced Engineering Materials* **8**, 973–980 (2006).
65. B. Phull, in *ASM Handbook, Volume 13A, Corrosion*, edited by S. D. Cramer and B. S. Covino, Jr. ASM International, Materials Park, OH, 2003, pp. 575–616.
66. A. Eliezer and F. Witte, The role of biological environments on magnesium alloys as biomaterials, in *Preceedings of the 7th International Conference on Magnesium Alloys and Their Applications*, Weinheim, Germany, edited by K.-U. Kainer. Wiley-VCH, Weinheim, Germany, 2006, pp. 822–827.
67. P. N. Ross, J.-A. MacCulloch, and R. J. Esdaile, in *Proceedings of the 20th International Die Casting Congress and Exposition*, Cleveland, Ohio, North American Die Casting Association, Rosemont, IL, 1999, pp. 213–218.
68. G. F. Sager, R. H. Brown, and R. B. Mears, *Tests for Determining Susceptibility to Stress-Corrosion Cracking*. American Society for Testing and Materials, Philadelphia, PA, 1945 pp. 255–272.
69. D. Sprowls, in *Stress Corrosion Cracking*, edited by R. H. Jones. ASM International, Materials Park, OH, 1992, pp. 336–415.
70. E. J. Meletis and R. F. Hochman *Corrosion* **40**, 39–45 (1984).
71. S. P. Lynch and P. Trevena, *Corrosion* **44**, 113–124 (1988).
72. K. Ebtehaj, D. Hardie, and R. N. Parkins, *Corrosion Science* **28**, 811–829 (1988).
73. R. S. Stampella, R. P. M. Procter, and V. Ashworth, *Corrosion Science* **24**, 325–341 (1984).
74. W. K. Miller, Stress corrosion cracking of magnesium die casting alloys, in *Materials Stability and Environmental Degradation*, Vol. 125, edited by A. Barkatt, E. D. Verink, L. R. Smith and R. Leslie, Materials Research Society, Pittsburgh, PA, 1988, pp. 253–259.
75. N. Winzer, G. Song, A. Atrens, W. Dietzel, C. Blawert, and K.-U. Kainer, Evaluation of Mg SCC using LIST and SSRT, in *Proceedings of the 7th International Conference on Magnesium Alloys and Their Applications*, Weinheim, Germany, edited by K.-U. Kainer. Wiley-VCH, Weinheim, Germany, 2006, pp. 715–720.
76. N. Winzer, A. Atrens, W. Dietzel, V. S. Raja, G. Song, and K.-U. Kainer, *Materials Science and Engineering*, **488**, 339–351 (2008).
77. R. N. Parkins and Y. Suzuki, *Corrosion Science* **23**, 577 (1983).
78. W. Dietzel and K.-H. Schwalbe, *Materialprüfung* **28**, 368–372 (1986).
79. W. Dietzel and K. H. Schwalbe, in *Research and Engineering Applications*, edited by R. D. Kane. ASTM, Philadelphia, PA, 1993, pp. 134–148.
80. A. Atrens, C. C. Brosnan, S. Ramamurthy, A. Oehlert, and I. O. Smith, *Measurement Science and Technology* **4**, 1281–1292 (1993).
81. N. Winzer, A. Atrens, W. Dietzel, G. Song, and K.-U. Kainer, *Materials Science and Engineering* **A472**, 97–106 (2008).

Part Six

Bibliography, International Units, and Abbreviations

Appendix 1

Corrosion and Prevention Books, Data, and ASTM Standards

A1.1. SOME RECOMMENDED BOOKS ON CORROSION

Cramer, S. D. and Covino, B. S. Jr., *Metals Handbook, Volume 13A, Corrosion: Fundamentals, Testing, and Protection, and Volume 13B, Corrosion: Materials*. ASM International, Materials Park, OH, 2003.

Evans, U. R., *The Corrosion and Oxidation of Metals*. Edward Arnold, London, 1960.

Fontana, M. G. and Greene, N. D., *Corrosion Engineering*, McGraw-Hill, New York, 1978.

Landrum, R. J., *Fundamentals of Designing for Corrosion Control*. NACE International, Houston, TX, 1989.

Marcus, P. and Oudar, J.,editors, *Corrosion Mechanisms in Theory and Practice*, Marcel Dekker, New York, 1995.

National Association of Corrosion Engineers, *Corrosion Testing Made Easy Series*. NACE International, Houston, TX, 1985.

Piron, D. L., *The Electrochemistry of Corrosion*. NACE International, Houston, TX, 1991.

Pourbaix, M., *Atlas of Electrochemical Equilibrium in Aqueous Solutions*. NACE International, Houston, TX, 1974.

Revie, R.W., editor, *Uhlig's Corrosion Handbook*, 2nd edition. Wiley, Hoboken, NJ, 2000.

Roberge, P. R., *Handbook of Corrosion Engineering*. McGraw-Hill, New York, 1999.

Sastri, V. S., *Corrosion Inhibitors, Principles and Applications*. Wiley, London, 1998.

Sastri, V. S., Ghali E., and Elboujdaini, M., *Corrosion Prevention and Protection*, Wiley, London, 2007.

Schutze, M., editor, *Corrosion and Environmental Degradation*, Vol. 1 and 2. Wiley-VCH, Weinheim, Germany, 2000.

Shreir, L.L., Jarman, R.A., and Burstein, G. T. *Corrosion, Metal/Environment Reactions*, 3rd edition, Vol. 1 and 2, Butterworth-Heinemann, London, 1995.

Uhlig, H. H. and Revie, R. W., *Corrosion and Corrosion Control*, 3rd edition. Wiley, Hoboken, NJ, 1985.

A1.2. BIBLIOGRAPHY OF CORROSION DATA FOR PERFORMANCE OF MATERIALS

These are some frequently consulted corrosion data sources that are related to the selection of materials (design) and/or case histories.

Baboian, R., editor, *NACE Corrosion Engineer's Reference Book*, 3rd edition. National Association of Corrosion Engineers, Houston, TX, 2002.

Craig, B. D. and Anderson, D. S., editors, *Handbook of Corrosion Data*, 2nd edition. ASM International, Materials Park, OH, 1995.

Corrosion Resistance of Aluminum and Magnesium Alloys: Understanding, Performance, and Testing By Edward Ghali

De Renzo, D. J., editor, *Corrosion Resistant Materials Handbook*, 4th edition. Noyes Data Corporation, NJ, 1985.

Graver, D. L., editor, *Corrosion Data Survey*, 6th edition, NACE International, Houston, TX, 1985.

Kreysa, G. and Schutze, M.,editors, *Corrosion Handbook, Volume 8B, Corrosive Agents and Their Interaction with Materials*. Wiley-VCH, Weinheim, Germany, 2007.

McEvily, A. J., editor, *Atlas of Stress-Corrosion Cracking and Corrosion Fatigue Curves*. ASM International, Materials Park, OH, 2000.

McNaughton, K. J., Materials engineering 1, selecting materials for process equipment, in *Chemical Engineering Magazine*. McGraw-Hill, New York, 1980.

Schweitzer, A. S., *Corrosion Resistant Tables, Metals, Plastics, Nonmetallics, and Rubbers*, 5th edition. Marcel Dekker, New York, 2004.

A1.3. ASTM STANDARDS

All ASTM standards referred to in this book are not detailed in the list of references at the end of every chapter. The reader is then invited to consider the following for detailed description of these important frequently used standards in North America. The address is: ASTM International, 100 Barr Harbor Drive, PO Box C700, West Conshohocken, PA, 19428-2959 USA.

ASTM B117-03, Standard practice for operating salt spray (fog) testing apparatus, in *Annual Book of ASTM Standards*. ASTM international, West Conshohocken, PA, 2009.

ASTM D2803-09, Standard guide for testing filiform corrosion resistance of organic coatings on metal, in *Annual Book of ASTM Standards*. ASTM International, West Conshohocken, PA, 2009.

ASTM G73-93, Standard practice for liquid impingement erosion testing, in *Annual Book of ASTM Standards*. ASTM International, West Conshohocken, PA, 2009.

Appendix 2

International Units and Some Equations

The units employed are those known as the International System of Units.

A2.1. CONSTANTS, CONVENTIONS, AND KEY EQUATIONS

A2.1.1. Constants

$F = 9.6487 \times 10^4\ \mathrm{C \cdot mol^{-1}}$

$R = 8.3143\ \mathrm{J.\,mol^{-1} \cdot K^{-1}} = 8.3143 \times 10^7\ \mathrm{erg \cdot mol^{-1}\,K^{-1}} = 1.9872\ \mathrm{cal\,mol^{-1} \cdot K^{-1}}$

$= 82.054\ \mathrm{mL \cdot atm \cdot mol^{-1} \cdot K^{-1}} = 62.361\ \mathrm{L \cdot mm \cdot mol^{-1} \cdot K^{-1}}$

$1\ \mathrm{J} = 1\ \mathrm{kg} \cdot m^2 \cdot \mathrm{s^{-2}} = 1\ \mathrm{V \cdot C} = 1\ \mathrm{W \cdot s} = 0.23901\ \mathrm{cal} = 10^7\ \mathrm{erg} = 9.869\ \mathrm{mL \cdot atm}$

$1\ \mathrm{atm} = 101325\ \mathrm{N \cdot m^{-2}(Pa)} = 760\ \mathrm{torr(mm)}$
$= 1.01325 \times 10^6\ \mathrm{dyne \cdot cm^{-2}} = 1.01325\ \mathrm{bar}$

$1\ \mathrm{L} = 1000.028\ \mathrm{cm^3}$

$1\ \mathrm{\mathring{A}} = 10^{-8}\ \mathrm{cm}$

1 year = 365.24 days

$\pi = 3.1415926536$

$\ln 10 = 2.302585093$

$g_s = 9.80665\ \mathrm{m \cdot s^{-2}}$

$N = 6.02252 \times 10^{23}\ \mathrm{molecule \cdot mol^{-1}}$

$1\ \mathrm{eV} = 1.60210 \times 10^{-19}\ \mathrm{J}$

1 horsepower = 746 W

$1\ \mathrm{W} = 1\ \mathrm{V \cdot A}$

$1\ \mathrm{C} = 1\ \mathrm{A \cdot s}$

$1\ \mathrm{V} = 1\ \mathrm{kg \cdot m^2 \cdot s^{-3} \cdot A^{-1}} = 1 \cdot \mathrm{J} \cdot A^{-1} \cdot \mathrm{s^{-1}}$

Corrosion Resistance of Aluminum and Magnesium Alloys: Understanding, Performance, and Testing By Edward Ghali

A2.1.2. Conventions

For a galvanic cell, a single vertical bar represents a phase boundary; a dashed vertical bars (¦) represents a junction between miscible liquids; and double dashed vertical bars (¦¦) represent a liquid junction in which the liquid junction potential has been eliminated by some more or less accurate extrapolation or other procedure.

A2.1.3. Key Equations

Faraday Quantitative Law

$$m = \frac{1}{F} \times \frac{A}{n} \times I \times t$$

where $F = 96{,}487$ coulombs, I is in amperes, t is in seconds, m and A are in grams, and n, is the number of electrons exchanged in the electrochemical equation. For example, $E°(Zn^{2+}/Zn)$ corresponds to $E°(Zn^{2+}(aq) + 2e^- \rightarrow Zn)$.

Solution Properties

$$PV = nRT,$$

where n = number of moles.

$$\text{Ionic strength} = \mu = \frac{1}{2}\sum_i m_i z_i^2$$

where m_i is the molar ion concentration and Z is the valence of the ion in the solution.

For an electrolyte $A_xB_y \Leftrightarrow xA^{y+} + yB^{x-}$, the average coefficient of activity is

$$\gamma_\pm = (\gamma_+^x \gamma_-^y)^{1/(x+y)}$$

$$a_{A^-} = \gamma_- |A^-|$$

$$A_xB_y = xA^{z+} + yB^{z-}$$

$\log\gamma_\pm = -Az_+z_-\sqrt{\mu}$; $A = 0.509$ for H_2O at 25 °C (Debye–Hückel limiting law).

$$\text{Coefficient or degree of dissociation} = \alpha_- = \left|\frac{A^-|}{|A^-| + |BA|}\right.$$

$$K = \frac{a_{B^+} \times a_{A^-}}{a_{BA}} = \frac{C\alpha^2}{1 - \alpha} \quad \text{when } \alpha_+ = \alpha_-$$

$$K_{\text{thermodynamic}} = \gamma_\pm^2 \times K_{\text{dissociation}}$$

where the concentration is considered arbitrarily equal to the activity of the ion in the dissociation constant.

$$\text{Tampon capacity} = \beta = \frac{\Delta B}{\Delta \text{pH}}$$

$$\text{pH} = \text{p}K + \log\frac{[\text{conjugated base}]}{[\text{acid}]}$$

Conductance and Conductivity

$R = \rho \times \frac{l}{A}$; resistivity ρ can be expressed in $\Omega \cdot$ cm.

$k = \frac{1}{\rho} = nC\Lambda_c \times 10^{-3}$; conductivity k can be expressed in $\Omega^{-1} \cdot \mathrm{cm}^{-1} = \mathrm{mho} \cdot \mathrm{cm}^{-1}$

$$\text{Force of resistance} = F_r = 6\pi\eta r v = kv$$

where η is the coefficient of viscosity, r is the radius of the ion, v the velocity of the ion in cm·s^{-1}, and k is a constant of proportionality.

Electromotive force $= F_m = zeE$ ($z =$ valence of the ion, $e =$ charge of the electron and $E =$ electric field in $\mathrm{V \cdot cm^{-1}}$)

Velocity $v = \frac{neE}{k} = uE$, where u is the mobility (the velocity for $E = 1$ V/cm)

Mobility of the ion $= u_+ = \frac{\lambda_+}{F} = \frac{\lambda_+}{Ne}$

$\alpha = \frac{\Lambda_c}{\Lambda_0}$

Equivalent conductance of the electrolyte at infinite dilution or zero concentration is given by

$$\Lambda_\infty = \Lambda_0 = \lambda_0^+ + \lambda_0^-$$

$$\Lambda_c = \alpha(\lambda_+ + \lambda_-) = \alpha F(u_+ + u_-)$$

where Λ, λ are the ion equivalent conductance (s), expressed in $\Omega^{-1} \cdot \mathrm{cm}^2 \cdot \mathrm{equivalent}^{-1}$. Transference or transport numbers of cations and anions are $t_+ + t_- = 1$:

$$t_+ = \frac{zC_1\lambda_1}{zC_1\lambda_{1+} + zC_2\lambda_{2+} + \cdots + zC_i\lambda_{i+}}$$

Variation of the concentration of the catholyte for inert metallic electrodes (e.g., platinum electrode) is

$$\Delta n_c = \frac{q}{F} \times t_-, \quad \text{where } q \text{ is the quantity of electricity in coulombs}$$

Electrode Potential

$$E = E^\circ - \frac{RT}{nF} \ln \frac{\text{Products activities}}{\text{Reactants activities}} \quad \text{(Nernst equation)}$$

At 25 °C, $RT/F = 0.0256926$ V, $(RT/F)\ln 10 = 0.0591594$ V

$$E = E^\circ - \frac{0.0592}{n} \log \frac{\text{Products}}{\text{Reactants}} \quad \text{at 25 °C and atmospheric pressure}$$

Free enthalpy $= \Delta G = -nFE = \Delta H - T\Delta S$

$\Delta G^\circ = -RT \ln K$

$$E^\circ = \frac{RT}{nF} \ln K$$

$$\text{EMF} = E_{\text{oxidation}} + E_{\text{reduction}}$$

$$E_x = E_s \times \frac{R_x}{R_s} \quad (s = \text{standard})$$

A2.2. EXAMPLES OF REFERENCE ELECTRODES AND METALLIC AND IONIC REDUCTION REACTIONS

Calomel: $Hg_2Cl_2 \rightarrow Hg$	$E^\circ = 0, 2679$ V, $E_{(\text{KClsat})} = 0.2420$ V, $E(\text{KCl}_{1\ \text{N}}) = 0.2810$ V, $E(\text{KCl}_{0.1\ \text{N}}) = 0.3335$ V
$AgCl \rightarrow Ag$	$E^\circ = 0.222$ V, $E(\text{KCl}_{0.1\ \text{N}}) = 0.288$ V
Cu^{2+}/Cu	$E^\circ = 0.337$ V, $E(\text{CuSO}_{4\ \text{sat}}) = 0.316$ V
$Hg_2SO_4 \rightarrow Hg$	$E^\circ = 0.615$ V; if K_2SO_4 is saturated $E = 0.64$ V
$Fe^{3+} \rightarrow Fe^{2+}$	$E^\circ = 0.771$ V

A2.3. ELECTROCHEMICAL CELLS AND THEIR POTENTIALS

Pt, H_2 I HC II AgCl, Ag	$E = E^\circ - \frac{RT}{F} \ln \frac{a_{H^+} \times a_{Cl^-}}{\sqrt{p_{H_2}}}$
$Zn/Zn^{2+} \vdots Cd^{2+}/Cd$	$E = E^\circ - \frac{RT}{2F} \ln \frac{a_{Zn^{2+}}}{a_{Cd^{2+}}}$
Pt, H_2 $(P_1) \vdots H^+ \vdots (P_2)$ H_2, Pt, $P_1 > P_2$	$E = \frac{RT}{nF} \ln \frac{P_1}{P_2}$
Pt, H_2/HCl $(a_1) \vdots$ HCl (a_2)/H_2, Pt, $a_1 < a_2$	$E = 2t - \frac{RT}{nF} \ln \frac{a_{2\pm}}{a_{1\pm}}$
Junction potential, $n = 1$	$E_j = (2t_- - 1) \frac{RT}{F} \ln \frac{a_{2\pm}}{a_{1\pm}}$
Pt, H_2/HCl(a_1)/AgCl, Ag I Ag, AgCl/HCl(a_2)/H_2, Pt, $a_1 < a_2$	$E = \frac{2RT}{F} \ln \frac{a_{2\pm}}{a_{1\pm}}$

A2.4. STANDARD ELECTRODE POTENTIAL OF CATIONS ($T = 25\ ^\circ C$)

Metal	E°	Metal	E°
$Li+/Li$	−3.045	Tl^{+}/Tl	−0.336
Cs^{+}/Cs	−2.923	Co^{2+}/Co	−0.277
Rb^{+}/Rb	−2.925	Ni^{2+}/Ni	−0.250
K^{+}/K	−2.925	Mo^{3+}/Mo	−0.200
Ra^{2+}/Ra	−2.916	In^{+}/In	−0.139
Ba^{2+}/Ba	−2.905	Sn^{2+}/Sn	−0.136
Sr^{2+}/Sr	−2.89	Pb^{2+}/Pb	−0.126
Ca^{2+}/Ca	−2.866	Fe^{3+}/Fe	−0.037
Na^{+}/Na	−2.713	$2D^{+}/D_2$	−0.003
La^{3+}/La	−2.52	$2H^{+}/H_2$	−0.000
Ce^{3+}/Ce	−2.483	Sb^{3+}/Sb	+0.1
Mg^{2+}/Mg	−2.363	Bi^{3+}/Bi	+0.215
Y^{3+}/Y	−2.372	Ge^{3+}/Ge	+0.24
Sc^{3+}/Sc	−2.077	As^{3+}/As	+0.3
Th^{4+}/Th	−1.90	Cu^{2+}/Cu	+0.337
Be^{2+}/Be	−1.847	Co^{3+}/Co	+0.4
U^{3+}/U	−1.80	Ru^{2+}/Ru	+0.455
Hf^{4+}/Hf	−1.70	Cu^{+}/Cu	+0.520
Al^{3+}/Al	−1.663	Te^{4+}/Te	+0.568
Ti^{2+}/Ti	−1.63	Tl^{3+}/Tl	+0.71
Zr^{4+}/Zr	−1.539	$Hg_2{}^{2+}/2Hg$	+0.799
U^{4+}/U	−1.4	Ag^{+}/Ag	+0.799
Mn^{2+}/Mn	−1.168	Rh^{3+}/Rh	+0.799
V^{2+}/V	−1.175	Pb^{4+}/Pb	+0.80
Cb^{3+}/Cb	−1.1	Os^{2+}/Os	+0.85
Cr^{2+}/Cr	−0.913	Hg^{2+}/Hg	+0.854
Zn^{2+}/Zn	−0.763	Pd^{2+}/Pd	+0.987
Cr^{3+}/Cr	−0.744	Ir^{3+}/Ir	+1.156
Ga^{3+}/Ga	−0.529	Pt^{2+}/Pt	+1.188
Ga^{2+}/Ga	−0.45	Ag^{2+}/Ag	+1.369
Fe^{2+}/Fe	−0.440	Au^{3+}/Au	+1.498
Cd^{2+}/Cd	−0.403	$Au+/Au$	+1.692
In^{3+}/In	−0.342		

A2.5. THE PERIODIC TABLE

1	2	3	4	5	6	7	8	9	10	11	12	13	14	15	16	17	18
1 **H** 1.0079																	2 **He** 4.0026
3 **Li** 6.941	4 **Be** 9.0122											5 **B** 10.811	6 **C** 12.011	7 **N** 14.007	8 **O** 15.999	9 **F** 18.998	10 **Ne** 20.180
11 **Na** 22.990	12 **Mg** 24.305											13 **Al** 26.982	14 **Si** 28.086	15 **P** 30.974	16 **S** 32.065	17 **Cl** 35.453	18 **Ar** 39.948
19 **K** 39.098	20 **Ca** 40.078	21 **Sc** 44.956	22 **Ti** 47.867	23 **V** 50.942	24 **Cr** 51.996	25 **Mn** 54.938	26 **Fe** 55.845	27 **Co** 58.933	28 **Ni** 58.693	29 **Cu** 63.546	30 **Zn** 65.409	31 **Ga** 69.723	32 **Ge** 72.64	33 **As** 74.922	34 **Se** 78.96	35 **Br** 79.904	36 **Kr** 83.798
37 **Rb** 85.468	38 **Sr** 87.62	39 **Y** 88.906	40 **Zr** 91.224	41 **Nb** 92.906	42 **Mo** 95.94	43 **Tc** (98)	44 **Ru** 101.07	45 **Rh** 102.91	46 **Pd** 106.42	47 **Ag** 107.87	48 **Cd** 112.41	49 **In** 114.82	50 **Sn** 118.71	51 **Sb** 121.76	52 **Te** 127.60	53 **I** 126.90	54 **Xe** 131.29
55 **Cs** 132.91	56 **Ba** 137.33	57-71 *	72 **Hf** 178.49	73 **Ta** 180.95	74 **W** 183.84	75 **Re** 186.21	76 **Os** 190.23	77 **Ir** 192.22	78 **Pt** 195.08	79 **Au** 196.97	80 **Hg** 200.59	81 **Tl** 204.38	82 **Pb** 207.2	83 **Bi** 208.98	84 **Po** (209)	85 **At** (210)	86 **Rn** (222)
87 **Fr** (223)	88 **Ra** (226)	89-103 #	104 **Rf** (261)	105 **Db** (262)	106 **Sg** (266)	107 **Bh** (264)	108 **Hs** (270)	109 **Mt** (268)	110 **Ds** (281)	111 **Rg** (272)	112 **Uub** (285)	113 **Uut** (284)	114 **Uuq** (289)	115 **Uup** (288)	116 **Uuh** (291)		118 **Uuo** (294)

* Lanthanide series	57 **La** 138.91	58 **Ce** 140.12	59 **Pr** 140.91	60 **Nd** 144.24	61 **Pm** (145)	62 **Sm** 150.36	63 **Eu** 151.96	64 **Gd** 157.25	65 **Tb** 158.93	66 **Dy** 162.50	67 **Ho** 164.93	68 **Er** 167.26	69 **Tm** 168.93	70 **Yb** 173.04	71 **Lu** 174.97
# Actinide series	89 **Ac** (227)	90 **Th** 232.04	91 **Pa** 231.04	92 **U** 238.03	93 **Np** (237)	94 **Pu** (244)	95 **Am** (243)	96 **Cm** (247)	97 **Bk** (247)	98 **Cf** (251)	99 **Es** (252)	100 **Fm** (257)	101 **Md** (258)	102 **No** (259)	103 **Lr** (262)

Source: Wieser, M.E., Atomic weight of the elements 2005 (IUPAC Technical Report). *Pure Applied Chemistry* **78** (11), 2051–2066 (2006).

Appendix 3

Abbreviations and Symbols

A	Arrhenius frequency factor
Ac	Alternating current
AES	Auger electron spectroscopy
AFM	Atomic force microscopy
AIDE	Adsorption-induced dislocation emission
AMC	Aluminum matrix composite
ATP	Adenosine triphosphate
AS	Artificial seawater
ASM	American Society of Metals
ASTM	American Society for Testing and Materials
AWWA	American Water Works Association
bcc	Body-centered cubic
BEI	Backscattered electron image
CAFE	Corporate average fuel economy
CCD	Corrosion current density
CCT	Cyclic corrosion testing
CCURBs	Corrosion control using regenerative biofilms
C_{dl}	Capacitor
CE	Chunk effect
CERT	Constant extension rate test
CF	Corrosion fatigue
c_I	Ion molar concentration
CIT	Corrosion inhibition tape
COD	Crack opening displacement
CoV	Coefficient of variance
CPE	Constant phase element
CTE	Coefficient of thermal expansion
CV	Crack velocity
D	Density of the oxide film
d	Density of the metal
da/dN	Fatigue crack propagation rate as a function of number of cycles
DASs	Data acquisition Systems
dc	Direct current
DCPD	Direct current potential drop
DHC	Delayed hydride cracking
DRA	Discontinuously reinforced aluminum
EAC	Environmentally assisted cracking
E_b	Breakdown potential

EBSD	Electron backscatter diffraction
EC	Electrical conductor
E_c or E_{corr}	Corrosion potential
EDP	Energy distribution plot
EDS	Energy dispersive spectroscopy
EHE	Environmental hydrogen embrittlement
EIC	Environmentally induced corrosion
EIMs	Electrochemical impedance measurements
EIR	Electrochemical ion reduction
EIS	Electrochemical impedance spectroscopy
ENMs	Electrochemical noise measurements
E_p	Passive potential (in some anodic polarization data) or pitting potential
E_{pit}	Pitting potential
Epm	Equivalents per million
EPS	Exopolysaccharidic substances
$E_{protection}$	Potential below which pitting does not occur
ΔE	Variation of the activation energy
f	Fiber
fcc	Face-centered cubic
FCG	Fatigue crack growth
FCP	Fatigue crack propagation
FESEM	Field-emission scanning electron microscopy
FFT	Fast Fourier transform
FGMs	Functionally graded materials
FRA	Frequency response analyzer
FSW	Friction stir welding
GDOES	Glow discharge optical emission spectroscopy
HAZ	Heat affected zone
HBSS	Hank's balanced salt solution
HIC	Hydrogen-induced cracking
HE	Hydrogen embrittlement
HEDE	Hydrogen enhanced decohesion
HELP	Hydrogen-enhanced localized plasticity
HER	Hydrogen evolution rate
HESS	Hydrogen electrode same solution
HPDC	High-pressure die casting
HSAAs	High-strength aluminum alloys
i	Current density per cm^2
IACS	International Anneated Copper Standard
IAP	Ion activity product
i_c	Corrosion current per cm^2
I_c	Total corrosion current
IGC	Intergranular corrosion
i_{cp}	Critical corrosion current necessary for passivation
i_p	Corrosion current of the metal in the passive region
IPN	Interpenetrating polymer network
IOZ	Inorganic zinc
$\mathbf{k_c}$	Rate constant or kinetic constant of the cubic rate law
k_l	Rate constant or kinetic constant of the linear rate law
k_g	Rate constant or kinetic constant of the logarithmic rate law
K_{max}	Maximum stress intensity
k_p	Rate constant or kinetic constant of the parabolic rate law
ΔK	Stress-intensity range

L	Liter (mL = 10^{-3} L)
LBW	Laser butt welding
LCA	Local cell action
LEISS	Low-energy ion scattering spectroscopy
LIST	Linearly increasing stress test
LME	Liquid metal embrittlement
LMIE	Liquid-metal-induced embrittlement
LSI	Langelier saturation index or Larson skold index
MA	Mechanical alloying
MAC	Mechanically assisted cracking
MAO	Microarc oxidation
MAP	Mechanical alloying processing
MECIC	Mechanically influenced corrosion
MEEDA	Monoethanolethylenediamine
METIC	Metallurgically influenced corrosion
MgSt	Magnesium stearate
MIC	Microbiologically influenced corrosion
MICI	Microbiologically induced corrosion inhibition
MMC	Metal matrix composite
MPC	Microplasmic ceramic
MSE	Mercury sulfate electrode
n	Number of metal atoms in the oxide molecule ($n = 2$ for Al_2O_3)
NDE	Nondestructive evaluation, negative difference effect
NDI	Nondestructive inspector
NDTs	Nondestructive techniques
n-type	Negative carrier
OCP	Open circuit potential
p	Pressure
PAA	Phosphoric acid anodization
PBR	Pilling–Bedworth ratio
P/M	Powder metallurgy
p_{O_3}	Partial pressure of ozone
p_{O_2}	Partial pressure of oxygen
PSBs	Persistent slipbands
PSI	Puckorius Scaling Index
p-type	Positive carrier
PVC	Pigment volume concentration
PWAA	Post-weld artificial aging
PZC	Potential of zero charge
R	Gas constant or ratio indicates the volume of oxide formed from a unit volume of metal
RBS	Rutherford backscattering spectroscopy
RH	Relative humidity
rms	Root mean square
R_p	Polarization resistance
RS	Rapidly solidified
RSI	Ryznar stability index
RSP	Rapid solidification process
SAD	Stress-assisted dissolution
SCC	Stress-corrosion cracking
SCE	Saturated calomel electrode
SECPM	Scanning electrochemical potential microscope
SEM	Scanning electron microscopy

SHS	Self-autopropagating high-temperature synthesis
SIMS	Secondary ion mass spectroscopy
SKP	Scanning Kelvin probe
SL	Saturation level
SLPR	Self-linear polarization resistance
SRB	Sulfate-reducing bacteria
SRET	Scanning reference electrode technique
SSRT	Slow strain rate tensile or slow strain rate test
SVEP	Scanning vibrating electrode probe
SVET	Scanning vibrating electrode technique
T	Temperature or absolute temperature
t	Time
TGSCC	Transgranular stress-corrosion cracking
THPS	Tetrakis-hydroxymethyl phosphonium sulfate
T_m	Melting temperature
TMAZ	Thermomechanically affected zone
TVC	Total viable count
TWI	The Welding Institute, UK
μ	Ionic force
UTS	Ultimate tensile stress
W	Molecular weight of the oxide film
w	Atomic weight of the metal
WBE	Wire beam electrode
WE	Working electrode
x	Oxide film thickness (or mass gain proportional to oxide film thickness).
x_o	Constant
XPS	X-ray photoelectron spectroscopy
XRD	X-ray diffraction
YS	Yield strength
z_I	Ion valence
ZRA	Zero-resistance ammeter

Index

Corrosion Resistance of Aluminum and Magnesium Alloys: Understanding, Performance, and Testing By Edward Ghali